Wastewater Engineering
Treatment and Resource Recovery

Fifth Edition

Metcalf & Eddy I AECOM

Revised by

George Tchobanoglous
Professor Emeritus of Civil and
Environmental Engineering
University of California at Davis

H. David Stensel
Professor of Civil and Environmental
Engineering
University of Washington, Seattle

Ryujiro Tsuchihashi
Wastewater Technical Leader, AECOM

Franklin Burton
Consulting Engineer
Los Altos, CA

Contributing Authors:

Mohammad Abu-Orf
North America Biosolids Practice
Leader, AECOM

Gregory Bowden
Wastewater Technical Leader, AECOM

William Pfrang
Wastewater Treatment Technology
Leader, AECOM

Mc
Graw
Hill
Education

WASTEWATER ENGINEERING: TREATMENT AND RESOURCE RECOVERY,
FIFTH EDITION (VOLUME 2)
International Edition 2014

10 09 08 07 06 05 04 03 02
20 15 14
CTP   SLP

All credits appearing on page or at the end of the book are considered to be an extension of the copyright page.

When ordering this title, use ISBN 978-1-259-01079-8 or MHID 1-259-01079-1 (Vol 1 & 2)

www.mhhe.com

George Tchobanoglous is Professor Emeritus in the Department of Civil and Environmental Engineering at the University of California, Davis. He received a B.S. degree in civil engineering from the University of the Pacific, an M.S. degree in sanitary engineering from the University of California at Berkeley, and a Ph.D. from Stanford University in 1969. Dr. Tchobanoglous' research interests are in the areas of wastewater treatment and reuse, wastewater filtration, UV disinfection, aquatic wastewater management systems, wastewater management for small and decentralized wastewater management systems, and solid waste management. He has authored or co-authored over 500 technical publications including 22 textbooks and 8 reference works. The textbooks are used in more than 225 colleges and universities, by practicing engineers, and in universities worldwide both in English and in translation. His books are famous for successfully bridging the gap between academia and the day-to-day world of the engineer. He is a Past President of the Association of Environmental Engineers and Science Professors. Among his many honors, in 2003 Professor Tchobanoglous received the Clarke Prize from the National Water Research Institute. In 2004, he received the Distinguished Service Award for Research and Education in Integrated Waste Management from the Waste-To-Energy Research and Technology Council. In 2004, he was also inducted into the National Academy of Engineering. In 2005, he was awarded an honorary Doctor of Engineering from the Colorado School of Mines. In 2007, he received the Frederick George Pohland Medal awarded by AAEE and AEESP. In 2012 he was made a WEF Fellow. He is a registered Civil Engineer in California.

H. David Stensel is a Professor in the Civil and Environmental Engineering Department at the University of Washington, Seattle, WA. Prior to his academic positions, he spent 10 years in practice developing and applying industrial and municipal wastewater treatment processes. He received a B.S. degree in civil engineering from Union College, Schenectady, NY, and M.E. and Ph.D. degrees in environmental engineering from Cornell University. His principal research interests are in the areas of wastewater treatment, biological nutrient removal, sludge processing methods, resource recovery, and biodegradation of micropollutants. He is a Past Chair of the Environmental Engineering Division of ASCE, has served on the board of the Association of Environmental Engineering Professors and on various committees for ASCE and the Water Environment Federation. He has authored or coauthored over 150 technical publications and a textbook on biological nutrient removal. Research recognition honors include the ASCE Rudolf Hering Medal, the Water Environment Federation Harrison Prescott Eddy Medal twice, and the Bradley Gascoigne Medal. In 2013, he received the Frederick George Pohland Medal awarded by AAEE and AEESP. He is a registered professional engineer, a diplomate in the American Academy of Environmental Engineers and a life member of the American Society of Civil Engineers and the Water Environment Federation.

Ryujiro Tsuchihashi is a technical leader with AECOM. He received his B.S. and M.S. in civil and environmental engineering from Kyoto University, Japan, and a Ph.D. in environmental engineering from the University of California, Davis. The areas of his expertise include wastewater/water reclamation process evaluation and design, evaluation and assessment of water reuse systems, biological nutrient removal, and evaluation of greenhouse gas emission reduction from wastewater treatment processes. He was a co-author of the textbook "Water Reuse: Issues, Technologies and Applications," a companion textbook to this textbook. He is a technical practice coordinator for AECOM's water reuse leadership team. Ryujiro Tsuchihashi is a member of the Water Environment Federation, American Society of Civil Engineer, and International Water Association, and has been an employee of AECOM for 10 years, during which he has worked on various projects in the United State, Australia, Jordan, and Canada.

Franklin Burton served as vice president and chief engineer of the western region of Metcalf & Eddy in Palo Alto, California for 30 years. He retired from Metcalf & Eddy in 1986 and has been in private practice in Los Altos, California, specializing in treatment technology evaluation, facilities design review, energy management, and value engineering. He received his B.S. in mechanical engineering from Lehigh University and an M.S. in civil engineering from the University of Michigan. He was co-author of the third and fourth editions of the Metcalf & Eddy textbook "Wastewater Engineering: Treatment and Reuse." He has authored over 30 publications on water and wastewater treatment and energy management in water and wastewater applications. He is a registered civil engineer in California and is a life member of the American Society of Civil Engineers, American Water Works Association, and Water Environment Federation.

Mohammad Abu-Orf is AECOM's North America biosolids practice leader and wastewater director. He received his B.S. in civil engineering from Birzeit University, West Bank, Palestine and received his M.S. and Ph.D. in civil and environmental engineering from the University of Delaware. He worked with Siemens Water Technology and Veolia Water as biosolids director of research and development. He is the main inventor on five patents and authored and co-authored more than 120 publications focusing on conditioning, dewatering, stabilization and energy recovery from biosolids. He was awarded first place for Ph.D. in the student paper competition by the Water Environment Federation for two consecutive years in 1993 and 1994. He coauthored manuals of practice and reports for the Water Environment Research Foundation. He served as an editor of the Specialty Group for Sludge Management of the International World Association for six years and served on the editorial board of the biosolids technical bulletin of the Water Environment Federation. Mohammad Abu-Orf has been an employee of AECOM for 6 years.

Gregory Bowden is a technical leader with AECOM. He received his B.S. in chemical engineering from Oklahoma State University and a Ph.D. in chemical engineering from the University of Texas at Austin. He worked for Hoechst Celanese (Celanese AG) for 10 years as a senior process engineer, supporting wastewater treatment facility operations at chemical production plants in North America. He also worked as a project manager in the US Filter/Veolia North American Technology Center. His areas of expertise include industrial wastewater treatment, biological and physical/chemical nutrient removal technologies and biological process modeling. Greg Bowden is a member of the Water Environment Federation and has been an AECOM employee for 9 years.

William Pfrang is a Vice-President of AECOM and Technical Director of their Metro-New York Water Division. He began his professional career with Metcalf & Eddy, Inc., as a civil engineer in 1968. During his career, he has specialized in municipal wastewater treatment plant design including master planning, alternative process assessments, conceptual, and detailed design. Globally, he has been the lead engineer for wastewater treatment projects in the United States, Southeast Asia, South America, and the Middle East. He received his B.S. and M.S. in civil engineering from Northeastern University. He is a registered professional engineer, a member of the American Academy of Environmental Engineers, and the Water Environment Federation. William Pfrang has been an employee of the firm for over 40 years.

Contents

Appendixes

Indexes

10 Anaerobic Suspended and Attached Growth Biological Treatment Processes

WORKING TERMINOLOGY

Term	Definition
Ammonia toxicity	Free ammonia (NH_3), at high enough concentrations, is considered toxic to acetoclastic methanogenic organisms.
Anaerobic expanded bed process	An anaerobic upflow process in which the medium, usually silica sand, is expanded but not fluidized by the upward liquid velocity.
Anaerobic fluidized bed process	An anaerobic upflow process which operates in a mode similar to the expanded bed, but the medium is fully fluidized by the upward velocity of the fluid.
Anaerobic granular sludge	Dense 0.50 to 4.0 mm particles in anaerobic upflow reactors containing fermentation, hydrogenotrophic, and methanogenic organisms in close proximity.
Anaerobic sequencing batch reactor	An anaerobic suspended growth process with reaction and solids-liquid separation in the same vessel, much like that for aerobic sequencing batch reactors (SBR).
Anaerobic sludge blanket processes	Influent wastewater is distributed at the bottom of an anaerobic reactor and travels in an upflow mode through a sludge blanket zone that typically contains dense granular biomass particles.
Anaerobic suspended growth processes	A mixed anaerobic reactor containing a suspension of fermentation and methanogenic organisms and feed particulate matter.
Anaerobic treatment processes	Any of a number of biological treatment processes carried out in the absence of oxygen.
Attached growth anaerobic process	Anaerobic treatment processes in which the biomass responsible for treatment is attached to some type of medium. The medium may be fixed, expanded or fluidized. Where fixed medium is used, the flow can either be downflow or upflow.
Covered anaerobic lagoon process	Covered liner earthen lagoon typically used for high-strength industrial wastewaters, such as meat processing wastewaters.
Expanded granular sludge blanket process (EGSB)	An anaerobic sludge blanket process with higher upflow velocities than the UASB process.
Hydrogen sulfide (H_2S)	A malodorous gas toxic gas formed under anaerobic conditions. Sulfur containing compounds in wastewater serve as electron acceptors for sulfate-reducing bacteria, which consume organic compounds and produce hydrogen sulfide (H_2S).
Membrane separation anaerobic treatment process	A suspended growth anaerobic process in which a synthetic membrane is used to separate the treated wastewater from the solids to achieve complete effluent suspended solids removal.
Methane (CH_4)	Methane and carbon dioxide along with cell biomass are the principal carbon end products of anaerobic conversion processes.
Organic loading rate (OLR)	The mass rate of organic substrate (COD) addition per unit volume of an anaerobic reactor.
Solids retention time (SRT)	The average period of time in which solids remain in a biological reactor.
Upflow anaerobic sludge blanket (UASB) reactor	The name of the first and most common of the anaerobic sludge blanket processes (see above).

Anaerobic biological reactions involve specialized bacteria and archaea that use a variety of electron acceptors in the absence of molecular oxygen for energy production. They are used in a number of different anaerobic processes in wastewater treatment. These include processes for nitrate/nitrite reduction to nitrogen gases, fermentation processes to produce volatile fatty acids for use in enhanced biological phosphorus removal, anaerobic contacting for acetate and propionate uptake in enhanced biological phosphorus removal, anaerobic oxidation of organic compounds in municipal and industrial wastewaters, anaerobic digestion of waste sludge, and anaerobic digestion of other organic wastes.

Anaerobic wastewater treatment processes include suspended growth, upflow and downflow attached growth, fluidized bed attached growth, upflow sludge blanket, lagoon, suspended growth with membrane separation, and many other proprietary processes. The purpose of this chapter is to describe and present typical design loadings and treatment process capabilities for the principal anaerobic processes, excluding conventional anaerobic digestion processes which are considered in Chap. 13. Before considering the individual anaerobic treatment processes, it will be helpful to consider the rational for the use of anaerobic treatment processes, the evolution of anaerobic treatment technologies, a brief review of the principal public and commercial processes, and general considerations for the application of anaerobic treatment processes.

10–1   THE RATIONALE FOR ANAEROBIC TREATMENT

The rationale for and interest in the use of anaerobic treatment processes can be explained by considering the advantages and disadvantages of these processes. The principal advantages and disadvantages of anaerobic treatment are listed in Table 10–1 and are discussed below.

Advantages of Anaerobic Treatment Processes

Anaerobic treatment processes have been used as an alternative to aerobic treatment for applications varying from low to extremely high strength wastes. Of the advantages cited in Table 10–1, energy considerations, lower biomass yield, less nutrients required, higher volumetric loadings, and effective pretreatment are examined briefly in the following discussion. The subjects are also considered in Chap. 13, 14, and 17.

Energy Considerations.   Anaerobic processes may be net energy producers instead of energy users as is the case for aerobic processes. The potential net energy production that can be achieved with anaerobic treatment depends on the strength of the wastewater, the operating temperature, and whether energy recovery is practiced. An energy balance comparison between anaerobic and aerobic treatment for various wastewater strengths is presented in Sec. 10–4.

Table 10–1

Advantages and disadvantages of anaerobic processes compared to aerobic processes

Advantages	Disadvantages
1. Less energy required	1. Longer startup time to develop necessary biomass inventory
2. Less biological sludge production	2. May require alkalinity addition
3. Less nutrients required	3. May require further treatment with an aerobic treatment process to meet discharge requirements
4. Methane production, a potential energy source	
5. Smaller reactor volume required	4. Biological nitrogen and phosphorus removal is not possible
6. Elimination of off-gas air pollution	5. Much more sensitive to the negative effect of lower temperatures on reaction rates
7. Able to respond quickly to substrate addition after long periods without feeding	6. May be more susceptible to upsets due to toxic substances or wide feeding changes
8. Effective pretreatment process	7. Potential for odor production and corrosiveness of gas
9. Potential for lower carbon footprint	

Lower Biomass Yield. Because the energetics of anaerobic processes result in lower biomass production by a factor of about 6 to 8 times, sludge processing and disposal costs are greatly reduced. Given the major environmental and monetary issues associated with the use or disposal of biomass produced from aerobic processes as discussed in Chap. 14, the fact that less sludge is produced in anaerobic treatment is a significant advantage for anaerobic treatment.

Less Nutrients Required. Because many industrial wastewaters lack sufficient nutrients, the lower cost for nutrient addition due to less biomass production from anaerobic treatment is a clear benefit.

Higher Volumetric Organic Loadings. Anaerobic processes generally have higher volumetric organic loading rates than aerobic processes so that smaller reactor volumes and less space may be required for treatment. Organic loading rates of 3.2 to 32 kg COD/m^3·d may be used for anaerobic processes, compared to 0.5 to 3.2 kg COD/ m^3·d for aerobic processes (Speece, 1996).

Effective Pretreatment Process. Often, anaerobic treatment processes are used in combination with aerobic treatment processes to achieve specific treatment goals. A common application in the wastewater field is the treatment of high strength waste before discharge to a municipal wastewater treatment facility. A typical example is shown on Fig. 10–1. Gas, recovered from the covered anaerobic section of the large anaerobic lagoon is used to produce electricity.

Disadvantages of Anaerobic Treatment Processes

Potential disadvantages also exist for anaerobic processes as reported in Table 10–1. Operational considerations, the need for alkalinity addition, and the need for further treatment are highlighted further in the following discussion.

Operational Considerations. The major concerns with anaerobic processes is their longer start-up time (months for a anaerobic versus days for aerobic processes), their sensitivity to possible toxic compounds, operational stability, the potential for odor production, and corrosiveness of the digester gas. However with proper wastewater characterization and process design these problems can be avoided and/or managed. Operational process knowledge and skill is also needed to maintain anaerobic process stability by

Figure 10–1

Combined anaerobic lagoon/ aerobic pond wastewater treatment system: (a) aerial view of large treatment ponds with covered anaerobic pretreatment lagoon, shown in circled area. The white dots are the plumes from large turbine type floating aerators (coordinates 37.9788 S, 144.6417 E). (b) View from inlet with gas recovery facilities in the foreground, floating membrane cover, and large surface floating aerators in the background.

(a)

(b)

proper control of the feed, temperature, and pH to maintain a balance between volatile fatty acid production by the acidogens and the capacity of the methanogenic organisms.

Need For Alkalinity Addition. The most significant negative factor that can affect the economics of anaerobic versus aerobic treatment is the possible need to add alkalinity. Alkalinity concentrations of at least 2000 to 3000 mg/L as $CaCO_3$ may be needed in anaerobic processes to maintain an acceptable pH with the characteristic high gas phase CO_2 concentration. If this amount of alkalinity is not in the influent wastewater or cannot be produced by the degradation of proteins and amino acid, a significant cost may be incurred to purchase alkalinity, which can affect the overall economics of the process.

Need For Further Treatment. Anaerobic wastewater treatment processes may have effluent BOD concentrations ranging from 50 to 150 mg/L due to the presence of residual volatile fatty acids and dispersed solids. The process can be followed by an aerobic process for effluent polishing to utilize the benefits of both processes. Series reactors of anaerobic-aerobic processes have been shown feasible for treating municipal wastewaters in both temperate and warmer climates resulting in less energy needs and less sludge production (Lew et al., 2003; Chong et al., 2012). More recently, as discussed subsequently, a number of integrated combined single tank anaerobic-aerobic reactors have also been developed.

Summary Assessment

In general, for municipal wastewaters with lower concentrations of degradable COD, lower temperatures, higher effluent quality needs, and nutrient removal requirements, aerobic processes are favored at present. For industrial wastewaters with much higher degradable COD concentrations and elevated temperatures, anaerobic processes may be more economical. In the future, with further developments in anaerobic treatment processes, it is anticipated that their use will become more widespread in a variety of applications, because of the overwhelming advantages from energy savings and less sludge production.

10–2 DEVELOPMENT OF ANAEROBIC TECHNOLOGIES

The earliest engineered anaerobic technologies were designed for and applied to the treatment of wastewater. At the time of their development in the late 1800s and early 1900s, a community's wastewater was an unhealthy combination of untreated sanitary wastes, animal manure, and various local industrial discharges. The historical development of the early anaerobic technologies, the application of anaerobic treatment for sludges, use of anaerobic treatment for high strength wastes, and some thoughts on the future are considered in this section. The types of anaerobic technologies currently available are considered in the following section.

Historical Developments in Liquefaction

Some of the noteworthy developments in the early stages of anaerobic treatment technology are summarized in Table 10–2. Schematic diagrams of some of the early treatment process developments are illustrated on Fig. 10–2. The early developments in anaerobic treatment were focused on liquefying the solids in wastewater to reduce or eliminate the need for sludge management and make the effluent suitable for subsequent treatment and or reuse for irrigation. The automatic scavenger developed by Mouras in France and patented in the 1880s [see Fig. 10–2(a)] was perhaps the first purposeful attempt to liquefy

Table 10–2

Important milestones in the development of anaerobic technologies[a]

Period	Event
Early developments	
1852	Henry Austin in England designed and built a tank that allows accumulation of solid matter layer on the bottom and scum layer on the top, with liquid draw off between layers. Closely resembles modern septic tanks (Kinnicutt et al., 1913).
1881	Jean-Louis Mouras obtained a French patent for an "automatic scavenger—an automatic and odorless cesspit" to treat wastewater anaerobically, a development that became one of the earliest known domestic wastewater treatment systems [see Fig. 10–1(a)]. Perhaps the first purposeful attempt to liquefy sludge. (Moigno, 1881,1882; Kinnicutt et al., 1913).
1887	The first Dortmund tank was designed and built by Kniebuhler in Germany. The advantage of the Dortmund tank was that sludge could be removed without stopping the flow. Dortmund tanks are still being built today [see Fig. 10–1(b)] (Kinnicutt et al., 1913).
1887	Lawrence Experimental Station established on the bank of the Merrimac River at Lawrence, MA. In 1890, the first report of the work at the Lawrence Station was published, which Winslow considered the "the most important single document in the history of sewage treatment" (Winslow, 1938).
1887	A sand bed, upflow anaerobic filter was constructed by the Lawrence Experimental Station to treat domestic wastewater and was operated for about 14 y (McCarty, 2001).
1891	Scott-Moncrieff of England constructed a two-tiered tank (empty lower level sludge compartment and upper level upflow anaerobic rock filter) to treat domestic wastewater anaerobically followed by an aerobic coke tray trickling filter [see Fig. 10–1(c)] (Kinnicutt et al., 1913). The complete treatment system comprised of the anaerobic rock filter and aerobic treatment unit is perhaps the first hybrid system ever built (McCarty, 2001).
Developments in the treatment of wastewater sludges	
1895	In Exeter, England, Donald Cameron installed a water-tight, covered basin to treat wastewater anaerobically, naming the device a "septic tank" [see Fig. 10–1(d)] (Kinnicutt et al., 1913).
1899	H. W. Clark at the Lawrence Experimental Station, noted that sludge should be fermented in a separate tank. Sludge lagoons had been used up to that time and later (Imhoff, 1938; Winslow, 1938).
1904	Travis developed a two story tank for the liquefaction of sludge [see Fig. 10–1(e)]. In the original design about one sixth of the flow passed through the lower chamber) (Kinnicutt et al., 1913). In one configuration effluent was directed to anaerobic filters [see Fig. 10–1(f)].
1906	Dr. Karl Imhoff of Germany patented a wastewater treatment device (the Imhoff tank) that anaerobically treated sewage while separating solids from the liquid phase prior to discharge [see Fig. 10–1(g)]. The Imhoff tank was based on earlier work by Travis, but avoided the flow through the liquefaction chamber (Imhoff, 1938).
1909–1913	Production of combustible gas by methane fermentation of strawboard was demonstrated at working scale in Netherlands in 1909. The gas was utilized for power generation. Gas utilization occurred at treatment plants for a diary and slaughterhouse in 1912, and at a wastewater treatment plant in 1914. Both fixed and floating gasholders were used (Kessener, 1938).
1914	Early experiments were conducted on the collection and heating of gas at Emschergenossenschaft in Germany (Imhoff, 1938).
1915	Experiments in heating digesters were done to increase gas production in the Netherlands (Kessener, 1938).
1927	The Ruhrverband, a German water management association, constructed a separate heated digestion tank at Essen—Rellinghausen to anaerobically digest treatment plant sludge and utilize the generated biogas for power and heating (Imhoff, 1938).
1929	A long rectangular digester with paddles on a horizontal shaft for seeding, stirring, and scum destruction was built in the Netherlands (Kessener, 1938).

(continued)

| **Table 10–2** (Continued)

Period	Event
1930 and 1932	Buswell of the Illinois State Water Survey started a series of reports on the fundamentals of anaerobic digestion of solids (Buswell and Neave, 1930; Buswell and Boruff, 1932).
1950	Stander of South Africa demonstrated the full-scale process benefits obtained by separating anaerobic solids externally and returning them to the digester tank (Stander, 1950; Standar and Snyders, 1950).
Early 1950s	Morgan and Torpey both demonstrated through their research the increased process performance achievable by adding mixing to an anaerobic biosolids digester (Morgan, 1954; Torpey, 1954).

Developments in the treatment of high strength wastes

Period	Event
1955	Schroepfer and others were involved in the first full-scale application of the anaerobic contact process at a meat packing plant in Minnesota (Schroepfer et al., 1955).
1969	Young and McCarty developed the anaerobic filter process to make attached growth (fixed-film) anaerobic biomass available for high strength wastewater treatment (Young and McCarty, 1969; Young, 1991).
1978	Grethlein experimented with the anaerobic treatment of wastewater using a septic tank and an external cross-flow membrane (Grethlein, 1978).
Late 1970s, early 1980	Lettinga developed the upflow anaerobic sludge (UASB) process [see Fig. 10–1(h)] using waste from the sugar beet industry (Lettinga et al., 1980). His work on anaerobic processes has been instrumental in the development of many commercial anaerobic technologies, especially for the treatment of high strength wastes (see Table 10–3).
1980	Switzenbaum and Jewell developed the anaerobic fluidized bed process, particularly useful in the treatment of high strength wastewater (Switzenbaum and Jewell, 1980).

[a] Adapted in part from Totzke (2012), McCarty (2001), Metcalf & Eddy (1915).

sludge. In the Dortmund tank [see Fig. 10–2(b)], sludge could liquefy and residual matter could be removed without stopping the process. In the Scott Moncrieff tank and filter [see Fig. 10–2(c)], the first tank was known as the liquefying tank. Referring to Fig. 10–2(c) wastewater entered the first compartment, which served as both a grease trap and sedimentation tank, and from there flowed into the space below the anaerobic rock filter. The wastewater then flowed up through the anaerobic rock filter. Bacteria attached to the rocks brought about the liquefaction of the colloidal material in the wastewater passing through the bed. Wastewater from above the stone bed then flowed and was distributed with tilting buckets over the uppermost series of nine perforated trays contain coke. The reactor was defined as anaerobic and the coke tray filter as aerobic. The time for the wastewater to pass through the nine trays was about 10 min.

Treatment of Wastewater Sludges

The first tank to be identified as a *septic tank* was built by Donald Cameron in England in 1895 [see Fig. 10–2(d)]. The next significant development in anaerobic technology was the Travis hydrolytic tank [see Fig. 10–2(e)] developed by W. O. Travis. As early as 1899, W. H. Clark, a chemist at the Lawrence laboratory, had suggested that sludge should be fermented in a separate tank. Travis partially implemented Clark's idea in the hydrolytic tank with the exception that in the original design a portion of the influent wastewater flowed through the liquefaction chamber. The Travis hydrolytic tank was followed by a series of hydrolyzing chambers containing an early form of inclined settlers to bring about additional treatment [see Fig. 10–2(f)]. It is interesting to note that the Imhoff tank, shown on Fig. 10–2(g), was basically an improvement of the Travis hydrolytic tank. K. Imhoff, who knew of Clark's idea, also recognized that the effluent quality was deteriorated by allowing a portion of the flow to pass through the liquefying chamber of the hydrolytic tank.

Figure 10–2

Evolution of anaerobic technology for wastewater treatment in process schematics: (a) Mouras automatic scavenger (1881), (b) Dortmund tank (1887), (c) Scott-Moncrieff two-tiered tank (1891), (d) Cameron septic tank (1895, first use of the name septic tank), (e) Travis two story hydrolytic tank (1904), (f) Travis two story hydrolytic tank with hydrolyzing chambers (1904), (g) Imhoff tank (1906), and (h) upflow anaerobic sludge blanket reactor (1980).

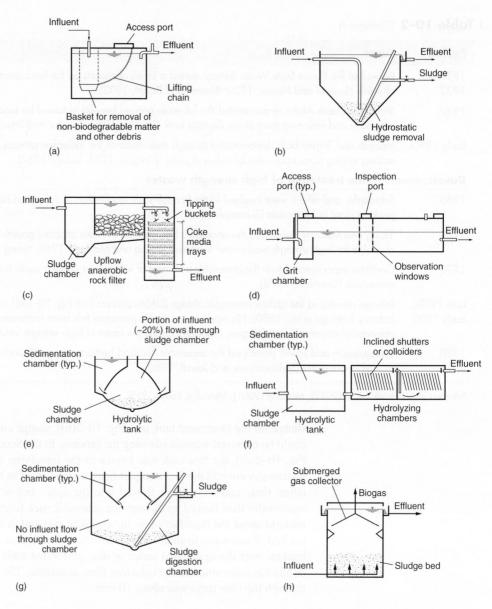

Imhoff tanks are still in common use at the present time. To increase gas production, experimental work on the collection, combustion, and utilization of digester gas was conducted in both Germany and the Netherlands in the early 1900s. The first heated digester constructed in Germany in 1927 evolved from early experimental work.

Treatment of High Strength Wastes

Following the stock market crash of the 1920s and the depression of the 1930s the rate of development in the anaerobic treatment of wastes slowed. Interest in wastewater treatment increased following the Second World War and considerable work was carried out in the 1950s by Stander (1950), Morgan (1954), and Torpey (1954). Concomitant with the work of Stander, Morgan, Torpey, and others in the 1950s, the principles of anaerobic degradation began to be applied to the private marketplace, a unique repository of high-strength, readily biodegradable wastes. Important technological developments in the anaerobic treatment of high strength wastes include the work of Schroepfer in the treatment of meat packing waste with the anaerobic contact process (Schroepher et al., 1955); the development of the anaerobic filter by Young and McCarty in the late 1960s (Young and McCarty, 1969; Young, 1991); and the development of the upflow anaerobic sludge blanket treatment process by Lettinga and his associates at Wageningen University in the Netherlands in the 1970s (Lettinga et al., 1980). Lettinga's work was motivated by the publications of Young and McCarty and McCarty.

The upflow anaerobic sludge blanket (UASB) reactor developed by Lettinga in 1980s [see Fig. 10–2(h)], is the most significant modern development in anaerobic process technology. Lettinga's work on the UASB and other anaerobic processes has been instrumental in the development of the many commercial anaerobic technologies discussed in the following section. Since these seminal advances, numerous researchers, consulting engineers, and manufacturers have pushed the development of anaerobic technologies to the point where there are many available suppliers and thousands of full-scale applications (Totzke, 2012).

Future Developments

In the evolution of anaerobic technologies it is interesting to note the significant amount of physical experimentation that took place in the late 1800s and early 1900s. One of the most interesting examples is the glass model [essentially the same as Fig. 10–2(f)] of the hydrolytic tank constructed by Travis to observe what was occurring in the process. The inclined glass plates, called *colloiders,* in the second and third chamber were used to remove colloidal material by interception and sedimentation, a precursor of the modern day lamella or tube settler described in Chap. 5. It should also be noted that the use of rock, coke, lath and slate filters are cited often in the literature from the late 1800s through the early 1900s. From this brief review, it is clear that many of the early process developments or variants thereof are in use today. Also, it is important to recognize that just as much or more research is going on today, but the primary focus is on understanding the fundamental process biochemistry, molecular microbiology, and physiology.

10–3 AVAILABLE ANAEROBIC TECHNOLOGIES

Currently, anaerobic treatment technology can be classified as public or proprietary and commercial. Public technologies are those technologies available to any experienced designer. Proprietary and commercial technologies are those technologies supplied as a complete package with little or no input from the designer. The available types of anaerobic technologies and their application are discussed below.

Types of Anaerobic Technologies

The principal types of anaerobic technologies used for the treatment of wastes, with the exception of the conventional complete-mix digester used for the treatment of municipal sludge's, are illustrated and described in Table 10–3. The digestion of municipal sludge is considered in detail in Chap. 13. The technologies identified in Table 10–3 are listed in their approximate order of application prevalence. In reviewing the technologies in Table 10–3, the lineage to Lettinga's UASB [see Table 10–3(b)] is apparent. The expanded granular sludge blanket [EGSB, see Table 10–3(c)] process is a modified UASB process with effluent recycle and higher upflow velocity to fluidize the granules within the reactor. The EGSB process was introduced by Lettinga and coworkers for the treatment of low strength brewery wastewater but it can handle both lower and higher

Table 10–3

Description of the principal types of commercially available anaerobic treatment technologies[a]

Process	Description
(a) Low loaded anaerobic lagoon system (ANL) 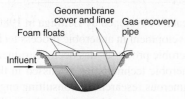	Generally unmixed reactor system employing suspended/flocculating anaerobic biomass and settled anaerobic solids with hydraulic retention times of 20 to 50 d and average SRTs of 50 to 100 d. Can handle a wider range of wastes including solids and soluble wastewaters. Designed for a total chemical oxygen demand (COD) loading of less than 2 kg/m³·d. Systems can be covered with synthetic membranes for gas collection.
(b) Upflow anaerobic sludge blanket (UASB) 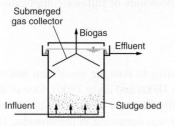	An upflow reactor with a bottom sludge bed and dense, granular anaerobic bio-mass with good mixing provided by the upflow velocity and biogas generation. The reactor effective anaerobic sludge concentration may be in the range of 35 to 40 kg/m³. A gas-liquid-solid separator at the top separates granular solids from the effluent and collects biogas. The sludge blanket has SRTs in excess of 30 d with hydraulic retention times in the range of 4 to 8 h. Designed for a COD loading of 5 to 20 kg/m³·d. Upflow velocities can vary from 1 to 6 m/h and reactor heights of 5 to 20 m have been used.
(c) Expanded granular sludge blanket (EGSB) 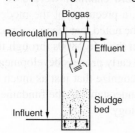	The EGSB is a commonly used modification of the UASB by employing a higher upflow velocity, a greater height to diameter ratio, and recirculation of effluent. Upflow velocities may be the range of 4 to 10 m/h and reactor heights up to 25 m have been used. The higher velocity provides a more efficient reactor for treatment of soluble substrates by improving mixing, reducing dead volume, and increasing diffusion rates from the bulk liquid to the granular biofilm. Was originally developed to treat low strength wastes but has been used for high strength as well and at low temperatures of 10°C. Organic loading rates as high as 35 kg/m³·d have been used. Not as effective as the UASB for colloidal and particulate solids capture.
(d) Internal circulation UASB (IC) 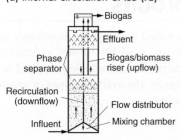	An IC reactor consists of two stacked UASB reactors in series, each with a gas separator at the top. The system uses a down comer pipe from a top chamber to the bottom inlet and a riser pipe from the first gas separator to induce recirculation and high upflow velocities in the lower granular sludge blanket reactor. Gas pro-duced from the lower reactor is captured in the first separator and creates a gas lift for water and biosolids in the riser pipe. The gas is separated (released) from the biosolids in the chamber above the second reactor gas separator. From there the biosolids/water mixture enters the down comer pipe to provide internal recirculation to the bottom compartment. The high recirculation ratio results in high upflow velocities in the bottom chamber, 8 to 20 times higher than in conventional UASB unit, to provide good mixing and a very efficient reactor operation. The upper reactor provides a second stage anaerobic treatment for more efficient overall COD removal and the lower upflow velocity and lower biogas production rate improves effluent solids and biomass capture. Reactor heights of up to 25 m have been used.

(continued)

Process	**Description**
(e) Fluidized bed (FB) 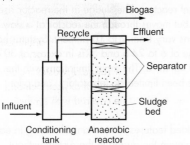	These systems are based on the development of a dense anaerobic biomass on small size (0.10 to 0.30 mm) inert particles of fine sand, basalt, pumice, or plastic. The particles are kept in suspension and mixed by a high upward velocity. The higher velocities leads to what is called a fluidized bed with 25 to 300 percent bed expansion and the expanded fluidized bed refers to operation at lower velocity with 15 to 25 percent bed expansion. These reactors are applicable for soluble wastes or easily degraded small particulates, such as whey. Upflow velocities may be in the range of 10 to 20 m/h and COD loadings of 20 to 40 kg/m³·d have been used.
(f) Anaerobic contact process (ANCP) 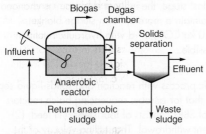	A completely mixed reactor system employing suspended anaerobic biomass, a mixing/flocculator degassing chamber, liquid-solids separation, and solids recycle so that the SRT is longer than the hydraulic retention time. Designed for a COD loading of in the range of 2 to 5 kg/m³·d.
(g) Anaerobic filter (ANF) 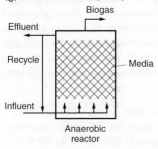	An anaerobic filter (ANF) system is an unmixed reactor system employing fixed film anaerobic biomass attached to supporting media, so that a large anaerobic biomass and long SRT can be maintained to allow treatment at hydraulic retention times in the range of 1 to 3 d and designed for a COD loading of 5 to 20 kg/m³·d. It is available in upflow (ANFU) and downflow (ANFD) configurations.
(h) Anaerobic hybrid process (ANHYB) 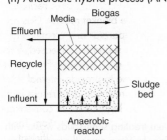	A combination of stand-alone anaerobic technologies employing a combination of an upflow anaerobic sludge blanket reactor and anaerobic filter to provide a high biomass concentration and high volumetric organic removal rates in the lower portion and further removal of volatile fatty acids and capture of suspended solids in the upper anaerobic filter portion.
(i) Anaerobic membrane process (ANMBR) 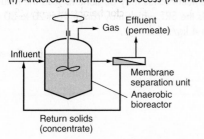	A mixed reactor system employing suspended/flocculating anaerobic biomass and a synthetic membrane solids-liquid separation with solids recycle to provide a long SRT with the short hydraulic retention time. Designed for a COD loading of 5 to 15 kg/m³·d.

(continued)

Process	Description
(j) Anaerobic baffled reactor (ABR) 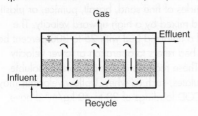	Baffles are used to direct the flow of wastewater in an upflow mode through a series of upflow anaerobic sludge blanket reactors. The sludge in the reactor rises and falls with gas production and flow, but moves through the reactor at a slow rate. Reactor volatile solids concentrations vary from 2 to 10 percent. Systems have been operated with τ values in the range of 6 to 24 h and SRTs in excess of 30 d. Designed for a COD loading of 5 to 10 kg/m^3·d. The main limitations with the ABR process are that many studies have been limited do laboratory- and pilot-scale treatment units.
(k) Anaerobic migrating blanket reactor (AMBR) 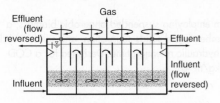	Process is similar to the ABR with the added features of mechanical mixing in each stage and an operating approach to maintain the sludge in the system without resorting to packing or settlers for additional solids capture. When a significant quantity of solids accumulates in the last stage, the influent feed point is changed to the effluent side, which helps to maintain a more uniform sludge blanket. Organic loading rates from 1.0 to 3.0 kg COD/m^3·d with hydraulic retention times ranging from 4 to 12 h are possible.
(l) Anaerobic sequencing batch reactor (ANSBR) 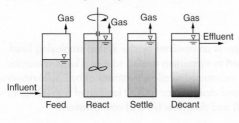	A mixed suspended growth anaerobic process with reaction and solids-liquid separation in the same vessel, much like that for aerobic sequencing batch reactors (SBRs) (see Chap. 8). The operation of SBRs consists of four steps: (1) feed, (2) react, (3) settle, and (4) decant/effluent withdrawal. The settling velocity of the sludge during the settle period before decanting the effluent is critical. Settling times used are about 30 min. After sufficient operating time, a dense granulated sludge develops that improves the liquid-solids separation. At τ values from 6 to 24 h, the SRT may range from 50 to 200 d, respectively. At 25°C, 92 to 98 percent COD removal was achieved at volumetric organic loadings of 1.2 to 2.4 kg COD/m^3·d. At 5°C, COD removal ranged from 85 to 75 percent for COD loadings from 0.9 to 2.4 kg/m^3·d, respectively.
(m) Continuously stirred tank anaerobic reactor (ANCSTR) 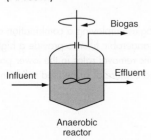	A completely mixed reactor system treating semi-solids wastes with suspended anaerobic biomass. The reactor detention time equals the SRT, which may range from 15 to 30 d, with resulted COD loadings typically less than 4 kg/m^3·d.
(n) Plug flow anaerobic system (ANPF) 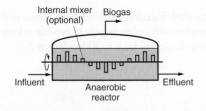	Generally an unmixed rectangular reactor system treating semi-solids waste with high (10 to 18 percent) total solids concentration. In some cases the rectangular reactor is slightly inclined. Recycle of effluent solids may be done to seed the influent feed. The feed retention time equals the SRT, which may range from 20 to 30 d with COD loadings generally less than 4 kg/m^3·d.

Adapted from Nicolella et al. (2000), Totzke (2012), Tauseef et al. (2013).

strength wastewaters (Kato et al., 1999). Similarly, the internal recycle (IC) reactor [see Table 10–3(d)] is essentially two UASB reactors in series with internal recycle and it has also been successful for the treatment of low and very high strength wastewaters. The principal advantages in the development of EGSB and IC processes has been to increase the volumetric organic loading and treatment efficiency. Other anaerobic processes have been developed to treat wastes with specific characteristic (e.g., colloidal and particulate wastes).

Combined Processes.   Process additions to the upflow granular sludge blanket have been made to improve treatment performance to approach or meet secondary treatment levels. These include hybrid anaerobic processes and a combined anaerobic-aerobic process. Hybrid processes typically involve two stages of anaerobic treatment, such as the one shown in Table 10–3(h) which involves an anaerobic sludge blanket process in the lower portion of the reactor followed by anaerobic attached growth process for polishing in the upper portion. A sequential anaerobic-aerobic process was reported as early as 1992 (Garuti et al., 1992) to provide a secondary effluent quality for anaerobic treatment in warm climates. More recently the aerobic process has been incorporated as an integral part of a combined anaerobic-aerobic process in a single tank (Tauseef, et al., 2013). Although there are a number of combined aerobic-anaerobic process the focus of this chapter is on the anaerobic processes.

Commercial Technologies.   Compiling a list of commercially available anaerobic technologies is beyond the scope of this book task, and is complicated by the wide range of technical descriptions, trade names, and research descriptions that are available in the literature and in the marketplace. Most of the technologies listed in Table 10–3 can be defined by (1) the organic loading criteria, (2) the methods used to condition the influent waste (e.g., dilution, pH adjustment, nutrient addition), (3) the method used to introduce and distribute the influent into the reactor; (4) the method used to contact the waste to be treated with the biomass, (5) the method used to retain and separate the anaerobic biomass, a key factor in a successful operation of an anaerobic treatment process, (6) the characteristics of the biological reactor, (7) the gas management system, and (8) the ultimate management of the residual solids. Because so many new processes are coming on the market it is important to review the current literature. Pilot-plant testing is recommended for new designs and for wastewater with no full-scale treatment experience. Fundamental considerations for anaerobic treatment are presented and discussed in the following section.

Application of Anaerobic Technologies

The technologies described in Table 10–3 are used for (1) the treatment of high-strength wastes from a variety of industries, (2) the pretreatment of high-strength wastes, (3) the treatment of domestic wastewater in combination with other aerobic processes, and (4) treatment of domestic wastewater. Of these applications, most of the activity has been for the treatment of high-strength and specialized industrial wastes, but there are also a number of installations for anaerobic treatment of domestic wastewater.

Applications for High Strength Wastes.   In the last 25 to 30 y, the number of industrial anaerobic installations worldwide has increased by nearly an order-of-magnitude and is as of 2013 close to 4750 (see Table 10–4). A representative listing of the industries in which anaerobic technologies are now used is presented in Table 10–5. Different types of anaerobic processes may be employed for the same type of industrial wastewater as indicated in Table 10–6. Because of their relatively small footprint, the EGSB and

Table 10–4

Estimated number of anaerobic process installations by type as of mid 2013[a] (exclusive of conventional anaerobic digesters used for wastewater sludges)

Process category	Installation
Anaerobic lagoon	>50,000
UASB	2000
EGSB	1500
Anaerobic contact process	500
Anaerobic filters (Upflow and downflow)	250
Anaerobic hybrid process	200
Anaerobic membrane process	50
Other	250

[a] Adapted from Totzke (2012).

IC processes have become quite popular for the treatment of industrial wastewaters. Typical views of EGSB and IC reactors are shown on Fig. 10–3. The use of UASB process for the treatment of high strength waste and domestic wastewater is considered below.

To meet current water quality requirements for most effluent discharges, an anaerobic process used for the treatment of high strength wastes may need to be followed by an aerobic process. A variety of aerobic processes have been used for post treatment of effluent from high rate anaerobic industrial pretreatment processes. These include conventional or biological nutrient removal activated sludge, sequencing batch reactors. trickling filters, biological aerated filters, rotating biological contactors, and wetlands (Chong et al., 2012). An example is shown on Fig. 10–4 in which a UASB reactor, located at a municipal wastewater treatment, is used to treat high strength food processing wastewater from a dedicated pipeline from the industry. The UASB effluent is directed to a biological nutrient removal activated sludge process receiving the municipal wastewater. The residual volatile fatty acids in the UASB effluent are expected to help the performance of the enhanced biological phosphorus removal process in the municipal plant.

Applications for Domestic Wastewater. In many parts of the world, especially in less developed countries with warm climates, anaerobic treatment may be the

Table 10–5

Representative examples of wastewater sources treated by anaerobic processes

Food and brewage industry	
Alcohol distillation	Slaughterhouse and meatpacking
Breweries	Soft drink beverages
Dairy and cheese processing	Starch production
Food processing	Sugar processing
Fish and seafood processing	Vegetable processing
Fruit processing	

Other applications	
Chemical manufacturing	Landfill leachate
Contaminated groundwater	Pharmaceuticals
Domestic wastewater	Pulp and paper

Table 10–6

Relative use of UASB, EGSB, and anaerobic contact process technologies in various applications[a]

Industry	UASB	EGSB	Anaerobic contact
Food and beverage industry			
Breweries	305	210	1
Candy	22	13	
Dairy and cheese processing	36	16	14
Food processing	61	29	8
Fruit	18	29	3
Meat/poultry/fish processing	8	1	11
Soft drink beverage	253	97	49
Starch production	59	30	13
Sugar processing	55	18	78
Vegetable processing	108	63	12
Yeast production	26	37	9
Other applications			
Chemical manufacturing	39	87	9
Pulp and paper	101	225	37
Miscellaneous	95	29	15
Total	1186	884	259

[a] Adapted from Totzke (2012).

most attractive option for domestic wastewater treatment. The USAB process is the most common anaerobic process used for treatment of domestic wastewater. Following its development for the treatment of high-strength wastewaters, it was successfully demonstrated for treating domestic wastewater at 25°C in a tropical climate during the early 1980s in Columbia (Gomec, 2010). The first full-scale facility was installed in

(a)

(b)

(c)

Figure 10–3

Various anaerobic reactors used for the treatment of high strength industrial wastewaters: (a) view of expanded granular sludge blanket (EGSB) treatment system. From right to left, the three tanks are (i) conditioning tank, (ii) EGSB reactor [see Table 10-3(c)], and (iii) sulfide oxidation tank (courtesy of Robert Pharmer of Pharmer Engineering); (b) view of EGSB reactor at cheese factory; and (c) view of internal circulation UASB [see Table 10-3(d)] (courtesy of Paques, BV).

Figure 10–4

Installation of upflow anaerobic sludge blanket (UASB) reactor for the pretreatment of industrial waste: (a) schematic of settling unit in UASB reactor (adapted from Biothane, BV), (b) view of settler as delivered to the construction site, (c) view of top of settler (Note the effluent weir on the right hand side of the settler, and (d) settler in position to be placed in UASB reactor. (Photographs courtesy of City of Yakima, WA.)

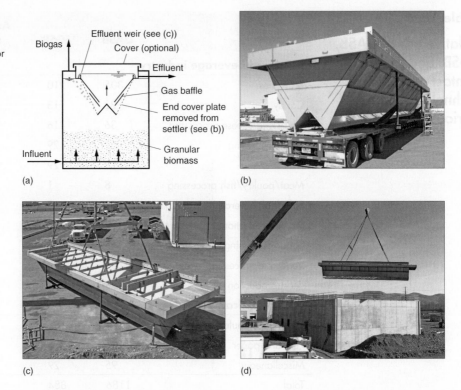

1989 in Kanpur, India, and is still in operation with a treatment capacity of 5000 m³/d. As of 2006, Aiyuk et al. (2006) reported that over 200 facilities were in operation worldwide. Many of these are in warm climates, but the UASB process had been applied for the treatment of domestic wastewaters at temperatures as low as 10°C. In a summary of UASB installations by Gomec (2010), 9 out of 35 facilities were operating at temperatures below 15°C. The hydraulic retention times in 75 percent of these facilities ranged from 2 to 10 h. A low energy wastewater treatment system employing a UASB reactor with a trickling filter treating the UASB effluent is shown on Fig. 10–5. The trickling filter effluent is used for crop irrigation following sedimentation. In the future, it is anticipated that many more combined anaerobic/aerobic treatment systems will be used and new ones developed, based on energy concerns and evolving treatment objectives.

The high rate anaerobic treatment process can be an attractive alternative for domestic wastewater treatment because of the relatively small size and low construction costs, low excess sludge production, low energy demand, and potential for biogas recovery. For countries with low water use and more concentrated wastewater, the advantages for the process increase. However, where a secondary effluent treatment quality is required, the anaerobic treatment process must be followed by an aerobic system for post treatment to further remove colloids, suspended solids and soluble BOD (Lew et al., 2002). Another emerging alternative is the use of synthetic membranes with a suspended growth anaerobic treatment process operated with a long SRT [see Table 10–3(i)]. The membrane separation can result in greater than 99 percent suspended solids removal to meet secondary treatment levels (Visvanathan and Abeynayaka, 2012). A number of processes that may be used after anaerobic treatment for nutrient removal, such as the Anammox process, are currently under investigation.

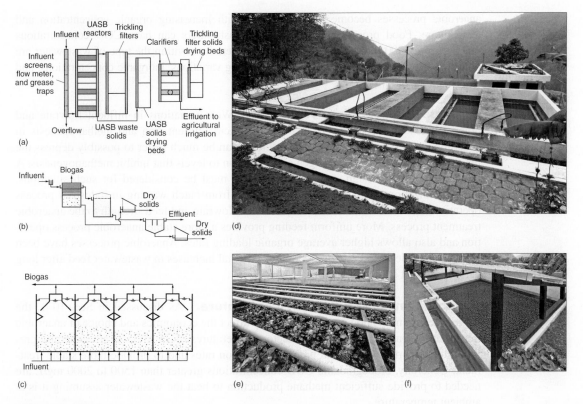

Figure 10–5

Combined UASB/trickling filter treatment process (a) schematic of the treatment system, (b) schematic of flow diagram, (c) cross-section through UASB reactor, (d) view of UASB reactor with four compartments, (e) view of gravity fed trickling filter, and (f) view of sludge drying beds. Settled trickling filter effluent is used to irrigated agricultural fields. (Photographs courtesy of S. Oakley and H. Leverenz, Coordinates: 14.7722 N, 91.1917 W; another similar plant can be seen at coordinates: W 14.7646 N, 91.1797 W.)

10–4 FUNDAMENTAL CONSIDERATIONS IN THE APPLICATION OF ANAEROBIC TREATMENT PROCESSES

The type of wastewater and its characteristics are important in the evaluation, design, and implementation of anaerobic processes. The wastewater characteristics affect the economics of selecting an anaerobic process over an aerobic treatment process, the type of anaerobic treatment process preferred, and the operational costs and concerns for using an anaerobic treatment process. The focus of the discussion in this section is on the effect of the wastewater characteristics on (1) important anaerobic process design issues, (2) the need for pretreatment and alkalinity and/or nutrient addition, and (3) the gas production and amount of energy that can be gained by treating the wastewater with an anaerobic treatment process. The topics discussed apply to the granular sludge blanket, suspended growth, attached growth, and membrane separation anaerobic processes presented in Table 10–3.

Characteristics of the Wastewater

As shown in Table 10–5, a wide variety of wastewaters have been treated by anaerobic processes. Due to the energy savings by eliminating aeration and minimal sludge production, anaerobic processes become more attractive with increasing organic concentration and temperature. Food processing and distillery wastewaters can have COD concentrations ranging from 3000 to 30,000 g/m³. Other considerations within the type of wastewater are potential toxic streams, daily flowrate and loading variations, inorganic concentrations, and seasonal load variations.

Flowrate and Loading Variations. Wide variations in influent flowrate and organic loads can upset the balance between acid fermentation and methanogenesis in anaerobic processes. The acidogenic reactions can be much faster to possibly depress the pH and increase VFA and hydrogen concentration to levels that inhibit methanogenesis. A more conservative design or flow equalization must be considered for such situations. Flow equalization tanks can store wastewaters from batch wasting in industrial process operations and then allow a more uniform feed flowrate and feed strength to the anaerobic treatment process. More uniform feeding provides a more stable anaerobic process operation and also allows higher average organic loading rates. Anaerobic processes have been shown to be able to respond quickly to incremental increases in wastewater feed after long periods without substrate addition.

Organic Concentration and Temperature. As discussed in Sec. 10–1, the wastewater strength and temperature greatly affect the economics and choice of anaerobic treatment over aerobic treatment. Reactor temperatures of 25 to 35°C are generally preferred to support more optimal biological reaction rates and to provide more stable treatment. Generally, biodegradable COD concentrations greater than 1500 to 2000 mg/L are needed to provide sufficient methane production to heat the wastewater assuming it is at ambient temperature.

Anaerobic treatment can be applied at lower temperatures and has been sustained at 10 to 20°C in granular sludge blanket, suspended growth, and attached growth reactors. At the lower temperatures slower reaction rates occur and longer SRTs, larger reactor volumes, and lower organic COD loadings are needed (Banik et al., 1996; Collins et al., 1998; and Alvarez et al., 2008).

When higher SRTs are needed, the solids loss in an anaerobic reactor can become a critical limiting factor. Anaerobic reactors generally produce more dispersed, less flocculent solids than aerobic systems with effluent TSS concentrations for suspended growth processes in the 100 mg/L range.

For dilute wastewaters, the effluent TSS concentration will limit the possible SRT of the process and treatment potential. Either a lower treatment performance occurs or it is necessary to operate the reactor at a higher temperature. Thus, the method to retain solids in the anaerobic reactor is important in the overall process design and performance. The maximum SRT possible is when biomass lost via the effluent is equal to the solids produced, which can be calculated from Eq. 8–20.

$$Q(\text{VSSe}) = \frac{QY_H(b\text{CODr})}{1 + b_H(\text{SRT})} + \frac{(f_d)(b_H)QY_H(b\text{CODr})\text{SRT}}{1 + b_H(\text{SRT})} \tag{10-1}$$

Where VSSe = effluent volatile suspended solids, g/m³

bCODr = COD degraded in anaerobic reactor, g/m³

Solving for SRT as a function of effluent volatile suspended solids concentration, biodegradable COD removed, and growth and decay coefficients yields

$$\text{SRT} = \frac{Y_H(b\text{CODr}) - \text{VSSe}}{(b_H)(\text{VSSe}) - (f_d)(b_H)(Y_H)(b\text{CODr})} \tag{10-2}$$

Figure 10–6

Estimate of biomass SRT in an anaerobic reactor as a function of the amount of COD degraded and effluent biomass VSS concentration. (Based on assumptions of a synthesis yield coefficient of 0.08 g VSS/g CODr and specific endogenous decay rate coefficient of 0.03 g VSS/g VSS·d from Table 10–13 and an f_d value of 0.10.)

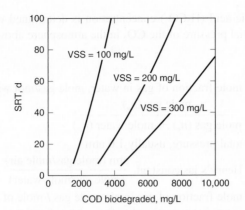

The maximum SRT as a function of bCOD removed and effluent VSS concentration is shown on Fig. 10–6. To achieve an SRT of 40 d, which would be desirable at a temperature of about 30°C, the amount of COD that must be degraded increases from about 2400 mg/L for an effluent VSS concentration of 100 mg/L to about 7400 mg/L if the effluent VSS concentration is 300 mg/L. For operation at lower temperatures much higher SRT values are needed, which requires a very low effluent VSS concentrations for weaker wastewaters or a higher biodegradable COD in the influent.

Fraction of Non-Dissolved Organic Matter.   The composition of the wastewater in terms of its particulate and soluble fractions affects the type of anaerobic reactor selected and its design. Wastewaters with high solids concentrations are treated more appropriately in suspended growth and UASB reactors than by upflow or downflow attached growth processes. Where more particulate conversion is required, longer SRT values may be needed if solids hydrolysis is the rate limiting step as compared to acid fermentation or methanogenesis in anaerobic treatment. In such cases, it may be appropriate to use a two phase anaerobic treatment process with hydrolysis and acid fermentation in a sludge bed or stirred reactor, followed by UASB reactor or other type of reactor for methanogenesis (Shuizhou and Zhou, 2005; Alvarez et al., 2008).

Wastewater Alkalinity.   During anaerobic treatment gas bubbles containing a high CO_2 content (25 to 35 percent) are produced in the liquid from fermentation reactions, which results in an elevated dissolved CO_2 concentration in the liquid. A high alkalinity concentration, in the range from 2000 to 4000 mg/L as $CaCO_3$, is typically needed to offset the dissolved carbonic acid and maintain the pH at or near neutral. For wastewaters low in alkalinity, proteins or amino acids can be metabolized and deaminated during anaerobic treatment to produce alkalinity as $NH_4(HCO_3)$. Where lacking, the purchase of alkalinity for pH control is necessary, which can have a significant impact on the economics of anaerobic treatment.

The relationship between pH and alkalinity as outlined in Appendix F is controlled by the bicarbonate chemistry as follows:

$$\frac{[HCO_3^-][H^+]}{[H_2CO_3]} = K_{a1} \tag{10–3}$$

where K_{a1} = first acid dissociation constant, which is a function of ionic strength and temperature

The carbonic acid (H_2CO_3) concentration is determined using Henry's law [Eq. (2–46)] and the partial pressure of the CO_2 in the atmosphere above the water.

$$x_g = \frac{P_T}{H} p_g \qquad (2\text{–}46)$$

where x_g = mole fraction of gas in water, mole gas/mol water

$$= \frac{\text{mole gas } (n_g)}{\text{mole gas } (n_g) + \text{mole water } (n_w)}$$

P_T = total pressure, usually 1.0 atm

H = Henry's las constant, $\dfrac{\text{atm (mole gas/mole air)}}{\text{(mole gas/mole water)}}$

P_g = mole fraction of gas in air, mole gas / mole of air (Note: The mole fraction of a gas is proportional to the volume fraction)

Once the carbonic acid concentration is known, the bicarbonate (HCO_3^-) alkalinity needed to maintain the required pH can be estimated. The use of the above equations is illustrated in Example 10–1.

EXAMPLE 10–1   **Alkalinity and pH in Anaerobic Process**   Determine the alkalinity required in kg $CaCO_3$/d to maintain a pH value of 7.0 in an anaerobic suspended growth process at 35°C, with a 30 percent CO_2 content in the gas phase. The influent wastewater flowrate is 2000 m^3/d and the alkalinity is 400 mg/L as $CaCO_3$. At 35°C, the Henry's constant for CO_2, computed using Eq. (2–28) and the data given in Table 2–8, is 2092 atm and the value of K_{a1} is 4.85×10^{-7} (see Appendix F).

Solution
1. Determine the concentration of HCO_3^- required to maintain the pH at or near a value of 7.0.

   a. Determine the concentration of H_2CO_3 using Eq. (2–46)

   $$x_{H_2CO_3} = \frac{P_T}{H} P_g = \frac{(1 \text{ atm})(0.30)}{2092 \text{ atm}} = 1.434 \times 10^{-4}$$

   Because one liter of water contains 55.6 moles [1000 g/(18 g/mole)], the mole fraction of H_2CO_3 is equal to

   $$x_{H_2CO_3} = \frac{\text{mole gas } (n_g)}{\text{mole gas } (n_g) + \text{mole water } (n_w)}$$

   $$1.434 \times 10^{-4} = \frac{[H_2CO_3]}{[H_2CO_3] + (55.6 \text{ mole/L})}$$

   Because the number of moles of dissolved gas in a liter of water is much less than the number of moles of water,

   $$[H_2CO_3] \approx (1.434 \times 10^{-4})(55.6 \text{ mole/L}) \approx 7.97 \times 10^{-3} \text{ mole/L}$$

   b. Determine the concentration of HCO_3^- required to maintain the pH at or near a value of 7.0 using Eq. (10–1)

   $$[HCO_3^-] = \frac{(4.85 \times 10^{-7})(7.97 \times 10^{-3} \text{ mole/L})}{(10^{-7} \text{ mole/L})}$$

   $$= 0.03863 \text{ mole/L}$$

   $$HCO_3^- = 0.03863 \text{ mole/L } (61 \text{ g/mole})(10^3 \text{ mg/1 g}) = 2356 \text{ mg/L}$$

2. Determine the amount of alkalinity required per day

$$\text{Equivalents of } HCO_3^- = \frac{(2356 \text{ g/L})}{(61 \text{ g/eq})} = 0.03863$$

$$1 \text{ eq. } CaCO_3 = \frac{\text{m.w.}}{2} = \frac{(100 \text{ g/mole})}{2} = 50 \text{ g } CaCO_3/\text{eq}$$

$$\text{Alkalinity as } CaCO_3 = (0.03863 \text{ eq/L}) (50 \text{ g/eq})(10^3 \text{ mg/1 g})$$

$$= 1931 \text{ mg/L as } CaCO_3$$

$$\text{Alkalinity needed} = (1931 - 400) \text{ mg/L}$$

$$= 1531 \text{ mg/L as } CaCO_3$$

Daily alkalinity addition:

$$= (1531 \text{ g/m}^3) (2000 \text{ m}^3/\text{d}) (1 \text{ kg}/10^3 \text{ g})$$

$$= 3062 \text{ kg/d}$$

Comment Based on the results of the above analysis, it is clear that the quantity of alkalinity required can be significant, and, as a consequence, a significant cost can be incurred.

The results of similar calculations to those presented in Example 10–1 for different temperatures and gas phase CO_2 concentrations are reported in Table 10–7. The data presented in Table 10–7 were derived using the bicarbonate dissociation constants given in Table F–2 in Appendix F and Henry's constants derived from the data given in Table 2–8 in Chap. 2. The values presented in Table 10–7 can be used to estimate the alkalinity requirements. For wastewaters with a higher total dissolved solids concentration and ionic strength, the alkalinity requirements will generally be much greater.

Nutrients. Though anaerobic processes produce less sludge and thus require less nitrogen and phosphorus for biomass growth, many industrial wastewaters may lack sufficient nutrients. Thus, the addition of nitrogen and/or phosphorus may be needed.

Macronutrients. The presence of the iron, nickel, cobalt, and molybdenum at trace concentrations are needed for the growth of methanogenic bacteria in anaerobic processes (Demirel and Scherer, 2011). The addition of trace metals has been shown in a number of cases to increase COD removal efficiency in anaerobic processes including granular

Table 10–7

Estimated minimum alkalinity as CaCO₃ required to maintain a pH of 7.0 as a function of temperature and percent carbon dioxide during anaerobic digestion

Temperature, °C	Gas phase CO₂, %			
	25	**30**	**35**	**40**
20	2040	2449	2857	3265
25	1913	2295	2678	3061
30	1761	2113	2465	2817
35	1609	1931	2253	2575
40	1476	1771	2066	2362

Table 10–8

Toxic and inhibitory inorganic compounds and concentrations harmful to methanogenesis in anaerobic processes[a]

Substance	Moderately inhibitory concentration, mg/L	Strongly inhibitory concentration, mg/L
Na^+	3500–5500	8000
K^+	2500–4500	12,000
Ca^{2+}	2500–4500	8000
Mg^{2+}	1000–1500	3000
Ammonia-nitrogen	1500–3000	3000
Sulfide, S^{2-}	200	200
Copper, Cu		0.5 (soluble)
		50–70 (total)
Chromium, Cr(VI)		3.0 (soluble)
		200–250 (total)
Chromium, Cr(III)		180–420 (total)
		2.0 (soluble)
Nickel, Ni		30.0 (total)
Zinc, Zn		1.0 (soluble)

[a] From Parkin and Owen (1986).

sludge reactors (Osuna et al., 2003; Fermoso et al., 2008) and a suspended growth process for the digestion of food waste (Evans et al., 2012). Digester studies by Takashima et al. (2011) led to a recommendation of having the following ratios available for iron, nickel, cobalt, and zinc in mg/g COD removed, respectively for efficient anaerobic degradation: 0.20, 0.0063, 0.017, and 0.049 in a mesophilic system and 0.45, 0.049, 0.054, and 0.24 in a thermophilic system.

The exact amounts of trace metals needed can vary for different wastewaters and, thus, successive trials are used to assess their benefit for high rate anaerobic processes.

Inorganic and Organic Toxic Compounds. Proper waste analysis and treatability studies are needed to assure that a chronic or serious transient toxicity does not exist for wastewater treated by anaerobic processes. At the same time, the presence of a toxic substance does not mean the process cannot function. Some toxic compounds inhibit anaerobic methanogenic reaction rates, but with a high biomass inventory and low enough loading, the process can be sustained. Toxic and inhibitory inorganic and organic compounds of concern for anaerobic processes are presented in Tables 10–8 and 10–9, respectively. Acclimation to toxic concentrations has also been shown (Speece, 1996) but it may be necessary to apply pretreatment steps to prevent toxicity problems in the anaerobic degradation process.

Pretreatment of Wastewater

Wastewater pretreatment requirements depend on the waste source, the type of anaerobic treatment process employed, and the need to prevent an anaerobic treatment process failure or unstable operation. Pretreatment considerations include screening, solids conditioning or reduction, pH and temperature adjustment, nutrient addition, and fats, oil, and grease (FOG) control, and toxicity reduction.

Table 10–9

Toxic and inhibitory organic compounds and concentrations harmful to methanogenesis in anaerobic processes[a]

Compound	Concentration resulting in 50% activity, mM
1-Chloropropene	0.1
Nitrobenrene	0.1
Acrolein	0.2
1-Chloropropane	1.9
Formaldehyde	2.4
Lauric acid	2.6
Ethyl benzene	3.2
Acrylonitrile	4
3-Chlorol- 1, 2-propandiol	6
Crotonaldehyde	6.5
2-Chloropropionic acid	8
Vinyl acetate	8
Acetaldehyde	10
Ethyl acetate	11
Acrylic acid	12
Catechol	24
Phenol	26
Aniline	26
Resorcinol	29
Propanol	90

[a] From Parkin and Owen (1986).

Screening. Some form of screening is normally used to removed objectionable material that could cause interference with the flow distribution in a granular sludge reactor, mixing problems in suspended growth reactors, or plugging of attached growth reactors. Fine screening, in the range of 2 to 3 mm, should be considered for anaerobic membrane reactors to prevent membrane fouling problems.

Solids Conditioning or Reduction. A solids conditioning pretreatment step may be considered in the processing of wastes that are high in solids content and/or lignin material, such as agriculture waste or certain pulp and paper mill streams, to enhance methane production and anaerobic degradation reaction rates. These may include mechanical, chemical, thermal, and biological processes or a combination of them (Sambusiti et al., 2013). Solids conditioning technology is addressed in Chap. 13 for anaerobic digestion of wastewater treatment plant sludges.

Solids reduction by a two-step process with solids removal or solids removal and hydrolysis in the first step before downstream granular sludge and attached growth anaerobic processes can be beneficial to COD removal performance and operational stability. High suspended solids concentration in the influent to a granular sludge anaerobic process may cause clogging and channeling in the sludge blanket to reduce treatment effectiveness. In addition the adsorption and incorporation of suspended and colloidal solids may impair the sludge granulation process and the density of the granules. Minimal influent solids concentrations are desired for the operation of packed bed anaerobic treatment processes to prevent plugging and channeling of the wastewater flow. One option is to provide gravity settling for solids removal and thickening with the settled solids treated in a separate sludge digester and another option is to provide solids reduction by anaerobic hydrolysis in the first step.

A two-step processes with solids contact and hydrolysis in the first step was proposed by van Haandel and Lettinga (1994) and has been used for both industrial and domestic wastewater treatment with downstream UASB, EGSB, and packed bed anaerobic processes (Seghezzo et al., 1998). A hydrolysis upflow sludge blanket (HUSB) reactor or alternatively referred to as an upflow anaerobic solids removal (UASR) reactor provides solids capture and hydrolysis prior to a downstream anaerobic process. It has been demonstrated in domestic wastewater treatment applications at temperatures from 14 to 26°C and with hydraulic retention times from 3 to 10 h (Alvarez et al., 2008; Zeeman et al., 1997). A greater solids accumulation occurs under cold temperature operation due to the slower hydrolysis rates and may require solids wasting and further treatment elsewhere.

pH Adjustment. The ability to operate the anaerobic process with little variation in the reactor pH and temperature leads to a more stable operation and better process efficiency. Based on knowledge of the wastewater characteristics and the anaerobic reactor operating conditions the amount of alkalinity, if needed, can be determined and provided in the operational protocols. For more dilute wastewaters with lower gas production rates relative to the wastewater flowrates, the percent CO_2 can be minimal compared to the amount when treating a wastewaters with high COD concentrations, such that the alkalinity demands are none or modest. The influent alkalinity should be controlled to maintain reactor operating pH values between 6.8 and 7.8 for which stable methanogenic activity occurs (Leitão et al., 2006).

Temperature Adjustment. Operation at constant temperature provides a more stable and more efficient process performance. A decrease in the reactor temperature is more detrimental than an increase. A sudden temperature drop of 1 to 2°C results in a slower acetate uptake rate by the methanogenic bacteria and accumulation of VFAs. Depending on the system alkalinity and buffer capacity, this could result in a pH decrease that could further slow down the methanogenic activity leading to a path of digester instability and potential failure if not adjusted soon enough. Sudden temperature drops can also affect the integrity of granular sludge in UASB, EGSB, and IC reactors.

Nutrient Addition. Many high strength wastewaters, such as those from food processing, brewery, beverage, and distillery operations, will require the addition of nitrogen and phosphorus to support growth of the anaerobic bacteria. The amount needed is higher during start up due to more rapid growth and less supply of nutrients within the reactor by endogenous decay. An influent COD:N:P ratio of 600:5:1 is recommended during start up and 300:5:1 during long term operation (Annachhatre, 1996).

Fats, Oil, and Grease (FOG) Control. The presence of FOG in wastewaters fed to anaerobic processes can be of concern for two reasons: (1) inhibition of methanogenesis due to inhibition by long chain fatty acids (LCFAs) and (2) sludge flotation due to the hydrophobicity and lower density of FOG components. The inhibition is thought to be due to sorption of LCFAs on the methanogenic organism's cell wall and membranes to interfere with cell substrate transport functions (Hanaki et al., 1981). Further, the presence of FOG can also be detrimental to the integrity of granular sludge particles in UASB and EGSB reactors and can cause fouling of the synthetic membranes used in anaerobic membrane reactors.

Anaerobic biodegradation of FOG has been demonstrated in the treatment of rendering waste with sustained FOG loadings of less than 1.0 kg FOG/m³·d, but unacceptable solids loss occurred due to flotation at higher loadings (Jeganathan et al., 2006). Upon initial exposure, FOG inhibition of methanogenic activity occurs at much lower loadings until FOG degradation occurs following an acclimation period (Evans et al., 2012). Thus, intermittent or variable FOG loadings can cause methanogenic inhibition and unstable operation of the anaerobic process. Source control and dissolved air flotation pretreatment must be considered for wastewater with high FOG concentration.

Toxicity Reduction. As shown in Tables 10–8 and 10–9 a wide range of inorganic and organic substances can be toxic to anaerobic processes, including certain heavy metals, high dissolved solids, chlorinate organic compounds, high nitrogen concentration from ammonia, amino acids, and or proteins, and industrial chemical products. Control of toxicity for anaerobic processes requires a careful evaluation of the wastewater and its sources. Source control is a first step and where necessary pretreatment steps for toxicity reduction may include for example dilution, air stripping, and chemical precipitation processes. Air stripping was found for example to be effective for the reduction of high ammonia concentration toxicity in piggery wastes (Zhang and Jahng, 2010).

A two step process in the anaerobic treatment system may be used to remove or biodegrade toxic substances before a final methanogenic treatment step. A first step solids removal/hydrolysis or acid phase anaerobic treatment may remove the toxin to a sufficient level before exposure of the more sensitive methanogenic bacteria to the toxic compound (Lettinga and Hulshoff Pol, 1991).

Expected Gas Production

Higher strength wastewaters will produce a greater amount of methane per volume of liquid treated to provide a relatively higher amount of energy to raise the liquid temperature if needed. Gas composition and volume relationships for methane are discussed below.

Gas Composition. Anaerobic degradation of organic substances results in the production of a gaseous product with the conversion of carbon to methane (CH_4) and carbon dioxide (CO_2), nitrogen to ammonia (NH_3), and sulfur to hydrogen sulfide (H_2S). The energy value of the gas produced is proportional to its methane content. Buswell and Boruff (1932) were the first to recognize that the composition of the gaseous product is a function of the organic compound type and composition. Buswell and Mueller (1952) later developed a molar stoichiometric relationship between the carbon, hydrogen, and oxygen in an organic compound to the volume of methane, and carbon dioxide produced in anaerobic degradation. Their relationship was further modified to include the organic compound nitrogen and sulfur content and the volume of ammonia and hydrogen sulfide produced (Parkin and Owen, 1986; Tchobanoglous et al., 2003):

$$C_v H_w O_x N_y S_z + \left(v - \frac{w}{4} - \frac{x}{2} + \frac{3y}{4} + \frac{z}{2}\right)H_2O \rightarrow$$

$$\left(\frac{v}{2} + \frac{w}{8} - \frac{x}{4} - \frac{3y}{8} - \frac{z}{4}\right)CH_4 + \left(\frac{v}{2} - \frac{w}{8} + \frac{x}{4} + \frac{3y}{8} + \frac{z}{4}\right)CO_2 + yNH_3 + zH_2S$$

(10–4)

The gaseous ammonia (NH_3) that is formed will react with the carbon dioxide to form the ammonium ion and bicarbonate according to the following relationship.

$$NH_3 + H_2O + CO_2 \rightarrow NH_4^+ + HCO_3^- \tag{10-5}$$

The reaction given by Eq. (10–5) is representative of the formation of alkalinity under anaerobic conditions, due to the conversion of organic compounds containing proteins, peptides or amino acids (i.e., nitrogen). The expected mole fractions of methane and carbon dioxide, and hydrogen sulfide are given by the following three expressions, respectively. Ammonia gas production from Eq. (10–4) is not included as most of it will remain in solution as ammonium bicarbonate. In general, the mole fraction of hydrogen sulfide will be somewhat less because of metal complexation/precipitation.

$$f_{CO_2} = \frac{4v - w + 2x + 2z}{8(v + z)} \tag{10-6}$$

$$f_{CH_4} = \frac{4v + w - 2x - 2z}{8(v + z)} \tag{10-7}$$

$$f_{H_2S} = \frac{z}{(v + z)} \tag{10-8}$$

Using approximate formulas for lipid, carbohydrate, and protein compounds of $C_{18}H_{33}O_2$, $C_6H_{10}O_5$, and $C_{11}H_{24}O_5N_4$, the percent methane in the anaerobic process gas calculated from Eq. (10–7) are 70, 50, and 66, respectively. These values compare well to respective values of 70, 50, and 68 percent in studies by Li et al. (2002). For carbohydrate, starch, and FOG wastes, alkalinity will be a problem due to the lack of ammonia production.

Volume of Methane Gas.

As derived by Eq. (7–142) in Sec. 7–14 in Chap. 7, the amount of methane (CH_4) produced from anaerobic oxidation of COD is equal to 0.35 L CH_4/g COD at standard conditions (0°C and 1 atm). The quantity of methane at other than standard conditions is determined by using the universal gas law [Eq. (2–44)] to determine the volume of gas occupied by one mole of CH_4 at the temperature in question.

$$V = \frac{nRT}{P} \tag{2-44}$$

Where V = volume occupied by the gas, L, m³

n = moles of gas, mole

R = universal gas law constant, 0.08205 atm·L/g mole·K

= 0.08205 atm·L/g mole·K

T = temperature, °K (273.15 + °C)

P = absolute pressure, atm

Thus, for 35°C the volume occupied by one mole of CH_4 is:

$$V = \frac{(1 \text{ mole})(0.082056 \text{ atm·L/g mole·K})(273.15 + 35)}{1.0 \text{ atm}} = 25.29 \text{ L/mole}$$

Because the COD of one mole of CH_4 is equal to 64 g, the amount of CH_4 produced per unit of COD converted under anaerobic conditions at 35°C is equal to 0.40 L as determined below.

(25.29 L/mole)/(64 g COD/mole CH_4) = 0.40 L CH_4/g COD

The total gas production rate is commonly estimated by calculating the methane production rate and dividing it by the fraction of methane in the gas phase, typically assuming a value in the range of 60 to 65 percent methane.

Table 10–10

Comparison of energy balance for aerobic and anaerobic processes. Wastewater at 20°C, 100 m³/d and 10,000 g/m³ COD concentration

Energy	Value, kJ/d	
	Anaerobic	**Aerobic**
Aeration		-1.9×10^6
Methane produced	12.5×10^6	
Increase wastewater temperature to 30°C	-5.3×10^6	
Net energy, kJ/d	7.2×10^6	-1.9×10^6

Energy Production Potential

In contrast to aerobic biological treatment very little electrical energy is used for anaerobic biological treatment, being limited to energy for pumping and tank mixing, depending on the anaerobic process. The methane produced in the anaerobic process is often used to heat the wastewater or sludge where required. In some cases for larger installations, the methane may be used in reciprocating engine or microturbine equipment to produce electricity. When the heat generated during electrical generation is captured and used for heating, the energy producing process is called cogeneration.

An energy balance comparison for aerobic and anaerobic treatment of a high strength wastewater at 20°C is presented in Table 10–10. For the conditions given, the aerobic process requires 1.9×10^6 kJ/d of energy. On the other hand, the anaerobic process produces a total of 12.5×10^6 kJ/d. Of the total energy produced anaerobically, about 5.3×10^6 kJ/d is required to raise the wastewater temperature from 20 to 30°C, assuming an 80 percent energy utilization efficiency after accounting for losses in the boiler and heat exchanger and reactor heat losses. Thus, the potential net energy production that can be achieved with anaerobic treatment is on the order of 7.2×10^6 kJ/d, or about 3.8 times the energy required for aerobic treatment. General considerations used for the energy balance are shown as follows.

A large portion of the energy demand for an aerobic activated sludge process is for aeration to supply oxygen and can be calculated as follows:

$$E_{\text{AER}}(\text{kJ/d}) = Q(\text{CODr})(A_n)(3600 \text{ kJ/kWh})/\text{AOTE} \tag{10–9}$$

Where E_{AER} = daily energy demand for oxygen transfer, kJ/d
$\quad\quad\quad Q$ = wastewater flowrate, m³/d
$\quad\quad$ CODr = COD removed by biodegradation, kg/m³
$\quad\quad\quad A_n$ = net oxygen required, kg O_2/kg CODr
$\quad$ AOTE = actual oxygen transfer efficiency, kg O_2/kWh

The value for A_n can be calculated for a given activated sludge design SRT using Eq. (8–67). The anaerobic process produces energy in the form of methane but may use some portion of the methane produced to heat the wastewater to a more optimal operating temperature, typically in the range of 30 to 35°C. The net energy production, accounting for the energy from methane production and energy used for heating is as follows:

$$E_{\text{ANAER}}, \text{kJ/d} = (Q)(\text{CODr})\left(\frac{0.35 \text{ m}^3 \text{ CH}_4}{\text{kg CODr}}\right)\left(\frac{35{,}846 \text{ kJ}}{\text{m}^3 \text{ CH}_4}\right)$$

$$-(Q)(\Delta T)(C_p)\left(\frac{10^3 \text{ kg}}{\text{m}^3 \text{ H}_2\text{O}}\right)\left(\frac{1}{\text{Eff}_{\text{heat}}}\right) \tag{10–10}$$

Where: E_{ANAER} = net energy available, kJ/d

ΔT = temperature increase for influent wastewater, °C

C_p = specific heat of water, 4.2 kJ/°C·kg

Eff_{heat} = fraction of heat available after losses from vessel and heat exchanger

The effect of the wastewater strength on the net energy consumption or production between aerobic and anaerobic treatment is illustrated in Example 10–2. The same assumptions used to generate the comparison in Table 10–10 are used in the example.

EXAMPLE 10–2   Comparison of Energy Consumption and Production in Aerobic and Anaerobic Treatment   Compare the energy needed for aeration for treatment of a high strength wastewater by an activated sludge aerobic treatment process to the net energy production for an anaerobic treatment process. The net energy production for anaerobic treatment accounts for the energy produced by methane generation minus the energy used to heat the wastewater from 20 to 30°C.

a. Express the net energy as a positive or negative production in kJ/m³ of wastewater treated and tabulate the results for aerobic and anaerobic treatment to achieve a COD biodegradation of 3800, 4200, 6000, 8000, and 10,000 mg/L. Include the volume of methane produced in m³/d for a wastewater flowrate of 400 m³/d.

b. Using similar calculations as for the above, prepare a graph that can be used to illustrate the net energy for anaerobic treatment in terms of kJ/m³ for three wastewater conditions that require heating from 25 to 30°C, 20 to 30°C and 10 to 30°C. Use an influent wastewater COD biodegradation range of 1500 to 10,000 mg/L. Where the net energy production is negative limit the Y axis to –5000 kJ/m³.

The following assumptions apply:

1. A_n = 0.80 g O_2/g COD removal
2. Aeration actual oxygen transfer efficiency (OTE) = 1.52 kg O_2/kWh
3. Net heat loss for methane utilization for heating = 20 percent
4. Heat capacity of water = 4.2 kJ/°C·kg
5. Energy content of methane at standard conditions = 38,846 kJ/m³
6. Methane production rate at standard conditions = 0.35 m³ CH_4/kg CODr (ignore COD to biomass, which is about 3 to 4 percent of total COD degraded)

Solution

1. Determine the energy needed for aeration per unit flow in the aerobic treatment process using Eq. 10–9 for a CODr of 3800 mg/L (3.8 kg/m³).

$$\frac{E_{AER}}{Q} = (CODr)(A_n)(3600 \text{ kJ/kWh})/AOTE$$

$$\frac{E_{AER}}{Q} = (3.8 \text{ kg COD/m}^3)(0.8 \text{ kg O}_2/\text{kg COD})\left[\frac{1}{(1.52 \text{ kg O}_2/\text{kWh})}\right](3600 \text{ kJ/kWh})$$

$$= 7200 \text{ kJ/m}^3$$

The aeration energy needed for the other CODr removal levels is directly proportional to the COD concentration and is tabulated in the summary table in Step 4a.

2. Determine the net energy produced per unit flow from methane production and wastewater heating using Eq. (10–10) for a CODr = 3800 mg/L for heating the wastewater from 20 to 30°C (ΔT = 10°C). With a 20 percent heat loss, the methane utilization efficiency for heating is 80 percent.

a. Net energy for a CODr = 3800 mg/L

$$\frac{E_{ANAER}}{Q} = (3.8 \text{ kg COD/m}^3)(0.35 \text{ m}^3 \text{ CH}_4\text{/kg COD})(35,846 \text{ kJ/m}^3 \text{ CH}_4)$$

$$-(10°C)(4.2 \text{ kJ/°C·kg})(10^3 \text{ kg/m}^3 \text{ H}_2\text{O})\left(\frac{1}{0.80}\right)$$

$$= -4825 \text{ kJ/m}^3$$

b. Net energy for a CODr = 4200 mg/L

$$\frac{E_{ANAER}}{Q} = (4.2 \text{ kg COD/m}^3)(0.35 \text{ m}^3 \text{ CH}_4\text{/kg COD})(35,846 \text{ kJ/m}^3 \text{ CH}_4)$$

$$-(10°C)(4.2 \text{ kJ/°C·kg})(10^3 \text{ kg/m}^3 \text{ H}_2\text{O})\left(\frac{1}{0.80}\right)$$

$$= 194 \text{ kJ/m}^3$$

Similarly for a CODr of 6000, 8000 and 10,000 mg/L, the specific net energy production is 22,777, 47,869, and 72,961 kJ/m³, respectively.

3. Determine the volume of methane produced for a flowrate of 400 m³/d as a function of temperature using the ideal gas law.

a. Methane volume for a CODr = 3800 mg/L at a temperature of 30°C
From the ideal gas law, $V_2 = (V_1/T_1) T_2$

$$\text{CH}_4 \text{ production at } 30°C = (\text{CODr})Q(0.35 \text{ m}^3 \text{ CH}_4\text{/kg COD})\left(\frac{273.15°C + 30°C}{273.15°C}\right)$$

$$= (3.8 \text{ kg/m}^3)(400 \text{ m}^3\text{/d})(0.35 \text{ m}^3 \text{ CH}_4\text{/kg COD})(1.1098)$$

$$= 590.4 \text{ m}^3 \text{ CH}_4\text{/d}$$

b. Methane volume for a CODr = 4200 mg/L at a temperature of 30°C

$$\text{CH}_4 \text{ production at } 30°C = (\text{CODr})Q(0.35 \text{ m}^3 \text{ CH}_4\text{/kg COD})\left(\frac{273.15°C + 30°C}{273.15°C}\right)$$

$$= (4.2 \text{ kg/m}^3)(400 \text{ m}^3\text{/d})(0.35 \text{ m}^3 \text{ CH}_4\text{/kg COD})(1.1098)$$

$$= 652.6 \text{ m}^3 \text{ CH}_4\text{/d}$$

Similarly for a CODr of 6000, 8000 and 10,000 mg/L, methane production is 930, 1240, and 1550 m³/d, respectively.

4. Prepare a summary table and a graphic plot to illustrate the results.

a. Prepare summary table

CODr, mg/L	Aeration energy, kJ/m³	Net anaerobic treatment energy, kJ/m³	Methane production, m³/d
3800	−7200	−4830	590
4200	−7960	190	650
6000	−11,370	22,780	930
8000	−15,160	47,870	1240
10,000	−18,950	72,960	1550

b. Similar calculations used in Step 2a were prepared for $\Delta T = 5°C$ and 10°C. The net energy produced from an anaerobic process as a function of the wastewater temperature increase needed to bring the temperature up to 30°C and the amount of COD degraded is shown on the following plot.

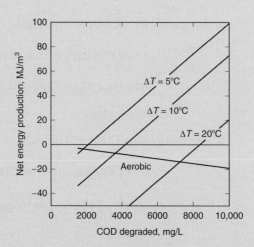

Comment The wastewater strength and temperature is important for comparing energy balances for aerobic and anaerobic processes, where the wastewater temperature must be increased. Based on the results presented in the summary table, both the aerobic and anaerobic processes require energy input for the degradation of a wastewater with a COD of 3800 mg/L COD. It can be shown that if the amount of COD degraded is 3640 mg/L. the energy input is equal for both systems. At lower COD concentrations, the aerobic process requires less energy and the anaerobic process has less methane production.

The effect of the influent wastewater temperature and the amount of COD that can be degraded on the net energy produced by anaerobic treatment is shown in the above figure. A net positive energy production for the anaerobic process requires biodegrading 2100, 4200, and 8400 mg/L COD if it is necessary to raise the influent wastewater temperature 5, 10, and 20°C, respectively. However, heat recovery from the anaerobic effluent stream can modify these values. Even at a break even net energy production, the lower biomass yield discussed below is still a major advantage offered by anaerobic treatment. Consideration should also be given to whether alkalinity addition is necessary for the anaerobic process.

Sulfide Production

Oxidized sulfur compounds, such as sulfate, sulfite, and thiosulfate, may be present in significant concentrations in various industrial wastewaters and to some degree in municipal wastewaters. These compounds can serve as electron acceptors for sulfate-reducing bacteria, which consume organic compounds in the anaerobic reactor and produce hydrogen sulfide (H_2S). For example, using methanol as the electron donor and an f_s value of 0.05 (see Sec. 7–4 in Chap. 7), the overall reaction for the reduction of sulfate to H_2S can be illustrated by the following expression:

$$0.119\ SO_4^{2-} + 0.167CH_3OH + 0.010CO_2 + 0.003NH_4^+ + 0.003\ HCO_3^- +$$

$$0.178H^+ = 0.003C_5H_7NO_2 + 0.060H_2S + 0.060HS^- + 0.331H_2O \qquad (10\text{–}11)$$

From Eq. (10–11), the amount of COD used for sulfate reduction is 0.89 g COD/g sulfate which is higher than the reported value of 0.67 g COD/g sulfate reduced (Arceivala, 1998) and is due to the lower biomass yield coefficient associated with methanol oxidation. Based on the following stoichiometry for H_2S oxidation, 2 moles of oxygen are required per mole of H_2S, as was the case for methane oxidation,

$$H_2S + 2O_2 \rightarrow H_2SO_4 \qquad (10\text{–}12)$$

Thus, the amount of H_2S produced per unit COD is the same as that for methane (0.40 L H_2S/g COD used at 35°C).

Hydrogen sulfide is malodorous and corrosive to metals. In contrast to methane, H_2S is highly soluble in water with a solubility of 2650 mg/L at 35°C, for example.

The concentration of oxidized sulfur compounds in the influent wastewater to an anaerobic treatment process is important as high concentrations can have a negative effect on anaerobic treatment. Sulfate reducing bacteria compete with the methanogenic bacteria for COD, and, thus, can decrease the amount of methane gas production. While low concentrations of sulfide (less than 20 mg/L) are needed for optimal methanogenic activity, higher concentrations can be toxic Speece (1996, 2008). Methanogenic activity has been decreased by 50 percent or more at H_2S concentrations ranging from 50 to 250 mg/L (Arceivala, 1998). A comprehensive evaluation of the dynamics of competition between sulfate-reducing and methanogenic bacteria and toxicity effects is given in Maillacheruvu et al. (1993).

Because unionized H_2S is considered more toxic than ionized sulfide, pH is important in determining H_2S toxicity. The degree of H_2S toxicity is also complicated by the type of anaerobic biomass present (granular versus dispersed), the particular methanogenic population, and the feed COD/SO$_4$ ratio. With higher COD concentrations, more methane gas is produced to dilute the H_2S and transfer more of it to the gas phase. Hydrogen sulfide exists in aqueous solution as either the hydrogen sulfide gas (H_2S), the ion (HS^-) or as the sulfide ion (S^{2-}), depending on the pH of the solution, in accordance with the following equilibrium reactions:

$$H_2S \rightleftarrows HS^- + H^+ \tag{10-13}$$

$$HS^- \rightleftarrows S^{2-} + H^+ \tag{10-14}$$

Applying the law of mass action [Eqs. (10–13)] to Eq. (10–14) yields

$$\frac{[HS^-][H^+]}{[H_2S]} = K_{a1} \text{ and } \frac{[S^{2-}][H^+]}{[HS^-]} = K_{a2} \tag{10-15}$$

where K_{a1} = first acid dissociation constant, 1×10^{-7}
K_{a2} = second acid dissociation constant, $\sim 10^{-19}$ (value uncertain)

The percentage H_2S as a function of the pH can be determined using the following relationship:

$$H_2S, \% = \frac{[H_2S](100)}{[H_2S] + [HS^-]} = \frac{100}{1 + [HS^-]/H_2S} = \frac{100}{1 + [H^+]/K_{a1}} \tag{10-16}$$

Dissociation constants for hydrogen sulfide as a function of temperature are presented in Table 10–11, along with values for ammonia. As illustrated on Fig. 10–7, at a pH value of 7, at 30°C, about 60 percent of the total H_2S is present as gaseous H_2S.

Table 10-11 Acid equilibrium constants for ammonia (NH₃) and hydrogen sulfide (H₂S)	Temperature, °C	$K_{NH_3} \times 10^{10}$, mole/L	$K_{H_2S} \times 10^7$, mole/L
	0	7.28	0.262
	10	6.37	0.485
	20	5.84	0.862
	25	5.62	1.000
	30	5.49	1.480
	40	5.37	2.440

Figure 10–7

Fraction of hydrogen sulfide in H₂S form as function of pH.

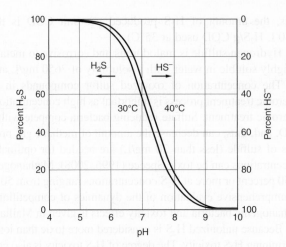

Ammonia Toxicity

Ammonia toxicity may be of concern for anaerobic treatment of wastewaters containing high concentrations of ammonium or proteins and/or amino acids, which can be degraded to produce ammonium. Free ammonia (NH_3), at high enough concentrations, has been found to be toxic to aceticlastic methanogens (Angelidaki and Arhing, 1994; Steinhaus et al., 2007; Lu et al., 2008), but the hydrogenotrophic organisms appear to be less sensitive (Sprott and Patel, 1986). As described in Chap. 2, ammonia is a weak acid and dissociates in water to form ammonium (NH_4^+) and hydroxyl ions. The amount of free ammonia is a function of temperature and pH. Dissociation constants for NH_3 as a function of temperature are given in Table 10–11. Based on the constants in Table 10–11, the free ammonia content decreases from 1.8 to 1.7 percent at a pH of 7.5 as the temperature increases from 20 to 35°C. At a pH of 7.8, the range is from 3.5 to 3.3 percent. Thus, the total ammonium- plus ammonia-N concentration (TAN) that can be tolerated is a function of the free ammonia toxicity concentration. Reported values for the threshold of NH_3-N toxicity range from 100 mg/L (McCarty, 1964) to 250 mg/L (Garcia and Angenent, 2009; Wilson et al., 2012). The higher values have been observed with more acclimation time, which may be due to acclimation by the aceticlasts or changes in the anaerobic process population with a shift to organisms more tolerant of free ammonia. These observations suggest that TAN concentrations of 3000 to 7000 mg/L may be tolerated in anaerobic treatment processes. With long-term acclimation non-inhibitory TAN concentrations of about 4000 mg/L have been reported by Garcia and Angement (2009) and 5000 to 8000 mg/L by van Velsen (1977) and Parkin and Miller (1982).

10–5 DESIGN CONSIDERATIONS FOR IMPLEMENTATION OF ANAEROBIC TREATMENT PROCESSES

Design of anaerobic treatment processes involves identification of process elements, determination of organic loading rate and other design parameters, and considerations for the conditions specific to anaerobic processes. When proprietary processes are considered, engineers must identify the quality and quantity of the waste stream to be treated, as well as the treatment goals. The purpose of this section is to provide a brief review of design parameters and considerations necessary to implement anaerobic treatment processes. The topics covered are (1) treatment efficiency needed, (2) general process design parameters and anaerobic degradation kinetics, and (3) other implementation issues.

Treatment Efficiency Needed

Anaerobic treatment processes are capable of very high COD conversion efficiency to methane production with minimal biomass production and relatively short hydraulic retention time reactors compared to aerobic treatment. Operation at high SRT values, greater than 20 to 50 d is possible with granular sludge, packed bed, and anaerobic membrane reactors, to provide maximum conversion of solids at temperatures above 25°C. However, the ability to meet secondary effluent discharge standards by anaerobic treatment alone is limited by high effluent suspended solids (50 to 150 mg/L) and a residual volatile fatty acid concentration that are common for anaerobic processes. These issues are magnified for anaerobic treatment at temperatures below 20°C. Some form of aerobic treatment is necessary to provide effluent polishing, either attached growth or suspended growth processes (Chong et al., 2012). For high strength wastewaters the combination of anaerobic and aerobic treatment can be very economical in terms of both capital and operating costs (Obayashi et al., 1981).

General Process Design Parameters

In the introduction to this chapter it was noted that the anaerobic digestion of wastewater treatment plant sludges is considered in Chap. 13. The material presented below is provided to serve as general guidance for comparing commercially available anaerobic technologies. Important design parameters used to size anaerobic processes include organic loading rate, hydraulic loading rate, and SRT. Biological kinetics and stoichiometric relationships between substrate removal and biomass growth are also useful for determining excess solids production and soluble substrate concentrations.

Organic Loading Rate.

The volumetric organic loading rate (OLR) is the key design parameter used to determine the reactor volume for granular sludge and attached growth anaerobic treatment processes. Generally, organic loading rates for anaerobic treatment processes are significantly higher than those used for aerobic processes. Organic loading rates used in anaerobic processes can vary from 1 to 50 kg COD/m³·d, whereas typical volumetric organic loading rates used for aerobic processes will vary from 0.5 and 3.2 kg COD/m³·d.

Organic loading rates are affected by the type of anaerobic process used, type of wastewater, and temperature. A range of organic loading rates are included with the process descriptions for commonly used anaerobic processes in Table 10–3. The wide range of organic loading rates reported for a given anaerobic treatment technology is illustrated in a comparison of organic loading rates for the UASB, EGSB, and anaerobic contact processes on Fig. 10–8. The values for the UASB and EGSB processes are of a similar

Figure 10–8

Range of organic loading rates reported for anaerobic treatment of various wastewaters with UASB, EGSB, and anaerobic contactor processes. (Data from Tauseef et al., 2013.)

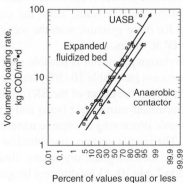

Table 10–12

Recommended UASB reactor SRTs for stable operation for treatment of domestic wastewater (van Lier et al., in Henze et al., 2011)

Temperature, °C	SRT, d
35	25
30	30
25	60
20	100
15	140

magnitude and generally higher than that used for anaerobic contactor reactors. The use of higher loadings for the granular sludge systems are in part due to the ability to carry a very high biomass concentration in the reactor as a result of the high biomass density developed in the granular sludge. The mixing regime for the EGSB and IC granular sludge systems also promote higher mass transfer of soluble substrate into the biofilm. The lower loading rates for the anaerobic contactor process may also be related to the more frequent application of this process for wastewaters with high solids concentration. The solids can limit the sludge concentration possible in the anaerobic reactor to thus require a larger reactor volume for a sufficient SRT for the necessary solids destruction, which then results in a lower organic loading rate.

Solids Retention Time. The solids retention time is a fundamental parameter for all anaerobic processes. The performance of an anaerobic treatment process is related to SRT in three ways: (1) from the basic biological kinetics discussed in Sec. 7–6, soluble substrate concentration is lower with increased SRT; (2) a higher SRT provides a greater mass of methanogenic organisms in the reactor, which is then better able to handle variations in process operating conditions to maintain a stable balance between volatile fatty acid production and utilization; and (3) the hydrolytic destruction efficiency of solids fed to an anaerobic reactor is increased at higher SRTs. At 30°C, SRT values greater than 20 d are needed for anaerobic processes for effective treatment performance. At lower temperatures much longer SRT values are needed. Recommended SRT values as a function of temperature for stable treatment of domestic wastewater by a UASB process are given in Table 10–12. Lower SRTs can be used for wastewaters that contain mainly soluble substrates.

Because of the nature of the biomass accumulation in granular sludge and attached growth anaerobic systems, it is impossible to predict the biomass concentration and SRT accurately. In addition, the substrate removal rate is not just a function of SRT as in a completely-mixed suspended growth reactor, but is also a function of substrate diffusion, mass transfer, and biofilm characteristics. Thus, the organic loading rate is the primary design parameter used for these systems, with the recognition that longer SRTs are commensurate with lower OLR values.

The use of SRT as design parameter is feasible for completely-mixed reactors such as the anaerobic contact process [see Table 10–3(f)] and the anaerobic membrane reactor [see Table 10–3(i)]. In such cases, the effect of the SRT on the effluent soluble substrate concentrations can be determined using Eq. (7–56) and appropriate biokinetic coefficients. The final step in anaerobic processing of organic material is the conversion of acetate and hydrogen to methane. The reactor acetate concentration is affected by the SRT and biokinetics of the acetoclastic methanogenic organisms. Biokinetic information for methanogenesis is summarized in Table 10–13 for different temperatures. Values for synthesis yield and specific endogenous decay coefficient coefficients for fermentation and methanogenesis

Table 10–13

Summary of biokinetic coefficients and methane stoichiometry for anaerobic treatment reactions[a]

Parameter	Unit	Value Range	Value Typical
Synthesis yield, Y_H			
Fermentation	g VSS/g COD	0.06–0.12	0.10
Methanogenisis	g VSS/g COD	0.02–0.06	0.04
Overall combined	g VSS/g COD	0.05–0.10	0.08
Decay coefficient, b_H			
Fermentation	g/g·d	0.02–0.06	0.04
Methanogenisis	g/g·d	0.01–0.04	0.02
Overall combined	g/g·d	0.02–0.04	0.03
Maximum specific growth rate, μ_m			
35°C	g/g·d	0.03–0.38	0.35
30°C	g/g·d	0.22–0.28	0.25
25°C	g/g·d	0.18–0.24	0.20
Half-velocity constant, K_S	mg COD/L	60–500	120
Methane			
Production at std. conditions	m³/kg CODr	—	0.35
Content of gas	%	60–70	65
Energy content at std.conditions	kJ/m³	—	38,846

[a] Tchobanoglous et al. (2003); Batstone et al. (2002).

are also included in Table 10–13. The methanogenic biokinetic values shown are based on *Methanosaeta* as the dominant organisms for acetoclastic methanogenesis, which are very common. However, under certain conditions, such as highly loaded anaerobic reactors, *Methanosarcina* may dominate. It has been shown that the maximum growth rate of *Methanosarcina* is about 2.5 times that for *Methanosaeta* at 35°C and its half-velocity coefficient for acetate utilization is about 3.5 times greater (Conklin et al., 2006).

The information given in Table 10–13 can also be used to approximate the amount of biomass and excess sludge production for completely-mixed anaerobic reactors, and for other types of anaerobic processes, if the SRT values can be estimated The use of SRT for the design of an anaerobic contact process is illustrated in Example 10–4 in Sec. 10–6.

Hydraulic Retention Time. The hydraulic retention time is directly related to the anaerobic process organic loading rates and wastewater strength:

$$\tau = S_o/\text{OLR} \tag{10-17}$$

Wastewaters with higher substrate concentration will require a longer hydraulic retention time for a given design OLR.

Process Implementation Issues

In addition to the process design consideration discussed above a number of other issues must be addressed when considering the implementation of anaerobic treatment technologies. Important issues include: solids separation, temperature management, corrosion control, and odor management. Additional issues are identified in Table 10–14.

Table 10–14

Design issues for anaerobic treatment of industrial wastewaters[a]

Design issue	Remarks
Flow equalization	
Pretreatment	TSS removal, FOG removal
Anaerobic process	
Corrosion control	Material selection
Temperature control	Preheating of waste stream, reactor heating
Chemical addition	Alkalinity control, nutrient, struvite control
Odor control	Odor removal processes (Chap. 16)
Sludge wasting	
Gas collection	Gas storage, cleaning, utilization (Chap. 17)

[a] Adapted, in part, from Totzke (2012).

Liquid-Solids Separation.   Solids separation is a critical element in the application of anaerobic processes. With respect to energy intensity, liquid sludge separation processes can vary from simple evaporation and/or decanting in sludge lagoons to centrifugation and electro-dewatering, depending on the application. A range of dewatering options for digested wastewater sludge, as well as other types of sludge, is considered in Sec. 14–2 in Chap. 14. Other solids separation processes that are an integral part of one or more of the processes identified in Table 10–3 are conventional sedimentation and membranes. Where sedimentation is used some form of degasification will be needed.

Temperature Management.   Anaerobic microorganisms responsible for the anaerobic treatment processes are in general temperature sensitive and anaerobic processes are often maintained at a significantly higher temperature than the ambient temperature, in the range of 30 to 36°C, or higher. Heating of the influent and the reactor may be achieved by using a heater directly in the reactor vessel, or using heat exchangers to preheat the influent. Heat in the effluent flow is used commonly to preheat the influent through a heat exchanger. Digester heating is considered in detail in Sec. 13–9 in Chap. 13.

Corrosion Control.   Anaerobic treatment usually generates hydrogen sulfide and H_2S, both of which can causes corrosion of various materials. Hydrogen sulfide in the anaerobic conditions can be converted biologically to sulfuric acid in the presence of moisture, also resulting in corrosion from sulfuric acid (U.S. EPA, 1991). Appropriate materials must be used for the equipment exposed to the anaerobic condition and the pipes to convey biogas containing hydrogen sulfide. Hydrogen sulfide removal in the gas phase is practiced commonly for odor control, and it is considered in Chap. 16. The use of materials such corrosion resistant concrete, fiber-reinforced plastic (FRP), polyvinyl chloride (PVC), high density polyethylene (HDPE) and a variety of similar materials and coatings in parts of the process exposed to corrosive gases must be considered to minimize the potential risk of corrosion and minimize the use of chemicals for corrosion control.

Odor Management.   Much of the early impetus for contained anaerobic process applications was related to complying with waste discharge regulations and minimizing odor complaints. With development of techniques to contain and remove odorous gases, odor has become less of the impetus but the odor control unit is a critical process element for the anaerobic treatment processes. As with the issue of corrosion, hydrogen sulfide is the most common odorous compound found in anaerobic processes, although a variety of mercaptan based odors have also been reported. The threshold values for a variety of odors are given in Sec. 16–3 in Chap. 16.

The principal methods used for the control of odors from anaerobic processes include the use of compost filters (biofilters), direct oxidation, carbon adsorption, acid scrubbing, and biotrickling filters. Relatively inexpensive, compost filters are generally used for smaller plants. In some wastewater treatment plants, odorous gases are oxidized by introducing them into sparged aeration diffusers in the activated sludge process. Views of carbon adsorption, acid scrubbing, and biotrickling filter facilities are shown on Fig. 9–9 in Chap. 9. All of these options for odor control are considered in greater detail in Sec. 16–3 in Chap. 16.

10–6 PROCESS DESIGN EXAMPLES

With the exception of the anaerobic lagoon process and anaerobic digesters used for wastewater treatment sludge, the three most common types of commercial anaerobic processes are: (1) the upflow anaerobic sludge blanket process, (2) the expanded granular sludge blanket process, and (3) the anaerobic contact process (see Table 10–4). The complete mix process which corresponds to anaerobic digestion of wastewater treatment plant sludges is considered separately in Chap. 13, along with sludge processing in Chap. 14. The purpose of this section is to illustrate the design of the upflow anaerobic sludge blanket process and the anaerobic contact process. It should be noted that because most commercial anaerobic processes illustrated in Table 10–3 are proprietary, there is little opportunity for individual process design as the treatment units are provided as a package.

Upflow Anaerobic Sludge Blanket Process

The UASB process and its evolution to the EGSB and IC processes have been described in Sec. 10–3. Some commercial and domestic installations have been illustrated previously on Figs. 10-3, 10–4, and 10–5. Some additional UASB and UASB type installations are illustrated on Fig. 10–9. The purpose of this section is to discuss key process elements

(a)

(b)

(c)

Figure 10–9

Granular sludge blanket anaerobic treatment systems: (a) UASB process in modular steel tanks, before insulation and hookup (courtesy of Robert Pharmer of Pharmer Engineering); (b) view of hybrid reactor equipped with internal packing above the granular sludge blanket [see Table 10-3(h)]. The exterior physical appearance of a UASB reactor without and with internal packing is similar; and (c) combined anaerobic-aerobic single-tank reactor treatment system (courtesy of Paques BV).

and process design considerations for granular sludge treatment systems. Important topics addressed are (1) the development and maintenance of the granular sludge, (2) physical design components, and (3) process design considerations.

Process Description.   The heart of the UASB, EGSB, and IC, and similar processes is the development and maintenance of the dense granular sludge particles, which results in a high reactor biomass concentration. Because of the high biomass concentration these anaerobic processes can be operated at high organic loading rates. The granular sludge particle size is generally in the range of 1 to 2 mm but may range from 0.10 to 8 mm depending on the waste treated and hydraulic and gas shear forces. Particle densities are in the range of 1.0 to 1.05 g/L with settling velocities of 15 to 50 m/h (Henze et al., 2011). Because of the granulated floc formation, the solids concentration at the bottom of the reactor may range from 50 to 100 kg/m³. Above the sludge blanket, a more diffused layer forms containing particles with lower settling velocities. The solids concentration in this layer may range from 10 to 30 kg/m³ (Aiyuk et al., 2006).

From studies on the microbial composition of the granules it appears that the surface is made up of coccid bacteria, while rod-shaped *Methanosaeta* are dominant in the interior and provide a filamentous structural backbone for the granulation (O'Flaherty et al., 2006). The physical characteristics of the granules provide a complex microbial ecology with methane-producing organisms in close proximity to hydrogen- and acetic acid-producers. A specific methane production activity of 0.10 g COD/VSS·d in the granular particles has been reported by Seghezzo et al. (2001).

Development of Granular Sludge.   Schmidt and Ahring (1996) describe a four-step process for the development of granular sludge: (1) attachment of cells to an uncolonized inert material or other cells, (2) initial adsorption of other colloidal or bacteria particles by reversible physiochemical forces, (3) irreversible attachment of microbial organisms due to microbial extracellular polymers, and (4) multiplication of cells from substrate diffusion into the granular structure. The substrate removal kinetics and mass transfer characteristics of the granular particles are similar to that described for biofilms in Sec. 7–7.

The development of a granular sludge bed can take many months, but this is normally avoided today by seeding with granular sludge wasted from other UASB reactors. The sludge blanket development is more rapid at higher temperatures (above 20°C) and with the presence of readily degradable soluble COD in the feed. The high upflow velocities washout unattached organisms and favor the growth of the dense granular particles.

Impact of Wastewater Characteristics.   The development and maintenance of granular sludge is affected by the wastewater characteristics. Granulation is very successful with high carbohydrate or sugar wastewaters, but less so with wastewaters high in protein resulting in a more fluffy floc instead (Thaveesri et al., 1994). Other factors affecting the development of granulated solids are pH, divalent cations and nutrient addition (Annachhatre, 1996). The pH should be maintained near 7.0 and a recommended COD/N/P ratio during startup is 300:5:1, while a lower ratio, 600:5:1, can be used during steady-state operation. Studies have shown that within certain concentrations, ferrous iron and calcium can enhance granular sludge formation; about 300 mg/L for Fe^{2+} and about 250 mg/L for Ca^{2+} (Yu et al., 2000; Yu et al., 2001).

The presence of suspended solids in the wastewater can adversely affect granular sludge formation and density, and the affects are more pronounced at lower temperatures

Table 10–15

Recommended design considerations for the gas-solids separators for UASB reactors[a]

1. The slope of the settler bottom, i.e., the inclined wall of the gas collector, should be between 45 to 60°.
2. The surface area of the apertures between the gas collectors should not be smaller than 15 to 20 percent of the total reactor surface area.
3. The height of the gas collector should be between 1.5 to 2 m at reactor heights of 5 to 7 m.
4. A liquid gas interface should be maintained in the gas collector in order to facilitate the release and collection of gas bubbles and to combat scum layer formation.
5. The overlap of the baffles installed beneath the apertures should be 100 to 200 mm to avoid upward flowing gas bubbles entering the settler compartment.
6. Generally scum layer baffles should be installed in front of the effluent weirs.
7. The diameter of the gas exhaust pipes should be sufficient to guarantee the easy removal of the biogas from the gas collection cap, particularly also in the case where foaming occurs.
8. Anti-foam spray nozzles should be installed in the upper part of the gas cap where the treatment of the wastewater is accompanied by heavy foaming.

[a] Adapted from Malina and Pohland (1992).

due to slower solids hydrolysis rates and solids accumulation (Letting and Hulshoff-Pol, 1991; Elmitwalli et al., 2002). For wastewaters with higher influent suspended solids and with lower temperatures, a two-step process, with a UASB reactor at a lower upflow velocity for solids capture and hydrolysis prior to a final UASB reactor, may provide more stable performance and operation flexibility. At certain feed solids concentrations (above 6 g TSS/L), an anaerobic contract process may be more appropriate.

Physical Design Considerations. The main physical design considerations are the feed inlet, gas separation, gas collection, and effluent withdrawal. The gas separation and effluent withdrawal functions are done with specifically designed gas-solids separators. Specific design features of these components are provided by the suppliers of proprietary upflow granular sludge anaerobic treatment processes. Gas-solids separator designs are located at the top of the liquid layer overlying the sludge blanket zone. Designs are employed to direct and trap the rising gas bubbles into hoods or collection zones [see Table 10–3(b) and (c)], and to capture the effluent from quiescent zones that allow solids settling and return to the reactor process volume below. Methods applied to improve effluent solids capture include a UASB-hybrid design with plastic packing at the top of the reactor (Tauseef et al., 2013) and the use of lamella-type plates in the effluent settling zone (Gomec, 2010). Design considerations for gas-solids separators are listed in Table 10–15.

The feed inlet designs must provide uniform flow distribution across the bottom of the granular sludge treatment process column to avoid channeling or formation of dead zones. The distribution is even more critical for designs with lower organic loading rates, as there would be less gas production to help mix the reactor contents. As shown in Table 10–16, the recommended bottom area served by individual feed pipes is a function of the bottom sludge layer density and the organic loading rate.

Process Design Considerations. Specific process designs will be influenced by the type of upflow granular sludge process used and the experience with the particular wastewater by the anaerobic treatment system supplier. Selection of the upflow granular sludge process reactor volume will be controlled by either (1) the allowable upflow velocity

Table 10–16

Guidelines for area served by feed inlet pipes for UASB reactors as a function of sludge density and organic loading rate[a]

Sludge type	COD loading, kg/m³·d	Area per feed inlet, m²
Dense flocculent sludge, (> 40 kg TSS/m³)	< 1.0	0.5–1
	1–2	1–2
	> 2	2–3
Medium flocculent sludge, (20 to 40 kg TSS/m³)	< 1–2	1–2
	> 2	2–5
Granular sludge	1–2	0.5–1.0
	2–4	0.5–2.0
	> 4	> 2.0

[a] Adapted from Lettinga and Hulshoff Pol (1991).

or (2) the organic loading rate (OLR). Limiting values for these parameters are affected by temperature and the type of wastewater being treated.

Upflow Velocity. The upflow velocity, based on the influent flowrate, is a critical design parameter. Design upflow velocities recommended for UASB reactors are shown in Table 10–17. Upflow velocities would be much higher in EGSB and IC reactors, which are more likely applied for higher strength industrial wastewaters. When the UASB reactor is applied for domestic wastewater treatment or for wastewater with higher influent solids concentrations, a lower velocity is needed to better retain the solids to provide sufficient time for solids capture and reduction by hydrolysis. The maximum allowable upflow velocity determines the cross-sectional area of the reactor, which is the feed rate divided by the upflow velocity:

$$A = \frac{Q}{v} \qquad (10\text{–}18)$$

where v = maximum design upflow superficial velocity, m/h
 A = reactor cross-section area, m²
 Q = influent flowrate, m³/h

The reactor process volume is equal to the cross-sectional area times the process reactor height (H) where V_v = reactor volume controlled by the maximum upflow velocity, m³.

$$V_v = H(A) \qquad (10\text{–}19)$$

The total reactor height is greater than this, as additional depth is added at the top for the gas-solids separator zone. The process volume must be large enough so that the allowable

Table 10–17

Design upflow velocities and reactor heights recommended for UASB reactors[a]

Wastewater type	Upflow velocity, m/h		Reactor height, m	
	Range	Typical	Range	Typical
COD near 100% soluble	1.0–3.0	1.5	6–10	8
COD partially soluble	1.0–1.25	1.0	3–7	6
Domestic wastewater	0.8–1.0	0.7	3–5	5

[a] Adapted from Lettinga and Hulshoff Pol (1991).

Figure 10–10

Effect of temperature on organic loading rate for UASB process (Henze et al., 2011).

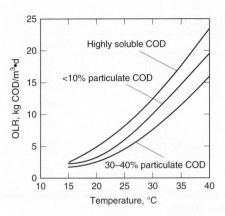

reactor organic loading rate is not exceeded. The reactor design is ultimately controlled by either the maximum allowable upflow velocity or by the required organic loading rate. For weaker wastewaters the upflow velocity will be the controlling parameter.

Organic Loading Rate.   As discussed in Sec. 10–5, a wide range of organic loading rates have been applied for upflow granular sludge processes, depending on the wastewater characteristics, reactor type, and temperature. In addition, Lettinga and Hulshoff-Pol (1991) point out that higher organic loadings can be used if higher effluent TSS concentrations are acceptable and also that higher loadings can be used for wastewaters with a lower fraction of particulate COD. As shown in Table 10–3, organic loading rates for UASB systems may range from 5 to 15 kg COD/ m³·d and from 10 to 40 kg COD/ m³·d for EGSB systems. An example of the effect of temperature on organic loading rate is shown on Fig. 10–10. For a highly soluble wastewater the organic loading rate is reduced by a factor of 5.0 for operation at 15 versus 30°C and by a factor of 4.3 for a wastewater with 30–40% particulate COD.

The reactor process volume is related to the organic loading rate as follows:

$$V_{OLR} = \frac{Q(S_o)}{OLR} \tag{10-20}$$

where V_{OLR} = reactor process volume controlled by the organic loading rate, m³. The process design for a UASB treatment process is illustrated in Example 10–3.

EXAMPLE 10–3   UASB Treatment Process Design   Determine the following for a UASB treatment process used to treat a sugar beet wastewater, with the characteristic given below, to achieve 90 percent COD removal:

 a. Reactor process volume

 b. Process hydraulic retention time

 c. Reactor dimensions

 d. Reactor SRT

 e. Daily sludge production rate in kg VSS/d

 f. Excess sludge waste volume in m³/d

 g. Methane gas production rate in m³/d

 h. Total gas production rate in m³/d

 i. Energy available from methane production in kJ/d

 j. Alkalinity requirements

Wastewater characteristics:

Item	Unit	Value
Flowrate	m^3/d	500
COD	g/m^3	12,000
TSS	g/m^3	600
nbVSS	g/m^3	500
Alkalinity	g/m^3 as $CaCO_3$	500
Temperature	°C	25

Use the design parameters given below and typical values from Table 10–13:
1. From Table 10–14
   $Y_H = 0.08$ g VSS/g COD
   $b_H = 0.03$ g VSS/g VSS·d,
2. $f_d = 0.10$ g VSS cell debris/g VSS biomass decay
3. Methane production at 0°C = 0.35 L CH_4/g COD
4. Energy content of methane at 0°C = 38,846 kJ/m^3
5. Percent methane in gas phase = 65%
6. Height of reactor process volume = 8 m
4. Height of clear zone above the sludge blanket = 0.50 m
8. Height of gas-solids separator = 2.5 m
9. Reactor length:width ratio = 2.0
10. Maximum reactor upflow velocity = 1.0 m/h
11. Average solids concentration in process volume = 30 kg VSS/m^3

Based on data from the treatment of sugar beet wastewater in other UASB facilities, 90 percent COD removal at 25°C, can be achieved with a design organic loading rate of 8.0 kg COD/m^3·d. The wastewater is mainly soluble containing carbohydrate compounds, and a granular sludge is expected. Assume an effluent VSS concentration of 120 g/m^3.

Solution
1. Determine the reactor process volume.
   a. Determine the reactor volume based on the maximum upflow velocity Eq. (10–18) and Eq. (10–19)

   $$A = \frac{Q}{v} = \frac{(500 \text{ m}^3/\text{d})}{(1.0 \text{ m/h})(24 \text{ h/d})} = 20.8 \text{ m}^2$$

   $$V_v = A(H) = 20.8 \text{ m}^2(8 \text{ m}) = 166.7 \text{ m}^3$$

   b. Determine the reactor volume based on the organic loading rate. From Eq. (10–20)

   $$V_{OLR} = \frac{Q S_o}{OLR} = \frac{(500 \text{ m}^3/\text{d})(12 \text{ kg COD/m}^3)}{(8.0 \text{ kg COD/m}^3 \cdot \text{d})} = 750 \text{ m}^3$$

   The organic loading rate controls the reactor volume design.
2. Determine the process hydraulic retention time.

   $$\frac{V}{Q} = \frac{750 \text{ m}^3}{(500 \text{ m}^3/\text{d})} = 1.5 \text{ d}$$

3. Determine the reactor dimensions.
   a. Reactor area = $(L)(W) = 2W(W) = 2W^2$

$$\text{Area} = \frac{V}{H} = \frac{750 \text{ m}^3}{8 \text{ m}} = 93.75 \text{ m}^2$$

$2W^2 = 93.75 \text{ m}^2$, $W = 6.85 \text{ m}$, $L = 13.7 \text{ m}$

b. Total reactor height

H_T = processing hgt + clear zone hgt + separator hgt

$H_T = 8 \text{ m} + 0.5 \text{ m} + 2.5 \text{ m} = 11 \text{ m}$

Reactor dimensions = 13.7 m $\times$ 6.85 m $\times$ 11 m

4. Determine the reactor SRT.
   a. From Eq. (7–56), $X(V) = P_x$ SRT
   b. From Eq. (8–20),

$$P_x = \frac{Q(Y_H)(S_o - S)}{1 + b_H(\text{SRT})} + \frac{f_d b_H (Q)(Y_H)(S_o - S)(\text{SRT})}{1 + b_H(\text{SRT})} + (\text{nbVSS})Q$$

   c. Substituting Eq. (8–20) into Eq. (7–56),

$$X_{\text{VSS}}(V) = \frac{Q(Y_H)(S_o - S)(\text{SRT})[1 + f_d b_H(\text{SRT})]}{1 + b_H(\text{SRT})} + (\text{nbVSS})Q(\text{SRT})$$

$S_o - S = 0.90 \, S_o = 0.90(12{,}000 \text{ g COD/m}^3)$

$S_o - S = 10{,}800 \text{ g COD/m}^3$

$(30{,}000 \text{ g VSS/m}^3)(750 \text{ m}^3) =$

$$\frac{(500 \text{ m}^3/\text{d})(0.08 \text{ g VSS/g COD})(10{,}800 \text{ g COD/m}^3)(\text{SRT})[1 + 0.10(0.03 \text{ g/g·d})(\text{SRT})]}{1 + (0.03 \text{ g/g·d})(\text{SRT})}$$

$+ \, 500 \text{ g VSS/m}^3 \, (500 \text{ m}^3/\text{d})\text{SRT}$

Solving: SRT = 50.2 d

5. Determine the daily sludge production rate from Eq. (7–56).

$$P_{X,\text{VSS}} = \frac{X_{\text{VSS}}(V)}{\text{SRT}}$$

$$= \frac{(30{,}000 \text{ g VSS/m}^3)(750 \text{ m}^3)(1 \text{ kg}/10^3 \text{ g})}{50.2 \text{ d}}$$

$P_{X,\text{VSS}} = 448.2 \text{ kg VSS/d}$

6. Determine the excess sludge daily waste volume.

$$P_{X,\text{VSS}} = Q(X_e) + (X)Q_W$$

$$Q_W = \frac{P_{X,\text{VSS}} - Q(X_e)}{X}$$

$$= \frac{(448{,}200 \text{ g VSS/d}) - (500 \text{ m}^3/\text{d})(0.120 \text{ g VSS/m}^3)}{(30{,}000 \text{ g VSS/m}^3)}$$

$Q_W = 14.9 \text{ m}^3/\text{d}$

7. Determine the methane gas production rate by COD balance.

COD removal = methane COD + biomass COD

$P_{X,\text{bio}} = P_{X,\text{VSS}} - \text{nbVSS}(Q)$

$$P_{X, \text{bio}} = 448{,}200 \text{ g VSS/d} - 500 \text{ g VSS/m}^3 \ (500 \text{ m}^3/\text{d})$$

$$P_{X, \text{bio}} = 448{,}200 \text{ g VSS/d} - 250{,}000 \text{ g VSS/d}$$

$$= 198{,}200 \text{ g VSS/d}$$

Methane COD = COD removed − biomass COD

CH_4 COD/d

$$= 500 \text{ m}^3/\text{d}(10{,}800 \text{ g COD/m}^3) - 1.42 \text{ g COD/g VSS } (198{,}200 \text{ g VSS/d})$$

CH_4 COD = 5,118,556 g CH_4 COD/d

At standard conditions, methane production rate =

$$(5{,}118{,}556 \text{ g } CH_4 \text{ COD/d})(0.35 \text{ L } CH_4/\text{g COD})(1 \text{ m}^3/10^3 \text{ L})$$

$$= 1719.5 \text{ m}^3 \ CH_4/\text{d at } 0°C$$

methane production rate at 25°C =

$$(1719.5 \text{ m}^3 \ CH_4/\text{d})\frac{(273.15 + 25)°C}{273.15°C} = 1955 \text{ m}^3 \ CH_4/\text{d}$$

8. Determine the total gas production rate; Percent methane = 65%

$$\text{Total gas production rate} = \frac{(1955 \text{ m}^3 \ CH_4\text{d})}{(0.65 \text{ m}^3 \ CH_4/\text{m}^3 \text{ gas})} = 3008 \text{ m}^3 \text{ gas/d}$$

9. Energy content of methane production

Energy = $(38{,}846 \text{ kJ/m}^3)(1719.5 \text{ m}^3 \ CH_4/\text{d}) = 66.8 \times 10^6 \text{ kJ/d}$

10. Determine alkalinity requirements.

Assume pH = 7.0

From Table 10–7 at pH = 7.0, T = 25°C,

percent CO_2 = 35%, alkalinity = 2678 g/m^3 as $CaCO_3$

Influent alkalinity = 500 g/m^3 as $CaCO_3$

Alkalinity needed = $(2678 - 500)$g/m^3 as $CaCO_3$

$$= 2178 \text{ g/m}^3 \text{ as } CaCO_3$$

Alkalinity in kg/d = $(2178 \text{ g/m}^3)(500 \text{ m}^3/\text{d})(1 \text{ kg}/10^3 \text{ g})$

$$= 1098 \text{ kg/d}$$

11. Summary of results.

Parameter	Unit	Value
Reactor process volume	m^3	750.0
Reactor total height	m	11.0
Reactor LxW	m	13.7 × 6.85
Hydraulic retention time	d	1.5

(*continued*)

Parameter	Unit	Value
SRT	d	50.2
Excess waste sludge	m^3/d	14.9
Total gas production rate	m^3/d	3008
Methane production rate	m^3/d	1955
Energy production rate	kJ/d	66.8×10^6
Alkalinity needed as $CaCO_3$	kg/d	1089

Comment   A significant amount of energy is produced daily in the form of methane gas. If the gas can be used as an energy source by the industrial facility, it could help to offset the cost of adding a considerable amount of alkalinity to maintain the anaerobic reactor pH near 7.

Anaerobic Contact Process

The anaerobic contact process is a completely-mixed reactor system with suspended/flocculating anaerobic biomass and a liquid-solids separation step for biomass capture and recycle.

Process Description.   The process flow diagram for the ANCP is illustrated in Table 10–3(f). As shown, the anaerobic contact process overcomes the disadvantages of equal values of hydraulic retention time and SRT in the completely mixed process. Biomass is separated and returned to the completely mixed or contact reactor so that the process SRT is longer than the hydraulic retention time. By separating the hydraulic retention time and SRT, the anaerobic reactor volume can be reduced. Gravity separation is the most common approach for solids separation and thickening prior to sludge recycle, however, the success is dependent on the settling properties of the anaerobic reactor solids.

Because the reactor sludge contains gas produced in the anaerobic process and gas production can continue in the separation process, solids-liquid separation can be inefficient and unpredictable. Various methods have been used to minimize the effect of trapped gas bubbles in the sludge settling step. These include gas stripping by agitation or vacuum degasification, inclined plate separators, and the use of coagulant chemicals. Clarifier hydraulic application rates range from 0.5 to 1.0 m/h. Practical reactor MLVSS concentrations are 4000 to 8000 mg/L (Malina and Pohland, 1992). Organic loading rates are typically $\leq$ 4.0 kg $COD/m^3 \cdot d$ and SRTs range from 15 to 30 d as indicted in Table 10–3(f).

Design Considerations for Anaerobic Contact Process.   The anaerobic contact process may be designed in a manner similar to completely mixed aerobic activated sludge processes, because the hydraulic regime and biomass concentration can be reasonably defined. The design procedure is as follows:

1. Select an SRT to achieve a given effluent concentration and percent COD removal.
2. Determine the daily solids production and mass of solids in the system to maintain the desired SRT.
3. Select the expected solids concentration in the reactor and determine the reactor volume.
4. Determine the gas production rate.
5. Determine the amount of excess sludge wasted and the nutrient needs.
6. Check the volumetric organic loading rate.
7. Determine alkalinity needs.

These same design considerations apply to other types of anaerobic treatment processes, with the main difference being the need to rely on using organic loading rates instead of SRT to size the reactor volume.

Information in Table 10–13 provides a summary of kinetic coefficients and other design values that may be used to design an anaerobic contactor process treating wastewater that is comprised of mostly soluble biodegradable COD. For wastewater with high solids concentrations, the design methods presented in Chap. 13 for anaerobic digestion may be applicable. Laboratory treatability studies or pilot plant testing is normally recommended for wastewater containing soluble and particulate constituents that are significantly different from past experience with anaerobic degradation of other wastewaters. The design of an anaerobic contact process is presented in Example 10–4.

EXAMPLE 10–4   **Suspended Growth Anaerobic Contact Reactor Process**   Determine the reactor volume and hydraulic retention time, the gas production rate and energy available, the solids production rate, and the amount of alkalinity and nutrient addition needed for an anaerobic contact process [see Table 10–3(f)] treating the following wastewaters to achieve 90 percent COD removal.

Wastewater characteristics:

Item	Unit	Value
Flowrate	m³/d	500
COD	g/m³	6000
Soluble COD	g/m³	4000
COD/VSS ratio	g/g	1.8
Degradable fraction of VSS	%	80
Nitrogen	g/m³	10
Phosphorus	g/m³	20
Alkalinity	g CaCO₃/m³	500
Temperature	°C	25

Design assumptions:
1. Effluent VSS concentration = 150 g/m³.
2. Factor of safety for design SRT = 3.0
3. f_d = 0.15 g VSS cell debris /g VSS biomass decay
4. At SRT ≥ 30, >99% of degradable VSS is transformed
5. MLVSS = 6000 g/m³
6. Settling rate = 24 m/d
7. Gas composition = 65% CH_4 and 35% CO_2
8. Use kinetic coefficients and methane production assumptions from Table 10–13.
9. Biomass nutrient content = 12% N and 2.4% P

Solution

1. Determine design SRT at 25°C.

   At 90 percent COD removal the effluent COD is:

   = (1.0 – 0.90) (6000 mg/L) = 600 g/m³

   The assumed effluent VSS concentration equals 150 g/m³.

Effluent COD from VSS = (150 g/m³L) 1.8 g COD/g VSS

= 270 g/m³

Allowable effluent soluble COD = (600 − 270) g/m³

= 330 g/m³

Solving for SRT in Eq. (7–46) and substituting $\mu_{max}=Y_H k$;

$$S = \frac{K_s[1 + b_H(SRT)]}{SRT(\mu_{max} - b_H) - 1}$$

$$SRT = \left[\frac{\mu_{max}(S)}{K_s + S} - b_H\right]^{-1}$$

Use kinetic coefficients from Table 10–13,

$\mu_{max} = 0.20$ g/g·d

$K_s = 120$ g/m³

$b_H = 0.03$ g/g·d

$$SRT = \left[\frac{(0.20 \text{ g/g·d})(330 \text{ g COD/m}^3)}{(120 + 330)\text{g COD m}^3} - (0.03 \text{ g/g·d})\right]^{-1} = 8.6 \text{ d}$$

With a safety of 3.0

Minimal design SRT = 3.0 (8.6) = 25.7 d

Use SRT = 30 d to complete degradable VSS transformation

2. Determine sludge production rate.

Calculate nondegraded VSS concentration

Nonsoluble COD = (6000 − 4000) g/m³

= 2000 g/m³

Nonsoluble COD as VSS = (2000 g/m³ COD) / (1.8 g COD/g VSS)

= 1110 g/m³ VSS

Degradable fraction of VSS = 0.8 (given)

Nondegraded VSS = 0.20 (1110) = 222 g VSS/m³

Use Eq. (8–20) to determine solids production:

$$P_{X,VSS} = \frac{Q(Y_H)(S_o - S)}{1 + b_H(SRT)} + \frac{f_d b_H(Q)(Y_H)(S_o - S)(SRT)}{1 + b_H(SRT)} + (nbVSS)Q$$

$S_o - S$ = COD degraded

= Influent COD − nondegraded VSS COD − effluent soluble COD

= 6000 g COD/m³ − 222 g VSS/m³ − 330 g COD/m³ = 5270 g COD/m³

Use following coefficients from Table 10–14 and assume $f_d = 0.15$

$Y_H = 0.08$ g VSS/g COD

$b_H = 0.03$ g/g·d

$$P_{X,VSS} = \frac{Q(Y_H)(S_o - S)[1 + f_d b_H (\text{SRT})]}{1 + b_H(\text{SRT})} + (\text{nbVSS})Q$$

$$P_{X,VSS} = \frac{(500 \text{ m}^3/\text{d})(0.08 \text{ g VSS/g COD})(5270 \text{ g COD/m}^3)[1 + 0.15(0.03 \text{ g/g·d})(30.0 \text{ d})]}{[1 + (0.03 \text{ g/g·d})(30.0 \text{ d})]}$$

$$+ (222 \text{ g VSS/m}^3)(500 \text{ m}^3/\text{d})$$

$P_{X,VSS} = 125{,}925 \text{ g VSS/d} + 111{,}000 \text{ g VSS/d} = 236{,}925 \text{ g VSS/d}$

3. Determine reactor volume and τ.
    a. Determine the volume using Eq (7–56).

    $$V = \frac{(P_{X,VSS})\text{SRT}}{\text{MLVSS}} = \frac{(236{,}925 \text{ g VSS/d})(30 \text{ d})}{6000 \text{ g/m}^3} = 1184.6 \text{ m}^3$$

    b. Determine the hydraulic detention time, τ.

    $$\tau = \frac{V}{Q} = \frac{1184.6 \text{ m}^3}{(500 \text{ m}^3/\text{d})} = 2.4 \text{ d}$$

4. Determine the methane and total gas production rate and energy production rates.
    a. Determine the methane gas production rate.

    From Table 10–13, 0.35 m³ CH₄/kg COD at 0°C

    COD removal = methane COD + biomass COD

    biomass COD = (1.42 g COD/g VSS)($P_{x,bio}$)

    $P_{x,bio}$ = first term in $P_{X,VSS}$ calculation = 125,925 g VSS/d

    Methane COD = COD removed − biomass COD

    CH₄ COD/d

    = 500 m³/d (5270 g COD/m³) − 1.42 g COD/g VSS (125,925 g VSS/d)

    = 2,456,186 g CH₄ COD/d

    At standard conditions, methane production rate =

    (2,456,186 g CH₄ COD/d)(0.35 L CH₄/g COD)(1 m³/10³ L)

    = 859.7 m³ CH₄/d at 0°C

    Methane production rate at 25°C =

    $$(859.7 \text{ m}^3 \text{ CH}_4/\text{d})\frac{(273.15 + 25)°\text{C}}{273.15°\text{C}} = 938.3 \text{ m}^3 \text{ CH}_4/\text{d}$$

    b. Determine the total gas production rate.

    Gas composition = 65% methane (given)

    $$\text{Total gas production rate at 25°C} = \frac{(938.3 \text{ m}^3 \text{ CH}_4/\text{d})}{(0.65 \text{ m}^3 \text{ CH}_4/\text{m}^3 \text{ gas})} = 1443.6 \text{ m}^3 \text{ gas/d}$$

    (Note gas rate = 1443.6/500 = 2.9 times liquid flowrate.)

c. Determine the energy production rate.

From Table 10–13, energy content of methane = 38,846 kJ/m^3 at 0°C.

Energy production rate = (859.7 m^3 CH$_4$/d)(38,846 kJ/m^3) = 33.4 × 10^6 kJ/d

5. Determine nutrient requirements.

Biomass production rate = 125,925 gVSS/d

Given: biomass N = 12% and P = 2.4% of VSS

N required = (125,925)(0.12) = 15,111 g/d

P required = (125,925)(0.024) = 3022 g/d

Influent nutrients:

N = (10 g/m^3)(500 m^3/d) = 5000 g/d

P = (20 g/m^3)(500 m^3/d) = 10,000 g/d

There is sufficient phosphorus in the influent, but nitrogen must be added.

N addition = (15,111 − 5000) g N/d

$\qquad$ = 10,111 g N/d

$\qquad$ = 10.1 kg N/d

6. Determine alkalinity requirement.

From Table 10–7 at pH = 7.0, T = 25°C, percent CO$_2$ = 35, alkalinity = 2678 g/m^3 as CaCO$_3$

Influent alkalinity = 500 g/m^3 as CaCO$_3$

Alkalinity needed = (2678 − 500) g/m^3 as CaCO$_3$

$\qquad$ = 2178 g/m^3 as CaCO$_3$

As NaHCO$_3$ = $\left[\dfrac{(2178 \text{ g as CaCO}_3/\text{m}^3)}{(50 \text{ mg/meq CaCO}_3)} \right]$(84 mg NaHCO$_3$/meq) = 3659 g NaHCO$_3$/m^3

NaHCO$_3$/d = (3659 g/m^3)(500 m^3/d)(1 kg/10^3 g) = 1830 kg/d

7. Determine clarifier diameter.

(Assume degasifier used before clarifier)

$$\text{Area} = \frac{(Q, \text{ m}^3/\text{d})}{(\text{settling rate, m/d})} = \frac{(500 \text{ m}^3/\text{d})}{(24 \text{ m/d})} = 20.83 \text{ m}^2$$

Diameter = 5.2 m

Comments　A considerable amount of energy (i.e., kJ) is generated by the production of methane (CH$_4$). The methane could be used to heat the anaerobic process, which would provide more rapid degradation and, thus, reduce the anaerobic bioreactor size.

Use of Simulation Models

The relatively straightforward design procedures described above can be used to obtain a reasonable estimate of reactor volume requirements, effluent soluble bCOD concentration, and gas production. However, as discussed in Sec. 8–5, in Chap. 8, for aerobic activated sludge processes, computer aided mechanistic dynamic simulation models have been developed and applied for anaerobic reactors. The most common is the model developed for anaerobic sludge digestion, termed ADM1, by an International Water Association task group (Batstone et al., 2002a). ADM1 has also been applied to other wastes with high solids contents such as pig slurry wastes (Girault et al., 2011) and waste mixtures of biodegradable soluble and particulate COD (Batstone et al., 2002b; Fezzani and Cheikh, 2008).

The ADM1 model follows the fate of feed COD according to the schematic given on Fig. 7–26 and also accounts for inert nondegradable soluble COD and volatile solids. Changes in concentration of biodegradable solid and dissolved COD components of the waste and intermediate degradation products are determined by the application of biokinetic equations along the various degradation pathways that include the type of microorganisms involved and effects of pH and temperature. The particulate COD is considered to be a homogeneous mixture of carbohydrates, proteins, and lipids. Kinetic relationships described a disintegration rate of particles to a production rate of carbohydrate, protein and lipid components. Another set of equations is used to describe the rate of hydrolysis of carbohydrates, proteins and lipids to sugars, amino acids, and long chain fatty acids (LCFA), respectively. Monod-based biokinetic equations are then applied for acidogenesis of sugars and amino acids to volatile fatty acids (VFA) and hydrogen and acetogenesis of LCFA and VFAs to acetate. Separate Monod-based kinetics are used to describe methanogenesis by acetate and hydrogen utilizing organisms.

The model also includes physicochemical processes to (1) calculate the reactor pH as a function of alkalinity and VFA concentrations and gas phase carbon dioxide concentration and (2) gas-liquid transfer of carbon dioxide, methane and hydrogen sulfide generated in the process. The model is applied as a series of differential equations encompassing 32 dynamic concentration state variables. Application of the model is most useful for dynamic simulations to analyze changes in VFA and hydrogen concentrations following transient feed and loading variations to evaluate conditions that can result in an imbalance between VFA production rates and methanogenesis acetate utilization rates that could lead to digester instability (Straub et al., 2006).

10–7 CODIGESTION OF ORGANIC WASTES WITH MUNICIPAL SLUDGE

Codigestion refers to an anaerobic digestion process in which different types of wastes from at least two different sources are combined and treated in a common anaerobic reactor. The main practice of codigestion occurs in municipal anaerobic sludge digesters. If there is excess capacity in municipal facility anaerobic digesters, codigestion of other wastes in the community can be an attractive means for increasing the methane production and energy available for the facility or for other community uses such as gas powered vehicles. Typical applications of codigestion have been for processing FOG wastes and food wastes. Codigestion applications with municipal sludge digesters is discussed in Chap. 13.

Benefits of Codigestion. A major benefit of codigestion is the ability to turn a waste product into a source of energy, while at the same time reducing the region carbon footprint by the replacement of other fuels with methane from the anaerobic conversion of the waste material and curtailing the carbon dioxide release from the waste decomposition in other ways (Rosso and Stenstrom, 2008). A variety of wastes are available in local communities for processing in codigestion. Examples are given in the list of wastes shown in Table 10–18 from a modest sized community that were found to be highly amenable for codigestion (Muller et al., 2009).

Table 10–18

Example of types of highly biodegradable wastes evaluated for a municipal codigestion application[a]

Description of wastes	Comment
Flower and vegetable wastes	Requires nutrients and alkalinity
Blood product from animal processing	High in nitrogen
Dissolved air flotation sludge from rendering plant	High in nitrogen
Brown grease from grease traps	Difficult to degraded alone, requires nutrients and alkalinity
Chili, soup and salad dressing production wastes	Requires nutrients and alkalinity
Confectionary sugar wastes	Requires nutrients and alkalinity
Beer, wine, soda and juice production wastes	Requires nutrients and alkalinity

[a] Muller et al. (2009).

Codigestion can also be an attractive alternative for various food processing operations in lieu of constructing their own onsite anaerobic treatment process. The advantages include economies of scale for the facility, the elimination of onsite operational requirements, and the elimination of operating cost for the addition of alkalinity and nutrients for high carbohydrate wastes. Without a sufficient amount of protein or amino acids in the waste to produce ammonium bicarbonate at a level needed to maintain a proper pH in anaerobic treatment, the cost for alkalinity addition can be prohibitive.

Operation of Digestion Process. Alkalinity production in municipal anaerobic digesters is normally sufficient due to the degradation of organic nitrogen in primary sludge and waste activated sludge. In some cases, such as oily wastewaters (Jeganathan et al., 2006), the waste is difficult to treat by itself, but can be handled within proper proportions in municipal anaerobic digesters. The economical impact to the municipal facility must also be considered and includes costs for feed stock management, storage, and pretreatment such as screening or heating, and increased operational requirements for the codigestion operation.

PROBLEMS AND DISCUSSION TOPICS

10–1 The alkalinity concentration in an anaerobic suspended growth reactor operated at 30°C, is 2200, 2600, or 2800 mg as $CaCO_3$/L (value to be selected by instructor). Assuming equilibrium between the liquid and gas phase with a CO_2 content in the gas phase of 35 percent, determine the reactor pH.

10–2 An industrial wastewater with a flowrate of 4000 m³/d has a soluble degradable COD concentration of 10,000, 5000, and 2500 mg/L (value to be selected by instructor), 20°C temperature, and 200 mg/L alkalinity concentration as $CaCO_3$. Determine and compare the net operating costs or revenue for anaerobic versus aerobic treatment based on the following key parameters and assumptions (labor and maintenance costs are omitted here) for each:

Anaerobic process:
Anaerobic operating cost items are related to raising the liquid temperature, and adding alkalinity, versus the revenue from methane production. The following assumptions apply:

1. Reactor temperature = 35°C

2. Heat exchanger recovery efficiency for raising liquid temperature = 80 percent

3. COD removal efficiency = 95 percent

4. CO_2 of gas phase = 35 percent and pH = 7.0

5. Value of methane=$5/10^6 kJ$

6. Alkalinity is provided as $NaHCO_3$ at $0.90/kg

Aerobic

Major aerobic treatment operating cost items are energy for aeration and sludge processing and disposal. The following assumptions apply:

1. COD removal efficiency = 99 percent

2. gO_2/g COD removal = 1.2

3. Actual aeration efficiency = 1.2 kgO_2/kWh

4. Electricity costs = $0.08/kWh

5. Net sludge production = 0.3 g TSS/g COD removed

6. Sludge processing/disposal cost = $0.10/kg dry solids

10–3 A wastewater has a daily average flowrate of 1000, 2000, or 3000 m^3/d (value to be selected by instructor) and 4000 mg/L of an organic substance with the following approximate composition: $C_{50}H_{75}O_{20}N_5S$. For anaerobic treatment at 95 percent degradation determine (a) the alkalinity production in mg/L as $CaCO_3$; and (b) the approximate mole fraction of CO_2, CH_4, and H_2S in the gas phase.

10–4 An industrial wastewater has an average flowrate of 2000 m^3/d, an influent COD concentration of 4000, 6000, or 8000 mg/L (value to be selected by instructor), and influent sulfate concentration of 500 mg/L. The percent of COD degraded in an anaerobic treatment process at 35°C is 95 percent, and 98 percent of the sulfate is reduced. Determine (a) the amount of methane produced in m^3/d; (b) the amount of methane produced in m^3/d, if the sulfate reduction is not accounted for; and (c) the amount of H_2S in the gas phase at a reactor pH value of 7.0.

10–5 A suspended growth anaerobic reactor is operated at an SRT of 30 d at a temperature of 30°C. On a given day, the methane gas production rate (m^3/d) decreases by 30 percent. List at least four possible causes that should be investigated and briefly explain the mechanism behind each one.

10–6 A 100 percent soluble industrial wastewater is to be treated by an anaerobic contact process consisting of a mixed covered reactor, a degasifier, and gravity settling. The effluent TSS concentration from the clarifier is 120 mg/L. For the following wastewater characteristics and design assumptions, determine and compare the following design parameters for treatment at 25 and 35°C to meet an effluent soluble COD concentration of ≤50 mg/L.

a. The design SRT, d

b. The reactor volume, m^3

c. The reactor hydraulic detention time τ, d

d. The methane gas production rate, m^3/d

e. The total gas production rate, m^3/d

f. The amount of solids and to be manually wasted daily, kg/d

g. The nitrogen and phosphorus requirements, kg/d.

Wastewater characteristics:

Parameter	Unit	Value
Flowrate	m^3/d	2000
Degradable COD	mg/L	

(continued)

(*Continued*)

Parameter	Unit	Value
Wastewater 1		4000
Wastewater 2		6000
Wastewater 3		8000
Percent sCOD	%	100
Alkalinity	mg/L as $CaCO_3$	500

Note: Wastewater 1, 2 or 3 to be selected by instructor.

Other design assumptions:

1. Reactor MLSS concentration = 5000 mg/L
2. Factor of safety for SRT = 1.5
3. VSS/TSS ratio = 0.85
4. f_d = 0.15 g VSS cell debris/g VSS biomass decay
5. Gas phase methane = 65 percent
6. Nitrogen content of biomass = 0.12 g N/g VSS5
7. Phosphorus content of biomass = 0.02 g P/g VSS
8. Use the appropriate kinetic coefficients and design information provided in Table 10–13

10–7 An anaerobic process is being considered for the treatment of a soluble industrial wastewater at 30°C. A design SRT of 30 d is required to provide the desired level of 95 percent soluble COD degradation. An effluent VSS concentration of 100, 150 or 200 mg/L (value to be selected by instructor) from biomass growth is assumed. Using the appropriate kinetic coefficient values from Table 10–13, determine the influent COD concentration that must be present to allow operation at a 30-d SRT, if all the biomass wasted is via the effluent solids losses.

10–8 Design a single UASB reactor to treat an industrial wastewater at 30°C with the following wastewater characteristics and using the assumptions given below. Assume 97 percent degradation of the soluble COD, 60 percent particulate COD degradation, and an effluent VSS concentration of 200 mg/L. Using the given information, determine:

1. The reactor liquid volume, m³
2. The reactor area (assume a circular reactor will be used), m²
3. The reactor area diameter and total height, m
4. The hydraulic retention time, d
5. The average SRT, d
6. The amount of solids to be manually wasted daily, kg VSS/d
7. The methane gas production rate, m³/d
8. The energy value of the gas, kJ/d
9. The alkalinity requirement, kg as $CaCO_3$/d

Design Assumptions:

1. Kinetic coefficients from Table 10–13.
2. f_d = 0.15 g VSS/g VSS biomass decayed
3. Maximum organic loading rate = 6.0 kg COD/m³·d
4. Maximum upflow velocity = 0.50 m/h
5. pH = 7.0
6. CO_2 in gas phase = 35 percent
7. Process reactor liquid height = 8 m
8. Average solids concentration in process reactor = 50 g VSS/L

Wastewater characteristics:

Parameter	Unit	Value
Flowrate	m³/d	500
Total bCOD	mg/L	
Wastewater 1		6000
Wastewater 2		7000
Wastewater 3		8000
Particulate COD	Percent	40
Particulate COD/VSS ratio	g/g	1.8
Particulate VSS/TSS ratio	g/g	0.85
Alkalinity	mg/L as CaCO₃	300

Note: Wastewater 1, 2, or 3 to be selected by instructor.

10-9 A domestic wastewater is to be treated using the UASB process at 25°C. The wastewater characteristics are given in the following table. Determine: (a) the reactor hydraulic retention time (hours); (b) the COD loading rate (kg COD/m³·d); and (c) the process reactor liquid height (m) and diameter (m). What effluent BOD and TSS concentration may be expected from the UASB reactor? Describe an aerobic secondary treatment process you would select to add after the UASB process to meet an effluent BOD concentration of 20 mg/L or less. Would alkalinity have to be added to the UASB reactor to maintain the pH near 7.0? Explain the basis for your answer.

Wastewater characteristics:

Parameter	Unit	Value
Flowrate	m³/d	
Wastewater 1		3000
Wastewater 2		4000
Wastewater 3		5000
COD	mg/L	450
BOD	mg/L	180
TSS	mg/L	180
Alkalinity	mg/L as CaCO₃	150

Note: Wastewater 1, 2, or 3 to be selected by instructor.

10-10 A brewery wastewater with a flowrate of 1000 m³/d and COD (mainly soluble) of 4000 mg/L is to be treated at 35°C in a 4 m upflow attached growth anaerobic reactor, which contains cross-flow plastic packing, with the aim of 90 percent COD removal. Assume that the attached growth SRT is 30 d. Determine (a) the reactor volume (m³) and dimensions; (b) the methane gas production rate (m³/d); and (c) the effluent TSS concentration (mg/L).

10-11 An industrial wastewater has a degradable COD concentration of 8000 mg/L and 4000 mg/L VSS concentration with 50 percent of the VSS degradable. Briefly critique the compatibility of the following processes for treatment of this wastewater and describe the potential impact of the influent solids on the reactor operation and performance.

Processes:
UASB
Anaerobic fluidized bed reactor
Anaerobic baffled reactor
Upflow packed bed reactor
Downflow attached growth reactor
Anaerobic covered lagoon

10–12 From the literature within the past three years, identify and summarize an application of an anaerobic membrane process. Include a description of the wastewater treated, the reactor design, the organic loading rate, the temperature, the membrane fouling control strategy, the membrane flux rate over time, the reactor solids concentration, the membrane cleaning method or restoration method, and any significant operating and performance issues.

10–13 From a review of the literature summarize the wastewater type and characteristics, system design and operating conditions and the treatment performance of a UASB, EGSB, or anaerobic contact process (to be selected by the instructor).

REFERENCES

Aiyuk, S., I. Forrez, D. K. Lieven, A. van Haandel, and W. Verstraete (2006) "Anaerobic, and a Complementary Treatment of Domestic Sewage in Regions with Hot Climates—A Review," *Bioresource Technol.*, **97**, 17, 2225–2241.

Alvarez, J. A., E. Armstrong, M. Gomez, and M. Soto (2008) "Anaerobic Treatment of Low-Strength Municipal Wastewater by a Two-Stage Pilot Plant Under Psychrophilic Conditions," *Bioresource Technol.*, **99**, 4, 7051–7062.

Angelidaki, I., and B. K. Ahring, (1994) "Anaerobic Thermophilic Digestion of Manure at Different Ammonia Loads – Effect of Temperature," *Water Res.* **28**, 3, 727–731.

Annachhatre, A. P. (1996) "Anaerobic Treatment of Industrial Wastewaters," *Resources, Conversation, and Recycling*, **16**, 1–4, 161–166.

Arceivala, S. J. (1998) *Wastewater Treatment for Pollution Control*, 2nd ed., Tata McGraw-Hill Publishing Company Limited, New Delhi.

Banik, G. C., and R. R. Dague (1996) "ASBR Treatment of Dilute Wastewater at Psychrophilic Temperatures," *Proceedings of the WEF 81ˢᵗ ACE*, Chicago, IL.

Batstone, D. J., J. Keller, I. Angelidaki, S. V. Kalyuzhnyi, S. G. Pavlostathis, A. Rossi, W. T. M. Sanders, H. Siegrist, and V. A. Vavilin (2002a) *The IWA Anaerobic Digestion Model No. 1 (ADM1); Scientific and Technical Report 13*, IWA Publishing, London.

Batstone, D. J., J. Keller, I. Angelidaki, S. V. Kalyuzhnyi, S. G. Pavlostathis, A. Rossi, W. T. M. Sanders, H. Siegrist, and V. A. Vavilin (2002b). "The IWA Anaerobic Digestion Model No 1 (ADM1)," *Water Sci. Technol.*, **45**, 10, 65–73.

Buswell, A. M. and S. L. Neave (1930) *Illinois State Water Survey Bulletin No. 32.*

Buswell, A. M., and C. B. Boruff (1932) "The Relationship Between Chemical Composition of Organic Matter and the Quality and Quantity of Gas Production During Digestion," *Sewage Works J.*, **4**, 3, 454–460.

Buswell, A. M., and H. F. Mueller (1952) "Mechanism of Methane Fermentation," *Ind. Eng. Chem.*, **44**, 3, 550–552.

Chong, S., T. K. Sen, A. Kayaalp, and H. M. Ang (2012) "The Performance Enhancements of Upflow Anaerobic Sludge Blanket (UASB) Reactors for Domestic Sludge Treatment—A State-of-the Art Review," *Water Res.*, **46**, 11, 3434–3470.

Collins, A. G., T. L. Theis, S. Kilambi, L. He, and S. G. Pavlostathis (1998) "Anaerobic Treatment of Low-Strength Domestic Wastewater Using an Anaerobic Expanded Bed Reactor," *J. Environ. Eng.*, **124**, 7, 652–655.

Conklin, A., H. D. Stensel, and J. F. Ferguson (2006) "The Growth Kinetics and Competition Between *Methanosarcina* and *Methanosaeta* in Mesophilic Anaerobic Digestion," *Water Environ. Res.*, **78**, 5, 486–496.

Demirel, B. and P. Scherer (2011) "Trace Element Requirements of Agricultural Biogas Digesters During Biological Conversion of Renewable Biomass to Methane," *Biomass and Bioenergy*, **35**, 3, 992–998.

Elmitwalli, T. A., K. L. T. Oahn, G. Zeeman, and G. Lettinga (2002) "Treatment of Domestic Sewage in a Two-Step Anaerobic Filter/Anaerobic Hybrid System at Low Temperature," *Water Res.*, **36**, 9 2225–2232.

Evans, P. J., J. Amadori, D. Nelsen, D. Parry, and H. D. Stensel (2012) "Factors Controlling Stable Anaerobic Digestion of Food Waste and FOG," *Proceedings of the WEF 85th ACE*, New Orleans, LA.

Fermoso, F. G., G. Collins, J. Bartacek, V. O'Flaherty, and P. N. L. Lens (2008) "Role of Nickel in High Rate Methanol Degradation in Anaerobic Granular Sludge Bioreactors," *Biodegradation*, **19**, 5, 725–737.

Fezzani, B., and R. B. Cheikh (2008) "Implementation of IWA Anaerobic Digestion Model No. 1 (ADM1) for Simulating the Thermophilic Anaerobic Codigestion of Olive Mill Wastewater with Olive Mill Solid Waste in a Semi-Continuous Tubular Digester," *Chem. Eng. J.*, **141**, 1–3, 75–88.

Garcia, M. L. and L. T. Angenent (2009) "Interaction Between Temperature and Ammonia in Mesophilic Digesters for Animal Waste Treatment," *Water Res.*, **43**, 9, 2373–2382.

Girault, R., P. Rousseau, J. P. Steyer, N. Bernet, and F. Beline (2011) "Combination of Batch Experiments with Continuous Reactor Data for ADM1 Calibration: Application to Anaerobic Digestion of Pig Slurry," *Water Sci. Technol.*, **63**, 11, 2575–2582.

Gomec, C. Y. (2010) "High Rate Anaerobic Treatment of Domestic Wastewater at Ambient Operating Temperatures: A Review on Benefits and Drawbacks," *J. Environ. Sci. Health Part A*, **45**, 10, 1169–1184.

Grethlein, H. E. (1978) "Anaerobic Digestion and Membrane Separation of Domestic Wastewater," *J. WPCF*, **50**, 4, 754–763.

Hanaki, K., T. Matsuo, and M. Nagase (1981). "Mechanism of Inhibition Caused by Long-Chain Fatty-Acids in Anaerobic-Digestion Process." *Biotechnol. Bioeng.* **23**, 7, 1591–1610.

Henze, M., M. C. M. van Loosdrecht, G. A. Ekama, and D. Brdjanovic (2011) *Biological Wastewater Treatment Principles, Modeling and Design*, IWA Publishing, London.

Imhoff, K. (1938) "Sedimentation and Digestion in Germany," in L. Pease (ed.) *Modern Sewage Disposal*, Federation of Sewage Works Associations, New York.

Jeganathan, J., G. Nakhla, and A. Bassi (2006) "Long-Term Performance of High-Rate Anaerobic Reactors for the Treatment of Oily Wastewater," *Environ. Sci. Technol.*, **40**, 20, 6466–6472.

Kato, M. T., S. Rebac, and G. Lettinga, (1999) "Anaerobic Treatment of Low-Strength Brewery Wastewater in Expanded Granular Sludge Bed Reactor," *Appl. Biochem. Biotechnol.*, **76**, 1, 15–32.

Kessener, H. J. N. H. (1938) "Sewage Treatment in the Netherlands," in L. Pease (ed.) *Modern Sewage Disposal*, Federation of Sewage Works Associations, New York.

Kinnicutt L. P., C. E. A. Winslow, and R. W. Pratt (1913) *Sewage Disposal*, John Wiley & Sons, Inc., New York.

Leitão, R. C., A. C. van Haandel, G. Zeeman, and G. Lettinga (2006) "The Effects of Operational and Environmental Variations on Anaerobic Wastewater Treatment Systems: A Review," *Bioresource Technol.*, **97**, 9, 1105–1118.

Lettinga, G., A. F. M. Van Velsen, S. W. Hobma, W. J. de Zeeuw, and A. Klapwijk (1980) "Use of the Upflow Sludge Blanket (USB) Reactor Concept for Biological Wastewater Treatment," *Biotechnol. Bioeng.*, **22**, 4, 699–734.

Lettinga, G., and L. W. Hulshoff Pol (1991) "UASB-Process Designs for Various Types of Wastewaters," *Water Sci. Technol.*, **24**, 8, 87–107.

Lew, B., M. Belavski, S. Admon, S. Tarre, and M. Green (2003) "Temperature Effect on UASB Reactor Operation for Domestic Wastewater Treatment and Temperate Climate Regions," *Water Sci. Technol.*, **48**, 3, 25–30.

Li, Y. Y., H. Sasaki, K. Yamashita, K. Saki, and K. Kamigochi (2002) "High-Rate Methane Fermentation of Lipid-Rich Food Wastes by a High-Solids Codigestion Process," *Water Sci. Technol.*, **45**, 12, 143–150.

Lu, F., M. Chen, P. J. He, and L. M. Shao (2008) "Effects of Ammonia on Acidogenesis of Protein-Rich Organic Wastes," *Environ. Eng. Sci.* **25**, 1, 114–122.

Maillacheruvu, K. Y., G. F. Parkin, C. Y. Peng, W. C. Kuo, Z. I. Oonge, and V. Lebduschka (1993) "Sulfide Toxicity in Anaerobic Systems Fed Sulfate and Various Organics," *Water Environ. Res.*, **65**, 2, 100–109.

Malina, J. F., and F. G. Pohland (1992) *Design of Anaerobic Processes for the Treatment of Industrial and Municipal Wastes*, Water Quality Management Library, Vol. 7, CRC Press, Boca Raton, FL.

McCarty, P. L. (1964) "Anaerobic Waste Treatment Fundamentals: I. Chemistry and Microbiology; II. Environmental Requirements and Control; III. Toxic Materials and Their Control; IV. Process Design," *Public Works*, **95**, 9, 107–112; 10, 123–126; 11, 91–94; 12, 95–99.

McCarty, P. L. (2001) "The Development of Anaerobic Treatment and Its Future," *Water Sci. Technol.*, **44**, 8, 159–156.

Metcalf, L., and H. P. Eddy (1915) *American Sewerage Practice, III, Disposal of Sewage* (1st ed.), McGraw-Hill Book Company, Inc., New York.

Moigno, A. F. (1881) "Mouras' Automatic Scavenger," *Cosmos*, 622.

Moigno, A. F. (1882) "Mouras' Automatic Scavenger," *Cosmos*, 97.

Morgan, P. W. (1954) "Studies of Accelerated Digestion of Sewage Sludge," *Sewage Ind. Wastes*, **26**, 4, 462–478.

Muller, C. D., H. L. Gough, D. Nelson, J. F. Ferguson, H. D. Stensel, and P. Randolph (2009) "Investigating the Process Constraints of the Addition of Codigestion Substrates to Temperature Phased Anaerobic Digestion," *Proceedings of the WEF 82nd ACE,* October, 13, 2009. Orlando, FL.

Nicolella, C., M. C. M. van Loosdrecht, and J. J. Heijnen (2000) "Wastewater Treatment with Particulate Biofilm Reactors," *J. Biotechnol.*, **80**, 1, 1–33.

Obayashi, A. W., E. G. Kominek, and H. D. Stensel, (1981) "Anaerobic Treatment of High Strength Industrial Wastewater," *Chem. Eng. Prog.*, **77**, 4, 68–73.

O'Flaherty V, G. Collins, and T. Mahony (2006) "The Microbiology and Biochemistry of Anaerobic Bioreactors with Relevance to Domestic Sewage Treatment," *Rev. in Environ. Sci. and Biotechnol.*, **5**, 1, 39–55.

Osuna, M. B., M. H. Zandvoort, J. M. Iza, G. Lettinga, and P. N. L. Lens (2003) "Effects of Trace Element Addition on Volatile Fatty Acid Conversions in Anaerobic Granular Sludge Reactors," *Environ. Technol.* **24**, 5, 573–587.

Parkin, G. F., and S. W. Miller (1982) "Response of Methane Fermentation to Continuous Addition of Selected Industrial Toxicants," *Proceedings of the 37th Purdue Industrial Waste Conference*, Lafayette, IN.

Parkin, G. F., and W. E. Owen (1986) "Fundamentals of Anaerobic Digestion of Wastewater Sludges," *J. Environ. Eng.*, **112**, 5, 867–920.

Rosso, D., and M. K. Stenstrom (2008) "The Carbon-Sequestration Potential of Municipal Wastewater Treatment," *Chemosphere*, **70**, 8, 1468–1475.

Sambusiti, C., F. Monlau, E. Ficara, H. Carrere, and F. Malpei (2013) "A Comparison of Different PreTreatments to Increase Methane Production from Two Agricultural Substrates," *Applied Energy*, **104**, 62–70.

Schmidt, J. E., and B. K. Ahring (1996) "Granular Sludge Formation in Upflow Anaerobic Sludge Blanket (UASB) Reactors," *Biotechnol. Bioeng.*, **49**, 3, 229–246.

Schroepfer, G. J., W. Fullen, A. Johnson, N. Ziemke, and J. Anderson (1955) "The Anaerobic Contact Process as Applied to Packinghouse Wastes," *Sewage Ind. Wastes*, **27**, 4, 460–486.

Seghezzo, L., G. Zeeman, J. B. van Lier, H. V. M. Hamelers, and G. Lettinga (1998) "A Review: The Anaerobic Treatment of Sewage in UASB and EGSB Reactors," *Bioresource Technol.*, **65**, 3, 175–190.

Shuizhou, K., and S. Zhou (2005) "Applications of Two-Phase Anaerobic Degradation in Industrial Wastewater Treatment," *Int. J. Environment and Pollution*, **23**, 1, 65–80.

Speece, R. E. (1996) *Anaerobic Biotechnology for Industrial Wastewaters*, Archae Press, Nashville, TN.

Speece, R. E. (2008) *Anaerobic Biotechnology and Odor/Corrosion Control for Municipalities and Industries,* Fields Publishing, Inc., Nashville, TN.

Sprott, G. D., and G. B. Patel (1986) "Ammonia Toxicity in Pure Cultures of Methanogenic Bacteria," *Syst. Appl. Microbiol.* **7**, 2–3, 358–363.

Stander, G. J. (1950) "Effluents from Fermentation Industries, Part IV, A New method for Increasing and Maintaining Efficiency in the Anaerobic Digestion of Fermentation Effluents," *J. Inst. Sewage Purification*, **4**, 438.

Stander, G. J., and R. Snyders (1950) "Effluents from Fermentation Industries, Part V, Re-Inoculation as an Integral Part of the Anaerobic Digestion Method of Purification of Fermentation effluents," *J. Inst. Sewage Purification*, **4**, 447.

Steinhaus, B., M. L. Garcia, A. Q. Shen, and L. T. Angenent (2007) "A Portable Anaerobic Microbioreactor Reveals Optimum Growth Conditions for the Methanogen Methanosaeta Concilii," *Appl. Environ. Microbiol.* **73**, 5, 1653–1658.

Straub, A. J., A. S. Q. Conklin, J. F. Ferguson, and H. D. Stensel (2006) "Use of the ADM1 to Investigate the Effects of Acetoclastic Methanogenic Population Dynamics on Mesophilic Digester Stability" *Water Sci. Technol.*, **54**, 4, 59–66.

Switzenbaum, M. S., and W. J. Jewell (1980) "Anaerobic-Attached Film Expanded-Bed Reactor Treatment," *J. WPCF*, **52**, 7, 1953–1965.

Takashima, M., K. Shimada, and R. E. Speece (2011) "Minimum Requirements for Trace Metals (Iron, Nickel, Cobalt, and Zinc) in Thermophilic and Mesophilic Methane Fermentation from Glucose," *Water Environ. Res.* **83**, 4, 339–346.

Tauseef, S. M., T. Abbasi, and S. A. Abbasi (2013) "Energy Recovery from Wastewater with High Rate Anaerobic Digesters," *Renewable and Sustainable Energy Reviews,* **19**, 704–741.

Tchobanoglous, G., H. D. Stensel, and F. L. Burton (2003) *Wastewater Engineering: Treatment and Reuse*, 4th ed., Metcalf & Eddy, Inc., McGraw-Hill, New York.

Thaveesri, J., K. Gernaey, B. Kaonga, G. Boucneau, and W. Verstraete (1994) "Organic and Ammonium Nitrogen and Oxygen in Relation to Granular Sludge Growth in Lab-Scale UASB Reactors," *Water Sci. Technol.*, **30**, 12, 43–53.

Torpey, W. N. (1954) "High Rate Digestion of Concentrated Primary and Activated Sludge," *Sewage Ind. Wastes*, **26**, 4, 479–496.

Totzke, D. (2012) "2012 *Anaerobic Treatment Technology Overview*, Applied Technologies, Inc., Brookford, WI.

U.S. EPA (1991) *Hydrogen Sulfide Corrosion in Wastewater Collection and Treatment Systems*, EPA 430/09-91-010, Report to Congress, Office of Water, U.S. Environmental Protection Agency, Washington, DC.

van Haandel, A. C. and G. Lettinga (1994) *Anaerobic Sewage Treatment: A Practical Guide for Regions with a Hot Climate*. John Wiley & Sons, Chichester, UK.

van Velsen, A. F. M. (1977) "Anaerobic Digestion of Piggery Waste," *Netherlands J. Agri. Sci.*, **25**, 3, 151–169.

Visvanathan, C. and A. Abeynayaka (2012) "Developments and Future Potentials of Anaerobic Membrane Bioreactors (AnMBRs)," *Membrane Water Treat.*, **3**, 1, 1–23.

Winslow, C. E. A. (1938) "Pioneers of Sewage Disposal in New England," in L. Pease (ed.) *Modern Sewage Disposal*, Federation of Sewage Works Associations, New York.

Wilson, C. A., J. Novak, I. Takacs, B. Wett, and S. Murthy (2012) "The Kinetics of Process Dependent Ammonia Inhibition of Methanogenesis from Acetic Acid," *Water Res.*, **46**, 19, 6247–6256.

Young, J. C. and P. L. McCarty (1969) "The Anaerobic Filter for Waste Treatment," *J. WPCF*, **41**, 5, Research Supplement to: **41**, 5, Part II, R160–R173.

Young, J. C. (1991) "Factors Affecting the Design and Performance of Upflow Anaerobic Filters," *Water Sci. Technol.*, **24**, 8, 133–155.

Yu, H. Q., H. H. P. Fang, and J. H. Tay (2000) "Effects of Fe^{2+} on Sludge Granulation in Upflow Anaerobic Sludge Blanket Reactors," *Water Sci. Technol.*, **41**, 12, 199–205.

Yu, H. Q., J. H. Tay, and H. H. P. Fang (2001) "The Roles of Calcium in Sludge Granulation During UASB Reactor Start-Up," *Water Res.*, **35**, 4, 1052–1060.

Zeeman, G., W. T. M. Sanders, K. Y. Wang, and G. Lettinga (1997) "Anaerobic Treatment of Complex Wastewater and Waste Activated Sludge–Application of an Upflow Anaerobic Solid Removal (UASR) Reactor for the Removal and Pre-Hydrolysis of Suspended COD," *Water Sci. Technol.*, **35**, 10, 121–128.

Zhang, L., and D. Jahng (2010) "Enhanced Anaerobic Digestion of Piggery Wastewater by Ammonia Stripping: Effects of Alkali Types," *J. Hazard. Mater.*, **182**, 1–3, 536–543.

11

Separation Processes for Removal of Residual Constituents

WORKING TERMINOLOGY

Term	Definition
Absorption	The process by which atoms, ions, molecules, and other constituents are transferred from one phase and are distributed uniformly in another phase (see also adsorption).
Activated carbon	A substance used commonly in adsorption processes for the removal of trace constituents from water and odor compounds from air. Activated carbon is derived from an organic base material, prepared using a high temperature pyrolysis process and activated at high temperature in the presence of steam resulting in properties conducive to mass transfer.

Term	Definition
Adsorption	The process by which atoms, ions, molecules, and other constituents are transferred from one phase and accumulate on the surface of another phase (see also absorption).
Backwash	The process of removing solids accumulated on or in a filtration medium by applying air and/or clean water in the opposing flow direction.
Brine	Concentrated liquid waste stream containing elevated concentrations of total dissolved solids
Depth filtration	The removal of particulate matter suspended from a liquid by passing the liquid through a granular medium such as sand or anthracite coal.
Electrodialysis (ED)	A process that moves ions (charged molecular species) from one solution to another by employing an electrical potential as the driving force and using a semipermeable membrane as a separator.
Flux	The mass or volume rate of transfer through the membrane surface, usually expressed as $m^3/m^2{\cdot}h$ or $L/m^2{\cdot}h$ $(gal/ft^2{\cdot}d)$.
Fouling	The accumulation of solid matter on the surface of or within the pores of a membrane that impedes the flow of permeate through the membrane.
Gas stripping	A process to remove a volatile constituent from a liquid phase, such as in the removal of ammonia from water in a packed column using air as the gas phase.
Ion exchange	A process used for the removal of dissolved ionic constituents where ions of a given species are displaced from a solid phase material by ions of a different species from solution.
Isotherm	A function used to relate the amount of a given constituent adsorbed from water per concentration of adsorbent at a given temperature.
Membrane	A device, usually made of an organic polymer, that allows the passage of water and certain constituents, but rejects others above a certain physical size or molecular weight.
Microfiltration (MF)	A membrane separation process used typically to remove particulate material from the feed water; microfiltration pore sizes range approximately from 0.05 to 2 μm.
Nanofiltration (NF)	A pressure-driven membrane separation process used to remove colloidal and dissolved material as small as approximately 0.001 μm.
Residuals	Waste streams produced by wastewater treatment processes. For depth and surface filtration, the residual waste stream is filter waste washwater. For membrane systems, residual waste streams include waste washwater, concentrate, and chemical cleaning wastes.
Reverse osmosis (RO)	The rejection of dissolved constituents by preferential diffusion using a pressure-driven, semipermeable membrane.
Semipermeable membrane	A membrane that is permeable to some components in a feed solution and impermeable to other components.
Separation processes	Physical and chemical processes used in water reclamation that bring about treatment by the isolation of particular constituents. The isolated constituents are concentrated into a waste stream that must be managed.
Sodium adsorption ratio (SAR)	A measure of the sodicity of the soil; the SAR is the ratio of the sodium cation to the calcium and magnesium cations.
Surface filtration	The removal of particulate matter suspended in a liquid by passing the liquid through a thin septum, usually a cloth or metal medium.
Synthetic organic compounds (SOCs)	Compounds of synthetic origin used extensively in industrial processes and contained in numerous manufactured consumer products. The presence of SOCs in drinking water as well as reclaimed water is of concern due to toxicity and unknown effects.
Ultrafiltration (UF)	A membrane separation process similar to MF except the membrane pore sizes can range from approximately 0.005 to 0.1 μm. Generally, UF membranes are able to achieve higher levels of separation than MF, particularly for bacteria and viruses.

The effluent from conventional secondary treatment contains varying amounts of residual suspended, colloidal, and dissolved constituents. Suspended and colloidal matter can reduce the effectiveness of downstream disinfection processes or make the effluent unsuitable for discharge or reuse. Dissolved constituents may range from relatively simple inorganic ions, such as calcium, potassium, sulfate, nitrate, and phosphate to an ever-increasing number of highly complex synthetic organic compounds. Research is ongoing to determine (1) the environmental effects of potentially toxic and biologically active substances found in wastewater and (2) how these substances can be removed by both conventional and advanced wastewater treatment processes. In recent years, the effects of many of these substances on the environment have become understood more clearly. As a result, wastewater treatment requirements are becoming more stringent in terms of limiting effluent concentrations of many of these substances.

To meet new treatment requirements, many of the existing secondary treatment facilities will have to be retrofit and new advanced wastewater treatment facilities will have to be constructed. The purpose of this chapter is to present an introduction to the unit processes used for the removal and/or treatment of residual particulate, colloidal, and dissolved constituents in treated wastewater. However, before discussing the individual unit processes, it will be helpful to review the need for additional wastewater treatment and the reasons that specific constituents are of concern.

11-1 NEED FOR ADDITIONAL WASTEWATER TREATMENT

Residual constituents found in secondary effluent can be grouped into four broad categories: (1) organic and inorganic suspended and colloidal particulate matter, (2) dissolved organic constituents, (3) dissolved inorganic constituents, and (4) biological constituents. Constituents within each category are reported in Table 11-1, along with the reasons for their removal. The potential impacts of the residual constituents identified in Table 11-1 will vary considerably depending on local conditions. The list of constituents presented in Table 11-1 is not meant to be exhaustive; rather, it is meant to highlight that a wide variety of substances must be considered in establishing and meeting discharge requirements. Also, based on the accumulation of scientific knowledge concerning the impacts of the residual constituents found in secondary effluent, derived from laboratory studies and environmental monitoring, it is anticipated that many of the treatment methods now classified as tertiary or advanced will be considered conventional within the next 10 to 20 years. For example, effluent filtration has become more commonplace within the past 20 years.

11-2 OVERVIEW OF TECHNOLOGIES USED FOR REMOVAL OF RESIDUAL PARTICULATE AND DISSOLVED CONSTITUENTS

Over the past 20 years, a wide variety of treatment technologies have been studied, developed, and applied for the removal of the residual constituents found in secondary and tertiary effluent. The unit processes used for the removal of residual constituents from water may be classified as (1) mass transfer separation processes and (2) chemical and biological transformation processes.

Separation Processes Based on Mass Transfer

The removal of constituents by the transfer of mass from one phase to another or by the concentration of mass within a phase is accomplished with various unit processes.

Table 11–1

Typical residual constituents found in treated wastewater effluents and reasons that additional treatment may be required

Residual constituent	Effect and/or need for additional treatment
Inorganic and organic suspended and colloidal particulate matter	
Suspended solids	• Can impact disinfection by shielding organisms • May cause sludge deposits or interfere with receiving water clarity • May affect effluent turbidity
Colloidal solids	• May affect effluent turbidity
Organic matter (particulate)	• May shield bacteria during disinfection, may deplete oxygen resources
Dissolved organic matter	
Total organic carbon	• May deplete oxygen resources
Refractory organics	• Toxic to humans; carcinogenic
Volatile organic compounds	• Toxic to humans; carcinogenic; form photochemical oxidants
Pharmaceutical compounds	• Impacts to aquatic species (e.g., endocrine disruption)
Surfactants	• Cause foaming and may interfere with coagulation
Dissolved inorganic matter	
Ammonia	• Increases chlorine demand • Can be converted to nitrates and, in the process, can deplete oxygen resources • With phosphorus, can lead to the development of undesirable aquatic growth • Toxic to fish
Nitrate	• Can stimulate algal and other aquatic growth • Can cause methemoglobinemia in infants (blue babies)
Phosphorus	• Can stimulate algal and other aquatic growth • Increases chemical requirements • Interferes with lime-soda softening
Calcium and magnesium	• Increase hardness and total dissolved solids • Can affect sodium adsorption ratio
Chloride and sulfate	• Can impart salty taste
Total dissolved solids	• Interfere with agricultural and industrial processes • Can interferes with coagulation
Biological	
Bacteria	• Can cause disease
Protozoan cysts and oocysts	• Can cause disease
Viruses	• Can cause disease

The principal mass transfer processes used for the separation (removal) of residual constituents are summarized in Table 11–2. It is important to note that a key characteristic of most separation processes is the generation of a waste stream that will require subsequent management (e.g., processing, disposal, reuse). The particular waste stream generated will depend on the type and effectiveness of the separation process used. For example,

Table 11-2

Unit processes based on mass transfer used for the removal of particulate and dissolved constituents in wastewater treatment and water reclamation[a]

Unit process	Phase	Application
Absorption	Gas → liquid	Aeration, O_2 transfer, SO_2 scrubbing, chlorination, chlorine dioxide and ammonia addition, ozonation
Adsorption	Gas → solid / Liquid → solid	Removal of inorganic and organic compounds using activated carbon, activated alumina, granular ferric hydroxide, or other adsorbent material
Distillation	Liquid → gas	Demineralization of water, concentrating of waste brines
Electrodialysis	Liquid → liquid	Removal of dissolved species, removal of salts
Filtration, depth	Liquid → solid	Removal of particulate material
Filtration, surface	Liquid → solid	Removal of particulate material
Flotation	Liquid → solid	Removal of particulate constituents
Gas stripping	Liquid → gas	Removal of NH_3 and other volatile inorganic and organic chemicals
Ion exchange	Liquid → solid	Demineralization of water, removal of specific constituents, softening
Microfiltration ultrafiltration	Liquid → liquid	Removal of particulate and colloidal species
Nanofiltration	Liquid → liquid	Removal of dissolved and colloidal species; softening
Precipitation, chemical	Liquid → solid	Removal of particulate and dissolved species; softening
Reverse osmosis	Liquid → solid	Removal of dissolved constituents
Sedimentation	Liquid → solid	Removal of particulate constituents

[a] Adapted in part from Crittenden et al. (2012).

adsorption results in a medium saturated with removed constituents, chemical precipitation produces a sludge containing both the precipitated constituents as well as the chemical(s) added to cause the precipitation, and reverse osmosis produces a liquid waste (brine) containing concentrated rejected constituents. In many cases, the management of waste streams resulting from separation processes, as discussed in Chap. 15, can present a significant technological challenge and cost.

Transformation Based on Chemical and Biological Processes

The second group of processes used for the removal of residual constituents make use of chemical and biological reactions to transform or destroy trace constituents in water, typically through oxidation and reduction reactions. Conventional chemical oxidants that have been used for constituent transformation include hydrogen peroxide, ozone, chlorine, chlorine dioxide, and potassium permanganate. Chemical oxidation processes that utilize hydroxyl radical species, referred to as advanced oxidation processes (AOPs), or photons generated through UV photolysis, are particularly effective for the transformation and destruction of trace constituents, often resulting in the complete mineralization of trace constituents to carbon dioxide and mineral acids. Chemical treatment processes including advanced oxidation and photolysis are considered in Chap. 6. Chemical disinfection of wastewater is considered separately in Chap. 12. Biological treatment and conversion processes are considered in Chaps. 7 through 10.

Application of Unit Processes for Removal of Residual Constituents

Information on the application of the unit processes identified in Table 11–2 is presented in Table 11–3. Selection of a given unit process or combination thereof depends on (1) the use to be made of the treated effluent; (2) the constituent(s) of concern; (3) the compatibility of the various operations and processes; (4) the available means for management of any process residuals; and (5) the environmental and economic feasibility of the various systems. Specific factors that should be considered in the selection of treatment processes were identified and discussed previously in Table 4–2 in Chap. 4. It should be noted that in some situations, economic feasibility may not be a controlling factor in the design of advanced wastewater treatment systems, especially where specific constituents must be removed to protect the environment and/or to meet discharge requirements. Because of the variations in performance observed in the field, bench-scale and pilot-plant testing is recommended for the development of local treatment performance data and design criteria. Representative performance data for the processes identified in Table 11–3 are presented in the discussion of the individual technologies that follows and in the indicated sections in other chapters.

11–3 UNIT PROCESSES FOR THE REMOVAL OF RESIDUAL PARTICULATE AND DISSOLVED CONSTITUENTS

The principal unit processes used for the removal of residual particulate matter discussed in this chapter include (1) depth filtration (passing the liquid through a filter bed comprised of a granular or compressible filter medium); (2) surface filtration (the removal of particulate material suspended in a liquid by mechanical sieving by passing the liquid through a thin septum); and (3) membrane filtration (passing the liquid through porous material to exclude particles ranging in size from 0.005 to 2.0 μm). Each of these processes is illustrated and described in Table 11–4. Flotation (attaching air bubbles to particulate matter to provide buoyancy so the particles can be removed by skimming) is included in Table 11–4 for completeness, but has been considered previously in Chap. 5.

The principal unit processes used for the removal of dissolved constituents, as discussed in this chapter, include (1) reverse osmosis (passing the liquid through semipermeable membranes to exclude particles ranging in size from 0.0001 to 0.001 μm), (2) electrodialysis (transport of ionic species through an ion-selective membranes), (3) adsorption (the accumulation of constituents on a solid phase), (4) gas stripping (transfer of a constituent from a liquid to a gas phase), (5) ion exchange (the exchange of ionic species), and (6) distillation (constituents are separated by evaporation).

Table 11–3

Application of the unit processes for the removal of residual particulate and dissolved constituents found in treated wastewater effluents[a]

	Unit process (Section discussed)					
Residual constituent	**Depth filtration (11–4)**	**Surface filtration (11–5)**	**Micro and ultra-filtration (11–6)**	**Reverse osmosis (11–6)**	**Electro-dialysis (11–7)**	**Adsorption (11–8)**
Inorganic and organic suspended and colloidal particulate matter						
Suspended solids	✔	✔	✔	✔		✔
Colloidal solids	✔	✔	✔			✔
Dissolved organic matter						
Total organic carbon				✔	✔	✔
Refractory organics				✔	✔	✔
Volatile organic compounds				✔	✔	✔
Dissolved inorganic matter						
Ammonia[a]				✔	✔	
Nitrate[a]				✔	✔	
Phosphorus[a]	✔[b]			✔	✔	
Totals dissolved solids				✔	✔	
Biological						
Bacteria		✔	✔	✔		
Protozoan cysts and oocysts	✔	✔	✔	✔		✔
Viruses			✔	✔		

[a] The biological removal of nitrogen and phosphorus is considered in Chaps. 7 through 10.

[b] Phosphorous removal is accomplished in a two-stage filtration process.

[c] Some carryover can occur.

Typical Process Flow Diagrams

Typical treatment process flow diagrams that incorporate the unit processes discussed above are illustrated on Fig. 11–1. The combination of unit processes will depend on the treatment objective. For example, in the flow diagram shown on Fig. 11–1(b), electrodialysis is used to remove salts to reduce the total dissolved solids in the treated effluent. On Fig. 11(d), a number of unit processes have been combined to produce a potable water. Where reverse osmosis is used, some form of membrane filter will be used upstream to mitigate the effects of particulate matter that tends to clog the membrane. On Fig. 11–1(f), two stages of reverse osmosis have been combined to produce water suitable for use in

Table 11–3 (Continued)

Unit process (Section discussed)						
Gas stripping (11–9)	Ion exchange (11–10)	Distillation (11–11)	Chemical precipitation (6-3, 4, 5)	Chemical oxidation (6–7)	Advanced oxidation processes (6–8)	Photolysis (6–9)
	✔	✔	✔			
	✔	✔	✔			
	✔	✔		✔	✔	✔
		✔		✔	✔	✔
✔		✔c		✔	✔	✔
✔	✔	✔				
	✔	✔				
		✔	✔			
		✔				
		✔	✔			
		✔				
		✔		✔	✔	✔

high-pressure boilers. Clearly, a wide variety of treatment process flow diagrams can be developed, depending on the specific requirements. Other examples of process flow diagrams are presented and discussed throughout this chapter.

Process Performance Expectations

It is important to know what typical mean effluent constituent values can be expected and the variability in those values for a given unit process. Information on the constituent values and variability is of importance in meeting effluent requirements and in the selection of technologies that might be used to further process the treated effluent. Typical mean

Table 11–4

Description of commonly used processes for the removal of residual suspended and colloidal solids

Unit process	Description
(a) Depth filtration 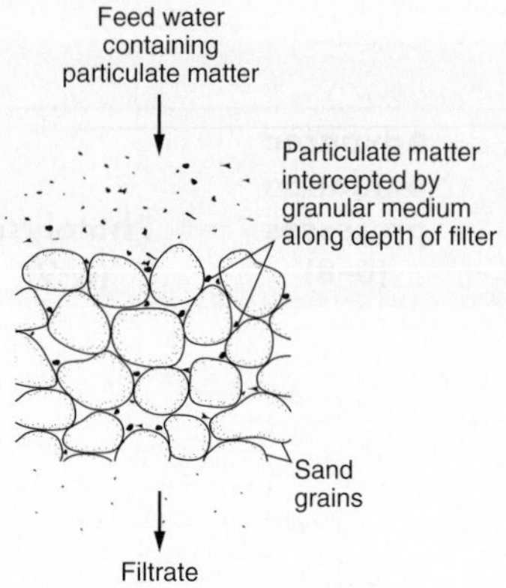 Feed water containing particulate matter Particulate matter intercepted by granular medium along depth of filter Sand grains Filtrate	Depth filtration was developed originally for the treatment of surface water for potable uses and later adapted for wastewater treatment applications. Depth filtration is used in to achieve supplemental removal of suspended solids (including particulate BOD) from wastewater effluents for the following purposes: (1) to allow more effective disinfection; (2) as a pretreatment step for subsequent treatment steps such as carbon adsorption, membrane filtration, or advanced oxidation; and (3) to remove chemically precipitated phosphorus.
(b) Surface filtration 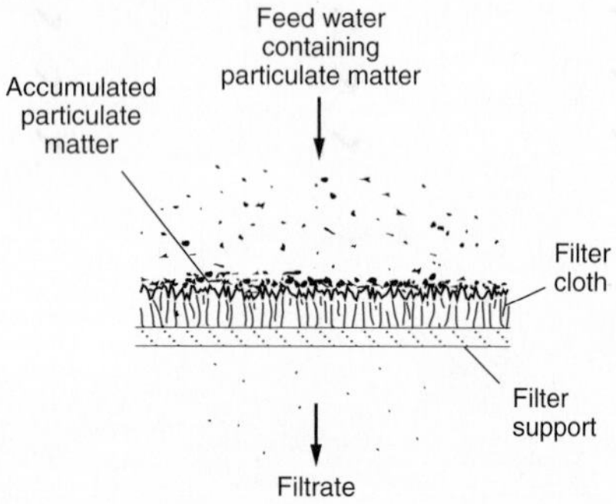 Feed water containing particulate matter Accumulated particulate matter Filter cloth Filter support Filtrate	Surface filtration is used to remove the residual suspended solids from secondary effluents and stabilization pond effluents, and as an alternative to depth filtration. Surface filtration, a relatively new technology, involves a sieving action similar to a kitchen colander.
(c) Membrane filtration 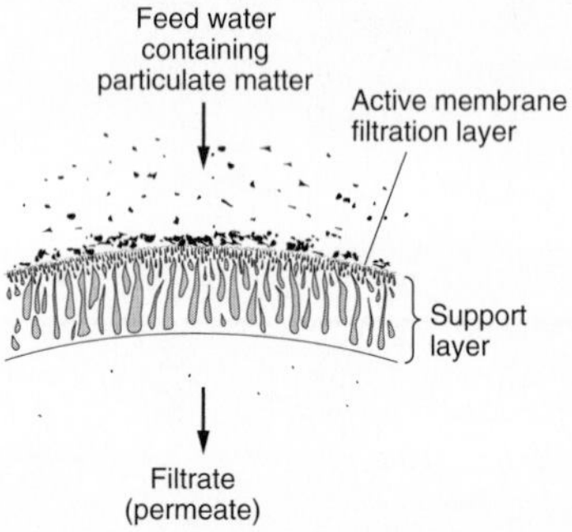 Feed water containing particulate matter Active membrane filtration layer Support layer Filtrate (permeate)	Membrane filtration with microfiltration (MF) and ultrafiltration (UF) membranes is being used increasingly for water and wastewater applications. Microfiltration and UF membrane filters are also surface filtration devices but are differentiated on the basis of the sizes of the pores in the filter medium; the pore size can vary from 0.005 to 2.0 μm. In water reuse applications, MF and UF usually follow biological treatment and are used to remove particulates, including pathogens; organic matter; and some nutrients, not removed by secondary clarification. Product water from MF and UF may be used directly for a variety of reuse applications (after disinfection) or used as pretreated feed water for further treatment by nanofiltration (NF) or reverse osmosis (RO).
(d) Dissolved air flotation 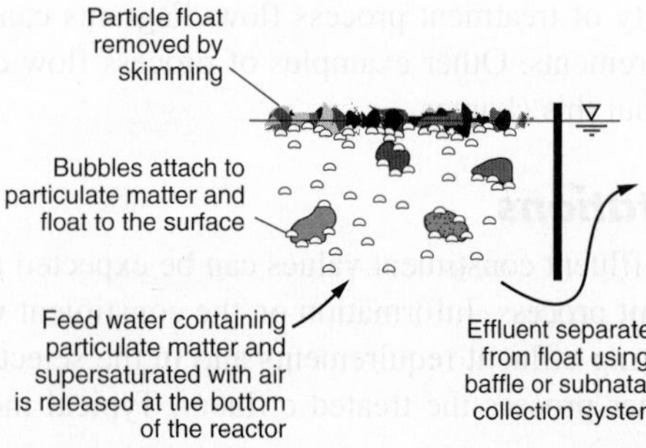 Particle float removed by skimming Bubbles attach to particulate matter and float to the surface Feed water containing particulate matter and supersaturated with air is released at the bottom of the reactor Effluent separated from float using baffle or subnatant collection system	Dissolved air flotation is a gravity separation process in which gas bubbles attach to solid particles to cause the density of the bubble-solid agglomerates to be lighter than water. For water reuse applications, DAF has been used principally for treating pond effluents containing algae and for low density particles that are difficult to remove by gravity sedimentation, as a replacement for conventional primary sedimentation, and as a pretreatment step for depth or surface filtration. Dissolved air flotation is considered in Sec. 5–7 in Chap. 5.

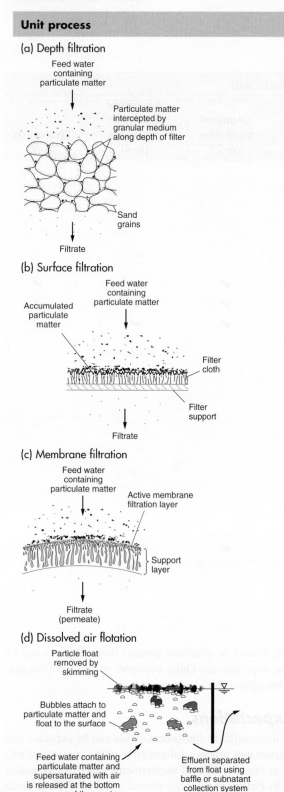

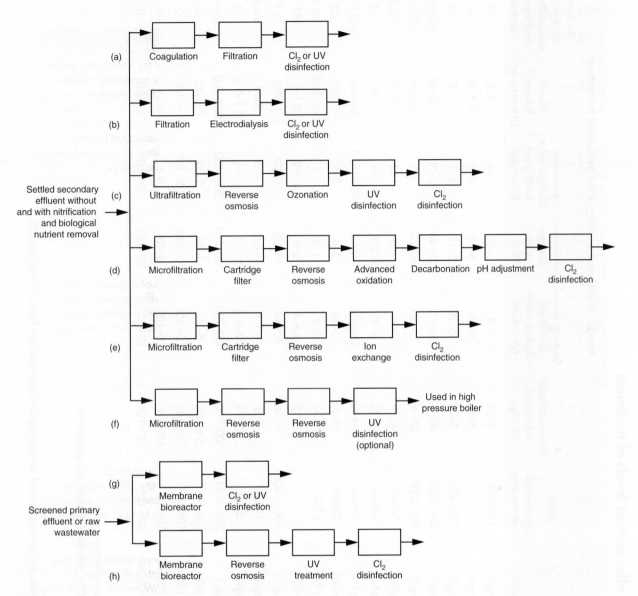

Figure 11-1

Typical process flow diagrams for wastewater treatment employing advanced treatment processes. All of the flow diagrams have been used at one time or another. For example, in flow diagram number (d) advanced oxidation is used for the destruction of NDMA. In flow diagram number (e) ion exchange is used for the removal of nitrate.

effluent constituent values that can be achieved with depth, surface, and membrane filtration following various forms of biological treatment are reported in Table 11–5. The variability observed in the performance of various particulate removal processes with respect to TSS and turbidity in the treated effluent is discussed in the sections which deal with depth, surface, and membrane filtration.

Table 11-5
Typical range of effluent quality after various levels of treatment

Constituent	Unit	Untreated wastewater[a]	Conventional activated sludge[b]	Conventional activated sludge with filtration[b]	Activated sludge with BNR[c]	Activated sludge with BNR and filtration[c]	Membrane bioreactor	Activated sludge with microfiltration and reverse osmosis
				Range of effluent quality after indicated treatment				
Total suspended solids (TSS)	mg/L	130–389	5–25	2–8	5–20	1–4	<1–5	≤1
Colloidal solids	mg/L		5–25	5–20	5–10	1–5	0.5–4	≤1
Biochemical oxygen demand (BOD)	mg/L	133–400	5–25	<–5–20	5–15	1–5	<1–5	≤1
Chemical oxygen demand (COD)	mg/L	339–1016	40–80	30–70	20–40	20–30	<10–30	≤2–10
Total organic carbon (TOC)	mg/L	109–328	20–40	15–30	10–20	1–5	<0.5–5	0.1–1
Ammonia nitrogen	mg N/L	14–41	1–10	1–6	1–3	1–2	<1–5	≤0.1
Nitrate nitrogen	mg N/L	0–trace	5–30	5–30	<2–8	1–8	<8[d]	≤1
Nitrite nitrogen	mg N/L	0–trace	0–trace	0–trace	0–trace	0.001–0.1	0–trace	≤0.001
Total nitrogen	mg N/L	23–69	15–35	15–35	3–8	2–5	<10[d]	≤1
Total phosphorus	mg P/L	3.7–11	3–10	3–8	1–2	≤2	<0.3^e–5	≤0.5
Turbidity	NTU		2–15	0.5–4	2–8	0.3–2	0.1–1	0.01–1
Volatile organic compounds (VOCs)	μg/L	<100–>400	10–40	10–40	10–20	10–20	10–20	≤1
Metals	mg/L	1–2.5	1–1.5	1–1.4	1–1.5	1–1.5	trace	trace
Surfactants	mg/L	4–10	0.5–2	0.5–1.5	0.1–1	0.1–1	0.1–0.5	≤1
Totals dissolved solids (TDS)	mg/L	374–1121	500–700	500–700	500–700	500–700	500–700	≤5–40
Trace constituents[f]	μg/L	10–50	5 to 40	5–30	5–30	5–30	0.5–20	~0.1
Total coliform	No./100 mL	10^6–10^{10}	10^4–10^5	10^3–10^5	10^4–10^5	10^4–10^5	<100	~0
Protozoan cysts and oocysts	No./100 mL	10^1–10^5	10^1–10^2	0–10	0–10	0–1	0–1	~0
Viruses	PFU/100 mL^g	10^1–10^4	10^1–10^3	10^1–10^3	10^1–10^3	10^1–10^3	10^0–10^3	~0

a From Table 3–18 in Chap. 3.
b Conventional activated sludge treatment includes nitrification.
c BNR is defined as biological nutrient removal for the removal of nitrogen and phosphorus.
d With anoxic stage.
e With coagulant addition.
f For example, fire retardants, personal care products, and prescription and non-prescription drugs (see also Table 2–16 in Chap. 2).
g PFU = plaque forming units.

11–4 INTRODUCTION TO DEPTH FILTRATION

Depth filtration with a non-compressible filter medium or media is one of the oldest unit process used in the treatment of potable water and is commonly used for the filtration of effluents from wastewater treatment processes, especially in low-level nutrient removal and water reuse applications. Depth filtration is used most commonly to (1) achieve supplemental removals of residual suspended solids (including particulate BOD and phosphorus), (2) reduce the mass discharge of solids, and (3) perhaps more importantly, as a conditioning step that will allow for the effective disinfection of the filtered effluent, especially with UV disinfection (see Chap. 12). Single- and two-stage filtration is used to remove chemically precipitated phosphorus. The relationship between depth filtration and other forms of filtration is illustrated on Fig. 11–2. In the past, depth filtration was used almost exclusively for effluent filtration. However, with the development of modern surface filtration technologies, as discussed in Sec. 11–5, depth filtration is no longer the dominant filtration technology.

To introduce the subject of depth filtration, the purpose of this section is to present (1) a general introduction to the depth filtration process, (2) an introduction to filter clean-water hydraulics, and (3) an analysis of the filtration process. The types of filters that are available and issues associated with their selection and design, including a discussion of the need for pilot-plant studies, are considered in the following section.

Description of the Filtration Process

The basics of the depth filtration can be understood by considering the (1) physical features of a conventional granular medium depth filter, (2) characteristics of the filter-medium, (3) the process by which suspended material is removed from the liquid, and (4) the backwash process in which material retained within the filter is removed.

Physical Features of a Depth Filter. The general features of a conventional granular medium depth filter are illustrated on Fig. 11–3. As shown, the filtering medium (sand in this case) is supported on a gravel layer, which, in turn, rests on the filter underdrain system. The water to be filtered enters the filter from an inlet channel. Filtered water

Figure 11–2

Classification of filtration processes used in wastewater management. Note: Intermittent and recirculating porous medium filters are used for small systems and are not considered in this text.

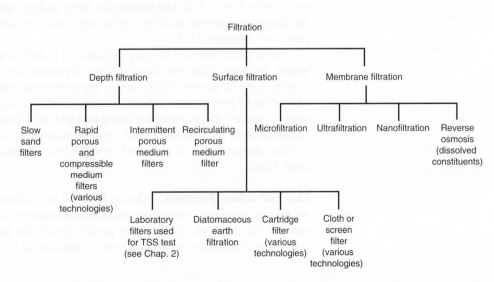

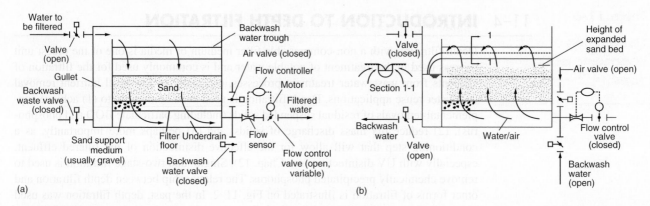

Figure 11–3

General features and operation of a conventional rapid granular medium depth filter: (a) flow during filtration cycle, and (b) flow during backwash cycle. (From Tchobanoglous and Schroeder, 1985.)

is collected in the underdrain system which is also used to reverse the flow to backwash the filter. Filtered water typically is disinfected before being discharged to the environment. If the filtered water is to be reused, it can be discharged to a storage reservoir or to the reclaimed water distribution system.

Characteristics of the Filter Medium. Grain size is the principal filter-medium characteristic that affects the filtration operation including the removal of suspended and colloidal material, the clear-water headloss, and the buildup of headloss during the filter run. If the size of the filtering medium is too small, much of the driving force will be wasted in overcoming the frictional resistance of the filter bed. If the size of the medium is too large, many of the small particles in the influent will pass directly through the filter bed. Thus, the selection of media size must balance the need for a target filtered water quality with an acceptable rate of filter headloss development. The size distribution of the filter material is usually determined by sieve analysis using a series of decreasing sieve sizes. The designation and size of opening for U.S. sieve sizes are given in Table 11–6. The results of a sieve analysis are usually analyzed by plotting the cumulative percent passing a given sieve size on arithmetic-log or probability-log paper (see Example 11–1).

The effective size of a filtering medium, $d_{10,}$ is defined as the 10 percent size based on weight. For sand, it has been found that the 10 percent size by weight corresponds approximately to the 50 percent size by count. The uniformity coefficient (UC) is defined as the ratio of the 60 percent size to the 10 percent size (UC = d_{60}/d_{10}). Sometimes it is advantageous to specify the 99 percent passing size and the 1 percent passing size to define the gradation curve for each filter medium more accurately. Additional information on filter medium characteristics is presented in the following section dealing with the design of depth filters.

The Filtration Process. During filtration in a conventional downflow depth filter, wastewater containing suspended and colloidal material is applied to the top of the filter bed [see Fig. 11–3(a)]. As the water passes through the filter bed, the suspended matter (measured as turbidity) in the wastewater is removed by a variety of removal mechanisms,

Table 11–6

Designation and size of opening of US sieve sizes[a]

Sieve size or number	Size of opening	
	in.	mm
3/8 in.	0.375[a]	9.51[b]
1/4 in.	0.250[a]	6.35[b]
4	0.187	4.76
6	0.132	3.36
8	0.0937	2.38
10	0.0787[a]	2.00[b]
12	0.0661	1.68
14	0.0555[a]	1.41[b]
16	0.0469	1.19
18	0.0394[a]	1.00[b]
20	0.0331	0.841
25	0.0280[a]	0.710[b]
30	0.0234	0.595
35	0.0197[a]	0.500[b]
40	0.0165	0.420
45	0.0138[a]	0.350[b]
50	0.0117	0.297
60	0.0098[a]	0.250[b]
70	0.0083	0.210
80	0.0070[a]	0.177[b]
100	0.0059	0.149
140	0.0041	0.105
200	0.0029	0.074

[a] Adapted from ASTM (2001b).

[b] Size does not follow the ratio $(2)^{0.5}$.

as described below. With the passage of time, as material accumulates within the interstices of the granular medium, the headloss through the filter starts to build up beyond the initial value, as shown on Fig. 11–4.

Headloss and Turbidity Considerations. After some period of time, the operating headloss or effluent turbidity reaches some predetermined headloss or turbidity value, and the filter must be cleaned. Under ideal conditions, the time required for the headloss buildup to reach the preselected terminal value should correspond to the time when the turbidity or suspended solids in the effluent reach the preselected terminal value for acceptable quality. Turbidity breakthrough occurs when the interstitial spaces within the filter bed fill to a point where the shearing force of the liquid passing through the filter exceeds the strength of the bond formed between the material being filtered and the accumulated material. At breakthrough, accumulated material will dislodge only to be replaced by new material so

Figure 11–4

Definition sketch for length of filter run based on: (a) headloss buildup and (b) effluent turbidity breakthrough.

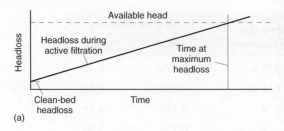

(a)

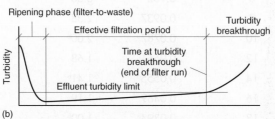

(b)

Note: Ripening period is the time required to reach an acceptable effluent turbidity value after the backwash cycle is completed. Backwash water during this period is typically returned to the process or plant inflow.

that an equilibrium condition is maintained. In actual practice, one or the other event will govern the backwash cycle.

Particle Removal Mechanisms. The principal particle-removal mechanisms, believed to contribute to the removal of material within a granular medium filter, are identified and described in Table 11–7. The major removal mechanisms (the first five listed in Table 11–7) are illustrated pictorially on Fig. 11–5. Straining has been identified as the principal mechanism that is operative in the removal of suspended solids during the filtration of settled secondary effluent from biological treatment processes (Tchobanoglous and Eliassen, 1970; Tchobanoglous, 1988).

Other mechanisms including interception, impaction, and adhesion are also operative even though their effects are small and, for the most part, masked by the straining action. The removal of the smaller particles found in wastewater (see Fig. 11–5) must be accomplished in two steps involving (1) the transport of the particles to or near the surface where they will be removed and (2) the removal of particles by one or more of the operative removal mechanisms. This two-step process has been identified as transport and attachment (O'Melia and Stumm, 1967).

Conventional down-flow filters, dual- and multi-media and deep-bed mono-medium depth filters (see Fig. 11–6) were developed to allow the suspended solids in the liquid to be filtered to penetrate further into the filter bed, and thus use more of the solids-storage capacity available within the filter bed. The deeper penetration of the solids into the filter bed also permits longer filter runs because the buildup of headloss is reduced. By comparison, in shallow mono-medium beds, most of the removal occurs in the upper few millimeters of the bed.

Backwash Process. The end of the filter run (filtration phase) is reached when the suspended solids in the effluent start to increase (breakthrough) beyond an acceptable level, or when a limiting headloss occurs across the filter bed (see Fig. 11–4). Once either of these conditions is reached, the filtration phase is terminated, and the filter must be cleaned (backwashed) to remove the material (suspended solids) that has accumulated within the granular medium filter bed. Backwashing is accomplished by reversing the flow

Table 11–7

Principal mechanisms and phenomena contributing to removal of material within a granular medium depth filter

Mechanism/phenomenon	Description
1. Straining	
a. Mechanical	Particles larger than the pore space of the filtering medium are strained out mechanically.
b. Chance contact	Particles smaller than the pore space are trapped within the filter by chance contact.
2. Sedimentation or impaction	Heavy particles that do not follow the flow streamlines settle on the filtering medium within the filter.
3. Interception	Many particles that move along in the streamline are removed when they come in contact with the surface of the filtering medium.
4. Adhesion	Particles become attached to the surface of the filtering medium as they pass by. Because of the force of the flowing water, some material is sheared away before it becomes firmly attached and is pushed deeper into the filter bed. As the bed becomes clogged, the surface shear force increases to a point at which no additional material can be removed. Some material may break through the bottom of the filter, causing the sudden appearance of turbidity in the effluent.
5. Flocculation	Flocculation can occur within the interstices of the filter medium. The larger particles formed by the velocity gradients within the filter are then removed by one or more of the above removal mechanisms.
6. Chemical adsorption	
a. Bonding	
b. Chemical interaction	Once a particle has been brought in contact with the surface of the filtering medium or with other particles, either one of these mechanisms, or both, may be responsible for holding it there.
7. Physical adsorption	
a. Electrostatic forces	
b. Electrokinetic forces	
c. van der Waals forces	
8. Biological growth	Biological growth within the filter will reduce the pore volume and may enhance the removal of particles with removal mechanisms 1 through 5.

Figure 11–5

Removal of suspended particulate matter within a granular filter by: (a) straining, (b) sedimentation or inertial impaction, (c) interception, (d) adhesion, and (e) flocculation with subsequent removal by one or more of the previous mechanisms. (Adapted from Tchobanoglous and Schroeder, 1985.)

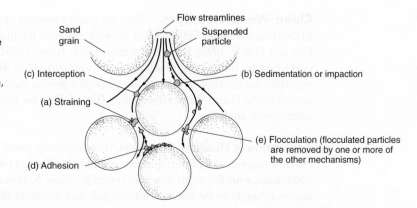

Flow streamlines
Sand grain
Suspended particle
(c) Interception
(b) Sedimentation or impaction
(a) Straining
(e) Flocculation (flocculated particles are removed by one or more of the other mechanisms)
(d) Adhesion

Figure 11–6

Schematic diagram of filter beds illustrating potential increase in storage capacity: (a) single medium, (b) dual media, and (c) multi-media.

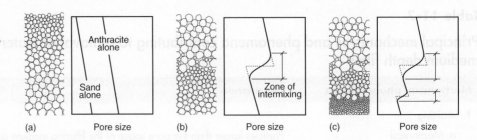

(a)   Pore size   (b)   Pore size   (c)   Pore size

through the filter [see Fig. 11–3(b)]. A sufficient flow of washwater is applied until the granular filtering medium is fluidized (expanded), causing the particles of the filtering medium to abrade against each other. The backwash water flow requirements are related to water temperature and the desired bed expansion during high-rate washing. Unless the filter bed is cleaned properly, fine material, grease, and bacterial slimes can accumulate within the bed, ultimately leading to the formation of mudballs.

The suspended matter retained within the filter is removed by the shear forces created by backwash water as it moves up through the expanded bed and by abrasion as the grains of the filter medium interact with each other. The material that has accumulated within the bed is then washed away. Surface washing with water and air scouring are often used in conjunction with the water backwash to enhance the cleaning of the filter bed. Air scour in particular will often reduce the required backwash duration and, therefore, will reduce the amount of washwater required. Following backwashing, the practice of filter-to-waste is often used to prevent any residual backwash particles from entering the filtered water. In most wastewater treatment plant flow diagrams, the washwater containing the suspended solids that are removed from the filter is returned either to the primary settling facilities or to the biological treatment process. Increasingly, especially at larger treatment plants, separate treatment facilities are provided to remove the solids from the backwash water.

Filter Hydraulics

During the past 60 years considerable effort has been devoted to the modeling of the filtration process. The models fall into two general categories: those models used to predict the clean-water headloss through a granular medium filter bed (*clean water* referring to water absent of suspended particles that would otherwise create headloss) and the filter backwash expansion, and those models used to predict the performance of filters for the removal of suspended solids. Headloss and backwash hydraulics are considered in the following discussion.

Clean-Water Headloss.   Over the years a number of equations have been proposed to describe the flow of clean-water through a porous medium (Darcy, 1856; Hazen, 1905; Fair and Hatch, 1933; Kozeny-Carman, 1937; Rose, 1945; Ergun, 1952). The equations developed by these early researchers are summarized in Table 11–8. In most cases, the equations for the flow of clean water through a porous medium are derived from a consideration of the Darcy-Weisbach equation [Eq. (5–78)] for flow in a closed conduit and from dimensional analysis.

Application of Headloss Equations. The equations given in Table 11–8 apply to different flow regimes. All of the equations presented in Table 11–8 can be used for laminar flow conditions, with Reynolds numbers typically below 6. However, only the Rose and Ergun equations apply to the laminar, transitional, and turbulent flow regimes. The equations for

Table 11–8
Formulas used to compute the clear-water headloss through a granular porous medium

Equation	No.	
Hazen (Hazen, 1905) $$h = \frac{1}{C}\left(\frac{60}{T+10}\right)\frac{L}{d_{10}^2}v_h$$	(11–1)	C = coefficient of compactness (varies from 600 for very closely packed sands that are not quite clean to 1200 for very uniform clean sand)
Fair-Hatch (Fair and Hatch, 1933) $$h = kvS^2\frac{(1-\alpha)^2}{\alpha^3}\frac{L}{d^2}\frac{v_s}{g}$$	(11–2)	C_d = coefficient of drag d = grain size diameter, m (ft) d_g = geometric mean diameter between sieve sizes d_1 and d_2, $\sqrt{d_1 d_2}$, mm (in.)
$$h = kv\frac{(1-\alpha)^2}{\alpha^3}\frac{Lv_s}{g}\left(\frac{6}{\psi}\right)^2\Sigma\frac{p}{d_g^2}$$	(11–3)	d_{10} = effective medium size diameter, mm (in.) f = friction factor
Kozeny-Carman[a] (Carman, 1937) $$h = \frac{k\,\mu\,(1-\alpha)^2}{g\,\rho\;\alpha^3}(S_v)^2 L\,v_s\;(k=5)$$	(11–4)	g = acceleration due to gravity, 9.81 m/s^2 (32.2 ft/s^2) h = headloss, m (ft) k = filtration constant, 5 based on sieve openings, 6 based on size of separation
Rose (Rose, 1945, 1949) $$h = \frac{1.067}{\psi}C_d\frac{1}{\alpha^4}\frac{L}{d}\frac{v_s^2}{g}$$	(11–5)	L = depth of filter bed or layer, m (ft) N_R = Reynolds number p = fraction of particles (based on mass) within adjacent sieve sizes
$$h = \frac{1.067}{\psi}\frac{Lv_s^2}{\alpha^4 g}\Sigma C_d\frac{p}{d_g}$$	(11–6)	S = shape factor (varies between 6.0 for spherical particles and 6/ψ for nonspherical particles)
$$C_d = \frac{24}{N_R} + \frac{3}{\sqrt{N_R}} + 0.34$$	(11–7)	S_v = specific surface area (A_p/V_p) is equal to 6/d for spheres and 6/$d\psi$ for nonspherical particles
$$N_R = \frac{\psi d v_s \rho}{\mu}$$	(11–8)	T = temperature, °C [°F in Eq. (11–10)] v_h = superficial (approach) filtration velocity, m/d (ft/d)
Ergun (Ergun, 1952) $$h = \frac{f}{\psi}\frac{(1-\alpha)}{\alpha^3}\frac{L}{d}\frac{v_s^2}{g}$$	(11–9)	v_s = superficial (approach) filtration velocity, m/s (ft/s) α = porosity μ = viscosity, N·s/m^2 (lb·s/ft^2) ν = kinematic viscosity, m^2/s (ft^2/s)
$$f = 150\frac{(1-\alpha)}{N_R} + 1.75$$	(11–10)	ρ = density, = kg/m^3 (slug/ft^3, lb·s^2/ft^4) ψ = sphericity, often identified as ϕ in the literature (1.0 for spheres, 0.94 for worn sand, 0.81 for sharp sand, 0.78 for angular sand, 0.70 for crushed coal and sand)
N_R = See Eq.(11–8)		

[a] Although known as the Kozeny-Carman equation, Blake (1922) should also be credited with its development.

the transitional or turbulent flow regimes are important because many of the newer filters are deeper with larger filter media and operate at higher filtration rates. In the Rose equation [Eq. (11–5)], use of the coefficient of drag, C_d [Eq. (11–7)], makes it possible to capture the affect of varying flow regimes from viscous to inertial. Similarly, in the Ergun friction equation [Eq. (11–10)], the first term accounts for viscous energy losses and the second term accounts for inertial energy losses. Based on a review of the literature, Trussell and Chang (1999) proposed some different coefficients for sand and anthracite for use in the Ergun equation [Eq. (11–10)].

The summation term in Eqs. (11–3) and (11–6) is included to account for the stratification that occurs in filters. To account for stratification, the mean size of the material retained between successive sieve sizes is assumed to correspond to the mean size of the successive sieves (see Table 11–6), assuming that the particles retained between sieve sizes are substantially uniform (Fair and Hatch, 1933). The Ergun equation [Eq. (11–9)] can also be applied to successive layers in a stratified filter.

Sphericity, Specific Surface Area, and Shape Factor. In applying the equations given in Table 11–8, some confusion exists over the definition of sphericity, ψ, specific surface area, S_v, and shape factor, S. The sphericity factor is defined as the ratio of the surface area of a sphere with the same volume as a given particle to the surface area of the particle and is given by the following formula (Wadell, 1935).

$$\psi = \frac{\pi^{1/3}(6V_p)^{2/3}}{A_p} \tag{11–11}$$

where ψ = sphericity, dimensionless
  V_p = equivalent volume sphere, L^3 (m^3)
  A_p = actual surface area of particle, L^2 (m^2)

Thus, for a spherical particle, the sphericity factor is equal to 1.0. Typically, sphericity factors can be applied to discrete particles and can vary from 1.0 for spheres to 0.70 for crushed sand. Further, because sphericity is difficult to measure, typical values are derived from experimental observations (Carman, 1937).

The specific surface area, S_v, defined as the area to volume ratio, is given by the following expressions for spherical and nonspherical particles.

For spherical particles

$$S_v = \frac{A_p}{V_p} = \frac{\pi d^2}{(\pi d^3/6)} = \frac{6}{d}, \text{ and} \tag{11–12a}$$

For nonspherical (irregular) particles

$$S_v = \frac{A_p}{V_p} = \frac{6}{\psi d} \tag{11–12b}$$

where S_v = specific surface area, m, mm
  A_p = surface area of filter medium particle, m^2, mm^2
  V_p = volume of filter medium particle, m^3, mm^3
  d = diameter of filter medium particle, m, mm
  ψ = sphericity, dimensionless

In the literature, the number 6 that appears in the above equations has been identified as a shape factor S for spherical particles and $6/\psi$ for nonspherical particles [see Eq. (11–2) in Table 11–8] (Fair et al., 1968). Computation of the clean-water headloss through a filter is illustrated in Example 11–1.

EXAMPLE 11–1

Determination of Clean-water Headloss in a Granular Medium Filter Determine the effective size, the uniformity coefficient, and the clean-water headloss in a filter bed composed of 0.75 m of uniform sand with the size distribution given below for a filtration rate of 160 L/m²·min. Assume that the operating temperature is 20°C. Use the Rose equation [Eq. (11–6)] given in Table 11–8 for computing the headloss. Assume the porosity of the sand in the various layers is 0.40 and use a value of 0.85 for the sphericity factor for sand.

Sieve size or number	Percent of sand retained	Cumulative percent passing	Geometric mean size[a], mm
6–8	0	100	
8–10	1	99	2.18
10–12	3	96	1.83
12–18	16	80	1.30
18–20	16	64	0.92
20–30	30	34	0.71
30–40	22	12	0.50
40–50	12		0.35

[a] Using sieve size data from Table 11–6, the geometric mean size $= \sqrt{d_1 d_2}$

Solution

1. Determine the effective size and the uniformity coefficient of the sand. Plot the cumulative percent passing versus the corresponding sieve size. Two different methods of plotting the data are presented below.

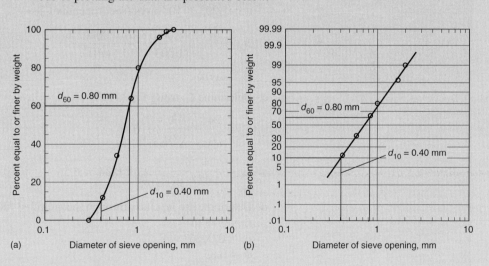

a. The effective size, d_{10}, read from the graphs is 0.40 mm
b. The uniformity coefficient is

$$\text{UC} = \frac{d_{60}}{d_{10}} = \frac{0.80 \text{ mm}}{0.40 \text{ mm}} = 2.0$$

2. Determine the clean-water headloss using Eq. (11–6).

$$h = \frac{1.067}{\psi} \frac{L v_s^2}{\alpha^4 g} \sum C_d \frac{p}{d_g}$$

a. Set up computation table to determine the summation term in Eq. (11–6)

Sieve size or number	Fraction of sand retained	Geometric mean size, mm	Reynolds number	C_d	$C_d\left(\frac{p}{d}\right)$, m^{-1}
8–10	0.01	2.18	4.93	6.56	30
10–12	0.03	1.83	4.15	7.60	124
12–18	0.16	1.30	2.93	10.28	1268
18–20	0.16	0.92	2.08	13.99	2441
20–30	0.30	0.71	1.60	17.71	7509
30–40	0.22	0.50	1.13	24.38	10,729
40–50	0.12	0.35	0.80	33.73	11,459
Sum					33,560

b. Determine the Reynolds number for each geometric mean as illustrated below.

$$N_R = \frac{\psi d v_s \rho}{\mu} = \frac{\psi d v_s}{\nu}$$

$d = 2.18$ mm

$$v_s = \left(\frac{160 \, L}{m^2 \cdot min}\right)\left(\frac{1 \, m^3}{1000 \, L}\right)\left(\frac{1 \, min}{60 \, s}\right) = 0.00267 \text{ m/s}$$

$\nu = 1.003 \times 10^{-6}$ m^2/s (see Appendix C)

$$N_R = \frac{(0.85)(0.00218 \, m)(0.00267 \, m/s)}{(1.003 \times 10^{-6} \, m^2/s)}$$

$N_R = 4.93$

c. Determine C_d using Eq. (11–7)

$$C_d = \frac{24}{N_R} + \frac{3}{\sqrt{N_R}} + 0.34$$

$$C_d = \frac{24}{4.93} + \frac{3}{\sqrt{4.93}} + 0.34 = 6.56$$

d. Determine the headloss through the stratified filter bed using Eq. (11–6).

$L = 0.75$ m
$v_s = 0.00267$ m/s
$\psi = 0.85$
$\alpha = 0.40$
$g = 9.81$ m/s^2

$$h = \frac{1.067(0.75 \, m)(0.00267 \, m/s)^2}{(0.85)(0.4)^4(9.81 \, m/s^2)} (33,560/m)$$

$h = 0.90$ m

Comment   Given that the Reynolds numbers are less than 6 (i.e., 4.93 or less), most of the headloss in this example is due to viscous forces operative in the laminar flow region, which is reflected by the first term in the C_d relationship [Eq. (11–7)]. As larger media are used and the flowrate increases, the second and third terms of the C_d relationship will have a greater impact. Although many equations have been proposed over the years, the Rose equation has proven to be quite satisfactory for estimating the clear-water headloss in granular medium filter beds for a variety of flow regimes.

Backwash Hydraulics.   To understand what happens during the backwash operation it will be helpful to refer to Fig. 11–7 in which the pressure drop across a packed bed is illustrated as the upward backwash velocity through it increases. Between points A and B, the bed is stable, and the pressure drop and Reynolds number N_R are related linearly. At point B, the pressure drop essentially balances the weight of the filter. Between points B and C the bed is unstable, and the particles adjust their position to present as little resistance to flow as possible. At point C, the loosest possible arrangement is obtained in which the particles are still in contact. Beyond point C, the particles begin to move freely but collide frequently so that the motion is similar to that of particles in hindered settling. Point C is referred to as the "point of fluidization." By the time point D is reached, the particles are all in motion, and, beyond this point, increases in N_R result in very small increases in ΔP as the bed continues to expand and the particles move in more rapid and more independent motion. Ultimately, the particles will stream with the fluid, and the bed will cease to exist at point E.

To expand a filter bed comprised of a uniform filter medium hydraulically, the headloss must equal the buoyant mass of the granular medium in the fluid. Mathematically this relationship can be expressed as

$$h = L_e(1 - \alpha_e)\left(\frac{\rho_m - \rho_w}{\rho_w}\right) \qquad (11\text{–}13)$$

where h = headloss required to expand the bed
$\quad L_e$ = the depth of the expanded bed
$\quad \alpha_e$ = the expanded porosity
$\quad \rho_m$ = density of the medium
$\quad \rho_w$ = density of water

Figure 11–7

Schematic diagram illustrating the fluidization of a filter bed. (Adapted from Foust et al., 1960.)

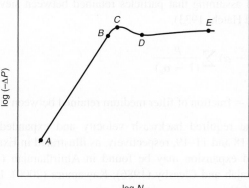

Because the individual particles are kept in suspension by the drag force exerted by the rising fluid it can be shown from settling theory (see Sec. 5–5 in Chap. 5) that

$$C_D A_p \rho_w \frac{v^2}{2} \phi(\alpha_e) = (\rho_m - \rho_w) g v_p \qquad (11\text{--}14)$$

where v = face velocity of backwash water, m/s

$\phi(\alpha_e)$ = correction factor to account for the fact that v is the velocity of the backwash water and not the particle-settling velocity v_p

other terms are as defined previously.

From experimental studies (Fair, 1951; Richardson and Zaki, 1954) it has been found that the expanded bed porosity can be approximated using the following relationships, assuming the Reynolds number is approximately one.

$$\phi(\alpha_e) = \left(\frac{v_s}{v}\right)^2 = \left(\frac{1}{\alpha_e}\right)^9 \qquad (11\text{--}15)$$

Thus

$$\alpha_e = \left(\frac{v}{v_s}\right)^{0.22} \qquad (11\text{--}16)$$

or

$$v = v_s \alpha_e^{4.5} \qquad (11\text{--}17)$$

where v_s = settling velocity of particle

However, because the volume of the filtering medium per unit area remains constant, $(1 - \alpha)L$ must be equal to $(1 - \alpha e)L_e$ so that

$$\frac{L_e}{L} = \frac{1 - \alpha}{1 - \alpha_e} = \frac{1 - \alpha}{1 - (v/v_s)^{0.22}} \qquad (11\text{--}18)$$

Where the filter medium is stratified, the smaller and lighter particles in the upper layers expand first. To expand the entire bed, the backwash velocity must be sufficient to lift the largest and heaviest particle. To account for filter bed stratification, Eq. (11–18) is modified assuming that particles retained between sieve sizes are substantially uniform (Fair and Hatch, 1933).

$$\frac{L_e}{L} = (1 - \alpha) \sum \frac{p}{(1 - \alpha_e)} \qquad (11\text{--}19)$$

Where p = fraction of filter medium retained between sieve sizes

Thus, the required backwash velocity and expanded depth can be estimated using Eqs. 11–18 and 11–19, respectively, as illustrated in Example 11–2. Additional details on filter bed expansion may be found in Amirtharajah (1978), Cleasby and Fan (1982), Dharmarajah and Cleasby (1986), Kawamura (2000), Leva (1959), and Richardson and Zaki (1954).

EXAMPLE 11–2   **Determination of Required Backwash Velocities for Filter Cleaning**   A stratified sand bed with the size distribution given below is to be backwashed at a rate of 0.75 m³/m²·min. Determine the degree of expansion and whether the proposed backwash rate will expand all of the bed. Assume the following data are applicable:

Sieve size or number	Percent of sand retained	Geometric mean size[a], mm
8–10	1	2.18[b]
10–12	3	1.83
12–18	16	1.30
18–20	16	0.92
20–30	30	0.71
30–40	22	0.50
40–50	12	0.35

[a] Based on sieve sizes given in Table 11–6.
[b] $2.18 = \sqrt{2.38 \times 2.0}$.

1. Granular medium = sand
2. Specific gravity of sand = 2.65
3. Depth of filter bed = 0.90 m
4. Temperature = 20°C

Solution

1. Set up computation table to determine the summation term in Eq. (11–19).

$$\frac{L_e}{L} = (1 - \alpha)\sum \frac{p}{(1 - \alpha_e)}$$

Sieve size or number	Percent of sand retained[a]	Geometric mean size, mm	v_s, m/s	v/v_s	α_e	$p/(1 - \alpha_e)$
8–10	1	2.18	0.304	0.041	0.496	1.98
10–12	3	1.83	0.270	0.046	0.509	6.11
12–18	16	1.30	0.210	0.060	0.538	34.62
18–20	16	0.92	0.157	0.080	0.573	37.51
20–30	30	0.71	0.123	0.102	0.605	75.97
30–40	22	0.50	0.085	0.146	0.655	63.81
40–50	12	0.35	0.055	0.227	0.722	43.15
Summation						263.15

[a] For ease of computation, the percentage value is used instead of the decimal fractional value.

   a. Determine the particle settling velocity using Fig. 5–20 in Chap. 5. Alternatively the particle settling velocity can be computed as illustrated in Example 5–5. The settling velocity values from Fig. 5–20 are entered in the computation table.

   b. Determine the values of v/v_s and enter the computed values in the computation table.

   The backwash velocity is
   $v = 0.75$ m/min = 0.0125 m/s

c. Determine the values of α_e and enter the computed values in the computation table.

$$\alpha_e = \left(\frac{v}{v_s}\right)^{0.22} = \left(\frac{0.0125}{0.304}\right)^{0.22} = 0.496$$

d. Determine the values for column 7 and enter the computed values in the computation table.

$$\frac{p}{1-\alpha_e} = \frac{0.01}{1-0.496} = 0.02$$

2. Determine the expanded bed depth using Eq. (11–19).

$$\frac{L_e}{L} = (1-\alpha)\sum\frac{p}{(1-\alpha_e)}\left(\frac{1}{100}\right)$$

$$L_e = (0.9\ \text{m})(1-0.4)(263.15)(1/100) = 1.42\ \text{m}$$

3. Because the expanded porosity of the largest size fraction (0.496) is greater than the normal porosity of the filter material, the entire filter bed will be expanded.

Comment   The expanded depth needs to be known to establish the minimum height of the washwater troughs above the surface of the filter bed. In practice, the bottom of the backwash water troughs is set from 50 to 150 mm (2 to 6 in) above the expanded filter bed and filter expansion of 30 to 50 percent is used commonly. The width and depth of the troughs should be sufficient to handle the volume of backwash water used to clean the bed, with a minimum freeboard of 600 mm (24 in) above the top of the trough.

Modeling the Filtration Process

The modeling of the filtration process involves the development of equations to describe the (1) removal of suspended solids with time and distance within the filter bed and (2) buildup of headloss as suspended solids are removed from the liquid passing through the filter.

Removal of Suspended Solids.   In general, the mathematical modeling of the time-space removal of particulate matter within the filter is based on a consideration of the equation of continuity, together with an auxiliary rate equation. The equation of continuity for the filtration operation may be developed by considering a suspended solids mass balance for a section of filter of cross-sectional area A, and of thickness Δz, measured in the direction of flow as illustrated on Fig. 11–8. Following the approach outlined in Chap. 1, the resulting equation is:

$$-v\frac{\partial C}{\partial z} = \frac{\partial q}{\partial t} + \alpha(t)\frac{\partial \overline{C}}{\partial t} \qquad (11-20)$$

where v = filtration velocity, L/m²·min
$\partial C/\partial z$ = change in concentration of suspended solids in fluid stream with distance, g/m³·m
$\partial q/\partial t$ = change in quantity of solids deposited within the filter with time, g/m³·min
$\alpha(t)$ = average porosity as a function of time
$\partial \overline{C}/\partial t$ = change in average concentration of solids in pore space with time, g/m³·min

Figure 11–8

Definition sketch for the analysis of the filtration process.

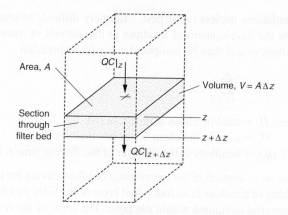

In Eq. (11–20), the first term represents the difference between the mass of suspended solids entering and leaving the section, the second term represents the time rate of change in the mass of suspended solids accumulated within the interstices of the filter medium, and the third term represents the time rate of change in the suspended solids concentration in the pore space within the filter volume.

Because the quantity of fluid contained within the bed in a flowing process is usually small compared with the volume of liquid passing through the bed, the materials balance equation can be written as:

$$-v\frac{\partial C}{\partial z} = \frac{\partial q}{\partial t} \tag{11–21}$$

This equation is the one most commonly found in the literature dealing with filtration theory.

To solve Eq. (11–21), an additional independent equation is required. The most direct approach is to derive a relationship that can be used to describe the change in concentration of suspended matter with distance, such as

$$\frac{\partial C}{\partial z} = \phi(V_1, V_2, V_3 \ldots) \tag{11–22}$$

in which V_1, V_2, and V_3 are the variables governing the removal of suspended matter from solution. An alternative approach is to develop a complementary equation in which the pertinent process variables are related to the amount of material retained (accumulated) within the filter at various depths. In equation form, this may be written as

$$\frac{\partial q}{\partial t} = \phi(V_1, V_2, V_3 \ldots) \tag{11–23}$$

Using one or the other of the above expressions, [Eq. (11–21) or (11–22)], a number of solutions have been proposed for the continuity equation [Eq. (11–21)] (Caliskaner and Tchobanoglous, 2000).

Headloss Development. In the past, the most commonly used approach to determine headloss in a clogged filter was to compute it with a modified form of the equations used to evaluate the clear-water headloss (see Table 11–8). In all cases, the difficulty encountered in using these equations is that the porosity must be estimated for various degrees of clogging. Unfortunately, the complexity of this approach renders most of these formulations useless or, at best, extremely difficult to use. An alternative approach is to relate the development of headloss to the amount of material removed by the filter. The headloss would then be computed using the expression

$$H_t = H_o + \sum_{i=i}^{n} (h_i)_t \tag{11-24}$$

where H_t = total headloss at time t, m (ft)
$\qquad H_o$ = total initial clean-water headloss, m (ft)
$\qquad (h_i)_t$ = headloss in the ith layer of the filter at time t, m (ft)

From an evaluation of the incremental headloss curves for uniform sand and anthracite, the buildup of headloss in an individual layer of the filter was found to be related to the amount of material contained within the layer. The form of the resulting equation for headloss in the ith layer is

$$(h_i)_t = a(q_i)_t^b \tag{11-25}$$

where $(q_i)_t$ = amount of material deposited in the ith layer at time t, mg/cm³
$\qquad a, b$ = constants

In this equation, it is assumed that the buildup of headloss is only a function of the amount of material removed. The application of these modeling equations may be found in the 3rd and 4th editions of this textbook.

11–5   DEPTH FILTRATION: SELECTION AND DESIGN CONSIDERATIONS

The ability to select and design filter technologies must be based on (1) knowledge of the types of filters that are available, (2) a general understanding of their performance characteristics, and (3) an appreciation of the process variables controlling depth filtration. Important design considerations for effluent filtration systems include (1) filter influent wastewater characteristics, (2) design and operation of the biological treatment process, (3) type of filtration technology to be used, (4) available flow control options, (5) type of filter backwashing system to be employed, (6) necessary filter appurtenances, and (7) filter control systems and instrumentation (not considered in this textbook). An understanding of issues related to effluent filtration with chemical addition, the type of filter problems encountered in the field, and the importance of pilot plant studies is also necessary. These subjects are presented and discussed in this section.

Available Filtration Technologies

The principal types of depth filters that have been used for the filtration of wastewater are described in Table 11–9. As shown in Table 11–9, the filters can be classified in terms of their operation as semi-continuous or continuous. Filters that must be taken offline periodically to be backwashed are classified operationally as semi-continuous. Filters in which the filtration and backwash operation occurs simultaneously are classified as continuous. Within each of these two classifications there are a number of different types of filters depending on bed depth (e.g., shallow, conventional, and deep bed), the type of filtering medium used (mono-medium, dual-, and multi-media), whether the

Table 11-9
Comparison of principal types of granular and synthetic medium filters

Type of filter	Type of filter operation	Filter bed details[a]		Typical direction of flow	Backwash operation	Flowrate through filter	Solids storage location	Type of design	Remarks
		Type	Filtering medium						
Conventional	Semi-continuous	Mono-medium (stratified or unstratified)	Sand or anthracite	Downward	Batch	Constant/variable	Surface and upper bed	Individual	Rapid headloss buildup
Conventional	Semi-continuous	Dual-media (stratified	Sand and anthracite	Downward	Batch	Constant/variable	Internal	Individual	Dual-media design used to extend length of filter run
Conventional	Semi-continuous	Multi-media (stratified)	Sand, anthracite, and garnet	Downward	Batch	Constant/variable	Internal	Individual	Multi-media design used for particle depth penetration
Deep bed	Semi-continuous	Mono-medium (stratified or unstratified)	Sand or anthracite	Downward	Batch	Constant/variable	Internal	Individual	Deep bed used to store solids and extend length of filter run
Deep bed	Semi-continuous	Mono-medium (stratified)	Sand	Upward	Batch	Constant	Internal	Proprietary	Deep bed used to store solids and extend length of filter run
Deep bed	Semi-continuous	Mono-medium (unstratified)	Sand	Upward	Continuous	Constant	Internal	Proprietary	Sand bed moves in countercurrent direction to fluid flow
Pulsed bed	Semi-continuous	Mono-medium (stratified)	Sand	Downward	Batch	Constant	Surface and upper bed	Proprietary	Air pulses used to break up surface mat and increase run length
Fuzzy filter	Semi-continuous	Mono-medium (unstratified)	Synthetic fiber	Upward	Batch	Constant	Internal	Proprietary	Perforated plate is used to retain the filter medium during backwash
Traveling bridge	Continuous	Mono-medium (stratified)	Sand	Downward	Semi-continuous	Constant	Surface and upper bed	Proprietary	Individual filter cells backwashed sequentially
Traveling bridge	Continuous	Dual-media (stratified)	Sand and anthracite	Downward	Semi-continuous	Constant	Surface and upper bed	Proprietary	Individual filter cells backwashed sequentially
Pressure filters	Semi-continuous	Mono medium or dual media	Sand and/or anthracite	Downward	Batch	Constant/variable	Surface and upper bed	Individual and proprietary	Used for small plants

[a] For filter bed depths, see Tables 11–15 and 11–16.

1145

filtering medium is stratified or unstratified, the type of operation (downflow or upflow), and the method used for the management of solids (i.e., surface or internal storage). For the mono-medium and dual-media semi-continuous filters, a further classification can be made based on the driving force (e.g., gravity or pressure) although most of the filters used commonly in wastewater applications are gravity flow. Another important distinction that must be noted for the filters identified in Table 11–9 is whether they are proprietary or individually designed.

The five types of depth filters used most commonly for wastewater filtration at larger treatment plants [greater than 1000 m³/d (0.25 Mgal/d)] are (1) conventional downflow filters (mono-medium, dual-, and multi-media), (2) deep-bed downflow filters, (3) deep-bed upflow continuous-backwash filters, (4) synthetic medium filters, (5) the pulsed bed filter, and (6) traveling bridge filters. A two-stage deep-bed filtration system which incorporates phosphorus removal is also used. Pressure filters, which operate in the same manner as gravity filters, are used at smaller plants. Many of the filters are proprietary and are supplied by the manufacturer as a complete unit. Each of these eight filter types is described in greater detail in Table 11–10. Views of several different types of filter installations are shown on Fig. 11–9.

Performance of Different Types of Depth Filters

The critical question associated with the selection of any depth filter is whether it will perform as anticipated. Performance of depth filters can be assessed from a review of the (1) hydraulic loading rate, (2) removal of turbidity and total suspended solids, (3) variability of turbidity and TSS removal, (4) removal of different particle sizes, (5) removal of microorganisms, and (6) backwash water requirements.

Hydraulic Loading Rate. The principal operational considerations for a depth filter are the volume of water produced in a given time period at a specified quality and the volume of washwater used to clean the filter. The volume of water produced is related to the development of headloss and filter performance, typically measured in terms of turbidity (see Fig. 11–4). The objective of a balanced filter design is to have the limiting headloss and turbidity breakthrough occur at or near the same time. In small plants, the water filtered during the ripening period is wasted during the filter-to-waste step (usually returned to the plant inflow). In large plants with many filters, the filter to waste cycle is often omitted. Chemical addition has also been used to extend the time to turbidity breakthrough and to achieve a variety of other treatment objectives including the removal of specific contaminants such as phosphorus, metal ions, and humic substances. Chemicals used commonly in effluent filtration include a variety of organic polymers, alum, and ferric chloride. It should be noted that the use of filter aid chemicals will normally result in more rapid headloss development and overuse of these chemicals can result in mudball formation.

Both the volume of water filtered and the rate at which headloss increases are related to the hydraulic loading rate (HLR). Typical operating characteristics, including hydraulic loading rates, for depth filters are reported in Table 11–11. Also reported in Table 11–11 are the filtration rates allowed by the California Department of Public Health for various filters in reuse applications. Because of the wide variation in the allowable rates, pilot plant studies are recommended.

Removal of Turbidity and Total Suspended Solids. The results of long-term testing of seven different types of pilot-scale filters on the effluent from the same activated sludge process (SRT > 8 d), without chemical addition, are shown on Fig. 11–10.

Table 11–10

Description of commonly used depth filters for reclaimed water applications[a]

Filter type	Description
(a) Conventional downflow 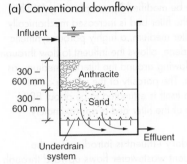	Wastewater containing suspended matter is applied to the top of the filter bed. Mono-medium, dual-, or multi-media filter materials are used. Typically sand or anthracite is used as the filtering material in single-medium filters. Dual-media filters usually consist of a layer of anthracite over a layer of sand. Other combinations include (1) activated carbon and sand, (2) resin beads and sand, and (3) resin beads and anthracite. Multi-media filters typically consist of a layer of anthracite over a layer of sand over a layer of garnet or ilmenite. Other combinations include (1) activated carbon, anthracite, and sand, (2) weighted, spherical resin beads, anthracite, and sand, and (3) activated carbon, sand, and garnet.
(b) Deep-bed downflow 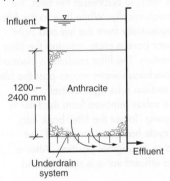	The deep-bed downflow filter is similar to the conventional downflow filter with the exception that the depth of the filter bed and the size of the filtering medium (usually anthracite) are greater than the corresponding values in a conventional filter. Because of the greater depth and larger medium size (i.e., sand or anthracite), more solids can be stored within the filter bed and the run length can be extended. The maximum size of the filter medium used in these filters depends on the ability to backwash the filter. In general, deep-bed filters are not fluidized completely during backwashing. To achieve effective cleaning, air scour plus water is used in the backwash operation.
(c) Deep-bed upflow continuous backwash 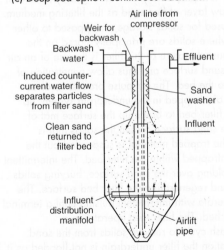	Wastewater to be filtered is introduced into the bottom of the filter where it flows upwards through a series of riser tubes and is distributed evenly into the sand bed through the open bottom of an inlet distribution hood. The water then flows upward through the downward moving sand. Clean filtrate exits from the sand bed, overflows a weir, and is discharged from the filter. At the same time sand particles, along with trapped solids, are drawn downward into the suction of an airlift pipe which is positioned in the center of the filter. A small volume of compressed air, introduced into the bottom of the airlift, draws sand, solids, and water upward through the pipe by creating a fluid with a density less than one.
	Impurities are scoured (abraded) from the sand particles during the turbulent upward flow. Upon reaching the top of the airlift, the dirty slurry spills over into the central reject compartment. A steady stream of clean filtrate flows upward, countercurrent to the movement of sand, through the washer section. The upflow liquid carries away the solids and reject water. Because the sand has a higher settling velocity than the removed solids, the sand is not carried out of the filter. The sand is cleaned further as it moves down through the washer. The cleaned sand is redistributed onto the top of the sand bed, allowing for a continuous uninterrupted flow of filtrate and reject water.

(continued)

Table 11-10 (Continued)	
Filter type	**Description**

(d) Synthetic medium (Fuzzy filter)

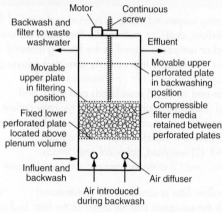

A synthetic medium filter, developed originally in Japan, is used for reclaimed water filtration. Unusual features of the filter are: (1) the porosity of the filter bed can be modified by compressing the filter medium and (2) the size of the filter bed is increased mechanically to backwash the filter. The filter medium, a highly porous synthetic material made of polyvaniladene, allows the influent to flow through the medium as opposed to flowing around the filtering medium, as in sand and anthracite filters. The porosity of the uncompacted quasi-spherical filter medium itself is estimated to be about 88 to 90 percent, and the porosity of the filter bed is approximately 94 percent.

In the filtering mode, secondary effluent is introduced in the bottom of the filter. The influent wastewater flows upward through the filter medium, retained by two porous plates, and is discharged from the top of the filter. To backwash the filter, the upper porous plate is raised mechanically. While flow to the filter continues, air is introduced sequentially from the left and right sides of the filter below the lower porous plate, causing the filter medium to move in a rolling motion. The filter medium is cleaned by the shearing forces as the backwash water moves past the filter and by abrasion as the filter medium rubs against itself. Backwash water containing the solids removed from the filter is diverted for subsequent processing. To put the filter back into operation after the backwash cycle has been completed, the raised porous plate is returned to its original position. After a short flushing cycle, the filtered effluent valve is opened, and filtered effluent is discharged.

(e) Pulsed-bed (PBF)

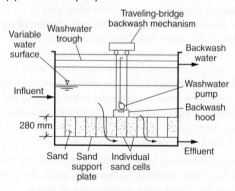

The pulsed bed filter is a proprietary downflow gravity filter with an unstratified shallow layer of fine sand as the filtering medium. The shallow bed is used for solids storage, as opposed to other shallow-bed filters where solids are principally stored on the sand surface. An unusual feature of this filter is the use of an air pulse to disrupt the sand surface and thus allow penetration of suspended solids into the bed. The air pulse process involves forcing a volume of air, trapped in the underdrain system, up through the shallow filter bed to break up the surface mat of solids and renew the sand surface. When the solids mat is disturbed, some of the trapped material is suspended but the most of solids are entrapped within the filter bed. The intermittent air pulse causes a folding over the sand surface, burying solids within the medium and regenerating the filter bed surface. The filter continues to operate with intermittent pulsing until a terminal headloss limit is reached. The filter then operates in a conventional backwash cycle to remove solids from the sand. During normal operation the filter underdrain is not flooded as it is in a conventional filter.

(continued)

Table 11–10 (*Continued*)	
Filter type	**Description**

(f) Traveling bridge

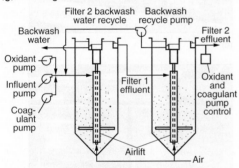

The traveling bridge filter is a proprietary continuous downflow, automatic backwash, low-head, granular medium depth filter. The bed of the filter is divided horizontally into long independent filter cells. Each filter cell contains approximately 280 mm (11 in.) of medium.

Treated wastewater flows through the medium by gravity and exits to the clearwell plenum via a porous plate, polyethylene underdrain. Each cell is backwashed individually by an overhead, traveling bridge assembly, while all other cells remain in service. Water used for backwashing is pumped directly from the clearwell plenum up through the medium and deposited in a backwash trough.

During the backwash cycle, wastewater is filtered continuously through the cells that are not being backwashed. The backwash mechanism includes a surface wash pump to assist in breaking up of the surface matting and "mudballing" in the medium. Because the backwashing operation is performed on an "as needed" basis, the backwash cycle is termed semi-continuous.

(g) Two-stage

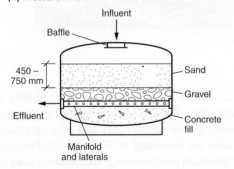

A proprietary two-stage filtration process is used for the removal of turbidity, total suspended solids, and phosphorus. Two deep-bed upflow continuous backwash filters are used in series to produce a high quality effluent. A large size sand diameter is used in the first filter to increase the contact time and to minimize clogging. A smaller sand size is used in the second filter to remove residual particles from the first stage filter. The waste washwater from the second filter which contains small particles and residual coagulant is recycled to the first filter to improve floc formation within the first stage filter and the influent to waste ratio. Based on full scale installations the reject rate has been found to be less than 5 percent. Phosphorus levels equal to or less than 0.02 mg/L have been achieved in the final filter effluent.

(h) Pressure filters

Pressure filters operate in the same manner as gravity filters and are used at smaller plants. The only difference is that, in pressure filters, the filtration operation is carried out in a closed vessel under pressurized conditions achieved by pumping. Pressure filters normally are operated at higher terminal headlosses, resulting in longer filter runs and reduced backwash requirements. If however, they are not backwashed on a regular basis, problems have been experienced with the formation of mudballs.

Long-term data from other water reclamation plants are also shown. The principal conclusions to be reached from an analysis of the data presented on Fig. 11–10 are that (1) given a high quality filter influent (turbidity less than 5 to 7 NTU) all of the filters tested are capable of producing an effluent with an average turbidity of 2 NTU or less without chemical addition; (2) when the influent turbidity is greater than about 7 to 10 NTU,

Figure 11–9

Views of typical filtration installations: (a) view of empty conventional gravity filter without underdrain system (see Fig. 11–20), but with washwater troughs in place; (b) typical traveling bridge filter (empty) with individual cells exposed; (c) deep-bed denitrifying filter; (d) continuous backwash upflow filters (courtesy Austep, Italy); (e) Fuzzy filter installation comprised of six filters; and (f) bank of small pressure filters used at small wastewater treatment plants. Additional information on these filters is presented in Tables 11–9 and 11–10.

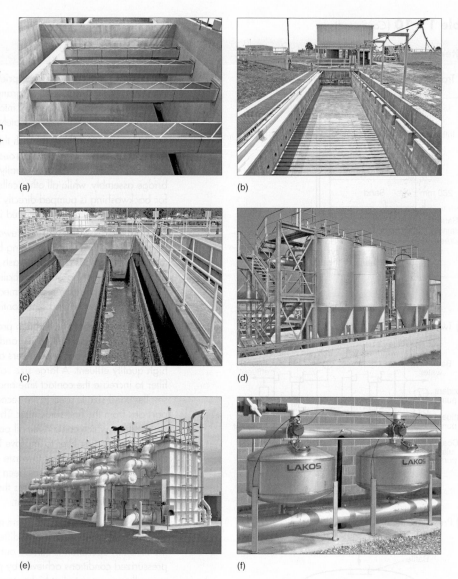

(a) (b)

(c) (d)

(e) (f)

chemical addition is required with all of the filters to achieve an effluent turbidity of 2 NTU or less; and (3) effluent quality is directly related to influent quality if chemical addition is not used. Typical values of effluent quality for turbidity and total suspended solids for depth filtration using granular media are presented in Table 11–12. For comparison, comparable data for other advanced filtration processes (e.g. membranes) used for the removal of particulate matter are presented in Table 11–31 in Sec 11–7.

Keeping in mind the limitations associated with turbidity measurements, the following two relationships can be used to approximate TSS values from measured turbidity values.

Settled secondary effluent

$$\text{TSS, mg/L} = (2.0 \text{ to } 2.4) \times (\text{turbidity, NTU}) \qquad (11\text{–}26)$$

Table 11–11

Comparison of operational characteristics for selected depth filters when filtering settled activated sludge effluent

| Type of filter | Filter bed details[a] | | Typical operational filtration rate | | Maximum filtration rate approved by CDPH[c,d] | | Backwash percentage |
	Type	Filtering medium[b]	gal/ft²·min	m³/m²·min	gal/ft²·min	m³/m²·min	
Conventional shallow	Mono-medium	S	2–6	0.08–0.24	5	0.20	4–8
Conventional	Dual-media	S and A	2–6	0.08–0.24	5	0.20	4–8
Conventional	Multi-media	S, A, and G	2–6	0.08–0.24	5	0.20	4–8
Deep bed	Mono-medium	S	5–8	0.20–0.33	5	0.20	4–8
Deep bed	Mono-medium	A	5–8	0.20–0.33	5	0.20	4–8
Deep bed, upflow	Mono-medium	S	4–6	0.16–0.15	5	0.20	8–15
Fuzzy filter	Mono-medium	SM	15–40	0.60–1.60	40	1.60	2–5
Pulsed bed	Mono-medium	S	2–6	0.08–0.24	5	0.08	4–8
Traveling bridge	Mono-medium	S	2–5	0.08–0.2	2	0.08	4–8
Traveling bridge	Dual-media	S and A	2–5	0.08–0.2	2	0.08	4–8
Pressure filters	Mono medium or dual media	S and A, A	2–6	0.08–0.24	5	0.20	4–8

[a] For filter bed depths, see Tables 11–15 and 11–16.
[b] S =sand, A = anthracite, G = garnet, SM = synthetic medium.
[c] California Department of Public Health.
[d] For Title 22 wastewater reuse applications.

Figure 11–10

Performance data for seven different types of depth filters used for wastewater applications tested using the effluent from the same activated sludge plant at filtration rate 160 L/m²·min (4 gal/ft²·min) with the exception of the Fuzzy Filter which was operated at 800 L/m²·min (20 gal/ft²·min).

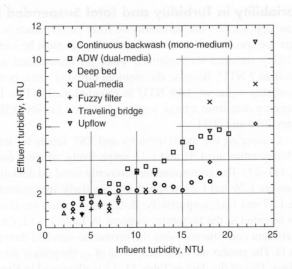

Table 11–12

Typical range of effluent quality variability observed from particulate removal processes

Particulate removal process	Unit	Typical range of effluent values	Geometric standard deviation, s_g[a] Range	Typical
Depth filtration following activated sludge process				
Turbidity	NTU	0.5–4	1.2–1.4	1.25
TSS	mg/L	2–8	1.3–1.5	1.4
Depth filtration following activated sludge with BNR				
Turbidity	NTU	0.3–2	1.2–1.4	1.25
TSS	mg/L	1–4	1.3–1.5	1.35
Surface filtration following activated sludge process				
Turbidity	NTU	0.5–2	1.2–1.4	1.25
TSS	mg/L	1–4	1.3–1.5	1.25

[a] s_g = geometric standard deviation; $s_g = P_{84.1}/P_{50}$.

Filter effluent

TSS, mg/L = (1.3 to 1.6) × (turbidity, NTU)   (11–27)

Using the above approximations, turbidity values of 5 to 7 NTU in the settled secondary effluent, which is the influent to the filter, correspond to TSS concentrations varying from about 10 to 17 mg/L, and an effluent turbidity of 2 NTU corresponds to TSS concentrations varying from 2.8 to 3.2 mg/L.

Variability in Turbidity and Total Suspended Solids Removal.   In water reuse applications the variability of filter performance is of critical importance because there are specific effluent turbidity limits that must be met consistently. For example, the turbidity standard for reclaimed water for unrestricted use in California is equal to or less than 2 NTU. Because the required turbidity value is written without a decimal point, a turbidity value of 2.49 NTU is reported as 2 NTU. The variability observed in the operating data from a large water reclamation facility is illustrated on Fig. 11–11, for the years 2010 and 2011.

Comparing the mean turbidity and TSS values for the two different years, the TSS/turbidity ratios are 1.51 and 1.32, respectively, which is consistent with the range given in Eq. (11–27). The corresponding geometric standard deviations, s_g, for turbidity for the two years are 1.26 and 1.23, respectively. Similarly, the geometric standard deviations for TSS are 1.37 and 1.42, respectively. Both sets of values are consistent with the range of s_g values reported in the literature, as given in Table 11–12. Characterization of the variability in effluent constituents using the geometric standard deviation, s_g, is discussed in Appendix D. The greater the numerical value of s_g, the greater the observed range in the measured values. Use of the data in Table 11–12 is illustrated in Example 11–3.

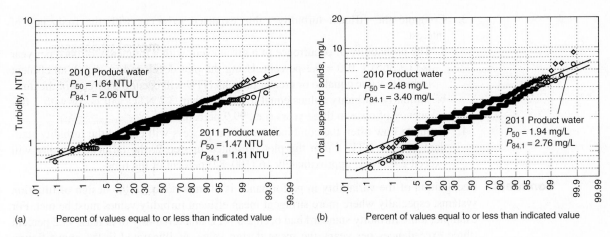

(a) Percent of values equal to or less than indicated value

(b) Percent of values equal to or less than indicated value

Figure 11–11

Probability distributions for filter performance for the filtration of settled activated sludge effluent from a large water reclamation facility: (a) turbidity and (b) total suspended solids.

EXAMPLE 11–3 **Evaluation of the Effluent Variability of an Activated Sludge Process with Mono-medium Filtration** An activated sludge process with mono-medium filtration has been designed to have a mean effluent turbidity value of 2 NTU. Determine the maximum turbidity value that is expected to occur with a frequency of (a) once per year and (b) once every three years. If the effluent turbidity standard is 2.49 NTU, estimate how often the process will exceed the turbidity limit.

Solution

1. Select an s_g value from Table 11–12 that corresponds to the effluent turbidity for an activated sludge with filtration process. From Table 11–12, use the typical s_g value of 1.25.
2. Determine the probability distribution of the effluent turbidity values.
   a. Using the s_g value, compute the turbidity value corresponding to the plotting position on $P_{84.1}$ (see Appendix D).

$$P_{84.1} = s_g \times P_{50} = 1.25 \times 2 \text{ NTU} = 2.5 \text{ NTU}$$

   b. Estimate the distribution of effluent turbidity values by plotting the $P_{84.1}$ and P_{50} values. As the effluent turbidity values are expected to follow a log normal distribution, a straight line can be drawn through the $P_{84.1}$ and P_{50} values, as shown on the following plot.

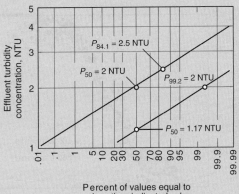

Percent of values equal to
or less than indicated value

3. Compute the effluent turbidity value expected to occur with the frequency of interest.
    a. The probability of occurrence of a given event with a frequency of once per year is (1 / 365) x 100 = 0.3 percent. Using the plot developed in step 2, an effluent turbidity value equal to or greater than 3.5 NTU will occur 0.3 percent of the time.
    b. Similarly, turbidity values equal to greater than 3.7 NTU will occur with a frequency of once in three years (i.e., 99.9 percent).
4. Estimate how often the combined treatment process will exceed the turbidity standard of 2.49 NTU. From the plot presented in step 2, the effluent turbidity will exceed 2.49 NTU approximately 16 (100–84) percent of the time.

Comment   Recognition of the variability in performance is of importance in the design of filtration systems, especially where more stringent mean effluent turbidity values must be met. For example, it the turbidity standard had been 2.0 NTU at a reliability of at least 99.2 percent (three exceedances per year), the mean design value, as illustrated in the above figure, would have to be about 1.17 NTU, assuming that the geometric standard deviation remained constant and was equal to 1.25. To reach a mean turbidity value of 1.17 NTU would, in most cases, require the addition of chemicals, although many plants with deep secondary clarifiers reach these values consistently without chemical addition.

Removal of Different Particle Sizes.   Although all of the filters shown on Fig. 11–10 can produce an effluent with an average turbidity of two or less with a suitable secondary effluent, the filtered effluent particle size distribution is different for each of the filters. Typical data on the removal of particle sizes from activated sludge effluent using depth filtration are shown on Fig. 11–12. As shown, the particle removal rate is essentially independent of the filtration rate in the range from 100 to about 260 L/m²·min. It is significant to note that most depth filters will pass some particles with diameters greater than 15 to 20 μm.

Depending on the quality of the settled secondary effluent, chemical addition has been used to improve the performance of effluent filters, with respect to turbidity. An example of the change in the distribution of particle sizes in the effluent from an activated sludge process following depth filtration without and with chemical coagulation is illustrated on

Figure 11–12

Particle size removal efficiency for a depth filter for effluent from an activated sludge plant at two different filtration rates.

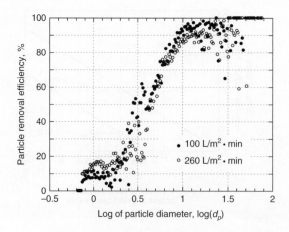

Figure 11–13

Effect of the use of chemicals on filter particle size removal performance (courtesy of K. Bourgeous, 2005): (a) original data as collected and, (b) the original data, plotted functionally according to the power law (see Example 2–4 in Chap. 2).

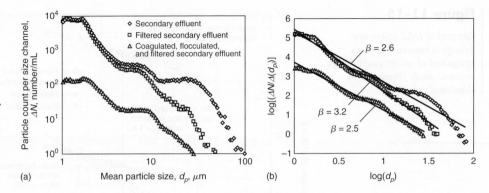

(a)

(b)

Fig. 11–13. The original data, as collected, are shown on Fig. 11–13(a). The original data, plotted functionally according to the power law (see Example 2–4 in Chap. 2), are presented on Fig. 11–13(b). As shown on Fig. 11–13(a), filtration alone only affected the larger particles, whereas with chemical coagulation all of the particles were affected more-or-less uniformly. As shown on Fig. 11–13(b), even though the number of particles in each size range was reduced by an order of magnitude, a significant number of particles remains in each size range.

Removal of Microorganisms. Where chemicals are not used, the removal of coliform bacteria and viruses from biologically treated secondary effluent is on the order of 0 to 1.0 and 0 to 0.5 logs, respectively. The degree of removal depends on the solids retention time (SRT) at which the biological process is operated. For example, as shown on Fig. 11–14, as the SRT is increased, fewer of the particles have one or more associated coliform bacteria. Typical data on the removal of the bacteriophage MS2 are illustrated on Fig. 11–15. As shown, the mean removal of MS2 across the effluent filters is about 0.3 logs. However, what is of more interest is the distribution of the removal data. Based on the distribution shown on Fig. 11–15, which is also typical for the removal of coliform organisms, allowing a disinfection credit of one log of removal for filtration in water reuse applications may not be protective of public health. Where chemicals are used, the data on the removal for microorganisms is confounded statistically. In general, it is not possible to separate the effect of chemical addition from the performance of the filter.

Backwash Water Requirements. The amount of backwash water needed to clean the filter bed, expressed as a percentage of plant flow, will depend on the characteristics of filter influent and the design of the filter bed. Typical backwash water percentages for

Figure 11–14

Number of particles with one or more associated coliform organisms as a function of the solids retention time for the activated sludge process. (Adapted from Darby, et al., 1999.)

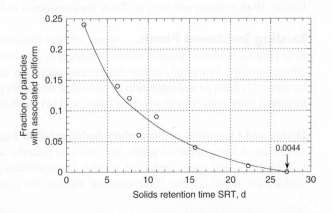

Figure 11-15

Removal of MS2 coliphage through a treatment process comprised of an activated sludge process, depth filtration, and chlorine disinfection.

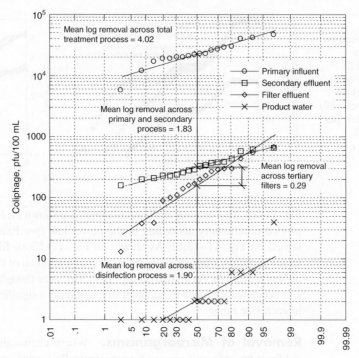

depth filters shown in Table 11–11 will vary from 4 to 15 percent. By comparison, the backwash percentages for surface filters, discussed in the following section, will typically vary from 1 to 4 percent. When adding effluent filtration to an existing plant, the impact of the return backwash water on plant hydraulics must be evaluated carefully. In many locations, the percentage of backwash water will control filter selection. The volume of backwash water required to wash a filter is related to filter area and desired degree of backwash bed expansion, the size of the media, and water temperature.

Considerations Related to Design and Operation of Treatment Facilities

The design and operation of the treatment facilities can have a significant impact on the performance of depth filters. Issues related to the design and operation of wastewater treatment facilities will depend on whether filtration is being considered for an existing or new facility. Both existing and new facilities are considered in the following discussion.

Existing Treatment Plants.   Where effluent filtration must be added to or retrofitted to an existing treatment plant it will be important to consider the factors that can impact the effluent quality including (1) the design and operation of the secondary sedimentation facilities, (2) the type and operation of biological treatment process, (3) the operation of the treatment plant with respect to return flows, and (4) the potential use of flow equalization.

Design and Operation of Secondary Sedimentation Facilities.   The concentration and variability of the suspended solids and colloidal material in the effluent from sedimentation facilities will vary depending on the characteristics of the biological solids to be settled; the depth of the sedimentation basins; the use of energy dissipation inlets (see Fig. 8–54); the presence of sludge density currents, as described in Chap. 5; short circuiting caused by a variety of factors and the placement of the effluent weirs. Typically, the TSS concentration in the effluent from activated sludge and trickling filter plants varies between 6 and 30 mg/L. Corresponding turbidity values, as discussed previously, can vary from 3 to 15 NTU. It should be noted that properly designed deep secondary sedimentation tanks 6 to 7 m (19.5 to 23 ft) can produce an average settled effluent with turbidity of 2 NTU or less consistently, and in some cases less than 1 NTU. Depending on the concentration and variability of the suspended solids in the effluent from the secondary sedimentation it may be appropriate to consider a filter that can continue to function even when heavily loaded. Both downflow and upflow deep-bed coarse medium filters, the fuzzy filter, and the pulsed-bed filter have been used in such applications.

Type and Operation of Biological Treatment Process. The three principal types of biological treatment processes used for secondary treatment are suspended growth, as exemplified by the activated sludge process; attached growth processes, as exemplified by the trickling filter process; and pond processes. The floc strength and filterability of the effluents from each of these processes will vary with the mode of operation. For example, the strength of the biological floc from the activated sludge process will vary with the mean cell-residence time, increasing with longer mean cell residence time. The increased strength derives in part from the production of extracellular polymers as the mean cell residence time is lengthened. At extremely long mean cell residence times (15 d and longer), it has been observed that the floc strength will decrease due to floc breakup. It has also been observed that the residual suspended solids remaining after secondary clarification from treatment processes with extremely long mean cell residence times are far more difficult to filter and as a result can impact the performance of disinfection systems (Emerick, 2012). Further, the residual floc from the chemical precipitation of biologically processed wastewater may be considerably weaker than the residual biological floc before precipitation. If there is uncertainty about the characteristics of the effluent from the secondary sedimentation facilities, pilot plant studies should be conducted, as discussed subsequently.

Management of Return Flows. Currently, in most treatment plants, return flows from sludge thickeners, sludge dewatering (e.g., centrifuges, and belt presses), sludge stabilization (e.g., digester supernatant), and sludge drying facilities are returned to the wastewater treatment plant headworks for reprocessing. In many instances, these return flows contain constituents that deteriorate overall plant performance (e.g., nitrogenous compounds, colloidal material and total dissolved solids). Unfortunately, these return flows can impact the overall performance of the biological treatment process especially with respect to the removal of nitrogen and colloidal material. Flow equalization facilities and or separate systems for the treatment of return flows, as discussed in Chap. 15, are now being installed at a number of treatment plants that need to meet more stringent discharge requirements. (Tchobanoglous et al., 2011).

Flow Equalization. Flow equalization is a method used to improve the performance and variability of the downstream treatment processes and to reduce the size and cost of treatment facilities (see extended discussion in Sec. 3–7 in Chap. 3). In advanced wastewater treatment, the principal benefits include (1) reduced variability of incoming water quality; (2) enhanced performance at constant flow operation, especially for membrane processes; and (3) reduced wear and tear on membranes due to fluctuating flows and loads (Tchobanoglous et al., 2003). If full plant flow equalization is not feasible, consideration should be given to flow equalization of the return flows.

New Treatment Plants. For new wastewater treatment plants all of the factors discussed above should be considered. However, extra care should be devoted to the design of the secondary settling facilities. With properly designed settling facilities resulting in an effluent with low TSS (typically 3 to 4 mg/L or less) and turbidity (less than 1 to 2 NTU), the decision on what type of filtration system is to be used is often based on plant-related variables, such as the space available, duration of filtration period (seasonal versus year-round), the time available for construction, and costs. The most important influent characteristics in the filtration of treated secondary effluents are the suspended-solids concentration, particle size and distribution, and floc strength.

Selection of Filtration Technology

In selecting a filter technology, important factors that must be considered include (1) the required effluent quality, (2) the influent wastewater characteristics, (3) type of filter to be used: proprietary or individually designed, (4) the filtration rate, (5) filtration driving force, (6) number and size of filter units, (7) the backwash water requirements, (8) the need for pilot-plant studies, and (9) system redundancy. Each of these issues is described in Table 11–13. The importance of the influent characteristics has been discussed previously. The backwash requirements, the need for chemical addition, and the need to conduct pilot plant studies are considered in the following discussion because of the impact these factors can have on the selection of a depth filtration process.

Table 11–13

Important factors in selecting filter technology for effluent filtration applications[a]

Factor	Remarks
Required effluent quality	Usually fixed regulatory requirement, depending on the final use of the effluent.
Influent wastewater characteristics	The required effluent quality will impact the selection process, as some filters are more able to withstand periodic shock loadings. For example, wider variations in effluent quality would be expected where shallow clarifiers are used. More predictable effluent quality can be expected from deep clarifiers. In recent designs employing deep clarifiers (5 to 6 m side water depths), effluent turbidity values of less than 2 NTU are achieved consistently.
Type of filter: proprietary vs. individually designed	Currently available filter technologies are either proprietary or individually designed. With proprietary filters, the manufacturer is responsible for providing the complete filter unit and its controls, based on basic design criteria and performance specifications. In individually designed filters, the design engineer is responsible for working with several suppliers in developing the design of the system components. Contractors and suppliers then furnish the materials and equipment in accordance with the engineer's design.
Filtration rate	The filtration rate affects the areal size of the filters that will be required. For a given filter application, the rate of filtration depends primarily on floc strength and the size of the filtering medium. For example, if the strength of the floc is weak, high filtration rates tend to shear the floc particles and carry much of the material through the filter. Filtration rates generally in the range of 80 to 330 L/m²·min will not affect the effluent quality when filtering settled activated sludge effluent (see Table 11–11).
Filtration driving force	Either the force of gravity or an applied pressure force can be used to overcome the frictional resistance to flow offered by the filter bed. Gravity filters of the type discussed in Table 8-5 are used most commonly for the filtration of treated effluent at large plants. Pressure filters operate in the same manner as gravity filters and are used at smaller plants. In pressure filters, the filtration operation is carried out in a closed vessel under pressurized conditions achieved by pumping.

(continued)

Table 11–13 (Continued)	
Factor	**Remarks**
Number and size of filtration units	The number of filter units generally should be kept to a minimum to reduce the cost of piping and construction, but it should be sufficient to assure that (1) backwash flowrates do not become excessively large and (2) that when one filter unit is taken out of service for backwashing, the transient loading on the remaining units is not excessive. Transient loadings due to backwashing are not an issue with filters that backwash continuously. To meet redundancy requirements, a minimum of two filters should be used. The sizes of the individual filter units should be consistent with the sizes of equipment available for use as underdrains, washwater troughs, and surface washers. Typically, width-to-length ratios for individually designed gravity filters vary from 1:1 to 1:4. A practical limit for the surface area on an individual depth filter (or filter cell) is about 100 m² (1075 ft²), although larger filters units have been built. For proprietary filters, use standard sizes that are available from manufacturers. The surface area of a depth filter is based on the peak filtration and peak plant flowrates. The allowable peak filtration rate is usually established on the basis of regulatory requirements. Operating ranges for a given filter type are based on past experience, the results of pilot-plant studies, manufacturers recommendations, and regulatory constraints.
Backwash water requirements	As noted in Table 8–4, depth filters operate in either a semi-continuous or continuous mode. In semi-continuous operation, the filter is operated until the effluent quality starts to deteriorate or the headloss becomes excessive at which point the filter is taken out of service and backwashed to remove the accumulated solids. With filters operated in the semi-continuous mode, provision must be made for the backwash water needed to clean the filters. Typically, the backwash water is pumped from a filtered water clearwell or obtained by gravity from an elevated storage tank. The backwash storage volume should be sufficient to backwash each filter every 12 h. For filters that operate continuously such as the upflow filter and the traveling bridge filter, the filtering and backwashing phases take place simultaneously. In the traveling bridge filter, the backwash operation can either be continuous or semi- continuous as required. For filters that operate continuously, there is no turbidity breakthrough or terminal headloss.
Chemical addition	Need for chemical addition is site specific. Depending on the final use of the effluent, provision of chemical dosing facilities may be mandated by local and/or state regulations.
Pilot-plant studies	Because of the many variables involved, pilot-plant studies are often conducted when filtration facilities are to be added to an existing facility. For new plants, pilot-plant studies can be conducted at treatment plants of similar design.
System redundancy	System redundancy is related to uninterruptible power and the need to provide standby capacity for routine maintenance. Most water reclamation plants in continuous service have emergency storage and onsite power generation to operate process equipment. In general, one standby filter, as a minimum, is recommended for standby service. Where the provision of standby facilities is not possible due to space or other limitations, the filters and related piping should be sized to handle periodic overloads during maintenance periods.

ᵃ Adapted, in part, from Tchobanoglous et al. (2003).

Filter Backwash Water Requirements. Methods commonly used for backwashing granular medium filter beds are considered subsequently. In general, depth filters require more backwash water as compared to surface filters. For existing plants with limited hydraulic capacity, the percentage of backwash water that can be processed may be the limiting factor in the selection of a filter technology. Because it is impossible to predict a priori what percentage backwash water will be needed, pilot-plant studies must be conducted to resolve the issue.

Effluent Filtration with Chemical Addition. Depending on the quality of the settled secondary effluent, chemical addition has been used to improve the performance of effluent filters. Chemical addition has also been used to achieve specific treatment objectives including the removal of specific contaminants such as phosphorus, metal ions, and humic substances. The removal of phosphorus by chemical addition is considered in Chap. 6. To control eutrophication, the contact filtration process is used in many parts of the country to remove phosphorus from wastewater treatment plant effluents which are discharged to sensitive water bodies. The two-stage filtration process, described in Table 11–10, has proven to be very effective, achieving phosphorus levels of 0.2 mg/L or less in the filtered effluent. Chemicals commonly used in effluent filtration include a variety of organic polymers, alum, and ferric chloride. Use of organic polymers and the effects of the chemical characteristics of the wastewater on alum addition are considered in the following discussion.

Use of Organic Polymers. Organic polymers are typically classified as long chain organic molecules with molecular weights varying from 10^4 to 10^6. With respect to charge, organic polymers can be cationic (positively charged), anionic (negatively charged), or nonionic (no charge). Polymers are added to settled effluent to bring about the formation of larger particles by bridging as described in Chap. 6. Because the chemistry of the wastewater has a significant effect on the performance of a polymer, the selection of a given type of polymer for use as a filter aid generally requires experimental testing (e.g., jar testing).

Common test procedures for polymers involve adding an initial dosage (usually 1.0 mg/L) of a given polymer and observing the effects. Depending upon the effects observed, the dosage should be increased by 0.5 mg/L increments or decreased by 0.25 mg/L increments (with accompanying observation of effects) to obtain an operating range. After the operating range is established, additional testing can be done to establish the optimum dosage. Great care must be taken to insure that the polymer is well dispersed before reaching the filter to avoid the formation of mudballs.

A recent development is the use of lower molecular weight polymers that are intended to serve as alum substitutes. When these polymers are used, the dosage is considerably higher (≥ 10 mg/L) than with higher molecular weight polymers (0.25 to 1.25 mg/L). As with the mixing of alum, the initial mixing step is critical in achieving maximum effectiveness of a given polymer. In general, mixing times of less than 1 second with G values of >2500 s^{-1} are recommended (see Table 5–9 in Chap. 5). It should be noted that, as a practical matter, as treatment plants get larger it is difficult to achieve mixing times less than one second unless multiple mixing devices are used.

Effects of Chemical Characteristics of Wastewater on Alum Addition. As with polymers, the chemical characteristics of the treated wastewater effluent can have a significant impact on the effectiveness of aluminum sulfate (alum) when it is used as an aid to filtration. For example, the effectiveness of alum is dependent on pH (see Fig. 6–9 in Chap. 6). Although Fig. 6–9 was developed for water treatment applications, it has been found to apply to most wastewater effluent filtration uses with minor variations. As shown on Fig. 6–9, the approximate regions in which the different phenomena associated with particle removal in conventional sedimentation and filtration processes are operative are plotted as a function of the alum dose and the pH of the treated effluent after alum has been added. For example, optimum particle removal by sweep floc occurs in the pH range of 7 to 8 with an alum dose of 20 to 60 mg/L. Generally, for many wastewater effluents that have high pH values (e.g. 7.3 to 8.5), low alum dosages in the range of 5 to 10 mg/L will not be effective. To operate with low alum dosages, pH control will generally be required.

Need for Bench-Scale and Pilot-Plant Studies. Although the information presented earlier in this section and previously in Sec. 11–3 will help the reader understand the nature of the filtration operation as it is applied to the filtration of treated wastewater, it must be stressed that there is no generalized approach to the design of full-scale filters. The principal reason is the inherent variability in the characteristics of the influent suspended solids to be filtered. For example, changes in the degree of flocculation of the suspended solids in the secondary settling facilities will significantly affect the particle sizes and their distribution in the effluent, which in turn will affect the performance of the filter. Further, because the characteristics of the effluent suspended solids will also vary with the organic loading on the process as well as with the time of day, filters must be designed to function under a rather wide range of operating conditions. The best way to ensure that the filter configuration selected for a given application will function properly is to conduct pilot-plant studies (see Fig. 11–16).

Because of the many variables that can be analyzed, care must be taken not to change more than one variable at a time so as to confound the results in a statistical sense. Bench-scale and pilot-plant testing should be carried out at several intervals, ideally throughout a full year, to assess seasonal variations in the characteristics of the effluent to be filtered. All test results should be summarized and evaluated in different ways to ensure their proper analysis. Because the specific details of each test program will be different, no generalization on the best method of analysis can be given.

Design Considerations for Granular Medium Filters

As noted in Table 11–9, the currently available filter technologies are either proprietary or individually designed. With proprietary filters, the manufacturer is responsible for providing the complete filter unit and its controls, based on basic design criteria and performance specifications. In individually designed filters, the designer is responsible for working with several suppliers in developing the design of the system components. Contractors and suppliers then furnish the materials and equipment in accordance with the engineer's design.

Because granular medium filters are still designed individually, important design considerations for depth filters are summarized in Table 11–14. Although some of the factors listed in Table 11–14 have been discussed previously and other design details are beyond the scope of this textbook, it is nevertheless important to consider the selection of the type of filter bed and filter medium(s) used in depth filters, the backwashing operation, and filter appurtenances.

Figure 11-16

Views of filtration pilot plant: (a) filter columns fed from the source and (b) instrumentation used to monitor filter performance including turbidity and particle size counting.

Table 11–14

Design considerations for granular medium filters for effluent for filtration

Variable	Significance
1. Required effluent quality	Usually fixed regulatory requirement, depending on the final use of the effluent.
2. Influent wastewater characteristics	Considered in previous section.
3. Filter medium characteristics	Affects clean-water headloss, particulate matter removal efficiency, and headloss buildup.
a. Effective size, d_{10}	
b. Uniformity coefficient, UC	
c. Type, grain shape, density, and composition	
4. Filter-bed characteristics	Porosity affects the amount of solids that can be stored within the filter. Bed depth affects
a. Bed depth	initial headloss, length of run. Degree of intermixing will affect performance of filter bed.
b. Porosity	
c. Stratification	
d. Degree of medium intermixing	
5. Filtration rate	Used in conjunction with variables 2, 3, and 4 to compute clean-water headloss. Maximum rate typically specified by regulatory agency (see Table 11–11).
6. Flowrate control	The principal methods now used to control the rate of flow through downflow gravity filters may be classified as (1) constant-rate filtration with fixed head, (2) constant-rate filtration with variable head, and (3) variable-declining-rate filtration. Other control methods are also in use.
7. Allowable headloss	Design variable, depends on whether driving force will be gravity or applied pressure.
8. Backwashing system	Methods commonly used for backwashing granular medium filter beds operated in the semi-continuous mode include (1) water backwash with auxiliary surface washwater agitation, (2) water backwash with auxiliary air scour, and (3) combined air-water backwashing. With the first two methods, fluidization of the granular medium is necessary to achieve effective cleaning of the filter bed at the end of the run. With the third method, fluidization is not necessary. Typical backwash flowrates required to fluidize various filter beds are reported in Table 11–11.
9. Backwash requirements	Affects size of filter piping and pipe gallery.
10. Filter appurtenances	Filter appurtenances include: (1) the underdrain system used to support the filtering materials, collect the filtered effluent, and distribute the backwash water and air (where used); (2) the washwater troughs used to remove the spent backwash water from the filter; and (3) the surface washing systems used to help remove attached material from the filter medium.

Filter Bed Configuration and Filter Medium. Important considerations in individually designed depth filters are the selection of the type of filter bed and the corresponding media characteristics.

Selection of Filter Bed Configuration. The principal types of non-proprietary filter bed configurations now used for wastewater filtration may be classified according to the number of filtering media that are used as mono-medium, dual-media, or multi-media beds (see Fig. 11–6). In conventional downflow filters, the distribution of grain sizes for each medium after backwashing is from small to large. Typical design data for mono-medium, and dual- and multi-media filters are presented in Tables 11–15 and 11–16, respectively.

Table 11–15

Typical design data for depth filters with mono-medium[a]

Characteristic	Unit	Value	
		Range	**Typical**
Shallow bed (stratified)			
Anthracite			
Depth	mm	300–500	400
Effective size	mm	0.8–1.5	1.3
Uniformity coefficient	unitless	1.3–1.8	≤1.5
Filtration rate	m³/m²·min	0.08–0.24	
Sand			
Depth	mm	300–360	330
Effective size	mm	0.45–0.65	0.45
Uniformity coefficient	unitless	1.2–1.6	≤1.5
Filtration rate	m³/m²·min	0.08–0.24	
Conventional (stratified)			
Anthracite			
Depth	mm	600–900	750
Effective size	mm	0.8–2.0	1.3
Uniformity coefficient	unitless	1.3–1.8	≤1.5
Filtration rate	m³/m²·min	0.08–0.40	
Sand			
Depth	mm	500–750	600
Effective size	mm	0.4–0.8	0.65
Uniformity coefficient	unitless	1.2–1.6	≤1.5
Filtration rate	m³/m²·min	0.08–0.24	
Deep-bed (unstratified)			
Anthracite			
Depth	mm	900–2100	1500
Effective size	mm	2–4	2.7
Uniformity coefficient	unitless	1.3–1.8	≤1.5
Filtration rate	m³/m²·min	0.08–0.40	
Sand			
Depth	mm	900–1800	1200
Effective size	mm	2–3	2.5
Uniformity coefficient	unitless	1.2–1.6	≤1.5
Filtration rate	m³/m²·min	0.08–0.40	
Fuzzy filter			
Depth	mm	600–1080	800
Effective size	mm	25–30	28
Uniformity coefficient	unitless	1.1–1.2	1.1
Filtration rate	m³/m²·min	0.60–1.60	

[a] Adapted in part from Tchobanoglous (1988) and Tchobanoglous et al. (2003).

Note: m³/m²·min × 24.5424 = gal/ft²·min.

Table 11–16

Typical design data for dual- and multi-media depth filters[a]

Characteristic	Unit	Value[b] Range	Typical
Dual-media			
Anthracite ($\rho = 1.60$)			
Depth	mm	360–900	720
Effective size	mm	0.8–2.0	1.5
Uniformity coefficient	unitless	1.3–1.6	≤1.5
Sand ($\rho = 2.65$)			
Depth	mm	180–360	360
Effective size	mm	0.4–0.8	0.65
Uniformity coefficient	unitless	1.2–1.6	≤1.5
Filtration rate	m³/m²·min	0.08–0.40	0.20
Multi-media			
Anthracite (top layer of quad-media filter, $\rho = 1.60$)			
Depth	mm	240–600	480
Effective size	mm	1.3–2.0	1.6
Uniformity coefficient	unitless	1.3–1.6	≤1.5
Anthracite (second layer of quad-media filter, $\rho = 1.60$)			
Depth	mm	120–480	240
Effective size	mm	1.0–1.6	1.1
Uniformity coefficient	unitless	1.5–1.8	1.5
Anthracite (top layer of tri-media filter, $\rho = 1.60$)			
Depth	mm	240–600	480
Effective size	mm	1.0–2.0	1.4
Uniformity coefficient	unitless	1.4–1.8	≤1.5
Sand ($\rho = 2.65$)			
Depth	mm	240–480	300
Effective size	mm	0.4–0.8	0.5
Uniformity coefficient	unitless	1.3–1.8	≤1.5
Garnet ($\rho = 4.2$)			
Depth	mm	50–150	100
Effective size	mm	0.2–0.6	0.35
Uniformity coefficient	unitless	1.5–1.8	≤1.5
Filtration rate	m³/m²·min	0.08–0.40	0.20

[a] Adapted from Tchobanoglous (1988) and Tchobanoglous et al. (2003).

[b] Anthracite, sand, and garnet sizes selected to limit the degree of intermixing. Use Eq. (11–28) for other values of density, ρ.

Note: m³/m²·min × 24.5424 = gal/ft²·min.

Figure 11–17

Typical particle size distribution ranges for sand and anthracite used in dual medium depth filters. Note that for sand the 10 percent size by weight corresponds approximately to the 50 percent size by count.

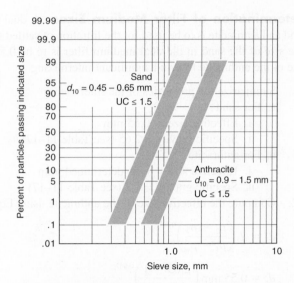

Selection of Filter Medium. Once the type of filter to be used has been selected, the next step is to specify the characteristics of the filter medium, or media, if more than one is used. Typically, this process involves the selection of the grain size as specified by the effective size, d_{10}, uniformity coefficient, UC, the 90 percent size, the specific gravity, solubility, hardness, and depth of the various materials used in the filter bed. Typical particle size distribution ranges for sand and anthracite filtering material are shown on Fig. 11–17. The 90 percent size designated, d_{90}, as read from a grain size analysis is used commonly to determine the required backwash rate for depth filters. The physical properties of filter materials used in depth filters are summarized in Table 11–17.

To avoid extensive intermixing of the individual mediums in multi-media filter beds, the settling rate of the filter media comprising the dual- and multi-media filters must have essentially the same settling velocity. Some intermixing is unavoidable, and the degree of intermixing in the dual- and multi-media beds depends on the density and size differences of the various media. The following relationship can be used to establish the appropriate sizes (Kawamura, 2000).

$$\frac{d_1}{d_2} = \left(\frac{\rho_2 - \rho_w}{\rho_1 - \rho_w} \right)^{0.667} \tag{11–28}$$

where d_1, d_2 = effective size of filter medium
ρ_1, ρ_2 = density of filter medium
ρ_w = density of water

The application of Eq. (11–28) is illustrated in Example 11–4.

Table 11–17

Typical properties of filter materials used in depth filtration[a]

Filter material	Specific gravity	Porosity, α	Sphericity
Anthracite	1.4–1.75	0.56–0.60	
Sand	2.55–2.65	0.40–0.46	0.75–0.85
Garnet	3.8–4.3	0.42–0.55	0.75–0.85
Ilmenite	4.5	0.40–0.5	
Fuzzy filter medium		0.87–0.89	

[a] Adapted in part from Cleasby and Logsdon (1999).

EXAMPLE 11-4 **Determination of Filter Medium Sizes** A dual media filter bed comprised of sand and anthracite is to be used for the filtration of settled secondary effluent. If the effective size of the sand in the dual medium filter is to be 0.55 mm, determine the effective size of the anthracite to avoid significant intermixing.

Solution

1. Summarize the properties of the filter media
   a. For sand
      i. Effective size = 0.55 mm
      ii. Specific gravity = 2.65 (see Table 11–17)
   b. For anthracite
      i. Effective size = to be determined, mm
      ii. Specific gravity = 1.7 (see Table 11–17)
2. Compute the effective size of the anthracite using Eq. (11–28)

$$d_1 = d_2 \left(\frac{\rho_2 - \rho_w}{\rho_1 - \rho_w} \right)^{0.667}$$

$$d_1 = 0.55 \text{ mm} \left(\frac{2.65 - 1}{1.7 - 1} \right)^{0.667}$$

$$d_1 = 0.97 \text{ mm}$$

Comment Another approach that can be used to assess whether intermixing will occur is to compare the fluidized bulk densities of the two adjacent layers (e.g., upper 450 mm sand and lower 100 mm of anthracite).

Filter Flowrate Control. The principal methods now used to control the rate of flow through downflow gravity filters may be classified as (1) constant-rate filtration with fixed head, (2) constant rate filtration with variable head, and (3) variable-declining-rate filtration. A variety of other control methods are also in use (Cleasby and Logsdon, 1999; Kawumura, 2000).

Constant Rate Filtration with Fixed Head. In constant-rate filtration with fixed head [see Fig. 11–18(a)], the flow through the filter is maintained at a constant rate. Constant-rate filtration systems are either influent controlled or effluent controlled. Pumps or weirs are used for influent control whereas an effluent modulating valve that can be operated manually or mechanically is used for effluent control. In effluent control systems, at the beginning of the run, a large portion of the available driving force is dissipated at the valve, which is almost closed. The valve is opened as the headloss builds up within the filter during the run. Because the required control valves are expensive and because they have malfunctioned on a number of occasions, alternative methods of flowrate control involving pumps and weirs have been developed and are coming into wider use.

Constant Rate Filtration with Variable Head. In constant-rate variable head filtration head [see Fig. 11–18(b)], the flow through the filter is maintained at a constant rate. Pumps or weirs are used for influent control. When the head or effluent turbidity reaches a preset value, the filter is backwashed.

Variable Rate Filtration with Fixed or Variable Head. In variable-declining-rate filtration [see Fig. 11–18(c)], the rate of flow through the filter is allowed to decline as the

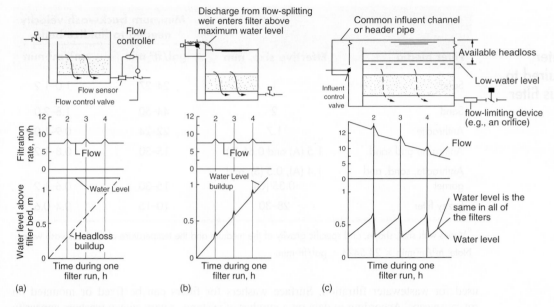

Figure 11–18

Definition sketch for filter operation: (a) fixed head, (b) variable head, and (c) variable-flow variable-head. Curves for filters in (a), (b), and (c) are for the operation of one filter in a bank of four filters. The numbers represent the filter that is backwashing during the filter run. In practice, the time before backwashing will not be the same for all of the filters. (Adapted from Tchobanoglous and Schroeder, 1985.)

rate of headloss builds up with time. Declining-rate filtration systems are either influent controlled or effluent controlled. When the rate of flow is reduced to the minimum design rate, the filter is removed from service and backwashed.

Filter Backwashing Systems.   Methods commonly used for backwashing granular-medium filter beds operated in the semi-continuous mode include (1) water backwash only, (2) water backwash with auxiliary surface water-wash agitation, (3) water backwash with auxiliary air scour, and (4) combined air-water backwashing. With the first three methods, fluidization of the granular medium is necessary to achieve effective cleaning of the filter bed at the end of the run. With the fourth method, fluidization is not necessary.

Water Backwash Only.   In the past, the most common method used to clean a filter of accumulated material was to backwash it with filtered water. Based on experimental studies, it has been found that the optimum cleaning of a conventional filter bed occurs when the expanded porosity of the bed is in the range of 0.65 to 0.70 (Amirtharajah, 1978). At this degree of expansion, it has been found that the shearing action of the rising backwash water and particle abrasion is most effective in removing the accumulated material from the filtering medium. Approximate backwash water flowrates required to fluidize various filter beds are reported in Table 11–18. To reduce the potential for the formation of mud balls and to enhance the removal of accumulated material either surface washers or air scour, as described below, are now used in conjunction with the water backwash.

Water Backwash with Auxiliary Surface Wash.   Surface washers (see Fig. 11–19) are often used to provide the shearing force required to clean the grains of the filtering medium

Table 11–18

Approximate backwash water flowrates required to fluidize various filter beds at 20°C

Filter media	Effective size, mm	Minimum backwash velocity needed to fluidize beda	
		gal/ft²·min	m³/m²·min
Sand	1	24–27	1.0–1.2
Sand	2	44–50	1.8–2.0
Anthracite	1.7	22–24	0.9–1.0
Anthracite and sand	1.5 (A) and 0.65 (S)	15–30	0.8–1.2
Anthracite, sand, and garnet	1.4 (A), 0.5 (S), and 0.35 (G)	15–30	0.6–1.2
Fuzzy filter	28–30	10–15	0.4–0.6

ᵃ Varies with size, shape, and specific gravity of the medium and the temperature of the backwash water.
Note: m³/m²·min × 24.5424 = gal/ft²·min.

used for wastewater filtration. Surface washers for filters can be fixed or mounted on rotary sweeps. According to data on a number of systems, rotary sweep washers appear to be the most effective. Operationally, the surface washing cycle is started about 1 or 2 min before the water backwashing cycle is started. Both cycles are continued for about 2 min, at which time the surface wash is terminated. Water usage for a single-sweep surface backwashing system varies from 0.02 to 0.04 m³/m²·min (0.5 to 1.0 gal/ft²·min) and from 0.06 to 0.08 m³/ m²·min (1.5 to 2.0 gal/ft²·min) for a dual-sweep surface backwashing system.

Water Backwash with Auxiliary Air Scour. The use of air to scour the filter provides a more vigorous washing action than water alone. Operationally, the water level above the filter bed is lowered to within 150 mm (6 in.) of the top of the media and air is usually applied for 3 to 4 min before the low-rate water backwashing cycle begins. In some

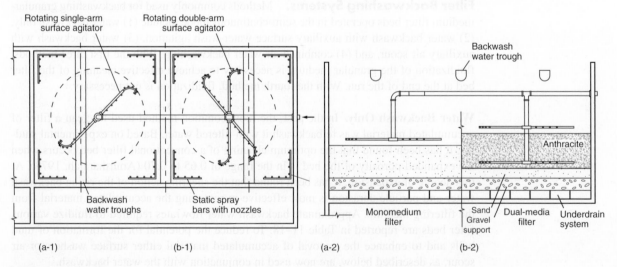

Figure 11–19

Surface washing facilities used to clean conventional granular medium filters: (a-1) and (a-2) single arm washer for a mono-medium filter and (b-1) and (b-2) a dual arm washer for a dual-medium filter.

Table 11–19

Approximate air and water flowrates used to backwash various filter beds at 20°C

Filter media	Medium characteristics		Backwash sequence	Backwash rates, m³/m²·min	
	Effective size, mm	Uniformity coefficient		Air	Water
Sand	1	1.4	1st–air + water	0.8–1.3	0.25–0.3
			2nd–water		0.5–0.6
Sand	2	1.4	1st–air + water	1.8–2.4	0.4–0.6
			2nd–water		0.8–1.2
Anthracite	1.7	1.4	1st–air + water	1.0–1.5	0.35–0.5
			2nd–water		0.6–0.8
Sand and	0.65(S)	1.4	1st–air	0.8–1.6	
anthracite[b]	1.5(A)	1.4	2nd–air + water	0.8–1.6	0.3–0.5
			3rd–water		0.6–0.9

[a] Adapted in part from Dehab and Young (1977) and Cleasby and Logsdon (2000).

[b] Dual medium filter bed is fluidized.

Note: m³/m²·min × 24.5424 = gal/ft²·min

m³/m²·min × 3.2808 = ft³/ft²·min.

systems, air is also injected during the first part of the low-rate water-washing cycle, referred to as combined or concurrent air scour (see below). Typical air flowrates range from 0.9 to 1.6 m³/ m²·min (3 to 5 ft³/ft²·min). A typical operating sequence for a dual media filter is given in Table 11–19. Also, as noted in Table 11–19, a water wash at the end of the air/water backwash cycle is used at the end to purge the filter of any residual air which could cause air binding. The reduced washwater requirements for the air-water backwash system can be appreciated by comparing the values given in Table 11–18 with those given in Table 11–19.

Combined Air-Water Backwash. The combined air-water backwash system is used in conjunction with the single-medium unstratified filter bed. Operationally, air and water are applied simultaneously for several minutes. The specific duration of the combined backwash varies with the design of the filter bed. Ideally, during the backwash operation, the filter bed should be agitated sufficiently so that the grains of the filter medium move in a circular pattern from the top to the bottom of the filter as the air and water rise up through the bed. Some typical data on the quantity of water and air required are reported in Table 11–19. At the end of the combined air-water backwash, a 2- to 3-min water a low-rate backwash at sub-fluidization velocities [typically 0.2 m³/ m²·min (5 gal/ft²·min)] is used to remove any air bubbles that may remain in the filter bed. This step is required to eliminate the possibility of air binding within the filter. Normally, a high-rate washing step follows the low-rate washing. High-rate washing is typically conducted at [0.6 − 0.8 m³/ m²·min (15−20 gal/ft²·min)]. Air scour combined with high-rate washing is not conducted because of the possibility of excessive expansion of media resulting in media loss into the washwater troughs.

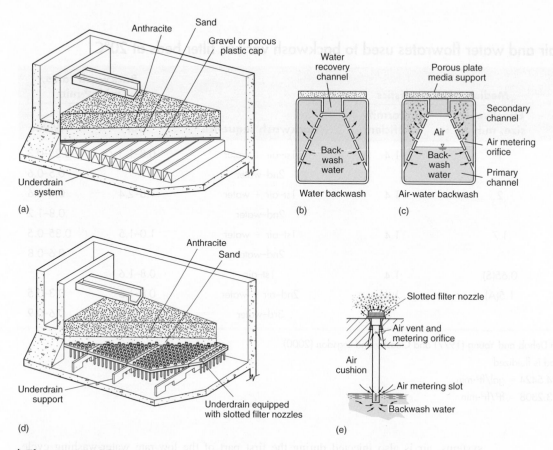

Figure 11–20

Typical underdrain systems used for granular media filters: (a) underdrain system used with gravel or porous plastic cap support for filter media, (b) underdrain system shown in (a) during water backwash, (c) underdrain system shown in (a) during air-water backwash, (d) underdrain system equipped with slotted air-water nozzles used without gravel support layer, and (e) air-water nozzle used in underdrain system shown in (d) without gravel support layer [(b) and (c) adapted from Leopold; (e) adapted from Infilco-Degremont, Inc.].

Filter Appurtenances. The principal filter appurtenances are as follows: (1) the underdrain system used to support the filtering materials, collect the filtered effluent, and distribute the backwash water and air (where used); (2) the washwater troughs used to remove the spent backwash water from the filter; and (3) the surface washing systems or air scour blower used to help remove attached material from the filter medium.

Underdrain Systems. The type of underdrain system to be used depends on the type of backwash system. In conventional water backwashed filters without air scour, it is common practice to place the filtering medium on a support consisting of several layers of graded gravel. The design of a gravel support for a granular medium is delineated in the AWWA Standard for Filtering Material B100–96 (AWWA, 1996). Typical underdrain systems are shown on Fig. 11–20. The gravel is not intended to aid in filtration. Rather, the purpose of the gravel is to prevent media from entering the underdrain. Air scour for filters with a gravel support system can be challenging, and would normally be comprised of the air scour piping grid laid on top of the gravel itself. Disruption of gravel support will

Figure 11–21

Details of baffle systems developed to minimize loss of filtering medium during backwash operation: (a) section through dual-baffle system and . (b) more elaborate baffle with two wings and side baffles.

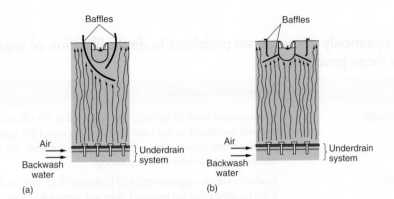

lead to potential media passage to the underdrain or media plugging the underdrain apertures. As an alternative to gravel, newer filter underdrains are often equipped with a porous, HDPE direct media retention plate, typically 25mm (1 in.) thick and fastened to the top of each underdrain block. This arrangement precludes the need for gravel and therefore can allow for a deeper filter bed.

Washwater Troughs. Washwater troughs are constructed of fiberglass, plastic, or sheet metal or of concrete with adjustable weir plates. The particular design of the trough will depend to some extent on the other equipment to be used in the design and construction of the filter. Loss of filter material during backwashing is a common operating problem. To reduce this problem, baffles can be placed on the under side of the washwater troughs as shown on Fig. 11–21.

Operational Problems with Depth Filters. The principal operational problems encountered in wastewater filtration with depth filters are (1) turbidity breakthrough; (2) mudball formation; (3) buildup of emulsified grease; (4) development of cracks and contraction of the filter bed; (5) loss of filter medium or media, by mechanical and operational means; and (6) gravel mounding. Because these problems can affect both the performance and operation of a filter system, care should be taken in the design phase to provide the necessary facilities to minimize their impact. These issues are considered further in Table 11–20. Because of the inherent variability in the wastewater characteristics and their potential impact upon filter design and operation, the best way to ensure the filter configuration selected for a given application will function properly is to conduct pilot plant studies representative of the range of operating conditions.

11–6 SURFACE FILTRATION

Surface filtration, as shown in Table 11–4, involves the removal of particulate material suspended in a liquid by mechanical sieving by passing the liquid through a thin septum (i.e., filter material). The mechanical sieving action is similar to a kitchen colander. Membrane filters, microfiltration, and ultrafiltration, discussed in Sec. 11–6, are also surface filtration devices but are differentiated on the basis of the sizes of the pores in the filter medium. Surface filter mediums typically have openings in the size range from 5 to 30 μm or larger; in microfiltration and ultrafiltration, the pore size can vary from 0.05 to 2.0 μm for MF and 0.005 to 0.1 μm for UF.

Surface filtration has been used in several applications including (1) as a replacement for depth filtration to remove residual suspended solids from secondary effluents, (2) for the

Table 11–20

Summary of commonly encountered problems in depth filtration of wastewater and control measures for those problems

Problem	Description/control
Turbidity breakthrough[a]	Unacceptable levels of turbidity are recorded in the effluent from the filter, even though the terminal headloss has not been reached. To control the buildup of effluent turbidity levels, chemicals and polymers have been added to the filter. The point of chemical or polymer addition must be determined by testing.
Mudball formation	Mudballs are an agglomeration of biological floc, dirt, and the filtering medium or media. If the mudballs are not removed, they will grow into large masses that often sink into the filter bed and ultimately reduce the effectiveness of the filtering and backwashing operations. The formation of mudballs can be controlled by auxiliary washing processes such as air scour or water surface wash concurrent with, or followed by, water wash.
Buildup of emulsified grease	The buildup of emulsified grease within the filter bed increases the headloss and thus reduces the length of filter run. Both air scour and water surface wash systems help control the buildup of grease. In extreme cases, it may be necessary to steam clean the bed or to install a special washing system.
Development of cracks and contraction of filter bed	If the filter bed is not cleaned properly, the grains of the filter bed filtering medium become coated. As the filter compresses cracks develop, especially at the sidewalls of the filter. Ultimately, mudballs may develop. This problem can be controlled by adequately backwashing and scouring.
Loss of filter medium or media (mechanical)	In time, some of the filter material may be lost during backwashing and through the underdrain system (where the gravel support has been upset and the underdrain system has been installed improperly). The loss of the filter material can be minimized through the proper placement of washwater troughs and underdrain system. Special baffles have also proved effective.
Loss of filter medium or media (operational)	Depending on the characteristics of the biological floc, grains of the filter material can become attached to it, forming aggregates light enough to be floated away during the backwashing operations. The problem can be minimized by the addition of an auxiliary air and/or water scouring system.
Gravel mounding	Gravel mounding occurs when the various layers of the support gravel are disrupted by the application of excessive rates of flow during the backwashing operation. A gravel support with an additional 60 to 75 mm (2 to 3 in.) layer of high density material, such as ilmenite or garnet, can be used to overcome this problem.

[a] Turbidity breakthrough does not occur with filters that operate continuously.

removal of suspended solids and algae from stabilization pond effluents, and (3) as a pretreatment operation before microfiltration or UV disinfection. Surface filtration is gaining in popularity because of the high quality effluent produced, smaller footprint, low backwash rates, and reduced maintenance requirements. Information on surface filtration technologies, their performance, and design considerations is presented and discussed in this section.

Available Filtration Technologies

The principal types of surface filtration devices are identified and described in Table 11–21. With the exception of the inclined surface and cartridge filters, all of the other surface filters have been used for the filtration of secondary effluent. Some of the surface filters have also been used for the filtration of algae for lagoon effluents. The inclined surface

Table 11–21

Description of some surface filters used in effluent filtration applications

Type	Description
(a) Cloth Media Filter (CMF) 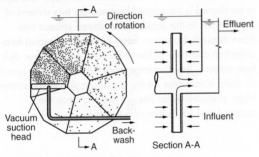	The CMF, marketed under the trademark AquaDisk® by Aqua-Aerobic Systems, also consists of several disks mounted vertically in a tank. Each disk is comprised of six equal segments. Operationally, water flows by gravity from the exterior of the disks through the filter medium to an internal collection system. Typically, two types of filter cloth are used: (1) a needle felt cloth made of polyester or (2) synthetic pile fabric cloth. A vacuum system is. Vacuum suction heads, located on either side of the disk, are used to remove the accumulated solids by drawing filtrate water from the filtrate header back through the cloth media while the disk is rotating. Solids are removed when a predetermined increase in headloss has occurred.
(b) Diamond Cloth Media Filter (DCMF) 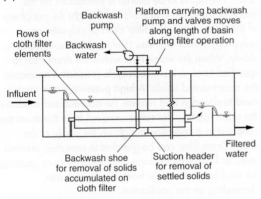	The DCMF, marketed under the trademark AquaDisk® by Aqua-Aerobic Systems, consists of cloth filter elements, which have a diamond shaped cross section. The filter elements are cleaned by a vacuum sweep which moves back and forth along the length of the filter, when a predetermined increase in headloss has occurred. Solids that settle to the bottom of the reactor below the filter element are removed periodically by a vacuum header. Using a diamond shape for the filter, it is possible to increase the cloth filter surface area per unit of aerial surface area. Because higher volumes for filtered water can be produced per unit area, the DCFM is used in new installations and as a replacement for existing sand filters.
(c) Discfilter® (DF) 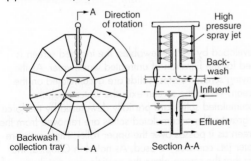	The DF, developed by Hydrotech and marketed in the U.S. by Veolia Water Systems, consists of a series of disks comprised of two vertically mounted parallel disks that are used to support the filter cloth Each disk is connected to a central feed tube. The cloth screen material used can be of either polyester or Type 304 or 316 stainless steel. Accumulated solids are removed from the screen by high-pressure water jets. The filter mechanism can be furnished with a self-contained tank or for installation in a concrete tank. In cold climates or where odor control is a consideration, an enclosure can be provided for the disks.

(continued)

Table 11–21 (Continued)	

Type	Description
(d) Ultrascreen® 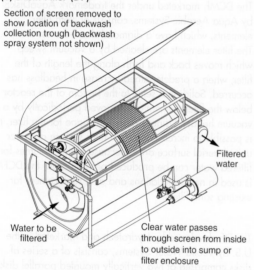	The Ultrascreen® developed by Nova Water technologies that consists of two continuously rotating circular screens of woven stainless steel mesh. The liquid to be filtered is introduced between the two screens at right angles to the screens. The filtered effluent that flows through the screen flows out and is directed to a collection chamber below the screen. Unlike other disk type screens, the discharge from the screen is by gravity as there is no water on the discharge side of the screen. High-pressure water jets are used to clean the accumulated solids from the screens.
(e) Drum Filter (DF)[a] 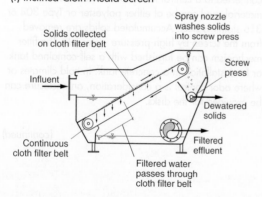	As the name implies the drum filter is in the shape of a drum. The liquid to be filtered is introduced on the inside of the drum and flows out thought the periphery of the drum, through a filter cloth of polyester or polypropylene or stainless steel, as the drum rotates slowly. When the water level within the drum rises to specified level, a backwash cycle is initiated to remove the accumulated solids. A high pressure water spray is used to dislodge and remove the accumulated solids as the drum rotates. The solids removed from the drum are collected in a collection trough on the inside of the drum. Drum filter can be installed in concrete, stainless steel, or fiberglass tankage. The range of pore openings for the filter cloth range from 10 μm to 1 mm, depending on the application.
(f) Inclined Cloth Media Screen	Developed by M2 Renewables, the inclined screen is used for the filtration of untreated wastewater. As the moving screen rotates, solids are accumulated on the screen. When the screen exits the water pool, the accumulated solids are partially dewatered by the force of gravity. The accumulated solids are removed from the screen as it passes over the upper roller. High-pressure water jets can also be used. As noted in Chap. 5, the fact that the screen alters the particle size distribution of the solids to be treated and has a relatively small footprint is significant relative to conventional primary clarification.

(continued)

Table 11–21 (Continued)	
Type	**Description**
(g) Cartridge Filter Filter element spacer Door opens for filter maintenance Replaceable cartridge filter elements Cartridge filter housing Feed water Filtered effluent Outlet of filter elements set in collection manifold	Most cartridge filters are usually polypropylene wound cartridges from 800 to 1000 mm in length housed inside a vertical or horizontal stainless steel or fiberglass vessel. They are employed in a number of different applications, typically to protect downstream applications. In advanced water treatment, they are employed to remove contaminates found in the chemicals added to control scaling in reverse osmosis membranes. Cartridge filters are not considered further in the discussion of surface filtration, but are included because they are used as a pretreatment step for reverse osmosis. Pleated cartridge filters are used almost exclusively to concentrate virus from treated wastewater for analysis.

ᵃ Courtesy of Xylem.

filter [see Table 11–21(f)], discussed in Sec. 5–9 in Chap. 5, is used for the filtration of untreated wastewater following coarse and intermediate screening. Cartridge filters [see Table 11–21(g)], are used for pretreatment prior to membrane filtration, particularly where reverse osmosis (RO) is used.

Description of the Surface Filtration Process

The key features of surface filters are (1) the filter configuration, (2) the filter medium, (3) the method used to introduce the liquid to be filtered, (4) the method used to clean the filtering medium, and (5) the impact of the accumulation of solids has on process performance.

Filter Configurations. Surface filters are available in a variety of configurations. The most common type of surface filter is comprised of a series of disks attached to a central shaft. The individual disks are made up of two filtering surfaces attached to a metal support frame as shown in Table 11–21 (a) and (c). The diamond cloth-media filter (DCMF), a relatively recent development, is shown schematically in Table 11–21(b) and pictorially on Fig. 11–22. The drum filter for effluent filtration [see Table 11–21(e)] is a relatively recent development, although drum filters are used in a number of filtering applications. Operation of the different surface filters depends on how the liquid to be filtered is applied to the filtering medium and how the filtering material is cleaned of accumulated material.

Filter Materials. The filter material used in surface filters can be categorized as two dimensional and three dimensional. Two dimensional mediums are typically made of synthetic fabrics of different weaves and woven metal fabrics (most commonly stainless steel). The most common weave for synthetic materials is known as a plain weave, which is similar to broadcloth. Stainless steel weaves can include plain weave, twilled weave, and Dutch weave wire meshes. Three dimensional filtering mediums include polyester needle felt cloth and synthetic pile fabric cloth.

Flow Path for Liquid to be Filtered. The flow path can also be used to classify surface filters. Basically, two methods are used to apply the liquid to be filtered to the

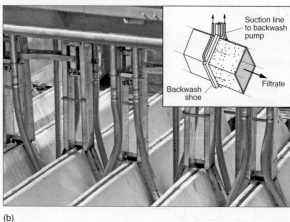

(a)   (b)

Figure 11–22

Views of Diamond Cloth Media Filter: (a) view of filters installed in an existing sand filter basin and (b) view of backwash shoe.

filtering medium. In the first, the liquid enters the feed tank and flows from the outside through the filter cloth into a receiving area (out-in) [see Table 11–21(a), (b) and (f)]. In the second method, the liquid to be filtered is introduced into the annular volume between the two filtering surfaces and flows outward through the filtering medium into the collection vessel (in-out) [see Table 11–21 (c), (d), and (e)]. In either case, solids accumulate on the surface in the direction of flow. The direction of flow affects the method to be used for the removal of the accumulated material, the submergence (i.e. the active filter area), and the overall depth of the unit.

Cleaning the Filter Medium.   Two types of methods are used to remove the accumulated material removed from the fluid: (1) vacuum removal and (2) intermittent and/or continuous high-pressure spray washing. The vacuum removal system is used for surface filters where the flow is from the outside in whereas the high-pressure water spray nozzles are used where the flow is from the inside to the outside.

Vacuum Removal.   When the headloss through the CMF reaches a predetermined set point, the disks are cleaned. As the disks rotate, solids are removed from both sides of the disk by liquid vacuum suction heads, located on either side of each disk, which draws filtered water from the filtrate header back through the cloth media while the disk is rotating. This reversal of flow removes particles that have become entrapped on the surface and within the cloth medium. The diamond cloth filter is also cleaned by a vacuum sweep which moves back and forth along the length of the filter. Solids that settle to the bottom of the reactor below the filter element are removed periodically by a vacuum header.

Over time, particles will accumulate in the cloth medium that can not be removed by a typical backwash. This accumulation of particles leads to increased headloss across the filter, an increase in the backwash suction pressure, and shorter run times between backwashes. When the backwash suction pressure or operating time reaches predetermined setpoints, a high pressure spray wash is initiated automatically. The high pressure spray wash flushes the particles that have become lodged inside the cloth filter media in 2 rev of the disk. The time interval between high pressure spray washes is a function of the feed water quality.

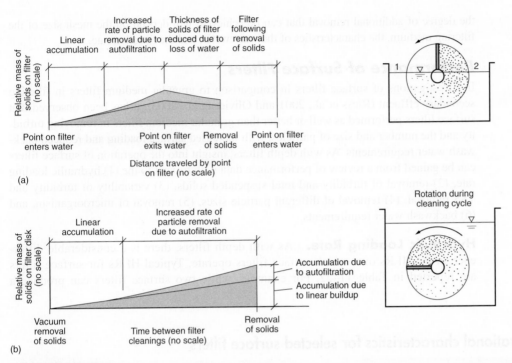

Figure 11–23

Generalized schematic of the operation of disc type surface filters based on following a point with
time on the filter cloth between cleaning events, including the removal of particles by size exclusion
and finer particles due to autofiltration: (a) for surface filters that are partially submerged and (b) for
submerged filter with vacuum removal of accumulated solids.

High-Pressure Water Sprays. In surface filters where the flow is from inside to outside,
high-pressure water sprays are used to remove the material that accumulates on the inside
of the disk. Most high-pressure water spray wash systems can operate in either an intermittent or continuous backwash mode. In the intermittent mode, high-pressure backwash
spray jets are activated only when headloss through the filter reaches a preset level or time.
Once activated, washwater is sprayed through the filter material from the outside as the
disk rotates. The accumulated solids that are dislodged fall into a collection trough. When
operating in a continuous backwash mode, the production of filtered water and backwashed occur simultaneously. The location and configuration of the high-pressure spray
nozzles and solids collection trough are manufacturer specific.

Impact of the Accumulation of Solids on Process Performance. The
removal of particulate matter with surface filters is illustrated schematically on Fig. 11–23
for partially and completely submerged surface filters. For partially submerged filters [see
Fig. 11–23(a)], the accumulation of solids on the filter medium occurs between point 1, where
the clean filter comes in contact with the fluid to be filtered, and point 2, where the filter leaves
the fluid. For the completely submerged filter [see Fig. 11–23(b)], the accumulation of solids
occurs with time until the backwash headloss is reached, at which time the filter is cleaned. In
both cases, the accumulated material on the surface of the filter begins to act as a filter. The
filtering action of the accumulated material is known as *autofiltration*. Autofiltration can be
used to explain why material of a smaller size than the pore size of the filter medium can be
removed by surface filtration. For both types of surface filters, the onset of autofiltration, and

the degree of additional removal that can be achieved, will depend on the mesh size of the filtering medium, the characteristics of the wastewater, and the filtration rate.

Performance of Surface Filters

In investigations of surface filters in comparison to granular medium filters in filtering secondary effluent (Riess et al., 2001 and Olivier et al., 2003), it has been observed that surface filters performed as well or better than granular medium filters in removing turbidity and the number and size of particles, with increased surface loading and reduced backwash water requirements. As with depth filters, insight into the operation of surface filters can be gained from a review of performance data with respect to the (1) hydraulic loading rate, (2) removal of turbidity and total suspended solids, (3) variability of turbidity and TSS removal, (4) removal of different particle sizes, (5) removal of microorganism, and (6) backwash water requirements.

Hydraulic Loading Rate. As with depth filters, there is a considerable difference in the HLRs over which surface filters operate. Typical HLRs for surface filters are reported in Table 11–22. For example, while two surface filters can produce a

Table 11–22

Comparison of operational characteristics for selected surface filters

Parameter	Unit	Cloth media filter®	Diamond cloth media filter®	Diskfilter®	Ultrascreen®	Drumfilter®
Typical hydraulic loading rate	m³/m²·min	0.08–0.20	0.08–0.20	0.08–0.20	0.20–0.65	0.08–0.26
	gal/ft²·min	2–5	2–5	2–5	5–16	2–6.5
Peak HLR	m³/m²·min	0.26	0.26	0.24	0.65	0.26
	gal/ft²·min	6.5	6.5	6	16	6.5
CDPH[a] allowable average HLR	m³/m²·min	—	—	—	0.32	—
	gal/ft²·min	—	—	—	8	—
CDPH allowable peak HLR	m³/m²·min	0.24	0.24	0.24	0.65	—
	gal/ft²·min	6	6	6	16	—
Influent TSS concentration	mg/L	5–20	5–20	5–20	5–20	5–20
Filter material	Type	Nylon and/or Polyester	Nylon and/or Polyester	Polyester or stainless steel	Stainless steel	Polyester or stainless steel
Nominal pore size of screen	μm	5–10	5–10	10–40	10–20	10–40
Direction of flow		out–in	out–in	in–out	in–out	in–out
Submergence	%	100	100	60–70	45	60–70
Headloss	mm	50–300	50–300	75–300	650	300
Disk diameter	m	0.90 or 1.80	na	1.75–3.0	1.6	
Backwash requirement	% of throughput	2–5	2–5	2–4	2–4	2–4

[a] CDPH = California Department of Public Health.

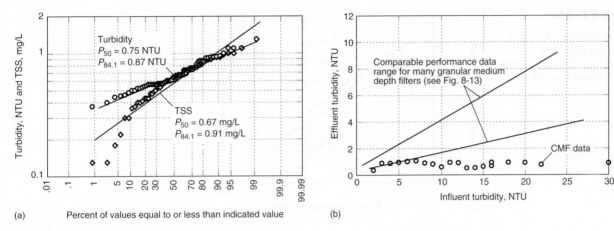

(a)   Percent of values equal to or less than indicated value   (b)

Figure 11–24

Performance data for cloth-media filter for secondary effluent: (a) effluent turbidity as a function of influent turbidity at a filtration rated of 176 L/min·m² and (b) effluent probability distributions for turbidity and TSS.

filtered effluent with turbidity of 2 NTU or less filtering the same effluent, the HLRs can vary by more that a factor of four or five. As with depth filters, the HLR and the backwash water requirements for surface filters will impact cost and the carbon footprint significantly.

Removal of Turbidity and Total Suspended Solids. To evaluate performance capabilities of surface filtration, a CMF pilot plant was tested using secondary effluent from an extended aeration activated sludge process with a solids retention time greater than 15 d. Effluent TSS and turbidity values from the activated sludge process ranged from 3.9 to 30 mg/L and 2 to 30 NTU, respectively. Based on a long-term study, it was found, as shown on Fig. 11–24(a), that both TSS and turbidity values of the filtered effluent were less than 1, 92 percent of the time (Riess et al., 2001). The performance of the CMF as compared to depth filters, all tested with the same activated sludge effluent, is shown on Fig. 11–24(b). As shown, the effluent turbidity from the CMF remained constant over a range of influent turbidity values that tested up to 30 NTU. The degree of removal of TSS from settled activated sludge effluent with surface filters, as with depth filters, will depend on the SRT at which the activated sludge process is operated. Similar results have been reported for other surface filtration technologies.

Variability in Turbidity and Total Suspended Solids Removal. The performance variability of surface filters is of critical importance where specific effluent turbidity limits must be met consistently. The variability observed in the operating data for surface filters as reported in Table 11–12, presented previously, is similar to that observed for depth filters. However it should be noted that average turbidity and TSS values tend to be lower.

Removal of Different Particle Sizes. In comparative testing with a granular medium filter, the surface filter consistently out-performed the granular medium filter in respect to particle removal (see Fig. 11–25). The particle size reduction also had a

Figure 11–25

Comparison of particle sizes in effluent from secondary treatment, granular medium filter, and cloth media filter (Olivier et al., 2003).

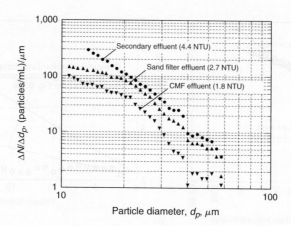

significant impact on the inactivation of total coliform bacteria when used with UV disinfection (Olivier et al., 2002, 2003).

Removal of Microorganisms. Where chemicals are not used, the removal of coliform bacteria and viruses from biologically treated secondary effluent is on the order of 0 to 1.0 and 0 to 0.5 logs, respectively, similar to the values observed for depth filters.

Backwash Water Requirements. The amount of backwash water needed to clean surface filters, expressed as a percentage of plant flow, will depend on the characteristics of filter influent and the design of the surface filter. Typical backwash water percentages for surface filters as shown in Table 11–22 will vary, from 1 to 4 percent.

Design Considerations

Pilot studies are recommended in developing design and operating parameters for new installations. Useful data for design includes (1) the variability of the characteristics of the feed water to be treated and (2) the amount of backwash water required for normal operation. The backwash water requirements are a function of the TSS in the feed water and the solids loading on the filters. If the secondary treatment system is effective in TSS removal, the volume of backwash water can be reduced substantially.

Because cloth-media surface filtration is a relatively new technology, little long-term data are available on the life of the filter cloth. Where surface filtration is being considered, performance should be evaluated from operating installations using a similar type of cloth medium. One operating advantage cited for cloth-media filters is that the filter cloth can be removed and washed in a heavy-duty washing machine.

Pilot Plant Studies

As with granular media filtration, discussed above, there is no generalized approach to the design of full-scale filters for the treatment of wastewater. The discussion presented in the previous section on pilot plant testing also applies to the cloth filter. Typical cloth filter test facilities are illustrated on Fig. 11–26. It should be noted that the single disk shown on Fig. 11–26(b) is full sized. In a larger installation, a number of disks would be arranged on the center shaft.

Figure 11–26

View of cloth filter pilot test filters. It should be noted that the cloth filters shown are full size.

(a)

(b)

11–7 MEMBRANE FILTRATION PROCESSES

Filtration, as defined in Sections 11–3 and 11–5, involves the separation (removal) of particulate and colloidal matter from a liquid. In membrane filtration the range of particle sizes is extended to include dissolved constituents (typically from 0.0001 to 1.0 μm). The role of the membrane, as shown in Table 11–4, is to serve as a selective barrier that will allow the passage of certain constituents and will retain other constituents found in the liquid. To introduce membrane technologies and their application, the following subjects are considered in this section: (1) membrane process terminology, (2) membrane classification, (3) membrane configurations, (4) application of membrane technologies, and (5) the need for pilot-plant studies. The disposal of concentrated waste streams is considered at the end of this section. Electrodialysis, also a membrane process, used typically for the removal of dissolved constituents, is considered separately in Sec 11–7 following the discussion of the application of pressure driven membranes.

Membrane Process Terminology

Terms used commonly in the membrane technology field include: *feed water*, *permeate*, and *retentate*. These terms are illustrated on Fig. 11–27. The influent water to supplied to the membrane system for treatment is known as the *feed water*. The liquid that has passed through the membrane is known as the *permeate*. The portion of the feed water that does not pass through the membrane is known as the *retentate* (also referred to as concentrate, reject, or waste stream). Flux, the rate at which permeate flows through the membrane expressed as L/m²·h or L/m²·d, is the principal measure of membrane performance. Flux is synonymous with the concept of filter hydraulic loading rate presented in the discussion of depth and surface filtration.

Figure 11–27

Definition sketch for operation of a membrane process.

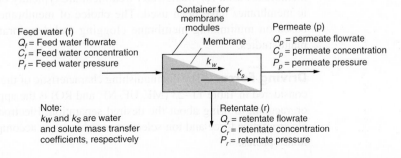

Container for membrane modules
Membrane

Feed water (f)
Q_f = Feed water flowrate
C_f = Feed water concentration
P_f = Feed water pressure

k_w
k_s

Permeate (p)
Q_p = permeate flowrate
C_p = permeate concentration
P_p = permeate pressure

Note:
k_w and k_S are water and solute mass transfer coefficients, respectively

Retentate (r)
Q_r = retentate flowrate
C_r = retentate concentration
P_r = retentate pressure

Membrane Process Classification

Membrane processes include microfiltration (MF), ultrafiltration (UF), nanofiltration (NF), reverse osmosis (RO), and electrodialysis (ED). Membrane processes can be classified in a number of different ways including (1) membrane configuration, (2) the type of material from which the membrane is made, (3) the nature of the driving force, (4) the separation mechanism, and (5) the nominal size of the separation achieved. Each of these methods of classifying membrane processes is considered in the following discussion. The general characteristics of membrane processes including typical operating ranges are reported in Table 11–23. The focus of the following discussion is on pressure driven membrane processes used for the removal of residual TSS, colloidal matter and dissolved solids. Pressure driven membranes are further defined as "low pressure," which includes MF and UF, and "high pressure," which includes NF and RO.

Membrane Configurations. In the membrane field, the term *module* is used to describe a complete unit comprised of the membrane elements (or modules), the pressure support structure for the membranes, the feed inlet and outlet permeate and retentate ports, and an overall support structure. The principal types of membrane modules used for wastewater treatment are (1) tubular, (2) hollow fine-fiber, and (3) spiral wound. Plate and frame and pleated cartridge filters are also available, but are used more commonly in industrial applications.

Definition sketches for the various membranes and detailed descriptions are presented in Table 11–24. There are two basic flow patterns with membranes: (1) outside-in [see Fig. 11–28(a)] and (2) inside-out [see Fig. 11–28(b)]. In most wastewater treatment applications where hollow fiber and membrane sheets are used, the flow is pattern is outside-in. With an outside-in flow pattern, the membrane can be backwashed with air, water, or a combination of both. The outside-in flow pattern is also used for feed water solutions with higher TSS and turbidities.

Membrane Materials. Most commercial membranes are produced as tubular, fine hollow fibers, or flat sheets. In general, three types of membranes are produced: symmetric, asymmetric and thin film composite (TFC) (see Fig. 11–29). As shown on Fig. 11–29(a) and (b), symmetric membranes are the same throughout. Symmetric membranes can vary from microporous to nonporous (so called dense). Asymmetric membranes [see Fig. 11–29(c)] are cast in one process and consist of a very thin (less than 1 μm) layer and a thicker (up to 100 μm) porous layer that adds support and is capable of high water flux.

Thin-film composite membranes [see Fig. 11–29(d)] are made by bonding a thin cellulose acetate, polyamide, or other active layer (typically 0.15 to 0.25 μm thick) to a thicker porous substrate, which provides stability. As reported in Table 11–23 membranes can be made from a number of different organic and inorganic materials. The membranes used for wastewater treatment are typically organic, although some ceramic membranes have been used. The choice of membrane and system configuration is based on minimizing membrane clogging and deterioration, typically based on pilot plant studies.

Driving Force. The distinguishing characteristic of the first four membrane processes considered in Table 11–23 (MF, UF, NF, and RO) is the application of hydraulic pressure, or vacuum, to bring about the desired separation. Electrodialysis involves the use of an electromotive force and ion selective membranes to accomplish the separation of charged ionic species.

Table 11-23
General characteristics of membrane processes

Membrane process	Membrane driving force	Typical separation mechanism	Typical pore size, μm	Typical operating range, μm	Materials (arranged alphabetically)	Configuration
Microfiltration	Hydrostatic pressure difference or vacuum in open vessels	Sieve	Macropores (> 50 nm)	0.07–2.0	Acrylonitrile, ceramic (various materials), polypropylene (PP), polysulfone (PS), polytetrafluorethylene (PTFE), polyvinylidene fluoride (PVDF), nylon	Spiral wound, hollow fiber, plate and frame
Ultrafiltration	Hydrostatic pressure difference or vacuum in open vessels	Sieve	Mesopores (2–50 nm)	0.008–0.2	Aromatic polyamides, ceramic (various materials) cellulose acetate (CA), polypropylene (PP), polysulfone (PS), polyvinylidene fluoride (PVDF), Teflon	Spiral wound, hollow fiber, plate and frame
Nanofiltration	Hydrostatic pressure difference in closed vessels	Sieve + solution/diffusion + exclusion	Micropores (<2 nm)	0.0009–0.01	Cellulosic, aromatic polyamide, polysulfone (PS), polyvinylidene fluoride (PVDF), thin-film composite (TFC)	Spiral wound, hollow fiber, thin film composit
Reverse osmosis	Hydrostatic pressure difference in closed vessels	Solution/diffusion + exclusion	Dense (<2 nm)	0.0001–0.002	Cellulosic, aromatic polyamide, thin-film composite (TFC)	Spiral wound, hollow fiber, thin film composite
Electrodialysis	Electromotive force	Ion exchange	Ion exchange	0.0003–0.002	Ion exchange resin cast as a sheet	Plate and frame

Table 11–24

Description of commonly used membrane types

Type	Description
(a) Tubular 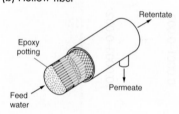 Tubular membranes · Retentate · Permeate · Feed water · Plastic mesh flow spacer	In the tubular configuration the membrane is cast on the inside of a support tube. A number of tubes (either singly or in a bundle) is then placed in an appropriate pressure vessel. The feed water is pumped through the feed tube and product water is collected on the outside of the tubes. The retentate continues to flow through the feed tube. These units are used generally for water with high suspended solids or plugging potential. Tubular units are the easiest to clean, which is accomplished by circulating chemicals and pumping a "foamball" or "spongeball" through to mechanically wipe the membrane. Tubular units produce at a low product rate relative to their volume, and the membranes are generally expensive. The internal diameter of the Tubes will vary from 6 to 40 mm in diameter with lengths as long as 3.66 m (12 ft).
(b) Hollow fiber Retentate · Epoxy potting · Permeate · Feed water	The hollow fiber membrane module consists of a bundle of hundreds to thousands of hollow fibers. The entire assembly is inserted into a pressure vessel. The feed can be applied to the inside of the fiber (inside-out flow) or the outside of the fiber (outside-in flow). Hollow fiber membrane modules are commonly used in membrane bioreactors (MBRs) as described in Chap. 7. Typical inside and outside diameters of the individual hollow fine fibers are about 35 to 45 and 90 to 100 μm, respectively. The typical length of a bundle of fibers is about 1.2 m (4 ft) long. A 100 mm (4 in.) diameter bundle may contain up to 650,000 individual fibers, although most contain fewer fibers per bundle. Fiber bundles can vary from 100 to 200 mm (4 to 8 in.) in diameter. Depending on the size of the bundle up to 7 bundles can be placed in a single pressure vessel [see Fig. 11–28(b)].
(c) Spiral wound 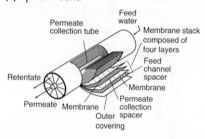 Feed water · Permeate collection tube · Membrane stack composed of four layers · Feed channel spacer · Retentate · Membrane · Permeate collection spacer · Permeate · Membrane · Outer covering	In the spiral wound membrane, a flexible permeate spacer is placed between two flat membrane sheets. The membranes are sealed on three sides. The open side is attached to a perforated pipe. A flexible feed spacer is added and the flat sheets are rolled into a tight circular configuration. Thin film composites are used most commonly in spiral wound membrane modules. The term spiral derives from the fact that the flow in the rolled-up arrangement of membranes and support sheets follows a spiral flow pattern. The diameter of spiral wound elements will typically vary from 100 to 200 mm (4 to 8 in.), however, up to 300 mm (12 in.) diameter elements have been used. The active length of the membrane element is typically about 0.9 m (3 ft) between glue lines, although element lengths varying from 150 mm (6 in.) up to 1.5 m (5 ft) have been used. Operationally, from 2 to 6 membrane elements are included in a single pressure vessel [see Fig. 11–28(c)]. Six membrane elements are normally used for reverse osmosis. As an example, the membrane surface area in a pressure vessel containing four 100 mm (4 in.) diameter by 0.9 m (3 ft) long membrane elements is about 8.33 m² (90 ft²).
(d) Plate and frame 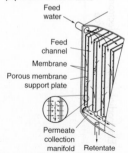 Feed water · Feed channel · Membrane · Porous membrane support plate · Permeate collection manifold · Retentate	Plate and frame membrane modules are comprised of a series of flat membrane sheets and support plates. The water to be treated passes between the membranes of two adjacent membrane assemblies. The plate supports the membranes and provides a channel for the permeate to flow out of the unit. Typically, the dimensions of individual plates that comprise the plate and frame filter are about 20 by 40 mm (7.5 × 15 in.). The packing density of plate and frame units will vary between 100 to 400 m²/m³.

Figure 11–28

Definition sketch for types of membrane operation:
(a) definition sketch for hollow fine-fiber membranes with flow from the outside to the inside of the fiber, (b) definition sketch for hollow fine fiber membranes with flow from the inside to the outside, and (c) spiral wound membranes in a containment vessel.

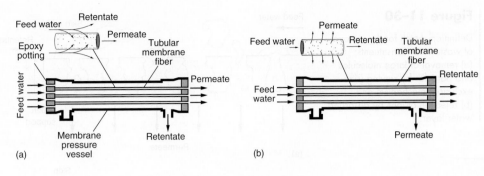

(a)

(b)

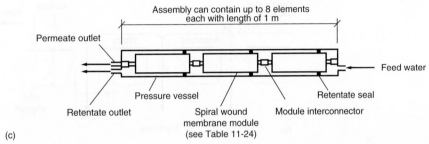

(c)

Removal Mechanisms.

The separation of particles in MF and UF is accomplished primarily by straining (sieving, i.e., physical size exclusion), as shown on Fig. 11–30(a). In NF and RO, in addition to straining, small particles are rejected by the water layer adsorbed on the surface of the membrane which is known as a *dense* membrane [see Fig. 11–30(b)]. Some ionic species such as sodium (Na^+) and chloride (Cl^-) can be transported across the membrane by diffusion through the pores of the macromolecule comprising the membrane. Typically NF can be used to reject constituents as small as 0.001 μm whereas RO can reject particles as small as 0.0001 μm.

Size of Separation.

The pore sizes in membranes are identified as macropores (> 50 nm), mesopores (2 to 50 nm), and micropores (< 2 nm). Because the pore sizes in RO membranes are so small, the membranes are defined as dense. The classification of membrane processes on the basis of the size of separation is shown on Fig. 11–31 and in Table 11–23. Referring to Fig.11–31, it can be seen that there is considerable overlap in the sizes of particles removed, especially between NF and RO. Nanofiltration is used most commonly in water softening operations in place of chemical precipitation.

Membrane Containment Vessels

Two types of containment vessels are used with membrane modules: (1) pressurized and (2) submerged.

Figure 11–29

Types of membrane construction:
(a) microporous symmetric membrane, (b) nonporous (dense) symmetric membrane,
(c) asymmetric membrane, and
(d) thin film composite (TFC), sometimes identified as an asymmetric membrane.

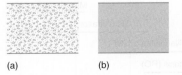

(a)

(b)

(c)

(d)

Figure 11–30

Definition sketch for the removal of wastewater constituents: (a) removal of large molecules and particles by sieving (size exclusion) mechanism and (b) rejection of ions by adsorbed water layer.

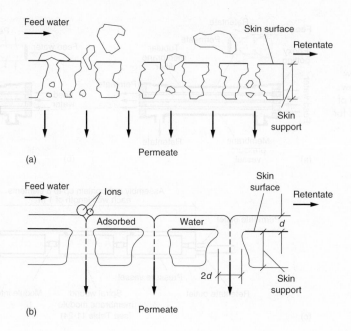

Pressurized. The primary purpose of the pressure vessel (or tube) is to support the membrane module or modules and keep the feed water and permeate (product water) isolated. The vessel must also be designed to prevent leaks and pressure losses to the outside, minimize the buildup of salt or fouling, and permit easy replacement of the membrane module. Microfiltration and ultrafiltration modules are generally 100 to 300 mm in

Figure 11–31

Comparison of the size of the constituents found in wastewater and the operating size ranges for membrane technologies. The operating size range for conventional depth filtration is also shown.

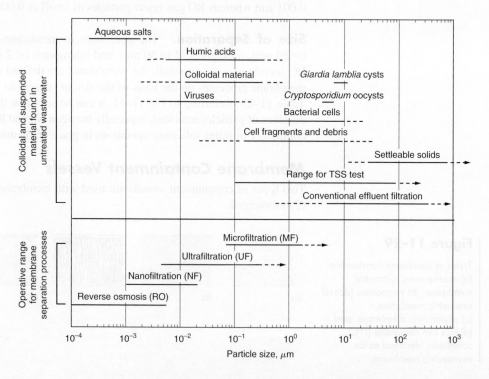

Figure 11–32

Views of various membrane installations: (a) pressurized microfiltration and (b) pressurized ultrafiltration membranes for the filtration of settled secondary effluent, (c) vacuum microfiltration membrane modules in open vessel, (d) membrane module used in open vessel shown in (c), (e) typical cartridge filter used before reverse osmosis, and (f) one bank of a large reverse osmosis installation used to treat activated sludge effluent following microfiltration, chemical addition, and cartridge filtration. Each bank of RO modules is designed to treat 19,000 m³/d (5 Mgal/d). The capacity of the entire facility is 265,000 m³/d (70 Mgal/d).

(a)　　　　　　　　　　(b)

(c)　　　　　　　　　　(d)

(e)　　　　　　　　　　(f)

diameter, 0.9 to 5.5 m long, and a single module is placed in a pressurized vessel arranged in racks or skids. Each module must be piped individually for feed and permeate water. Typical pressurized MF membrane modules are shown on Figs. 11–32(a) and (b). The modules for NF and RO are 100 to 300 mm in diameter, 0.9 to 5.5 m long, and from two to eight modules are placed in a single pressurized vessel arranged in racks, either horizontally or vertically [see Figs. 11–32 (d) and (f)]. Vertical placement helps reduce the number of pipes and fittings and the total plant footprint.

In the pressurized configuration, pumps are used to pressurize the feed water and circulate it through the membrane [see Figs. 11–33(a) and (c)]. Centrifugal pumps can be used for MF, UF, and NF; positive displacement pumps or high-pressure turbine pumps are necessary for RO. Depending on the operating pressure and the characteristic of the feed water, a variety of materials have been used including plastic and fiberglass tubes and plumbing components. Steel pressure tubes are required for some reverse osmosis applications, and stainless steel is required for seawater and brackish water having high TDS.

Submerged (Vacuum) Type. In the submerged system, the membrane elements are immersed in a feed water tank as shown on Fig. 11–32(c). The permeate is withdrawn

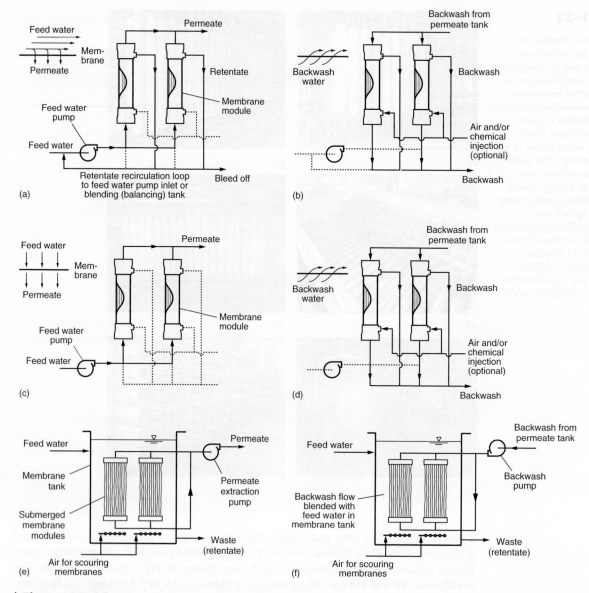

Figure 11-33

Definition sketch for membrane systems: (a) pressurized cross-flow membrane system (see insert), (b) backwashing pressurized cross-flow system, (c) pressurized dead-end flow [see insert, see also Fig. 8–32(a)] membrane system, (d) backwashing pressurized direct flow system, (e) submerged membrane with vacuum draw-off [see also Fig. 11-32(c)], and (f) backwashing submerged system.

through the membrane by applying a vacuum, usually from the suction of a centrifugal pump [see Fig. 11–33(e)]. Transmembrane pressure developed by the permeate pump causes clean water to be extracted through the membrane. Net positive suction head (NPSH) limitations of the permeate pump restrict the submerged membranes to a maximum transmembrane pressure of about 50 kPa, and they operate typically at a transmembrane pressure of 20 to 40 kPa (−28 to −100 kPa vacuum).

Operational Modes for Pressurized Configurations

Two different operational modes are used with pressurized microfiltration and ultrafiltration units: (1) cross-flow and (2) dead-end.

Cross-Flow Mode. In the *cross-flow* mode [see Fig. 11–33(a) and insert] the feed water is pumped more-or-less tangentially to the membrane. The accumulation of particulate matter on the surface of the membrane can be controlled by the shear force of the fluid velocity. The differential pressure across the membrane causes a portion of the feed water to pass through the membrane. Water that does not pass through the membrane is recirculated back to the membrane after blending with influent feed water or is recirculated to a blending (or balancing) tank. In addition, a portion of the water that did not pass through the membrane is bled off for separate processing and disposal [see Fig. 11–33(a)]. It should be noted that cross-flow is the flow pattern in spiral-wound membranes.

Dead-End Mode. In the second configuration, known as *dead-end* (also known as direct-feed or perpendicular feed) [see Fig. 11–33(c) and in the insert], there is no cross-flow (or liquid waste stream) during the permeate production mode. All of the water applied to the membrane passes through the membrane. Particulate matter that cannot pass through the membrane pores is retained on the membrane surface. Dead-end filtration is most effective when the concentration of particulate matter is low or where the accumulated material dose not cause a rapid headloss buildup. Dead-end filtration is used both for pretreatment and where the filtered water is to be used directly.

Membrane Cleaning. As constituents in the feed water accumulate on the membranes (often termed membrane fouling), the pressure builds up on the feed side, the membrane flux (i.e., flow through membrane) starts to decrease, and the percent rejection of certain water quality constituents [see Eq. (11–34)] may actually increase (see Fig. 11–34). When the flux has deteriorated to a given level, the membrane modules are taken out of service and backwashed and periodically cleaned chemically [see Figs. 11–33(b), (d), and (f)]. It should be noted that the quantity of waste washwater produced during membrane cleaning is typically less from pressurized as compared to submerged vacuum systems.

Figure 11–34

Definition sketch for the performance of a membrane filtration system as function of time with and without proper cleaning.

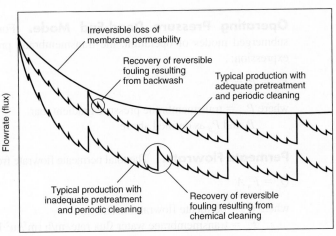

Chemical cleaning is used to restore the membrane performance to near its initial permeability. Two chemical cleaning methods are used typically: (1) clean-in-place (CIP) and (2) chemically enhanced backwash (CEB). In the CIP method, cleaning agents are used to soak and pass across the microfiber elements within the membrane module. In the CEB method, chemical cleaning agents are added to backwash water at selected backwash intervals (typically based on pressure buildup). Sometimes the CEB cleaning method is used for regular operation and when the membrane performance has deteriorated to a given level the CIP cleaning method is used.

A certain irreversible loss of membrane permeability will occur during process operation (see Fig. 11–34). The degree of irreversible permeability loss depends on the membrane material and operating conditions, including (1) long-term aging of the membrane material, (2) mechanical compaction and deformation from high operating pressures, (3) hydrolysis reactions related to solution pH, and (4) reactions with specific constituents in the feed water.

Process Analysis for MF and UF Membranes

Referring to Fig. 11–27, the process analysis for MF and UF membranes involves consideration of the operating pressure, permeate flow, the degree of recovery, and degree of rejection. Constituent and flowrate mass balances are used to assess the performance of the membranes.

Operating Pressure Cross-Flow Mode. For the cross-flow mode of operation, the transmembrane pressure is given by the following expression:

$$P_{tm} = \left(\frac{P_f + P_r}{2}\right) - P_p \qquad (11\text{--}29)$$

where P_{tm} = transmembrane pressure gradient, bar (Note: 1 bar = 10^5 Pa)
P_f = inlet pressure of the feed water, bar
P_r = pressure of the retentate, bar
P_p = pressure of the permeate, bar

The overall pressure drop across the filter module for the cross-flow mode of operation is given by

$$P = P_f - P_p \qquad (11\text{--}30)$$

where P = pressure drop across the module, bar
P_f and P_p as defined above

Operating Pressure Dead-End Mode. For the dead end pressurized and submerged modes of operation, the transmembrane pressure is given by the following expression:

$$P_{tm} = P_f - P_p \qquad (11\text{--}31)$$

where P_{tm} = transmembrane pressure gradient, bar
P_f and P_p as defined above

Permeate Flowrate. The total permeate flowrate from a membrane system is given by

$$Q_p = F_w A \qquad (11\text{--}32)$$

where Q_p = permeate flowrate, m³/h
F_w = transmembrane water flux rate, m/h (m³/m²·h)
A = membrane area, m²

As would be expected, the transmembrane water flux rate is a function of the quality and temperature of the feed water, the degree of pretreatment, the characteristics of the membrane, and the system operating parameters. Note that membrane area (A) is not the cross sectional area of the membrane module, but rather the active surface area of the membrane material. For example, a standard 200 mm diameter 1020 mm long (8-in. × 40-in.) RO module contains 37 m² (400 ft²) of membrane surface area.

Recovery. Recovery, r, expressed as a percentage, is defined as the ratio of the net water produced to the total water applied during a filter run as follows:

$$r,\% = \frac{V_p}{V_f} \times 100 \tag{11-33}$$

where V_p = net volume of the permeate, m³
V_f = volume of water fed to the membrane, m³

In computing the net volume of permeate, the amount of backwash water used must also be taken into consideration.

Rejection. Rejection, R, expressed as a percentage or as a dimensionless fraction, is a measure of the amount of material removed from the feed water. It should be noted that there is a difference in the recovery, r, (which refers to the water) and rejection, R, (which refers to the solute). Rejection, R, is given by the following expression.

$$R,\% = \frac{C_f - C_p}{C_f} \times 100 = \left(1 - \frac{C_p}{C_f}\right) \times 100 \tag{11-34}$$

where C_f = feed water concentration, g/m³, mg/L
C_p = permeate concentration, g/m³, mg/L

Log Reduction. Another commonly used approach to express the rejection is as log rejection, LR, as given below.

$$LR = -\log(1 - R) = \log\left(\frac{C_f}{C_p}\right) \tag{11-35}$$

where R is the dimensionless form of Eq. (11-34).

Materials Mass Balance. The corresponding flowrate and constituent mass balance equations for the pressurized cross-flow membrane are:

Flowrate balance: $Q_f = Q_p + Q_r$ \qquad (11-36)

Constituent mass balance: $Q_f C_f = Q_p C_p + Q_r C_r$ \qquad (11-37)

where Q_f = feed water flowrate, m³/h, m³/s
Q_r = retentate flowrate, m³/h, m³/s
C_r = retentate concentration, g/m³, mg/L

Figure 11–35

Three modes of membrane operation with respect to membrane flux and transmembrane pressure (TMP): (a) constant flux, (b) constant pressure, and (c) non restricted flux and pressure. (Adapted from Bourgeous et al., 1999.)

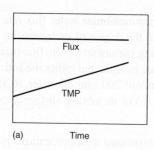

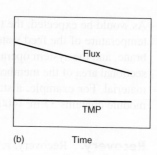

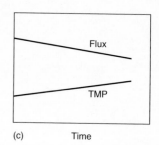

(a)　　　Time　　　　　(b)　　　Time　　　　　(c)　　　Time

Operating Strategies for MF and UF Membranes

Operating strategies for membranes are developed based on a consideration of operating pressures and flux rates. Three different operating strategies can be used to control the operation of a membrane process with respect to flux and the transmembrane pressure (TMP). The three modes, illustrated on Fig. 11–35, are (a) constant flux in which the flux rate is fixed and the TMP is allowed to vary (increase) with time, (b) constant TMP in which the TMP is fixed and the flux rate is allowed to vary (decrease) with time, and (c) both the flux rate and the TMP are allowed to vary with time. Traditionally, the constant flux mode of operation has been used. However, based on the results of a study with various wastewater effluents (Bourgeous et al., 1999), the mode in which both the flux rate and the TMP are allowed to vary with time may be the most effective mode of operation. It should be noted that the diagrams on Fig. 11–35 do not reflect the irreversible permeability loss, as described previously. Regardless of the operating strategy an important issue with membranes systems is fiber breakage. The impact of fiber breakage is examined in Example 11–5.

EXAMPLE 11–5   **Impact of Broken Fibers on Membrane Filter Effluent Quality**   Membrane filtration is used to treat secondary effluent for reuse applications. The effluent from the wastewater treatment plant, which serves as the influent to the membrane filter installation, has an effluent turbidity of 5 NTU and contains a heterotrophic plate count (HPC) of 10^6 microorganisms/L. The effluent from the membrane filters typically contains less than 10 microorganisms/L and a turbidity of about 0.2 NTU. Using this information, what is the log rejection for microorganisms under normal operation with no broken fibers? If it is assumed that 6 out of 6000 (0.1 percent) membrane fibers have been broken during operation, determine the impact on the effluent microorganism count and turbidity. For the following analysis, neglect the water lost during the backwashing cycle.

Solution

1. Calculate the log rejection for microorganisms with no broken fibers using Eq. (11–35).

$$LR = \log\left(\frac{C_f}{C_p}\right) = \log\left[\frac{(10^6 \text{ org/L})}{(10 \text{ org/L})}\right] = 5.0$$

2. Determine the log rejection for microorganisms assuming that 6 fibers have been broken.

   a. Prepare a mass balance diagram for the condition with the broken fibers.

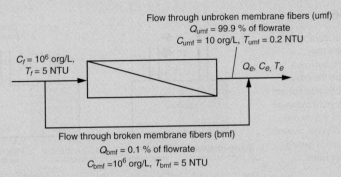

Flow through unbroken membrane fibers (umf)
Q_{umf} = 99.9 % of flowrate
C_{umf} = 10 org/L, T_{umf} = 0.2 NTU

C_f = 10⁶ org/L, T_f = 5 NTU

Q_e, C_e, T_e

Flow through broken membrane fibers (bmf)
Q_{bmf} = 0.1 % of flowrate
C_{bmf} = 10⁶ org/L, T_{bmf} = 5 NTU

b. Write mass balance equation for microorganisms in the effluent from the membrane and solve for effluent microorganism concentration.

$$C_e = \frac{C_{umf}Q_{umf} + C_{bmf}Q_{bmf}}{Q_e}$$

$$= \frac{(10 \text{ org/L})(0.999) + (10^6 \text{ org/L})(0.001)}{1} = 1010 \text{ org/L}$$

c. Calculate the log rejection for microorganisms for the condition with the broken fibers.

$$R_{log} = \log\left(\frac{C_p}{C_f}\right) = \log\left[\frac{(10^6 \text{ org/L})}{(1010 \text{ org/L})}\right] = 3.0$$

3. Calculate the impact on turbidity assuming that 6 fibers have been broken. Use the mass balance equation developed in Step 2 and solve for the effluent turbidity.

$$T_e = \frac{T_{umf}Q_{umf} + T_{bmf}Q_{bmf}}{Q_e}$$

$$= \frac{(0.2 \text{ NTU})(0.999) + (5 \text{ NTU})(0.001)}{1} = 0.205 \text{ NTU}$$

Comment This example is used to demonstrate that a few broken fibers can have a significant impact on the microorganism count in the effluent (1010 versus 10/L) and the log removal (5 versus 3.0 log) and essentially no impact on the effluent turbidity (0.2 versus 0.205 NTU, the difference is not measurable). For this reason, turbidity alone cannot be used as a surrogate measure for bacterial quality, and disinfection of microfiltration effluent will be required to protect public health in sensitive applications. The use of turbidity monitoring is often accompanied with the practice of pressure decay testing and particle counting for monitoring membrane integrity.

Process Analysis for Reverse Osmosis

Process analysis for reverse osmosis involves consideration of the membrane water and mass flux rate, permeate recovery ratio, rejection factor, and the corresponding mass balance analysis. To understand the details of the process analysis for reverse osmosis, it will be helpful to first review the fundamental basis for the reverse osmosis process.

Figure 11–36

Definition sketch for reverse osmosis: (a) osmosis (the differential pressure between solutions is less than the osmotic pressure), (b) osmotic equilibrium (the differential pressure between solutions is equal to the osmotic pressure), and (c) reverse osmosis (the applied pressure is greater than the osmotic pressure).

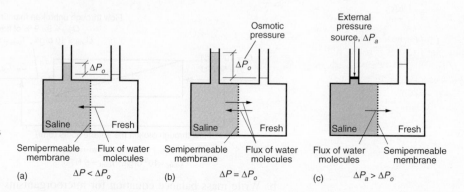

The fundamental basis is presented below followed by the expressions used for the analysis of the process.

Fundamental Basis for Reverse Osmosis Process. When two solutions having different solute concentrations are separated by a semipermeable membrane, a difference in chemical potential will exist across the membrane (see Fig. 11–36). Water will tend to diffuse through the membrane from the lower-concentration (higher-potential) side to the higher-concentration (lower-potential) side; this phenomenon is called forward osmosis [see Fig. 11–26(a)]. In a system having a finite volume, flow continues until the pressure difference balances the chemical potential difference. This balancing pressure difference is termed the osmotic pressure and is a function of the solute characteristics and concentration and temperature. If a pressure gradient, opposite in direction and greater than the osmotic pressure, is imposed across the membrane, flow from the more concentrated to the less concentrated region will occur; this phenomenon is termed *reverse osmosis* [see Fig. 11–36(c)].

Membrane Flux and Area Requirements. A number of different models have been developed to determine the membrane surface area and the number of stages (arrays) required (see Fig. 11–37). The basic equations used to develop the various models are as follows.

Feed Water Flux Rate. Referring to Fig. 11–27, the flux of water through the membrane is a function of the pressure gradient:

$$F_w = k_w(\Delta P_a - \Delta \Pi) = \frac{Q_p}{A} \tag{11–38}$$

where F_w = feed water flux rate, $L/m^2 \cdot h$

k_w = mass transfer coefficient for water flux (involving temperature, membrane characteristics, and solute characteristics), $L/m^2 \cdot h \cdot bar$

ΔP_a = average applied pressure gradient, bar (Note: Number 1 bar = 10^5 Pa)

$$= \left(\frac{P_f + P_c}{2}\right) - P_p$$

$\Delta \Pi$ = osmotic pressure gradient, bar

$$= \left(\frac{\Pi_f - \Pi_c}{2}\right) - \Pi_p$$

Figure 11-37

Typical process flow diagrams: (a) depth or surface filtration with nanofiltration and (b) combined microfiltration or ultrafiltration with reverse osmosis.

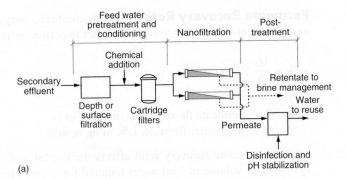

(a)

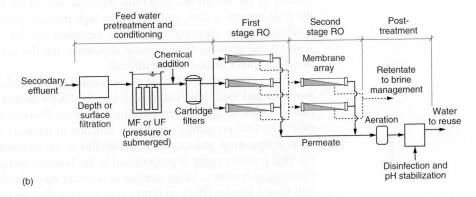

(b)

P_f = inlet pressure of feed water, bar
P_r = pressure of retentate, bar
P_p = pressure of permeate, bar
Π_f = osmotic pressure of feed water, bar
Π_r = osmotic pressure of retentate, bar
Π_p = osmotic pressure of permeate, bar
Q_p = permeate flowrate, L/h
A = membrane area, m^2

Mass (Solute) Flux Rate. Some solute passes through the membrane in all cases. Solute flux can be described adequately by an expression of the form

$$F_s = k_s \, \Delta C_s = \frac{(Q_p)(10^{-3}\,\text{m}^3/\text{L})C_p}{A} \tag{11-39}$$

where F_s = mass flux of solute, g/m^2·h
k_s = mass transfer coefficient for solute, m/h
ΔC = solute concentration gradient across membrane, g/m^3

$$= \left(\frac{C_f + C_r}{2}\right) - C_p$$

C_f = solute concentration in feed water, g/m^3
C_r = solute concentration in retentate (concentrate), g/m^3
C_p = solute concentration in permeate, g/m^3
Q_p = permeate flowrate, L/h

Permeate Recovery Ratio. The permeate recovery ratio, r, expressed as a percentage, represents the conversion of feed water to permeate (product water), and is defined as

$$r, \% = \frac{Q_p}{Q_f} \times 100 \tag{11-40}$$

where Q_p = permeate flowrate, L/h, m³/h, or m³/s
Q_f = feed water flowrate, L/h, m³/h, or m³/s

The permeate recovery ratio affects the capital and operating cost of a membrane system. The volume of feed water required for a given permeate capacity is determined directly by the design recovery ratio. Also, the size of the feed water system, capacity of the pretreatment system, and size of the high pressure pumps and supply piping are also functions of the recovery ratio. With increased recovery, the feed water flowrate, is reduced, the pressure may increase somewhat, but the brine will be more concentrated which can make disposal more difficult.

An example of the effect of the permeate recovery ratio on feed pressure, power consumption and feed flow is shown on Fig. 11-38 for an RO system operating at recovery rates between 60 and 90 percent. The feed water flowrate depends only on the recovery ratio. The feed pressure is a complex function of recovery ratio, feed water salinity feed water temperature, and specific permeate flux of the membrane. The power requirement of the high pressure pump is proportional to the flowrate and pressure. In the usual range of operating parameters, for an increase in recovery ratio, the decrease in feed water flowrate will have a greater effect on power consumption than an increase in feed water pressure (Wilf, 1998). For RO, higher operating pressures are desirable because the degree of separation and the quality of the product are improved.

Rejection Factor. Rejection (or retention), R, expressed as percentage or as a dimensionless fraction, is a measure of the amount of solute or solid that is retained or does not pass through the membrane; it is calculated using the following expression.

$$R, \% = \left(\frac{C_f - C_p}{C_f}\right) \times 100 = \left(1 - \frac{C_p}{C_f}\right) \times 100 \tag{11-41}$$

where C_f = concentration in the feed water, g/m³
C_p = concentration in the permeate, g/m³

Figure 11-38

Effect of permeate recovery on feed pressure, feed flowrate, and power consumption.

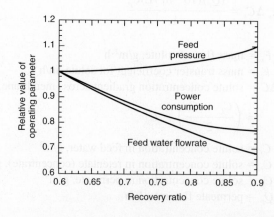

The rejection efficiency of reverse osmosis membranes for specific species can range from 85 to 99.5 percent and is quoted by the manufacturer for a standard set of feed conditions. When the rejection of microorganisms is considered, it is more convenient to express the rejection as log rejection as given by Eq. (11–35), repeated here for convenience.

$$LR = -\log(1 - R) = \log\left(\frac{C_f}{C_p}\right) \tag{11-42}$$

Materials Mass Balances. The flowrate and constituent mass balance equations given previously for microfiltration and ultrafiltrations also apply to NF and RO systems. The equations are

Flowrate balance: $Q_f = Q_p + Q_r$ $\qquad$ (11–43)

Constituent mass balance: $Q_f C_f = Q_p C_p + Q_r C_r$ $\qquad$ (11–44)

where Q_r = retentate flowrate, m³/h, m³/s
$\qquad C_r$ = retentate concentration, g/m³
$\qquad$ Other terms are as defined previously

Use of the above equations to estimate the required surface area for TDS reduction is illustrated in Example 11–6.

EXAMPLE 11–6 **Determination of Membrane Area Required for Demineralization** A brackish water having a TDS concentration of 3000 g/m³ is to be desalinized using a thin film composite membrane having a solvent (water) mass transfer coefficient k_w of 9 × 10⁻⁹ s/m (9 × 10⁻⁷ m/s·bar) and a solute (i.e., TDS) mass transfer coefficient k_i of 6 × 10⁻⁸ m/s. The product water is to have a TDS of no more than 200 g/m³. The feed water flowrate is 0.010 m³/s. The net operating pressure ($\Delta P_a - \Delta P$) will be 2500 kPa (2.5 × 10⁶ kg/m·s²). Assume the recovery rate, r, will be 90 percent and that all of the water is to be processed through the membrane unit to remove other constituents in addition to TDS. Estimate the rejection rate, R, and the concentration of the retentate.

Solution

1. The problem involves determination of the membrane area required to produce 0.009 m³/s (0.9 × 0.010 m³/s) of water with a TDS concentration equal to or less than 200 g/m³. If the estimated permeate TDS concentration is well below 200 g/m³ and TDS is the only constituent of concern, blending of feed water and permeate can be used to reduce the required membrane area.

2. Estimate the membrane area using Eq. (11–38) and the water mass transfer coefficient. The estimated area may need to be adjusted based on the solute mass transfer rate.

$$F_w = k_w(\Delta P_a - \Delta P)$$

$$= (9 \times 10^{-9} \text{ s/m})(2.5 \times 10^6 \text{ kg/m·s}^2) = 2.25 \times 10^{-2} \text{ kg/m}^2\text{·s}$$

$$Q_p = F_w \times A, \; Q_p = r\,Q_f, \; Q_p = 0.9\,Q_f$$

$$A = \frac{(0.9 \times 0.010 \text{ m}^3/\text{s})(10^3 \text{ kg/m}^3)}{(2.25 \times 10^2 \text{ kg/m}^2\text{·s})} = 400 \text{ m}^2$$

3. Estimate the permeate TDS concentration using Eq. (11–39) and the estimated area.

$$F_i = k_i\,\Delta C_i = \frac{Q_p C_p}{A}$$

Substituting for ΔC_i and solving for C_p yields

$$C_p = \frac{k_i[(C_f + C_r)/2]A}{Q_p + k_iA}$$

Assume $C_c \approx 10C_f$ (Note: If the estimated C_r value and the computed value of C_r, as determined below, are significantly different, the value of C_p must be recomputed)

$$C_p = \frac{(6 \times 10^{-8}\,\text{m/s})[(3\,\text{kg/m}^3 + 30\,\text{kg/m}^3)/2](400\,\text{m}^2)}{(0.01\,\text{m}^3/\text{s}) + (6 \times 10^{-8}\,\text{m/s})(400\,\text{m}^2)} = 0.044\,\text{kg/m}^3$$

4. Estimate the rejection rate using Eq. (11–41)

$$R,\% = \frac{C_f - C_p}{C_f} \times 100$$

$$R = \frac{(3.0\,\text{kg/m}^3) - (0.044\,\text{kg/m}^3)}{(3.0\,\text{kg/m}^3)} \times 100 = 98.5\%$$

5. Estimate the retentate TDS using Eq. (11–41).

$$C_r = \frac{Q_f C_f - Q_p C_p}{Q_r}$$

$$C_r = \frac{(1.0\,\text{L})(3.0\,\text{kg/m}^3) - (0.9\,\text{L})(0.044\,\text{kg/m}^3)}{(0.1\,\text{L})} = 29.6\,\text{kg/m}^3$$

The estimated value of C_r used in Step 3 (30 kg/m³) is ok.

Comment If the permeate TDS concentration were significantly below 200 g/m³, blending of feed and permeate could be used to reduce the required membrane area. In this example blending cannot be used.

Membrane Fouling

Membrane fouling is, perhaps, the most important consideration in the design and operation of membrane systems as it affects pretreatment needs, cleaning requirements, operating conditions, cost, and performance. Membrane fouling will occur depending on the site-specific physical, chemical, and biological characteristics of the feed water, the type of membrane, and operating conditions. As reported in Table 11–25, four general forms of fouling can occur: (1) particulate fouling, due to a buildup of the constituents in the feed water on the membrane surface, (2) precipitation of inorganic salts resulting in the formation of inorganic scales, (3) organic fouling due to the presence of organic matter, and (4) biological fouling due to the presence of microorganisms in the feed water. Any or all of these forms of fouling can occur simultaneously and over time. In addition, membranes can be damaged by the presence of chemical substances that can react with the membrane. Typical wastewater constituents in that can cause membrane fouling are also presented in Table 11–25.

Table 11–25

Typical constituents in wastewater that cause membrane fouling and other constituents that can cause damage to the membranes[a]

Type of fouling	Responsible wastewater constituents	Remarks
Particulate fouling	Organic and inorganic colloids Emulsified oils Clays and silts Silica Iron and manganese oxides Oxidized metals Metal salt coagulant products Powdered activated carbon	Particulate fouling can be reduced by cleaning the membrane at regular intervals
Scaling (precipitation of supersaturated salts)	Barium sulfate Calcium carbonate Calcium fluoride Calcium phosphate Strontium sulfate Silica	Scaling can be reduced by limiting salt content, by pH adjustment, and by other chemical treatments such as the addition of antiscalants
Organic fouling	Natural organic matter (NOM) including humic and fulvic acids, proteins and polysccharides Emulsified oils Polymers used in treatment process	Effective pretreatment can be used to limit organic fouling
Biofilm fouling	Dead microorganisms Living microorganisms Polymers produced by microorganisms	Biofilms are formed on membrane surface by colonizing bacteria
Damage to membrane	Acids Bases pH extremes Free chlorine Free oxygen	Membrane damage can be limited by controlling the amount of these substances in the feed water. The extent of the damage depends on the nature of the membrane selected

[a] In many cases all four types of fouling can occur simultaneously.

Particulate Fouling. Particulate fouling is caused by the presence of particulate matter in the feed water. To protect RO and NF membrane systems from large particulate fouling, cartridge filters [see Table 11–21(g)] are used ahead of the membrane feed pumps. These are normally 5 to 15 μm pore size woven filters designed to prevent fouling from relatively large particles. Nonetheless, much smaller particles can also damage and foul RO and NF modules. As noted in Table 11–25, particulate constituents can include organic and inorganic colloids, emulsified oils, clays and silts, silica, and metal oxides and salts. Of these constituents, silica has proven to be one of the more problematic in wastewater. Silica $(SiO_2)_n$ in wastewater can be found in a variety of different forms including reactive, colloidal, and particulate silica depending on the chemical characteristics of the feed water.

Figure 11–39

Modes of membrane fouling: (a) pore narrowing, (b) pore plugging, and (c) gel/cake formation caused by concentration polarization.

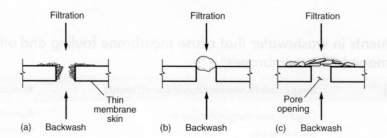

Three accepted mechanisms resulting in resistance to flow due to the accumulation of material (see Fig. 11–39) are (1) pore narrowing, (2) pore plugging, and (3) gel/cake formation caused by concentration polarization (Ahn et al., 1998). The mechanisms of pore plugging and pore narrowing will only occur when the particulate matter in the feed water is smaller than the pore size or the molecular weight cutoff. Pore plugging occurs when particles the size of the pores become stuck in the pores of the membrane. Pore narrowing consists of solid material attaching to the interior surface of the pores, which results in a narrowing of the pores. It has been hypothesized that once the pore size is reduced, concentration polarization is amplified further causing an increase in fouling (Crozes et al., 1997).

Gel/cake formation, caused by concentration polarization, occurs when the majority of the solid matter in the feed is larger than the pore sizes or molecular weight cutoff of the membrane. Concentration polarization can be described as the build up of matter close to or on the membrane surface that causes an increase in resistance to solvent transport across the membrane. Some degree of concentration polarization will always occur in the operation of a membrane system. The formation of a gel or cake layer, however, is an extreme case of concentration polarization where a large amount of matter has actually accumulated on the membrane surface forming a gel or cake layer.

Scaling. As chemical constituents in the feed water are removed at the surface of the membrane, their concentration increases locally. When the concentrations of the individual constituents increase beyond their solubility limits, a variety of different types of salts can be precipitated, depending on the chemical characteristics and temperature of the feed water. Chemical precipitation is especially critical in RO units used for desalination, because of the high initial salt concentration in seawater. The chemical scale that forms on the membrane surface is of importance because it can reduce the water permeability of the membrane and potentially cause irreversible damage to the membrane.

Organic Fouling. Most treated wastewater contains a variety of organic matter in varying concentrations. As noted in Table 11–25, organic foulants can include NOM that was present originally in the water supply, NOM produced during biological treatment, emulsified oils, and organic polymers that may have been used in the wastewater treatment process including polymers used as filter aids in tertiary treatment and polymers recycled to the treatment process from dewatering activities. Because these polymeric materials are sticky, they can accumulate on the membrane surface and accelerate fouling by forming stable organic/inorganic particulate matter, which can reduce the water permeability of the membrane.

Biological Fouling. Effluent from biological treatment systems presents a special problem, as the membranes are susceptible to fouling because of the biological activity that can occur. Because the concentration of organic matter and nutrients is elevated at the membrane surface, conditions are favorable for the growth of microorganisms. As microorganisms begin to colonize on the membrane surface, the water permeability of the membrane will be reduced. When a membrane process is operated intermittently, the water permeability of the membrane can be reduced further if the microorganisms start to grow into the membrane pores. The growth of microorganisms is also of concern because of the production of extracellular polymers which can interact with other foulants, as described previously.

Control of Membrane Fouling

Typically, three approaches are used to control membrane fouling: (1) pretreatment of the feed water, (2) membrane backflushing, and (3) chemical cleaning of the membranes. Pretreatment is used to reduce the TSS, colloidal material, and bacterial content of the feed water. Often the feed water will be conditioned chemically to limit chemical precipitation within the units. Many proprietary pretreatment chemicals such as anti-scalants, biocides, and scale inhibitors are available in the marketplace to control NF and RO fouling. In low pressure membrane operations (MF and UF), the method of eliminating the accumulated material from the membrane surface is backflushing with water and/or air. Chemical treatment is used to remove constituents that are not removed during conventional backwashing. Chemical precipitates can be removed by altering the chemistry of the feed water and by chemical treatment. Damage of the membrane due to deleterious constituents typically cannot be reversed. The need for pretreatment and pretreatment options for NF and RO are discussed below.

Assessing Need for Pretreatment for NF and RO. To assess the treatability of a given wastewater with NF and RO membranes, a variety of fouling indexes have been developed over the years. The three principal indices are the silt density index (SDI), the modified fouling index (MFI), and the mini plugging factor index (MPFI). Fouling indexes are determined from simple membrane tests. The sample must be passed through a 0.45 μm Millipore filter with a 47-mm internal diameter at 210 kPa (30 $lb_f/in.^2$) gauge to determine any of the indexes. The time to complete data collection for these tests varies from 15 min to 2 h, depending on the fouling nature of the water.

Silt Density Index. The most widely used index is the SDI (DuPont, 1977; ASTM, 2002). The SDI is defined as follows.

$$SDI = \frac{100[1 - (t_i/t_f)]}{t} \tag{11–45}$$

where t_i = time to collect the initial sample of 500 mL
t_f = time to collect final sample of 500 mL
t = total time for running the test

The silt density index is a static measurement of resistance which is determined by samples taken at the beginning and end of the test. The SDI does not measure the rate of change of resistance during the test. Recommended SDI values are reported in Table 11–26. The calculation of the SDI is demonstrated in Example 11–7.

Modified Fouling Index. The modified fouling index (MFI) is determined using the same equipment and procedure used for the SDI, but the volume is recorded every 30 s over a

Table 11–26

Recommended values for fouling indexes[a]

Membrane process	Fouling index		
	SDI	MFI, s/L²	MPFI, L/s²
Nanofiltration	0–2	0–10	0–1.5 × 10⁻⁴
Reverse osmosis hollow fiber	0–2	0–2	0–3 × 10⁻⁵
Reverse osmosis spiral wound	0–3[b]	0–2	0–3 × 10⁻⁵

[a] Adapted in part from Taylor and Wiesner (1999) and AWWA (1996).

[b] Although a value of 3 is acceptable, the trend is to lower the upper limit to a value of 2 or less.

15-min filtration period (Schippers and Verdouw, 1980). Derived from a consideration of cake filtration, the MFI is defined as follows:

$$\frac{1}{Q} = a + \text{MFI} \times V \tag{11–46}$$

where Q = average flowrate, L/s

a = constant (intercept of linear portion of curve)

MFI = modified fouling index, s/L²

V = volume, L

The value of the MFI is obtained as the slope of the straight-line portion of the curve obtained by plotting the inverse of the flowrate versus the cumulative volume [see Fig. 11–40(a)].

Mini Plugging Factor Index. The mini-plugging factor index (MPFI) is a measure of the change in flowrate as a function of time, as illustrated on Fig. 11–40(b) (Taylor and Jacobs, 1996). The equipment used for the MPFI test is the same as that used for the SDI and MFI tests. The MPFI is defined as the slope of the linear portion of the flowrate versus time curve [see Fig. 11–40(b)], which is ascribed to cake fouling. In equation form, the MPFI is expressed as follows:

$$Q = (\text{MPFI})t + a \tag{11–47}$$

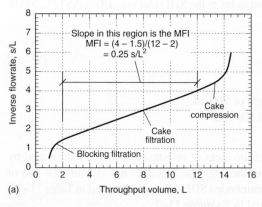
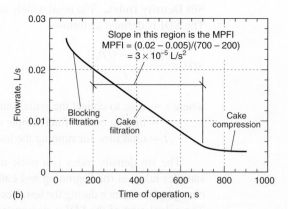

(a)

(b)

Figure 11–40

Typical plot to determine fouling indexes: (a) modified fouling factor (MFI) and (b) mini plugging factor index (MPFI).

where Q = average flowrate at 30 s intervals, L/s
MPFI = mini plugging factor index, L/s^2
t = time, s
a = constant (intercept of linear portion of curve)

Typical values for the MPFI are reported in Table 11–26. Because the MFI is based on throughput volume, it is thought to be a more sensitive index than the MPFI for characterization of fouling.

EXAMPLE 11–7 Silt Density Index for Reverse Osmosis Determine the silt density index for a proposed feed water from the following test data. If a spiral wound RO membrane is to be used, will pretreatment be required?

Test run time = 30 min
Initial 500 mL = 2 min
Final 500 mL = 10 min

Solution

1. Calculate the SDI using Eq. (11–45).

$$SDI = \frac{100[1 - (t_i/t_f)]}{t}$$

$$SDI = \frac{100[1 - (2/10)]}{30} = 2.67$$

2. Compare the SDI to the acceptable criteria.
Calculated SDI value of 2.67 is less than 3 (see Table 11–26); therefore, further pretreatment would not be needed normally.

Comment As a practical matter, because the SDI value is close to 3.0 it may be prudent to consider some form of pretreatment to prolong the filtration cycle.

Limitations of Fouling Indexes. The SDI and MFI fouling indexes described above, and others currently in use, have serious limitations including (1) the fact that a dead-end test is used to gather data to predict the fouling performance of a cross-flow membrane, (2) the test is conducted with a 0.45 μm filter which does not capture the effect of smaller colloidal particles, (3) the test is not representative of cake filtration, which occurs in cross-flow, (4) the test does not measure the propensity for scale formation, and (5) the test is conducted under conditions of constant pressure with variable flux, where the opposite operational mode is normally used in practice. It should be noted that several other indexes, using MF or UF membranes in place of the Millipore filter, to reflect the effect of smaller colloidal material and large dissolved organic material on fouling, are currently under development.

Pretreatment for NF and RO. A very high quality feed is required for efficient operation of a nanofiltration or reverse osmosis unit. Membrane elements in the reverse osmosis unit can be fouled by colloidal matter and dissolved constituents in the feed water. The pretreatment options identified in Table 11–27 have been used singly and in combination. The effectiveness of the treatment options can be assessed with one or more of the indexes discussed previously. Regular chemical cleaning of the membrane elements (about once a month) is also necessary to restore and maintain the membrane flux.

Application and Performance of Membranes

With evolving health concerns and the development of new and lower cost membranes, the application of membrane technologies in the field of environmental engineering has increased dramatically within the past 5 years. The increased use of membranes is expected to continue well into the future. In fact, the use of conventional filtration technology, such as described in Sections 11–4 and 11–5, may be a thing of the past within 10 to 15 years, especially in light of the need to remove resistant organic constituents of concern. The principal applications of the various membrane technologies in wastewater treatment are reported in Table 11–28. Application of the membrane technologies for the removal of specific constituents from wastewater is given in Table 11–29. Each of the membrane technologies is considered further in the following discussion.

Microfiltration. Microfiltration membranes are the most numerous on the market and are the least expensive. The use of membranes for biological treatment is currently one of the most important uses of membranes in wastewater treatment. In advanced treatment applications, microfiltration has been used, most commonly as a replacement for depth filtration to reduce turbidity, remove residual suspended solids, and reduce microorganisms for effective disinfection and as a pretreatment step for reverse osmosis (see Fig. 11–41). Typical operating information for microfiltration including size range, operating pressures, and flux rate are presented in Table 11–30. Typical performance data are reported in Table 11–31. Corresponding variability data are presented in Table 11–32. Care should be used in applying the performance data reported in Table 11–30 as it has been found that the performance of MF is to a large extent site specific, especially with respect to fouling.

Ultrafiltration. Ultrafiltration (UF) membranes are used for many of the same applications as described above for microfiltration. Some UF membranes with small pore sizes have also been used to remove dissolved compounds with high molecular weight, such as colloids, proteins, and carbohydrates. The membranes do not remove sugar or salt. The major distinction between UF and MF is that UF can remove viruses whereas MF cannot.

Figure 11–41

Typical process flow diagram for the production of potable water employing filter screens, open vessel microfiltration, cartridge filters, revers osmosis, UV advanced oxidation, decarbonation, and lime stabilization. (Adapted from Orange County Water District, CA.)

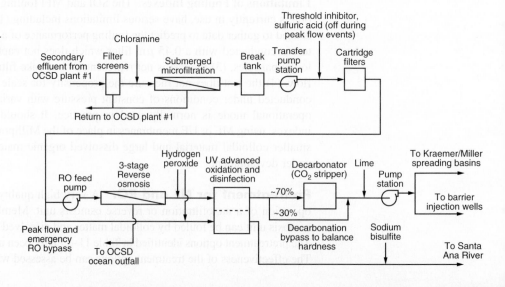

Table 11–27

Methods of pretreatment for nanofiltration and reverse osmosis systems

Material to be removed	Method of pretreatment	Description or discussion
Particulate matter and colloidal	Depth or surface filtration, microfiltration, or ultrafiltration	Particulate matter can be removed by various methods of filtration. Fouling agents may pass through these filtration systems, thus the potential for membrane fouling should be verified by pilot testing.
Particulate matter	Cartridge filter	Cartridge filters are pressure-driven filters with pore sizes varying from 5 to 15 μm and are commonly installed ahead of RO membranes. Cartridge filters provide a final level of protection against the intrusion of relatively large solids into the reverse osmosis system. When antiscalants are used, cartridge filters are used to remove the colloidal material and impurities found in antiscalent chemicals. The filters do not remove dissolved substances. Generally, the pressure drop across a clean cartridge filter is between 0 and 35 kPa. As the solids accumulate and the pressure drop reaches a threshold range of 70 to 80 kPa, the cartridge has to be removed and replaced (Paranjape et al., 2003).
Microorganisms	Disinfection	Disinfection of the feed water may be accomplished using either chlorine, ozone, or UV irradiation to limit bacterial activity. Ultrafiltration can also be used to reduce the number of microorganisms.
Scale formation	pH adjustment	To inhibit scale formation, the pH of the feed water is adjusted (usually with sulfuric acid) within the range from 4.0 to 7.5. A low pH enhances conversion of carbonate into bicarbonate species, which are much more soluble. Cellulose acetate RO membranes have an optimum pH range of 5 to 7 as they are prone to hydrolysis below a pH of 5. Newer polyamide RO membranes can be used over a broader pH range of 2 to 11 (Paranjape et al., 2003).
	Antiscalants	Antiscalants are compounds that either prevent scale formation entirely or permit formation of scales that can be removed easily during cleaning. Certain antiscalants, however, may increase the fouling caused by humic acids (Richard et al., 2001).
Iron and manganese	Ion exchange or chemical treatment	Removal of iron and manganese will decrease scaling potential. The exclusion of oxygen may be necessary to prevent oxidation of iron and manganese.
Sparingly soluble salts	Chemical treatment	Sparingly soluble salts such as silica can be removed by chemical treatment for industrial purposes, i.e., removal of silica may be required to prevent precipitation on heat exchangers. Chemical treatment may include the addition of aluminum and iron oxides, zinc chloride, magnesium oxide, ozone (when ozone-resistant membranes are used), and ultra-high lime clarification. Lime clarification, however, may not be as effective as other pretreatment methods in removing materials that foul RO membranes thus resulting in more frequent cleaning of the membranes (Gagliardo, 2000).

Ultrafiltration is used typically in industrial applications for the production of high purity process rinse water. Typical operating and performance data are presented in Tables 11–30 and 11–31, respectively. Performance variability data are presented in Table 11–32.

Nanofiltration. Nanofiltration, also known as "loose" RO or low pressure RO, can reject particles as small as 0.001 μm. Nanofiltration is used for the removal of selected dissolved constituents from wastewater such as the multivalent metallic ions responsible for hardness (i.e., calcium and magnesium). For this reason, NF is the preferred membrane

Table 11–28

Typical applications for membrane technologies in wastewater treatment[a]

Applications	Description
Microfiltration and Ultrafiltration	
Aerobic biological treatment	Membranes are used to separate the treated wastewater from the active biomass in an activated sludge process. The membrane separation unit can be internal (immersed in the bioreactor) or external to the bioreactor (see Fig. 8–2 in Chap. 8). Such processes are known as membrane bioreactor (MBR) processes
Anaerobic biological treatment	Membrane is used to separate the treated wastewater from the active biomass in an anaerobic complete-mix reactor
Membrane aeration biological treatment	Plate and frame, tubular, and hollow membranes are used to transfer pure oxygen to the biomass attached to the outside of the membrane. Such processes are known as membrane aeration bioreactor (MABR) processes
Membrane extraction biological treatment	Membranes are used to extract degradable organic molecules from inorganic constituents such as acids, bases, and salts from the waste stream for subsequent biological treatment [see Fig. 11–47(b)]. Such processes are known as extractive membrane bioreactor (EMBR) processes
Pretreatment for effective disinfection	Membranes are used to remove residual suspended solids from settled secondary effluent or from the effluent from depth or surface filters to achieve effective disinfection with either chlorine or UV radiation for reuse applications.
Pretreatment for nanofiltration and reverse osmosis	Microfilters are used to remove residual colloidal and suspended solids as a pretreatment step for additional processing
Nanofiltration	
Effluent reuse	Used to treat prefiltered effluent (typically with microfiltration) for indirect potable reuse applications such as groundwater injection. Credit is also given for disinfection when using nanofiltration
Wastewater softening	Used to reduce the concentration of multivalent ions contributing to hardness for specific reuse applications
Reverse osmosis	
Effluent reuse	Used to treat prefiltered effluent (typically with microfiltration) for indirect potable reuse applications such as groundwater injection. Credit is also given for disinfection when using reverse osmosis
Effluent dispersal	Reverse osmosis processes have proven capable of removing sizable amounts of selected compounds such as NDMA.
Two-stage treatment for boiler use	Two-stages of reverse osmosis are used to produce water suitable for high pressure boilers.

[a] Adapted in part from Stephenson et al. (2000).

for membrane softening. The advantages of nanofiltration over lime softening include the production of a product water that meets the most stringent reuse water quality requirements. Because both inorganic and organic constituents and bacteria and viruses are removed, disinfection requirements are minimized. Typical operating and performance data are presented in Tables 11–30 and 11–33, respectively. Performance variability data are presented in Table 11–32.

Reverse Osmosis.   Worldwide, reverse osmosis (RO) is used primarily for desalination (Voutchkov, 2013). In wastewater treatment, RO is used for the removal of dissolved constituents from wastewater, remaining after advanced treatment with depth filtration or

Table 11-29

Application of membrane technologies for the removal of specific constituents found in wastewater[a]

Constituent	Membrane technology				Comments
	MF	UF	NF	RO	
Biodegradable organics		✔	✔	✔	
Hardness			✔	✔	
Heavy metals			✔	✔	
Nitrate			✔	✔	
Priority organic pollutants		✔	✔	✔	
Synthetic organic compounds			✔	✔	
TDS			✔	✔	
TSS	✔	✔			TSS removed during pretreatment for NF and RO
Bacteria	✔[b]	✔[b]	✔	✔	Used for membrane disinfection. Removed as pretreatment for NF and RO with MF and UF
Protozoan cysts and oocysts and helminth ova	✔	✔	✔	✔	
Viruses			✔	✔	Used for membrane disinfection

[a] Specific removal rates will depend on the composition and constituent concentrations in the treated wastewater.

[b] Variable performance, depending on the membrane nominal pore size and operating conditions.

microfiltration. The membranes exclude ions, but require high pressures to produce deionized water. The Orange County Water District flow diagram involving the use of reverse osmosis for the production of potable water for groundwater recharge is shown on Fig. 11–41. Typical operating information for reverse osmosis used for wastewater including operating pressures and flux rate rates is reported in Table 11–30. Corresponding performance data and variability data are presented in Tables 11–34 and 11–32. As noted above, care should be used in applying the performance data reported in Table 11–34 as it has been found that the performance of RO is also site specific, especially with respect to fouling (see Table 11–25). Important process design considerations for NF and RO membranes are reported in Table 11–35.

Depending on the level of dissolved solids removal, the product water from NF and RO processes may be corrosive to equipment and piping. Typical postreatments will involve the addition of chemical to adjust the stability of the treated water, in some reuse applications, the removal or addition of gases (see Fig. 11–41), and the addition of chemicals to meet disinfection requirements and to control the growth of microorganisms in pipelines. The types and use of chemicals to stabilize NF and RO product water are considered in Chap. 6. In some cases, blending with other waters, especially in potable reuse applications, may be appropriate.

Membrane Energy Consumption. Typical product recovery and energy consumption values for various membrane systems are presented in Table 11–30 for the processing of wastewater. Corresponding data for seawater desalination are also given for the

Table 11-30

Typical operating characteristics of membrane technologies used in wastewater treatment applications and for desalination[a]

Membrane technology	Product recovery[b], %	Operating pressure[c]		Rate of flux		Energy consumption[c]	
		lb/in.²	kPa	gal/ft²·d	L/m²·h	kWh/10³ gal	kWh/m³
Wastewater with TDS from 800 to 1200 mg/L							
Microfiltration (vacuum type)	85–95	−3−−14	−28−−100	15–25	25–42	0.75–1.1	0.2–0.3
Microfiltration (pressure type)	85–95	5–30	34–200	24–35	40–60	0.75–1.1	0.2–0.3
Ultrafiltration	85–95	10–35	68–350	24–35	40–60	0.75–1.1	0.2–0.3
Nanofiltration	85–90	100–200	700–1400	8–12	14–20	1.5–1.9	0.4–0.5
Reverse osmosis (without energy recovery)	80–85	125–230	800–1900	8–12	14–20	1.9–2.5	0.5–0.65
Reverse osmosis (with energy recovery)[d]	80–85	125–230	800–1900	8–12	14–20	1.7–2.3	0.46–0.6
Electrodialysis	75–95			20–25	33–42	4.2–8.4	1.1–2.2
Seawater with TDS of about 35,000 mg/L							
Ultrafiltration (pretreatment)	85–95	10–35	68–350	24–47	40–80	0.75–1.1	0.2–0.3
Reverse osmosis (without energy recovery)[e]	30–55	700–1000	4800–6900	8–12	14–20	34–45	9–12
Reverse osmosis (with turbine/pump energy recovery)	30–55	700–1000	4800–6900	8–12	14–20	19–26	5–7
Reverse osmosis (with pressure exchange energy recovery)	30–55	700–1000	4800–6900	8–12[f]	14–20	9.5–15	~2.5–4

[a] Adapted in part from Patel (2013), Voutchkov (2013), Wetterau (2013).

[b] Cross-flow mode [see Fig, 11–33(a)]. In dead end mode [see Fig, 11–33(c)] all of the water passes through the membrane.

[c] The operating pressure and energy consumption will vary with the influent water quality and temperature of the feed water.

[d] Overall total energy reduction will vary from 6 to 12 percent, depending on the energy recovery device (ERD) and process configuration.

[e] At 50 percent recovery, the minimum theoretical energy required is 1.06 kWh/m³; the corresponding practical limit is about 1.56 kWh/m³ (Elimelech and Phillip, 2007).

[f] Flux rate with open intake is in the range from 12–17 L/m²·h (7–10 gal/ft²·d).

Note:
$$kPa \times 0.1450 = lb_f/in.^2$$
$$L/m^2 \cdot h \times 0.5890 = gal/ft^2 \cdot d$$
$$kWh/m^3 \times 3.785 = kWh/10^3 \, gal$$
$$Bar = 100 \, kPa.$$

Table 11–31

Expected performance of microfiltration and ultrafiltration membranes on secondary effluent

Constituent	Rejection	Value	
		Microfiltration	**Ultrafiltration**
TOC	%	45–65	50–75
BOD	%	75–90	80–90
COD	%	70–85	75–90
TSS	%	95–98	96–99.9
TDS	%	0–2	0–2
NH_3-N	%	5–15	5–15
NO_3-N	%	0–2	0–2
PO_4^-	%	0–2	0–2
SO_4^{2-}	%	0–1	0–1
Cl^-	%	0–1	0–1
Total coliform[a]	log	2–5	3–6
Fecal coliform[a]	log	2–5	3–6
Protozoa[a]	log	2–5	>6
Viruses[a]	log	0–2	2–7[b]

[a] The reported values reflect observed practice and integrity concerns (see Example 8–4 in Chap. 8) and also a wide range of performance differences between membranes, as given in following footnote.

[b] The low and corresponding mean removal values for four different UF membranes treating the same water were 2.5, 4.0, 5.3, and 6.1 and 3.8, 5.0, 6.5, and 7.5, respectively (Sakaji, R. H., 2006).

purpose of comparison. The impact of water quality on energy consumption can be seen by comparing the energy requirement for wastewater with TDS ~1000 mg/L (~0.6 kWh/m³) versus seawater with TDS ~35,000 mg/L (~10.5 kWh/m³, without energy recovery). The importance of energy recovery, as discussed below, especially in seawater desalination, is also clearly evident.

In reviewing the information presented in Table 11–30, it is important to note that the reported operating pressure values for all of the membrane processes are considerably lower than comparable values of ten years ago. It is anticipated that operating pressures will continue to go down as new membranes and operating techniques are developed, but at a considerably lower rate as compared to the last ten years. At the present time, where the use of membranes is being considered, special attention must be devoted to the characteristics of the wastewater to be processed.

Energy Recovery from Nanofiltration and Reverse Osmosis. Because NF and RO in particular produce a high-pressure retentate flowrate , especially in seawater desalination, various methods have been developed or are under development to recover the energy lost in depressurizing the retentate flowrate. Energy recovery devices (ERDs) are designed to recover energy from the retentate flowrate and transfer it to the feed water to reduce the overall process energy (see Fig. 11–42 on page 1213). Typical devices that have been used operate on the following principles.

- Reverse running pumps
- Pelton wheel turbines

Table 11-32

Typical effluent quality variability observed with processes used for the removal of from dissolved constituents from reclaimed wastewater

Particulate removal process	Unit	Range of effluent values[a]	Geometric standard deviation, s_g[b]	
			Range	Typical
Microfiltration				
Turbidity	NTU	0.1–0.4	1.1–1.4	1.3
TSS	mg/L	0–1	1.3–1.9	1.5
Ultrafiltration				
Turbidity	NTU	0.1–0.4	1.1–1.4	1.3
TSS	mg/L	0–1	1.3–1.9	1.5
Nanofiltration				
TDS	mg/L	50–100	1.3–1.5	1.4
TOC	mg/L	1–5	1.2–1.4	1.5
Turbidity	NTU	0.01–0.1	1.5–2.0	1.75
Reverse osmosis[c]				
TDS	mg/L	25–50	1.3–1.8	1.6
TOC	mg/L	0.1–1	1.2–2.0	1.8
Turbidity	NTU	0.01–0.1	1.2–2.2	1.8
Electrodialysis				
TDS	mg/L	na	1.2–1.75	1.5

[a] Typical effluent values are not given for the processes because they will vary widely and depend on the operating conditions and water quality requirements.

[b] s_g = geometric standard deviation; $s_g = P_{84.1}/P_{50}$.

[c] Because measured effluent values are typically near the constituent detection limits, the error in the detection method can contribute to the observed effluent variability.

- Hydraulic turbocharger
- Isobaric energy recovery – piston type
- Isobaric energy recovery – rotary type
- Pressure amplifying pump

Pumps, Turbines, and Hydraulic Turbochargers. The reverse running pumps (i.e., Francis turbines), Pelton wheel turbines, and hydraulic turbochargers are adaptations of well know hydraulic machinery applied for the recovery of energy from NF and RO installations. Functionally, as shown on Fig. 11–42(a), the Pelton wheel turbine shaft is coupled to the motor used to drive the pump that pressurizes the feed water. The operation of the hydraulic turbocharger is similar to the Pelton turbine, with the exception that a pump impeller is mounted on the same shaft as the turbine, and a motor is not used.

Isobaric Devices. Isobaric energy recovery devices (also known as flow work exchangers) [see Fig. 11–42(b)] utilize the principles of positive displacement and isobaric

Table 11–33

Typical rejection rates for NF and "loose" RO membranes used to treat wastewater

Constituent	Unit	Rejection rate	
		Nanofiltration	Loose RO
Total dissolved solids	%	40–60	
Total organic carbon	%	90–98	
Color	%	90–96	
Hardness	%	80–85	
Sodium chloride	%	10–50	70–95
Sodium sulfate	%	80–95	80–95
Calcium chloride	%	10–50	80–95
Magnesium sulfate	%	80–95	95–98
Nitrate	%	80–85	85–90
Fluoride	%	10–50	
Arsenic (+5)	%	<40	
Atrazine	%	85–90	
Proteins	log	3–5	3–5
Bacteria[b]	log	3–6	3–6
Protozoa[b]	log	> 6	> 6
Viruses[b]	log	3–5	3–5

[a] Adapted in part from www.gewater.com and Wong (2003).

[b] Theoretically all microorganisms should be removed. The reported values reflect integrity concerns (see Example 8–4 in Chap. 8).

chambers to transfer energy from a high pressure stream (the brine stream in the case of NF and RO) to a low pressure incoming feed water (Stover, 2007). The dual work exchange energy recovery (DWEER)® is an alternating piston driven device with two isobaric chambers. The PX® technology combines an isobaric positive displacement device with a centrifugal ERD; the transfer of energy is accomplished without the use of a piston. Isobaric ERDs, because of their ease of operation and flexibility, are used extensively throughout the world in desalination installations and are replacing and/or reducing the use of centrifugal type ERDs.

Performance of Energy Recovery Devices.   Because of the relatively low feed water pressures used in the RO treatment of wastewater, energy recovery efficiency is relatively low. Typical recoveries for Pelton wheels and isobaric devices vary from 25 to 45 and 45 to 65 percent, respectively. The overall process energy reduction is about 6 to 12 percent, depending on the device and process configuration. By comparison, in seawater desalination, recovery efficiencies as high as 95 percent have been achieved, depending on the ERD (Voutchkov, 2013). The overall total energy reductions that can be achieved in seawater desalination will vary from 30 to 75 percent. On a relatively small scale, the Clark® pressure amplifying pump employs two opposing cylinders and pistons on a single rod to pressurize the feed water in conjunction with a small feed pump.

Table 11–34

Typical performance for reverse osmosis treatment[a]

Constituent	Unit	Rejection rate
Total dissolved solids	%	90–98
Total organic carbon	%	90–98
Color	%	90–96
Hardness	%	90–98
Sodium chloride	%	90–99
Sodium sulfate	%	90–99
Calcium chloride	%	90–99
Magnesium sulfate	%	95–99
Nitrate	%	84–96
Fluoride	%	90–98
Arsenic (V)	%	85–95
Atrazine	%	90–96
Proteins	log	4–7
Bacteria[b]	log	4–7
Protozoa[b]	log	>7
Viruses[b]	log	4–7

[a] Adapted in part from www.gewater.com and Wong (2003).

[b] Theoretically all microorganisms should be removed. The reported values reflect integrity concerns (see Example 8–4 in Chap. 8).

Forward Osmosis: An Emerging Membrane Technology

The membrane processes discussed in this section are based on reversing the natural osmosis process through the addition of a driving force greater than the osmotic pressure to produce purified water. Although not used commonly for water purification, a number of alternative processes based on the utilization of the osmotic pressure are under development. Processes that utilize the natural osmotic pressure [see Fig. 11–43(a)] are termed forward osmosis (FO) or direct osmosis (DO). In the FO process, illustrated on Fig. 11–43(b), water from the feed solution permeates through the membrane to dilute a more concentrated solution, known by a variety of names including draw solution, osmotic agent, and driving agent. The draw solution is the name used most commonly for this solution.

The principal requirement for the draw solution is that its osmotic pressure must be greater than that of the feed solution. Another requirement for the draw solution is that it must be easy to reconcentrate after being diluted by the water from the feed solution. A solution of sodium chloride (NaCl) has been used as the draw solution because it can be reconstituted easily by reverse osmosis without the problems associated with scaling. A draw solution comprised of multivalent ions has been used where high rejection is required. Based on the results of preliminary testing, advantages of the FO process include minimal pressure requirements and high rejection for a variety of constituents. The FO process may also result in less membrane clogging, but more research is needed to define the controlling conditions that will minimize clogging.

Table 11–35

Process design considerations for NF and RO[a]

Design consideration	Discussion
Feed water characterization	Complete characterization of the feed water is essential for identifying constituents that produce a high potential for membrane fouling. The effect of residual suspended solids in the influent to the membranes especially should be evaluated
Pretreatment	Pretreatment must be evaluated to extend membrane life, and issues such as flow equalization, pH control, chemical treatment, and residual solids removal should be considered
Flux rate	Flux rate influences system costs by establishing the filter area, affecting polarization control, and affecting membrane life
Recovery	Recovery rate affects solute rejection, membrane performance, and brine generation volumes
Membrane fouling	Parameters should be developed based on pilot plant testing. Acid, antiscalants, and biocides are used to control membrane fouling, as are staging and operational conditions
Membrane cleaning	Cleaning procedures and frequency need to be established
Membrane life	The principal economic consideration that governs successful application of membrane technology
Operating and maintenance costs	High pressure systems require significant energy costs, high capital costs for high pressure pumps, and high maintenance costs associated with equipment wear. After membrane replacement, energy is the next major operating expense
Recycle flows	Provisions for recycling a portion of the product water should be included as an operating consideration to control membrane velocity, influent concentration, and equalizing influent flow variations
Retentate and backwash disposal	Retentate and backwash characteristics need to be considered especially, if chemicals are used in pretreatment or membrane cleaning and large volumes of waste require disposal

[a] Adapted in part from Celenza (2000).

For example, potential applications in the wastewater management field include concentration of dilute industrial wastewater, concentration of RO brines, concentration of digester supernatant, and the concentration of the return flows from sludge thickening processes. By reducing the volume of RO brines that must be processed,

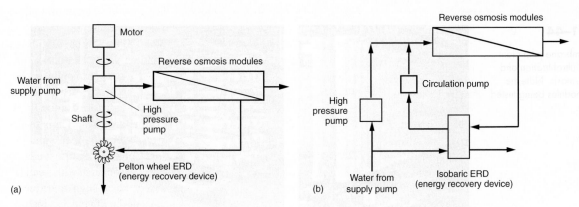

(a)   (b)

Figure 11–42

Application of energy recovery devices in conjunction with reverse osmosis: (a) Pelton wheel and (b) isobaric piston type.

Figure 11–43

Application of forward osmosis: (a) definition sketch for forward osmosis (the differential pressure between solutions is less than the osmotic pressure) and (b) flow diagram for the application of forward osmosis. Note: the draw solution is far more concentrated than the feed water.

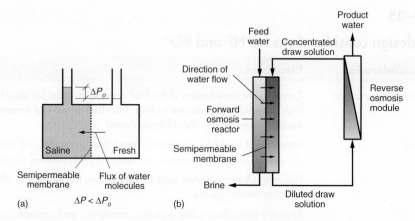

(a) $\Delta P < \Delta P_o$

(b)

technologies such as evaporation and crystallization become technically and economically feasible. An excellent review article has been prepared by Cath and his associates (Cath et al., 2006).

Pilot-Plant Studies for Membrane Applications

Because every wastewater is unique with respect to its constituent characteristics, it is difficult to predict *a priori* how a given membrane process will perform. As a result, the selection of the best membrane for a given application is based usually on the results of pilot studies. Membrane fouling indexes (see Table 11–26) can be used to assess the need for pretreatment. In some situations, manufacturers of membranes will provide a testing service to identify the most appropriate membrane for a specific feed water. Typical pilot scale facilities used to evaluate the performance of NF and RO treatment processes are shown on Fig. 11–44.

The elements that comprise a pilot plant include (1) the pretreatment system; (2) tankage for flow equalization and cleaning; (3) pumps for pressurizing the membrane, recirculation, and backflushing with appropriate controls; (4) the membrane test module; (5) facilities for monitoring the performance of the test module; and an appropriate membrane backflushing system. Typical membrane operating parameters and water quality

Figure 11–44

Views of membrane pilot plant test units: (a) ultrafiltration and (b) reverse osmosis. Note the membrane modules being tested are full scale.

(a)

(b)

Table 11–36

Typical operating parameters and water quality measurements used for pilot testing membrane facilities[a]

Membrane operating parameters
Pretreatment requirements including chemical dosages
Transmembrane flux rate correlated to operating time
Transmembrane pressure
Recovery
Washwater requirements
Recirculation ratio
Cleaning frequency including protocol and chemical requirements
Posttreatment requirements

Typical water quality measurements

Turbidity	Heterotrophic plate count
Particle counts	Other bacterial indicators
Total organic carbon	Specific constituents that can limit recovery such as silica, barium, calcium, fluoride, strontium, and sulfate
Nutrients	
Heavy metals	
Organic priority pollutants	Biotoxicity
Total dissolved solids	Fouling indexes
pH	
Temperature	

[a] Tchobanoglous et al. (2003).

measurements are presented in Table 11–36. Additional specific parameters selected for evaluation will depend on the final use of the product water.

Management of Retentate

Management of the retentate produced by membrane processes represents the major problem that must be dealt with in their applications. Methods that can be used for the treatment and disposal of the retentate are reported in Table 11–37. While small facilities can dispose of small quantities of retentate by blending with other wastewater flows, this approach is not suitable for large facilities. The retentate from NF and RO facilities will contain hardness, heavy metals, high molecular weight organics, microorganisms, and often hydrogen sulfide gas. The pH is usually high due to the concentration of alkalinity, which increases the likelihood of metal precipitation in disposal wells. As a result, most of the large-scale desalination facilities are located along coastal regions, both in the United States and in other parts of the world. For inland locations, long transmission lines to coastal regions are being considered. While controlled evaporation is technically feasible, because of high operating and maintenance costs, this approach is used where no other alternatives are available and the value of product water is high. The quality and quantity of the concentrated retentate produced from nanofiltration, reverse osmosis and electrodialysis can be estimated using simplified recovery and rejection computations as illustrated in Example 11–8.

Table 11–37

Treatment methods and disposal options for concentrated brine solutions from membrane processes

Disposal option	Description
Treatment options	
Concentration by using multistage membrane arrays	Concentration of brine stream
Falling film evaporators	Thicken and concentrate brine streams
Crystallizers	Concentration of brine stream into a crystalized form for processing or disposal
Forward osmosis	Concentration of brine stream
Membrane distillation	Concentration of brine stream
Solar evaporators	Thicken and concentrate brine streams
Spray dryers	Concentration of brine stream
Vapor compression evaporators	Concentration of brine stream
Disposal options	
Deep well injection	Depends on whether subsurface aquifer is brackish water or is otherwise unsuitable for domestic uses.
Discharge to wastewater collection system	This option is only suitable for very small discharges such that the increase in TDS is not significant (e.g., less than 20 mg/L).
Evaporation ponds	Large surface area required in most areas with the exception of some southern and western states.
Land application	Land application has been used for some low concentration brine solutions.
Ocean discharge	The disposal option of choice for facilities located in the coastal regions of the United States. Typically, a brine line, with a deep ocean discharge, is used by a number of dischargers. Combined discharge with power plant cooling water has been used in Florida. For inland locations, truck, rail hauling or pipeline is needed for transportation.
Surface water discharge	Discharge of brines to surface waters is the most common method of disposal for concentrated brine solutions.

EXAMPLE 11–8 **Estimate Quantity and Quality of Waste Streams from a Reverse Osmosis Facility** Estimate quantity and quality of the retentate and the total quantity of water that must be processed, from a reverse osmosis facility that is to produce 4000 m³/d of water to be used for industrial cooling operations. Assume that both the recovery and rejection rates are equal to 90 percent and that the TDS concentration of the feed steam is 400 mg/L.

Solution

1. Determine the flowrate of the concentrated retentate and the total amount of water that must be processed.

   a. Combining Eqs. (11–36) and (11–40) results in the following expression for the retentate flowrate.

   $$Q_r = \frac{Q_p(1 - r)}{r}$$

b. Determine the retentate flowrate.

$$Q_r = \frac{(4000 \text{ m}^3/\text{d})(1 - 0.9)}{0.9} = 444 \text{ m}^3/\text{d}$$

c. Determine the total amount of water that must be processed to produce 4000 m³/d of RO water. Using Eq. (11–43) the required amount of water is

$$Q_f = Q_p + Q_r = 4000 \text{ m}^3/\text{d} + 444 \text{ m}^3/\text{d} = 4444 \text{ m}^3/\text{d}$$

2. Determine the concentration of the permeate. The permeate concentration is obtained by writing Eq. (11–41) in decimal form as follows:

$$C_p = C_f(1 - R) = 400 \text{ mg/L} \, (1 - 0.9) = 40 \text{ mg/L}$$

3. Determine the concentration of the retentate. The required value is obtained by solving Eq. (11–44).

$$C_r = \frac{Q_f \, C_f - Q_p \, C_p}{Q_r}$$

$$C_r = \frac{(4444 \text{ m}^3/\text{d})(400 \text{ mg/L})C_f - (4000 \text{ m}^3/\text{d})(40 \text{ mg/L})}{(444 \text{ m}^3/\text{d})}$$

$$C_r = 3643 \text{ mg/L}$$

Comment A variety of concentration methods are currently under investigation to reduce the volume of the retentate that must be treated.

11–8 ELECTRODIALYSIS

Electrodialysis (ED) is an electrochemical separation process in which mineral salts and other ionic species are transported through ion-selective membranes from one solution to another under the driving force of a direct current (DC) electric potential. As compared to NF and RO, which transport pure water through the membrane leaving the salts behind, with ED salt is gradually stripped from solution leaving a dilute solution behind containing particulate matter and neutral species not removed by the ED process. The salt transferred through the membrane then forms the concentrate. A typical flow diagram employing electrodialysis for the control of dissolved solids is shown on Fig. 11–45.

Description of the Electrodialysis Process

The key to the ED process is the ion selective membranes that are essentially ion exchange resins cast in sheet form. Ion exchange membranes that allow passage of positively charged ions such as sodium and potassium are called cation membranes. Membranes that allow passage of negatively charged ions such as chloride and phosphate are called anion membranes. To demineralize a solution using ED, cation and anion membranes are arranged alternately between plastic spacers in a stacked configuration with a positive electrode (anode) at one end and a negative electrode (cathode) at the other (see Fig. 11–46). When a DC voltage is applied, the electrical potential created becomes the driving force to move ions, with the membranes forming barriers to the ions of opposite charge. Therefore, anions attempting to migrate to the anode will pass through the adjacent anion membrane but will be stopped by the first cation membrane they encounter. Cations trying to

Figure 11–45

Typical process flow diagram employing electrodialysis for the removal of total dissolved solids (TDS) from secondary effluent.

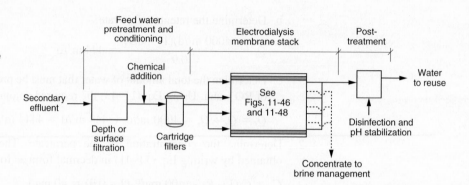

migrate to the cathode will pass through the cation membrane but will be stopped by the anion membrane. The membranes, therefore, form ion diluting compartments and ion concentrating compartments (www.gewater.com).

An ED assembly, known as a *stack*, consists of multiple cell pairs located between an anode and a cathode. A set of adjacent components consisting of a diluting compartment spacer, an anion membrane, a concentrating compartment spacer, and a cation membrane is called a *cell pair*. Electrolysis stacks can contain as many as 600 cell pairs. Feed water (filtered wastewater) is pumped through the stack assembly. Typical flux rates are from 35 to 45 L/m²·h. Dissolved solids removals vary with the (1) wastewater temperature, (2) amounts of electric current passed, (3) type and amount of ions, (4) permeability/selectivity of the membrane, (5) fouling and scaling potential of the feed water, (6) feed water flowrates, and (7) number and configuration of stages.

Electrodialysis Reversal

In the early 1970s, the electrodialysis reversal (EDR) process was introduced. An EDR unit operates on the same principle as ED technology, except that both the product and concentrate channels are identical in construction (see Fig. 11–47). The same membranes are used to provide a continuous self-cleaning ED process that uses periodic reversal of the DC polarity to allow systems to run at high recovery rates. Polarity reversal causes the concentrating

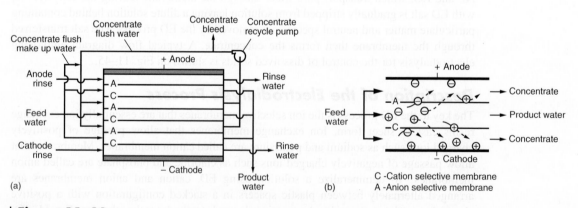

Figure 11–46

Conventional electrodialysis: (a) schematic of electrodialysis membrane stack with anode and cathode rinse and (b) schematic illustration of ion migration within the membrane stack. Note: The conventional electrodialysis process has been largely replaced the electrodialysis reversal (EDR) process (see Fig 11–47).

Figure 11–47

Schematic of electrodialysis reversal (EDR) process: (a) negative polarity and (b) positive polarity. Because the polarity is reversed, the anode and cathode rinse shown on Fig. 11–46 is not needed.

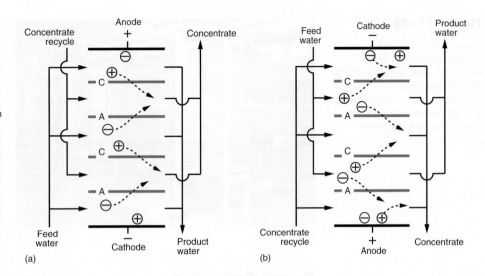

and diluting flow to switch after every cycle. Any fouling or scaling constituents are removed when the process reverses, sending fresh product water through the compartments filled previously with concentrated waste. The reversal process is useful in breaking up and flushing out scales, slimes, and other deposits in the cells before they buildup. Product water is not collected during a short interval of time following reversal.

EDR systems are able to reduce dissolved ions in feed water containing 10,000 to 12,000 mg/L of total dissolved solids, but because of energy requirements are ideally suited for the treatment of brackish water in the range from 800 to 5000 mg/L. As a rule of thumb, it takes about 1 to 1.2 kWh/m³ to remove a kilogram of salt (see Table 11–38). Typical removal rates can range from 50 to 94 percent removal (www.gewater.com). A view of an EDR installation and an exposed membrane stack are shown on Fig. 11–48. The EDR facility shown on Fig. 11–48 is used to remove TDS from a portion (sidestream) of the reclaimed water produced at the North City plant in San Diego, CA. The treated water with a reduced TDS concentration is blended back into the main flow which has a TDS concentration that varies from 1200 to 1300 mg/L to produce a final reclaimed water with a TDS equal to or less than 1000 mg/L to meet contractual agreements with the users of the reclaimed water.

Table 11–38

Typical operating parameters for electrodialysis units

Parameter	Unit	Range
Flux rate	m³/m²·d	0.8–1.0
Water recovery	%	75–90
Concentrate flowrate	% of feed	12–20
TDS removal	%	50–94
CD/N (current density to normality) ratio	(mA/cm²)/(g-eq/L)	500–800
Membrane resistance, Ω	ohms	4–8
Current efficiency	%	85–95
Energy consumption[a]	kWh/m³	1.5–2.6
Approximate energy per kg of salt removed	kWh/m³·kg	1–1.2

[a] Based on treating reclaimed water with a TDS concentration in the range from 1000 to 2500 mg/L. Not recommended for TDS concentration values beyond 10,000 to 12,000 mg/L.

Figure 11–48

Electrodialysis reversal process used to remove TDS from reclaimed water at the North City plant in San Diego, CA: (a) view of full scale electrodialysis facility and (b) view of electrodialysis membrane stack with cover removed.

(a)

(b)

Power Consumption

The ED/EDR process uses electric power to transfer ions through the membranes and to pump water through the system. Two, or sometimes three, pumping stages are used typically.

Power Requirements for Ion Transfer. The current required for ED can be estimated using Faraday's laws of electrolysis. Because one Faraday of electricity will cause one gram equivalent of a substance to migrate from one electrode to another, the number of gram equivalents removed per unit time is given by:

$$\text{Gram-eq/unit time} = Q_p(N_{\text{inf}} - N_{\text{eff}}) = Q_p\Delta N = Q_p N_{\text{inf}} E_r \qquad (11–48)$$

$$\text{where gram/eq} = \frac{\text{Mass of solute, g}}{\text{Equivalent weight of solute, g}}$$

Q_p = product water flowrate, L/s
N_{inf} = normality of influent (feed), g-eq/L
N_{eff} = normality of effluent (product), g-eq/L
ΔN = change in normality between the influent and effluent, g-eq/L
E_r = efficiency of salt removal, % (expressed as a decimal)

The corresponding expression for the current for a stack of membranes is given by:

$$i = \frac{FQ_p(N_{\text{inf}} - N_{\text{eff}})}{nE_c} = \frac{FQ_p N_{\text{inf}} E_r}{nE_c} \qquad (11–49)$$

where i = current, A, ampere
F = Faraday's constant, 96,485 A·s/g-eq
n = number of cell pairs in the stack
E_c = current efficiency, % (expressed as a decimal)

In the analysis of the ED process, it has been found that the capacity of the membrane to pass an electrical current is related to the current density (CD) and the normality (N) of the feed solution. Current density is defined as the current in milliamperes that flows through a square centimeter of membrane perpendicular to the current direction. Normality corresponds to the concentration of a solution based on the number of gram equivalent weights of a solute per liter of solution. A solution containing one gram of equivalent weight per liter is referred to as *one normal* (1 N). The relationship between current density and the solution normality is known as the *current density to normality* (CD/N) ratio.

High values of the CD/N ratio are indicative that there is insufficient charge to carry the current. When high ratios exist, a localized deficiency of ions may occur on the surface of the membrane, causing a condition called *polarization*. Polarization should be avoided as it results in high electrical resistance leading to excessive power consumption. In practice, CD/N ratios will vary from 500 to 800 when the current density is expressed as mA/cm^2. The resistance of an ED unit used to treat a particular water must be determined experimentally. Once the resistance, R, and the current flow, i, are known, the power required can be computed using Ohm's law as follows:

$$P = E \times i = R(i)^2 \tag{11–50}$$

where P = power, W
E = voltage, V
  = $R \times i$
R = resistance, Ω
i = current, A

The application of the above relationships is considered in Example 11–9.

EXAMPLE 11–9 **Determine Power Requirements and Membrane Area for ED Treatment of Reclaimed Water** Determine the power and area required to reduce the TDS content of 4000 m^3/d of treated wastewater to be used for industrial cooling water. Assume the following data apply.

1. Number of cell pairs in stack = 500
2. Influent TDS concentration = 2500 mg/L (~ 0.05 g-eq/L)
3. TDS removal efficiency, E_r = 50%
4. Product water flowrate = 90% of feed water
5. Current efficiency, E_c = 90%
6. CD/N ratio = (500 mA/cm^2)/(g-eq/L)
7. Resistance = 5.0 Ω

Solution
1. Calculate the current using Eq. (11–49).

$$i = \frac{FQ_pN_{inf}E_r}{nE_c}$$

Q_p = (4000 m^3/d)(10^3 L/1 m^3)/(86,400 s/d) = 46.3 L/s

$$i = \frac{(96,485 \text{ A·s/g-eq})(46.3 \text{ L/s})(0.05 \text{ g-eq/L})(0.5)}{(500)(0.90)}$$

i = 248 A

2. Determine the power required using Eq. (11–50).

$P = R(i)^2$
$P = (5.0 \ \Omega)(248 \text{ A})^2 = 307{,}520 \text{ W} = 308 \text{ kW}$

3. Determine the power requirement per m^3 of treated water.

$$\text{Power consumption} = \frac{(308 \text{ kW}) (24 \text{ h/d})}{(4000 \text{ m}^3/\text{d})(0.9)} = 2.05 \text{ kWh/m}^3$$

4. Determine the required surface area per cell pair. The area is given by

$$A = \frac{i, \text{current}}{CD, \text{current density}}$$

a. Determine the current density from the CN/N ratio:

$$CD = [(500 \text{ mA/cm}^2)/(\text{g-eq/L})](0.05 \text{ g-eq/L}) = 25 \text{ mA/cm}^2$$

b. The required area is:

$$\text{Area} = \frac{i}{CD} = \frac{(248 \text{ A})(1000 \text{ mA/A})}{(25 \text{ mA/cm}^2)} = 9920 \text{ cm}^2 = 0.99 \text{ m}^2$$

Comment   The actual performance will have to be determined from pilot tests. The computed value for the power required per unit volume, 2.05 kWh/m³, is within the range of values reported in Table 11–38 (1.1 to 2.6 kWh/m³) for water with 1000 to 2500 mg/L TDS.

Power Requirements for Pumping.   For pumping, the power requirements depend on the concentrate recirculation rate, the need for both product and waste pumping for discharge, and the efficiency of the pumping equipment (USBR, 2003).

Operating Considerations

The ED process may be operated in either a continuous or a batch mode. The units can be arranged either in parallel to provide the necessary hydraulic capacity or in series to obtain the desired degree of demineralization. A typical three-stage, two-line ED flow diagram is shown on Fig. 11–49. The ED process should be protected from particulate fouling by a 10 micron cartridge filter [see Table 11–21(g) and Fig. 11–45].

A single electrodialysis stack can remove from 25 to 60 percent of the TDS, depending on the feed water characteristics. Further desalting requires that two or more stacks be used in series (USBR, 2003). A portion of the resulting concentrate is recycled to improve system performance. Makeup water, usually about 10 percent of the feed volume, is required to wash the membranes continuously. A portion of the concentrate flowrate is recycled to maintain nearly equal flowrates and pressures on both sides of each membrane. Typical operating parameters for the electrodialysis process are reported in Table 11–38.

Operating Issues.   Problems associated with the ED process for wastewater treatment include chemical precipitation of salts with low solubility on the membrane surface and clogging of the membrane by the residual colloidal organic matter in wastewater treatment plant effluents. To reduce membrane fouling, some form of filtration may be necessary. With a properly designed plant, membrane cleaning should be infrequent. However, for both ED and EDR systems, clean-in-place (CIP) systems are provided normally to

Figure 11–49

Schematic diagram for a three-stage, two-line electrodialysis process.

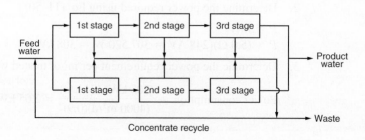

circulate either hydrochloric acid solution for mineral scale resolution or sodium chloride solution with pH adjustment for organics removal (USBR, 2003).

Membrane and Electrode Life.   Membranes for ED and EDR applications have a life of about 10 years before they are replaced. Effective and timely cleaning-in place extends the membrane life and improves product quality and power consumption. Cation membranes typically last longer than anion membranes because anion membranes are particularly susceptible to oxidation by chlorine and other strong oxidants (USBR, 2003). With the development of the EDR process and new electrode design, the life of anode and cathode electrodes is typically 2 to 3 years. Anode life is typically less than cathode life. Electrodes can be reconditioned (USBR, 2003).

Electrodialysis Versus Reverse Osmosis

In a recently completed study, two advanced treatment processes were compared to reduce the salinity of reclaimed water from a TDS concentration of 750 ± 50 mg/L to 500 mg/L or less (Adham et al., 2004). The two advanced treatment processes evaluated were (1) MF followed by RO and (2) EDR. The study was conducted for a period of about six months. Based on the results of the side-by-side testing, it was found that the EDR process with cartridge prefiltration was more cost effective than the combined MF/RO process. Some of the advantages and disadvantages cited for each advanced treatment process are reported in Table 11–39. As more of the potential applications of EDR are currently under investigation, the current literature should be consulted.

Table 11–39

Comparison of advantages and disadvantages of electrodialysis and reverse osmosis for desalination[a]

Advantages	Disadvantages
Electrodialysis (EDR)	
• Minimal pretreatment may be required (cartridge filtration is recommended)	• Limited to 50 percent salt rejection for a single membrane stack (stage)
• Operates at a low pressure	• Requires larger footprint to produce similar quantity and quality of water if multiple staging is used
• Process is much quieter because high pressure pumps are not required	• Electrical safety requirements
• Antiscalant is not required	• Less experience for wastewater demineralization in the U.S.
• Membrane life expectancy is longer because foulants are removed continuously during the reversal process	• Not as effective at removing microorganisms and many anthropogenic organic contaminants
• Requires less maintenance than RO due to reversal process	
Reverse osmosis	
• RO membranes provide a barrier to microorganisms and many anthropogenic organic contaminants (for the treated portion of the water produced)	• Requires high pressure to achieve high salt rejection
• More demonstrated experience for wastewater demineralization	• Requires pretreatment processes to minimize scaling and fouling
• RO membranes can remove more than 90 percent of TDS	• Requires chemical addition for MF & RO fouling control
• Source water blending will reduce size of systems	• More routine maintenance may be required to maintain performance
• Flexibility to provide higher quality water, if desired	

[a] Adapted from Adham et al. (2004).

11–9 ADSORPTION

In wastewater treatment adsorption is used for the removal of substances that are in solution by accumulation of those substances on a solid phase. Adsorption is considered to be a mass transfer operation as a constituent is transferred from a liquid phase to a solid phase (see Table 11–2). The *adsorbate* is the substance that is being removed from the liquid phase at the interface. The *adsorbent* is the solid, liquid, or gas phase onto which the adsorbate accumulates. Although adsorption is used at the air-liquid interface in the flotation process (see Sec. 5–8), only the case of adsorption at the liquid-solid interface is considered in this section. Activated carbon is the primary adsorbent used in adsorption processes. The basic concepts of adsorption are presented in this section along with elements of design and limitations of the adsorption process.

Applications for Adsorption

Adsorption treatment of wastewater is usually thought of as a polishing process for water that has already received normal biological treatment. Adsorption has been used for the removal of refractory organic constituents; residual inorganic constituents such as nitrogen, sulfides, and heavy metals; and odor compounds from wastewater. Under optimum conditions, it appears that adsorption can be used to reduce the effluent COD to less than 10 mg/L. In water reclamation applications adsorption is used for (1) the continuous removal of organics and (2) as a barrier against the breakthrough of organics from other unit processes. In some cases, adsorption is used for the control of precursors that may form toxic compounds during disinfection.

Representative compounds that are readily and poorly adsorbed onto activated carbon are listed in Table 11–40. As shown in Table 11–40, activated carbon is known to have a low adsorption affinity for low molecular weight polar organic compounds. If biological activity is low in the carbon contactor or in other biological unit processes, low molecular weight polar organic compounds may be difficult to remove with activated carbon.

Types of Adsorbents

Treatment with adsorbent materials involves either (1) passing a liquid to be treated through a bed of adsorbent material held in a reactor/contactor (either fixed or fluidized) or (2) blending the adsorbent material into a unit process followed by sedimentation or filtration for removal of the spent adsorbent. The principal types of adsorbents include activated carbon, granular ferric hydroxide (GFH), and activated alumina. Carbon-based adsorbents are used most commonly for wastewater adsorption because of their relatively low cost. Other adsorbents that may prove to be effective with further research include manganese greensand, manganese dioxide, hydrous iron oxide particles, and iron oxide coated sand. Regardless of the adsorbent selected for a particular application, pilot testing will be necessary for determination of process performance and design parameters. The characteristics of materials used for adsorption are summarized in Table 11–41.

Activated Carbon. Activated carbon is derived by subjecting an organic base material, such as wood, coal, almond, coconut, or walnut hulls to a pyrolysis process followed with activation by exposure to oxidizing gases such as steam and CO_2 at high temperatures. The resulting carbon structure is porous, as illustrated on Fig. 11–50, on page 1226, with a large internal surface area. The resulting pore sizes are defined as follows:

Macropores > 500 nm
Mesopores > 20 nm and < 500 nm
Micropores < 20 nm

Table 11–40

Readily and poorly adsorbed organics on activated carbon[a]

Readily adsorbed organics	Poorly adsorbed organics
Aromatic solvents	Low-molecular weight ketones, acids, and aldehydes
Benzene	Sugars and starches
Toluene	Very high molecular weight or colloidal organics
Nitrobenzenes	Low-molecular weight aliphatics
Chlorinated aromatics	
PCBs	
Chlorophenols	
Polynuclear aromatics	
Acenaphthene	
Benzopyrenes	
Pesticides and herbicides	
DDT	
Aldrin	
Chlordane	
Atrazine	
Chlorinated non-aromatics	
Carbon tetrachloride	
Chloroalkyl ethers	
Trichloroethene	
Chloroform	
Bromoform	
High-molecular weight hydrocarbons	
Dyes	
Gasoline	
Amines	
Humics	

[a] From Froelich (1978).

The surface properties, pore size distribution, and regeneration characteristics that result are a function of both the initial material used and the preparation procedure, therefore many variations are possible. The two size classifications of activated carbon are granular activated carbon (GAC), which has a diameter greater than 0.1 mm (~140 sieve) and is used in pressure or gravity filtration, and powdered activated carbon (PAC), which typically has a diameter of less than 0.074 mm (200 sieve) and is added directly to the activated sludge process or solids contact processes.

Granular Ferric Hydroxide. Granular ferric hydroxide (GFH) is manufactured from a ferric chloride solution by neutralization and precipitation with sodium hydroxide. The adsorption capacity of GFH depends on water quality parameters, including pH, temperature, and other constituents in the water. Constituents that have been removed using GFH include arsenic, chromium, selenium, copper, and other metals. The process performance is reduced by suspended solids and precipitated iron and manganese, and by

Table 11–41

Comparison of various adsorbent materials[a]

| Parameter | Unit | Activated carbon | | Activated alumina | Granular ferric hydroxide |
		Granular (GAC)	Powdered (PAC)		
Total surface area	m^2/g	700–1300	800–1800	280–380	250–300
Bulk density	kg/m^3	400–500	360–740	600–800	1200–1300
Particle density, wetted in water	kg/L	1.4–1.5	1.3–1.4	3.97	1.59
Particle size range	μm	100–2400	5–50	290–500	150–2000
Effective size	mm	0.6–0.9	na		
Uniformity coefficient	UC	≤ 1.9	na		
Mean pore radius	Â	16–30	20–40		
Iodine number		600–1100	800–1200		
Abrasion number	minimum	75–85	70–80		
Ash	%	≤ 10	≤ 6		
Moisture as packed	%	2–4	2–4		

[a] Specific values will depend on the source material used for the production of activated carbon.

constituents that compete for adsorption sites, including organic matter and other ions (e.g., phosphate, silicate, sulfate). While GFH adsorbents can be effective from a performance standpoint for removal of specific constituents (e.g., arsenic), the cost associated with GFH the GFH process is often prohibitive for large systems. The adsorption capacity of GFH media is reduced significantly following regeneration; thus after reaching capacity, GFH adsorbents are typically disposed of in a landfill and replaced with new media. However, because GFH is not regenerated, the costs associated with management of the waste regenerant can be avoided, making the process viable in some situations, especially where the waste regenerant must be handled as a hazardous waste.

Figure 11–50

Definition sketch for the adsorption of an organic constituent onto an activated carbon particle.

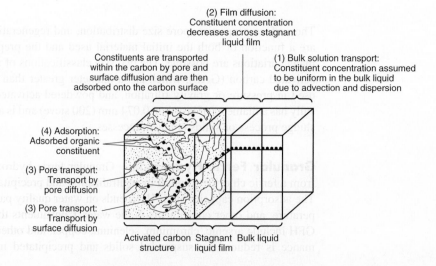

Activated Alumina. Activated alumina is derived from a naturally occurring mineral processed from bauxite that has been treated to remove molecules of water from its crystalline structure. Activated alumina is used in drinking water treatment for the removal of arsenic and fluoride (Clifford, 1999) and may have application in water reclamation for specific constituents. Activated alumina can be regenerated with a strong-base followed by a strong-acid. The regeneration of activated alumina and subsequent waste management issues result in significant operation and maintenance costs. As mentioned for GFH, pH (best performance at pH of 5.5 to 6), temperature, and competing constituents will affect the performance of activated alumina adsorption. The use of powdered activated alumina coupled with membranes (microfiltration and ultrafiltration) may also be a promising treatment process.

Fundamentals of Adsorption Processes

The adsorption process, as illustrated on Fig. 11–50, takes place in four, more or less definable steps: (1) bulk solution transport, (2) film diffusion transport, (3) pore and surface transport, and (4) adsorption (or sorption). The adsorption step involves the attachment of the material to be adsorbed to the adsorbent at an available adsorption site (Snoeyink and Summers, 1999). Additional details on the physical and chemical forces involved in the adsorption process may be found in Crittenden et al. (2012). Adsorption can occur on the outer surface of the adsorbent and in the macropores, mesopores, micropores, and submicropores, but the surface area of the macro and mesopores is small compared with the surface area of the micropores and submicropores and the amount of material adsorbed there is usually considered negligible.

Because the adsorption process occurs in a series of steps, the slowest step in the series is identified as the rate-limiting step. When the rate of adsorption equals the rate of desorption, equilibrium has been achieved, and the capacity of the adsorbent has been reached. The theoretical adsorption capacity for a given adsorbent for a particular contaminant can be determined by developing adsorption isotherms, as described below. Because activated carbon is the most common adsorbent used in advanced wastewater treatment applications, the focus of the following discussion is on activated carbon.

Development of Adsorption Isotherms

The quantity of adsorbate that can be taken up by an adsorbent is a function of both the characteristics and concentration of adsorbate and the temperature. The characteristics of the adsorbate that are of importance include: solubility, molecular structure, molecular weight, polarity, and hydrocarbon saturation. Generally, the amount of material adsorbed is determined as a function of the concentration at a constant temperature, and the resulting function is called an adsorption isotherm. Adsorption isotherms are developed by exposing a given amount of absorbate in a fixed volume of liquid to varying amounts of activated carbon. Typically, more than ten containers are used, and the minimum time allowed for the samples to equilibrate where powdered activated carbon is used is seven days. If activated carbon is used, it is usually in the powdered form (as opposed to granular) to minimize adsorption times.

Mass Balance. If a mass balance is performed for a batch reactor into which a quantity of powdered activated carbon has been added (see Fig. 11–51), the resulting expression at equilibrium at the completion of the mass transfer process is given by

    1. General word statement:

| Amount of reactant adsorbed within the system boundary | = | initial amount of reactant within the system boundary | − | final amount of reactant within the system boundary | (11–51) |

Figure 11–51

Definition sketch for mass balance of carbon adsorption.

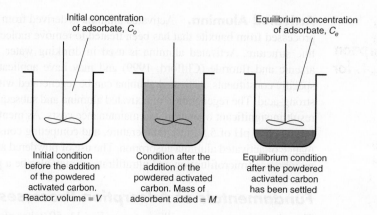

Initial concentration of adsorbate, C_o

Equilibrium concentration of adsorbate, C_e

Initial condition before the addition of the powdered activated carbon. Reactor volume = V

Condition after the addition of the powdered activated carbon. Mass of adsorbent added = M

Equilibrium condition after the powdered activated carbon has been settled

2. Simplified word statement:

$$
\begin{array}{c} \text{Amount} \\ \text{adsorbed} \end{array} = \begin{array}{c} \text{initial amount} \\ \text{of adsorbate} \\ \text{present} \end{array} - \begin{array}{c} \text{final amount} \\ \text{of adsorbate} \\ \text{present} \end{array} \tag{11–52}
$$

3. Symbolic representation at equilibrium (refer to Fig. 11–51):

$$
q_e M = V C_o - V C_e \tag{11–53}
$$

where q_e = adsorbent phase concentration after equilibrium, mg adsorbate/g adsorbent

M = mass of adsorbent, g

V = volume of liquid in the reactor, L

C_o = initial solution concentration of adsorbate, mg/L

C_e = final solution equilibrium concentration of adsorbate after adsorption has occurred, mg/L

Equation (11–53) can be written as follows:

$$
q_e = -\frac{V}{M}(C_e - C_o) \tag{11–54}
$$

The adsorbent phase concentration data computed using Eq. (11–54) are then used to develop adsorption isotherms as described below.

Freundlich Isotherm. Equations used to describe the experimental isotherm data were developed by Freundlich, Langmuir, and Brunauer, Emmet, and Teller (BET isotherm) (Shaw, 1966). Of the three, the Freundlich isotherm is used most commonly to describe the adsorption characteristics of the activated carbon used in water and wastewater treatment. Derived empirically in 1912, the Freundlich isotherm is defined as follows:

$$
\frac{x}{m} = K_f C_e^{1/n} \tag{11–55}
$$

where x/m = mass of adsorbate adsorbed per unit mass of adsorbent, mg adsorbate/g activated carbon

K_f = Freundlich capacity factor, (mg absorbate/g activated carbon) × (L water/mg adsorbate)$^{1/n}$ = (mg/g)(L/mg)$^{1/n}$

C_e = equilibrium concentration of adsorbate in solution after adsorption, mg/L

$1/n$ = Freundlich intensity parameter

Table 11–42

Freundlich adsorption isotherm constants for selected organic compounds[a,b]

Compound	pH	$K(mg/g)(L/mg)^{1/n}$	$1/n$
Benzene	5.3	1.0	1.6–2.9
Bromoform	5.3	19.6	0.52
Carbon tetrachloride	5.3	11	0.83
Chlorobenzene	7.4	91	0.99
Chloroethane	5.3	0.59	0.95
Chloroform	5.3	2.6	0.73
DDT	5.3	322	0.50
Dibromochloromethane	5.3	4.8	0.34
Dichlorobromomethane	5.3	7.9	0.61
1, 2–Dichloroethane	5.3	3.6	0.83
Ethylbenzene	7.3	53	0.79
Heptachlor	5.3	1,220	0.95
Hexachloroethane	5.3	96.5	0.38
Methylene chloride	5.3	1.3	1.16
N-Dimethylnitrosamine	na	6.8×10^{-5}	6.60
N-Nitrosodi-n-propylamine	na	24	0.26
N-Nitrosodiphenylamine	3–9	220	0.37
PCB	5.3	14,100	1.03
PCB 1221	5.3	242	0.70
PCB 1232	5.3	630	0.73
Phenol	3–9	21	0.54
Tetrachloroethylene	5.3	51	0.56
Toluene	5.3	26.1	0.44
1, 1, 1–Trichloroethane	5.3	2–2.48	0.34
Trichloroethylene	5.3	28	0.62

[a] Adapted from Dobbs and Cohen (1980) and LaGrega et al. (2001).

[b] The adsorption isotherm constants reported in this table are meant to be illustrative of the wide range of values that will be encountered for various organic compounds. It is important to note that the characteristics of the activated carbon used as well as the analytical technique used for the analysis of the residual concentrations of the individual compounds will have a significant effect on the coefficient values obtained for specific organic compounds.

The constants in the Freundlich isotherm can be determined by plotting $\log (x/m)$ versus $\log C_e$ and making use of the linear form of Eq. (11–55) rewritten as

$$\log\left(\frac{x}{m}\right) = \log K_f + \frac{1}{n}\log C_e \qquad (11\text{–}56)$$

The U.S. EPA (1980) has developed adsorption isotherms for a variety of toxic organic compounds, some of which are presented in Table 11–42. As shown in Table 11–42, the variation in the Freundlich capacity factor for the various compounds is extremely wide (e.g., 14,100 for PCB to 6.8×10^{-5} for N-Dimethylnitrosamine). Because of the wide variation, the Freundlich capacity factor must be determined for each new compound. Application of the Freundlich adsorption isotherm is illustrated in Example 11–10.

EXAMPLE 11–10   **Activated Carbon Required to Treat a Wastewater**   As a result of effluent chlorination, the amount of chloroform formed was found to be 0.12 mg/L. How much powdered activated carbon will be required to treat an effluent flowrate of 4000 m³/d to reduce the chloroform concentration to 0.05 mg/L? The Freundlich adsorption isotherm coefficients for chloroform are: $K_f = 2.6$ (mg/g)(L/mg)$^{1/n}$ and $1/n = 0.73$.

Solution

1. Combine Eqs. (11–54) and (11–55) to obtain an expression for V/M as follows:

$$q_e = \frac{V}{M}(C_e - C_o)$$

$$q_e = \frac{x}{m}K_f C_e^{1/n}$$

$$-\frac{V}{M} = \frac{K_f C_e^{1/n}}{(C_e - C_o)}$$

2. Substitute the isotherm coefficients and solve for M/V:

$$-\frac{V}{M} = \frac{K_f C_e^{1/n}}{(C_e - C_o)} = \frac{2.6(0.05)^{0.73}}{0.05 - 0.12} = -4.17 \text{ L/g}$$

$M/V = 1/4.17 = 0.24$ g/L

3. Determine the amount of carbon required to treat 4000 m³/d.

$$\text{PAC required} = \frac{(0.24 \text{ g/L})(4000 \text{ m}^3/\text{d})(10^3 \text{ L/1 m}^3)}{(10^3 \text{ g/1 kg})} = 960 \text{ kg/d}$$

Comment   Due to the cost and the amount of PAC required to treat the effluent to reduce the residual chloroform to 0.05 mg/L, carbon adsorption is a poor choice for the removal of residual chloroform.

Langmuir Isotherm.   Derived from rational considerations, the Langmuir adsorption isotherm is defined as:

$$\frac{x}{m} = \frac{abC_e}{1 + bC_e} \tag{11–57}$$

where x/m = mass of adsorbate adsorbed per unit mass of adsorbent, mg adsorbate/ g activated carbon
  a, b = empirical constants
  C_e = equilibrium concentration of adsorbate in solution after adsorption, mg/L

The Langmuir adsorption isotherm was developed by assuming (1) a fixed number of accessible sites are available on the adsorbent surface, all of which have the same energy, and

(2) adsorption is reversible. Equilibrium is reached when the rate of adsorption of molecules onto the surface is the same as the rate of desorption of molecules from the surface. The rate at which adsorption proceeds is proportional to the driving force, which is the difference between the amount adsorbed at a particular concentration and the amount that can be adsorbed at that concentration. At the equilibrium concentration, this difference is zero.

Correspondence of experimental data to the Langmuir equation does not mean that the stated assumptions are valid for the particular system being studied because deviations from the assumptions can have a canceling effect. The constants in the Langmuir isotherm can be determined by plotting $C_e/(x/m)$ versus C_e and making use of the linear form of Eq. (11–57) rewritten as:

$$\frac{C_e}{(x/m)} = \frac{1}{ab} + \frac{1}{a}C_e \tag{11-58}$$

For the case where the Langmuir adsorption isotherm best represents a set of experimental isotherm data, a plot of $C_e/(x/m)$ vs. C_e will be linear, with a slope of $1/a$ and a y-intercept of $1/(ab)$. Application of the Langmuir adsorption isotherm is illustrated in Example 11–11.

EXAMPLE 11-11   **Analysis of Activated Carbon Adsorption Data**   Determine which isotherm equation (i.e., Freundlich and Langmuir) best fits the isotherm coefficients for the following GAC adsorption test data. Also determine the corresponding coefficients for the isotherm equation. The liquid volume used in the batch adsorption tests was 1 L. The initial concentration of the adsorbate in solution was 3.37 mg/L. Equilibrium was obtained after 7 d.

Mass of GAC, m, g	Equilibrium concentration of adsorbate in solution, C_e, mg/L
0.0	3.37
0.001	3.27
0.010	2.77
0.100	1.86
0.500	1.33

Solution

1. Derive the values needed to plot the Freundlich and Langmuir adsorption isotherms using the batch adsorption test data.

Adsorbate concentration, mg/L					
C_o	C_e	$C_o - C_e$	m, g	x/m,[a] mg/g	$C_e/(x/m)$
3.37	3.37	0.00	0.000	—	—
3.37	3.27	0.10	0.001	100	0.0327
3.37	2.77	0.60	0.010	60	0.0462
3.37	1.86	1.51	0.100	15.1	0.1232
3.37	1.33	2.04	0.500	4.08	0.3260

[a] $\dfrac{x}{m} = \dfrac{(C_o - C_e)V}{m}$

2. Plot the Freundlich and Langmuir adsorption isotherms using the data developed in Step 1 and determine which isotherm best fits the data.

   a. The required plots are given below.

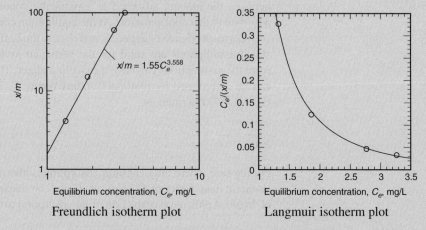

Freundlich isotherm plot   Langmuir isotherm plot

   b. From the above plots, the experimental data are best represented by the Freundlich isotherm. Because the plot for the Langmuir isotherm is curvilinear, use of the Langmuir adsorption isotherm is inappropriate.

3. Determine the Freundlich adsorption isotherm coefficients.

   a. When x/m versus C_e is plotted on log-log paper, the intercept on the x/m axis when $C_e = 1.0$ is the value of K_f and the slope of the line is equal to $1/n$. Thus, $x/m = 1.55$, and $K_f = 1.55$

   b. When $x/m = 1.0$, $C_e = 0.89$, and $1/n = 3.6$

   c. The form of the resulting isotherm is $\dfrac{x}{m} = 1.55 C_e^{3.6}$

   d. The Freundlich adsorption isotherm equation may also be determined using a power-type best fit through the data.

Adsorption of Mixtures

In the application of adsorption in water reclamation, mixtures of organic compounds in reclaimed water are always encountered. Typically, there is a depression of the adsorptive capacity of any individual compound in a solution of many compounds, but the total adsorptive capacity of the adsorbent may be larger than the adsorptive capacity with a single compound. The amount of inhibition due to competing compounds is related to the size of the molecules being adsorbed, their adsorptive affinities, and their relative concentrations. It is important to note that adsorption isotherms can be determined for a heterogeneous mixture of compounds including total organic carbon (TOC), dissolved organic carbon (DOC), chemical oxygen demand (COD), dissolved organic halogen (DOH), UV absorbance, and fluorescence (Snoeyink and Summers, 1999). The adsorption from mixtures is considered further in Crittenden et al. (1985, 1987a, 1987b, 1987c,) and Sontheimer and Crittenden (1988).

Adsorption Capacity

The adsorptive capacity of a given adsorbent is estimated from isotherm data as follows. If isotherm data are plotted, the resulting isotherm will be as shown on step 2 of Example 11–11. As shown on Fig. 11–52, the adsorptive capacity of the carbon can be estimated by

Figure 11–52

Plot of Freundlich isotherm used for determination of breakthrough adsorption capacity.

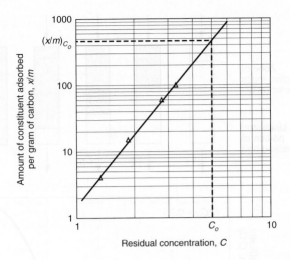

extending a vertical line from the point on the horizontal axis corresponding to the initial concentration C_o, and extrapolating the isotherm to intersect this line. The value at the point of intersection $[(x/m)_{C_o}]$ can be read from the vertical axis. The value represents the amount of constituent adsorbed per unit weight of carbon when the carbon is at equilibrium with the initial concentration of constituent, C_o. The equilibrium condition generally exists in the upper section of a carbon bed during column treatment, and it, therefore, represents the ultimate capacity of the carbon for a particular reclaimed water. The value of the breakthrough adsorption capacity $(x/m)_b$ can be determined using the small-scale column test described later in this section. Typically, breakthrough is said to have occurred when the effluent concentration reaches 5 percent of the influent value. Exhaustion of the adsorption bed is assumed to have occurred when the effluent concentration is equal to 95 percent of the influent concentration. A number of equations have been developed to describe the breakthrough curve including those by Bohart and Adams (1920) and Crittenden et al. (1987a).

Mass Transfer Zone.   The area of the GAC bed in which sorption is occurring is called the mass transfer zone (MTZ), as shown on Fig. 11–53. After the water containing the constituent to be removed passes through a region of the bed whose depth is equal to the MTZ, the concentration of the contaminant in the water will have been reduced to its minimum value. No further adsorption will occur within the bed below the MTZ. As the top layers of carbon granules become saturated with organic material, the MTZ will move down in the bed until breakthrough occurs. The volume of a given water processed until breakthrough and exhaustion is designated as V_{BT} and V_E, respectively, as shown on Fig. 11–53. The length of the MTZ is typically a function of the hydraulic loading rate applied to the column and the characteristics of the activated carbon. In the extreme, if the loading rate is too great the length of the MTZ will be larger than the GAC bed depth, and the adsorbable constituents will not be removed completely by the carbon. At complete exhaustion, the effluent concentration is equal to the influent concentration.

Breakthrough Curve.   In addition to the applied hydraulic loading rate, the shape of the breakthrough curve will also depend on whether the applied liquid contains nonadsorbable and biodegradable constituents. The impact of the presence of nonadsorbable and biodegradable organic constituents on the shape of the breakthrough curve is illustrated on

Figure 11–53

Typical breakthrough curve for activated carbon showing movement of mass transfer zone (MTZ) with throughput volume.

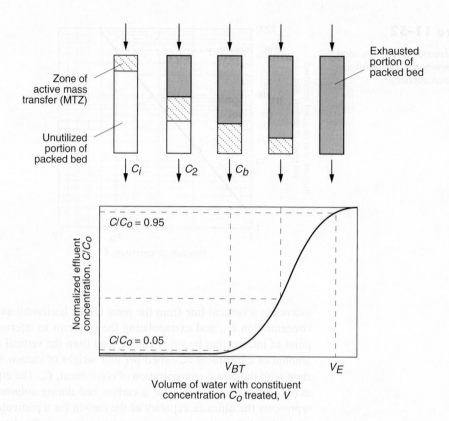

Fig. 11–54. As shown on Fig. 11–54, if the liquid contains nonadsorbable constituents, the nonadsorbable constituents will appear in the effluent as soon as the carbon column is put into operation (ignoring the short period of time for hydraulic conductivity). If adsorbable and biodegradable constituents are present in the applied liquid, the breakthrough curve will not reach a C/C_o value of 1.0 but will be depressed, and the observed C/C_o value will depend on the biodegradability of the influent constituents because biological activity continues even though the adsorption capacity has been utilized. If the liquid contains nonadsorbable and biodegradable constituents, the observed breakthrough curve will not

Figure 11–54

Impact of the presence of adsorbable, nonadsorbable, and biodegradable organic constituents on the shape of the activated carbon breakthrough curve. (Adapted from Snoeyink and Summers, 1999.)

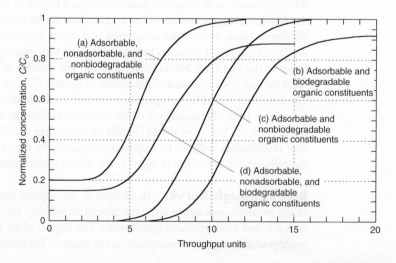

Figure 11–55

Activated carbon contactor configurations: (a) series and (b) parallel operation.

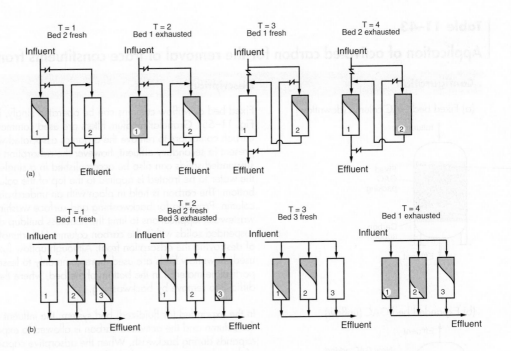

start at zero and will not terminate at a value of 1.0 (Snoeyink and Summers, 1999). The above effects are observed commonly in wastewater adsorption applications, especially with respect to the removal of COD.

In practice, the only way to use the capacity at the bottom portion of the carbon adsorption column is to have two or more columns in series and switch them as they are exhausted (lead-lag), or to use multiple columns in parallel so that breakthrough in one column does not affect the combined effluent quality. This mode of operation is referred to as the carousel technique, where multiple columns can be rotated so that only one column reaches exhaustion at a given time. The arrangement of adsorption columns in series and parallel configurations is shown on Fig. 11–55(a) and (b), respectively. A minimum of two parallel or series carbon contactors is recommended for design. Multiple units permit one or more units to remain in operation while one unit is taken out of service for removal and regeneration of spent carbon, or for maintenance. The optimum flowrate and bed depth, as well as the operating capacity of the carbon, must be established to determine the dimensions and the number of columns necessary for continuous treatment. These parameters can be determined from dynamic column tests, as discussed below.

Adsorption Contactors. Several types of activated carbon contactors are used for trace constituent removal, including fixed and expanded beds, addition of PAC to the activated sludge process, and separate mixed systems with subsequent carbon separation, as summarized in Table 11–43. A typical pressurized, down-flow carbon contactor is shown on Fig. 11–56. The sizing of carbon contactors is based on a number of factors, as summarized in Table 11–44 for a downflow packed bed contactor. For the case where the mass transfer rate is fast and the mass transfer zone is a sharp wave front, a steady-state mass balance around a fixed bed carbon contactor may be written as:

Accumulation = inflow − outflow − amount adsorbed

Table 11–43

Application of activated carbon for the removal of trace constituents from wastewater

Configuration	Description
(a) Fixed bed GAC column (downflow) 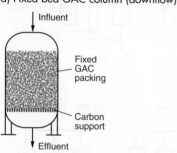	Fixed-bed downflow columns can be operated singly, in series, or in parallel (see Fig. 11–55). Granular medium filters are used commonly upstream of the activated carbon contactors to remove the organics associated with the suspended solids present in secondary effluent, however the adsorption of organics and filtration of suspended solids can also be accomplished in a single step. In the downflow design, the water to be treated is applied to the top of the column and withdrawn at the bottom. The carbon is held in place with an underdrain system at the bottom of the column. Provision for backwashing and surface washing is often provided in wastewater applications to limit the headloss buildup due to the removal of particulate suspended solids within the carbon column. Unfortunately, backwashing has the effect of destroying the adsorption front. Although upflow fixed-bed reactors have been used, downflow beds are used more commonly to lessen the chance of accumulating particulate material in the bottom of the bed, where the particulate material would be difficult to remove by backwashing.
(b) Expanded bed GAC (upflow) 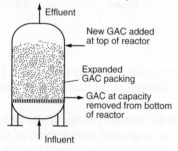	In the expanded (or fluidized) bed system, the influent is introduced at the bottom of the column and the activated carbon is allowed to expand, much as a filter bed expands during backwash. When the adsorptive capacity of the carbon at the bottom of the column is exhausted, the bottom portion of carbon is removed, and an equivalent amount of regenerated or virgin carbon is added to the top of the column. In such a system, headloss does not build up with time after the operating point has been reached. In general, expanded bed upflow contactors may have more carbon fines in the effluent than downflow contactors because bed expansion leads to the creation of fines as the carbon particles collide and abrade, and allows the fines to escape through passageways created by the expanded bed. While not used commonly, continuous backwash moving-bed and pulsed-bed carbon contactors have been used (see Table 8–4 for filter configurations).
(c) Activated Sludge with PAC addition 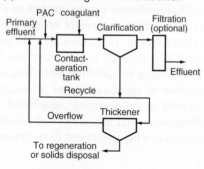	The use of powdered activated carbon with the activated sludge process, where activated carbon is added directly to the aeration tank, results in simultaneous biological oxidation and physical adsorption. A feature of this process is that it can be integrated into existing activated sludge systems at nominal capital cost. The addition of powdered activated carbon has several process advantages, including: (1) system stability during shock loads, (2) reduction of refractory priority pollutants, (3) color and ammonia removal, and (4) improved sludge settleability. In some industrial waste applications where nitrification is inhibited by toxic organics, the application of powdered activated carbon may reduce or limit this inhibition.
(d) Mixed PAC contactor with gravity separation 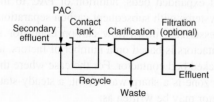	Powdered activated carbon can be applied to the effluent from biological treatment processes in a separate contacting basin. The contactor can operate in a batch or continuous flow mode. In the batch mode, after a specified amount of time for contact, the carbon is allowed to settle to the bottom of the tank, and the treated water is then removed from the tank. The continuous flow operation consists of a basin divided for contacting and settling. The settled carbon may be recycled to the contact tank. Because carbon is very fine, a coagulant, such as a polyelectrolyte, may be needed to aid in the removal of the carbon particles, or filtration through rapid sand filters may be required. In some treatment processes, PAC is used in conjunction with chemicals used for the precipitation of specific constituents.

(continued)

Table 11–43 (Continued)

Configuration	**Description**
(e) Mixed PAC contactor with membrane separation 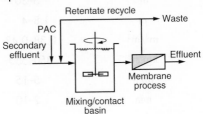	The removal of trace constituents in a complete mix or plug flow contactor may be combined with separation by micro or ultrafiltration membranes. The PAC is added to the secondary effluent by continuous or pulse addition, followed by concentration of the PAC on the membrane. When the headloss across the membrane reaches a given value, a backwash cycle is initiated. The backwash containing the PAC retentate may be wasted or recycled to the contact basin. A number of full-scale plants have used this process (Snoeyink et al., 2000, Anselme et al., 1997).

$$0 = QC_o t - QC_e t - m_{GAC} q_e \qquad (11\text{--}59)$$

where Q = volumetric flowrate, L/h

C_o = initial concentration of adsorbate, mg/L

t = time, h

C_e = final equilibrium concentration of adsorbate, mg/L

m_{GAC} = mass of adsorbent, g

q_e = adsorbent phase concentration after equilibrium, mg adsorbate/g adsorbent

From Eq. (11–59), the adsorbent usage rate is defined as

$$\frac{m_{GAC}}{Qt} = \frac{C_o - C_e}{q_e} \qquad (11\text{--}60)$$

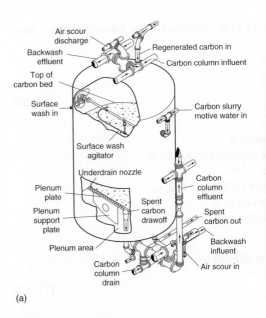

(a)

(b)

Figure 11–56

Activated carbon contactors: (a) illustration of typical pressure vessel contactor and (b) view of typical granular activated carbon contactors operated in parallel, used for the treatment of filtered secondary effluent.

Parameter	Symbol	Unit	Value
Volumetric flowrate	V	m³/h	50–400
Bed volume	V_b	m³	10–50
Cross-sectional area	A_b	m²	5–30
Carbon depth	D	m	1.8–4
Void fraction	α	m³/m³	0.38–0.42
GAC density	ρ	kg/m³	350–550
Approach velocity	v_f	m/h	5–15
Effective contact time	t	min	2–10
Empty bed contact time	EBCT	min	5–30
Operation time	t	d	100–600
Throughput volume	V_L	m³	10–100
Specific throughput	V_{sp}	m³/kg	50–200
Bed volumes[b]	BV	m³/m³	2000–20,000

[a] Adapted from Sontheimer et al. (1988).

[b] Total volume of water processed expressed in terms of the reactor bed volume.

If it is assumed that the mass of the adsorbate in the pore space is small compared to the amount adsorbed, then the term $QC_e t$ in Eq. (11–60) can be neglected without serious error and the adsorbent usage rate is given by:

$$\frac{m_{\mathrm{GAC}}}{Qt} \approx \frac{C_o}{q_e} \tag{11–61}$$

To quantify the operational performance of GAC contactors, the following terms have been developed and are used commonly.

1. Empty bed contact time (EBCT)

$$EBCT = \frac{V_b}{Q} = \frac{A_b D}{v_f A_b} = \frac{D}{v_f} \tag{11–62}$$

where EBCT = empty bed contact time, h
V_b = volume of contactor occupied by GAC, m³
Q = volumetric flowrate, m³/h
A_b = cross-sectional area of GAC filter bed, m²
D = depth of GAC in contactor, m
v_f = linear approach velocity, m/h

2. Activated carbon density.
The density of the activated carbon is defined as

$$\rho_{\mathrm{GAC}} = \frac{m_{\mathrm{GAC}}}{V_b} \tag{11–63}$$

where ρ_{GAC} = density of granular activated carbon, g/L
m_{GAC} = mass of granular activated carbon, g
V_b = volume of contactor occupied by GAC, L

3. Specific throughput, expressed as m^3 of water treated per gram of carbon:

$$\text{Specific throughput, m}^3/\text{g} = \frac{Qt}{m_{GAC}} = \frac{V_b t}{\text{EBCT} \times m_{GAC}} \qquad (11\text{–}64)$$

Using Eq. (11–63), Eq. (11–64) can be written as

$$\text{Specific throughput} = \frac{V_b t}{\text{EBCT}(\rho_{GAC} \times V_b)} = \frac{t}{\text{EBCT} \times \rho_{GAC}} \qquad (11\text{–}65)$$

4. Carbon usage rate (CUR) expressed as gram of carbon per m^3 of water treated:

$$\text{CUR, g/m}^3 = \frac{m_{GAC}}{Qt} = \frac{1}{\text{Specific throughput}} \qquad (11\text{–}66)$$

5. Volume of water treated for a given EBCT, expressed in liters, L:

$$\text{Volume of water treated, m}^3 = \frac{\text{Mass of GAC for given EBCT}}{\text{GAC usage rate}} \qquad (11\text{–}67)$$

6. Bed life, expressed in days, d:

$$\text{Bed life, d} = \frac{\text{Volume of water treated for given EBCT}}{Q} \qquad (11\text{–}68)$$

The application of these terms is illustrated in Example 11–12.

EXAMPLE 11–12 Estimation of Activated Carbon Adsorption Breakthrough Time A fixed-bed activated carbon adsorber has a fast mass transfer rate and the mass transfer zone is essentially a sharp wave front. Assuming the following data apply, determine the carbon requirements to treat a flowrate of 1000 L/min, and the corresponding bed life.

1. Compound to be treated = Trichloroethylene (TCE)
2. Initial concentration, C_o = 1.0 mg/L
3. Final concentration C_e = 0.005 mg/L
4. GAC density = 450 g/L
5. Freundlich capacity factor, K_f = 28 $(\text{mg/g})(\text{L/mg})^{1/n}$ (see Table 11–42)
6. Freundlich intensity parameter, $1/n$ = 0.62 (see Table 11–42)
7. EBCT = 10 min

Ignore the effects of biological activity within the column.

Solution

1. Estimate the GAC usage rate for TCE. The GAC usage rate is estimated using Eq. (11–60) and Eq. (11–55).

$$\frac{m_{GAC}}{Qt} = \frac{C_o - C_e}{q_e} = \frac{C_o - C_e}{K_f C_o^{1/n}}$$

$$= \frac{(1.0\,\text{mg/L}) - (0.005\,\text{mg/L})}{28\,(\text{mg/g})(\text{L/mg})^{0.62}\,(1.0\,\text{mg/L})^{0.62}}$$

$$= 0.036\ \text{g GAC/L}$$

2. Determine the mass of carbon required for a 10 min EBCT.

 The mass of GAC in the bed = $V_b \rho_{GAC}$ = (EBCT)(Q)(ρ_{GAC})

 Carbon required = 10 min (1000 L/min) (450 g/L) = 4.5×10^6 g

3. Determine the volume of water treated using a 10 min EBCT.

$$\text{Volume of water treated} = \frac{\text{Mass of GAC for given EBCT}}{\text{GAC usage rate}}$$

$$\text{Volume of water treated} = \frac{4.5 \times 10^6 \, \text{g}}{(0.036 \, \text{g GAC/L})} = 1.26 \times 10^8 \, \text{L}$$

4. Determine the bed life.

$$\text{Bed life} = \frac{\text{Volume of water treated for given EBCT}}{Q}$$

$$\text{Bed life} = \frac{1.26 \times 10^8 \, \text{L}}{(1000 \, \text{L/min})(1440 \, \text{min/d})} = 87.5 \, \text{d}$$

Comment   In this example, the full capacity of the carbon in the contactor was utilized based on the assumption that two columns in series are used. If a single column is used, then a breakthough curve must be used to arrive at the bed life. The Freundlich isotherm parameters, K and $1/n$, are a function of the initial concentration, the actual carbon that is used, as well as the water quality (temperature, pH). Equilibrium isotherms are used to determine these parameters for the conditions of interest.

Small Scale Column Tests

Over the years, a number of small scale column tests have been developed to simulate the results obtained with full scale reactors. One of the early column tests was the high-pressure minicolumn (HPMC) technique developed by Rosene et al. (1980), and later modified by Bilello and Beaudet (1983). In the HPMC test procedure, a high-pressure liquid chromatography column loaded with activated carbon is used. Typically the HPMC test procedure is used to determine the capacity of activated carbon for the adsorption of volatile organic compounds. The principal advantage of the HPMC test procedure is that it allows for the rapid determination of the GAC adsorptive capacity under conditions similar to those encountered in the field.

 An alternative procedure known as the *rapid small-scale column test* (RSSCT) has been developed by Crittenden et al. (1986, 1987d, 1991). The test procedure allows for the scaling of data obtained from small columns (see Fig. 11–57) to predict the performance of pilot or full-scale carbon columns. In developing the procedure, mathematical models were used to define the relationships between the breakthrough curve for small and large columns. In adsorption columns, the mass transfer mechanisms that are responsible for the spreading of the mass transfer zone are (1) dispersion, (2) film diffusion, and (3) intraparticle diffusion. Two different design relationships were developed, one for constant diffusivity and one for proportional, or non-constant, diffusivity. In the constant diffusivity model, it is assumed that dispersion is negligible because the hydraulic loading rate is high in the RSSCT, and that mass transfer occurs as a result of film diffusion. Further, it is

Figure 11–57

Schematic of column used for rapid small scale column testing (RSSCT) to develop data for pilot or full scale carbon columns.

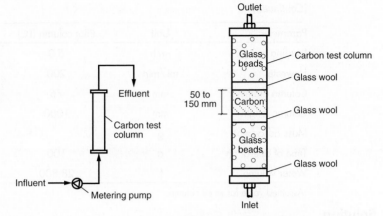

Column to particle diameter = 20:1 (or greater)
Column diameter = 20 to 40 mm
Column length = 300 mm

assumed that the intraparticle diffusivity is the same for both the small and large columns. In the proportional diffusivity model, it is assumed that dispersion is negligible because the hydraulic loading rate is high in the RSSCT, and that mass transfer occurs as a result of intraparticle diffusion. The relationships for the two cases can be generalized as follows:

$$\frac{\text{EBCT}_{SC}}{\text{EBCT}_{LC}} = \left(\frac{d_{SC}}{d_{LC}}\right)^{2-x} = \frac{t_{SC}}{t_{LC}} \tag{11–69}$$

$$\frac{v_{SC}}{v_{LC}} = \frac{d_{LC}}{d_{SC}} \tag{11–70}$$

where d_{SC} = diameter of particle in small-scale column, mm
$\quad d_{LC}$ = diameter of particle in large-scale column, mm
$\quad t_{SC}$ = time in small-scale column, min
$\quad t_{LC}$ = time in large-scale column, min
$\quad v_{SC}$ = superficial velocity in small-scale column, m/h
$\quad v_{LC}$ = superficial velocity in large-scale column, m/h

For constant and proportional diffusivity, the value of x in the exponent in Eq. (11–69) is 0 and 1, respectively. The application of the above equations is illustrated in Example 11–13.

EXAMPLE 11–13 **Comparison of Rapid Small-scale Column Test Parameters to Pilot Scale Parameters** Determine the corresponding parameters for a RSSCT based on the following data proposed for a pilot scale column. Assume that film diffusion is the controlling mechanism.

Parameter	Unit	Pilot column (LC)	RSSCT (SC)
Particle diameter	mm	0.5	0.1
Carbon density	g/L	450	450
EBCT	min	10	

(continued)

(*Continued*)

Parameter	Unit	Pilot column (LC)	RSSCT (SC)
Loading rate	m/h	5.0	
Flowrate	mL/min	200	
Column diameter	mm	75	10[a]
Column length	mm	1000	
Mass adsorbent	g		
Time of operation	d	100	
Water volume	L	28,800	

[a]Assumed value for small column.

Solution

1. Estimate the EBCT for the RSSCT.

$$EBCT_{SC} = EBCT_{LC}\left(\frac{d_{SC}}{d_{LC}}\right)^2$$

$$EBCT_{SC} = 10 \text{ min}\left(\frac{0.1}{0.5}\right)^2 = 0.4 \text{ min}$$

2. Estimate the loading rate for the RSSCT.

$$v_{SC} = v_{LC}\frac{d_{LC}}{d_{SC}}$$

$$v_{SC} = 5 \text{ m/h}\frac{0.5}{0.1} = 25 \text{ m/h}$$

3. Estimate flowrate for the RSSCT.

$$A = \frac{\pi}{4}d_{SC}^2 = \frac{\pi}{4}(10 \text{ mm})^2 = 78.5 \text{ mm}^2$$

$$Q_{SC} = (v_{SC})(A)$$

$$Q_{SC} = \frac{(25 \text{ m/h})(10^3 \text{ mm/1 m})(78.5 \text{ mm}^2)}{(60 \text{ min/h})(10^3 \text{ mm}^3/1 \text{ mL})} = 32.7 \text{ mL/min}$$

4. Estimate column length for the RSSCT.

$$L_{SC} = \frac{Q_{SC} \times EBCT_{SC}}{A} = \frac{(32,700 \text{ mm}^3/\text{min})(0.4 \text{ min})}{78.5 \text{ mm}^2} = 166.7 \text{ mm}$$

5. Estimate mass of adsorbent required for the RSSCT.

$$M_{SC} = EBCT_{LC}\left(\frac{d_{SC}}{d_{LC}}\right)^2 (Q_{SC})(\rho_{SC})$$

$$M_{SC} = 10 \text{ min}\left(\frac{0.1 \text{ mm}}{0.5 \text{ mm}}\right)^2 \left[\frac{(32.7 \text{ mL/min})(450 \text{ g/L})}{(10^3 \text{ mL/1 L})}\right] = 5.9 \text{ g}$$

6. Estimate time of operation for the RSSCT.

$$t_{SC} = t_{LC}\frac{EBCT_{SC}}{EBCT_{LC}}$$

$$t_{SC} = 100\left(\frac{0.4 \text{ min}}{10 \text{ min}}\right) = 4 \text{ d}$$

7. Estimate volume of water required for the RSSCT.

$$v_W = Q_{SC} \times t_{SC}$$

$$v_W = \frac{(32.7 \text{ mL/min})(4 \text{ d})(1440 \text{ min/d})}{(10^3 \text{ mL/1 L})} = 188.4 \text{ L}$$

8. Summarize the findings for the RSSCT.

Parameter	Unit	Pilot column	RSSCT
Particle radius	mm	0.5	0.1
Carbon density	g/L	450	450
EBCT	min	10	0.4
Loading rate	m/h	5.0	25.0
Flowrate	mL/min	200	32.7
Column diameter	mm	75	10[a]
Column length	mm	1000	166.7
Mass adsorbent	g		5.9
Time of operation	d	100	4
Water volume	L	28,800	188.4

[a]Assumed value for small column.

Comment The time savings in conducting the RSSCT versus the pilot column is apparent. Furthermore, many more tests can be conducted to test alternative configurations and carbon types. Often RSSCT's are performed in advance of piloting to narrow the list of the most appropriate carbon media for use in piloting.

Analysis of Powdered Activated Carbon Contactor

For a powdered activated carbon (PAC) application, the isotherm adsorption data can be used in conjunction with a materials mass balance analysis to obtain an approximate estimate of the amount of carbon that must be added as illustrated below. Here again, because of the many unknown factors involved, column and bench scale tests are recommended to develop the necessary design data. If a mass balance is written around the contactor (i.e., a batch reactor) after equilibrium has been reached, the resulting expression is given by Eq. (11–54), as derived previously. Estimation of the of powdered activated carbon (PAC) dose for adsorption is illustrated in Example 11–14.

EXAMPLE 11–14 **Estimation of Powdered Activated Carbon (PAC) Adsorption Dose and Cost** A treated wastewater with a flowrate of 1000 L/min is to be treated with PAC to reduce the concentration of residual organics measured as TOC from 5 to 1 mg/L. The Freundlich adsorption isotherm parameters were developed as discussed previously. Assuming the following data apply, determine the PAC requirements to treat the wastewater flow. If PAC costs $0.50/kg, estimate the annual cost for treatment, assuming the PAC will not be regenerated.

1. Compound = mixed organics
2. Initial concentration, C_o = 5.0 mg/L
3. Final concentration, C_e = 1.0 mg/L
4. GAC density = 450 g/L
5. Freundlich capacity factor, K_f = 150 $(mg/g)(L/mg)^{1/n}$
6. Freundlich intensity parameter, $1/n$ = 0.5

Solution

1. Estimate the PAC dose based on the isotherm data. The PAC dose can be estimated by writing Eq. (11–54) as follows:

$$\frac{m}{V} = \frac{(C_o - C_e)}{q_e} = \frac{(C_o - C_e)}{K_f C_e^{1/n}}$$

Substituting the given values in the expression yields:

$$\frac{m}{V} = \frac{(5 \text{ mg/L} - 1 \text{ mg/L})}{150(mg/g)(L/mg)^{0.5}(1.0 \text{ mg/L})^{0.5}} = 0.0267 \text{ g/L}$$

2. Estimate the annual cost for the PAC treatment.

Annual cost =

$$= \frac{(0.0267 \text{ g/L})(1000 \text{ L/min})(1440 \text{ min/d})(365 \text{ d/y})(\$0.50/kg)}{(10^3 \text{ g/1 kg})}$$

Annual cost = $7008/y

Comment For small wastewater flows, it is not usually cost effective to plan for carbon regeneration.

Activated Sludge Powdered Activated Carbon Treatment

Powdered activated carbon treatment (PACT), a proprietary process, is described in Table 11–43 along with other applications. The dosage of powdered activated carbon and the mixed liquor-powdered activated carbon suspended solids concentration are related to the SRT as follows:

$$X_p = \frac{X_i \text{ SRT}}{\tau} \tag{11–71}$$

where X_p = equilibrium powdered activated carbon-MLSS content, mg/L
X_i = powdered activated carbon dosage, mg/L
SRT = solids retention time, d
τ = hydraulic retention time, d

Carbon dosages typically range from 20 to 200 mg/L. With higher SRT values, the organic removal per unit of carbon is enhanced, thereby improving the process efficiency. Reasons cited for this phenomenon include (1) additional biodegradation due to decreased toxicity, (2) degradation of normally nondegradable substances due to increased exposure time to the biomass through adsorption on the carbon, and (3) replacement of low molecular weight compounds with high molecular weight compounds, resulting in improved adsorption efficiency and lower toxicity.

Carbon Regeneration

In many situations, the economical application of activated carbon depends on an efficient means of regenerating and reactivating the carbon after its adsorptive capacity has been reached. Regeneration is the term used to describe all of the processes that are used to recover the adsorptive capacity of the spent carbon, exclusive of reactivation. Typically, some of the adsorptive capacity of the carbon (about 4 to 10 percent) is lost in the regeneration process, while a loss of 2 to 5 percent is expected during the reactivation process, and a 4 to 8 percent loss of carbon is assumed due to attrition, abrasion, and mishandling. In general, regenerated activated carbon is not used in reclaimed water applications because of the potential for residual constituents, not removed in the regeneration process, to desorb and contaminate the reclaimed water. Additional details on carbon reactivation and regeneration may be found in Sontheimer and Crittenden (1988).

Adsorption Process Limitations

The adsorption process in water reuse applications is limited by (1) the logistics involved with transport of large volumes of adsorbent materials, (2) the area requirements for the carbon contactors, and (3) the production of waste adsorbent that can be difficult to regenerate and may need to be disposed of as hazardous waste due to the presence of toxic constituents. In particular, PAC contributes directly to the residuals solid loading and must be considered in terms of the impact on residuals handling. Further, the regeneration of some adsorbents is not feasible, resulting in potentially high media replacement costs. Process monitoring and control is essential, as the performance of carbon contactors will be affected by variations in pH, temperature, and flowrate.

11–10 GAS STRIPPING

Gas stripping involves the mass transfer of a gas from the liquid phase to the gas phase. The transfer is accomplished by contacting the liquid containing the gas that is to be stripped with a gas (usually air) which does not contain the gas initially. The removal of dissolved gases from wastewaters by gas (usually air) stripping has received considerable attention, especially for the removal of ammonia and odorous gases and volatile organic compounds (VOCs). Early work on the air stripping of ammonia from wastewater was conducted at Lake Tahoe, CA (Culp and Slechta, 1966; Slechta and Culp, 1967). The removal of VOCs by aeration is considered in Sec. 16–4 in Chap. 16.

The purpose of this section is to introduce the fundamental principles involved in gas stripping and to illustrate the general application of these principles. A design procedure is also presented. The material presented in this section is applicable to the removal of ammonia (NH_3), carbon dioxide (CO_2), oxygen (O_2), hydrogen sulfide (H_2S), and a variety of VOCs. The focus of the discussion in this section is on the analysis of facilities designed specifically for the removal of gaseous constituents as opposed to the removal of odorous gases (see Section 16–3, Chap. 16) and VOCs in aeration systems designed for the biological treatment of wastewater (see Sec. 16–4 in Chap. 16).

Analysis of Gas Stripping

Important factors that must be considered in the analysis of gas stripping include (1) the characteristics of the compound(s) to be stripped, (2) the type of contactor to be used and the required number of stages, (3) the materials mass balance analysis of the stripping tower, and (4) the required physical features and dimensions of the required stripping tower.

Characteristics of the Compound(s) to be Stripped.   As noted above, the removal of volatile dissolved compounds by stripping involves contacting the liquid with a gas that does not contain the compound initially. The compound that is to be stripped will come out of solution and enter the gas phase to satisfy the Henry's law equilibrium as discussed in Chap. 2. Compounds such as benzene, toluene, and vinyl chloride which have Henry's law constants greater than 500 atm are especially amenable to stripping. Compounds with Henry's law constants values greater than 0.1 atm are classified as volatile and are considered amenable to stripping. Compounds with Henry's law constants between 0.001 and 0.1 atm are classified as semi-volatile and are marginally amenable to stripping. Compounds with Henry's law constants less than 0.001 atm are essentially not amenable to stripping.

The air stripping of ammonia from wastewater requires that the ammonia be present as a gas. Ammonium ions in wastewater exist in equilibrium with gaseous ammonia, as given by Eq. (2–38):

$$NH_4^+ \rightleftharpoons NH_3 + H^+ \tag{2–38}$$

As the pH of the wastewater is increased above 7, the equilibrium is shifted to the right and the ammonium ion is converted to ammonia, which may be removed by gas stripping. The amount of lime required to raise the pH of wastewater to 11 as a function of the alkalinity is given on Fig. 6–12 in Chap. 6.

Methods Used to Contact Phases.   In practice, two methods are used to achieve contact between phases so that mass transfer can occur: (1) continuous contact and (2) staged contact. As shown on Fig. 11–58, three flow patterns are used in practice: (1) cocurrent, (2) countercurrent, and (3) cross-flow. In addition, the contact packing may be fixed or mobile (Crittenden, 1999). The most common flow pattern in mass transfer operations is the countercurrent mode, in which the liquid to be stripped is pumped to the top of the tower and sprayed over the packing surface. Air is introduced into the bottom of the tower (i.e., counter current) and either blown or sucked up through the packing material. The packing material is used to distribute the applied liquid in a thin film to enhance the stripping process. In the cross-flow stripper, not used commonly, air is introduced along the side. One of the most critical issues in the design and operation of stripping towers is maintaining uniform airflow across the packing surface. To achieve a more uniform

Figure 11–58

Typical water and air flow patterns for gas stripping towers: (a) countercurrent flow, (b) co-current flow, and (c) cross-flow.

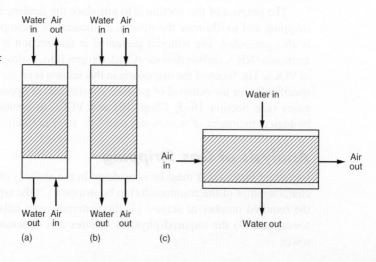

(a)   (b)   (c)

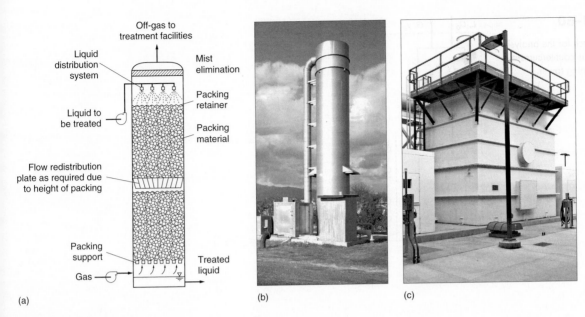

Figure 11–59

Typical examples of stripping towers: (a) schematic of packed bed stripping tower used for the removal of volatile gases from water, (b) view of stripping tower shown schematically in (a), and (c) typical stripping tower used for the removal of carbon dioxide from water following reverse osmosis treatment.

distribution of both air and water flow through the tower, packing is placed in individual stages on flow redistribution plates within the tower. A variety of packing materials are used in stripping towers including Raschig rings (cylinders), Berl saddles, or other proprietary plastic packings. Although a wide range of packing sizes is available, the most common size range is from 25 to 50 mm. A schematic and photograph of a typical gas stripping tower is shown on Fig. 11–59.

Mass Balance Analysis for a Continuous Stripping Tower. A steady-state materials balance for the lower portion of a countercurrent continuous stripping tower used for the removal of a dissolved gas from wastewater (see Fig. 11–60) is given by

1. General word statement:

$$\begin{array}{c}\text{Moles of solute} \\ \text{entering in} \\ \text{liquid stream}\end{array} + \begin{array}{c}\text{moles of solute} \\ \text{entering in} \\ \text{gas stream}\end{array} = \begin{array}{c}\text{moles of solute} \\ \text{leaving in} \\ \text{liquid stream}\end{array} + \begin{array}{c}\text{moles of solute} \\ \text{leaving in} \\ \text{gas stream}\end{array} \qquad (11\text{–}72)$$

2. Simplified word statement:

$$\text{inflow} = \text{outflow} \qquad (11\text{–}73)$$

3. Symbolic representation (refer to Fig. 11–61):

$$LC + Gy_o = LC_e + Gy \qquad (11\text{–}74)$$

where L = liquid flowrate, moles per unit time
C = concentration of solute in liquid at point within the tower, moles of solute per mole of liquid
G = gas flowrate, moles per unit time

Figure 11–60

Definition sketch for the analysis of a continuous countercurrent flow gas stripping tower.

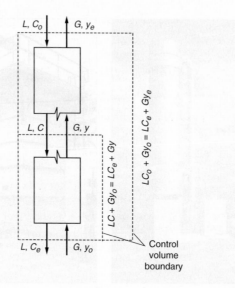

y_o = concentration of solute in gas entering the bottom of the tower, moles of solute per mole of solute-free gas

C_e = concentration of solute in liquid leaving the bottom of the tower, moles of solute per mole of liquid

y = concentration of solute at a point within the tower, moles of solute per mole of solute-free gas

Figure 11–61

Operating lines for various gas stripping conditions: (a) general case, (b) condition when $y_o = 0$, (c) condition when $y_o = 0$ and y_e is in equilibrium with C_o, the constituent concentration in the incoming water, and (d) condition when $y_o = 0$, $C_e = 0$, and y_e is in equilibrium with C_o.

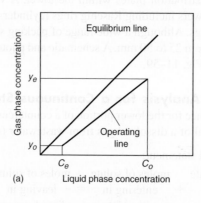

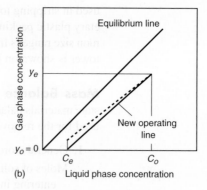

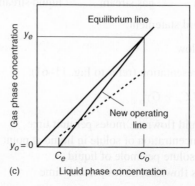

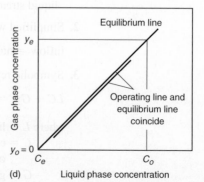

Figure 11–62

Equilibrium curves for ammonia in water as a function of temperature based on Henry's law.

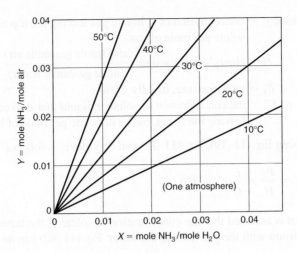

Combining terms, Eq. (11–74) can be written as:

$$(y_o - y) = L/G(C_e - C) \tag{11–75}$$

If the overall tower is considered, Eq. (11–74) can be written as:

$$LC_o + Gy_o = LC_e + Gy_e \tag{11–76}$$

Combining terms, Eq. (11–76) can be written as:

$$(y_o - y_e) = L/G(C_e - C_o) \tag{11–77}$$

where C_o = concentration of solute in liquid entering at the top of the tower, moles of solute per mole of liquid

y_e = concentration of solute in gas leaving the top of the tower, moles of solute per mole of gas

Because Eq. (11–77) is derived solely from a consideration of the equality of input and output, it holds regardless of the internal equilibria that may control the mass transfer. Equation (11–77) represents the equation of a straight line with slope L/G which passes through the point (C_o, y_e) and point (C_e, y_o). The line passed through these two points [see Fig. 11–61(a)] is known as the *operating line* and represents the conditions at any point within the column. The equilibrium line is based on Henry's law. For example, equilibrium lines defined by Henry's law for ammonia as a function of temperature are presented on Fig. 11–62. It should be noted that when a gas is being stripped from solution, the operating line will lie below the equilibrium line. If a gas is being absorbed into solution the operating line will lie above the equilibrium line.

If it is assumed that the air entering the bottom of the tower contains no solute (i.e, $y_o = 0$), then Eq. (11–77) can be written as

$$y_e = L/G(C_o - C_e) \tag{11–78}$$

The new operating line for the condition defined by Eq. (11–78) is shown on Fig. 11–61(b). Using Henry's law [see Eq. (2–46)], y_e is defined as follows

$$y_e = \frac{H}{P_T} C'_o \tag{11–79}$$

where y_e = concentration of solute in gas leaving the top of the tower, moles of solute per mole of gas

H = Henry's law constant, $\dfrac{\text{atm (mole gas/mole air)}}{\text{(mole gas/mole water)}}$

P_T = total pressure, usually 1.0 atm

C_o' = the concentration of solute in liquid that is in equilibrium with the gas leaving the tower, moles of solute per mole of liquid

Using Eq. (11–79), Eq. (11–78) can be written as follows:

$$C_o' = \frac{P_T}{H} \times \frac{L}{G}(C_o - C_e) \tag{11-80}$$

If it is assumed that the concentration of solute in the liquid entering the tower is in equilibrium with the gas leaving the tower, Eq. (11–80) can be written as:

$$\frac{G}{L} = \frac{P_T}{H} \times \frac{(C_o - C_e)}{C_o} \tag{11-81}$$

The operating line for the condition defined by Eq. (11–81) is shown on Fig. 11–61(c). The value of G/L (air to liquid ratio) defined by Eq. (11–81) represents the minimum amount of air that can be used for stripping for the given conditions (i.e, $y_o = 0$ and $y_e = HC_o/P_T$). In practice, from one and a half to three times the theoretical minimum air to liquid ratio is used to achieve effective stripping of most constituents. The application of this relationship is illustrated in Example 11–15.

If it is assumed further that the liquid leaving and the air entering the bottom of the tower contains no solute, then Eq. (11–81) can be written as

$$\frac{G}{L} = \frac{P_T \times C_o}{H \times C_o} = \frac{P_T}{H} \tag{11-82}$$

The value of G/L for this condition corresponds to the equilibrium line defined by Henry's law [see Fig. 11–61(d)], and represents the theoretical minimum amount of air that can be used for stripping for the given conditions (i.e., $y_o = 0$, $C_e = 0$, and $y_e = HC_o/P_T$). The range of air-to-liquid ratios for stripping ammonia from wastewater as a function of temperature is plotted on Fig. 11–63. The theoretical ratio is derived by assuming the process

Figure 11–63

Air requirements for ammonia stripping as function of temperature.

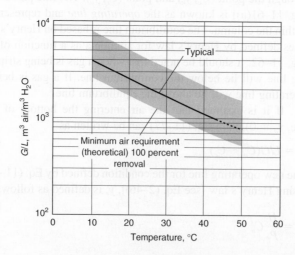

to be 100 percent efficient with a stripping tower of infinite height—obviously unachievable in practice. Computation of the theoretical air to water ratio is illustrated in Example 11–15.

EXAMPLE 11–15   **Air Requirements for Ammonia Stripping**   Determine the theoretical amount of air required at 20°C to reduce the ammonia concentration from 40 to 1 mg/L in a treated wastewater with a flowrate of 4000 m³/d. Assume that the Henry's constant for ammonia at 20°C is 0.75 atm (see Table 2–7 in Chap. 2) and the air entering the bottom of the tower does not contain any ammonia.

Solution

1. Determine the influent and effluent mole fractions of ammonia in the liquid using Eq. (2–3).

$$x_B = \frac{n_B}{n_A + n_B}$$

where x_B = mole fraction of solute B
n_B = number of moles of solute B
n_A = number of moles of solute A

$$C_o = \frac{[(40 \times 10^{-3})/17]}{[55.5 + (40 \times 10^{-3})/17]} = 4.24 \times 10^{-5} \, \text{mole NH}_3/\text{mole H}_2\text{O}$$

$$C_e = \frac{[(1 \times 10^{-3})/17]}{[55.5 + (1 \times 10^{-3})/17]} = 1.06 \times 10^{-6} \, \text{mole NH}_3/\text{mole H}_2\text{O}$$

2. Determine the mole fraction of ammonia in the air leaving the tower using Eq. (11–79).

$$y_e = \frac{H}{P_T} C_o$$

$$H = \left[\frac{(0.75 \, \text{atm})(\text{mole NH}_3/\text{mole air})}{(\text{mole NH}_3/\text{mole H}_2\text{O})} \right] = (0.75 \, \text{atm}) \left(\frac{\text{mole H}_2\text{O}}{\text{mole air}} \right)$$

$$y_e = \frac{H}{P_T} \times C_o = \frac{0.75 \, \text{atm}}{1.0 \, \text{atm}} \left(\frac{\text{mole H}_2\text{O}}{\text{mole air}} \right) \times (4.24 \times 10^{-5}) \, \text{mole NH}_3/\text{mole H}_2\text{O}$$

$$= 3.18 \times 10^{-5} \frac{\text{mole NH}_3}{\text{mole air}}$$

3. Determine the gas to liquid ratio using Eq. (11–81) rearranged as follows:

$$\frac{G}{L} = \frac{P_T}{H} \times \frac{(C_o - C_e)}{C_o} = \frac{(C_o - C_e)}{y_e}$$

$$\frac{G}{L} = \frac{(4.24 \times 10^{-5} - 0.106 \times 10^{-5})(\text{mole NH}_3/\text{mole H}_2\text{O})}{(3.18 \times 10^{-6} \, \text{mole NH}_3/\text{mole air})} = 1.3 \frac{\text{mole air}}{\text{mole H}_2\text{O}}$$

4. Convert the moles of air and water to liters of air and water.

For air at 20°C:

1.3 mole × 24.1 L/mole = 31.33 L

For water:

(1.0 mole H$_2$O)(18 g/mole)(1L/1000 g) = 0.018 L

$$\frac{G}{L} = \frac{31.33 \text{ L air}}{0.018 \text{ L water}} = 1741 \text{ L/L} = 1741 \text{ m}^3/\text{m}^3$$

6. Determine the total quantity of air required based on ideal conditions.

$$\text{Air required} = \frac{(1741 \text{ m}^3/\text{m}^3)\,(4000 \text{ m}^3/\text{d})}{(1440 \text{ min/d})} = 4835 \text{ m}^3/\text{min}$$

Comment   The procedure followed to determine the height of the stripping tower is illustrated in Example 11–14 presented later in this section. Also, it should be noted that the ammonium must first be converted to ammonia gas for this process to be effective. Steam stripping of ammonia is considered in Chap. 15.

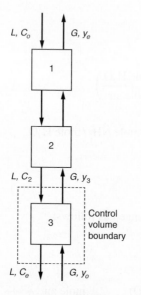

Figure 11–64

Definition sketch for the analysis of a three stage countercurrent flow gas stripping tower.

Mass Balance Analysis of a Multistage Stripping Tower.   In the analysis of stripping towers, reference is often made to the number of ideal stages required for stripping. The analysis for the number of stages is analogous to the simulation of plug flow with a series of complete-mix reactors as detailed in Chap. 1. Separate stages are used to improve performance of stripping towers. Equilibrium conditions are assumed in each stage of the tower. A steady-state materials balance for the lower portion of a countercurrent staged stripping tower (see Fig. 11–64) is given by:

inflow = outflow

$$LC_e + Gy_3 = LC_2 + Gy_o \tag{11–83}$$

or

$$(y_3 - y_o) = L/G(C_2 - C_e) \tag{11–84}$$

If an overall mass balance is performed around the tower, the resulting equations are the same as derived above for the continuous stripping tower [see Eq. (11–76)].

In the 1920s McCabe and Theile (1925) developed a graphical procedure for determining the required number of ideal stages. The method is illustrated on Fig. 11–65 for a stripping tower comprised of three stages. The operating line for the three stages is shown on Fig. 11–65. The number of ideal stages required for stripping of a constituent is obtained as follows. The point C_o, y_e represents the air leaving and the water entering the top of the stripping column. The composition of the liquid in equilibrium with the constituent concentration in the air is found by extending a horizontal line from the

Figure 11–65

Operating line for three-stage countercurrent flow gas stripping tower.

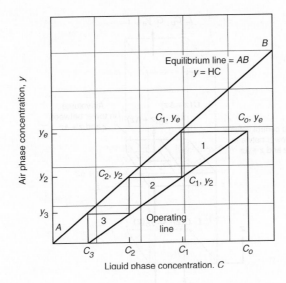

point C_o, y_e to the point C_1, y_e. From the point C_1, y_e the value of y_2, the air entering stage 1 from stage 2, is obtained from the equation of the operating line. If a materials mass balance is performed between stages 1 and 2, the resulting expression for y_2 is:

$$y_2 = \frac{L}{G}C_1 + \frac{Gy_e - LC_o}{G}$$

(11–85)

The value of y_2 is obtained by drawing a vertical line from point C_1, y_e to the operating line at point C_1, y_2 as shown on Fig. 11–65. In a similar manner the value of C_2 is obtained by drawing a horizontal line from the point C_1, y_2 to the equilibrium line. This procedure is repeated until the point C_n, y_{n+1} is reached. The number of ideal stages is typically a fractional number (e.g., 4.2, 5.6, etc.). In practice, the number of stages is rounded to the next whole number.

Determination of Height of Stripping Tower Packing. The purpose of the following analysis is to illustrate how the height of a stripping tower packing is determined, based on an analysis of the mass transfer occurring within the tower. A mass balance performed on the liquid phase within the stripping tower shown on Fig. 11–66 is as follows.

Simplified word statement

Accumulation = inflow − outflow + generation

$$\frac{\partial C}{\partial t}\Delta V = LC|_z - LC|_{z+\Delta z} + r_V \Delta V$$

(11–86)

where $\partial C/\partial t$ = change in concentration of constituent C with time, g/m³·s

ΔV = differential volume, m³

Δz = differential height, m

L = liquid volumetric flowrate, m³/s

C = concentration of constituent C, g/m³

r_V = rate of mass transfer of constituent C per unit volume per unit time, g/m³·s

Figure 11–66

Definition sketch for the analysis of mass transfer within a stripping tower. Note: the packing material is not shown. (Adapted from Hand et al., 1999.) (LD)

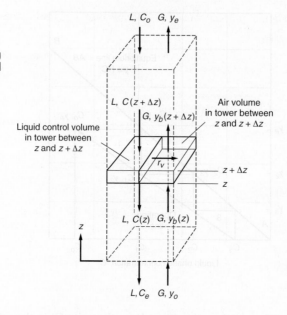

Substituting area times the differential height ($A\Delta z$) for the differential volume (ΔV) and writing the differential form for the term $LC|_{z+\Delta z}$ in Eq. (11–86) results in the following expression:

$$\frac{\partial C}{\partial t} A\Delta z = LC - L\left(C + \frac{\Delta C}{\Delta z}\Delta z\right) + r_V A\Delta z \tag{11–87}$$

Simplifying Eq. (11–87) and taking the limit as Δz approaches zero yields

$$\frac{\partial C}{\partial t} = -\frac{L}{A}\frac{\partial C}{\partial z} + r_V \tag{11–88}$$

The rate of mass transfer as described in Chap. 5 [see Eq. (5–57)]:

$$r_V = K_L a(C_b - C_s) \tag{11–89}$$

where r_V = rate of mass transfer of constituent C per unit volume per unit time, g/m³·s
$K_L a$ = volumetric mass transfer coefficient which depends on water quality characteristics and temperature, 1/s
C_b = concentration of constituent C in liquid bulk phase at time t, g/m³
C_S = concentration of constituent C in liquid in equilibrium with gas as given by Henry's law, g/m³

Assuming steady-state conditions within the tower ($\partial C/\partial t = 0$) and substituting for r_V, Eq. (11–88) can now be written as:

$$\frac{dC_b}{dz} = \frac{K_L a A}{L}(C_b - C_s) \tag{11–90}$$

The height of the tower can be obtained by integrating the above expression:

$$\int_o^z dz = \frac{L}{K_L aA} \int_{C_e}^{C_o} \frac{dC_b}{(C_b - C_s)} \tag{11–91}$$

To integrate the right hand side of the above equation, a relationship must be found between C_b and C_s because C_s is changing continuously throughout the height of the tower. From Henry's law, the value of C_s is given by:

$$C_s = \frac{P_T}{H} y \tag{11–92}$$

Substituting a modified form of Eq. (11–78) for y in Eq. (11–92) yields:

$$C_s = \frac{P_T}{H} \times \frac{L}{G}(C_b - C_e) \tag{11–93}$$

If Eq. (11–93) is substituted into Eq. (11–91) and the resulting expression is integrated, the following expression is obtained (Hand et al., 1999)

$$Z = \frac{L}{K_L aA} \left(\frac{C_o - C_e}{C_o - C_e - C_o'} \right) \ln \left(\frac{C_o - C_o'}{C_e} \right) \tag{11–94}$$

where Z = height of stripping tower packing, m

$$C_o' = \frac{P_T}{H} \times \frac{L}{G}(C_o - C_e), \text{ given previously [see Eq. (11–80)]}$$

It should be noted that if $C_o' = C_o$, Eq. (11–80) is the same as Eq. (11–81).

Design Equations for Stripping Towers. Utilizing the above equations, a number of process models and design equations have been developed. Equations that can be used to determine the height of a stripping tower are as follows:

$$Z = HTU \times NTU \tag{11–95}$$

where Z = height of stripping tower packing material, m

HTU = height of a transfer unit, m

NTU = number of transfer units

The height of a transfer unit is defined as:

$$HTU = \frac{L}{K_L aA} \tag{11–96}$$

where L = liquid volumetric flowrate, m³/s

$K_L a$ = volumetric mass transfer coefficient, 1/s

A = cross-sectional area of tower, m²

The HTU is a measure of the mass transfer characteristics of the packing material. The number of transfer units is defined as:

$$\text{NTU} = \left(\frac{C_o - C_e}{C_o - C_e - C_o'}\right) \ln \left(\frac{C_o - C_o'}{C_e}\right) \tag{11-97}$$

Substituting Eq. (11–93) in Eq (11–97) yields:

$$\text{NTU} = \left(\frac{S}{S-1}\right) \ln \left[\frac{(C_o/C_e)(S-1)+1}{S}\right] \tag{11-98}$$

where S is known as the stripping factor and is defined as:

$$S = \frac{G}{L} \times \frac{H}{P_T} \tag{11-99}$$

A value of $S = 1$ corresponds to the minimum amount of air required for stripping. When $S > 1$ the amount of air is in excess and complete stripping is possible given a tower of infinite height. When $S < 1$, there is insufficient air for stripping. In practice, stripping factors vary from 1.5 to 5.0.

Values for $K_L a$ for specific compounds are best obtained from pilot plant studies or by using empirical correlations such as given in Chap. 16, and repeated here for convenience as Eq. (11–100). It should be noted that many other relationships have been proposed in the literature (Sherwood and Hollaway, 1940 and Onda et al., 1968).

$$K_L a_{\text{VOC}} = K_L a_{O_2} \left(\frac{D_{\text{VOC}}}{D_{O_2}}\right)^n \tag{11-100}$$

where $K_L a_{\text{VOC}}$ = system mass transfer coefficient, 1/h
$K_L a_{O_2}$ = system oxygen mass transfer coefficient, 1/h
D_{VOC} = diffusion coefficient of VOC in water, cm^2/s
D_{O_2} = diffusion coefficient of oxygen in water, cm^2/s
n = coefficient (0.5 for stripping towers)

Air and water temperature are significant factors in the design of stripping towers because of their effect on air and water viscosities, Henry's law constants, and volumetric mass transfer coefficients. The effect of temperature on the Henry's law constant is illustrated on Fig. 11–62. The value of $K_L a$ can be adjusted for temperature effects using Eq. (1–44) with a theta value of 1.024.

Design of Stripping Towers

In its simplest form a stripping tower consists of a tower (usually circular cross-section), a support plate for the packing material, a distribution system for the liquid to be stripped, located above the packing material, and an air supply located at the bottom of the stripping tower (see Fig. 11–59). The process design variables include (1) the type of packing material, (2) the stripping factor, (3) the cross-sectional area of the tower, and (4) the height of the packing material in the stripping tower. The cross-sectional area will depend on the pressure drop through the packing. Representative design values for stripping of VOCs and ammonia are presented in Table 11–45. The significant difference in the amount of air required for stripping is a clear illustration of the importance of the Henry's law constant.

Table 11–45

Typical design parameters for stripping towers for the removal of VOC and ammonia[a]

Item	Symbol	Unit	VOC removal[b]	Ammonia removal[c]
Liquid loading rate		L/m²·min	600–1,800	40–80
Air to liquid ratio[d]	G/L	m³/m³	20–60:1	2,000–6,000:1
Stripping factor	S	unitless	1.5–5.0	1.5–5.0
Allowable air pressure drop,	ΔP	(N/m²)/m	100–400	100–400
Height to diameter ratio	H/D	m/m	≤10:1	≤10:1
Packing depth[e]	D	m	1–6	2–6
Factor of safety	SF	%D, %H	20–50	20–50
Wastewater pH	pH	unitless	5.5–8.5	10.8–11.5
Approximate packing factors				
Pall rings, Intalox saddles				
12.5 mm[f]	C_f	1/m	180–240	180–240
25 mm[f]	C_f	1/m	30–60	30–60
50 mm[f]	C_f	1/m	20–25	20–25
Berl saddles, Raschig rings				
12.5 mm[f]	C_f	1/m	300–600	300–600
25 mm[f]	C_f	1/m	120–160	120–160
50 mm[f]	C_f	1/m	45–60	45–60

[a] Adapted in part from Eckert (1970, 1975), Kavanaugh and Trussell (1980), and Hand (1999).

[b] Typical data for VOCs with Henry's law constants greater than 500 atm (mole H_2O/mole air).

[c] Ammonia with a Henry's law constant of 0.75 atm (mole H_2O/mole air) is considered only marginally strippable, which accounts for the low loading rate and high air to liquid ratio.

[d] Ratio is highly temperature dependent.

[e] For packing depths greater than 5 to 6 m, redistribution of the liquid flow is recommended.

[f] Size of packing material.

The headloss through the packing is determined using a generalized gas pressure drop relationships such as plotted on Fig. 11–67 (Eckert, 1975). The pressure drop is expressed in Newton per square meter per meter of depth (N/m²)/m. The upper line on Fig. 11–67 labeled *approximate flooding* represents the condition that occurs when the amount of water and air applied are so great that the pore spaces fill to the point where water starts to flood within the tower. The units for the x and y axis are as follows

X axis:

$$x = \frac{L'}{G'}\left(\frac{\rho_G}{\rho_L - \rho_G}\right)^{1/2} \approx \frac{L'}{G'}\left(\frac{\rho_G}{\rho_L}\right)^{1/2} \qquad (11\text{–}101)$$

Y axis:

$$y = \frac{(G')^2(C_f)(\mu_L)^{0.1}}{(\rho_G)(\rho_L - \rho_G)} \qquad (11\text{–}102)$$

Figure 11–67

Generalized pressure drop curves for packed bed stripping towers. Note the curves in this plot have been converted to metric units from US customary units in which the original curves were plotted. (Adapted from Eckert, 1975.)

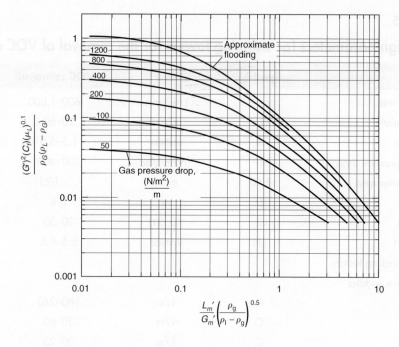

Which can be rewritten as follows:

$$G' = \left[\frac{(\text{value from } y \text{ axis})(\rho_G)(\rho_L - \rho_G)}{(C_f)(\mu_L)^{0.1}} \right]^{1/2}$$ (11–103)

where L' = liquid loading rate, kg/m²·s
G' = gas loading rate, kg/ m²·s
ρ_G = density of gas, kg/m³
ρ_L = density of liquid, kg/m³
C_f = packing factor for packing material, 1/m
μ_L = viscosity of liquid, kg/m·s

The packing factor C_f depends on the type and size of the packing. Typical ranges for packing factors that can be used for preliminary assessments are reported in Table 11–35. For more detailed design calculations, current values should be obtained from manufacturers.

To use Fig. 11–67, select a value for G'/L' and compute the corresponding x value. Enter the plot at the computed value of x and move vertically upward to a preselected pressure drop line. Move horizontally from the point of intersection to the y axis and note the value on the y axis. Using the y-axis value, determine the gas loading rate, G', using Eq. (11–103) and the corresponding liquid loading rate, L'. To determine the required cross-sectional area, the liquid flowrate is divided by the liquid loading rate.

A generalized analysis procedure is as follows:

1. Select a packing material and its corresponding packing factor for use in Eq. (11–101).
2. Select several stripping factors for successive trials (e.g., 2.5, 3.0, 4.0, etc.).
3. Select acceptable pressure drop ΔP (typically a function of the packing material selected).
4. Determine the cross-sectional area of the tower, based on the allowable pressure drop, using the data presented on Fig. 11–67 or other appropriate relationships.

5. Determine the height of the transfer units using Eq. (11–96). To apply Eq. (11–96) the value of $K_L a$ must be known or estimated using Eq. (11–100).
6. Determine the number of transfer units using Eq. (11–98).
7. Determine the height of the stripping tower packing material using Eq. (11–95).
8. Determine the total height of the stripping tower. To account for the entrance plenum and exit gas collection system, an additional 2 to 3 m is added to the computed height of the packing to obtain the overall height of the stripping tower.

The design procedure outlined above is illustrated in Example 11–16. Representative design values for stripping towers are given in Table 11–45. To evaluate the stripping process more thoroughly, any of the commercially available software packages can be used.

EXAMPLE 11–16   **Determination of Height of Stripping Tower for the Removal of Ammonia**   Determine the diameter and height of the stripping tower required to treat the wastewater in Example 11–15. The ammonia concentration in a treated wastewater from a flow of 4000 m³/d is to be reduced from 40 to 1 mg/L. Assume that the Henry's law constant for ammonia at 20°C is 0.75 atm, and the air entering the bottom of the tower does not contain any ammonia. Assume the $K_L a$ value for ammonia is 0.0125 s⁻¹.

Solution

1. Select a packing material. Assume a packing factor of 20 for 50 mm Pall rings (see Table 11–45).
2. Select a stripping factor. Assume a stripping factor of 3.
3. Select an acceptable pressure drop. Assume a pressure drop of 400 (N/m²)/m (see Table 11–45).
4. Determine the cross-sectional area of the stripping tower using the pressure drop plot given on Fig. 11–67.
   a. Determine the value of the ordinate value for a stripping factor of 3.

$$S = \frac{G}{L} \times \frac{H}{P_T} = \frac{G \text{ mole air}}{L \text{ mole water}} \times \frac{0.75 \text{ atm}}{1.0 \text{ atm}} = \frac{G \text{ mole air}}{L \text{ mole water}} \times 0.75$$

$$S = 0.75 \times \left(\frac{G \text{ mole air}}{L \text{ mole water}}\right)\left(\frac{28.8 \text{ g}}{\text{mole air}}\right)\left(\frac{\text{mole water}}{18 \text{ g}}\right) = 1.2 \frac{G \text{ g}}{L \text{ g}} = 1.2 \frac{G' \text{ kg}}{L' \text{ kg}}$$

$$\frac{L'}{G'} = \frac{(1.2 \text{ kg/kg})}{3} = 0.4$$

$$\frac{L'}{G'}\left(\frac{\rho_G}{\rho_L - \rho_G}\right)^{1/2} \approx \frac{L'}{G'}\left(\frac{\rho_G}{\rho_L}\right)^{1/2} = (0.4 \text{ kg/kg})\left[\frac{(1.204 \text{ kg/m}^3)}{(998.2 \text{ kg/m}^3)}\right]^{1/2} = 0.0139$$

   b. For an abscissa value of 0.0139 and a pressure drop of 400 (N/m²)/m, the ordinate value from Fig. 11–67 is 0.3.
   c. Using an ordinate value 0.3, determine the loading rate using Eq. (11–103). From Appendix C, $\mu_L = 1.002$ kg/m·s

$$G' = \left[\frac{(\text{value from } y \text{ axis})(\rho_G)(\rho_L - \rho_G)}{(C_f)(\mu_L)^{0.1}} \right]^{1/2}$$

$$G' = \left[\frac{(0.3)(1.204)(998.2 - 1.204)}{(20)(1.002)^{0.1}} \right]^{1/2} = 4.24 \text{ kg/m}^2 \cdot \text{s}$$

$L' = 0.4 \, G' = 0.4 \times 4.24 \text{ kg/m}^2 \cdot \text{s} = 1.70 \text{ kg/m}^2 \cdot \text{s}$

d. Solve for the diameter of the tower, using the loading rate determined in Step 4.

$$D = \left[\frac{4}{3.14} \times \frac{(4000 \text{ m}^3/\text{d})(998.2 \text{ kg/m}^3)}{(4.24 \text{ kg/m}^2 \cdot \text{s})} \times \frac{1 \text{ d}}{86{,}400 \text{ s}} \right]^{1/2} = 3.73 \text{ m}$$

5. Determine the height of the transfer unit using Eq. (11–96).

$$\text{HTU} = \frac{L}{K_L a A}$$

$$\text{HTU} = \frac{L}{K_L a A} = \left[\frac{(4000 \text{ m}^3/\text{d})}{(0.0125/\text{s})[(3.14/4)(3.73)^2]} \times \frac{1 \text{ d}}{86{,}400 \text{ s}} \right] = 0.34 \text{ m}$$

6. Determine the number of transfer units using Eq. (11–98).

$$\text{NTU} = \left(\frac{S}{S - 1} \right) \ln \left[\frac{(C_o/C_e)(S - 1) + 1}{S} \right]$$

$$\text{NTU} = \left(\frac{3}{3 - 1} \right) \ln \left[\frac{(40/1)(3 - 1) + 1}{3} \right] = 4.94$$

7. Determine the theoretical height of the stripping tower packing using Eq. (11–95).

$Z = \text{HTU} \times \text{NTU} = 0.34 \times 4.94 = 1.68 \text{ m}$

8. Determine the total height of the stripping tower.

Add 3 m to obtain the total height of the stripper

Hstripper = Hpacking, m + 3 m = 1.68 m + 3 m = 4.68 m

Comment   In this example, the value of $K_L a$ for ammonia was known. Quite often the required $K_L a$ value must be determined in the field, using pilot scale facilities. Alternatively Eq. (11–100) can be used to estimate a value for $K_L a$. In some cases, data from the literature or from manufacturers may be used to obtain preliminary sizing. Because of the relatively low Henry's law constant, a large surface area is required to achieve the desired ammonia removal. Further, because of the relatively large radius that is required, it is likely that two stripping towers, each with a radius and area of about 2.6 and 5.3 m², respectively would be used to optimize air flow. To optimize the design, various stripping ratios must be evaluated. Optimization is best accomplished using one of the commercially available stripping tower software packages. Because the Henry's law constant for ammonia is so low (0.75 atm), ammonia is often not considered amenable or suitable for stripping at ambient temperatures and is seldom done. Steam stripping of the concentrated ammonia present in return flows from sludge processing is considered in Chap. 15.

Figure 11–68

Typical flow diagram for the air stripping of ammonia from wastewater.

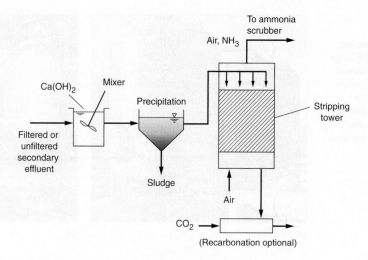

Air Stripping Applications

As noted previously, air stripping is used to remove a variety of gaseous constituents including VOCs, carbon dioxide (CO_2), oxygen (O_2), hydrogen sulfide (H_2S), and ammonia (NH_3). The removal and treatment of VOCs in aeration systems is considered in Sec. 16–4 in Chap. 16. The removal and treatment of odorous compounds (e.g., H_2S) is considered in Sec. 16–3 in Chap. 16. Air stripping has been used for the removal of ammonia from various wastewaters including untreated and treated wastewater, digester supernatant (also by steam stripping), and for the recovery of ammonia from return flows. The recovery of ammonia from return flows is considered in detail in Chap. 15. Ammonia stripping from wastewater is considered briefly below.

A typical flow diagram for the removal of ammonia from wastewater by air stripping is shown on Fig. 11–68. In most cases where ammonia stripping has been tried with wastewater, a number of operating problems have developed, the most serious being (1) maintaining the required pH for effective stripping, (2) calcium carbonate scaling within the tower and feed lines, and (3) poor performance during cold weather operation. Maintaining the required pH is a control problem that can be managed with multiple sensors. The amount and nature (soft to extremely hard) of the calcium carbonate scale formed varies with the characteristics of the wastewater and local environmental conditions and cannot be predicted a priori. Under conditions of icing, the liquid-air contact geometry in the tower is altered, which further reduces the overall efficiency. The best solution for cold weather conditions is to enclose the stripper. For the reasons cited above and cost, ammonia stripping from wastewater is seldom done. However, ammonia stripping is done on concentrated return flows resulting from the treatment of biosolids. The use of ammonia stripping for the recovery of nitrogen in the form of ammonium sulfate is considered in Chap. 15.

11–11 ION EXCHANGE

Ion exchange is a unit process in which ions of a given species are displaced from an insoluble exchange material by ions of a different species in solution. The most widespread use of this process is in domestic water softening, where sodium ions from a cationic exchange resin replace the calcium and magnesium ions in the treated water, thus reducing the hardness. Ion exchange has been used in wastewater applications for the removal of nitrogen, heavy metals, and total dissolved solids.

Figure 11–69

Two examples of full scale ion exchange installations: (a) large downflow packed-bed columns, and (b) ion exchange canisters on a rotating platform. The canisters are rotated so that one canister can be regenerated while the others are in operation.

(a)   (b)

Ion exchange processes can be operated in a batch or continuous mode. In a batch process, the resin is stirred with the water to be treated in a reactor until the reaction is complete. The spent resin is removed by settling and subsequently is regenerated and reused. In a continuous process, the exchange material is placed in a bed or a packed column, similar to the one shown previously on Fig. 11–56(a), and the water to be treated is passed through it. Continuous ion exchangers are usually of the downflow, packed-bed column type. Wastewater enters the top of the column under pressure, passes downward through the resin bed, and is removed at the bottom. When the resin capacity is exhausted, the column is backwashed to remove trapped solids and is then regenerated. Two examples of commercial ion exchange reactors are shown on Fig. 11–69.

Ion Exchange Materials

Naturally occurring ion exchange materials, known as zeolites, are used for water softening and ammonium ion removal. Zeolites used for water softening are complex aluminosilicates with sodium as the mobile ion. Ammonium exchange is accomplished using a naturally occurring zeolite, clinoptilolite. Synthetic aluminosilicates are manufactured, but most synthetic ion exchange materials are resins or phenolic polymers. Five types of synthetic ion exchange resins are in use: (1) strong-acid cation, (2) weak-acid cation, (3) strong-base anion, (4) weak-base anion, and (5) heavy-metal selective chelating resins. The properties of these resins are summarized in Table 11–46.

Most synthetic ion exchange resins are manufactured by a process in which styrene and divinylbenzene are copolymerized. The styrene serves as the basic matrix of the resin and divinylbenzene is used to cross link the polymers to produce an insoluble tough resin. Important properties of ion exchange resins include exchange capacity, particle size, and stability. The exchange capacity of a resin is defined as the quantity of an exchangeable ion that can be taken up. The exchange capacity of resins is expressed as eq/L or eq/kg (meq/L or meq/g). The particle size of a resin is important with respect to the hydraulics of the ion exchange column and the kinetics of ion exchange. In general, the rate of exchange is proportional to the inverse of the square of the particle diameter. The stability of a resin is important to the long-term performance of the resin. Excessive osmotic swelling and shrinking, chemical degradation, and structural changes in the resin caused by physical stresses are important factors that may limit the useful life of a resin.

Table 11–46

Classification of ion exchange resins[a]

Type of resin	Characteristics
Strong-acid cation resins	Strong-acid resins behave in a manner similar to a strong-acid, and are highly ionized in both the acid (R–SO_3H) and salt (R–SO_3Na) form, over the entire pH range.
Weak-acid cation resins	Weak-acid resins have a weak-acid functional group (–COOH), typically a carboxylic group. These resins behave like weak organic acids that are weakly dissociated.
Strong-base anion resins	Strong-base resins are highly ionized having strong-base functional groups such as (OH^-); and can be used over the entire pH range. These resins are used in the hydroxide (OH^-) form for water deionization.
Weak-base anion resins	Weak-base resins have weak-base functional groups in which the degree of ionization is dependent on pH.
Heavy-metal selective chelating resins	Chelating resins behave like weak-acid cation resins, but exhibit a high degree of selectivity for heavy-metal cations. The functional group in most of these resins is EDTA, and the resin structure in the sodium form is R–EDTA–Na.

[a] Adapted in part from Ford (1992).

Typical Ion Exchange Reactions

Typical ion exchange reactions for natural and synthetic ion exchange materials are given below.

For natural zeolites used in water softening (Z):

$$ZNa_2 + \begin{bmatrix} Ca^{2+} \\ Mg^{2+} \\ Fe^{2+} \end{bmatrix} \rightleftarrows Z \begin{bmatrix} Ca^{2+} \\ Mg^{2+} \\ Fe^{2+} \end{bmatrix} + 2Na^+ \tag{11-104}$$

For synthetic resins (R):
Strong-acid cation exchange:

$$RSO_3H + Na^+ \rightleftarrows RSO_3Na + H^+ \tag{11-105}$$

$$2RSO_3Na + Ca^{2+} \rightleftarrows (RSO_3)_2Ca + 2Na^+ \tag{11-106}$$

Weak-acid cation exchange:

$$RCOOH + Na^+ \rightleftarrows RCOONa + H^+ \tag{11-107}$$

$$2RCOONa + Ca^{2+} \rightleftarrows (RCOO)_2Ca + 2Na^+ \tag{11-108}$$

Strong-base anion exchange:

$$RR_3'NOH + Cl^- \rightleftarrows RR_3'NCl + OH^- \tag{11-109}$$

Weak-base anion exchange:

$$RNH_3OH + Cl^- \rightleftarrows RNH_3Cl + OH^- \tag{11-110}$$

$$2RNH_3OCl + SO_4^{2-} \rightleftarrows (RNH_3)_2SO_4 + 2Cl^- \tag{11-111}$$

Exchange Capacity of Ion Exchange Resins

Reported exchange capacities vary with the type and concentration of regenerant used to restore the resin (see table 11–47.) Typical synthetic resin exchange capacities are in the range of 2 to 10 eq/kg of resin. Zeolite cation exchangers used for water softening have exchange capacities of 0.05 to 0.1 eq/kg. Exchange capacity is measured by placing the resin in a known form. A cationic resin would be washed with a strong-acid to place all of the exchange sites on the resin in the H^+ form or washed with a strong NaCl brine to place all of the exchange sites in the Na^+ form. A solution of known concentration of an exchangeable ion (e.g., Ca^{2+}) can then be added until exchange is complete and the amount of exchange capacity can be measured, or in the acid case, the resin is titrated with a strong- base. Determination of the capacity of an ion exchange resin based on titration is illustrated in Example 11–17.

Exchange capacities for resins often are expressed in terms of grams $CaCO_3$ per cubic meter of resin (g/m^3) or gram equivalents per cubic meter ($g\text{-}eq/m^3$). Conversion between these two units is accomplished using the following expression:

$$\frac{1 \text{ g-eq}}{m^3} = \frac{(1\text{ g-eq})\left(\dfrac{100\text{ g CaCO}_3}{2\text{ g-eq}}\right)}{m^3} = 50\text{ g CaCO}_3/m^3 \tag{11–112}$$

Calculation of the required resin volume for an ion-exchange process is also illustrated in Example 11–17.

EXAMPLE 11–17 **Determination of Ion Exchange Capacity for a New Resin** A column study was conducted to determine the capacity of a cation exchange resin. In conducting the study, 0.1 kg of resin was washed with NaCl until the resin was in the R-Na form. The column was then washed with distilled water to remove the chloride ion (Cl^-) from the intersticies of the resin. The resin was then titrated with a solution of calcium chloride ($CaCl_2$), and the concentrations of chloride and calcium were measured at various throughput volumes. The measured concentrations of Cl^- and Ca^{2+} and the corresponding throughput volumes are as given below. Using the data given below, determine the exchange capacity of the resin and the mass and volume of a resin required to treat 4000 m^3 of water containing 18 mg/liter of ammonium ion (NH_4^+). Assume the density of the resin is 700 kg/m^3.

Throughput volume, L	Constituent, mg/L	
	Cl^-	Ca^{2+}
2	0	0
3	trace	0
5	7	0
6	18	0
10	65	0
12	71	trace
20	71	13
26	71	32
28	71	38
32	$C_o = 71$	$C_o = 40$

Solution

1. Prepare a plot of the normalized concentrations of Cl^- and Ca^{2+} as a function of the throughput volume. The required plot is given below.

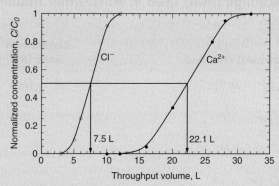

2. Determine the exchange capacity.
   The exchange capacity (EC) of the resin in meq/kg is:

$$EC = \frac{VC_o}{R}$$

where V = throughput volume between the Cl^- and Ca^{2+} breakthrough curves at $C/C_o = 0.5$
   C_o = initial calcium concentration in meq/L
   R = amount of resin in kg

$$EC = \frac{(22.1\ L - 7.5\ L)\left[\dfrac{(40\ mg/L)}{(20\ mg/meq)}\right]}{0.1\ kg\ of\ resin} = 292\ meq/kg\ of\ resin$$

3. Determine the mass and volume of resin required to treat 4000 m^3 of water containing 18 mg/liter of ammonium ion NH_4^+.
   a. Determine the meq of NH_4^+.

$$NH_4^+,\ meq/L = \frac{(18\ mg/L\ as\ NH_4^+)}{(18\ mg/meq)} = 1\ meq/L$$

   b. The required exchange capacity is equal to

$$(1.0\ meq/L)(4000\ m^3)(10^3\ L/m^3) = 4 \times 10^6\ meq$$

   c. The required mass of resin is

$$R_{mass},\ kg = \frac{4 \times 10^6\ meq}{(292\ meq/kg\ of\ resin)} = 13{,}700\ kg$$

   d. The required volume of resin is

$$R_{vol},\ m^3 = \frac{13{,}700\ kg\ of\ resin}{(700\ kg/m^3)} = 19.6\ m^3$$

Comment In practice, because of leakage and other operational and design limitations, the required volume of resin will usually be about 1.1 to 1.4 times that computed on the basis of exchange capacity. Also, the above computation is based on the assumption that the entire capacity of the resin is utilized.

Table 11–47

Characteristics of ion exchange resins used in wastewater treatment processes[a]

Resin Type	Acronym	Fundamental reaction[b]	Regenerant ions (X)	pK	Exchange capacity, meq/mL	Constituents removed
Strong-acid cation	SAC	$n[RSO_3^-]X^+ + M^{n+} \rightleftarrows$ $[nRSO_3^-]M^{n+} + nX^+$	H^+ or Na^+	<0	1.7 to 2.1	H^+ form: any cation; Na^+ form: divalent cations
Weak-acid cation	WAC	$n[RCOO^-]X^+ + M^{n+} \rightleftarrows$ $[nRCOO^-]M^{n+} + nX^+$	H^+	4 to 5	4 to 4.5	Divalent cations first, then monovalent cations until alkalinity is consumed
Strong-base anion (type 1)	SBA–1[c]	$n[R(CH_3)_3N^+]X^- + A^{n-} \rightleftarrows$ $[nR(CH_3)_3N^+]A^{n-} + nX^-$	OH^- or Cl^-	>13	1 to 1.4	OH^- form: any anion; Cl^- form: sulfate, nitrate, perchlorate, etc.
Stong-base anion (type 2)	SBA–2[d]	$n[R(CH_3)_2(CH_3CH_2OH)N^+]X^- + A^{n-} \rightleftarrows$ $[nR(CH_3)_2(CH_3CH_2OH)N^+]A^{n-} + nX^-$	OH^- or Cl^-	>13	2 to 2.5	OH^- form: any anion; Cl^- form: sulfate, nitrate, perchlorate, etc.
Weak-base anion	WBA	$[R(CH_3)_2N]HX + HA \rightleftarrows$ $[R(CH_3)_2N]HA + HX$	OH^-	5.7 to 7.3	2 to 3	Divalent anions first, then monovalent anions until strong-acid is consumed

[a] From Crittenden et al. (2005).
[b] Term within the brackets represents the solid phase of the resin.
[c] Greater chemical stability than SBA–1.
[d] Greater regeneration efficiency and capacity than SBA–2.

Ion Exchange Chemistry

The chemistry of the ion exchange process may be represented by the following equilibrium expression for the reaction of constituent A on a cation exchange resin and constituent B in solution.

$$nR^-A^+ + B^{n+} \rightleftarrows R_n^-B^{n+} + nA^+ \tag{11–113}$$

where R^- is the anionic group attached to an ion exchange resin and A and B are cations in solution. The generalized form of the equilibrium expression for the above reaction is

$$\frac{[A^+]_S^n[R_n^-B^{n+}]_R}{[R^-A^+]_R^n[B^{n+}]_S} = K_{A^+ \rightarrow B^{n+}} \tag{11–114}$$

where $K_A{}^+ \to B^{n+}$ = selectivity coefficient
$[A^+]_S$ = concentration of A in solution
$[R^-A^+]_R$ = concentration A on the exchange resin

The reactions for the removal of sodium (Na^+) and calcium (Ca^{2+}) ions from water using a strong-acid synthetic cationic exchange resin R, and the regeneration of the exhausted resins with hydrochloric acid (HCl) and sodium chloride (NaCl) are as follows:

Reaction:

$$R^-H^+ + Na^+ \rightleftarrows R^-Na^+ + H^+ \tag{11–115}$$

$$2R^-Na^+ + Ca^{2+} \rightleftarrows R_2^-Ca^{2+} + 2Na^+ \tag{11–116}$$

Regeneration:

$$R^-Na^+ + HCl \rightleftarrows R^-H^+ + NaCl \tag{11–117}$$

$$R_2^-Ca^{2+} + 2NaCl \rightleftarrows 2R^-Na^+ + CaCl_2 \tag{11–118}$$

The corresponding equilibrium expressions for sodium and calcium are as follows:

For sodium:

$$\frac{[H^+][R^-Na^+]}{[R^-H^+][Na^+]} = K_{H^+ \to Na^+} \tag{11–119}$$

For calcium:

$$\frac{[Na^+]^2[R^-Ca^{2+}]}{[R^-Na^+]^2[Ca^{2+}]} = K_{Na^+ \to Ca^{2+}} \tag{11–120}$$

The selectivity coefficient depends primarily on the nature and valence of the ion, the type of resin and its saturation, and the ion concentration in wastewater and typically is valid over a narrow pH range. In fact, for a given series of similar ions, exchange resins have been found to exhibit an order of selectivity or affinity for the ions. Approximate selectivity coefficients for cationic and anionic resins are given in Tables 11–48 and 11–49,

Table 11–48

Approximate selectivity coefficients scale for cations on 8 percent cross-linked strong-acid ion exchange resins[a]

Cation	Selectivity coefficient	Cation	Selectivity coefficient
Li^+	1.0	Co^{2+}	3.7
H^+	1.3	Cu^{2+}	3.8
Na^+	2.0	Cd^{2+}	3.9
$NH_4{}^+$	2.6	Be^{2+}	4.0
K^+	2.9	Mn^{2+}	4.1
Rb^+	3.2	Ni^{2+}	3.9
Cs^+	3.3	Ca^{2+}	5.2
Ag^+	8.5	Sr^{2+}	6.5
Mg^{2+}	3.3	Pb^{2+}	9.9
Zn^{2+}	3.5	Ba^{2+}	11.5

[a] Adapted from Bonner and Smith (1957), see also Slater (1991).

Table 11–49

Approximate selectivity coefficients for anions on strong-base ion exchange resins[a]

Anion	Selectivity coefficient	Anion	Selectivity coefficient
HPO_4^{2-}	0.01	BrO_3^-	1.0
CO_3^{2-}	0.03	Cl^-	1.0
OH^- (Type I)	0.06	CN^-	1.3
F^-	0.1	NO_2^-	1.3
SO_4^{2-}	0.15	HSO_4^-	1.6
CH_3COO^-	0.2	Br^-	3.0
HCO_3^-	0.4	NO_3^-	3.0–4.0
OH^- (Type II)	0.05–0.65	I^-	18.0

[a] Adapted from Peterson (1953) and Bard (1966).

respectively. The use of the selectivity coefficients given in these tables is illustrated in Example 11–18.

Typical selectivity series for synthetic cationic and anionic exchange resins are as follows.

$$Li^+ < H^+ < Na^+ < NH_4^+ < K^+ < Rb^+ < Ag^+ \tag{11–121}$$

$$Mg^{2+} < Zn^{2+} < Co^{2+} < Cu^{2+} < Ca^{2+} < Sr^{2+} < Ba^{2+} \tag{11–122}$$

$$OH^- < F^- < HCO_3^- < Cl^- < Br^- < NO_3^- < ClO_4^- \tag{11–123}$$

In practice, the selectivity coefficients are determined by measurement in the laboratory and are valid only for the conditions under which they were measured. At low concentrations, the value of the selectivity coefficient for the exchange of monovalent ions by divalent ions is, in general, larger than the exchange of monovalent ions by monovalent ions. This fact has, in many cases, limited the use of synthetic resins for the removal of certain substances in wastewater, such as ammonia in the form of the ammonium ion. There are, however, certain natural zeolites that favor NH_4^+ or Cu^{2+}.

Anderson (1975), in a classic paper, developed a method that can be used to evaluate the effectiveness of a proposed ion exchange process using strong ionic resins. In the development proposed by Anderson, it is assumed that at 100 percent leakage, the effluent concentration of a constituent is equal to the influent concentration (i.e., equilibrium has been reached). The equilibrium condition can be assumed to be either the limiting operating exchange capacity of the resin or the capacity corresponding to the maximum regeneration level that can be attained. Using this assumption, Eq. (11–114) is converted from concentration units to units of equivalent fractions by making the following substitutions:

$$X_{A^+} = \frac{[A^+]_S}{C} \text{ and } X_{B^+} = \frac{[B^+]_S}{C} \tag{11–124}$$

$$X_{A^+} + X_{B^+} = 1 \tag{11–125}$$

where X_A^+ and X_B^+ are the equivalent fractions of A and B in solution and C is the total cationic or anionic concentration in solution.

$$\overline{X}_{A^+} = \frac{[R^-A^+]_R}{\overline{C}} \text{ and } \overline{X}_{B^+} = \frac{[R^-B^+]_R}{\overline{C}} \tag{11–126}$$

Figure 11–70

Distribution curves for a single monovalent ion A between the solution and the resin for different values of the selectivity coefficient.

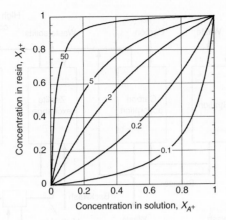

$$\overline{X}_{A^+} + \overline{X}_{B^+} = 1 \tag{11–127}$$

Where $\overline{X}_A$ and $\overline{X}_B$ are the equivalent fractions of A and B in the resin and $\overline{C}$ is the total ionic concentration in the resin (i.e., the total resin capacity in eq/L). Substituting the above terms into Eq. (11–114) and simplifying results in the following expression:

$$\frac{\overline{X}_{B^+} X_{A^+}}{\overline{X}_{A^+} X_{B^+}} = K_{A^+ \rightarrow B^+} \tag{11–128}$$

Substituting for X_{A^+} and $\overline{X}_{A^+}$ in Eq. (11–128) results in:

$$\frac{\overline{X}_{B^+}}{1 - \overline{X}_{B^+}} = (K_{A^+ \rightarrow B^+})\left(\frac{X_{B^+}}{1 - X_{B^+}}\right) \tag{11–129}$$

It should be noted that Eq. (11–129) is only valid for exchanges between monovalent ions on fully ionized exchange resins. The distribution of a single monovalent ion A between the solution and the resin for different values of the selectivity coefficient is presented on Fig. 11–70. The distribution curves can be used to assess the effectiveness of a resin for the removal of a given ion, based on the selectivity coefficient.

The following three attributes of Eq. (11–129) were identified by Anderson (1975).

1. The term $\overline{X}_B/(1 - \overline{X}_B)$ corresponds to the state of the resin in an exchange column when the influent and effluent concentrations are the same.
2. The term $\overline{X}_B$ corresponds to the extent to which the resin can be converted to the B^+ form when the resin is in equilibrium with a solution of composition X_B.
3. The term $\overline{X}_B$ also corresponds to the maximum extent of regeneration that can be achieved with a regenerant composition of X_B.

The corresponding equation for exchanges between monovalent and divalent ions on a fully ionized exchange resin is:

$$\frac{\overline{X}_{B^{+2}}}{(1 - \overline{X}_{B^{+2}})^2} = (K_{A^+ \rightarrow B^{+2}})\left(\frac{\overline{C}}{C}\right)\frac{X_{B^{+2}}}{(1 - X_{B^{+2}})^2} \tag{11–130}$$

The application of these equations is illustrated in Example 11–18.

Figure 11-71

Typical flow diagram for the removal of ammonium by zeolite exchange. Note: the ammonium removed is recovered by high pH air stripping and acid scrubbing.

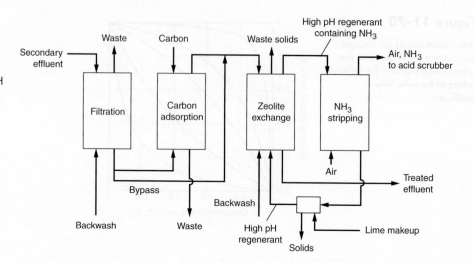

Application of Ion Exchange

As noted previously, ion exchange has been used in wastewater applications for the removal of nitrogen, heavy metals, and total dissolved solids.

For Nitrogen Control. For nitrogen control, the ions typically removed from the waste stream are ammonium, NH_4^+, and nitrate, NO_3^-. The ion that the ammonium displaces varies with the nature of the solution used to regenerate the bed. Although both natural and synthetic ion exchange resins are available, synthetic resins are used more widely because of their durability. Some natural resins (zeolites) have found application in the removal of ammonium from wastewater. Clinoptilolite, a naturally occurring zeolite, has proven to be one of the best natural exchange resins. In addition to having a greater affinity for ammonium ions than other ion exchange materials, it is relatively inexpensive when compared to synthetic media. One of the novel features of this zeolite is the regeneration system employed. Upon exhaustion, the zeolite is regenerated with lime $[Ca(OH)_2]$ and the ammonium ion removed from the zeolite is converted to ammonia because of the high pH. A flow diagram for this process is shown on Fig. 11–71. The stripped liquid is collected in a storage tank for subsequent reuse. A problem that must be solved is the formation of calcium carbonate precipitates within the zeolite exchange bed and in the stripping tower and piping appurtenances. As indicated on Fig. 11–71, the zeolite bed is equipped with backwash facilities to remove the carbonate deposits that form within the filter.

When using conventional synthetic ion exchange resins for the removal of nitrate, two problems are encountered. First, while most resins have a greater affinity for nitrate over chloride or bicarbonate, they have a significantly lower affinity for nitrate as compared to sulfate, which limits the useful capacity of the resin for the removal of nitrate. The impact of the presence of sulfate on the nitrate removal capacity of conventional resins is illustrated in Example 11–18. Second, because of the lower affinity for nitrate over sulfate, a phenomenon known as *nitrate dumping* can occur. Nitrate dumping occurs when an ion exchange column is operated past the nitrate breakthrough, at which point the sulfate in the feed water will displace the nitrate on the resin causing a release of nitrate. To overcome the problems associated with low affinity and nitrate breakthrough, new types of resins have been developed within which the affinities for nitrate and sulfate have been exchanged. When significant amounts of sulfate are present (i.e., typically greater than 25 percent of the total of the sum of the sulfate and nitrate expressed in meq/L), the use of nitrate selective resins is advantageous. Because the performance of nitrate selective resins

Figure 11–72

Typical ion exchange test columns (a) ion exchange columns used to study the removal of nitrate from water which has been processed with reverse osmosis and (b) bench scale ion exchange columns. (Courtesy of David Hand.)

(a)

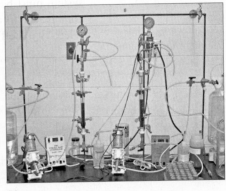

(b)

will vary with the composition of the treated wastewater, pilot testing will usually be required (McGarvey et al., 1989; Dimotsis and McGarvey, 1995). Typical ion exchange test columns used to study the removal of nitrate from water which has been processed with reverse osmosis are shown on Fig. 11–72.

EXAMPLE 11–18

Ion Exchange for the Removal of Nitrate without and with Sulfate Present in the Water
Nitrate is to be removed from two different treated wastewaters with the compositions given below. For the purpose of illustration, assume a conventional ion exchange resin will be used.

Wastewater A

Cation	Conc., mg/L	mg/meq	meq/L	Anion	Conc., mg/L	mg/meq	meq/L
Ca^{2+}	82.2	20.04	4.10	HCO_3^-	305.1	61.02	5.00
Mg^{2+}	17.9	12.15	1.47	SO_4^{2-}	0.00	48.03	0.00
Na^+	46.4	23.00	2.02	Cl^-	78.0	35.45	2.20
K^+	15.5	39.10	0.40	NO_3^-	50.0	62.01	0.81
		Σcations	7.99			Σanions	8.01

Wastewater B

Cation	Conc., mg/L	mg/meq	meq/L	Anion	Conc., mg/L	mg/meq	meq/L
Ca^{2+}	82.2	20.04	4.10	HCO_3^-	220.0	61.02	3.61
Mg^{2+}	17.9	12.15	1.47	SO_4^{2-}	79.2	48.03	1.65
Na^+	46.4	23.00	2.02	Cl^-	78.0	35.45	2.20
K^+	15.5	39.10	0.40	NO_3^-	50.0	62.01	0.81
		Σcations	7.99			Σanions	8.27

Determine the maximum amount of water that can be processed per liter of a strong-base anion exchange resin with an exchange capacity of 2.0 eq/L.

Solution:
Wastewater A

1. Estimate the selectivity coefficient (see Table 11–49). To apply Eq. (11–129) the system must be reduced to two components. For this purpose, HCO_3^- and Cl^- are combined into a single component. Using a selectivity value of 4 for nitrate, the selectivity coefficient is estimated as follows:

$$K_{HCO_3^- \to NO_3^-} = \frac{4.0}{0.4} = 10.0$$

$$K_{Cl^- \to NO_3^-} = \frac{4.0}{1.0} = 4.0$$

$$K_{[(HCO_3^-)(Cl^-)] \to NO_3^-} = 7.0 \text{ (estimated)}$$

2. For the equilibrium condition ($C_e/C_o = 1.0$), estimate the nitrate equivalent fraction in solution.

$$X_{NO_3^-} = \frac{0.81}{8.01} = 0.101$$

3. Compute the equilibrium resin composition using Eq. (11–129).

$$\frac{\overline{X}_{B^+}}{1 - \overline{X}_{B^+}} = (K_{A^+ \to B^+})\left(\frac{X_{B^+}}{1 - X_{B^+}}\right)$$

$$\frac{\overline{X}_{NO_3^-}}{1 - \overline{X}_{NO_3^-}} = 7.0\left(\frac{0.101}{1 - 0.101}\right)$$

$$\overline{X}_{NO_3^-} = 0.44$$

Thus, 44 percent of the exchange sites on the resin can be used for the removal of nitrate.

4. Determine the limiting operating capacity of the resin for the removal of nitrate.

Limiting operating capacity = (2 eq/L of resin)(0.44) = 0.88 eq/L of resin

5. Determine the volume of water that can be treated during a service cycle.

$$Vol = \frac{\text{(nitrate removal capacity of resin, eq/L of resin)}}{\text{(nitrate in solution, eq/L of water)}}$$

$$= \frac{(0.88 \text{ eq/L of resin})}{(0.81 \times 10^{-3} \text{ eq/L of water})} = 1086 \frac{\text{L of water}}{\text{L of resin}}$$

Solution:
Wastewater B

1. Estimate the selectivity coefficient (see Table 11–49). To apply Eq. (11–130) the system must be reduced to two components. For this purpose, HCO_3^-, Cl^-, and NO_3^- are combined into a single monovalent component. The selectivity coefficient is estimated as follows:

$$K_{HCO_3^- \to SO_4^{2-}} = \frac{0.15}{0.4} = 0.4$$

$$K_{Cl^- \to SO_4^{2-}} = \frac{0.15}{1.0} = 0.15$$

$$K_{NO_3^- \to SO_4^{2-}} = \frac{0.15}{4.0} = 0.04$$

$$K_{[(NO_3^-)(HCO_3^-)(Cl^-)] \to SO_4^{2-}} = 0.2 \text{ (estimated)}$$

2. For the equilibrium condition ($C_e/C_o = 1.0$), estimate the sulfate equivalent fraction in solution.

$$X_{SO_4^{2-}} = \frac{1.65}{8.27} = 0.2$$

3. Compute the equilibrium resin composition using Eq. (11–130).

$$\frac{\overline{X}_{B^{2-}}}{(1 - \overline{X}_{B^{2-}})^2} = (K_{A^- \to B^{2-}}) \left(\frac{C}{C} \right) \left[\frac{X_{B^{2-}}}{(1 - X_{B^{2-}})^2} \right]$$

$$\frac{\overline{X}_{SO_4^{2-}}}{(1 - \overline{X}_{SO_4^{2-}})^2} = 0.2 \frac{2}{0.00827} \left[\frac{0.2}{(1 - 0.2)^2} \right]$$

$$\overline{X}_{SO_4^{2-}} = 0.77, \text{ determined by successive trials}$$

Thus, 77 percent of the exchange sites on the resin will be in the divalent form at equilibrium. The relative amount of NO_3^- can be estimated by assuming that the remaining 23 percent of the resin sites are in equilibrium with a solution of NO_3^-, HCO_3^-, and Cl^- with the same relative concentration as the feed.

The equivalent fraction of nitrate in the solution will then be

$$X_{NO_3^-} = \frac{0.81}{6.62} = 0.12$$

The selectivity coefficient for the monovalent system is estimated:

$$K_{HCO_3^- \to NO_3^-} = \frac{4.0}{0.4} = 10.0$$

$$K_{Cl^- \to NO_3^-} = \frac{4.0}{1.0} = 4.0$$

$$K_{[(HCO_3^-)(Cl^-)] \to NO_3^-} = 7.0 \text{ (estimated)}$$

Compute the equilibrium resin composition using Eq. (11–129).

$$\frac{\overline{X}'_{B^+}}{1 - \overline{X}'_{B^+}} = (K_{A^+ \to B^+}) \left(\frac{\overline{X}'_{B^+}}{1 - \overline{X}'_{B^+}} \right)$$

$$\frac{\overline{X}'_{NO_3^-}}{1 - \overline{X}'_{NO_3^-}} = 7.0 \left(\frac{0.12}{1 - 0.12} \right)$$

$$\overline{X}'_{NO_3^-} = 0.5$$

The fraction of the total resin capacity in the nitrate form is then computed.

$$\overline{X}_{NO_3^-} = (1 - \overline{X}_{SO_4^{2-}})(\overline{X}'_{NO_3^-}) = (0.23)(0.5) = 0.115$$

4. Determine the limiting operating capacity of the resin for the removal of nitrate.

Limiting operating capacity = (2 eq/L of resin)(0.115) = 0.23 eq/L of resin.

5. Determine the volume of water that can be treated during a service cycle.

$$\text{Vol} = \frac{\text{(nitrate removal capacity of resin, eq/L of resin)}}{\text{(nitrate in solution, eq/L of water)}}$$

$$= \frac{\text{(0.23 eq/L of resin)}}{(0.81 \times 10^{-3} \text{ eq/L of water})} = 284 \frac{\text{L of water}}{\text{L of resin}}$$

Comment   As illustrated in this problem, the ionic composition of the wastewater can have a significant effect on the amount of water that can be treated per unit volume of resin, especially where nitrate is to be removed. Because the sulfate is more than 25 percent of the sum of the sulfate and nitrate, the use of a nitrate selective resin would be advantageous in this application. The approximate nature of these calculations also demonstrates the importance of conducting pilot plant tests to establish actual throughput volumes.

Removal of Heavy Metals.   Metal removal may be required as a pretreatment before discharge to a municipal sewer system. Because of the potential accumulation and toxicity of these metals, it is desirable to remove them from wastewater effluents before release to the environment. Ion exchange is one of the most common forms of treatment used for the removal of metals. Facilities and activities that may discharge wastewater containing high concentrations of metals include metal processing, electronics industries (semiconductors, printed circuit boards), metal plating and finishing, pharmaceuticals and laboratories, and vehicle service shops. High metal concentrations can also be found in leachate from landfills, and stormwater runoff.

Where industries produce effluents with widely fluctuating metal concentrations, flow equalization may be required to make ion exchange feasible. The economic feasibility of using ion exchange processes for metal removal greatly improves when the process is used for the removal and recovery of valuable metals. Because it is now possible to manufacture resins for specific applications, the use of resins that have a high selectivity for the desired metal(s) also improves the economics of ion exchange.

Materials used for the exchange of metals include zeolites, weak and strong anion and cation resins, chelating resins, and microbial and plant biomass. Biomass materials are generally more abundant, and therefore, less expensive when compared to other commercially available resins. Natural zeolites, clinoptilolite (selective for Cs), and chabazite (mixed metals background Cr, Ni, Cu, Zn, Cd, Pb) have been used to treat wastewater with mixed metal backgrounds (Ouki and Kavannagh, 1999). Chelating resins, such as aminophosphonic and iminodiacetic resins, have been manufactured to have a high selectivity for specific metals such as Cu, Ni, Cd, and Zn.

Ion exchange processes are highly pH dependent. Solution pH has a significant impact on the metal species present and the interaction between exchanging ions and the resin. Most metals bind better at higher pH due to less competition from protons for sites. Operating and wastewater conditions determine selectivity of the resin, pH, temperature, other ionic species, and chemical background. The presence of oxidants, particles, solvents, and polymers may affect the performance of ion exchange resins. The quantity and quality of regenerate produced and subsequently requiring management must also be considered.

Removal of Total Dissolved Solids.   For the reduction of the total dissolved solids, both anionic and cationic exchange resins must be used (see Fig. 11–73). The wastewater is

Figure 11-73

Typical flow diagram for the removal of hardness and for the complete demineralization of water using ion exchange resins.

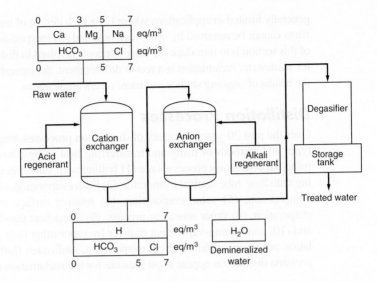

first passed through a cation exchanger where the positively charged ions are replaced by hydrogen ions. The cation exchanger effluent is then passed over an anionic exchange resin where the anions are replaced by hydroxide ions. Thus, the dissolved solids are replaced by hydrogen and hydroxide ions that react to form water molecules.

Total dissolved solids removal can take place in separate exchange columns arranged in series, or both resins can be mixed in a single reactor. Wastewater application rates range from 0.20 to 0.40 m³/m²·min (5 to 10 gal/ft²·min). Typical bed depths are 0.75 to 2.0 m (2.5 to 6.5 ft). In reuse applications, treatment of a portion of the wastewater by ion exchange, followed by blending with wastewater not treated by ion exchange, would possibly reduce the dissolved solids to acceptable levels. In some situations, it appears that ion exchange may be as competitive as reverse osmosis.

Operational Considerations

To make ion exchange economical for advanced wastewater treatment, it would be desirable to use regenerants and restorants that would remove both the inorganic anions and the organic material from the spent resin. Chemical and physical restorants found to be successful in the removal of organic material from resins include sodium hydroxide, hydrochloric acid, methanol, and bentonite. To date, ion exchange has had limited application because of the extensive pretreatment required, concerns about the life of the ion exchange resins, and the complex regeneration system required.

High concentrations of influent TSS can plug the ion exchange beds, causing high headlosses and inefficient operation. Resin binding can be caused by residual organics found in biological treatment effluents. Some form of chemical treatment and clarification is required before ion exchange demineralization. This problem has been solved partially by prefiltering the wastewater or by using scavenger exchange resins before application to the exchange column.

11-12 DISTILLATION

Distillation is a unit process in which the components of a liquid solution are separated by vaporization and condensation. Along with reverse osmosis, distillation can be used to control the buildup of salts in critical reuse applications. Because distillation is expensive, its application is generally limited to applications where (1) a high degree of treatment is required, (2) contaminants cannot be removed by other methods, and (3) inexpensive heat is available. The purpose of this section is to introduce the basic concepts involved in distillation. As the use of distillation for wastewater reclamation is a recent development, the current literature must be consulted for the results of ongoing studies and more recent applications.

Distillation Processes

Over the past 20 years, a variety of distillation processes, employing a variety of evaporator types and methods of using and transferring heat energy, have been evaluated or used. The principal distillation processes are (1) boiling with submerged tube heating surface, (2) boiling with long-tube vertical evaporator, (3) flash evaporation, (4) forced circulation with vapor compression, (5) solar evaporation, (6) rotating-surface evaporation, (7) wiped-surface evaporation, (8) vapor reheating process, (9) direct heat transfer using an immiscible liquid, and (10) condensing-vapor-heat transfer by vapor other than steam. Of these types of distillation processes, multiple-effect evaporation, multistage flash evaporation, and vapor-compression distillation appear most feasible for the reclamation of municipal wastewater.

Multiple-Effect Evaporation Distillation.   In multiple-effect evaporation distillation systems, several evaporators (boilers) are arranged in series, each operating at a lower pressure than the preceding one. In a three-stage, vertical-tube evaporator (see Fig. 11–74), preheated influent water to be demineralized is introduced into the first evaporation stage where it is evaporated with steam contained within heat exchange tubes. The vapor from the first stage enters the second stage where it is condensed within the evaporation tubes. Water from the first stage is the feed water for the second stage. The process is repeated in the next n^{th} stages. Heated vapor from the last stage is used to heat the influent feed water. In an alternative arrangement, the feed water for the second and subsequent stages is the preheated influent. Water which does not evaporate is taken off as brine at each stage. If air entrainment is kept low, almost all of the nonvolatile contaminants can be removed in a single evaporation step. Volatile contaminants, such as ammonia gas and low-molecular weight organic acids, may be removed in a preliminary evaporation step, but if their concentration is so small that their presence in the final product is not objectionable, this step with its added cost can be eliminated.

Multistage Flash Evaporation Distillation.   Multistage flash evaporation distillation systems have been used commercially in desalination for many years. In the multistage flash process (see Fig. 11–75), the influent wastewater is first treated to remove TSS and deaerated before being pumped through heat transfer units in the several stages of the

Figure 11–74

Schematic of multiple-effect evaporation distillation process.

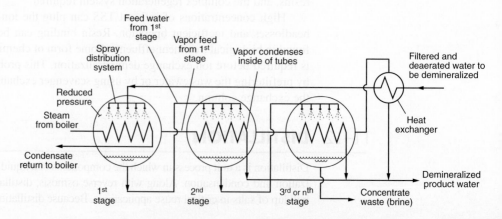

Figure 11-75

Schematic of multistage flash evaporation distillation process.

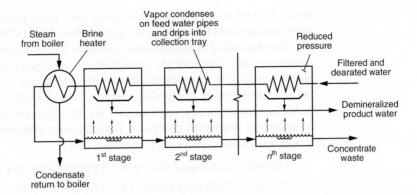

distillation system, each of which is maintained at a lower pressure. Vapor generation or boiling caused by reduction in pressure is known as *"flashing."* As the water enters each stage through a pressure reducing nozzle, a portion of the water is flashed to form a vapor. In turn, the flashed water vapor condenses on the outside of the condenser tubes and is collected in trays (see Fig. 11–75). As the vapor condenses, its latent heat is used to preheat the wastewater that is being returned to the main heater where it will receive additional heat before being introduced to the first flashing stage. When the concentrated wastewater reaches the lowest pressure stage, it is pumped out. Thermodynamically, multistage flash evaporation is less efficient than ordinary evaporation. However, by combining a number of stages in a single reactor, external piping is eliminated and construction costs are reduced.

Vapor Compression Distillation.   In the vapor compression process an increase in pressure of the vapor is used to establish the temperature difference for the transfer of heat. The basic schematic of a vapor compression distillation unit is shown on Fig. 11–76. After initial heating of the wastewater, the vapor pump is operated so that the vapor under higher pressure can condense in the condenser tubes, at the same time causing the release of an equivalent amount of vapor from the concentrated solution. Heat exchangers can conserve heat from both the condensate and the waste brine. The only energy input required during operation is the mechanical energy for the vapor pump. Hot concentrated wastewater must be discharged at intervals to prevent the buildup of excessive concentrations of salt in the boiler.

Performance Expectations in Reclamation Applications

The principal issues with the application of the distillation processes for wastewater reclamation are the carryover of volatile constituents found in treated wastewater and the degree of subsequent cooling and treatment that may be required to renovate the distilled

Figure 11-76

Schematic of vapor compression distillation process.

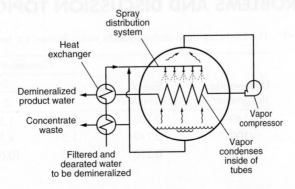

water. Typical water quality performance data for a multiple-effect distillation process have been reported for a pilot scale unit by Rose et al. (1999).

The theoretical thermodynamic minimum energy required to raise the temperature of wastewater and to provide the latent heat of vaporization is about 2260 kJ/kg. Typically, about 1.25 to 1.35 times the latent heat of vaporization will be required. Unfortunately, because of the many irreversibilities in an actual distillation processes, the thermodynamic minimum energy requirements are of little practical relevance in the practical evaluation of distillation processes. Typical energy requirements for the three distillation processes discussed above are (Voutchkov, 2013):

Multiple-effect evaporation: 5.7–7.8 kWh/m³ (23–30 kWh/10³ gal)

Multistage flash evaporation: 12.7–15.0 kWh/m³ (48–57 kWh/10³ gal)

Vapor compression: 8–12 kWh/m³ (30–45 kWh/10³ gal)

These values for distillation can be compared to the values given in Table 11–30 for reverse osmosis: 9–12 kWh/m³ (34–45 kWh/10³ gal) without energy recovery and <3–4 kWh/m³ (<11–32 kWh/10³ gal) with energy recovery. Based on this energy comparison it is clear why reverse osmosis used in seawater distillation has largely replaced distillation for seawater desalination.

Operating Problems

The most common operating problems encountered include scaling and corrosion. Due to temperature increases, inorganic salts come out of solution and precipitate on the inside walls of pipes and equipment. The control of scaling due to calcium carbonate, calcium sulfate, and magnesium hydroxide is one of the most important design and operational considerations in distillation desalination processes. Controlling the pH minimizes carbonate and hydroxide scales. Most inorganic solutions are corrosive. Cupronickel alloys are used most commonly in sea water desalination. Other metals that are used include aluminum, bronze, titanium, and monel.

Disposal of Concentrated Waste

All distillation processes reject part of the influent wastewater. Hence, all of these processes have concentrated wastewater disposal problems. The permissible maximum concentration in the wastewater depends on the solubility, corrosion, and vapor pressure characteristics of the wastewater. Therefore, the waste concentration is an important consideration in process optimization. Disposal of concentrated wastewater brines is essentially the same problem encountered with the membrane processes discussed in Sec. 11–5.

PROBLEMS AND DISCUSSION TOPICS

11-1 The following sieve analysis results were obtained for four different stock sands.

US sieve size designation[a]	Size of opening, mm	Cumulative weight passing, %.			
		Sand sample			
		1	2	3	4
140	0.105	0.4	1.5	0.1	5.0
100	0.149	1.5	4.1	0.8	11.1
70	0.210	4.0	10.0	2.5	20.0

(continued)

(*Continued*)

US sieve size designation[a]	Size of opening, mm	Cumulative weight passing, %. Sand sample			
		1	2	3	4
50	0.297	9.5	21.0	8.2	32.0
40	0.420	18.5	40.6	18.5	49.5
30	0.590	31.0	61.0	32.0	62.3
20	0.840	49.0	78.3	58.1	78.3
16	1.190	63.2	90.0	76.3	88.5
12	1.680	82.8	96.0	90.0	94.4
8	2.380	89.0	99.0	96.7	97.8
6	3.360	98.0	99.9	99.0	99.0
4	4.760	100.0	100.0	100.0	100.0

[a] Note: sieve size number 18 has an opening size of 1.0 mm.

a. For sand sample 1, 2, 3, or 4 (to be selected by instructor) determine the geometric mean size, the geometric standard deviation, the effective size, and the uniformity coefficient for the stock sand.

b. It is desired to produce from the stock sand a filter sand with an effective size of 0.45 mm and a uniformity coefficient of 1.6. Estimate the amount of stock sand needed to obtain one ton of filter sand.

c. What U.S. Standard sieve size should be used to eliminate the excess coarse material?

d. If the sand remaining after sieving in Part c above is placed in a filter, what backwash rise rate would be needed to eliminate the excess fine material?

e. What depth of sieved material would have to be placed in the filter to produce 600 mm of usable filter sand?

f. On log-probability paper, plot the size distribution of the modified sand. Check against the required distribution and sizes.

g. Determine the headloss through 600 mm of the filter sand specified in part (b) for a filtration rate of 160 L/m ^{2}·min (4 gal/ft^2·min). Assume the sand is stratified and that the maximum and minimum sand sizes are 1.68 (sieve size 12) and 0.297 mm (sieve size 50), respectively. Assume also that $T = 20°C$, $\alpha = 0.4$ for all layers, and $\psi = 0.75$.

An excellent discussion of the procedures involved in developing a usable filter sand from a stock filter sand may be found in Fair et al. (1968).

11–2 Determine the sphericity and specific surface area of two of the following filter mediums (filter medium to be selected by instructor).

Item	Filter medium				
	1	2	3	4	5
Shape	Ellipsoid	Icosahedron	Cube	Rod	Isosceles tetrahedron
Dimensions	1 mm × 1.2 mm × 2 mm	Each face, 0.5 mm × 0.5 mm × 0.5 mm	1 mm × 1 mm × 1 mm	0.5 mm diameter × 2 mm in length	1.2 mm × 1.2 mm × 1.2 mm × 2.5 mm

11-3 Using the equations developed by Kozeny, Rose, and Ergun, given in Table 11–8, compare the headloss through a 600 mm sand bed. Assume that the sand bed is composed of spherical unsized sand with a diameter of 0.45, 0.55, or 0.6 mm (to be selected by instructor), the porosity of the sand is 0.40, the filtration rate is 240 L/m ^{2}·min (5 gal/ft^2·min) and the temperature is 15°C. The kinematic viscosity at 15°C is equal to 1.139×10^{-6} m^2/s (see Table C–1, Appendix C).

11-4 Using the equation developed by Rose, determine the headloss through a 750-mm sand bed for a filtration rate of 300 L/m^2·min. Assume that the sand bed is composed of spherical unsized sand with a diameter of 0.40, 0.45, 0.55, or 0.6 mm (size to be selected by instructor), the porosity of the sand is 0.40, and the temperature of the water is 10°C. The kinematic viscosity at 10°C is equal to 1.306×10^{-6} m^2/s (see Table C-1, Appendix C).

11-5 Solve Problem 11–4 assuming the bed is stratified. Assume that the given sand sizes correspond to the effective size (d_{10}) and that the uniformity coefficient for all of the sand sizes is equal to 1.5.

11-6 If a 0.3-m layer of uniform anthracite is placed on top of the sand bed in Prob. 11–3, determine the ratio of the headloss through the anthracite to that of the sand. Assume that the grain-size diameter of the anthracite is 2.0 mm and porosity is 0.50. Will intermixing occur?

11-7 Given the particle size distribution 1, 2, 3, or 4 (to be selected by instructor), determine the effective size (d_{10}) and uniformity coefficient UC, and clean-water headloss through a stratified bed 600 mm deep. If a layer of anthracite is to be added over 600 mm of sand determine the effective size required to minimize intermixing. Assume the filtration rate is 160 L/m^2·min, $\phi = 0.85$, and $\alpha = 0.4$.

| | Percent of sand retained | | | |
| | Particle size distribution | | | |
Sieve number	1	2	3	4
6–8	2	0	1	0.1
8–10	8	0.1	2	0.7
10–14	10	0.5	4	1.2
14–20	30	7.4	13	10
20–30	26	32	20	24
30–40	14	30	20	29
40–60	8	25	23	25
Pan	2	5	17	10

11-8 Four stratified sand beds with the size distribution given below are to be backwashed at a rate of 0.75 m^3/m^2·min. Determine the degree of expansion and whether the proposed backwash rate will expand all of the filter bed (to be selected by instructor). Assume the following data are applicable: sand specific gravity is 2.65, depth of the filer bed is 0.90 m, and the temperature is 20°C.

| | Percent of sand retained | | | |
| | Stratified bed number | | | |
Sieve size or number	1	2	3	4
8–10	10	2	0.1	0
10–12	10	4	0.5	0
12–18	30	14	4.4	0

(continued)

(*Continued*)

Sieve size or number	Percent of sand retained			
	Stratified bed number			
	1	2	3	4
18–20	10	8	7	1
20–30	34	40	48	28
30–40	5	22	30	41
40–50	1	9	9	27
Pan		1	1	3

11-9 Given the following granular medium filter effluent turbidity data collected at four different wastewater treatment plants, estimate the mean, the geometric standard deviation, s_g, and the probability of exceeding a turbidity reading of 2.5 NTU (treatment plant to be selected by instructor).

Turbidity, NTU			
Treatment plant			
1	2	3	4
1.7	1.7	1.0	1.2
1.8	1.1	1.8	1.4
2.2	0.9	1.5	1.5
2.0	1.4	1.1	1.6
	1.3	1.7	1.7
		1.3	1.9
			2.0
			2.1

11-10 Assuming the data for treatment plants 1 and 2 in Problem 11–9 were collected at the same treatment plant at different times, what is the impact of using all of the data given for 1 and 2 as one data set versus using the individual data sets? In general, what are the advantages or disadvantages of collecting more turbidity samples?

11-11 Gravity filters are to be used to treat 16,000, 20,000, or 24,000 m³/d, to be selected by instructor, of settled effluent at a filtration rate of 200 L/m²·min (5 gal/ft²·min). The filtration rate with one filter taken out of service for backwashing is not to exceed 240 L/m²·min (6 gal/ft²·min). Determine the number of units and the area of each unit to satisfy these conditions. If each filter is backwashed for 30 min every 24 h at a wash rate of 960 L/m²·min (24 gal/ft²·min), determine the percentage of filter output used for washing if the filter is out of operation for a total of 30 min/d. What would be the total percentage of filter output used for backwashing if a surface washing system that requires 40 L/m²·min (1 gal/ft²·min) of filtered effluent is to be installed?

11-12 Using the performance data given in the following table for a microfiltration membrane, determine (water to be selected by instructor) the rejection and log rejection for each microorganism group.

Microorganism	Microorganisms concentration, org/mL			
	Water 1		Water 2	
	Feed water	Permeate	Feed water	Permeate
HPC	6.5×10^7	3.3×10^2	8.6×10^7	1.5×10^2
Total coliform	3.4×10^6	100	5×10^5	60
Enteric virus	7×10^3	6.6×10^3	2.0×10^3	9.1×10^2

11-13 A hollow-fiber membrane system with inside to outside flow is operated in a cross-flow arrangement. Each module contains 6000 fibers that have an inside diameter of 1.0 mm and a length of 1.25 m. Using this information determine

a. the feed water flowrate at the entrance to the module needed to achieve a cross-flow velocity of 1 m/s within the membrane fibers.

b. the permeate flowrate if the permeate flux of 100 L/m²·h is maintained.

c. the retentate cross-flow velocity at the exit from the membrane fibers.

d. the ratio of velocity of flow through the membrane surface to the average cross-flow velocity within an individual membrane fiber.

e. the ratio of permeate flowrate to feed water flowrate.

This problem was adapted from Crittenden et al., 2012.

11-14 Membrane filtration, operated in a dead-end mode, is used to treat secondary effluent. If the heterotrophic microorganism plate count (HPC) in the effluent increased from 5 org/L under normal operation to 200 org/L after an extended period of operation, estimate the number of broken fibers for the following conditions. The influent flowrate and organism count are 4000 m³/d and 6.7×10^7 org/L, respectively. The membrane bundle contains 5000 individual fibers. If the influent and effluent turbidity values under normal operation are 4 and 0.25 NTU respectively, estimate the increase in the effluent turbidity assuming the increase could be measured.

11-15 Contrast the advantages and disadvantages between depth filtration, surface filtration, and microfiltration. Cite a minimum of three recent articles (after 2000).

11-16 Four different waters are to be desalinized by reverse osmosis using a thin film composite membrane. For water 1, 2, 3, or 4 (water to be selected by instructor), determine the required membrane area, the rejection rate, and the concentration of the retentate.

		Water			
Item	Unit	1	2	3	4
Flowrate	m³/d	4000	5500	20,000	10,000
Influent TDS	g/m³	2850	3200	2000	2700
Permeate TDS	g/m³	200	500	400	225
Flux rate coefficient k_w	m/s·bar[a]	1.0×10^{-6a}	1.0×10^{-6}	1.0×10^{-6}	1.0×10^{-6}
Mass transfer rate coefficient, k_i	m/s	6.0×10^{-8}	6.0×10^{-8}	6.0×10^{-8}	6.0×10^{-8}
Net operating pressure	kPa	2750	2500	2800	3000
Recovery	%	88.0	90.0	89	86

[a] 1.0×10^{-6} m/s·bar = 1.0×10^{-8} s/m.

11-17 Using the data given below, determine the recovery and rejection rates for one of the following reverse osmosis units (unit to be selected by instructor).

		Reverse osmosis unit			
Item	Unit	1	2	3	4
Feed water flowrate	m³/d	4000	6000	8000	10,000
Retentate flowrate	m³/d	350	600	7500	9000
Permeate TDS	g/m³	65	88	125	175
Retentate TDS	g/m³	1500	2500	1850	2850

11-18 Using the data given below, determine the flux rate coefficient and the mass transfer rate coefficient.

Item	Unit	Reverse osmosis unit			
		1	2	3	1
Flowrate, Q_f	m³/d	4000	5500	20,000	10,000
Influent TDS, C_f	g/m³	2500	3300	5300	2700
Permeate TDS, C_p	g/m³	20	50	40	23
Net operating pressure, ΔP	bar	28	25	28	30
Membrane area	m²	1600	1700	9600	5500
Recovery, r	%	88.0	90.0	89	86

11-19 Estimate the SDI for the following filtered wastewater samples. If the water is to be treated with reverse osmosis will additional treatment be required?

Test run time, min	Volume filtered, mL			
	Wastewater sample number			
	1	2	3	4
2	315	480	180	500
5	575	895	395	700
10	905	1435	710	890
20	1425	2300	1280	1150

11-20 Calculate the modified fouling index (MFI) for the effluent from a microfiltration process (water sample to be selected by instructor) using the following experimental data:

Time, min	Volume filtered, L		Time, min	Volume filtered, L	
	Water sample			Water sample	
	1	2		1	2
0			3.5	6.78	7.17
0.5	1.50	1.50	4.0	7.48	8.03
1.0	2.50	2.50	4.5	8.08	8.87
1.5	3.45	3.48	5.0	8.57	9.67
2.0	4.36	4.40	5.5		10.34
2.5	5.22	5.37	6.0		10.97
3.0	6.03	6.28	6.5		11.47

11-21 Determine the cost (based on the current price of electricity) to treat a flowrate of 2500 m³/d with a TDS concentration of 1300 g/m³ and a cation and anion concentration of 0.13 g-eq/L using an electrodialysis unit. Assume the following typical values of operation for the electrodialysis unit.

Product flowrate = 90% of the feed water flowrate
Efficiency of salt removal = 50%
The current efficiency = 90%
Resistance = 5.0 ohms
Number of cell pairs in the stack = 350, 400, 450 (to be selected by instructor)
Assume an energy cost of $0.13/kWh and 24 h/d operation.

11-22 Review and cite three current articles (within the last five years) dealing the disposal of nanofiltration, reverse osmosis, or electrodialysis brine. What types of process combinations are being proposed? What are the critical issues that stand out in your mind?

11–23 A wastewater is to be treated with activated carbon to remove residual COD. The following data were obtained from a laboratory adsorption study in which 1 g of activated carbon was added to a beaker containing 1 L of wastewater at selected COD values. Using these data, determine the more suitable isotherm (Langmuir or Fruendlich) to describe the data (sample to be selected by instructor).

Initial COD, mg/L	Equilibrium COD, mg/L			
	Wastewater sample number			
	1	2	3	4
140	5	10	0.4	5
250	12	30	0.9	18
300	17	50	2	28
340	23	70	4	36
370	29	90	6	42
400	36	110	10	50
450	50	150	35	63

11–24 Using the following isotherm test data, determine the type of model that best describes the data and the corresponding model parameters. Assume that a 1 L sample volume was used for each of the isotherm experiments.

Mass of GAC, mg	Equilibrium concentration of adsorbate in solution, C_e, μg/L			
	Test number			
	1	2	3	4
0	5.8	26	158.2	25.3
0.001	3.9	10.2	26.4	15.89
0.01	0.97	4.33	6.8	13.02
0.1	0.12	2.76	1.33	6.15
0.5	0.022	0.75	0.5	2.1

11–25 Using the results from Problem 11–23, determine the amount of activated carbon that would be required to treat a flowrate of 4800 m³/d to a final COD concentration of 2 mg/L if the COD concentration after secondary treatment is equal to 30 mg/L.

11–26 Design a fixed-bed activated carbon process using the following data. Determine the number of contactors, mode of operation, carbon requirements, and corresponding bed life. Ignore the effects of biological activity within the column.

Parameter	Unit	Compound			
		Chloroform	Heptachlor	Methylene chloride	NDMA
Flowrate	m³/d	4000	4500	5000	6000
C_o	ng/L	500	50	2000	200
C_e	ng/L	50	10	10	10
GAC density	g/L	450	450	450	450
EBCT	min	10	10	10	10

11–27 Referring to the data presented in Table 11–23, prepare a list of the top 5 most and least readily adsorbable substances.

11–28 Using the results from Prob. 11–13, determine the amount of activated carbon that would be required to treat a flowrate of 5000 m³/d to a final COD concentration of 20 mg/L if the COD concentration after secondary treatment is equal to 120 mg/L.

11-29 Using the following carbon adsorption data (sample number to be selected by instructor) determine the Freundlich capacity factor (mg absorbate/g activated carbon) and Freundlich intensity parameter, $1/n$.

Carbon dose, mg/L	Residual concentration, mg/L					
	Sample number					
	1	2	3	4	5	6
0	25.9	9.20	9.89	27.5	20.4	9.88
5	17.4	7.36	9.39	24.8	19.3	7.95
10	13.2	6.86	8.96	24.2	18.6	7.02
25	10.2	3.86	7.83	18.9	16.1	3.66
50	3.6	1.13	5.81	11.8	12.2	0.98
100	2.5	0.22	4.45	2.3	6.7	0.25
150	2.1	0.18	2.98	1.1	3.1	0.09
200	1.4	0.11	2.01	0.9	1.1	0.04

11-30 Determine the theoretical air flowrate required to remove the following compounds in a stripping tower at the indicated concentrations (compound and water to be selected by instructor). Also estimate the height of the stripping tower for a water flowrate of 3000 m³/d. Values of the Henry's law constant may be found in Table 16–12.

Compound	K_La, s^{-1}	Concentration μg			
		Water 1		Water 2	
		Influent	Effluent	Influent	Effluent
Clorobenzene	0.0163	100	5	120	7
Chloroethene	0.0141	100	5	150	5
TCE[a]	0,0176	100	5	180	10
Toluene	0.0206	100	5	200	15

[a] Henry's law constant for TCE is equal to 0.00553 m³·atm/mole

11-31 A quantity of sodium-form ion exchange resin (5 g) is added to a water containing 2 meq of potassium chloride and 0.5 meq of sodium chloride. Calculate the residual concentration of potassium if the exchange capacity of the resin is 4.0 meq/g of dry weight and the selectivity coefficient is equal to 1.46.

11-32 Determine the exchange capacity for one of the following resins (resin to be selected by instructor). How much resin would be required to treat a flowrate of 4000 m³/d to reduce the concentration of calcium (Ca^{2+}) from 125 to 45 mg/L? Assume the mass of resin used to obtain the data given in the table is 0.1 kg.

Throughput volume, L	Resin 1		Resin 2	
	Cl^-	Ca^{2+}	Cl^-	Ca^{2+}
0	0	0	0	0
5	2	0	2	0
10	8	0	13	0
15	44	0	29	0
20	65	0	45	0
25	70	0	60	1
30	71	0	69	8
35	71	6	71	17
40	71	20	71	27

(continued)

(Continued)

Throughput volume, L	Resin 1		Resin 2	
	Cl⁻	Ca²⁺	Cl⁻	Ca²⁺
45		34	71	35
50		39		39
55		40		40
60		40		40

11-33 Determine the exchange capacity for one of the resins given in Problem 11–32 (resin to be selected by instructor). How much resin would be required to treat a flowrate of 5500 m^3/d to reduce the concentration of magnesium, Mg^{2+}, from 115 to 15 mg/L?

11-34 Four different wastewaters have been reported to have the following ionic composition data. Estimate the selectivity coefficient and determine the amount of wastewater (1, 2, 3, or 4, to be selected by instructor) that can be treated by a strong-base ion exchange resin, per service cycle, for the removal of nitrate. Assume the resin has an ion exchange capacity of 1.8 eq/L.

Cation	Conc., mg/L	Anion	Concentration, mg/L			
			Wastewater sample number			
			1	2	3	4
Ca^{2+}	82.2	HCO_3^-	304.8	152	254	348
Mg^{2+}	17.9	SO_4^{2-}	0	0	0	0
Na^+	46.4	Cl^-	58.1	146.3	124	60
K^+	15.5	NO_3^-	82.5	90	21.5	42

11-35 Four different wastewaters have been reported to have the following ionic composition data. Estimate the selectivity coefficient and determine the amount of wastewater (1, 2, 3, or 4, to be selected by instructor) that can be treated by a strong-base ion exchange resin, per service cycle, for the removal of nitrate. Assume the resin has an ion exchange capacity of 2.5 eq/L.

Cation	Conc., mg/L	Anion	Concentration, mg/L			
			Wastewater sample number			
			1	2	3	4
Ca^{2+}	82.2	HCO_3^-	321	180	198.5	69
Mg^{2+}	17.9	SO_4^{2-}	65	36.5	124	136
Na^+	46.4	Cl^-	22	95	56	87
K^+	15.5	NO_3^-	46	93	34.5	97

11-36 For each compound in the following list, which of the treatment methods discussed in this chapter, if any, are suitable for use to reduce the concentration from 100 to 10 $\mu g/L$?

Benzene

Chloroform

Dieldrin

Heptachlor

N-Nitrosodimethylamine

Trichloroethylene (TCE)

Vinyl chloride

REFERENCES

Adham, S., T. Gillogly, G. Lehman, E. Rosenblum, and E. Hansen, (2004) *Comparison of Advanced Treatment Methods for Partial Desalting of Tertiary Effluents, Desalination and Water Purification Research and Development,* Report No. 97, Agreement No. 99-FC-81–0189, U.S. Department of the Interior, Bureau of Reclamation, Denver, CO.

Ahn, K. H., J. H. Y. Song Cha, K. G. Song, and H. Yoo (1998) "Application of Tubular Ceramic Membranes For Building Wastewater Reuse," *Proceedings IAWQ 19 th International Conference,* Vancouver, p. 137.

Amirtharajah, A. (1978) "Optimum Backwashing of Sand Filters," *J. Environ. Eng. Div., ASCE,* **104**, EE5, 917–932.

Anderson, R. E. (1975) "Estimation Of Ion Exchange Process Limits By Selectivity Calculations," in I. Zwiebel and N. H. Sneed (eds), *Adsorption and Ion Exchange, AICHE Symposium Series,* **71**, 152, 236–242.

Anderson, R. E. (1979) "Ion Exchange Separations," in P. A. Scheitzer (ed), *Handbook of Separation Techniques For Chemical Engineers,* McGraw-Hill, New York.

ASTM (2001a) *C136–01 Standard Test Method for Sieve Analysis of Fine and Coarse Aggregates,* American Society for Testing and Materials, Philadelphia, PA.

ASTM (2001b) *E11–01 Standard Specification for Wire Cloth and Sieves for Testing Purposes,* American Society for Testing and Materials, Philadelphia, PA.

ASTM (2002) *D4189–95 Standard Test Method for Silt Density Index (SDI) of Water,* American Society for Testing and Materials, Philadelphia, PA.

AWWA (1996) *AWWA Standard for Filtering Material, B100–96,* American Water Works Association, Denver, CO.

Bard, A. J. (1966) *Chemical Equilibrium,* Harper & Row, Publishers, New York.

Bilello, L. J., and B. A. Beaudet (1983) "Evaluation of Activated Carbon by the Dynamic Minicolumn Adsorption Technique," in M. J. McGuire and I. H. Suffet (eds.), *Treatment of Water by Granular Activated Carbon,* American Chemical Society, Washington, DC.

Blake F. C., (1922) "The Resistance of Packing to Fluid Flow," *Trans. Am. Inst. Chem. Eng.,* **14**, 415–421.

Bohart, G. S., and E. Q. Adams (1920) "Some Aspects of the Behavior of Charcoal with Respect to Chlorine," *J. Am. Chem. Soc.,* **42**, 3, 523–529.

Bonner, O. D., and L. L. Smith (1957) "A Selectivity Scale For Some Divalent Cations on Dowex 50," *J. Physical Chem.,* 61, 3, 326–329.

Bourgeous, K., G. Tchobanoglous, and J. Darby (1999) "Performance Evaluation of the Koch Ultrafiltration (UF) Membrane System for Wastewater Reclamation," Center For Environmental And Water Resources Engineering, Report No. 99–2, Department of Civil and Environmental Engineering, University of California, Davis, Davis, CA.

Caliskaner, O., and G. Tchobanoglous (2000) "Modeling Depth Filtration of Activated Sludge Effluent Using a Synthetic Compressible Filter Medium," Presented at the 73rd Annual Conference and Exposition on Water Quality and Wastewater Treatment, Water Environment Federation, Anaheim, CA.

Carman, P. C. (1937) "Fluid Flow Through Granular Beds," *Trans. Inst. Chem. Engrs.,* London, **15**, 150–166.

Cath, T. Y., A. E. Childress, and M. Elimelech (2006) "Forward Osmosis: Principles, Applications, and Recent Developments," *J. Mem. Sci.,* **281**, 9, 70–87.

Celenza, G. (2000) *Specialized Treatment Systems, Industrial Wastewater Process Engineering, vol. III,* Technomic Publishing Co., Inc., Lancaster, PA.

Cleasby, J. L., and K. Fan (1982) "Predicting Fluidization and Expansion of Filter Media," *J. Environ. Eng. Div., ASCE,* **107**, EE3, 455–472.

Cleasby, J. L., and G. S. Logsdon (1999) "Granular Bed and Precoat Filtration," Chap. 8, in R. D. Letterman (ed), *Water Quality and Treatment: A Handbook of Community Water Supplies,* 5th ed., American Water Works Association, McGraw-Hill, New York.

Crittenden, J. C., R. R. Trussell, D. W. Hand, K. J. Howe, and G. Tchobanoglous (2005) *Water Treatment: Principles and Design,* 2nd ed., John Wiley & Sons, Inc., Hoboken, NJ.

Crittenden, J. C., R. R. Trussell, D. W. Hand, K. J. Howe, and G. Tchobanoglous (2012) *Water Treatment: Principles and Design,* 3rd ed., John Wiley & Sons, Inc., New York.

Crittenden, J. C., P. Luft, D. W. Hand, J. L. Oravitz, S. W. Loper, and M. Art (1985) "Prediction of Multicomponent Adsorption Equilibria Using Ideal Adsorption Solution Theory," *Environ. Sci, Technol.,* **19**, 11, 1037–1043.

Crittenden, J. C., J. K. Berrigan, and D. W. Hand (1986) "Design of Rapid Small Scale Adsorption Tests for a Constant Diffusivity," *J. WPCF,* **58**, 4, 312–319.

Crittenden, J. C., D. W. Hand, H. Arora, and B. W. Lykins, Jr. (1987a) "Design Considerations for GAC Treatment of Organic Chemicals," *J. AWWA,* **79**, 1, 74–82.

Crittenden, J. C., T F. Speth, D. W. Hand, P. J. Luft, and B. W. Lykins, Jr. (1987b) "Multicomponent Competition in Fixed Beds," *J. Environ. Eng. Div., ASCE,* **113**, EE6, 1364–1375.

Crittenden, J. C., P. J. Luft, and D. W. Hand (1987c) "Prediction of Fixed-Bed Adsorber Removal of Organics in Unknown Mixtures," *J. Environ. Eng. Div., ASCE,* **113**, 3, 486–498.

Crittenden, J. C., J. K. Berrigan, and D. W. Hand (1987d) "Design of Rapid Fixed-bed Adsorption Tests for Nonconstant Diffusivities," *J. Environ. Eng. Div., ASCE,* **113**, 2, 243–259.

Crittenden, J. C., P. S. Reddy, H. Arora, J. Trynoski, D. W. Hand, D. L. Perram, and R. S. Summers (1991) "Predicting GAC Performance With Rapid Small-Scale Column Tests," *J AWWA,* **83**, 1, 77–87.

Crittenden, J. C., K. Vaitheeswaran, D. W. Hand, E. W. Howe, E. M. Aieta, C. H. Tate, M. J. Mcgurie, and M. K. Davis (1993) "Removal of Dissolved Organic Carbon Using Granular Activated Carbon," *Water Res.,* **27**, 4, 715–721.

Crittenden, J. C. (1999) *Class Notes,* Michigan Technological University, Houghton, MI.

Crozes, G. F., J. G. Jacangelo, C. Anselme, and J. M. Laine (1997) "Impact of Ultrafiltration Operating Conditions on Membrane Irreversible Fouling," *J. Membr. Sci.,* **124**, 63–76.

Culp, G. L., and A. Slechta (1966) *Nitrogen Removal From Sewage,* Final Progress Report, U. S. Public Health Service Demonstration Grant 29–01.

Dahab, M. F., and J. C. Young (1977) "Unstratified-Bed Filtration of Wastewater," *J. Environ. Eng. Div., ASCE,* **103**, 1, 21–36.

Darby, J., R. Emerick, F. Loge, and G. Tchobanoglous (1999) "The Effect of Upstream Treatment Processes on UV Disinfection Performance," Project 96-CTS-3, *Water Environment Research Foundation,* Washington DC.

Darcy, H. (1856) *Les fontaines publiques de la ville de Dijon* (in french), Victor Dalmont, Paris.

Dharmarajah, A. H., and J. L. Cleasby (1986) "Predicting the Expansion of Filter Media," *J. AWWA,* **78**, 12, 66–76.

Dimotsis, G. L., and F. McGarvey (1995) "A Comparison of a Selective Resin with a Conventional Resin for Nitrate Removal," IWC, No. 2.

Dobbs, R. A., and J. M. Cohen (1980) *Carbon Adsorption Isotherms for Toxic Organics,* EPA-600/8–80–023, U. S. Environmental Protection Agency, Washington, DC.

Dupont (1977) "Determination of the Silt Density Index," Technical Bulletin No. 491, Dupont de Nemours and Co., Wilmington, DE.

Eckert, J. S. (1970) "Selecting the Proper Distillation Column Packing," *Chem. Eng. Prog.,* **66**, 3, 39–44.

Eckert, J. S. (1975) "How Tower Packings Behave," *Chem. Eng.,* **82**, 4, 70–76.

Elimelech, M., and W. A. Phillip (2007) "The Future of Seawater Desalination: Energy, Technology, and the Environment," *Science,* **333**, 712–717.

Emerick, R. (2012) Personal communication.

Ergun, S. (1952) "Fluid Flow through Packed Columns." *Chem. Eng. Prog.,* **48**, 2, 89–94.

Fair, G. M. (1951) "The Hydraulics of Rapid Sand Filters," *J. Inst. Water Eng.,* **5**, 171–213.

Fair G. M., and L. P. Hatch (1933) "Fundamental Factors Governing the Streamline Flow of Water Through Sand," *J. AWWA,* **25**, 11, 1551–1565.

Fair, G. M., J. C. Geyer, and D. A. Okun (1968) *Water and Wastewater Engineering,* Vol. 2, Wiley, New York.

Ford, D. L. (1992) *Toxicity Reduction: Evaluation and Control,* Technomic Publishing Company, Inc., Lanchester, PA.

Foust, A. S., L. A. Wenzel, C. W. Clump, L. Maus, and L. B. Andersen (1960) *Principles of Unit Operations,* John Wiley & Sons, Inc., New York.

Hand, D. W., J. C. Crittenden, D. R. Hokanson, and J. L. Bulloch (1997) "Predicting the Performance of Fixed-bed Granular Activated Carbon Adsorbers," *Water Sci. Technol.,* **35**, 7, 235–241.

Hand, D. W., D. R. Hokanson, and J. C. Crittenden (1999) "Air Stripping And Aeration," Chap. 6, in R. D. Letterman, ed., *Water Quality And Treatment: A Handbook of Community Water Supplies,* 5th ed., American Water Works Association, McGraw-Hill, New York.

Hazen, A. (1905) *The Filtration of Public Water-Supplies,* 3rd ed, John Wiley & Sons, New York.

Kavanaugh, M. C., and R. R. Trussell (1980) "Design of Stripping Towers to Strip Volatile Contaminants From Drinking Water," *J. AWWA,* **72**, 12, 684–692.

Kozeny, J. (1927) "Uber Grundwasserbewegung," *Wasserkraft und Wasserwirtschaft,* **22**, 5, 67–70, 86–88.

Kawamura, S. (2000) *Intergrated Design And Operation of Water Treatment Facilities,* 2nd ed., John Wiley & Sons, Inc., New York.

LaGrega, M. D., P. L. Buckingham, and J. C. Evans (2001) *Hazardous Waste Management,* McGraw-Hill Book Company, Boston, MA. Reissued in 2010 by Wayland Press, Inc., Long Grove, IL.

Leva, M. (1959) *Fluidization,* McGraw-Hill Book Company, Inc. New York, NY.

McCabe, W. L., and E. W. Thiele (1925) "Graphical Design of Fractionating Columns," *Ind. Eng. Chem.,* **17**, 605–611.

McGarvey, F. B. Bachs, and S Ziarkowski (1989) "Removal of Nitrates from Natural Water Supplies," Presented at the American Chemical Society Meeting, Dallas TX.

Olivier, M., and D. Dalton (2002) "Filter Fresh: Cloth-media Filters Improve a Florida Facility's Water Reclamation Efforts," *Water Environ. Technol.,* 14, 11, 43–45.

Olivier, M., J. Perry, C. Phelps, and A. Zacheis (2003) "The Use of Cloth Media Filtration Enhances UV Disinfection through Particle Size Reduction," 2003 WateReuse Symposium, WateReuse Association, Alexandria, VA.

O'Melia, C. R., and W. Stumm (1967) "Theory of Water Filtration," *J AWWA,* **59**, 11, 1393–1412.

Onda, K., H. Takeuchi, and Y. Okumoto (1968) "Mass Transfer Coefficients Between Gas and Liquid Phases in Packed Columns," *J. Chem, Eng.,* Japan, **1**, 1, 56–62.

Ouki, S. K., and M. Kavanagh (1999) "Treatment of Metals-Contaminated Watewaters by Use of Natural Zeolites," *Water Sci., Technol.,* **39**, 10–11, 115–122.

Patel, M. (2013) Personal Communication, Orange County Water District, Fountain Valley, CA.

Peterson, S. (1953) Annuals of the New York Academy of Science, vol. 57, p. 144.

Richardson, J. F., and W. N. Zaki (1954) " Sedimentation and Fluidisation: Part I, *Trans. Instn. Chem. Engrs.,* **32**, 35–53.

Riess, J., K. Bourgeous, G. Tchobanoglous, and J. Darby (2001) *Evaluation of the Aqua-aerobics Cloth Medium Disk Filter (CMDF) for Wastewater Recycling In California,* Center for Environmental and Water Resources Engineering, Report No. 01–2, Department of Civil and Environmental Engineering, University of California, Davis, CA.

Rose, H. E. (1945) "On the Resistance Coefficient-Reynolds Number Relationship for Fluid Flow through a Bed of Granular Material," *Proc. Inst. Mech. Engrs.,* **153**, 154–161, London.

Rose, H. E. (1949) "Further Researches in Fluid Flow through Beds of Granular Material," *Proc. Inst. Mech. Engrs.,* **160**, 493–503, London.

Rose, J., P. Hauch, D. Friedman, and T. Whalen (1999) "The Boiling Effect: Innovation for Achieving Sustainable Clean Water," *Water* 21, No. 9/10.

Rosene, M. R., R. T. Deithun, J. R. Lutchko, and W. J. Wayner, (1980) "High Pressure Technique for Rapid Screening of Activated Carbons for Use in Potable Water," in M. J. McGuire and I. H. Suffet (eds.) *Treatment of Water by Granular Activated Carbon,* American Chemical Society, Washington, DC.

Sakaji, R. H. (2006) "What's New for Membranes in the Regulatory Arena," presented at Microfiltration IV, National Water Research Institute, Anaheim/Orange County, Orange, CA.

Schippers, J. C., and Verdouw, J. (1980) "The Modified Fouling Index, a Method for Determining the Fouling Characteristics of Water," *Desalination,* **32**, 137–148.

Shaw, D. J. (1966) *Introduction to Colloid and Surface Chemistry,* Butterworth, London, England.

Sherwood, T. K., and F. A. Hollaway (1940) "Performance of Packed Towers-Liquid Film Data for Several Packings," *Trans. Am. Inst. Chem. Engrs.,* **36**, 39–70.

Slater, M. J. (1991) *Principles of Ion Exchange Technology,* Butterworth Heinemann, New York.

Slechta, A., and G. L. Culp (1967) "Water Reclamation Studies at the South Lake Tahoe Public Utility District," *J. WPCF,* **39**, 5, 787–814.

Snoeyink, V. L., and R. S. Summers (1999) "Adsorption Of Organic Compounds," Chap. 13, in R. D. Letterman (ed.), *Water Quality And Treatment: A Handbook of Community Water Supplies,* 5th ed., American Water Works Association, McGraw-Hill, New York.

Sontheimer, H., J. C. Crittenden, and R. S. Summers (1988) *Activated Carbon For Water Treatment,* 2nd ed., in English, DVGW-Forschungsstelle, Engler-Bunte-Institut, Universitat Karlsruhe, Germany.

Stephenson, T., S. Judd, B. Jefferson, and K. Brindle (2000) *Membrane Bioreactors for Wastewater Treatment,* IWA Publishing, London.

Stover, R. L. (2007) "Seawater Reverse Osmosis with Isobaric Energy Recovery Devices," *Desalination* **203** 168–175.

Taylor, J. S., and M. Wiesner (1999) "Membranes," Chap. 11, in R. D. Letterman, ed., *Water Quality And Treatment: A Handbook of Community Water Supplies,* 5th ed., American Water Works Association, McGraw-Hill, New York.

Taylor, J. S., and E. P. Jacobs (1996) "Reverse Osmosis and Nanofiltration," Chap. 9, in J. Mallevialle, P. E. Odendaal, and M R. Wiesner (eds.) *Water Treatment Membrane Processes,* American Water Works Association, published by McGraw-Hill, New York.

Tchobanoglous, G., and R. Eliassen (1970) "Filtration of Treated Sewage Effluent," *Journal San. Eng. Div., ASCE,* **96**, SA2, 243–265.

Tchobanoglous, G., and E. D. Schroeder (1985) *Water Quality: Characteristics, Modeling, Modification,* Addison-Wesley Publishing Company, Reading, MA.

Tchobanoglous, G. (1988) "Filtration of Secondary Effluent for Reuse Applications," Presented at the 61st Annual Conference of the WPCF, Dallas, TX.

Tchobanoglous, G., F. L. Burton, and H. D. Stensel (2003) *Wastewater Engineering: Treatment and Reuse,* 4th ed., McGraw-Hill, New York.

Tchobanoglous, G., H. Leverenz, M. H. Nellor, and J. Crook (2011) *Direct Potable Reuse: A Path Forward,* WateReuse Research and WateReuse California, Washington, DC.

Trussell, R. R., and M. Chang (1999) "Review of Flow Through Porous Media as Applied to Head Loss in Water Filters, *J. Environ. Eng., ASCE,* **125**, 11, 998–1006.

USBR (2003) *Desalting Handbook for Planners,* 3rd ed., Desalination Research and Development Program Report No. 72, United States Department of the Interior, Bureau of Reclamation.

Voutchkov, N. (2013) *Desalination Engineering Planning and Design,* McGraw-Hill Book Company, New York.

Wadell, H. (1935). "Volume, Shape and Roundness of Quartz Particles". *J. Geol.,* **43**, 3, 250–280.

Wetterau, G. D. (2013) Personal Communication, CDM Smith, Los Angeles, CA.

Wilf, M. (1998) "Reverse Osmosis Membranes for Wastewater Reclamation," In T. Asano (ed.) *Wastewater Reclamation and Reuse,* Chap. 7, pp. 263–344, Water Quality Management Library, vol. 10, CRC Press, Boca Raton, FL.

Wong, J. (2003) "A Survey of Advanced Membrane Technologies and Their Applications in Water Reuse Projects," *Proceedings of the 76th Annual Technical Exhibition & Conference,* Water Environment Federation, Alexandria, VA.

12 Disinfection Processes

WORKING TERMINOLOGY

Term	Definition
Absorbance	A measure of the amount of light of a specified wavelength that is absorbed by a solution and the constituents in the solution.
Breakpoint chlorination	A process whereby enough chlorine is added to react with all oxidizable substances in water such that if additional chlorine is added it will remain as free chlorine (see below, $HOCl + OCl^-$).
Chlorine residual, total	The concentration of free or combined chlorine in water, measured after a specified time period following addition. Combined chlorine residual is measured most commonly amperometrically.
Combined chlorine	Chlorine combined with other compounds [e.g., monochloramine (NH_2Cl), dichloramine ($NHCl_2$), and nitrogen trichloride (NCl_3), among others].
Combined chlorine residual	Chlorine residual comprised of combined chlorine compounds [e.g., monochloramine (NH_2Cl), dichloramine ($NHCl_2$), and nitrogen trichloride (NCl_3) and others].
CT	The product of disinfectant residual, C, expressed in mg/L and contact time, T, expressed in min. The term CT is used to assess the effectiveness of the disinfection process.
Dechlorination	The removal of residual chlorine from solution by a reducing agent such as sulfur dioxide or by reacting it with activated carbon.
Disinfection	The partial destruction and inactivation of disease-causing organisms from exposure to chemical agents (e.g., chlorine) or physical processes (e.g., UV radiation).
Disinfection byproducts (DBPs)	Chemicals that are formed with the residual organic matter found in treated wastewater as a result of the addition of a strong oxidant (e.g., chlorine or ozone) for the purpose of disinfection.
Dose response curve	The relationship between the degree of microorganism inactivation and the dose of the disinfectant.
Free chlorine	The total quantity of hypochlorous acid (HOCl) and hypochlorite ion (OCl^-) in solution.
Inactivation	Rendering microorganisms incapable of reproducing, and thus their ability to cause disease.
Irradiation	Exposure to penetrating UV radiation.
Natural organic matter (NOM)	Dissolved or particulate organic constituents that are typically derived from three sources: (1) the terrestrial environment (mostly humic materials), (2) the aquatic environment (algae and other aquatic species and their byproducts), and (3) the microorganisms in the biological treatment process.
Pasteurization	The process of heating food or water at a specified temperature and time for the purpose of killing microorganisms.
Pathogens	Microorganisms capable of causing diseases of varying severity.
Photoreactivation/dark repair	The ability of microorganisms to repair the damage caused by exposure to UV irradiation.
Radiation	Energy such as light, heat, and sound that can be transmitted over large distances without conductors or special conduits.
Reduction equivalent dose (RED)	The inactivation observed through the UV disinfection system as compared to the UV dose response derived from a collimated beam dose response study.
Sterilization	The total destruction of disease-causing and other organisms.
Total chlorine	The sum of the free and combined chlorine.
Transmittance	The ability of a solution to transmit light. Transmittance is related to absorbance.
Ultraviolet (UV) light	Electromagnetic radiation with a wavelength less than that of visible light in the range from 100 to 400 nm.
Ultraviolet (UV) irradiation	A disinfection process in which the exposure to UV radiation (or light) is used to inactivate microorganisms.

Because of the critical importance of the disinfection process in wastewater treatment and/or reuse applications, the purpose of this chapter is to introduce the reader to the important issues that must be considered in the disinfection of treated water with various disinfectants to render it safe for dispersal to the environment or for reuse in a variety of applications. The four categories of human enteric organisms found in wastewater that are of the greatest consequence in producing disease are (1) bacteria, (2) protozoan oocysts and cysts, (3) viruses, and (4) helminth ova. Diseases caused by these waterborne microorganisms have been discussed previously in Chap. 2. Disinfection, the subject of this chapter, is the process used to achieve a given level of destruction or inactivation of pathogenic organisms. Because not all the organisms present are destroyed during the process, the term *disinfection* is differentiated from the term *sterilization*, which is the destruction of all organisms.

To delineate the issues involved in disinfection the following topics are considered: (1) an introduction to the disinfectants used in wastewater, (2) general considerations in wastewater disinfection, (3) disinfection with chlorine and related compounds, (4) disinfection with chlorine dioxide, (5) dechlorination, (6) design considerations for chlorination and dechlorination facilities, (7) disinfection with ozone, (8) disinfection with other chemicals and combination of chemicals, (9) disinfection with UV irradiation, and (10) disinfection by pasteurization.

12–1 INTRODUCTION TO DISINFECTANTS USED IN WASTEWATER

Before discussing the details of the individual disinfection technologies and the practical aspects of disinfection that follow, it is appropriate to consider the characteristics of an ideal disinfectant, the major types of disinfection agents used for wastewater, and to provide a general comparison between disinfectants.

Characteristics for an Ideal Disinfectant

To provide a perspective on the disinfection of wastewater, it is useful to consider the characteristics of an ideal disinfectant as given in Table 12–1. As reported, an ideal disinfectant would have to possess a wide range of characteristics such as safe to handle and apply, stable in storage, toxic to microorganisms, nontoxic to higher forms of life, and soluble in water or cell tissue. It is also important that the strength or concentration of the disinfectant be measurable. The latter consideration is an issue with the use of ozone, where little or no residual may remain after disinfection, and UV and pasteurization disinfection where no residual is measurable.

Disinfection Agents and Methods

Disinfection is most commonly accomplished by the use of (1) chemical agents and (2) nonionizing radiation. Each of these techniques is considered briefly in the following discussion. Other methods of disinfection and or inactivation are mentioned for completeness.

Chemical Agents. Chlorine and its compounds, and ozone, are the principal chemical compounds employed for the disinfection of wastewater. Other chemical agents that have been used as disinfectants in different applications include (1) bromine, (2) iodine, (3) phenol and phenolic compounds, (4) alcohols, (5) heavy metals and related compounds,

Table 12-1
Characteristics of an ideal disinfectant[a]

Characteristic	Properties/response
Alteration of solution characteristics	Should be effective with minimum alteration of the solution characteristics such as increasing the total dissolved solids (TDS)
Availability	Should be available in large quantities and reasonably priced
Deodorizing ability	Should deodorize while disinfecting
Homogeneity	Solution must be uniform in composition
Interaction with extraneous material	Should not be absorbed by organic matter other than bacterial cells
Noncorrosive and nonstaining	Should not disfigure metals or stain clothing
Nontoxic to higher forms of life	Should be toxic to microorganisms and nontoxic to humans and other animals
Penetration	Should have the capacity to penetrate through particle surfaces
Safety	Should be safe to transport, store, handle, and use
Solubility	Must be soluble in water or cell tissue
Stability	Should have low loss of germicidal action with time on standing
Toxicity to microorganisms	Should be effective at high dilutions
Toxicity at ambient temperatures	Should be effective in ambient temperature range

[a] Adapted from Tchobanoglous et al. (2003).

(6) dyes, (7) soaps and synthetic detergents, (8) quaternary ammonium compounds, (9) hydrogen peroxide, (10) peracetic acid, (11) various alkalies, and (12) various acids. Highly acidic or alkaline water will destroy pathogenic bacteria, because water with a pH greater than 11 or less than 3 is relatively toxic to most microorganisms.

Non-Ionizing Radiation. In general, energy in the form of electromagnetic waves, heat, and acoustic waves that can be transmitted over large distances without conductors or special conduits is termed *radiation*. Electromagnetic waves include visible light, infrared light, microwaves, and radio waves. Ultraviolet light (UV) is the most common form of electromagnetic radiation used for the disinfection of treated wastewater. Heating water to the boiling point, for example, will destroy the major disease producing non-spore forming bacteria. Commonly used in the food processing industry, pasteurization in the wastewater field has received greater interest recently because of the availability of new equipment, the opportunity to utilize waste heat, and energy concerns with other disinfectants. Pasteurization of sludge is used extensively in Europe.

Ionizing Radiation. Radiation with sufficient energy to ionize atoms is termed *ionizing radiation*. Alpha particles, beta particles, gamma rays, X-ray radiation, and neutrons are generally considered to be forms of ionizing radiation. For example, gamma rays emitted from radioisotopes, such as cobalt 60, have been used to disinfect (sterilize) both water and wastewater. Although the use of a high-energy electron-beam device for the irradiation of wastewater or sludge has been studied extensively, there are no commercial devices or full-scale installations in operation.

Removal by Mechanical Means. Incidental removal of bacteria and other organisms also occurs by mechanical means during wastewater treatment. The removals accomplished are byproducts of the primary function of the treatment process (e.g., screening, sedimentation, filtration, etc.). The use of membrane filtration (e.g., microfiltration and ultrafiltration) has been recognized as a means to reduce pathogenic organisms for water reuse applications. The level of reduction is assessed by spiking a known concentration of an indicator organism and measuring the inactivation achieved. In a full-scale operation, surrogate parameters, such as pressure decay across the membrane and turbidity, are used to monitor the integrity of the membrane. Pathogen removal by membrane processes is discussed further in Chap. 11.

Mechanisms Used to Explain Action of Disinfectants

The five principal mechanisms that have been proposed to explain the action of disinfectants are (1) damage to the cell wall, (2) alteration of cell permeability, (3) alteration of the colloidal nature of the protoplasm within the cell, (4) alteration of the organism's DNA or RNA, and (5) inhibition of enzyme activity within the protoplasm. A comparison of the mechanisms of disinfection using chlorine, ozone, UV irradiation, and pasteurization is presented in Table 12–2. To a large extent, observed performance differences for the various disinfectants can be explained on the basis of the operative inactivation mechanisms.

Damage, destruction, or alteration of the cell wall by oxidizing chemicals, such as chlorine and ozone, results in cell lysis and death. Oxidizing chemicals can also alter the chemical arrangement of enzymes and inactivate the enzymes. Some oxidants can inhibit the synthesis of the bacterial cell wall. Exposure to UV irradiation can cause the formation of double bonds in the DNA of microorganisms as well as rupturing some DNA strands. When UV photons are absorbed by the DNA in bacteria and protozoa and the DNA and RNA in viruses, covalent dimers can be formed from adjacent thymines in DNA or uracils in RNA. The formation of double bonds disrupts the replication process so that the organism can no longer reproduce and is thus inactivated. When heat is applied, both the

Table 12–2

Mechanisms of disinfection using chlorine, ozone, UV, and pasteurization

Chlorine	Ozone	UV radiation	Pasteurization
1. Direct oxidation of cell wall allowing cellular constituents to flow out of the cell	1. Direct oxidation of cell wall allowing cellular constituents to flow out of the cell	1. Photochemical damage to RNA and DNA (e.g., formation of double bonds) within the cells of an organism	1. The structure of the enzymes within the cell is altered by heat (e.g. denatured), rendering them inoperative
2. Modification of cell wall permeability	2. Reactions with radical byproducts of ozone decomposition	2. The nucleic acids in microorganisms are the most important absorbers of the energy of light in the wavelength range of 240–280 nm	2. The structure of the proteins and fatty acids that make up the cell wall are damaged by heat, allowing contents of cell to escape
3. Alteration of the cell protoplasm	3. Damage to the constituents of the nucleic acids (purines and pyrimidines)	3. Because DNA and RNA carry genetic information for reproduction, damage of these substances can effectively inactivate the cell	3. The fluids within the cell can expand and rupture the cell wall, releasing the contents of the cell
4. Inhibition of enzyme activity	4. Breakage of carbon-nitrogen bonds leading to depolymerization		
5. Damage to the cell DNA and RNA			

nature of the enzymes in the cell protoplasm and the structure of the cell wall are altered, rendering the microorganism incapable of reproducing.

Comparison of Disinfectants

Using the criteria defined in Table 12–1 as a frame of reference and the issues discussed above, the disinfectants that have been used in wastewater applications are compared in Table 12–3. Additional details on the relative performance of the various disinfection technologies are presented in the following sections. In reviewing Table 12–3, important comparisons that should be noted include safety (e.g., chlorine gas versus sodium hypochlorite) and the increase in TDS (e.g., chlorine gas versus UV irradiation). These issues are also addressed in the subsequent sections.

12–2 DISINFECTION PROCESS CONSIDERATIONS

The purpose of this section is to present background material that will serve as a basis for the discussion of individual disinfectants considered in the following sections. Topics to be discussed include (1) an introduction to the physical facilities used for disinfection, (2) the factors that affect the performance of the disinfection process, (3) development of CT values (residual disinfectant concentration times time) for predicting disinfection performance, (4) application of CT values, (5) a comparison of the performance of alternative disinfection technologies, and (6) a review of the advantages and disadvantages of each disinfection technology. Costs, both capital and operation and maintenance, have not been provided other than in a general context. Costs are influenced by many site-specific factors and must be evaluated on a case-by-case basis.

Physical Facilities Used for Disinfection

In general, the disinfection is accomplished as a separate unit process in specially designed reactors. The purpose of the reactors is to maximize contact between the disinfecting agent and the liquid to be disinfected. The specific design of the reactor depends on the nature and action of the disinfecting agent. The types of reactors used are illustrated on Figs. 12–1 and 12–2 and described below briefly.

Chlorine and Related Compounds.   As shown on Figs. 12–1(a) and 12–1(b) baffled serpentine contact chambers or long pipelines are used for the application of diluted chlorine and related compounds. Both of these contact chambers are designed to perform as ideal plug-flow reactors. As will be discussed later, the efficacy of disinfection is affected by the degree to which the flow in these chambers is less than ideal. Views of full scale chlorine contact basins are shown on Figs. 12–2(a) and 12–2(b).

Ozone.   Ozone is typically applied by bubbling ozone gas through the liquid to be disinfected in a contact chamber [see Fig. 12–1(c)] or in a sidestream [see Fig. 12–1(d)] and then injected into an ozone contactor [see Fig. 12–2(d)]. Fine bubble diffusers are used to improve ozone transfer to the liquid. Eductors and Venturi injectors are used in sidestream designs. To limit the amount of short circuiting that can occur in a single contact chamber a series of baffled chambers are used [see Fig. 12–1(c)].

Ultraviolet Light (UV).   Both open [see Figs. 12–1(e) and (f)] and closed [see Fig. 12–1(g)] contact chambers (reactors) are used for UV disinfection. Open channel reactors are used commonly for low-pressure, low-intensity and low-pressure, high-intensity

Table 12–3
Comparison of technologies used for the disinfection of treated wastewater[a]

Characteristic[b]	Chlorine gas[c]	Sodium hypochlorite[c]	Combined chlorine	Chlorine dioxide	Ozone	UV radiation	Pasteurization
Availability/cost	Low	Moderately low	Moderately low	Moderately low	Moderately high	Moderately high	Moderate
Deodorizing ability	High	Moderate	Moderate	High	High	na[d]	na
Interaction with organic matter	Oxidizes organic matter	Oxidizes organic matter	Oxidizes organic matter	Oxidizes organic matter	Oxidizes organic matter	Absorbance of UV radiation	na
Corrosiveness	Highly corrosive	Corrosive	Corrosive	Highly corrosive	Highly corrosive	na	na
Toxic to higher forms of life	Highly toxic	Highly toxic	Toxic	Toxic	Toxic	Toxic	Toxic
Penetration into particles	High	High	Moderate	High	High	Moderate	High
Safety concern	High	Moderate to low	High to moderate[e]	High	Moderate	Low	Low
Solubility	High	High	High	High	Moderate	na	na
Stability	Stable	Slightly unstable	Slightly unstable	Unstable[f]	Unstable[f]	na	na
Effectiveness for							
Bacteria	Excellent	Excellent	Good	Excellent	Excellent	Good	Excellent
Protozoa	Fair to poor	Fair to poor	Poor	Good	Good	Excellent	Excellent
Viruses	Excellent	Excellent	Fair	Excellent	Excellent	Good	Good
Byproduct formation	THMs and HAAs[g]	THMs and HAAs[g]	Traces of THMs and HAAs, cyanogens, NDMA	Chlorite and chlorate	Bromate	None known in measurable concentrations	None known in measurable concentrations
Increases TDS	Yes	Yes	Yes	Yes	No	No	No
Use as a disinfectant	Common	Common	Common	Increasing slowly	Increasing slowly	Increasing rapidly	Increasing slowly

[a] Adapted in part from Tchobanoglous et al. (2003) and Crittenden et al. (2012).

[b] See Table 12–1 for a description of the characteristics of an ideal disinfectant.

[c] Free chlorine (HOCl and OCl⁻).

[d] na = not applicable.

[e] Depends on whether chlorine gas or sodium hypochlorite is used to combine with nitrogenous compounds.

[f] Must be generated as used.

[g] THMs = trihalomethanes and HAAs = haloacetic acids.

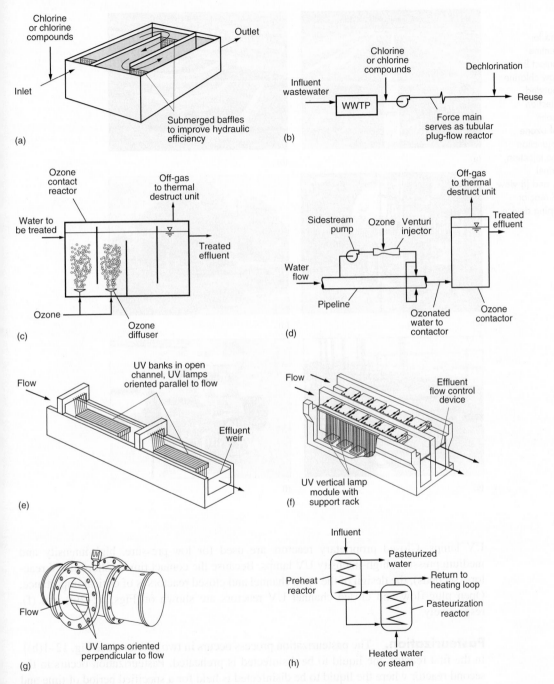

Figure 12-1

Types of reactors used to accomplish disinfection process: (a) plug-flow reactor in back-and-forth configuration, (b) force main which serves as a tubular plug-flow reactor, (c) multiple chamber inline ozone contactor, (d) sidestream ozone injection system, (e) UV irradiation in an open channel with two UV banks with flow parallel to UV lamps, (f) UV irradiation in an open channel with six UV banks with flow perpendicular to UV lamps, (g) UV irradiation in an closed reactor, and (h) reactors for pasteurization system.

Figure 12–2

Views of reactors used for disinfection: (a) serpentine plug-flow chlorine contact basin, (b) serpentine plug-flow chlorine contact basin with rounded corners and flow deflection baffles, (c) typical ozone generator, (d) view of ozone contactor used in conjunction with sidestream ozone injection, (e) view of open channel plug-flow UV reactor, and (f) view of closed channel UV reactor with manual lamp wiping device.

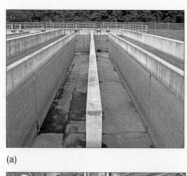
(a)

(b)

(c)

(d)

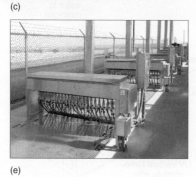
(e)

(f)

UV lamps. Closed proprietary reactors are used for low-pressure, high-intensity and medium pressure, high-intensity UV lamps. Because the contact time is short in UV reactors (seconds), the design of the open channel and closed reactors is of critical importance. Open plug-flow and closed channel UV reactors are shown on Figs. 12–2(e) and (f), respectively.

Pasteurization.   The pasteurization process occurs in two reactors [see Fig. 12–1(h)]. In the first reactor, the liquid to be disinfected is preheated. Pasteurization occurs in the second reactor where the liquid to be disinfected is held for a specified period of time and temperature.

Factors Affecting Performance

In applying disinfection agents or physical processes, the following factors must be considered: (1) contact time and hydraulic efficiency of contact chambers, (2) concentration of the disinfectant, (3) intensity and nature of physical agent or means, (4) temperature, (5) types of organisms, (6) nature of suspending liquid (e.g., unfiltered or filtered secondary effluent), and (7) the upstream treatment processes. The subjects introduced in this

Figure 12–3

Log inactivation of dispersed microorganisms as a function of time in a batch reactor using increasing disinfectant dosages.

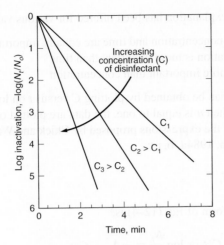

section are considered further in the subsequent sections dealing with the individual disinfectants.

Contact Time. Perhaps one of the most important factors in the disinfection process is contact time. Once the disinfectant has been added, the time of contact before the effluent is to be discharged or reused is of paramount importance. As shown on Fig. 12–1, disinfection reactors are designed to ensure that an adequate contact time is provided. The hydraulic efficiency of disinfection reactors is considered in Sec. 12–6.

Working in England in the early 1900s, Harriet Chick observed that for a given concentration of disinfectant, the longer the contact time, the greater the kill (see Fig. 12–3). This observation was first reported in the literature in 1908 (Chick, 1908). In differential form, Chick's law is

$$\frac{dN_t}{dt} = -KN_t \tag{12–1}$$

where dN_t/dt = the rate of change in the number (concentration) of organisms with time
$\quad\quad K$ = inactivation rate constant, T^{-1}
$\quad\quad N_t$ = number of organisms at time t
$\quad\quad t$ = time

If N_o is the number of organisms when t equals 0, Eq. (12–1) can be integrated to

$$\ln \frac{N_t}{N_o} = -Kt \tag{12–2}$$

The value of the inactivation rate constant, K, in Eq. (12–2) can be obtained by plotting $-\ln(N_t/N_o)$ versus the contact time t, where K is the slope of the resulting line of best fit.

Concentration of Chemical Disinfectant. Also working in England in the early 1900s, Herbert Watson reported that the inactivation rate constant was related to the concentration as follows (Watson, 1908):

$$K = \Lambda C^n \tag{12–3}$$

where K = inactivation rate constant, T^{-1}, base e
$\quad\quad \Lambda$ = coefficient of specific lethality, units vary with the value of n
$\quad\quad C$ = concentration of disinfectant, mg/L
$\quad\quad n$ = empirical constant related to dilution, dimensionless

The following explanation has been offered for various values of the dilution constant n:

$n = 1$, both the concentration and time are equally important
$n > 1$, concentration is more important than time
$n < 1$, time is more important than concentration

The value of n can be obtained by plotting C versus t on log-log paper for a given level of inactivation. When n is equal to one, the data are plotted on log-arithmetic paper.

Combining the expressions proposed by Chick and Watson in differential form yields (Haas and Karra, 1984a, b, c):

$$\frac{dN_t}{dt} = -\Lambda C^n N_t \tag{12-4}$$

The integrated form of Eq. (12–4) is:

$$\ln \frac{N_t}{N_o} = -\Lambda_{\text{base }e} C^n t \quad \text{or} \quad \log \frac{N_t}{N_o} = -\Lambda_{\text{base }10} C^n t \tag{12-5}$$

If n is equal to one, a reasonable assumption based on past experience (Hall, 1973), Eq. (12–5) can be written as follows:

$$\log \frac{N_t}{N_o} = -\Lambda_{\text{base }10}(CT) = -\Lambda_{\text{base }10}(D) \tag{12-6}$$

Where C = residual concentration of disinfectant, mg/L
T = contact time in the reactor, min
D = *germicidal dose* for a given degree of inactivation, mg·min/L

The concept of dose (concentration times time) is significant as the performance of the disinfectants, as discussed subsequently, is based on the concept (Morris, 1975). This concept has also been adopted by the U.S. EPA in establishing guidelines for the disinfection of public water supplies (see "Development of the CT Concept for Predicting Disinfection Performance" later in the chapter).

EXAMPLE 12–1 **Determination of the Coefficient of Specific Lethality Based on the Chick-Watson Expression** Using the microorganism inactivation data given below, determine the coefficient of specific lethality of the chemical disinfecting agent using Eq. (12–6).

C, mg/L	Time, min	Number of organisms, Number/100 mL
0	0	1.00×10^8
4.0	2	1.59×10^7
4.0	4.5	1.58×10^6
4.0	8	2.01×10^4
4.0	11.5	3.16×10^3

Solution

1. To determine the coefficient of lethality prepare a plot of $\log[N/N_o]$ as a function of CT and fit a linear trend line through the data.

a. Determine the values of log[N/N_o] and CT. The required data table is shown below.

C, mg/L	Time, min	Number of organisms, N/100 mL	CT, mg·min/L	log(N/N_o)
0	0	1.00×10^8	0	0
4.0	2	1.59×10^7	8	−0.8
4.0	4.5	1.58×10^6	18	−1.8
4.0	8	2.01×10^4	32	−3.7
4.0	11.5	3.16×10^3	46	−4.5

b. Prepare a plot of log(N/N_o) as a function of CT. The required plot is shown below.

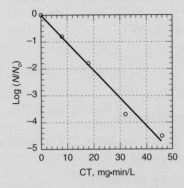

2. Determine the coefficient of specific lethality. The slope of the line in the above plot corresponds to the coefficient of specific lethality, $-\Lambda_{CW}$ (base 10). From the plot

$$-\Lambda_{CW} \text{ (base 10)} = \frac{-5 - 0}{49 - 0}$$

$$\Lambda_{CW} \text{(base 10)} = 0.102 \, \text{L/mg·min}$$

Check, when CT = 46,

$$\log \frac{N_t}{N_o} = -\Lambda_{\text{base 10}} CT = -0.102(46) = -4.69 \text{ versus } - 4.5 \, \text{OK}$$

Temperature.   The effect of temperature on the rate of kill with chemical disinfectants can be represented by a form of the van't Hoff-Arrhenius relationship. Increasing the temperature results in a more rapid kill. In terms of the coefficient of specific lethality, Λ, the effect of temperature is given by the following relationship, repeated here from Chap. 1.

$$\ln \frac{\Lambda_1}{\Lambda_2} = \frac{E(T_2 - T_1)}{RT_1 T_2} \tag{12–7}$$

where Λ_1, Λ_2 = coefficient of specific lethality at temperatures T_1 and T_2, respectively
   E = activation energy, J/mole
   R = universal gas constant, 8.3144 J/mole·K

Typical values for the activation energy for various chlorine compounds at different pH values are given in Sec. 12–3. The effect of temperature is considered in Example 12–2.

EXAMPLE 12–2 **Effect of Temperature on Disinfection Times** Estimate the time required for a 99 percent kill for a chlorine dosage of 0.05 mg/L at a temperature of 20°C. Assume the activation energy is equal to 26,800 J/mole (from Table 12–12 in Sec. 12–3). The following coefficients were developed for Eq. (12–5) at 5°C using a batch reactor.

$$\Lambda = 10.5 \text{ L/mg·min}$$

$$n = 1$$

Solution

1. Estimate the time required at 5°C for a 99 percent kill using Eq. 12–5.

$$\log \frac{N_t}{N_o} = -10.5\, CT$$

$$\log \frac{10}{100} = -(10.5 \text{ L/mg·min})(0.05 \text{ mg/L})T$$

$$T = \frac{-6.91}{(-10.5)(0.05)} = 13.2 \text{ min at 5°C}$$

2. Estimate the time required at 20°C using the van't Hoff-Arrhenius equation [Eq. (12–7)].

$$\ln \frac{\Lambda_1}{\Lambda_2} = \frac{E(T_2 - T_1)}{RT_1 T_2}$$

$$\ln \frac{10.5}{\Lambda_2} = \frac{(26{,}800 \text{ J/mole})(278 - 293)\text{K}}{(8.3144 \text{ J/mole·K})(293)(298)}$$

$$\ln \frac{10.5}{\Lambda_2} = -0.594$$

$$\ln \frac{10.5}{\Lambda_2} = e^{-0.594} = -0.552$$

$$\Lambda_2 = 19.0 \text{ L/mg·min}$$

$$T = \frac{-6.91}{(-19.0)(0.05)} = 7.27 \text{ min at 20°C}$$

Intensity and Nature of Non-Ionizing Radiation. As noted earlier, irradiation with ultraviolet light (UV) is used commonly for the disinfection of water. It has been found that the effectiveness of UV disinfection is a function of the average UV intensity, expressed as milliwatts per square centimeter (mW/cm^2). When the exposure time is considered, the dose of UV to which the microorganisms in the liquid are exposed to is given by the following expression.

$$D = I_{avg} \times t \qquad (12\text{–}8)$$

Where D = UV dose, mJ/cm^2 (Note: $mJ/cm^2 = mW·s/cm^2$)

I_{avg} = average UV intensity, mW/cm^2

t = time, s

The UV dose is expressed in mJ/cm^2 (millijoule per square centimeter) which is equivalent to $mW \cdot s/cm^2$. Thus, the concept of dose can also be used to define the effectiveness of UV light in a manner analogous to that used for chemical disinfectants, as well as when heat is used as in pasteurization.

Types of Organisms. The effectiveness of various disinfectants is influenced by the type, nature, and condition of the microorganisms. For example, viable growing (vegetative) bacteria cells are often killed or inactivated more easily than older cells that have developed a slime (polymer) coating. Bacteria that are able to form spores enter this protective state when stressors, such as increased temperature or a toxic agent, is applied. Bacterial spores are extremely resistant, and many of the chemical disinfectants normally used have little or no effect on them. Similarly, many of the viruses and protozoa of concern respond differently to each of the chemical disinfectants. In some cases, other disinfecting agents, such as heat or UV irradiation, may have to be used for effective disinfection. The inactivation of different types of microorganism groups is considered in the following sections.

Nature of Suspending Liquid. In reviewing the development of the relationships developed by Chick and Watson for the inactivation of microorganisms, as cited above, it is important to note that most of the tests were conducted in batch reactors using distilled or buffered water, under laboratory conditions. In practice, the nature of the suspending liquid must be evaluated carefully. Three constituents found in wastewater are significant: (1) inorganic constituents that can react with the disinfectant, (2) organic matter including both natural organic material (NOM) and other organic compounds, and (3) suspended material. The NOM found in treated wastewater will react with most oxidizing disinfectants and reduce their effectiveness or result in greater dosages to effect disinfection. The NOM is derived from three sources: (1) the terrestrial environment (mostly humic materials), (2) the aquatic environment (algae and other aquatic species and their byproducts), and (3) the microorganisms in the biological treatment process. The source of the other organic compounds is from the constituents discharged to the collection system. The presence of suspended matter will also reduce the effectiveness of disinfectants by absorption of the disinfectant and by shielding the entrapped bacteria.

Because of the interactions that can occur between the disinfecting agent and the wastewater constituents, departures from the Chick-Watson rate law [Eqs.(12–5) and (12–6)] are common as shown on Fig. 12–4. As shown on Fig. 12–4(a), there can be a lag or shoulder effect in which constituents in the suspending liquid react initially with the disinfectant rendering the disinfectant ineffective followed by a log-linear portion. The tailing effect in which large particles shield the organisms to be disinfected is shown on Fig. 12–4(b). The combined effects of lag and tailing are illustrated on Fig. 12–4(c). In general, Eq. (12–5) as applied to wastewater fails to account for the variable, heterogeneous characteristics of wastewater.

Effect of Upstream Treatment Processes. The extent to which upstream processes remove NOM, other organic matter, and suspended matter will greatly influence the disinfection process. Incidental removals of bacteria and other organisms are also achieved by mechanical and biological means during wastewater treatment. Typical removal efficiencies for various treatment operations and processes are reported in Table 12–4. The first and last four operations listed are essentially physical. The actual removal accomplished is a byproduct of the primary function of the process.

Another factor that impacts the performance of both chlorine and UV disinfection for unfiltered effluents (especially when coliform bacteria are used as the regulatory indicator) is the number of particles with associated coliform bacteria. It has been observed that for

Figure 12–4

Departures observed from the Chicks' law: (a) lag or shoulder effect in which the disinfectant reacts first with constituents in the suspending liquid after which the response is log-linear (i.e., first order kinetics), (b) log-linear response followed by tailing effect in which large particles shield the organisms to be disinfected following the inactivation of dispersed organisms, and (c) combined lag, log-linear, and tailing effects.

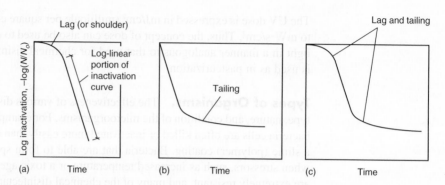

activated sludge plants the number of particles with associated coliform organisms is a function of the solids retention time (SRT). The relationship between the fraction of wastewater particles with one or more associated coliform organisms and the SRT is illustrated on Fig. 12–5. As illustrated, longer SRTs result in a decrease in the fraction of particles containing coliform bacteria. The use of deep final clarifiers (or other filtration methods) reduces the number of large particles that may shield bacteria [see Fig. 7–7(b) in Chap. 7]. In general, without some form of filtration, it is difficult to achieve extremely low coliform concentrations in the settled effluent from activated sludge plants operated at low SRT values (e.g., 1 to 2 d).

Development of the CT Concept for Predicting Disinfection Performance

Although the disinfection models discussed above are useful for analyzing disinfection data, they are difficult to use to predict disinfection performance over a wide range of operating conditions. In the water treatment field, before the adoption of the Surface Water Treatment Rule (SWTR) (circa 1989) and before the importance of *Cryptosporidium* as a

Table 12–4

Removal or destruction of total coliform by different treatment processes

Process	Removal	
	Percent	log[a]
Coarse screens	0–5	~0
Fine screens	10–20	0–0.1
Grit chambers	10–25	0–0.1
Plain sedimentation	25–75	0.1–0.6
Chemical precipitation	40–80	0.2–0.7
Trickling filters	90–95	1–1.3
Activated sludge	90–98	1–1.7
Depth filtration	—	0.25–1
Microfiltration	—	2–4[b]
Ultrafiltration	—	2–5[b]
Reverse osmosis	—	2–6[b]

[a] The log-reduction credit allowed by regulatory agencies for these processes will vary from state to state.

[b] Depends on the characteristics and configuration of the membrane.

Figure 12–5

Fraction of particles in settled wastewater with one or more associated coliform organisms as function of the solids retention time. (From Emerick et al., 1999.)

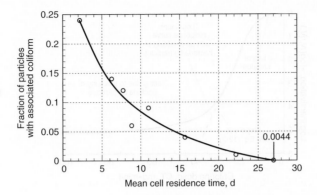

causative agent in waterborne disease outbreaks was recognized, meeting water quality requirements was quite straightforward. Chlorine and its compounds were generally used to inactivate coliform bacteria to meet the drinking water standards in effect at that time.

In developing the rationale for the first SWTR, the U.S. EPA needed some way to ensure the safety of public water supplies that were unfiltered (e.g., New York City, San Francisco, Seattle). Based on ongoing research, the U.S. EPA determined that four logs of virus and three logs of *Giardia* reduction would be required by means of disinfection. Recognizing that guidance was required on how to achieve adequate disinfection, the U.S. EPA undertook an evaluation of the most commonly used disinfectants for the disinfection of viruses and *Giardia* cysts. In conducting their evaluation, the U.S. EPA adopted the CT concept (the product of the residual disinfectant concentration, C, in mg/L times and the contact time, T, in min), derived from the simplified Chick-Watson model [see Eq. (12–6)], as a measure of performance. The CT values obtained, typically in laboratory bench scale studies, are used as a surrogate measure of disinfection effectiveness. Thus, if a given CT value is achieved it could be assumed generally that disinfection requirements had been met. Bauman and Ludwig (1962) were among the first, if not the first, to suggest use of the CT concept in a paper published in 1962. The CT concept was not picked up again in a meaningful way until 1980 when it was used by the Safe Drinking Water Committee of the National Research Council in its evaluation of the disinfection literature (NRC, 1980; Hoff, 1986).

Although *Cryptosporidium* had been identified at the time the SWTR was adopted in 1989, CT values for *Cryptosporidium* were not included because it would have delayed adoption of the SWTR. It has since been found that many pathogens, including *Cryptosporidium*, remain intact and viable while in the presence of various disinfectants at concentrations that are sufficient to inactivate most other pathogens. Based on ongoing work, the U.S. EPA has now published extensive tables of CT values for a variety of disinfectants, microorganisms, and operating conditions (U.S. EPA, 2003a, 2006). In addition, corresponding UV dose values have also been published for *Cryptosporidium, Giardia,* and viruses. From a practical standpoint, the utility of the CT or UV dose approach can be appreciated as it is relatively easy to measure the residual concentration of the disinfectant or the UV intensity and the exposure contact time. With respect to the contact time, the t_{10} value (the contact time during which no more than 10 percent in the influent water has passed through the process–see discussion in Sec. 12–3) is used commonly in the field of water treatment for disinfectants other than UV irradiation.

Application of the CT Concept to Wastewater Disinfection

Use of the CT concept to control the disinfection process is now becoming more common in the wastewater field. In some states, the CT value and the chlorine contact time are

Figure 12–6

Typical disinfection dose-response curve obtained with wastewater containing oxidizable constituents and suspended solids. Both lag and tailing effects are evident.

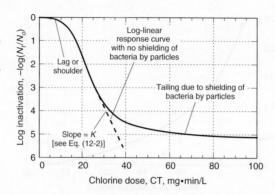

specified in regulatory requirements. For example, the California Department of Public Health (CDPH) California requires a minimum CT value of 450 mg·min/L (based on combined chlorine residual) and a modal contact time of 90 min at peak flow for certain water reclamation applications. It is assumed, based on past testing, that a minimum CT value of 450 mg·min/L will produce a four-log inactivation of poliovirus. As the use of the CT concept becomes more common in the wastewater field, a number of past limitations must be considered in the application of this concept for regulatory purposes. In the past, most of the CT values reported in the literature are obtained using (1) complete-mix batch reactors (i.e., ideal plug flow conditions) in a laboratory setting under controlled conditions, (2) discrete organisms grown in the laboratory in pure culture, (3) a buffered fluid for the suspension of the discrete organisms, and (4) an absence of particulate matter.

Further, many of the CT values reported in the literature were based on older analytical techniques. As a consequence, CT values used for regulatory purposes often do not match what is observed in the field. Referring to Fig. 12–6, it can be seen that in the tailing region, the residual concentration of microorganisms is essentially independent of the CT value. In addition, some compounds present in treated wastewater will (1) react with the chlorine and its compounds, (2) be measured as combined residual, and (3) have no disinfection properties (see Sec. 12–3). In a similar manner, dissolved constituents, such as metals and humic acids, will reduce the effectiveness of UV disinfection. Thus, it is difficult to develop standardized CT or UV dose values suitable for all conditions encountered in wastewater treatment. Clearly, as discussed subsequently, site-specific testing is required to establish the appropriate disinfectant dose.

Performance Comparison of Disinfection Technologies

A general comparison of the germicidal effectiveness of the disinfection technologies based on Eq. (12–6) by classes of organisms is presented in Table 12–5. Additional information is presented in the sections dealing with the individual technologies. It is important to note that the values given in these tables are only meant to serve as a guide in assessing the effectiveness of these technologies. The CT values also vary with both temperature and pH. Because the characteristics of each wastewater and the degree of treatment will significantly impact the effectiveness of the various disinfection technologies, site-specific testing must be conducted to evaluate the effectiveness of alternative disinfection technologies and to establish appropriate dosing ranges.

Advantages and Disadvantages of Alternative Disinfection Technologies. The general advantages and disadvantages of using chlorine, chlorine dioxide, ozone, and UV for the disinfection of wastewater are summarized in Table 12–6. In most wastewater

Table 12–5

Relative CT values for various levels of inactivation of bacteria, viruses, *Cryptosporidium*, and *Giardia lamblia* cysts in filtered secondary effluent (pH ~7.5, ~20°C)[a,b]

Disinfectant	Unit	Inactivation			
		1–log	2–log	3–log	4–log[c]
Bacteria[d]					
Chlorine (free)	mg·min/L	0.4–0.6	0.8–1.2	1.2–1.8	1.6–2.4
Chloramine	mg·min/L	50–70	95–150	140–220	200–300
Chlorine dioxide	mg·min/L	0.4–0.6	0.8–1.2	1.2–1.8	1.6–2.4
Ozone	mg·min/L	0.005–0.01	0.01–0.02	0.015–0.03	0.02–0.04
UV radiation	mJ/cm^2	10–15	20–30	30–45	40–60
Virus					
Chlorine (free)	mg·min/L		1.5–1.8	2.2–2.6	3–3.5
Chloramine	mg·min/L		370–400	550–600	750–800
Chlorine dioxide	mg·min/L		5–5.5	9–10	12.5–13.5
Ozone	mg·min/L		0.25–0.3	0.35–0.45	0.5–0.6
UV radiation[e]	mJ/cm^2		40–50	60–75	80–100
Protozoa (*Cryptosporidium*)[f]					
Chlorine (free)	mg·min/L	2000–2600	4000–5000		
Chloramine	mg·min/L	4000–5000	8000–10,000		
Chlorine dioxide	mg·min/L	120–150	235–260	350–400	
Ozone	mg·min/L	4–4.5	8–8.5	12–13	
UV radiation	mJ/cm^2	2.5–3	6–7	12–13	
Protozoa (*Giardia lamblia* cysts)[g]					
Chlorine (free)	mg·min/L	20–30	45–55	70–80	
Chloramine	mg·min/L	400–450	800–900	1100–1300	
Chlorine dioxide	mg·min/L	5–5.5	9–11	15–16	
Ozone	mg·min/L	0.25–0.3	0.45–0.5	0.75–0.8	
UV radiation	mJ/cm^2	2–2.5	5.5–6.6	11–13	

[a] Adapted in part from AWWA (1991), Baumann and Ludwig (1962), Crittenden et al. (2012), Hoff (1986), Code of Federal Regulations – Title 40 (40 CFR 141.2), Maguin et al. (2009), Montgomery (1985), Roberts et al. (1980), Sung (1974), U.S. EPA (1999b).

[b] Reported CT values are highly temperature and pH sensitive. Disinfection rates will increase by a factor of 2 to 3 for each 10°C increase in temperature.

[c] The range of CT values for 4-log removal is for the linear portion of the dose-response curve (see Fig. 12–6). Depending on the particle size distribution resulting from the filtration of secondary effluent, much higher CT values may be needed to achieve a 4-log removal.

[d] The reported CT values are for total coliform. Significantly lower CT values have been reported for fecal coliform and *E coli*.

[e] With the exception of adenovirus which requires a much higher UV dose (as high as 160–200 mJ/cm^2 for 4-log inactivation).

[f] The data for *Cryptosporidium* inactivation with free or combined chlorine are extremely variable. Values of CT greater than 10,000 mg·min/L have been reported for 99 percent inactivation with chloramines. Clearly, free or combined chlorine is not an effective disinfectant for *Cryptosporidium*. Further, *Cryptosporidium* oocysts will, in general, require even higher CT values.

[g] Based primarily on the results of infectivity studies.

Note: Because there is such a wide variability in the susceptibility of different microorganism groups as well as within a microorganism group to the different disinfection technologies, a wide range of dosage values has been reported in the literature. Thus, the data presented in this table are only meant to serve as general guide to the relative effectiveness of the different disinfection technologies and are not for a specific microorganism.

Table 12–6

Advantages and disadvantages of chlorine, chlorine dioxide, ozone, and UV for the disinfection of treated wastewater[a]

Advantages	Disadvantages
Free and combined chlorine species	
1. Well established technology	1. Hazardous chemical that can be a threat to plant workers and the public; thus, strict safety measures must be employed especially in light of the Uniform Fire Code
2. Effective disinfectant	2. Relatively long contact time required as compared to other disinfectants
3. Chlorine residual can be monitored and maintained	3. Combined chlorine is less effective in inactivating some viruses, spores, and cysts at low dosages used for coliform organisms
4. Combined chlorine residual can also be provided by adding ammonia	4. Residual toxicity of treated effluent must be reduced through dechlorination
5. Germicidal chlorine residual can be maintained in long transmission lines	5. Forms trihalomethanes and other DBPs including NDMA[b] (see Table 12–16)
6. Availability of chemical system for auxiliary uses such as odor control, dosing RAS, and disinfecting plant water systems	6. Releases volatile organic compounds from chlorine contact basins
	7. Oxidizes iron, magnesium, and other inorganic compounds (consumes disinfectant)
7. Oxidizes sulfides	8. Oxidizes a variety of organic compounds (consumes disinfectant)
8. Capital cost is relatively inexpensive, but cost increases considerably if conformance to Uniform Fire Code regulations is required	9. Increases TDS level of treated effluent
	10. Increases chloride content of treated effluent
	11. Acid generation; pH of the wastewater can be reduced if alkalinity is insufficient
9. Available as calcium and sodium hypochlorite that are considered to be safer than chlorine gas	12. Chemical scrubbing facilities may be required to meet Uniform Fire Code regulations
	13. Formal risk management plan may be required
10. Hypochlorite can be generated onsite	14. Not effective disinfectant for *Cryptosporidium*
Chlorine dioxide	
1. Effective disinfectant for bacteria, *Giardia* and viruses	1. Unstable, must be produced onsite
2. More effective than chlorine in inactivating most viruses, spores, cysts and oocysts	2. Oxidizes iron, magnesium, and other inorganic compounds (consumes disinfectant)
	3. Oxidizes a variety of organic compounds
3. Biocidal properties not influenced by pH	4. Forms DBPs (i.e., chlorite and chlorate), limiting applied dose
4. Under proper generation conditions, halogen-substituted DBPs are not formed	5. Potential for the formation of halogen-substituted DBPs
	6. Decomposes in sunlight
5. Oxidizes sulfides	7. Can lead to the formation of odors
6. Provides residuals	8. Increases TDS level of treated effluent
	9. Operating costs can be high (e.g., must test for chlorite and chlorate)

(continued)

Table 12-6 (Continued)	
Advantages	**Disadvantages**
Ozone	

Ozone	
1. Effective disinfectant	1. Ozone residual monitoring and recording requires more operator time than chlorine residual monitoring and recording
2. More effective than chlorine in inactivating most viruses, spores, cysts and oocysts	2. No residual effect
3. Biocidal properties not influenced by pH	3. Less effective in inactivating some viruses, spores, cysts at low dosages used for coliform organisms
4. Shorter contact time than chlorine	4 Forms DBPs (see Table 12–15)
5. Oxidizes sulfides	5. Oxidizes iron, magnesium, and other inorganic compounds (consumes disinfectant)
6. Requires less space	6. Oxidizes a variety of organic compounds (consumes disinfectant)
7. Contributes dissolved oxygen	7. Off-gas requires treatment
8. At higher dosages than required for disinfection, ozone reduces the concentration of trace organic constituents	8. Safety concerns
	9. Highly corrosive and toxic
	10. Energy intensive
	11. Relatively expensive
	12. Highly operational and maintenance sensitive
	13. Has been shown to control the growth of filamentous microorganisms, but more expensive than chlorine

UV	
1. Effective disinfectant	1. No immediate measure of whether disinfection was successful
2. Requires no hazardous chemicals	2. No residual disinfectant
3. No residual toxicity	3. Less effective in inactivating some viruses, spores, and cysts at low dosages used for coliform organisms
4. More effective than chlorine in inactivating most viruses, spores, and cysts	4. Energy intensive
5. No formation of DBPs at dosages used for disinfection	5. Hydraulic design of UV system is critical
6. Does not increase TDS level of treated effluent	6. Capital cost is relatively expensive, but price is coming down as new and improved technology is brought to the market
7. At very high dosages, effective in the destruction of resistant organic constituents such as NDMA	7. Large number of UV lamps required where low-pressure, low-intensity systems are used
8. Improved safety as compared to use of chemical disinfectants	8. Acid washing to remove scale from quartz sleeves may be required for any technology
9. Requires less space than chlorine disinfection	9. Lacks a chemical system that can be adapted for auxiliary uses such as odor control, dosing RAS, and disinfecting plant water systems
10. At higher UV dosages than required for disinfection, UV radiation can be used to reduce the concentration of trace organic constituents of concern such as NDMA (see Sec. 10–8 in Chap. 10)	10. Fouling of UV lamps
	11. Lamps require routine periodic replacement
	12. Lamp disposal is problematic due to presence of mercury

[a] Adapted in part from Crites and Tchobanoglous (1998), U.S. EPA (1999b), and Hanzon et al. (2006).

[b] DBPs = disinfection byproducts.

treatment applications, the choice of disinfectant has usually between chlorine and UV. Recently, however, with concerns regarding trace constituents of concern, a renewed interest has developed in the use of ozone. Deciding factors in the selection of a disinfectant are commonly (1) economic evaluation, (2) public and operator safety, (3) environmental effects, and (4) ease of operation (Hanzon et al., 2006). Other treatment objectives are also important in the selection of a disinfectant. Potential concerns with pesticides, trace constituents of concern, endocrine disruptors, and similar compounds may influence the choice of disinfectants. Each disinfectant offers varying treatment performance with regard to these potential concerns.

12–3 DISINFECTION WITH CHLORINE

Chlorine, of all the chemical disinfectants, is the one used most commonly throughout the world. Specific topics considered in this section include a brief description of the characteristics of the various chlorine compounds, a review of chlorine chemistry and breakpoint chlorination, an analysis of the performance of chlorine as a disinfectant and the factors that may influence the effectiveness of the chlorination process, a discussion of the formation of disinfection byproducts (DBPs), and a consideration of the potential impacts of the discharge of DBPs to the environment. Disinfection with chlorine dioxide and dechlorination are considered in the following two sections, respectively. Chlorination facilities are considered in Sec. 12–6.

Characteristics of Chlorine Compounds

The principal chlorine compounds used at water reclamation plants are chlorine (Cl_2), sodium hypochlorite (NaOCl), and chlorine dioxide (ClO_2). Calcium hypochlorite [$Ca(OCl)_2$], another chlorine compound, is used in very small treatment plants due to its ease of operation. Many large cities have switched from chlorine gas to sodium hypochlorite because of the safety concerns and regulatory requirements related to the handling and storage of pressurized liquid chlorine (see Table 12–3). The characteristics of Cl_2, NaOCl, and $Ca(OCl)_2$ are considered below. The characteristics of chlorine dioxide and its use as a disinfectant are discussed in the following section.

Chlorine.   The general properties of chlorine (Cl_2) are summarized in Table 12–7. Chlorine can be present as a gas or a pressurized liquid. Chlorine gas is greenish yellow in color and about 2.48 times as heavy as air. Liquid chlorine is amber colored and about 1.44 times as heavy as water. Unconfined liquid chlorine vaporizes rapidly to a gas at standard temperature and pressure with one liter of liquid yielding about 450 liters of gas. Chlorine is moderately soluble in water, with a maximum solubility of about 1 percent at 10°C (50°F).

Although the use of chlorine for the disinfection has been of great significance from a public health perspective in both potable water supplies and treated wastewater, serious concerns have been raised about its continued use. Important concerns include the following:

1. Chlorine is a highly toxic substance that is transported by rail and truck, both of which are prone to accidents.
2. Chlorine is a highly toxic substance that potentially poses health risks to treatment plant operators, and the general public, if released by accident.
3. Because chlorine is a highly toxic substance, stringent requirements for containment and neutralization must be implemented as specified in the Uniform Fire Code (UFC).

Table 12–7

Properties of chlorine, chlorine dioxide, and sulfur dioxide[a]

Property	Unit	Chlorine (Cl_2)	Chlorine dioxide (ClO_2)	Sulfur dioxide (SO_2)
Molecular weight	g	70.91	67.45	64.06
Boiling point (liquid)	°C	−33.97	11	−10
Melting point	°C	−100.98	−59	−72.7
Latent heat of vaporization	kJ/kg	253.6	27.28	376.0
Liquid density at 15.5°C	kg/m^3	1422.4	1640[b]	1396.8
Solubility in water at 15.5°C	g/L	7.0	70.0[b]	120
Specific gravity of liquid at 0°C (water = 1)	s.g.	1.468		1.486
Vapor density at 0°C and 1 atm	kg/m^3	3.213	2.4	2.927
Vapor density compared to dry air at 0°C and 1 atm	unitless	2.486	1.856	2.927
Specific volume of vapor at 0°C and 1 atm	m^3/kg	0.3112	0.417	0.342
Critical temperature	°C	143.9	153	157.0
Critical pressure	kPa	7811.8		7973.1

[a] Adapted in part from U.S. EPA (1986), White (1999).

[b] At 20°C.

4. Chlorine reacts with the organic constituents in wastewater to produce odorous compounds.

5. Chlorine reacts with the organic constituents in wastewater to produce byproducts, many of which are known to be carcinogenic and/or mutagenic.

6. Residual chlorine in treated effluent is toxic to aquatic life.

7. The discharge of chloro-organic compounds has long-term effects on the environment that are not known.

Sodium Hypochlorite. Sodium hypochlorite (NaOCl) (i.e., liquid bleach), is only available as an aqueous solution and usually contains 12.5 to 17 percent available chlorine at the time it is manufactured. Sodium hypochlorite can be purchased in bulk or manufactured onsite; however, the solution decomposes more readily at high concentrations and is affected by exposure to light and heat. A 16.7 percent solution stored at 26.7°C (80°F) will lose 10 percent of its strength in 10 d, 20 percent in 25 d, and 30 percent in 43 d. It must, therefore, be stored in a cool location in a corrosion resistant tank. Another disadvantage of sodium hypochlorite is the chemical cost. The purchase price may range from 150 to 200 percent of the cost of liquid chlorine. The handling of sodium hypochlorite requires special design considerations because of its corrosiveness, the presence of chlorine fumes, and gas binding and caking in chemical feed lines. Several proprietary systems are available for the generation of sodium hypochlorite from sodium chloride (NaCl) or seawater. These systems are electric power intensive and result in a very dilute solution, a maximum of 0.8 percent as chlorine. Onsite generation systems have been used only on a limited basis, typically at relatively large plants, due to their complexity and high power cost.

Calcium Hypochlorite. Calcium hypochlorite [Ca(OCl)$_2$], is available commercially in either a dry or a wet form. In dry form it is available as an off-white powder or as granules, compressed tablets, or pellets. Calcium hypochlorite granules or pellets are readily soluble in water, varying from about 21.5 g/100 mL at 0°C (32°F) to 23.4 g/100 mL at 40°C (104°F). Because of its oxidizing potential, calcium hypochlorite should be stored in a cool, dry location away from other chemicals in corrosion resistant containers. With proper storage conditions the granules are relatively stable. Calcium hypochlorite is more expensive than liquid chlorine, loses its available strength when stored, and because it must be dissolved before being used, it is difficult to handle for large installations. In addition, calcium hypochlorite can clog metering pumps, piping, and valves as it tends to crystallize readily. Calcium hypochlorite is used most commonly at small installations in a dry form as tablets, where handling is relatively easy for plant operators.

Chemistry of Chlorine Compounds

The reactions of chlorine in water and the reaction of chlorine with ammonia are presented below.

Chlorine Reactions in Water. When chlorine in the form of Cl$_2$ gas is added to water, two reactions take place: *hydrolysis* and *ionization*.

Hydrolysis may be defined as the reaction in which chlorine gas combines with water to form hypochlorous acid (HOCl).

$$Cl_2 + H_2O \rightarrow HOCl + H^+ + Cl^- \tag{12-9}$$

The equilibrium constant, K_H, for this reaction is

$$K_H = \frac{[HOCl][H^+][Cl^-]}{[Cl_2]} = 4.5 \times 10^{-4} \text{ at } 25°C \tag{12-10}$$

Because of the magnitude of the equilibrium constant, large quantities of chlorine can be dissolved in water.

Ionization of hypochlorous acid to hypochlorite ion (OCl$^-$) may be defined as

$$HOCl \rightleftarrows H^+ + OCl^- \tag{12-11}$$

The ionization constant, K_i, for this reaction is

$$K_i = \frac{[H^+][OCl^-]}{[HOCl]} = 3 \times 10^{-8} \text{ at } 25°C \tag{12-12}$$

The variation in the value of K_i with temperature is reported in Table 12–8.

The total quantity of HOCl and OCl$^-$ present in water is called the *free chlorine*. The relative distribution of these two species (see Fig. 12–7) is very important because the killing efficiency of HOCl is many times that of OCl$^-$. The percentage distribution of HOCl at various temperatures can be computed using Eq. (12–13) and the data in Table 12–8.

$$\frac{[HOCl]}{[HOCl] + [OCl^-]} = \frac{1}{1 + [OCl^-]/[HOCl]} = \frac{1}{1 + [K_i]/[H^+]} = \frac{1}{1 + K_i 10^{pH}} \tag{12-13}$$

Hypochlorite Reactions in Water. Free chlorine can also be added to water in the form of hypochlorite salts. Both sodium and calcium hypochlorite hydrolyze to form hypochlorous acid (HOCl) as follows:

$$NaOCl + H_2O \rightarrow HOCl + NaOH \tag{12-14}$$

Table 12–8

Values of the ionization constant of hypochlorous acid at different temperatures[a]

Temperature, °C	$K_i \times 10^8$, mole/L
0	1.50
5	1.76
10	2.04
15	2.23
20	2.62
25	2.90
30	3.18

[a] Computed using equation from Morris (1966).

$$Ca(OCl)_2 + 2H_2O \rightarrow 2HOCl + Ca(OH)_2 \tag{12-15}$$

The ionization of hypochlorous acid was discussed previously [see Eq. (12–11)].

Chlorine Reactions with Ammonia. Untreated wastewater contains nitrogen in the form of ammonia, ammonium, and various combined organic forms (see Table 2–6 in Chap. 2). The effluent from most treatment plants also contains significant amounts of nitrogen, usually in the form of ammonia, ammonium, or nitrate, if the plant is designed to achieve nitrification. As noted in Chap. 2 the relative distribution between ammonia and ammonium will depend on the pH. Because hypochlorous acid is a very active oxidizing agent, it will react readily with ammonia (used here for the purpose of illustration) in water to form three types of chloramines in successive reactions:

$$NH_3 + HOCl \rightarrow NH_2Cl \text{ (monochloramine)} + H_2O \text{ [see also Eq. (12–19)]} \tag{12-16}$$

$$NH_2Cl + HOCl \rightarrow NHCl_2 \text{ (dichloramine)} + H_2O \tag{12-17}$$

$$NHCl_2 + HOCl \rightarrow NCl_3 \text{ (nitrogen trichloride)} + H_2O \tag{12-18}$$

These reactions are very dependent on the pH, temperature, and contact time, and on the ratio of chlorine to ammonia (White, 1999). The two species that predominate, in most cases, are monochloramine (NH_2Cl) and dichloramine ($NHCl_2$). The ratio of dichloramine to monochloramine as a function of the ratio of chlorine to ammonia at various pH values is presented in Table 12–9. The amount of nitrogen trichloride present is negligible up to chlorine-to-nitrogen ratios of 2.0. As will be discussed subsequently, chloramines also

Figure 12–7

Distribution of hypochlorous (HOCl) acid and hypochlorite ion (OCl⁻) in water as a function of pH at 0 and 20°C.

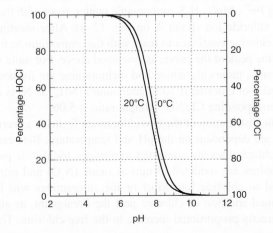

Table 12–9

Ratio of dichoramine to monochloramine under equilibrium conditions as a function of pH and applied molar dose ratio of chlorine to ammonium[a]

Molar Ratio Cl₂:NH₄⁺	pH			
	6	**7**	**8**	**9**
0.1	0.13	0.014	1E-03	0.000
0.3	0.389	0.053	5E-03	0.000
0.5	0.668	0.114	0.013	1E-03
0.7	0.992	0.213	0.029	3E-03
0.9	1.392	0.386	0.082	0.011
1.1	1.924	0.694	0.323	0.236
1.3	2.700	1.254	0.911	0.862
1.5	4.006	2.343	2.039	2.004
1.7	6.875	4.972	4.698	4.669
1.9	20.485	18.287	18.028	18.002

[a] From U.S. EPA (1986).

serve as disinfectants, although they are slow-reacting. When chloramines are the only disinfectants, the measured residual chlorine is defined as *combined chlorine residual* as opposed to free chlorine in the form of hypochlorous acid and hypochlorite ion.

Breakpoint Reaction with Chlorine

The maintenance of a residual (free or combined) for the purpose of disinfection is complicated because free chlorine not only reacts with ammonium, as noted previously, but also is a strong oxidizing agent. The term *breakpoint chlorination* is the term applied to the process whereby enough chlorine is added to react with all oxidizable substances such that if additional chlorine is added it will remain as free chlorine. The main reason for adding enough chlorine to obtain a free chlorine residual is that effective disinfection can usually then be assured. The amount of chlorine that must be added to reach a desired level of residual is called the *chlorine demand*. Breakpoint chlorination chemistry, acid generation, and the buildup of dissolved solids are considered in the following discussion.

Breakpoint Chlorination Chemistry. The stepwise phenomena that result when chlorine is added to water containing oxidizable substances and ammonium can be explained by referring to Fig. 12–8. As chlorine is added, readily oxidizable substances, such as Fe^{2+}, Mn^{2+}, H_2S, and organic matter, react with the chlorine and reduce most of it to the chloride ion (point A on Fig. 12–8). After meeting this immediate demand, the added chlorine continues to react with the ammonium to form chloramines between point A and the peak of the curve, as discussed above. For mole ratios of chlorine to ammonium less than 1, monochloramine and dichloramine are formed. At the peak of the curve, the mole ratio of chlorine (Cl_2) to ammonium (NH_4^+ as N) is equal to one [see Eq. (12–16)]. The corresponding Cl_2/NH_4^+ weight ratio is 5.06.

The distribution of the two chloramine forms is governed by their rates of formation, which are dependent on the pH and temperature. Between the peak and the breakpoint, some chloramines are converted to nitrogen trichloride [see Eq. (12–18)], the remaining chloramines are oxidized to nitrous oxide (N_2O) and nitrogen (N_2), and the chlorine is reduced to chloride ion. Most of the chloramines will be oxidized at the breakpoint. Continued addition of chlorine past the breakpoint, as shown on Fig. 12–8, will result in a directly proportional increase in the free chlorine. Theoretically, the weight ratio of

Figure 12–8

Generalized breakpoint chlorination curve. The upper portion of the diagram represents residual chlorine as a function of the amount of chlorine added to wastewater containing ammonium. The lower portion represents the fate of ammonium and chloramines during the breakpoint chlorination process. The dashed line reflects the fact that along with the formation of chloramines, some destruction of the chloramines occurs simultaneously before the peak is reached.

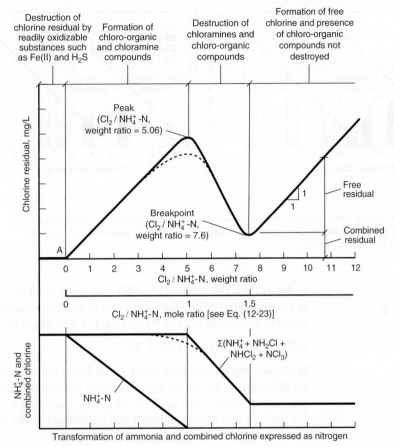

chlorine to ammonium nitrogen at the breakpoint is 7.6 to 1 (see Example 12–3) and the mole ratio is equal to 1.5 to 1 [see Eq. (12–23)].

Possible reactions to account for the appearance of N_2 and N_2O and the disappearance of chloramines during breakpoint chlorination are as follows (Saunier, 1976; Saunier and Selleck, 1976):

$$NH_4^+ + HOCl \rightarrow NH_2Cl + H_2O + H^+ \qquad (12\text{–}19)$$

$$NH_2Cl + HOCl \rightarrow NHCl_2 + H_2O \qquad (12\text{–}20)$$

$$NHCl_2 + H_2O \rightarrow NOH + 2HCl \qquad (12\text{–}21)$$

$$NHCl_2 + NOH \rightarrow N_2 + HOCl + HCl \qquad (12\text{–}22)$$

The overall reaction, obtained by summing Eqs. 12–19 through 12–22, is given as:

$$2NH_4^+ + 3HOCl \rightarrow N_2 + 3H_2O + 3HCl + 2H^+ \qquad (12\text{–}23)$$

Occasionally, serious odor problems have developed during breakpoint-chlorination operations because of the formation of nitrogen trichloride and related compounds. The presence of additional compounds that will react with chlorine, such as organic nitrogen, may greatly alter the shape of the breakpoint curve, as shown on Fig. 12–9. The formation of disinfection byproducts is considered later in this section.

Acid Generation. The addition of chlorine gas produces acid. When chlorine is added to water the hydrolysis reaction results in the formation of a strong acid (HCl) as given

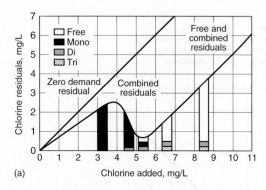
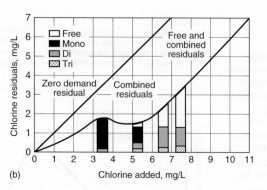

Figure 12–9

Curves of chlorine residual versus chlorine dosage for wastewater: (a) for wastewater containing ammonia nitrogen and (b) for wastewater containing nitrogen in the form of ammonia and organic nitrogen. (Adapted from White, 1999.)

by Eq. (12–9). The reaction of HOCl with ammonium also results in the formation of acid as given by Eq. (12–23). The total moles of hydrogen that must be neutralized can be determined by combining Eq. (12–9) with Eq. (12–23), which results in the following expression:

$$2NH_4^+ + 3Cl_2 \rightarrow N_2 + 6HCl + 2H^+ \tag{12–24}$$

In practice, the hydrochloric acid formed during chlorination [see Eq. (12–16)] reacts with the alkalinity of the wastewater, and under most circumstances, there is a slight pH drop. Stoichiometrically, 14.3 mg/L of alkalinity, expressed as $CaCO_3$, are required for each 1.0 mg/L of ammonium nitrogen that is oxidized in the breakpoint-chlorination process (see Example 12–3).

Buildup of Total Dissolved Solids (TDS). In addition to the formation of hydrochloric acid, the chemicals added to achieve the breakpoint reaction also contribute an incremental increase in the TDS. As shown in Eq. (12–24), 6 moles of HCl and 2 moles of H^+ are formed, while 2 moles of NH_4^+ are removed from solution. In situations where the level of total dissolved solids may be critical with respect to water reuse applications, this incremental buildup from breakpoint chlorination should always be checked. The TDS contribution for each of several chemicals that may be used in the breakpoint reaction is summarized in Table 12–10. The magnitude of the possible buildup of TDS is illustrated

Table 12–10

Effects of chemical addition on total dissolved solids in breakpoint chlorination[a]

Chemical addition	Increase in total dissolved solids per unit of NH_4^+ consumed
Breakpoint with chlorine gas	6.2 : 1
Breakpoint with sodium hypochlorite	7.1 : 1
Breakpoint with chlorine gas—neutralization of all acidity with lime (CaO)	12.2 : 1
Breakpoint with chlorine gas—neutralization of all acidity with sodium hydroxide (NaOH)	14.8 : 1

[a] From U.S. EPA (1986).

in Example 12–3 in which the use of breakpoint chlorination is considered for the seasonal control of nitrogen.

EXAMPLE 12–3 **Analysis of Disinfection Process for Nitrified Secondary Effluent with Free Chlorine** Estimate the daily required chlorine dosage, the required alkalinity, if alkalinity needs to be added, and the resulting buildup of TDS when breakpoint chlorination is used achieve disinfection with free chlorine. Assume that the following data apply to this problem:

1. Plant flowrate = 3800 m³/d
2. Secondary effluent characteristics
   a. BOD = 20 mg/L
   b. Total suspended solids = 25 mg/L
   c. Residual NH_3–N = 2 mg/L
   d. Alkalinity = 150 mg/L as $CaCO_3$
3. Required free chlorine residual concentration for disinfection = 0.5 mg/L
4. Any alkalinity added is in the form of lime (CaO)

Solution

1. Determine the molecular weight ratio of hypochlorous acid (HOCl), expressed as Cl_2, to ammonium (NH_4^+), expressed as N, using the overall reaction for the breakpoint reactions given by Eq. (12–23).

$$2NH_4^+ + 3HOCl \rightarrow N_2 + 3H_2O + 3HCl + 2H^+$$

2(18)   3(52.45)
2(14)   3(2 × 35.45)

Molecular weight ratio is:

$$\frac{Cl_2}{NH_4^+\text{-N}} = \frac{3(2 \times 35.45)}{2(14)} = 7.60$$

2. Estimate the required Cl_2 dosage.
   a. Determine the Cl_2 dosage needed to reach the breakpoint using the molecular ratio developed in Step 1.

   $$Cl_2 = (2 \text{ g/m}^3)(7.6 \text{ g/g}) = 15.2 \text{ g/m}^3$$

   b. Determine the required Cl_2 dosage including the free residual.

   $$Cl_2/d = (3800 \text{ m}^3/d)[(15.2 + 0.5) \text{ g/m}^3](1 \text{ kg}/10^3 \text{ g}) = 59.9 \text{ kg/d}$$

3. Determine the alkalinity required.
   a. The total number of moles of H^+ that must be neutralized per mole of NH_4^+ oxidized is given by Eq. (12–24), which has been divided by 2.

   $$NH_4^+ + 1.5Cl_2 \rightarrow 0.5N_2 + 3HCl + H^+$$

   b. When using lime to neutralize the acidity, the required alkalinity ratio is computed as follows:

   $$2CaO + 2H_2O \rightarrow 2Ca^{2+} + 4OH^-$$

   $$\text{Required alkalinity ratio} = \frac{2(100 \text{ g/mole of } CaCO_3)}{(14 \text{ g/mole of } NH_4^+ \text{ as N})} = 14.3$$

c. The required alkalinity is

$$Alk = \frac{[(14.3\,mg/L\,alk)/(mg/L\,NH_4^+)](2\,mg/L\,NH_4^+)(3800\,m^3/d)}{(10^3\,g/kg)}$$

$$= 108.7\,mg/L\;as\;CaCO_3$$

4. Determine whether sufficient alkalinity is available to neutralize the acid during breakpoint chlorination.

Because the available alkalinity (150 mg/L) is greater than the required alkalinity (108.7 mg/L), alkalinity will not have to be added to complete the reaction.

5. Determine the increment of TDS added to the secondary effluent. Using the data reported in Table 12–10, the TDS increase per mg/L of ammonia consumed when CaO is used to neutralize the acid formed is equal to 12.2 to 1.

TDS increment added = 12.2(2) mg/L = 24.4 mg/L

Comment The ratio computed in Step 1 will vary somewhat, depending on the actual reactions involved. In practice, the actual ratio typically has been found to vary from 8:1 to 10:1. Similarly, in Step 3, the stoichiometric coefficients will also depend on the actual reactions involved. In practice, it has been found that about 15 mg/L of alkalinity are required because of the hydrolysis of chlorine. In Step 5, it should be noted that although breakpoint chlorination can be used to control nitrogen, it may be counter productive if in the process the treated effluent is rendered unusable for other applications because of the buildup of total dissolved solids, and the potential formation of disinfection byproducts.

Effectiveness of Free and Combined Chlorine as Disinfectants

In view of the renewed interest in public health, environmental water quality, and water reclamation, the effectiveness of the chlorination process is of great concern. Numerous tests have shown that when all the physical parameters controlling the chlorination process are held constant, the germicidal efficiency of disinfection, as measured by the survival of "discrete bacteria," depends primarily on the form of the chlorine residual and time (i.e., CT).

Relative Effectiveness of Free and Combined Chlorine. Generalized data on the relative germicidal effectiveness of combined and free chlorine for the disinfection of different microorganisms were presented previously in Table 12–5 in terms of the required CT values to achieve various levels of inactivation. A comparison of the relative germicidal efficiency of hypochlorous acid (HOCl), hypochlorite ion (OCl⁻), and monochloramine (NH_2Cl), based on the work of Butterfield et al. (1943), is presented on Fig. 12–10. As shown on Fig. 12–10, for a given contact time or chlorine residual, the germicidal efficiency of HOCl is 100 times that of OCl⁻ and more than 400 times that of NH_2Cl. However, because of the equilibrium relationship that exists between HOCl and OCl⁻ ion (see Fig. 12–7), maintenance of the proper pH is extremely important in achieving effective disinfection. It should be noted, however, that given an adequate contact time, monochloramine is nearly as effective as free chlorine in achieving disinfection. In addition to the data for the chlorine compounds given on Fig. 12–10, corresponding CT values have been added for the purpose of comparison. As shown, the

Figure 12–10

Comparison of the germicidal efficiency of hypochlorous (HOCl) acid and hypochlorite ion (OCl⁻), and monochloramine for 99 percent destruction of *E. coli* at 2 to 6°C with CT values added for the purpose of comparison. (From Butterfield et al., 1943.)

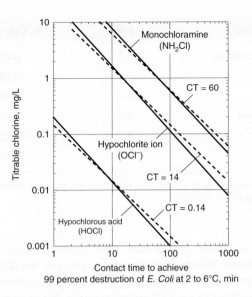

disinfection data presented on Fig. 12–10 can be represented quite well with the CT relationship.

Coefficient of Specific Lethality. Another parameter that can be used to assess the relative effectiveness of the various forms of chorine is the coefficient of specific lethality, Λ. Utilizing the data from Table 12–5 and Fig. 12–10 as well as numerous literature sources, coefficients of specific lethality, computed for various microorganism groups and disinfecting agents, are summarized in Table 12–11. It is important to note that the data presented in Table 12–11 were derived primarily using batch reactors operated under

Table 12–11

Relative coefficients of specific lethality, Λ, for the inactivation of bacteria, viruses, *Cryptosporidium*, and *Giardia lamblia* cysts in filtered secondary effluent with various disinfectants (pH ~7.5, ~20°C)[a]

Disinfectant	Unit	Coefficient of specific lethality[b], $\Lambda_{(base\ 10)}$			
		Total coliform[c]	Virus	Protozoa *Cryptosporidium*	Protozoa *Giardia lamblia* cysts
Chlorine (free)	L/mg·min	2	1.2	0.00044	0.04
Chloramine	L/mg·min	0.016	0.0052	0.00022	0.0024
Chlorine dioxide	L/mg·min	2	0.38	0.008	0.2
Ozone	L/mg·min	44	7.27	0.24	4.21
UV radiation	cm²/mJ	5.7	0.0215	0.31	0.33

[a] Based on Eq. (12–6).
[b] The coefficient of specific lethality values are for the linear portion of the dose (CT) response curve.
[c] The reported coefficient of specific lethality values for fecal coliform and *E coli* are quit different.

Table 12–12

Activation energies for aqueous chlorine and chloramines at normal temperatures[a]

Compound	pH	E, Cal/mole	E, J/mole
Aqueous chlorine	8.5	6400	26,800
	9.8	12,000	50,250
	10.7	15,000	62,810
Chloramines	7.0	12,000	50,250
	8.5	14,000	58,630
	9.5	20,000	83,750

[a] Adapted from Fair et al. (1948) who developed the reported values using the data developed by Butterfield et al. (1943).

controlled conditions, and, as such, are of limited use other than for the purpose of illustrating the relative differences in the effectiveness of the different disinfectants for different organism groups. As shown, there are significant differences in the effectiveness of the various disinfectants for each organism group. For example, free chlorine is very effective for the inactivation of bacteria and viruses, but less so for *cryptosporidium* and *Giardia* cysts.

Effect of pH. The importance of pH and temperature on the disinfection process with chlorine and chloramines was investigated by Butterfield and his associates in 1943 (Butterfield et al., 1943; Wattie and Butterfield, 1944). Based on the results published by Butterfield et al. (1943), Fair and Geyer (1954) determined the activation energy values reported in Table 12–12 for the disinfection of *E. coli* in clean water. Reviewing the data in Table 12–12, it is important to note the magnitude of the activation energy as a function of pH. As the pH increases, the value of the activation energy increases which corresponds to a reduced effectiveness which is consistent with the data presented on Fig. 12–10.

Effect of Temperature. Temperature also has a significant impact on the coefficient of specific lethality, Λ. As a rule of thumb, it has been found that for each 10°C increase in temperature (identified as the Q_{10} temperature coefficient in biological and chemical engineering literature) there is a 2 to 2.5 times increase in the coefficient of specific lethality. Thus, when referring to the CT values given in Table 12–5 and the coefficient of specific lethality values in Table 12–11, it is important to note that the given values are for a pH and temperature of approximately 7.5 and 20°C, respectively.

Measurement and Reporting of Disinfection Process Performance

To provide a framework in which to consider the effectiveness of disinfection and the factors that affect the disinfection of treated wastewater, it is appropriate to consider how the effectiveness of the chlorination process is now assessed and how the results are analyzed. When using chlorine for the disinfection, the principal parameters that can be measured, apart from environmental variables such as pH and temperature, are the number of organisms and the form of the chlorine residual (i.e., combined or free or both) remaining after a specified period of time.

Number of Organisms Remaining. The coliform group of bacteria can be determined using a number of different techniques (Standard Methods, 2012). The membrane filter technique or the most probable number (MPN) procedure as discussed in

Chap. 2 are used commonly. The organisms remaining can also be determined by the plate-count procedure using an agar mixture as the plating medium. Either the standard "pour-plate" method or the "spread-plate" method can be used. The plates should be incubated at 37°C (98.6°F) because this temperature results in the optimum growth of *E. coli*, and the colonies should be counted after a 24-h incubation period.

Measurement of Chlorine Residual. The principal methods used to measure the free and combined chlorine residual include (1) the DPD (N,N-diethyl-p-phenylenediamine) colorimetric method, (2) DPD titration method, (3) the Iodometric titration method, and (4) the amperometric titration method. Of these methods, the DPD colorimetric method is currently the most widely used because it can be used to differentiate between free and combined chlorine species. Both field hand-held and continuous online residual analyzers are available. In the DPD method, appropriate chemicals, typically preformed in packets, are added to a sample containing chlorine. The red color resulting from the presence of chlorine is measured with a spectrophotometer or filter photometer. The initial color is due to free chlorine. Additional chemicals are added to obtain total residual chlorine (free and combined). Additional details on these chlorine analysis methods may be found in Harp (2002) and Standard Methods (2012).

Reporting of Results. Disinfection process results are reported in terms of the number of organisms and the chlorine residual remaining after a specified period of time. When the results are plotted it is common practice to plot the logs of removal versus the corresponding CT value as shown previously on Fig. 12–6.

Factors that Affect Disinfection of Wastewater with Chlorine Compounds

The purpose of the following discussion is to explore the important factors that affect the disinfection efficiency of chlorine compounds in actual wastewater applications. These include the following:

1. Initial mixing
2. Chemical characteristics of the water to be disinfected
3. NOM content
4. Impact of particles and particle associated microorganisms
5. Characteristics of the microorganisms
6. Contact time

Each of these factors are discussed in more detail below.

Issues related to the design of chlorine contact basins not included in this chapter include (1) basin configuration, (2) the use of baffles and guide vanes, (3) number of chlorine contact basins, (4) precipitation of solids in chlorine contact basins, (5) solids transport velocity, and (6) a procedure for predicting disinfection performance. These subjects are considered in detail elsewhere (Tchobanoglous et al., 2003).

Initial Mixing. The importance of initial mixing on the disinfection process cannot be overstressed. It has been shown that the application of chlorine in a highly turbulent regime ($N_R \geq 10^4$) results in kills two orders of magnitude greater than when chlorine is added separately to a conventional rapid-mix reactor under similar conditions. Although the importance of initial mixing is well delineated, the optimum level of turbulence is not known. Examples of mixing facilities designed to achieve the rapid mixing of chlorine with the water are presented later in Sec. 12–6 (see Fig. 12–22).

Based on recent findings, questions have now been raised about the form in which the chlorine compounds are added. In some plants where chlorine injectors are used, there is concern over the practice of using chlorinated wastewater for the chlorine injection water as opposed to clean water. The concern is that if nitrogenous compounds are present in the wastewater, a portion of the chlorine that is added reacts with these compounds, and by the time chlorine solution is injected, it is in the form of monochloramine or dichloramine. The formation of chloramines can be a problem if adequate retention time is not available in the chlorine contact basin as combined chlorine requires a longer contact time. Again, it should be remembered that although both HOCl and NH$_2$Cl are both effective as disinfecting compounds, the contact time required is significantly different for the same residual concentration (see Fig. 12–10).

The formation of disinfection byproducts (DBPs) is another major concern with the use of free chlorine. When wastewater is exposed to free chlorine, competing reactions, such as the formation of chloramines (free chlorine and ammonia), and DBPs can occur. The predominant reaction depends on the applicable kinetic rates for the various reactions. The formation and control of DBPs is discussed later in this section.

Chemical Characteristics of Wastewater. It has often been observed that, for treatment plants of similar design with exactly the same effluent characteristics measured in terms of BOD, COD, and nitrogen, the effectiveness of the chlorination process varies significantly from plant to plant. To investigate the reasons for this observed phenomenon and to assess the effects of the compounds present in the chlorination process, Sung (1974) studied the characteristics of the compounds in untreated and treated wastewater. Among the more important conclusions derived from Sung's study are the following:

1. In the presence of interfering organic compounds, the total chlorine residual cannot be used as a reliable measure for assessing the bactericidal efficiency of chlorine.
2. The degree of interference of the compounds studied depended on their functional groups and their chemical structure.
3. Saturated compounds and carbohydrates exert little or no chlorine demand and do not appear to interfere with the chlorination process.
4. Organic compounds with unsaturated bonds may exert an immediate chlorine demand, depending on their functional groups. In some cases, the resulting compounds may titrate as chlorine residual and yet may possess little or no disinfection potential.
5. Compounds with polycyclic rings containing hydroxyl groups and compounds containing sulfur groups react readily with chlorine to form compounds which have little or no bactericidal potential, but which still titrate as chlorine residual.
6. To achieve low bacterial counts in the presence of interfering organic compounds, additional chlorine and longer contact times are required.

From the results of Sung's work, it is easy to see why the efficiency of chlorination at plants with the same general effluent characteristics can be quite different. Clearly, it is not the value of the BOD or COD that is significant, but the nature of the organic compounds that make up the measured values. Thus, the nature of the treatment process used in any plant also has an effect on the chlorination process. The impact of wastewater characteristics on chlorine disinfection is presented in Table 12–13. The presence of oxidizable compounds such as humics and iron causes the inactivation curve to have a lag or shoulder affect as shown on Fig. 12–6. In effect, the added chlorine is being utilized in the oxidation of these substances and is not available for the inactivation of microorganisms.

Because more wastewater treatment plants are now removing nitrogen, operational problems with chlorine disinfection are now reported more frequently. In treatment plants

Table 12–13

Impact of wastewater constituents on the use of chlorine for wastewater disinfection

Constituent	Effect
BOD, COD, TOC, etc.	Organic compounds that comprise the BOD and COD can exert a chlorine demand. The degree of interference depends on their functional groups and their chemical structure
NOM (natural organic matter)	Reduces effectiveness of chlorine by forming chlorinated organic compounds that are measured as chlorine residual, but are not effective for disinfection
Oil and grease	Can exert a chlorine demand
TSS	Shields embedded bacteria
Alkalinity	No or minor effect
Hardness	No or minor effect
Ammonia	Combines with chlorine to form chloramines
Nitrite	Oxidized by chlorine, formation of N-nitrosodimethylamine (NDMA)
Nitrate	Chlorine dose is reduced because chloramines are not formed. Complete nitrification may lead to the formation of NDMA due the presence of free chlorine. Partial nitrification, especially diurnal swings in nitrification, may lead to difficulties in establishing the proper chlorine dose
Iron	Oxidized by chlorine
Manganese	Oxidized by chlorine
pH	Affects distribution between hypochlorous acid and hypochlorite ion
Industrial discharges	Depending on the constituents, may lead to a diurnal and seasonal variations in the chlorine demand

where the effluent is nitrified completely, the chlorine added to the water is present as free chlorine, after satisfying any immediate and nitrogenous (see Example 12–3) chlorine demand. In general, the presence of free chlorine will reduce significantly the required chlorine dosage. However, the presence of free chlorine may lead to the formation of disinfection byproducts including N-nitrosodimethylamine (NDMA). In treatment plants that do not nitrify completely, partially nitrify, or move in and out of nitrification diurnally, control of the chlorination process is especially difficult because of the variation in the effectiveness of the chlorine compounds. Some of the chlorine is used to satisfy the demand of the residual nitrite and/or ammonia. Because of the uncertainties involved in knowing to what degree the plant is nitrifying at any point in time, the chlorine dosage that is added is based on the dosage required if the disinfection is to be accomplished by combined chlorine compounds, resulting in excessive chlorine use.

Impact of Particles Found in Treated Wastewater. Another factor that must be considered is the presence of suspended solids in the water to be disinfected. As shown previously on Fig. 12–6, when suspended solids are present, the disinfection process is controlled by two different mechanisms. The log-linear bacterial inactivation that is observed initially, after the shoulder effect, is of individual free swimming bacteria and bacteria in small clumps. The straight line portion of the bacterial inactivation can be described using Eq. (12–2). In the curved portion of the curve the bacterial kill is controlled by the presence of suspended solids. The slope of the curved portion of the curve is a function of (1) the particle size distribution and (2) the number of particles with associated coliform organisms. Further, as noted previously, if particles contain

Figure 12–11

Inactivation of MS2 coliphage and polivirus in buffer and treated wastewater effluent with combined chlorine. (From BioVir Laboratories, 2001.)

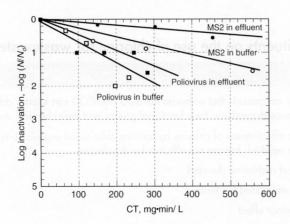

significant numbers of organisms, the organisms can provide protection to other organisms embedded within the particle by limiting the penetration of chlorine through diffusion. Unfortunately, the observed variability caused by the presence of particles often is masked by the addition of excess chlorine to overcome both chemical and particle effects.

Characteristics of the Microorganisms.   Other important variables in the chlorination process are the type, characteristics, and age of the microorganisms. For a young bacterial culture (1 d old or less) with a free chlorine dosage of 2 mg/L, only 1 min was needed to reach a low bacterial number. When the bacterial culture was 10 d old or more, approximately 30 min was required to achieve a comparable reduction for the same applied chlorine dosage. It is likely that the resistance offered by the polysaccharide sheath, which microorganisms develop as they age, accounts for this observation. In the activated sludge treatment process, the operating solids retention time (SRT), which to some extent is related to the age of the bacterial cells in the system, will, as discussed previously, affect the performance of the chlorination process. Some recent data on the disinfection of bacteriophage MS2 and poliovirus are shown on Fig. 12–11. As shown on Fig. 12–11, it is clear that a CT value of 450 mg·min/L, as used by the State of California, does not result in a four-log reduction of virus, when the measured residual chlorine is combined chlorine (i.e., mono- and dichloramine). Clearly, site-specific testing is required to establish the appropriate chlorine dose.

Some representative data on the effectiveness of chlorine for the inactivation of *E. coli* and three enteric viruses are reported on Fig. 12–12. Because of newer analytical techniques that have been developed, the data presented on Fig 12–12 are only meant to illustrate the differences in the resistances of different organisms. From the available evidence on the viricidal effectiveness of the chlorination process, it appears that chlorination beyond the breakpoint to obtain free chlorine is required to kill many of the viruses of concern. Where breakpoint chlorination is used, it is necessary to dechlorinate the treated water before discharge to the environment or reuse in sensitive applications to reduce any residual toxicity that may remain after chlorination. Based on the use of integrated cell culture-PCR techniques (see Chap. 2), it has been reported that the inactivation of poliovirus may require five times more chlorine than thought previously (Blackmer et al., 2000).

Contact Time.   Along with the residual concentration of the disinfectant, contact time is of critical importance in the design and operation of chlorination facilities.

Figure 12–12

Concentration of chlorine as HOCl required for 99 percent kill of *E. coli* and three enteric viruses at 0 to 6°C (Butterfield et al., 1943).

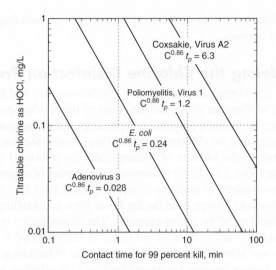

Coxsakie, Virus A2
$C^{0.86} t_p = 6.3$

Poliomyelitis, Virus 1
$C^{0.86} t_p = 1.2$

E. coli
$C^{0.86} t_p = 0.24$

Adenovirus 3
$C^{0.86} t_p = 0.028$

Titratable chlorine as HOCl, mg/L

Contact time for 99 percent kill, min

The principal design objective for chlorine contact basins is to ensure that some defined percentage of the flow remains in the chlorine contact basin for the design contact time to ensure effective disinfection. The mean contact time is usually specified by the regulatory agency and may range from 30 to 120 min; contact times of 15 to 90 min at peak flow are common. To be assured that a given percentage of the flow remains in the chlorine contact basin for a given period of time, the most common approach is to use long plug-flow, around-the-end type of contact basins (see Fig. 12–13). For example, for water reuse applications the CDPH requires a CT value of 450 mg·min/L based on a modal contact time of 90 min at peak flow. In other states, the t_{10} is used in the

Figure 12–13

Views of chlorine contact basins: (a) and (b) serpentine plug-flow chlorine contact basins with flow deflection baffles, (c) plug-flow chlorine contact basin with rounded corners, and (d) plug-flow basin with inlet diffuser.

(a)

(b)

(c)

(d)

CT relationship (see subsequent discussion on assessing the performance of chlorine contact basins).

Modeling the Chlorine Disinfection Process

When considering the disinfection of both secondary and filtered secondary effluent, both the lag or shoulder effect and the effect of the residual particles (see Fig. 12–6) must be considered. As noted previously, depending on the constituents in the wastewater, a shoulder region may be observed in which there is no reduction in the number of organisms as the result of the addition of a disinfectant. As additional chlorine is added beyond some limiting value, a log linear reduction in the number of organisms is observed with increased chlorine dosages. If particles (typically greater than 20 μm) are present, the disinfection curve starts to diverge from the log linear form and a tailing region is observed due to particle shielding of the microorganisms. The tailing region is of importance as more restrictive standards are to be achieved (e.g., 23 MPN/100 mL). It is interesting that the tailing region was identified in an early report on the chlorination of treated wastewater (Enslow, 1938). Further, because large particles have little effect on turbidity (see Chap. 2), effluents with low measured turbidity values can still be difficult to disinfect, due the presence of undetected large particles (Ekster, 2001; see also discussion of turbidity in Chap. 8).

The Collins-Selleck Model. In the early 1970s, Collins conducted extensive experiments on the disinfection of various wastewaters (Collins, 1970; Collins and Selleck, 1972). Using the batch reactor whose contents were well stirred, Collins and Selleck found that the reduction of coliform organisms in a chlorinated primary treated effluent followed a linear relationship when plotted on log-log paper (see Fig. 12–14). The equation developed to describe the observed results is

$$\frac{N}{N_o} = \frac{1}{(1 + 0.23\,CT)^3} \qquad (12\text{--}25)$$

Note that the form of the equation developed by Collins accounts for the shoulder effect and for tailing. A number of other models have been proposed including an empirical model proposed by Gard (1957), Hom (1972), which was subsequently rationalized by Haas and Joffe (1994) and Rennecker et al. (1999).

Figure 12–14

Coliform survival in a batch reactor as a function of amperometric chlorine residual and contact time (temperature range 11.5 to 18°C). (From Collins, 1970; Collins and Selleck, 1972.)

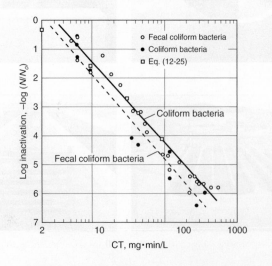

Figure 12-15

Definition sketch for the application of Eq. (12–27).

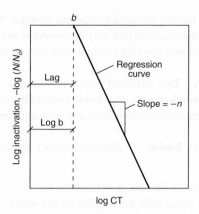

The Refined Collins-Selleck Model.   A refinement of the original Collins model for the disinfection of secondary effluent in which a shoulder effect and tailing is observed, as proposed by White (1999), is

$$N/N_o = 1 \text{ for CT} < b \tag{12–26}$$

$$N/N_o = [(\text{CT})/b]^{-n} \text{ for CT} > b \tag{12–27}$$

Where C = residual concentration of chemical agent at the end of time t, mg/L
  T = contact time, min
  n = slope of inactivation curve
  b = value of x-intercept when $N/N_o = 1$ or log $(N/N_o) = 0$ (see Fig. 12–15)

Typical values for the coefficients n and b for secondary non-nitrified effluent for coliform and fecal coliform organisms are 2.8 and 4.0 and 2.8 and 3.0, respectively (Roberts et al., 1980; White, 1999; Black & Veatch Corporation, 2010). However, because of the variability of the chemical composition of wastewater and the variable particle size distribution, it is recommended that the constants be determined for the wastewater in question.

Effluent from Membrane Processes.   The most important characteristic of these effluents is that they do not contain particles that can shield microorganisms. Depending on the type of membrane process used (microfiltration, ultrafiltration, nanofiltration, or reverse osmosis), moderate to significant reductions in the number of microorganisms present will also be observed (see discussion in Chaps. 8, 9, and 11). For these effluents, the Chick-Watson model, as given by Eq. (12–6), or, if a shoulder exists, the Collins-Selleck relationship, can be used to model the disinfection process with chlorine. Typically, the shoulder effect is reduced considerably, especially with reverse osmosis effluent.

Required Chorine Dosages for Disinfection

The required chemical dosage for disinfection can be estimated by considering (1) the initial chlorine demand of the treated wastewater, (2) the allowance needed for decay during the chlorine contact time, and (3) the required chlorine residual concentration determined using Eq. (12–27) for the organism under consideration (e.g., bacteria, virus, or protozoan oocysts and cysts).

Initial Chlorine Demand.   The chlorine dosage required to meet the initial demand depends on the constituents in the water (see Fig. 12–15). It is important to remember that the chlorine added to meet the initial demand, due to inorganic compounds, is reduced to the chloride ion and will not be measured as chlorine residual. Also, chlorine that combines with humic and other organic material that may be present is not effective as a disinfectant, but is nevertheless measured as a chlorine residual contributing to the lag term, b, in Eq. (12–27).

Chlorine Decay. Two different cases of chlorine decay have to be considered: (1) chlorine decay, which occurs in chlorine contact basins and (2) chlorine decay, which occurs in long effluent transmission lines and in water reuse distribution systems.

In Chlorine Contact Basins. In chlorine contact basins the principal factors affecting chlorine decay are

1. Chemical reactions that occur in the bulk liquid.
2. Reactions that occur with the biofilm on the walls of the chlorine contact basin.
3. Photooxidation that occurs in uncovered chlorine contact basins.

In addition to the rapid reactions that occur initially, slower addition and substitution type reactions can occur with the residual organic matter in the bulk of the treated effluent as it moves through the chlorine contact basin (Gang et al., 2003). Similarly, addition and substitution chemical reactions can occur with the biofilms attached to the wall of the chlorine contact basin. The nature of the reactions is site specific, but biofilms are almost always present, especially in uncovered chlorine contact basins. To reduce the decay due to UV oxidation observed in open uncovered chlorine contact basins, a variety of floating and fixed covers have been added to existing contact basins (see Fig. 12–16). Typical decay values for chlorine residual are on the order of 2 to 4 mg/L for a contact time of about one hour.

In Transmission and Distribution Piping. The principal factors affecting chlorine decay in transmission and distribution piping are

1. Chemical reactions that occur in the bulk liquid flowing within the pipe.
2. Reactions that occur with the biofilm on the walls of the piping system.
3. Chemical reactions due to the release of constituents under anaerobic conditions that will react with chlorine.

The reactions that occur in the bulk fluid and in the biofilm on the pipe wall are as described above for the chlorine contact basin. Under anaerobic conditions that often occur in long transmission lines, bacteria on the pipe walls will convert the sulfate present in the treated effluent to form sulfide, which in turn forms hydrogen sulfide, which readily reacts with any chlorine that may be present. There are many articles in the literature that deal with the modeling of chlorine decay in pipelines. In general, the decay process is modeled

Figure 12–16

Typical examples of covered back-and-forth chlorine contact basins: (a) basins covered with inexpensive floating tarp, and (b) basins covered with specially designed polypropylene cover.

(a)　　　　　　　　　　　　　　　　(b)

Table 12–14

Typical chlorine dosages, based on combined chlorine unless otherwise indicated, required to achieve different total coliform disinfection standards for various wastewaters based on a 30 min contact time with a decay factor of 0.6[a]

Type of wastewater	Initial total coliform count, MPN/100 mL	Chlorine dose, mg/L			
		Effluent standard, MPN/100 mL			
		1000	200	23	≤2.2
Raw wastewater	10^7–10^9	16–30			
Primary effluent	10^7–10^9	8–12	18–24		
Trickling filter effluent	10^5–10^6	6–7.5	12–15	18–22	
Activated sludge effluent	10^5–10^6	5.5–7.5	10–13	13–17	
Filtered activated sludge effluent	10^4–10^6	2.5–3.5	5.5–7.5	10–13	13–17
Nitrified effluent[b]	10^4–10^6		0.02–0.03	0.03–0.04	0.04–0.05
Filtered nitified effluent[b]	10^4–10^6		0.02–0.03	0.03–0.04	0.04–0.05
Microfiltration effluent[b]	10^1–10^3			0.02–0.03	0.03–0.04
Reverse osmosis[b]	~ 0				0.01–0.02
Septic tank effluent	10^7–10^9	16–30	30–60		
Intermittent sand filter effluent	10^2–10^4	1–2	2–4	3–6	4–8

[a] The combined chlorine values are based on the assumption that the added chlorine only combines with ammonia to form monochloramine. The reported values are independent of the chlorine dose required to meet the immediate chlorine demand.

[b] Based on free chlorine. The reported values are independent of the chlorine dose required to reach the breakpoint (see Example 12–3).

using either first or second-order decay models. The computer model EPANET 2, developed by the U.S. EPA to simulate the hydraulic and water quality behavior within pressurized pipe networks, has also been used to study the decay of chlorine in pipelines (Rossman, 2000). It is critical that decay be considered in long transmission and distribution piping in determining the required residual that may be needed.

Required Chlorine Residual. Typical chlorine dosage values for various for residual total coliform concentrations, based on a contact time of 30 min, are reported in Table 12–14. It should be noted that the dosage values given in Table 12–14 are only meant to serve as a guide for the initial estimation of the required chlorine dose. As noted above, site-specific testing is required to establish the appropriate chlorine dose. Estimation of the required chlorine dose is illustrated in Example 12–4.

EXAMPLE 12–4 **Estimate the Required Chlorine Dose for a Typical Non-nitrified Secondary Effluent** Estimate the chlorine dose needed to disinfect a filtered non-nitrified secondary activated sludge effluent using the refined Collins-Selleck model assuming a shoulder effect exists and that the following conditions apply. Check the computed summer combined residual using Eq. (12–6).

1. Effluent total coliform count before disinfection = $10^7/100$ mL
2. Required summer effluent total coliform count = 23/100 mL
3. Required winter effluent total coliform count = 240/100 mL
4. Immediate summer or winter effluent chlorine demand, not including the shoulder effect = 2 mg/L
5. Chlorine demand due to decay in chlorine contact tanks during the summer months (May–October) = 2.5 mg/L
6. Chlorine demand due to decay in chlorine contact tanks during the winter months (November–April) = 1.5 mg/L
7. Required chlorine contact time = 30 min
8. Use the typical values given in the above discussion for the coefficients $n = 2.8$ and $b = 4.0$
9. The coefficient of specific lethality for summer conditions = 0.024 L/mg·min (base 10)

Solution

1. Estimate the required combined chlorine residual using the refined Collins-Selleck Model, Eq. (12–27) and the given coefficients.

   $$N/N_o = (CT/b)^{-n}$$

   a. Summer

   $$23/10^6 = (CT/4.0)^{-2.8}$$

   $$(23/10^6)^{-\frac{1}{2.8}} = (CT/4.0)$$

   $$(45.3)4 = C(30)$$

   $$C = 6.0 \text{ mg/L}$$

   b. Winter

   $$240/10^6 = (CT/4.0)^{-2.8}$$

   $$(19.6)4 = C(30)$$

   $$C = 2.6 \text{ mg/L}$$

2. The required chlorine dosage is
   a. Summer

   Chlorine dosage = 2.0 mg/L + 2.5 mg/L + 6.0 mg/L = 10.5 mg/L

   b. Winter

   Chlorine dosage = 2.0 mg/L + 1.5 mg/L + 2.6 mg/L = 6.1 mg/L

3. Determine the required chlorine summer dose using Eq. (12–6).
   a. Solve Eq. (12–6) for the combined residual.

   $$\log \frac{N_t}{N_o} = -\Lambda_{\text{base 10}}CT = \log \frac{23}{10^6} = (-0.024)(C)(30)$$

   $$C = \frac{-4.64}{-(0.024\,\text{L/mg·min})(30\,\text{min})} = 6.4 \text{ mg/L}$$

   b. The required chlorine doses computed with the two methods are similar.

Comment   The chlorine dosage increases significantly as the effluent standards become more stri gent. In the above computation, it was assumed that the filtered effluent to be disinfected remained in the chlorine contact tank for the full 30 minutes. Thus, it is clear that the proper design of a plug-flow chlorine contact basin is critical to the effective use of chlorine as disinfectant. The design of chlorine contact basins is considered in Sec. 12–6.

Formation and Control of Disinfection Byproducts (DBPs)

In the early 1970s, it was found that the use of oxidants, such as chlorine and ozone, in water treatment for disinfection; for the control of tastes, odors, and color removal; and other in-plant uses resulted in the production of undesirable DBPs (Rook, 1974; Bellar and Lichtenberg, 1974). The DBPs measured most frequently and with the highest concentration are trihalomethanes (THMs) and haloacetic acids (HAAs), resulting from chlorination. In addition to trihalomethanes and haloacetic acids, a variety of other DBPs are also produced. The principal DBPs that have been identified are reported in Table 12–15. Many of the compounds identified in Table 12–15 have also been identified in treated effluent that has been disinfected using chlorine, chloramines, chlorine dioxide, and ozone.

Concerns with DBPs.   Formation of DBPs is of great concern in effluent dispersal to the environment and for indirect and direct potable reuse because of the potential long-term (chronic) impact of these compounds on public health and the environment. Chloroform, for example, is a well-known animal carcinogen and many of the haloforms are also thought to be animal carcinogens. In addition, many of these compounds have been classified as probable human carcinogens. Still others of these compounds are known to cause chromosomal aberrations and sperm abnormalities. Recognizing the many unknowns and the potential public health and environmental risks associated with these compounds, the U.S. EPA has moved aggressively to control their formation in drinking water.

Formation of DBPs Using Chlorine for Disinfection.   Trihalomethanes (THMs) and other DBPs are formed as a result of a series of complex reactions between free chlorine and a group of organic acids known collectively as humic acids. The reactions lead to the formation of single carbon molecules that are often designated as CHX_3, where X is either a chlorine (Cl^-) or bromine (Br^-) atom. For example, the chemical formula for chloroform is $CHCl_3$.

The rate of formation of DBPs is dependent on a number of factors, including

1. Presence of organic precursors
2. Free chlorine concentration
3. Bromide concentration
4. pH
5. Temperature
6. Time

The type and concentration of the organic precursor affects both the rate of the reaction and extent to which the reaction is completed.

The presence of free chlorine was thought to be necessary for the THM formation reaction to proceed, but it appears that THMs can form in the presence of combined chlorine (chloramines), but at a much reduced rate. It is important to note that initial mixing can affect the formation of THMs because of the competing reactions between chlorine and ammonia,

disinfection byproducts formed during application of chlorine, chloramine,
rine dioxide in natural waters[a]

Class	Byproduct	Chemical agent	Molecular formula
Trihalomethanes	Chloroform	Chlorine	$CHCl_3$
	Bromodichloromethane	Chlorine	$CHBrCl_2$
	Dibromochloromethane	Chlorine	$CHBr_2Cl$
	Bromoform	Chlorine, ozone	$CHBr_3$
	Dichloroiodomethane	Chlorine	$CHICl_2$
	Chlorodiiodomethane	Chlorine	CHI_2Cl
	Bromochloroiodomethane	Chlorine	$CHBrICl$
	Dibromoiodomethane	Chlorine	$CHBr_2I$
	Bromodiiodomethane	Chlorine	$CHBrI_2$
	Triiodomethane	Chlorine	CHI_3
Haloacetic acids	Monochloroacetic acid	Chlorine	$CH_2ClCOOH$
	Dichloroacetic acid	Chlorine	$CHCl_2COOH$
	Trichloroacetic acid	Chlorine	CCl_3COOH
	Bromochloroacetic acid	Chlorine	$CHBrClCOOH$
	Bromodichloroacetic acid	Chlorine	$CBrCl_2COOH$
	Dibromochloroacetic acid	Chlorine	$CBr_2ClCOOH$
	Monobromoacetic acid	Chlorine	$CH_2BrCOOH$
	Dibromoacetic acid	Chlorine	$CHBr_2COOH$
	Tribromoacetic acid	Chlorine	CBr_3COOH
Haloacetonitriles	Trichloroacetonitrile	Chlorine	$CCl_3C\equiv N$
	Dichloroacetonitrile	Chlorine	$CHCl_2C\equiv N$
	Bromochloroacetonitrile	Chlorine	$CHBrClC\equiv N$
	Dibromoacetonitrile	Chlorine	$CHBr_2C\equiv N$
Haloketones	1,1-Dichloroacetone	Chlorine	$CHCl_2COCH_3$
	1,1,1-Trichloroacetone	Chlorine	CCl_3COCH_3
Aldehydes	Formaldehyde	Ozone, chlorine	$HCHO$
	Acetaldehyde	Ozone, chlorine	CH_3CHO
	Glyoxal	Ozone, chlorine	$OHCCHO$
	Methyl glyoxal	Ozone, chlorine	CH_3COCHO
Aldoketoacids	Glyoxylic acid	Ozone	$OHCCOOH$
	Pyruvic acid	Ozone	$CH_3COCOOH$
	Ketomalonic acid	Ozone	$HOOCCOCOOH$
Carboxylic acids	Formate	Ozone	$HCOO^-$
	Acetate	Ozone	CH_3COO^-
	Oxalate	Ozone	$OOCCOO^{2-}$

(continued)

Table 12-15 (Continued)			
Class	**Byproduct**	**Chemical agent**	**Molecular formula**
Oxyhalides	Chlorite	Chlorine dioxide	ClO_2^-
	Chlorate	Chlorine dioxide	ClO_3^-
	Bromate	Ozone	BrO_3^-
Nitrosamines	N-nitrosodimethylamine	Chloramines	$(CH_3)_2NNO$
Cyanogen Halides	Cyanogen chloride	Chloramines	$ClCN$
	Cyanogen bromide	Chloramines	$BrCN$
Misc.	Chloral hydrate	Chlorine	$CCl_3CH(OH)_2$
Trihalonitromethanes	Trichloronitromethane (Chloropicrin)	Chlorine	CCl_3NO_2
	Bromodichloronitromethane	Chlorine	$CBrCl_2NO_2$
	Dibromochloronitromethane	Chlorine	CBr_2ClNO_2
	Tribromonitromethane	Chlorine	CBr_3NO_2

ᵃ Adapted from Krasner (1999), Krasner et al. (2001), and Thibaud et al. (1987).

and between chlorine and humic acids. If bromide is present, it can be oxidized to bromine by free chlorine. In turn the bromine ion can combine with the organic precursors to form THMs, including bromodichoromethane, dibromochoromethane, and bromoform. The rate of formation of THMs has been observed to increase with both pH and temperature. Additional details on the formation of THMs may be found in U.S. EPA (1999a).

Although chloramines, as discussed above, produce THMs at reduced rates, they can, nevertheless, produce other DBP compounds that are of concern. Other DBPs that are produced when treated wastewater is disinfected with chloramines include N-nitrosodimethylamine (NDMA), a member of a class of compounds known as nitrosoamines, cyanogen chloride, and cyanogen bromide (see Table 12–15). As a class of compounds nitrosoamines are among the most powerful carcinogens known (Snyder, 1995). The compounds in this class have been found to produce cancer in every species of laboratory animal tested.

One pathway leading to the formation of NDMA can be illustrated with the following two reactions:

$$NO_2^- + HCl \rightarrow HNO_2 + Cl^- \tag{12-28}$$
nitrite anion   hydrochloric acid   nitrous acid   chloride ion

$$HNO_2 + CH_3{-}NH{-}CH_3 \rightarrow CH_3{-}\overset{\displaystyle NO}{\underset{\displaystyle |}{N}}{-}CH_3^- \tag{12-29}$$
nitrous acid   dimethylamine   N-nitrosodimethylamine

The concern in biological wastewater treatment facilities is that some nitrite may leak through the process. While the concentration of nitrite may be too low to measure by conventional means, concentrations of NDMA as low as 1 or 2 ng/L are being measured and the CDPH notification level for groundwater recharge is 10 ng/L. Based on a limited number of test locations, it has been observed that the concentrations of NDMA in the incoming wastewater can be quite variable, with concentrations as high as 6000 ng/L being measured.

In addition to the formation of NDMA as outlined above, it appears the addition of chloramines for disinfection can serve to amplify the concentration of any NDMA that may be present in the treated effluent before disinfection. In a series of studies conducted by the Los Angeles County Sanitation Districts (Jalali et al., 2005), it was found that chloramination increased the concentration of NDMA in treated effluent following disinfection by tenfold.

Other DBPs resulting from the use of chloramines as disinfectants in treated effluents include cyanogen chloride and cyanogen bromide, where bromides are present (see Table 12–15). In very large quantities, cyanogen chloride is used in tear gas, in fumigant gases, and as a reagent in the formation of other compounds. In the body, cyanogen chloride is metabolized rapidly to cyanide. Because there is limited information on the toxicity of low-level concentrations of cyanogen chloride, proposed guidelines are based on cyanide. The cyanogen compounds are of concern and they are now beginning to be regulated in effluent discharge permits. The current NPDES permit limit for cyanide is 5 mg/L.

Control of DBP Formation Using Chlorine for Disinfection. The principal means of controlling the formation of THMs and other related DBPs is to avoid the direct addition of free chlorine. Based on the evidence to date, it appears that the use of chloramines generally does not lead to the formation of THMs in amounts that would be of concern relative to current standards. As discussed previously, other DBPs may be produced that are of equal concern, but for other reasons (see following discussion). It is important to note that if chloramines are to be used for disinfection, the chloramine solution must be prepared with a potable water supply containing little or no ammonia (i.e., *treated plant effluent should not be used*). If the formation of DBPs is of concern due to the presence of specific organic precursors (i.e., humic materials), the practice of breakpoint chlorination cannot be used. Further, if humic materials are present consistently, it may be appropriate to investigate alternative means of disinfection such as UV irradiation.

The control of DBPs produced when chloramines are used (by reducing direct reactions of organics with residual free chlorine) can be more challenging as chloramination may form other DBPs. With respect to NDMA it appears that with proper control and operation of the biological treatment process, the potential for the formation or amplification of this compound can be reduced. Removals of 50 to 70 percent have been reported for NDMA when using reverse osmosis employing thin film composite membranes (see Chap. 11). The use of UV irradiation has also proven to be effective in the control of NDMA. Where the formation of NDMA and cyanogen chloride is a persistent concern, a number of wastewater agencies have switched to UV irradiation for disinfection. In the study cited above (Jalali et al., 2005), it was also found that there was no net change in the total cyanide (CN^-) concentration in the treated effluent due to UV irradiation. The use of sequential chlorination to control the formation of NDMA is considered in Sec. 12–8.

Environmental Impacts of Disinfection with Chlorine

The environmental impacts associated with the use of chlorine and chlorine compounds as a disinfectant in wastewater applications include the discharge of DBPs and the regrowth of microorganisms.

Discharge of DBPs.   It has been shown that many of the DBPs can cause environmental impacts at very low concentrations. The occurrence of DBPs and compounds such as NDMA raises serious questions about the continued use of free chlorine for disinfection.

Regrowth of Microorganisms.   In many locations, a regrowth of microorganisms has been observed in receiving water bodies and in long transmission pipelines following dechlorination of treated effluent disinfected with chlorine. The regrowth of microorganisms is not unexpected as it is well known that a number of microorganisms survive the disinfection process. It has been hypothesized that regrowth (also known as aftergrowth) results, in part, because (1) the amount of organic matter and available nutrients in treated wastewater is sufficient to sustain the limited number of organisms remaining after disinfection, (2) predators such as protozoa are absent, (3) there are favorable temperatures, and (4) disinfectant residuals are ineffective. Because regrowth is an especially important issue in transmission lines used for the transport of reclaimed water, a suitable chlorine residual (on the order of 1 to 2 mg/L, depending on local conditions) should be maintained in the pipeline to control regrowth (a common practice in water distribution systems). In very long pipelines, it may be necessary to add additional chlorine at intermediate points along the length of the pipeline.

12–4   DISINFECTION WITH CHLORINE DIOXIDE

Chloride dioxide (ClO_2), another bactericide, is equal to or greater than chlorine in disinfecting power. Chlorine dioxide has proven to be an effective virucide, being more effective in achieving inactivation of viruses than chlorine. A possible explanation is that because chlorine dioxide is absorbed by peptone (a protein), and that viruses have a protein coat, adsorption of ClO_2 onto this coating could cause inactivation of the virus. In the past, ClO_2 did not receive much consideration as a wastewater disinfectant due to its high costs; sodium chlorite feed stock is about ten times as expensive as chlorine on a weight basis.

Characteristics of Chlorine Dioxide

Chlorine dioxide (ClO_2) is, under atmospheric conditions, a yellow to red unpleasant smelling, irritating, unstable gas with a high specific gravity. Because chlorine dioxide is unstable and decomposes rapidly, it is usually generated onsite before its application. Chlorine dioxide is generated by mixing and reacting a chlorine solution in water with a solution of sodium chlorite ($NaClO_2$) according to the following reaction:

$$2NaClO_2 + Cl_2 \rightarrow 2ClO_2 + 2NaCl \tag{12–30}$$

Based on Eq. (12–30), 1.34 mg sodium chlorite reacts with 0.5 mg chlorine to yield 1.0 mg chlorine dioxide. Because technical grade sodium chlorite is only about 80 percent pure, about 1.68 mg of the technical grade sodium chlorite is required to produce 1.0 mg of chlorine dioxide. Sodium chlorite may be purchased and stored as a liquid (generally a 25 percent solution) in refrigerated storage facilities. The properties of chlorine dioxide were presented previously in Tables 12–3 and 12–7.

Chlorine Dioxide Chemistry

The active disinfecting agent in a chlorine dioxide system is free dissolved chlorine dioxide (ClO_2). At the present time, the complete chemistry of chlorine dioxide in an aqueous environment is not understood completely. Because ClO_2 does not hydrolyze in a manner similar to the chlorine compounds discussed in the previous section, the oxidizing power of ClO_2 is often referred to as *equivalent available chlorine*. The definition of the term equivalent available chlorine is based on a consideration of the following oxidation half reaction for ClO_2:

$$ClO_2 + 5e^- + 4H^+ \rightarrow Cl^- + 2H_2O \tag{12-31}$$

As shown in Eq. (12–31), the chlorine atom undergoes a 5 electron change in its conversion from chlorine dioxide to the chloride ion. Because the weight of chlorine in ClO_2 is 52.6 percent and there is a 5 election change, the equivalent available chlorine content is equal to 263 percent as compared to chlorine. Thus, ClO_2 has 2.63 times the oxidizing power of chlorine. The concentration of ClO_2 is usually expressed in g/m^3. On a molar basis, one mole of ClO_2 is equal to 67.45 g, which is equivalent to 177.25 g (5×35.45) of chlorine. Thus, 1 g/m^3 of ClO_2 is equivalent to 2.63 g/m^3 of chlorine.

Effectiveness of Chlorine Dioxide as a Disinfectant

Chlorine dioxide has an extremely high oxidation potential, which probably accounts for its potent germicidal powers. Because of its extremely high oxidizing potential, possible bactericidal mechanisms may include inactivation of critical enzyme systems or disruption of protein synthesis. It should be noted, however, that when ClO_2 is added to water it is often reduced to the chlorite ion (ClO_2^-), a weak disinfectant, according to the following reaction. The formation of ClO_2^- may help to explain the variability that is sometimes observed in the performance of ClO_2 as a disinfectant.

$$ClO_2 + e^- \rightarrow ClO_2^- \tag{12-32}$$

Based on the coefficient of specific lethality as reported in Table 12–11, the effectiveness of ClO_2 with respect to bacteria is similar to that of free chlorine. However, there are some differences, depending on the microorganism group and members within each group. Chlorine dioxide appears to be more effective than free chlorine in the inactivation of protozoan cysts.

Modeling the Chlorine Dioxide Disinfection Process

As discussed previously in Sec. 12–3, the models that have been developed to describe the disinfection process with chlorine can also be used, with appropriate caution, for chlorine dioxide. As with chlorine, the shoulder effect and the effect of the residual particles must be considered. Further, the differences between (1) secondary and filtered secondary effluent and (2) microfiltration and reverse osmosis effluent must also be considered.

Required Chlorine Dioxide Dosages for Disinfection

The required chlorine dioxide dosage will depend on the pH and the specific organism under investigation. Relative CT values for chlorine dioxide are given in Table 12–5, presented previously in Sec. 12–2, and values of the coefficient of specific lethality are given in Table 12–11. Because the data on chlorine dioxide in the literature are limited, site-specific testing is recommended to establish appropriate dosage ranges although the values given in Table 12–5 can be used as a starting point.

Byproduct Formation and Control

The formation of DBPs is of great concern with the use of chlorine dioxide. The formation and control of DBPs with chlorine dioxide is considered in the following discussion.

Formation of DBPs Using Chlorine Dioxide for Disinfection. The principal DBPs formed when chlorine dioxide is used as a disinfectant are chlorite (ClO_2^-) and chlorate (ClO_3^-), both of which are potentially toxic at low concentrations. The principal sources of the chlorite ion are from the process used to generate the chlorine dioxide and from the reduction of chlorine dioxide. As given by Eq. (12–30), all of the $NaClO_2$ reacts with chlorine to form ClO_2. Unfortunately, on occasion some unreacted chlorite ion can escape from the reactor where the chlorine dioxide is being generated and find its way into the water that is being treated. The second source of chlorite is from the reduction of chlorine dioxide as discussed above [see Eq. (12–32)]. The chlorate ion can be derived from the oxidation of chlorine dioxide, from the impurities in the sodium chlorite feed stock, and from the photolytic decomposition of chlorine dioxide.

The chlorine dioxide residuals and other end products are believed to degrade more quickly than chlorine residuals, and, therefore, may not pose as serious a threat to aquatic life as chlorine residuals. An advantage in using chlorine dioxide is that it does not react with ammonia to form the potentially toxic chlorinated DBPs. It has also been reported that halogenated organic compounds are not produced to any appreciable extent.

Control of DBP Formation Using Chlorine Dioxide for Disinfection. The formation of chlorite can be controlled by careful management of the feedstock or increasing the chlorine dose beyond the stoichiometric amount. Treatment methods for the removal of the chlorite ion involve reducing the chlorite ion to the chloride ion using either ferrous iron or sulfite. Granular activated carbon (GAC) can also be used to absorb trace amounts of chlorite. At the present time there are no cost-effective methods for the removal of the chlorate ion. The control of the chlorate ion depends primarily on the effective management of the facilities used for the production of chlorine dioxide (White, 1999; Black and Veatch Corporation, 2010).

Environmental Impacts

The environmental impacts associated with the use of chlorine dioxide as a wastewater disinfectant are not well known. It has been reported that the impacts are less adverse than those associated with chlorination. Chlorine dioxide does not dissociate or react with water as does chlorine. However, because chlorine dioxide is normally produced from chlorine and sodium chlorite, free chlorine may remain in the resultant chlorine dioxide solution (depending on the process) and impact the receiving aquatic environment, as does chlorine and its byproducts.

12–5 DECHLORINATION

Chlorination is one of the most commonly used methods for the destruction of pathogenic and other harmful organisms that may endanger human health. As noted in the previous sections, however, certain organic constituents in wastewater interfere with the chlorination process. Many of these organic compounds may react with the chlorine to form toxic compounds that can have long-term adverse effects on the beneficial uses of the waters to which they are discharged or reused. To minimize the effects of these potentially toxic chlorine residuals on the environment, dechlorination of treated effluent is necessary. Dechlorination may be accomplished by reacting the residual chlorine with a reducing agent such as sulfur dioxide or sodium bisulfite or by adsorption on and reaction with activated carbon.

Dechlorination of Treated Wastewater with Sulfur Dioxide

Where effluent toxicity requirements are applicable, or where dechlorination is used as a polishing step following the breakpoint chlorination process for the removal of ammonia nitrogen, sulfur dioxide (SO_2) is used most commonly for dechlorination. Sulfur dioxide is available commercially as a liquefied gas under pressure in steel containers. Sulfur dioxide is handled in equipment very similar to standard chlorine systems. When added to water, sulfur dioxide reacts to form sulfurous acid (H_2SO_3), a strong reducing agent. In turn, the sulfurous acid dissociates to form HSO_3^- that will react with free and combined chlorine, resulting in formation of chloride and sulfate ions. Sulfur dioxide gas successively removes free chlorine, monochloramine, dichloramine, nitrogen trichloride, and poly-n-chlor compounds as illustrated in Eqs. (12–33) through (12–38).

Reactions between sulfur dioxide and free chlorine:

$$SO_2 + H_2O \rightarrow H_2SO_3 \tag{12–33}$$

$$HOCl + H_2SO_3 \rightarrow HCl + H_2SO_4 \tag{12–34}$$

$$SO_2 + HOCl \rightarrow HCl + H_2SO_4 \tag{12–35}$$

Reactions between sulfur dioxide and monochloramine, dichloramine, and nitrogen trichloride are:

$$NH_2Cl + H_2SO_3 + H_2O \rightarrow NH_4Cl + H_2SO_4 \tag{12–36}$$

$$NHCl_2 + 2H_2SO_3 + 2H_2O \rightarrow NH_4Cl + 2H_2SO_4 + HCl \tag{12–37}$$

$$NCl_3 + 3H_2SO_3 + 3H_2O \rightarrow NH_4Cl + 3H_2SO_4 + 2HCl \tag{12–38}$$

For the overall reaction between SO_2 and chlorine [Eq. (12–35)], the stoichiometric amount of SO_2 required per mg/L of chlorine residual is 0.903 mg/L. In practice, as reported in Table 12–16, it has been found that about 1.0 to 1.2 mg/L of sulfur dioxide will be required for the dechlorination of 1.0 mg/L of chlorine residue (expressed as Cl_2). Because the reactions of sulfur dioxide with chlorine and chloramines are nearly instantaneous, contact time is not usually a factor and contact chambers are not used, but rapid and positive mixing at the point of application is an absolute requirement.

The ratio of free chlorine to the total combined chlorine residual before dechlorination will determine whether the dechlorination process is partial or proceeds to completion. If the ratio is less than 85 percent, it can be assumed that significant organic nitrogen is present and that it will interfere with the dechlorination of free residual chlorine.

Table 12–16

Typical information on the quantity of dechlorinating compound required for each mg/L of residual chlorine

Dechlorinating compound			Quantity, mg/(mg/L) residual	
Name	**Formula**	**Molecular weight**	**Stoichiometric amount**	**Range in use**
Hydrogen peroxide	H_2O_2	34.01	0.48	0.5–0.7
Sodium bisulfite	$NaHSO_3$	104.06	1.46	1.5–1.7
Sodium metabisulfite	$Na_2S_2O_5$	190.10	1.34	1.4–1.6
Sodium sulfite	Na_2SO_3	126.04	1.78	1.8–2.0
Sodium thiosulfate	$Na_2S_2O_3$	112.12	0.56	0.6–0.9
Sulfur dioxide	SO_2	64.09	0.903	1.0–1.2

In most situations, sulfur dioxide dechlorination is a very reliable unit process, provided that the precision of the combined chlorine residual monitoring service is adequate. Excess sulfur dioxide dosages should be avoided, not only because of the chemical wastage, but also because of the oxygen demand exerted by the excess sulfur dioxide. The relatively slow reaction between excess sulfur dioxide and dissolved oxygen is given by the following expression:

$$HSO_3^- + 0.5O_2 \rightarrow SO_4^{2-} + H^+ \tag{12–39}$$

The result of this reaction is a reduction in the dissolved oxygen in the water, a corresponding increase in the measured BOD and COD, and a possible drop in the pH. All these effects can be eliminated by proper control of the dechlorination system.

Dechlorination of Treated Wastewater with Sodium Based Compounds

Sodium based chemicals that have been used for dechlorination include sodium sulfite (Na_2SO_3), sodium bisulfite ($NaHSO_3$), sodium metabisulfite ($Na_2S_2O_5$), sodium thiosulfate ($Na_2S_2O_3$), and hydrogen peroxide (H_2O_2). When these chemicals are used for dechlorination, the following reactions occur. The stoichiometric weight ratios of these compounds needed per mg/L of residual chlorine are given in Table 12–15, along with the range of values used in practice.

Sodium Sulfite. Reactions between sodium sulfite and free chlorine residual and combined chlorine residual, as represented by monochloramine:

$$Na_2SO_3 + Cl_2 + H_2O \rightarrow Na_2SO_4 + 2HCl \tag{12–40}$$

$$Na_2SO_3 + NH_2Cl + H_2O \rightarrow Na_2SO_4 + Cl^- + NH_4^+ \tag{12–41}$$

Sodium Bisulfite. Reaction between sodium bisulfite and free chlorine residual and combined chlorine residual, as represented by monochloramine:

$$NaHSO_3 + Cl_2 + H_2O \rightarrow NaHSO_4 + 2HCl \tag{12–42}$$

$$NaHSO_3 + NH_2Cl + H_2O \rightarrow NaHSO_4 + Cl^- + NH_4^+ \tag{12–43}$$

Sodium Metabisulfite. Reactions between sodium metabisulfite and free chlorine residual and combined chlorine residual, as represented by monochloramine:

$$Na_2S_2O_5 + Cl_2 + 3H_2O \rightarrow 2NaHSO_4 + 4HCl \tag{12–44}$$

$$Na_2S_2O_5 + 2NH_2Cl + 3H_2O \rightarrow Na_2SO_4 + H_2SO_4 + 2Cl^- + 2NH_4^+ \tag{12–45}$$

Sodium Thiosulfate and Related Compounds. Often used as a dechlorinating agent in analytical laboratories, the use of sodium thiosulfate ($Na_2S_2O_3$) in full scale water reclaimation treatment plants is limited for the following reasons. It appears the reaction of sodium thiosulfate with residual chlorine is stepwise, creating a problem with uniform mixing. The ability of sodium thiosulfate to remove residual chlorine is a function of the pH (White, 1999; Black and Veatch Corporation, 2010). The reaction with residual chlorine is only stoichiometric at a pH value of 2, making prediction of the required dose impossible in wastewater applications. As reported in Table 12–16, the stoichiometric weight ratio of sodium thiosulfate per mg/L of residual chlorine is 0.556. Although not in common use, calcium thiosulfate (CaS_2O_3) ascorbic acid ($C_6H_8O_6$), and sodium ascorbate ($C_6H_7NaO_6$) have all been used at full scale for dechlorination.

Dechlorination with Hydrogen Peroxide

Hydrogen peroxide (H_2O_2) has also been used for dechlorination. Unlike sulfur dioxide and the sodium-based compounds discussed above, hydrogen peroxide does not result in an increase in the total dissolved solids as it only adds oxygen to the water. When hydrogen peroxide is used for dechlorination, the following reaction occurs.

$$H_2O_2 + Cl_2 \rightarrow O_2 + 2HCl \tag{12–46}$$

The stoichiometric weight ratio of hydrogen peroxide needed per mg/L of residual chlorine is 0.48. Because the reaction between hydrogen peroxide and chlorine compounds is so rapid, other inorganic and organic compounds generally do not interfere with the reaction. The optimal pH range is about 8.5 at which the reaction occurs instantaneously, although there is no upper limit. In the past, hydrogen peroxide has not been used for dechlorination because it is difficult to handle.

Dechlorination with Activated Carbon

Both combined and free residual chlorine can be removed by means of adsorption on and reaction with activated carbon. When activated carbon is used for dechlorination, the following reactions occur once chlorine or chlorine compounds have been adsorbed.
Reactions with free chlorine residual:

$$C + 2Cl_2 + 2H_2O \rightarrow 4HCl + CO_2 \tag{12–47}$$

Reactions with combined residual as represented by mono- and dichloramine:

$$C + 2NH_2Cl + 2H_2O \rightarrow CO_2 + 2NH_4^+ + 2Cl^- \tag{12–48}$$

$$C + 4NHCl_2 + 2H_2O \rightarrow CO_2 + 2N_2 + 8H^+ + 8Cl^- \tag{12–49}$$

Granular activated carbon is used in either a gravity or pressure filter bed. If carbon is to be used solely for dechlorination, it must be preceded by an activated carbon process for the removal of other constituents susceptible to removal by activated carbon. In treatment plants where granular activated carbon is used to remove organics, either the same or separate beds can also be used for dechlorination.

Because granular carbon in column applications has proved to be very effective and reliable, activated carbon should be considered where dechlorination is required. However, this method is quite expensive. It is expected that the primary application of activated carbon for dechlorination will be in situations where high levels of organic removal are also required.

Dechlorination of Chlorine Dioxide with Sulfur Dioxide

Where treated wastewater is disinfected with chlorine dioxide, dechlorination can be achieved using sulfur dioxide. The reaction that takes place in the chlorine dioxide solution can be expressed as

$$SO_2 + H_2O \rightarrow H_2SO_3 \tag{12–50}$$

$$5H_2SO_3 + 2ClO_2 + H_2O \rightarrow 5H_2SO_4 + 2HCl \tag{12–51}$$

Based on Eq. (12–51), it can be seen that 2.5 mg of sulfur dioxide will be required for each mg of chlorine dioxide residual (expressed as ClO_2). In practice, 2.7 mg SO_2/mg ClO_2 would normally be used.

12–6   DESIGN OF CHLORINATION AN[ ] FACILITIES

The chemistry of chlorine in water and wastewater has [ ] tions, along with an analysis of how chlorine functions [ ] erations in the implementation of chlorination and dech[ ] purposes include (1) estimation of the chlorine dosag[ ] (3) dosage control, (4) injection and initial mixing, ([ ]...inc contact basin design, (6) assessing the hydraulic performance of existing chlorine contact basins, (7) outlet control and chlorine residual measurement, (8) chlorine storage facilities, (9) chemical containment and neutralization facilities, and (10) dechlorination facilities. These topics are considered in the following discussion.

Sizing Chlorination Facilities

To aid in the design and selection of the required chlorination facilities and equipment, it is important to know the uses, including dosage ranges, to which chlorine and its compounds have been applied. Chlorination capacities for disinfection are generally selected to meet the specific design criteria of the state or other regulatory agencies controlling the receiving body of water. In any case, where the residual in the effluent is specified or the final number of coliform bacteria is limited, onsite testing is preferred to determine the dosage of chlorine required. Typical chlorine dosages for disinfection have been given previously in Table 12–14. Typical chlorine dosages for applications other than disinfection are given in Table 12–17. A range of dosage values is given because they will vary depending on the characteristics of the wastewater. In the absence of more specific data, the maximum values given in Tables 12–14 and 12–17 can be used as a guide in sizing chlorination equipment. The sizing of chlorination facilities is illustrated in Example 12–5.

Table 12–17

Typical dosages for various chlorination applications in wastewater collection and treatment

Application	Dosage range, mg/L
Collection:	
Corrosion control (H$_2$S)	2–9[a]
Odor control	2–9[a]
Slime growth control	1–10
Treatment:	
BOD reduction	0.5–2[b]
Digester and Imhoff tank foaming control	2–15
Digester supernatant oxidation	20–140
Ferrous sulfate oxidation	–[c]
Filter fly control	0.1–0.5
Filter ponding control	1–10
Grease removal	2–10
Sludge bulking control	1–10

[a] Per mg/L of H$_2$S.

[b] Per mg/L of BOD$_5$ destroyed.

[c] $6FeSO_4 \cdot 7H_2O + 3Cl_2 \rightarrow 2Fe_2(SO_4)_3 + 42H_2O$.

EXAMPLE 12–5 **Sizing of Chlorination Facilities** Determine the capacity of a chlorinator for a treatment plant with an average wastewater flowrate of 1000 m³/d (0.26 Mgal/d). The peak daily factor for the treatment plant is 3.0 and the maximum required chlorine dosage (set by state regulations) is to be 20 mg/L.

Solution

1. Determine the capacity of the chlorinator at peak flow.

$$Cl_2, kg/d = (20 \text{ g/m}^3)(1000 \text{ m}^3/d)(3)(1 \text{ kg}/10^3 \text{ g})$$

$$= 60 \text{ kg/d}$$

Use the next largest standard size chlorinator: two 90 kg/d (200 lb/d) units with one unit serving as a spare. Although the peak capacity will not be required during most of the day, it must be available to meet the chlorine requirements at peak flow. Best design practice calls for the availability of a standby chlorinator.

2. Estimate the daily consumption of chlorine. Assume an average dosage of 10 mg/L.

$$Cl_2, kg/d = (10 \text{ g/m}^3)(1000 \text{ m}^3/d)(1 \text{ kg}/10^3 \text{ g})$$

$$= 10 \text{ kg/d}$$

Comment In sizing and designing chlorination systems, it is also important to consider the low flow/dosage requirements. The chlorination system should have sufficient turndown capability for these conditions so that excessive chlorine is not applied.

Disinfection Process Flow Diagrams

Process flow diagrams and equipment used to inject (feed) chlorine, hypochlorite, dry calcium hypochlorite, and chlorine dioxide into wastewater are illustrated and discussed below.

Flow Diagram for Chlorine. Chlorine may be applied directly as a gas or in an aqueous solution. Typical chlorine/sulfur dioxide chlorination/dechlorination process flow diagrams are shown on Fig. 12–17. The difference in the two diagrams shown on Fig. 12–17 is in the method of introducing and mixing the chlorine solution with the wastewater. Chlorine can be withdrawn from storage containers either in liquid or gas form. If withdrawn as a gas, the evaporation of the liquid in the container results in frost formation that restricts gas withdrawal rates to 18 kg/d (40 lb/d) for 68 kg (150 lb) cylinders and 205 kg/d (450 lb/d) for 0.9-tonne (1-ton) containers at 21°C (70°F). Evaporators are used normally where the maximum rate of chlorine gas withdrawal from a 0.9-tonne (1-ton) container must exceed approximately 180 kg/d (400 lb/d). Although multiple ton cylinders can be connected to provide more than 180 kg/d (400 lb/d), the use of an evaporator conserves space. Evaporators are almost always used when the total dosage exceeds 680 kg/d (1500 lb/d). Chlorine evaporators are available in sizes ranging from 1818 to 4545 kg/d (4000 to 10,000 lb/d) capacities; chlorinators are available normally in sizes ranging from 227 to 4545 kg/d (500 to 10,000 lb/d).

Flow Diagram for Liquid Hypochlorite Solutions. A typical sodium hypochlorite/sodium bisulfite chlorination/dechlorination process flow diagram is shown on Fig. 12–18. For small treatment plants, the most satisfactory means of feeding sodium or calcium hypochlorite is through the use of low capacity proportioning pumps. Generally, pumps are available in capacities up to 450 L/d (120 gal/d), with adjustable stroke for any reduced values. Large capacities or multiple units are available from some of

Figure 12–17

Schematic flow diagrams for chlorination/dechlorination: (a) using a chlorine injector system and (b) using a molecular chlorine vapor induction system.

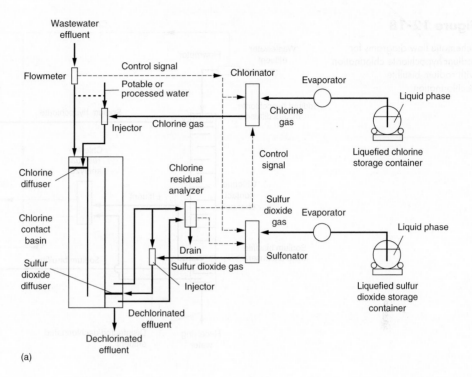

(a)

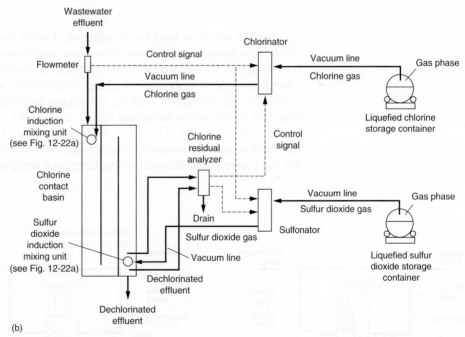

(b)

the manufacturers. The pumps can be arranged to feed at a constant rate, or they can be provided with variable speed and with analog signals for varying the feed rate. The stroke length can also be controlled.

Flow Diagram for Dry Calcium Hypochlorite Feed System.

For small wastewater flowrates up to about 400 m³/d (10^5 gal/d) chlorine in the form of dry calcium

Figure 12–18

Schematic flow diagrams for sodium hypochlorite chlorination with sodium bisulfite dechlorination.

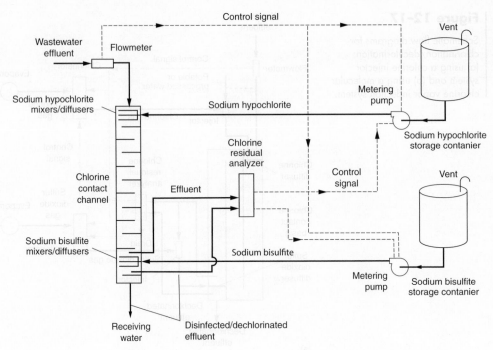

hypochlorite tablets is used for disinfection. Two of the most common types of tablet chlorinators (nonpressurized and pressurized) are shown on Fig. 12–19. The schematic flow diagrams for the two tablet chlorinators are essentially the same; a sidestream of water is diverted from the main discharge line, chlorine at relatively high concentrations is added to the sidestream, and the chlorinated sidestream is discharged back into the main flow by means of a pump [see Fig. 12–19(a)] or by reducing the pressure in the main discharge line [see Fig. 12–19(b)]. As shown on Fig. 12–19(a), in the nonpressurized tablet chlorinator the sidestream contacts the bottom surface of the chlorine tablets that rest on a screen. The chlorine tablets have been designed to dissolve at a more or less

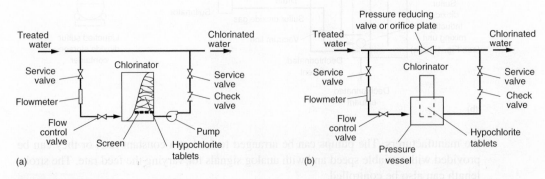

Figure 12–19

Schematic flow diagrams for calcium hypochlorite tablet chlorinators: (a) nonpressurized (adapted from PPG Industries, Inc.) and (b) pressurized (adapted from PPG Industries, Inc.).

Figure 12–20

Typical flow diagram for the addition of chlorine dioxide.

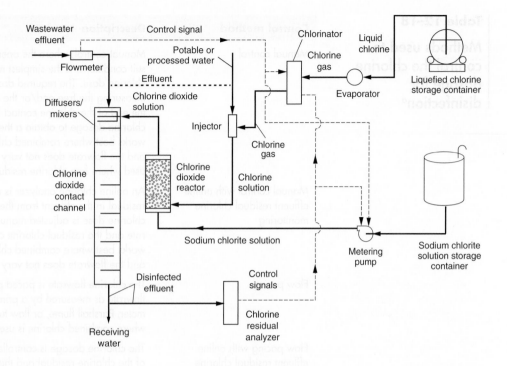

constant rate, releasing a controlled amount of chlorine. The amount of chlorine added is dependent on the flowrate through the tablet chlorinator. The tablet chlorinator shown on Fig. 12–19(b) is pressurized and hypochlorite is released as water flows over the hypochlorite tablets. Dry calcium hypochlorite tablets, typically 75 mm (3 in.) in diameter, contain about 65 to 70 percent available chlorine. For small treatment plants, the use of chlorine tablets eliminates the hazards associated with handling chlorine cylinders. Further, because there are no moving parts, tablet chlorinators are simple to operate and maintain.

Flow Diagram for Chlorine Dioxide. The chlorine dioxide, generated onsite, is present in an aqueous solution which is applied in same manner as that used for typical chlorination systems. A schematic process flow diagram of a typical chlorine dioxide installation is shown on Fig. 12–20.

Dosage Control

The control of the chlorine dosage can be accomplished in a number of different ways depending on the disinfection objective. The principal control methods are summarized in Table 12–18. The specific control method to be used will depend on the variability of the influent flowrate, the presence of unoxidized constituents that can react with chlorine, the pH of the wastewater, and whether combined or free or a combination of the two forms of chlorine will be used for disinfection. Dosage control is easiest where combined chlorine is used as the disinfectant. Dosage control has also become more difficult due to the impact of climate change resulting in short duration, high intensity rainfall events. Because of the rapid increase in the influent flowrate observed at some treatment plants, due to increased stormwater runoff, the carryover (washout) of solids from the secondary sedimentation basins, especially where shallow basins are used, has further complicated dosage control

Table 12–18

Methods used to control the chlorine dosage for disinfection[a]

Control method	Description
Manual control	Manual control, where the operator changes the feed rate to suit conditions, is the simplest method for controlling the chlorine dose. The required dosage is usually determined by measuring the free and/or the combined chlorine residual at the end of the chlorine contact basin and adjusting the chlorine dosage to obtain a the desired residual. This method works best where combined chlorine is used for disinfection, and the flowrate does not vary too rapidly, but can also be used where free chlorine residual is used.
Manual control with online effluent residual chlorine monitoring	An online chlorine analyzer is used to monitor the chlorine residual in the effluent from the chlorine contact basin. The chlorine dose is adjusted manually based on the plant flowrate and the residual chlorine concentrations. This method works best where combined chlorine is used for disinfection, and the flowrate does not vary too rapidly.
Flow pacing	The chlorine flowrate is paced proportional to the wastewater flowrate as measured by a primary meter such as a magnetic meter, Parshall flume, or flow tube. This method works best where combined chlorine is used for disinfection.
Flow pacing with online effluent residual chlorine monitoring	The chlorine dosage is controlled by automatic measurement of the chlorine residual and the wastewater flowrate. An automatic analyzer with signal transmitter and recorder is required.
Flow pacing with online effluent residual chlorine monitoring and automatic control	The control signals obtained from the wastewater flowmeter and the residual monitor are fed to a programmable logic controller (PLC) to provide more precise control of chlorine dosage and residual. This method works best where combined chlorine is used for disinfection.
Flow pacing with online residual chlorine monitoring after initial demand and automatic control	In this method, the chlorine residual is measured a short distance downstream from the point of chlorine addition. The readings from the wastewater flowmeter and the residual chlorine monitor are fed to a PLC to provide more precise control of chlorine dosage and residual. This method works best where combined chlorine is used for disinfection.
Flow pacing with online free and combined residual chlorine monitoring and automatic control	This approach is used for the disinfection of nitrified effluents with free chlorine where a variable residual ammonia must be removed to reach the breakpoint. The free and combined chlorine residual concentrations along with the readings from the wastewater flowmeter are fed to a PLC to provide more precise control of chlorine dosage. This approach is complex as the PLC must be programmed to recognize the difference between free and combined chlorine residuals and be able to interpret the data with respect to the chemistry of the breakpoint reaction. Data from an online ammonia analyzer, now available for field use, can also be integrated with the other data fed to the PLC to optimize the disinfection process with free chlorine.

[a] Adapted in part from Kobylinski et al. (2006).

because of the increase in chlorine demand needed to disinfect microorganisms embedded in floc particles. With manual control it is difficult, if not impossible, to maintain a constant CT, system monitoring and control is resource intensive, manual system is resource intensive; and chemical usage is higher due to residual variability. With automated systems, the ability to maintain the online analyzers is of critical importance if the benefits of automation are to be realized (Hurst, 2012).

Where free chlorine is to be used as the disinfectant, dosage control is more difficult, especially if the concentration of residual ammonia in the effluent to be disinfected is somewhat variable. As noted in Table 12–18, both free and combined chlorine residual monitors can be used in conjunction with readings from the influent wastewater flowmeter(s) to provide input to a programmable logic controller (PLC). More recently, online ammonia analyzers have been developed that can be used in conjunction with a PLC and inputs from chlorine residual monitors and flowrate measuring devices. Because the three online monitors must be maintained for complete automated control of the disinfection process, such systems are best employed at larger treatment plants with adequate staff.

Injection and Initial Mixing

As pointed out previously in Section 12–3, other things being equal, effective mixing of the chlorine solution with the wastewater, the contact time, and the chlorine residual are the principal factors involved in achieving effective bacterial kill. The addition of chlorine solution is often accomplished with a diffuser, which may be a plastic pipe with drilled holes through which the chlorine solution can be distributed into the path of wastewater flow (see Fig. 12–21). Unfortunately, the use of diffusers for adding chlorine is not very effective. To optimize the performance of disinfection systems, the chlorine should be introduced and mixed as rapidly as possible (ideally in less than a second). Techniques that can be used to mix chlorine in a fraction of a second were introduced and discussed in Chap. 5. Effective devices for mixing chlorine with the wastewater within a fraction of a second are illustrated on Fig. 12–22.

Chlorine Contact Basin Design

The principal design objective for chlorine contact basins is to ensure that some defined percentage of the flow remains in the chlorine contact basin for the design contact time to ensure effective disinfection. The contact time is usually specified by the regulatory agency and may range from 30 to 120 min; contact times of 15 to 90 min at peak flow are common. For example, for reuse applications the CDPH of the State of California requires a CT value of 450 mg·min/L with a modal contact time of 90 min at peak flow. Issues related to the design and analysis of chlorine contact basins considered in the following discussion include (1) basin configuration, (2) the use of baffles and guide vanes, (3) number of chlorine contact basins, (4) precipitation of solids in chlorine contact basins, (5) solids transport velocity and (6) a procedure for predicting disinfection performance.

Chlorine Contact Basin Configuration. To be assured that a given percentage of the flow will remain in the chlorine contact basin for a given period of time, the most common approach is to use long plug-flow, around-the-end type of contact basins (see Fig. 12–13) or a series of interconnected basins or compartments. Plug-flow chlorine contact basins that are built in a serpentine fashion (e.g., folded back and forth) to conserve space require special attention in their design to eliminate the formation of hydraulic dead zones that will reduce the hydraulic detention times. Length-to-width ratios (L/W) of at least 20 to 1 (preferably 40 to 1) and the use of baffles and guide vanes, as described below,

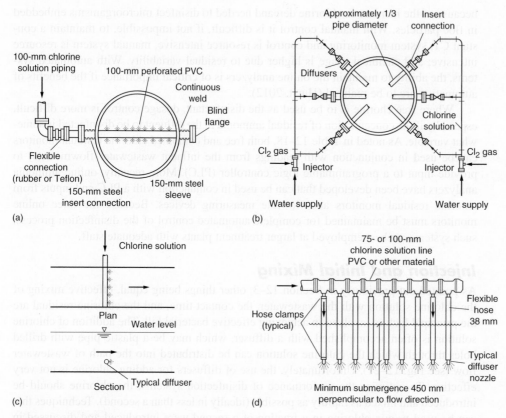

Figure 12–21

Typical diffusers used to inject chlorine solution: (a) across the pipe diffuser (b) diffuser system for large conduits, (c) single across the channel diffuser, and (d) typical hanging nozzle type chlorine diffuser for open channels. (Adapted from White, 1999.)

will help to minimize short circuiting. In some small plants, chlorine contact basins have been constructed of large diameter sewer pipe. The design of a chlorine contact basin based on dispersion is considered in Example 12–6.

Use of Baffles and Deflection Guide Vanes. To improve the hydraulic performance of chlorine contact basins, it has become common practice to use either submerged baffles, deflection guide vanes, or combinations of the two. Submerged baffles are used to break up density currents caused by temperature gradients, to limit short circuiting, and to minimize the effect of hydraulic dead spaces. The location of the baffles is critical in improving the performance of chlorine contact basins. A typical placement of baffles, and the effect on the corresponding tracer response curves, is illustrated on Fig. 12–23. As shown on Fig. 12–23, the addition of baffles improves the hydraulic performance of the chlorine contact basin significantly. The open area in submerged baffles will typically vary from 6 to 10 percent of the cross-sectional area of flow. The headloss through each baffle can be estimated using the following expression:

$$h = \frac{1}{2g}\left(\frac{Q}{Cna}\right)^2$$

(12–52)

Figure 12–22

Typical mixers for the addition of chlorine: (a) Water Champ® induction mixer can be mounted horizontally, as shown, or vertically, depending on the basin configuration, (b) inline static mixer, (c) inline turbine mixer, and (d) inline injector pump type. For additional types of chlorine mixers see Fig. 5–12 in Chap. 5.

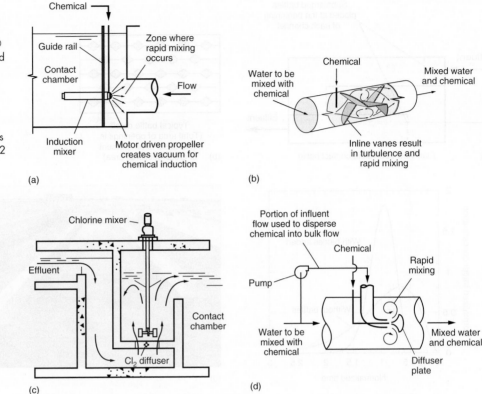

where h = headloss, m

   g = acceleration due to gravity, 9.81 m/s^2
   Q = discharge through chlorine contact basin channel, m^3/s
   C = discharge coefficient, unitless (typically about 0.8)
   n = number of openings
   a = area of individual opening, m^2

An alternative approach that has been used to improve the performance of chlorine contact basins is through the addition of deflection guide vanes, as shown on Fig. 12–24. The placement and number of vanes will depend on the layout of the chlorine contact basin. Two or three guide vanes are used most commonly. The beneficial effect of adding guide vanes was studied extensively by Louie and Fohrman (1968).

Number of Chlorine Contact Basins.   For most treatment plants, two or more contact basins should be used to meet reliability and redundancy requirements to facilitate maintenance and cleaning. Provisions should also be included for draining and scum removal. Vacuum type cleaning equipment may be used as an alternative to draining the basin for removal of accumulated solids. Bypassing the contact basin for maintenance should only be practiced on rare occasions, with the approval of regulatory agencies. If the time of travel in the outfall sewer at the maximum design flowrate is sufficient to equal or exceed the required contact time, it may be possible to eliminate the chlorine contact chambers, provided regulatory authorities agree.

Precipitation in Chlorine Contact Basins.   A problem often encountered in the operation of chlorine contact basins is the formation and precipitation of a light flocculent

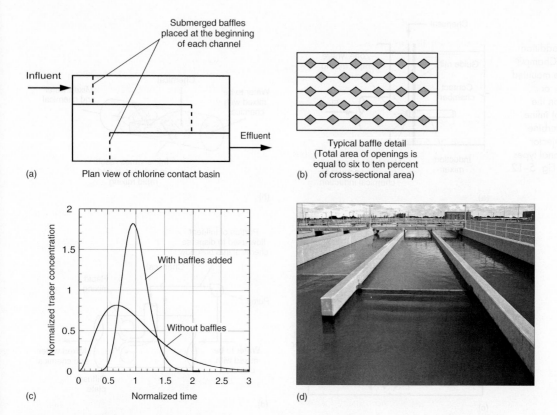

Submerged baffles placed at the beginning of each channel

Influent

Effluent

(a) Plan view of chlorine contact basin

Typical baffle detail
(Total area of openings is equal to six to ten percent of cross-sectional area)

(b)

(c)

(d)

Figure 12–23

Baffling in chlorine contact basins: (a) placement of baffles in chlorine contact tank at the beginning of each channel (or pass) is critical (adapted from Crittenden et al., 2005), (b) typical submerged baffle detail (adapted from Kawamura, 2000), (c) effect of the use of baffles in chlorine contact basins (adapted from Hart, 1979), and (d) view of chlorine contact tank with submerged wooden baffles placed at the beginning and end of each channel.

Figure 12–24

Chlorine contact basin with flow deflection vanes: (a) schematic and (b) photograph of empty chlorine contact basin designed with guide vanes.

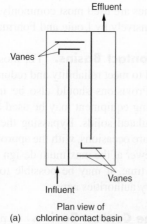

Effluent

Vanes

Influent

Vanes

Plan view of
chlorine contact basin

(a)

(b)

material. The principal cause of the formation and precipitation of floc is the lowering of the pH that results from the addition of chlorine. The problem occurs most frequently where alum is used for phosphorus removal in separate precipitation facilities or is added before the effluent filters. For a variety of reasons including high pH and inadequate initial mixing, not all of the alum added will react completely to form a floc that can be removed by precipitation or filtration. However, when the pH is lowered in the chlorine contact basin due to the addition of chlorine, some of the unreacted alum may form a floc. Oxidation of organic material in the treated effluent can also result in the formation of precipitates. Thus, in addition to meeting reliability and redundancy requirements, a minimum of two chlorine contact basins is necessary to allow one basin to be removed from service so that the accumulated solids can be removed from the basins.

Solids Transport Velocity. The horizontal velocity at minimum flow in a chlorine contact basin should, in theory, be sufficient to scour the bottom or to limit the deposition of sludge solids that may have passed through the settling tank. To limit excessive deposition, horizontal velocities should be at least 2 to 4.5 m/min (6.5 to 15 ft/min). In general, it will be difficult to achieve such velocities and simultaneously meet stringent dispersion requirements (see Example 12–6). If floc particles form, it will generally be impossible to avoid the accumulation of a sludge layer in the chlorine contact basins, another reason at least two chlorine contact basins should be used.

EXAMPLE 12–6 **Design of a Chlorine Contact Basin Based on Dispersion** Design a chlorine contact basin for secondary effluent with an average flowrate of 4000 m³/d. The estimated peaking factor is 2.0. The detention time at peak flow is to be 90 min. A minimum of two parallel channels must be used for redundancy requirements. The dimensions of the chlorine contact basin should be such to achieve a dispersion number equal to or less 0.015 at peak flow. Also check the dispersion number at average flow. What will happen if the low flow drops to 33 percent of the average flow in the early morning hours? Based on the resulting calculations will solids deposition occur, requiring periodic draining and cleaning of the chlorine contact basin?

Solution
1. Assume some trial cross-sectional dimensions for the chlorine contact basin and determine the corresponding length and flow velocity.
   a. Assumed dimensions

   Width = 2 m (6.6 ft)

   Depth = 3 m (9.8 ft)

   Number of parallel channels = 2

   b. Determine required length

   $$L = \frac{(2 \times 4000 \text{ m}^3/\text{d})}{(2)(1440 \text{ min/d})} \times (90 \text{ min}) \times \frac{1}{(2 \text{ m} \times 3 \text{ m})} = 41.7 \text{ m}$$

   c. Check velocity at peak flow

   $$v = \frac{(2 \times 4000 \text{ m}^3/\text{d})}{(2)(1440 \text{ min/d})(60 \text{ s/min})} \times \frac{1}{(2 \text{ m} \times 3 \text{ m})} = 0.0077 \text{ m/s}$$

2. Determine the coefficient of dispersion using Eq. (I–14) from Appendix I and the dispersion number using Eq. (I–9) from Appendix I for the chlorine contact basin.

    a. Compute the coefficient of dispersion

$$D = 1.01\nu N_R^{0.875}$$

       i. Compute the Reynolds number

$$N_R = 4\nu R/\nu$$

        ν = velocity in open channel, LT^{-1}, (m/s)

        R = hydraulic radius = area/wetted perimeter, L, (m)

        ν = 0.0077 m/s

        ν = 1.003×10^{-6} m²/s (at 20°C)

$$N_R = \frac{(4)(0.0077 \text{ m/s})[(2.0 \text{ m} \times 3.0 \text{ m})/(2 \times 3.0 \text{ m} + 2.0 \text{ m})]}{(1.003 \times 10^{-6} \text{ m}^2/\text{s})} = 23{,}031$$

      ii. Determine the coefficient of dispersion

$$D = 1.01(1.003 \times 10^{-6} \text{ m}^2/\text{s})(23{,}031)^{0.875} = 6.648 \times 10^{-3} \text{ m}^2/\text{s}$$

    b. Determine the dispersion number

$$d = \frac{D}{\nu L} = \frac{Dt}{L^2} = \frac{(0.006648 \text{ m}^2/\text{s})(90 \text{ min} \times 60 \text{ s/min})}{(41.4 \text{ m}^2)} = 0.0206$$

Because the computed dispersion number (0.0206) is greater than the desired value (0.015), an alternative design must be evaluated. For the alternative design, assume three parallel channels will be used.

3. Assume new trial cross-sectional dimensions for the chlorine contact basin and determine the new length and flow velocity.

    a. Assumed dimensions

    Width = 1.25 m (5.0 ft)

    Depth = 3 m (9.8 ft)

    Number of parallel channels = 3

    b. Determine required length

$$L = \frac{(2 \times 4000 \text{ m}^3/\text{d})}{(3)(1440 \text{ min/d})} \times (90 \text{ min}) \times \frac{1}{(1.25 \text{ m} \times 3 \text{ m})} = 44.4 \text{ m}$$

    c. Check velocity at peak flow

$$\nu = \frac{(2 \times 4000 \text{ m}^3/\text{d})}{(3)(1440 \text{ min/d})(60 \text{ s/min})} \times \frac{1}{(1.25 \text{ m} \times 3 \text{ m})} = 0.0082 \text{ m/s}$$

4. Check the dispersion number for the chlorine contact basin.

    a. Compute the Reynolds number

$$N_R = 4\nu R/\nu$$

    ν = 0.0082 m/s

    ν = 1.003×10^{-6} m²/s (at 20°C)

$$N_R = \frac{(4)(0.0082 \text{ m/s})[(1.25 \text{ m} \times 3.0 \text{ m})/(2 \times 3.0 \text{ m} + 1.25 \text{ m})]}{(1.003 \times 10^{-6} \text{ m}^2/\text{s})} = 16{,}915$$

b. Compute the coefficient of dispersion

$$D = 1.01 \nu N_R^{0.875}$$

$$D = 1.01 \times 1.003 \times 10^{-6} \text{ m}^2/\text{s } (16{,}915)^{0.875} = 5.07 \times 10^{-3} \text{ m}^2/\text{s}$$

c. Determine the dispersion number

$$d = \frac{D}{\nu L} = \frac{Dt}{L^2} = \frac{(0.00507 \text{ m}^2/\text{s})(90 \text{ min} \times 60 \text{ s/min})}{(44.4 \text{ m})^2} = 0.0139$$

Because the computed dispersion number (0.0139) is smaller than the desired value (0.015), the proposed design is acceptable.

5. Check the dispersion number for the chlorine contact basin at average flow.
   a. Compute the Reynolds number

   $$N_R = 4\nu R/\nu$$

   $$\nu = 0.0082/2 = 0.0041 \text{ m/s}$$

   $$\nu = 1.003 \times 10^{-6} \text{ m}^2/\text{s}$$

   $$N_R = \frac{(4)(0.0041 \text{ m/s})[(1.25 \text{ m} \times 3.0 \text{ m})/(2 \times 3.0 \text{ m} + 1.25 \text{ m})]}{(1.003 \times 10^{-6} \text{ m}^2/\text{s})} = 8{,}457$$

   b. Determine the coefficient of dispersion

   $$D = 1.01 \, \nu N_R^{0.875}$$

   $$D = 1.01(1.003 \times 10^{-6} \text{ m}^2/\text{s})(8{,}457)^{0.875} = 2.77 \times 10^{-3} \text{ m}^2/\text{s}$$

   c. Determine the dispersion number

   $$d = \frac{D}{\nu L} = \frac{Dt}{L^2} = \frac{(0.00277 \text{ m}^2/\text{s})(90 \text{ min} \times 60 \text{ s/min})}{(44.4 \text{ m})^2} = 0.0076$$

   d. Because the velocity is reduced at average flow, the computed dispersion number is equivalent to about 66 complete-mix reactors in series.

Comment Under all flow conditions, deposition of residual suspended solids would be expected in the chlorine contact basin, especially so at low flow.

Predicting Disinfection Performance. An extremely important issue in the design of chlorine contact basins is being able to predict the performance of the proposed design. To predict performance, the actual residence time that a given molecule of the fluid spends in the reactor must be known. The residence time in the reactor can be determined using some of the analytical techniques presented previously in Chap. 1 and Appendix I. The pertinent equations from Appendix I are repeated here for convenience. In Appendix I, it was noted that the Peclet number divided by 2 is equal to the number of reactors in series. The relationship of the Peclet number to the dispersion number is

$$P_e = \frac{\nu L}{D} = \frac{1}{d} \tag{12–53}$$

For complete mix-reactors in series, the normalized residence time distribution curve, $E(\theta)$, where θ is equal to t/τ for n reactors in series, as derived in Appendix I, is given by

$$E(\theta) = \frac{n}{(n-1)!}(n\theta)^{n-1}e^{-n\theta} \tag{12–54}$$

Further, the fraction of tracer, $F(\theta)$, that has been in the reactor for less than time t is defined as follows:

$$F(\theta) = \int_0^t E(\theta)d\theta \approx \sum_0^n E(\theta)\Delta\theta \qquad (12\text{-}55)$$

Thus, for a given dispersion number, the Peclet number can be used to determine the corresponding number of complete-mix reactors in series needed to achieve that dispersion number. Knowing the number of reactors in series, the value of $E(\theta)$ can be computed for various values of θ. The value of $F(\theta)$ can then be determined by summing the area under the $E(\theta)$ curve. The amount of flow that has been in the reactor for less than time θ can now be determined. Coupling normalized microorganism inactivation dose response data, obtained using a batch reactor, with the normalized detention time data, the actual performance for the chlorine contact basin can be estimated using what is known as a segregated flow model (SFM).

In the SFM approach, it is assumed that each block of fluid that enters a chlorine contact basin does not interact with other blocks of water. Thus, each block of water corresponds to an ideal plug flow reactor, each having a different residence time as defined by the value of $E(\theta)$, as given above. The reduction in organisms that would occur in each block of water can then be estimated for the period of time the block of water has remained in the chlorine contact basin. The overall performance is obtained by summing the results for each block of water. The SFM approach can be described as follows (Fogler, 1999):

Word statement

$$
\begin{array}{ccc}
\begin{array}{c}
\text{Mean reduction in} \\
\text{number of} \\
\text{microorganisms} \\
\text{spending between} \\
\text{time } t \text{ and } t + dt \text{ in the} \\
\text{chlorine contact basin}
\end{array}
&=&
\begin{array}{c}
\text{number of} \\
\text{microorganisms} \\
\text{remaining after spending} \\
\text{time } t \text{ in the chlorine} \\
\text{contact basin based on} \\
\text{batch test results}
\end{array}
\times
\begin{array}{c}
\text{fraction of flow} \\
\text{that remained} \\
\text{in the chlorine} \\
\text{contact basin} \\
\text{between time } t \\
\text{and } t + dt
\end{array}
\end{array}
\qquad (12\text{-}56)
$$

In equation form,

$$d\overline{N} = N(\theta) \times E(\theta)dt \qquad (12\text{-}57)$$

The values of $N(\theta)$ and $E(\theta)$ are obtained from batch disinfection and tracer or dispersion prediction studies. Application of the above equations for predicting the hydraulic performance and the effluent microorganism concentration using the SFM are illustrated in Example 12–7 using the data from Example 12–6.

EXAMPLE 12–7   **Estimation of Performance of a Chlorine Contact Basin**   Using the design information from Example 12–6, determine the fraction of flow that has not remained in the chlorine contact basin for the full hydraulic detention time. Determine how much larger the chlorine contact basin must be to be assured that 90 percent of the flow remains in the chlorine contact basin for the full design hydraulic detention time. Using the following normalized dose response data for an enteric virus, based on a τ value of 90 min and combined chlorine residual of 6 mg/L, estimate the performance of the chlorine contact basin in terms of the residual number of organisms remaining in the effluent.

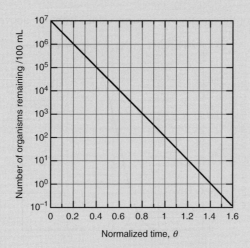

Normalized time, θ

Solution

1. Determine the number of complete mix reactors in series.
   a. From Example 12–6, the dispersion number at peak flow is

   $$d = 0.0139$$

   b. Using Eq. (12–53), the number of complete-mix reactors in series is

   $$\text{Number of reactors in series} = \frac{P_e}{2} = \frac{1}{2d} = \frac{1}{(2)\,0.0139} = 36$$

2. Determine the percentage of the flow that has been in the chlorine contact basin for less than the hydraulic detention time.
   a. Set up a computation table and compute $E(\theta)$ using Eq. (12–54) and the data given above.

   $$E(\theta) = \frac{n}{(n-1)!}(n\theta)^{n-1}e^{-n\theta}$$

Normalized time, θ	$E(\theta)$	$E(\theta) \times \Delta\theta \times 100$	Cumulative percent, $F(\theta)$
0.30	0.0000	0.000	0.000
0.40	0.0000	0.000	0.001
0.50	0.0046	0.046	0.046
0.60	0.0737	0.737	0.783
0.70	0.4435	4.435	5.218
0.80	1.2976	12.976	18.193
0.90	2.1878	21.878	40.071
1.00	2.3881	23.881	63.952
1.10	1.8337	18.337	82.290
1.20	1.0531	10.531	92.821
1.30	0.4739	4.739	97.560
1.40	0.1733	1.733	99.293
1.50	0.0530	0.530	99.822
1.60	0.0139	0.139	99.961
1.70	0.0031	0.032	99.992
1.80	0.0006	0.006	99.999
1.90	0.0001	0.001	100.00
2.00	0.0000	0.000	100.00

b. Plot the cumulative percent values from the above table.

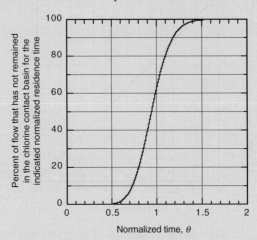

c. From the computation table and graph given above, the percentage of the flow that has been in the chlorine contact basin for less than the hydraulic residence time is 64 percent. In fact, about 18 percent of the flow has left the chlorine contact basin before 80 percent of the nominal hydraulic detention time has elapsed.

3. Estimate how much larger the chlorine contact basin must be to be assured that 90 percent of the flow remains in the chlorine contact basin for the full hydraulic detention time. From the above graph, the size of the chlorine contact basin would have to be increased by a factor of 1.2.

4. Estimate the performance of the chlorine contact basin.

a. Set up a computation table to determine the number of organisms remaining in the effluent from the chlorine contact basin. The SFM approach, described above, will be used for this analysis. In effect, flow in each time period is treated as a batch reactor for the time interval it has remained in the reactor. The corresponding concentration of microorganisms leaving in any given volume of liquid is taken from the normalized dose response curve. The computation table for the application of the SFM is given below. The data in columns (1) and (3) are from the computation table prepared in Step 2 above, except that the data in column (3) are divided by 100. The data in column (2) are from the normalized dose response curve obtained as part of the process analysis for the design of the chlorine contact basin.

Normalized time, θ (1)	Number of organisms remaining, $N(\theta)$ MPN/100 mL (2)	$E(\theta) \times \Delta\theta$ (3)	Number of organisms remaining in effluent, ΔN MPN/100 mL (4)
0.30	300,000	0.00000	0.000
0.40	100,000	0.00000	0.00
0.50	30,000	0.00046	13.80
0.60	10,000	0.00737	73.70
0.70	3,000	0.04435	133.05
0.80	1,000	0.12976	129.76
0.90	300	0.21878	65.63

(continued)

(*Continued*)

Normalized time, θ (1)	Number of organisms remaining, $N(\theta)$ MPN/100 mL (2)	$E(\theta) \times \Delta\theta$ (3)	Number of organisms remaining in effluent, ΔN MPN/100 mL (4)
1.00	100	0.23881	23.88
1.10	30	0.18337	5.50
1.20	10	0.10531	1.05
1.30	3	0.04739	0.14
1.40	1	0.01733	0.02
1.50	0.3	0.00530	—
1.60	0.1	0.00139	—
1.70	0.03	0.00032	—
1.80	0.01	0.00006	—
1.90	0.003	0.00001	—
2.00	0.001	0.00000	—
Total		1.00000	446.53

b. The number of organisms in the effluent leaving the chlorine contact basin is:

Organisms in effluent $N = \sum[N(\theta) \times E(\theta)\Delta\theta] = 447$ MPN/100 mL

c. By comparison, if it was assumed that the basin had performed as an ideal plug-flow reactor, then the organism concentration in the effluent would have been estimated to be 100 MPN/100 mL.

Comment The SFM method of analysis used to determine the number of organisms in the effluent is useful for estimating the performance of reactors with varying amounts of dispersion such as chlorine contact basins.

Assessing the Hydraulic Performance of Existing Chlorine Contact Basins

To be assured that a chlorine contact basin performs properly, most regulatory agencies request that tracer studies be conducted to determine the hydraulic characteristics of the chlorine contact basin. The types of tracers that have been used, the conduct of tracer tests, and analysis of tracer data are reviewed briefly below.

Compounds Used as Tracers. Tracers of various types are used commonly to assess the hydraulic performance of reactors used for wastewater disinfection. Dyes and chemicals that have been used successfully in tracer studies include congo red, fluorescein, fluosilicic acid (H_2SiF_6), hexafluoride gas (SF_6), lithium chloride (LiCl), Pontacyl Brilliant Pink B, potassium, potassium permanganate, rhodamine WT, sodium floride (NaF), and sodium chloride (NaCl). Pontacyl Brilliant Pink B (the acid form of rhodamine WT) is especially useful in the conduct of dispersion studies because it is not readily adsorbed onto surfaces. Because fluorescein, rhodamine WT, and Pontacyl Brilliant Pink B can be detected at very low concentrations using a flourometer, they are the dye tracers used most commonly in the evaluation of the performance of wastewater treatment facilities.

Figure 12-25

Schematic of setup for the conduct of a tracer study of a plug-flow chlorine contact basin using either a slug of tracer added to flow or a continuous input of tracer. The tracer response curve is measured continuously.

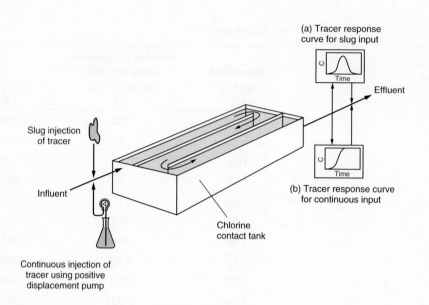

Conduct of Tracer Tests. In tracer studies, a tracer (i.e., a dye, most commonly) is introduced into the influent end of the reactor or basin to be studied (see Fig. 12–25). The time of its arrival at the effluent end is determined by collecting a series of grab samples for a given period of time or by measuring the arrival of a tracer using instrumental methods (see Fig. 12–25). The method used to introduce the tracer controls the type of response observed at the downstream end. Two types of tracer input methods are used, the choice depending on the reactor influent and effluent configurations. The first method involves the injection of a quantity of tracer (sometimes referred to a pulse or slug of tracer) over a short period of time. Initial mixing is usually accomplished with a static mixer or an auxiliary mixer. With the slug injection method it is important to keep the initial mixing time short relative to the detention time of the reactor being measured. The measured output is as described on Fig. 12–25(a). In the second method, a continuous step input of tracer is introduced until the effluent concentration matches the influent concentration. The measured response is as shown on Fig. 12–25(b). It should also be noted that another response curve can be measured after the dye injection has ceased and the dye in the reactor is flushed out.

Analysis of Tracer Test Response Curves. Tracer response curves, measured using a slug or continuous injection of a tracer, are known as C (concentration versus time) and F (fraction of tracer remaining in the reactor versus time) curves, respectively. The fraction remaining is based on the volume of water displaced from the reactor by the step input of tracer. The generalized results of three different dye tracer tests are shown on Fig. 12–26. As shown on Fig. 12–26, each of the three basins is subject to differing amounts of short circuiting. Length-to-width ratios (L/W) of at least 20 to 1 (preferably 40 to 1) and the use of baffles and guide vanes helps to minimize short circuiting. In some small plants, chlorine contact basins have been constructed of large diameter sewer pipe. The beneficial effect of using submerged baffles to improve the hydraulic efficiency of serpentine chlorine contact basins is illustrated on Fig. 12–23.

Tracer curves, such as shown on Figs. 12–25 and 12–26, are used to assess the hydraulic efficiency of chlorine contact basins. Parameters used to assess the hydraulic efficiency of chlorine contact basins are summarized in Table 12–19 and are illustrated on Fig. 12–27. As discussed previously, the mean, modal, and t_{10} times have been used to define the

Figure 12–26

Typical chlorine contact basin tracer response curves for three different basins with the same hydraulic detention time. The degree of short circuiting is illustrated clearly by the shape of the tracer curve.

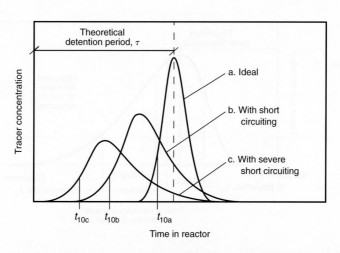

Table 12–19

Various terms used to describe the hydraulic performance of chlorine contact basins[a]

Term	Definition
τ	Theoretical hydraulic residence time (V, volume/Q, flowrate).
t_i	Time at which tracer first appears.
t_p	Time at which the peak concentration of the tracer is observed (mode).
t_g	Mean time to reach centroid of the residence time distribution (RTD) curve (see Appendix H).
t_{10}, t_{50}, t_{90}	Time at which 10, 50, and 90 percent of the tracer has passed through the reactor.
t_{90}/t_{10}	Morrill Dispersion Index, MDI (Morrill, 1932).
1/MDI	Volumetric efficiency as defined by Morrill (1932).
t_i/τ	Index of short circuiting. In an ideal plug-flow reactor, the ratio is 1, and approaches 0 with increased mixing.
t_p/τ	Index of modal retention time. Ratio will approach 1 in a plug-flow reactor, and 0 in a complete-mix reactor. For values of the ratio greater than or less than 1 the flow distribution in the reactor is not uniform.
t_g/τ	Index of average retention time. A value of 1 would indicate that full use is being made of the volume. A value of the ratio greater than or less than 1 indicates the flow distribution is not uniform.
t_{50}/τ	Index of mean retention time. The ratio t_{50}/τ is a measure of the skew of the RTD curve. A value of t_{50}/τ of less than 1 corresponds to an RTD curve that is skewed to the left. Similarly, for values greater than 1.0 the RTD curve is skewed to the right.
$t/\tau = \theta$	Normalized time, used in the development of the normalized RTD curve.
$\tau_{\Delta c} \approx \dfrac{\sum t_i C_i \Delta t_i}{\sum C_i \Delta t_i}$	Expression used to determine the mean hydraulic residence time, τ, if the concentration versus time tracer response curve is defined by a series of discrete time step measurements, where t_i is time at ith measurement C_i is concentration at ith measurement, and Δt_i is time increment about C_i.
$\sigma_{\Delta c}^2 \approx \dfrac{\sum t_i^2 C_i \Delta t_i}{\sum C_i \Delta t_i} - (\tau_{\Delta c})^2$	Expression used to determine variance for a concentration versus time tracer response curve, which is defined by a series of discrete time step measurements.

[a] Adapted from Morrill (1932), Fair and Geyer (1954), and U.S. EPA (1986).

Figure 12–27

Definition sketch for the parameters used in the analysis of concentration versus time tracer response curves.

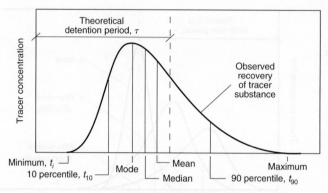

contact time in the CT relationship. The analysis of a tracer response curve is illustrated in Example 12–8. Additional details on the analysis of tracer response curves may be found in Appendix I and in Crittenden et al. (2012).

EXAMPLE 12–8   Analysis of Tracer Data for an Existing Chlorine Contact Basin   The following tracer data have been gathered during a tracer test of a chlorine contact basin. During the tracer test, the total chlorine residual measured at the tank outlet was 4.0 mg/L. Using these data determine the mean hydraulic residence time (HRT), the variance, and the t_{10} time. Determine the CT values corresponding to the mean HRT and the t_{10} time. To further assess the performance of the chlorine contact basin, determine the Morrill Dispersion Index (MDI) and the corresponding volume efficiency (1/MDI) as defined in Table 12–19.

Time, min	Tracer concentration, μg	Time, min	Tracer concentration, μg
0.0	0.0000	144	9.333
16	0.000	152	16.167
40	0.000	160	20.778
56	0.000	168	19.944
72	0.000	176	14.111
88	0.000	184	8.056
96	0.056	192	4.333
104	0.333	200	1.556
112	0.556	208	0.889
120	0.833	216	0.278
128	1.278	224	0.000
136	3.722		

Solution

1. Determine the mean hydraulic residence time and variance for the tracer response data using equations given in Table 12–19.

a. Set up the required computation table. In setting up the computation table given below, the Δt value was omitted as it appears in both the numerator and in the denominator of the equations used to compute the residence time and the corresponding variance.

Time, t, min	Conc., C, μg	$t \times C$	$t^2 \times C$	Cumulative conc., μg	Cumulative percentage
88	0.000	0.000	0		
96	0.056	5.338	512.41	0.05	0.05
104	0.333	34.663	3604.97	0.39[a]	0.38[b]
112	0.556	62.227	6969.45	0.94	0.92
120	0.833	99.996	11,999.52	1.78	1.74
128	1.278	163.558	20,935.48	3.06	2.99
136	3.722	506.219	68,845.81	6.78	6.63
144	9.333	1343.995	193,535.31	16.11	15.75
152	16.167	2457.384	373,522.37	32.28	31.58
160	20.778	3324.480	531,916.80	53.06	51.91
168	19.944	3350.592	562,899.46	73.00	71.41
176	14.111	2483.536	437,102.34	87.11	85.22
184	8.056	1482.230	272,730.39	95.17	93.10
192	4.333	831.994	159,742.77	99.50	97.34
200	1.556	311.120	62,224.00	101.06	98.87
208	0.889	184.891	38,457.37	101.94	99.73
216	0.278	60.005	12,961.04	102.22	100.00
224	0.000	0.000			
Total	102.222	16,702.229	2,757,959.48		

[a] $0.056 + 0.333 = 0.39$.

[b] $(0.39/102.222) \times 100 = 0.38$.

b. Determine the mean hydraulic residence time.

$$\tau_{\Delta c} = \frac{\sum t_i C_i \Delta t_i}{\sum C_i \Delta t_i} = \frac{16,702.23}{102.22} = 163.4 \text{ min} = 2.7 \text{ h}$$

c. Determine the variance.

$$\sigma_{\Delta c}^2 = \frac{\sum t_i^2 C_i \Delta t_i}{\sum C_i \Delta t_i} - (\tau_{\Delta c})^2 = \frac{2,757,959.48}{102.22} - (163.4)^2 = 280.5 \text{ min}^2$$

$$\sigma_{\Delta c} = 16.7 \text{ min}$$

d. Determine the t_{10} time using the cumulative percentage values. Because of the short time interval, a linear interpolation method can be used.

$$(15.75\% - 6.63\%)/(144 \text{ min} - 136 \text{ min}) = 1.14\%/\text{min}$$

$$t_{10} = 136 + (10\% - 6.63\%)/ (1.14\%/\text{min}) = 139.0 \text{ min}$$

e. Identify the mean hydraulic residence and t_{10} times on the tracer curve.

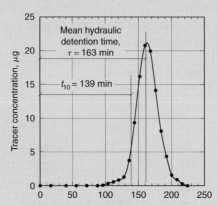

2. Another technique that can be used to obtain the above times is to plot the cumulative concentration data on log-probability paper. Such a plot is also useful for determining the MDI. The required plot is given below.

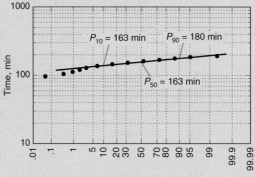

Cumulative concentration percentage

The mean hydraulic retention and t_{10} times are read directly from the above plot.

$t_{50} = 163$ min

$t_{10} = 139$ min

3. Determine the corresponding CT values for the mean HRT and the t_{10} time determined above in Step 1.

CT (modal) = (4.0 mg/L)(163.4 min) = 654 mg·min/L

CT (t_{10}) = (4.0 mg/L)(139 min) = 556 mg·min/L

4. Determine the MDI and the corresponding volume efficiency using the expressions given in Table 12–19 and the values from the plot given in Step 2 above.
    a. The Morrill Dispersion Index is

$$\text{Morrill Dispersion Index, MDI} = \frac{P_{90}}{P_{10}} = \frac{180}{139} = 1.30$$

c. The corresponding volumetric efficiency for the chlorine contact basin is

$$\text{Volumetric efficiency, \%} = \frac{1}{\text{MDI}} = \frac{1}{1.3} \times 100 = 77\%$$

Comment The variance computed in Step 1 is useful in assessing the dispersion in the chlorine contact basin (see Example I–1 in Appendix I; Crittenden et al., 2012). The CT values, based on the modal and t_{10} times, exceed the CT value of 450 mg·min/L required in California. It is important to note that if the tracer curve is very skewed, it may not be possible to achieve effective disinfection, especially if the t_{10} value is used. Thus, the design of a chlorine contact basin to achieve near plug flow is of critical importance. The MDI value (1.30) is characteristic of a chlorine contact basin with low dispersion. A MDI value below 2.0 has been established by the U.S. EPA as an effective design (U.S. EPA, 1986). Similarly, the volumetric efficiency is high, signifying near-ideal plug flow with a small amount of axial dispersion.

Outlet Control and Chlorine Residual Measurement

The flow at the end of the contact basin may be metered by means of a V-notch or rectangular weir or a Parshall flume. Control devices for chlorination in direct proportion to the flowrate may be operated from these meters or from the main plant flowmeter. Final determination of the success of a chlorine contact basin must be based on samples taken and analyzed to correlate chlorine residual and the MPN of coliform or other indicator organisms. When the chlorine residual is used for chlorinator control, chlorine residual sample pumps should be located at the front end of the first pass of the contact basin after rapid mixing to allow time for the initial demand to be met. Chlorine residual measurements should also be taken at the contact basin outlet to ensure compliance with the regulatory agency requirements. In the event that no chlorine contact basin is provided and the effluent pipeline is used for contact, the sample can be obtained at the point of chlorination, held for the theoretical detention time, and the residual determined. The sample is then dechlorinated and subsequently analyzed for bacteria using standard laboratory procedures.

Chlorine Storage Facilities

Storage and handling facilities for chlorine can be designed with the aid of information developed by The Chlorine Institute. Although all the safety devices and precautions that must be designed into the chlorine handling facilities are too numerous to mention, the following are fundamental:

1. Chlorine gas is toxic and very corrosive. Adequate exhaust ventilation with intakes at floor level should be provided because chlorine gas is heavier than air. The ventilation system should be capable of at least 60 air changes per hour with the exhaust directed vertically upwards.
2. Chlorine storage and chlorinator equipment rooms should be walled off from the rest of the plant and should be accessible only from the outdoors. A fixed glass viewing window should be included in an inside wall to check for leaks before entering the equipment rooms. Fan controls should be located at the room entrance. Air purifying respirators or self-contained breathing apparatus (SCBA) should also be located nearby in protected but readily accessible locations.
3. Temperatures in the scale and chlorinator areas should be controlled to avoid freezing.
4. Dry chlorine liquid and gas can be handled in black steel piping, but chlorine solution is highly corrosive and should be handled in Schedule 80 polyvinylchloride (PVC) piping.
5. Adequate storage of standby cylinders should be provided. The amount of storage should be based on the availability and dependability of the supply and the quantities used. Cylinders in use are set on scales and the loss of weight is used as a positive record of chlorine dosage.

6. Chlorine cylinders should be protected from direct sunlight in warm climates to prevent overheating of the full cylinders.
7. In larger systems, chlorine residual analyzers should be provided for monitoring and control purposes to prevent the under- or over-dosing of chlorine.
8. The chlorine storage and feed facilities should be protected from fire hazards. In addition, chlorine leak detection equipment should be provided and connected to an alarm system.

Chemical Containment Facilities

In 1991, the International Conference of Building Officials revised Article 80: Hazardous Materials of the Uniform Fire Code (UFC). The revisions were extensive and covered a variety of issues. The provisions of the new code apply to new facilities and to old facilities as well, if it is determined that they constitute a distinct hazard to life or property. The new code provisions are contained in the following divisions apply to the chemicals used for disinfection: I General Provisions, II Classification by Hazard, III Storage Requirements, and IV Dispensing, Use, and Handling. The classification of hazardous materials used for wastewater disinfection are summarized in Table 12–20. Storage requirements include provisions for spill control and containment, ventilation, treatment, and storage. Emergency scrubbing systems, usually using a caustic solution, are also required to neutralize leaking chlorine and sulfur dioxide gas. Many of the same topics contained in the storage requirements section also apply to the dispensing, use, and handling. Hazardous material management, provision for standby power, security, and alarms are among the additional topics covered. It is extremely important to review current UFC regulations in the design of new facilities and in the refurbishing of existing facilities. Furthermore, the U.S. EPA and many states, as well as OSHA, have implemented chemical safety regulations requiring formal hazard reviews, air dispersion modeling of release scenarios, and emergency response preparedness.

Dechlorination Facilities

Dechlorination of chlorinated effluents is accomplished most commonly using sulfur dioxide. Where granular activated carbon is used for the removal of residual organic material, the carbon can also be used for the dechlorination of chlorinated effluents.

Sulfur Dioxide. The principal elements of a sulfur dioxide dechlorination system include the sulfur dioxide containers, scales, sulfur dioxide feeders (sulfonators), solution

Table 12–20

Classification of hazardous materials used in wastewater disinfection

Category	Typical chemicals
Physical Hazards	
Compressed gases	Oxygen, ozone, chlorine, ammonia, sulfur dioxide
Oxidizers	Oxygen, ozone, chlorine, hydrogen peroxide, acids, chlorine
Health Hazards	
Highly toxic material	Chlorine, chlorine dioxide, ozone, acids, bases
Corrosives	Acids, bases, chlorine, sulfur dioxide, ammonia, hypochlorite, sodium bisulfite
Other health hazards—irritants, suffocating, etc.	Chlorine, sulfur dioxide, ammonia

injectors, diffuser, mixing chamber, and interconnecting piping. For facilities requiring large withdrawal rates of SO_2, evaporators are used because of the low vaporization pressure of 240 kN/m^2 at 21°C (35 $lb_f/in.^2$ at 70°F). Common sulfonator sizes are 216, 864, and 3409 kg/d (475, 1900, and 7500 lb/d). The key control parameters of this process are (1) proper dosage based on precise (amperometric) monitoring of the combined chlorine residual and (2) adequate mixing at the point of application of sulfur dioxide.

Sodium Bisulfite.   Sodium bisulfite is available as white powder, a granular material, or as a liquid. The liquid form is used most commonly for dechlorination at wastewater treatment facilities. Although available in solution strengths up to 44 percent, a 25 percent solution is most typical to minimize viscosity increases during cold weather. In most applications, a diaphragm type pump is used to meter the sodium bisulfite. The reaction between sodium bisulfite and chlorine residual was presented previously [see Eqs. (12–42 and 12–43)]. Based on Eq. (12–42), each mg/L of chlorine residual requires about 1.46 mg/L of sodium bisulfite and 1.38 mg/L of alkalinity as $CaCO_3$ will be consumed.

Granular Activated Carbon.   The common method of activated carbon treatment used for dechlorination is downflow through either an open or enclosed vessel. The activated carbon system, while significantly more costly than other dechlorination approaches, may be appropriate when activated carbon is being used as an advanced wastewater treatment process. Typical hydraulic loading rates and contact times for activated carbon columns used for dechlorination are 3000 to 4000 $L/m^2 \cdot d$ and 15 to 25 min, respectively.

12–7 DISINFECTION WITH OZONE

Although historically used primarily for the disinfection of water, recent advances in ozone generation and solution technology have made the use of ozone economically more competitive for the disinfection of treated wastewater. Further, interest in the use of ozone for disinfection has also been renewed because of its ability to reduce or eliminate trace constituents. Ozone can also be used in water reuse applications for the removal of soluble refractory organics, in lieu of the carbon adsorption process. The characteristics of ozone, the chemistry of ozone, the generation of ozone, an analysis of the performance of ozone as a disinfectant, and the application of the ozonation process are considered in the following discussion.

Ozone Properties

Ozone is an unstable gas produced when oxygen molecules dissociate into atomic oxygen. Ozone can be produced by electrolysis, photochemical reaction, and radiochemical reaction by electrical discharge. Ozone is often produced by ultraviolet light and lightning during a thunderstorm. The electrical discharge method is used for the generation of ozone in water and wastewater disinfection applications. Ozone is a blue gas at normal room temperatures, and has a distinct odor. Ozone can be detected at concentrations of 2×10^{-5} to 1×10^{-4} g/m^3 (0.01 to 0.05 ppm_v, by volume). Because ozone has an odor, it can usually be detected by the human olfactory system before health concerns develop. The stability of ozone in air is greater than it is in water, but in both cases is on the order of minutes. Gaseous ozone is explosive when the concentration reaches about 240 g/m^3 (20 percent weight in air). The properties of ozone are summarized in Table 12–21. The solubility of ozone in water is governed by Henry's law. Typical values of Henry's constant for ozone are presented in Table 12–22.

Table 12–21

Properties of ozone[a]

Property	Unit	Value
Molecular weight	g	48.0
Boiling point	°C	−111.9 ± 0.3
Freezing point	°C	−192.5 ± 0.4
Latent heat of vaporization at 111.9°C	kJ/kg	14.90
Liquid density at −183°C	kg/m³	1574
Gaseous density at 0°C and 1 atm	g/mL	2.154
Solubility in water at 20.0°C	mg/L	12.07
Vapor pressure at −183°C	kPa	11
Vapor density compared to dry air at 0°C and 1 atm	unitless	1.666
Specific volume of vapor at 0°C and 1 atm	m³/kg	0.464
Critical temperature	°C	−12.1
Critical pressure	kPa	5532.3

[a] Adapted in part from Rice (1996), U.S. EPA (1986), White (1999).

Ozone Chemistry

Some of the chemical properties displayed by ozone may be described by its decomposition reactions which are thought to proceed as follows:

$$O_3 + H_2O \rightarrow HO_3^+ + OH^- \tag{12–58}$$

$$HO_3^+ + OH^- \rightarrow 2HO_2 \tag{12–59}$$

$$O_3 + HO_2 \rightarrow HO\cdot + 2O_2 \tag{12–60}$$

$$HO\cdot + HO_2 \rightarrow H_2O + O_2 \tag{12–61}$$

The dot ($\cdot$) that appears next to the hydroxyl and other radicals is used to denote the fact that these species have an unpaired electron. The free radicals formed, HO_2 and $HO\cdot$, have great oxidizing powers and are active in the disinfection process. These free radicals also possess the oxidizing power to react with other impurities in aqueous solutions.

Table 12–22

Values of Henry's constant for ozone[a]

Temperature, °C	Henry's constant, atm/mole fraction
0	1940
5	2180
10	2480
15	2880
20	3760
25	4570
30	5980

[a] U.S. EPA (1986).

Table 12–23

Impact of wastewater constituents on the use of ozone for wastewater disinfection

Constituent	Effect
BOD, COD, TOC, etc.	Organic compounds that comprise the BOD and COD can exert an ozone demand. The degree of interference depends on their functional groups and their chemical structure
NOM (natural organic matter)	Affects the rate of ozone decomposition and the ozone demand
Oil and grease	Can exert an ozone demand
TSS	Increases ozone demand and shielding of embedded bacteria
Alkaliniy	No or minor effect
Hardness	No or minor effect
Ammonia	No or minor effect, can react at high pH
Nitrite	Oxidized by ozone
Nitrate	Can reduce effectiveness of ozone
Iron	Oxidized by ozone
Manganese	Oxidized by ozone
pH	Effects the rate of ozone decomposition
Industrial discharges	Depending on the constituents, may lead to a diurnal and seasonal variations in the ozone demand
Temperature	Affects the rate of ozone decomposition

Effectiveness of Ozone as a Disinfectant

Ozone is an extremely reactive oxidant and it is generally believed that bacterial kill through ozonation occurs directly because of cell wall disintegration (cell lysis). The impact of the wastewater characteristics on ozone disinfection is reported in Table 12–23. The presence of oxidizable compounds will cause the ozone inactivation curve to have a shoulder affect as discussed previously for chlorine (see Fig. 12–6). Tailing will also occur in the presence of residual floc particles.

Ozone is also a very effective viricide and is generally believed to be more effective than chlorine. The relative germicidal effectiveness of ozone for the disinfection of different microorganisms was presented previously in Table 12–5. Ozonation does not increase dissolved solids and disinfection effectiveness is not affected by ammonium ion. Although ozone is not necessarily impacted by water pH, ozone residual is more stable in acidic environments and less stable in waters with caustic pH. Therefore, it is typically easier to achieve disinfection when pH is reduced than when pH is greater than neutral. For these reasons, ozonation is considered as an alternative to either chlorination or hypochlorination, especially where dechlorination may be required and high purity oxygen facilities are available at the treatment plant.

Modeling the Ozone Disinfection Process

In practice, an ozone contactor will be comprised of three or more compartments or chambers (see Fig. 12–31 in discussion of ozone reactor characteristics). Water depth is typically 4.6 to 6 m (15 to 20 ft). Ozone is typically added to first or the first and second of the compartments and the remaining compartments serve as contact compartments. The detention time in the first compartment, used to meet the immediate ozone demand (i.e., peroxidation), is short, typically 2 to 4 min. Contact time in the subsequent compartments will vary from 3 to 10 min depending on the rate of ozone utilization.

Over the years a number of different mathematical relationships have been developed to model the disinfection process with ozone. The most common of these is Eq. (12–6) repeated here for convenience.

$$\log \frac{N_t}{N_o} = -\Lambda_{\text{base 10}}(\text{CT}) \tag{12-6}$$

Values for the coefficient of specific lethality, Λ, are given in Table 12–11. Of the values given in Table 12–11, those for viruses, *Cryptosporidium,* and *Giardia* cysts are the most reliable as they can be derived from the published U.S. EPA CT tables (U.S. EPA, 2003a).

Because ozone is sparingly soluble, bench and/or pilot-scale studies (see Fig 12–28), using the same retention time as the full-scale reactor, are conducted to assess (1) the immediate ozone demand, (2) the amount of ozone that can be transferred to the liquid, and (3) the ozone decay profile along the reactor. The information gathered is used to determine the CT value and the level of inactivation that can be expected in the full-scale reactor. The amount of ozone that is utilized or transferred to the liquid is computed using the following expression:

$$\text{Ozone dose mg/L} = \frac{Q_g}{Q_l}(C_{g,\text{in}} - C_{g,\text{out}}) \tag{12-62}$$

where Q_g = gas flowrate, L/min
$\quad\quad Q_l$ = liquid flowrate, L/min
$\quad\quad C_{g,\text{in}}$ = concentration of ozone in feed gas, mg/L
$\quad\quad C_{g,\text{out}}$ = concentration of ozone in off-gas, mg/L

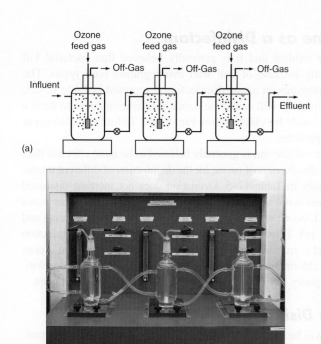

Figure 12–28

Typical ozone test reactors: (a) schematic of setup for bench-scale ozone testing, (b) view of bench-scale ozone test reactors, and (c) pilot-scale ozone test reactor.

The bench and/or pilot-scale reactors are operated in a continuous mode at various ozone concentrations. Once steady-state has been reached, both the water and ozone dosing are stopped, and the ozone decay is observed with time. The continuous operation simulates the compartments where ozone is being added and accounts for the immediate demand of the wastewater. The decay curve is used to estimate the residual ozone concentrations in the downstream compartments. Analysis of bench-scale ozone test data is illustrated in Example 12–9. Computation of the CT for an ozone contactor is illustrated in Example 12–10.

EXAMPLE 12–9 **Estimate the Immediate Ozone Requirement for a Typical Secondary Effluent** Estimate the immediate ozone demand from the following bench scale steady-state ozone test data collected at 20°C. Determine the first-order equation for the corresponding decay data.

The steady-state test results are

Test	Ozone dose, mg/L	Ozone residual, mg/L
1	5	1.5
2	8	5.0
3	10	7.5
4	13	10.3
5	18	17.5

The corresponding decay data are

Time	Ozone residual, mg/L
0	4.02
4	2.58
7	1.72
10	1.28

Solution

1. Plot the bench scale steady-state data and determine the immediate ozone demand.
   a. The required plot is given below.

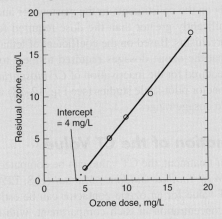

b. Determine the immediate ozone demand.

The immediate ozone demand corresponds to the intercept on the x-axis. From the above plot the value is equal to 4 mg/L.

2. Plot the bench scale steady-state decay data and determine an appropriate first-order equation.

a. The required plot is given below.

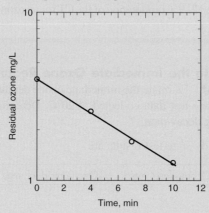

b. The corresponding first-order equation is

$$C_{\text{residual ozone}} = (4.0 \text{ mg/L})e^{-0.116t}$$

where t = contact time, min

Comment The existence of an immediate ozone demand is the reason why the first compartment where ozone is added is generally not considered in establishing the CT value for an ozone contactor. The CT value can be determined from the decay curve by considering the reactor as a whole or by considering each compartment individually.

Required Ozone Dosages for Disinfection

The required ozone dosage for disinfection can be estimated by considering (1) the initial ozone demand, based on the results of a bench scale test as illustrated above, and (2) the corresponding decay curve. Computation of the CT value for an existing ozone reactor is illustrated in Example 12–10. The ozone dosage required to meet the initial demand depends on the constituents in the wastewater and is site specific and, in most cases, will be significantly greater than the dose required for disinfection of the coliform group of microorganisms. Based on the coefficient of lethality, Λ, values given in Table 12–11, it is clear that the ozone dosages required to meet total coliform standards are a fraction of those required for the inactivation of *Cryptosporidium* and *Giardia lamblia*. In most cases, bench and/or pilot-scale studies (see Fig. 12–28) will need to be conducted to establish the required dosage ranges.

Estimation of the CT Value

In water treatment, the CT value can be computed in four different ways as defined in the *LT2ESWTR Toolbox Guidance Manual* (U.S. EPA, 2010). The t_{10} approach is as follows. The CT value for an ozone contactor can be estimated as the summation of the average ozone concentration in each compartment, with the exception of the first compartment, times the detention time in each compartment. The time in each compartment is based on the t_{10} time (see Table 12–19) measured across all of the compartments divided by the proportional volume of each compartment. The first compartment is omitted, as noted above, as it is used to meet the immediate ozone demand and does not contribute to disinfection. Additional details may be found in the *LT2ESWTR Toolbox Guidance Manual* (U.S. EPA, 2010).

EXAMPLE 12–10 **Estimate the CT Value for an Ozone Contactor and Corresponding Log Reduction in Cryptosporidium** Estimate the CT value for the ozone contactor shown below and the log-reduction of *Cryptosporidium* that can be achieved. If the log reduction is greater than 2, estimate the number of reaction compartments that would be needed to achieve a 2-log reduction. The detention time of each compartment of the ozone contactor is 3 min. Assume the decay curve developed in Example 12–9 is applicable and that the t_{10}/t ratio is 0.6. The observed ozone concentration at the end of the first compartment is 4 mg/L.

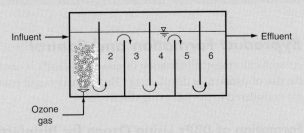

Solution

1. Using the decay curve from Example 12–9, determine the residual ozone concentration at the end of each compartment starting with compartment 2.

Compartment no.	Ozone residual, mg/L
2	2.82[a]
3	1.99
4	1.41
5	0.99
6	0.88

[a] $C = (4 \text{ mg/L})e^{-0.116 \times t}$

$\quad = (4 \text{ mg/L})e^{-0.116 \times 3} = 2.82 \text{ mg/L}$

2. Using the data from Step 1, determine the CT value for the ozone contactor, noting that the theoretical detention time in each reactor is 3 min and the t_{10}/t ratio is 0.6.

$$CT = \sum_{i=2}^{b} C_i T_i = [(2.82 + 1.99 + 1.41 + 0.99 + 0.88) \text{ mg/L}](3 \text{ min} \times 0.6)$$

$$= 14.6 \text{ mg·min/L}$$

3. Estimate the log reduction that can be achieved for *Cryptosporidium* using the information given in Table 12–5.
   From Table 12–5, the estimated log reduction that can be achieved is 3 plus logs.

4. Check the log reduction that can be achieved using Eq. 12–6.
    a. The coefficient of specific lethality for cryptosporidium from Table 12–11 is 0.256 L/mg·min (base 10).
    b. The log reduction using Eq. (12–6) is

$$\log \frac{N_t}{N_o} = -\Lambda_{\text{base}10}\text{CT} = (0.256\ \text{L/mg·min})(13.7\ \text{mg·min/L}) = 3.74$$

5. Estimate the number of compartments that would be needed to achieve a 2-log reduction in *Cryptosporidium*.
    a. Assuming only one reaction compartment is needed, determine the CT value.
    CT = (2.82 mg/L)(3 min × 0.6) = 5.1 < 8.25 (from Table 12–5), hence an additional compartment is needed.
    b. Assuming two reaction compartments are needed, determine the CT value.
    CT = [(2.82 + 1.99) mg/L](3 min × 0.6) = 8.6 > 8.25, hence two compartments should be used.

Byproduct Formation and Control

As with chlorine, the formation of unwanted DBPs is one of the problems associated with the use of ozone as a disinfectant. The formation and control of DBPs when using ozone are considered in the following discussion.

Formation of DBPs Using Ozone for Disinfection.   One advantage of ozone is that it does not form chlorinated DBPs such as THMs and HAAs (see Table 12–14). Ozone does, however, form other DBPs (see Table 12–24) including aldehydes, various acids, and aldo- and keto-acids when significant amounts of bromide are not present. In the presence of bromide, the following DBPs may also be produced: bromoform, brominated acetic acid, bromopicrin, brominated acetonitriles, cyanogen bromide, and bromate (see Table 12–24). On occasion, hydrogen peroxide can also be generated. The specific amounts and the relative distribution of compounds depend on the nature of the precursor compounds that are present. Because the chemical characteristics of wastewater vary from location to location, pilot testing will be required to assess the effectiveness of ozone as a disinfectant.

Control of DBP Formation Using Ozone for Disinfection.   Because the nonbrominated compounds appear to be readily biodegradable, they can be removed by passage through a biologically active filter, by soil application, or by other biologically active processes. The removal of inorganic, brominated DBPs is more complex. Bench and pilot-scale testing is recommended to determine if brominated DBPs will be problematic. If it is expected that brominated DBPs will remain problematic, it may be appropriate to investigate an alternative means of disinfection such as by UV irradiation.

Environmental Impacts of Using Ozone

It has been reported that ozone residuals can be acutely toxic to aquatic life (Ward and DeGraeve, 1976). Several investigators have reported that ozonation can produce some toxic mutagenic and/or carcinogenic compounds. These compounds are usually unstable, however, and are present only for a matter of minutes in the ozonated water. White (1999) has reported that ozone destroys certain harmful refractory organic substances, such as humic

Table 12–24

Representative disinfection byproducts resulting from the ozonation of wastewater containing organic and selected inorganic constituents[a]

Class	Representative compounds
Acids	Acetic acids
	Formic acid
	Oxalic acid
	Succinic acid
Aldehydes	Acetaldehyde
	Formaldehyde
	Glyoxal
	Methyl glyoxal
Aldo- and ketoacids	Pyruvic acid
Brominated byproducts[b]	Bromate ion
	Bromoform
	Brominated acetic acids
	Bromopicrin
	Brominated acetonitriles
	Cyanogen bromide
Other	Hydrogen peroxide

[a] Adapted, in part, from U.S. EPA (1999a, 2002).

[b] The bromide ion must be present to form brominated byproducts.

acid (precursor of trihalomethane formation) and malathion. Whether toxic intermediates are formed during ozonation depends on the ozone dose, the contact time, and the nature of the precursor compounds. White (1999) has also reported that ozone treatment ahead of chlorination for disinfection purposes reduces the likelihood for the formation of THMs.

Ozone residual quenching is still required to meet OSHA indoor and outdoor ambient air quality standards. Ozone quenching of the off-gas is also required prevention or limit the corrosion of downstream piping and equipment. Where required, hydrogen peroxide, sodium bisulfite, and calcium thiosulfate have been used to quench residual ozone.

Other Benefits of Using Ozone

An additional benefit associated with the use of ozone for disinfection is that the dissolved oxygen concentration of the effluent will be elevated to near saturation levels as ozone rapidly decomposes to oxygen after application. The increase in oxygen concentration may eliminate the need for reaeration of the effluent to meet required dissolved oxygen water quality standards.

Ozone Disinfection Systems Components

A complete ozone disinfection system, as illustrated on Fig. 12–29, is comprised of the following components: (1) facilities for the preparation of the feed gas, (2) power supply, (3) the ozone generation facilities, (4) two alternative types of facilities for contacting the ozone with the liquid to be disinfected (inline or sidestream), and (5) facilities for the destruction of ozone in the off-gas (Rice, 1996; Rakness, 2005). Additional details on the design of ozone systems and related components may be found in a recently published book by Rakness (2005).

Figure 12-29

Schematic flow diagram for complete ozone disinfection system with alternative air sources. (Adapted from U.S. EPA, 1986.)

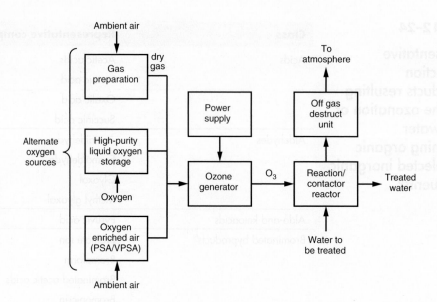

Preparation of Feed Gas.   Ozone can be generated using air, high purity oxygen, or oxygen enriched air. If air is used for ozone generation, it must conditioned by removing the moisture and particulate matter before being introduced into the ozone generator. The following steps are involved in conditioning the air: (1) gas compression, (2) air cooling and drying, and (3) air filtration. If high purity oxygen is used, the conditioning steps are not required. The liquid oxygen (LOX) supply is stored onsite and is either generated onsite or trucked in as needed. In the oxygen enriched air system, high purity oxygen is generated onsite with a vacuum pressure swing adsorption (AVPAS) system or pressure swing adsorption (PSA) system for smaller treatment plants. Both oxygen generation systems have facilities for adsorbing moisture, which can damage the ozone generator dielectrics, and for the removal of hydrocarbons and nitrogen to enhance the purity of the oxygen. The choice of feed gas is influenced by the local cost of high purity oxygen.

Power Supply.   The major requirement for power is for the production of ozone from oxygen. Additional power is required for preparation of the feed gas, contacting the ozone, destroying the residual ozone, and for the controls, instrumentation, and monitoring facilities. The energy requirements for the major components are reported in Table 12–25.

Table 12-25

Typical energy requirements for the application of ozone

Component	kWh/lb ozone	kWh/kg ozone
Air preparation (compressor and dryers)	2–3	4.4–6.6
Ozone generation		
Air feed	6–9	13.2–19.8
Pure oxygen	3–6	6.6–13.2
Ozone contacting	1–3	2.2–6.6
All other uses	0.5–1	1.2–2.2

Figure 12–30

Schematic detail of the generation of ozone. (Adapted from U.S. EPA, 1986.)

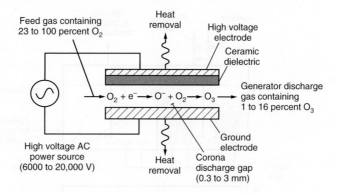

Ozone Generation.

Because ozone is chemically unstable, it decomposes to oxygen very rapidly after generation, and thus must be generated onsite. The most efficient method of producing ozone today is by electrical discharge. Ozone is generated either from air or high purity oxygen when a high voltage is applied across the gap of narrowly spaced electrodes (see Fig. 12–30). The high energy corona created by this arrangement dissociates one oxygen molecule, which reforms with two other oxygen molecules to create two ozone molecules. The gas stream generated by this process from air will contain about 1 to 3 percent ozone by weight, and from pure oxygen about 8 to 12 percent ozone. Ozone concentrations up to 12 percent can now be generated with the latest medium frequency ozone generators.

Inline Ozone Contact/Reaction Reactors.

The concentration of ozone generated from either air or pure oxygen is so low that the transfer efficiency to the liquid phase is an extremely important economic consideration. To optimize ozone dissolution, deep and covered contact chambers are normally used. Two four-compartment ozone contact reactors are shown schematically on Fig. 12–31 without and with chimneys. The chimneys shown on Fig. 12–31(b) are used to enhance the countercurrent flow within the reactor. The chimneys also provide locations for ozone residual sampling.

Ozone is introduced by means of porous diffusers or injectors into the bottom of the first and second, and in some cases, the third chamber. Fast ozone reactions occur in the first chamber. The combined water-ozone mixture then enters the second chamber where slower reactions occur. Disinfection generally occurs in the second chamber. The third and fourth chambers are used to complete the slow reactions and to allow the ozone to decompose. The first and second chambers are identified as the *reaction* chambers. The third and fourth chambers, without ozone addition, are known as the *contact* chambers. The number of chambers used will depend on the treatment objectives.

Sidestream Ozone Contact/Reaction System.

With the ability to generate higher concentrations of ozone (e.g., 10 to 12 percent), sidestream injection of ozone (see Fig. 12–32) is now a viable alternative to the use of porous diffusers in deep tanks as described above. As shown on Fig. 12–32, the ozone injection system is independent of the ozone contactor. The ozone is injected under pressure through a Venturi injector. Two sidestream configurations are used: (1) one with the inclusion of a degas vessel and (2) one without. The purpose of the degas vessel is to minimize the DO level in the water which has been ozonated and (2) to minimize the number of gas bubbles in the downstream pipe which serves as a reactor. The pipeline into which the ozonated water is injected also serves as a reactor prior to the discharge into the contactor (Rakness, 2005).

Figure 12–31

Schematic of typical four compartment ozone contactors: (a) without chimneys and (b) with chimneys. The chimneys in (b) are used to enhance the counter current flow through the reactor.

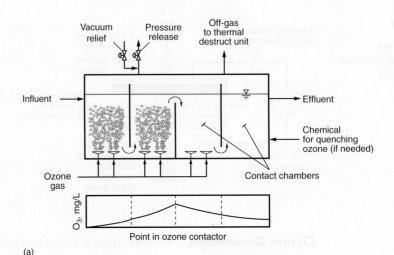

(a)

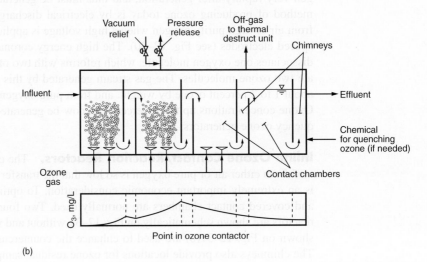

(b)

Destruction of Off-Gases. The off-gases from the contact chamber and the degas vessel must be treated to destroy any remaining ozone as it is an extremely irritating and toxic gas. Off-gas is destroyed to a concentration of <0.1 ppm$_v$. The product formed by destruction of the remaining ozone is pure oxygen which can be recycled if pure oxygen is being used to generate the ozone.

12–8 OTHER CHEMICAL DISINFECTION METHODS

Because of the concerns over the effectiveness of disinfection processes and concern over the formation of DBPs, ongoing research is continuing into the evaluation of alternative disinfection methods. The use of peracetic acid, peroxone, sequential chlorination, and combined disinfection processes are introduced and considered briefly in this section. Pasteurization, a physical method, is considered in Sec. 12–10. Because research on these and other disinfection methods is ongoing, current conference proceedings and literature must be consulted for the latest findings.

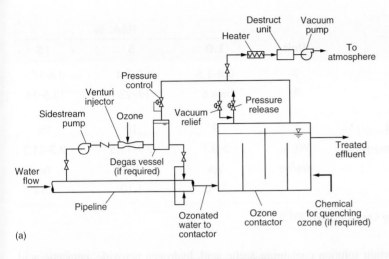

(a)

(b)

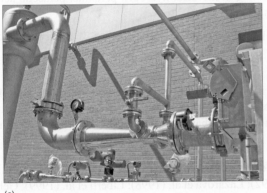

(c)

(d)

Figure 12–32

Sidestream ozone injection for disinfection: (a) typical schematic for sidestream injection system (adapted from Rakness, 2005), (b) view of degas vessel (venturi injector located on back right), (c) Venturi injector used in conjunction with degas vessel shown in (b) (photos (b) and (c) courtesy of Glenn Hunter, Process Applications, Inc.), and (d) view of sidestream injection system located above enclosed ozone contactor, including venturi injectors (left side), degas vessels (center), and destruct units (right).

Peracetic Acid

In the late 1980s, the use of peracetic acid (PAA) was proposed as a wastewater disinfectant. Peracetic acid, made up of acetic acid and hydrogen peroxide, has been used for many years as a disinfectant and sterilizing agent in hospitals. Peracetic acid is also used as a bactericide and fungicide, especially in food processing. Interest in the use of PAA as a wastewater disinfectant arises from considerations of safety and the possibility that its use will not result in the formation of DBPs. The use of PAA is considered briefly in this section as an example of the continuing search for alternative disinfectants to replace chlorine.

Peracetic Acid Chemistry and Properties. Commercially available PAA, also known as ethaneperoxide acid, peroxyacetic acid, or acetyl hydroxide, is only available as a

Table 12–26

Properties of various peracetic acid (PAA) formulations[a]

Property	Unit	PAA, %		
		1.0	**5**	**15**
Weight PAA	%	0.8–1.5	4.5–5.4	14–17
Weight hydrogen peroxide	%	min 6	19–22	13.5–16
Weight acetic acid	%	9	10	28
Weight available oxygen	Wt, %	3–3.1	9.9–11.5	9.3–11.1
Stabilizers	Yes/no	Yes	Yes	Yes
Specific gravity		1.10	1.10	1.12

[a] Adapted from Solvay Chemicals, Inc. (2013).

quaternary equilibrium solution containing acetic acid, hydrogen peroxide, peracetic acid, and water. The pertinent reaction is as follows.

$$CH_3CO_2H + H_2O_2 \rightleftarrows CH_3CO_3H + H_2O \tag{12–63}$$

Acetic   Hydrogen   Peracetic
acid   peroxide   acid

The undissociated PAA (CH_3CO_3H) is considered to be the biocidal form in the equilibrium mixture, however, the hydrogen peroxide may also contribute to the disinfection process. Hydrogen peroxide is also more stable than PAA. The properties of PAA are summarized in Table 12–26.

Effectiveness of Peracetic Acid as a Disinfectant. The effectiveness of PAA has been studied by Lefevre et al. (1992); Lazarova et al. (1998); Liberti et al. (1999); Gehr (2000, 2006); Koivunen (2005b); and Gehr et al. (2003), among others. A recent review was published by Kitis (2004). The findings to date are mixed concerning the bactericidal effectiveness of PAA, as well as the impact of wastewater characteristics on the effectiveness of PAA, especially when used alone. When combined with UV the effectiveness of PAA appears to be enhanced significantly (see discussion of combined disinfectants presented below). It has been hypothesized that the principal means by which disinfection is accomplished by PAA may be by the release of hydroxyl radicals (HO•) and the active oxygen resulting from secondary reactions (Caretti and Lubello, 2003). The current literature must be consulted for more information on the application of PAA.

In a report by the U.S. EPA (1999b), PAA was included among a total of 5 possible disinfectants for use on combined sewer overflows (CSOs). Based on data for disinfection of secondary treatment plant effluents, it was suggested that PAA be strongly considered for CSO disinfection. Among the desirable attributes listed are absence of persistent residuals and byproducts, not affected by pH, short contact time, and high effectiveness as a bactericide and viricide.

Formation of Disinfection Byproducts. Based on the limited data available, the principal end products identified were CH_3COOH (acetic acid or vinegar), O_2, CH_4, CO_2, and H_2O, none of which are considered toxic in the concentrations typically encountered.

Use of Peroxone as a Disinfectant

Peroxone is the combination of ozone (or ultraviolet light) with hydrogen peroxide. When hydrogen peroxide is added with ozone, it will quench ozone residual. In reacting with dissolved ozone (or reacting with photons of light from a UV process) it becomes an advanced oxidation process (AOP), which is characterized by the prolific generation of hydroxyl radicals (OH·). These OH· are stronger and less selective than ozone or other oxidants in regard to destroying target synthetic or naturally occurring micropollutants and microorganisms. The use of peroxone as an advanced oxidation process is discussed in Chap. 6.

Sequential Chlorination

Developed by the Sanitation Districts of Los Angeles County, *sequential chlorination* is a two-step disinfection process. In the first step, chlorine is added to nitrified filtered effluent to produce a free chlorine residual (FCR). As noted previously, free chorine, because of its high germicidal effectiveness, rapidly inactivates both bacteria and viruses. It also reacts with NDMA precursors to make them less available for subsequent NDMA formation. In the second step, ammonia and additional chlorine, if needed, are added to form chloramines, which provide additional bacterial and viral disinfection and minimize the formation of THMs. At the lowest free chlorine CT values tested (2 to 4 mg·min/L), it was possible to achieve an average of more than 6-log MS2 bacteriophage inactivation. With respect to DBP formation, the levels of THMs increased while the corresponding levels of NDMA decreased as compared to conventional chloramination. The sequential chlorination process has been developed to provide an alternative to the prescriptive CT value of 450 mg·min/L at a minimum modal contact time of 90 min required in the CDPH reclamation criteria [CCR, Section 60301.230(a)] (Maguin et al., 2009; Friess et al., 2013).

Combined Chemical Disinfection Processes

Interest in the sequential or simultaneous use of two or more disinfectants has increased within the last few years, especially in the water supply field. Reasons for the increased interest in the use of multiple disinfectants include (U.S. EPA, 1999a):

- The use of less reactive disinfectants, such as chloramines, has proven to be quite effective in reducing the formation of DBPs, and more effective for controlling biofilms in the distribution system.

- Regulatory and consumer pressure to produce water that has been disinfected to achieve high levels of inactivation for various pathogens has forced both the water and wastewater industry to search for more effective disinfectants. To meet more stringent disinfection standards, higher disinfectant doses have been used which, unfortunately, have resulted in the production of increased levels of DBPs.

- Based on the results of recent research it has been shown that the application of sequential disinfectants is more effective than the additive effect of the individual disinfectants. When two (or more) disinfectants are used to produce a synergistic effect by either simultaneous or sequential application to achieve more effective pathogen inactivation, the process is referred to as interactive disinfection (U.S. EPA, 1999a).

Currently, extensive research is being conducted on these processes. Some examples for the use of combined and/or sequential application of disinfectants are presented in Table 12–27. Because the application of multiple disinfectants is, at present, site specific, depending on the microorganism, the disinfection technologies employed, and other non-disinfection process objectives, the current literature must be reviewed to assess the suitability and effectiveness of combined disinfection technologies. For example, in Australia, combined use of chlorination and UV is becoming the norm in water-reuse applications.

Table 12-27

Effectiveness of combined disinfectants and processes for water and wastewater treatment[a,b]

Combined disinfectants	Response	Reference
In water treatment		
Ozone (O_3), UV, and chloramines replaced chlorination	Increase in CT credits by as much as 3 to 5 logs	Malley (2005)
Sequential sonification and chlorine	Increase in effectiveness over use of sonification or chlorine alone	Plummer and Long (2005)
Ozone and free chlorine, ozone and monochloramine, chlorine dioxide and free chlorine, chlorine dioxide, chlorine, and monochloramine	Synergistic response observed in the inactivation of *C. parvum* oocysts	Li et al. (2001)
Sequential UV and chlorine for inactivation of adenoviruses	Increase in effectiveness over use of UV or chlorine alone	Sirikanchana et al. (2005)
In wastewater treatment		
Free chlorine and combined chlorine	Reduced CT times for the inactivation of virus	Maguin et al. (2009), Friess et al. (2013)
Peracetic acid (PAA) and UV	Increase in effectiveness over use of UV or PAA alone	Chen et al. (2005), Lubello et al. (2002)
PAA and UV and PAA and ozone	Increase in effectiveness over use of PAA and UV alone	Caretti and Lubello (2005)
PAA and hydrogen peroxide (H_2O_2), H_2O_2 and UV, and H_2O_2 and O_3	No improved effectiveness	Caretti and Lubello (2005), Lubello et al. (2002)
Ozone, PAA, H_2O_2 and copper (Cu)	PAA and H_2O_2 alone had no effect, addition of 1 mg/L Cu had a dramatic effect	Orta de Velasque et al. (2005)
PAA/UV and H_2O_2/UV	PAA/UV had synergistic effects, whereas H_2O_2/UV did not	Koivunen (2005a)
Ultrasound and UV	Increase in effectiveness over use of UV alone	Blume et al. (2002) see also Blume and Neis (2004)

[a] Adapted in part from Gehr (2006).

[b] Additional combinations are reviewed in U.S. EPA (1999a).

12-9 ULTRAVIOLET (UV) RADIATION DISINFECTION

The germicidal properties of ultraviolet (UV) light sources have been used in a wide variety of applications since the use of UV was pioneered in the early 1900s, having been discovered first in the 1880s. First used on high quality water supplies, the use of UV light as a wastewater disinfectant evolved during the 1990s with the development of new lamps, ballasts, and ancillary equipment. With the proper combination of UV dose and water quality, UV irradiation has proven to be an effective disinfectant for bacteria, protozoa, and viruses in both unfiltered and filtered secondary effluent, while not contributing to the formation of toxic byproducts. In several cases, UV has even been proven to be effective at disinfecting primary effluents. To develop an understanding of the application of UV for the disinfection of wastewater the following topics are considered in this section: (1) source of UV radiation, (2) UV system configurations, (3) the germicidal effectiveness of UV irradiation, (4) modeling the UV disinfection process, (5) estimating the UV dose,

Figure 12–33

Definition sketch for ultraviolet (UV) radiation disinfection: (a) identification of the ultraviolet radiation portion of the electromagnetic spectrum, (b) identification of the germicidal portion of the UV radiation spectrum, and (c) UV radiation spectra for both low-pressure low-intensity and medium-pressure high-intensity UV lamps and the relative UV absorption for DNA superimposed over spectra of the UV lamps.

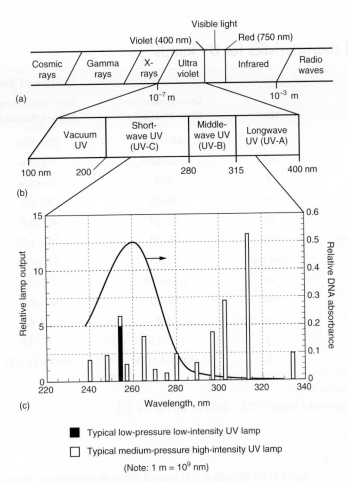

■ Typical low-pressure low-intensity UV lamp

□ Typical medium-pressure high-intensity UV lamp

(Note: 1 m = 10^9 nm)

(6) UV disinfection guidelines, (7) analysis of a UV disinfection system, (8) operational issues with UV systems, and (9) the environmental impacts of disinfection with UV radiation.

Source of UV Radiation

The portion of the electromagnetic spectrum in which UV radiation occurs, is between 100 and 400 nm [see Fig. 12–33(a)]. The UV radiation range is characterized further according to wavelength as longwave (UV-A), also known as near-ultraviolet irradiation, middlewave (UV-B), and shortwave (UV-C), also known as far UV. The germicidal portion of the UV radiation band is between about 220 and 320 nm, principally in the UV-C range. The UV wavelengths between 255 to 265 nm are considered to be most effective for microbial inactivation [see Fig. 12–33(c)]. Most commonly, UV radiation is produced by striking an electric arc between two electrodes in specially designed lamps containing liquid mercury, as well as other gas mixtures. The energy generated by the excitation of the liquid mercury causes it to vaporize. Mercury in its gaseous form excites electrons in the lamps thus producing photons of UV light.

When used for water and wastewater disinfection, quartz sleeves are most often used to isolate the UV lamps from direct water contact and to control the lamp wall temperature by buffering the effluent temperature extremes to which the UV lamps are exposed, thereby maintaining a fairly uniform UV lamp output. In another less common configuration, the

Table 12–28

Typical operational characteristics for UV lamps

Item	Unit	Type of lamp		
		Low-pressure low-intensity	Low-pressure high-intensity	Medium-pressure high-intensity
Power consumption	W	40–100	200–500ᵃ	1000–13,000
Lamp current	mA	350–550	Variable	Variable
Lamp voltage	V	220	Variable	Variable
Germicidal output/input	%	30–50	35–50	15–20ᵇ
Lamp output at 254 nm	W	25–27	60–400	100–2000
Lamp operating temperature	°C	35–50	100–150	600–800
Pressure	mm Hg	0.007	0.01–0.8	10^2–10^4
Lamp length	m	0.75–1.5	1.8–2.5	0.3–1.2
Lamp diameter	mm	15–20	Variable	Variable
Sleeve life	y	4–6	4–6	1–3
Ballast life	y	10–15	10–15	3–5
Estimated lamp life	h	8000–12,000	9000–15,000	3000–8000

ᵃ Up to 1200 W in very high output lamp.
ᵇ Output in the most effective germicidal range (~255 – 265 μm, see Fig. 12–33).

water to be disinfected is passed through proprietary plastic tubes that are themselves surrounded by UV lamps. The output of UV disinfection systems also decreases with time due
to a reduction in the electron pool within the UV lamp, deterioration of the electrodes, and
the aging of the quartz sleeve. Lamps with other gas mixtures and without electrodes, as
described below, are also used to generate UV light.

Types of UV Lamps

The principal electrode-type lamps used to produce UV light fall into three categories
based on the lamp's internal operating parameters: *low-pressure low-intensity, low-
pressure high-intensity,* and *medium-pressure high-intensity* systems. Comparative
information on the operational characteristics of these three types of UV lamps is presented in Table 12–28. In the brief discussion of these types of UV lamps presented
below, it is important to note that UV lamp technology is changing rapidly. It is, therefore, imperative that current manufacturers' literature be consulted when designing a UV
disinfection facility. The ballasts used in conjunction with UV lamps are also discussed
briefly.

Low-Pressure Low-Intensity UV Lamps.   Low-pressure low-intensity mercury-
argon electrode type UV lamps [see Fig. 12–34(a)] are used to generate a broad spectrum
of essentially monochromatic radiation in the UV-C region with an intense peak at a wavelength of 253.7 nm (essentially 254 nm) and a lesser peak at about 184.9 nm. The peak at
254 nm is close to the 260 nm wavelength considered to be most effective for microbial

Figure 12–34

Typical examples of UV lamps:
(a) low-pressure low-intensity,
(b) medium-pressure high-intensity
lamps with cleaning device, and
(c) schematic illustration of the
electrode-less microwave driven
UV lamp [see Fig. 12–36(d)].

(a)

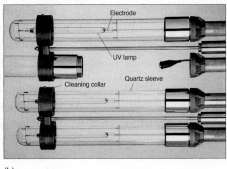

(b)

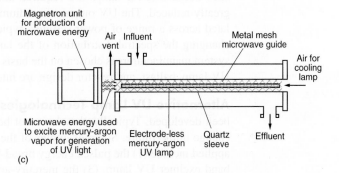

(c)

inactivation. Compared to power input, approximately 30 to 50 percent of the lamp's UV energy output is monochromatic at 254 nm, making it an efficient choice for disinfection processes. Also, approximately 85 to 88 percent of the lamp's output is monochromatic at 254 nm, making it an efficient choice for disinfection processes. Because there is an excess of liquid mercury in the low-pressure low-intensity UV lamp, the mercury vapor pressure is controlled by the coolest part of the lamp wall. If the lamp wall does not remain relatively near the optimum temperature of 40°C, some of the mercury will condense back to its liquid state thereby decreasing the number of mercury atoms available to release photons of UV; hence UV output declines.

Low-Pressure High-Intensity UV Lamps. Low-pressure high-intensity UV lamps are similar to the low-pressure low-intensity lamps [see Fig.12–34(a)] with the exception that a mercury-indium amalgam is used in place of liquid mercury. Use of the mercury amalgam allows greater UV-C output, typically from 2 to 10 times the output of conventional low-intensity lamps. Similar to low-pressure low-intensity lamps, low-pressure high-intensity lamps are very efficient at converting lamp input power into UV light. Compared to power input, approximately 35 to 50 percent of the lamp's UV energy output is monochromatic at 254 nm. Low-pressure high-intensity UV lamps operate at temperatures of 100 – 150°C. The amalgam in the low-pressure high-intensity UV lamps is used to maintain a constant level of mercury atoms, and, thus, provides greater stability over a broad temperature range. The UV output of low-pressure high-intensity lamps can be modulated between 30 and 100 percent. The range of modulation varies between different lamps. Current manufacturer's literature should be reviewed for lamp specifications as new low-pressure high-intensity lamps are being developed continuously.

Medium-Pressure High-Intensity UV Lamps. A number of medium-pressure high-intensity UV lamps have been developed over the last decade. Medium-pressure high-intensity UV lamps, which operate at temperatures of 600 to 800°C and vapor pressures of 10^2 to 10^4 mm Hg, generate polychromatic irradiation [see Fig. 12–33(c)]. Medium-pressure high-intensity UV lamps [see Fig. 12–34(b)] generate approximately 20 to 50 times the total UV-C output of low-pressure high-intensity UV lamps. Although the UV output of medium-pressure high-intensity lamps is significantly higher compared to low-pressure low-and high-intensity lamps, their efficiency is much lower. Compared to power input, only 15 to 20 percent of the lamp UV energy output is within the germicidal UV range. The use of medium-pressure high-intensity lamps is limited primarily to potable water supplies, large wastewater facilities, stormwater overflows, or on space-limited sites because fewer lamps are required and the footprint of the disinfection system is greatly reduced. The UV output of medium-pressure high-intensity lamps can be modulated across a range of power settings (typically 30 to 100 percent) without significantly changing the spectral distribution of the lamp. The particular UV lamp selected by UV system manufacturers is chosen on the basis of an integrated design approach in which the UV lamp, ballast, and reactor design are interdependent.

Alternative UV Lamp Technologies. A number of alternative technologies have been developed. Typically, they have not been used at the municipal scale, but this may change in the future. Some examples of the types of lamps that are being developed and applied include (1) the pulsed energy broad-band xenon lamp (pulsed UV), (2) the narrow band excimer UV lamp, (3) the mercury-argon electrode-less microwave powered high-intensity UV lamp, and (4) UV light emitting diodes (LED) lamp.

The pulsed UV lamp produces polychromatic light at high levels of radiation. It is estimated that the radiation produced by the pulsed UV lamp is 20,000 times as intense as sunlight at sea level. The disinfection effectiveness provided by pulsed UV lamps has been researched in some detail (O'Brien et al., 1996; EPRI, 1996; Mofidi et al., 2001). Narrow band excimer lamps produce essentially monochromatic light in three wavelengths: 172, 222, and 308 nm depending on the gas used in the lamp. Gases that have been used for the purpose include xenon (Xe), xenon chloride (XeCl), krypton (Kr), and krypton chloride (KrCl). In the microwave powered UV lamp, UV light is generated by striking a mercury-argon filled electrode-less UV lamp with microwave energy generated with a magnetron [see Fig. 12–34(c)]. Because the lamp does not contain electrodes, longer lamp life (3 to 5 y) is claimed, though no third-party certification has been completed. Based on preliminary results, it appears that the UV LED lamps currently under development will compete directly with conventional UV technologies. At present, there is no LED lamp technology that can compete with high output UV lamps.

As noted above, because developments in UV technology are occurring at such a rapid pace, it is essential that the current literature be consulted when designing UV disinfection systems. Note that in most cases, emerging technologies do not have a proven track record of cost-effective, reliable performance.

Ballasts for UV Lamps. A ballast is a type of transformer that is used to limit the current to a lamp. Because UV lamps are arc discharge devices similar to fluorescent lamps, the more current in the arc, the lower the resistance becomes. Without a ballast to limit current, the lamp would destroy itself. Thus, matching the lamp and ballast is of critical importance in the design of UV disinfection systems. Three types of ballasts are used: (1) standard (core coil), (2) energy efficient (core coil), and standard electronic (solid-state). In general, electronic ballasts are about 10 percent more energy efficient than magnetic ballasts. Electronic ballasts are now used most commonly for controlling the UV lamps used for disinfection.

UV Disinfection System Configurations

In addition to the type of lamp used, UV systems for the disinfection of wastewater can also be classified according to whether the flow occurs in open or closed channels. Each of these system configurations is described below.

Open Channel Disinfection Systems. The principal components of low-pressure low- and high-intensity open channel UV systems used for the disinfection of wastewater are illustrated on Fig. 12–35. As shown, lamp placement can be horizontal and parallel to the flow [see Fig. 12–35(a)], vertical and perpendicular to the flow [see Fig. 12–35(c)], or inclined (e.g. diagonal) to the flow [see Fig. 12–35(e)]. Each module contains a specified number of UV lamps encased in quartz sleeves. The total number of lamps is specific to each application, but the number of lamps in each module depends on the channel and overall system configuration and lamp manufacturer. The lamp spacing is manufacturer and lamp type specific and can range from 75 mm (3 in.) to 150 mm (6 in.). The inclined lamp UV systems are a relatively recent development. Stated advantages include the use of longer lamps with higher output, which reduces the total required number of lamps; improved system hydraulics and performance; and ease of installation, maintenance, and operation.

Figure 12–35

Isometric cut-away and photographic views of typical open channel UV disinfection systems: (a) horizontal lamp system parallel to flow (adapted from Trojan Technologies, Inc.), (b) view of one UV bank of a horizontal lamp system removed for cleaning, (c) vertical lamp system perpendicular to flow (adapted from Infilco Degremont, Inc.), (d) vertical lamp module removed from channel for cleaning, (e) inclined (45°) lamp system (adapted from Xylem, Inc.), and (f) view of inclined lamp UV system with lamps elevated out of the channel (courtesy of Xylem, Inc.).

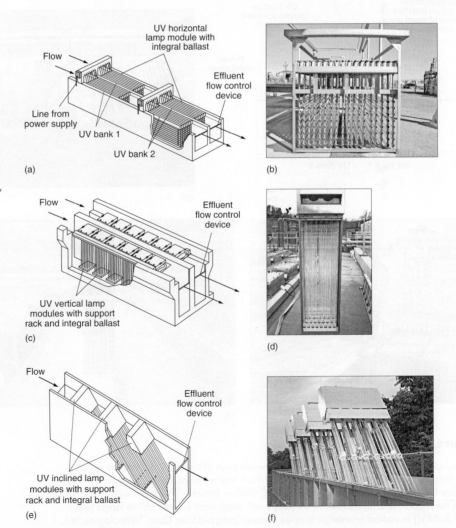

An extended serpentine fixed sharp-crested weir, automatic level controlled adjustable weir, or a weighted flap gate (not recommended) is used to control the depth of flow through each disinfection channel. Proper level control is essential to (1) maintain submergence of the lamps at all times, (2) prevent short circuiting by ensuring that the water level above the top lamp is not too high, and (3) adequately seal the channel to prevent undisinfected water from bleeding through to the effluent channel when a bank of lamps is out of service. An inadequate level control device can often be the cause of poor UV disinfection performance.

Each channel typically contains two or more banks of UV lamps in series, and each bank is comprised of a number of modules (or racks of UV lamps). It is important to note that a standby bank or channel is normally provided for system reliability. The design flowrate is usually divided equally among a number of open channels. Typical examples of horizontal and vertical low-pressure low-high-intensity UV disinfection systems are shown on Fig. 12–35(c) through 12–35(f), respectively. A typical medium-pressure UV disinfection system is shown on Figs. 12–36(a) and 12–36(b). The lamps are arranged in

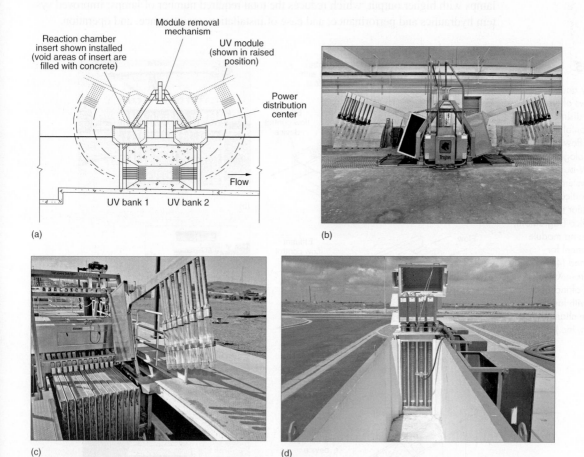

(a)

(b)

(c)

(d)

Figure 12–36

Typical examples of medium-pressure and microwave open channel UV disinfection systems:
(a) schematic view through UV reactor (adapted from Trojan Technologies), (b) typical medium pressure UV system installed in open channel, (c) medium pressure UV system with one lamp module out of the reactor, and (d) microwave UV lamps with magnetron located above lamps [see Fig. 12–34(c)] in vertical orientation in open channel (adapted from Quay Technologies, Ltd).

modules and are positioned in a reactor with a fixed geometry [see Fig. 12–36(c)]. The lamp cleaning sleeves can be seen on Fig. 12–36(c). Vertical mercury-argon electrode-less microwave powered high-intensity UV lamps are shown on Fig. 12–36(d).

Closed Channel Disinfection Systems. A number of low- and medium-pressure high-intensity UV disinfection systems are designed to operate in closed channels or pipes. Two UV system configurations are used. In the first configuration, the direction of flow is perpendicular to the placement of the lamps, as shown on Fig. 12–37(a). In the second configuration, the direction of flow is parallel to the UV lamps [see Fig. 12–37(b)]. Because high-intensity UV lamps operate at a lamp wall temperature of between 600 to 800°C, the UV output of these lamps is unaffected by the effluent temperature. A typical medium-pressure UV disinfection reactor is shown on Figs. 12–37(c) and 12–37(d). A closed system pulsed UV reactor is shown on Fig. 12–37(f).

Figure 12–37

Views of medium-pressure high-intensity closed inline UV disinfection systems: (a) schematic of closed reactor with flow perpendicular to UV lamps, (b) schematic of closed reactor with flow parallel to UV lamps, (c) view through inline UV reactor (courtesy of Trojan Technologies, Inc.) (d) view of installed UV system, (e) close up of small inline UV system with manual cleaning device, and (f) view of pulsed UV reactor.

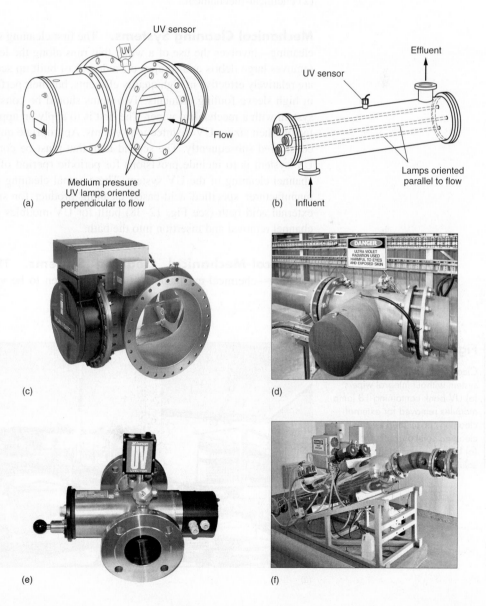

Quartz Sleeve Cleaning Systems

In UV disinfection systems, various physical and chemical characteristics of the water result in fouling of the quartz sleeves that encase each UV lamp. During operation of a UV lamp in wastewater, factors such as interfacial temperature, reactor hydraulics, and the quartz microstructure and topography allow attachment of inorganic debris and organic films or greases onto the protective quartz sleeve surrounding the lamp. These deposits absorb UV light and decrease the intensity of UV light penetration into the wastewater. The decrease in UV intensity leads to a decrease in UV dose resulting in reduction in disinfection performance. Fouling has been found to be complex and is difficult to predict. It also tends to be site-specific, owing mainly to the chemical and biological nature of the liquid matrix being treated by the UV system. To overcome quartz sleeve fouling, the majority of UV systems on the market have in-situ sleeve cleaning systems. These cleaning systems, as discussed below, can be divided into 2 categories: (1) mechanical, and (2) chemical-mechanical.

Mechanical Cleaning Systems. The first cleaning system category—mechanical cleaning—involves the use of a wiper that runs along the length of the quartz sleeve and removes large debris and scrapes off a degree of built up scaling. These types of systems are relatively effective in high quality effluents, but their performance can be compromised in high sleeve fouling effluents. Two items should be considered when designing a UV system with a mechanical wiper. The first is to apply an appropriate quartz sleeve fouling factor when sizing UV disinfection systems. Appropriate quartz sleeve fouling factors are considered subsequently. The second item that must be considered during design of the UV system is to include provisions for periodic (period of time is site specific) out-of-channel cleaning of the UV system. The external cleaning can be done manually with a manufacturer specified acid-based cleaning product for smaller UV systems or in an external acid bath (see Fig. 12–38) built for UV modules and with a crane for module channel removal and insertion into the bath.

Chemical-Mechanical Cleaning Systems. The second cleaning system category—chemical-mechanical—has been proven to be very effective at removing all

Figure 12–38

Cleaning UV disinfection system without integral wipers: (a) UV bank containing 18 lamp modules removed for external cleaning, positioned over cleaning solution bath and (b) UV bank placed in cleaning solution bath.

(a)   (b)

scaling from quartz sleeves and maintaining a near 100 percent UV light output through the life of the system. Chemical-mechanical cleaning systems typically utilize two wipers that contain a small volume of an acidic gel (either phosphoric or citric acid based). This chemical gel remains in a canister surrounding each quartz sleeve and is replaced annually. Although a chemical-mechanical cleaning system has been shown to be more effective than mechanical cleaning alone, a quartz sleeve fouling factor should be applied to UV system sizing in any case.

Mechanism of Inactivation by UV Irradiation

Ultraviolet light is a physical rather than a chemical disinfecting agent. The mechanism of inactivation and photoreactivation are important concepts to understand as they are some of the fundamental principles of UV disinfection.

Inactivation Mechanisms. UV radiation penetrates the cell wall of the microorganism and is absorbed by the nucleic acids (DNA and RNA), which guide the development of all living organisms. Damage to the nucleic acid interferes with normal cell processes such as cell synthesis and cell division. Deoxyribonucleic acid (DNA) controls the structure, while ribonucleic acid (RNA) controls the metabolic processes. Typically, DNA is a double-stranded helical structure with four nucleotides: adenine, guanine, thymine, and cytosine, while RNA is a single-stranded structure with the nucleotides adenine, guanine, uracil, and cytosine.

Exposure to UV radiation damages DNA by manipulating adjacent thymine molecules as illustrated on Fig. 12–39. The process of forming double bonds is known as dimerization. Cytosine-cytosine and cytosine-thymine dimers can also be formed. Thus, organisms rich in thymine such as protozoans *C. parvum* and *G. lamblia* tend to be more sensitive to UV radiation (see Table 12–5) (Mofidi et al., 2001; Mofidi et al., 2002). Uracil and cytosine are the corresponding molecules in RNA. Viruses contain either DNA or RNA, which is either single or double stranded. Adenovirus contains double-stranded DNA, which is considered as a possible explanation for its high sensitivity to UV light (Sommer et al., 2001). Exposure to UV radiation can also cause more severe damage, such as breaking chains, cross-linking DNA with itself, and cross-linking DNA with other proteins. (Crittenden et al., 2005). In general, UV irradiation must form a significant number of bonds or other damage to the cell to be effective, which is the case with the doses delivered in properly sized UV disinfection systems.

Figure 12–39

Formation of double bonds in microorganisms exposed to ultraviolet radiation.

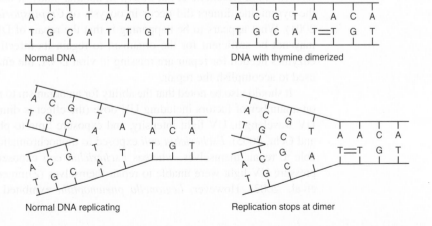

Microbial Growth Phase and Resistance to UV Irradiation. In addition to the mechanisms discussed above, it has been found that the growth phases of organisms can inherently provide protection against disinfectants. As noted in Chap. 7 there are four readily identifiable stages that cells undergo: lag growth phase, exponential growth phase, stationary growth phase, and death phase. The effect that different phases of growth cycle may have on UV susceptibility of *E. coli* has been evaluated by Modifi et al. (2002). It appears that when cell DNA is not actively dividing, bacteria may be more resistant to UV disinfection. Thus, if naturally occurring bacteria are not stressed by environmental factors, they may exhibit a similar spectrum of resistance to UV disinfection. Based on these findings, knowledge of the cell growth stage is of importance in establishing bacterial dose-response relationships.

Microbial Repair Following UV Irradiation. Because some organisms are able to maintain some metabolic activities after being exposed to UV radiation, they may be able to repair the damage caused by the exposure. Many organisms in nature have evolved mechanisms for reversing UV damage. Two different types of mechanisms are involved: (1) photoreactivation and (2) dark repair.

Photoreactivation. Photoreactivation involves specific enzymes that can repair sections of damaged DNA after being energized by exposure to light. The mechanism of photoreactivation, first discovered in 1949 for *Streptomyces griseus* by Kelner (1949) and for bacteriophage by Dulbecco (1949), was demonstrated to be enzyme-catalyzed (Rupert, 1960). The enzyme responsible for DNA repair is named *photolyase*. Photoreactivation can be described as the two-step enzymatic reaction between photolyase and its substrate, pyrimidine dimers (Friedberg et al., 1995). The first step is for photolyase to recognize any dimers (see Fig. 12–39) and specifically bind them to form an enzyme-substrate complex. The first step is light-independent and, therefore, can occur even under dark conditions. The enzyme-dimer complex is stable and goes through the second repair step in which the dimers are broken utilizing the energy of light at wavelengths between 310 and 490 nm. The second step is dependent only on light input.

For example, the *E. coli* photolyase has a round shape with a hole inside, which recognizes and structurally binds to the pyrimidine dimers sticking out from the genome DNA. Once the pyrimidine dimers are repaired (i.e., broken) and the structure is changed, the bind is loosened and the enzyme leaves the dimer (Friedberg et al., 1995). In the case of pathogenic parasites, the effects of photoreactivation are unclear. Based on infectivity studies, it was reported that the oocysts of *Cryptosporidium parvum* did not undergo photoreactivation (Rochelle et al., 2004). In another study, it has been reported that repair of the pyrimidine dimers did occur in oocysts of *Cryptosporidium parvum* (Oguma et al., 2001). What appears to be happening is that the repair of DNA following UV irradiation may not be sufficient for the organism to regain its infectivity. Although the necessary enzymes needed for repair are missing in viral DNA, the enzymes of the host cell can be used to accomplish the repair.

It should also be noted that the ability for an organism to repair itself appears to depend on a number of factors including UV dose (the effect is diminished at higher UV doses), UV wavelength, UV light intensity, and exposure time to photoreactivating light (Martin and Gehr, 2005). *Escherichia coli* exposed to monochromatic low-pressure UV light were able to repair themselves, whereas *Escherichia coli* exposed to polychromatic medium-pressure UV light were unable to repair themselves (Zimmer and Slawson, 2002; Oguma et al., 2002). However, *Legionella pneumophila* exhibited very high photoreactivation ability after exposure to either low-pressure or medium-pressure UV light (Oguma et al., 2004). From a review of some recent published findings, it appears that if effluent that has undergone UV disinfection is subsequently kept in the dark for approximately 3 hours, the regrowth potential is reduced significantly (Martin and Gehr, 2005). Clearly, more research needs to be done to understand what is causing the effect observed with medium-pressure UV light.

Dark Repair. In the early 1960s it was found that UV-induced DNA damage could be repaired without light (Hanawalt et al., 1979). Dark repair appears to be accomplished by two mechanisms: (1) excision repair and (2) recombination repair. In excision repair, enzymes remove the damaged section of DNA, and in recombination repair, the damaged DNA is regenerated using a complementary strand of DNA. Although the necessary enzymes needed for repair are missing in viral DNA, the enzymes of the host cell can be used to accomplish the repair. Contrary to photoreactivation, with high specificity to pyrimidine dimers, dark repair can act on various kinds of damage in the genome. Dark repair is a rather slow process compared to photoreactivation.

Germicidal Effectiveness of UV Irradiation

The overall effectiveness of the UV disinfection process depends on a number of factors including (1) the chemical characteristics of the wastewater to be irradiated, (2) the presence of particles, (3) the characteristics of the microorganisms, and (4) the physical characteristics of the UV disinfection system. Before considering these subjects, it is appropriate to consider the definition of UV dose to provide a frame of reference for the discussion of the factors affecting UV disinfection. The material presented below will also be useful in assessing the modeling of the UV process that is considered subsequently.

Definition of UV Dose. The effectiveness of UV disinfection is based on the UV dose to which the microorganisms are exposed. The UV dose, D, as defined previously, is given by Eq. (12–8), which is repeated here for convenience.

$$D = I_{avg} \times t \tag{12-8}$$

where D = UV dose, mJ/cm^2 (note $mJ/cm^2 = mW \cdot s/cm^2$)
   I_{avg} = average UV intensity, mW/cm^2
   t = exposure time, s

Note that the UV dose term is analogous to the dose term used for chemical disinfectants (i.e., CT). As given by Eq. (12–8), the UV dose can be varied by changing either the intensity or exposure time. Additional details on the measurement of UV dose may be found in Linden and Mofidi (2003) and Jin et al. (2006).

Effect of Chemical Constituents in Wastewater. The constituents in wastewater can have a significant impact on the average UV intensity. The impact is measured in term of absorbance and transmittance. The reduction in UV intensity with distance is defined by Beers-Lambert Law, repeated here for convenience from Chap. 2.

$$\log\left(\frac{I}{I_o}\right) = -\varepsilon(\lambda)Cx = k(\lambda)x = [A(\lambda)/x]x \tag{2-19}$$

where I = light intensity at distance x from the light source, mW/cm^2
   I_o = light intensity at light source, mW/cm^2

$\varepsilon(\lambda)$ = molar absorptivity (also known as the extinction coefficient) of the light-absorbing solute at wavelength λ, L/mole·cm

C = concentration of light-absorbing solute, mole/L

x = light path-length, cm

$k(\lambda)$ = the absorptivity, cm^{-1}

$A(\lambda)$ = absorbance, dimensionless

Although the absorbance, $A(\lambda)$, is dimensionless, it is often reported in units of cm^{-1}, which corresponds to absorptivity $k(\lambda)$. If the length of the light path is 1 cm, absorptivity is equal to the absorbance. In UV practice, it is more common to use transmittance, which is defined as

$$\text{Transmittance, } T, \% = \left(\frac{I}{I_o}\right) \times 100 \qquad (2\text{–}21)$$

Typical absorbance and transmittance values for wastewater after several different treatment processes are presented in Table 12–29.

Dissolved constituents impact UV disinfection either directly via absorbance (increasing absorbance serves to attenuate UV light to a larger degree) or via fouling of UV lamp sleeves such that a reduced intensity is applied to the bulk liquid medium. The effects of constituents found in effluent from different wastewater treatment processes are reported in Table 12–30. One of the most perplexing problems encountered in the application of UV disinfection for wastewater disinfection is the variation typically observed in the absorbance (or transmittance) at treatment plants. Often, the variations in transmittance are caused by industrial discharges, which can lead to diurnal as well as seasonal variations. Common industrial impacts are related to the discharge of inorganic and organic dyes, wastes containing metals, and complex organic compounds.

Of the inorganic compounds that affect transmittance, iron is considered to be the most important with respect to UV light absorbance because dissolved iron can absorb UV light directly. Organic compounds containing double bonds and aromatic functional groups can also absorb UV light. Absorbance values for a variety of compounds found in wastewater are given in Table 12–31. From a review of the information presented in Table 12–31, it is clear that the presence of iron in wastewater can have a significant impact on the use of UV. If iron salts are used within the treatment process, the economic benefits of changing to another chemical (e.g., alum) should be evaluated to determine whether the cost savings of having a smaller UV system are greater than the capital and operating costs of switching to the new chemical. It is also important to note that stormwater inflows can cause wide variations, especially when humic materials from terrestrial sources are present. In general, the solution to the problem of varying transmittance levels may require monitoring of industrial

Table 12–29	Type of wastewater	Absorbance, a.u./cm	Transmittance[a],%
Absorbance and transmittance values for various wastewaters at a wavelength of 254 nm	Primary	0.70 to 0.30	20 to 50
	Secondary	0.35 to 0.15	45 to 70
	Nitrified secondary	0.25 to 0.10	56 to 79
	Filtered secondary	0.25 to 0.10	56 to 79
	Microfiltration	0.10 to 0.04	79 to 91
	Reverse osmosis	0.05 to 0.01	89 to 98

[a] T, % = 10$^{-A(\lambda)}$ × 100.

Table 12–30

Impact of wastewater constituents on the use of UV radiation for wastewater disinfection

Constituent	Effect
BOD, COD, TOC, etc.	No or minor effect, unless humic materials comprise a large portion of the BOD
NOM (natural organic matter)	Strong absorbers of UV radiation
Oil and grease	Can accumulate on quartz sleeves of UV lamps, can absorb UV radiation
TSS	Absorption of UV radiation, can shield embedded bacteria
Alkalinity	Can impact scaling potential. Also affects solubility of metals that may absorb UV light
Hardness	Calcium, magnesium and other salts can form mineral deposits on quartz tubes, especially at elevated temperatures
Ammonia	No or minor effect
Nitrite	No or minor effect
Nitrate	No or minor effect
Iron	Strong absorber of UV radiation, can precipitate on quartz tubes, can become embedded in suspended solids and shield bacteria by absorption
Manganese	Strong absorber of UV radiation
pH	Can affect solubility of metals and carbonates
TDS	Can impact scaling potential and the formation of mineral deposits
Industrial discharges	Depending on the constituents (e.g., dyes), may lead to a diurnal and seasonal variations in the transmittance
Stormwater inflow	Depending on the constituents, may lead to short term as well as seasonal variations in the transmittance

Table 12–31

UV Absorbance of water and common chemicals found in wastewater

Compound	Form or designation	Molar absorption coefficient, L/mole·cm	Threshold concentration, mg/L
Ferric iron	$Fe[III]$	3069	0.057
Ferrous iron	$Fe[II]$	466	9.6
Hypochlorite ion	OCl^-	29.5	8.4
N-nitrosodimethylamine	NDMA	1974	
Nitrate	NO^{3-}	3.4	
Natural organic matter	NOM	80 to 350	
Ozone	O_3	3250	0.071
Zinc	Zn^{2+}	1.7	187
Water	H_2O	6.1×10^{-6}	

Figure 12–40

Particle interactions that effect the effectiveness of UV disinfection including microorganism shading; light scattering, reflection, and refraction; and incomplete penetration.

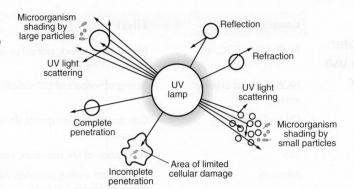

discharges, the implementation of source control programs, and correcting sources of infiltration. In some cases, biological treatment will mitigate the influent variations. In some extreme situations, the conclusion may be that UV disinfection is not practical.

Where the implementation of UV disinfection is being assessed, it is useful to install online transmittance monitoring equipment to document the variations that occur in the transmittance with time. Alternatively, a bench-top photometer may be used to sample UV transmittance manually. Sampling with a photometer will provide "snapshots" of UV transmittance as opposed to constant measurements from an online monitor. If enough "snapshots" are taken, fairly accurate UV transmittance trending data can be determined, which may be equal in accuracy to the online monitoring data.

Effect of Particles. The presence of particles in the wastewater to be irradiated can also impact the effectiveness of UV disinfection (Qualls et al., 1983; Parker and Darby, 1995; Emerick et al., 1999). The manner in which particles can affect UV performance is illustrated on Fig. 12–40. Many organisms of interest in wastewater (e.g., coliform bacteria) occur both in a dispersed state (i.e., not bound to other objects) and a particle-associated state (i.e., bound to other objects such as other bacteria or cellular debris). Coliform bacteria are of particular importance because of the central role they play in discharge permits [i.e., coliform bacteria are used as indicators for the presence of other pathogenic organisms (see Chap. 2) and their inactivation is assumed to correlate with the inactivation of other pathogenic organisms]. Dispersed coliform bacteria are inactivated readily because they are exposed fully to the average UV light intensity as compared to particle-embedded microorganisms (see Fig. 12–40). Treatment process related disinfection problems, when disinfecting unfiltered effluent, usually result from the influence of particle associated organisms (see also Fig. 12–5). In fact, coliform bacteria can associate with particles to such a degree that they are completely shielded from UV light resulting in a residual coliform bacteria concentration post UV irradiation.

It has been observed in activated sludge effluents that a minimum particle size (on the order of 10 μm) governs the ability to shield coliform bacteria from UV light (Emerick et al., 2000). Due to the inherent porous nature of activated sludge particles, particles smaller than that critical size are unable to reduce the applied intensity and thus embedded organisms are inactivated in a manner similar to dispersed organisms. Particles greater than the critical size can either reduce the applied UV intensity, leading to a reduced inactivation rate for organisms associated with the particle, or shield coliform bacteria. Particle size does not appear to be a governing factor once the critical size is exceeded because coliform bacteria are located randomly within particles, are not typically located in the most shielded regions within particles, and common enumeration techniques typically exclude the larger sized particles.

Table 12–32

Typical UV dosages required to achieve different effluent total coliform disinfection standards for various wastewaters

Type of wastewater	Initial coliform count, MPN/100 mL	UV dose, mJ/cm²			
		Effluent standard, MPN/100 mL			
		1000	200	23	≤2.2
Raw wastewater	10^7–10^9	20–50			
Primary effluent	10^7–10^9	20–50			
Trickling filter effluent	10^5–10^6	20–35	25–40	40–60	90–110
Activated sludge effluent	10^5–10^6	20–30	25–40	40–60	90–110
Filtered activated sludge effluent	10^4–10^6	20–30	25–40	40–60	80–100
Nitrified effluent	10^4–10^6	20–30	25–40	40–60	80–100
Filtered nitified effluent	10^4–10^6	20–30	25–40	40–60	80–100
Microfiltration effluent	10^1–10^3	5–10	10–15	15–30	40–50
Reverse osmosis	~0	—	—	—	5–10
Septic tank effluent	10^7–10^9	20–40	25–50		
Intermittent sand filter effluent	10^2–10^4	10–20	15–25	25–35	50–60

Characteristics of the Microorganisms.   The effectiveness of the UV disinfection process depends on the characteristics of the microorganisms as well as the microorganism group. Typical values for the disinfection of coliform organisms with UV light for various wastewaters are reported in Table 12–32. It should be noted that the dosage values given in Table 12–32 are only meant to serve as a guide for the initial estimation of the required UV dose. The range of the reported values reflects the variable nature of wastewater. The relative effectiveness of UV irradiation for disinfection of representative microorganisms of concern is in wastewater is reported in Table 12–33. As with the values given in Table 12–5, the values given in Table 12–33 are only meant to serve as a guide in assessing the relative UV dose required for different microorganisms. Knowledge concerning the required UV dose for specific pathogen inactivation is changing continuously as improved methods of analysis are applied. For example, before infectivity studies were conducted, it was thought that UV irradiation at reasonable dosage values (i.e., less than 200 mJ/cm²) was not effective for the inactivation of *Cryptosporidium parvum* and *Giardia lamblia*. However, based on infectivity studies, it has been found that both of these protozoans are inactivated with extremely low UV dosage values (typically in the range of 5 to 15 mJ/cm²) (Linden et al., 2001; Mofidi et al., 2001; Mofidi et al., 2002). The current literature should be consulted to obtain the most contemporary information regarding required UV dosages for the inactivation of specific microorganisms.

Impact of System Characteristics.   Problems with the application of Eq. (12–8) for use in the design of UV disinfection reactors are associated with (1) inaccurate knowledge of the average UV intensity and (2) the exposure time associated with all of the pathogens passing through a UV disinfection system. In practice, field-scale UV disinfection

Table 12–33

Estimated relative effectiveness of UV radiation for the disinfection of representative microorganisms of concern in wastewater

Organism	Dosage relative to total coliform dosage
Bacteria	
Escherichia coli (E.coli)	0.6–0.8
Fecal coliform	0.9–1.0
Pseudomonas aeruginosa	1.5–2.0
Salmonella typhosa	0.8–1.0
Streptococcus fecalis	1.3–1.4
Total coliform	1.0
Vibrio cholerae	0.8–0.9
Viruses	
Adenovirus	6–8
Coxsackie A2	1.2–1.2
MS-2 bacteriophage	2.2–2.4
Polio type 1	1.0–1.1
Rotavirus SA 11	1.4–1.6
Protozoa	
Acanthamoeba castellanii	6–8
Cryptosporidium parvum	0.4–0.5
Cryptosporidium parvum oocysts	1.3–1.5
Giardia lamblia	0.3–0.4
Giardia lamblia cysts	0.3–0.4

ᵃ Relative doses based on discrete non clumped single organisms in suspension. If the organisms are clumped or particle associated, the relative dosages have no meaning.

reactors have dose distributions resulting from both the internal intensity profiles and exposure time distribution. The internal intensity profiles are a reflection of the nonhomogeneous placement of lamps within the system, lack of ideal radial mixing within the system, the scattering/absorbing effects of particulate material, and the absorbance of the liquid medium. The distribution associated with exposure time is a reflection of non-ideal hydraulics leading to longitudinal mixing.

One of the most serious problems encountered with UV disinfection systems in both open and closed channel systems is achieving a uniform velocity field in the approach and exit to and from the UV banks. Achieving a uniform velocity field can be especially difficult when UV systems are retrofitted into existing open channels, such as converted chlorine contact basins. A second equally serious problem with UV system hydraulics is even flow distribution between channels whether the channels are new or are a retrofit of existing structures. An uneven flow split can lead to overdosing in one channel and most importantly, underdosing in the other(s) therefore compromising disinfection performance. Ensuring ideal flow distribution to multiple channels and uniform velocities within those channels is critical. To optimize the hydraulic performance of UV disinfection systems, computational fluid dynamics (CFD) modeling should be considered.

Estimating UV Dose

The first step in assessing the performance of a UV disinfection system is to determine the UV dose needed to inactivate the challenge microorganism to a level prescribed by the treatment plant discharge permit and/or is protective of public health in water reuse applications. Three methods have been used to estimate the UV dose. In the first method, an average UV dose is determined by assuming an average system UV intensity and exposure time. The average UV intensity is estimated using a computational procedure known as the point source summation (PSS) method (U.S. EPA, 1992). Over the past decade, the PSS method has been used less frequently by designers due to its failure to account for system-specific hydraulics (i.e., ideal hydraulic behavior is assumed in the PSS that never occurs in field-scale disinfection systems). At present, this method should not be used to determine UV dose.

The second method involves the use of CFD to integrate both the distribution of UV intensities and velocity profiles within the reactor to obtain a distribution of UV doses within a system (Batchley et al., 1995). Although the CFD method is promising, its use is limited at the present time (2013) because (1) the methodology is not standardized, (2) the methodology has been unable to predict disinfection performance adequately, and (3) the reporting of a distribution of UV doses, even if accurate, is problematic for UV disinfection system specification. In the third, and most widely used method, the UV dose is determined using a collimated beam bioassay. Use of the bioassay approach in designing UV disinfection systems is discussed below.

Determination of UV Dose by Collimated Beam Bioassay.

The most common and industry accepted procedure for determining the required UV dose for the inactivation of challenge microorganisms involves the use of a collimated beam and a small reactor (i.e., a Petri dish) to which a known UV dose is applied. Typical collimated beam devices are shown on Fig. 12–41. Use of a monochromatic low-pressure low-intensity

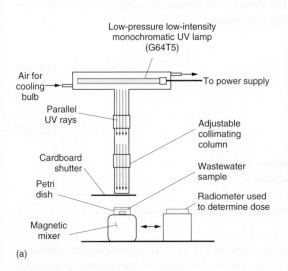

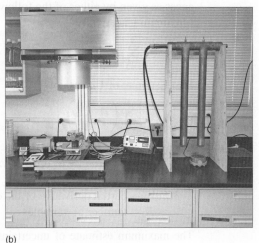

(a)   (b)

Figure 12–41

Collimated beam device used to develop dose-response curves for UV disinfection: (a) schematic and (b) view of two different types of collimated beam devices. The collimated beam on the left is of European design; the collimated beam on the right is of the type shown schematically in (a).

lamp in the collimated beam apparatus allows for accurate characterization of the applied UV intensity. Use of a batch reactor allows for accurate determination of exposure time. The applied UV dose, as defined by Eq. (12–8), can be controlled by simply varying the exposure time and maintaining a constant, known UV intensity. Because the geometry is fixed, the depth-averaged UV intensity within the Petri dish sample (i.e., the batch reactor) can be computed using the following relationship.

$$D = I_o t(1-R)P_f \left[\frac{(1-10^{-k_{254}d})}{2.303(k_{254}d)}\right]\left(\frac{L}{L+d}\right) \tag{12–64}$$

$$D = I_o t(1-R)P_f \left[\frac{(1-e^{-2.303 k_{254}d})}{2.303(k_{254}d)}\right]\left(\frac{L}{L+d}\right) \tag{12–65}$$

where D = average collimated beam UV dose ($I_o \times t$), mJ/cm^2

$\quad I_o$ = incident UV intensity averaged over the surface of the sample before and after irradiating sample, mW/cm^2

$\quad t$ = exposure time, s

$\quad R$ = reflectance at the air water interface at 254 nm

$\quad P_f$ = Petri dish factor

$\quad k_{254}$ = absorbance of sample, absorptivity, a.u./cm (base 10)

$\quad d$ = depth of sample, cm

$\quad L$ = distance from lamp centerline to liquid surface, cm

The term $(1-R)$ on the right side of Eq. (12–64) accounts for the reflectance at the air water interface. The value of R is typically about 2.5 percent. The term P_f accounts for the fact that the UV intensity may not be uniform over the entire area of the Petri dish. The value of P_f is typically greater than 0.9. The term within the brackets is the depth averaged UV intensity within the Petri dish and is based on the Beers-Lambert Law (see Example 2–5, Chap. 2). The final term is a correction factor for the height of the UV light source above the sample. The application of Eq. (12–64) is illustrated in Example 12–11.

The uncertainty of the computed UV dose can be estimated using the sum of the variances as given by either of the following expressions:

Maximum Uncertainty

$$U_D = \pm \sum_{n=1}^{N}\left|U_{V_n}\frac{\partial D}{\partial V_n}\right| \tag{12–66}$$

Best Estimate of Uncertainty

$$U_D = \pm \left[\sum_{n=1}^{N}\left(U_{V_n}\frac{\partial D}{\partial V_n}\right)^2\right]^{1/2} \tag{12–67}$$

where U_D = uncertainty of UV dose value, mJ/cm^2

$\quad U_{V_n}$ = uncertainty or error in variable n

$\quad V_n$ = variable n

$\quad \partial D/\partial V_n$ = partial derivative of the expression with respect to the variable V_n

$\quad N$ = number of variables

The maximum estimate of uncertainty as given by Eq. (12–66) represents the condition where every error will be a maximum value. The best estimate of uncertainty, as given by Eq. (12–67), is used most commonly because it is unlikely that every error will be a maximum at the same time and the fact that some errors may cancel each other. The application of Eq. (12–67) is illustrated in Example 12–11. Knowledge of the average UV intensity and exposure time allows calculation of the average applied UV dose using Eq. (12–8). The UV dose is then correlated to the microorganism inactivation results as discussed below.

EXAMPLE 12–11   **Determination of UV Dose Delivered in Collimated Beam Test** The following measurements were made to establish the UV dose using a collimated beam. Using these data determine the average UV dose delivered to the sample and best estimate of the uncertainty associated with the measurement.

$I_o = 5 \pm 0.35$ mW/cm^2 (accuracy of meter $\pm 7\%$)

$t = 60 \pm 1$ s

$R = 0.025$ (assumed to be the correct value)

$P_f = 0.94 \pm 0.02$

$k_{254} = 0.065 \pm 0.005$ cm^{-1}

$d = 1 \pm 0.05$ cm

$L = 40 \pm 0.5$ cm

Solution

1. Using Eq. (12–64) estimate the UV dose delivered by the collimated beam.

$$D = I_o t (1 - R) P_f \left[\frac{(1 - 10^{-k_{254}d})}{2.303(k_{254}d)} \right] \left(\frac{L}{L + d} \right)$$

$$= (5 \times 60)(1 - 0.025)(0.94) P_f \left[\frac{(1 - 10^{-0.065 \times 1})}{2.303(0.065 \times 1)} \right] \left(\frac{40}{40 + 1} \right)$$

$$= (300)\,(0.975)\,(0.94)\,(0.928)\,(0.976) = 249 \text{ mJ/cm}^2$$

2. Determine the best estimate of uncertainty for the computed UV dose. The uncertainty of the computed dose can be estimated using Eq. (12–67). The procedure is illustrated for one of the variables and summarized for the remaining variables.

    a. Find the variability in the measured UV dose due to the variability of the measured time t. The partial derivative of the expression used in step one with respect to t is

$$U_t \frac{\partial D}{\partial t} = U_t \left\{ I_o (1 - R) P_f \left[\frac{(1 - 10^{-k_{254}d})}{2.303(k_{254}d)} \right] \left(\frac{L}{L + d} \right) \right\}$$

$$U_t \frac{\partial D}{\partial t} = (1) \left\{ 5(1 - 0.025)(0.94) \left[\frac{(1 - 10^{-0.065 \times 1})}{2.303(0.065 \times 1)} \right] \left(\frac{40}{40 + 1} \right) \right\}$$

$$= 4.15 \text{ mJ/cm}^2$$

$$U_{D,t} = \pm \left[\left(U_t \frac{\partial D}{\partial t} \right)^2 \right]^{1/2} = \pm[(4.15 \text{ mJ/cm}^2)^2]^{1/2} = \pm 4.15 \text{ mJ/cm}^2$$

Percent $= 100 \, U_{D,t}/D = 100(4.15/249) = 1.67\%$

    b. Similarly, for the remaining variables, the corresponding values of the partial derivatives are as given below:

$U_{D,I_o} = 17.44$ mJ/cm^2 and 7.0%

$U_{D,P_f} = 5.30$ mJ/cm^2 and 2.13%

$U_{D,k_{254}} = 1.40$ mJ/cm^2 and 0.56%

$D_{D,d} = 1.21$ mJ/cm^2 and 0.49%

$U_{D,L} = 0.076$ mJ/cm^2 and 0.03%

c. The best estimate of uncertainty using Eq. (12–67) is

$$U_D = \pm [(4.15)^2 + (17.44)^2 + (5.30)^2 + (1.40)^2 + (1.21)^2 + (0.076)^2]^{1/2}$$

$$= \pm 18.8 \text{ mJ/cm}^2$$

$$\text{Percent} = (100 \times 18.8)/249.0 = 7.55 \text{ percent}$$

3. Based on the above uncertainty computation the most likely UV dose is

$$D = 249 \pm 19 \text{ mJ/cm}^2$$

Comment   Based on the best estimate of uncertainty, the most conservative estimate of the UV dose that can be delivered consistently, based on the collimated beam test, is 230 mJ/cm^2 $(249 - 19)$. The maximum uncertainty would correspond to the summation of the individual errors and would be ± 30 mJ/cm^2.

Bioassay Testing.   To assess the degree of microbial inactivation that can be achieved at a given UV dose, the concentration of microorganism is determined before and after exposure in a collimate beam (see Fig. 12–41). Microorganisms inactivation is measured using the most probable number (MPN) procedure or the membrane filtration test for bacteria, a plaque count procedure for viruses, or an animal infectivity procedure for protozoa. To verify the accuracy of the laboratory collimated beam dose-response test data, the collimated beam test must be repeated to obtain statistical significance. To be assured that stock solution of the challenge microorganisms is mono-dispersed, the laboratory inactivation test data must fall within an accepted set of quality control limits. Quality control limits proposed by the National Water Research Institute (NWRI, 2003) and the U.S. EPA (2003b) for Bacteriophage MS2 are as follows.

NWRI

Upper bound: $-\log_{10}(N/N_o) = 0.040 \times D + 0.64$ (12–68a)

Lower bound: $-\log_{10}(N/N_o) = 0.033 \times D + 0.20$ (12–68b)

U.S. EPA

Upper bound: $-\log_{10}(N/N_o) = -9.6 \times 10^{-5} \times D^2 + 4.5 \times 10^{-2} \times D$ (12–69a)

Lower bound: $-\log_{10}(N/N_o) = -1.4 \times 10^{-4} \times D^2 + 7.6 \times 10^{-2} \times D$ (12–69b)

where D = UV dose, mJ/cm^2

As will be illustrated in Example 12–12, the bounds proposed by the U.S. EPA are more lenient as compared to those used by NWRI. Similar bounding curves have been proposed for *B. subtilus* (U.S. EPA, 2003b). The NWRI guidelines are used for water reuse applications.

EXAMPLE 12–12   Verification of Laboratory Procedures for Bacteriophage MS2 Response   The following collimated beam test results were obtained for a stock solution of bacteriophage MS2 which is to be used to test a UV reactor. These results are used to verify that the laboratory test results are acceptable and define the dose response equation.

Dose, mJ/cm²	Surviving concentration, phage/mL	Log survival,[a] log (phage/mL)	Log inactivation
0	1.00E + 07	7.000	0.000
20	1.12E + 06	6.049	0.951[b]
40	7.41 + 04	4.870	2.130
60	1.95E + 04	4.290	2.710
80	4.37E + 03	3.640	3.360
100	1.02E + 03	3.009	3.991
120	7.08E + 01	1.850	5.150

[a] The rule followed in the log transformation of a number is to retain in the mantissa the same number of significant figures as in the number that is being transformed.

[b] Sample calculation: Log inactivation = 7.000 − 6.049 = 0.951.

Solution

1. Plot the collimated beam test results and compare to the quality control range expressions provided in the NWRI [Eqs. 12–68(a) and 12–68(b)] and U.S. EPA [Eqs. 12–69(a) and 12–69(b)] UV Guidelines. The results are plotted on the figure given below.

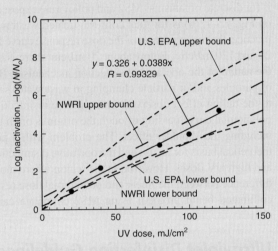

2. As shown in the above plot all of data points fall within the acceptable range as defined by both NWRI and U.S. EPA.
3. Define the dose response relationship. Based on a linear regression analysis, the UV dose response relationship is

$$\text{UV dose} = \frac{\log \text{inactivation} - 0.326}{0.0389}$$

Comment

In general, when conducting bioassay testing, the initial concentration of MS2 should be 2-log higher than the number of logs of inactivation to be achieved. Irradiated samples should be diluted so that number of plaque forming units per plate is between 20 and 200 (NWRI, 2012).

Reporting and Using Bioassay Collimated Beam Test Results.

The results of collimated beam bioassays are reported in the form of a dose response curve as developed in Example 12–12 and shown on Fig. 12–42. The inactivation curve shown on Fig. 12–42(a)

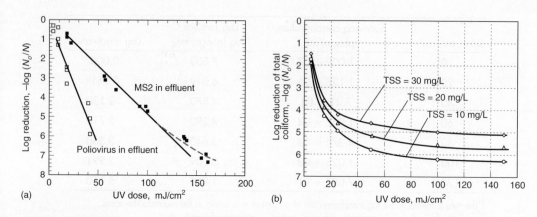

Figure 12–42

Typical dose response curves for UV disinfection developed from data obtained using a collimated beam device: (a) for dispersed microorganisms (Cooper et al., 2000) and (b) wastewater containing varying concentrations of TSS.

is for discrete organisms (MS2 and poliovirus) exposed to UV light, whereas the curve shown on Fig. 12–42(b) is for total coliform treated wastewater containing particulate material. In practice, the linear portion of the dose response curve for MS2 coliphage typically is between 20 and 120 mJ/cm². Below about 20 mJ/cm² there is uncertainty in the measurements and in the nature of the operative disinfection mechanism. Beyond about 120 mJ/cm² the presence of particles and/or particle clumping in wastewater samples causes a shoulder effect similar to the tailing effect observed with chlorine disinfection (see Fig. 12–6). In the literature, a polynomial curve passing through the origin is often used to fit all of the dose response data including the shoulder effect. The problem with a polynomial curve fit is that there is no theoretical basis for its use and the operative disinfection mechanisms are not the same at low and high UV doses. However, in the region where most UV reactors are tested, there is little difference between the linear and polynomial dose response curves. Additional details on the collimated beam protocol using MS2 coliphage can be found in the NWRI Guidelines (2012).

Ultraviolet Disinfection Guidelines

The National Water Research Institute and the American Water Works Association Research Foundation published "Ultraviolet Disinfection Guidelines for Drinking Water and Wastewater Reclamation" (NWRI, 1993; NWRI and AWWARF, 2000; NWRI, 2003; NWRI, 2012). The following elements are considered in the UV guidelines: (1) reactor design, (2) reliability design, (3) monitoring and alarm design, (4) the field commissioning test, (5) performance monitoring, and (6) an engineering report for unrestricted effluent reuse applications. Some of the items may not be applicable when utilizing UV disinfection for less demanding applications.

The guidelines that cover reclaimed water are similar to those that cover drinking water systems. The primary difference is that recommended (or mandatory) doses are provided for reclaimed water systems, whereas there is no mention of recommended doses for non-reclaimed wastewater applications. For reclaimed water systems, the recommended design UV doses for various effluents are 100 mJ/cm² for media filtration or equivalent effluent, 80 mJ/cm² for membrane filtration effluent, and 50 mJ/cm² for reverse osmosis effluent. The different dose requirements reflect the different virus density concentrations expected within each type of treatment process effluent. For example, the dosage of 100 mJ/cm² for media filtration effluent is intended to provide 5 logs of poliovirus inactivation with a factor of safety of about 2.

In addition to differing dose recommendations as a function of effluent quality, there are differing design transmittance recommendations. For granular medium and other types of filtration, microfiltration, and reverse osmosis effluents, the design transmittance's are 55, 65, and 90 percent, respectively. The differing transmittance values are based on field observations made to date, though site-specific variation does occur and should be accounted for. All UV disinfection systems installed for either drinking water or unrestricted reuse applications must undergo validation testing prior to their installation. Although the guidelines do not apply to the disinfection of non reclaimed wastewater, the general design issues addressed are applicable. The IUVA Manufacturer's Council has published a "low dose" bioassay approach (IUVA, 2011).

Relationship of UV Guidelines to UV System Design

The design of a UV disinfection system involves a number of issues including (1) determination of the UV dose required, based on bioassay testing, for adequate inactivation of the challenge (target) microorganism(s), (2) selection of manufacturer-specific validated UV disinfection reactors or systems, (3) determination of process operational parameters and UV system configuration (e.g., the number of lamps per module, modules per bank, banks per channel, and the overall number of channels) and, in some circumstances, (4) conduct of a spot-check bioassay test on the full-scale system to check compliance with the required UV design performance. For reclaimed wastewater applications, the first issue is addressed directly in the UV guidelines as discussed above. For the majority of applications which are not disinfecting for reclaiming water, an appropriate dose must be selected. Guidance on dose selection is provided in Table 12–33. Beyond this guidance, collimated beam tests should be performed on the actual wastewater to determine an appropriate dose to use for UV system design. A final resource for dose selection is information from UV equipment manufacturers, who maintain an extensive databases detailing dose requirements for varying disinfection limits, varying solids contents, and varying plant processes.

The general procedure for validating a UV reactor and some important guidance on design aspects are also included in the guidelines. Because of their fundamental importance in understanding the application of UV disinfection systems, these issues are discussed in the text and illustrated in the examples that follow.

Validation of UV Reactor or System Performance

Validation testing consists of quantifying the level of inactivation of a virus surrogate (e.g., Bacteriophage MS2) by the UV disinfection reactor or system as a function of a number of process variables such as flowrate, transmittance, sensor settings, water level (where appropriate), and power settings. To quantify the inactivation achieved through the UV disinfection system, the UV dose response of the challenge microorganism to be used is determined using a collimated beam illustrated on Fig. 12–43. The inactivation observed through the UV disinfection reactor or system is compared to the UV dose response to establish a term called *reduction equivalent dose* (RED) or *delivered dose*, which corresponds to the UV dose delivered by the UV disinfection system. It should be noted that the RED is specific to the challenge organism and the test conditions.

In the past, validation testing was done once a UV system was installed and operational. To avoid unnecessary testing and the risk that an installed system does not perform adequately, validation testing is now typically completed by UV equipment manufacturers at test centers in the United States or at selected treatment facilities around the world. The manufacturers then provide design engineers with design information on which to base the

Figure 12–43

Schematic illustration of the application of biodosimetry as used to determine the performance of a test or full scale VU reactor.

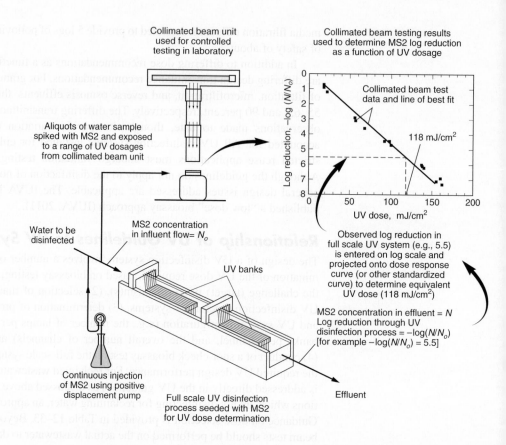

Collimated beam unit used for controlled testing in laboratory

Aliquots of water sample spiked with MS2 and exposed to a range of UV dosages from collimated beam unit

Collimated beam testing results used to determine MS2 log reduction as a function of UV dosage

Collimated beam test data and line of best fit

118 mJ/cm^2

Log reduction, $-\log(N/N_o)$

UV dose, mJ/cm^2

Water to be disinfected

MS2 concentration in influent flow= N_o

UV banks

Observed log reduction in full scale UV system (e.g., 5.5) is entered on log scale and projected onto dose response curve (or other standardized curve) to determine equivalent UV dose (118 mJ/cm^2)

MS2 concentration in effluent = N
Log reduction through UV disinfection process = $-\log(N/N_o)$
[for example $-\log(N/N_o) = 5.5$]

Continuous injection of MS2 using positive displacement pump

Full scale UV disinfection process seeded with MS2 for UV dose determination

Effluent

design of a full-scale installation. The process flow diagram used for testing both open and closed UV reactors is illustrated on Fig. 12–44. In general, validation testing of UV disinfection equipment, using the setup shown on Fig. 12–44(a), consists of the following steps:

1. Selection of representative test water for use in the validation testing of the disinfection system.

2. Selection of the configuration of the UV disinfection system to be tested (i.e., 1, 2, 3, etc. UV banks in series). If the power to the UV lamps cannot be turned down to simulate the end of life lamp performance for a portion of the testing, then aged UV lamps must be used in the test.

3. Hydraulic performance testing of the UV disinfection system is done to verify the uniformity of the approach and exit velocities.

4. Quantification of the inactivation of the test organism (e.g., MS2) through the UV test reactor [see Fig. 12–44(b)] as a function of hydraulic loading rate and other variables.

5. Simultaneous with the field testing, a collimated beam test is conducted on the test water to determine the inactivation response of the viral test organism as a function of applied UV dose. The laboratory test data must fall within the area bounded by Eqs. (12–68a) and (12–68b) or Eqs. (12–69a) and (12–69b) given previously.

6. Assign UV doses to the pilot reactor or system based on the standardized dose response relationship (see NWRI, 2012). In the past, the dose response relationship developed from the collimated beam test was used.

7. Based on the assigned UV dose and the operative control parameters, manufacturers will develop design equations for the test reactor or system.

The steps required in conducting a validation test are illustrated in Example 12–13.

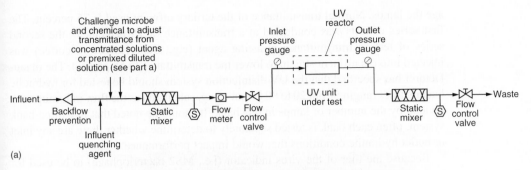

(b)

Figure 12–44

Validation testing of UV reactors: (a) schematic of the experimental test setup and (b) view of large closed UV reactor undergoing validation testing.

EXAMPLE 12–13 **Analysis of Pilot Test Results Used to Validate Performance of UV Reactor or Disinfection System** A manufacturer has supplied a pilot scale UV disinfection system whose performance is to be validated as a function of lamp hydraulic loading rate and water quality only. Other important variables such as power variation and water level variation are not included in this example. Operational curves are to be based on flowrate alone and flowrate and transmittance. For this test, the manufacturer chose to make use of a four-lamp per bank pilot facility with three banks in series to achieve the total applied dose. Each bank of lamps is hydraulically independent of subsequent banks. The engineer and owner are interested in knowing the range of flowrates and water quality over which the UV system can deliver a UV dose of 80 mJ/cm², before any design correction factors are applied. Assume the MS2 UV dose response curve given in Example 12–12 will be used for the analysis of the test results. The test program and the results of the field tests are as follows.

Solution
1. Develop test program.
   The testing was conducted on tertiary effluent from a local water reclamation facility. New lamps were placed in the pilot facility, as it would take more than a year to

age the lamps. Normal transmittance of the tertiary effluent used is 75 percent. The first series of tests was conducted at a transmittance of 75 percent. In the second series of tests a transmittance reducing agent (e.g., SuperHume® or coffee) was injected into the effluent stream to lower the transmittance to 55 percent. The manufacturer has specified that the UV disinfection system should be tested for hydraulic loading rates ranging from 20 to 80 L/min·lamp, calculated as the flow in L/min·bank divided by the number of lamps in one bank. It should be noted that in a three-bank system, often each bank is tested separately to determine whether there are any inlet or outlet hydraulic conditions that would impact performance.

Because the titer of the virus indicator (i.e., MS2 bacteriophage) to be used for performance testing was approximately 1×10^{11} phage/mL, it was decided to test the system under the conditions outlined in the following table.

Hydraulic loading rate, L/min·lamp (1)	Flowrate, L/min·bank (2)	Virus titer Concentration, phage/mL (3)	Virus titer injection flowrate, L/min (4)	Approximate resulting virus concentration in process flow, phage/mL (5)
20	80	1E+11	0.008	1E+7
40	160	1E+11	0.016	1E+7
60	240	1E+11	0.024	1E+7
80	320	1E+11	0.032	1E+7

Notes on column entries:

(1) Desired range to be tested as specified by the manufacturer.

(2) The pilot system contained three banks with 12 lamps total; however, the hydraulic loading rate is only based upon the flowrate through one bank, which makes the calculation more similar to a velocity determination. Thus, at a hydraulic loading rate of 20 L/min·lamp, the process flowrate is equal to 80 L/min·bank [(20 L/min·lamp)(4 lamps/bank)].

(3) Provided by the laboratory.

(4) It was desired to obtain a virus titer in the process flow of about 1×10^7 phage/mL. Therefore, at 80 L/min, the solution containing the virus had to be injected at a rate of 0.008 L/min to obtain the desired initial titer.

2. Test results at 75 percent transmittance.

In conducting the test, each flowrate was tested randomly with respect to order. Three distinct replicate samples were collected per flowrate. An inlet and outlet sample (i.e., that contained the concentration of phage prior to any inactivation) was collected with each process replicate.

a. The inlet test results at 75 percent transmittance are as follows:

Flowrate, L/min·lamp	Replicate	Inlet concentration, phage/mL	Log-transformed inlet conc., log(phage/mL)[a]	Average log-transformed inlet conc., log(phage/mL)
20	1	5.25E+06	6.720	
20	2	1.00E+07	7.000	6.927
20	3	1.15E+07	7.061	
40	1	1.00E+07	7.000	
40	2	1.23E+07	7.090	7.067
40	3	1.29E+07	7.111	

(continued)

(*Continued*)

Flowrate, L/min·lamp	Replicate	Inlet concentration, phage/mL	Log-transformed inlet conc., log(phage/mL)[a]	Average log-transformed inlet conc., log(phage/mL)
60	1	1.23E+07	7.090	
60	2	1.05E+07	7.021	7.030
60	3	9.55E+06	6.980	
80	1	1.23E+07	7.090	
80	2	1.20E+07	7.079	7.023
80	3	7.94E+06	6.900	

[a] The rule followed in the log transformation of a number is to retain in the mantissa the same number of significant figures as in the number that is being transformed.

b. The outlet test results at 75 percent transmittance, based on triplicate samples, are as follows. Only the average log-transformed outlet concentration values from the 75 percent transmittance test are given. The procedure followed in obtaining these values was the same as illustrated above for the inlet test results.

Flowrate, L/min·lamp	Number of banks	Average log-transformed outlet conc., log(phage/mL)
20	2[a]	2.233
40	3	1.832
60	3	3.232
80	3	3.591

[a] Notice that at the low flowrate investigated (20 L/min·lamp), only two operational banks were investigated rather than 3. Only two banks were tested because three operational banks resulted in no detectable viruses in the effluent. Because the banks were hydraulically independent, it is allowed under the UV Guidelines to investigate the inactivation for only two banks and extrapolate to performance expected for additional banks of lamps.

3. Test results at 55 percent transmittance.
   For the purposes of this example, assume the average log-transformed inlet concentration values from the 75 percent transmittance test apply to the 55 percent transmittance test. Only the average of the triplicate log-transformed outlet concentration values from the 55 percent transmittance test are given. The procedure followed in obtaining these values was the same as illustrated above for the 75 percent test.

Flowrate, L/min·lamp	Number of banks	Average log-transformed outlet conc., log(phage/mL)
20	3	1.703
40	3	3.987
60	3	4.662
80	3	4.997

4. Using the test data and the given information develop the necessary UV regression equation for 75 percent transmittance based on **flowrate** only.

   a. Set up computation table to determine the UV dose based on the test results. Using the measured phage data determine the corresponding UV dose based on the log-linear regression expression developed in Example 12–12.

Flowrate, L/min·lamp	Average phage concentration, log(phage/mL)			Assigned UV dose, mJ/cm^2	Log transformed values	
	Inlet	Outlet	Diff.		Flowrate	UV dose
20	6.927	2.233	7.041[a]	172.6[b]	1.301	2.237
40	7.067	1.832	5.235	126.2	1.602	2.101
60	7.03	3.232	3.798	89.3	1.778	1.951
80	7.023	3.591	3.432	79.8	1.903	1.902

   [a] The inactivation for this flowrate was extrapolated from the two-bank results. Because the system is a three-bank system, the inactivation for three banks is 150 percent greater than the inactivation observed with two operational banks [7.041 = (6.927 − 2.233) × 1.5].

   [b] Sample calculation. Using the linear regression expression derived from the collimated beam test in Example 12–12, the equivalent UV dose at a flowrate of 20 L/min·lamp is:

   $$\text{UV dose} = \frac{\log \text{ inactivation} - 0.326}{0.0389}$$

   $$\text{UV dose, mJ/cm}^2 = \frac{7.041 - 0.326}{0.0389} = 172.6$$

   b. Develop the UV operational design equation.

   i. Use a linear regression analysis to develop a regression equation based on water flowrate. Other equations are possible, depending on the control strategy (e.g., flowrate and transmittance, flowrate, transmittance, and power setting).

   ii. To complete a regression analysis, the flowrate and UV dose data must first be log transformed. The data are log transformed to develop a linear relationship that can be used with the linear dose response curve developed using the collimated beam (see Example 12–12). The log transformed data are presented in columns 6 and 7 in the table developed in Step 4.

   iii. Using the UV dose (column 7) as the dependent variable and the flowrate (column 6) as the independent variable, the following results are obtained using the linear regression analysis program in Excel or other statistical analysis program.

   Model Parameters

Source	Value
Intercept	2.997
X1	−0.577

   iv. The equation for UV dose as a function of flowrate, based on the regression analysis, is

   $$\log (\text{UV dose}) = 2.997 - 0.577 \text{ (log flowrate) or}$$

   $$\text{UV dose, mJ/cm}^2 = (10^{2.997})[(\text{flowrate})^{-0.577}]$$

   where the unit for flowrate is L/min·lamp

   Note: The above dose equation is the delivered dose based upon three UV banks operating in series and with a UV transmittance (UVT) of 75 percent.

If the test had been conducted with one UV bank, the UV dose for two or three banks would be obtained by multiplying the regression equation for one UV bank by 2 or 3, respectively.

c. Plot the regression equation for UV dose versus the UV lamp hydraulic loading rate based on the results of the single variable (i.e., flowrate) linear regression analysis.

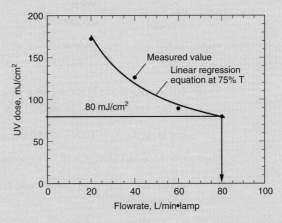

d. Determine the range of flowrates over which the UV disinfection system will deliver a UV dose of 80 mJ/cm². From the plot given above, the range of flows is up to 80 L/min·lamp.

5. Using the test data and the given information develop the necessary UV design curve based on **flowrate and transmittance**.

a. Set up computation table to determine the UV dose based on the test results. Using the measured phage data determine the corresponding UV dose based on the linear regression expression developed in Example 12–12.

Flowrate, L/min·lamp	T, %	Average phage concentration, log(phage/mL) Inlet	Average phage concentration, log(phage/mL) Outlet	Average phage concentration, log(phage/mL) Diff.	Assigned UV dose, mJ/cm²	Log transformed values Flowrate	Log transformed values Transmittance	Log transformed values UV dose
20	75	6.927	2.233	7.041	172.6	1.301	1.875	2.237
40	75	7.067	1.832	5.235	126.2	1.602	1.875	2.101
60	75	7.03	3.232	3.798	89.3	1.778	1.875	1.951
80	75	7.023	3.591	3.432	79.8	1.903	1.875	1.902
20	55	6.927	1.703	5.224	125.9	1.301	1.740	2.100
40	55	7.067	3.987	3.08	70.8	1.602	1.740	1.850
60	55	7.03	4.662	2.368	52.5	1.778	1.740	1.720
80	55	7.023	4.997	2.026	43.7	1.903	1.740	1.640

b. Develop the operational design equation.

i. Use a linear regression analysis to develop an operational equation based on water flowrate and transmittance.

ii. To complete a regression analysis, the flowrate and UV dose data must first be log transformed. The log transformed data are presented in columns 7, 8, and 9 in the above table.

iii. Using the UV dose (column 9) as the dependent variable and the flowrate (column 7) and transmittance (column 8) as the independent variables, the following results are obtained using the linear regression analysis program in Excel or other statistical analysis program.

Model Parameters

Source	Value
Intercept	0.097
X1	−0.673
X2	1.631

iv. The equation for UV dose as a function of flowrate and transmittance, based on the linear regression analysis, is

$$\log (\text{UV dose}) = 0.097 - 0.673 \,(\log \text{flowrate}) + 1.631 \,(\log \text{transmittance}) \text{ or}$$

$$\text{UV dose, mJ/cm}^2 = (10^{0.097})[(\text{flowrate})^{-0.673}][(\text{transmittance})^{1.631}]$$

where the units for flowrate and transmittance are L/min·lamp and percent, respectively.

Note: The above dose equation is the delivered UV dose based upon three banks operating in series and with UVT varying from 55 to 75 percent. If the UV validation test had been conducted with one bank, the UV dose for two or three banks would be obtained by multiplying the regression equation for one bank by 2 or 3, respectively.

c. Based on the results of the multiple variable (e.g., flowrate and transmittance) linear regression analysis, plot the curves of UV dose versus the UV lamp hydraulic loading rate for 75 and 55 transmittance, The required curves are shown on the following plot. It should be noted that the curve resulting from the multiple linear regression analysis for 75 percent transmittance is not exactly the same as that derived from the single variable (i.e., flowrate) linear regression analysis developed in Step 5. The reason for the difference is that the regression analysis with two variables must cover a significantly broader range of values as compared to a single variable regression analysis.

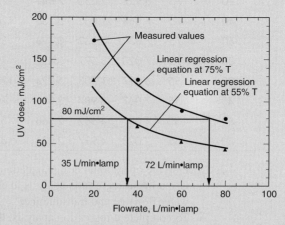

d. Determine the range of flowrates over which the UV disinfection system will deliver a UV dose of 80 mJ/cm². From the plot given above, the range of flowrates is up to 72 L/min·lamp at 75 percent transmittance and up to 35 L/min·lamp at 55 percent transmittance.

Comment   When the lamps are new and the protective quartz sleeves are clean, it may not be necessary to operate all three banks, depending on the actual UV dose requirements of the full scale disinfection system.

Factors Affecting UV System Design

Factors that affect the minimum number of UV lamps necessary for disinfection are (1) the UV lamp hydraulic loading rate based on the equipment validation test, (2) the level of confidence desired in meeting the permit requirements, and (3) the aging and fouling characteristics of the UV lamp/quartz sleeve assembly (discussed below). The validation of UV equipment has been considered in Example 12–13.

Confidence Level in Meeting Permit Limits.

With respect to the level of confidence desired in the system performance it should be noted that the linear regression equations developed in Example 12–13 correspond to the line of best fit with half of the data points lying above and half lying below the predicted curve. Because some of the actual data points lie below the regression equation, a factor of safety must be used to account for the observed variability. One approach is to determine the confidence interval (CI) of the regression equation. Another is to develop a prediction interval (PI) based on the regression equation. The difference between the CI and the PI is as follows. The upper and lower CI for the regression analysis represents the interval in which the true average measurement is likely to lie if the procedure were repeated many times. Stated differently, a 75 percent confidence interval will contain the true mean value (not estimated value from data measurements) 75 percent of the time the interval is calculated. The upper and lower PI represents the interval in which a given percentage of new observations, independent of those used to develop the regression equation, will lie. Because there is more uncertainty in future measurements, the interval between the upper and lower PI limits is greater than that for the confidence interval, which is based on repeating the procedure an infinite number of times. The development of the PI is illustrated in Example 12–14. The relationship between the CI and PI and the regression equation are illustrated on Fig. 12–45.

Figure 12–45

Definition sketch illustrating the relationship between (1) the values measured in the UV reactor validation test, (2) the linear regression equation based on the measured values, (3) the lower 75 percent confidence interval (CI) based on the linear regression, (4) the lower 75 percent prediction interval (PI) based on the linear regression, and (5) the design curve based on the lower 75 percent PI with a combined correction factor for lamp aging and fouling.

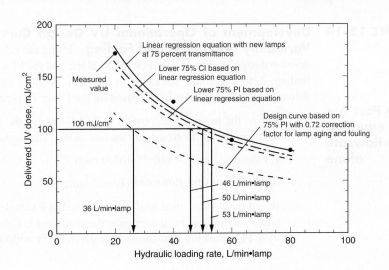

Lamp Aging. As UV lamps age, the UV output from the lamps decreases. Aging factors range from 0.5 (NWRI default value in the absence of validation data to support a higher factor) to 0.98. However, manufacturers may gain approval of a higher (less conservative) factor if sufficient data are available to support an increase. As discussed previously, UV reactor validation is conducted with new lamps. Thus, when sizing a UV system, the validated performance must be de-rated by the lamp aging factor. For example, if 100 lamps are required to deliver a certain dose at a specific UVT, the UV system would need to be sized with 200 lamps so that when the lamps have aged and are producing 50 percent UV output compared to when they were new, the dose critical for disinfection is still being delivered. Lamp aging factors vary widely from vendor to vendor and from lamp to lamp. Aging factors range from 0.5 to 0.98.

Quartz Sleeve Fouling Factor. The correction factor for quartz sleeve fouling will vary between 0.7 and 0.95 depending on the cleaning system that is employed. For applications with high UVT, low solids and little iron present in the effluent, it is generally appropriate to accept a UV system vendor's validation of a higher factor (as high as 0.95 for mechanical cleaning systems). For chemical-mechanical cleaning system the fouling factor can be as high as 0.95 and should be validated independently of the UV system manufacturer. For any combination of low UVT, high solids, and high iron concentration in the effluent, a 0.8 (or lower) factor should be applied when sizing a UV system regardless of whether a UV system manufacturer has validation of a higher factor.

Application of Design Factors in UV System Sizing. With the variation of lamp aging factors from 0.5 to 0.98 and quartz sleeve fouling factors from 0.7 to 0.95, the combined design or correction factor can range from 0.35 to 0.94. The correction factor is applied to the PI to obtain the final design curve for the disinfection system. The importance of these correction factors can be assessed from a review of the plot given on Fig. 12–45. The design curve given on Fig. 12–45 is based on the lower 75 percent PI with a combined correction factor of 0.72 for lamp aging and fouling (0.72 factor based on an lamp aging factor of 0.9 and a quartz sleeve fouling factor of 0.8). Clearly, if the UV system had been designed on the basis of the manufacturer's design curve, the system would be undersized with respect to lamp aging and fouling. Determination of confidence and prediction intervals and the development of design equations taking into account lamp aging and fouling are illustrated in Example 12–14.

EXAMPLE 12–14	**Development of Operational UV Design Curves Taking into Account Variability and Aging and Fouling** Using the information from Example 12–13, develop design equations based on the PI alone and the PI with a factor for lamp aging and fouling. Also, determine the range of hydraulic loading rates over which the system can deliver a UV dose of 80 mJ/cm², based on the lower 75 percent PI with new lamps.
Solution Part A— Design equation based on flowrate alone	1. Define the lower 75 percent CI and PI limits for the regression equation developed in Example 12–13 based on flowrate alone. The regression equation is UV dose, mJ/cm² $= (10^{2.997})[(\text{flowrate})^{-0.577}]$ where the unit for flowrate is L/min·lamp Based on the statistical analysis presented in Example 12–13, the assigned UV dose based on the field measurements, the predicted UV dose, the lower 75% CI, and the 75% PI values log transformed are given in the following table:

	Log-transformed values			
Flowrate, L/min·lamp	Assigned UV dose, mJ/cm²	Predicted UV dose, mJ/cm²	Predicted, 75% CI UV dose, mJ/cm²	Predicted, 75% PI UV dose, mJ/cm²
20	2.237	2.247	2.209	2.191
40	2.101	2.073	2.052	2.027
60	1.951	1.972	1.948	1.924
80	1.902	1.900	1.869	1.848

Although the lower 75 percent CI and PI values are obtained with a standard statistical program as given above, the procedure for determining these values is illustrated below for a linear regression expression with one variable–in this case, flowrate per lamp.

a. The lower 75 percent CI and PI intervals for the predicted mean response can be obtained using the following expressions:

  i. Confidence interval

$$\text{UV dose}_{75\%} = y_p - t_{\alpha/2}S\sqrt{\frac{1}{n} + \frac{(x - \bar{x})^2}{SS_{xx}}}$$

  ii. Prediction interval

$$\text{UV dose}_{75\%} = y_p - t_{\alpha/2}S\sqrt{1 + \frac{1}{n} + \frac{(x - \bar{x})^2}{SS_{xx}}}$$

  where y_p = the predicted UV dose computed using the regression equation given above, mJ/cm²

  $t_{\alpha/2}$ = 1.706 which corresponds to the value of the t-distribution based on a 75% prediction level with $n - 2$ degrees of freedom

  S = sample variance

$$S = \sqrt{\frac{\Sigma(y - y_p)^2}{n - 2}}$$

  y = assigned UV dose from field measurements, mJ/cm²
  y_p = predicted UV dose, mJ/cm²
  n = number of sample pairs
  x = flowrate, L/min·lamp
  $\bar{x}$ = average flowrate, L/min·lamp
  SS_{xx} = the sample corrected sum of squares

$$SS_{xx} = \sum_{1}^{n}(x - \bar{x})^2$$

b. Compute the values needed to determine the confidence intervals. Set up two computation tables, one for the UV dose and another for flowrate.

x	y	y_p	$(y - y_p)$	$(y - y_p)^2$
1.301	2.237	2.247	−0.010	0.000100
1.602	2.101	2.073	0.028	0.000784
1.778	1.951	1.972	−0.021	0.000441
1.903	1.902	1.900	0.002	0.000004
				0.001329

x	$(x - \bar{x})^a$	$(x - \bar{x})^2$
1.301	−0.345	0.119025
1.602	−0.044	0.001936
1.778	0.132	0.017424
1.903	0.257	0.066049
6.584		0.204434

$^a \bar{x} = 6.584/4 = 1.646$

i. Solve for the sample variance, S.

$$S = \sqrt{\frac{0.001329}{4 - 2}} = 0.025778$$

ii. Solve for sample corrected sum of squares, SS_{xx}.

$$SS_{xx} = \sum_{1}^{n}(x - \bar{x})^2 = 0.204434$$

iii. Solve for the lower 75 percent CI at a flowrate of 40 L/min·lamp.

$$\text{UV dose}_{75\% \text{CI}} = y_p - t_{\alpha/2}S\sqrt{\frac{1}{n} + \frac{(x - \bar{x})^2}{SS_{xx}}}$$

The value of y_p, computed using the regression equation given above, is equal to 2.073. Thus,

$$\text{UV dose}_{75\% \text{CI}} = 2.073 - (1.706)(0.024434)\sqrt{\frac{1}{4} + \frac{0.001936}{0.204434}}$$

$$\text{UV dose}_{75\% \text{CI}} = 2.073 - (1.706)(0.024434)(0.509382) = 2.052$$

iv. Solve for the lower 75 percent PI at a flowrate of 40 L/min·lamp.

$$\text{UV dose}_{75\% \text{PI}} = y_p - t_{\alpha/2}S\sqrt{1 + \frac{1}{n} + \frac{(x - \bar{x})^2}{SS_{xx}}}$$

The value of y_p, computed using the regression equation given above, is equal to 2.073. Thus,

$$\text{UV dose}_{75\% \text{CI}} = 2.073 - (1.706)(0.024434)\sqrt{1 + \frac{1}{4} + \frac{0.001936}{0.204434}}$$

$$\text{UV dose}_{75\% \text{PI}} = 2.073 - (1.706)(0.024443)(1.122248) = 2.026$$

The CI and PI values computed manually are essentially the same as the values obtained from the linear regression analysis program as given in Step 1. Values computed manually may not be exact due to rounding errors that are magnified when dealing with log-transformed values.

2. Correct the lower 75 percent PI values for lamp aging and fouling.

a. An overall correction factor of 0.72 is assumed for account for lamp aging and fouling. The correction factor for lamp aging is 0.9, based on the manufacturers recommendation. The corresponding fouling factor is 0.8. Note: the design engineer must decide if additional factors of safety may be required, depending on local conditions.

The UV dose based on lamp aging and fouling is given in following table in which the log-transformed values have been transformed back to arithmetic form.

Flowrate, L/min·lamp	Assigned UV dose, mJ/cm²	Predicted UV dose,[a] mJ/cm²	Predicted UV dose at 75% PI, mJ/cm²	Correction factor for lamp aging and fouling[b]	Design UVdose, mJ/cm²
20	172.6	176.5	155.3	0.72	111.8
40	126.2	118.3	106.5	0.72	76.7
60	89.3	93.7	84.0	0.72	60.5
80	79.8	79.4	70.5	0.72	50.8

[a] From regression equation.

[b] Correction factor = Lamp aging factor (0.9) × fouling factor (0.8).

b. The measured values, the linear regression equation, the 75 percent PI curve, and the design curve based on the 75 percent PI and taking into account lamp aging and fouling are plotted on the following graph for 75 percent transmittance.

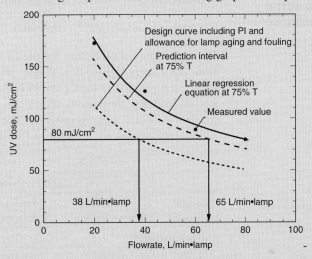

c. Determine the range of flowrates over which the UV disinfection system will deliver a UV dose of 80 mJ/cm², based on the lower 75 percent PI with new lamps. From the plot given above, the range of flowrates is up to 65 L/min·lamp at 75 percent transmittance.

3. Develop the design equations for the 75 percent PI curve, and the design curve based on the 75 percent PI and taking into account lamp aging and fouling for a transmittance value of 72 percent. The required equations can be be obtained by noting the ratio of the predicted PI UV dose to the predicted UV dose and the ratio of the design UV dose to the predicted UV dose as illustrated in the following table:

Predicted UV dose,[a] mJ/cm²	Predicted UV dose at 75% PI, mJ/cm²	Design UV dose,[b] mJ/cm²	Ratio, PI/predicted UV dose	Ratio, Design/predicted UV dose
176.5	155.3	111.8	0.88	0.63
118.3	106.5	76.7	0.90	0.65
93.7	84.0	60.5	0.90	0.65
79.4	70.5	50.8	0.89	0.63

[a] From regression equation.

[b] Design equation based on PI with correction factor for lamp aging and fouling.

The ratios in the above table are not exact, because the prediction interval at the extremes of the range is greater than for the centermost values. Use of the ratio for the extremes of the range is conservative. Thus, the pertinent equations are

Design equation based on 75 percent PI

$$\text{UV dose, mJ/cm}^2 = (10^{2.997})[(\text{flowrate})^{-0.577}](0.88)$$

Design equation based on 75 percent PI and including lamp aging and fouling

$$\text{UV dose, mJ/cm}^2 = (10^{2.997})[(\text{flowrate})^{-0.577}](0.63)$$

where the unit for flowrate is L/min·lamp

**Solution Part B—
Design equation
based on flowrate
and transmittance**

1. Define the lower 75 percent CI and PI limits for the regression equation developed in Example 12–13 based on flowrate and transmittance. The regression equation is

$$\text{UV dose, mJ/cm}^2 = (10^{0.097})[(\text{flowrate})^{-0.673}][(\text{transmittance})^{1.631}]$$

The computational procedure for the CI and PI values is similar to that illustrated above for a linear regression with one variable. In a multiple linear regression analysis, the computation of the CI and PI is more complicated because more three terms are involved. For this reason, the CI and PI values are usually determined using a standard statistical analysis program. The lower 75 percent CI and PI values for the regression equation are summarized in the following table for 75 and 55 percent transmittance.

	Log-transformed values			
Flowrate, L/min·lamp	Assigned UV dose, mJ/cm²	Predicted UV dose, mJ/cm²	Predicted, 75% CI UV dose, mJ/cm²	Predicted, 75% PI UV dose, mJ/cm²
75% transmittance				
1.30	2.237	2.280	2.248	2.227
1.60	2.101	2.077	2.056	2.029
1.78	1.951	1.959	1.936	1.910
1.90	1.902	1.875	1.847	1.824
55% transmittance				
1.30	2.100	2.060	2.028	2.006
1.60	1.850	1.857	1.835	1.809
1.78	1.720	1.739	1.715	1.690
1.90	1.640	1.654	1.627	1.604

2. Correct the lower 75 percent PI values for lamp aging and fouling.
   a. To account for lamp aging and fouling a correction factor of 0.72 will be applied. Note: the design engineer must decide if additional factors of safety may be required, depending on local conditions.

   The UV dose based on lamp aging and fouling is given in the following table in which the log-transformed values have been transformed back to arithmetic form.

Flowrate, L/min·lamp	Assigned UV dose, mJ/cm²	Predicted UV dose, mJ/cm²	Predicted UV dose at 75% PI, mJ/cm²	Correction factor for lamp aging and fouling	Design UV dose, mJ/cm²
75% transmittance					
20	172.6	190.5	168.5	0.72	121.4
40	126.2	119.5	107.0	0.72	77.0
60	89.3	91.0	81.3	0.72	58.5
80	79.8	74.9	66.6	0.72	48.0
55% transmittance					
20	125.9	114.7	101.5	0.72	73.1
40	70.8	72.0	64.4	0.72	46.4
60	52.5	54.8	49.0	0.72	35.3
80	43.7	45.1	40.1	0.72	28.9

b. The linear regression equation, the 75 percent PI curve, and the design curve taking into account lamp aging and fouling based on the lower 75 percent PI are plotted on the following graph for transmittance values of 75 and 55 percent.

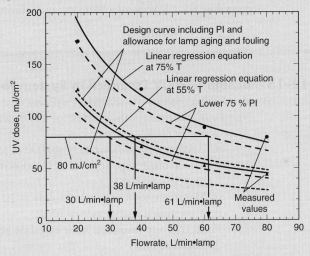

c. Determine the range of flowrates over which the UV disinfection system will deliver a UV dose of 80 mJ/cm², based on the lower 75 percent PI with new lamps. From the plot given above the range of flowrates is up to 61 L/min·lamp at 75 percent transmittance and up to 30 L/min·lamp at 55 percent transmittance. The range of flowrates per lamp at 75 percent transmittance with aged and fouled lamps is up to 38 L/min·lamp.

3. Develop the design equations for the 75 percent PI curve, and the design curve based on the 75 percent PI and taking into account lamp aging and fouling for transmittance values varying from 55 to 75 percent. The required equations, developed as outlined above in Part 1, Step 4, are:

Design equation based on 75 percent PI.

$$\text{UV dose, mJ/cm}^2 = (10^{0.198})[(\text{flowrate})^{-0.674}][(\text{transmittance})^{1.5713}](0.88)$$

Design equation based on 75 percent PI and including lamp aging and fouling.

$$\text{UV dose, mJ/cm}^2 = (10^{0.198})[(\text{flowrate})^{-0.674}][(\text{transmittance})^{1.5713}](0.64)$$

where the units for flowrate and transmittance are L/min·lamp and percent, respectively.

Comment   In both cases presented above and in previous sections discussing the impact of lamp aging and quartz sleeve fouling, the allowance made for these correction factors is significant as compared to the lower 75 percent PI. Thus, in evaluating UV disinfection systems, the selection of appropriate and validated lamp aging and fouling factors is of critical importance.

Selection and Sizing of a UV Disinfection System

Factors that affect the selection and sizing of a UV disinfection system include the selection and sizing of the UV disinfection reactor or system based on the UV design curve which takes into account the confidence and/or prediction interval associated with the hydraulic loading rate as determined in the equipment validation test and the lamp aging and fouling correction factors as illustrated in Example 12–13. The selection and sizing procedure for a UV disinfection system is illustrated in Example 12–15.

EXAMPLE 12–15   **Design of a UV Disinfection System for Secondary Effluent**   Design a UV disinfection system for secondary effluent that will deliver a minimum design dose of 30 mJ/cm². Assume for the purpose of this example that the following data apply:

1. Wastewater characteristics
   a. Average design flowrate = 40,000 m³/d = 27,778 L/min
   b. Maximum design flowrate = 100,000 m³/d = 69,444 L/min (peak hour flow with recycle streams)
   c. Maximum total suspended solids = 20 mg/L
   d. Minimum transmittance = 65%
2. Fecal coliform discharge limit based on geometric mean
   200 FC/100 mL
3. System characteristics
   a. Horizontal lamp configuration
   b. From a validation study conducted on a single UV bank using the procedure described in Example 12–12, the following equation was developed based on the 75 percent PI with a lamp aging and fouling factor allowance of 72 percent.

   $$\text{UV dose, mJ/cm}^2 = (10^{-2.428})[(\text{flowrate})^{-0.650}][(\text{transmittance})^{3.126}](0.64)$$

   where the units for flowrate and transmittance are L/min·lamp and percent, respectively
   c. System headloss coefficient = 0.75 (manufacturer specific)
   d. Lamp/sleeve diameter = 23 mm
   e. Cross-sectional area of quartz sleeve = 4.15 × 10⁻⁴ m²
   f. Lamp spacing = 75 mm (center to center)
   g. One standby UV bank will be required per channel

Solution

1. Determine the flowrate per lamp using the design equation based on the test conducted on a single UV bank. Based on the UV design equation for a dose of 30 mJ/cm², the corresponding flowrate per lamp is 258 L/min·lamp.

$$\text{Flowrate, L/min·lamp} = \left\{ \frac{30}{(10^{-2.428})[(65)^{3.126}](0.64)} \right\}^{-(1/0.650)} = 258$$

2. Specify the flowrate range per UV channel assuming three channels will be in operation during peak flow conditions.
   i. Up to 24,000 L/min, use one channel
   ii. From 24,000 to 48,000 L/min, split the flow between two channels such that each channel receives up to 24,000 L/min.
   iii. From 48,000 to 72,000 L/min, split the flow between three channels such that each channel receives up to 24,000 L/min.

3. Determine the number of lamps required per bank.
   At 24,000 L/min, the total number of required lamps is:

$$\text{Lamps required, Lamps/bank} = \frac{(24,000 \text{ L/min·bank})}{(258 \text{ L/lamp·min})} = 93 \text{ lamps/bank}$$

4. Configure the UV disinfection system.
   Typically, 2, 4, 8, or 16 lamps per module are available. Using an 8 lamp module, 12 modules are required per bank for a total of 96 lamps per bank.

5. Determine the total number of lamps per channel including standby.

$$\text{Total number of lamps per channel} = (2 \text{ banks/channel})(96 \text{ lamps/bank})$$
$$= 192 \text{ lamps/channel}$$

6. Determine total number of lamps.

$$\text{Total number of lamps per channel} = (3 \text{ channel})(192 \text{ lamps/channel})$$
$$= 586 \text{ lamps/channel}$$

7. Check whether the headloss for the selected configuration is acceptable.
   a. Determine the channel cross-sectional area.

$$\text{Cross sectional area of channel} = (12 \times 0.075 \text{ m})(8 \times 0.075 \text{ m})$$
$$= 0.54 \text{ m}^2$$

   b. Determine the net channel cross-sectional area by subtracting the cross sectional area of the quartz sleeves (4.15×10^{-4} m²/lamp).

$$A_{\text{channel}} = 0.54 \text{ m}^2 - [(12 \times 8) \text{ lamps/bank}] \times (4.15 \times 10^{-4} \text{ m}^2/\text{lamp})$$
$$= 0.50 \text{ m}^2$$

   c. Determine the maximum velocity in the channel.

$$v_{\text{channel}} = \frac{(24,000 \text{ L/min·channel})(0.001 \text{ m}^3/\text{L})(1 \text{ min}/60\text{s})}{0.5 \text{ m}^3} = 0.8 \text{ m/s}$$

   d. Determine the headloss per UV channel.

$$h_{\text{channel}} = 0.75 \frac{v^2}{2g}$$

$$h_{channel} = \frac{(0.75)(0.80 \text{ m/s})^2(1000 \text{ mm/m})}{2(9.81 \text{ m/s}^2)} \ (2 \text{ banks}) = 49.0 \text{ mm}$$

Note that 2 banks were used to determine system headloss. Use of two banks includes one redundant bank of lamps in each channel. The clear spacing between quartz sleeves is 52 mm (75 mm − 23 mm) and the headloss should not exceed this value.

8. Summarize the system configuration.

   System utilizes three channels, each channel containing two banks of lamps in series, one operational bank and one redundant bank. Each bank contains 12 modules, each of which contains 8 lamps.

Comment   The majority of UV disinfection systems have the ability to turn banks of lamps on and off and vary power (and therefore UV output) to the banks that remain on. Turning UV lamps on and off is done automatically in response to varying flowrates and water quality (UVT). Varying the output of a UV system based on flow is accomplished by connecting a plant flowrate signal to the UV system's programmable logic controller (PLC). UVT can be manually entered based on readings taken from a bench-top photometer or based on continous readings from an online transmittance monitor.

Use of Spot-Check Bioassay to Validate UV System Performance

A spot-check-bioassay (SCB) test procedure has been developed to validate the performance of a newly installed and operational UV disinfection system. The test involves making a minimum of eight spot-check viral assays to demonstrate that the full-scale UV reactor performance complies with the design intent. Because new lamps are installed, the 75 percent PI is used as a reference. The CHPH has approved the use of the SCB test procedure to assess compliance of a full-scale disinfection reactor or system with the design intent. As implemented by CDPH, seven of the eight bioassay tests results must lie above the lower 75 percent PI predicted values. The rationale is that the percent ratio of seven out of eight is 87.5, which corresponds to the lower prediction interval. If more than one out of eight SCB test bioassays is below the PI curve, it is usually a clear indication that something may be wrong with the installation (e.g., poor inlet and outlet flow distribution, poor channel geometry, poor alignment, inappropriate weir placement, inappropriate flow control devices, inappropriate power settings, as well as other site conditions). If the installation site features can be corrected, they should be corrected, and the system should be retested. If the installation site features cannot be corrected, the UV system should be derated. The SCB test procedure along with the procedure for derating the UV system is illustrated in Example 12–16.

The SCB test procedure is similar to the procedure followed for UV reactor validation, as delineated in Example 12–14, Part A, with the exception that a wide range of operating conditions is evaluated. For example, consider a system comprised of two channels, each containing 4 banks of UV lamps. For such a system a typical test program might include four tests conducted under the following conditions:

1. Maximum flowrate per lamp, minimum transmittance
2. Average flowrate per lamp, minimum transmittance
3. Maximum power setting, minimum transmittance
4. Minimum flowrate per lamp, minimum transmittance, first operational UV bank in sequence (i.e., 1, 2, 3, and 4)

Four additional tests could be conduced under the following conditions:

5. Ambient transmittance, maximum flowrate per lamp
6. Ambient transmittance, intermediate ballast output settings (60, 70, 80, or 90 percent)
7. Ambient transmittance, intermediate flowrates
8. Ambient transmittance, with the last operational UV bank in sequence (i.e., 1, 2, 3, and 4)

The goal of test 1 is to check performance under worst case conditions. The goal of test 2 is to check performance under typical flowrates and worst case water quality conditions. The goal of test 3 is to check performance under worst case water quality conditions at maximum power. The goal of tests 4 and 8 is to determine whether bank placement has an impact on operational performance. The goal of test 5 is to check performance under maximum flowrate per lamp and typical water quality conditions. The goal of test 6 is to check performance under different power settings. The goal of test 7 is to evaluate the performance of the UV system at various intermediate operating conditions. It should be noted that any number of test sequences can be used, as long as a wide range of operating conditions is evaluated.

EXAMPLE 12–16   Conduct of Spot-check Bioassay to Validate Performance of Full-scale UV Disinfection System   A spot-check bioassay is to be conducted to validate the performance of a newly installed and operational UV disinfection system at a wastewater treatment plant. The UV system, validated in Example 12–14, is comprised of two channels, each containing four banks of lamps. Each bank contains 4 UV lamps oriented parallel to flow. The UV system was validated over a range of flowrates from 20 to 80 L/min·lamp at a transmittance of 75 percent. Based on the 75 percent PI, the UV system with new lamps will deliver a UV dose of 100 mJ/cm² up to a flowrate of 44 L/min·lamp, as shown in the following plot:

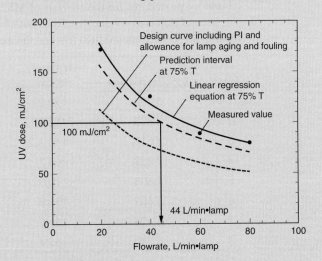

The regression equation is

$$\text{UV dose, mJ/cm}^2 = (10^{2.997})[(\text{flowrate})^{-0.577}]$$

The corresponding equation for the 75 percent PI is

UV dose, mJ/cm^2 = $(10^{2.997})[(\text{flowrate})^{-0.577}](0.88)$

Solution

1. Determine the maximum required flowrate for one channel.

   Max. flowrate = (4 lamp/bank)(44 L/min·lamp·bank) = 176 L/min

2. Determine the minimum required flowrate for one channel.

   Min. flowrate = (4 lamp/bank)(20 L/min·lamp·bank) = 80 L/min

3. Develop the test conditions.
   A minimum of eight spot-check viral assays must be conducted to demonstrate that full-scale UV reactor performance complies with the design intent.
   a. Bioassay test conditions.

Test No.	UVT, %	Operational banks	Flowrate, L/min	Hydraulic loading rate, L/min-lamp	Power setting, %
1	75	1, 2, 3	176	44	100
2	75	2, 3, 4	176	44	100
3	75	1, 3, 4	176	44	100
4	75	1, 2, 4	176	44	100
5	75	1, 2, 3	140	35	100
6	75	2, 3, 4	120	30	100
7	75	1, 3, 4	100	25	100
8	75	1, 2, 4	80	20	100

   b. Conduct spot-check bioassays.
      i. The first step is to conduct a quality assurance test to demonstrate that the laboratory procedures for the analysis of MS2 are valid (see Example 12–8).
      ii. The second step is to conduct the field spot-check bioassay tests. The log inactivation achieved from the field test and the assigned UV dose are as follows:

Test[a]	Log$_{10}$ inactivation	UV Dose[b], mJ/cm^2
1	5.002	120.2
2	4.803	115.1
3	4.617	110.3
4	4.438	105.7
5	4.605	110.0
6	5.609	135.8
7	6.406	156.3
8	6.760	165.4

   [a] See above table for operating conditions.
   [b] The UV dose is based on the following equation:

   $$\text{UV dose} = \frac{\log \text{inactivation} - 0.326}{0.0389}$$

4. Compare the SCB test results to the values obtained from the linear regression equation and the PI equation.

   a. The two comparisons are presented in the following table:

Test	UV dose, mJ/cm²			Ratio spot-check/ predicted from regression equation	Ratio spot-check/ predicted PI from regression equation
	Predicted from regression equation	PI predicted from regression equation	Measured from spot check		
1	111.9	98.4	120.2	1.07	1.22
2	111.9	98.4	115.1	1.03	1.17
3	111.9	98.4	110.3	0.99	1.12
4	111.9	98.4	105.7	0.94	1.07
5	127.7	112.3	110.0	0.86	0.98
6	139.5	122.8	135.8	0.97	1.11
7	155.0	136.4	156.3	1.01	1.15
8	176.3	155.2	165.4	0.94	1.07

   b. In comparing the SCB data to the regression equation it can be seen that the distribution of values is as would be expected, with a more or less equal distribution of values above and below the value obtained from the regression analysis.

   c. Based on the performance ratio, seven of the eight test results are above the predicted value of the PI, thus, the operation of the full-scale UV disinfection system is consistent with the design intent, as required by CDPH.

5. System adjustments for poor performance. In a situation in which more than one of the eight SCB test values lies below the PI curve, the following steps should be taken:

   a. Review the features of the installation, as discussed above, that may be leading to poor performance, correct any of the problems commonly encountered, and conduct a new SCB test.

   b. If the new test results are the same as the previous test, the UV system target UV dose set point must be adjusted or the system must be derated.

      i. Where the regulatory agency prefers not to modify the system dose equation, a site-specific target dose can be developed. The site-specific target dose can be computed using the following expression:

$$\text{Target UV dose} = \frac{\text{Design equation UV dose}}{\text{7th lowest 75\% PI spot check ratio}}$$

      The 7th lowest spot check ratio, based on the PI, is obtained as shown in column 5 in the above table. In this example the value is 0.94.

      ii. Alternatively, the system target UV dose can be derated using the following expression:

$$\text{UV dose}_{Adj} = \text{Design equation} \times \text{7th lowest 75\% PI spot check ratio}$$

      As above, 7th lowest spot check ratio, based on the PI, is 0.94.

Comment   To avoid having to derate a UV system or change the target UV dose, it is imperative that careful attention be devoted to the design, installation, and operation of the UV disinfection system and its appurtenant facilities, especially the outlet control structure.

Troubleshooting UV Disinfection Systems

Problems associated with UV disinfection systems are related primarily to the inability to achieve permit limits. Some issues that must be considered when diagnosing problems associated with UV disinfection systems are discussed below.

UV Disinfection System Hydraulics. Perhaps one of the most serious problems encountered in the field is erratic or reduced inactivation performance due to poor system hydraulics. The most common hydraulic problems are related to (1) the creation of density currents that can cause the incoming water to move along the bottom or top of the UV lamp banks resulting in short circuiting, (2) inappropriate entry and exit conditions that can lead to the formation of eddy currents which ultimately create uneven velocity profiles that induce short circuiting, (3) the creation of dead spaces or zones within the reactor resulting in short circuiting, and (4) uneven flow distribution in systems with multiple channels leading to overloading in certain channels and underloading in others. The occurrence of short circuiting, channel overloading, and dead zones reduces the average contact time and leads to a decrease in UV dose and therefore compromises disinfection. When designing a UV system, the use of CFD modeling may be warranted to ensure hydraulic issues that could negatively impact disinfection performance are accounted for.

The principal hydraulic design features that can be used to improve system hydraulics in open channels include the use of (1) submerged perforated diffusers at the inlet of UV channel(s), (2) corner fillets in rectangular open channel systems with horizontal lamp placement, and (3) flow deflectors in open channel systems with vertical lamp placement. In rare cases, power input to mix the incoming flow may be necessary. Some of these corrective measures for open channel UV disinfection systems are illustrated on Fig. 12–46. Submerged perforated baffles should have an open area of about 4 to 6 percent of the cross-sectional area of the flow channel. Similar to open-channel systems, closed vessel UV systems may require similar design features to improve hydraulics through the reactors. Again, the use of CFD modeling may be of great value in studying the effect of various physical interventions in bringing about a more uniform approach velocity flow field (Sotirakos et al., 2013).

Biofilms on Walls of UV Channels and on UV Equipment. Another serious problem encountered with UV disinfection systems is the development of biofilms on the exposed surfaces of the UV reactor. The problem is especially serious in open channel systems covered with standard grating. It has been found that if the UV channels are exposed to any light, even very dim light, biofilms (typically fungal and filamentous bacteria) will develop on the exposed surfaces. The problem with biofilms is that they can harbor and effectively shield bacteria. When the clumped biofilms break away from the attachment surface, bacteria can be shielded as the clumps pass through the disinfection system. The best control measure is to completely cover the UV channels. Further, all concrete channels should be lined or coated to avoid the formation of bacterial colonies in the crevices and rough spots found in poured concrete. In addition, the channels can be cleaned and disinfected occasionally using hypochlorite, peracetic acid (see Sec. 12–8), or another suitable cleaning agent/disinfectant.

It should be noted that biofilm development can also occur in closed UV systems, but the severity is usually less, with the exception of UV systems in which medium-pressure high-intensity UV lamps are employed. Because medium-pressure high-intensity UV lamps emit some light in or near the visible light range (see Fig. 12–33) they can stimulate the growth of microorganisms on exposed surfaces. In some cases, growths approaching 300 mm in length have been found attached to the lamp support structure. The amount of

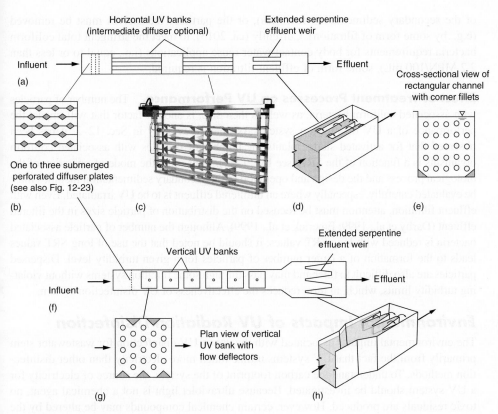

Figure 12–46

Typical examples of physical features that can be used to improve the performance of horizontal and vertical open UV reactors: (a) plan view of horizontal lamp UV system in channel with lining or coating, (b) perforated submerged diffuser plates to promote plug flow, (c) use of delta wings for enhanced internal mixing (courtesy of Calgon Carbon), (d) and (h) extended serpentine effluent weirs to promote plug flow, (e) elimination of dead space with corner fillets, (f) plan view of vertical lamp UV system in channel with lining or coating, and (g) use of baffle diffusers for enhanced internal mixing.

light emitted in the visible light range will vary with each type of lamp (i.e., manufacturer). Removal of these growths with a suitable disinfectant must be conducted on a periodic basis.

Overcoming the Impact of Particles by Increasing UV Intensity.

It was thought at one time that the impact of particles on the performance of UV disinfection systems could be overcome by increasing the UV intensity. Unfortunately, it has been found that increasing the UV intensity tenfold has little effect on reducing the number of surviving particle-associated coliform bacteria because the absorption of UV irradiation by particles in wastewater is typically 10,000 times or more greater than the bulk liquid medium. Particles essentially block the transmission of UV light. Particles larger than a certain critical size (a function of the size of the target organism) will effectively shield the embedded microorganisms (Emerick et al., 1999; Emerick et al., 2000). Because the effectiveness of UV disinfection is governed primarily by the number of particles containing coliform bacteria, to improve the performance of a UV disinfection system, either the number of particles with associated coliform bacteria must be reduced (e.g., by modifying the treatment process mode of operation or by adding polymer to improve the performance of the secondary sedimentation facilities), or the particles themselves must be removed (e.g., by some form of filtration). Currently (ca. 2013), to meet the stringent total coliform bacteria requirements for body contact water reuse applications (i.e., equal to or less than 2.2 MPN/100 mL), some form of effluent filtration is required.

Effect of Treatment Processes on UV Performance. The number of particles with associated coliform bacteria, as well as their size, is another factor that will impact the performance of a UV disinfection system. As noted previously in Sec. 12–2, it has been observed that for activated sludge plants the number of particles with associated coliform organisms is a function of the SRT (see Fig. 12–5). Thus, both the mode of operation of the biological process and the design and operation of the secondary sedimentation facilities must be evaluated carefully, especially where an unfiltered effluent is to be UV irradiated. Even with effluent filtration, attention must be focused on the distribution of particle sizes in the filtered effluent (Darby et al., 1999; Emerick et al., 1999). Although the number of particle associated bacteria is reduced with long SRT values, it should be noted that the use of long SRT values leads to the formation of a larger number of particles for a given turbidity level. Dispersed particles are also difficult to filter and may pass through some filtration systems without violating turbidity limits, which, in turn, reduces the effectiveness of UV disinfection system.

Environmental Impacts of UV Radiation Disinfection

The environmental impacts associated with the use of UV disinfection for wastewater stem primarily from the fact that UV systems utilize much more electricity than other disinfection methods. To understand the carbon footprint of the system, the source of electricity for a UV system should be investigated. Because ultraviolet light is not a chemical agent, no toxic residuals are produced. However, certain chemical compounds may be altered by the ultraviolet irradiation. On the basis of the evidence to date, it appears that the compounds formed are harmless or are broken down into more innocuous forms at the dosages used for the disinfection of wastewater and reclaimed water (20 to 100 mJ/cm^2). Photooxidation, which does alter the structure of compounds, occurs above about 400 kJ/cm^2 range. Thus, the disinfection of wastewater with ultraviolet light is not considered to have any adverse environmental impacts. The impacts associated with some of the new very high-energy lamps, which may operate in the kilojoule range, is not known at present (ca. 2013).

12–10 DISINFECTION BY PASTEURIZATION

The process of heating food or water at a specified temperature and time for the purpose of killing microorganisms is known as *pasteurization*. The process was first demonstrated on April 20, 1862 in France by Pasteur and Bernard in response to a request by Emperor Napoleon III to save France's wine industry from what were called "diseases of wine" (Lewis and Heppell, 2000). The major contribution made by Pasteur was to define the exact time and temperature required to kill specific microorganisms, without affecting the taste of the wine. From the early beginnings, pasteurization is now used universally in the food industry to control pathogenic microorganisms. It is important to note that pasteurization is not intended to kill all microorganisms as compared to sterilization. Rather, it is intended to reduce the viable number of microorganisms present. Description of the pasteurization process, reported performance data, and regulatory requirements are discussed below.

Description of the Pasteurization Process

The operation of a pasteurization process is illustrated schematically on Fig. 12–47. As shown, effluent to be disinfected is introduced into the preheat reactor where heat from the

Figure 12–47

Definition sketch for the pasteurization process for the wastewater. (Adapted from Salveson et al., 2011.)

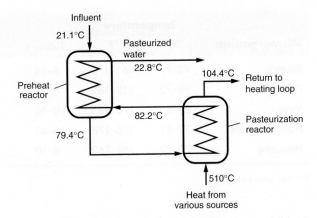

disinfected effluent is used to preheat the incoming effluent. The preheated effluent is then directed to the pasteurization reactor where heat from an external source is used to heat the preheated effluent to the desired temperature and where it is retained for a prescribed period of time. The external heat source can be from turbine exhaust, engine exhaust, waste gas burner exhaust, hot water, or from another suitable fluid. The reported temperature values shown on Fig. 12–47 are site specific and will depend on local conditions and the design of the heat exchange equipment.

Operationally, three different types of pasteurization are in use: (1) batch, (2) HTST (high-temperature short time) and (3) UHT (ultra-high temperature). Because of the large volumes required, batch pasteurization is only suitable for very small operations. Typical operational ranges for the three types of pasteurization are summarized in Table 12–34. The continuous flow HTST pasteurization process is used in most industrial operations and is the form used for the disinfection of treated wastewater. General operational data for the HTST process to achieve four-log inactivation of specific organism groups is presented in Table 12–35. The UHT pasteurization process, also known as flash pasteurization, is only used in more specialized applications.

Thermal Disinfection Kinetics

The disinfection performance of the pasteurization process depends on both temperature and holding time. As noted in Table 12–2, and as compared to other disinfection methods,

Table 12–34

General operating ranges for pasteurization technologies[a]

Pasteurization technology	Temperature		Time	Comments
	°C	**°F**		
Batch	62–64	144–147	30–35 min	Inactivates most vegetative bacterial calls including streptococci, staphylococci, and mycobacterium tuberculosis
High-temperature short time (HTST)	72–75	161–165	8–30s	Same effect as batch, but at much shorter times
Ultra-high temperature (UHT)	135–140	275–285	<1–5s	Lethal for most bacterial cells at even shorter times than HTST

[a] Adapted in part from Toder (2012), Hudson et al. (2003), Sorqvist (2003).

Table 12–35

General operating ranges for HTST pasteurization to achieve approximately 4-log inactivation of selected microorganisms

Microorganism	Temperature °C	°F	Time, s	Comments
Bacteria	72–77	161–170	6–16	
Protozoa	70–72	158–162	8–16	Essentially complete inactivation
Virus	80–85	176–185	10–30	
MS2 Coliphage	79–81	175–178	15–40	
Helminths	70–72	158–162	8–10	Essentially complete destruction

Various sources.

the high temperature is needed to denature the enzymes in the cell protoplasm and to alter the structure of the cell wall. The holding time is needed to complete the reaction within the cell and with the cell wall constituents. In a manner similar to chemical disinfection it has been observed that if the temperature is increased the time required to inactivate a given microorganism will decrease (Pflug et al., 2001).

First Order Kinetics. The microbial disinfection of microorganisms, at a specific temperature, can be modeled as a first order reaction as follows:

$$\frac{dN}{dt} = -kN \tag{12-70}$$

where N = the number of organisms surviving after time t
k = reaction rate constant, base e
t = the exposure time

The integrated form of Eq. (12–70) in base 10 is

$$N_t = N_o 10^{-Kt} \tag{12-71}$$

where N_t = the number of organisms remaining at time t
N_o = the initial number of organisms
K = reaction rate constant, base 10, Note $K = 0.4343k$
t = time, s, min, h

The reaction rate constant, K, is given by

$$K = (\log N_o - \log N)/t \tag{12-72}$$

Heat Resistance Parameters. Two parameters, D and Z, are used commonly to describe the effectiveness of the pasteurization process (Goff, 2012; Pflug et al., 2001). The term D is a measure of the heat resistance of a given microorganism and corresponds to the time required to achieve 1-log of inactivation (i.e., 90 percent) at a given temperature T, as illustrated on Fig.12–48(a). Also known as the *decimal reduction time*, D, is given by the following expression:

$$D = 1/K = t/(\log N_o - \log N_t) \tag{12-73}$$

Using Eq. (12–73), Eq. (12–72) can be written as

$$N_t = N_o 10^{-t/D} \tag{12-74}$$

Figure 12–48

Definition sketches for the pasteurization process: (a) plot to determine the time, D, at constant temperature required to reduce the concentration of microorganisms by 1-log and (b) plot to determine, Z, the temperature increase required to reduce the D value by 1-log.

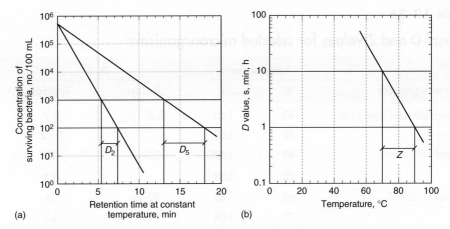

Each microorganism will have a specific value of D. It should be noted that the K and D values are only for a specific temperature. For example, at temperature T, an organism having a D value of 6 is less resistant than an organism with a D_T value of 10 [see Fig. 12–48(a)].

The term, Z, reflects the temperature dependence of D as illustrated on Fig. 12–48(b). The Z value corresponds to the temperature required to achieve 1-log change in the D value. The Z value is given by the following expression:

$$Z = (T_2 - T_1)/[\log(D_1) - \log(D_2)] \text{ or} \tag{12–75}$$

$$\log\left(\frac{D_1}{D_2}\right) = \frac{(T_2 - T_1)}{Z} \tag{12–76}$$

When D_1 and D_2 differ by one log, Eq. (12–76) reduces to

$$Z = (T_2 - T_1) \tag{12–77}$$

Using these two parameters, the heat resistance of a single microorganism can be quantified. Typical D and Z values for selected microorganisms are reported in Table 12–36. The values given in Table 12–36 are derived from a number of sources. The most extensive evaluation of D and Z values has been by Sorqvist (2003). Typical Z values for bacteria are in the range from 5 to 10. The application of the terms D and Z is illustrated in Example 12–17.

Non-Linear Inactivation. Although the equations for D and Z assume a more or less log-linear response, a number of researchers have observed both shoulder and tailing effects as described previously (see also Fig. 12–4). In general, the shoulder effect reflects the condition where the microorganism is less effected by temperature initially as compared to the subsequent linear inactivation phase. The tailing effect corresponds a period of reduced thermal destruction in which the response is less rapid than that that observed in the linear inactivation phase. In both cases, the effects are not understood completely (Hiatt, 1964). In addition, a number of other microorganisms such as spore forming bacteria *(B. anthracis and B. cereus)* exhibit a nonlinear response. Inactivation of spore-forming bacteria requires extremely high temperatures, especially if short contact times are to be used.

Table 12–36

Typical D and Z values for selected microorganisms[a]

Microorganism	Temperature		D, s		Z, °C	
	°C	**°F**	**Range**	**Typical**	**Range**	**Comp.[b]**
Campylobacter jejuni/coli	60	140	6.5–10	8.2	2.8–8.0	5.5
	70	158		0.12		
E. coli	60	140	35–42	40	3.2–9.2	5.0
	70	158		0.4		
E. coli 0157:H7	60	140	23–26	24	4.3–9.8	4.8
	70	158		0.2		
Enterococcus fecalis	60	140	360–480	415	2.2–14.2	6.0
	70	158		9.0		
Listeria monocytogenes	60	140	81–93	87	4.3–11.5	6.1
	70	158		2.0		
MS2 coliphage	70			14		10
	80			1.4		
Salmonella spp.	60	140	23–26	24	3.3–9.5	5.6
	70	158		0.4		
Staphylococcus aureus	60	140		54		10.5
	70	158		6		
Streptococcus	60	140		24		7.7
	70	158		1.2		
Total, coliform	60	140	42–60	50		7.9
	70	160		2.7		
Yersinia enterocolitica	60	140	24–37	30	4.0–13.7	6.6
	70	158		0.9		

[a] Adapted from Hudson et al. (2003), Sorqvist (2003), Salveson (2012).

[b] Computed Z values based on the typical D values given in column 5.

EXAMPLE 12–17   Estimate Pasteurization Operating Conditions Some new pasteurization equipment is to be installed to replace some existing equipment for the disinfection of an unknown strain of bacteria. The current equipment operates at a temperature of 65°C (150°F). The corresponding D and Z values for the bacteria are 10 s and 12°C, respectively. The initial bacterial count is 10^6 org/100 mL. If the new equipment is to operate at 77°C (170°F) for 4 s, what level of inactivation can be achieved? If the new equipment cannot be installed, what time would be required at a temperature of 65°C (150°F) to achieve 4 log removal?

Solution

1. Determine D_T at 77°C using Eq. (12–76).

    $$\log\left(\frac{10.0 \text{ s}}{D_2}\right) = \frac{(77 - 65)}{12} = 1.0$$

    $$\left(\frac{10.0 \text{ s}}{D_2}\right) = \text{inverse } \log(1.0) = 10.0$$

    $$D_2 = 10.0 \text{ s}/10.0 = 1.0 \text{ s}$$

2. Determine the degree of inactivation using Eq. (12–73).

    $$\log\left(\frac{N}{N_o}\right) = -4\text{s}/1.0 \text{ s} = -4$$

    Four-log inactivation can be achieved

3. Determine the time required to achieve a 4-log inactivation using the existing equipment. The required time can be estimated using Eq. (12–73).

    $$\log\left(\frac{N}{N_o}\right) = -t/D_T$$

    $$\log\left(\frac{N}{N_o}\right) = -4 = -t/10$$

    $$t = 4 \times 10 = 40 \text{ s}$$

Germicidal Effectiveness of Pasteurization

The literature on the thermal destruction of microorganisms in the food processing industry is vast. However, the literature on the disinfection of treated effluent is relatively limited. In a study completed recently, the use of UV and pasteurization were evaluated at pilot scale as possible replacements for the existing chlorine disinfection system. The secondary effluent was filtered. In the studies on pasteurization, conducted using proprietary equipment, contact times and temperatures were varied depending on the test organism (Salveson et al., 2011). Based on the test results for MS2 coliphage, it was found that within the range of times tested, contact time did not appear to be significant up to a temperature of about 73°C (163°F) (i.e., non-linear response). The temperature and time required to achieve a 4-log reduction of MS2 was on the order 80°C (176°F) and 7.7 s, respectively. These values are consistent with the values approved by the CDPH for pasteurization.

Regulatory Requirements

The CDPH has approved the use of the pasteurization process for the disinfection of treated effluent for Title 22 reuse applications. To achieve a 4-log virus reduction (based on MS2), the CDPH has set the temperature at 82°C (180°F) at a contact time of 10 s. For nonreuse applications where virus reduction is not needed, it may be possible to reduce the temperature to 74°C (165°F) and the time to 8 s. It is anticipated that these values will be revised as additional operational data are collected.

Application of Pasteurization for Disinfection

Pasteurization has been studied at a number of locations and has been found to have the lowest cost when compared to other disinfection technologies, and especially where waste heat is available (Salveson et al., 2011). A typical heat balance for the pasteurization process was shown previously on Fig. 12–47. The recovery of the heat from wastewater is another possibility that can be considered in the application of the pasteurization process. The survival of bacterial indicator species and bacteriophages in sludge and wastewater after thermal treatment has been studied (Moce-Uivina et al., 2003). The pasteurization of biosolids is considered further in Chap. 14.

PROBLEMS AND DISCUSSION TOPICS

12–1 Assuming Chick's law applies, determine the inactivation rate constant for total coliform for one of the following four treated effluents (sample to be selected by instructor). The effluent temperature was 20°C. If the activation energy for the disinfection reaction is 52 kJ/mole, determine the inactivation rate constant at 12°C.

Log of organisms remaining	Time, min			
	Effluent sample			
	1	2	3	4
8	0.0	0.0	0.0	0.0
7	1.8	5.5	3.8	2.6
6	3.6	11.5	8.0	5.5
5	5.6	17.5	12.3	8.0
4	7.4	23.5	16.5	11.0
3	9.2	20.9	20.9	13.9

12–2 Using the rate constant developed in Problem 12–1, determine the chlorine dose required to achieve a 99.99 percent inactivation of total coliform in 60 min at 15 and 25°C.

12–3 The following combined chlorine disinfection data were obtained in a series of laboratory tests performed on three different filtered activated sludge effluents:

Combined chlorine CT, mg·min/L	Residual fecal coliform count, no./100 mL		
	Test		
	1	2	3
0	10^6	10^6	10^6
50	10,000	199,500	316,000
100	10,200	31,600	63,000
200	126	800	4000
300	1	25	280
400		1	20
			1

a. Using these data, determine the value of the coefficient of specific lethality in Eq. (12–6) and the CT value to achieve a residual coliform count of 200/100 mL and 1000/100 mL.

b. Using the following data, determine the required volume in m^3 of a chlorine contact chamber designed to provide 60-min contact at the average winter flowrate. Using the equations developed in Part a, determine the minimum dosage required in mg/L to give the required kill for one of the test results given above (test condition to be selected by instructor). Assuming that the yearly chlorine requirement can be computed on the basis of the average flowrate for each of the two 6-mo periods, determine the minimum yearly chlorine requirement in kilograms.

Item	Unit	May-Oct	Nov-Apr
Average flowrate	m³/d	20,000	26,000
Peak daily flowrate	m³/d	40,000	52,000
Maximum permissible fecal coliform count in effluent	MPN/100 mL	200	1000

12-4 The following data were obtained for several filtered wastewater effluents. Using these data estimate the coefficients for the refined Collins-Selleck model [Eq. (12–27)] for wastewater number (to be selected by instructor).

	Time min			
	Wastewater number			
$-\log (N/N_o)$	1	2	3	4
1	2.1	6.9	2.9	3.5
2	4	15	5.9	8.1
3	7.1	36	12.3	18
4	13.6	80	24	40
5	21.5	190	55.5	90
6	42.3	430	115	200

Using the derived values, estimate the inactivation that could be achieved with a CT value of 30, 60, or 120 mg·min/L (value to be selected by instructor).

12-5 A consultant has proposed using chorine dosages of 15 and 8, 20 and 10, 30 and 20 mg/L during the summer and winter, respectively for effluent disinfection. If the effluent total coliform count before disinfection is $10^7/100$ mL, estimate the final total coliform counts that can be achieved during the summer and winter with one of the dosage sets (to be selected by instructor).

1. Demand due to decay during chlorine contact = 2.0 mg/L

2. Required chlorine contact time = 45 min

3. Use the typical values given below for the coefficients.

   $b = 4.0$ and $n = 2.8$

12-6 The chlorine residuals measured when various dosages of chlorine were added to four different wastewater effluents are given below. For one of the effluents (to be selected by instructor), determine: (a) the breakpoint dosage and (b) the design dosage to obtain a free chlorine residual of 1, 2, or 3.5, mg/L (value to be selected by instructor).

	Residual, mg/L			
	Effluent number			
Dosage, mg/L	1	2	3	4
0	0	0	0	0
1	0	1	0	0
2	1	2	1	1
3	0.2	3	2	2
4	1	4	2.3	2.9
5	2	4.3	1.2	3.4
6	3	3.6	0.9	2.7
7		2.3	1.7	1.2

(continued)

(Continued)

Dosage, mg/L	Residual, mg/L			
	Effluent number			
	1	2	3	4
8		0.7	2.7	1.2
9		0.7	3.7	2.1
10			1.7	3.1
11			2.8	4.1

12-7 Estimate the daily required chlorine dosage, the required alkalinity, if alkalinity will have to be added, and the resulting buildup of total dissolved solids when breakpoint chlorination is used for the seasonal control of nitrogen. Assume that the following data apply to this problem:

1. Plant flowrate = 4800 m³/d

2. Effluent characteristics

    a. BOD = 15 mg/L

    b. Total suspended solids = 15 mg/L

    c. NH_3-N = 1, 1.25, or 1.5 mg/L (value to be selected by instructor)

    d. Alkalinity = 125, 145, or 165 mg/L as $CaCO_3$ (value to be selected by instructor)

12-8 Review the current literature and prepare an assessment of the use of chlorine gas versus sodium hypochlorite for the disinfection of treated wastewater. A minimum of 3 recent (after 2000) articles and/or reports should be cited in your assessment.

12-9 The following data were obtained from dye tracer studies of five different chlorine contact basins. Using these data, determine the mean hydraulic residence time and the corresponding variance, the t_{10} time, and the Morrill Dispersion Index and the volumetric efficiency for one of the basins (to be selected by instructor). How would the performance of the basin selected for analysis be classified according to the U.S. EPA guidelines?

Time, min	Tracer concentration, μg				
	Basin number				
	1	2	3	4	5
0	0.0	0.0	0.0	0.0	0.0
10	0.0	0.0	0.0	0.0	0.0
20	3.5	0.1	0.1	0.0	0.0
30	7.6	2.1	2.1	0.0	0.7
40	7.8	7.5	10.0	0.3	4.0
50	6.9	10.1	12.0	1.8	9.0
60	5.9	10.2	10.2	4.5	12.5
70	4.8	9.7	8.0	8.0	11.5
80	3.8	8.1	6.0	11.0	8.8
90	3.0	6.0	4.3	11.0	5.5
100	2.4	4.4	3.0	9.0	3.0
110	1.9	3.0	2.1	4.3	1.8
120	1.5	1.9	1.5	2.0	0.8
130	1.0	1.0	1.0	1.0	0.4
140	0.6	0.4	0.5	0.2	0.1

(continued)

(Continued)

Time, min	Tracer concentration, μg				
	Basin number				
	1	2	3	4	5
150	0.3	0.1	0.1	0.0	0.0
160	0.1	0.0	0.0	0.0	0.0
170	0.0	0.0	0.0	0.0	0.0

12-10 Using the following dose response data for an enteric virus and the tracer data for four different chlorine contact basins, determine for one of the basins (to be selected by instructor) the expected effluent microorganism concentration based on the t_{10} and mean hydraulic residence times. Also estimate the chlorine residual that would be required to achieve 4 logs of removal with the existing basins.

Dose response data for enteric viruses

CT, mg/L·min[a]	Number of organisms remaining
0	10^7
100	$10^{6.2}$
200	$10^{5.4}$
400	$10^{3.8}$
600	$10^{2.1}$
800	$10^{0.6}$
1000	10^{-1}

[a] Combined chlorine residual = 6.0.

Tracer data for chlorine contact basins

Time, min	Tracer concentration, mg/L			
	Chlorine contact basin			
	1	2	3	4
0	0.0	0.0	0.0	0.0
10	0.0	0.0	0.0	0.0
20	0.0	0.0	0.0	0.0
30	0.1	0.0	0.0	0.0
40	2.0	0.0	0.0	0.0
50	7.3	1.1	0.1	0.0
60	7.0	7.0	1.3	0.1
70	5.2	7.3	8.0	1.5
80	3.3	5.7	8.5	7.5
90	1.7	4.2	6.2	8.0
100	0.7	2.9	2.9	5.5
110	0.2	1.7	1.3	3.5
120	0.0	0.9	0.4	1.8
130		0.3	0.0	0.9
140		0.1		0.3
150		0.0		0.1
160				0.0
τ, min	80	85	90	100

12-11 Determine the amount of sulfur dioxide (SO_2), sodium sulfite (Na_2SO_3), sodium bisulfite ($NaHSO_3$), sodium metabisulfite ($Na_2S_2O_5$), and activated carbon (C) that would be required per year to dechlorinate treated effluent containing a combined chlorine residual of 5.0, 6.5, 7.0, or 7.7 mg/L as Cl_2 (residual to be selected by instructor) from a plant with an average flowrate of 1500, 3300, 4600, or 7500 m^3/d (flowrate to be selected by instructor).

12-12 Estimate the immediate ozone demand and the first order decay equation for wastewater number (to be selected by instructor) using the following bench-scale steady-state and decay test data collected at 25°C. If the coefficient of specific lethality for the inactivation of a newly discovered microorganism with ozone is 0.15 L/mg·min, estimate the degree of inactivation that could be achieved at 15°C using an ozone contactor with 4 compartments following the injection of ozone. The theoretical detention time in each compartment is 3 min. Assume the activation energy for ozone for the new microorganism is 48 kJ/mole.

Test	Ozone dose, mg/L	Ozone residual, mg/L Wastewater number			
		1	2	3	4
1	6		2.4	1.0	3.3
2	10	1.1	4.9	5.9	7.0
3	14	6.9	7.4	10.5	10.3
4	18	12.2	10.0	15.5	14.0
5	20	15.0	11.1	18.0	15.7

The corresponding decay data are:

Time, min	Ozone residual, mg/L Wastewater number			
	1	2	3	4
0	3.8	2.8	2.0	3.25
5	2.25	1.4	1.37	2.3
10	1.35	0.72	0.95	1.65
15	0.82	0.37	0.67	1.19
20	0.50	0.19	0.46	0.84

12-13 Estimate the immediate ozone demand and the first order decay equation using the following bench-scale steady-state test data collected at 20°C. If a four-compartment ozone contactor, similar to the one shown on Fig. 12–31(a), is used, estimate the log reduction in *Cryptosporidium* that can be achieved at 5°C. Assume the activation energy for ozone for *Cryptosporidium* is 54 kJ/mole.

The steady-state test results are as follows.

Test	Ozone dose, mg/L	Ozone residual, mg/L
1	5	1.5
2	8	5
3	10	7.5
4	13	10.3
5	18	17.5

The corresponding decay data are:

Time, min	Ozone residual, mg/L
0	5
4	3
7	2.5
10	2

12-14 Given the following ozone decay test data, estimate the number of compartments that would be required in an ozone contactor to achieve 3-log reduction in *Cryptosporidium* based on test number (to be selected by instructor). Assume the theoretical detention time in each compartment is 3 min and the t_{10}/t for the reactor is 0.65.

	Ozone residual, mg/L			
	Test number			
Time, min	1	2	3	4
0	3.3	1.5	3.2	2.8
2	3.0		2.75	
4		1.0		2.1
6	2.0		1.8	
10		0.65		1.8
12	1.5		0.9	
16		0.3		1.6

12-15 Review the current literature and prepare an assessment of the use of ozone for the disinfection of treated wastewater. A minimum of three articles and/or reports dating back to 1995 should be cited in your assessment.

12-16 Review the current literature and prepare an assessment of the use of peracetic acid alone or in combination with other disinfectants. A minimum of three articles and/or reports dating back to 2000 should be cited in your assessment.

12-17 Given the following measurements and data for a collimated beam test, determine the average UV dose delivered to the sample and best estimate of the uncertainty associated with the measurement.

$I_m = 10 \pm 0.5$ mW/cm^2 (accuracy of meter $\pm 7\%$)

$t = 30 \pm 1$ s

$R = 0.025$ (assumed to be the correct value)

$P_f = 0.94 \pm 0.02$

$\alpha = 0.065 \pm 0.005$ cm^{-1}

$d = 1 \pm 0.05$ cm

$L = 48 \pm 0.5$ cm

12-18 If the intensity of the UV irradiation measured at the water surface in a Petri dish is 12 mW/cm^2, determine the average UV intensity to which a sample will be exposed if the depth of water in the Petri dish is 10, 22, 14, 15, or 16 mm (water depth to be selected by instructor).

12-19 If the intensity of UV irradiation measured at the water surface in a Petri dish in Problem 12–18 is 8 mW/cm^2, and that the computed UV dose was based on a water depth of 10 mm. What would be the effect if the actual water depth in the Petri dish were 20 mm?

12-20 Determine the mean, the standard deviation, the 75 percent confidence interval, and the 75 percent prediction interval for following MS2 bacteriophage inactivation data, (test to be selected by instructor) obtained using a collimated beam device. What UV dose would be required to achieve a 4-log inactivation of MS2 based on the lower prediction interval of 75 percent?

| | Log reduction, $-\log N/N_o$ | | | | |
| | Test number | | | | |
Applied UV dose, mJ/cm²	1	2	3	4	5
20	0.9	1.7	1.4	1.1	1
40	1.7	3.3	2.6	2.2	1.8
60	2.4	5.2	4.1	3	2.8
80	3.5	6.5	5.1	4.3	3.7
100	4.3			5.5	4.7
120	4.9			6.2	5.4

12-21 In the latest edition of the NWRI UV Guidelines (NWRI, 2012), the standard dose response curve that should be used to evaluate reactor performance is given as

$$UV\ dose = \frac{\log\ inactivation - 0.5464}{0.0368}$$

The dose response curve used in this chapter is:

$$UV\ dose = \frac{\log\ inactivation - 0.326}{0.0389}$$

Compare the two curves for UV dosages of 20, 40, 60, 80, 100, and 120 mJ/cm² (dosages to be selected by instructor). Is the difference significant? Why is it reasonable to specify a standard curve that all equipment manufacturers should use to evaluate the performance of their equipment?

12-22 A UV reactor comprised of two banks with 4 lamps per bank was tested on two different reclaimed waters (1 and 2) at four flowrates using MS2 bacteriophage as the test organism. The transmittance for both wastewaters was 65 percent. The hydraulic loading rates were varied from 50 to 200 L/min·lamp. In conducting the test, each flowrate was tested randomly with respect to order. The measured inlet and outlet phage concentrations are as follows:

| Flowrate, L/min | Replicate | Wastewater 1, phage/mL | | Wastewater 2, phage/mL | |
		Inlet	Outlet	Inlet	Outlet
200	1	9.65×10^6	3.80×10^1	1.05×10^7	2.19×10^2
200	2	1.00×10^7	3.98×10^1	6.98×10^6	1.54×10^2
200	3	1.15×10^7	3.72×10^2	1.15×10^7	1.70×10^2
400	1	1.00×10^7	1.95×10^3	1.00×10^7	3.75×10^2
400	2	1.29×10^7	1.55×10^3	1.23×10^7	3.62×10^2
400	3	9.55×10^6	1.77×10^3	1.12×10^7	3.08×10^2
600	1	1.23×10^7	1.12×10^3	1.20×10^7	1.32×10^4
600	2	1.05×10^7	9.33×10^3	1.05×10^7	1.05×10^4
600	3	1.25×10^6	8.91×10^3	9.55×10^6	9.95×10^3
800	1	1.13×10^7	4.79×10^4	1.03×10^7	5.95×10^4
800	2	1.08×10^7	8.35×10^4	1.19×10^7	1.00×10^5
800	3	8.95×10^6	6.61×10^4	1.11×10^7	7.68×10^4

Using the given data, for water 1 or 2 (water to be selected by instructor), develop design equations based on (1) the regression analysis, (2) the 75 percent prediction interval, and

(3) the 75 percent prediction interval taking into account lamp aging and fouling for a transmittance value of 72 percent. What is the maximum flowrate per lamp over which the UV system will deliver a dose of 50 mJ/cm^2? Assume the MS2 UV dose response curve given in Example 12–12 will be used for the analysis of the inactivation test results.

12-23 The following MS2 bacteriophage inactivation data were obtained for filtered wastewater with a transmittance of 55 percent at 254 nm with a UV pilot test unit comprised of 6 UV lamps in a single UV bank operated at various ballast settings. The reported inactivation test results are the average of triplicate samples. Determine the maximum flowrate, expressed as L/min·lamp, over which the UV disinfection system will deliver a dose of 100 mJ/cm^2 at a ballast setting of 100 percent, 80 mJ/cm^2 at a ballast setting of 80 percent, or 50 mJ/cm^2 at a ballast setting of 50 percent (ballast setting to be selected by instructor) taking into account lamp fouling and aging. Assume the lamp aging and fouling factor is 60 percent based on the regression equation developed for each ballast setting.

Flowrate, L/min	Ballast output, %	Log$_{10}$ MS2 bacteriophage inactivation
180	100	7.7559
180	80	6.7445
180	50	5.4219
400	100	6.3555
400	80	5.383
400	50	5.383
560	100	5.5775
560	80	4.7606
560	50	3.5547
732	100	5.0718
732	80	4.2549
732	50	3.2046

12-24 Using the data from Problem 12–23, develop the design curve based on the regression analysis taking into account flowrate and ballast settings. Plot the original data, the regression curve, and the 75 percent prediction interval.

12-25 Review the current literature and prepare an assessment of the use of low-pressure low-intensity versus low-pressure high-intensity UV disinfection systems for the disinfection of filtered secondary effluent. A minimum of 3 articles and/or reports dating back to 2005 should be cited in your assessment.

12-26 Using the D and Z values given in Table 12–36, determine whether the CDPH pasteurization requirements of 82°C for 10 s is sufficient to achieve a 4-log reduction in MS2 coliphage.

12-27 The following data were obtained from a pilot-plant pasteurization test. Using these data determine the D and Z values. If the temperature were increased to 68°C, how long would it take to achieve a 4-log reduction?

Temp., °C	Observed log reduction at indicated time		
	3 s	7 s	10 s
60	0.25	0.4	0.5
65	0.55	1.32	1.8
70	1.80	4.35	6.00

REFERENCES

Baumann, E. R., and D. D. Ludwig (1962) "Free Available Chlorine Residuals for Small Nonpublic Water Supplies," *J. AWWA*, **54**, 11, 1379–1388.

Bellar, T. A., and J. J. Lichtenberg, (1974) "Determining Volatile Organics at Microgram-per-Litre Levels by Gas Chromatography," *J. AWWA*, **66**, 12, 739–744.

Bill Sotirakos, B., K. Bircher, and A. Salveson (2013) "Development, Challenges and Validation of a High-Efficiency UV System for Water Reuse and Low Effluent Quality Wastewater," WEAO 2013 Technical Conference, Toronto, Ontario, Canada.

Black & Veatch Corporation (2010) *White's Handbook of Chlorination and Alternative Disinfectants*, 5th. ed., John Wiley & Sons, Inc., Hoboken, New Jersey.

Blackmer, F., K. A. Reynolds, C. P. Gerba, and I. L. Pepper (2000) "Use of Integrated Cell Culture-PCR to Evaluate the Effectiveness of Poliovirus Inactivation by Chlorine," *Appl. Environ. Microbiol.*, **66**, 5, 2267–2268.

Blatchley, E. R. et al. (1995) "UV Pilot Testing: Intensity Distributions and Hydrodynamics," *J. Environ. Eng. ASCE*, **121**, 3, 258–262.

Blume, T., I. Martinez, and U. Neis (2002) "Wastewater Disinfection Using Ultrasound and UV Light," in U. Neis (ed.) *Ultrasound in Environmental Engineering II*, ISSN 0724–0783, ISBN 3-930400-47-2.

Blume, T., and U. Neis (2004) "Combined Acoustical-Chemical Method for the Disinfection of Wastewater," *Chemical Water and Wastewater Treatment, Vol. VIII*, 127–135, Proceedings of the 11th Gothenburg Symposium, Orlando, FL.

Butterfield, C. T., E. Wattie, S. Megregian, and C. W. Chambers (1943) "Influence of pH and Temperature on the Survival of Coliforms and Enteric pathogens When Exposed to Free Chlorine," *U.S. Public Health Service Report*, **58**, 51, 1837–1866.

Caretti, C., and C. Lubello (2003) "Wastewater Disinfection with PAA and UV Combined Treatment: a Pilot Plant Study," *Water Res.*, **37**, 10, 2365–2371.

Chen, D., X. Dong, and R. Gehr (2005) "Alternative Disinfection Mechanisms for Wastewaters Using Combined PAA/UV Processes," in *Proceedings of WEF, IWA and Arizona Water Pollution Control Association Conference, Disinfection 2005*, Mesa, AZ.

Chick, H. (1908) "Investigation of the Laws of Disinfection," *J. Hygiene,* British, **8**, 92–158.

Collins, H. F. (1970) "Effects of Initial Mixing and Residence Time Distribution on the Efficiency of the Wastewater Chlorination Process," paper presented at the California State Department of Health Annual Symposium, Berkeley and Los Angeles, CA, May 1970.

Collins, H. F., and R. E. Selleck (1972) "Process Kinetics of Wastewater Chlorination," *SERL Report* 72–5, Sanitary Engineering Research Laboratory, University of California, Berkeley, CA.

Cooper, R. C., A. T. Salveson, R. Sakaji, G. Tchobanoglous, D. A. Requa, and R. Whitley (2000) "Comparison Of The Resistance of MS2 And Poliovirus to UV And Chlorine Disinfection," Presented at the California Water Reclamation Meeting, Santa Rosa, CA.

Crittenden, J. C., R. R. Trussell, D. W. Hand, K. J. Howe, and G. Tchobanoglous (2012) *Water Treatment: Principles and Design*, 3rd ed., John Wiley & Sons, Inc., New York.

Darby, J., R. Emerick, F. Loge, and G. Tchobanoglous (1999) "The Effect of Upstream Treatment Processes on UV Disinfection Performance," Project 96-CTS-3, *Water Environment Research Foundation*, Alexandria, VA.

Dulbecco, R. (1949) "Reactivation of Ultraviolet Inactivated Bacteriophage by Visible Light," *Nature* **163**, 949–950.

Ekster, A. (2001) Personal communication.

Emerick, R. W., F. J. Loge, D. Thompson, and J. L. Darby (1999) "Factors Influencing Ultraviolet Disinfection Performance Part II: Association of Coliform Bacteria with Wastewater Particles," *Water Environ. Res.*, **71**, 6, 1178–1187.

Emerick, R. W., F. Loge, T. Ginn, and J. L. Darby (2000) "Modeling the Inactivation of Particle-Associated Coliform Bacteria," *Water Environ. Res.*, **72**, 4, 432–438.

Enslow, L. H. (1938) "Chlorine in Sewage Treatment Practice," Chap. VIII, in L. Pearse (ed.) *Modern Sewage Disposal*, Federation of Sewage Works Associations, New York.

EPRI (1996) *UV Disinfection for Water and Wastewater Treatment*, Report CR-105252, Electric Power Research Institue, Inc., Report prepared by Black & Veatch, Kansas City, MO.

Fair, G. M., J. C. Morris, S. L. Chang, I. Weil, and R. P Burden (1948) "The Behavior of Chlorine as a Water Disinfectant," *J. AWWA*, **40**, 10, 1051–1056.

Fair, G. M., and J. C. Geyer (1954) *Water Supply and Waste-Water Disposal*, John Wiley & Sons, Inc. New York.

Friedberg, E. R., G. C. Walker, and W. Siede (1995) *DNA Repair and Mutagenesis,* WSM Press, Washington, DC.

Friess, P. L., C-C. Tang, S-J. Huitric, P. Ackerman and N. Munakata (2013) *Demonstration of Sequential Chlorination for Tertiary Recycled Water Disinfection at the San Jose Creek East Water Reclamation Plant,* Final Report, The Sanitation Districts of Angeles County, Whittier, CA.

Gang, D. C., T. E. Clevenger, and S. K. Banerji (2003) "Modeling Chlorine Decay in Surface Water," *J. Environ. Inform.*, **1**, 1, 21–27.

Gard, S. (1957) *Chemical Inactivation of Viruses, in CIBA Foundation Symposium on the Nature of Viruses*, Little Brown and Company, Boston, MA.

Gehr, R. (2000) Seminar Lecture Notes, Universidad Autonoma Metropolitana, Mexico City, Mexico.

Gehr, R. (2006) Seminar Notes, Presented at III Simposio Internacional en Ingenieria y Ciencias para la Sustenabilidad Ambiental, Universidad Autonoma Metropolitana, Mexico City, Mexico.

Gehr, R., M. Wagner, P. Veerasubramanian, and P. Payment (2003) "Disinfection Efficiency of Peracetic Acid, UV and Ozone after Enhanced Primary Treatment of Municipal Wastewater," *Water Res.* **37**, 19, 4573–4586.

Goff, D. (2012) "Thermal Destruction of Microorganisms," *Dairy Science and Technology Education*, University of Guelph, Canada, www.foodsci.uoguelph.ca/dairyedu/home.html.

Haas, C., and S. Karra, (1984a) "Kinetics of Microbial Inactivation by Chlorine-I Review of Results in Demand-Free Systems," *Water Res.*, **18**, 11, 1443–1449.

Haas, C., and S. Karra (1984b) "Kinetics of Microbial Inactivation by Chlorine-II. Review of Results in Systems with Chlorine Demand," *Water Res.*, **18**, 11, 1451–45.

Haas, C., and S. Karra (1984c) "Kinetics of Wastewater Chlorine Demand Exertion," *J. WPCF*, **56**, 2, 170–182.

Haas, C. N., and J. Joffe (1994) "Disinfection Under Dynamic Conditions: Modification of Hom's Model for Decay," *Environ. Sci. Technol.*, **28**, 7, 1367–1369.

Hall, E. L. (1973) "Quantitative Assessment of Disinfection Interferences," *Water Treat. Exam.*, **22**, 153–174.

Hanawalt, P. C., P. K. Cooper, A. K. Ganesan, and C. A. Smith (1979) "DNA Repair in Bacteria and Mammalian Cells," *Annu. Rev. Biochem.*, **48**, 783–836.

Hanzon, B., J. Hartfelder, S. O'Connell, and D. Murray (2006) "Disinfection Deliberation," *WE&T, Water Environ. Fed.,* **18**, 2, 57–62.

Harp, D. L. (2002) *Current Technology of Chlorine Analysis for Water and Wastewater*, Technical Information Series—Booklet No.17, Lit. no. 7019, Hach Company, Loveland, CO.

Hart, F. L. (1979) "Improved Hydraulic Performance of Chlorine Contact Chambers," *J. WPCF*, **51**, 12, 2868–2875.

Hiatt, C. W. (1964) "Kinetics of the Inactivation of Viruses," *Bacteriol. Rev.*, **28**, 2, 150–163.

Hoff, J. C. (1986) *Inactivation of Microbial Agents by Chemical Disinfectants,* EPA-600/2-86-067, Water Engineering Research Laboratory, U.S. Environmental Protection Agency, Cincinnati, OH.

Hom, L. W. (1972) "Kinetics of Chlorine Disinfection in an Eco-System," *J. Environ. Eng. Div.* ASCE, **98**, SA1, 183–194.

Hudson, A., T. Wong, and R. Lake (2003) *Pasteurization of Dairy Products: Times, Temperatures and Evidence for Control of Pathogens*, Institute of Environmental Science & Research Limited, Christchurch, New Zealand.

IUVA (2011) "Uniform Protocol for Wastewater UV Validation Applications," *IUVE News*, **13**, 2 26–33.

Jalali, Y., S. J. Huitric, J. Kuo, C. C. Tang, S. Thompson, and J. F. Stahl (2005) "UV Disinfection of Tertiary Effluent and Effect on NDMA and Cyanide," paper presented at WEF Technology 2005, San Francisco, CA.

Jin, S., A. A. Mofidi, and K. G. Linden (2006) "Polychromatic UV Fluence Measurement Using Chemical Actinometry, Biodosimetry, and Mathematical Techniques," *J. Environ. Eng. Div.,* ASCE **132**, 8, 831–841.

Kawamura, S. (2000) *Integrated Design and Operation of Water Treatment Facilities,* 2nd ed., Wiley Interscience, New York.

Kelner, A. (1949) "Effect of Visible Light on the Recovery of Streptomyces Griseus Conidia from Ultra-violet Irradiation Injury," *Proc. Nat. Acad. Sci.,* **35**, 73–79.

Kitis, M. (2004) "Disinfection of Wastewater With Peracetic Acid: A Review," *Environ. Int.,* **30**, 1, 47–55.

Kobylinski, E. A., G. L. Hunter, and A. R. Shaw (2006) "On Line Control Strategies for Disinfection Systems: Success and Failure," *Proceedings of the 2006 WEFTEC Conference,* 6371–6394, WEF, Dallas, TX.

Koivunen, J. (2005a) "Inactivation of Enteric Microorganisms with Chemical Disinfectants, UV Irradiation and Combined Chemical/UV Treatments," *Water Res.,* **39**, 8, 1519–1526.

Koivunen, J. (2005b) "Peracetic Acid (PAA) Disinfection of Primary, Secondary and Tertiary Treated Municipal Wastewaters," Water Res., 39, 18, 4445–4453.

Krasner, S. W. (1999) "Chemistry of Disinfection By-Product Formation," in P. C. Singer, (ed.), *Formation and Control of Disinfection By-Products in Drinking Water,* AWWA, Denver, CO.

Krasner, S. W., S. Pastor, R. Chinn, M. J. Sclimenti, H. S. Wienberg, S. D. Richardson, and A. D. Thruston, Jr. (2001) "The Occurrence of a New Generation of DBPs (Beyond the ICR)," paper presented at the AWWA Water Quality Technology Conference, Nashville, TN.

Lazarova, V., M. L. Janex, L. Fiksdal, C. Oberg, I. Barcina, and M Ponimepuy (1998) "Advanced Wastewater Disinfection Technologies: Short and Long Term Efficiency," *Water Sci. Technol.,* **38**, 12, 109–117.

Lefevre, F., J. M. Audic, and F. Ferrand (1992) "Peracetic Acid Disinfection of Secondary Effluents Discharged Off Coastal Seawater," *Water Sci. Technol.,* **25**, 12, 155–164.

Lewis, M. J., and N. J. Heppell (2000) *Continuous Thermal Processing of Food: Pasteurization and UHT Sterilization,* Aspen Publishers, Inc., A Wolters Kluwer Company, Gaithersburg, MD.

Li, H., G. R. Finch, D. W. Smith, and M. Belosevic (2001) *Sequential Disinfection Design Criteria for Inactivation of Cryptosporidium Oocysts in Drinking Water,* AWWA Research Foundation and the American Water Works Association, Denver, CO.

Liberti, L., A. Lopez, and M. Notarnicola (1999) "Disinfection with Peracetic Acid for Domestic Sewage Re-Use in Agriculture," *J. Water Environ. Mgmt.,* (Canadian), **13**, 8, 262–269.

Linden, K., G. Shin, and M. Sobsey (2001) "Comparative Effectiveness of UV Wavelengths For the Inactivation of *Cryptosporidium Parvum* Oocysts in Water," *Water Sci. Technol.,* **43**, 12, 171–174.

Linden, K. G., and A. A. Mofidi (2003) Disinfection Efficiency and Dose Measurement for Polychromatic UV Systems, American Water Works Association Research Foundation, Denver, CO.

Louie, D., and M. Fohrman (1968) "Hydraulic Model Studies of Chlorine Mixing and Contact Chambers," *J. WPCF,* **40**, 2 174–184.

Lubello, C., C. Caretti, and R. Gori (2002) "Comparison Between PAA/UV and H_2O_2/UV Disinfection for Wastewater Reuse," *Water Sci. Technol.: Water Supply*; **2**, 1, 205–212.

Maguin, S. R., P. L. Friess, S-J. Huitric, C-C. Tang, J. Kuo, and N. Munakata (2009) "Sequential Chlorination: A New Approach for Disinfection of Recycled Water," *Environ. Eng: App. Res. Pract.,* **9**, 2–11.

Malley, J. P. (2005) "A New Paradigm For Drinking Water Disinfection," Presented at the 17th World Ozone Congress, International Ozone Association, Strasbourg, Germany.

Martin, N., and R. Gehr (2005) "Photoreactivation Following Combined Peracetic Acid-UV Disinfection of a Physicochemical Effluent," Presented at the Third International Congress on Ultraviolet Technologies, IUVA, Whistler, BC, Canada.

Moce-Uivina, L., M. Muniesa, H. Pimienta-Vale, F. Lucena, and J. Jofre (2003) "Survival of Bacterial Indicator Species and Bacteriophages after Thermal Treatment of Sludge and Sewage," *App. Environ. Microbiol.,* **69**, 3, 1452–1456.

Mofidi, A. A., H. Baribeau, P. A. Rochelle, R. De Leon, B. M. Coffey, and J. F. Green, (2001) "Disinfection of *Cryptosporidium Parvum* with Polychromatic Ultraviolet Light." *J. AWWA,* **93**, 6, 95–109.

Mofidi, A. A., E. A. Meyer, P. M. Wallis, C. I. Chou, B. P. Meyer, S. Ramalingam, and B. M. Coffey (2002) "Effect of Ultraviolet Light on Giardia lamblia and Giardia muris Cysts as Determined by Animal Infectivity." *J. Water Res.,* **36**, 2098–2108.

Morrill, A. B. (1932) "Sedimentation Basin Research and Design," *J. AWWA,* **24**, 9, 1442–1458.

Morris, J. C. (1966) "The Acid Ionization Constant of HOCl from 5°C to 35°C," *J. Phys. Chem.* **70**, 12, 3798–3806.

Morris, J. C. (1975) "Aspects of the Quantitative Assessment of Germicidal Efficiency," Chap. 1, in J. D. Johnson (ed.), *Disinfection: Water and Wastewater,* Ann Arbor Science Publishers, Inc., Ann Arbor, MI.

NRC (1980) "The Disinfection of Drinking Water" in *Drinking Water and Health, Vol. 2.* Safe Drinking Water Committee, Board on Toxicology and Environmental Health Hazards, Assembly of Life Sciences, National Research Council, The National Academies Press, Washington, DC.

NWRI (1993) *UV Disinfection Guidelines for Wastewater Reclamation in 33California and UV Disinfection Research Needs Identification,* National Water Research Institute, Prepared for the California Department of Health Services. Sacramento, CA.

NWRI and AWWARF (2000) *Ultraviolet Disinfection Guidelines for Drinking Water and Wastewater Reclamation,* NWRI-00–03, National Water Research Institute and American Water Works Association Research Foundation, Fountain Valley, CA.

NWRI (2003) *Ultraviolet Disinfection Guidelines for Drinking Water and Water Reuse,* 2nd ed., National Water Research Institute, Fountain Valley, CA.

NWRI (2012) *Ultraviolet Disinfection Guidelines for Drinking Water and Water Reuse,* 3rd ed., Updated Edition, National Water Research Institute, Fountain Valley, CA.

O'Brien, W. J., G. L. Hunter, J. J. Rosson, R. A. Hulsey, and K. E. Carns (1996) "Ultraviolet System Design: Past, Present, and Future, Proceedings Disinfecting Wastewater for Discharge & Reuse," *Water Environment Federation,* Alexandria, VA.

Oguma, K., H. Katayama, H. Mitani, S. Morita, T. Hirata, and S. Ohgaki (2001) "Determination of Pyrimidine Dimers in *Escherichia Coli* and *Cryptosporidium Parvum* During Ultraviolet Light Inactivation, Photoreactivation and Dark Repair," *Appl. Environ. Microbiol.,* **67**, 4630–4637.

Oguma, K., H. Katayama, and S. Ohgaki (2002) "Photoreactivation of *Escherichia coli* after Low- or Medium-Pressure UV Disinfection Determined by an Endonuclease Sensitive Site Assay," *Appl. Environ. Microbiol.,* **68**, 12, 6029–6035.

Oguma, K., H. Katayama, and S. Ohgaki. (2004) "Photoreactivation of *Legionella Pneumophilia* after Inactivation by Low- or Medium-Pressure Ultraviolet Lamp," *Water Res.,* **38**, 11, 2757–2763.

Orta de Velasquez, M. T., I. Yanez-Noguez, N. M. Rojas-Valencia and C.l. Lagona-Limon (2005) "Ozone in the Disinfection of Municipal Wastewater Compared with Peracetic Acid, Hydrogen Peroxide, and Copper after Advanced Primary Treatment," Presented at the 17th International Ozone Association World Congress & Exhibition, Strasbourg, France.

Parker, J. A., and J. L. Darby (1995) "Particle-Associated Coliform in Secondary Effluents: Shielding From Ultraviolet Light Disinfection," *Water Environ. Res.,* **67**, 7, 1065–1075.

Pflug, I. J., R. G. Holcomb, and M. M. Gomez. (2001) "Principles of the Thermal Destruction of Microorganisms," in S.S. Block (ed.) *Disinfection, Sterilization, and Preservation,* Lippincott Williams & Wilkins, Philadelphia, PA.

Plummer, J. D., and S. C. Long (2005) "Enhancement of Chlorine Inactivation with Chemical Free Sonication," Presented at the Water Quality Technology Conference, Quebec City, Canada.

Qualls, R. G., M. P. Flynn, and J. D. Johnson (1983) "The Role of Suspended Particles in Ultraviolet Disinfection," *J. WPCF,* **55**, 10, 1280–1285.

Qualls, R. G., and J. D. Johnson (1985) "Modeling and Efficiency of Ultraviolet Disinfection Systems," *Water Res., * **19**, 8, 1039–1046.

Rakness, K. L. (2005) *Ozone in Drinking Water Treatment: Process Design, Operation and Optimization,* American Water Works Association, Denver, CO.

Rennecker, J., B. Marinas, J. Owens, and E. Rice (1999) "Inactivation of Cryptosporidium Parvum Oocysts with Ozone," *Water Res., * **33**, 11, 2481–2488.

Rice, R. G. (1996) *Ozone Reference Guide,* Prepared for the Electric Power Research Institute, Community Environment Center, St. Louis, MO.

Roberts, P. V., E. M. Aieta, J. D. Berg, and B. M. Chow (1980) "Chlorine Dioxide for Wastewater Disinfection: A Feasibility Evaluation," Technical Report No. 21, Civil Engineering Department, Stanford University, Stanford, CA.

Rochelle, P. A., A. A. Mofidi, M. M., Marshall, S. J. Upton, B. Montelone, K. Woods, and G. DiGiovanni (2004) *An Investigation of UV Disinfection and Repair in Cryptosporidium Parvum,* AWWA Research Foundation, Denver, CO.

Rook, J. J. (1974) "Formation of Haloforms During the Chlorination of Natural Water," *Water Treat. Exam., * **23**, 2, 234–243.

Rossman, L. A. (2000) *EPANET 2 Users Manual,* EPA/600/R-00/057, U.S. Environmental Protection Agency, Cincinnati, OH.

Rupert, C. S. (1960) "Photoreactivation of Transforming DNA by an Enzyme from Baker's Yeast," *J. Gen. Physiol., * **43**, 573–595.

Salveson, A., N. Gael, and G. Ryan (2011) "Not Just for Milk Anymore: Pasteurization for Disinfection of Wastewater and Reclaimed Water," *Water Environ. Tech., * **23**, 3, 43–45.

Salveson, A. (2012) Personal Communication on Inactivation of Total Coliform by Pasteurization, Carollo Engineers, Walnut Creek, CA.

Saunier, B. M. (1976) *Kinetics of Breakpoint Chlorination and Disinfection,* Ph.D. Thesis, University of California, Berkeley, CA.

Saunier, B. M., and R. E. Selleck (1976) "The Kinetics of Breakpoint Chlorination in Continuous Flow Systems," Paper presented at the American Water Works Association Annual Conference, New Orleans, LA.

Severin, B. F., M. T. Suidan, and R. S. Engelbrecht (1983) "Kinetic Modeling of UV Disinfection of Water," *Water Res., * British, **17**, 11, 1669–1678.

Sinikanchana, K., J. Shisler, and B. Marinas (2005) "Sequential Inactivation of Adenoviruses by UV and Chlorine," presented at 2005 Water Quality Technology Conference and Exposition, American Water Works Association, Denver CO.

Snyder, C. H. (1995) *The Extraordinary Chemistry of Ordinary Things,* 2nd ed., John Wiley & Sons, Inc., New York.

Solvay Chemicals, Inc. (2013) Proxitane™ WW-12 Peracetic Acid Technical Data Sheet, Houston, TX.

Sommer, R., W. Pribil, S. Appelt, P. Gehringer, H. Eschweiler, H. Leth, A. Cabal, and T. Haider (2001) "Inactivation of Bacteriophages in Water by Means of Non-Ionizing (UV-253.7 nm) and Ionizing (Gamma) Radiation: A Comparative Aproach," *Water Res., * **35**, 13, 3109–3116.

Sorqvist, S. (2003) "Heat Resistance in Liquids of *Enterococcus* spp., *Listeria* spp., *Escherichia coli, Yersinia Enterocolitica, Salmonella* spp. and *Campylobacter* spp.," *Acta Vet. Scand, * **44**, 1–2, 1–19.

Sung, R. D. (1974) *Effects of Organic Constituents in Wastewater on the Chlorination Process,* Ph.D. thesis, Department of Civil Engineering, University of California, Davis, CA.

Tchobanoglous, G., F. L. Burton, and H. D. Stensel (2003) *Wastewater Engineering: Treatment and Reuse,* 4th ed., Metcalf and Eddy, Inc., McGraw-Hill Book Company, New York.

Thibaud, H, J. De Laat, and M. Dore (1987) "Chlorination of Surface Waters: Effect of Bromide Concentration on the Chloropicrin Formation Potential," paper presented at the Sixth Conference on Water Chlorination: Environmental Impact and Health Effects, Oak Ridge, Tennessee.

Toder, K. (2012) *Online Textbook of Bacteriology,* http://textbookofbacteriology.net

U.S. EPA (1986) *Design Manual, Municipal Wastewater Disinfection,* EPA/625/1-86/021, U.S. Environmental Protection Agency, Cincinnati, OH.

U.S. EPA (1992) *User's Manual for UVDIS, Version 3.1, UV Disinfection Process Design Manual*, EPA G0703, Risk Reduction Engineering Laboratory, U.S. Environmental Protection Agency, Cincinnati, OH.

U.S. EPA (1999a) *Alternative Disinfectants and Oxidants Guidance Manual*, EPA 815-R-99-014, U.S. Environmental Protection Agency, Cincinnati, OH.

U.S. EPA (1999b) *Combined Sewer Overflow Technology Fact Sheet, Alternative Disinfection Methods*, EPA832-F-99-033, U.S. Environmental Protection Agency, Cincinnati, OH.

U.S. EPA (2003a) *EPA Guidance Manual*, Appendix B. CT Tables, LT1ESWTR, Disinfection Profiling and Benchmarking, U.S. Environmental Protection Agency, Washington, DC.

U.S. EPA (2003b) *Ultraviolet Disinfection Guidance Manual*, Draft, U.S. Environmental Protection Agency, Office of Water, Washington, DC.

U.S. EPA (2006) *National Primary Drinking Water Regulations: Long Term 2 Enhanced Surface Water Treatment Rule*, LT2ESWTR, Federal Register, **71**, 3, 654–786.

U.S. EPA (2010) *Long Term 2 Enhanced Surface Water Treatment Rule Toolbox Guidance Manual*, EPA 815-R-09-016, Office of Water, U.S. Environmental Protection Agency, Washington, DC.

Wagner, M., D. Brumelis, and R. Gehr (2002) "Disinfection of Wastewater by Hydrogen Peroxide or Peracetic Acid: Development of Procedures for Measurement of Residual Disinfectant and Application to a Physicochemically Treated Municipal Effluent," *Water Environ. Res.*, **74**, 33, 33–50.

Wang, H., X. Shao, and R. A. Falconer (2003) "Flow and Transport Simulation Models for Prediction of Chlorine Contact Tank Flow-Through Curves," *Water Environ. Res.*, **75**, 5, 455–471.

Ward, R. W., and DeGraeve (1976) *Disinfection Efficiency and Residual Toxicity of Several Wastewater Disinfectants*, EPA-600/2-76-156, U.S. Environmental Protection Agency, Cincinnati, OH.

Watson, H. E. (1908) "A Note On The Variation of the Rate of Disinfection With Change in the Concentration of the Disinfectant," *J. Hygiene* (British), **8**, 536.

Wattie, E., and C. T. Butterfield, (1944) "Relative resistance of *Escherichia Coli* and Eber-thella Typhosa to Chlorine and Chloramines," *U.S. Public Health Service Report*, **59**, 52, 1661–1671.

White, G. C. (1999) *Handbook of Chlorination and Alternative Disinfectants*, 4th ed., John Wiley & Sons, Inc., New York (see also Black & Veatch Corporation, 2010).

Zimmer, J. L., and R. M. Slawson (2002) "Potential Repair of *Escherichia coli* DNA following Exposure to UV Radiation from Both Medium- and Low-Pressure UV Sources Used in Drinking Water Treatment," *Appl. Environ. Microbiol.*, **68**, 7, 3293–3299.

13

Processing and Treatment of Sludges

WORKING TERMINOLOGY

Term	Definition
Acid-gas digestion	A modified process of anaerobic digestion where the acid phase hydrolysis is separated from the gas producing phase for increased volatile solids reduction.
Aerobic digestion	Biological stabilization process operated in the presence of oxygen in which the biodegradable matter in primary and secondary sludge is oxidized to carbon dioxide and other end products.
Anaerobic digestion	Biological stabilization process operated in the absence of oxygen in which the biodegradable matter in primary and secondary sludge is converted to methane, carbon dioxide, and other end products.
Autothermal thermophilic aerobic digestion (ATAD)	An aerobic digestion process in which the microbes generate enough heat to maintain temperatures in the thermophilic range. When maintained for enough time to meet 40CFR 503 requirements, the process results in biosolids that are relatively pathogen free and meet Class A standards.
Biosolids	Sludge from wastewater treatment processes that has been stabilized to meet the criteria in the U.S. EPA's 40 CFR 503 regulations and, therefore, can be used beneficially.
Class A biosolids	Biosolids that contain less than 1000 most probable number (MPN)/g of fecal coliforms and less than 3 MPN/4g of Salmonella bacteria and meet one of six stabilization alternatives given in 40 CFR 503. The material also must meet the pollutant limits and vector attraction reduction requirements set forth in 40 CFR 503.
Class B biosolids	Biosolids that contain less than 2 million colony-forming units (CFU) or most probable number (MPN) of fecal coliforms per gram of dry biosolids. The material also must meet the pollutant limits and vector attraction reduction requirements set forth in 40 CFR 503.
Digestion	The process of biologically degrading organic matter in sludge, thereby reducing the concentrations of volatile solids and pathogens.
Disposition	Disposition is a term used to reflect disposal of biosolids or sludge for either beneficial or non-beneficial use due to the value of the material.
Dissolved air flotation	A clarification process in which small air bubbles become attached to flocculated material, float to the surface, and are removed by skimming. Heavier solids which settle are removed by mechanical scrapers.
Dual digestion	A two-stage digestion process wherein the first stage is aerobic thermophilic digestion and the second stage is mesophilic anaerobic digestion. High-purity oxygen has also been used for the first stage.
Grit	Sand, gravel, cinders, other heavy inorganic materials and also organic matter such as eggshells, bone chips, seeds, and coffee grounds.

Term	Definition
Humus	Sludge removed from trickling filters.
Mesophilic anaerobic digestion	Anaerobic digestion that occurs in a temperature range of 30 to 38°C (85 to 100°F).
Methanogenesis	The metabolic conversion of organic acids or hydrogen and carbon dioxide to methane.
Screenings	The material removed from a screening device.
Scum	Buoyant materials (e.g., grease, food waste, paper, and foam) often found floating on the surface of primary and secondary clarifiers and thickeners.
Sidestream	A portion of the wastewater flow that has been diverted from the main treatment process flow for specialized treatment.
Solids	A term often used as a replacement for sludges that have not been stabilized by physical, chemical or biological treatment. The term solids is not used as a substitute for sludge in this chapter. The mass of dry material in sludge is referred to as the solids content.
Sludge	Any material (i.e., sludge) produced during primary, secondary, or advanced wastewater treatment that has not undergone any process to reduce pathogens or vector attraction.
Stabilization	A treatment process designed to reduce the number of pathogens in sludge and to reduce the attraction of vectors as defined in the requirements of 40 CFR 503.
Thermal hydrolysis	A thermal conditioning process, utilizing high-pressure steam for pretreating dewatered sludge prior to anaerobic digestion, that hydrolyzes and reduces the viscosity of the sludge.
Thermophilic anaerobic digestion	Anaerobic digestion that occurs in a temperature range of 50 to 57°C (122 to 135°F).

The constituents removed and/or produced in wastewater treatment plants include screenings, grit, scum, sludge, and biosolids. The sludge and biosolids (formerly collectively called sludge) resulting from wastewater treatment processes are usually in the form of a liquid or semisolid liquid, which typically contains from 0.25 to 12 percent solids by weight, depending on the operations and processes used. In the United States, the term *biosolids*, as defined by the Water Environment Federation (WEF 2010a), refers to any sludge that has been stabilized to meet the criteria in the U.S. Environmental Protection Agency's 40 CFR 503 regulations and, therefore, can be used beneficially. The term *sludge* is only used before beneficial use criteria (discussed in Sec. 14–2) have been achieved. The term *sludge* is generally used in conjunction with a process descriptor, such as *primary sludge*, *enhanced primary sludge*, *waste activated sludge*, and *secondary sludge*. Although the terms solids and has been used as a substitute for sludge, to avoid confusion only the terms *sludge*, as defined above, and *biosolids* are used in this chapter and book.

Of the constituents removed by treatment, sludge is by far the largest in volume, and its processing, reuse, and disposition present perhaps the most complex problem in the field of wastewater treatment. For this reason, two chapters have been devoted to this subject. The disposition of grit and screenings is discussed in Chap. 5. The problems of dealing with sludge are complex because (1) sludge is composed largely of the substances responsible for the offensive character of untreated wastewater; (2) the portion of sludge produced from biological treatment requiring disposition is composed of the organic matter contained in the wastewater but in another form, and it, too, will decompose and become offensive; and (3) only a small part of sludge is solid matter.

The purpose of this chapter is to describe the principal processes and methods used for sludge processing and treatment as identified in Table 13–1. Resource recovery

Table 13–1

Sludge handling and processing methods

Handling or processing method	Function	See Sec.
Pumping	Transport of sludge and biosolids	13–4
Preliminary operation		13–5
Grinding	Particle size reduction	13–5
Screening	Removal of fibrous material	13–5
Degritting	Grit removal	13–5
Blending	Homogenization of sludge	13–5
Storage	Flow equalization	
Thickening		
Gravity thickening	Volume reduction	13–6
Flotation thickening	Volume reduction	13–6
Centrifugation	Volume reduction	13–6
Gravity belt thickening	Volume reduction	13–6
Rotary drum thickening	Volume reduction	13–6
Stabilization		
Alkaline stabilization	Stabilization	13–8
Anaerobic digestion	Stabilization, mass reduction, resource recovery	13–9
Aerobic digestion	Stabilization, mass reduction	13–10
Composting	Stabilization, product recovery	14–5
Heat drying	Stabilization, volume reduction, resource recovery	14–3
Conditioning	Improve dewatering	14–1
Dewatering		
Centrifuge	Volume reduction	14–2
Belt filter press	Volume reduction	14–2
Rotary press	Volume reduction	14–2
Screw press	Volume reduction	14–2
Filter press	Volume reduction	14–2
Advanced dewatering	Volume reduction and stabilization	14–2
Drying beds	Volume reduction	14–2
Reed beds	Storage and volume reduction	14–2
Lagoons	Storage and volume reduction	14–2
Advanced Thermal Oxidation	Volume and mass reduction, resource recovery	14–4
Application of biosolids to land	Beneficial use and disposition	14–10
Conveyance and storage	Transport and storage of sludge and biosolids	14–6

methods, also identified in Table 13–1, and the beneficial use of the biosolids is discussed in Chapter 14. To understand the various sludge handling and processing methods, the first two sections of this chapter are devoted to a discussion of the sources, characteristics, and quantities of sludge; the current regulatory environment; and a presentation of representative sludge-treatment process flow diagrams. Because the pumping of sludge is a fundamental part of wastewater treatment plant design, a separate discussion (Sec. 13–4) is devoted to sludge and scum pumping. The preliminary processing of sludge is discussed in Secs. 13–5 and 13–6. Stabilization of sludge is introduced in Sec. 13–7 and is divided into three subsequent sections for more detailed discussion: alkaline stabilization, anaerobic digestion, and aerobic digestion (see Secs. 13–8 through 13–10). Composting, also used for sludge stabilization after dewatering, is considered in Chap. 14.

13–1 SLUDGE SOURCES, CHARACTERISTICS, AND QUANTITIES

To design sludge processing, treatment, and disposition facilities properly, the sources, characteristics, and quantities of the sludge to be handled must be known. The method of primary and secondary treatment of wastewater has a significant impact on quantity and quality of the sludge produced. For example, using membrane bioreactors in secondary treatment produces sludge that is difficult to dewater and digest anaerobically, as compared to using a conventional waste activated sludge process. Stringent regulations for producing high quality effluent have an impact on the process used for secondary treatment, which in turn impact the quantity and quality of biosolids produced from sludge. For example, using biological nutrient removal (BNR) systems to meet stringent nutrient effluent quality produces lesser amount of sludge, but a sludge that is more difficult to process downstream by dewatering or digestion. The purpose of this section is to present background data and information on these topics that will serve as a basis for the material to be presented in the subsequent sections of this chapter.

Sources

The sources of sludge in a treatment plant vary according to the type of plant and its method of operation. The principal sources of sludge and the types generated are reported in Table 13–2. For example, in a complete mix activated sludge process, if the wasting of sludge is accomplished from the mixed liquor line or aeration chamber, the activated sludge

Table 13–2

Sources of sludge from conventional wastewater treatment plants[a]

Unit operation or process	Types of sludge	Remarks
Preaeration	Grit and scum	In some plants, scum removal facilities are not provided in preaeration tanks. If the preaeration tanks are not preceded by grit removal facilities, grit deposition may occur in preaeration tanks.
Primary sedimentation	Primary and scum	Quantities of sludge and scum depend upon the nature of the collection system and whether industrial wastes are discharged to the system.
Biological treatment	Secondary and scum	Suspended solids are produced by the biological conversion of BOD. Some form of thickening may be required to concentrate the waste sludge stream from biological treatment.
		Provision for scum removal from secondary settling tanks is a requirement of the U.S. EPA.

[a] The coarse material removed by screening and grit during preliminary treatment are considered in Chap. 5.

settling tank is not a source of sludge. On the other hand, if wasting is accomplished from the activated sludge return line, the activated sludge settling tank constitutes a source of sludge. Processes used for thickening, digesting, conditioning, and dewatering of sludge produced from primary and secondary settling tanks also constitute sources.

Characteristics

To treat and reuse the sludge produced from wastewater treatment plants in the most effective manner, it is important to know the characteristics of the sludge that will be processed. The characteristics vary depending on the origin of the sludge, the amount of aging that has taken place, and the type of processing to which the sludge has been subjected (see Table 13–3).

General Composition. Typical data on the chemical composition of sludges are reported in Table 13–4. Many of the chemical constituents, including nutrients, are important in considering the ultimate disposition of the processed sludge and the liquid removed during processing. The measurement of pH, alkalinity, and organic acid content

Table 13–3
Characteristics of sludge and biosolids produced during wastewater treatment

Type	Description
Scum/grease	Scum consists of the floatable materials skimmed from the surface of primary and secondary settling tanks. Scum may contain grease, vegetable and mineral oils, animal fats, waxes, soaps, food wastes, vegetable and fruit skins, hair, paper and cotton, cigarette tips, plastic materials, condoms, grit particles, and similar materials. The specific gravity of scum is less than 1.0 and usually around 0.95.
Primary sludge	Sludge from primary settling tanks is usually gray and slimy and, in most cases, has an extremely offensive odor. Primary sludge can be readily digested under suitable conditions of operation.
Sludge from chemical precipitation	Sludge from chemical precipitation with metal salts is usually dark in color, though its surface may be red if it contains much iron. Lime sludge is grayish brown. The odor of chemical sludge may be objectionable, but is not as bad as primary sludge. While chemical sludge is somewhat slimy, the hydrate of iron or aluminum in it makes it gelatinous. If the sludge is left in the tank, it undergoes decomposition similar to primary sludge, but at a slower rate. Substantial quantities of gas may be given off and the sludge density increased by long residence times in storage.
Activated sludge	Activated sludge generally has a brown flocculant appearance. If the color is dark, the sludge may be approaching a septic condition. If the color is lighter than usual, there may have been underaeration with a tendency for the sludge to settle slowly. Sludge in good condition has an inoffensive "earthy" odor. The sludge tends to become septic rapidly and then has a disagreeable odor of putrefaction. Activated sludge digests well aerobically, but not anaerobically.
Trickling filter sludge	Humus sludge from trickling filters is brownish, flocculant, and relatively inoffensive when fresh. It generally undergoes decomposition more slowly than other undigested sludges. When trickling filter sludge contains many worms, it may become inoffensive quickly. Trickling filter sludge digests readily.
Aerobically digested biosolids	Aerobically digested biosolids are brown to dark brown and have a flocculant appearance. The odor of aerobically digested sludge is not offensive; it is often characterized as musty. Well digested aerobic sludge dewaters easily on drying beds.
Anaerobically digested biosolids	Anaerobically digested biosolids are dark brown to black and contain an exceptionally large quantity of gas. When thoroughly digested, they are not offensive, the odor being relatively faint and like that of hot tar, burnt rubber, or sealing wax.

ª The characteristics of the coarse material removed by screening and grit during preliminary treatment are considered in Chap. 5.

Table 13–4

Typical chemical composition of untreated primary and activated sludge[a]

Item	Untreated primary sludge		Untreated activated sludge	
	Range	Typical	Range	Typical
Total dry solids (TS),%	1–6	3	0.4–1.2	0.8
Volatile solids (% of TS)	60–85	75	60–85	70
Grease and fats (% of TS)	5–8	6	5–12	8
Protein (% of TS)	20–30	25	32–41	36
Nitrogen (N, % of TS)	1.5–4	2.5	2.4–5	3.8
Phosphorus (P_2O_5, % of TS)	0.8–2.8	1.6	2.8–11	5.5
Potash (K_2O, % of TS)	0–1	0.4	0.5–0.7	0.6
Cellulose (% of TS)	8–15	10	—	
Iron (not as sulfide)	2–4	2.5	—	
Silica (SiO_2, % of TS)	15–20	—	—	
pH	5–8	6	6.5–8	7.1
Alkalinity (mg/L as $CaCO_3$)	500–1500	600	580–1100	790
Organic acids (mg/L as HAc)	200–2000	500	1100–1700	1350
Energy content, kJ/kg VSS	23,000–29,000	25,000	19,000–23,000	20,000

[a] Adapted, in part, from U.S. EPA (1979).

Note: kJ/kg × 0.4303 = Btu/lb.

is important in process control of anaerobic digestion. The content of heavy metals, pesticides, and hydrocarbons has to be determined when incineration and land application methods are contemplated. The thermal content of sludge is important where a thermal reduction process such as incineration or gasification is considered.

Specific Constituents. Biosolids characteristics that affect their suitability for application to land and for beneficial use include organic content (usually measured as volatile solids), nutrients, pathogens, metals, and toxic organics. The fertilizer value of the biosolids, which should be evaluated where they are to be used as a soil conditioner, is based primarily on the content of nitrogen, phosphorus, and potassium (potash). Typical nutrient values of wastewater biosolids as compared to commercial fertilizers are reported in Table 13–5. In most land application systems, biosolids provide sufficient nutrients for

Table 13–5

Comparison of nutrient levels in commercial fertilizers and wastewater biosolids

Product	Nutrients, %		
	Nitrogen	Phosphorus	Potassium
Fertilizers for typical agricultural use[a]	5	10	10
Typical values for stabilized wastewater biosolids (based on TS)[b]	3.3	2.3	0.3

[a] The concentrations of nutrients may vary widely depending upon the soil and crop needs.

[b] The concentrations of nutrients may vary depending on wastewater nutrient removal requirements.

Table 13–6

Typical metal content in wastewater solids[a]

Metal	Range of dry solids, mg/kg[b]
Arsenic[c]	1.18–49.2
Cadmium[c]	0.21–11.8
Chromium[c]	6.74–1160
Cobalt	0.87–290
Copper[c]	115–2580
Iron	1575–299,000
Lead[c]	5.81–450
Manganese	34.8–14,900
Mercury[c]	0.17–8.3
Molybdenum[c]	2.51–132
Nickel	7.44–526
Selenium[c]	1.1–24.7
Tin	7.5–522
Zinc[c]	216–8550

[a] US EPA (2009).

[b] Because of the wide range of values there is no typical value.

[c] Metals currently regulated under 40 CFR 503.

plant growth. In some applications, the phosphorus and potassium content may be low and require augmentation.

Trace elements are those inorganic chemical elements that, in very small quantities, can be essential or detrimental to plants and animals. The term "heavy metals" is used to denote several of the trace elements present in sludge and biosolids. Concentrations of heavy metals may vary widely, as indicated in Table 13–6. Due to successful implementation of pretreatment programs, quality of sludge and biosolids in terms regulated heavy metals improved significantly. For the application of biosolids to land, concentrations of heavy metals may limit the application rate and the useful life of the application site (see Sec. 14–10).

Quantities

Data on the quantities of sludge produced from various processes and operations are presented in Table 13–7. Although the data in Table 13–7 are useful as presented, it should be noted that the quantity of sludge produced would vary widely. Corresponding data on expected sludge concentrations from various processes are given in Table 13–8.

Quantity Variations.   The quantity of semi-solid and solid material entering the wastewater treatment plant daily may be expected to fluctuate over a wide range. To ensure capacity capable of handling these variations, the following items must be considered in the design of sludge processing and disposition facilities (1) the average and maximum rates of sludge production, and (2) the potential storage capacity of the treatment units within the plant. The variation in daily quantity of sludge that may be expected in large cities is shown

Table 13-7

Typical data for the physical characteristics and quantities of sludge produced from various wastewater treatment operations and processes

Treatment operation or process	Specific gravity of solids	Specific Gravity of Sludge	Dry solids lb/10³ gal		Dry solids kg/10³ m³	
			Range	Typical	Range	Typical
Primary sedimentation	1.4	1.02	0.9–1.4	1.25	110–170	150
Activated sludge	1.25	1.05	0.6–0.8	0.7	70–100	80
Trickling filter	1.45	1.025	0.5–0.8	0.6	60–100	70
Extended aeration	1.3	1.015	0.7–1.0	0.8ᵃ	80–120	100ᵃ
Aerated lagoon	1.3	1.01	0.7–1.0	0.8ᵃ	80–120	100ᵃ
Filtration	1.2	1.005	0.1–0.2	0.15	12–24	20
Algae removal	1.2	1.005	0.1–0.2	0.15	12–24	20
Chemical addition to primary tanks for phosphorous removal						
Low lime (350–500 mg/L)	1.9	1.04	2.0–3.3	2.5ᵇ	240–400	300ᵇ
High lime (800–1600 mg/L)	2.2	1.05	5.0–11.0	6.6ᵇ	600–1300	800ᵇ
Suspended growth nitrification	—	—	—	—	—	—ᶜ
Suspended growth denitrification	1.2	1.005	0.1–0.25	0.15	12–30	18
Roughing filters	1.28	1.02	—	—ᵈ	—	—ᵈ

ᵃ Assuming no primary treatment.

ᵇ Solids in addition to that normally removed by primary sedimentation.

ᶜ Negligible.

ᵈ Included in biosolids production from secondary treatment processes.

on Fig. 13–1. The curve is characteristic of large cities having a number of large collection lines laid on flat slopes; even greater variations may be expected at small plants.

A limited quantity of sludge may be stored temporarily in the sedimentation and aeration tanks. Where digestion tanks with varying levels are used, their large storage capacity provides a substantial dampening effect on peak digested sludge loads. In sludge treatment systems where digestion is used, the design is usually based on maximum monthly loadings to provide a minimum of 15 d residence time during these maximum month loadings. Thus, based on average daily loadings, digesters have some sludge storage capacity. Where digestion is not used, the sludge treatment processes should be designed based on the inherent storage capacity available in the sludge handling system. For example, the mechanical dewatering system following gravity thickening could be based on the maximum 1 or 3 d sludge production. Certain components of the sludge processing system, such as sludge pumping and thickening, are sized to handle the maximum-day conditions.

Volume-Mass Relationships. The volume of sludge depends mainly on its water content and only slightly on the character of the solid matter. A 10 percent sludge, for example, contains 90 percent water by weight. If the solid matter is composed of fixed

Table 13-8

Expected solids concentrations from various treatment operations and processes

Operation or process application	Solids concentration, % dry solids	
	Range	Typical
Primary settling tank		
Primary sludge	1–6	3
Primary sludge to a cyclone degritter	0.5–3	1.5
Primary sludge and waste activated sludge	1–4	2
Primary sludge and trickling filter humus	4–10	5
Primary sludge with iron addition for phosphorus removal	0.5–3	2
Primary sludge with low lime addition for phosphorus removal	2–8	4
Primary sludge with high lime addition for phosphorus removal	4–16	10
Scum	3–10	5
Secondary settling tank		
Waste activated sludge with primary settling	0.5–1.5	0.8
Waste activated sludge without primary settling	0.8–2.5	1.3
High purity oxygen activated sludge with primary settling	1.3–3	2
High purity oxygen activated sludge without primary settling	1.4–4	2.5
Trickling filter humus	1–3	1.5
Rotating biological contactor waste sludge	1–3	1.5
Gravity thickener		
Primary sludge only	3–10	5
Primary sludge and waste activated sludge	2–6	3.5
Primary sludge and trickling filter humus	3–9	5
Dissolved air flotation thickener		
Waste activated sludge with polymer addition	4–6	5
Waste activated sludge without polymer addition	3–5	4
Centrifuge thickener (waste activated sludge only)	4–8	5
Gravity belt thickener (waste activated sludge with polymer addition)	3–6	5
Anaerobic digester		
Primary sludge	2–5	4
Primary sludge and waste activated sludge	1.5–4	2.5
Primary sludge and trickling filter humus	2–4	3
Aerobic digester		
Primary sludge only	2.5–7	3.5
Primary sludge and waste activated sludge	1.5–4	2.5
Waste activated sludge only	0.8–2.5	1.3

Figure 13–1

Peak sludge load as a function of the average daily load.

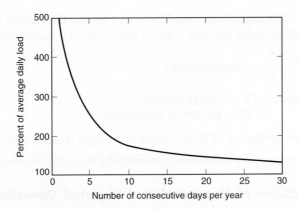

(mineral) and volatile (organic) solids, the specific gravity of all of the solid matter can be computed using Eq. (13–1).

$$\frac{W_s}{S_s \rho_w} = \frac{W_f}{S_f \rho_w} + \frac{W_v}{S_v \rho_w} \qquad (13-1)$$

where W_s = weight of solids
S_s = specific gravity of solids
ρ_w = density of water
W_f = weight of fixed solids (mineral matter)
S_f = specific gravity of fixed solids
W_v = weight of volatile solids
S_v = specific gravity of volatile solids

Therefore, if one-third of the solid matter in a sludge containing 90 percent water is composed of fixed mineral solids with a specific gravity of 2.5, and two-thirds is composed of volatile solids with a specific gravity of 1.0, then the specific gravity of all solids S_s would be equal to 1.25, as follows:

$$\frac{1}{S_s} = \frac{0.33}{2.5} + \frac{0.67}{1.0} = 0.82$$

$$S_s = \frac{1}{0.82} = 1.25$$

If the specific gravity of the water is taken to be 1.0, the specific gravity of the sludge S_{sl} is 1.02, as follows:

$$\frac{1}{S_{sl}} = \frac{0.1}{1.25} + \frac{0.9}{1.0} = 0.98$$

$$S_{sl} = \frac{1}{0.98} = 1.02$$

The volume of sludge may be computed with the following expression:

$$V = \frac{M_s}{\rho_w S_s P_s} \qquad (13-2)$$

where V = volume, m³
M_s = mass of dry solids, kg
ρ_w = specific weight of water, 10^3 kg/m³
S_{sl} = specific gravity of the sludge
P_s = percent solids expressed as a decimal

For approximate calculations for a given solids content, it is simple to remember that the volume varies inversely with the percent of solid matter contained in the sludge as given by

$$\frac{V_1}{V_2} = \frac{P_2}{P_1} \quad \text{(approximate)}$$

Where V_1, V_2 = sludge volumes
$\qquad$ P_1, P_2 = percent of solid matter

The application of these volume and weight relationships is illustrated in Example 13–1.

EXAMPLE 13–1   **Volume of Untreated and Digested Dewatered Sludge**   Determine the liquid volume before and after digestion and dewatering and the percent reduction for 500 kg (dry basis) of primary sludge with the following characteristics:

	Primary	Digested and dewatered
Solids, %	5	20
Volatile matter, %	80	60 (destroyed)
Specific gravity of fixed solids	2.5	2.5
Specific gravity of volatile solids	≈1.0	≈1.0

Solution

1. Compute the average specific gravity of all the solids in the primary sludge using Eq. (13–1).

    $$\frac{1}{S_s} = \frac{0.2}{2.5} + \frac{0.8}{1.0} = 0.88$$

    $$S_s = \frac{1}{0.88} = 1.14 \quad \text{(primary solids)}$$

2. Compute the specific gravity of the primary sludge.

    $$\frac{1}{S_{sl}} = \frac{0.05}{1.14} + \frac{0.95}{1} = 0.99$$

    $$S_{sl} = \frac{1}{0.99} = 1.01$$

3. Compute the volume of the primary sludge using Eq. (13–2).

    $$V = \frac{500\,\text{kg}}{(10^3\,\text{kg/m}^3)(1.01)(0.05)}$$

    $$= 9.9\,\text{m}^3$$

4. Compute the percentage of volatile matter after digestion total volatile solids after digestion.

    $$\text{Volatile matter, \%} = \frac{\text{total VS after digestion}}{\text{total TS after digestion}} \times 100$$

    $$= \frac{(\text{VS Primary})M_s(1 - \text{VSR})}{M_s - M_s(\text{VS Primary})(\text{VSR})} \times 100$$

    $$= \frac{(0.8)(500\,\text{kg})(1 - 0.6)}{500\,\text{kg} - 500\,\text{kg}(0.8)(0.6)} \times 100 = 61.5\%$$

5. Compute the average specific gravity of all the solids in the digested sludge using Eq. (13–1).

$$\frac{1}{S_s} = \frac{0.385}{2.5} + \frac{0.615}{1.0} = 0.769$$

$$S_s = \frac{1}{0.769} = 1.30 \text{ (digested solids)}$$

6. Compute the specific gravity of the digested sludge (S_{ds}).

$$\frac{1}{S_{ds}} = \frac{0.20}{1.3} + \frac{0.80}{1} = 0.95$$

$$S_{ds} = \frac{1}{0.95} = 1.05$$

7. Compute the volume of digested sludge using Eq. (13–2).

$$V = \frac{500 \text{ kg} - 500\text{kg}(0.8)(0.6)}{(10^3 \text{ kg/m}^3)(1.05)(0.20)}$$

$$= 1.2\,\text{m}^3$$

8. Determine the percentage reduction in the sludge volume after digestion.

$$\text{Reduction} = \frac{(9.9 - 1.2)\,\text{m}^3}{9.9\,\text{m}^3} \times 100 = 87.8\%$$

13–2 REGULATIONS FOR THE REUSE AND DISPOSITION OF SLUDGE IN THE UNITED STATES

In selecting the appropriate methods of sludge processing, reuse, and disposition, consideration must be given to the appropriate regulations. In the United States, regulations (40 CFR Part 503) were promulgated in 1993 by the U.S. Environmental Protection Agency (U.S. EPA) that established pollutant numerical limits and management practices for the reuse and disposition of sludge generated from the processing of municipal wastewater and septage (Federal Register, 1993). The regulations were designed to protect public health and the environment from any reasonably anticipated adverse effects of pollutants contained in the biosolids.

The regulations addressed by 40 CFR Part 503 cover specifically (1) land application of biosolids, (2) surface disposition of biosolids, (3) pathogen and vector reduction in treated biosolids, and (4) incineration. Each of these subjects is discussed below. The regulations directly affect selection of many of the processes used for sludge treatment, especially for sludge stabilization, i.e., alkaline stabilization, anaerobic digestion, aerobic digestion, and composting. In some cases, to achieve compliance, appropriate treatment requirements or methods are stipulated by the regulations. Additional discussion regarding regulations for applying biosolids on land is provided in Sec. 14–8.

Land Application

Land application relates to biosolids reuse and includes all forms of applying bulk or bagged biosolids to land for beneficial uses at agronomic rates, i.e., rates designed to provide the amount of nitrogen needed by crop or vegetation while minimizing the amount that passes below the root zone. The regulations establish two levels of biosolids quality with respect to heavy metals concentrations—pollutant ceiling and pollutant concentrations ("high" quality biosolids); two levels of quality with respect to pathogen densities— Class A and Class B; and two types of approaches for meeting vector attraction—biosolids processing or use of physical barriers. Vector attraction reduction decreases the potential for spreading infectious disease by vectors such as rodents, insects, and birds.

Surface Disposition

The surface disposition part of the Part 503 regulations applies to (1) dedicated surface disposition sites; (2) monofills, i.e., sludge-only landfills; (3) piles or mounds; and (4) impoundments or lagoons. Disposition sites and sludge placed on those sites for final disposition are addressed in the surface disposition rules. Surface disposition does not include placement of sludge for storage or treatment purposes. Where surface disposition sites do not have a liner or leachate collection system, limits are established for pollutants such as arsenic and nickel and vary based on the distance of the active surface disposition site boundary from the site property line (see Federal Register, 1993).

Pathogen and Vector Attraction Reduction

The 40 CFR Part 503 regulations divide the quality of biosolids into two categories, referred to as Class A and Class B (see requirements in Table 13–9,). Class A biosolids must meet specific criteria to ensure they are safe to be used by the general public and for nurseries, gardens, and golf courses. Class B biosolids have lesser treatment requirements than Class A, and typically are used for application to agricultural land and daily cover in a landfill.

In addition to meeting the requirements in Table 13–9, Class A biosolids must meet one of the following criteria:

- A fecal coliform density of less than 1000 MPN/g total dry solids
- Salmonella sp. density of less than 3 MPN/4 g total dry solids (3 MPN/4 g TS)

Bulk biosolids applied to lawns and home gardens or sold or given away in bags or other containers must meet the Class A criteria for pathogen reduction (see Table 13–9) and one of several vector attraction reduction processing options (see Table 13–10). Alternatively, biosolids can be treated by a prescribed process that reduces pathogens beyond detectable levels.

Class B pathogen requirements are the minimum level of pathogen reduction for land application and surface disposition. The only exception to achieving at least Class B level occurs when the sludge is placed in a surface disposition facility that is covered daily. Biosolids that do not qualify as Class B cannot be land applied. To meet Class B requirements, biosolids must be treated by a process that reduces but does not eliminate pathogens (see PSRP, also discussed below), or that must be tested to meet fecal coliform limits of less than 2.0×10^6 MPN/g TS or less than 2.0×10^6 CFU/g TS.

To meet pathogen and vector attraction reduction requirements, two levels of preapplication treatment are required and have been defined by the U.S. EPA as Processes to Further Reduce Pathogens (PFRP) and Processes to Significantly Reduce Pathogens (PSRP). These processes are defined in Tables 13–11 and 13–12. Because PSRPs reduce but do not eliminate pathogens, PSRP-treated biosolids still have the potential to transmit disease. Because PFRPs reduce pathogens below detectable levels, there are no pathogen-related restrictions for land application. Minimum frequency of monitoring, record-keeping, and reporting requirements must be met, however.

Table 13–9

Pathogen reduction alternatives[a]

Class A	Description
Alternative 1	Thermally Treated Sewage Sludge: Use one of four time-temperature regimes.
Alternative 2	Sewage Sludge Treated in a High pH-High Temperature Process: Specifies pH, temperature, and air-drying requirements.
Alternative 3	For Sewage Sludge Treated in Other Processes: Demonstrate that the process can reduce enteric viruses and viable helminth ova. Maintain operating conditions used in the demonstration.
Alternative 4	Sewage Sludge Treated in Unknown Processes: Demonstration of the process is unnecessary. Instead, test for pathogens—*Salmonella* sp. bacteria, enteric viruses, and viable helminth ova—at the time the sewage sludge is used or disposed, or is prepared for sale or give-away in a bag or other container for application to the land, or when prepared to meet the requirements in 503.10(b), (c), (e), or (f).
Alternative 5	Use of PFRP: Sewage sludge is treated in one of the processes to further reduce pathogens (PFRP).
Alternative 6	Use of a Process Equivalent to PFRP: Sewage sludge is treated in a process equivalent to one of the PFRPs, as determined by the permitting authority.

Class B	Description
Alternative 1	Monitoring of Indicator Organisms: Test for fecal coliform density as an indicator for all pathogens at the time of sewage sludge use or disposal.
Alternative 2	Use of PSRP: Sewage sludge is treated in one of the processes to significantly reduce pathogens (PSRP).
Alternative 3	Use of Processes Equivalent to PSRP: Sewage sludge is treated in a process equivalent to one of the PSRPs, as determined by the permitting authority.

[a] From US EPA (1992).

[b] In addition to meeting the requirements in one of the six alternatives listed below, fecal coliform or *Salmonella* spp. bacterial levels must meet specific densities at the time of sewage sludge use or disposal, when prepared for sale or give-away in a bag or other container for application to the land, or when prepared to meet the requirements in 503.10(b), (c), (e), or (f).

[c] The requirements in one of the three alternatives below must be met in addition to Class B site restrictions.

Incineration

Originally, the definition of a nonhazardous solid waste material included sludges from wastewater treatment and other secondary material being discarded. However, wastewater sludges were eventually exempted from being defined under the nonhazardous solid waste ruling and were instead covered under 40 *CFR* 503 regulations.

Resource Conservation and Recovery Act. Under the Resource Conservation and Recovery Act, in the Identification of Non-Hazardous Secondary Materials that are Solid Waste (published by the U.S. EPA on March 21, 2011), sludge is defined somewhat differently. If the secondary material (e.g., sludge) is being discarded, it falls under the solid waste definition. As such, the burning of sludges in sewage sludge incinerators (SSIs) would be considered discarding, and, thus, would fall under the "other discarded material" part of the solid waste definition. Accordingly, SSIs would be regulated under Sec. 129 of

Table 13–10

Vector attraction reduction[a]

Requirement	What is required?	Most appropriate for:
Option 1 503.33(b)(1)	At least 38 percent reduction in volatile solids during biosolids treatment	Biosolids processed by: Anaerobic biological treatment Aerobic biological treatment Chemical oxidation
Option 2 503.33(b)(2)	Less than 17 percent additional volatile solids loss during bench–scale anaerobic batch digestion of the biosolids for 40 additional d at 30 to 37°C (86 to 99°F)	Only for anaerobically digested biosolids
Option 3 503.33(b)(3)	Less than 15 percent additional volatile solids reduction during bench–scale aerobic batch digestion for 30 additional d at 20°C (68°F)	Only for aerobically digested biosolids with 2 percent or less solids—e.g., biosolids treated in extended aeration plants
Option 4 503.33(b)(4)	SOUR at 20°C (68°F) is 1.5 mg O_2/h·g total biosolids solids	Biosolids from aerobic processes (should not be used for composted sludges). Also for biosolids that has been deprived of oxygen for longer than 1 to 2 h
Option 5 503.33(b)(5)	Aerobic treatment of the biosolids for at least 14 d at over 40°C (104°F) with an average temperature of over 45°C (113°F)	Composted biosolids (Options 3 and 4 are likely to be easier to meet for biosolids from other aerobic processes)
Option 6 503.33(b)(6)	Addition of sufficient alkali to raise the pH to at least 12 at 25°C (77°F) and maintain a pH of 12 for 2 h and a pH of 11.5 for 22 more h	Alkali-treated biosolids (alkalies include lime, fly ash, kiln dust, and wood ash)
Option 7 503.33(b)(7)	Percent solids of 75 percent prior to mixing with other materials	Biosolids treated by an aerobic or anaerobic process (i.e., biosolids that do not contain unstabilized sludge generated in primary wastewater treatment)
Option 8 503.33(b)(8)	Percent solids of 90 percent prior to mixing with other materials	Biosolids that contain unstabilized sludge generated in primary wastewater treatment (e.g., any heat-dried sludges)
Option 9 503.33(b)(9)	Biosolids is injected into soil so that no significant amount of biosolids is present on the land surface 1 hour after injection, except Class A biosolids which must be injected within 8 h after the pathogen reduction process	Liquid biosolids applied to the land. Domestic septage applied to agricultural land, a forest, or a reclamation site
Option 10 503.33(b)(10)	Biosolids is incorporated into the soil within 6 h after application to land. Class A biosolids must be applied to the land surface within 8 h after the pathogen reduction process, and must be incorporated within 6 h after application	Biosolids applied to the land. Domestic septage applied to agricultural land, forest, or a reclamation site

[a] From U.S. EPA (1992).

the Clean Air Act, instead of how they have been regulated historically, under Sec. 112 of the Clean Air Act through Part 503 biosolids regulations.

Clean Air Act. Section 129 of the Clean Air Act requires the U.S. EPA to develop standards for solid waste combustion processes. As a result, the U.S. EPA was required to develop new source performance standards (NSPSs) and emission guidelines (EGs) for sewage sludge incineration units (SSIs). The new standards and emission guidelines were finalized by the U.S. EPA on February 21, 2011 and published in the Federal Register on March 21, 2011.

Table 13–11

Regulatory definition of processes to further reduce pathogens (PFRP)[a]

Process	Definition
Composting	Using either within-vessel or static aerated pile composting, the temperature of the biosolids is maintained at 55°C or higher for 3 d. Using windrow composting, the temperature of the wastewater sludge is maintained at 55°C or higher for 15 d or longer. During this period, a minimum of five windrow turnings is required.
Heat drying	Dewatered biosolids are dried by direct or indirect contact with hot gases to reduce the moisture content to 10 percent or lower. Either the temperature of biosolids particles exceed 80°C or the wet bulb temperature of the gas stream in contact with the biosolids as the biosolids leave the dryer exceeds 80°C.
Heat treatment	Liquid biosolids are heated to a temperature of 180°C or higher for 30 min.
Thermophilic aerobic digestion	Liquid biosolids are agitated with air or oxygen to maintain aerobic conditions, and the MCRT is 10 d at 55 to 60°C.
Beta ray irradiation	Biosolids are irradiated with beta rays from an accelerator at dosages of at least 1.0 megarad (Mrad) at room temperature (approximately 20°C).
Gamma ray irradiation	Biosolids are irradiated with gamma rays from certain isotopes such as 60 Cobalt or 135 Cesium at dosages of at least 1.0 Mrad at room temperature (approximately 20°C).
Pasteurization	The temperature of the biosolids is maintained at 70°C or higher for at least 30 min.

[a] Federal Register (1993).

Table 13–12

Regulatory definition of processes to significantly reduce pathogens (PSRP)[a]

Process	Definition
Aerobic digestion	Biosolids are agitated with air or oxygen to maintain aerobic conditions for a MCRT and temperature between 40 d at 20°C and 60 d at 15°C.
Air drying	Biosolids are dried on sand beds or on paved or unpaved basins for a minimum of 3 mo. During 2 of the 3 mo, the ambient average daily temperature exceeds 0°C.
Anaerobic digestion	Biosolids are treated in the absence of air between an MCRT of 15 d at temperatures of 35 to 55°C and an MCRT of 60 d at a temperature of 20°C. Times and temperatures between these endpoints may be calculated by linear interpolation.
Composting	Using either within-vessel, static aerated pile, or windrow composting, the temperature of the biosolids is raised to 40°C or higher for 5 d. For 4 h during the 5 d period, the temperature in the compost pile should exceed 55°C.
Lime stabilization	Sufficient lime is added to raise the pH of the biosolids to pH 12 and maintained for 2 h of contact.

[a] Federal Register (1993).

Emission Guidelines and New Source Performance Standards. The new rule requires facilities to meet the maximum achievable control technology (MACT) limits. The MACT standards for existing units were based on the best performing 12 percent of the existing units, while MACT standards for new or "modified" units are based on the "best controlled similar unit." MACT standards have been set for nine pollutants. These pollutants are: cadmium (Cd), lead (Pb), mercury (Hg), particulate matter (PM), carbon monoxide (CO), hydrogen chloride (HCl), sulfur dioxide (SO_2), nitrogen oxides (NOx), and dioxins and furans (PCDD/PCDF).

The new SSI rules and emission guidelines contain standards for existing and new multiple-hearth furnaces (MHFs) and fluidized bed incinerators (FBIs). This rule stipulates that all SSIs will require Title V operating permits, annual operator training, annual stack testing and/or continuous emissions monitoring systems, recordkeeping requirements, and establishment of operating limits. For new or modified SSIs, the proposed MACT limits are set as a composite of the best emissions performance from the best SSIs tested. For new SSIs, owners or operators are required to conduct a siting analysis prior to construction. This analysis would include site specific analysis of air pollution control alternatives to minimize the environmental and health impacts to the maximum extent practicable. Details regarding air pollution controls are discussed in Sec. 14–6 in Chap.14.

Clean Water Act. Under the 503 regulations, the requirements for biosolids incineration are still applicable. There is some overlap between the 503 regulations and the Clean Air Act Section 129 MACT based EGs and NSPSs, but they differ significantly due to the different approaches used to develop each set of rules. The 503 regulations are based on a risk based approach and aim to avoid adverse impacts. Limits are based partially on maximum allowable concentrations of pollutants within the feed biosolids coupled with stack monitoring of total hydrocarbons (or carbon monoxide) and operational standards to ensure good combustion and emissions performance. However, the MACT rules are technology based and set the limits relative to the best performing incinerator units within their class (ie. MHF or FBI). These limits are expressed as maximum concentrations of pollutants in the incinerator flue gases. The methods of measurement and media in which the concentrations are measured are entirely different and are not directly comparable to the 503 regulations. In practical application, the MACT emission requirements are much more stringent than previously required under Part 503 and they will generally dictate the required emission performance levels for both new and existing incinerators. For the time being, both sets of rules apply with overlapping requirements leading to duplicate sampling by operators of both the feed biosolids and the flue gases.

13–3 SLUDGE PROCESSING FLOW DIAGRAMS

A generalized flow diagram incorporating the unit operations and processes to be discussed in this chapter and chapter 14 is presented on Fig. 13–2. As shown, an almost infinite number of combinations are possible. In practice, the most commonly used process flow diagram for sludge processing involves biological treatment. Typical flow diagrams incorporating biological processing are presented on Fig. 13–3. Thickeners may be used depending upon the source of sludge and the method of sludge stabilization, dewatering, and disposition. Following biological digestion, any of the several methods shown may be used to dewater the sludge; the choice depends on economic evaluation, beneficial use requirements, and local conditions. In instances where biological stabilization is not used, dewatered sludge undergoes thermal decomposition, in either multiple-hearth or fluidized-bed incinerators. Furthermore, unstabilized dewatered cake can be dried, alkaline stabilized, or hauled to a landfill.

Figure 13–2

Generalized sludge processing
flow diagram.

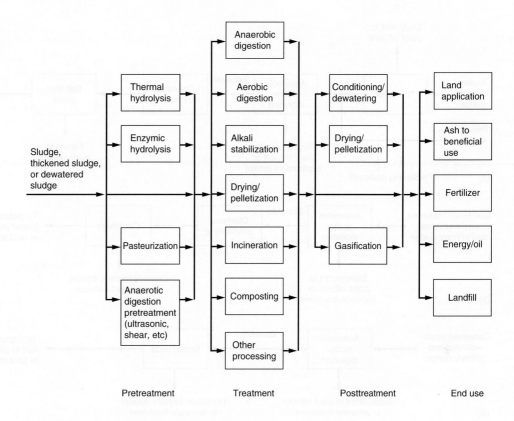

13–4 SLUDGE AND SCUM PUMPING

Sludge produced in wastewater treatment plants must be conveyed from point to point in the plant in conditions ranging from a watery sludge or scum to a thick sludge. Sludge may also be pumped off-site for long distances for treatment and disposition. For each type of sludge and pumping application, a different type of pump may be needed (see Table 13–13).

Pumps

Pumps used most frequently to convey sludge include the plunger, progressive cavity, hose, solids handling centrifugal (screw centrifugals and traditional "non-clog designs), recessed impeller, diaphragm, high-pressure piston diaphragm, and rotary lobe types. Other types of pumps such as hydraulic piston slurry pumps have also been used to pump sludge. Chopper pumps are used extensively for pumping scum containing rags, plastics, and other fibrous materials that require shredding. The advantages and disadvantages of each type of pump are summarized in Table 13–14.

Plunger Pumps. Plunger pumps [see Fig. 13–4(a)] have been used frequently for sludge applications, especially primary sludges, and have proved to be quite satisfactory. The advantages of plunger pumps are as follows:

1. Pulsating action of simplex and also duplex pumps tends to concentrate the sludge in the hoppers ahead of the pumps and resuspend solids in pipelines when pumping at low velocities.
2. They are suitable for suction lifts up to 3 m (10 ft) and are self-priming.

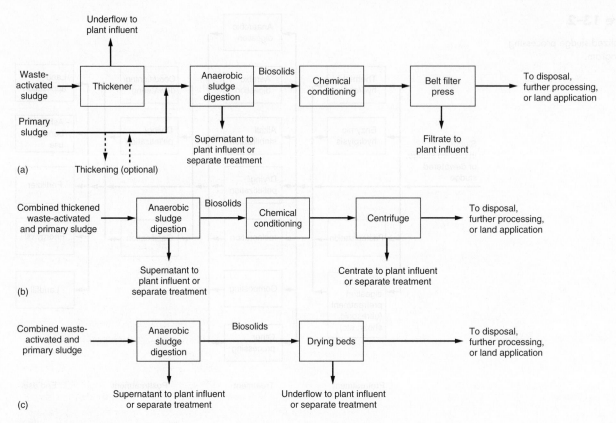

Figure 13-3

Typical sludge treatment flow diagrams with biological digestion and three different sludge dewatering processes: (a) belt filterpress, (b) centrifuge, (c) drying bed. In some plants, flows that are to be returned to the headworks are stored in equalization basins for return to the treatment process during the early morning hours when the plant load is reduced.

3. Low pumping rates can be used with large port openings.
4. Positive delivery is provided unless some object prevents the ball checks valves from seating.
5. They have constant but adjustable capacity, regardless of large variations in pumping head.
6. Discharge pressure limitations are approximately 10 to 11 bar (150 to 165 $lb_f/in.^2$).
7. Heavy sludge concentrations may be pumped if the equipment is designed for the load conditions.

Plunger pumps come with one, two, or three plungers (called simplex, duplex, or triplex units) with capacities of 2.5 to 3.8 L/s (40 to 60 gal/min) per plunger, and larger models are available. Pump speeds should be between 40 and 50 strokes per min. Because grease accumulations in sludge lines cause a progressive increase in head with use, heavier duty pumps should be designed for a minimum head of 6.9 bar (100 $lb_f/in.^2$). Capacity is decreased in constant speed pumps by shortening the stroke of the plunger; however, the pumps seem to operate more satisfactorily at or near full stroke. For this reason, many pumps are provided with variable-speed drives for speed control of capacity. A plunger pump differs from a centrifugal or recessed impeller pump in that its

Table 13–13

Application of pumps to types of sludge and biosolids[a]

Type of sludge or solids	Applicable pump	Comment
Ground screenings	Pumping screenings should be avoided	Pneumatic ejectors may be used.
Grit	Torque flow centrifugal	The abrasive character of grit and the presence of rags make grit difficult to handle. Hardened casings and impellers should be used for torque flow pumps. Pneumatic ejectors may also be used.
Scum	Plunger, progressive cavity, diaphragm, centrifugal, chopper	Scum is often pumped by the sludge pumps; valves are manipulated in the scum and sludge lines to permit this. In larger plants separate scum pumps are used. Scum mixers are often used to ensure homogeneity prior to pumping. Pneumatic ejectors may also be used.
Primary sludge	Plunger, centrifugal torque flow, diaphragm, progressive cavity, rotary lobe, chopper, hose	In most cases, it is desirable to obtain as concentrated a sludge as practicable from primary sedimentation tanks, usually by collecting the sludge in hoppers and pumping intermittently, allowing the sludge to collect and consolidate between pumping periods. The character of untreated primary sludge will vary considerably, depending on the characteristics of the solids in the wastewater, and the types of treatment units and their efficiency. Where biological treatment follows, the quantity of sludge from (1) waste activated sludge (2) humus sludge from settling tanks following trickling filters, (3) overflow liquors from digestion tanks, (4) and centrate or filtrate return from dewatering operations will also affect the sludge characteristics. In many cases, the character of the sludge is not suitable for the use of conventional nonclog centrifugal pumps. Where sludge contains rags, chopper pumps may be used.
Chemical precipitation	Same as for primary sludge	The precipitate may contain large amounts of inorganic constituents depending on the type and amount of chemicals used.
Trickling filter humus	Nonclog and torque flow centrifugal, progressive cavity, plunger, diaphragm	Humus is usually of homogenous character and can be easily pumped.
Return or waste activated sludge	Nonclog and torque flow centrifugal, progressive cavity, diaphragm	Sludge is dilute and contains only fine solids so that nonclog pumps may be used. For nonclog pumps, slow speeds are recommended to minimize the breakup of flocculent particles.
Thickened or concentrated sludge	Plunger, progressive cavity, diaphragm, high pressure piston, rotary lobe, hose	Positive displacement pumps are most applicable for concentrated sludge because of their ability to generate movement of the sludge mass. Torque flow pumps may be used but may require the addition of flushing or dilution facilities.
Digested biosolids	Plunger, torque flow centrifugal, progressive cavity, diaphragm, high pressure piston, rotary lobe	Well digested biosolids are homogenous, containing 2 to 5 percent total solids and a quantity of gas bubbles. Poorly digested biosolids may be difficult to handle. If good screening and grit removal are provided, nonclog centrifugal pumps may be considered.

[a] Adapted in part from U.S. EPA (1979).

Table 13–14

Advantages and disadvantages of various types of sludge pumps[a]

Type of Pump	Advantages	Disadvantages
Plunger	• Can pump heavy sludge concentrations (up to 15 percent) • Self-priming and can handle suction lifts up to 3 m (10 ft) • Constant but adjustable capacity regardless of variations in head • Cost-effective choice for flowrates up to 30 L/s (500 gal/min) and heads up to 60 m (200 ft) • Pulsating action of simplex and duplex pumps sometimes helps to concentrate sludge in hoppers ahead of pumps and resuspended solids in pipelines when pumping at low velocities • High Pressure Capacity	• Low efficiency • High maintenance if operated continuously • Depending on downstream processes, pulsating flow may not be acceptable
Progressing Cavity	• Provides a relatively smooth flow • Pumps greater than 3 L/s (50 gal/min) capacity can pass solids of about 20 mm (0.8 in.) in size • Easily controlled flowrates • Minimal pulsation • Relatively simple operation • Stator/rotor tends to act as a check valve, thus preventing backflow through pump. An external check valve may not be required	• Stator will burn out if pump is operated dry; needs a run dry protection system • Smaller pumps usually require grinders to prevent clogging • Power cost escalates when pumping heavy sludge • Grit in sludge may cause excessive stator wear • Seals and water required typically
Diaphragm	• Pulsating action may help to concentrate sludge in hoppers ahead of pumps and resuspend solids in pipelines when pumping at low velocities • Self-priming with suction lifts up to 3 m (10 ft) • Can pump grit with relatively minimum wear • Relatively simple operation	• Depending on downstream processes, pulsating flow may not be acceptable • Requires a source of compressed air • Operation may be excessively noisy • Low head and efficiency • High maintenance if operated continuously
Centrifugal Nonclog (mixed flow)	• Has high volume and excellent efficiency for activated sludge pumping applications • Relatively low cost	• Not recommended for other sludge pumping applications because of potential clogging due to rags and other debris
Recessed Impeller	• Because of recessed impeller design, pump can pass large solids and grit • Can pump digested sludges up to approximately 4%	• Low efficiency-about 5 to 20 percent lower than standard nonclog pumps • Limited to raw sludge with solid concentrations of 2.5 percent or less • Abrasion-resistant impellers cannot be trimmed to modify pumping characteristics
Chopper	• Reduces clogging of pump suction • May eliminate need for grinder or comminutor • Can handle higher sludge concentrations than nonclog pumps	• Relatively low efficiency-efficiency ranges from about 40 to 60 percent • Requires a level of maintenance similar to grinders
Rotary Lobe	• Provides a relatively smooth flow • Does not require a check valve in most applications with low to moderate discharge static heads • Able to run dry for short period of time without significant damage Low speed and low maintenance	• Because of close tolerances between rotating lobes, grit will cause excessive wear, thus reducing pumping efficiency • Fluid pumped must act as a lubricant • Cost of pumping increases with volume
Peristalic Hose	• Has self-priming capabilities • Because it is a positive-displacement pump, it is capable of metering flow • Relatively simple to maintain • Can pump sludge with abrasive grit	• Depending on downstream processes, pulsating flow may not be acceptable • High starting torque (two to three times running torque) • Replacement hoses may be expensive
High-pressure pistion	• Can be used to pump thickened sludge long distances • Can pump at rates of 30 L/s (500 gal/min) at pressures up to 13,800 kPa (2000 $lb_f/in.^2$) • Can run dry without major damage • Unobstructed internal flow path; can pass large solids	• High Capital Cost • Requires skilled maintenance personnel

[a] Adapted in part from WEF (2010a).

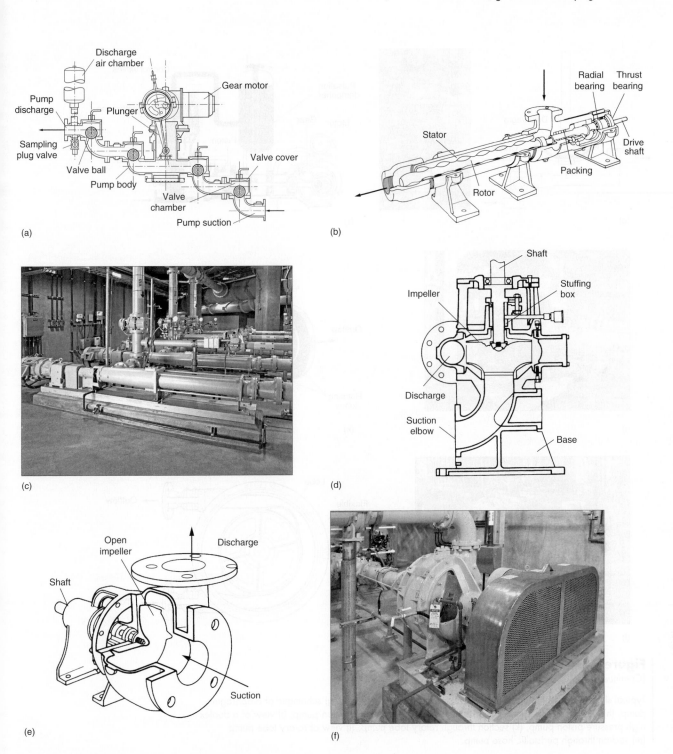

Figure 13-4

Typical sludge and scum pumps used in wastewater treatment plants: (a) plunger pump, (b) progressive cavity pump, (c) view of progressive cavity pump installation (d) section through nonclog centrifugal pump, (e) section through torque flow pump, (f) view of belt driven torque flow pump.
(Figure continues on next page.)

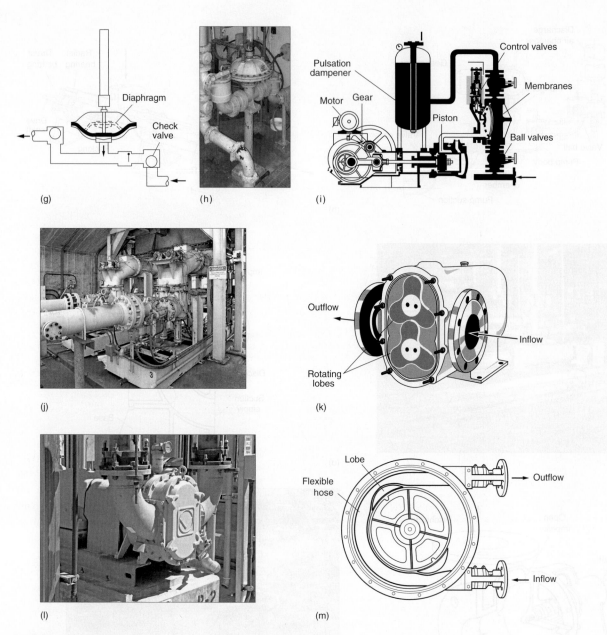

Figure 13–4
(Continued)

Typical sludge and scum pumps used in wastewater treatment plants: (g) schematic of diaphragm pump, (h) view of diaphragm pump, (i) schematic of high pressure piston pump, (j) view of a duplex high pressure piston pump, (k) section through rotary lobe pump, (l) view of rotary lobe pump (m) section through peristaltic hose pump.

discharge is pulsing due to the action of a piston; consequently, the actual flow while sludge is moving in the pipeline is greater than average pumping capacity. The headloss calculations, therefore, must be based on the peak pulsating flow rather than the design flow. The factors given in Table 13–15 can be used to account for the actual peak pulsating or instantaneous flow.

Table 13–15

Factors for computing peak pulsating flowrate when using plunger pumps

Type of plunger pump	Actual pulsating peak flowrate
Simplex	3.1 × design flowrate
Duplex	1.55 × design flowrate
Triplex	1.2 × design flowrate

Progressive Cavity Pumps. The progressive cavity pump [see Fig. 13–4(b) and (c)] has been used successfully on almost all types of sludges. The pump is composed of a single-threaded rotor that operates with a minimum of clearance in a double-threaded helix elastomer stator. A volume or "cavity" moves progressively from suction to discharge when the rotor turns. The pump is self-priming at suction lifts up to 8.5 m (28 ft), but it must not be operated dry or it will burn out the elastomer stator. Progressive cavity pumps are available in capacities up to 126 L/s (2000 gal/min) and may be operated at discharge heads of 48 bar (720 $lb_f/in.^2$) with sludge. This type of pump requires oversizing to meet system conditions over the life of the of equipment. For example, if a 9.5 L/s (150 gal/min) pump is required the pump selection should be sized for an additional 50 percent or a pump sized for 14.25 L/s (225 gal/min). Speed for sludge applications should be limited to approximately 250 rev/min. For sludges and for systems feeding dewatering equipment, a grinder normally precedes these pumps. The pumps are expensive to maintain because of wear of the rotors and stators, particularly in primary sludge pumping applications where grit is present. For primary sludge applications consideration should be given to recessed impeller pumps. Advantages of the pumps are (1) the flowrates are controlled easily using variable speed drives, (2) pulsation is minimal, and (3) operation is relatively simple.

Centrifugal Pumps. Centrifugal pumps of solids handling or "non-clog" design [see Fig. 13–4(d)] are commonly used to pump activated sludge. In centrifugal pumping applications, the problem is choosing the proper number and capacity to accommodate the typical wide range of flowrates required. At any given speed, centrifugal pumps operate well only if the pumping head is within a relatively narrow range; the variable nature of sludge, however, causes pumping heads to change. The selected pumps must have sufficient clearance to pass the solids without clogging and have a small enough capacity to avoid pumping a sludge diluted by large quantities of wastewater overlying the sludge blanket. Throttling the discharge to reduce the capacity is impractical because of frequent stoppages; hence it is absolutely essential that these pumps be equipped with variable-speed drives. Centrifugal pumps of special design: recessed impeller and "chopper" type pumps have been used for pumping primary sludge.

Recessed impeller pumps [see Figs. 13–4(e) and (f)] have impellers that are fully recessed and are very effective in conveying sludge and higher sludge concentrations than the solids handling centrifugal pumps. The size of particles that can be handled is limited only by the diameter of the suction or discharge openings. The rotating impeller develops a vortex in the sludge so that the main propulsive force is the liquid itself. Most of the fluid does not actually pass through the vanes of the impeller, thereby minimizing abrasive contact; however, pumps used in sludge service are recommended to have nickel or chrome abrasion-resistant volute and impellers. The pumps can operate only over a narrow head range at a given speed, so the system operating conditions must be evaluated carefully. Variable speed control is recommended where the pumps are expected to operate over a wide range of head conditions. For high-pressure applications, multiple pumps may be used, connected together in series.

Chopper-type pumps have a cutter knife attached to a non-clog impeller that agitates and breaks up large solids that tend to block the pump suction. Incoming sludge is chopped by sharpened impeller blades that turn across the cutter bar. Chopper pumps are manufactured in sizes up to 380 L/s (6000 gal/min) in both horizontal and vertical dry pit configurations as well as submersible configurations.

Slow-speed centrifugal and mixed-flow pumps are commonly used for returning activated sludge to the aeration tanks. Screw pumps are also being used for this service especially where pumps are required to have a large turndown. Screw centrifugal pumps tend to have less clogging issues for these applications.

Diaphragm Pumps. Diaphragm pumps use a flexible membrane that is pushed and pulled to contract and enlarge an enclosed cavity [see Fig. 13–4(g) and (h)]. Flow is directed through this cavity by check valves, which may be either ball or flap type. The capacity of a diaphragm pump is altered by changing either the length of the diaphragm stroke or the number of strokes per minute. Pump capacity can be increased and flow pulsations smoothed out by providing two pump chambers and using both strokes of the diaphragm for pumping. Diaphragm pumps are relatively low capacity and low head; the largest available air diaphragm pump delivers 14 L/s (220 gal/min) against 15 m (50 ft) of head.

High-Pressure Piston Diaphragm Pumps. High-pressure piston pumps are used in high-pressure applications such as pumping sludge long distances. Several types of piston pumps have been developed for high-pressure applications and are similar in action to plunger pumps. The high-pressure piston pumps use separate power pistons or membranes or diaphragms to separate the drive mechanisms from contacting the sludge. A schematic of a piston pump is shown on Fig 13–4(i). A view of a duplex piston pump is shown on Fig. 13–4(j). Advantages of these types of pumps are (1) they can pump relatively small flowrates at high pressures, up to 13.8 bar (note 1 bar = 100 kPa) (200 lb$_f$/in.^{2}), (2) large solids up to the discharge pipe diameter can be passed, (3) a range of sludge concentrations can be handled, and (4) the pumping can be accomplished in a single stage. The pumps, however, are very expensive.

Rotary Lobe Pumps. Rotary lobe pumps [see Fig. 13–4(k) and (l)] are positive displacement pumps in which two rotating synchronous lobes push the fluid through the pump. Rotational speed and shearing stresses are low. For sludge pumping, lobes are made of hard metal or hard rubber. This type of pump requires oversizing to meet system conditions over the life of the of equipment. For example if a 9.5 L/s (150 gal/min) pump is required the pump selection should be sized for an additional 50 percent or a pump sized for 14.25 L/s (225 gal/min). Speed for sludge applications should be limited to approximately 250 to 300 rev/min depending on the abrasiveness of the sludge. An advantage cited for the rotary lobe pump is that lobe replacement is less costly than rotor, and stator replacement for progressive cavity pumps and the space required for installation is less. Rotary lobe pumps, like other positive-displacement pumps, must be protected against pipeline obstructions.

Hose Pumps. Peristaltic hose pumps [see Fig. 13–4(m)] have also been used for pumping sludge. The pump works by alternately compressing and relaxing a specially designed resilient but reinforced hose. The hose is compressed between the inner wall of the pump housing and the compression shoes on the rotor. A lubricant is used to reduce heat and wear on the hose. The pumped sludge only comes in contact with the inner wall of the hose, which cushions entrained abrasives during compression. The pumps are available in capacities ranging from 36 to 1250 L/min (10 to 330 gal/min). As a positive-displacement pump, the pump output is directly proportional to speed at either high or low discharge pressures. The primary disadvantages of the hose pump are the pulsating flow, hose wear, and the relatively high cost of hose replacement.

Headloss Determination

The headloss encountered in the pumping of sludge depends on the flow properties (rheology) of sludge, the pipe diameter, and the flow velocity. It has been observed that headlosses increase with increased solids content, increased volatile content, and lower temperatures. When the percent volatile matter multiplied by the percent solids exceeds 600, difficulties may be encountered in pumping sludge.

Water, oil, and most other fluids are "Newtonian," which means that the pressure drop is proportional to the velocity and viscosity under laminar flow conditions. As the velocity increases past a critical value, the flow becomes turbulent. Dilute sludges such as unconcentrated activated and trickling-filter sludges behave similar to water. Concentrated wastewater sludges, however, are non-Newtonian fluids. The pressure drop under laminar conditions for non-Newtonian fluids is not proportional to flow, so the viscosity is not a constant. Special procedures may be used to determine headloss under laminar-flow conditions, and the velocity at which turbulent flow begins. In this section both the simplified approach of calculating headloss and a method using the sludge rheology will be discussed.

The headloss in pumping unconcentrated activated and trickling-filter sludges may be from 10 to 25 percent greater than for water. Primary, digested, and concentrated sludges at low velocities may exhibit a plastic-flow phenomenon in which a definite pressure is required to overcome resistance and start flow. The resistance then increases approximately with the first power of the velocity throughout the laminar range of flow, which extends to about 1.1 m/s (3.5 ft/s), the lower critical velocity. Above the higher critical velocity at about 1.4 m/s (4.5 ft/s), the flow may be considered turbulent. In the turbulent range, the losses for well-digested sludge may be more than two to three times the losses for water. The losses for primary and concentrated sludges, especially those conditioned with polymer, and scum may be considerably greater. The risk of underestimating the headloss also increases as the piping distance and sludge concentration increases. Where possible, particularly in long-distance sludge pumping, hydraulic studies should be conducted to confirm the ranges of headloss characteristics.

Simplified Headloss Computations. Relatively simple procedures are used to compute headloss for short sludge pipelines. The accuracy of these procedures may be adequate, especially at sludge solids concentrations less than 3 percent by weight. To determine the headloss, the factor k is obtained from Fig. 13–5(a) for a given solids content and type of sludge. The headloss when pumping sludge is computed by multiplying the headloss of water, determined by using the Darcy-Weisbach, Hazen-Williams, or Manning equations, by k. The values given on Fig. 13–5(a) should be used only when (1) velocities are at least 0.8 m/s (2.5 ft/s), (2) velocities do not exceed 2.4 m/s (8 ft/s), (3) thixotropic behavior is not considered, and (4) the pipe is not obstructed by grease or other materials.

Another approximate method makes use of empirical multiplication factor charts [see Fig. 13–5(b)]. The approximate method involves only velocity and percent solids consideration. Usually, the consistency of untreated primary sludge changes during pumping. At first, the most concentrated sludge is pumped. When most of the sludge has been pumped, the pump must handle a dilute sludge that has essentially the same hydraulic characteristics as water. The change in characteristics causes a centrifugal pump to operate farther out

Figure 13-5

Headloss multiplication factors:
(a) for different sludge types and
concentrations (b) for different
pipeline velocities and sludge
concentrations.

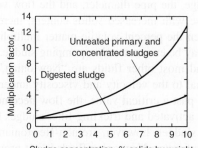

(a)

Note: Multiply loss with clean water by k to estimate
friction loss under laminar conditions (see text).

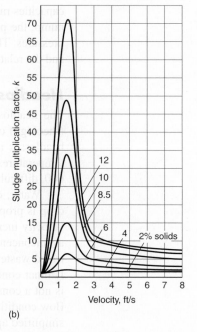

(b)

on its head-capacity curve, beyond the areas of best efficiency. The pump motor should be sized for the additional load, and a variable-speed drive should be considered to reduce the flow under changing sludge characteristics. If the pump motor is not sized for the maximum load when pumping water at top speed, it is likely to be overloaded or damaged if the overload devices do not function or are set too high.

To determine the operating speeds and motor power required for a centrifugal pump handling sludge, system curves should be computed (1) for the most dense sludge anticipated with design friction factor, (2) for average conditions, and (3) for water with a new pipe friction factor to cover the full anticipated range of the pumping system. The curves should be plotted on a graph of the pump curves for a range of available speeds. The maximum and minimum speeds required of a particular pump are obtained from the intersection of the pump head-capacity curves with the system curves at the desired capacity. Where the maximum speed head-capacity curve intersects the system curve for water determines the power required. In constructing the system curves for sludge for velocities from 0 to 1.1 m/s (3.5 ft/s), the headloss can be considered constant at the figure computed for 1.1 m/s (3.5 ft/s). The intersection of the pump curves with the system curve for average conditions can be used to estimate hours of operations, average speed, and power costs.

Because the usual flow formulas cannot be used in the plastic and laminar range, judgment and experience must be relied upon. In this range, capacities will be small, and plunger, progressive cavity, or rotary-lobe pumps should be used with ample head and capacity as recommended previously.

Application of Rheology to Headloss Computations.

For pumping sludge over long distances, an alternative method of computing headloss characteristics has been developed based on the flow properties of the sludge. A method of computing headloss for laminar flow conditions was derived originally by Babbitt and Caldwell (1939), based on the results of experimental and theoretical studies. Additional studies have been performed for the transition from laminar to turbulent flow (Mulbarger et al., 1981;

U.S. EPA, 1979) and are summarized in Sanks et al. (1998). Long-distance pumping of mixtures of untreated (raw) primary and secondary sludge is discussed by Carthew et al. (1983). The approach used in those studies for turbulent flow, which is of critical importance for long pipelines, is described below. For laminar and transitional flow, computational procedures described in Sanks et al. (1998) are recommended.

As stated previously, water, oil, and most other common fluids are "Newtonian," which means the pressure drop is directly proportional to the velocity and viscosity under laminar-flow conditions. As the velocity increases past a critical value, the flow becomes turbulent. The transition from laminar to turbulent flow depends on the Reynolds number, which is inversely proportional to the fluid viscosity. Wastewater sludge, however, is a non-Newtonian fluid. The pressure drop under laminar conditions is not proportional to flow, so the viscosity is not a constant. The precise Reynolds number at which turbulent-flow characteristics are encountered is uncertain for sludges.

Sludge has been found to behave much like a Bingham plastic, a substance with a straight-line relationship between shear stress and flow only after flow begins. A Bingham plastic is described by two constants: the yield stress s_y and the coefficient of rigidity η. Typical ranges of values for yield stress and coefficient of rigidity are shown on Figs. 13–6(a) and (b). If the two constants can be determined, the pressure drop over a wide range of velocities can be obtained using ordinary equations for water and the use of Fig. 13–6(c). As observed on Figs. 13–6(a) and (b), published data quantifying yield stress and the coefficient of rigidity values for wastewater sludges are highly variable. Pilot studies should be conducted to determine the rheological data for specific applications. Procedures for developing yield stress and the coefficient of rigidity using a pipeline viscometer and rotational viscometer are also given by Carthew et al. (1983).

Two dimensionless numbers can be used to determine the pressure drop due to friction for sludge: Reynolds number and Hedstrom number. Reynolds number is calculated by using the following expression:

$$N_R = \frac{\rho v D}{\eta} \qquad \text{SI units} \tag{13-3a}$$

$$N_R = \frac{\gamma v D}{\eta} \qquad \text{U.S. customary units} \tag{13-3b}$$

where N_R = Reynolds number, dimensionless
ρ = density of sludge, kg/m^3
γ = specific weight of sludge, lb/ft^3
v = average velocity, m/s (f/s)
D = diameter of pipe, m (ft)
η = coefficient of rigidity, kg/m·s (lb/ft·s)

Hedstrom number, which is reviewed by Hill et al. (1986), is calculated as follows:

$$H_e = \frac{D^2 s_y \rho}{\eta^2} \qquad \text{SI units} \tag{13-4a}$$

$$H_e = \frac{D^2 s_y g_c \gamma}{\eta^2} \qquad \text{U.S. customary units} \tag{13-4b}$$

where H_e = Hedstrom number, dimensionless
s_y = yield stress, N/m^2 (lb$_f$/ft^2)
g_c = 32.2 lb$_m$·ft/lb$_f$·s^2
Other terms are as defined previously.

Figure 13-6

Curves for computing pipeline headloss by the sludge rheology method: (a) yield stress vs. percent sludge solids, (b) coefficient of rigidity vs. percent sludge solids, and (c) friction factor for sludge analyzed as a Bingham plastic. (Adapted from Carthew et al., 1983.)

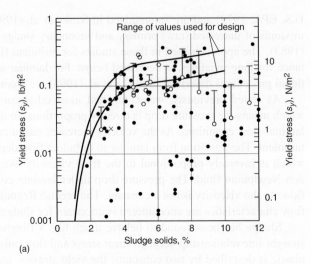

(a)

Note: lb/ft^2 × 47.8803 = N/m^2

LEGEND

o - Raw primary sludge
× - Secondary sludge
• - Digested sludge
⊥ - Median + standard deviation

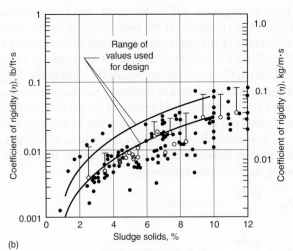

(b)

Note: lb/ft^2·s × 1.488 = kg/m·s

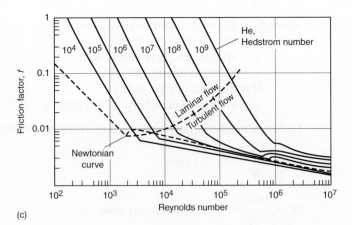

(c)

Using the calculated Reynolds number and the Hedstrom number, the friction factor f can be determined from Fig. 13–6(c). The pressure drop for turbulent conditions can then be calculated from the following relationship:

$$\Delta p = \frac{2f\rho Lv^2}{D} \qquad \text{SI units} \tag{13-5a}$$

$$\Delta p = \frac{2f\gamma Lv^2}{g_c D} \qquad \text{U.S. customary units} \tag{13-5b}$$

where Δp = pressure drop due to friction, N/m^2 (lb_f/ft^2)
 f = friction factor [from Fig. 13–6(c)]
 L = length of pipeline, m (ft)
Other terms are as defined previously.

In using Eqs. (13–3), (13–4), and (13–5), it should be noted that the Reynolds number is not the same as the Reynolds number based on viscosity. In plastic flow, an effective viscosity may be defined, but it is variable and can be much greater than the coefficient of rigidity. Consequently, the two Reynolds numbers can differ greatly. The friction factor f will usually differ significantly from the f values reported in standard hydraulic texts for clear water, which may be four times the values used on Fig. 13–6(c). These equations apply to the entire range of laminar and turbulent flows, except that Fig. 13–6(c) does not allow for pipe roughness. To allow for pipe roughness, if customary water formulas for headloss result in a higher pressure drop than computed with Eq. (13–5), then roughness is dominant, the flow is fully turbulent, and the pressure drop given by the water headloss formula will be reasonably accurate. A safety factor on the order of 1.5 is recommended for worst-case design conditions (Mulbarger et al., 1981). The use of Eqs. (13–3), (13–4), and (13–5) is illustrated in Example 13–2.

EXAMPLE 13–2 **Computation of Headloss Using Sludge Rheology** Calculate the headloss in a 250-mm-diameter pipeline 10,000 m long conveying untreated (raw) sludge at an average flowrate 0.04 m³/s. Determine also if the flow is turbulent. By testing, the following sludge rheology data were found:

Yield stress $s_y = 1.3$ N/m²

Coefficient of rigidity $\eta = 0.035$ kg/m·s

Specific gravity = 1.01

Solution

1. Calculate the pipeflow velocity.
   a. Determine the pipe cross-sectional area.

   $$A = \pi \times \frac{D^2}{4} = 3.14\frac{(0.25 \text{ m})^2}{4} = 0.49 \text{ m}^2$$

   b. Determine velocity.

   $$v = \frac{Q}{A} = \frac{(0.04 \text{ m}^3/\text{s})}{0.049 \text{ m}^2} = 0.82 \text{ m/s}$$

2. Compute sludge specific weight.

   $$\rho = 1000 \text{ kg/m}^3 \times 1.01 = 1010 \text{ kg/m}^3$$

3. Compute Reynolds number using Eq. (13–3).

$$N_R = \frac{\rho v D}{\eta} = \frac{(1010 \text{ kg/m}^3)(0.82 \text{ m/s})(0.25 \text{ m})}{(0.035 \text{ kg/m·s})} = 5.92 \times 10^3$$

4. Compute Hedstrom number using Eq. (13–4).

$$H_e = \frac{D s_y \rho}{\eta^2} = \frac{(0.25 \text{ m})^2 (1.3 \text{ N/m}^2)(1010 \text{ kg/m}^3)}{(0.035 \text{ kg/m·s})^2} = 6.70 \times 10^4$$

5. Determine friction factor f from Fig. 13–6(c) using the computed Reynolds and Hedstrom numbers.

$$f = 0.007$$

Note, on Fig. 13–6c, that the flow is in the turbulent zone.

6. Compute pressure drop using Eq. (13–5).

$$\Delta p = \frac{2 f \rho L v^2}{D} = \frac{2(0.007)(1010 \text{ kg/m}^3)(10{,}000 \text{ m})(0.82 \text{ m/s})^2}{0.25 \text{ m}}$$

$$= 380{,}309 \text{ kg/m·s}^2 \ (\text{N/m}^2 \text{ or Pa})$$

Convert to meters of water.

$$\Delta p = \frac{380{,}309 \text{ kg/m·s}^2}{(10^3 \text{ kg/m}^3)(9.81 \text{ m/s}^2)} = 38.8 \text{ m}$$

Comment   In this example, only one set of rheology data was used. In actual design, test data should be used for a range of probable conditions so that a family of headloss curves can be developed for the range of operating conditions. In addition, appropriate safety factors should be used for worst-case conditions. Comparison of the headloss to the headloss for water using the Hazen-Williams formula is left as a homework problem.

Sludge Piping

In wastewater treatment plants, conventional sludge piping should not be smaller than 150 mm (6 in.) in diameter although smaller-diameter glass-lined pipe has been used successfully. Sludge piping may not need to be larger than 200 mm (8 in.), unless the velocity exceeds 1.5 to 1.8 m/s (5 to 6 ft/s), in which case the pipe is sized to maintain that velocity. Gravity sludge withdrawal lines should not be less than 200 mm (8 in.) in diameter. It is common practice to install a number of cleanouts in the form of plugged tees or crosses instead of elbows so that the lines can be rodded if necessary. Pump connections should not be smaller than 100 mm (4 in.) in diameter.

A liberal number of hose gates should be installed in the piping, and an ample supply of high-pressure flushing water should be available for clearing stoppages. The flushing water should be plant effluent. The flushing water system should have a capacity of not less than 0.010 m³/s (150 gal/min) at 500 kN/m² ($\sim$70 lb$_f$/in.²). In large plants with larger piping, a greater capacity should be available, and the available pressure should be increased to 700 kN/m² (100 lb$_f$/in.²).

Grease has a tendency to coat the inside of piping used for transporting primary sludge and scum. Grease accumulation is more of a problem in large plants than in small ones. The coating results in a decrease in effective diameter and a large increase in pumping head. For this reason, low capacity positive-displacement pumps are designed for heads greatly in excess of the theoretical head. Centrifugal pumps, with their larger capacity, usually pump a more dilute sludge, often containing some wastewater, and head buildup due to grease accumulations appears to occur more slowly. In some plants, provisions have been made for melting the grease by circulating hot water, steam, or digester supernatant through the main sludge lines.

In treatment plants, friction losses are low because the pipe runs are short; consequently, there is little difficulty in providing an ample safety factor. In the design of long sludge lines, however, special design features should be considered including (1) providing two pipes unless a single pipe can be shut down for several days without causing problems; (2) providing for external corrosion and pipe loads; (3) adding facilities for applying dilution water for flushing the line; (4) providing means to insert a pipe cleaner; (5) including provisions for steam injection, especially in cold climates and where excessive grease accumulation occurs; (6) providing air relief and blowoff valves for the high and low points, respectively, and (7) considering the potential effects of waterhammer. A discussion of waterhammer in force mains is provided in the companion volume to this text (Metcalf & Eddy, 1981).

13–5 PRELIMINARY SLUDGE PROCESSING OPERATIONS

Grinding, degritting, blending, and storage of sludge is necessary to provide a relatively constant, homogeneous feed to subsequent processing facilities. Blending and storage can be accomplished either in a single unit designed to do both or separately in other plant components. Screening of raw sludge or digested biosolids is sometimes required in reuse applications for the removal of plastics, rags, and other material. Each of these preliminary operations is discussed in this section.

Grinding

Sludge grinding is a process in which large and stringy material contained in sludge is cut or sheared into small particles to prevent clogging or wrapping around rotating equipment. A typical sludge grinder installation is shown on Fig. 13–7. Some of the processes that must be preceded by sludge grinders and the purposes of grinding are reported in Table 13–16. Grinders historically have required high maintenance, but newer designs of slow-speed grinders have been more durable and reliable. These designs include improved bearings and seals, hardened steel cutters, overload sensors, and mechanisms that reverse the cutter rotation to clear obstructions or shut down the unit if the obstruction cannot be cleared.

Figure 13–7

Typical inline sludge grinder:
(a) side view, (b) end view, and
(c) view of typical installation
[(a) and (b) adapted from
Franklin Miller].

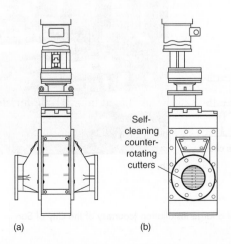

Self-
cleaning
counter-
rotating
cutters

(a)   (b)   (c)

Table 13-16

Operations or processes requiring the grinding of sludge and biosolids

Operation or process	Purpose of grinding
Pumping with progressive cavity pumps	Prevent clogging and reduce wear
Solid bowl centrifuges	Prevent clogging. Large solid bowl units generally can handle larger particles and may not require sludge or biosolids grinding
Belt filter press	Prevent clogging of the sludge or biosolids distribution system, prevent warping of rollers, and provide more uniform dewatering

Screening

Because raw wastewater screens can allow significant quantities of solid material to pass through, sludge screening is an alternative to grinding. Screening is advantageous in that nuisance material is removed from the sludge stream. Step screens, shown on Fig. 5–4(c) in Chap. 5, can be used for the removal of fine solids from septage and primary sludge. Screen openings normally range from 3 to 6 mm (0.12 to 0.24 in.), although openings up to 10 mm (0.4 in.) can be used.

Another type of sludge screen is an inline screen that can be installed in a pipeline (see Fig. 13–8). The screen removes material by passing the flow stream through a screen with 3 to 10-mm (0.12 to 0.4-in.) openings although 5 mm (0.2 in.) is the typical size for wastewater sludges. Material captured by the screen moves by a screw conveyor into a press or compaction zone where it is dewatered and compacted. Material is ejected from the press zone when sufficient solids build up to overcome the force on the unit's discharge cone. Screening solids concentrations range from 30 to 50 percent. Allowable operating pressure is reported to be 100 kPa (14 $lb_f/in.^2$) (Arakaki et al., 1998). The screened sludge is diluted and may require thickening.

Degritting

In some plants where separate grit removal facilities are not used ahead of the primary sedimentation tanks, or where the grit removal facilities are not adequate to handle peak

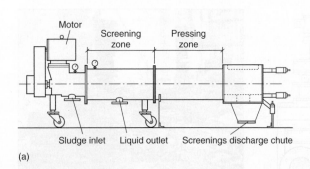

(a)   (b)

Figure 13-8

Sludge screenings press: (a) schematic and (b) view of a large installation (courtesy of the City of San Diego, CA).

Table 13–17

Grit removal efficiency using cyclone degritters for primary sludge[a]

Primary sludge concentration, % total solids	Mesh of removal[b]
1	150
2	100
3	65
4	28–35

[a] For a 300 mm (12 in.) hydrocyclone at 42 kN/m^2 (6 lb$_f$/in.^{2} gage) at 13 L/s (200 gal/min).

[b] About 95 percent or more of indicated particle size is removed.

Note: Normal design range is for 1 to 1.5 percent sludge.

flows and peak grit loads, it may be necessary to remove the grit before further processing of the sludge. Where further thickening of the primary sludge is desired, a practical consideration is sludge degritting. The most effective method of degritting sludge is through the application of centrifugal forces in a flowing system to achieve separation of the grit particles from the organic sludge. Such separation is achieved through the use of cyclone degritters, which have no moving parts. The sludge is applied tangential to a cylindrical feed section, thus imparting a centrifugal force. The heavier grit particles move to the outside of the cylinder section and are discharged through a conical feed section. The organic sludge is discharged through a separate outlet. The efficiency of the cyclone degritter is affected by pressure and by the concentration of the organics in the sludge. To obtain effective grit separation, the sludge must be relatively dilute, 1 to 2 percent TS. As the sludge concentration increases, the particle size that can be removed decreases. The general relationship between sludge concentration and effectiveness of removal for primary sludges is shown in Table 13–17.

Blending

Sludge is generated in primary, secondary, and advanced wastewater treatment processes. Primary sludge consists of settleable solids carried in the raw wastewater. Secondary sludge consists of biological solids as well as additional settleable solids. Sludge produced in the advanced wastewater may consist of biological and chemical solids. Sludge is blended to produce a uniform mixture to downstream operations and processes. Uniform mixtures are most important in short-detention-time systems, such as sludge dewatering, heat treatment, and incineration. Provision of well-blended sludge with consistent characteristics to these treatment units will enhance greatly plant operability and performance.

Sludge from primary, secondary, and advanced processes can be blended in several ways:

1. *In primary settling tanks.* Secondary or tertiary sludges can be returned to the primary settling tanks where they will mix and co-settle with the primary sludge.
2. *In pipes.* Blending in pipes requires careful control of sludge sources and feed rates to ensure the proper blend. Without careful control, wide variations in sludge consistency may be expected.
3. *In sludge-processing facilities with long detention times.* Aerobic and anaerobic digesters (complete-mix type) can blend the feed sludges uniformly.
4. *In a separate blending tank.* This practice provides the best opportunity to control the quality of the blended sludges.

In treatment plants of less than 0.05 m³/s (1 Mgal/d) capacity, blending is accomplished usually in the primary settling tanks. In large facilities, optimum efficiency is achieved by separately thickening sludges before blending.

Storage

Storage should be provided to minimize fluctuations in the rate of sludge and biosolids production and to allow sludge to accumulate during periods when subsequent processing facilities are not operating, e.g., night shifts, weekends, and periods of unscheduled equipment downtime. Sludge and biosolids storage is particularly important in providing a uniform feed rate ahead of the following processes: mechanical dewatering, lime stabilization, heat drying, and thermal reduction.

Short-term sludge and biosolids storage may be accomplished in wastewater settling tanks or in thickening tanks. Long-term sludge and biosolids storage may be accomplished in stabilization processes with long detention times, e.g., aerobic and anaerobic digestion, or in specially designed separate tanks. In small installations, sludge is usually stored in the settling tanks and digesters. In large installations that do not use aerobic and anaerobic digestion, sludge is often stored in separate blending and storage tanks. Such tanks may be sized to retain the sludge for a period of several hours to a few days. If sludge or biosolids is stored longer than 2 to 3 d, it will deteriorate, become odorous, and be more difficult to dewater. The determination of the required storage volume is illustrated in Example 13–3. Sludge or biosolids is often aerated to prevent septicity and to promote mixing. Mechanical mixing may be necessary to assure complete blending of the sludge. Chlorine, iron salts, potassium permanganate, and hydrogen peroxide have been used with limited success to limit or control septicity and to control the odors from sludge storage and blending tanks. In cases where sludge storage occurs in enclosed tanks, ventilation should be provided along with appropriate odor-control technologies such as chemical scrubbers or biofilters (see Chap. 15).

EXAMPLE 13–3   **Determination of Volume Required for Sludge Storage**   Assume that the yearly average rate of sludge production from an activated sludge treatment plant is 12,000 kg/d. Develop a curve of sustained sludge mass loading rates that can be used to determine the size of sludge-storage facilities required with various downstream sludge-processing units. Then, using the developed curve, determine the volume required for sludge storage, assuming that sludge accumulated for 7 d is to be processed in 5 working d, and that sludge accumulated for 14 d is to be processed in 10 working d. Note that the 5- and 10-d work periods correspond to 1 and 2 wk, respectively, assuming that certain sludge-processing facilities, such as belt-filter presses, will not be operated on the weekends.

Solution
1. Develop a curve of sustained sludge mass loadings.
   a. Because no information is specified, it will be assumed that the sustained sludge production will mirror the sustained BOD plant loadings given on Fig. 3–13(a) and used in Example 3–7.
   b. Set up an appropriate computation table and compute the values necessary to plot the curve.

Length of sustained peak, d	Peaking factor[a]	Peak sludge mass loading, kg/d	Total sustained loading, kg[b]
(1)	(2)	(3)	(4)
1	2.4	28,800	28,800
2	2.1	25,200	50,400
3	1.9	22,800	68,400
4	1.8	21,600	86,400
5	1.7	20,400	102,000
10	1.4	16,800	168,000
15	1.3	15,600	234,000
365	1.0	12,000	

[a] From Fig. 3–13(a).

[b] Total mass produced for the corresponding sustained period given in col. 1.

    c. Plot the sustained sludge loading curve (see following figure).

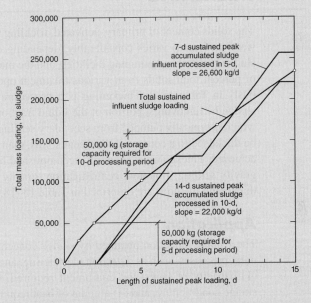

2. Determine the sludge storage volume required for the stated operating conditions.
    a. Determine the daily rate at which sludge must be processed to handle the 7-d sustained peak (from figure) in 5 working d.

$$kg/d = \frac{133,000}{5\,d} = 26,600\,kg/d$$

    b. Determine the daily rate at which sludge must be processed to handle the 14-d sustained peak (from figure) in 10 working d.

$$kg/d = \frac{220,000}{10\,d} = 22,000\,kg/d$$

c. Assuming that the sludge storage facilities are empty on Friday just before the weekend, plot on the figure the average daily rate at which sludge must be processed during the 5- and 10-d periods.
d. From the figure, the required storage capacity in pounds of sludge is
   i. Capacity based on 5 working d = 50,000 kg
   ii. Capacity based on 10 working d = 50,000 kg

Comment The downstream processing equipment can now be sized using the daily rate at which sludge must be processed. For example, if the number of kilograms per hour that can be processed with a belt-filter press is known, then the size and number of units can be computed from the number of shifts to be used per day and the assumed value of the actual working hours per shift. In sizing equipment, a trade-off analysis should always be performed between the cost of storage and processing facilities versus labor costs (for both one shift and two shifts) to determine the most cost-effective combination.

13–6 THICKENING

The solids content of primary, activated, trickling filter, or mixed sludge (i.e., primary plus waste activated) varies considerably, depending on the characteristics of the sludge, the sludge removal and pumping facilities, and the method of operation. Representative values of percent total solids from various treatment operations or processes were shown previously in Table 13–8. Thickening is a procedure used to increase the solids content of sludge by removing a portion of the liquid fraction. To illustrate, if waste activated sludge which is typically pumped from secondary settling tanks with a content of 0.8 percent, can be thickened to a content of 4 percent solids, then a five-fold decrease in sludge volume is achieved. Thickening is generally accomplished by physical means, including co-settling, gravity settling, flotation, centrifugation, gravity belt, and rotary drum. Typical sludge-thickening methods are described in Table 13–18.

Application

The volume reduction obtained by sludge concentration is beneficial to subsequent treatment processes, such as digestion, dewatering, and drying from the following standpoints: (1) capacity of tanks and equipment required, (2) quantity of chemicals required for sludge conditioning, and (3) amount of heat required by digesters and amount of auxiliary fuel required for heat drying.

For large facilities where sludge must be transported a significant distance, such as to a separate plant for processing, a reduction in sludge volume may result in a reduction of pipe size and pumping costs. For smaller facilities, the requirements of a minimum practicable pipe size and minimum velocity may necessitate pumping of significant volumes of wastewater in addition to sludge, thereby diminishing the value of volume reduction. Volume reduction is very desirable when liquid sludge is transported by tank trucks for direct application to land as a soil conditioner.

Sludge thickening is achieved at all wastewater treatment plants in some manner—in the primary clarifiers, in sludge digestion facilities, or in specially designed separate units. If separate units are used, the recycled flows are returned normally to the wastewater treatment facilities. In treatment plants of less than 4000 m^3/d ($\sim$1 Mgal/d) capacity, separate sludge thickening is seldom practiced. In small plants, gravity thickening is accomplished in the primary settling tank or in the sludge-digestion units, or both. In larger treatment

Table 13–18

Occurrence of thickening methods in sludge processing

Method	Type of sludge	Frequency of use and relative success
Gravity, co-settling in clarifier	Primary and waste activated	Occasional use; may negatively impact the effectiveness of the primary clarifier
Gravity, thickening in separate tank	Untreated primary	Commonly used with excellent results. Sometimes used with hydrocyclone degritting of sludge
	Untreated primary and waste activated	Often used. For small plants, generally satisfactory results with sludge concentrations in the range of 4 to 6 percent. For large plants, results are marginal. Can be odorous in warm weather
	Waste activated	Seldom used; poor solids concentration (2 to 3 percent)
Dissolved air flotation	Untreated primary and waste activated	Limited use; results similar to gravity thickeners
	Waste activated	Commonly used, but use is decreasing because of high operating cost; good results (3.5 to 5 percent solids concentration)
Solid bowl centrifuge	Waste activated	Often used in medium to large plants; good results (4 to 6 percent solids concentration)
Gravity belt thickener	Waste activated	Often used; good results (3 to 6 percent solids concentration)
Rotary drum thickener	Waste activated	Limited use; good results (5 to 9 percent solids concentration)

facilities, the additional costs of separate sludge thickening are often justified by the improved control over the thickening process and the higher concentrations attainable.

Description and Design of Thickeners

The following discussion is intended to introduce the reader to the operations used for the thickening of sludges. Because most of the equipment is mechanical, the primary concern is with its proper application to meet a given treatment objective rather than with the theory of mechanical design. In designing thickening facilities, it is important to (1) provide adequate capacity to meet peak demands and (2) prevent septicity, with its attendant odor problems, during the thickening process. The six methods of thickening discussed in this section are (1) co-settling thickening, (2) gravity, (3) dissolved air flotation, (4) centrifugal, (5) gravity belt, and (6) rotary drum.

Co-settling Thickening.
Primary clarifiers are often used to thicken sludge for downstream processing. To thicken sludge, a sludge blanket must be created to consolidate the sludge without allowing the clarified water to be pulled through. Often, sludge retention times of 12 to 24 h or more are maintained in clarifiers to achieve thickened sludge concentration levels in the clarifier underflow. Excessive retention of sludge in the clarifier can cause septic conditions and gasification, and reduce the levels of TSS and BOD removal. Typical effects of sludge blanket retention on TSS removal are illustrated on Fig. 13–9.

Figure 13–9

Effect of sludge blanket retention time on TSS removal for co-thickening of primary sludge (Albertson and Walz, 1997).

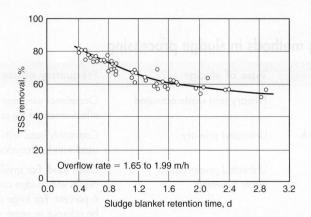

Overflow rate = 1.65 to 1.99 m/h

Successful thickening of sludge in primary clarifiers has been achieved by a combination of the following: (1) using one clarifier in a bank of clarifiers for co-settling thickening; dilute sludge underflow (less than 1 percent solids) from the other clarifiers is discharged to the thickening clarifier, (2) maintaining the sludge inventory for about 6 to 12 h, and (3) providing for the addition of coagulating chemicals such as polymer and ferric chloride to condition the sludge to enhance settling. The need for chemical addition depends upon the clarifier overflow rates. Underflow sludge concentrations on the order of 3 to over 5 percent have been reported (Albertson and Walz, 1997). By controlling the sludge blanket within the above sludge retention parameters, clarifier removal rates are enhanced, and sludge thickening is achieved. A schematic diagram of the co-settling thickening system is shown on Fig. 13–10.

Gravity Thickening.   Gravity thickening is one of the most common methods used and is accomplished in a tank similar in design to a conventional sedimentation tank. Normally, a circular tank is used, and dilute sludge is fed to a center feed well. The feed sludge is allowed to settle and compact, and the thickened sludge is withdrawn from the conical tank bottom. Conventional sludge collecting mechanisms with deep trusses (see Fig. 13–11) or vertical pickets stir the sludge gently, thereby opening up channels for water to escape and promoting densification. The supernatant flow that results is drawn off and returned to either the primary settling tank, the influent of the treatment plant, or a return flow treatment process. The thickened sludge is pumped to the digesters or dewatering equipment as required; thus, storage space must be provided for the sludge. As indicated in Table 13–18,

Figure 13–10

Schematic diagram of a sludge co-thickening system.

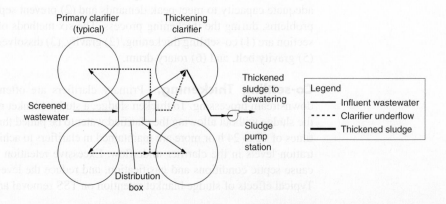

Figure 13–11

Schematic diagram of a gravity thickener: (a) plan and (b) section.

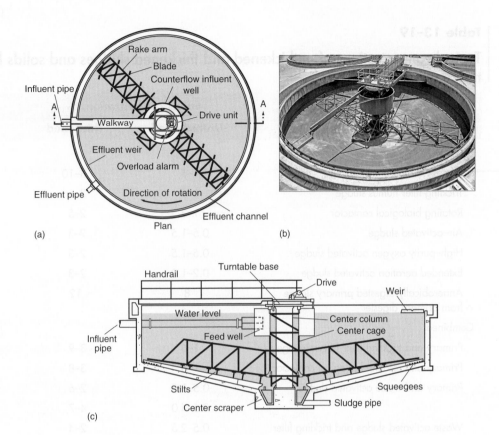

(a)

(b)

(c)

gravity thickening is most effective on primary sludge. Gravity thickeners are designed on the basis of solids loading and thickener overflow rate. Typical solids loadings based on existing data are reported in Table 13–19. Recommended maximum hydraulic overflow rates range from 15.5 to 31 $m^3/m^2 \cdot d$ (380 to 760 $gal/ft^2 \cdot d$) for primary sludges, 4 to 8 $m^3/m^2 \cdot d$ (100 to 200 $gal/ft^2 \cdot d$) for waste activated sludge, and 6 to 12 $m^3/m^2 \cdot d$ (150 to 300 $gal/ft^2 \cdot d$) for combined primary and waste activated sludge (WEF, 1980). High hydraulic loadings can cause excessive solids carryover. Conversely, low hydraulic loadings can cause septic conditions and odors, and floating sludge can result.

Provisions for dilution water and occasional chlorine addition are frequently included to improve process performance by maintaining the hydraulic loading. Polymer addition is frequently provided. To maintain aerobic conditions in gravity thickeners, especially when wastewater is warm (22 to 28°C), provisions should be included for adding up to 24 to 30 $m^3/m^2 \cdot d$ (600 to 750 $gal/ft^2 \cdot d$) of dilution water (final effluent) to the thickening tank. The dilution water may also remove certain soluble organic and inorganic compounds that consume large amounts of conditioning chemicals used in dewatering. Dilution water that is part of supernatant returned and recycled to the liquid process must be considered in process design.

Because the thickening characteristics of wastewater sludge can vary considerably, it is desirable to design a thickening facility using criteria based on a testing program. Testing programs that can be used include batch settling tests, bench-scale settling tests, and pilot-scale testing. The latter method is recommended wherever possible because data can be obtained from a variety of operating parameters. Test methods are described in WEF (2010a).

Table 13–19

Typical concentrations of unthickened and thickened sludges and solids loadings for gravity thickeners[a]

Type of sludge or biosolids	Solids concentration, %		Solids loading	
	Unthickened	Thickened	lb/ft²·d	kg/m²·d
Separate				
Primary sludge	1–6	3–10	20–30	100–150
Trickling filter humus sludge	1–4	3–6	8–10	40–50
Rotating biological contactor	1–3.5	2–5	7–10	35–50
Air–activated sludge	0.5–1.5	2–3	4–8	20–40
High–purity oxygen activated sludge	0.5–1.5	2–3	4–8	20–40
Extended aeration activated sludge	0.2–1.0	2–3	5–8	25–40
Anaerobically digested primary sludge from primary digester	8	12	25	120
Combined				
Primary and trickling filter humus sludge	1–6	3–9	12–20	60–100
Primary and rotating biological contactor	1–6	3–8	10–18	50–90
Primary and waste activated sludge	0.5–1.5	2–6	5–14	25–70
	2.5–4.0	4–7	8–16	40–80
Waste activated sludge and trickling filter humus sludge	0.5–2.5	2–4	4–8	20–40
Anaerobically digested primary and waste activated sludge	4	8	14	70
Chemical sludge:				
High lime	3–4.5	12–15	24–60	120–300
Low lime	3–4.5	10–12	10–30	50–150
Iron	0.5–1.5	3–4	2–10	10–50

[a] Adapted from WEF (2010a).

In operation, a sludge blanket is maintained on the bottom of the thickener to aid in concentrating the sludge. An operating variable is the sludge volume ratio, which is the volume of the sludge blanket held in the thickener divided by the volume of the thickened sludge removed daily. Values of the sludge volume ratio normally range between 0.5 and 20 d; the lower values are required during warm weather. Alternatively, sludge blanket depth should be measured. Blanket depths may range from 0.5 to 2.5 m (2 to 8 ft); shallower depths are maintained in the warmer months.

EXAMPLE 13–4 **Design a Gravity Thickener for Combined Primary and Waste Activated Sludge** Design a gravity thickener for a wastewater treatment plant having primary and waste activated sludge with the following characteristics:

Type of sludge	Specific gravity	Solids, %	Flowrate, m^3/d
Average design conditions:			
Primary sludge	1.03	3.3	400
Waste activated	1.005	0.2	2250
Peak design conditions:			
Primary sludge	1.03	3.4	420
Waste activated	1.005	0.23	2500

Solution

1. Compute the dry solids at peak design conditions.
    a. Primary sludge

    $$kg/d \text{ dry solids} = (420 \, m^3/d)(1.03)(0.034 \, g/g)(10^3 \, kg/m^3)$$
    $$= 14,708 \, kg/d$$

    b. Waste activated sludge

    $$kg/d \text{ dry solids} = (2500 \, m^3/d)(1.005)(0.0023 \, g/g)(10^3 \, kg/m^3)$$
    $$= 5779 \, kg/d$$

    c. Combined sludge mass = 14,708 + 5779 = 20,487 kg/d
    d. Combined sludge flowrate = 2,500 + 420 = 2,920 m^3/d

2. Compute solids concentration of the combined sludge, assuming the specific gravity of the combined sludge is 1.02.

    $$\% \text{ solids} = \frac{(20,487 \, kg/d)}{(2920 \, m^3/d)(1.02)(10^3 \, kg/m^3)} \times 100 = 0.69\%$$

3. Compute surface area based on solids loading rate. Because the sludge concentration is between 0.5 and 1.5%, select a solids loading rate of 50 $kg/m^2{\cdot}d$ from Table 13–19.

    $$\text{Area} = \frac{(20,487 \, kg/d)}{(50 \, kg/m^2{\cdot}d)} = 409.7 \, m^2$$

4. Compute hydraulic loading rate.

    $$\text{Hydraulic loading} = \frac{(2920 \, m^3/d)}{409.7 \, m^2} = 7.13 \, m^3/m^2{\cdot}d$$

5. Compute diameter of thickener; assume two thickeners.

    $$\text{Diameter} = \sqrt{\frac{4 \times 409.7 \, m^2}{2 \times \pi}} = 16.15 \, m$$

Comment
The hydraulic loading rate of 7.13 $m^3/m^2{\cdot}d$ at peak design flow is at the lower end of the recommended rate. To prevent septicity and odors, dilution water should be provided. Calculation of the dilution water requirements for average design flow is a homework problem. The thickener size of 16.15 m is within the maximum size of 20 m customarily recommended by thickener equipment manufacturers for use in municipal wastewater treatment. In actual design, round the thickener diameter to the nearest 0.5 m, or, in this case, 16 m.

Flotation Thickening. In dissolved air flotation, air is introduced into a solution that is being held at an elevated pressure. A typical unit used for thickening waste activated sludge is shown on Fig. 13–12. When the solution is depressurized, the dissolved air is released as finely divided bubbles carrying the sludge to the top, where it is removed.

Figure 13–12

Typical dissolved air flotation unit used for thickening waste activated sludge: (a) cross-section through typical circular flotation unit, (b) view inside covered circular flotation unit, and (c) view inside building containing rectangular flotation units.

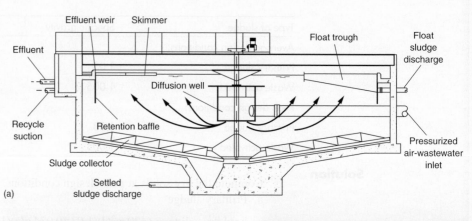

(a)

(b)

(c)

Flotation thickening is used most efficiently for waste sludges from suspended-growth biological treatment processes, such as the activated sludge process or the suspended growth nitrification process. Other sludges such as primary sludge, trickling filter humus, aerobically digested sludge, and sludges containing metal salts from chemical treatment have been flotation thickened. In locations where freezing is a problem or where odor control is of concern, flotation thickeners are normally enclosed in a building.

The float solids concentration that can be obtained by flotation thickening of waste activated sludge is influenced primarily by the air-to-solids ratio, sludge characteristics (in particular the sludge volume index, SVI), solids loading rate, and polymer application. Although float solids concentrations have ranged historically between 3 and 6 percent by weight, float solids concentration is difficult to predict during the design stage without bench-scale or pilot-plant testing. The air-to-solids ratio is probably the most important factor affecting performance of the flotation thickener, and is defined as the weight ratio of air available for flotation to the solids to be floated in the feed stream. The air-to-solids ratio at which float solids are maximized varies from 2 to 4 percent. The SVI is also important because better thickening performance has been reported when the SVI is less than 200, using nominal polymer dosages. At high SVIs, the float concentration deteriorates and high polymer dosages are required.

Higher loadings can be used with dissolved air flotation thickeners than are permissible with gravity thickeners, because of the rapid separation of solids from the wastewater. Flotation thickeners typically are designed for the solids loadings given in Table 13–20. For design without the benefit of pilot studies, the minimum loadings should be used. The higher solids loadings generally result in lower concentrations of thickened sludge.

Table 13–20

Typical solids loadings for dissolved air flotation units[a,b]

Type of sludge	Loading, lb/ft²·h		Loading, kg/m²·h	
	Without chemical addition	With chemicals	Without chemical addition	With chemicals
Air activated sludge:				
Mixed Liquor	0.25–0.6	Up to 2	1.2–3	Up to 10
Settled	0.5–0.8	Up to 2	2.4–4	Up to 10
High purity oxygen activated sludge	0.6–0.8	Up to 2	3–4	Up to 10
Trickling filter humus sludge	0.6–0.8	Up to 2	3–4	Up to 10
Primary + air activated sludge	0.6–0.8	Up to 2	3–6	Up to 10
Primary + trickling filter humus sludge	0.83–1.25	Up to 2	4–6	Up to 10
Primary sludge only	0.83–1.25	Up to 2.5	4–6	Up to 12.5

[a] Adapted, in part, from U.S. EPA (1979) and WEF (2010a).
[b] Loading rates necessary to produce a minimum 4 percent solids concentration in the float.

Operational difficulties may arise when the solids loading rate exceeds approximately 10 kg/m²·h (2.0 lb/ft²·h). The increased amount of float created at high solids loading necessitates continuous skimming, often at high skimming speeds.

Primary tank effluent or plant effluent is recommended as the source of air-charged water rather than flotation tank effluent, except when chemical aids are used, because of the possibility of fouling the air-pressure system with solids. The use of polymers as flotation aids is effective in increasing the solids recovery in the floated sludge from 85 to 98 or 99 percent, and in reducing the recycle loads. Polymer dosages for thickening waste activated sludge are 2 to 5 kg of dry polymer per tonne of dry solids (4 to 10 lb/ton).

Centrifugal Thickening. Centrifuges are used both to thicken and to dewater sludges. As indicated in Table 13–18, their application in thickening is limited normally to waste activated sludge. Thickening by centrifugation involves the settling of sludge particles under the influence of centrifugal forces. The basic type of centrifuge used for sludge thickening is the solid bowl centrifuge (see Fig. 13–13).

The solid-bowl centrifuge consists of a long bowl, normally mounted horizontally and tapered at one end. Sludge is introduced into the unit continuously, and the solids concentrate on the periphery. An internal helical scroll, spinning at a slightly different speed, moves the accumulated sludge toward the tapered end where additional solids concentration occurs and the thickened sludge is discharged.

Under normal conditions, thickening can be accomplished by centrifugal thickening without polymer addition. Maintenance and power costs for the centrifugal thickening process, however, can be substantial. Therefore, the process is usually attractive only at facilities larger than 0.2 m³/s (5 Mgal/d), where space is limited and skilled operators are available, or for sludges that are difficult to thicken by more conventional means. Many systems are designed with standby polymer systems for use to improve system performance. Polymer dosages for thickening waste activated sludge range from 0 to 4 kg of dry polymer per 1 tonne of dry solids (0 to 8 lb/ton).

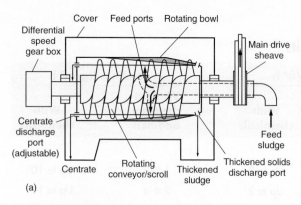

(a) (b)

Figure 13-13

Centrifuge used for sludge thickening: (a) schematic, and (b) scroll rotor removed for maintenance.

The performance of a centrifuge is often quantified by the concentration achieved in the thickened sludge product and the TSS recovery (sometimes termed "capture"). The recovery is calculated as the thickened dry solids as a percentage of the feed solids. Using the commonly measured solids concentrations, the recovery is calculated by the following expression (WEF, 2010a):

$$R = \frac{\text{TSS}_P(\text{TSS}_F - \text{TSS}_C)}{\text{TSS}_F(\text{TSS}_P - \text{TSS}_C)} \times 100 \tag{13-6}$$

where R = recovery, percent
TSS_P = total suspended solids concentration in thickened product, percent by weight
TSS_F = total suspended solids concentration in feed, percent by weight
TSS_C = total suspended solids concentration in centrate, percent by weight

For a constant feed concentration, the percent recovery increases as the concentration of solids in the centrate decreases. In concentrating sludge solids, recovery is important because with a higher recovery lesser amounts of biodegradable solids are returned to the treatment process for further treatment. In developing a mass balance for the treatment plant, return flows (also termed sidestream flows) from thickening, stabilization, and dewatering processes must be taken into account (see Sec. 14–7).

The principal operational variables include the following: (1) characteristics of the feed sludge (its water holding structure and the sludge volume index); (2) rotational speed; (3) hydraulic loading rate; (4) depth of the liquid pool in the bowl; (5) differential speed of the screw conveyor; and (6) polymer conditioning to improve the performance. Because the interrelationships of these variables will be different in each location, specific design recommendations are not available; in fact, bench-scale or pilot-plant tests are recommended.

Gravity Belt Thickening. The development of gravity belt thickeners stemmed from the application of belt presses for sludge dewatering. In belt-press dewatering, particularly for sludges having solids concentrations less than 2 percent, effective thickening occurred in the gravity drainage section of the press. The equipment developed for thickening consists of a gravity belt that moves over rollers driven by a variable-speed drive unit (see Fig. 13–14). The sludge is conditioned with polymer and fed into a feed/distribution box at one end, where the sludge is distributed evenly across the width of the moving belt. The water drains through the belt as the concentrating sludge is carried toward the discharge

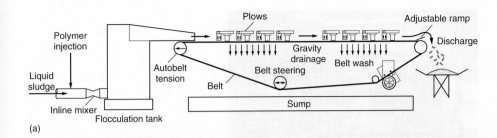

(a)

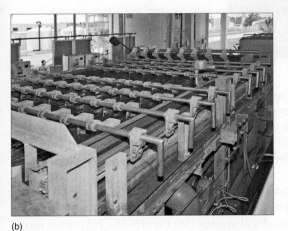

(b)

(c)

Figure 13–14

Gravity belt thickener: (a) schematic diagram (courtesy of Ashbrook Corporation), (b) top view of sludge plows used to aid the dewatering process, and (c) top view of gravity belt thickener viewed from the discharge end.

end of the thickener. The sludge is ridged and furrowed by a series of plow blades placed along the travel of the belt [see Fig. 13–14(b)], allowing the water released from the sludge to pass through the belt. After the thickened sludge is removed [see Fig. 13–14(c)], the belt travels through a wash cycle. The gravity-belt thickener has been used for thickening waste activated sludge, anaerobically and aerobically digested sludge, and some industrial sludges. Polymer addition is required. Testing is recommended to verify that the sludge can be thickened at typical polymer dosages.

Typical hydraulic loading rates for gravity-belt thickeners are given in Table 13–21. In lieu of pilot-plant data, a value of 800 L/m·min (64 gal/ft·min) is suggested as a design value;

Table 13–21

Typical hydraulic loading rates for gravity belt thickeners[a,b]

Belt size (effective dewatering width), m	Hydraulic loading range	
	gal/min	L/s
1.0	100–250	6.7–16
1.5	150–375	9.5–24
2.0	200–500	12.7–32
3.0	300–750	18–47

[a] Assumes 0.5 to 1.0 percent feed solids for municipal sludges. Variations in sludge density, belt porosity, polymer reaction rate, and belt speed will act to increase or decrease the rates of flow for any given size belt.

[b] Adapted from WEF (2010a).

Figure 13–15

Rotary drum thickener. (Courtesy of Parkson Corporation.)

the higher the feed rate, the greater the operator attention required to maintain stable operation. Solids loading rates range on the order of 200 to 600 kg/m·h (135 to 400 lb/ft²·h). Systems are often designed for a maximum of 4 to 7 percent thickened solids. Solids capture typically ranges between 90 and 98 percent (WEF, 2010a). Polymer dosages for thickening waste activated sludge range from 3 to 7 kg of dry polymer per tonne of dry solids (6 to 14 lb/ton).

Rotary Drum Thickening. Rotary media-covered drums are also used to thicken sludges. A rotary drum thickening system consists of a conditioning system (including a polymer feed system) and rotating cylindrical screens (see Fig. 13–15). Polymer is mixed with dilute sludge in the mixing and conditioning drum. The conditioned sludge is then passed to rotating screen drums, which separate the flocculated solids from the water. Thickened sludge rolls out the end of the drums, while separated water decants through the screens. Some designs also allow coupling of the rotary drum unit to a belt filter press for combination thickening and dewatering.

Rotary drum thickeners can be used as a prethickening step before belt-press dewatering and are typically used in small- to medium-sized plants for waste activated sludge thickening. The addition of large amounts of polymer for conditioning can be of concern because of floc sensitivity and shear potential in the rotating drum (WEF, 2010a). Rotary drum thickeners are available in capacities up to 24 L/s (400 gal/min). Typical performance data for rotary drum thickeners are given in Table 13–22.

Table 13–22

Typical Performance ranges for rotary drum thickeners for sludge and biosolids[a]

Type of feed	Feed, % TS	Water removed, %	Thickened solids, %	Solids recovery, %
Untreated sludge				
Primary	3.0–6.0	40–75	7–9	93–98
WAS[b]	0.5–1.0	70–90	4–9	93–99
Primary + WAS	2.0–4.0	50	5–9	93–98
Anaerobically digested biosolids	2.5–5.0	50	5–9	90–98
Aerobically digested biosolids	0.8–2.0	70–80	4–6	90–98

[a] WEF (2010a).

[b] WAS = waste activated sludge.

13–7 INTRODUCTION TO SLUDGE STABILIZATION

Sludge is stabilized to (1) reduce pathogens, (2) eliminate offensive odors, and (3) inhibit, reduce, or eliminate the potential for putrefaction. The success in achieving these objectives is related to the effects of the stabilization operation or process on the volatile or organic fraction of the sludge. Survival of pathogens, release of odors, and putrefaction occur when microorganisms are allowed to flourish in the organic fraction of the sludge. The means to eliminate these nuisance conditions is mainly related to the biological reduction of the volatile content and the addition of chemicals to the sludge or biosolids to render them unsuitable for the survival of microorganisms.

Stabilization is not practiced at all wastewater treatment plants, but it is used by an overwhelming majority of plants ranging in size from small to very large. In addition to the health and aesthetic reasons cited above, stabilization can result for volume reduction, production of usable gas (methane), and improved sludge dewaterability.

The principal methods used for stabilization of sludge are (1) alkaline stabilization, usually with lime; (2) anaerobic digestion; (3) aerobic digestion; and (4) composting. These processes are generally defined in Table 13–23. Each of the processes, with the exception of

Table 13–23

Description of sludge stabilization processes

Process	Description	Comments
Alkaline stabilization	Addition of an alkaline material, usually lime, to maintain a high pH level to effect the destruction of pathogenic organisms.	An advantage of alkaline stabilization is that a rich soil-like product results with substantially reduced pathogens. A disadvantage is that the product mass is increased by the addition of the alkaline material. Some alkaline stabilization processes are capable of producing a Class A sludge.
Anaerobic digestion	The biological conversion of organic matter by fermentation in a heated reactor to produce methane gas and carbon dioxide. Fermentation occurs in the absence of oxygen.	Methane gas can be used beneficially for the generation of heat or electricity. The resulting biosolids may be suitable for land application. The process requires skilled operation as it may be susceptible to upsets and recovery is slow.
Aerobic digestion	The biological conversion of organic matter in the presence of air (or oxygen), usually in an open-top tank.	Process is much simpler to operate than an anaerobic digester, but no usable gas is produced. The process is energy-intensive because of the power requirements necessary for mixing and oxygen transfer.
Autothermal thermophilic digestion	Process is similar to aerobic digestion except higher amounts of oxygen are added to accelerate the conversion of organic matter. Process operates at temperatures of 40 to 80 °C, autothermally in an insulated tank.	Process is capable of producing a Class A sludge. Skilled operators are required and the process is a high energy user (to produce air or oxygen).
Composting	The biological conversion of solid organic matter in an enclosed reactor or in windrows or piles.	A variety of sludge or biosolids can be composted. Composting requires the addition of a bulking agent to provide an environment suitable for biological activity. Volume of compost produced is usually greater than the volume of wastewater sludge being composted. Class A or Class B sludge can be produced. Odor control is very important as process is odorous.

Table 13–24

Relative degree of attenuation achieved with various sludge stabilization processes[a]

Process	Degree of attenuation		
	Pathogens	**Putrefaction**	**Odor potential**
Alkaline stabilization	Good	Fair	Fair
Anaerobic digestion	Fair	Good	Good
Advanced anaerobic digestion	Excellent	Good	Good
Aerobic digestion	Fair	Good	Good
Autothermal thermophilic digestion (ATAD)	Excellent	Good	Good
Composting	Good	Good	Fair to good

[a] Adapted in part from WEF (2010a).

composting which is considered in Chap. 14, is discussed in more detail in the following sections, and their ability to mitigate or stabilize the effects related to pathogens, putrefaction, and odors is given in Table 13–24. Heat treatment and the addition of oxidizing chemicals, processes that seldom are used in the United States for stabilization, are not included in this text. For information about these processes please refer to Metcalf & Eddy (1991).

When designing a stabilization process, it is important to consider the sludge quantity to be treated, the integration of the stabilization process with the other treatment units, and the objectives of the stabilization process. The objectives of the stabilization process are often affected by existing or pending regulations. If sludge is to be applied on land, pathogen reduction has to be considered. The effect of regulations on application of biosolids to land is discussed in Sec. 14–8.

13–8 ALKALINE STABILIZATION

A method used to eliminate nuisance conditions in sludge involves the use of an alkaline material to render the sludge unsuitable for the survival of microorganisms. In the lime stabilization process, lime is added to untreated sludge in sufficient quantity to raise the pH to 12 or higher. The high pH creates an environment that halts or substantially retards the microbial reactions that can otherwise lead to odor production and vector attraction. The sludge will not putrefy, create odors, or pose a health hazard so long as the pH is maintained at this level. However, high ammonia odor levels have been observed during lime stabilization. The process can also inactivate virus, bacteria, and other microorganisms present. Advantages and disadvantages of alkaline stabilization are summarized in Table 13–25.

Chemical Reactions in Lime Stabilization

The lime stabilization process involves a variety of chemical reactions that alter the chemical composition of the sludge. The following simplified equations are illustrative of the types of reactions that may occur (WEF, 2010a):

Calcium

$$Ca^{2+} + 2HCO_3^- + CaO \rightarrow 2CaCO_3 + H_2O \qquad (13\text{-}7)$$

Phosphorus

$$2PO_4^{3+} + 6H^+ + 3CaO \rightarrow Ca_3(PO_4)_2 + 3H_2O \qquad (13\text{-}8)$$

Table 13–25

Advantages and disadvantages of alkaline stabilization[a]

Advantages	Disadvantages
1. Well proven process	1. The resulting product is not suitable for use on all soil, especially high alkaline soils
2. Product is suitable for a variety of uses that are consistent with the EPA's national beneficial reuse policy	2. The volume of material to be managed and moved off-site is increased by approximately 15 to 50 percent in comparison with other stabilization techniques, such as digestion. The increased volume results in higher transportation costs when material is moved off-site
3. Simple technology requiring few special skills for reliable operation	
4. Easy to construct of readily available parts	
5. Small footprint	3. Potential for odor generation both at the processing and end use site due to ammonia and TMA release
6. Flexible operation, easily started and stopped	4. Potential for dust production
7. Can produce Class A or Class B biosolids	5. The nitrogen content in the final product is lower than that in several other biosolids products because of ammonia volatilization. In addition, available phosphorous can be reduced through the formation of calcium phosphate

[a] Adapted, in part, from EPA (2000).

Carbon dioxide

$$CO_2 + CaO \rightarrow CaCO_3 \qquad (13\text{–}9)$$

Reactions with organic contaminants:
Acids:

$$RCOOH + CaO \rightarrow RCOOCaOH \qquad (13\text{–}10)$$

Fats:

$$Fat + Ca(OH)_2 \rightarrow glycerol + fatty\ acids \qquad (13\text{–}11)$$

Other reactions also occur, such as the hydrolysis of polymers, especially polymeric carbohydrates and proteins, and the hydrolysis of ammonia from amino acids.

Initially, lime addition raises the pH of the sludge. Then, reactions occur such as those in the above equations. If insufficient lime is added, the pH decreases as the reactions take place. Therefore, excess lime is required.

Biological activity produces compounds, such as carbon dioxide and organic acids, that react with lime. If biological activity in the sludge being stabilized is not sufficiently inhibited, these compounds will be produced, reducing the pH and resulting in inadequate stabilization. Many odorous, volatile off-gases are also produced, especially ammonia, which require collection and treatment in odor-control systems such as chemical scrubbers or biofilters (see Chap. 16). Other odorous material such as trimethyl amine (TMA) is generated from the degradation of conditioning polymers during lime stabilization (Dentel et al., 2005).

Heat Generation

If quicklime, CaO (or any compound high in quicklime), is added to sludge, it initially reacts with water to form hydrated lime. This reaction is exothermic and releases approximately

64 kJ/g·mole (2.75 × 10⁴ Btu/lb·mole) (WEF, 2010a). The reaction between quicklime and carbon dioxide is also exothermic, releasing approximately 180 kJ/g·mole (7.8 × 10⁴ Btu/lb·mole). These reactions can result in substantial temperature rise (see discussion of lime posttreatment).

Application of Alkaline Stabilization Processes

Three methods of alkaline stabilization are commonly used: (1) addition of lime to sludge prior to dewatering, termed "lime pretreatment," (2) the addition of lime to sludge after dewatering, or "lime posttreatment," and (3) advanced alkaline stabilization technologies. Either hydrated lime, $Ca(OH)_2$, or quicklime is used most commonly for lime stabilization. Fly ash, cement kiln dust, and carbide lime have also been used as a substitute for lime in some cases.

Lime Pretreatment. Pretreatment (before dewatering) of liquid sludge with lime has been used for either (1) the direct application of liquid sludge to land, or (2) combining benefits of sludge conditioning and stabilization prior to dewatering. In the former case, large quantities of liquid sludge have to be transported to land disposition sites, which limits utilization of lime pretreatment of sludge to small treatment plants. When pretreatment is used prior to dewatering, dewatering has been accomplished using a pressure-type filter press and/or screw press. Lime pretreatment is seldom used with centrifuges or belt filter presses because of abrasive wear and scaling problems.

Lime pretreatment of liquid sludge requires more lime per unit weight of sludge processed than that necessary for dewatering. The higher lime dose is needed to attain the required pH because of the chemical demand of the liquid. In addition, sufficient contact time must be provided before dewatering so as to provide a high level of pathogen kill. The recommended design objective is to maintain the pH above 12 for about 2 h to ensure pathogen destruction (the minimum U.S. EPA criterion for lime stabilization), and to provide enough residual alkalinity so that the pH does not drop below 11 for several days. The lime dosage required varies with the type of sludge and solids concentration. Typical dosages are reported in Table 13–26. Generally, as the percent solids concentration increases, the required lime dose decreases. Testing should be performed for specific applications to determine the actual dosage requirements.

Table 13-26
Typical lime dosages for pretreatment sludge stabilization[a]

| Type of sludge | Solids concentration, % | | Lime dosage[b] | | | |
| | Range | Average | lb Ca(OH)₂/ton dry solids | | g Ca(OH)₂/ kg dry solids | |
			Range	Average	Range	Average
Primary	3–6	4.3	120–340	240	60–170	120
Waste activated	1–1.5	1.3	420–860	600	210–430	300
Septage	1–4.5	2.7	180–1020	400	90–510	200

[a] Adapted from WEF, (1995a).

[b] Amount of Ca(OH)₂ required to maintain a pH of 12 for 30 min.

Figure 13–16

Theoretical temperature increase in post-lime stabilized sludge using quicklime (Roediger, 1987).

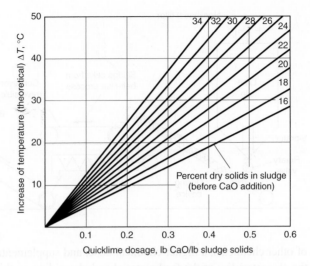

Note: In practice, higher temperature values are expected. 1.8 (°C) + 32 = °F

Because lime stabilization does not destroy the organics necessary for bacterial growth, the sludge must be treated with an excess of lime or used beneficially before the pH drops significantly. An excess dosage of lime may range up to 1.5 times the amount needed to maintain the initial pH of 12. Reasons for pH drop could be due to generation of carbon dioxide and organic acids from degradation of organic material in the sludge. The dissolution of carbon dioxide in the air could be another reason for pH decay. For additional details about pH decay following lime stabilization, WEF (1995a) is recommended.

Lime Posttreatment. In lime posttreatment, quicklime is mixed with dewatered sludge in a pugmill, paddle mixer, or screw conveyor to raise the pH of the mixture. Quicklime is used because the exothermic reaction of quicklime and water can raise the temperature of the mixture above 50°C, sufficient to inactivate worm eggs. The theoretical temperature increase by the addition of quicklime is illustrated on Fig. 13–16.

Lime posttreatment is more common than lime pretreatment and has several significant advantages when compared to lime pretreatment: (1) dry lime can be used; therefore, no additional water and equipment is needed for hydrated lime; (2) there are no special requirements for dewatering; and (3) scaling problems and associated maintenance problems of lime-sludge dewatering equipment are eliminated. Adequate mixing is critical for a posttreatment stabilization system so as to avoid pockets of putrescible material. A lime posttreatment stabilization system consists typically of a dry lime feed system, dewatered sludge cake conveyor, and a lime-sludge mixer (see Fig. 13–17). Good mixing is especially important to ensure contact between lime and small particles of sludge. When the lime and sludge are well mixed, the resulting mixture has a crumbly texture, which allows it to be stored for long periods or easily distributed on land by a conventional manure spreader. A potential disadvantage of lime posttreatment may be the release of odorous gases, specifically trimethyl amine (Novak, 2001; Dentel, 2005).

Advanced Alkaline Stabilization Technologies. Alkaline stabilization using materials other than lime is used by a number of municipalities. Most of the technologies that rely on additives, such as cement kiln dust, lime kiln dust, or fly ash, are modifications of conventional dry lime stabilization. The most common modifications include the addition

Figure 13–17

Typical lime posttreatment system. (From Roediger Pittsburgh.)

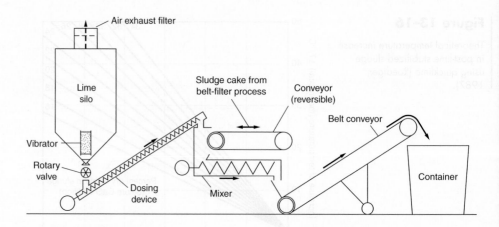

of other chemicals, a higher chemical dose, and supplemental drying. These processes alter the characteristics of the feed material and, depending on the process, may increase product stability, decrease odor potential, and provide product enhancement. To utilize these technologies, dewatered sludge is required.

Pasteurization may be accomplished by the exothermic reaction of quicklime with water to achieve a process temperature of 70°C and maintain it for more than 30 min. Other sources of energy can be used to aid in increasing the temperature generated from the exothermic chemical reaction. For example, the pasteurization system marketed by RDP Company uses electricity to generate heat for raising the temperature to the required degree. N-Viro International Corporation markets an advanced alkaline stabilization system combined with drying. To meet Class A biosolids criteria, the pasteurization reaction must be carried out under carefully controlled and monitored mixing and temperature conditions to ensure uniform treatment and inactivation of pathogens by the heat generated during the reaction. The process produces a soil-like material that is not subject to liquefaction under mechanical stress. Several other process variations of advanced alkaline stabilization are available, some of which are proprietary. Additional information may be found in WEF (2010a) and WEF (2012).

13–9 ANAEROBIC DIGESTION

Anaerobic digestion is among the oldest processes used for the stabilization of sludge. As described in Chap. 10, anaerobic digestion involves the decomposition of organic matter and reduction of inorganic matter (principally sulfate) in the absence of molecular oxygen. The major applications of anaerobic digestion are in the stabilization of concentrated sludges produced from the treatment of municipal and industrial wastewater. Great progress has been made in the fundamental understanding and control of the process, the sizing of tanks, and the design and application of equipment. Because of the emphasis on energy conservation and recovery and the desire to obtain beneficial use of wastewater biosolids, anaerobic digestion continues to be the dominant process for stabilizing sludge. Furthermore, anaerobic digestion of municipal wastewater sludge can, in many cases, produce sufficient digester gas to meet most of the energy needs for plant operation. An aerial view of a large digester installation is shown on Fig. 13–18.

In this section, a brief review is provided of process fundamentals followed by discussions of mesophilic anaerobic digestion, the most common basic process used; thermophilic digestion; and phased digestion. Phased digestion covers many of the new developments in anaerobic digestion.

Figure 13–18

Aerial view of several large anaerobic digesters at Boston, MA.

Process Fundamentals

As described in Chap. 7, the three types of chemical and biochemical reactions that occur in anaerobic digestion are hydrolysis; fermentation, also called acidogenesis (the formation of soluble organic compounds and short-chain organic acids); and methanogenesis (the bacterial conversion of organic acids into methane and carbon dioxide). Important environmental factors in the anaerobic digestion process are (1) solids retention time, (2) hydraulic retention time, (3) temperature, (4) alkalinity, (5) pH, (6) the presence of inhibitory substances, i.e., toxic materials, and (7) the bioavailability of nutrients and trace metals. The first three factors are important in process selection and are discussed in this section. Alkalinity is a function of feed solids and is important in controlling the digestion process. The effects of pH and inhibitory substances are discussed in Chaps. 7 and 10. The presence of nutrients and trace metals necessary for biological growth is described in Sec. 10–2 in Chap. 10.

Solids and Hydraulic Retention Times. Anaerobic digester sizing is based on providing sufficient residence time in well-mixed reactors to allow significant destruction of volatile suspended solids (VSS) to occur. Sizing criteria that have been used are (1) solids retention time SRT, the average time the solids are held in the digestion process, and (2) the hydraulic retention time τ, the average time the liquid is held in the digestion process. For soluble substrates, the SRT can be determined by dividing the mass of solids in the reactor (M) by the mass of solids removed daily (M/d). The hydraulic retention time τ is equal to the volume of liquid in the reactor (m^3) divided by the quantity of biosolids removed (m^3/d). For digestion systems without recycle, SRT $= \tau$.

The three reactions (hydrolysis, fermentation, and methanogenesis) are directly related to SRT (or τ). An increase or decrease in SRT results in an increase or decrease in the extent of each reaction. There is a minimum SRT for each reaction. If the SRT is less than the minimum SRT, bacteria cannot grow rapidly enough and the digestion process will fail eventually (WEF, 2010a).

Temperature. As discussed in Sec. 7–5, temperature not only influences the metabolic activities of the microbial population but also has a profound effect on such factors as gas transfer rates and the settling characteristics of biological sludges. In anaerobic digestion, temperature is important in determining the rate of digestion, particularly the rates of hydrolysis and methane formation. The minimum SRT required to achieve a given amount of VSS destruction is based on the design operating temperature. Most anaerobic digestion systems are designed to operate in the mesophilic temperature range, between 30 and 38°C (85 and 100°F). Other systems are designed for operation in the thermophilic temperature range of 50 to 57°C (122 to 135°F). Newly developed systems, as discussed in a latter part of this section, use a combination of mesophilic and thermophilic digestion in separate stages.

While selection of the design operating temperatures is important, maintaining a stable operating temperature is more important because the bacteria, especially the methane formers, are sensitive to temperature changes. Generally, temperature changes greater than 1°C/d affect process performance, and thus changes less than 0.5°C/d are recommended (WEF, 2010a).

Alkalinity. Calcium, magnesium, and ammonium bicarbonates are examples of buffering substances found in a digester. The digestion process produces ammonium bicarbonate from the breakdown of protein in the raw sludge feed; the others are found in the feed sludge. The concentration of alkalinity in a digester is, to a great extent, proportional to the solids feed concentration. A well-established digester has a total alkalinity of 2000 to 5000 mg/L.

The principal consumer of alkalinity in a digester is carbon dioxide, and not volatile fatty acids as is commonly believed (Speece, 2001). Carbon dioxide is produced in the fermentation and methanogenesis phases of the digestion process (see Sec. 7–12 in Chap. 7). Due to the partial pressure of gas in a digester, the carbon dioxide solubilizes and forms carbonic acid, which consumes alkalinity. The carbon dioxide concentration in the digester gas is, therefore, reflective of the alkalinity requirements. Volatile fatty acids are intermediate products from the acid phase of digestion and consume alkalinity. Volatile acids in digesters range from 50 to 300 mg/L. The ratio of volatile acids to the alkalinity is a parameter that is used to monitor the health of the digestion process and should be monitored closely. The volatile acids to alkalinity ratio for well-established digesters should fall between 0.05 to 0.25 with a 0.1 value indicating a good buffering capacity. Supplemental alkalinity can be supplied by the addition of sodium bicarbonate, lime, or sodium carbonate.

Description of Mesophilic Anaerobic Digestion Processes

The operation and physical facilities for mesophilic anaerobic digestion in single-stage high-rate, two-stage, and separate digesters for primary sludge and waste activated sludge are described in this section. Standard-rate, sometimes called low-rate, digestion is seldom used for digester design (because of the large tank volume required and the lack of adequate mixing) and is not covered in this text. For information about standard-rate digestion, the reader is referred to the third edition of this text (Metcalf & Eddy, 1991) and WEF (1998). The processes described below normally operate in the mesophilic range; high-rate digesters also operate in the thermophilic range. Thermophilic digestion is discussed at the end of the section.

Single-Stage High-Rate Digestion. Heating, auxiliary mixing, uniform feeding, and thickening of the feed stream characterize the single-stage high-rate digestion process. The sludge is mixed by one of many systems such as gas recirculation, pumping, or draft-tube mixers (separation of scum and supernatant does not take place), and sludge is heated to achieve optimum digestion rates [see Fig. 13–19(a)].

Uniform feeding is very important, and sludge should be pumped to the digester continuously or on a 30-min to 2-h time cycle to help maintain constant conditions in

Figure 13–19

Schematic d diagram of typical anaerobic digesters (a) high-rate and (b) two-stage.

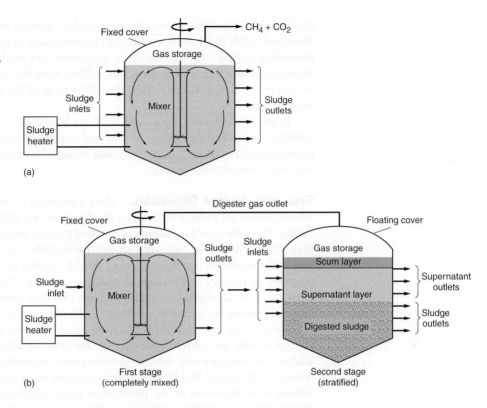

(a)

(b)

the reactor. In digesters fed on a daily cycle of 8 or 24 h, it is important to withdraw digested sludge from the digester before adding the feed sludge, because the pathogen kill is significantly greater when compared to using the feed sludge to displace the waste sludge (Speece, 2001). Because there is no supernatant separation in the high-rate digester, and the total solids are reduced by 45 to 50 percent and given off as gas, the digested sludge is about half as concentrated as the untreated sludge feed. Digestion tanks may have fixed roofs or floating covers (see subsequent discussion of digester types). Any or all of the floating covers may be of the gas holder type, which provides excess gas storage capacity. Alternatively, gas may be stored in a separate low-pressure gas holder or compressed and stored under pressure.

Two-Stage Digestion. Two-stage digestion, which was frequently used in the past, is seldom used in modern digester design. In two-stage digestion, a high-rate digester is coupled in series with a second tank [see Fig. 13–19(b)]. The first tank is used for digestion and is heated and equipped with mixing facilities. The second tank is usually unheated and used principally for storage. The tanks may be identical, in which case either one may be the primary. Tanks may have fixed roofs or floating covers, the same as single-stage digestion. In other cases, the second tank may be an open tank or a sludge lagoon. In the case of an open second digester some methane would escape if digestion continued leading to increased carbon footprint of the processes. Two-stage digestion of the type described above is seldom used, mainly because of the expense of building a large tank that is not fully utilized and because the second tank was of negligible benefit, operationally.

Because anaerobically digested biosolids may not settle well, the supernatant withdrawn from the second-stage tank may contain high concentrations of suspended solids. Reasons for poor settling characteristics include incomplete digestion in the primary digester (which generates gases in the secondary digester and causes floating solids) and fine-sized solids that have poor settling characteristics. Supernatant returned to the liquid processing system could cause upset conditions and might require separate treatment. Where two-stage digestion is used, return flows from the second tank must be accounted for in the solids mass balance. Less than 10 percent of the gas generated comes from the second stage.

In some installations, the second stage is a heated and mixed reactor to achieve further stabilization prior to dewatering or other subsequent processing. Additional discussion is provided later in this section on two-phase mesophilic digestion that provides more effective utilization of tank capacity.

Separate Sludge Digestion. Most wastewater treatment plants employing anaerobic digestion use a single digester for the digestion of a mixture of primary and biological sludge. The solid-liquid separation of digested primary sludge, however, is downgraded by even small additions of biological sludge, particularly activated sludge. The rate of reaction under anaerobic conditions is also slowed slightly. In separate sludge digestion, the digestion of primary and biological sludges is accomplished in separate tanks. Reasons cited for separate digestion include (1) the excellent dewatering characteristics of the digested primary sludge are maintained, (2) the digestion process is specifically tailored to the sludge being treated, and (3) optimum process control conditions can be maintained. Design criteria and performance data for the separate anaerobic digestion of biological sludges, however, are very limited. In some cases, especially where biological phosphorus removal is practiced, biological sludge is digested aerobically instead of anaerobically to prevent resolubilization of the phosphorus under anaerobic conditions. Separate sludge digestion is not currently a common practice at most plants.

Process Design for Mesophilic Anaerobic Digestion

Ideally, the design of anaerobic sludge digestion processes should be based on an understanding of the fundamental principles of biochemistry and microbiology discussed in Chap. 7 in Sec. 7–12. Because these principles have not been appreciated fully in the past, a number of empirical methods have also been used in the design of digesters. The purpose of the following discussion is to illustrate the various methods that have been used to design single-stage, high-rate digesters in terms of size. These methods are based on (1) solids retention time, (2) the use of volumetric loading factors, (3) volatile solids destruction, (4) observed volume reduction, and (5) loading factors based on population.

Solids Retention Time. Digester design based on SRT involves application of the principles discussed in Chaps. 7 and 10. To review briefly, the respiration and oxidation end products of anaerobic digestion are methane gas and carbon dioxide. The quantity of methane gas can be calculated using Eq. (13–12):

$$V_{CH_4} = (0.35)[(S_o - S)(Q)(1 \text{ kg}/10^3 \text{ g}) - 1.42P_x] \qquad (13\text{–}12)$$

where V_{CH_4} = volume of methane produced at standard conditions (0°C and 1 atm), m³/d
$\quad$ 0.35 = theoretical conversion factor for the amount of methane produced, m³, from the conversion of 1 kg of bCOD at 0°C (conversion factor at 35°C = 0.40, see Example 7–10 in Chap. 7)
$\quad Q$ = flowrate, m³/d
$\quad S_o$ = bCOD in influent, g/m³
$\quad S$ = bCOD in effluent, g/m³
$\quad P_x$ = net mass of cell tissue produced per day, kg/d

Table 13–27

Suggested solids retention times for use in the design of complete-mix anaerobic digesters[a]

Operating temperature, °C	SRT (minimum)	SRT$_{des}$
18	11	28
24	8	20
30	6	14
35	4	10
40	4	10

[a] From McCarty (1964) and (1968).

Note: 1.8 (°C) + 32 = °F.

The theoretical conversion factor for the amount of methane produced from the conversion of 1 g of bCOD is derived in Sec. 7–12 in Chap. 7. For a complete-mix high-rate digester without recycle, the mass of biological solids synthesized daily, P_x, can be estimated using Eq. (13–13).

$$P_x = \frac{YQ(S_o - S)(1 \text{ kg}/10^3 \text{ g})}{1 + b(\text{SRT})} \tag{13-13}$$

where Y = yield coefficient, g VSS/g bCOD

b = endogenous coefficient, d^{-1} (typical values range from 0.02 to 0.04)

SRT = solids retention time, d

other terms as defined previously

For a complete-mix digester, the SRT is the same as the hydraulic retention time τ.

Typical anaerobic reaction values for Y and b are given in Table 10–13 in Chap. 10 and range from 0.05 to 0.10 and 0.01 to 0.04, respectively. Typical values for SRT at various temperatures are reported in Table 13–27. In practice for high-rate digestion, however, values for SRTs range from 15 to 20 d. Grady, Daigger, and Lim (1999) observed that (1) a lower SRT limit of 10 days at a temperature of 35°C is sufficient to ensure an adequate safety factor against a washout of the menthanogenic population, and (2) incremental changes in volatile solids destruction are relatively small for SRT values above 15 d at 35°C. In selecting the design SRT for anaerobic digestion, peak hydraulic loading must be considered. The peak loading can be estimated by combining poor thickener performance with the maximum sustained plant loading expected during seven continuous days during the design period (U.S. EPA, 1979). The application of Eqs. (13–12) and (13–13) in the process design of a high-rate digester is illustrated in Example 13–5.

EXAMPLE 13–5   Estimating Single-stage, High-rate Digester Volume and Performance

Estimate the size of digester required for primary sludge from a primary clarifier designed for 38,000 m³/d (10 Mgal/d) of wastewater. Check the volumetric loading and the amount of gas produced. The influent wastewater BOD and TSS concentrations are 400 and 300 mg/L, respectively. The primary clarifier achieves 35 percent BOD removal and 50 percent TSS removal. Assume that the primary sludge contains about 95 percent moisture and has a specific gravity of 1.02. Other pertinent design assumptions are as follows:

1. The hydraulic regime of the reactor is complete-mix.
2. τ = SRT = 15 d at 35°C (see Table 13–27).

3. Efficiency of waste utilization (solids conversion) E = 0.70.
4. The sludge contains adequate nitrogen and phosphorus for biological growth.
5. $Y = 0.08$ kg VSS/kg bCOD utilized and $b = 0.03$ d^{-1}.
6. Constants are for a temperature of 35°C.
7. Digester gas is 65 percent methane.

Solution

1. Determine the daily sludge mass and volume using Eq. (13–2).

$$\text{Sludge mass} = \frac{(38{,}000 \text{ m}^3/\text{d})(300 \text{ g/m}^3)(0.5)}{(10^3 \text{ g}/1 \text{ kg})} = 5700 \text{ kg/d}$$

$$\text{Sludge volume} = \frac{(5700 \text{ kg/d})}{1.02(10^3 \text{ kg/m}^3)(0.05)} = 111.8 \text{ m}^3/\text{d}$$

2. Determine the bCOD loading.

bCOD loading = $(0.35)(400 \text{ g/m}^3)(38{,}000 \text{ m}^3/\text{d})(1 \text{ kg}/10^3 \text{ g}) = 5320 \text{ kg/d}$

3. Compute the digester volume.

$$\tau = \frac{V}{Q}$$

$V = Q\tau = (111.8 \text{ m}^3/\text{d})(15\text{d}) = 1677\text{m}^3$

4. Compute the volumetric loading.

$$\frac{(\text{kg bCOD/d})}{\text{m}^3} = \frac{(5320 \text{ kg/d})}{1677 \text{ m}^3} = 3.17 \text{ kg/m}^3\cdot\text{d}$$

5. Compute the quantity of volatile solids produced per day using Eq. (13–13).

$$P_x = \frac{YQ(S_o - S)(10^3 \text{ g/kg})^{-1}}{1 + b(\text{SRT})}$$

$S_o = 5320 \text{ kg/d}$

$S = 5320(1 - 0.70) = 1596 \text{ kg/d}$

$S_o - S = 5320 - 1596 = 3724 \text{ kg/d}$

$$P_x = \frac{(0.08)[(5320 - 1596)\text{kg/d}]}{1 + (0.03\text{d}^{-1})(15\text{d})} = 205.5$$

6. Compute the volume of methane produced per day at 35°C using Eq. (13–12) (conversion factor at 35°C = 0.40).

$$V_{\text{CH}_4} = (0.40)[(S_o - S)(Q)(10^3 \text{ g/kg})^{-1} - 1.42 \, P_x]$$

$$V_{\text{CH}_4} = (0.4 \text{ m}^3/\text{kg})[(5320 - 1596)\text{kg/d} - 1.42(205.5 \text{ kg/d})]$$

$$= 1373 \text{ m}^3\text{d}$$

7. Estimate the total gas production.

$$\text{Total gas volume} = \frac{1373}{0.65} = 2112 \text{ m}^3/\text{d}$$

Table 13–28

Typical design criteria for sizing mesophilic high-rate complete-mix anaerobic sludge digesters[a]

Parameter	U.S. customary units		SI units	
	Units	Value	Units	Value
Volume criteria				
Primary sludge	ft^3/capita	1.3–2.0	m^3/capita	0.03–0.06
Primary sludge + trickling filter humus sludge	ft^3/capita	2.6–3.3	m^3/capita	0.07–0.09
Primary sludge + activated sludge	ft^3/capita	2.6–4.0	m^3/capita	0.07–0.11
Solids loading rate[b]	lb VSS/ 10^3 ft^3·d	100–300	kg VSS/m^3·d	1.6–4.8
Solids retention time[b]	d	15–20	d	15–20

[a] Adapted, in part, from U.S. EPA (1979).

[b] Based on combined primary and secondary sludges digestion without any pretreatment methods.

Loading Factors. One of the most common methods used to size digesters is to determine the required volume based on a loading factor. Although a number of different factors have been proposed, the two most favored are based on (1) the mass of volatile solids added per day per unit volume of digester capacity and (2) the mass of volatile solids added to the digester each day per mass of volatile solids in the digester. Of the two, the first method is preferred. Loading criteria are based generally on sustained loading conditions (see Chap. 3), typically peak 2-wk or peak mo sludge production with provisions for avoiding excessive loadings during shorter periods. Typical design criteria for sizing mesophilic high-rate anaerobic digesters are given in Table 13–28. The upper limit of volatile solids loading rates is typically determined by the rate of accumulation of toxic materials, particularly ammonia, or washout of methane formers (WEF, 2010a).

Excessively low volatile solids loading rates can result in designs that are costly to build and are troublesome to operate. In a survey conducted by Speece (1988) of 30 digester installations in the United States, one of the most significant observations was the relatively low solids content in the sludge feed to the digesters. The average TSS in the sludge feed was 4.7 ± 1.6 percent and the average volatile solids content was 70 percent. The average VSS value in the digesters was a dilute 1.6 percent. Dilute sludge feed causes low volatile solids loading leading to starving conditions within the digester, resulting in the following adverse effects in digester operation: (1) reduced τ, (2) reduced VS destruction, (3) reduced methane generation, (4) reduced alkalinity, (5) increased volumes of digested biosolids and supernatant, (6) increased heating requirements, (7) increased dewatering capacity, and (8) increased hauling cost for liquid biosolids. As a cautionary note, a potential problem with ammonia toxicity could occur if the waste activated sludge is thickened too much. Thus, in planning the design and operation of anaerobic digesters, consideration should be given to optimizing volatile solids loading to effectively utilize digester capacity. The effect of solids concentration and hydraulic detention time on volatile solids loading is reported in Table 13–29.

Estimating Volatile Solids Destruction. The degree of stabilization obtained is often measured by the percent reduction in volatile solids. The reduction in volatile solids can be related either to the SRT or to the detention time based on the untreated sludge feed.

Table 13–29

Effect of sludge concentration and hydraulic detention time on volatile solids loading factors[a]

| Sludge concentration, % | Volatile solids loading factor | | | | | | | |
| | lb/ft³·d | | | | kg/m³·d | | | |
	10 d[b]	12 d	15 d	20 d	10 d	12 d	15 d	20 d
2	0.09	0.07	0.06	0.04	1.4	1.2	0.95	0.70
3	0.13	0.11	0.09	0.07	2.1	1.8	1.4	1.1
4	0.18	0.15	0.12	0.09	2.9	2.4	1.9	1.4
5	0.22	0.19	0.15	0.11	3.6	3.0	2.4	1.8
6	0.27	0.22	0.18	0.13	4.3	3.6	2.9	2.1
7	0.31	0.26	0.21	0.16	5.0	4.2	3.3	2.5
8	0.36	0.30	0.24	0.18	5.7	4.8	3.8	2.9

[a] Based on 70 percent volatile content of sludge, and a sludge specific gravity of 1.02 (concentration effects neglected).
[b] Hydraulic detention time, d.

The amount of volatile solids destroyed in a high-rate complete-mix digester can be roughly estimated by the following empirical equation (Liptak, 1974):

$$V_d = 13.7 \ln(\text{SRT}_{des}) + 18.9 \qquad (13\text{–}14)$$

where V_d = volatile solids destruction, %
   SRT_{des} = time of digestion, d (range 15 to 20 d)

The equation does not account for variation in the sludge feed to digestion and the digestion mixing and other operating conditions and should be used to obtain a rough estimate only, and it appears that the equation overestimates volatile solids destruction. Typical volatile solids destruction ranges as a function of SRT are provided in Table 13–30. Because the untreated sludge feed can be measured easily, this method is also used commonly. In plant operation, calculation of volatile solids reduction should be made routinely as a matter of record whenever sludge is drawn to processing equipment or drying beds. Alkalinity and volatile acids content should also be checked daily as a measure of the stability of the digestion process.

In calculating the volatile solids reduction, the ash content of the sludge is assumed to be conservative; that is, the number of pounds of ash going into the digester is equal to that being removed. Digester VSR can be calculated based on two different methods. The first method is the mass balance method, which is shown below.

$$R_{VSS} = \frac{M_{VS\,in\,feed} - M_{VS\,in\,digested\,sludge} - M_{VS\,in\,supernatant}}{M_{VS\,in\,feed}} \times 100 \qquad (13\text{–}15)$$

Table 13–30

Estimated volatile solids destruction in high-rate complete-mix mesophilic anaerobic digestion

Digestion time, d	Volatile solids destruction, %
30	50–65
20	50–60
15	45–50

where R_{VSS} = volatile solids destruction, %

$M_{VS\ in\ feed}$ = mass flowrate of volatiles in digester feed, kg/d

$M_{VS\ in\ digested\ sludge}$ = mass flowrate of volatiles out of digester, kg/d

$M_{VS\ in\ surpernatent}$ = mass flowrate of volatiles in digester decant stream, kg/d

It should be noted that in modern high rate digesters, there is no decant so the $M_{VS\ in\ surpernatent}$ number goes to zero. Digester VSR can also be calculated with the simplified Van Kleeck formula given below.

$$R_{VSS} = \frac{W_{VS\ in\ feed} - W_{VS\ in\ digested\ sludge}}{W_{VS\ in\ feed} - (W_{VS\ in\ digested\ sludge})(W_{VS\ in\ feed})} \times 100 \qquad (13\text{--}16)$$

where $W_{VS\ in\ feed}$ = Weight fraction of digested sludge volatile content per total dry solids

$W_{VS\ in\ digested\ sludge}$ = Weight fraction of volatiles out of digester per total dry solids

It should be noted that the Van Kleeck formula assumes that there is no supernatant withdrawal or accumulation of grit inside the digester so in practice the results may not be 100 percent accurate. A typical example calculation of volatile solids reduction is presented in Example 13–6.

EXAMPLE 13–6   **Determination of Volatile Solids Reduction**   From the following analysis of untreated and digested biosolids, determine the total volatile solids reduction achieved during digestion. It is assumed that (1) the weight of fixed solids in the digested biosolids equals the weight of fixed solids in the untreated sludge and (2) the volatile solids are the only constituents of the untreated sludge lost during digestion.

	Volatile solids, %	Fixed solids, %
Untreated sludge	68	32
Digested sludge	50	50

Solution

1. Determine the weight of the digested solids. Because the quantity of fixed solids remains the same, the weight of the digested solids based on 1.0 kg of dry untreated sludge, as computed below, is 0.64 kg.

    $$\text{Fixed solids in untreated sludge} = \frac{0.32\ kg}{(0.32 + 0.68)\ kg}\ 100 = 32\%$$

    Let X equal the weight of volatile solids after digestion. Then

    $$\text{Fixed solids after digestion} = \frac{0.32\ kg}{(0.32 + X)\ kg}\ 100 = 50\%$$

    $$\text{Weight of volatile solids after digestion,}\ X\ kg = \frac{0.32\ kg}{0.5} - 0.32 = 0.32\ kg$$

    Weight of digested solids = 0.32 kg + 0.32 kg = 0.64 kg

2. Determine the percent reduction in total and volatile suspended solids.
    a. Percent reduction of total suspended solids

    $$R_{TSS} = \frac{(1.0 - 0.64)\ kg}{1.0\ kg}\ 100 = 36\%$$

b. Percent reduction in volatile suspended solids using both methods
Using the mass balance method [Eq. (13–15)]

$$R_{VSS} = \frac{(0.68 - 0.32)\,kg}{0.68\,kg}\,100 = 52.9\%$$

Using the Van Kleeck method [Eq. (13–16)]

$$R_{VSS} = \frac{0.68 - 0.5}{0.68 - 0.5(0.68)}\,100 = 52.9\%$$

Population Basis. Digestion tanks are also designed on a volumetric basis by allowing a certain number of cubic meters per capita (cubic feet per capita). Detention times range from 10 to 20 d for high-rate digesters (U.S. EPA, 1979). These detention times are recommended for design based on total tank volume, plus additional storage volume if sludge is dried on beds and weekly sludge withdrawals are curtailed because of inclement weather.

Typical design criteria for heated anaerobic digesters based on population are shown in Table 13–28. The criteria are applied only where analyses and volumes of sludge to be digested are not available. The capacities shown in Table 13–28 should be increased 60 percent in a municipality where the use of food-waste grinders is universal and should be increased on a population-equivalent basis to allow for the effect of industrial wastes.

Selection of Tank Design and Mixing System

Most anaerobic digestion tanks are either cylindrical, conventional German design, or egg-shaped (see Fig. 13–20). The most common shape used in the United States is a shallow, vertical cylinder with a floating cover [see Fig. 13–20(a)] or fixed cover [see Fig. 13–20(b)]. Rectangular tanks were used in the past, but they experienced great difficulty in mixing the tank contents uniformly. German designers have worked on optimizing the shape of digesters, and two basic types have emerged: the conventional German digester and the egg-shaped digester. The conventional German digester [see Fig. 13–20(c)] is a deep cylindrical vessel with steeply sloped top and bottom cones (Stukenberg et al., 1992). The egg-shaped digester, shown on Fig. 13–20(d), is similar in appearance to an upright egg, and the design is sometimes modified to a sphere-cone shape. Egg-shaped tanks have been used extensively in Europe, especially in Germany, and are growing in popularity in the United States. Essentially all of the modern digester designs in the United States are of either the cylindrical or egg-shaped type. Cylindrical and egg-shaped digesters and the mixing systems used for each type of tank are discussed in the following paragraphs. Advantages and disadvantages of each type of digester are summarized in Table 13–31.

Proper mixing is one of the most important considerations in achieving optimum process performance. Various systems for mixing the contents of the digesters have been used; the most common types involve the use of (1) gas injection, (2) mechanical stirring, and (3) mechanical pumping. Some digester installations use a combination of gas mixing and recirculation by pumping. The advantages and disadvantages of the various mixing systems are summarized in Table 13–32; typical design parameters are shown in Table 13–33.

Cylindrical Tanks. Cylindrical sludge digesters are seldom less than 6 m (20 ft) or more than 38 m (125 ft) in diameter. The water depth should not be less than 7.5 m (25 ft) at the sidewall because of the difficulty in mixing shallow tanks, and the depth may be as

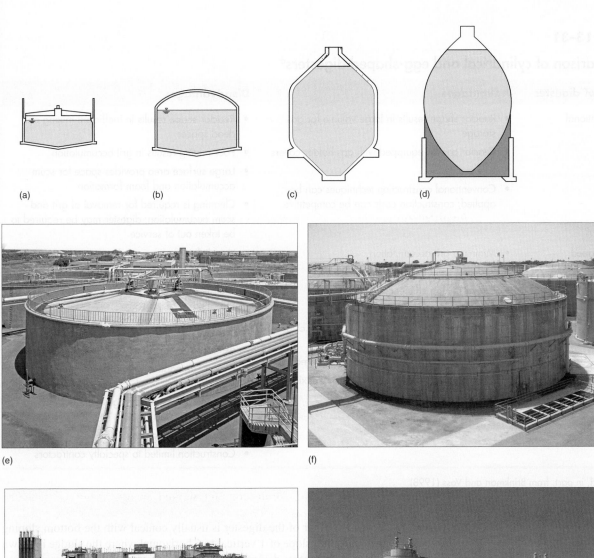

(a)   (b)   (c)   (d)

(e)   (f)

(g)   (h)

Figure 13–20

Typical shapes of anaerobic digesters: (a) and (e) cylindrical with floating cover, (b) and (f) cylindrical with fixed cover, (c), (g), and (h) conventional German design with reinforced concrete construction [note digesters in (g) and (h) are clad in a metal sheath], and (d) egg-shaped with steel shell (see also Figs. 13–24 and 13–25).

Table 13–31

Comparison of cylindrical and egg-shaped digesters[a]

Type of digester	Advantages	Disadvantages
Conventional	• Reactor shape results in large volume for gas storage • Reactor can be equipped with gas holder covers • Low profile • Conventional construction techniques can be applied; construction costs can be competitive	• Reactor shape results in inefficient mixing and dead spaces • Poor mixing results in grit accumulation • Large surface area provides space for scum accumulation and foam formation • Cleaning is required for removal of grit and scum accumulation; digester may be required to be taken out of service
Egg-shaped	• Minimum grit accumulation • Reduced scum formation • Higher mixing efficiency • More homogeneous biomass is obtained • Lower operating and maintenance costs; cleaning frequency significantly reduced • Smaller footprint; less land area is required • Foaming is minimized (except for gas mixing)	• Very little gas storage volume; external gas storage is required if as is recovered • High profile structures; may be aesthetically objectionable • Difficult access to top-mounted equipment; installation requires a high stair tower or an elevator • Greater foundation requirements and seismic considerations • Foaming of gas-mixed digester may be a problem in collecting gas • Higher construction costs • Construction limited to specialty contractors

[a] Adapted, in part, from Brinkman and Voss (1998).

much as 15 m (50 ft). The floor of the digester is usually conical with the bottom sloping to the center, with a minimum slope of 1 vertical to 6 horizontal where the sludge is drawn off (see Fig. 13–21). An alternative design uses a "waffle" bottom to minimize grit accumulation and to reduce the need for frequent digester cleaning (see Fig. 13–22).

Gas-injection systems used in cylindrical tanks are classified as unconfined or confined [see Figs. 13–23(a) and (b) on page 1518]. Unconfined gas systems collect gas at the top of the digesters, compress the gas, and then discharge the gas through a pattern of bottom diffusers or through a series of radially placed top-mounted lances. Unconfined gas systems mix the digester contents by releasing gas bubbles that rise to the surface, carrying and moving the sludge. These systems are suitable for digesters with fixed, floating, or gas holder covers. In confined gas systems, gas is collected at the top of the digesters, compressed, and discharged through confined tubes. Two major types of confined systems are the gas lifter and the gas piston [see Figs. 13–23(c) and (d)]. The gas lifter system consists of submerged gas pipes or lances inserted into an eductor tube or gas lifter. Compressed gas is released from the lances or pipes, and the gas bubbles rise, creating an air-lift effect. In the gas piston system, gas bubbles are released intermittently at the bottom of a cylindrical tube or piston. The bubbles rise and act like a piston, pushing the sludge to the surface. These systems are suitable for fixed, floating, or gas holder covers.

Mechanical stirring systems commonly use low-speed turbines or mixers [see Figs. 13–23(e) and (f)]. In both systems, the rotating impeller(s) displaces the sludge, mixing

Table 13–32

Summary of advantages and disadvantages of various anaerobic digester mixing systems

Type of mixer	Advantages	Disadvantages
All systems	• Increased rate of biosolids stabilization	• Corrosion and tear of ferrous metal piping and supports • Equipment wear by grit • Equipment plugging and operational interference by rags
Gas injection: Unconfined: Cover-mounted lances	• Lower maintenance and less hindrance to cleaning than bottom-mounted diffusers. Effective against scum buildup	• Corrosion of gas piping and equipment • High maintenance for compressor • Potential gas-seal problem • Compressor problems if foam gets inside, solids deposition • Plugging of gas lances • Entire tank contents are not mixed
Bottom-mounted diffusers	• Better movement of bottom deposits than cover-mounted lances	• Corrosion of gas piping and equipment • High maintenance for compressor • Potential gas-seal problem • Foaming. Incomplete mixing • Scum formation • Diffuser plugging • Bottom deposits can alter mixing patterns • Requires digester dewatering for maintenance
Confined: Gas lifters	• Better mixing and gas production and better movement of bottom deposits than cover mounted lances. Lower power requirements than cover mounted lances	• Corrosion of gas piping and equipment • High maintenance for compressor • Potential gas-seal problem • Corrosion of gas lifter • Lifter interferes with digester cleaning • Scum buildup • Does not provide good top mixing • Requires digester dewatering for maintenance if bottom mounted
Gas pistons	• Good mixing efficiency • Less susceptible to plugging due to rags or fibrous material • Provides surface agitation for management of scum layer • Can contain optional heating jacket	• Corrosion of gas piping and equipment • High maintenance for compressor • Potential gas-seal problem • Equipment internally mounted • Pistons interfere with digester cleaning • Requires digester dewatering for maintenance • Cannot operate at varying liquid levels

(continued)

Table 13–32 (Continued)		
Type of mixer	**Advantages**	**Disadvantages**
Mechanical stirring:		
Low-speed turbines	• Good mixing efficiency	• Wear of impellers and shafts
		• Bearing failures. Interferences of impellers with rags
		• Requires oversized gear boxes
		• Gas leaks at shaft seal
		• Long overhung loads
Low-speed mixers	• Breaks up scum layers	• Not designed to mix entire tank contents
		• Bearing and gear box failures
		• Impeller wear. Interference of impellers by rags
Linear motion mixing	• Lower energy consumption compared to other technologies	• Limited number of installations in the United States
	• High mixing efficiency	• May require maintenance platform and equipment hoist
	• Capable of operating at various liquid levels	• No redundancy
	• Suitable for retrofits	• Sole supplier
	• Equipment maintenance can be completed without taking digester out of service	
	• Minimal mechanical complexity	
Mechanical pumping:		
Internal draft tubes	• Good top-to-bottom mixing	• Sensitive to liquid level
	• Reversibility provides variable mixing dynamics	• Corrosion and wear of impeller
	• Provides surface agitation for management of scum layer	• Bearing and gear box failures
		• Requires oversized gear box
		• Structural modifications required for retrofit
External draft tubes	• Same as internal draft tube	• Same as internal draft tube
	• Draft tube maintenance easier than internal type	
	• Can contain optional heating jacket	
Pumps	• Good mixing control. Scum layer and sludge deposits can be recirculated. Pumps easier to maintain than compressors	• Impeller wear
	• Conductive for FOG addition	• Plugging of pumps by rags[a]
		• Bearing failures
		• High electrical consumption
		• Decreased efficiency at higher solids concentrations
		• High mixing energy may contribute to foaming

[a] Impact from plugging with rags can be minimized or eliminated by using chopper pumps.

the digester contents. Low-speed turbine systems usually have one cover-mounted motor with two turbine impellers located at different sludge depths. A low-speed mixer system usually has one cover-mounted mixer. Mechanical stirring systems are most suitable for digesters with fixed or floating covers. Another type of mechanical mixer that is new to the market (as of writing this book) is a linear motion mixer that consists of a ring-shaped disc

Table 13–33

Typical design parameters for anaerobic digester mixing systems[a]

Parameter	Type of mixing system	Typical values[b]	
		U.S. customary units	**SI units**
Unit power	Mechanical systems	0.025–0.04 hp/10^3 gal of digester volume	0.005–0.008 kW/m^3 of digester volume
Unit gas flow[c]	Gas mixing		
	Unconfined	4.5–5 ft^3/10^3 ft^3·min	0.0045–0.005 m^3/m^3·min
	Confined	5–7 ft^3/10^3 ft^3·min	0.005–0.007 m^3/m^3·min
Velocity gradient, G[d]	All	50–80 s^{-1}	50–80 s^{-1}
Turnover time of tank contents	Confined gas mixing and mechanical systems	20–30 min	20–30 min

[a] Adapted from U.S. EPA (1987).

[b] Actual design values may differ depending on the type of mixing system, manufacturer, and digestion process or function.

[c] Quantity of gas delivered by the gas injection system divided by the digester gas volume.

[d] See Eq. (5–3) in Chap. 5.

Figure 13–21

Typical cross section through a high-rate, gas mixed cylindrical digester.

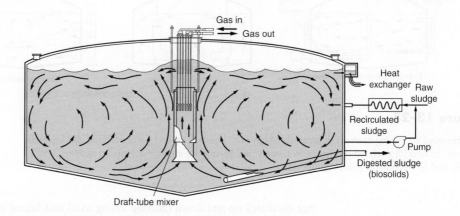

Figure 13–22

Typical waffle bottom anaerobic digester: (a) plan view (b) section.

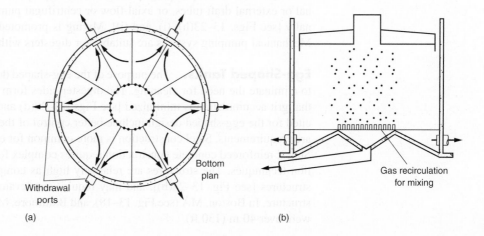

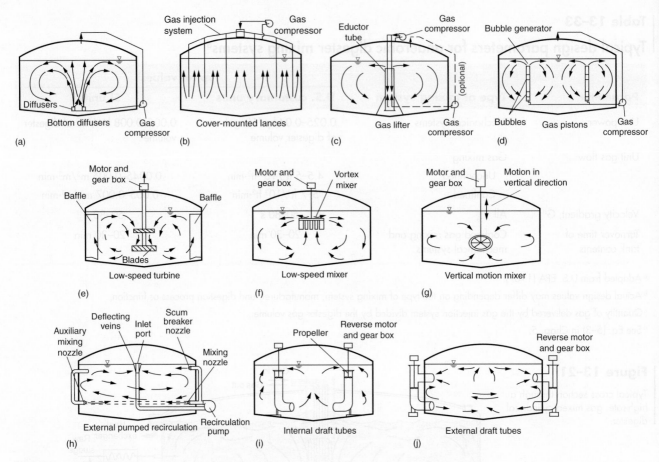

Figure 13–23

Devices used for mixing contents of anaerobic high rate digesters (a) and (b) unconfined gas injection systems; (c) and (d) confined gas injection systems, (e), (f), and (g) mechanical mixing systems; and (h), (i), and (j) mechanical pumping systems.

that oscillates up and down creating strong axial and lateral agitation needed for mixing of the digester contents [see Fig. 13–23(g)].

Most mechanical pumping systems consist of propeller-type pumps mounted in internal or external draft tubes, or axial-flow or centrifugal pumps and piping installed externally [see Figs. 13–23(h), (i), and (j)]. Mixing is promoted by the circulation of sludge. Mechanical pumping systems are suitable for digesters with fixed covers.

Egg-Shaped Tanks. The purpose of the egg-shaped design is to enhance mixing and to eliminate the need for cleaning. The digester sides form a steep cone at the bottom so that grit accumulation is minimized [see Figs. 13–20(d) and 13–24(a)]. Other advantages cited for the egg-shaped design include better control of the scum layer and smaller land-area requirements. Steel construction is more common for egg-shaped tanks in the United States; reinforced concrete construction requires complex formwork and special construction techniques. The structures are relatively high as compared to other treatment plant structures [see Fig. 13–24(b)], and may require an elevator for access to the top of the structure. In Boston, MA (see Fig. 13–18), and Baltimore, MD, the heights of the digesters were over 40 m (130 ft).

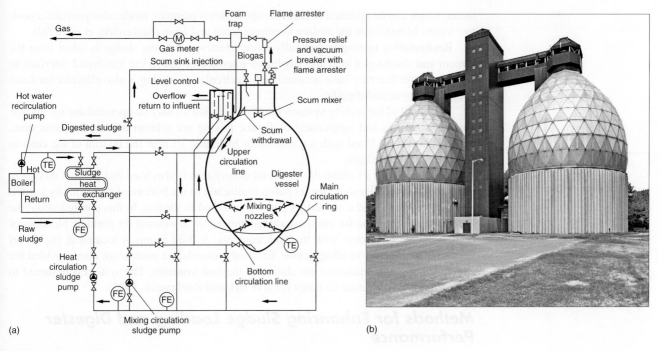

Figure 13–24

Egg-shaped anaerobic digester: (a) schematic diagram from Walker Process catalogue
(b) pictorial view.

Egg-shaped digester mixing systems are similar to those for cylindrical tanks and consist of unconfined gas mixing, mechanical draft-tube mixing, or pumped recirculation mixing (see Fig. 13–25). Gas mixing is considered by some to be relatively ineffective in mixing the digester contents below the level of the injection nozzles. The mechanical draft tube and pumped recirculation mixing systems, however, are considered able to provide sufficient energy to mix even the sludge in the bottom cone of the digester. The mechanical draft-tube

Figure 13–25

Mixing systems for egg-shaped anaerobic digesters. (From Stukenberg et al., 1992.)

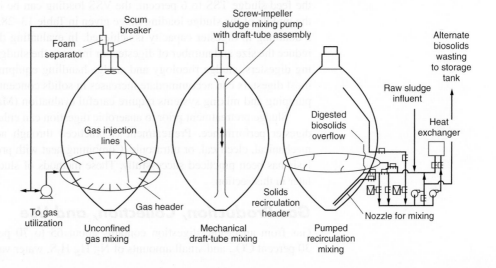

mixer, which can be operated in either an up- or down-pumping mode, also provides a positive means of mixing at the surface to control scum and foam (Stukenberg et al., 1992).

Recirculation mixing is generally more effective when the sludge is taken from the bottom and discharged near the gas-liquid interface or above the gas-liquid interface to break up scum that may have accumulated. Recirculation mixing is also effective for foam control in gas-mixed digesters.

Any or all of the mixing systems may be used, and all may be operated during any one day, although gas and mechanical draft-tube mixing are seldom used at the same time. Most digesters are fitted with a gas lance or hydraulic jet near the bottom of the cone to stir any accumulated grit.

A combination jet-pump draft-tube mixing system is also used that permits mixing in three zones of the digester. One jet pump is attached to the bottom of the centrally located vertical draft tube, and a second jet pump is attached to the top. In this configuration, the draft tube can function for pumping sludge upward or downward for periodic blending of bottom sludge and scum with the tank contents. A third pump is located at the vessel perimeter to create a swirling action. External recirculation pumps are also provided for sludge heating and additional circulation of the tank contents. The system is designed to circulate the tank volume 10 times per d (Clark and Ruehrwein, 1992).

Methods for Enhancing Sludge Loading and Digester Performance

Opportunities for enhancing the performance of anaerobic digesters include thickening the digester feed sludge or thickening a portion of the digesting sludge to increase the SRT. Recirculating a portion of the digested sludge and cothickening with untreated primary and waste sludge was reported originally by Torpey and Melbinger (1967). The solids concentration in the feed sludge improved and the performance of the digester, as measured by volatile solids destruction, increased significantly. The thickening system was installed at wastewater plants in New York City. In a study by Maco et al. (1998), the effects of thickening digested biosolids, either thickened separately or combined with prethickening of untreated sludge, increased the SRT of the digestion process and the production of biogas and decreased the hydraulic retention time.

The value of thickening the feed sludge to the digester is indicated by data presented in Table 13–29. For example, for a 15-d hydraulic retention time and an average TSS of 3 percent, the volatile solids loading factor in Table 13–29 is 1.4 kg/m^3·d. By improving the feed sludge TSS to 6 percent, the VSS loading can be increased to 2.9 kg/m^3·d, near the middle of the sludge loading range given in Table 13–28. In this hypothetical example, a doubling in digester capacity is achieved. In evaluating digested biosolids recycling to reduce the size and number of digesters or increase the sludge processing capacity of existing digesters, sludge rheology and sludge handling equipment require evaluation. While most digesters can accommodate increases in solids concentrations, the limits imposed by pumping and mixing systems require careful evaluation (Maco et al., 1998).

Sludge pretreatment prior to anaerobic digestion can enhance the solids loading and the digester performance. Pretreatment is practiced through adding energy in the form of mechanical, electrical, or ultrasonic. Combining heat with pressure to cause thermal hydrolysis has been practiced successfully. These methods of sludge pretreatment are presented later in this section.

Gas Production, Collection, and Use

Gas from anaerobic digestion contains about 65 to 70 percent CH_4 by volume, 25 to 30 percent CO_2, and small amounts of N_2, H_2, H_2S, water vapor, and other gases. Digester gas has a specific gravity of approximately 0.86 relative to air. Because production of gas is one of the best measures of the progress of digestion and because digester gas can be used as fuel, the designer should be familiar with its production, collection, and use.

Gas Production. The volume of methane gas produced during the digestion process can be estimated using Eq. (13–12), discussed previously. Total gas production is usually estimated from the percentage of volatile solids reduction. Typical values vary from 0.75 to 1.12 m^3/kg (12 to 18 ft^3/lb) of volatile solids destroyed. Gas production can fluctuate over a wide range, depending on the volatile solids content of the sludge feed and the biological activity in the digester. Excessive gas production rates sometimes occur during startup and may cause foaming and escape of foam and gas from around the edges of floating digester covers. In egg-shaped and shallow cylindrical digesters, foaming can clog the gas outlet unless foam control is provided. If stable operating conditions have been achieved and the foregoing gas production rates are being maintained, a well-digested sludge can be obtained.

Gas production can also be estimated crudely on a per capita basis. The normal yield is 15 to 22 m^3/10^3 persons·d (0.6 to 0.8 ft^3/person·d) in primary plants treating normal domestic wastewater. In secondary treatment plants, the gas production is increased to about 28 m^3/10^3 persons·d (1.0 ft^3/person·d).

Gas Collection. In cylindrical digesters, gas is collected under the cover of the digester. Three principal types of covers are used: (1) floating, (2) fixed, and (3) membrane. Floating covers fit on the surface of the digester contents and allow the volume of the digester to change without allowing air to enter the digester [see Fig. 13–26(a)]. Gas and air must not be allowed to mix, or an explosive mixture may result. Explosions have occurred in wastewater treatment plants. Gas piping and pressure-relief valves must include adequate flame traps. The covers may also be installed to act as gas holders for a limited storage of gas. High-rate digesters produce about two volumes of gas per volume of digester capacity/d (Speece 2001). Floating covers can be used for single-stage digesters or in the second stage of two-stage digesters.

Fixed covers provide a free space between the roof of the digester and the liquid surface [see Fig. 13–26(b)]. Gas storage must be provided so that (1) when the liquid volume is changed, gas, and not air, will be drawn into the digester; otherwise an overflow weir with a U-shaped trap needs to be provided to maintain a liquid seal, and (2) gas will not be lost by displacement. Gas can be stored either at low pressure in external gas holders that use floating covers or at high pressure in pressure vessels if gas compressors are used. Gas not used should be burned in a flare. Gas meters should be installed to measure gas produced and gas used or wasted.

Another development in gas holder covers for cylindrical tanks is the membrane cover [see Fig. 13–26(c)]. This cover consists of a support structure for a small center gas dome and flexible air and gas membranes. An air-blower system is provided to pressurize the air space between the two membranes and vary the air space volume. Only the gas membrane and the center gas dome are in contact with the digester contents. The gas membrane is made from a flexible polyester fabric.

In egg-shaped digesters, the volume available for gas storage is small. For efficient utilization of digester gas, supplemental external storage may be required.

Gas Pretreatment and Use. Methane gas at standard temperature and pressure (20°C and 1 atm) has a lower heating value of 35,800 kJ/m^3 (960 Btu/ft^3). Lower heating value is the heat of combustion less the heat of vaporization of any water vapor present. Because digester gas is only 65 percent methane, the lower heating value of digester gas

Figure 13-26

Types of anaerobic digester covers: (a) floating, (b) fixed, and (c) and (d) schematic and view of membrane gas cover.

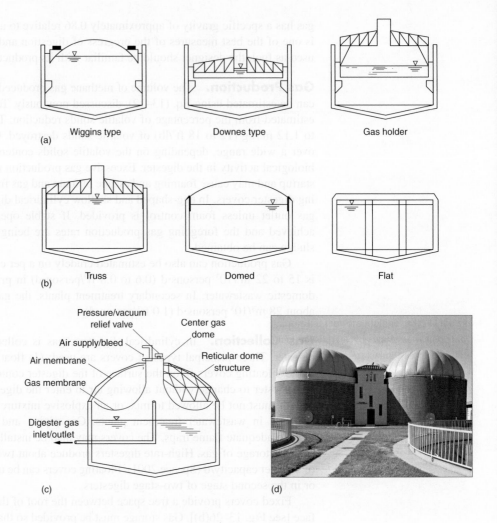

(a) Wiggins type   Downes type   Gas holder

(b) Truss   Domed   Flat

Pressure/vacuum relief valve
Center gas dome
Air supply/bleed
Air membrane
Reticular dome structure
Gas membrane
Digester gas inlet/outlet

(c)   (d)

is approximately 22,400 kJ/m³ (600 Btu/ft³). By comparison, natural gas, which is a mixture of methane, propane, and butane, has a heating value of 37,300 kJ/m³ (1000 Btu/ft³). Digester gas can be used as a fuel for boilers to provide heat for digestion and other parts of the plant if surplus heat is available. Digester gas can also be used in cogeneration and when purified can be used as a natural gas substitute.

Because digester gas contains hydrogen sulfide, nitrogen, particulates, and water vapor, the gas frequently has to be cleaned before use. These impurities can significantly impact the operation and performance of equipment that utilizes the gas, especially cogeneration equipment. Proper design of a digester gas use system requires careful consideration of possible pretreatment requirements. The primary constituents of digester gas that require pretreatment are moisture, hydrogen sulfide, mercaptans, and siloxanes.

Pretreatment for Removing Moisture. Digester gas is typically saturated with water vapor. Common practice and regulatory codes require designing for removal of the moisture that condenses in the digester gas piping system using sediment traps and drip traps. Piping should be provided with a minimal slope of 10 mm/m (1/8-in./ft) to low points at which sediment traps or drip traps are located. Consideration should be given to access these devices for removal of the collected condensate. Automatic drip traps are available but are not allowed in some jurisdictions. A significant amount of the moisture in the gas will be condensed simply through the cooling in the piping system at ambient temperatures, which can then be removed by the appropriate traps. Long runs of piping or travel through other pretreatment systems can be advantageous for reducing the moisture content of the gas.

Some gas use systems may require additional removal of moisture from the gas. Chilling of the gas (reducing the dew point) is commonly used in systems where very dry gas is required. This treatment step is usually associated with systems where activated carbon is supplied for removal of other contaminants or where the gas use system requires a very dry gas. As cooling of the gas is costly to operate, this system should only be included after careful consideration of the costs and benefits obtained through that level of treatment.

Pretreatment for Removing Hydrogen Sulfide. Hydrogen sulfide can damage piping systems and digester gas use equipment. The hydrogen sulfide combines with the condensate and forms a weak sulfuric acid. Piping materials such as stainless steel or lined ductile iron pipe should be considered to resist the corrosion and erosion that can be caused by the slightly acidic condensate. Cement lined ductile iron pipe should not be used as the cement lining is readily destroyed by the condensate. Additional wall thickness ductile iron pipe has been used successfully providing an allowance for some corrosion and erosion.

A significant amount of the hydrogen sulfide is removed along with the condensate. This removal may be adequate for some digester gas uses like boilers, although even then the boiler may incur additional maintenance due to the presence of the hydrogen sulfide. Several systems are commonly used for removal of hydrogen sulfide from digester gas. Historically, iron sponge impregnated into wood chips has been used. The sulfide combines with the iron to form iron sulfide, a solid. These systems can also include methods to regenerate the iron in place.

There are several other commercially available processes for removal of hydrogen sulfide that include a liquid phase oxidation process, biological scrubbers or treatment systems, chemical systems, activated carbon, adsorptive resins, and others. Most, aside from the biological systems, resins, and activated carbon, use some form of iron chemistry. Systems that remove moisture through chilling will also remove a significant amount of the hydrogen sulfide in the condensate. However, consideration of condensate handling must be assessed. If returned to the treatment plant, consideration needs to be given to the ultimate fate of the hydrogen sulfide.

Pretreatment for Removing Siloxanes. Siloxanes are silicon containing volatile organic compounds. Siloxanes are used as carriers or conditioners in antiperspirants, skin care products, deodorants, liquid soaps, and hair care products and are now ubiquitous in the environment and wastewater. They are hydrophobic and tend to attach to the sludge produced during treatment. Once in an anaerobic digester, the conditions of mixing and heating tend to cause the siloxanes to volatilize into the digester gas. Wherever digester gas is burned, such as in a boiler, engine, or turbine, silicon dioxide (SiO_2), an inert white powder, is formed. Silica dioxide is the base material for glass, sandpaper, and grinding tools. The silicon dioxide can build up in the combustion equipment causing significantly increased maintenance or total equipment failure.

Sampling and testing procedures for siloxanes has not been standardized. Also, the allowable levels of siloxanes for gas use equipment vary by type and manufacturer of equipment. A proper design requires careful assessment and consideration of the siloxane composition of the gas and required cleaning for use. Sampling and testing program would be best for obtaining design information.

Siloxanes can be removed primarily in condensate through chilling or adsorption on a media or activated carbon. Return of the condensate to the treatment plant has been

Table 13–34

Typical electricity and heat generation efficiency from various co-generation systems[a]

Co-generation system	Electricity generation efficiency, %	Heat recovery efficiency,%
Internal combustion engine	37–42	35–43
Lean burn internal combustion engine	30–38	41–49
Conventional turbine	26–34	40–52
Recuperated Turbine	36–37	30–45
Microturbine	26–30	30–37
Molten carbonate fuel cell	40–45	30–40
Phosphoric acid fuel cell	36–40	NA

[a] Adapted from U.S. EPA (2010).

determined to be acceptable as a significant amount of the siloxane will be removed from the plant through volatilization from an aeration tank and in the final solids disposition. Media are regenerated through steam or other processes.

Digester Gas Use in Cogeneration. Cogeneration is generally defined as a system for generating electricity and producing another form of energy (usually heat in the form of steam or hot water). Cogeneration systems are also known as combined heat and power (CHP) systems. Most common at wastewater facilities are internal combustion engines or microturbines connected to generators. Several larger facilities also use turbines. Fuel cells are also used to create electricity with the heat recovered for process uses. Typical efficiencies for these processes are provided in Table 13–34. A typical internal combustion engine is shown on Fig. 13–27.

Design of a cogeneration system must consider variations in gas production, use of all of the gas or allowing occasional flaring of excess gas, and redundancy. These considerations are also impacted by the cost of electricity. It will be cost-effective to provide a more robust system when the electrical savings are greater due to higher electrical rates. Gas storage may also be considered when large variations in gas production are anticipated

Figure 13–27

Internal combustion engines for cogeneration at the Back River Wastewater Treatment facility in Baltimore, MD.

(due to industrial loadings or feed of supplemental organic material to the anaerobic digesters) or to maximize electricity generation at facilities where time of day electrical charges are significantly higher than off peak rates. The level of gas cleaning required for the cogeneration systems can vary widely and also needs to be considered in the system design.

The design of a cogeneration system must also consider air pollution impacts. Several jurisdictions have significant restrictions on combustion of fuels. Permitting aspects for cogeneration systems can impact the cost-effectiveness or ability to implement a specific system or device. Fuel cells have the lowest air emissions but the highest cost of the systems listed in Table 13–34.

Digester Gas Use as a Natural Gas. Digester gas is purified to enable its use as a substitute for natural gas. Besides the removal of H_2S, siloxanes, and water vapor, the additional removal of carbon dioxide (CO_2) is required to upgrade the digester gas to natural gas pipeline quality. Purified digester gas can have a methane content of 95 percent plus and can be modified slightly to meet pipeline quality methane for sale or use as natural gas. Purified digester gas can also be compressed to become compressed natural gas (CNG) for use as vehicle fleet fuel in natural gas burning engines. CNG plants can be useful in areas where air pollution is an issue and natural gas fleet vehicles are already in operation. Use of CNG can also be a substantial cost savings in areas where gasoline prices are higher. Several technologies are available for gas purification. The common technologies for gas purification are water adsorption, chemical adsorption, pressure swing adsorption (PSA), and cryogenic separation.

Digester Heating

The heat requirements of digesters consist of the amount needed (1) to raise the incoming sludge to digestion tank temperatures, (2) to compensate for the heat losses through walls, floor, and roof of the digester, and (3) to make up the losses that might occur in the piping between the source of heat and the tank. The sludge in digestion tanks is heated by pumping the sludge through external heat exchangers and back to the tank.

Analysis of Heat Requirements. In computing the energy required to heat the incoming sludge to the temperature of the digester, it is assumed that the specific heat of most sludges is essentially the same as that of water. The assumption that the specific heats of sludge and water are essentially the same has proved to be acceptable for engineering computations. The heat loss through the digester sides, top, and bottom is computed using the following expression:

$$q = UA\Delta T \tag{13–17}$$

where q = heat loss, J/s (Btu/h)
U = overall coefficient of heat transfer, J/m²·s·°C (Btu/ft²·h·°F)
A = cross-sectional area through which the heat loss is occurring, m² (ft²)
ΔT = temperature drop across the surface in question, °C (°F)

In computing the heat losses from a digester using Eq. (13–17), it is common practice to consider the characteristics of the various heat transfer surfaces separately and to develop transfer coefficients for each one. The application of Eq. (13–17) in the computation of digester heating requirements is illustrated in Example 13–7.

Heat-Transfer Coefficients. Typical overall heat-transfer coefficients are reported in Table 13–35. As shown, separate entries are included for the walls, bottom, and top of the digester. Digestion tank walls may be surrounded by earth embankments that serve as

Table 13–35

Typical values for the overall coefficients of heat transfer for computing digester heat losses[a]

Item	U.S. customary, Btu/ft²·°F·h	SI units, W/m²·°C
Plain concrete walls (above ground)		
300 mm (12 in.) thick, not insulated	0.83–0.90	4.7–5.1
300 mm (12 in.) thick with air space plus brick facing	0.32–0.42	1.8–2.4
300 mm (12 in.) thick wall with insulation	0.11–0.14	0.6–0.8
Plain concrete walls (below ground)		
Surrounded by dry earth	0.10–0.12	0.57–0.68
Surrounded by moist earth	0.19–0.25	1.1–1.4
Plain concrete floors		
300 mm (12 in.) thick in contact with moist earth	0.5	2.85
300 mm (12 in.) thick in contact with dry earth	0.3	1.7
Floating covers		
With 35 mm (1.5 in.) wood deck, built–up roofing, and no insulation	0.32–0.35	1.8–2.0
With 25 mm (1 in.) insulating board installed under roofing	0.16–0.18	0.9–1.0
Fixed concrete covers		
100 mm (4 in.) thick and covered with built–up roofing, not insulated	0.70–0.88	4.0–5.0
100 mm (4 in.) thick and covered, but insulated with 25 mm (1 in.) insulating board	0.21–0.28	1.2–1.6
225 mm (9 in.) thick, not insulated	0.53–0.63	3.0–3.6
Fixed steel covers 6 mm (0.25 in.) thick	0.70–0.95	4.0–5.4

[a] Adapted in part from U.S. EPA (1979).

insulation, or they may be of compound construction consisting of approximately 300 mm (12 in.) of concrete, insulation, or an insulating air space, plus brick facing or corrugated aluminum facing over rigid insulation. The heat transfer from plain concrete walls below ground level and from floors depends on whether they are below the groundwater level. If the groundwater level is not known, it may be assumed that the sides of the tank are surrounded by dry earth and that the bottom is saturated earth. Because the heat losses from the tank warm up the adjacent earth, it is assumed that the earth forms an insulating blanket 1.5 to 3 m (5 to 10 ft) thick before stable ambient earth temperatures are reached. In northern climates, frost may penetrate to a depth of 1.2 m (4 ft). Therefore, the ground temperature can be assumed to be 0°C (32°F) at this depth and to vary uniformly above this depth to the design air temperatures at the surface. Below the frost depth, normal winter ground temperatures can be assumed, which are 5 to 10°C (10 to 20°F) higher at the base of the wall. Alternatively, an average temperature may be assumed for the entire wall below grade.

The loss of heat through the roof depends on the type of construction, the absence or presence of insulation and its thickness, the presence of air space (as with floating covers between the skin plate and the roofing), and whether the underside of the roof is in contact with sludge liquor or gas.

Radiation from roofs and aboveground walls also contributes to heat losses. At the temperatures involved, the effect is small and is included in the coefficients normally used, such as those given in the foregoing discussion. For the theory of radiant-heat transmission, the reader is referred to McAdams (1954). Heat requirements for a digester are determined in Example 13–7.

When external heaters are installed, the sludge is pumped at high velocity through the tubes while water circulates at high velocity around the outside of the tubes. The circulation promotes high turbulence on both sides of the heat transfer surface and results in higher heat transfer coefficients and better heat transfer. Another advantage of external heaters is that untreated cold sludge on its way into the digesters can be warmed, intimately blended, and seeded with sludge liquor before entering the tank. Heat exchangers require cleaning periodically to maintain heat transfer efficiency.

Digestion tanks have also been heated using internal heating systems. Some arrangements have included pipes mounted to the interior face of the digester wall and mixing tubes equipped with hot-water jackets. Because of inherent operating and maintenance problems with this type of heating system, internal heating is not recommended. Reported problems include caking of sludge on the heating surface and the inability to inspect or service the equipment unless the tank is dewatered (WEF, 1987).

EXAMPLE 13–7 Estimation of Digester Heating Requirements   A digester with a capacity of 90,700 wet kg/d (200,000 lb/d) of thickened sludge is to be treated by circulation of sludge through an external hot water heat exchanger. Assuming that the following conditions apply, find the heat required to maintain the required digester temperature. If all heat were shut off for 24 h, what would be the average drop in temperature of the tank contents?

1. Concrete digester dimensions:

    Diameter = 20 m

    Side depth = 7 m

    Middepth = 10 m

2. Heat-transfer coefficients:

    Dry earth embanked for entire depth, $U = 0.68$ W/m²·°C

    Floor of digester in moist earth, $U = 2.85$ W/m²·°C

    Fixed Concrete Insulated Roof exposed to air, $U = 1.5$ W/m²·°C

3. Temperatures:

    Air = −5°C

    Earth next to wall = 0°C

    Incoming sludge = 10°C

    Earth below floor = 5°C

    Sludge contents in digester = 35°C

4. Specific heat of sludge 4200 J/kg·°C

Solution

1. Compute the heat requirement for the sludge.

$$q = (90,700 \text{ kg/d})[(35 - 10)°\text{C}](4200 \text{ J/kg·°C})$$

$$= 95.2 \times 10^8 \text{ J/d}$$

2. Compute the area of the walls, roof, and floor.

   Wall area $= \pi(20)\ (7) = 439.6\ m^2$

   Floor area $= \pi(10)\ [10^2 + (10 - 7)^2]^{1/2} = 327.8\ m^2$

   Roof area $= \pi(10^2) = 314\ m^2$

3. Compute the heat loss by conduction using Eq. (13–17).

   $$q = UA\Delta T$$

   a. Walls:

   $$q = 0.68\ W/m^2{\cdot}°C\ (439.6\ m^2)(35 - 0°C)(86,400\ s/d) = 9.0 \times 10^8\ J/d$$

   b. Floor:

   $$q = 0.85\ W/m^2{\cdot}°C\ (268.2\ m^2)(32 - 5°C)(86,400\ s/d) = 5.32 \times 10^8\ J/d$$

   $$q = 2.85\ W/m^2{\cdot}°C\ (327.8\ m^2)(35 - 5°C)(86,400\ s/d) = 24.2 \times 10^8\ J/d$$

   c. Roof:

   $$q = 1.5\ W/m^2{\cdot}°C\ (314\ m^2)[35 - (-5°C)](86,400\ s/d) = 16.2 \times 10^8\ J/d$$

   d. Total losses:

   $$q_t = (9.0 + 24.2 + 16.2) \times 10^8\ J/d = 49.4 \times 10^8\ J/d$$

4. Compute the required heat-exchanger capacity.

   Capacity = heat required for sludge and heat required for digester

   $$= (95.2 + 49.4) \times 10^8\ J/d = 144.6 \times 10^8\ J/d$$

5. Determine the effect of heat shutoff.

   a. Digester volume $= \pi\left(\dfrac{D^2}{4}\right)h_s + \pi\left(\dfrac{D^2}{12}\right)h_c$

   $$= \pi\left(\frac{20^2}{4}\right)(7) + \pi\left(\frac{20^2}{12}\right)(10 - 7) = 2198 + 314$$

   $$= 2512\ m^3$$

   b. Weight of sludge $= (2512\ m^3)(10^3\ kg/m^3)$

   $$= 2.51 \times 10^6\ kg$$

   c. Drop in temperature $= \dfrac{(144.6 \times 10^8\ J/d)(1d)}{(2.51 \times 10^6\ kg)(4200\ J/kg{\cdot}°C)} = 1.37°C/d$

Heating Equipment. The contents of the digester can be heated by tube-in-tube, spiral-plate, or water-bath external heat exchangers. The tube-in-tube and spiral-heat exchangers are similar in design. A tube-in-tube exchanger consists of two concentric pipes, one containing the circulating sludge and the other containing hot water. Flow through the pipes is countercurrent. Spiral-plate heat exchangers [see Fig. 13–28(a) and (b)] are composed of two long strips of plate that are wrapped to form a pair of

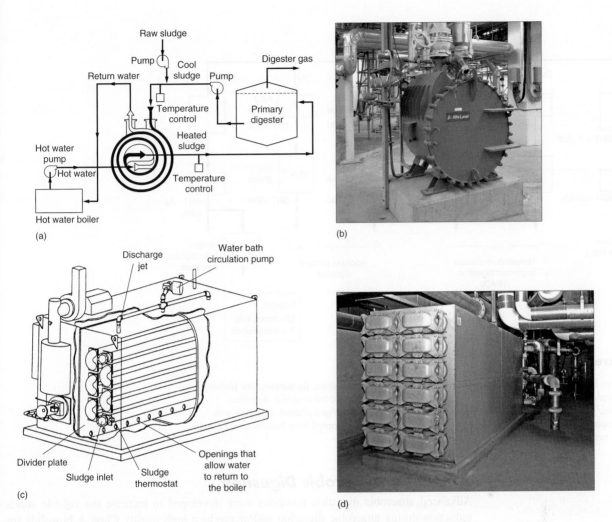

Figure 13–28

Heat exchangers used for heating digesting sludge: (a) schematic diagram of a spiral type, (b) view of a spiral type and (c) schematic of a water bath type heat exchanger, and (d) view of a water bath type heat exchanger.

concentric passages. The flow regime is also countercurrent. Water temperatures are kept generally below 68°C (154°F) to prevent caking of the sludge. Heat-transfer coefficients for external heat exchangers range from 0.9 to 1.6 W/m²·°C (WEF 2010a).

Operation of a water-bath heat exchanger involves circulation of the sludge through a heated water bath [see Fig. 13–28(c) and (d)]. The heat transfer rate is increased by pumping hot water in and out of the bath. Recirculation pumps allow the sludge feed to be heated before introduction to the digester.

Boilers and cogeneration systems are used typically to supply heat to the circulating water in the heat exchangers. Boilers can be fueled by digester gas; however, natural gas or fuel oil may be used as auxiliary fuel for times when sufficient digester gas is not available, such as for digester startup. If a cogeneration system is provided that uses digester gas to fuel an internal-combustion engine for generating electricity or powering pumps or blowers, heat from the engine jacket water can be used in the heat exchanger.

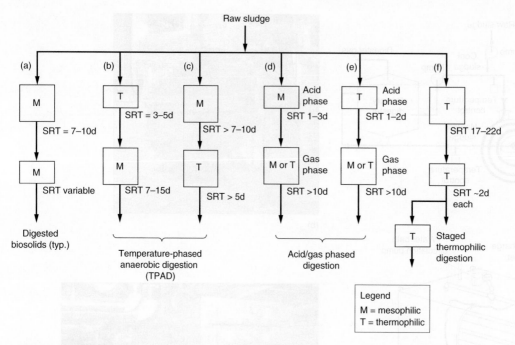

Figure 13–29

Options for staged anaerobic digestion: (a) staged mesophilic digestion, (b) temperature phased thermophilic-mesophilic digestion, (c) temperature phased mesophilic-thermophilic digestion, (d) acid/gas phased digestion with mesophilic acid-phase, (e) acid/gas phased digestion with thermophilic acid phase, and (f) staged thermophilic digestion. (Adapted from Schafer and Farrellm 2000 and Moen, 2000.)

Advanced Anaerobic Digestion

Advanced anaerobic digestion processes were developed to increase the volatile solids reduction during anaerobic digestion and/or produce high quality Class A biosolids for beneficial use of the generated biosolids. Among the advanced anaerobic digestion processes are the thermophilic digestion, staged thermophilic digestion, staged mesophilic digestion, acid/gas phase digestion and temperature phased digestion. These digestion options are shown on Fig. 13–29 and are discussed below. Typical SRTs are also noted on Fig. 13–29.

Thermophilic Anaerobic Digestion. Thermophilic digestion occurs at temperatures between 50 and 57°C (120 and 135°F), conditions suitable for thermophilic bacteria. Because biochemical reaction rates increase with temperature, doubling with every 10°C (18°F) rise in temperature until a limiting temperature is reached, thermophilic digestion is much faster than mesophilic digestion. Advantages and disadvantages of thermophilic digestion when compared to mesophilic digestion are provided in Table 13–36. Single-stage thermophilic digesters have been used only in limited applications; for municipal sludge treatment, they have been mainly used as the first stage of a temperature-phased anaerobic digestion process (Moen, 2000).

Although there may be greater reductions in pathogens in thermophilic digestion than in mesophilic digestion, U.S. federal regulations controlling land application of biosolids do not classify thermophilic digestion as a process to further reduce pathogens (PFRP). Both mesophilic and thermophilic digestion are classified as processes to significantly

Table 13–36

Advantages and disadvantages of thermophilic anaerobic digestion as compared to. mesophilic anaerobic digestion

Advantages	Disadvantages
1. Improved pathogen destruction, Class A sludge production is possible[a]	1. If not batch treatment process will require EPA certification for Class A PFRP
2. Reaction rate is increased which can reduce volume requirements (capital savings)	2. Increased thermal energy requirements[b]
3. May improve overall VSR and increase digester gas production	3. Biosolids may not dewater as well[c]
4. Components and design are essentially the same as conventional mesophilic digesters	4. Higher odor potential in dewatered cake[c]
	5. Increased ammonia concentration in dewatering sidestream
	6. Process may not be as stable
	7. More complex system due to heat recovery requirements
	8. May be more susceptible to foaming

[a] To meet Class A with thermophilic digestion, the process must incorporate batch thermophilic tanks that can meet the time/temperature requirements under Alternative 1. Otherwise, the process will require site specific testing under Alternative 3.

[b] Many plants incorporate a heat recovery loop where thermophilic sludge is cooled and the energy is utilized to preheat the raw sludge. Incorporating the heat recovery loop reduces thermal energy requirements but increases the cost and complexity of the digestion system.

[c] To improve dewatering and reduce odor potential many plants incorporate a mesophilic digestion stage prior to dewatering.

reduce pathogens (PSRP). Therefore, single-stage thermophilic digestion has significant limitations, as cited above.

Staged Thermophilic Digestion.   A staged thermophilic digestion process [see Fig. 13–29(f)] uses a large reactor followed by one or more smaller reactors to reduce pathogen short circuiting and achieve a Class A sludge. At the Annacis Island Wastewater Treatment Plant in Vancouver, BC, the first stage is followed by three subsequent stages. Volatile solids reductions for the digestion system are reported to be on the order of 63 percent (Schafer and Farrell, 2000b).

Staged Mesophilic Digestion.   Although digestion performed in two tanks coupled in series has been done in the past, little information is available about the operation of two-stage heated and mixed high-rate digesters. Researchers Torpey and Garber found that there were few benefits in volatile solids reduction and gas production in two series tanks as compared to a single-stage high-rate process (Torpey and Melbinger, 1967; Garber, 1982). More recent testing indicates that two-stage mesophilic digestion may produce more stable, less odorous biosolids that are easier to dewater (Schafer and Farrell, 2000a). Staged mesophilic digestion is shown on Fig. 13–29(a).

Acid/Gas Phased Digestion.   In the acid/gas (AG) digestion process, anaerobic digestion proceeds through the three distinct phases of digestion described earlier—hydrolysis, fermentation (acidogenesis), and methanogenesis—but the process is divided into two separate steps. In the first stage, known as the acid phase digester, solubilization of particulate matter occurs (hydrolysis), and volatile acids are formed (acidogenesis). The first stage is conducted at a pH of 6 or less and at a short SRT conducive to the production of high concentrations of volatile acids (> 6000 mg/L). The second stage, known as the gas phase, is conducted at a neutral pH and a longer SRT to suit the environmental conditions for the methane-generating bacteria and maximize gas production. Advantages of this method of digestion are (1) greater volatile solids reduction can be achieved, (2) digester foaming can be controlled, and (3) either stage can be operated at mesophilic or thermophilic temperatures [see Figs. 13–29(d) and (e)]. More than 30 full-scale plants using the AG process are in operation at the time of writing of this text (2012) (Wilson et al., 2008). Total volatile solids reductions range from 50 to 60 percent. Most acid/gas systems produce Class B biosolids and operate in the mesophilic range for both phases and termed AGMM. However, at the Belmont Wastewater Plant in Indianapolis, IN, where a thermophilic acid phase and mesophilic gas phase system was pilot tested, it was found that the process was effective in meeting Class A requirements for pathogen reduction (Schafer and Farrell, 2000a).

The design of an AG digestion system requires control of the organic loading to the acid phase to prevent the formation of methanogens. Control is provided through control of the detention time. Ideal detention time is between 1 to 2 d. Due to the low SRT, volatile solids leading rates for the acid phase digestion are an order of magnitude greater than conventional digestion and range from 24 to 40 kg VS/m³·d (1.5 to 2.5 lb/d·ft³) (WEF, 2012). The methane phase can then be a little as 10 d. Regulatory approval of this short a detention time maybe required to retain Class B digested sludge quality. The original concept for this process recommended a plug flow acid phase reactor. Current designs have used a tall cylindrical tank with level control to enable good mixing while maintaining the high organic loading. There will be very little gas produced in this process and the gas will contain little, if any, methane. This gas may be burned separately or combined with the methane phase digester gas, but is often wasted and not otherwise used. Counterintuitive to typical biological processes, operation requires increasing the loading (by decreasing the detention time) when the process begins to become less acidic and/or methane begins to be produced.

Temperature-Phased Digestion. Temperature-phased anaerobic digestion (TPAD), shown on Figs. 13–29(b) and (c), was developed in Germany and is an approach that incorporates the advantages of thermophilic digestion and mitigates the disadvantages through the addition of a mesophilic phase that enhances stabilization. The design of the temperature-phased process utilizes the advantage of the greater thermophilic digestion rate, which generally is four times faster than mesophilic digestion. The TPAD process has shown the capability for absorbing shock loadings better, as compared to single-stage mesophilic or thermophilic digestion. The process can operate in either of two modes, thermophilic-mesophilic or mesophilic-thermophilic. In the thermophilic-mesophilic mode, shown on Fig. 13–29(b), the thermophilic phase is designed to operate at 55°C (130°F) with a 3 to 5 d detention time. The mesophilic phase is designed to operate at 35°C (95°F) with a 10 d or greater detention. The total average detention time of 15 d compares to the typical 10 to 20 d range of the single-stage high-rate mesophilic digestion process. The volatile suspended solids (VSS) destruction efficiencies of the TPAD process are on the order of 15 to 25 percent greater than single-stage mesophilic digestion (Schafer and Farrell, 2000b).

Through greater hydrolysis and biological activity in the thermophilic phase, the system tends to have greater VSS destruction and gas production. Foaming is also reduced. The mesophilic phase provides additional VSS destruction and conditions the sludge for further processing. The main advantages of the mesophilic phase are (1) the destruction of odorous compounds (mostly fatty acids) that are common to the thermophilic digestion process and (2) the improved stability of the digestion operation. The process is also reported to be capable of meeting Class A sludge requirements (WEF, 2010a).

A second temperature-phased digestion process shown on Fig. 13–29(c) has a mesophilic stage that precedes the thermophilic stage. Limited results from full-scale and pilot testing show that the volatile solids reduction is greater than that from single-stage mesophilic digestion (Schafer and Farrell, 2000b). Design considerations for the temperature-phased anaerobic digestion process include selection of the heating and mixing systems to ensure proper temperature control of each stage, sizing of the gas-handling equipment to meet the greater gas production rates, and control of the pumping systems for digester feed and heating (WEF, 2010a).

Sludge Pretreatment for Anaerobic Digestion

Pretreatment of sludge prior to anaerobic digestion is used to increase the solids loading, increase the volatile solids reduction, increase biogas production and in some cases produce Class A biosolids. Sludge pretreatment results in increased hydrolysis through the application of some form of energy to the sludge. The form of pretreatment can be chemical, physical, electrical, or thermal. This section discusses two main categories of pretreatment: thermal hydrolysis and physical-chemical and electrical pretreatment.

Thermal Hydrolysis Pretreatment. Thermal hydrolysis (TH) is a thermal conditioning process that operates at lower temperatures in the range 150–200°C and functions as a pretreatment step before anaerobic digestion. The cited benefits of the process include (1) break down of longer organic polymer chains to shorter chain organic matter to increase digestion and gas production, (2) production of Class A product under the time and temperature stipulations of the EPA Part 503 biosolids rule, (3) enhancement of digested biosolids dewatering characteristics achieving in many cases greater than 30 percent cake solids with mechanical dewatering, (4) produce good quality product in terms of odor and texture, and (5) significant reduction of digestion volume through reducing treated sludge viscosity allowing higher solids concentrations to be pumped and mixed during digestion.

At present (ca. 2013), two commercially available TH processes are available; one is called Cambi™ and is provided by Cambi AS, Norway; the other is called Exelys™ and is provided by Veolia Water Systems, France. The two systems operate under the same treatment parameters; however, the first system is batch based, and the second system is a plug flow design that is a modification of Veolia's batch TH process (Biothelys™). The Cambi™ system is the most widely used system for thermal hydrolysis and is considered fully developed system; however, comparison with other systems may be warranted (Abu-Orf and Goss, 2012). The focus of the information in this section is mainly on the batch TH system as provided by Cambi™.

Description of the TH Batch System as Provided by Cambi™. The Batch TH process is a well established process outside North America with more than two dozen installations (2012) processing sludge from 3.3 to 250 dry tonne/d (3.6 DT/d to 275 DT/d). The process is gaining interest in North America with the first installation expected at Blue Plains Advanced Wastewater Treatment Plant in 2014. The process appears to be cost effective and meets the future needs of Blue Plains (Abu-Orf et al., 2009). As illustrated on Fig. 13–30, the thermal hydrolysis step consists of three basic units: the pulper, the reactor and the flash tank that defines a process train. In reality there is more to the process train than those three units. The sludge needs to pass through some type of screening process ahead of the TH to remove damaging materials from the sludge stream. Furthermore, the sludge must be pre-dewatered to slightly higher than 16 percent total solids. The dewatered sludge is then transferred to a silo or bin that is large enough to provide equalization,

Figure 13–30

Main components of batch thermal hydrolysis system.

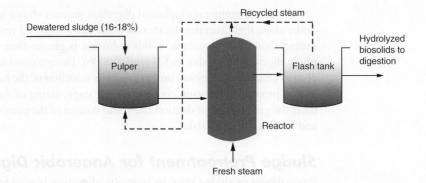

allowing the TH process to operate at a uniform flowrate. The sludge can then be transferred to the pulper tank via augers in the bottom of the storage bin and pumps. The concentration of solids to the pulper is diluted to a concentration between 14.5 and 16.5 percent TS using dilution water (typically plant effluent). The sludge is mixed in the pulper and preheated using steam that is recycled back from the flash tank. The preheated sludge is pumped to the reactor vessel where it will be heated using fresh stream to about 165°C (329°F) and a pressure of approximately 8–9 bar (120–130 lb$_f$/in.^{2}) gauge After the appropriate hydrolyzing time, the sludge is transferred from the reactor to the flash tank. The flash tank receives treated sludge with a solids concentration about 3 points less due to the steam injection and dilution.

The treated sludge temperature at this point is too hot to feed a mesophilic digester and requires cooling and dilution if needed. Again, the water used for dilution is required to lower the concentration of the treated sludge to between 8 and 12 percent for feeding into the digester. The temperature must also be reduced to about 42–44°C by heat transfer. The sludge at 8 to 12 percent TS is then transferred to the digestion facility. Finally, the digested biosolids are transferred to final dewatering before being distributed for beneficial reuse. The final product has very good stacking properties and is low in odor, making it very suitable for soil blending and for land application. In addition to the process train, there must be a source of fresh steam for heating the reactors, so there will be at least a boiler as part of the train. Usually, the biogas generated from the digestion process is processed via a combined heat and power (CHP) facility for producing electricity and a majority of the necessary steam for the TH process. The amount of waste heat from the CHP system typically provides 75–95 percent of the total steam energy requirements so a supplemental standby boiler that operates on digester gas or an auxiliary fuel source is required.

Reactor Operation Scheme. The TH reactors operate in a batch mode with a typical volume of 12 m^3 (424 ft^3) and receive approximately 7.6 m^3 (268 ft^3) (~70 percent of total volume) of sludge in each batch. Once the sludge is transferred, approximately one tonne of fresh steam is injected per one dry tonne of sludge (from Cambi™ Specifications). After the reaction period elapses, the sludge is released to the flash tank. When the steam is released from the reactors, it passes to the pulper and preheats the incoming sludge. The pulper and the flash tanks are usually sized twice the size of one reactor. The batch steps for a typical 90 min cycle for each reactor are shown in Table 13–37. The theoretical amount of steam required can be estimated by the following equation:

$$\frac{M_{steam}}{M_s} = \frac{\left(C_{PS} + \dfrac{C_{PW}}{W_s} - C_{PW}\right)(T_H - T_{raw})}{H - C_{PW}(T_H - T_{ref})}$$

(13–18)

Table 13-37

Steps for 90 min batch cycle operation

Step	Action	Time, min	Description
1	Fill	15	Fill reactor with 7.6 m³ of sludge
2	Steam injection	15	Inject steam in the Reactor
3	React	30	Hold reactor at 160°C and 620 kP (90 lb/in.²)
4	Steam out	15	Release steam to pulper
5	Empty	15	Transfer sludge to flash tank by pressure release.

where

M_{steam} = mass of live steam fed to the process

C_{PS} = specific heat of dry sludge fraction, 1.5 kJ/kg°C (0.36 Btu/lb·°F)

C_{PW} = specific heat of water, 4.18 kJ/kg·°C (1 Btu/lb·°F)

T_H = temperature out of the thermal hydrolysis system. For Cambi™, this temperature corresponds to the temperature out of the flash tank which is approximately 105–110°C (220–230°F)

T_{raw} = raw sludge temperature, typically 10–25°C (50–77°F)

H = enthalpy of steam which is approximately 2785 kJ/kg (1200 Btu/lb) at 12 bar (175 lb$_f$/in.²)

T_{ref} = reference temperature, typically 0°C (32°F)

For example, sludge at 10°C that is dewatered to 16 percent TS by weight would require the following amount of steam for the hydrolyzed sludge to reach a temperature of 110°C:

$$\frac{M_{stream}}{M_s} = \frac{\left[(1.5 \text{ kJ/kg} \cdot °C) + \dfrac{(4.18 \text{ kJ/kg} \cdot °C)}{0.16} - (4.18 \text{ kJ/kg} \cdot °C)\right][(110 - 10)°C]}{(2785 \text{ kJ/kg}) - (4.18 \text{ kJ/kg} \cdot °C)[(110 - 0)°C]}$$

$$= 1.0 \text{ kg steam/kg sludge}$$

It should be noted that Eq. (13–18) is based on some simplifying assumptions such as no heat loss and that all flash steam is recovered and condensed successfully in the pulper. In reality some losses would be expected, and sizing for a design should be conducted carefully with the vendor.

Reactor System Sizing. The overall system sizing is based on a single reactor capacity. Usually it is assumed that sludge is delivered to the reactor at 14.7 percent. The reactor capacity is based on the cycle time and the concentration of the sludge in the reactor. For a standard 90 min cycle as described in Table 13–37 and 7.6 m³/batch of sludge at a 14.7 percent solids concentration, the capacity of a single reactor is about 17.88 dry tonne/d (19.66 dry ton/d). Design size of the reactor is usually based on 95 percent reactor availability. The throughput capacity of train is dependent on the number of reactors within each train. Single reactor capacity can be increased by increasing the solids concentration (maximum would be 17 to 17.5 percent) or shortening the cycle time without compromising the reaction time that is necessary to achieve Class A biosolids according to Part 503 Regulations. In addition, Cambi™ is currently (2012) developing alternative reactor sizes allowing the system to be tailored to sizes at small to medium sized plants. An aerial view of a Cambi™ process installation in Davyhulme, Manchester UK, in which the thermal hydrolysis reactors are visible is shown on Fig. 13–31(a).

Figure 13–31

The Cambi™ process (a) Aerial view of installation in Davyhulme, Manchester, UK. The Cambi™ reactors are visible in the center of the photograph. The 8 × 7600 m³ digesters, shown on the lower right, which originally processed 40,000 tonne/y of dry solids now process 92,000 tonne/y with the installation of the Cambi™ thermal hydrolysis pretreatment process (courtesy of Cambi™) and (b) schematic of the operation of a six reactor process, based on a 90 min cycle time.

(a)

Time	Reactor 1	Reactor 2	Reactor 3	Reactor 4	Reactor 5	Reactor 6
15 min	Fill	Empty	Steam out	React	React	Steam in
15 min	Steam in	Fill	Empty	Steam out		React
15 min	React	Steam in	Fill	Empty	Steam out	
15 min		React	Steam in	Fill	Empty	Steam out
15 min	Steam out		React	Steam in	Fill	Empty
15 min	Empty	Steam out	React	React	Steam in	Fill

(b)

Reactor Sequence Operation. The preferred smallest Cambi process train will have two reactors; otherwise it will be a batch operation. With two reactors the train works as a continuous operation. Based on solids residence of 90 min within each reactor, a sketch portraying the sequential status of a six reactor Cambi™ process, as selected by the Blue Plains Advanced Wastewater Treatment Plant, is shown on Fig. 13-31(b) (Abu-Orf et al., 2009). Process staggering allows continuous operation of all major mechanical equipment.

As shown on Fig. 13-31(b), one reactor would always be filling, one would always be receiving steam, one would always be releasing steam, and one would always be emptying treated sludge. Based on this operating sequence the sludge feed pumps and the steam plant would always be operating and that only valves would be opening and closing to direct the flow to and from the reactors.

Physical, Chemical, and Electrical Pretreatment. These pretreatment processes are generally applied to sludge produced from secondary treatment process as these sludges typically do not digest well anaerobically. Pretreatment of these sludges is accomplished through application of ultrasonic waves, mechanical shear, electrical pulse, pressure drop, or electrical field. The application of these different treatment methods has resulted in various degrees of success in enhance sludge digestion. For the application of the pretreatment process to be practically effective the amount of sludge entering the digester from secondary treatment must be more or at least same as the sludge produced from primary treatment.

A description of six sludge pretreatment technologies is presented in Table 13–38. Pretreatment technologies use pulse power, pressure drop combined with mechanical

Table 13–38

Description of commercially available physical, chemical and electrical technologies for pretreatment of secondary sludges prior to anaerobic digestion[a]

Technology/ manufacturer	Description	Advantages	Disadvantages
OpenCel/OpenCel, USA	Pulsed Power Technology, as offered by Open Cel®, exposes biological sludge to high voltage bursts between 20 and 100 microseconds to lyses cell membranes	Relatively low energy requirements Imparts usable heat to secondary sludge Low space requirements Low pressure operation	Relatively new Limited number of installations in North America Offered by only one manufacturer
Crown Disintegration/ Siemens, USA	Pretreatment is applied to only a portion of the secondary sludge, which includes grinding and mixing followed by pressurization. The high pressure drop causes cavitation of the sludge and rupturing of the cell membrane	Similar technology offered by other manufacturers Low space requirements Approximately 20 installations worldwide, all in Europe. Small footprint	Relatively high energy requirements. No operating facilities in North America. High pressure system 1200 kPa (175 $lb_f/in.^2$) with wear parts
Sludge Squeezer/ Huber, USA	Technology imparts a high pressure drop to a portion of the secondary sludge in a two stage process. In the first stage the sludge flocs are mechanically ruptured. In the second stage the flocs are mixed into the sewage sludge via a hydrodynamic flow field and homogenized	Similar technology offered by other manufacturers Low space requirements Approximately 3 installations, all in Europe	Relatively high energy requirements No operating facilities in North America High pressure system 1200 kPa (175 $lb_f/in.^2$) with wear parts
MicroSludge/ MicroSludge, Canada	A portion of the secondary sludge is pretreated with lime to soften cell membrane, then undergoes grinding and mixing followed by pressurization up to 1200 kPa (175 $lb_f/in.^2$). When the pressure is released, the biological cells are exposed to high shear forces which is hypothesized to rupture the cell membranes. The process includes course and fine screens for the thickened sludge and conditioned sludge, respectively, as well as a gas liquid separator to release ammonia gas formed at high pH	Small footprint Reduction in dewatering cost (polymer and electrical) Benefit for digester heating from 45°F rise of processed sludge Reduced digester mixing energy due to reduced viscosity and volume	Relatively high energy requirements. Requires lime addition High pressure system 1200 kPa (175 $lb_f/in.^2$) with wear parts. No operating facilities in North America
Sonolyzer/ Ovivo, USA	Based on ultrasonic treatment of secondary sludges, which consists of applying high frequency sound waves to the sludge matrix, causing cavitation and disintegration of the cell membranes	Intensively studied over the last 15 y. >25 installations worldwide	Relatively high energy requirements. Only one manufacturer currently in the North America
Electrokinetic Disintegration/ Sud-Chemie AG, Germany	The sludge is run through a series of pipes containing an internal electrical high voltage field and as it moves, the cellular structure is weakened and cracked which allows the bacteria to more effectively digest the sludge	Optimizes digestion Increase gas production Increase settling of sludge Several installations in Europe	High energy requirement Operation with high voltage

[a] The information in this table is current ca. 2013.

Table 13–39

Factors favoring direct co-digestion of organic feedstocks (WEF, 2010b)[a]

Category	Description
Technical	• Remove nuisance wastes from the collection system, especially if a waste is causing stoppage, odor or damage.
	• Remove organic loadings and nuisance factors from headworks and liquid treatment train.
	• Increase use of existing digester capacity, especially with co-digestion of wastes that are synergistic with wastewater sludge in terms of increasing the volatile solids loading rate.
	• Improve knowledge of how to handle organic wastes.
	• Provide a reliable outlet for organic wastes.
Economical	• Develop a new revenue stream from tipping fees for organic wastes.
	• Produce more biogas for combined heat and power systems, or thermal dryer systems, or other beneficial uses.
	• Reduce cost of operation, maintenance, and odor control in the liquid treatment train, from headworks to final clarifiers.
	• Avoid or defer construction of additional liquid train treatment capacity.
	• Increase the throughput rate of the sludge processing train.
Environmental	• Earn carbon credits, where applicable.
	• Reduce land application of organic wastes that contribute to methane production rather than carbon sequestration.
	• Reduce emission of greenhouse gases, particularly methane, coincidental to increasing energy recovery from waste materials.

[a] Adapted from WEF (2010b).

disintegration or chemical treatment, ultrasonic waves, and electrokinetic disintegration. These technologies have different advantages and disadvantages (see Table 13–38). One important observation is that these pretreatment technologies are used more widely in Europe than in North America. The main reason for the increased usage of pretreatment options is that treatment facilities in Europe benefit greatly from the small to moderate increase in biogas production from applying these technologies through green energy credit and other incentives for generating additional green energy from biosolids.

Co-digestion with Other Organic Waste Material

Anaerobic digestion has been applied traditionally as a single substrate, single purpose treatment process and is commonly used in municipal, industrial and agricultural treatment facilities. Most municipal wastewater treatment plants have reported an excess digestion capacity of 15 to 30 percent (Hansen, 2006). These facilities may be able to process with their existing digester capacity a wide range of organic material with municipal sludge and increase their biogas production. The process of digesting more than one substrate is called co-digestion. The technical, economical, and environmental drivers for co-digesting organic wastes with municipal sludge are outlined in Table 13–39 (adapted from WEF, 2010b).

Co-digestion or "co-fermentation" is the simultaneous digestion of a mixture of two or more organic substrates, usually a primary substrate such as wastewater sludge together with lesser amounts of one or more secondary substrate, such as organic municipal solid waste (MSW), source separated organic waste, flotation scum layers, glycerin, and brown

Table 13–40

Biogas unit production and methane content[a]

Feedstock type	Gas yield per unit solids destroyed, m³/kg	Methane content
Fat	1.2–1.6	62–72
Scum	0.9–1.0	70–75
Grease	1.1	68
Protein	0.7	73

[a] Adapted from WEF (2010a).

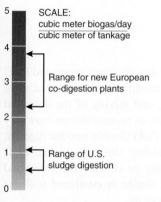

Figure 13–32

Gas production efficiency comparison. (Adapted from Schafer and Lekven, 2008.)

grease collected from grease traps. Yellow grease, having other commercial uses, is considered too valuable to be used as a feedstock for co-digestion. When blending organic substrates, the possible outcomes can be synergistic, antagonistic, or neutral based upon methane production that is greater than, less than, or the same as that observed when each material is digested alone. The successful co-digestion of any organic feedstock requires careful management (Zitomer et al., 2008).

Anaerobic digestion process appears to become more stable when a variety of substrates are co-digested causing improved digester performance (Braun and Wellinger, 2003; Schafer and Lekven, 2008; STOWA, 2006). The addition of certain organic substrate to the anaerobic digestion system can stimulate biological activity and improve digestion performance in terms of biogas production as illustrated on Fig. 13–32.

Co-digestion of Liquid, High Strength Organic Wastes.   Co-digestion of fats, oils, and grease (FOG) is the most common among high strength liquid organic wastes due to its observed high biogas yield. Other liquid organic wastes are suitable for co-digestion including whey from cheese production or residual glycerin from biodiesel production. Biogas production and methane content from the degradation of some organic wastes are presented in Table 13–40. Detailed waste characterization data for selected high-energy organic feedstocks for co-digestion is presented in Table 13–41.

FOG is mostly referred to as material that is collected in interceptors or grease traps of restaurants and cafeterias. Separately collected unprocessed FOG generally has a high residual water content. Brown grease (often also referred to as trap grease) is obtained after the residual water content of the collected FOG is removed. In general, unprocessed FOG is comprised of a brown grease content of around 10 percent by volume; the remaining 90 percent is water (NREL, 2008).

Table 13–41

Characteristics of selected high-energy organic feedstocks[a]

Component	Unit	Restaurant interceptor grease	Biodiesel glycerin	Polymer dewatered FOG	Lime dewatered FOG
Total solids	% solids	1.8–21.9	14.7	42.4	49.1
Volatile solids	% solids	1.2–21.6	14.0	40.9	37.4
Volatile solids/total solids	%	88.9–98.6	95.2	96.5	76.5
pH	—	4.3–4.8	8.4	4.0	6.5

[a] Adapted from WEF (2010a).

Figure 13–33

City of San Francisco waste water treatment plant FOG receiving station.

Considerations for FOG Co-digestion. WWTPs that co-digest pretreated FOG and other high strength (liquid) organic wastes had to make modifications to their existing facilities design to accommodate delivery, storage, dosing, and mixing of the delivered material (WEF, 2010b). The actual co-digestion program and its process design may vary from plant to plant. While some facilities feed the pretreated FOG directly into the digester, most co-digestion programs use one or more separate holding tanks for treated FOG storage. In this case the received FOG (see Fig. 13–33) may be stored temporarily and added to the digester through combining with the influent sludge or combined with the digested sludge through existing heat exchange recycling system.

Additional components of the pretreatment train may include a rock trap or screen for contaminate removal, chopper pump for tank mixing, tank air ventilation, or activated carbon for odor control. Some co-digestion programs may also be designed to receive only substrate that has already been pretreated off-site. A dosing pump automatically and steadily feeds the FOG into the anaerobic digester. The treated FOG addition should be increased gradually to minimize lag effects and increase process stability. The gradual increase in FOG addition should also prevent any sudden surge in gas production that may result in solids flotation or digester stratification (WEF, 2010b).

Once the biomass in the digester has been acclimated to the FOG addition, an increase in both overall biogas production as well as methane content has been reported. For a successful FOG co-digestion program and plant operation, adequate injection, effective digester heating and mixing (to avoid dead zones) were found to be important process factors. To take full advantage of the anticipated increase in digester gas production extra co-generation capacity for power and heat production may be required.

Co-digestion of Organic Solid Wastes. Anaerobic digestion technologies were commercially developed for source-separated organic waste (SSO) and the organic fraction of MSW (OFMSW) in Europe around 1990 and are now in use worldwide. As the technologies for the treatment of SSO and OFMSW become more developed, the number of plants that co-digest these material will increase. SSO and OFMSW have been found suitable for co-digestion with municipal solids with promising gas yields. The amount of biogas that can be produced from co-digesting SSO and other organic substrates can be found in Braun and Wellinger (2003).

East Bay Municipal Utility District (EBMUD) in California is leading the effort on co-digestion of organic solid wastes. Funded in part through an U.S. EPA Region 9 grant, bench-scale studies were conducted to determine biodegradability, methane gas production,

Table 13–42

Inputs and outputs of the Co-digestion Economic Analysis Tool (CoEAT) of EPA

Inputs (including financial data)	Outputs
1. Feedstock type and generation	1. Fixed and r...
2. Collection and Transportation	2. Savings from...
3. Processing	3. Capital invest...
4. Digestion infrastructure	4. Biogas produc...on and energy values
5. Disposal of biosolids	5. Methane reduction from landfills

and the required minimum mean residence time when feeding pulped SSO to the digesters. The results show that three to three and a half times more methane was produced from co-digestion when compared to digesting only sludge for a given digester volume. The EBMUD has developed and patented its in-house food-waste recycling process that has been operating since early 2000 and is accepting approximately 36 tonne/d (79,400 lb/d) of pretreated (crushed and screened) food scraps for co-digestion (Peck, 2008; Gray et al., 2008a; Gray et al., 2008b).

Cost-Effectiveness of Co-digestion. Cost effectiveness of co-digestion of organic wastes depends on many factors. The important factors for cost consideration include waste type, location and distance from plant, tipping fee, required on-site pretreatment, digestion capacity, method of beneficial use of the generated biogas, and electricity prices. Region 9 of the U.S. EPA developed a co-digestion economic analysis tool (CoEAT) based on the extensive research and experience of EBMUD. The CoEAT is designed for use by decision makers and is considered the initial step in assessing the economic feasibility of food waste co-digestion at wastewater treatment plants for the purpose of biogas production. The input and output information for the CoEAT program is presented in Table 13–42. The types of organic wastes considered includes residential food waste, commercial food waste, FOG, food processing waste (fruit, vegetables, breads, rendering byproducts), dairy waste—milk solids, and agricultural waste (fruit and vegetable trimmings).

13–10 AEROBIC DIGESTION

Aerobic digestion may be used to treat (1) waste activated sludge only, (2) mixtures of waste activated sludge or trickling-filter sludge and primary sludge, or (3) waste sludge from extended aeration plants. Aerobic digestion has been used primarily in plants of a size less than 0.2 m³/s (5 Mgal/d), but in recent years the process has been employed in larger wastewater treatment plants with capacities up to 2 m³/s (50 Mgal/d) (WEF, 2010a). In cases where separate sludge digestion is considered, aerobic digestion of biological sludge may be an attractive application. Advantages and disadvantages of conventional aerobic digestion as compared to anaerobic digestion are provided in Table 13–43.

As discussed in Sec. 13–2 and Table 13–12, aerobic digestion is one of the processes defined to meet PSRP requirements for Class B biosolids. To meet Class B requirements for pathogen reduction, the regulations state the solids retention times must be at least 40 d at 20°C and 60 d at 15°C. In many instances, plants that have facilities designed for SRTs less than 40 d and wish to meet the Class B requirements for pathogen reduction have had to add additional storage capacity or thickeners. If the design engineer uses aerobic digestion for stabilization and does not meet the above SRTs, it will be necessary to monitor the performance of the process to demonstrate that the pathogen reduction

Table 13–43

Advantages and disadvantages of aerobic digestion

Advantages	Disadvantages
1. Volatile solids reduction in a well-operated aerobic digester is approximately equal to that obtained anaerobically	1. High power cost is associated with supplying the required oxygen
2. Lower BOD concentrations in sidestreams than anaerobic digestion	2. Does not produce methane for energy recovery
3. Produces an odorless, humuslike, biologically stable end product	3. Aerobically digested biosolids produced have poorer mechanical dewatering characteristics than anaerobically digested biosolids
4. Allows recovery of the basic fertilizer values in the biosolids	4. The process is affected significantly by temperature, location, tank geometry, concentration of feed solids, type of mixing/aeration device, and type of tank material
5. Simple technology requiring few special skills for reliable operation	
6. Low capital cost for small facilities	5. Process consumes alkalinity
7. Easy to construct of readily available parts	
8. Suitable for digesting nutrient-rich waste activated sludges	
9. No risk for explosions	

ᵃ Adapted, in part, from WEF (2012).

criterion has been met. Monitoring is also required to demonstrate that the volatile solids reduction requirements are met for compliance to the vector attraction criterion (U.S. EPA, 2003).

Process Description

Aerobic digestion is similar to the activated sludge process. As the supply of available substrate (food) is depleted, the microorganisms begin to consume their own protoplasm to obtain energy for cell maintenance reactions. When energy is obtained from cell tissue, the microorganisms are said to be in the endogenous phase. Cell tissue is oxidized aerobically to carbon dioxide, water, and ammonia. In actuality, only about 75 to 80 percent of the cell tissue can be oxidized; the remaining 20 to 25 percent is composed of inert components and organic compounds that are not biodegradable. The ammonia is subsequently oxidized to nitrate as digestion proceeds. Nonbiodegradable volatile suspended solids will remain in final product from aerobic digestion. Considering the biomass wasted to a digester and the formula $C_5H_7NO_2$ is representative for cell mass of a microorganism, the biochemical changes in an aerobic digester can be described by the following equations:

Biomass destruction:

$$C_5H_7NO_2 + 5O_2 \rightarrow 4CO_2 + H_2O + NH_4HCO_3 \tag{13–19}$$

Nitrification of released ammonia nitrogen:

$$NH_4^+ + 2O_2 \rightarrow NO_3^- + 2H^+ + H_2O \tag{13–20}$$

Overall equation with complete nitrification:

$$C_5H_7NO_2 + 7O_2 \rightarrow 5CO_2 + 3H_2O + HNO_3 \tag{13–21}$$

Using nitrate nitrogen as electron acceptor (denitrification):

$$C_5H_7NO_2 + 4NO_3^- \rightarrow 5CO_2 + 2N_2 + NH_3 + 4OH^- \tag{13-22}$$

With complete nitrification/denitrification:

$$2C_5H_7NO_2 + 11.5O_2 \rightarrow 10CO_2 + N_2 + 7H_2O \tag{13-23}$$

As given by Eqs. (13–19) through (13–21), the conversion of organic nitrogen to nitrate results in an increase in the concentration of hydrogen ions and subsequently a decrease in pH if sufficient buffering capacity is not available in the sludge. Approximately 7 kg of alkalinity, expressed as $CaCO_3$, are destroyed per each kg of ammonia oxidized. Theoretically, approximately 50 percent of the alkalinity consumed by nitrification can be recovered by denitrification. If the dissolved oxygen is kept very low (less than 1 mg/L), however, nitrification will not occur. In practice, cycling of the aerobic digester between aeration and mixing has been found to be effective in maximizing denitrification while maintaining pH control. In situations where the buffering capacity is insufficient, resulting in pH values below 5.5, it may be necessary to install alkalinity feed equipment to maintain the desired pH.

Where activated or trickling-filter sludge is mixed with primary sludge and the combination is to be digested aerobically, direct oxidation of the organic matter in the primary sludge and oxidation of the cell tissue will both occur. Aerobic digesters can be operated as batch or continuous flow reactors (see Fig. 13–34). Three proven variations of the process are most commonly used: (1) conventional aerobic digestion, (2) high-purity oxygen aerobic digestion, and (3) autothermal aerobic digestion (ATAD). Aerobic digestion accomplished

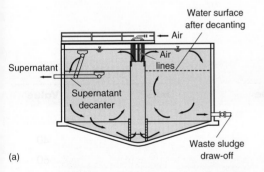
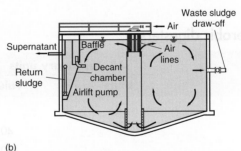

Figure 13–34

Examples of aerobic digesters (a) batch operation with air addition, (b) continuous operation with air addition, (c) view of empty aerobic digester with mechanical aerator, and (d) aerobic digester in lined earthen basin.

with air is the most commonly used process, so it is considered in greater detail in the following discussion.

Conventional Air Aerobic Digestion

Factors that must be considered in designing conventional aerobic digesters include temperature, solids reduction, tank volume, feed solids concentration, oxygen requirements, energy requirements for mixing, and process operation. Typical design criteria for aerobic digestion are presented in Table 13–44.

Temperature.   Because the majority of aerobic digesters are open tanks, digester liquid temperatures are dependent on weather conditions and can fluctuate extensively. As with all biological systems, lower temperatures retard the process while higher temperatures accelerate it. In considering the temperature effects, heat losses should be minimized by using concrete instead of steel tanks, placing the tanks below grade instead of above grade or providing insulation for above-grade tanks, and using subsurface instead of surface aeration. In extremely cold climates, consideration should be given to heating the sludge or the air supply, covering the tanks, or both. The design should provide for the necessary degree of sludge stabilization at the lowest expected liquid operating temperature and should provide the maximum oxygen requirements at the maximum expected liquid operating temperature.

Volatile Solids Reduction.   A major objective of aerobic digestion is to reduce the mass of the solids for disposition. This reduction is assumed to take place only with the

Table 13–44

Design criteria for aerobic digesters[a]

Parameter	U.S. customary units		SI units	
	Units	Value	Units	Value
SRT[b]	d		d	
At 20°C		40		40
At 15°C		60		60
Volatile solids loading	lb/ft³·d	0.1–0.3	kg/m³·d	1.6–4.8
Oxygen Requirements:				
Cell tissue[c]	lb O₂/lb VSS destroyed	~ 2.3	kg O₂/kg VSS destroyed	~ 2.3
BOD in primary sludge		1.6–1.9		1.6–1.9
Energy requirements for mixing				
Mechanical aerators	hp/10³ ft³	0.75–1.5	kW/10³ m³	20–40
Diffused air mixing	ft³/10³ ft³·min	20–40	m³/m³·min	0.02–0.040
Dissolved oxygen residual in liquid	mg/L	1–2	mg/L	1–2
Reduction volatile suspended solids	%	38–50	%	38–50

[a] Adapted, in part, from WEF (1995a); Federal Register (1993).

[b] To meet pathogen reduction requirements (PSRP) of 40 CFR Part 503 regulations.

[c] Ammonia produced during carbonaceous oxidation oxidized to nitrate.

biodegradable content of the sludge, although there may be some destruction of the nonorganics as well. Volatile solids reductions ranging from 35 to 50 percent are achievable by aerobic digestion. Optional criteria for meeting vector attraction requirements of 40 CFR Part 503 are (1) a minimum of 38 percent reduction in volatile solids during biosolids treatment or (2) less than a specific oxygen uptake rate (SOUR) of (1.5 mg O_2/h)/g of total sludge solids at 20°C (U.S. EPA, 1999).

The change in biodegradable volatile solids in a completely mixed digester can be represented by a first-order biochemical reaction at constant-volume conditions:

$$r_M = - k_d M \qquad\qquad (13-24)$$

where r_M = rate of change of biodegradable volatile solids (M) per unit of time (Δmass/time), MT^{-1}

k_d = reaction rate constant, T^{-1}

M = mass of biodegradable volatile solids remaining at time t in the aerobic digester, M

The time factor in Eq. (13–24) is the solid's retention time (SRT) in the aerobic digester. Depending on how the aerobic digester is being operated, time t can be equal to or considerably greater than the theoretical hydraulic residence time (t). Use of the biodegradable portion of the volatile solids is based on the fact that approximately 20 to 35 percent of the waste activated sludge from wastewater treatment plants with primary treatment is not biodegradable. The percentage of nonbiodegradable volatile solids in waste activated sludge from contact stabilization processes (no primary tanks) ranges from 25 to 35 percent (WEF, 2010a).

The reaction rate term k_d is a function of the sludge type, temperature, and solids concentration. Representative values for k_d may range from 0.05 d^{-1} at 15°C to 0.14 d^{-1} at 25°C for waste activated sludge. Because the reaction rate is influenced by several factors, it may be necessary to confirm decay coefficient values by bench-scale or pilot-scale studies.

Solids destruction is primarily a direct function of both basin liquid temperature and the SRT (sometimes referred to as sludge age), as indicated on Fig. 13–35. The data were derived from both pilot-and full-scale studies. The plot on Fig. 13–35 relates volatile solids reduction to degree-days (temperature times sludge age). Initially, as the degree-days increase, the rate of volatile solids reduction increases rapidly. As the degree-days approach 500, the curve begins to flatten. To produce well-stabilized biosolids, at least 550 degree-days are recommended for the aerobic digestion system (Enviroquip, 2000). The use of Fig. 13–35 is demonstrated in Example 13–8, the design of an aerobic digester.

Figure 13–35

Volatile solids reduction in an aerobic digester as a function of digester liquid temperature and digester sludge age.

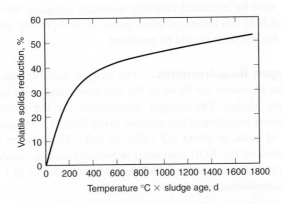

Tank Volume and Detention Time Requirements. The tank volume is governed by the detention time necessary to achieve the desired volatile solids reduction. In the past, SRTs of 10 to 20 d were the norm for the design of aerobic digestion systems (Metcalf & Eddy, 1991). To meet the pathogen reduction requirements of 40 CFR Part 503 regulations, the SRT criteria (see Table 13–44) in conventional aerobic digesters take precedence over the vector attraction criteria of 38 percent solids reduction for sizing the tank volume.

The digester tank volume can be calculated by Eq. (13–25) (WEF, 2010a):

$$V = \frac{Q_i(X_i + YS_i)}{X(k_d P_v + 1/\text{SRT})} \tag{13–25}$$

where V = volume of aerobic digester, m^3 (ft³)
Q_i = influent average flowrate to digester, m^3/d (ft³/d)
X_i = influent suspended solids, mg/L
Y = fraction of the influent BOD consisting of raw primary solids, (expressed as a decimal)
S_i = influent BOD, mg/L
X = digester suspended solids, mg/L
k_d = reaction rate constant, d^{-1}
P_v = volatile fraction of digester suspended solids (expressed as a decimal)
SRT = solids retention time, d

The term YS_i can be neglected if primary sludge is not included in the sludge load to the aerobic digester.

If the aerobic digestion process is operated in a complete-mix, staged configuration (two or three stages), the total SRT should be divided approximately equally among the stages. For more information on staging aerobic digestion, Enviroquip (2000) should be consulted.

Feed Solids Concentration. The concentration of the digester feed solids is important in the design and operation of the aerobic digester. If thickening precedes aerobic digestion, higher feed solids concentrations will result in higher oxygen input levels per digester volume, longer SRTs, smaller digester volume requirements, easier process control (less decanting in batch-operated systems), and subsequently increased levels of volatile solids destruction (WEF 2010a). However, feed solids concentrations greater than 3.5 to 4 percent may affect the ability of the mixing and aeration system in maintaining well-mixed tank contents with adequate dissolved oxygen levels necessary to support the biological process. At feed solids concentrations greater than 4 percent, the aeration equipment must be evaluated carefully to ensure adequate mixing and aeration are achieved. Also at feed solids concentration greater than 4 percent, provisions for removing the heat from the digesters should be practiced.

Oxygen Requirements. The oxygen requirements that must be satisfied during aerobic digestion are those of the cell tissue and, with mixed sludges, the BOD in the primary sludge. The oxygen requirement for the complete oxidation of cell tissue (including nitrification), computed using Eqs. (13–19) and (13–20), is equal to 7 mole/mole of cells, or about 2.3 kg/kg of cells. The oxygen requirement for the complete oxidation of the BOD contained in primary sludge varies from about 1.6 to 1.9 kg/kg destroyed. The oxygen residual should be maintained at 1 mg/L or above under all operating conditions.

Energy Requirements for Mixing. To ensure proper operation, the contents of the aerobic digester should be well mixed. In large tanks, multiple mixing devices should be installed to ensure good distribution of the mixing energy. Typical energy requirements for mixing are given in Table 13–44. In general, because of the large amount of air that must be supplied to meet the oxygen requirement, adequate mixing should be achieved; nevertheless, mixing power requirements should be checked, particularly when feed solids concentrations are greater than 3.5 percent. If polymers are used in the thickening process, especially for centrifuge thickening, a greater amount of unit energy may be required for mixing.

If fine-pore diffused air mixing is used, considerations for selecting the aeration system should include limitations of feed solids concentration on achieving good mixing. Recommendations on feed solids limitations should be obtained from manufacturers of aeration equipment. In addition, the potential for diffuser fouling should be evaluated, especially if the process operation requires decanting.

Process Operation. Depending on the buffering capacity of the system, the pH may drop to a low value of about 5.5 at long hydraulic detention times. The potential drop in pH is due to the increased presence of nitrate ions in solution and the lowering of the buffering capacity due to air stripping. Filamentous growths may also develop at low pH values. The pH should be checked periodically and adjusted if found to be excessively low. Dissolved oxygen levels and respiration rates should also be checked to ensure proper process performance.

Aerobic digesters that do not include prethickening should be equipped with decanting facilities for thickening the digested biosolids before discharge to subsequent operations. Operator control and visibility of the decanting operation are important design considerations. If the digester is operated so that the incoming sludge is used to displace supernatant and the biosolids are allowed to build up, the solids retention time will not be equal to the hydraulic retention time.

EXAMPLE 13–8 **Aerobic Digester Design** Design an aerobic digester to treat the waste sludge produced by the activated sludge treatment plant. Assume that the following conditions apply:

1. The amount of waste sludge to be digested is 2100 kg TSS/d.
2. The minimum and maximum liquid temperatures are 15°C for winter operation and 25°C for summer operation.
3. The system must achieve 40 percent volatile solids reduction in the winter.
4. The minimum SRT for winter conditions is 60 d.
5. Waste activated sludge is concentrated to 3 percent, using a dissolved air flotation thickener.
6. The specific gravity of the waste sludge is 1.03.
7. Sludge concentration in the digester is 70 percent of the incoming thickened sludge concentration.
8. The reaction rate coefficient k_d is 0.06 d^{-1} at 15°C.
9. Volatile fraction of digester TSS is 0.65.
10. No primary sludge is included in the influent to the digester.
11. Diffused-air mixing is used.
12. Air temperature in diffused air system = 20°C.

Solution

1. Compute the volatile solids reduction for winter conditions using Fig. 13–35 and compute the percent volatile solids reduction under summer (maximum) conditions.

   a. For winter conditions, the degree-days from Fig. 13–35 are 15°C × 60 d = 900 degree-days. From Fig. 13–35, the volatile solids reduction is 45 percent, which exceeds the winter requirements of 40 percent.

      To meet the pathogen reduction requirements, the SRT must be 60 d; therefore, the required volume is 68.0 m³/d × 60 d = 4080 m³.

   b. During the summer, the liquid temperature will be 25°C, and the degree-days will be 25 × 60 = 1500. From Fig. 13–35, the volatile solids reduction in the summer will be 50 percent.

2. Compute the winter and summer volatile solids reduction based on a total mass of volatile suspended solids.

   Total mass of VSS $(VSS_M) = (0.65)(2100 \text{ kg/d}) = 1365$ kg/d

   a. Winter: $1365 \times 0.45 = 614$ kg VSS_M reduced/d
   b. Summer: $1365 \times 0.50 = 682$ kg VSS_M reduced/d

3. Determine oxygen requirements (see Table 13–45 for oxygen requirements).
   a. Winter: $614 \times 2.3 = 1412$ kg O_2/d
   b. Summer: $682 \times 2.3 = 1569$ kg O_2/d

4. Compute the volume of air required per d at standard conditions. For the density of air, see Appendix B–1. Note that air is approximately 23.2% oxygen by weight.

   a. Winter: $V = \dfrac{1412 \text{ kg}}{(1.204 \text{ kg/m}^3)(0.232)} = 5055$ m³/d

   b. Summer: $V = \dfrac{1569 \text{ kg}}{(1.204 \text{ kg/m}^3)(0.232)} = 5617$ m³/d

   Assuming an oxygen transfer efficiency of 10 percent, the air flowrates are

   Winter: $q = \dfrac{(5055 \text{ m}^3/\text{d})}{(0.1)(1440 \text{ min/d})} = 35.1$ m³/min

   Summer: $q = \dfrac{(5617 \text{ m}^3/\text{d})}{(0.1)(1440 \text{ min/d})} = 39.0$ m³/min

5. Compute the volume of sludge to be disposed of per day using Eq. (13–2).

   $Q = \dfrac{2100 \text{ kg}}{(10^3 \text{ kg/m}^3)(1.03)(0.03)} = 68.0$ m³/d

6. Compute the air requirement per m³ of digester volume.

   $q = \dfrac{(39.0 \text{ m}^3/\text{min})}{4080 \text{ m}^3} = 0.0096$ m³/min · m³

7. Check the mixing requirements. Because the air requirement computed in Step 7 is below the range of values given in Table 13–44, mixing requirements will govern the design of the aeration system, unless separate mixing is provided.

Comment   The above example is based on a single-stage aerobic digester. If a two-stage or more digester were used, a significant reduction in tank volume is possible. In a multistage arrangement, the air distribution between tanks would vary based on the expected demand as most of the volatile solids reduction will occur in the first stage where the biomass is most active.

Table 13–45

Advantages and disadvantages of ATAD

Advantages	Disadvantages
1. High reaction rate and low retention time requirements when compared to other digestion processes	1. High odor potential
2. Simple operation	2. Potential for poor dewatering characteristics[b]
3. Greater reduction of bacteria and viruses are achieved as compared to mesophilic anaerobic digestion	3. Does not nitrify
4. Can meet Class A when the reactor is well mixed and maintained at or above 55°C	4. Requires upstream mechanical thickening
5. Fully enclosed reactors	5. Sidestreams contain high nutrient loads that may require further treatment
6. Lower energy requirements than conventional aerobic digestion	6. Foam control is necessary
	7. Many processes are proprietary
	8. Potential for erosion / corrosion

[a] Adapted, in part, from WEF (2012).

[b] Many newer designs now incorporate product cooling and an additional mesophilic digestion system which is reported to significantly improve dewatering characteristics.

Dual Digestion

Aerobic thermophilic digestion has also been used extensively in Europe as a first stage in the dual digestion process. The second stage is mesophilic anaerobic digestion. Dual digestion has also been tried in the United States using high-purity oxygen in the first stage. Residence times in the aerobic digester range typically from 18 to 24 h, and the reactor temperature ranges from 55 to 65°C. Typical residence time in the anaerobic digester is 10 d. The advantages of using aerobic thermophilic digestion in dual digestion are (1) increased levels of pathogen reduction, (2) improved overall volatile solids reduction, (3) increased methane gas generation in the anaerobic digester, (4) less organic material in and fewer odors produced by the stabilized sludge, and (5) equivalent volatile solids reductions can be achieved in one-third less tankage than a single-stage anaerobic digester. Prior hydrolysis in the aerobic reactor results in increased degradation during subsequent anaerobic digestion and gas production. Approximately 10 to 20 percent of the volatile solids is liquefied in the aerobic digester, while COD reduction is less than 5 percent. Provisions for foam suppression and odor control are required (Roediger and Vivona, 1998).

Autothermal Thermophilic Aerobic Digestion (ATAD)

Autothermal thermophilic aerobic digestion (ATAD), illustrated on Fig. 13–36, represents a variation of both conventional and high-purity oxygen aerobic digestion. In the ATAD process, the feed sludge is generally prethickened, and the reactors are insulated to conserve the heat produced from the oxidation of volatile solids during the digestion process. Thermophilic operating temperatures (generally in the range of 55 to 70°C) can be achieved without external heat input by using the heat released by the exothermic microbial oxidation process. The heat produced per kg of volatile solids destroyed is approximately 20,000 kJ. Because supplemental heat is not provided (other than the heat introduced by aeration and mixing), the process is termed autothermal.

Within the ATAD reactor, sufficient levels of oxygen, volatile solids, and mixing allow aerobic microorganisms to degrade organic matter to carbon dioxide, water, and nitrogen byproducts. The major advantages and disadvantages of ATAD are cited in Table 13–45.

Figure 13–36

Autothermal thermophilic aerobic
digester (ATAD) system:
(a) process flow diagram and
(b) detail of typical reactor.

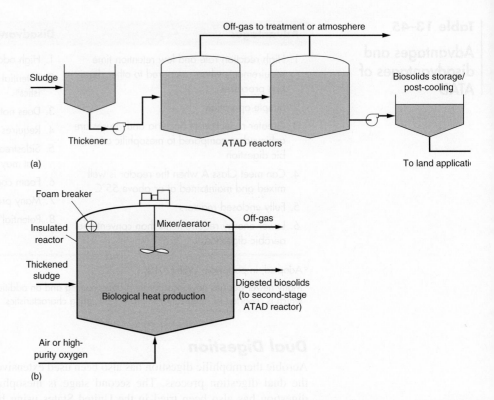

Because the ATAD system is capable of producing Class A biosolids, it is growing in popularity. In a partial survey of ATAD systems in USA, ~25 ATAD systems were in operation (Meckes, 2011).

Process Theory. The biochemical conditions in thermophilic aerobic digesters differ significantly from conventional aerobic digesters. Because of the high operating temperatures, nitrification is inhibited, and aerobic destruction of volatile solids occurs as described by Eq. (13–19) without the subsequent reactions described by Eqs. (13–20) through (13–23). Additionally, most, if not all, ATAD systems may be operating under microaerobic conditions where oxygen demand exceeds oxygen supply (Stensel and Coleman, 2000). Under microaerobic conditions, proteinaceous cellular material will undergo fermentation where protein is represented as peptone as described by Eq. (13–26) (Chu and Mavinic, 1998):

$$4CH_2NH_2COOH + 4H_2O \rightarrow 3CH_3COOH + 2(NH_4)_2CO_3 \tag{13–26}$$

Both Eqs. (13–19) and (13–26) result in the production of ammonia that reacts with water and carbon dioxide to form ammonium bicarbonate and ammonium carbonate to increase alkalinity. Because nitrification does not occur, the pH in the ATAD system will typically range from 8 to 9, higher than in conventional aerobic digesters. Ammonia-nitrogen produced will be present in the off-gas and in solution with concentrations of several hundred mg/L in each. Most of the ammonia nitrogen will be returned to the liquid process in side-streams from the odor control and dewatering facilities. The acetic acid (or acetate) produced by the fermentation of proteins is one of the volatile fatty acids. Acetic acid will be oxidized subsequently in the presence of sufficient dissolved oxygen as described by Eq. (13–27):

$$CH_3COOH + 2O_2 \rightarrow 2CO_2 + 2H_2O \tag{13–27}$$

Anaerobic conditions will occur at times in ATAD systems and will most likely take place in the pre-ATAD sludge holding facilities and in the first-stage ATAD reactors during and immediately after sludge transfers and batch feeding. Under anaerobic conditions, reduced sulfur compounds can be formed that can affect the design and performance of odor-control systems.

Process Design. ATAD systems are designed to have short hydraulic retention times within insulated reactors [see Fig. 13–36(b)]. As long as the ATAD system is well mixed and sufficient oxygen is provided, the temperature in the reactor will rise until a balance occurs; i.e., the heat lost equals the heat input from the exothermic reaction and mechanical energy input. The temperature will continue to rise until the process becomes oxygen mass-transfer-limited.

Factors that must be considered in designing an ATAD system include prethickening, number and type of reactors, postcooling/thickening, feed characteristics, detention time, feed cycle, aeration and mixing, temperature and pH, and foam and odor control. Nearly all of the ATAD systems currently installed in the United States utilize two or more reactors operated in series [see Fig. 13–36(a)]. Design considerations for ATAD systems are presented in Table 13–46; typical design criteria are summarized in Table 13–47.

ATAD systems must be designed to (1) transfer sufficient oxygen to meet the high demand of the reactors and (2) supply the required oxygen while minimizing the latent

Table 13–46
Typical design considerations for an ATAD system[a]

System component	Design consideration
Prethickening system	Thickening or blending facilities may be required to maintain an influent COD to the ATAD reactor greater than 40 g/L
Reactors	Number of reactors; a minimum of two enclosed insulated reactors in series equipped with mixing aeration, and foam control equipment
Screening	Fine screening 6 to 12 mm (0.25 to 0.5 in.) clear openings of raw wastewater or sludge feed stream should be provided for the removal of inert materials, plastics, and rags
Feed cycle	Continuous or batch processing is acceptable, except batch processing provides greater assurance in meeting Class A pathogen reduction requirements
Foam control	Foam suppression is required to ensure effective oxygen transfer and enhanced biological activity. Freeboard of 0.5 to 1.0 m (1.65 to 3.3 ft) is recommended
Post-ATAD storage/ dewatering	Postprocess cooling is necessary to achieve solids consolidation and to enhance dewaterability. A minimum of 20 d detention may be necessary unless heat exchangers are used for cooling the processed biosolids
Odor control	Because of high temperatures in the ATAD system, relatively high concentrations of ammonia are released. Reduced sulfur compounds also result, which can include hydrogen sulfide, carbonyl sulfide, methyl mercaptan, ethyl mercaptan, dimethyl sulfide, and dimethyl disulfide. Odor-control systems may include wet scrubbers, biofilters, or a combination of both (see Chap. 16)
Sidestreams	Liquid sidestreams from odor-control and dewatering systems, when returned to the liquid processing system, may contain constituents that could affect process performance unless accounted for or treated separately

[a] Adapted in part from WEF (2010a) and Stensel and Coleman (2000).

Table 13-47

Typical design parameters for autothermal aerobic digester (ATAD)[a]

Parameter	U.S. customary units			SI units		
	Units	Range	Typical	Units	Range	Typical
Reactor						
HRT	d	4–30	6–8	d	4–30	6–8
Volumetric loading						
TSS, 40 to 60 g/L	lb/10³ ft³·d	320–520		kg/m³·d	5–8.3	
VSS, 25 g/L	lb/10³ ft³·d	200–260		kg/m³·d	3.2–4.2	
Temperature						
Stage 1	°F	95–122	104	°C	35–50	40
Stage 2	°F	122–158	131	°C	50–70	55
Aeration and Mixing						
Mixer type		Aspirating				Aspirating
Oxygen transfer efficiency	lb O₂/kWh		4.4	kg O₂/kWh		2
Energy requirement	hp/10³ ft³	5–6.4		W/m³	130–170	

[a] Adapted, in part, from Stensel and Coleman (2000).

heat loss in the exhaust air. It is difficult to define the oxygen transfer rate in an ATAD system while using typical design procedures used for selecting and sizing aeration equipment for wastewater treatment processes. The oxygen transfer coefficient α (alpha) and the oxygen saturation coefficient β (beta) have not been quantified under the environmental conditions present in an ATAD reactor (Stensel and Coleman, 2000). Factors affecting oxygen transfer are the high temperatures (that would reduce α values) and the foam layer and low dissolved oxygen levels (that might increase oxygen transfer). Nearly all ATAD systems utilize a type of aspirating aerator to introduce oxygen into the reactors. The types include hollow-shaft propeller or turbine aerators, pumped venturi aspirators, and jet aspirators. With all air aspirating systems, the equipment provides both mixing and oxygen transfer. Typical energy requirements for mixing and aeration are given in Table 13–47.

Substantial amounts of foam are generated in the ATAD process as cellular proteins, lipids, and oil and grease materials are broken down and released into solution. The foam layer contains high concentrations of biologically active solids that provide insulation of the reactor and improved oxygen utilization. It is important, therefore, that the foam layer be managed and controlled effectively. Mechanical foam cutters are used most commonly for foam control, but other methods such as spray systems have been employed. A freeboard of 0.5 to 1.0 m is generally recommended for controlling the foam layer (Stensel and Coleman, 2000).

Where ATAD systems are followed by mechanical dewatering, post-ATAD storage is recommended to allow for cooling of the biosolids to improve dewatering performance. Post-ATAD storage coupled with long detention times in the final-stage ATAD reactors may further increase the reduction of volatile solids.

Process Control. The provisions of the 40 CFR Part 503 regulations applicable for meeting the Class A biosolids requirements with the ATAD process are complex because several alternative pathogen-reduction requirements are given. The basic requirements that need to be demonstrated are (1) fecal coliform densities are less than 1000 MPN/g of total solids (dry weight basis), or (2) *Salmonella spp.* bacteria concentrations are below detection limits of 3 MPN/4 g of total solids (dry weight basis). For compliance with these pathogen regulations for Class A biosolids, the withdrawal and feeding of the sludge to the reactors is performed on a batch basis. (In flow-through systems, it is possible that some pathogens might pass through.) Two or more reactors in a series configuration are used typically to ensure that all particles in the reactor are subjected to the time and temperature requirements and that no insufficiently treated biosolids are released to the environment. The ATAD pumping system is designed to withdraw and feed the daily amount of sludge in 1 h or less. The reactor is then isolated for the remaining 23 h each day at a minimum temperature of 55°C.

Improved ATAD Systems

Because many of the first generation ATAD systems suffered from poor performance, odor problems, and high dewatering costs a second generation of ATAD systems has been developed. The differences between the second ATAD generation and previous ATAD installations includes: (1) a single thermophilic stage instead of 2 to 3 stages of shorter detention time; (2) use of pressurized (blower) air for aeration instead of aspirated air; (3) sufficient aeration pressure to maintain aerobic conditions and reduce odor; (4) aeration control based on ORP; (5) non-mechanical foam control; (6) a mesophilic aeration stage following the thermophilic stage. The volume of the first stage thermophilic reactor is approximately two thirds of the total treatment volume. In general, the second generation ATAD systems are reported to provide high volatile solids reduction, a class A biosolids product without offensive odor, and good dewatering with high solids concentration without high chemical demand (Smith et al., 2012). In addition, using a water scrubber and photo catalyter oxidizer for off gas odor control has been reported to be effective (Smith et al., 2012).

The first municipal installation of the second generation ATAD was in 2002 (Scisson, 2009). As of 2012, the largest second generation ATAD began operating at Middletown, OH in May 2009 with a processing capacity of 15 tonne/d (16.5 ton/d) (Pevec, 2010). Based on the performance data from this installation, the system has operated with minimal odor, produces an average volatile solids reduction of about 57 percent, and with centrifuge dewatering cake solids of about 31 percent have been achieved.

High-Purity Oxygen Digestion

High-purity oxygen aerobic digestion is a modification of the aerobic digestion process in which high-purity oxygen is used in lieu of air. The resultant biosolids are similar to biosolids from conventional aerobic digestion. Influent sludge concentrations vary from 2 to 4 percent. Recycle flows are similar to those achieved by conventional aerobic digestion. High-purity oxygen aerobic digestion is particularly applicable in cold weather climates because of its relative insensitivity to changes in ambient air temperatures due to the increased rate of biological activity and the exothermal nature of the process.

While one variation of the high-purity aerobic digestion process uses open tanks, aerobic digestion is usually done in closed tanks similar to those used in the high-purity oxygen activated sludge process. Using closed tanks for high-purity oxygen aerobic digestion will generally result in higher operating temperatures because of the exothermic nature of the digestion process. Maintenance of these higher temperatures in the digester results in a significant increase in the rate of volatile suspended solids destruction.

Where covered tanks are used, a high-purity oxygen atmosphere is maintained above the liquid surface, and oxygen is transferred into the sludge via mechanical aerators. Where an open aeration tank is used, oxygen is introduced to the liquid sludge by a special diffuser that produces minute oxygen bubbles. The bubbles dissolve before reaching the air-liquid interface.

The major disadvantage of high-purity oxygen aerobic digestion is the increased cost associated with oxygen generation. As a result, high-purity oxygen aerobic digestion is cost-effective generally only when used in conjunction with the high-purity oxygen activated sludge system. Also, neutralization may be required to offset the reduced buffering capacity of the system.

PROBLEMS AND DISCUSSION TOPICS

13-1 The water content of waste activated sludge is reduced from 98 to 95 percent. What is the percent reduction in volume by the approximate method and by the more exact method, assuming that the solids contain 70 percent organic matter of specific gravity 1.00 and 30 percent mineral matter of specific gravity 2.00? What is the specific gravity of the 98 and the 95 percent slurry?

13-2 Consider an activated sludge treatment plant with a flowrate of 40,000 m³/d. The untreated wastewater contains 200 mg/L suspended solids. The plant provides 60 percent removal of the suspended solids in the primary settling tank. If the primary sludge alone is pumped, it will contain 5 percent solids. Assume that 400 m³/d of waste activated sludge containing 0.5 percent solids is to be transferred to the digester. If the waste activated sludge is thickened in a gravity belt thickener to 6 percent TS, calculate the thickened waste activated sludge volume. Calculate the total reduction in daily volume of biosolids pumped to the digester that can be achieved by thickening the waste activated sludge in a gravity belt thickener as compared with discharging the primary and waste activated sludge directly to the digester. Assume complete capture of the waste activated sludge in the gravity belt thickener.

13-3 For Example 13-4 for gravity thickening, calculate the amount of dilution water required at average design flow using the data provided to maintain a hydraulic loading rate of 12 m³/m²·d for the thickener size computed in the example.

13-4 Determine the required digester volume for the treatment of the sludge quantities specified in Example 13-5 using the (a) volatile solids loading factor, and (b) volumetric per capita allowance methods. Set up a comparison table to display the results obtained using the three different procedures for sizing digesters (two in this problem and one in Example 13-5). Assume the following data apply:

1. Volatile solids loading method
   a. Solids concentration = 5%
   b. Detention time = 15 d
   c. Loading factor = 2.4 kg VSS/m³·d
   d. Volatile solids concentration = 75%
2. Volumetric loading method
   a. Sewer basin population = 70,000
   b. Per capita contribution = 0.72 g/capita·d
   c. Volume required 50 m³/10³ capita·d

13-5 A wastewater treatment plant is planning to provide for separate anaerobic sludge digestion for its primary sludge. The plant receives an influent wastewater with the following characteristics:

Average flowrate = 8000 m³/d

Suspended solids removed by primary sedimentation 200 mg/L

Volatile matter in settled solids = 75%

Water in untreated sludge = 96%

Specific gravity of mineral solids = 2.60

Specific gravity of organic solids = 1.30

Using these data, determine (a) the required digester volume using an SRT of 20 d, and (b) the minimum digester capacity using the recommended loading parameters of kg VM/m³·d (kilograms of volatile matter per cubic meter per day).

13-6 A wastewater treatment plant currently dewaters on average 750 kg/d of primary and waste activated sludge to an average solid content of 22 percent TS. The plant currently uses post lime stabilization and on average mixes 300 kg/d of quicklime with their dewatered sludge in a pug mill. What is the theoretical temperature increase after adding quicklime? Discuss the advantages and disadvantage for the plant if they were to switch to anaerobic digestion for sludge stabilization in lieu of lime stabilization.

13-7 A digester is loaded at a rate of 300 kg COD/d. Using a waste-utilization efficiency of 75 percent, what is the volume of gas produced when SRT = 40 d? Assume $Y = 0.10$ and $b = 0.02$ d^{-1}.

13-8 Volatile acid concentration, pH, or alkalinity should not be used alone to control a digester. How should they be correlated to predict most effectively how close to failure a digester is at any time?

13-9 A digester is to be heated by circulation of sludge through an external hot water heat exchanger. Using the following data, find the heat required to maintain the required digester temperature:

1. U_x = overall heat-transfer coefficient, W/m²·°C

2. Wall above ground: $U_{air} = 0.85$, wall below ground: $U_{ground} = 1.2$, cover: $U_{cover} = 1.0$

3. Digester is a concrete tank with floating steel cover; diameter = 11 m and sidewall depth = 8 m, 4 m of which is above the ground surface. The tank walls and floor are 300 mm thick.

4. Sludge fed to digester = 15 m³/d at 14°C

5. Outside temperature = –15°C

6. Average ground temperature = 5°C

7. Sludge in tank is to be maintained at 35°C

8. Assume a specific heat of the sludge = 4200 J/kg·°C

9. Sludge contains 4% solids

10. Assume a cone-shaped cover with center 0.6 m above digester top, and a cone-shaped bottom with center 1.2 m below bottom edge.

13-10 A wastewater treatment plant is considering options for expanding their existing anaerobic digestion system to handle increased sludge production. The plant currently sends 25,000 kg/d of 5 percent thickened sludge to anaerobic digestion for stabilization. The plant currently has three 6200 m³ digesters but normally only operates two keeping the third as a redundant unit. The digesters are high rate complete mix digesters with no decanting.

a. Using one of the data sets as selected by the instructor determine how much additional digester volume would be required to maintain at least 15 d HRT at future build out with one digester out of service?

b. One option also being considered is adding a thermal hydrolysis process to increase the capacity of the existing anaerobic digesters to avoid building new digester volume. Assuming the solid content going to the digester is 9 percent (after thermal hydrolysis and dilution) would the existing digestion volume be sufficient to meet digestion requirements? What would be the theoretical steam requirement for the thermal hydrolysis system assuming the temperature out of the thermal hydrolysis process is 110°C?

Assume the specific gravity of the thickened sludge and diluted hydrolyzed sludge going to digestion is 1.03.

		Data set			
Item	Unit	1	2	3	4
Future sludge loading	kg/d	55,000	60,000	50,000	58,000
Raw sludge temperature	°C	10	15	20	12

13–11 A small wastewater plant currently utilizes aerobic digestion to stabilize their waste activated sludge prior to Class B liquid land application and they are looking at ways the system can be expanded to handle future loads. The plant currently sends waste activated sludge to a single stage complete mix aerobic digester with no decanting. Using one of the data sets to be selected by instructor, recommend possible options to consider in the upgrade if aerobic digestion is to be maintained. In the example use the following assumptions.

1. The winter and summertime liquid temperatures are 15 and 25 d, respectively

2. The system must be able to achieve >40% VSR in the winter and meet Class B requirements (SRT > 60 d at 15°C)

3. Specific Gravity of liquid waste activated sludge is 1.01

4. Air temperature in diffused air system is 20°C.

5. Assume a diffused air oxygen transfer efficiency of 10%

In the calculation be sure to comment on volume requirements, SRT, aeration requirements and mixing. Note any perceived advantage or disadvantage with options.

		Data set		
Item	Unit	1	2	3
Current sludge loading	kg/d	500	750	900
Waste activated sludge	% TS	1	0.8	1.3
Future sludge loading	kg/d	1500	2000	3000
Digester volume	m³	3000	5600	2400
Current blower size	m³/min	90	165	125

REFERENCES

Abu-Orf, M. M., S. Pound, R. Sobeck., E. Locke, L. Benson, W. Bailey, C. Peot, M. Sultan, J. Carr, S. Kharkar, S. Murthy, R. Derminassian, and G. Shih (2009) "DC WASA Adopts Thermal Hydrolysis for Anaerobic Digestion Pretreatment: Conceptual Design Details for the Largest Cambi™ System." *Proceedings WETEC 2009,* Water Environment Federation, Alexandria, VA.

Abu-Orf, M. M., and C. Goss (2012) "Comparing Thermal Hydrolysis Processes (Cambi™ and Exelys) for Solids Pretreatment Prior to Anaerobic Digestion" *Proceedings of the WEF Residuals and Biosolids Management Conference 2012,* Water Environment Federation, Alexandria, VA.

Albertson, O. E., and T. Walz (1997) "Optimizing Primary Clarification and Thickening," *Water Environ. Technol.,* **9**, 12, 41–45.

Arakaki, G., R. Vander Schaaf, S. Lewis, and G. Himaka (1998) "Design of Sludge Screening Facilities," *Proceedings of the 71st Annual Conference & Exposition,* vol. 2, Water Environment Federation, Alexandria, VA.

Babbitt, H., and D. H. Caldwell (1939) *Laminar Flow of Sludge in Pipes,* University of Illinois Bulletin 319, Urbana, IL.

Braun, R., and A. Wellinger (2003) *Potential of Co-digestion;* Report, International Energy Agency Bioenergy, Task 37, IEA Energy Technology Network, Comprised of a number of International Collaborators.

Brinkman, D., and D. Voss (1998) "Egg-Shaped Digesters—Are They All They're Cracked Up to Be?" *Proceedings of the 71st Annual Conference & Exposition,* vol. 2, Water Environment Federation, Alexandria, VA.

Carthew, G. A., C. A. Goehring, and J. E. van Teylingen (1983) "Development of Dynamic Headloss Criteria for Raw Sludge Pumping," *J. WPCF,* **55,** 5, 472–483.

Chu, A., and D. S. Mavinic (1998) "The Effects of Macromolecular Substrates and a Metabolic Inhibitor on Volatile Fatty Acid Metabolism in Thermophilic Aerobic Digestion," *Water Sci. Technol.,* (British.),. **38,** 2, 55–61.

Clark, S. E., and D. N. Ruehrwein (1992) "Egg-Shaped Digester Mixing Improvements," *Water Environ. Technol.,* **4,** 1.

Dentel, S. K., Chang, J. S., and Abu-Orf, M. M (2005) "Alkylamine Odors from Degradation of Flocculant Polymers in Sludges," Water Research, **39,** 14.

Enviroquip (2000) Aerobic Digestion Workshop, vol. III, Enviroquip, Inc., Austin, TX.

Federal Register (1993) 40 CFR Part 503, *Standards for the Disposal of Sewage Sludge.*

Garber, W. F. (1982) "Operating Experience with Thermophilic Anaerobic Digestion," *J. WPCF,* **54,** 8, 1170–1175.

Gray (Gabb), D. M. D., P. Suto, and J. Hake (2008a) *"Technical Process Considerations for Providing Community Organics Recycling with Municipal Wastewater Treatment Plant Anaerobic Digesters,"* Proceedings of WEFTEC 2008, Water Environment Federation, Alexandria, VA.

Gray (Gabb), D. M. D., P. Suto, and M. Chien (2008b) *"Producing green energy from post-consumer food wastes at a wastewater treatment plant using a innovative new process,"* Proceedings of the Water Environment Federation Sustainability 2008, Water Environment Federation, Alexandria, VA.

Grady, C. P. L., Jr., G. T. Daigger, and H. C. Lim (1999) *Biological Wastewater Treatment,* 2d ed., Marcel Dekker, New York.

Hill, R. A., P. E. Snoek, and R. L. Gandhi (1986) "Hydraulic Transport of Solids," in, by I. J. Karassik, W. C. Krutzsch, W. H. Fraser, and J. P. Medina (eds.), Pump Handbook McGraw-Hill, New York.

Kester, G. (2008) Fats, Oils, and Grease (FOG), Presented at the *BioCycle West Coast Conference,* San Diego, CA.

Liptak, B. G. (1974) *Environmental Engineers' Handbook,* Chilton Book Co., Radnor, PA.

Maco, R. S., H. D. Stensel, and J. F. Ferguson (1998) "Impacts of Solids Recycling Strategies on Anaerobic Digester Performance," *Proceedings of the 71st Annual Conference & Exposition,* Water Environment Federation, Alexandria, VA.

McAdams, W. H. (1954) Heat Transmission, 2d ed., McGraw-Hill, New York.

McCarty, P. L. (1964) "Anaerobic Waste Treatment Fundamentals," Parts 1, 2, 3, and 4, *Public Works,*. **95,**. 9, 107–112, 10, 123–126, 11, 91–94, 12, 95–99.

McCarty, P. L. (1968) "Anaerobic Treatment of Soluble Wastes," in E. F. Gloyna and W. W. Eckenfelder, Jr. (eds.), *Advances in Water Quality Improvement,* University of Texas Press, Austin, TX.

Meckes, M (2011). Survey of Thermophilic Anaerobic and Aerobic Digestion Systems in USA, USEPA-NRMRL, Cincinnati, OH, November, 2011.

Metcalf & Eddy, Inc. (1981) *Wastewater Engineering: Collection and Pumping of Wastewater,* McGraw-Hill, New York.

Metcalf & Eddy, Inc. (1991) *Wastewater Engineering: Treatment, Disposal, Reuse,* 3d ed., McGraw-Hill, New York.

Moen, G. (2000) *Comparison of Thermophilic and Mesophilic Digestion,* Master's Thesis, Department of Civil and Environmental Engineering, University of Washington, Seattle, WA.

Mulbarger, M. C., S. R. Copas, J. R. Kordic, and F. M. Cash (1981) "Pipeline Friction Losses for Wastewater Sludges," *J. WPCF,* **51**, 8, 1303–1313.

Novak, J. (2001) Personal Communication, Virginia Polytechnic Institute, Blacksburg, VA.

NREL (2008) *Urban Waste Grease Resource Assessment,* National Renewable Energy Laboratory, NREL/SR-570-26141 (http://www.nrel.gov/docs/fy99osti/26141.pdf) (July 23, 2011).

Peck, C. (2008) Food Waste Opportunity! Investigating the Anaerobic Digestion Process to Recycle Post-Consumer Food Waste; *Presentation at the California Resource Recovery Association 32nd Annual Conference,* San Francisco, CA.

Pevec, T., and E. S. John (2010) Largest Municipal ATAD – Class A Biosolids at Middletown WWTP," *Proceedings of the Residuals and Biosolids Conference 2010,* Water Environment Federation, Alexandria, VA.

Roediger, H. (1987) "Using Quicklime—Hygienization and Solidification of Dewatered Sludge," Water Environment Federation, Operations Forum, **4**, 4, 18–21.

Roediger, M., and M. A. Vivona (1998) "Processes for Pathogen Reduction to Produce Class A Solids," *Proceedings of the 71st Annual Conference & Exposition,* Water Environment Federation, pp. 137–148, Alexandria, VA.

Sanks, R. L., G. Tchobanoglous, D. Newton, B. E. Bosserman, and G. M. Jones (1998) *Pumping Station Design,* 2d ed., Butterworths, Stoneham, MA.

Schafer, P. L., and J. B. Farrell (2000a) "Turn Up the Heat," *Water Environ. Technol.,* **12**, 11, 27–32.

Schafer, P. L., and J. B. Farrell (2000b) "Performance Comparisons for Staged and High-Temperature Anaerobic Digestion Systems," *Proceedings of WEFTEC 2000,* Water Environment Federation, Alexandria, VA.

Schafer, P., and C. Lekven (2008) "Co-Digestion Issues That Wastewater Agencies are Facing" *Proceedings of WEFTEC 2008,* 6776–6780, Water Environment Federation: Alexandria, VA.

Scisson, J. P (2009) "As good as the hype: An overview of the second generation and performance," *Proceedings of the Residuals and Biosolids Conference 2009,* Water Environment Federation, Alexandria, VA.

Smith, J. E., Bizier, P., and Sobrados-Bernardos, L. (2012) "Global Development of the ATAD Process and Its Significant Achievements in Energy Recovery and Utilization" *Proceedings of the Residuals and Biosolids Conference 2012,* Water Environment Federation, Alexandria, VA.

Speece, R. E. (1988) "A Survey of Municipal Anaerobic Sludge Digesters and Diagnostic Activity Assays," *Water Res,* **22**, 3, 365–372.

Speece, R. E. (2001) Personal communication.

Stensel, H. D., and T. E. Coleman (2000) "Assessment of Innovative Technologies for Wastewater Treatment: Autothermal Aerobic Digestion (ATAD)," Preliminary Report, Project 96-CTS-1.

STOWA (Dutch acronym for the Foundation for Applied Water Research) (2006) *Co-digestion Sheet;* http://www.stowa-selectedtechnologies.nl/Sheets/Sheets/Co.Digestion.html (accessed July 4, 2009).

Stukenberg, J. R., J. H. Clark, J. Sandine, and W. Naydo (1992) "Egg-Shaped Digesters: from Germany to the U. S.," *Water Environ. Technol.,* **4**, 4, 42–51.

Torpey, W. N., and N. R. Melbinger (1967) "Reduction of Digested Sludge Volume by Controlled Recirculation," *J. WPCF,* **39**, 9, 1464–1474.

U.S. EPA (1979) *Process Design Manual Sludge Treatment and Disposal,* EPA 625/1-79-011, Office of Research and Development, U.S. Environmental Protection Agency, Washington, DC.

U.S. EPA (1987) *Design Information Report—Anaerobic Digester Mixing Systems,* EPA-68-03-3208, EPA/600/J-87/014, Water Engineering Research Laboratory, U.S. Environmental Protection Agency, Cincinnati, OH (see also *J. WPCF,* **59**, 3, 162–170).

U.S. EPA (1995) *Process Design Manual—Land Application of Sewage Sludge and Domestic Septage,* EPA/625/R-95/001, Center for Environmental Research Information, U.S. Environmental Protection Agency, Washington, DC.

U.S. EPA (2000) *BiosolidsTechnology Fact Sheet: Alkaline Stabilization of Biosolids,* EPA 832-F-00-052, U.S. Environmental Protection Agency, Washington, DC.

U.S. EPA (2003) *Control of Pathogens and Vector Attraction in Sewage Sludge,* EPA/625/R-92/013, Office of Research and Development, U.S. Environmental Protection Agency, Washington, DC.

U.S. EPA (2009) *Targeted National Survey – Overview Report,* EPA/822/R-08/014, U.S. Environmental Protection Agency, Washington, DC.

U.S. EPA (2010) *Evaluation of Combined Heat and Power Technologies for Wastewater Facilities,* EPA-832-R-10-006, U.S. Environmental Protection Agency, Washington, DC.

WEF (1980) *Sludge Thickening, Manual of Practice No. FD-1,* Water Environment Federation, Alexandria, VA.

WEF (1987) *Anaerobic Digestion, Manual of Practice No. 16,* 2d ed., Water Environment Federation, Alexandria, VA.

WEF (1995) *Wastewater Residuals Stabilization, Manual of Practice No. FD-9,* Water Environment Federation, Alexandria, VA.

WEF (1998) *Design of Municipal Wastewater Treatment Plants,* 4th ed., Manual of Practice no. 8, vol. 3, Chaps. 17–24, Water Environment Federation, Alexandria, VA.

WEF (2010a) *Design of Municipal Wastewater Treatment Plants,* 5th ed., Manual of Practice no. 8, vol. 3, Chaps. 20–27, Water Environment Federation, Alexandria, VA.

WEF (2010b) *Direct Addition of High-Strength Organic Waste to Municipal Wastewater Anaerobic Digesters,* Water Environment Federation, Alexandria, VA.

WEF (2012) *Solids Process Design and Management.* Water Environment Federation, Alexandria, VA.

Wilson, T. E., R. Kilian, and L. Potts (2008) "Update on 2-phase AG Systems," *Proceedings of WEF Residuals and Biosolids Conference 2008,* Water Environment Federation, Alexandria, VA.

Zitomer, D., P. Adhikari, C. Heisel, and D. Deneen (2008) Municipal Anaerobic Digesters for Codigestion, Energy Recovery, and Greenhouse Gas Reductions. *Water Environ. Res.,* **80, 3,** 229–237.

14

Biosolids Processing, Resource Recovery and Beneficial Use

WORKING TERMINOLOGY

Term	Definition
Aerated static pile composting	A method of composting where the sludge or biosolids and bulking agent are mixed and distributed in a long pile over a grid of air piping which provides the air for reaction.
Belt-filter press	A device that uses a series of porous moving belts revolving over a series of pulleys to drain water from sludge or biosolids.
Carbon footprint	A measure of the impact human activities have on the environment in terms of the amount of greenhouse gases produced, measured in units of carbon dioxide equivalent.
Centrifuge	A dewatering device that relies on centrifugal force to separate particles of varying density (e.g., water and solids).
Composting	A stabilization process that relies on the aerobic decomposition of organic matter in sludge and biosolids by bacteria and fungi.

Term	Definition
Conditioning	A chemical, physical, or biological process designed to improve the thickening or dewatering characteristics of sludge or biosolids.
Decanting	Separating liquid from settled sludge or biosolids by drawing or pouring off the upper layer of liquid after the sludge or biosolids have settled.
Dewatering	A process, usually by belt-filter press or centrifuge that removes a portion of the water contained in solids. Dewatering is distinguished from thickening in that the resulting dewatered cake may be handled as a solid, not as a liquid.
Dissolved-air flotation	A clarification process in which minute bubbles become attached to flocculated material, float to the surface, and are removed by skimming. Heavier solids settle and are removed by mechanical scrapers.
Filter press	A dewatering device in which water is forced from semi-solid materials under high pressure.
Fluidized-bed incinerator	A furnace that uses a high-temperature gas to fluidize solid particles (usually sand and waste sludge and biosolids) to produce and sustain combustion.
Gravity-belt thickener	A thickening device that uses a porous filter belt to promote gravity water drainage.
Heat drying	The application of heat to evaporate water and reduce the moisture content in biosolids below that achievable by conventional dewatering methods.
Humus	Sludge removed from trickling filters.
Incineration	The reduction of the volume of a solid by the thermal destruction of organic matter.
In-vessel composting	A method of composting, mainly proprietary, that occurs inside an enclosed vessel or container.
Mass balance	A method for analyzing physical systems based on the law of conservation of mass.
Multiple-hearth incinerator	An incinerator consisting of numerous hearths that is used for the thermal destruction of organic sludge or biosolids.
Reed bed	A treatment system in which biosolids are used to grow reeds, which in turn utilize the water, nitrogen, and other nutrients to stabilize and dewater the biosolids.
Rheology	The flow properties of a liquid (generally biosolids and sludge) that include elasticity, viscosity, and plasticity.
Rotary drum thickener	A rotating cylindrical screen used to thicken liquid streams of sludge and biosolids.
Rotary press	A sludge or biosolids dewatering device in which the material to be dewatered flows through a channel that is bound between two rotating screens; filtrate passes through the screens and the dewatered material continues through the channel.
Sidestream	A portion of the wastewater flow that has been diverted from the main treatment process flow for specialized treatment (see Chap. 15).
Sludge drying beds	Devices used for the dewatering and drying of sludge and biosolids in which a semi-solid solution is spread over a porous (e.g., sand) or impervious medium and allowed separate and air dry or decant.
Solids	A term often used as a replacement for sludges that have not been stabilized by physical, chemical or biological treatment. The term solids is not used as a substitute for sludge in this chapter. The mass of dry material in sludge is referred to as the solids content.
Thickener	A tank, vessel, or device where residuals or a slurry are concentrated by removing a portion of the water.
Windrow composting	A method of composting where sludge or biosolids are mixed with a bulking agent and arranged in windrows (long piles) that are turned over periodically and remixed mechanically.

Processes used to reduce organic content and to render the processed sludge suitable for reuse or final disposal were considered in Chap. 13. The focus of Chap. 14 is on the many processes used for recovery and beneficial use of biosolids. Before biosolids can be processed or used beneficially, they are typically dewatered to reduce the volume that must be handled. However, to achieve effective dewatering, sludges and biosolids must be conditioned for enhanced water removal. The conditioning of sludge and biosolids is discussed in Sec. 14–1. Commercially available dewatering methods are identified and discussed in Sec. 14–2. Drying the dewatered sludge or biosolids is for the purpose of removing more water and further stabilization; producing granular material that is beneficially used as a fertilizer or energy source is presented in Sec. 14–3. Thermal oxidation of the sludges for destroying harmful constituents and producing ash-like material that can also be used beneficially is discussed in Sec. 14–4. Composting for stabilizing the biosolids and producing fertilizer like material is discussed in Sec. 14–5. Conveyance and storage of biosolids are discussed in Sec. 14–6. The preparation of solids balances for treatment facilities is described in Sec. 14–7. Biosolids resource recovery and energy recovery are covered in Sec. 14–8. The application of biosolids to land and conveyance and storage of biosolids after processing is discussed in Sec. 14–9.

14–1 CHEMICAL CONDITIONING

Sludges and biosolids are conditioned expressly to improve their dewatering characteristics. For proper mechanical dewatering systems such as centrifugation, belt filter press, rotary press, screw press, and pressure filter press, as discussed in Sec. 14–2 , sludges and biosolids must be chemically conditioned. Chemical conditioning results in the flocculation (aggregation) of the sludge and biosolids to achieve efficient solid-liquid separation. Other conditioning methods such as heat treatment and freeze-thaw, have also been used to a limited extent or experimentally and are discussed in previous editions of this textbook. Chemical conditioning uses inorganic chemicals and water soluble polymers, or both. Inorganic conditioners, which include lime, ferric chloride, ferrous sulfate, aluminum sulfate, and aluminum chloride, are used mainly for recessed chamber filter press discussed in Sec. 14–2. However, in certain applications when biosolids are hard to dewater and require high polymer dosages, iron salts are often used (Abu-Orf et al., 2001). Polymers do not increase the dry solids content, while iron salts and lime can increase the dry solids content by 20 to 30 percent. A general discussion of polymers, the most used conditioning agents for mechanical dewatering, is the primary focus of this section. Following a general description of polymers, polymer characteristics, factors affecting conditioning, dosage determination, mixing, and polymer makeup and feeding are also considered.

Polymers

Water-soluble polymers are used most commonly for sludges and biosolids conditioning prior to mechanical dewatering. Polymers are also termed *organic polyelectrolytes* as they dissociate upon addition to water into negatively and positively charged species. Polymers are chains of individual monomer units and linked together in linear, branched, or structured configuration with functional groups located along the chains that determine the charge of the polymer. Because sludges and biosolids are mainly negatively charged, cationic polymers are used most commonly for conditioning. Cationic polyacrylamide (PAM) is the backbone of most, if not all, commercially available polymers for biosolids and sludge conditioning (WEF, 2012). These polymers upon dissociation release anions such as chloride leaving behind long chains of high molecular weight polymeric molecules that are positively charged. These long chains flocculate the suspended solids and the colloidal

Table 14–1

Distribution of polymer charge density and molecular weight[a]

Relative charge density and molecular weight	Charge density, mole %	Relative molecular weight
Very high	> 70–100	> 6,000,000–18,000,000
High	> 40–70	> 1,000,000–6,000,000
Medium	> 10–40	> 200,000–1,000,000
Low	< 10	< 200,000

[a] Adapted from WEF (2012).

material in the sludge biofloc resulting in better solids liquid separation, and is the reason polymers are called flocculants. The form of the cationic PAM can be dry, liquid, or emulsion. The choice of form of polymer to use depends on (1) achieving desired dewatering performance; (2) cost effectiveness; (3) space for storage and handling of the neat product; (4) requirements for polymer make down, aging, and feeding equipment; and (5) safety considerations.

Polymers are proprietary chemicals and vary according to electrical charge, charge density, molecular weight and molecular structure (WEF, 2012). Polymers that are used in sludge or biosolids conditioning are usually cationic, high charge density, and high molecular weight. Degrees of polymer charge density and molecular weight are reported in Table 14–1 (WEF, 2012). The molecular structure can be straight, branched, or structured. In high shear dewatering devices like centrifuges, branched or structured polymers are used. Polymer characterization methods to determine charge density and molecular weight are not commonly practiced in wastewater treatment plants. However, for large treatment facilities that spend significant amounts of money on polymers, onsite characterization is recommended to ensure receiving consistent performing product (Abu-Orf et al., 2009).

Factors Affecting Polymer Conditioning

The selection of the type and dosage of polymers depends on the properties of the sludge and biosolids, mixing conditions between the chemicals and the sludge, and dewatering devices to be used. Important sludge and biosolids properties include source, solids concentration, electrical charge of sludge, biopolymer content, and rheology (WEF, 2012). The probable range of polymer doses required can be estimated based on the source of sludge or biosolids (e.g., primary sludge, waste activated sludge, and digested biosolids). Increasing the volatile solids destruction in aerobic and anaerobic digestion increases the polymer demand (Novak et al., 2004). Solids concentrations will affect the polymer dosage and its dispersion. The anionic electrical charge of the sludge or biosolids is an important factor in determining the cationic amount of polymer used. The biopolymer (protein and polysaccharides) content in the sludge and biosolids has a significant impact on polymer dose and dewatering. A linear relationship was found between the colloidal biopolymer content in the biosolids and the optimum polymer dose for dewatering and the dewatering potential (Novak et al., 2004). The method of dewatering affects the selection of the conditioning chemical because of the differences in mixing equipment used by various vendors and the characteristics of particular methods of dewatering.

Polymer Dosage Determination

The polymer dosage required for conditioning can be determined in the laboratory and need to be verified in full-scale trials. Laboratory tests used for selecting polymer dosage include the Buchner funnel test for the determination of specific resistance of filtration

Figure 14–1

Buchner funnel test apparatus used for the determination of the specific resistance of sludges and biosolids.

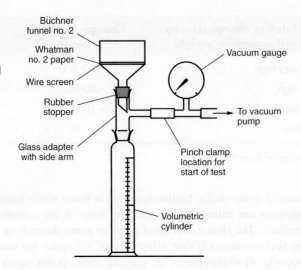

(see Fig. 14–1), capillary suction time test (CST) (Standard Methods 2012), and the standard jar test (ASTM, 2008). The Buchner funnel test is a method of testing sludge drainability or dewatering characteristics using various conditioning agents. The capillary suction test relies on gravity and the capillary suction of a piece of a standard thick filter paper to draw out water from a small sample of conditioned sludge or biosolids. The standard jar test, the easiest method to use, consists of testing standard volumes of sludge samples (usually 1 L) with different conditioner concentrations, followed by rapid mixing, flocculation, and settling using standard jar test apparatus. Other methods used successfully in laboratory and full scale for determination of optimum polymer dose for conditioning and dewatering include charge of the liquid stream (centrate or filtrate) as measured by streaming current detector (Abu-Orf and Dentel, 1997), viscosity of the liquid stream (Abu-Orf et al., 2003), and rheology of the conditioned sludge or biosolids (Abu-Orf and Ormeci, 2005). Laboratory or pilot-scale testing is recommended to determine the types of polymer and dose required, particularly for sludge and biosolids that may be difficult to dewater.

Mixing

The intimate and uniform mixing of sludge or biosolids and polymer is essential for proper conditioning. The intensity of the mixing must not break the flocculated material after it has formed, and the detention should be kept to a minimum so that sludge reaches the dewatering unit as soon after conditioning as possible. Mixing requirements vary depending on the dewatering method used. A separate mixing and flocculation tank is provided ahead of pressure filters; a separate flocculation tank may be provided for a belt filter press, or the polymer may be added directly to the sludge feed line of the belt filter press unit; and inline mixers are usually used with a centrifuge. It is generally desirable to provide at least two locations for the addition of conditioning chemicals.

In general, it has been observed that the type of sludge has the greatest impact on the quantity of chemical required. Difficult-to-dewater sludges that require larger doses of chemicals generally do not yield as dry a cake and have poorer quality of filtrate or centrate. Sludge types, listed in the approximate order of increasing conditioning chemical requirements, are as follows:

1. Untreated (raw) primary sludge
2. Untreated mixed primary and trickling filter sludge

   3. Untreated mixed primary and waste-activated sludge

   4. Anaerobically digested primary sludge

   5. Anaerobically digested mixed primary and waste activated sludge

   6. Aerobically digested mixed primary and waste activated sludge

   7. ATAD biosolids

   8. Aerobically digested waste activated sludge

   9. Untreated waste activated sludge

Typical levels of polymer addition for various types of sludges using belt-filter press centrifuge, rotary press, and screw press dewatering can be found in Sec. 14–2. Actual dosages in any given case may vary considerably from the indicated values. Polymer dosages will also vary greatly depending on the molecular weight, ionic strength, and activity levels of the polymers used. Manufacturers should be consulted for applicability and dosage information.

Conditioning Makeup and Feed

Chemicals are most easily applied and metered in the liquid form. Dissolving tanks are needed if the chemicals are received as dry powder. In most plants, these tanks should be large enough for at least one day's supply of chemicals and should be furnished in duplicate. In large plants, tankage sufficient for one shift is usually adequate. The tanks must be fabricated or lined with corrosion-resistant material. Polyvinyl chloride, polyethylene, and rubber are suitable materials for tank and pipe linings for acid solutions. Metering pumps must be corrosion-resistant. These pumps are generally of the positive displacement type with variable speed or variable stroke drives to control the flowrate.

14–2 DEWATERING

Dewatering is a physical unit operation used to separate the solid matter and water in the sludge or biosolids resulting in a high solids content stream called "cake" and a liquid stream. The liquid stream contains fine, low-density solids and a high concentration of nutrients when anaerobically digested sludge is dewatered and is typically returned to the wastewater treatment system or treated separately (sidestream treatment, see Chap. 15) to reduce nutrient loading to the main treatment system. For effective solids liquid separation, chemical conditioning is required. Increasing the solids content of sludge and biosolids is mainly practiced for one or more of the following reasons:

   1. The costs for trucking sludge and biosolids to the ultimate disposition site become substantially lower when the volume is reduced by dewatering.

   2. Dewatered sludge and biosolids are generally easier to handle than thickened or liquid sludge. In most cases, dewatered sludge may be shoveled, moved about with tractors fitted with buckets and blades, and transported by belt conveyors.

   3. Dewatering is required normally prior to the incineration of the sludge to increase the calorific value by removal of excess moisture.

   4. Dewatering is required before composting to reduce the requirements for supplemental bulking agents or amendments.

   5. Dewatering is required prior to thermal drying as it is cost effective to remove the water mechanically or by other means, compared to evaporating the water during drying.

   6. In some cases, removal of the excess moisture may be required to render biosolids odorless and nonputrescible.

   7. Dewatering is required prior to landfilling sludge and biosolids in monofills to reduce leachate production at the landfill site.

In mechanical dewatering devices, mechanically assisted physical means are used to dewater the sludge more quickly. Other dewatering devices rely on the application of electric and heat energy. Following a brief overview of dewatering, each of the major dewatering technologies is considered in this section.

Overview of Dewatering Technologies

To provide a perspective on the discussion of the individual dewatering technologies, it will be useful to first consider the fundamental principles of dewatering; some important factors in the selection of dewatering technologies, including the advantages and disadvantages of the various technologies; and the need for bench and pilot-sate testing.

Fundamental Principles of Dewatering. When considering the dewatering of sludge or biosolids, it is important to consider the various forms of water associated with the biosolids. In a relatively simplified overview, the four types of water associated with sludge, as proposed by Tsang and Vesiland (1990) and others and illustrated on Fig. 14–2(a), are (1) free water, (2) interstitial water, (3) surface water, and (4) bound water. Water not attached to particles that can be removed by gravitational forces, filtration, and centrifugation is known as *free water*. Water trapped within the sludge matrix is known as *interstitial water*. Water bound to the sludge particles by adsorption and adhesion is known as *surface water*. Intercellular and chemically bound water is known as *bound water*. The form of water that can be removed by the various dewatering technologies is illustrated on Fig. 14–2(b). Free and a portion of the interstitial water can be removed by physical means. Electro-dewatering can be used to remove interstitial and a portion of the bound water, depending on the adsorption forces. Thermal drying is required to remove the majority of bound water (Mahmoud et al., 2010).

Important Factors in Technology Selection. The selection of the dewatering device is determined by the type of sludge or biosolids to be dewatered, characteristics of the dewatered product, downstream processing, ultimate disposition, and the space available. The dewatering processes that are used commonly include centrifuges;

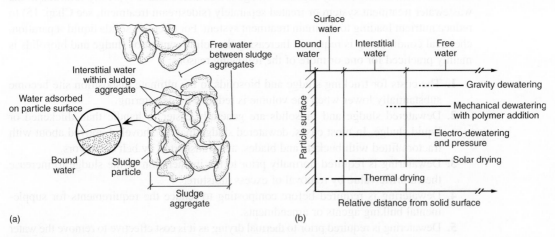

(a)   (b)

Figure 14–2

Overview of dewatering treated biosolids: (a) different forms of water associated with treated biosolids as proposed by Tsang and Vesiland (1990) and (b) potential operating regions for various technologies used for dewatering biosolids. The dashed portion represents the maximum potential achievable.

belt filter presses; rotary presses; screw presses; recessed-plate filter presses; electro-dewatering, a new innovative process; drying beds; and lagoons. Vacuum filtration, often used in the past for municipal sludge dewatering, has essentially been replaced by alternative mechanical dewatering equipment. The advantages and disadvantages of the various methods of sludge dewatering presented in this chapter are summarized in Table 14–2.

Table 14–2

Comparison of alternative methods for dewatering various types of sludge and biosolids[a]

Dewatering method	Advantages	Disadvantages
Solid bowl centrifuge	• Clean appearance, minimal odor problems, fast startup and shut down capabilities • Easy to install • Produces relatively dry sludge cake • Low capital cost-to-capacity ratio	• Scroll wear potentially a high maintenance problem • Requires grit removal and possibly sludge grinder in the feed stream • Skilled maintenance personnel required • Moderately high suspended solids content in centrate • Cannot observe dewatering zone to optimize/adjust performance
Belt filter press	• Low energy requirements • Relatively low capital and operating costs • Less complex mechanically and is easier to maintain • High pressure machines are capable of producing very dry cake • Minimal effort required for system shut down	• Hydraulically limited in throughput • Requires sludge grinder in feed stream • Very sensitive to incoming sludge feed characteristics • Short media life as compared to other devices using cloth media • Automatic operation generally not advised
Recessed plate filter press	• Highest cake solids concentration • Low suspended solids in filtrate • Simple operation • High solids capture rate	• Batch operation • High equipment cost • High labor cost • Special support structure requirements • Large floor area required for equipment • Skilled maintenance personnel required • Additional solids due to large chemical addition require disposal • Limitations on filter cloth life
Rotary Press	• Low speed 0.5 to 2.5 rev/min • Low noise < 68 dBA • Enclosed design contains odors and aerosols • Relatively low energy use drive motor ranges from 0.56 to 15 kW (0.75 to 20 hp) depending on size of unit • Overdosing polymer does not clog screen and hinder dewatering • Washwater only used during shut down of system • Low shearing force reduces odors in dewatered cake stockpile	• Relatively large footprint per unit volume of dewatering capacity • Capacity limitations will require multiple units for wastewater facilities treatment facilities > 19,000 m³/d (> 5 Mgal/d) • Cannot observe dewatering zone to optimize/adjust performance

(continued)

| **Table 14–2** (Continued) |

Dewatering method	Advantages	Disadvantages
Screw Press	• Low speed 0.3 to 1.5 rev/min • Low noise < 68 dBA • Enclosed design with hinged access doors contains odors and aerosols • Low energy use drive motor ranges from 0.37 to 3.7 kW (0.5 to 5 hp) depending on size of unit • Overdosing polymer does not clog screen and hinder dewatering • Low shearing force reduces odors in dewatered cake stockpile	• Capacity limitations will require multiple units for wastewater facilities treatment facilities 19,000 m³/d (> 5 Mgal/d) • Washwater required periodically throughout operating cycle • Cannot observe dewatering zone to optimize/adjust performance
Electro-dewatering	• Automatic operation • Good results for difficult sludge and biosolids • Mechanics are simple and easy to maintain • Odor improvement and pathogen kill on the sludge and biosolids • Some flexibility to incoming sludge characteristics • 3–5 times more energy efficient than dryers	• Batch operation • Moderate to high capital costs • Not particularly suited for larger plant 75,700 m³/d (20 Mgal/d) and above • Limited final dryness achievable (max 45 to 50 percent DS) • Difficult to predict performance without bench scale testing • New technology • Requires odor treatment for the process off gases • Require predewatering, range of feed between 10 and 25 percent • Operational cost sensitive to local electricity tariff
Sludge drying beds	• Lowest capital cost method where land is readily available • Small amount of operator attention and skill required • Low energy consumption • Little to no chemical consumption • Less sensitive to sludge variability • Higher solids content than mechanical methods	• Requires large area of land • Requires stabilized sludge • Design requires consideration of climatic effects • Sludge removal is labor intensive
Sludge lagoons	• Low energy consumption • No chemical consumption • Organic matter is further stabilized • Low capital cost where land is available • Least amount of skill required for operation	• Potential for odor and vector problems • Potential for groundwater pollution • More land intensive than mechanical methods • Appearance may be unsightly • Design requires consideration of climatic effects

ᵃ Adapted in part from U.S. EPA (2000).

For smaller plants where land availability is not a problem, drying beds or lagoons are generally used. Conversely, for larger facilities or facilities situated on constricted sites, mechanical dewatering devices are often chosen. Odor control is an important design consideration as the level of odor release varies based on the type of sludge and the mechanical equipment selected. High shear dewatering and conveyance equipment can increase odor release, especially from anaerobically digested sludge (WERF, 2003) (see also Sec. 16–4 in Chap. 16).

Need for Bench and Pilot-Scale Testing. When particular types of sludge or biosolids must be dewatered mechanically, it is often difficult or impossible to select the optimum dewatering device and polymer dosage without conducting bench-scale or pilot studies. Bench-scale testing is usually conducted by manufacturers of dewatering equipment to narrow down the types of polymers and doses to be used in pilot testing. Trailer-mounted, full-size equipment is available from several manufacturers for field-testing purposes. In selecting the type of mechanical dewatering to be used, it is important not to rely on published industry standard performance information and data. Side-by-side pilot testing should be undertaken to select the most cost-effective dewatering device, suitable for the treatment plant sludge or biosolids. With side-by-side testing, dewatering devices can be compared on the same sludge or biosolids, as it is well established that sludge characteristics vary seasonally, if not daily, at some plants. Pilot testing should be designed carefully to determine solids throughput, optimum polymer dose, percent cake solids, and percent solids recovery, which are important factors in comparing capital and operation costs of the various dewatering devices.

Centrifugation

The centrifugation process is used widely in industry for separating liquids of different density, thickening slurries, or removing solids. The process is applicable to the dewatering of wastewater sludges and has been used widely in both the United States and Europe. Solid-bowl centrifugal devices used for thickening sludge (see Sec. 13–6) may also be used for sludge and biosolids dewatering. In this section, standard solid-bowl and "high-solids" centrifuges are discussed. The high-solids centrifuge is a modification of the standard centrifuge.

Solid-Bowl Centrifuge. In the solid-bowl machine (see Figs. 14–3 and 14–4), biosolids or sludge is fed at a constant flowrate into the rotating bowl, where it separates into a dense cake containing the solids and a dilute liquid stream called "centrate." The centrate is returned to the wastewater treatment system or treated separately, if necessary.

Figure 14–3

View of typical solid bowl centrifuge dewatering installations.

(a)　　　　　　　　　　　　　　　　　　(b)

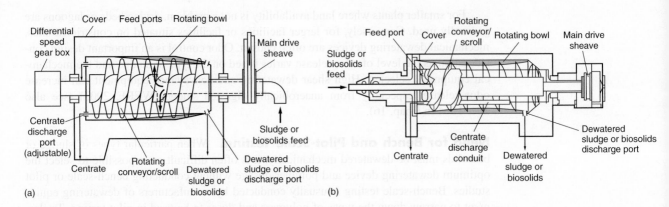

Figure 14–4

Schematic diagrams of two solid bowl centrifuge configurations for dewatering sludge: (a) countercurrent and (b) cocurrent.

The sludge cake is discharged from the bowl by a screw feeder into a hopper or onto a conveyor belt. Depending on the type of sludge or biosolids, solids concentration in the cake varies generally from 20 to 30 percent range. Cake concentrations above 25 percent are desirable for processing by incineration, drying, or by hauling to offsite processing, land application or disposition at a sanitary landfill.

Solid-bowl centrifuges are suitable generally for a variety of dewatering applications. Chemicals for conditioning are added to achieve the desired dewatering performance including cake solids and centrate quality and usually added to the feed line or within the bowl of the centrifuge. Dosage rates for conditioning with polymers vary from 1.0 to 25 g/ kg (2 to 50 lb/ton) of sludge (dry solids basis). Typical performance data for solid-bowl centrifuges are reported in Table 14–3.

Table 14–3

Typical dewatering performance for solid bowl centrifuges for various types of sludge and biosolids[a]

Type of feed	Feed solids, %	Cake solids, %	Polymer use, lb/ton dry TS	Polymer use, g/kg dry TS	Solids capture, %
Untreated sludge					
Primary	4–8	25–50	5–10	2.5–5	95+
Primary + WAS	3–5	25–35	5–16	2.5–8	95+
WAS	1–2	16–25	15–30	7.5–15	95+
Anaerobically digested biosolids					
Primary	2–5	25–40	8–12	4–6	95+
Primary + WAS	2–4	22–35	15–30	7.5–15	95+
Aerobically digested WAS	1–3	18–25	20–30	10–15	95+
ATAD biosolids	2–5	20–30	25–45	12.5–22.5	95+

[a] Adapted in part from U.S. EPA (2000) and feedback from centrifuge vendors.

High-Solids Centrifuge. High-solids (also called "high-torque") centrifuges are modified solid-bowl centrifuges that are designed to produce a dryer solids cake. These units have a slightly longer bowl length to accommodate a longer "beach" section, a lower differential bowl speed to increase residence time, and a modified scroll to provide a pressing action within the beach end of the unit. In some cases, the high-solids units are capable of achieving solids contents in excess of 30 percent in dewatering municipal wastewater sludges, although a higher polymer usage may be required.

Design Considerations. Centrifugation design is based on either helical or axial flow. Two basic designs of helical centrifuges are used: countercurrent flow and cocurrent flow [see Figs. 14–4 (a) and (b)]. The main difference in the designs is the location of the feed ports, removal of centrate, and internal flow patterns of the liquid and solid phases. In the countercurrent design, the feed slurry enters axially at the junction of the cylindrical conical section; solids travel to the conical end while the liquid phase moves in the opposite direction. In axial flow countercurrent centrifuges, flights are mounted on spokes in the path from the feed zone to the centrate dams. These flights reduce the flow velocity, resulting in enhanced separation. This design allows centrifuges to handle variations of the feed rate without needing to change the dams. As a result, axial flow is the most common type of centrifuge design (WEF, 2012). In the cocurrent design, the solid phase travels the full length of the bowl as does the liquid phase. Cocurrent centrifuge designs are seldom used because of maintenance problems (WEF, 2012).

Process Variables. Process variables affecting centrifuge performance, as measured by the sludge cake solids and TSS recovery, include feed flowrate, rotational speed, differential speed of the scroll, depth of the settling zone, conditioning dose, and the physicochemical properties of the suspended solids and suspending liquid. Important properties are particle size and shape, particle density, temperature, and liquid viscosity.

Selection of units for plant design is dependent on manufacturer's rating and performance data. Several manufacturers have portable pilot units, which can be used for field testing if sludges or biosolids are available. Unfortunately, biosolids or sludges from similar treatment processes but in different localities may differ markedly from each other. For this reason, pilot-plant tests as previously discussed should be conducted whenever possible before final design decisions are made.

Other Design Considerations. The area required for a centrifuge installation is less than that required for other dewatering devices of equal capacity, and the initial cost is lower. Higher power usage costs will partially offset the lower initial cost. Special consideration must also be given in providing sturdy foundations and soundproofing because of the vibration and noise that result from centrifuge operation. An adequate electric power source is required because large motors may be used.

Because centrifuges are enclosed, on-site odor generation may be better contained as compared to other types of dewatering systems. Ventilation of the centrifuge facility to control potential odors and moisture accumulation should be provided, however. On the other hand, cake solids produced from high solids centrifuges are more odorous compared to other dewatering devices, which could adversely affect beneficial use methods such as land application. Moreover, sudden increase in pathogen indicator organisms in cake solids after anaerobic digestion was observed after centrifugation (WERF, 2008a).

Combined Centrifuge Process. Technologies that combine centrifuge dewatering with flash air drying are available commercially. It is reported that these systems are able to process sludges or biosolids with total solids ranging from 2 to 7 percent and produce a

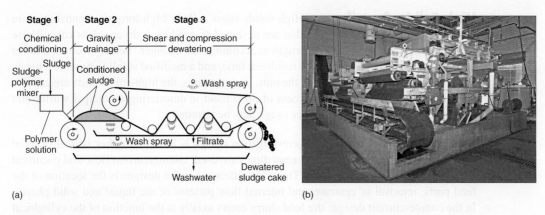

Stage 1 | **Stage 2** | **Stage 3**

Chemical conditioning | Gravity drainage | Shear and compression dewatering

Sludge-polymer mixer

Sludge

Conditioned sludge

Wash spray

Polymer solution

Wash spray   Filtrate

Washwater   Dewatered sludge cake

(a)   (b)

Figure 14–5

Belt press dewatering: (a) schematic of the three basic stages of belt press dewatering (b) view of a typical installation.

product with 60 to 90 percent dryness. Currently (2013), there is no installation in North America.

Belt-Filter Press

Belt-filter presses (BFPs) are continuous-feed dewatering devices that use the principles of gravity drainage and mechanically applied pressure to dewater chemically conditioned sludges or biosolids (see Fig. 14–5). The belt-filter press was introduced in the United States in the early 1970s and has become one of the predominant sludge dewatering devices. It has proven to be effective for almost all types of municipal wastewater sludge and biosolids.

Description.   In most types of BFPs, conditioned sludge or biosolids are first introduced on a gravity drainage section where it is allowed to thicken. In this section, a majority of the free water is removed by gravity. Following gravity drainage, pressure is applied in a low-pressure section, where the sludge is squeezed between opposing porous cloth belts. On some units, the low-pressure section is followed by a high-pressure section where the sludge is subjected to shearing forces as the belts pass through a series of rollers. The squeezing and shearing forces thus induce release of additional quantities of water from the sludge. Many vendors also offer systems with three belts which allow for independent control of the gravity section from the pressure section. The three belt system is ideal for dilute sludges. Improved dewatering can be achieved with this design as each zone can be optimized independently. The final dewatered sludge cake is removed from the belts by scraper blades.

System Operation and Performance.   A typical BFP system consists of sludge feed pumps, polymer feed equipment, belt filter press, sludge cake conveyor, and support systems (sludge feed pumps, washwater pumps, and compressed air). Most units do not use a conditioning tank. A schematic diagram of a typical two belt-filter press installation is shown on Fig. 14–6.

Many variables affect the performance of the BFP: sludge or biosolids characteristics, method and type of chemical conditioning, pressures developed, machine configuration (including gravity drainage), belt porosity, belt speed, and belt width. The BFP is sensitive to wide variations in sludge characteristics, resulting in improper conditioning and reduced dewatering efficiency. Blending facilities should be included in the system design where

Figure 14–6

Schematic diagram of a complete belt press dewatering system.

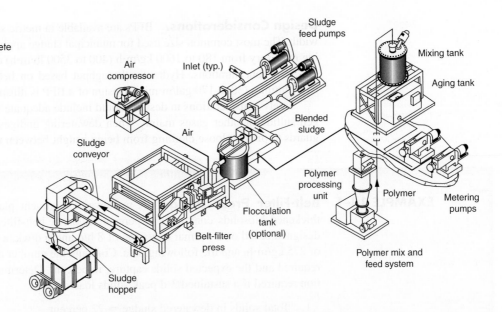

the sludge or biosolids characteristics are likely to vary widely. Based on actual operating experience, it has been found that the solids throughput is greater and the cake dryness is improved with higher solids concentrations in the feed sludge or biosolids. Typical BFP performance data for various types of sludge and biosolids are reported in Table 14–4.

Table 14–4

Typical dewatering performance for belt filter presses[a]

Type of feed	Dry feed solids, %	Loading per meter of belt length		Dry polymer[b], g/kg dry solids	Cake solids, %	
		L/min	kg/h		Typical	Range
Untreated sludge						
Primary	4–8	230–640	1130–1590	1.5–2.5	30	26–35
WAS	1–2	190–380	180–340	5–10	16	12–20
Primary plus WAS	3–5	150–450	340–820	3–5.5	23	15–25
Primary plus trickling filter	3–6	150–450	360–910	3–7	27	16–30
SBR	1–2	190–380	250–360	5–7.5	16	12–19
MBR	1–2	260–420	230–320	5.5–10	15	11–18
Anaerobically digested:						
Primary	2–5	230–610	680–910	2–5	28	24–35
WAS	2–3	110–340	230–410	4–10	20	13–23
Primary plus WAS	2–4	150–450	320–540	4–8.5	24	15–28
Aerobically digested WAS	1–3	150–340	250–410	6–10	18	12–22
ATAD	2–5	110–490	360–590	5–12.5	19	12–22

[a] Based on feedback from belt filter press vendors.

[b] Polymer needs based on high molecular weight polymer (100 percent strength, dry basis).

Design Considerations. BFPs are available in metric sizes from 0.5 to 3.0 m in belt width. The most common size used for municipal sludge applications is 2.0 m. Solids loading rates vary from 180 to 1600 kg/m·h (400 to 3500 lb/m·h) depending on the sludge type and feed concentrations. Hydraulic throughput based on belt width ranges from 110 to 640 L/m·min (30 to 170 gal/m·min). Design of a BFP is illustrated in Example 14–1.

Safety considerations in design should include adequate ventilation to remove hydrogen sulfide or other gases mainly when dewatering undigested sludges, and equipment guards to prevent loose clothing from being caught between the rollers.

EXAMPLE 14–1 **Belt-Filter Press Design** A wastewater treatment plant produces 75,000 L/d of thickened biosolids containing 3 percent solids. A belt-filter press installation is to be designed based on a normal operation of 8 h/d and 5 d/wk, a belt-filter press loading rate of 275 kg/m·h, and the following data. Compute the number and size of belt-filter presses required and the expected solids capture, in percent. Determine the daily hours of operation required if a sustained 3-d peak solids load occurs.

1. Total solids in dewatered sludge = 22 percent.
2. Total suspended solids concentration in filtrate = 900 mg/L = 0.09 percent.
3. Washwater flowrate = 90 L/min per m of belt width.
4. Specific gravities of sludge feed, dewatered cake, and filtrate are 1.02, 1.07, and 1.01, respectively.

Solution

1. Compute average weekly sludge production rate.

   Wet biosolids = (75,000 L/d)(7 d/wk)(10³ g/1 L)(1 kg/10³ g)(1.02)
   = 535,500 kg/wk
   Dry solids = (535,500 kg/wk)(0.03) = 16,065 kg/wk

2. Compute daily and hourly dry solids-processing requirements.

   $$\text{Daily rate} = (16{,}065 \text{ kg/wk})\left(\frac{1 \text{ wk}}{5 \text{ operating d}}\right)$$
   $$= 3213 \text{ kg/d}$$

   $$\text{Hourly rate} = \frac{(3213 \text{ kg/d})}{(8 \text{ h per operating d})}$$
   $$= 401.6 \text{ kg/h (8 h operating d)}$$

3. Compute belt-filter press size 3

   $$\text{Belt width} = \frac{(401.6 \text{ kg/h})}{(275 \text{ kg/m·h})} = 1.46 \text{ m}$$

   Use one 1.5-m belt-filter press and provide one identical size for standby.
4. Compute filtrate flowrate by developing solids balance and flow balance equations.
   a. Develop daily solids balance equation.

      Solids in sludge feed = solids in sludge cake + solids in filtrate
      3213 kg/d = (S kg/d)(0.22) + (F kg/d) × (0.0009)
      3213 kg/d = (0.22)(S) + (0.0009)(F)

      where S = sludge cake flowrate (wet), kg/d
      F = filtrate flowrate (wet), kg/d

b. Develop mass flowrate equation.

Sludge flowrate + washwater flowrate = filtrate flowrate + cake flowrate

Daily sludge flowrate = (535,500 kg/wk)/(5 d/wk) = 107,100 kg/d

Washwater flowrate = (90 L/min·m)(1.5 m)(60 min/h)(8 h/d)(1 kg/L)(1.0)
$$= 64,800 \text{ kg/d}$$

107,100 kg/d + 64,800 kg/d = 171,900 kg/d = $F + S$

c. Solve the mass balance and flowrate equations simultaneously.

First solve S in terms of F per the flowrate equation in 4.b.

$$S = 171,900 \text{ kg/d} - F$$

Next solve for F from the solids balance equation in 4.a.

$$3213 \text{ kg/d} = 0.22 \ (171,900 \text{ kg/d} - F) + 0.0009(F)$$
$$= 37,818 \text{ kg/d} - 0.2191(F)$$
$$F = 157,942 \text{ kg/d}$$
$$= (157,942 \text{ kg/d})/(1 \text{ kg/L})/(1.01) = 159,521 \text{ L/d}$$

5. Determine solids capture.

$$\text{Solid capture} = \frac{\text{solids in feed} - \text{solids in filtrate}}{\text{solids in feed}} \times 100$$

$$= \frac{[(3213 \text{ kg/d}) - (157,942 \text{ kg/d})(0.0009)]}{(3213 \text{ kg/d})} \times 100$$

$$= 95.6\%$$

6. Determine operating requirements for sustained peak biosolids load.

a. Determine peak 3-d load.

From Fig. 3–14 (b), the ratio of peak to average mass loading for 3 consecutive days is 2. The peak load is (75,000 L/d) (2) = 150,000 L/d.

b. Determine daily operating time requirements, neglecting sludge in storage.

Dry solids per day = (150,000 L/d)(1 kg/L)(1.02)(0.03)
$$= 4590 \text{ kg/d}$$

$$\text{Operating time} = \frac{(4590 \text{ kg/d})}{(275 \text{ kg/m·h})(1.5 \text{ m})} = 11.1 \text{ h/d}$$

The operating time can be accomplished by running the standby belt-filter press in addition to the duty press, or operating the duty press for an extended shift.

Comment   The value of sludge storage is important in dewatering applications because of the ability to schedule operations to suit labor availability most efficiently. Scheduling sludge dewatering operations during the day shift is also desirable if sludge has to be hauled off-site.

Rotary Press

The rotary press as a dewatering device for municipal sludges and biosolids has been adapted since 1994 with the first installation at the City of Montreal, Canada. Similar to centrifuges and BFPs, the rotary press feed solids are typically conditioned with cationic polymer that is injected into the feed upstream of the inlet to the dewatering unit. A typical process flow diagram for a rotary press system is shown on Fig. 14–7(a). Because rotary press systems are totally enclosed, they provide enhanced safety, odor containment, and relatively low noise levels compared with some other types of sludge or biosolids dewatering systems discussed in this textbook.

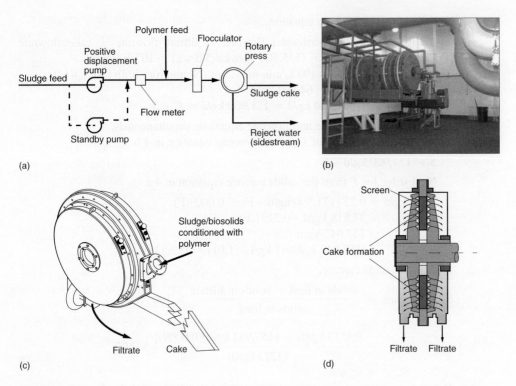

Figure 14–7

Rotary press for sludge dewatering: (a) schematic process flow diagram, (b) view of typical installation (courtesy of Scarborough Sanitart District, Scarborough ME), (c) view of a rotary press unit, and (d) cross-section through press (courtesy of Fournier Industries, Inc.).

Description. The rotary press is a slow-speed, enclosed, modular unit. The sludge or biosolids are fed into the unit at a relatively low pressure into the space between the two parallel filter screens. The sludge or biosolids are typically fed into a cylindrical coated carbon steel containment vessel that has a rectangular cross section. The solids within each module are retained between two parallel rotating stainless steel screens. The flocculated slurry material advances forward within the rotating channel that is formed between the parallel screens, and the filtrate passes out through sides of the screens. The sludge travelling around the channel (between the rotating screens) continues to be dewatered. The frictional force of the slow-moving screen assembly and the adjustable outlet restriction creates back pressure that forces additional filtrate out through the screen. The solids retained between the screens are dewatered, effectively forming an extrusion of relatively dry dewatered cake. Washwater is required intermittently to clean the screens and flush the solids from the unit during shutdown. A cutaway view of the typical rotary press is shown on Fig. 14–7(d).

Design Considerations. The design of the rotary press dewatering system is affected by a number of factors. These include quantity of sludge or biosolids, feed concentration, characteristics of sludge (digested or undigested, primary sludges or secondary sludges or both), safety, allowable noise levels, allowable odor levels, desired operating schedule, level of automation, space availability, reliability and redundancy requirements, washwater quality and availability, filtrate quality and treatment requirements, manufacturers' local service capabilities, and budgetary constraints.

The quantity of sludge or biosolids should be established at the beginning of the design process. The quantities should be developed for both current operation and projected future conditions. The characteristics of the sludges or biosolids that will be dewatered are an important consideration. The digested biosolids and raw secondary sludges are typically the most difficult to dewater effectively. Conversely, the presence of primary sludges tends to aid the dewatering process. The operating schedule at the facility, level of automation, reliability/redundancy requirements, and physical space availability need consideration when selecting the system and number of units. Washwater is used intermittently to flush the rotating screen assemblies during the shutdown sequence. The amount of washwater needed is relatively small, typically 190 L/min (50 gal/min) per channel. The time required for the flushing sequence is approximately 5 min per channel per d. All of the channels are normally flushed simultaneously, although sequential flushing can be used if the washwater supply has a capacity limitation. The units typically are offered with a single-channel, a dual-channel, a four-channel and a six-channel configuration. Special three- and five-channel units can also be provided, but the inlet configuration can create a flow imbalance due to the unequal number of channels on each side of the inlet connection. The screen diameter ranges from 460 to 1220 mm (18 to 48 in.) depending on the selected manufacturer. The effective dewatering area for various rotary press configurations is reported in Table 14–5.

The maximum hydraulic loading limit for a rotary press is approximately 8.5 m/h (3.5 gal/min·ft^2) at a total solids feed concentration of 3 percent. The maximum solids' loading is approximately 244 – 254 kg/h·m^2 (50 – 52 lb/h·ft^2). To optimize cake dryness and minimize polymer usage, rotary presses are normally operated below the maximum hydraulic and solids loading criteria. The typical average operating hydraulic loading rate for a rotary press system is approximately 2.4 m/h (1.0 gal/min·ft^2) at a total solids feed concentration of 4 percent and approximately 3.7 m/h (1.5 gal/min·ft^2) at a total solids feed concentration of 2 percent.

As discussed earlier, the types of sludges and biosolids to be dewatered have a significant impact on the actual allowable hydraulic loading, solids loading, polymer use, and dewatered cake moisture content. The characteristics and composition of sludge and biosolids vary from facility to facility so on-site pilot testing is recommended if strict performance and capacity criteria will be specified for the rotary press equipment.

System Operation and Performance. A typical rotary press system consists of a sludge feed pump, flow metering, polymer feed equipment, conditioning system (either inline or tank designs can be used), the rotary press, a dewatered cake conveyor (if direct discharge to a container or truck is not feasible), compressed air for valve control functions,

Table 14–5 Rotary press dewatering area[a]

| No. of Channels | Effective dewatering area, m^2 | | | |
	460 mm diameter screens	610 mm diameter screens	915 mm diameter screens	1220 mm diameter screens
1	0.23	0.40	0.96–1.00	1.75
2	NA	0.79	1.91–1.20	3.49
4	NA	NA	3.84–4.00	7.00
6	NA	NA	6.00	NA

[a] Based on feedback from rotary press vendors.

Table 14–6
Typical dewatering performance of rotary press[a]

Type of feed	Polymer use		Cake solids, % TS	Solids capture, %
	lb/ton dry TS	g/kg dry TS		
Untreated sludge				
Primary	4–12	2–6	28–45	95+
Primary plus WAS	15–20	7.5–10	20–32	92–98
WAS	20–35	12.5–17.5	13–18	90–95
Anaerobically digested biosolids				
Primary	15–20	7.5–10	22–32	90–95
Primary plus WAS	20–30	10–15	18–25	90–95
WAS	20–35	10– 17.5	12–17	85–90
Aerobically digested WAS	17–25	8.5–17.5	28–45	90–95

[a] Based on feedback from rotary press vendors.

electrical power for the drive components, and control/signal wiring for remote monitoring or operational control. A typical rotary press installation is shown on Fig. 14–7(b).

Rotary press process variables that can be adjusted to optimize performance or minimize operational costs include feed pump speed and flow, polymer concentration, polymer feed pump speed and flow, polymer mixing intensity, screen rotational speed, cake discharge outlet pressure, and washwater flushing frequency and duration. The performance of the unit is affected by each of these variables. In optimizing the performance of a rotary press, it is important to change one variable at a time so that the effect of each variable can be evaluated fully. The typical performance for a rotary press system varies from facility to facility and is affected by the type of solids feed, the process variables listed above, and the physical condition of the equipment and controls. Typical operating performance data for a rotary press system is reported in Table 14–6.

Screw Press

The screw press was adapted from industrial applications in 1990 for dewatering wastewater sludges or digested biosolids. The screw press feed solids are typically conditioned with cationic polymer that is injected into the feed upstream of the inlet to the dewatering unit. A typical process flow diagram for a screw press system is shown on Fig. 14–8(a).

Description. The screw press is a slow-speed, enclosed cylindrical unit. The sludges or biosolids are fed into the unit at a relatively low pressure into a stationary wedge wire screening basket with a rotating screw assembly that transfers the sludges or biosolids through the wedge wire screening basket. The flocculated material advances forward up along the rotating screw assembly. The filtrate passes out through the bottom and sides of the wedge wire screen. The sludges travelling along the screw continue to be dewatered. The frictional force of the slow moving screw assembly and the adjustable outlet restriction

Figure 14–8

Screw press for sludge dewatering: (a) schematic process flow diagram (courtesy of PW Tech, Inc.), (b) cutaway view (adapted from Huber Technology Inc.) and (c) view of inclined screw press.

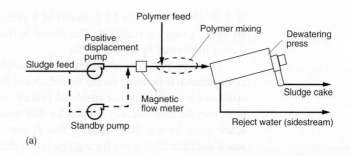

(a)

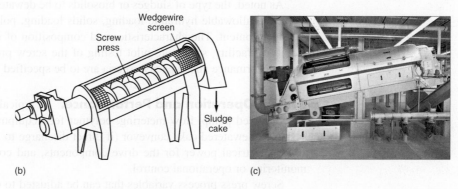

(b)   (c)

creates back pressure that forces additional filtrate out through the screen near the outlet end of the unit. The dewatered sludges form an extrusion of relatively dry dewatered cake. Washwater is required intermittently to clean the wedge wire screening assembly and flush the solids from the unit throughout the normal operating cycle. Units are available in either horizontal or inclined configuration depending on the manufacturers' preference. A cut-away view of a typical screw press is illustrated on Fig. 14–8(b). A view of a typical screen press is presented on Fig. 14–8(c).

Design Considerations. The design of the screw press dewatering system will be affected by a number of factors. These include quantity and feed concentration of sludges, volatile solids (VS) content, sludge characteristics (digested or undigested, primary sludges or secondary sludges or both), safety considerations, allowable noise and odor levels, facility operating schedule, extent of automation, building space availability, redundancy and reliability requirements, quality and availability of washwater, required filtrate quality and treatment, local service capabilities of the manufacturer, and financial constraints.

Similar to other types of dewatering systems the quantities of sludges or biosolids need to be established at the beginning of the design process. The quantities should be developed for both current operation, and projected future conditions. The characteristics of the material to be dewatered are an important consideration. Digested biosolids and raw secondary sludges are typically more difficult to effectively dewater than primary sludges. The screw press systems are low speed, totally enclosed units that provide enhanced safety, odor containment, and relatively low noise levels compared with some other types of dewatering systems discussed in this textbook. The schedule of operation, automation requirements, redundancy and reliability considerations, number of units, and building space availability need due consideration. The screw press intermittently utilizes washwater to flush the stationary wedge wire screen assembly regularly throughout the normal operating cycle. The amount of washwater needed is relatively small. Typically a flowrate of 7 to 45 L/min (2 to 12 gal/min) at a pressure between 2.8 to 5.5 bar-gauges (40 to 80 lb/in.²-gauge) is required. The screen is flushed for approximately 15 s every 10 min during its normal operating cycle.

The units typically are offered with a single screw, dual screw, inclined, or horizontal configuration depending on the manufacturer that is selected. Units are available in various sizes and a single unit is capable of processing up to 500 kg/h (1100 lb/h) although the units may be hydraulically limited for thin waste activated sludges (< 1 percent TS). Dual screw units have a flow capacity that is approximately twice the capacity of the single screw configuration units for a given screw diameter and screen length.

As noted, the type of sludges or biosolids to be dewatered has a significant impact on the actual allowable hydraulic loading, solids loading, polymer use, and dewatered cake moisture content. The characteristics and composition of sludges or biosolids vary from facility to facility, so on-site pilot testing of the screw press system is recommended if strict performance and capacity criteria are to be specified for the screw press equipment.

System Operation and Performance.   A typical screw press system consists of a sludge feed pump, flow metering, polymer feed equipment, inline solids conditioning system, a dewatered cake conveyor (if direct discharge to a container or truck is not feasible), electrical power for the drive components, and control/signal wiring for remote monitoring or operational control.

Screw press process variables that can be adjusted to optimize performance or minimize operational costs include feed pump speed and flow, polymer concentration, polymer feed pump speed and flow, polymer mixing intensity, screw rotational speed, cake discharge outlet pressure, and washwater flushing frequency and duration. The typical performance of for a screw press system varies from facility to facility and is impacted by the type of feed, the process variables listed above, and the physical condition of the equipment and controls. Typical operating performance data for a screw press system are presented in Table 14–7.

Table 14–7
Typical dewatering performance of screw press[a]

| Type of feed | Process parameter | | | |
| | Polymer use | | Cake solids, % TS | Solids capture, % |
	lb/ton dry TS	g/kg dry TS		
Untreated sludge				
Primary	8–20	4–10	30–40	90+
Primary plus WAS	10–20	5–10	25–35	90+
WAS	17–22	8.5–11	15–22	88–95
Anaerobically digested biosolids				
Primary	20–35	10–17.5	22–28	90+
Primary plus WAS	20–35	10–17.5	17–25	90+
WAS	17–35	8.5–17.5	15–25	88–95
Aerobically digested WAS	17–25	8.5–17.5	15–20	88–95

[a] Based on feedback from screw press vendors.

Filter Presses

In a filter press, dewatering is achieved by forcing the water from the sludge or biosolids under high pressure. Advantages and disadvantages of filter presses are included in Table 14–2. When requiring cake solids contents greater than 35 percent on a routine basis, use of filter press is often dictated as other mechanical dewatering devices cannot achieve this high solids content consistently.

Various types of filter presses have been used to dewater sludges or biosolids. The two types used most commonly are the fixed-volume and variable-volume recessed-plate filter presses.

Fixed-Volume, Recessed-Plate Filter Press.

The fixed-volume, recessed-plate filter press consists of a series of rectangular plates, recessed on both sides, that are supported face to face in a vertical position on a frame with a fixed and movable head [see Fig. 14–9(a)]. A filter cloth is hung or fitted over each plate. The plates are held together with sufficient force to seal them to withstand the pressure applied during the filtration process. Hydraulic rams or powered screws are used to hold the plates together.

In operation, chemically conditioned sludge or biosolids is pumped into the space between the plates, and pressure of 700 to 2100 kPa (100 to 300 $lb_f/in.^2$) is applied and maintained for 1 to 3 h, forcing the liquid through the filter cloth and plate outlet ports. The plates are then separated and the cake is removed. The filtrate normally is returned to the influent of the treatment plant. The cake thickness varies from about 25 to 38 mm (1 to 1.5 in.), and the moisture content varies from 45 to 70 percent. The filtration cycle time varies from 2 to 5 h and includes the time required to (1) fill the press, (2) maintain the press under pressure, (3) open the press, (4) wash and discharge the cake, and

Figure 14–9

Typical fixed-volume, recessed plate filter press used for dewatering sludge: (a) schematic of filter press, (b) and (c) views of a typical installations, and (d) cross section through a variable-volume recessed plate filter press.

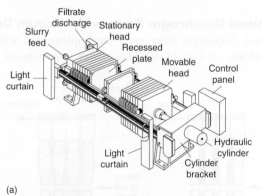

(a)

(b)

(c)

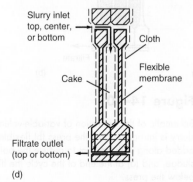

(d)

(5) close the press. Depending on the degree of automation incorporated into the machine, operator attention must be devoted to the filter press during feed, discharge, and wash intervals.

Variable-Volume, Recessed-Plate Filter Press.

Another type of filter press used for wastewater sludge dewatering is the variable-volume recessed-plate filter press, commonly called the "diaphragm press." This type of filter press is similar to the fixed-volume press except that a rubber diaphragm is placed behind the filter media, as shown on Fig. 14–9(d). The rubber diaphragm expands to achieve the final squeeze pressure, thus reducing the cake volume during the compression step. Generally about 10 to 20 min are required to fill the press and 15 to 30 min of constant pressure are required to dewater the cake to the desired solids content. Variable-volume presses are generally designed for 690 to 860 kN/m^2 (100 to 125 lb$_f$/in.^{2}) for the initial stage of dewatering followed by 1380 to 2070 kN/m^2 (200 to 300 lb$_f$/in.^{2}) for final compression. Variable-volume presses can handle a wide variety of sludges and biosolids with good performance results but require considerable maintenance (WEF, 2010).

Design Considerations.

Several operating and maintenance problems have been identified for recessed-plate filter presses ranging from difficulties in the chemical feed and sludge-conditioning system to excessive downtime for equipment maintenance. Features that should be considered in the design of a filter press installation include (1) adequate ventilation in the dewatering room (6 to 12 air changes per hour are recommended depending on the ambient temperature), (2) high-pressure washing systems, (3) an acid wash circulation system to remove calcium scale when lime is used, (4) a grinder ahead of the conditioning tank, (5) cake breakers or shredders following the filter press (particularly if the dewatered sludge is incinerated), and (6) equipment to facilitate removal and maintenance of the plates. Other design criteria are can be found in WEF, (2010).

Combined Diaphragm Press with Vacuum Drying.

Two components to the traditional diaphragm filter press are implemented with this combined process shown schematically on Fig. 14–10. First sludge is introduced into the press, and internal filter

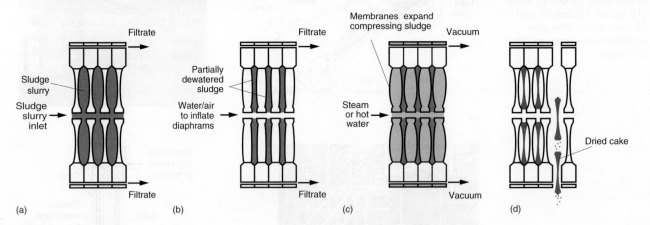

Figure 14–10

Schematic of the operation of variable-volume vacuum assisted recessed-plate filter press: (a) sludge slurry is introduced into the press (b) flexible membrane is expanded with air or water, (c) steam is added along with vacuum to reduce the boiling point of the water and, thus, further dewater the sludge, and (d) at the end of the cycle the filter press is open, and dried sludge is collected in the bin below the press.

plate diaphragms are inflated with air or water to squeeze out free water Next, hot water or steam is introduced to raise the temperature of the sludge while a vacuum is applied to lower the boiling point of water. Vaporized water in sludge is drawn out through the vacuum. At the end of the cycle, the press is opened to drop the dried sludge onto a receiving bin. There are several installations of this technology in North America and Europe (ca. 2013).

Electro-Dewatering

Electric field-assisted dewatering, also called electro-dewatering, is used to dewater municipal sludges and biosolids, industrial residuals, as well as less conventional biomass. The process can be used to dewater raw sludges or digested biosolids. The basic principle of operation, the commercial application of the principle, design considerations, and some performance data are presented and discussed below.

Description. In practice, because the bonding forces and the particle size of the sludge or biosolids, bound water cannot be removed readily by mechanical means. Electro-dewatering involves the application of a direct voltage to sludge or biosolids placed between two electrodes as illustrated on Fig. 14–11. Negatively charged sludge or biosolids move towards or gather at the positive electrode (anode). Positively charged water molecules move towards or gather at the negative electrode (cathode). At the negative electrode water moves through the filter cloth which covers the electrode. The filter cloth on the negative electrode does not clog, a common problem in mechanical filtration, as the sludge or biosolids particles are repelled by the negative electrode (Yoshida, 1993). Pressure is applied to the material to be dewatered to further accelerate the process and to allow the applied DC current to move more uniformly from the anode to the cathode [see Fig. 14–11(a)]. The advanced dewatering capability of electro-dewatering process as compared to other and other dewatering technologies was illustrated previously on Fig. 14–2(b).

Commercial Implementation of the Electro-Dewatering Process. In the most common commercial configurations encountered, a typical electro-dewatering system consists of a feeding module, the electro-dewatering unit, a high pressure wash system, a rectifier, a dewatered cake conveyor (if direct discharge to a container or truck is not feasible), compressed air for valve control functions, electrical power for the drive components and control/signal wiring for remote monitoring or operational control. A typical schematic of an electro-dewatering system is shown on Fig. 14–12(a).

Referring to Fig. 14–12(a), sludges or biosolids are fed in a hopper built as part of a feeding module. At the beginning of each treatment cycle, a conveyor system is actuated and

Figure 14–11

Definition sketch for the operation of electro-dewatering process. (Adapted from Mahmud et al., 2010.)

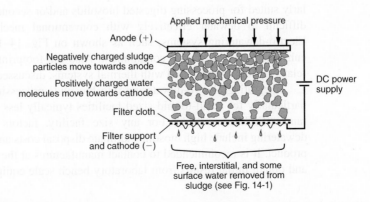

Applied mechanical pressure

Anode (+)

Negatively charged sludge particles move towards anode

Positively charged water molecules move towards cathode

Filter cloth

Filter support and cathode (−)

DC power supply

Free, interstitial, and some surface water removed from sludge (see Fig. 14-1)

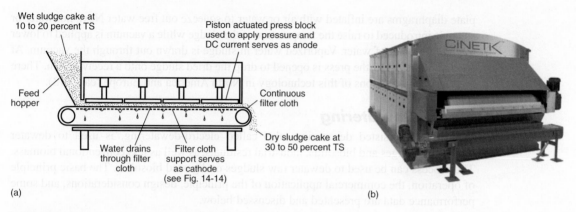

Figure 14-12

Electro-dewatering apparatus: (a) schematic diagram of operation and (b) view of discharge side of a linear electro-dewatering machine (courtesy of Ovivo, Inc.).

the feeding module extrudes a thin and uniform layer of the material to be dewatered on the filtering belt, to a predetermined thickness. Once the newly formed cake is moved into the treatment area as shown on Fig. 14–12 (a), power-blocks move down on the cake and apply a predetermined pressure. The applied pressure also allows DC current (generated by a rectifier) to flow through the cake and the filtering belt, between the anodes and the cathodes. The DC current is applied at predetermined levels for a controlled period of time until the required cake dryness is achieved, after which the power-blocks are lifted back to their upper position and the treated cake is discharged from the system. The filtering belt is cleaned using a high pressure wash system (filtrated process water or potable water) during its displacement. The high pressure wash system is also used daily to clean the cathodes and the equipment, during programmed automatic wash sequences and before a shutdown sequence. Machine controls and instrumentation ensure management of operating parameters such as pressure, voltage, current, treatment time, belt speed, and automatic wash cycles.

Design Considerations. Important design considerations for the electro-dewatering process are summarized in Table 14–8 and discussed briefly below. Specific design parameters are presented in Table 14–9. The quantity of sludge or biosolids should be established at the beginning of the design process. The quantities should be developed for both current operation and projected, future conditions. The characteristics of the sludge or biosolids that will be dewatered are the most important consideration. Electro-dewatering is particularly suited for processing digested biosolids and/or secondary sludge, typically the most difficult to dewater effectively with conventional mechanical dewatering processes. Electro-dewatering systems, such as shown on Fig. 14–12(b), are totally enclosed for enhanced safety and odor containment. Also the footprint is small and energy usage is relatively low as compared with thermal systems, discussed later in this chapter.

Integrating electro-dewatering into biosolids processing is worthy of consideration, particularly for small-to-mid-sized facilities typically less than 57,000 m³/d (15 Mgal/d) range (Eschborn, 2011). For any size facility, factors favoring integrating electro-dewatering include high biosolids ultimate disposal costs and a desire to produce a Class A product. It is recommended to contact manufactures at the early stages of design process and obtain initial results from laboratory bench scale equipment. Piloting to evaluate the

Table 14-8

Design considerations for electro-dewatering[a]

Item	Comment
Quantity of sludge and biosolids	The quantity of solids should be established at the beginning of the design process. The quantities should be developed for both current operation and projected, future conditions.
Characteristics of the feed sludge or biosolids	Important feed characteristics include conductivity, pH, particles sizes, concentration, ionic composition and polymer used. Although unit capacity can sometimes be reduced with a wetter feed solids, better treatment response is commonly observed with feeds at 12 to 20 percent TS.
Type of sludge or biosolids	Electro-dewatering is particularly suited for processing digested biosolids and/or secondary sludge which are difficult to dewater effectively by conventional mechanical means.
Sludge or biosolids pretreatment	Conventional mechanical dewatering equipment such as belt filter presses, centrifuges, screw presses or rotary presses are used upstream of the electro-dewatering.
Final dryness desired	Depends on the ultimate use of the dewatered material. Cake solids are typically in the range between 25 and 50 percent TS.
Expected volume reduction	Typically, volume reduction will vary from 50 to 75 percent.
Filtrate characteristics and treatment requirements	The filtrate contains relatively high concentrations of organics (BOD_5 and COD), suspended solids (TSS), ammonia and organic nitrogen (TKN). Although the flow of this filtrate is small, these constituents can have an impact on downstream treatment processes, so they should be characterized and considered during the design of electro-dewatering.
Washwater quality and availability	Filtered process water or potable water is used to avoid clogging the filter. The amount of water needed is relatively small.
Electricity cost	Three to five fold reduction in energy usage compared to thermal drying.
Plant related issues in selection of system and number of units	Safety considerations, allowable odor levels, facility operating schedule, level of automation, space availability, reliability and redundancy requirements, manufacturers local service capabilities, and budgetary constraints.

[a] Courtesy of OVIVO.

Table 14-9

Design parameters for various linear electro-dewatering models[a]

Item	Unit	Effective dewatering area, m²		
		4	**8**	**16**
Effective dewatering area	m²	4	8	16
	ft²	43	86	172
Footprint	m²	11.6	19.5	27.9
	ft²	125	210	300
Capacity at inlet	kg/h	270–600	545–1180	1090–2360
	lb/h	600–1320	1200–2600	2400–5200
Washwater	L/min	15.9	18.9	22.7
	gal/min	4.2	5	6

[a] Courtesy of OVIVO.

Table 14–10

Typical linear electro-dewatering performance for total solids and energy usage

Type of feed	Cake solids, % TS		Energy usage[a]	
	Inlet	Outlet	kWh/ton	kWh/kg
Untreated sludge				
Primary	22–24	29–49	110–260	0.12–0.29
WAS	13–17	28–43	150–270	0.17–0.30
	25	33–38	210–310	0.23–0.34
	16–20	32–43	230–310	0.25–0.34
Anaerobically digested biosolids	12–18	30–46	190–280	0.21–0.31
	20–23	32–48	165–260	0.18–0.29
Aerobically digested WAS	16–20	32–43	230–310	0.25–0.34

[a] Wet basis.

compatibility of the technology with the sludge or biosolids and to determine time required to achieve the desired cake dewatering along with electricity consumption is highly recommended, once satisfactory laboratory results are obtained.

Process Performance Data. Process variables that can be adjusted to optimize performance or minimize operational costs include cycle time (time of the batch treatment), applied voltage, applied current intensity, applied pressure, thickness of the formed cake (feed), sludge conductivity and ionic composition, polymer type and concentration (applied upstream, at the mechanical dewatering step). Typical operating performance data for linear electro-dewatering, which will vary facility to facility, is presented in Table 14–10. Thermal drying requires around 617–1200 kWh/m³ of water removed (Gazbar et al., 1994; Mujumdar, 2007) whereas application of linear electro-dewatering often results in a 3 to 5 fold reduction in energy usage. In areas where electricity costs are high, economic evaluation to verify the cost benefit from obtaining the desired cake solids is required.

Sludge Drying Beds

Drying beds used to be the most widely used method of sludge dewatering in the United States. Sludge drying beds are typically used to dewater digested biosolids and settled sludge from plants using the extended aeration activated-sludge treatment process without prethickening. After drying, the dried material is removed and either disposed of in a landfill or used as a soil conditioner. The advantages and disadvantages of drying beds are summarized in Table 14–11. Conventional sand drying beds are the most commonly used sludge drying beds. Other types of drying beds include paved drying beds, wedge-wire drying beds, and vacuum-assisted drying beds. Because paved and wedge-wire drying beds are not used commonly, they are not covered in this textbook. Vacuum-assisted sludge

Table 14–11

Advantages and disadvantages of drying beds

Advantages	Disadvantages
1. Low capital and operating cost alternative	1. Requires a large footprint
2. Simple operation with minimal operator attention required	2. Performance highly susceptible to weather conditions
3. High-solids content in the dried product	3. Removing sludge can be labor-intensive high potential for odors
	4. Can attract insects

drying beds were used in the past but are no longer commercially available and are not included in this text (WEF, 2012). Because conventional sand drying beds are used most extensively, more detailed discussion is provided for these units.

Conventional Sand Drying Beds. Conventional sand drying beds are generally used for small- and medium-sized communities. For cities with populations over 20,000, consideration should be given to alternative means of sludge dewatering. In larger municipalities, the initial cost, the cost of removing the sludge and replacing sand, and the large area requirements generally preclude the use of sand drying beds.

In a typical sand drying bed, sludge is placed on the bed in a 200 to 300 mm (8 to 12 in.) layer and allowed to dry. Sludge dewaters by gravity drainage through the sludge mass and supporting sand and by evaporation from the surface exposed to the air (see Fig. 14–13). Most of the water leaves the sludge by drainage; thus the provision of an adequate underdrainage system is essential. Drying beds are equipped with lateral drainage lines (perforated plastic pipe or vitrified clay pipe laid with open joints), sloped at a minimum of 1 percent and spaced 2.5 to 6 m (8 to 20 ft) apart. The drainage lines

Figure 14–13

Typical conventional sand drying bed: (a) plan and (b) section. Insert – view of sludge drying beds with sludge in various stages of dryness.

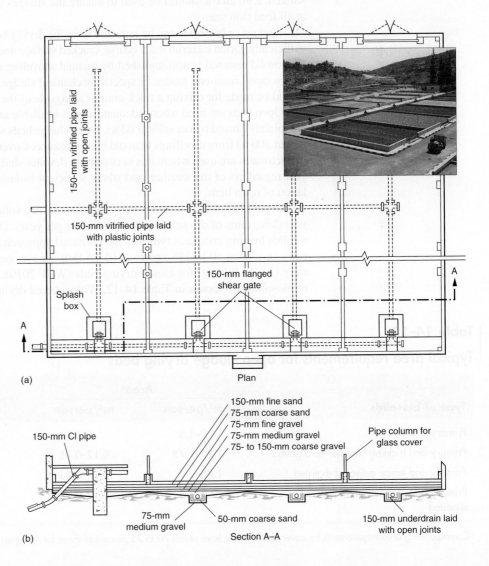

150-mm vitrified pipe laid with open joints

150-mm vitrified pipe laid with plastic joints

150-mm flanged shear gate

Splash box

A

A

(a)   Plan

150-mm fine sand
75-mm coarse sand
75-mm fine gravel
75-mm medium gravel
75- to 150-mm coarse gravel

150-mm CI pipe

Pipe column for glass cover

75-mm medium gravel

50-mm coarse sand

150-mm underdrain laid with open joints

(b)   Section A–A

should be adequately supported and covered with coarse gravel or crushed stone. The sand layer should range from 200 to 460 mm (9 to 18 in.) deep with an allowance for some loss from cleaning operations. Deeper sand layers generally retard the draining process. Sand should have a uniformity coefficient of not over 4.0 and an effective size of 0.3 to 0.75 mm.

The drying area is typically partitioned into individual beds 7.5 m wide although the actual width for a particular site should be set to accommodate the sludge removal method. The bed length can vary and beds can be up to 30 to 60 m (100–200 ft) long. The partitions are typically constructed of earthen embankments, wooden planks, concrete planks or reinforced concrete blocks. The outer wall is normally constructed with a 500–900 mm (20–36 in.) freeboard above the sand area and the wall usually extends to the underdrain gravel. Concrete foundation walls are required if the beds are to be covered.

Sludges or biosolids are typically fed to drying beds via an open channel or closed piping. Distribution boxes or valves are required to divert the sludge or biosolids flow into the bed selected. Splash plates are placed in front of the sludge or biosolids outlets to spread the material over the bed and to prevent erosion of the sand. If the feed pipe is pressurized, a 90 elbow should be used to ensure the sludges or biosolids hit the splash plate at all feed flowrates.

Sludges or biosolids can be removed from the drying bed after it has drained and dried sufficiently. Dried material has a coarse, cracked surface and is black or dark brown. Sludge or biosolid removal is accomplished by manual shoveling into wheelbarrows or trucks, or by a scraper, front-end loader, or special mechanical sludge removal equipment. Provisions should be made for driving a truck onto or alongside of the bed to facilitate loading.

Open beds are used where adequate area is available and sufficiently isolated to avoid complaints caused by occasional odors. Open sludge beds should be located at least 100 m (about 300 ft) from dwellings to avoid odor nuisance. Covered beds with greenhouse types of enclosures are used where it is necessary to dewater sludge continuously throughout the year regardless of the weather, and where sufficient isolation does not exist for the installation of open beds.

Drying bed solids loadings are computed on dry solids loading per square foot per year (kilograms of dry solids per square meter per year). Designing drying beds based on a solids loading criteria is typically the preferred approach and loading requirements typically vary from 50 to 125 kg/m²·y (10–25 lb/ft²·y) for open drying beds and 60–200 kg/m²·y (12–40 lb/ft²·y) for closed drying beds (WEF 2010). Typical data for various types of biosolids are shown in Table 14–12. With covered drying beds, more biosolids can be

Table 14–12

Typical area requirements for open sludge drying beds

Type of biosolids	Area[a]		Dry sludge loading rate	
	ft²/person	m²/person	lb/ft²·y	kg/m²·y
Primary digested	1.0–1.5	0.1	25–30	120–150
Primary and trickling filter humus digested	1.25–1.75	0.12–0.16	18–25	90–120
Primary and waste activated digested	1.75–2.5	0.16–0.23	12–20	60–100
Primary and chemically precipitated digested	2.0–2.5	0.19–0.23	20–33	100–160

[a] Corresponding area requirements for covered beds vary from about 70 to 75 percent of those for the open beds.

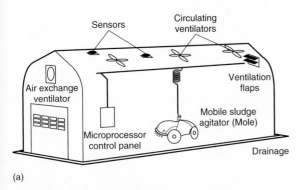

(a)

(b)

(c)

Figure 14–14

Solar sludge drying bed system employing mobile sludge agitator: (a) schematic (adapted from Parkson Corp.), (b) view of exterior of typical installation, and (c) view interior of a typical installation and mobile sludge mixer.

applied per year because of the protection from rain and snow. Polymer conditioning is also sometimes used to improve the performance of sludge drying beds.

Solar Drying Beds.   A method used to enhance the dewatering and drying of liquid, thickened, or dewatered biosolids is solar drying in covered drying beds (see Fig. 14–14). Solar drying systems are typically not used for undigested primary sludge due to concerns of odor and requirements to dry to greater than 90 percent to get Class A biosolids product instead of 75 percent drying for stabilized biosolids. The solar drying system, which is a sophisticated "greenhouse," consists of a rectangular base structure, translucent chamber, sensors to measure atmospheric drying conditions, air louvers, circulation fans, ventilation fans, a mobile electromechanical device that agitates and moves the drying biosolids, and a microprocessor that controls the drying environment [see Figs. 14–14(a) and (b)]. The system's main source of drying energy is solar radiation.

In most solar drying systems, mechanically dewatered biosolids are distributed in a greenhouse either manually or automatically. It is also possible to add liquid sludge directly to the greenhouse; however, the additional greenhouse area required typically outweighs the benefit from eliminating the mechanical dewatering step. Solar drying is best suited for tropical or arid environments however there are installations in both northern and mountainous climates.

During the drying cycle, the microprocessor evaluates the number of climatic variables, such as temperature, humidity and solar radiation, which initiates one or more operations that optimize the conditions inside the greenhouse. The greenhouse contains circulation fans and exhaust fans to provide convective drying and control of climatic conditions inside the greenhouse. The biosolids are periodically turned and aerated with varying devices depending on the manufacturer [see Fig. 14–14(c)]. In addition, low temperature waste heat can be used to

Figure 14–15

Cross section of a reed bed for dewatering and storage of biosolids.

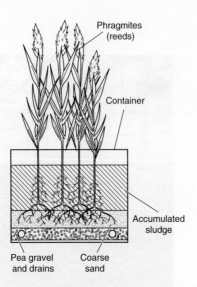

Phragmites (reeds)

Container

Accumulated sludge

Pea gravel and drains

Coarse sand

enhance solar drying and reduce area requirements for the greenhouse. Ultimately, it is possible to obtain a dry pelletized material with a solids content as high as 90 percent.

Reed Beds

Reed beds can be used for biosolids dewatering at treatment plants with capacities up to 0.2 m³/s (5 Mgal/d). A typical reed bed for biosolids dewatering, treatment, and storage is shown on Fig. 14–15. Reed beds are similar in appearance to subsurface flow constructed wetlands, which consist of channels or trenches filled with sand or rock to support emergent vegetation. The difference between reed beds used for biosolids application and subsurface flow wetlands is that the liquid biosolids are applied to the surface of the beds (as compared to subsurface application) and the filtrate flows through the gravel to underdrains.

Typically, reed beds are constructed of washed river-run gravel in the following layers: (1) a 250 mm (10 in.) deep drainage layer composed of 20 mm (0.8 in.) washed gravel, (2) a 250 mm (10 in.) deep layer composed of 4 to 6 mm (0.16 to 0.24 in.) washed gravel, and (3) a 100 to 150 mm (4 to 6 in.) layer of sand (0.4 to 0.6 mm). Sometimes an even coarser bottom layer is used. At least 1 m (3 ft) of freeboard above the sand layer is provided for a 10-y accumulation of sludge. Phragmites (reeds) are planted on 300 mm (12 in.) centers in the gravel layer just below the sand. Other wetland vegetation can be used, although reeds are the most popular. The first sludge application is made after the reeds are well established. Harvesting of the reeds is practiced typically in the winter by cutting the tops back to a level above the sludge blanket. Harvesting is necessary whenever the plant growth becomes too thick and restricts the even flow of biosolids. The harvested material can be composted, burned, or landfilled (Crites and Tchobanoglous, 1998).

The purpose of the plants is to provide a pathway for continuous drainage of water from the sludge layer. As the plants move back and forth due to wind currents, pathways are created for water to drain from the biosolids into the underdrains. The plants also absorb water from the sludge. Oxygen transfer to the plant roots assists in the biological stabilization and mineralization of the sludge. The reed bed system is a form of passive composting. The design loading rates for reed beds range from 30 to 60 kg/m²·y (6 to 12 lb/ft²·y). Loading rates as high as 100 kg/m²·y (20 lb/ft²·y) have been used, depending on the nature of the sludge and climatic conditions. The liquid sludge is applied intermittently, as in sand drying beds. The typical biosolids depth applied is 75 to 100 mm (3 to 4 in.) every week to 10 d (Crites and Tchobanoglous, 1998; Cooper et al., 1996).

Lagoons

Drying lagoons may be used as a substitute for drying beds for the dewatering of digested biosolids. Lagoons are not suitable for dewatering untreated sludges, limed sludges, or sludges with a high-strength supernatant because of their odor and nuisance potential. The performance of lagoons, like that of drying beds, is affected by climate; precipitation and low temperatures inhibit dewatering. Lagoons are most applicable in areas with high evaporation rates. Dewatering by subsurface drainage and percolation is limited by increasingly stringent environmental and groundwater regulations. If a groundwater aquifer used for a potable water supply underlies the lagoon site, it may be necessary to line the lagoon or otherwise restrict significant percolation.

Unconditioned digested biosolids are discharged to the lagoon in a manner suitable to accomplish an even distribution. Biosolids depths usually range from 0.75 to 1.25 m (2.5 to 4 ft). Evaporation is the principal mechanism for dewatering. Facilities for decanting of supernatant are usually provided, and the liquid is recycled to the treatment facility. Biosolids are removed mechanically, usually at a solids content of 25 to 30 percent. The cycle time for lagoons varies from several months to several years. Typically, biosolids are pumped to the lagoon for 18 months, and then the lagoon is rested for 6 mo. Solids loading criteria range from 36 to 39 kg/m^3·y (2.2 to 2.4 lb/ft^3·y) of lagoon capacity (U.S. EPA, 1987a). A minimum of two cells is essential, even in very small plants, to ensure availability of storage space during cleaning, maintenance, or emergency conditions.

14–3 HEAT DRYING

Heat drying involves the application of heat to evaporate water and to reduce the moisture content of biosolids below that achievable by conventional dewatering methods. The advantages and disadvantages for heat drying are summarized in Table 14–13.

Heat-Transfer Methods

The classification of dryers is based on the predominant method of transferring heat to wet sludge or biosolids. Heat can be transferred by convection, conduction, radiation, or a combination of two or more methods. Dryers using infrared radiation have been used mostly for demonstration testing and are not covered in detail in this text.

All drying systems follow the three stages of drying which are illustrated in the drying curve shown on Fig. 14–16. The three stages of drying are warm up stage, constant

Table 14–13

Advantages and disadvantages of heat drying

Advantages	Disadvantages
1. Proven process	1. Relatively high capital cost
2. Product is readily marketable and suitable for a variety of uses that are consistent with the EPA's national beneficial reuse policy	2. Large fuel requirements
	3. Payback from the sale of Class A production usually will not offset high operating costs
3. Small footprint	4. Potential to create dust
4. Reduced product transportation costs	5. Increased fire and explosion risk
5. Significant pathogen reduction (Class A)	6. Relatively complex system, requires highly trained operating staff
6. Improved storage capability	
7. Does not require chemical additives	7. Odor potential and potential for odorous and dusty end product
8. Enhances heat value of biosolids	

Figure 14–16

Three stages of drying:
(a) warming up, (b) constant rate,
and (c) falling rate.

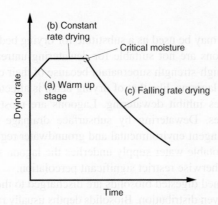

rate drying stage, and the falling rate drying stage. During the warm up stage, the solids and moisture are heated up to the process temperature by transferring sensible heat. During the constant rate drying period, the moisture evaporated from the surface of the sludge or biosolids is replaced with internal moisture at an equal rate thus the surface remains saturated and the temperature of the sludge or biosolids surface being dried during this period is approximately equal to the wet bulb temperature of the gas present in the dryer. The constant rate period is typically the longest period and most drying occurs during this phase. The falling rate period occurs when most of the free moisture is removed and the rate of drying becomes controlled by diffusion of internal water to the surface of the sludge or biosolids being dried. During this period, the sludge or biosolids surface temperature begins to approach the gas or air temperature. The moisture content at the transition of the constant rate to the falling rate period is known as the critical moisture content.

Convection. In convection (direct drying) systems, wet sludge or biosolids are contacted directly with the heat transfer medium, usually hot gases. Under equilibrium conditions of constant rate drying, mass transfer is proportional to (1) the area of wetted surface exposed, (2) the difference between water content of the drying air and saturation humidity at the wet-bulb temperature of the solid-air interface, and (3) other factors, such as velocity and turbulence of drying air expressed as a mass transfer coefficient. The heat-transfer rate for evaporation is determined by the following equation (WEF, 2010):

$$q_{conv} = h_c A(T_g - T_s) \tag{14–1}$$

where q_{conv} = convective heat transfer rate, kJ/h (Btu/h)
h_c = convection heat transfer coefficient, kJ/m^2·h·°C (Btu/ft^2·h·°F)
A = area of the heated surface, m^2 (ft^2)
T_g = gas temperature, °C (°F)
T_s = temperature at sludge/gas interface, °C (°F)

The convection heat transfer coefficient can be obtained from dryer manufacturers or from pilot studies; however, many manufacturers consider this proprietary information.

Conduction. In conduction (indirect) drying systems, a solid retaining wall separates the wet sludge from the heat transfer medium, usually steam, thermal oil or another hot fluid. Heat transfer for conduction is determined by the following equation (WEF, 2010):

$$q_{cond} = h_{cond}A(T_m - T_s) \tag{14–2}$$

where q_{cond} = conductive heat-transfer rate, kJ/h (Btu/h)

$\quad h_{\text{cond}}$ = conductive heat-transfer coefficient, kJ/m^2·h·°C (Btu/ft^2·h·°F)

$\quad\quad A$ = area of wetted surface exposed to gas, m^2 (ft^2)

$\quad\quad T_m$ = temperature of heating medium, °C (°F)

$\quad\quad T_s$ = temperature sludge at drying surface, °C (°F)

The conductive heat-transfer coefficient, a composite term, includes the effects of the heat-transfer surface films of the sludge and the medium. The conduction heat-transfer coefficient can be obtained from dryer manufacturers or from pilot studies.

Process Description

Heat dryers are classified as follows: direct, indirect, combined direct-indirect, and infrared. Direct and indirect dryers are described as they are the types most used commonly for municipal biosolids drying. Coal, oil, gas, infrared radiation, or dried sludge may be used as the means of supplying the energy for heat drying.

Direct Dryers.   Direct (convection) dryers that have been used for drying municipal wastewater sludges and biosolids are the flash dryer, rotary dryer, and fluidized-bed dryer. Although as many as 50 municipal flash dryers have been installed in the United States since 1940, only Houston, TX, is still known to use this technology at two of its wastewater treatment plants. Because of safety concerns, high energy requirements, high O&M, and limited interest from vendors to work in the wastewater market, the popularity of this technology has declined, thus flash drying is not discussed in detail in this text (WEF, 2010). Rotary dryers are now very commonly used for wastewater sludges or biosolids. Fluidized-bed drying is a relatively new application in the United States.

Rotary Dryer.   Rotary dryers have been used for the drying of raw primary sludge, waste-activated sludge, and digested biosolids from combined primary and WAS (see Fig. 14–17). In general, drying raw primary sludge is not recommended due to material handling concerns, odor and stability of the final product. A rotary dryer consists of a cylindrical steel shell that is rotated on bearings and usually is mounted with its axis at a slight slope from

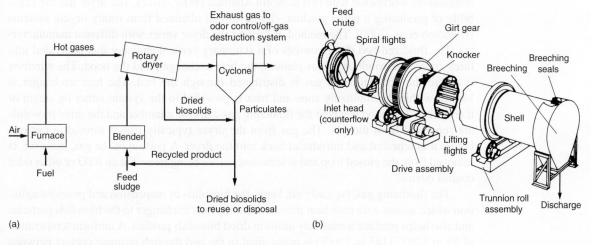

(a)

(b)

Figure 14–17

Rotary sludge dryer: (a) typical process flow diagram, and (b) isometric view of rotary dryer.

Figure 14–18

Cross-section through a fluidized-bed reactor. (From Andritz.)

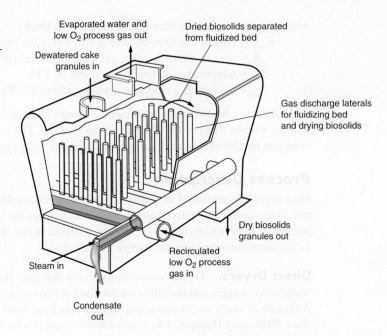

the horizontal [see Figs. 14–17(a) and (b)]. The feed sludge or biosolids is mixed with previously dried material in a blender located ahead of the dryer [see Fig. 14–17(a)]. The blended feed material has a moisture content of approximately 65 percent that improves its ability to move through the dryer without sticking. The mixture and hot gases are conveyed to the discharge end of the dryer. During conveyance, axial flights along the rotating interior wall pick up and cascade the sludge through the dryer. The product, which has a dry solids content of 90 to 95 percent, is screened and the oversize material passes through a crusher and then is transported to a recycle bin. The dried product is amenable to handling, storage, and marketing as a fertilizer or soil conditioner.

Fluidized-Bed Dryer. Fluidized-bed dryers, developed in Europe, were first adapted in the United States in the late 1990's (see Fig. 14–18). There are approximately 30 known installations worldwide with two in North America (WEF, 2012). The dryer has the capability of producing a pellet product, similar to that obtained from rotary drying systems (Holcomb et al., 2000). The method of feeding the dryer varies with different manufacturers. The fluidized-bed dryer consists of a stationary vertical chamber that is divided into three zones; the windbox or gas plenum, the heat exchanger, and the hood. The windbox is where the hot fluidizing gas is distributed through the bed. The heat exchanger is located within the fluidizing zone and heat is provided to the system either by steam or thermal oil. The hood is where the fluidizing gas exits the chamber and the dried biosolids are separated from the gas. The gas from the dryer typically passes through a cyclone before it is reheated and introduced back into the dryer. A portion of the gas, however, is removed from the closed loop and is scrubbed before being treated in an RTO or other odor control device.

The fluidizing gas, typically air, keeps the biosolids in suspension and provides agitation which assists with both heat transfer from the heat exchanger to the biosolids particles and also helps produce a relatively uniform dried biosolids product. A uniform temperature of 85 to 120°C (185 to 230°F) is maintained in the bed through intimate contact between the biosolids granules and the fluidizing air. Since fluid bed dryers operate at relatively low temperatures, they are suitable for low temperature waste heat recovery.

Figure 14–19

Example of belt dryer:
(a) schematic of operation with
three different heat sources, and
(b) view of enclosed belt dryer
(courtesy of SH+E Group U.S.).

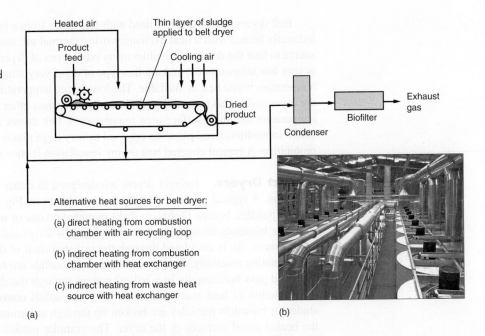

(a)

(b)

Alternative heat sources for belt dryer:

(a) direct heating from combustion
    chamber with air recycling loop

(b) indirect heating from combustion
    chamber with heat exchanger

(c) indirect heating from waste heat
    source with heat exchanger

Belt Dryers. At the time of writing this text, belt drying is still a relatively new type of sludge or biosolids dryer being implemented in the United States. However, there are up to seven belt dryers either in operation or under construction. Belt dryers are more widely used in Europe. Belt dryers are convection dryers that consist of a conveyor belt where biosolids are distributed in a thin layer (Fig. 14–19). The slowly moving belt transports the sludge or biosolids through the dryer while warm heated gases pass either through the belt and biosolids layer or pass across the biosolids layer to provide convective heat transfer. There are several manufactures currently available each with slightly different configurations and sludge or biosolids distribution methods. Some dryers contain metal mesh belts or perforated plates, while others contain fabric belts similar to what is used on belt filter presses. The dried product can be recycled and back mixed with dewatered cake, similar to rotary drum dryers, ensuring the biosolids being fed to the dryer are past the sticky phase and allowing for better distribution in the dryer. The sludge or biosolids fed to the dryer can also be extruded in ribbons (see Fig. 14–20) with a high drying surface area eliminating the need for back mixing systems. In the case of distributing the biosolids using extrusion method, belt drying cannot typically accept more than 30 percent cake solids.

Figure 14–20

Extruded biosolids for enhanced
heat drying: (a) view of extruded
biosolids strings (courtesy of
Kruger) and (b) view of biosolids
ropes being extruded (courtesy of
SH+E Group U.S.).

(a)

(b)

Belt dryers can be directly fired with flue gases from a hot gas furnace, or they can be indirectly heated with a heat exchanger using thermal oil, steam, or flue gas as the heating source to heat the drying air. Unlike many other types of dryers, belt dryers can also operate at very low temperatures, making this type of dryer very attractive for situations where low temperature waste heat is available. The low drying temperature and minimized agitation in the dryer makes this type of dryer inherently safer than other types of dryers. With the low temperature and high surface area requirements, belt dryers are typically better suited for small- to medium-sized plants as opposed to very large plants where they may become cost prohibitive. A typical covered belt drying installation is shown on Fig. 14–19(b).

Indirect Dryers. Indirect dryers are designed in either a horizontal or vertical configuration. A typical process flow diagram is shown on Fig. 14–21(a). Horizontal dryers employ paddles, hollow flights, or disks mounted on one or more rotating shafts to convey sludge or biosolids through the dryer [see Figs. 14–21(c) and (d)]. A heated medium, usually steam or oil, is circulated through the jacketed shell of the dryer and the hollow core of the rotating assembly. Dewatered sludge or biosolids are fed perpendicular to the dryer shaft and pass horizontally in a helical pattern through the dryer. The dryer performs the dual function of heat transfer and sludge or biosolids conveying. Drying occurs as the sludge or biosolids particles are broken up through agitation and come into contact with the heated metal surfaces in the dryer. The granular particles are both hot and abrasive.

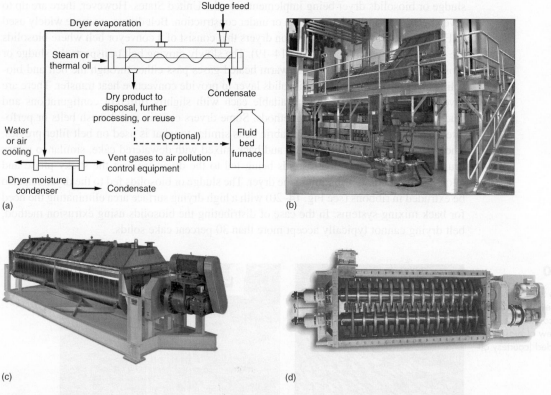

(a)

(b)

(c)

(d)

Figure 14–21

Indirect sludge dryer: (a) schematic process flow diagram, (b) view of typical dryer installation, (c) view of dryer without support facilities, and (d) view of interior paddle flights. [Figures (b), (c), and (d) courtesy of Komline-Sanderson.]

Figure 14–22

Cross section through a vertical indirect dryer (From Pelletech).

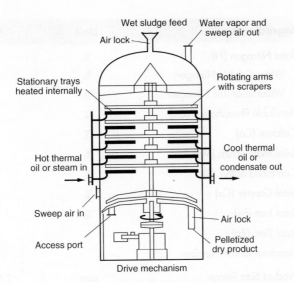

Moving parts, as in most dryers, will become abraded, and corrosion will accelerate the deterioration of the metal. The design of the agitator has to allow for efficient heat transfer, mixing of the sludge mass, and minimum fouling of the agitator. A weir at the discharge end ensures complete submergence of the heat-transfer surface in the biosolids being dried. The water vapor derived from the drying operation may be drawn off under a slightly negative pressure by an induced-draft fan located in the off-gas duct.

In vertical indirect dryers (see Fig. 14–22), sludge or biosolids contacts a metal surface heated by a medium, such as steam or oil, and the heat is conductively transferred to the sludge. The biosolids or sludge does not come in contact with the heating medium. Dewatered product (approximately 20 percent solids) mixed with recycled product is fed through the top inlet of a multistage dryer. Rotating arms move the sludge from the heated stationary tray to another in a rotating zigzag motion until it exits at the bottom as a dried, pelletized product. The rotating arms are equipped with adjustable scrapers that move and tumble the sludge in thin layers (20 to 30 mm) over the heated stationary trays. Dried product exits the dryer and is conveyed by a bucket elevator to a separation hopper. A portion of the properly sized pellets are cooled and conveyed to storage. Oversized particles are crushed and mixed with fines and a portion of the properly sized pellets and backmixed with the dewatered cake being fed to the dryer. Because of the backmixing process, this type of dryer produces a relatively uniform product similar to rotary drum dryers.

In indirect dryers, sludge or biosolids is dried to a specific level of dryness and discharged to a product conveyor for transfer to storage. Solids concentrations in the dryer product can range from 65 to over 95 percent depending on the ultimate use of the product.

Product Characteristics and Use

Dried biosolids are mainly used in a similar manner to commercial fertilizers, depending on the nutrient content of the material. Some of the dried material is well established as fertilizer and sold for profit to the producer. For example, OCEANGRO™, which is a dried pellet product from Ocean County Utilities Authority in New Jersey, is sold in bulk or in bags as a fertilizer to the agricultural community. The OCEANGRO™ guaranteed analysis is summarized in Table 14–14. The characteristics of the finished biosolids product depend on the type of sludge, the type of preprocessing, and the physical configuration of the drying surfaces. Raw primary sludge produces more of a fibrous, dusty, and odorous

Table 14–14

OCEANGROW™ guaranteed analysis

Ingredients	Unit	Value
Total Nitrogen (N)	%	5.0
Water Soluble Nitrogen	%	0.50
Water Insoluble Nitrogen[a]	%	4.50
Available Phosphate (P_2O_5)	%	5.00
Calcium (Ca)	%	2.50
Total Magnesium (Mg)	%	0.33
Combined Sulfur (S)	%	1.00
Total Copper (Cu)	%	0.04
Total Iron (Fe)	%	2.50
Total Zinc (Zn)	%	0.05
Standard Guide Number		150
Product Size Range	mm	1.5 to 2.5
Uniformity Index	%	60

[a] This product contains 4.50 percent slow release nitrogen.

material that is difficult to manage and pelletize. Digested sludge can be pelletized, depending on the type of dryer or downstream conditioning process, to produce an amorphous particle that can be easily handled and transported. Dryers with backmixing typically produce more uniform pellets but other downstream processes such as screening, conditioning agents or pelletizing can be added to dryers without backmixing if a more uniform product is required for marketing purposes. The best size range for marketing is approximately from 2 to 4 mm, but can vary depending on the specific market (WEF, 2012). To maximize marketing potential, screening for sizes smaller and larger than the selected size range will be required. The fines and oversized particles can be returned to blend with the incoming sludge, hence increasing the solids content entering the dryer but not changing the amount of moisture that must be evaporated.

Biosolids dried products can be used as an energy source depending on the characteristics of the sludge or biosolids that were dried. In recent years, dried biosolids have been used as a coal substitute for fuel in cement kilns. The use of dried product as an energy source is expected to increase, due to high energy costs and because the energy recovered from dried biosolids is considered renewable energy.

Product Transport and Storage

Although the granular product from heat drying is reasonably durable, long mechanical conveyors, such as screw conveyors, drag conveyors, and pneumatic conveyors that create an abrasive action could cause crumbling and dust formation. These types of conveyors, however, are commonly used at many sites. If product friability is a concern, open or folded belt conveyors may be preferred.

Upon exiting the dryer, the hot biosolids should be cooled to below 50°C (120°F) before placing in silos or storage vessels. The combination of initial heat plus heat that could be introduced by biological activity in the silo can cause smoldering or an open flame. This condition can occur where the drying operation is expected to pass through frequent start-and-stop cycles and drying may not be complete. In general, the product entering storage should be from 92 to 98 percent dry solids.

Fire and Explosion Hazards

With the fine particles and high levels of dryness in heat-dried sludge, hazards due to fire and explosion may exist in the dryer or when dried sludge is conveyed or stored. An organic dust suspended in air can rapidly combust if exposed to an ignition source. The heat of combustion can rapidly increase the volume and/or pressure of hot combustion products. If the pressure exceeds the rupture strength of the containing vessel, an explosion occurs. The phenomenon is called "deflageration," and deflageration explosions are the most serious concern when handling dried biosolids (Haug et al., 1993). Dried biosolids can also reheat if rewetted due to biological activity. In addition, the sludge being dried could contain a high level of fiber or grease content which could cause problems inside the dryer. Design considerations that are recommended for safety purposes and to prevent thermal events are given in Table 14–15.

Air Pollution and Odor Control

Two important control measures associated with heat drying of sludge are dust collection and odor control. All dryers produce some sort of offgas which results from the process gas and evaporated water that is continuously removed from the drying process. The offgas can contain

Table 14–15
Prevention measures to avoid dust hazards in heat drying[a]

Item	Prevention measure
Venting system (for processing, conveyance, and storage components)	Provide explosion relief vents.
	Size explosion vents for "worst case" explosion in an air atmosphere.
Temperature control	Use controls to prevent high temperatures which could cause unsafe conditions
	Temperature control can also be used to prevent the material from being over dried which could cause excess dust formation.
Water deluge	Include a water deluge or sprinkler system that reacts based on high temperature conditions. Water deluge can also be set to react to the presence of carbon monoxide or carbon dioxide which could indicate smoldering.
Nitrogen padding	Provide a nitrogen inerting atmosphere for all dried biosolids conveyance and processing facilities. Maintain oxygen levels below 5 percent by volume to reduce potential for self-heating and ignition of hot biosolids.
Electrical equipment	Design in accordance with appropriate National Fire Protection Association criteria. If dust is present, all equipment must be dust tight and electronic cabinets nitrogen purged. Motor control centers that contain sparking devices, such as starters and relays, must be located outside classified areas.
Ducts and vessels	Electrically bond and ground all conductive elements of the system that contact dried biosolids.
Maintenance	Keep areas clean to prevent accumulation of dust.
	Any vessel containing powder must have powder removed before opening or the powder must be cooled to ambient temperatures before safe entry clearance is given.
Product cooling	Ensure product is cooled to below 50°C (120°F) before being transported to product storage.
Miscellaneous	Eliminate or move outside all heat sources from classified areas. Equip electric motors located in a Class II, Division 2 area with Class F insulation to reduce "skin" temperatures.

[a] Adapted from Haug, et al. (1993) and WEF (2012).

high levels of odors due to volatilization and heat reactions. The method for dealing with dryer offgas varies depending on the dryer type, manufacturer and local site requirements.

Most dryers include wet scrubbers that have relatively high efficiencies and will condense the water evaporated from the sludge as well as remove some of the organic matter in the vent gas. Most wet scrubbers will also include mist eliminators to minimize carryover of water droplets. Some dryers also include cyclone separators, either on the process air circulation loop or upstream of the wet scrubber, to remove particles entrained in the air. The cyclones used typically have efficiencies of 75 to 80 percent and are suitable for vent gas temperatures up to 340 or 370°C (650 or 700°F). The final scrubbed off gas must then go through some sort of odor control treatment. The use of a thermal oxidizer is one common method used on many types of dryers to eliminate odors. Thermal oxidizers typically operate at temperatures greater than 815°C (1500°F) with a residence times of 0.75 to 1 s (WEF, 2012). The dryer offgas can also be sent to an incinerator (if one is onsite), or it can be mixed with combustion air for the dryer system's burner (if the dryer is indirectly fired). Other options that have been used for off-gas odor control include chemical scrubbers, biofilters, and directing the off-gas to diffusers in the aeration basin of the activated sludge process.

14-4 ADVANCED THERMAL OXIDATION

Incineration, referred to in this chapter as advanced thermal oxidation (ATO) of sludge and biosolids, involves the total conversion of organic solids to oxidized end products, primarily carbon dioxide, water, and ash. The major advantages and disadvantages of incineration are summarized in Table 14–16. ATO is used most commonly by medium to large sized plants with limited disposal or reuse options.

Sludges processed by ATO are usually first dewatered. It is normally unnecessary to stabilize sludge before incineration. In fact, such practice may be detrimental because stabilization, specifically aerobic and anaerobic digestion, decreases the volatile content of the sludge and consequently increases the requirement for an auxiliary fuel. When ATO is practiced for biosolids, it is desirable that the dewatered biosolids concentration be between 30 to 35 percent,

Table 14–16 Advantages and disadvantages of incineration	**Advantages**	**Disadvantages**
	1. Maximum volume reduction and end product stability thereby lessening disposal requirements	1. High capital and operating cost
	2. Maximum destruction of pathogens and toxic compounds	2. Complex system that requires highly trained operating staff
	3. Relatively small process footprint	3. Possible adverse environmental effects from air emission and ash
	4. Greatest control over final biosolids disposition	4. Disposal of residuals which may be classified as hazardous wastes if they exceed prescribed maximum pollutant concentrations
	5. Well suited for larger plants with larger disposal requirements	
	6. Energy recovery potential	5. Relatively lengthy implementation timeframe due to planning and public consultation requirements
	7. Well proven process	6. Air permitting can be tedious and difficult
		7. May not be feasible for a nonattainment area

to reduce the use of an auxiliary fuel and achieve autogenous oxidation. In Europe, ATO has been used with dewatered biosolids from thermally hydrolyzed sludges as discussed in Chap. 13. Sludges may be thermally oxidized separately or in combination with municipal solid wastes. The ATO processes considered in the following discussion include multiple-hearth incineration, fluidized-bed incineration, and coincineration with municipal solid waste. Before discussing these processes, some fundamental aspects of complete combustion are introduced.

Fundamental Aspects of Complete Combustion

Combustion is the rapid exothermic oxidation of combustible elements in fuel. ATO is complete combustion. The predominant elements in the carbohydrates, fats, and proteins composing the volatile matter of sludge are carbon, oxygen, hydrogen, sulfur, and nitrogen (C-O-H-S-N). Other major components include moisture and ash. The approximate percentages of these may be determined in the laboratory by a technique known as ultimate analysis (ASTM, 2009) and proximate analysis (ASTM, 2009). Heating (or calorific) value may be obtained using ASTM (2011).

Oxygen requirements for complete combustion of a material may be determined by knowing its constituents, assuming that carbon and hydrogen are oxidized to the ultimate end products CO_2 and H_2O. The formula becomes

$$C_aO_bH_cN_d + (a + 0.25c - 0.5b)O_2 \rightarrow aCO_2 + 0.5cH_2O + 0.5dN_2 \tag{14-3}$$

The theoretical quantity of air required will be 4.35 times the calculated quantity of oxygen because air is composed of 23 percent oxygen on a mass basis. To ensure complete combustion, sufficient excess air is required. Too little excess air may lead to poor emissions performance and may also limit the throughput of the process. Too much excess air may lead to high auxiliary fuel usage and unnecessarily large equipment sizing. In some jurisdictions, air pollution regulations may require a minimum flue gas oxygen concentration. The design range of excess air required depends on the ATO process configuration, the system design, and the characteristics of the biosolids. For some units operating with very dry, high heating value biosolids, air is also used to quench, control, and spread the heat released in the process to maintain below an upper temperature limit. In this case, the air required for cooling may far exceed the theoretical quantity of air.

A materials balance must be made to include the above compounds and the inorganic substances in the sludge, such as the inert material (ash), moisture, and the other constituents of the air (primarily the approximately 77 percent N_2 in the air that is supplied along with O_2 reaction in Eq. (14–3). The specific heat of each of these substances and of the products of combustion must be taken into account in determining the heat required for the incineration process.

Heat requirements will include the sensible heat Q_s in the ash, plus the sensible heat required to raise the temperature of the flue gases to 760°C (1400°F) or whatever temperature of operation is selected for complete oxidation, elimination of odors, and assurance of suitable environmental performance less the heat recovered in preheaters or recuperators. Heat loses to the ambient space around the ATO process is also a heat requirement. Latent heat must also be furnished to evaporate all of the moisture in the sludge. The total heat required may be expressed as

$$Q = \sum Q_S + Q_E + Q_L = \sum C_P W_S(T_2 - T_1) + W_W\lambda + Q_L \tag{14-4}$$

where Q = total heat, kJ (Btu)

$\quad Q_S$ = sensible heat in the ash, kJ (Btu)

$\quad Q_E$ = latent heat, kJ (Btu)

$\quad Q_L$ = heat loss

C_p = specific heat for each category of substance in ash and flue gases, kJ/kg°C (Btu/lb°F)

W_s = mass of each substance, kg (lb)

W_w = mass of water, kg (lb)

T_1, T_2 = initial and final temperatures

λ = latent heat of evaporation, kJ/kg (Btu/lb)

Reduction of moisture content of the sludge is the principal way to lower heat requirements, and the moisture content may determine whether additional fuel will be needed to support combustion. It is desirable both from a cost perspective and from an environmental sustainability perspective not to have to add additional fuel to the process. Therefore, it is desirable that the heat required to drive the process should come from the exothermic oxidation of the biosolids' volatile components, which is called "autogenous combustion."

The heating value of a sludge may be estimated by using Eq. (2–66) presented previously in Chap. 2 and repeated here for convenience:

$$\text{HHV (MJ/kg)} = 34.91\,\text{C} + 117.83\,\text{H} - 10.34\,\text{O} - 1.51\,\text{N} + 10.05\,\text{S} - 2.11\text{A} \qquad (2\text{–}66)$$

where, HHV = higher heating value, MJ/kg (Btu/lb = MJ/kg × 0.00043)

C = carbon, percent by weight expressed as a decimal (dry basis)

H = hydrogen, percent by weight expressed as a decimal (dry basis)

O = oxygen, percent by weight expressed as a decimal (dry basis)

N = nitrogen, percent by weight expressed as a decimal (dry basis)

S = sulfur, percent by weight expressed as a decimal (dry basis)

A = ash, percent by weight expressed as a decimal (dry basis)

The comparable expression proposed by the U.S. EPA (1979) and WEF (2010) is

$$\text{HHV (MJ/kg)} = 33.83\,\text{C} + 144.70\,(\text{H} - \text{O/8}) + 9.42\,\text{S} \qquad (14\text{–}5)$$

The fuel value of sludge ranges widely depending on the type of sludge and the volatile solids content. The fuel value of untreated primary sludge is the highest, especially if it contains appreciable amounts of grease and skimmings. Where kitchen food grinders are used, the volatile and thermal content of the sludge will also be high. Digested biosolids have significantly lower heating values than raw sludge. Typical heating values for various types of sludge and biosolids are reported in Table 14–17. The heating value for sludge is equivalent to that of some of the lower grades of coal. Computation of the heating value of biosolids is illustrated in Example 14–2.

Table 14–17

Typical heating values for various types of sludge and biosolids[a]

Type of sludge or biosolids	Btu/lb of total solids[b]		kJ/kg of total solid[b]	
	Range	Typical	Range	Typical
Raw primary	10,000–12,500	11,000	23,000–29,000	25,000
Activated	8500–10,000	9000	20,000–23,000	21,000
Anaerobically digested primary	4000–6000	5000	9000–14,000	12,000
Raw chemically precipitated primary	6000–8000	7000	14,000–18,000	16,000
Biological filter	7000–10,000	8500	16,000–23,000	20,000

[a] Adapted, in part, from WEF (1988).

[b] Lower value applies to plants with long solids retention time.

EXAMPLE 14–2 **Energy Content of Biosolids** A wastewater treatment plant is considering an ATO for handling of their biosolids. The plant currently dewaters waste activated sludge on belt filter presses prior to lime addition for Class B land application. A sample of dewatered sludge, before lime addition, was sent out for an ultimate analysis and the results are shown below.

Parameter	As received basis	Dry basis
Carbon	6.84	41.33
Hydrogen	0.94	5.66
Oxygen	3.71	22.41
Nitrogen	0.92	5.57
Sulfur	0.14	0.86
Ash	4.00	24.17
Moisture	83.45	0.00
HHV (MJ/kg)	2.96	17.88

Compare the lab results of the HHV on a dry basis to the theoretical HHV's calculated from Eq (2–66) and Eq (14–5).

Solution

1. Determine the energy content of the biosolids using Eq. (2–66).

    HHV (MJ/kg) = 34.91 C + 117.83 H − 10.34 O − 1.51 N + 10.05 S − 2.11 A

    HHV (MJ/kg) = 34.91 (41.33/100) + 117.83 (5.66/100) − 10.34 (22.41/100)
    $$\qquad\qquad - 1.51\,(5.57/100) + 10.05\,(0.86/100) - 2.11\,(24.17/100)$$

    HHV = 18.27 MJ/kg

    $$\text{Percent difference (measured vs. calculated)} = \left(\frac{17.88 - 18.27}{17.88} \right) 100$$

    $$= - 2.18\%$$

2. Determine the energy content of the biosolids using Eq. (14–5).

    HHV (MJ/kg) = 33.83 C + 144.70 (H − O/8) + 9.42 S

    HHV (MJ/kg) = 33.83 (41.33/100) + 144.70 (5.66/100 − (22.41/100)/8)
    $$\qquad\qquad + 9.42\,(0.86/100)$$

    HHV = 18.20 MJ/kg

    $$\text{Percent difference (measured vs. calculated)} = \left(\frac{17.88 - 18.20}{17.88} \right) 100$$

    $$= - 1.79\,\%$$

Comment The computed results from both Eq. (2–66) and Eq. (14–5) are very close to each other. There is also good agreement, within 3 percent, between the predicted and measured values. For theoretical calculations Eq. (2–66) is favored because it was derived from an analysis of hundreds of organic feed stocks and includes nitrogen and ash, whereas Eq. (14–5) is a modification of the well-known Dulong formula for coal. To design an ATO system, a detailed heat balance must be prepared. Such a balance must include the energy required to evaporate water in the sludge and heat losses from the process equipment, ductwork, stack, and ash as well as any heat recovery (such as combustion air preheating). Heat is obtained from the combustion of volatile matter in the sludge and from the burning of auxiliary fuels. If the process is autogenous, the auxiliary fuel is needed only for warming up the ATO or adjusting from a process upset should the desired temperature be too low or operation become unstable. Regardless of the feed biosolids characteristics, the design should include provisions for auxiliary heat for startup and for assuring complete oxidation at the desired temperature under all conditions. Fuels such as oil, natural gas, or excess digester gas are typically used for supplemental heating purposes.

To design an ATO system, a detailed heat balance must be prepared. Such a balance must include heat losses through the walls and pertinent equipment of the incinerator, as well as losses in the stack gases and ash. Approximately 4.0 to 5 MJ (4000 to 5500 Btu) are required to evaporate each kg (2.2 lb) of water in the sludge. Heat is obtained from the combustion of the volatile matter in the sludge and from the burning of auxiliary fuels. If the process is autogenous, auxiliary fuel is needed only for warming up the incinerator or adjusting from a process upset should the desired temperature be too low or operation become unstable. Regardless of the feed biosolids characteristics, the design should include provisions for auxiliary heat for startup and for assuring complete oxidation at the desired temperature under all conditions. Fuels such as oil, natural gas, or excess digester gas are suitable.

Multiple-Hearth Incineration

Multiple-hearth incineration is used to convert dewatered sludge cake to an inert ash. Because the process is complex and requires specially trained operators, multiple-hearth furnaces are normally used only in large plants. Multiple-hearth incinerators have been used at smaller facilities where land for the disposal of sludge is limited and at chemical treatment plants for the recalcining of lime sludges. As of 2011, about 70 percent of ATO installations in the United States utilized multiple-hearth units. These units are all fairly old, all having been installed in the mid and late twentieth century.

Process Description.   As shown on Fig. 14–23 , a multiple-hearth incinerator is set up as a counter flow process with a series of hearths (typically 7–11) where the sludge introduced into the unit is successively, dried, combusted, cooled, and discharged. Air and flue gases are the heat exchange fluids in the vessel and generally travel counter-current to the biosolids. The sludge cake is fed onto the top hearth and is slowly raked to the center by a series of teeth mounted on a set of rake (or "rabble") arms. The teeth direct the solids in a spiral pattern toward the center with each consecutive arm pushing a ridge of biosolids in toward the center and in the direction of rotation. The teeth also act to agitate the exposed surfaces of the biosolids and to mix the solids in an attempt to provide even exposure to the heat and air. From the center, sludge cake drops to the second hearth, where the rabble arms and teeth move it to the periphery. The sludge cake then drops to the third hearth and is again raked to the center. The hottest temperatures are typically on the middle hearths, where the sludge begins combustion after having evaporated enough water to raise the solids to a high enough temperature to sustain combustion. Preheated air is introduced to the lowest hearth and gains heat as it travels pass the sludge where the fixed carbon continues to burn off and the remaining ashes cool prior to discharge from the unit. The air is further heated by the sludge as the air rises past the middle hearths where combustion occurs and generates flue gases. The flue gases then cool as it gives up its heat to dry the incoming sludge on the top hearths.

Because air and flue gases are used as the primary heat exchange fluid in this counter flow system, proper air and flue gas flow, sufficient convection, and mixing is required to

Figure 14–23

Corss section through a typical multiple-hearth incinerator.

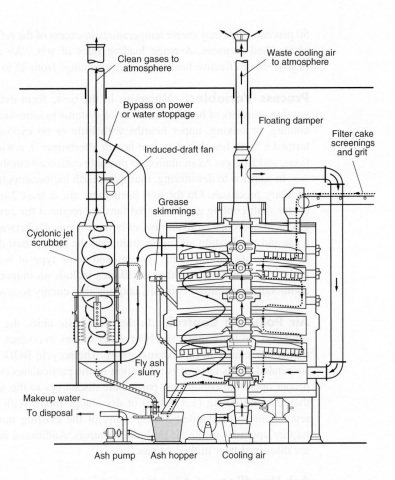

operate the system properly. The rabble tooth pattern is also important to provide even distribution, and frequent overturning of the biosolids on the hearths. Auxiliary fuel may be added at many different points on the unit. Typically, there are 3 to 4 burners around the circumference of a hearth and every second hearth is fitted with burners. By choosing the location and firing rate for these burners, the temperature and location of the combustion zone within the furnace can be changed to suit the material being combusted. The highest moisture content of the flue gas is found on the top hearths where sludge with the highest moisture content is heated and some water is vaporized. Cooling air is initially blown into the central column and hollow rabble arms to keep them from overheating. A large portion of this air, after passing out of the central column at the top, is recirculated to the lowest hearth as preheated combustion air.

Operational Controls. Operators may set feed rate, excess air, air injection location, burner location, burner firing rate, rake speed, etc. Given the large number of inter-related and codependent process variables for a multiple hearth furnace, consistent operation is sometimes challenging. These furnaces generally operate at relatively high levels of excess air (up to 100 percent) to avoid localized air/oxygen deficits and to provide sufficient buffer to ensure consistent, quality combustion.

Feed sludge must contain more than 15 percent solids because of limitations on the maximum evaporating capacity of the furnace. Auxiliary fuel is required usually when the feed sludge contains between 15 and 30 percent solids. Feed sludge containing more than

50 percent solids may create temperatures in excess of the refractory and metallurgical limits of standard furnaces. Average loading rates of wet cake are approximately 40 kg/m²·h (8 lb/ft²·h) of effective hearth area but may range from 25 to 75 kg/m²·h (5 to 15 lb/ft²·h).

Process Variables. Variations in the basic form exist with some furnaces having varying numbers of hearths, an array of systems to introduce auxiliary air for combustion, cooling, or mixing, upper hearths with little or no exposure to the biosolids (typically termed a "zero hearth" or an "on-hearth afterburner"), a wide degree of burner configurations, and flue gas recirculation to promote enhanced combustion controls.

In addition to dewatering, multiple hearth incinerators have a large amount of required ancillary processes. On the unit itself there are at least 2 or more fans for burner air and center shaft cooling air, additional fans are required for auxiliary combustion air and flue gas recirculation, a large number of burners (typically between 9 and 36 per unit) each with a gas and combustion air supply train, and the center shaft drive and gear reducer. In addition, there are ash handling systems and some type of wet or dry scrubber to meet air pollution requirements. Other ancillaries include an induced draft fan, bypass stack, any additional air pollution controls, and possibly energy recovery.

Air Pollution Control. In the most basic units, the air pollution controls are wet scrubbers. In these units, scrubber water comes in contact with and removes most of the particulate matter in the exhaust gases. The recycle BOD and COD is nil, and the total suspended solids content is a function of the particulates captured in the scrubber. Under proper operating conditions, particulate discharges to the air from wet scrubbers are less than 0.65 kg/10³ kg (1.3 lb/ton) of dry sludge input. With recent (2011) upgrades to the regulations in the United States, many of the existing multiple hearth incinerators will require upgrades to the air pollution controls. Additional details on air pollution controls are discussed later in this chapter.

Ash Handling. Ash handling may be either wet or dry. In the wet system, the ash falls into an ash hopper located beneath the furnace, where it is slurried with water from the exhaust gas scrubber. After agitation, the ash slurry is pumped to a lagoon or is dewatered mechanically. The effluent water from the ash lagoon or ash dewatering process can either be sent back into the wastewater treatment process or discharged back into the plant effluent and sent to the outfall. In the dry system, the ash is conveyed mechanically to a storage hopper for discharge into a truck for eventual disposal. The ash is usually conditioned with water. Ash density is about 5.6 kg/m³ (0.35 lb/ft³) dry and 880 kg/m³ (55 lb/ft³) wet.

Fluidized-Bed Incineration

The fluidized-bed incinerator used commonly for sludge incineration is a vertical, cylindrically shaped, refractory-lined steel shell that contains a sand bed (media) and fluidizing air orifices to produce and sustain combustion (see Fig. 14–24). As of 2011, about 30 percent of ATO installations in the United States utilized fluidized-bed technology. These units tend to be much newer than the multiple-hearth installations that they are typically replacing. As of 2012, all of the new full-scale units currently being installed are fluidized-bed incinerators.

Process Description. The fluidized-bed incinerator ranges in size from 2.7 to 9.1 m (9 to 30 ft) in diameter. The incinerator is generally composed of three areas. These are (from bottom to top): (1) the windbox, (2) the sand bed, and (3) the freeboard. The windbox acts as an air plenum for distributing the fluidizing air, and may also have one or more burners for startup. Above the windbox is the sand bed. When quiescent, the

Figure 14–24

Cross section through a typical fluidized-bed incinerator.

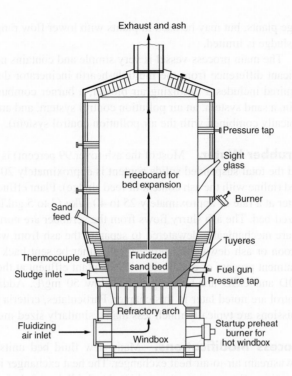

Labels on figure:
- Exhaust and ash
- Pressure tap
- Sight glass
- Freeboard for bed expansion
- Burner
- Sand feed
- Tuyeres
- Thermocouple
- Fluidized sand bed
- Fuel gun
- Pressure tap
- Sludge inlet
- Refractory arch
- Fluidizing air inlet
- Startup preheat burner for hot windbox
- Windbox

sand bed is approximately 0.8 m (2.5 ft) thick and rests on a brick dome, refractory-lined grid, or a steel plate that keeps the sand bed separated from the windbox. The sand bed support area contains orifices, called "tuyeres," through which the fluidizing air is injected into the incinerator at a pressure of 20 to 35 kN/m² (3 to 5 lb$_f$/in.²) by fluidizing air blowers. At low velocities, combustion gas "bubbles" appear within the fluidized bed. The main bed of suspended particles remains at a certain elevation in the combustion chamber and "boils" in place. Units that function in this manner are called "bubbling-bed" incinerators. The mass of suspended solids and gas, when active and at operating temperatures, expands to about double the at-rest volume. Sludge is mixed quickly within the fluidized-bed by the turbulent action of the bed. If required and auxiliary fuel injection ports are available, auxiliary fuel, such as oil, natural gas, or digester gas can be injected directly into the bed.

The minimum temperature needed in the sand bed prior to injection of sludge is approximately 700°C (1300°F). The temperature of the sand bed is controlled between 760 and 820°C (1400 and 1500°F). Evaporation of the water and combustion of the sludge solids takes place rapidly due to the high turbulence. The freeboard area, located above the bed, provides for residence time for the completion of combustion of the gaseous constituents. The flue gas (products of combustion) and ash leave the gas outlet through the top of the incinerator. No ash exits from the bed at the bottom of the incinerator as it is entrained with the flue gases.

Most fluid bed units that have been designed for autogenous operation should not require any auxiliary fuel after startup. The large thermal mass within the unit (i.e., the sand bed and refractory dome) helps to dampen fluctuations in heat input and also tends to retain the temperature in the unit if required to shut down for a limited duration. The large thermal mass tends to provide a stable, slow acting process that is robust, reliable, and controllable by most operators.

The fluidized-bed, though reliable, is complex and requires the use of trained personnel. Because fluidized-bed incinerators are complex, they are normally used in medium to large plants, but may be used in plants with lower flow ranges where land for the disposal of sludge is limited.

The main process vessel is very simple and contains no moving parts, which is a significant difference from the multiple-hearth incinerator design. The ancillary equipment required includes a fluidizing air blower, a burner combustion air blower, auxiliary fuel train, a sand system, an air pollution control system, and an ash handling system (which is typically combined with the air pollution control system).

Scrubber Water.   Most of the ash (over 99 percent) is captured in the scrubber water, and the total suspended solids content is approximately 20 to 30 percent of the dry solids feed (inline with the ash content of feed sludge). Plant effluent is typically used as scrubber water at a rate of approximately 25 to 40 L/kg (3 to 5 gal/lb) of dry solids feed to the fluidized-bed. The ash slurry flows from the scrubber are normally directed to an ash lagoon or are mechanically dewatered to separate the ash from water. The effluent from the ash lagoon or ash dewatering process can either be sent back to the head of the wastewater treatment process or combined with the plant effluent at the outfall. The concentrations of BOD and COD are low, typically below 50 mg/L. Additional details on air pollution control are noted later in this chapter. Particulates, criteria air contaminants, and other air emissions are typically much less than a similarly sized multiple-hearth incinerator.

Process Modifications.   Many new fluid bed units now incorporate at least one downstream air-to-air heat exchanger. The heat exchanger is used to transfer the heat from the exiting flue gases to the preheat the fluidizing air before it is introduced into the unit below the tuyeres. In this manner, heat is preserved within the fluid bed unit, despite the high volumetric flowrate of air through the bed. These units are called "hot windbox" fluid beds and are able to operate autogenously on biosolids with higher moisture contents. Some new units also incorporate energy recovery systems (separate from the fluidizing air preheat heat exchanger), which is discussed below.

A modification of the fluidized-bed incineration technology is the "circulating-bed" incinerator. In the circulating-bed unit, the reactor gas passes through the combustion chamber at much higher velocities, ranging from 3 to 8 m/s (10 to 25 ft/s). At these velocities, the bubbles in the fluidized bed disappear and streamers of solids and gas prevail. The entire mass of entrained particles flow up the reactor shaft to a particle separator, are deposited in storage momentarily, and are recirculated back to the primary combustion zone in the bottom of the reactor. Ash is removed continuously from the bottom of the bed. On turndown, the circulating bed becomes a bubbling bed.

Energy Recovery from Thermal Oxidation

Energy recovery from thermal oxidation processes is now an important component of the overall ATO process. The design of hot wind box fluidized-bed units incorporate at least one fluidizing air preheat heat exchanger as part of the main process to allow for highly efficient operation. Additional heat exchangers may be utilized to reheat the flue gases further downstream depending on the requirements of the air pollution control system. The reheating of the flue gases prior to stack tip discharge may also be desired for enhanced atmospheric dispersion or for aesthetic/plume suppression reasons. The heat recovery heat exchangers can also be used generate hot water, steam, or heated thermal oil for waste heat recovery or electricity production.

The quantity, quality, and form of the energy extracted from the ATO process depend on the heat and solids content of the dewatered sludge fed to the process. Undigested sludge with solids content greater than 25 percent offers good potential for energy recovery. One example

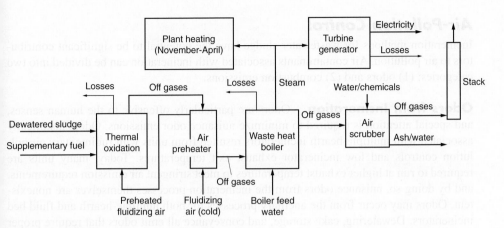

Figure 14–25

An example of an energy recovery process diagram.

of energy recovery is shown on Fig. 14–25. The hot off gases from thermal oxidation process is first introduced to an air preheater, then fed to a waste heat boiler that produces steam, which is fed to a steam turbine generator producing electricity. Depending on the electricity and fuel costs, the steam can be used to heat the buildings in the winter months instead of the turbine generator. Other uses for the steam extracted from the ATO process include process heating for digesters, heating FOG systems, or dewatering and drying processes.

Because generating and operating high pressure steam facilities require specialized boiler and pressure vessel operators, this application is not used widely. In this case, the hot off gases generated from the thermal oxidation process can be used to heat thermal oil or hot water that is used to generate electricity using the organic Rankin cycle. Other sources of waste heat (typically lower grade heat sources) which have been utilized at biosolids ATO facilities include capturing and recuperating heat lost from the exterior of the incinerator vessel via an air to glycol heat reclaim circuit, recovering heat from the scrubber water prior to the ash lagoon or ash dewatering with a water to water heat exchanger, and recovering heat from the residual temperature of the flue gas.

Coincineration with Municipal Solid Waste

Coincineration is the process of incinerating wastewater sludges with municipal solid wastes. Coincinerator types can vary but many consist of reciprocating grates that are designed for more of a solid fuel than the multiple heath and fluidized-bed incinerators discussed in the previous section. The major objective of coincineration is to reduce the combined costs of incinerating sludge and solid wastes. Coincineration is not practiced widely. The process has the advantages of producing the heat energy necessary to evaporate water from sludges, supporting combustion of solid wastes and sludge, and providing an excess of heat for steam generation, if desired, without the use of auxiliary fossil fuels. In properly designed systems, the hot gases from the process can be used to remove moisture from sludges to a content of 10 to 15 percent. Direct feeding of sludge cake containing 70 to 80 percent moisture over solid wastes on traveling or reciprocating grates has been found to be ineffective. For systems operating without heat recovery, a disposal ratio of 1 kg (2.2 lb) of dry wastewater sludge to 4.6 kg (11 lb) of solid wastes is fired in normal operation. In the case of the water-walled boiler with heat recovery, the ratio is approximately 1 kg of dry (industrial plant) sludge to 7 kg (17 lb) of solid wastes. Based on past experience in municipal solid-waste disposal, the application of coincineration will likely continue to proceed very slowly, despite the advantages to the community in combining the two waste-disposal functions.

Air-Pollution Control

Incineration methods for wastewater sludge have the potential to be significant contributors to air pollution. Air contaminants associated with incineration can be divided into two categories: (1) odors and (2) combustion emissions.

Odors from Incineration. Odors are particularly offensive to the human senses, and special attention is required to minimize nuisance odor emissions. Odors historically associated with multiple-hearth incinerators resulted from units with rudimentary air pollution controls and low incinerator exhaust exit temperatures. Today, many units are required to run at higher exhaust temperatures to meet stringent air emission requirements, and by doing so, nuisance odors from the incineration processes themselves are nonexistent. Odors may occur from the ancillary processes for both multiple-hearth and fluid bed incinerators. Dewatering, cake storage, and conveyance all emit odors that require proper capture, treatment, and discharge. Incinerators can generally use the foul air as combustion air in the ATO process resulting in cost effective odor elimination.

Emissions from Incineration. Combustion emissions vary depending upon the type of thermal reduction technology employed and the nature of the sludge and auxiliary fuel used in the combustion process. Combustion emissions of particular concern are the criteria air pollutants (i.e., particulate matter, carbon monoxide, oxides of nitrogen, and sulfur dioxide), acid gases (mainly hydrogen chloride), and hazardous air pollutants (mercury, cadmium, lead, dioxins and furans). Under Part 503 regulations in the United States, heavy metals (arsenic, beryllium, chromium, and nickel), and total hydrocarbons (or carbon monoxide) are also a concern for regulators. Regulations promulgated by the U.S. EPA in 2011 concerning incinerator emissions are discussed in Sec. 13–2 and are generally more strict than the rules in the European Union.

Generally, emission controls fall into three main categories. These are (1) source control, (2) combustion control, and (3) air pollution control equipment. For certain pollutants, namely the heavy metals, source control is one of the most cost effective methods of reducing air pollution from the incinerator stack.

Source Control. Source control programs and collection systems use bylaw enforcement, and directed control programs (such as dental amalgam separators for mercury reduction) are effective at reducing loadings into the downstream collection system, which eventually ends up in the biosolids.

Combustion Control. Nitrogen oxides, carbon monoxide, total hydrocarbons, certain volatile organic compounds, dioxin, and furans are all air pollutants that are formed during the ATO process and combustion controls are effective at reducing levels of these contaminants in the flue gases. Most of these pollutants will be decreased to acceptable levels with elevated temperatures in the incinerator exhaust (generally greater than 750°C) and sufficient residence time (greater than 1 s). However, increased levels of nitrogen oxides may form with highly elevated temperatures, so a careful balance must be maintained in the specific details of the incinerator design and operation.

Air Pollution Control. For both multiple-hearth and fluid bed incinerators, the typical (historically most common, as least in North America) type of air pollution control equipment has been discussed previously. However, advances in the level of air pollution control equipment are being driven by stricter legislation. In general because fluid bed ATO units produce cleaner emissions than multiple-hearths, the industry is moving towards fluid bed technology for new systems. Air pollution control equipment is used as the final step to reduce the remaining pollutant levels in the flue gases. Although the majority of systems in North America have historically been wet scrubbers, there are also "dry" systems which are more common in Europe and Asia and more closely resemble the equipment utilized on solid waste incinerators. For primarily fluid bed incinerators, ammonia or urea injection may be used for selective non-catalytic reduction of nitrogen oxides. This treatment is done early while the flue gases are hot as they exit from the incinerator.

Advanced wet systems are now being used which contain multiple throat venturi scrubbers for particulate and heavy metal removal, caustic wet scrubbing for residual acid gas and sulfur dioxide removal, wet electrostatic precipitation for enhanced fine particulate and metal removal and polishing. A packed bed of sulfur impregnated media is sometimes used for the removal of mercury. Unfortunately, the mercury removal process requires a significant quantity of water, which can result in large quantities of wastewater. The generation of wastewater is usually not a problem for biosolids incinerators, as they are most commonly co-located with a wastewater treatment plant. Wet ash handling is used similar to the traditional wet venturi scrubber system.

Dry systems typically employ some method of flue gas cooling (whether this is part of a waste heat recovery boiler or a quench tower) to temper the gases before proceeding to the downstream equipment. Powdered activated carbon is injecting into the flue gas stream for heavy metal removal, and dioxin and furan control. Lime may also be injected for sulfur dioxide and acid gas control. A bag house, employing filter bags, is used to filter the flue gases. For this type of system, the ash is dry and is sent to silos for truck loading and disposal.

Hybrid systems may be employed which incorporate components of wet and dry air pollution control trains depending on the requirements of the level of pollution control, the site constraints, and the availability of water.

14–5 COMPOSTING

Composting is a process in which organic material undergoes biological degradation to a stable end product. Sludge that has been composted properly is a nuisance-free, humus-like material. Approximately 20 to 30 percent of the volatile solids are converted to carbon dioxide and water. As the organic material in the sludge decomposes, the compost heats to temperatures in the pasteurization range of 50 to 70°C (120 to 160°F), and enteric pathogenic organisms are destroyed. Properly composted biosolids may be used as soil conditioners in agricultural or horticultural applications, subject to any limitations based on the constituents in the composed biosolids (WEF, 2010). Composting is a cost-effective and environmentally sound alternate for the stabilization of wastewater sludges. Composting can be practiced for dewatered sludges or dewatered digested biosolids. However, composting biosolids is preferred, due to the odor issues during the composting process itself and the quality of the compost product.

Although composting may be accomplished under anaerobic or aerobic conditions, essentially all municipal wastewater biosolids composting applications are under mostly aerobic conditions (composting is never completely aerobic). Aerobic composting accelerates material decomposition and results in the higher rise in temperature necessary for pathogen destruction. Aerobic composting also minimizes the potential for nuisance odors.

The anticipated daily production of biosolids from a wastewater-treatment facility will have a pronounced effect on the alternate composting systems available for use, as will the availability of land for the construction of the composting facility. Other factors affecting

Figure 14–26

Stages during composting as related to carbon dioxide respiration and temperature (Epstein, 1997.).

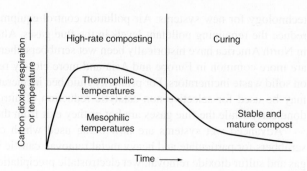

the type of composting system are the nature of the biosolids produced; stabilization, if any, of the biosolids prior to composting; the type of dewatering equipment and chemicals used. Biosolids that are stabilized by aerobic or anaerobic digestion prior to composting may result in reducing the size of the composting facilities by up to 40 percent.

Process Microbiology

The composting process involves the complex destruction of organic material coupled with the production of humic acid to produce a stabilized end product. The microorganisms involved fall into three major categories: bacteria, actinomycetes, and fungi. Although the interrelationship of these microbial populations is not fully understood, bacterial activity appears to be responsible for the decomposition of proteins, lipids, and fats at thermophilic temperatures, as well as for much of the heat energy produced. Fungi and actinomycetes are also present at varying levels during the mesophilic and thermophilic stages of composting and appear to be responsible for the destruction of complex organics and the cellulose supplied in the form of amendments or bulking agents.

Composting Process Stages

During the composting process, three separate stages of activity and associated temperatures are observed: mesophilic, thermophilic, and curing (cooling) (see Fig. 14–26). In the initial mesophilic stage, the temperature in the compost pile increases from ambient to approximately 40°C (104°F) with the appearance of fungi and acid-producing bacteria. As the temperature in the composting mass increases to the thermophilic range of 40 to 70°C (104 to 160°F), these microorganisms are replaced by thermophilic bacteria, actinomycetes, and thermophilic fungi. It is in the thermophilic temperature range that the maximum degradation and stabilization of organic material occur. The curing stage is characterized by a reduction in microbial activity, and replacement of the thermophilic organisms with mesophilic bacteria and fungi. During the curing stage, further evaporative release of water from the composted material will occur, as well as stabilization of pH and completion of humic acid formation.

Composting Process Steps

Most composting operations consist of the following basic steps (see Fig. 14–27): (1) preprocessing, the mixing of dewatered sludge with an amendment and/or a bulking agent; (2) high-rate decomposition, aerating the compost pile either by the addition of air, by mechanical turning, or by both; (3) recovery of the bulking agent (at the end of either the high-rate decomposition or curing phase, if practical); (4) further curing and storage, which allows further stabilization and cooling of the compost; (5) postprocessing,

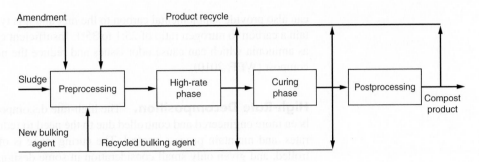

Figure 14–27

Generalized process diagram for composting showing inputs of sludge (feed substrate), amendments, and bulking agents (Haug, 1993.).

screening for the removal of nonbiogradable material such as metals and plastics or grinding for size reduction, and (6) final disposition. A portion of the final product is sometimes recycled to the preprocessing step to aid in conditioning the compost mixture.

Feed Stock Amendments. An amendment is an organic material added to the feed substrate primarily to, reduce moisture content, and increase the air voids for proper aeration. Amendments can also be used to increase the quantity of degradable organics in the mixture. Commonly used amendments are sawdust, straw, recycled compost, and rice hulls. A bulking agent is an organic or inorganic material that is used to provide structural support and to increase the porosity of the mixture for effective aeration. Wood chips are the most commonly used bulking agents and can be recovered and reused. The characteristics of bulking agents used most commonly are reported in Table 14–18. An amendment

Table 14–18

Characteristics of compost bulking agents used in the aerobic composting of sludge from wastewater treatment[a]

Bulking agent	Comments
Wood chips	May have to be purchased
	High recovery rate by screening
	Provides supplemental carbon source
Chipped brush	Possibly available as a waste material
	Low recovery rate by screening
	Provides supplemental carbon source
	Longer curing time of compost
Leaves and yard waste	Must be shredded
	Wide range of moisture content
	Readily available source of carbon
	Relatively low porosity
	Nonrecoverable
Shredded tires	Often mixed with other bulking agents
	Supplemental carbon is not available
	Nearly 100 percent recoverable
	May contain metals
Ground waste lumber	Possibly available as a waste material
	Often a poor source of supplemental carbon

[a] Adapted in part from WEF (2010).

can also provide supplemental carbon to the mix and it is typically recommended to maintain a carbon to nitrogen ratio of 25:1 to 35:1. Insufficient carbon can lead to nitrogen loss as ammonia which can cause odor issues and reduce the nutrient content of the resulting compost (WEF, 2010).

High-Rate Decomposition. The high-rate decomposition stage of composting has been more engineered and controlled due to the need to reduce odors, supply high aeration rates, and maintain process control. The curing stage is often less engineered, less controlled, and given only small consideration in some designs, however, the curing stage is an integral part of the system design and operation, and both stages need to be designed and operated properly to produce a mature compost product.

Recovery of the Bulking Agent. If practicable, the bulking agent should be recovered at the end of either the high-rate decomposition or curing phase. The most common method for recovery of bulking agent is screening with mesh size based on the physical size of the bulking agent.

Postprocessing. Postprocessing is often used to prepare the finished compost for marketing. Preparation includes conveying the finished compost from the active composting area to the curing, screening, and preparation areas. Trommel screens and belt shredders are used frequently; shredding can precede or follow curing. In some cases, double screening is preferable, especially for the horticultural market to meet product quality requirements. Particle size of the finished product for general use ranges typically from 6 to 25 mm (1/4 to 1 in.).

Composting Methods

The two principal methods of composting now in use in the United States may be classified as agitated or static. In the agitated method the material to be composted is agitated periodically to introduce oxygen, to control the temperature, and to mix the material to obtain a uniform product. In the static method, the material to be composted remains static and air is blown through the composting material. The most common agitated and static methods of composting are known as the windrow and static pile methods, respectively. Proprietary composting systems in which the composting operation is carried out in a reactor of some type are known as in-vessel composting systems.

Windrow. In a windrow system, a mixture of dewatered sludge and bulking agent is placed in windrows, which are typically from 1 to 2 m (3 to 6 ft) high and 2 to 4.5 m (6 to 14 ft) at the base [see Fig. 14–28(a)]. The windrows are turned and mixed periodically during the composting period using specialized equipment [see Fig. 14–28(b)]. Supplemental mechanical aeration is used in some applications. The composting period is about 21 to 28 d. Under typical operating conditions, the windrows are turned a minimum of five times while the temperature is maintained at or above 55°C. In windrow composting, aerobic conditions are difficult to maintain throughout the cross-sectional area of the windrow. Thus, the microbial activity within the pile may be aerobic, facultative, anaerobic, or various combinations thereof, depending on when and how often the pile is turned. Turning of the windrows is often accompanied by the release of

Figure 14–28

Composting systems: (a) view of compost windrows, (b) view of equipment for turning and grading windrows, (c) schematic of aerated static pile compost process, and (d) view of force-aerated static pile in compost bags for odor control.

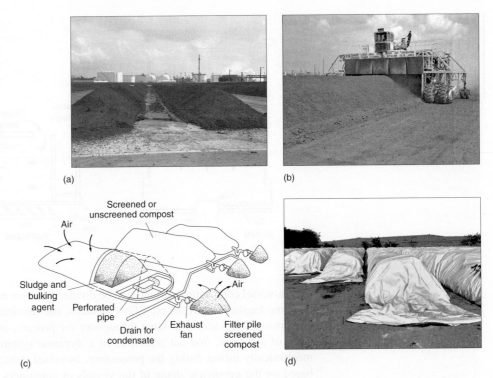

(a)

(b)

(c)

(d)

offensive odors. The release of odors occurs typically when anaerobic conditions develop within the windrow. Some windrow operations are covered or enclosed, similar to aerated static piles.

Aerated Static Pile. The aerated static pile system consists of a grid of aeration or exhaust piping over which a mixture of dewatered sludge and bulking agent is placed [see Fig. 14–28(c)]. An alternative version of the static pile composting method is carried out in an enclosed plastic bag with forced aeration [see Fig. 14–28(d)]. In a typical static pile system, the bulking agent consists of wood chips, which are mixed with the dewatered sludge by a pug-mill type or rotating-drum mixer or by movable equipment such as a front-end loader. Material is composted for 21 to 28 d and is typically followed by a curing period of 30 d or longer. Typical pile heights are generally about 2 to 2.5 m (6 to 8 ft). A layer of screened compost is often placed on top of the pile for insulation. Disposable corrugated plastic drainage pipe is commonly used for air supply and each individual pile is recommended to have an individual blower for more effective aeration control. Screening of the cured compost usually is done to reduce the quantity of the end product requiring ultimate disposal and to recover the bulking agent. For improved process and odor control, many facilities cover or enclose all or significant portions of the system.

In-Vessel Composting Systems. In-vessel composting is accomplished inside an enclosed container or vessel. Mechanical systems are designed to minimize odors and process time by controlling environmental conditions such as air flow, temperature, and oxygen concentration. The advantages of in-vessel composting systems are better process and odor control, faster throughput, lower labor costs, and smaller area requirements.

Figure 14–29

Plug flow in-vessel composting reactors: (a) unmixed vertical plug flow reactor and (b) unmixed tunnel (horizontal plug flow) reactor.

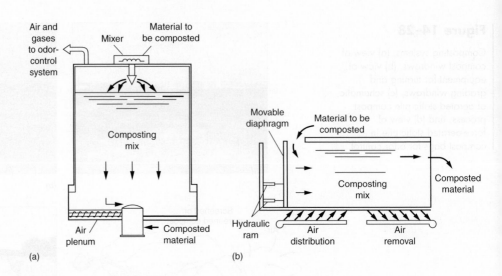

(a)

(b)

In-vessel composting systems can be divided into two major categories: plug flow and dynamic (agitated bed). In plug-flow systems, the relationship between particles in the composting mass stays the same throughout the process, and the system operates on the basis of a first-in, first-out principle. In a dynamic system, the composting material is mechanically mixed during the processing. In-vessel systems can be further categorized based on the geometric shape of the vessels or containers used. Examples of plug-flow reactors are shown on Fig. 14–29 and examples of dynamic-type systems are illustrated on Fig. 14–30.

Design Considerations

A number of factors, each of which must be evaluated to meet the specific requirements, must be considered in the design of a composting system (see Table 14–19). A design approach using a materials balance is particularly useful because the amount of each component (sludge or biosolids, bulking agent, and amendment) used during each phase of the process is determined. In a materials balance, the following parameters must be measured or calculated for each component: (1) total volume, (2) total wet weight, (3) total solids content (dry weight), (4) volatile solids content (dry weight), (5) water content

Figure 14–30

Dynamic (mixed) in-vessel composting units: (a) vertical reactor and (b) horizontal reactor.

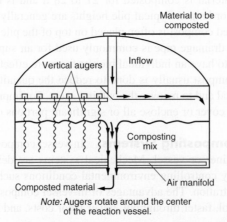

Note: Augers rotate around the center of the reaction vessel.

(a)

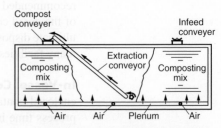

Note: Extraction conveyor either mixes the compost in the reactor or discharges compost to the compost conveyor.

(b)

Table 14–19

Design considerations for aerobic sludge composting processes[a]

Item	Comment
Type of sludge	Both sludge and biosolids can be composted successfully. Sludge has a greater potential for odors, particularly for windrow systems. Sludge, as compared to biosolids, has more energy available, will degrade more readily, and has a higher oxygen demand.
Amendments and bulking agents	Amendment and bulking agent characteristics, such as moisture content, particle size, and available carbon, affect the process and quality of product. Bulking agents should be readily available.
Carbon-nitrogen ratio	The initial C:N ratio should be in the range of 25:1 to 35:1 by weight. At lower ratios, ammonia is given off. Carbon should be checked to ensure it is readily biodegradable.
Volatile solids	The volatile solids of the composting mix should be greater than 30 percent of the total solids content. Dewatered sludge will usually require an amendment or bulking agent to adjust the solids content.
Air requirements	Air with at least 50 percent of the oxygen remaining should reach all parts of the composting material for optimum results, especially in mechanical systems.
Moisture content	Moisture content of the composting mixture should be not greater than 40 percent for static pile, windrow, and in-vessel composting.
pH control	The pH of the composting mixture should generally be in the range of 6 to 9. To achieve optimum aerobic decomposition, pH should remain at 7 to 7.5 range.
Temperature	For best results, temperature should be maintained between 50 and 55°C for the first few days and between 55 and 60°C for the remainder of the active composting period. If the temperature is allowed to increase beyond 65°C for a significant period of time, biological activity will be reduced.
Control of pathogens	If properly conducted, it is possible to kill all pathogens, weeds, and seed during the composting process. To achieve this level of control, the temperature must be maintained at levels required by the EPA 503 regulations for Class A (see Table 13–11).
Mixing and turning	To prevent drying, caking, and air channeling, material in the process of being composted should be mixed or turned on a regular schedule or as required. Frequency of mixing or turning will depend on the type of composting operation.
Heavy metals and trace organics	Heavy metals and trace organics in the sludge and finished compost should be monitored to ensure that the concentrations do not exceed the applicable regulations for end use of the product.
Site constraints	Factors to be considered in selecting a site include available area, access, proximity to treatment plant and other land uses, climatic conditions, and availability of buffer zone.

[a] Adapted in part from Tchobanoglous, et al. (1993).

(weight), (6) bulk density (wet weight/unit volume), (7) percent water content, and (8) percent volatile solids of the compost mix.

An important output of the materials balance is to determine the composition of the compost mix. The compost mix should be about 40 percent dry solids to ensure adequate composting in windrow and static-pile composting. In-vessel systems require similar solids requirements, but slightly lower values may be used, depending on the aeration system. Starting the compost process with solids higher than 40 percent soon results in a "dry mix." A dry mix is dusty, and sufficient biological activity and temperature levels are difficult to maintain. The addition of water will likely be required for the duration of the process; thus, provisions for a water supply should be made.

The effect of moisture content in the dewatered sludge on the compost mix is illustrated on Fig. 14–31. The moisture content of the sludge affects the wet weight of the mixture and the amount of amendment that has to be used. Using Fig. 14–31 (a), for

Figure 14–31

Effect of sludge solids content on compost mix and amendment quantities: (a) mix quantities versus sludge solids content, (b) amendment requirements versus sludge solids content. (U.S. EPA, 1989.)

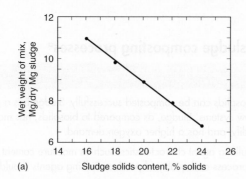

(a)

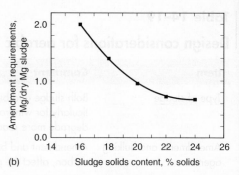

(b)

example, if the sludge cake contains 24 percent solids, the wet weight of the mix is about 6.7 Mg (tonne) per dry Mg of sludge. If the sludge solids content decreases to 16 percent, the wet weight increases to about 11 Mg per dry Mg of sludge. The additional moisture content would require larger materials-handling systems and larger reactors. The amendment requirements, as indicated on Fig. 14–31(b), would triple over the same range of sludge solids. In compost-system design, the type of sludge-dewatering system and the consistency of resulting product have to be evaluated carefully.

Co-composting with Municipal Solid Wastes

Co-composting of sludge and municipal solid wastes is a possible alternative where integrated waste-disposal facilities are considered. Mixing the sludge with the organic fraction of municipal solid waste or source separated yard wastes is beneficial because (1) sludge dewatering may not be required, and (2) the overall metals content of the composted material will be less than that of the composted sludge alone. Liquid treatment plant sludges typically have a solids content ranging from 3 to 8 percent. A 2-to-1 mixture of compostable municipal solid or yard wastes to sludge is recommended as a minimum. Both static and agitated compost systems have been tried (Tchobanoglous et al., 1993). Two examples of composting facilities in reactors with mechanical mixing, located within a building for odor control, are shown on Fig. 14–32.

Public Health and Environmental Issues

The principal public health and environmental issues concerning compost operations relate to exposure to pathogens and bioaerosols. Exposure to pathogens can occur during the composting process or through the use of the product if the composting process is not executed properly and the resulting product is not disinfected. The potential modes of infection for workers are (1) inhalation of aerosols containing airborne microorganisms,

Figure 14–32

Examples of compost reactors with mechanical mixing. The reactors are in enclosed buildings for odor control.

(a)

(b)

(2) dermal contact, or (3) oral through inadvertent contact of dust or contaminated food or through hand-to-mouth contact such as cigarette smoking. Ingestion of contaminated product or contamination by the product to cigarettes or food is the greatest potential source of pathogen invasion by workers or users (Epstein, 1997).

Compost bioaerosols are organisms or biological agents that can be dispersed through the air and affect human health. Bioaerosols can contain living organisms including bacteria, fungi, actinomycetes, arthropods, protozoa and microbial products such as endotoxin and microbial enzymes. During composting, bioaerosols are not only present in waste materials but also can be generated during the process. The level and type of bioaerosols are a function of feedstock. The two bioaerosols of greatest interest to worker health and the environment surrounding composting facilities are *Aspergillus fumigatus* and endotoxin.

A. fumigatus, a common fungus, is of concern to both worker health and populations surrounding composting facilities, as it can cause lung disease. Endotoxin, part of the cell wall of gram-negative bacteria that is released to the environment during the composting process, is of primary concern to workers in composting, recycling, and other solid-waste processing facilities. There is little evidence that exposure to airborne endotoxin causes toxic conditions. Most of the data, however, concerning worker illness is associated with composting municipal solid waste, principally in Europe. Proper ventilation, dust control, and use of dust masks reduce worker exposure to bioaerosols (Epstein, 1997).

14-6 SLUDGE AND BIOSOLIDS CONVEYANCE AND STORAGE

The sludges from primary and biological treatment processes are concentrated and stabilized by mechanical, biological, and thermal means and are reduced in volume in preparation for final disposal. Because the methods of conveyance and final disposal often determine the type of stabilization required and the amount of volume reduction that is needed, they are considered briefly in the following discussion.

Conveyance Methods

Biosolids may be transported long distances by (1) pipeline, (2) truck, (3) barge, (4) rail, or any combination of these four modes. Truck transportation, however, is the most common method used currently [see Fig. 14–33(a)]. To minimize the danger of spills, odors, and dissemination of pathogens to the air, liquid biosolids should be transported in closed vessels, such as tank trucks, railroad tank cars, or covered tank barges. Stabilized, dewatered biosolids can be transferred in open vessels, such as dump trucks, or in railroad gondolas. If biosolids are hauled long distances, the vessels should be covered. The method of transportation chosen and its costs are dependent on a number of factors,

Figure 14–33

Views of sludge transport vehicle and storage facilities: (a) typical side dump sludge transport vehicle, and (b) large bins used for the temporary storage for lime-stabilized sludge. Overhead bucket is used to distribute sludge in bins and to transport to trucks.

(a)   (b)

including (1) the nature, consistency, and quantity of biosolids to be transported; (2) the distance from origin to destination; (3) the availability and proximity of the transit modes to both origin and destination; (4) the degree of flexibility required in the transportation method chosen; and (5) the estimated useful life of the ultimate disposal facility.

Each transportation method contributes a minor air pollutant load, either directly or indirectly. A certain amount of air pollution is produced from the facility that generates electricity necessary for sludge pumping. The engines that move trucks, barges, and railroad cars also produce some air pollutants. On a mass (tonnage) basis, the transportation mode that contributes the lowest pollutant load is piping. Next, in sequence, are barging and unit train rail transportation. The highest pollutant load is from trucking. Other factors of environmental concern include traffic, noise, and construction disturbance.

Storage

It is often necessary to store biosolids that have been digested anaerobically before they are disposed of or used beneficially. Although sludge can be stored for short periods of time in clarifiers, biological systems, digesters or holding tanks, the available storage capacity may not be sufficient for long term storage. Storage of liquid biosolids can be accomplished in storage basins and lagoons, and storage of dewatered biosolids can be done on storage pads. An example of large temporary sludge storage bins for lime-stabilized biosolids is shown on Fig.14–33 (b). Drying beds, discussed previously, can also be used for storage.

Storage Basins and Lagoons. Biosolids stored in basins become more concentrated and are further stabilized by continued anaerobic biological activity. Long-term storage is effective in pathogen destruction. Depth of the biosolids storage basins may vary from 3 to 5 m (10 to 16 ft). Solids loading rates vary from about 0.1 to 0.25 kg VSS/m^2·d (20 to 50 lb VSS/10^3 ft^2·d) of surface area. If the basins are not loaded too heavily ($\leq$ 0.1 kg VSS/m^2·d), it is possible to maintain an aerobic surface layer through the growth of algae and by atmospheric reaeration. Alternatively, surface aerators can be used to maintain aerobic conditions in the upper layers. The number of basins to be used should be sufficient to allow each basin to be out of service for a period of about 6 months. Stabilized and thickened biosolids can be removed from the basins using a mud pump mounted on a floating platform or by mobile crane using a drag line. Biosolids concentrations as high as 35 percent solids have been achieved in the bottom layers of these basins.

Long-term storage of sludge and biosolids in lagoons is simple and economical if the treatment plant is in a remote location. A lagoon is an earthen basin into which untreated sludge, or digested biosolids are deposited. In lagoons with untreated sludge, the organic matter is stabilized by anaerobic and aerobic decomposition, which may give rise to objectionable odors. The stabilized biosolids settle to the bottom of the lagoon and accumulate, and excess liquid from the lagoon, if there is any, is returned to the plant for treatment. Lagoons should be located away from highways and dwellings to minimize possible nuisance conditions and should be fenced to keep out unauthorized persons. Lagoons should be relatively shallow, 1.25 to 1.5 m (4 to 5 ft), if they are to be cleaned by scraping. If the lagoon is used only for digested biosolids, the nuisances mentioned should not be a problem. If subsurface drainage and percolation are potential problems, the lagoon should be lined. Solids may be stored indefinitely in a lagoon and may be removed periodically after draining and drying.

Storage Pads. Where dewatered biosolids have to be stored prior to land application, sufficient storage area should be provided based on the number of consecutive days that biosolids hauling could occur without applying biosolids to land. Allowances also have to be made for paved access and for area to maneuver the biosolids hauling trucks, loaders, and application vehicles. The storage pads should be constructed of concrete and designed to withstand the truck loadings and biosolids piles. Provisions for leachate and stormwater collection and disposal also have to be included in the design of sludge storage pads.

14–7 SOLIDS MASS BALANCES

Sludge and biosolids processing facilities, such as thickening, digestion, and dewatering, produce waste streams that must be recycled to the treatment process or to treatment facilities designed specifically for the purpose. The recycled flows impose an incremental solids, hydraulic, organic, and nutrient load on the wastewater-treatment facilities that must be considered in the plant design. When the flows are recycled to the treatment process, they should be directed to the head of the plant and blended with the plant flow following preliminary treatment. Equalization facilities can be provided for the recycled flows so that their reintroduction into the plant flow will not cause a shock loading on the subsequent treatment processes. To predict the incremental loads imposed by the recycled flows, it is necessary to perform a materials mass balance for the treatment system.

Preparation of Solids Mass Balances

Typically, a materials mass balance is computed on the basis of average flow, average BOD and total suspended solids concentrations. To size certain facilities properly, such as sludge storage tanks and plant piping, it is also important to perform a materials mass balance for the maximum expected concentration of BOD and TSS in the untreated wastewater. However, the maximum concentrations will not usually result in a proportional increase in the recycled BOD and TSS. The principal reason is that the storage capacity in the wastewater and sludge-handling facilities tends to dampen peak solids loads. For example, for a maximum TSS load equal to twice the average value, the resulting peak solids loading to a dewatering unit may be only 1.5 times the average loading. Further, it has been shown that periods of maximum hydraulic loading typically do not correlate with periods of maximum BOD and TSS. Therefore, coincident maximum hydraulic loadings should not be used in the preparation of a materials mass balance for maximum organic loadings (see Chap. 5). The preparation of a mass balance is illustrated in Example 14–3.

Performance Data for Solids Processing Facilities

To prepare a materials mass balance, it is necessary to have information on the operational performance and efficiency of the various unit operations and processes that are used for the processing of waste sludge and biosolids. Representative data on the solids capture and expected solids concentrations for the most commonly used operations are reported in Tables 14–20 and 14–21. These data were derived from an analysis of the records from a number of installations throughout the United States. The wide variation that can occur in the reported values is apparent; thus, the values in Tables 14–20 and 14–21 should be used only if no other information is available. Wherever possible, local conditions and data should be used in performing the mass balance.

Impact of Return Flows and Loads

In addition to performance data for expected solids capture and constituent concentrations for the various process components, data for the expected concentrations of BOD and TSS in the return flows must also be included in preparing of mass balances. If the quantities

Table 14–20

Typical solids concentration and capture values for various sludge and biosolids processing methods

Operation	Solids concentration, %		Solids capture %	
	Range	Typical	Range	Typical
Gravity thickeners:				
Primary sludge only	3–10	5	85–92	90
Primary and waste activated	2–6	3.5	80–90	85
Flotation thickeners:				
With chemicals	4–6	5	90–98	95
Without chemicals	3–5	4	80–95	90
Centrifuge thickeners:				
With chemicals	4–8	5	90–98	95
Without chemicals	3–6	4	80–90	85
Belt-filter press:				
With chemicals	15–30	22	85–98	95
Filter press:				
With chemicals	20–50	36	90–98	95
Centrifuge dewatering:				
With chemicals	10–35	25	85–98	95

Table 14–21

Typical BOD and total suspended-solids (TSS) concentrations in the recycle flows from various sludge processes[a]

Operation	BOD, mg/L		Suspended Solids, mg/L	
	Range	Typical	Range	Typical
Gravity thickening supernatant:				
Primary sludge only	100–400	250	80–300	200
Primary + waste activated sludge	60–400	300	100–350	250
Flotation thickening subnatant	50–1200	250	100–2500	300
Centrifuge thickening centrate	170–3000	1000	500–3000	1000
Aerobic digestion supernatant	100–1700	500	100–10,000	3400
Anaerobic digestion (two-stage, high rate) supernatant	500–5000	1000	1000–11,500	4500
Centrifuge dewatering centrate	100–2000	1000	200–20,000	5000
Belt-filter press filtrate	50–500	300	100–2000	1000
Recessed-plate-filter press filtrate	50–250		50–1000	
Sludge lagoon supernatant	100–200		5–200	
Sludge drying bed underdrainage	20–500		20–500	

(continued)

Table 14–21 (*Continued*)	BOD, mg/L		Suspended Solids, mg/L	
Operation	**Range**	**Typical**	**Range**	**Typical**
Composting leachate		2000		500
Incinerator scrubber water	20–60		600–8000	
Depth filter washwater	50–500		100–1000	
Microscreen washwater	100–500		240–1000	
Carbon adsorber washwater	50–400		100–1000	

[a] Adapted, in part, from U.S. EPA (1987c) and WEF (2010).

and characteristics of recycled flows and loads are not accounted for properly, the facilities that receive them may be underdesigned significantly. The major impacts of return flows and measures that can mitigate these impacts are summarized in Table 14–22. Impact of returned flows on the overall treatment process is discussed in detail in Chap. 15.

Table 14–22

Major impacts and potential mitigation measures for return flows from sludge and biosolids-processing facilities[a]

Source of return flow	Impact	Process impacted	Mitigation measure
Sludge thickening	Effluent degradation by colloidal SS	Sedimentation	Add flocculent aid ahead of sedimentation tank
			Separately thicken primary and biological sludges
			Optimize gravity thickener dilution water
	Floating sludge	Sedimentation	Minimize gravity thickener detention time
			Remove sludge continuously and uniformly
	Odor release and septicity	Recycle point	Reduce gravity thickener detention time
			Return flows ahead of aerated grit chamber
			Provide odor containment, ventilation, and treatment (scrubber or biofilter)
		Biological	Return odorous flows to aeration tank
			Remove sludge continuously and uniformly
			Provide separate return flow treatment (with other recycle streams)
	Solids buildup	Sedimentation	Increase dewatering unit operation time
		Biological	Remove sludge continuously and uniformly
			Include recycle loads in mass balance analysis
Sludge dewatering	Effluent degradation by colloidal suspended solids	Sedimentation	Optimize dewatering units solids capture by improved sludge conditioning
			Add flocculent aid ahead of sedimentation tank
			Return centrate/filtrate to thickener
			Provide separate return flow treatment (with other recycle streams)
	Solids buildup	Sedimentation	Increase dewatering unit operation time

(*continued*)

Table 14–22 (Continued)

Source of return flow	Impact	Process impacted	Mitigation measure
		Biological	Remove sludge continuously and uniformly
			Reduce trickling-filter recycle rate
			Include recycle loads in mass balance analysis
Sludge stabilization	Effluent degradation by excessive BOD load	Biological	Optimize supernatant/decant removal, i.e., remove smaller amounts over a longer period of time, or reschedule removal to off-peak periods
			Provide separate return flow treatment
			Increase RBC speed
			Increase MLVSS in activated-sludge system (decrease F:M ratio)
			Increase dissolved oxygen level in activated-sludge process
	Effluent degradation by nutrients	Biological	Regulate digester supernatant/decant removal
			Thicken sludge before stabilization
			Provide separate return flow treatment
Washwater from depth filters	Hydraulic surges	Sedimentation	Provide backwash storage for flow equalization
			Schedule filter backwashing for off-peak periods

[a] Adapted, in part, from U.S. EPA (1987b).

EXAMPLE 14–3 **Preparation of a Solids Mass Balance for a Secondary Treatment Facility** Prepare a solids balance for the treatment flow diagram shown in the following figure, using an iterative computational procedure.

1. Definition of terms

   BOD_C = biochemical oxygen demand expressed as a concentration, g/m^3
   BOD_M = biochemical oxygen demand expressed as a mass, kg/d
   TSS_C = total suspended solids expressed as a concentration, g/m^3
   TSS_M = total suspended solids expressed as a mass, kg/d
   Assume for the purpose of this example that the following data apply:

2. Wastewater flowrates
   a. Average dry weather flowrate = 21,600 m^3/d
   b. Peak dry weather flowrate = 2.5(21,600 m^3/d) = 53,900 m^3/d

3. Influent characteristics
   a. BOD_C = 375 g/m^3
   b. TSS_C = 400 g/m^3 (assume VSS_c / TSS_c ratio in influent = 67%)
   c. TSS_C after grit removal = 360 g/m^3 (assume volatile fraction of the grit = 10%)

4. Sludge and biosolids characteristics
   a. Concentration of primary sludge = 6%
   b. Concentration of thickened waste-activated sludge = 4% (assume solids capture in the flotation thickeners = 90%)
   c. Total suspended solids in digested sludge = 5%
   d. For the purposes of this example, assume that the specific gravity of the solids from the primary sedimentation tank and the flotation thickener is equal to 1.0

     e. Fraction of the biological solids that are biodegradable = 65%

     f. The value of BOD_C can be obtained by multiplying the value of UBOD by a factor of 0.68 (corresponds to a k value of 0.23 d^{-1} in the BOD equation, see Chap. 2)

5. Effluent characteristics

     a. $BOD_C = 20$ g/m^3

     b. $TSS_C = 22$ g/m^3

6. Primary clarifier

     a. Assumes performance of 33% BOD removal and 70% TSS removal

     b. Assume VSS_C/TSS_C ratio in primary effluent going to the secondary treatment process = 85%

7. Secondary treatment

     a. Assume the mixed-liquor VSS_C/TSS_C ratio is 0.8

     b. Aeration tank volume, $V_r = 4700$ m^3

     c. $Y = 0.5$ kg/kg

     d. $b = 0.06$ d^{-1}

     e. SRT = 10 d

8. Sludge digestion

     a. Assume SRT = 20 d

     b. Assume VSR = 50%

     c. Assume digester gas production = 1.12 m^3/kg VSS destroyed

     d. Assume BOD_C in digester supernatant = 1000 g/m^3

     e. Assume TSS_C in digested sludge = 5%

9. Sludge dewatering

     a. Assume sludge cake = 22% solids

     b. Specific gravity of sludge = 1.06

     c. Solids capture = 93%

     d. Centrate $BOD_C = 2000$ mg/L

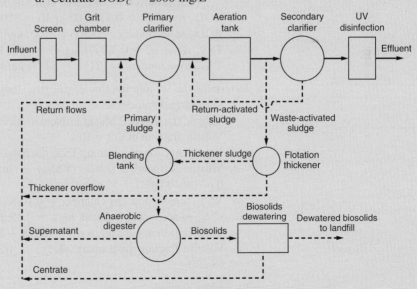

Solution

1. Convert the given constituent quantities to daily mass values.

     a. BOD_M in influent:

$$BOD_M = (21{,}600 \text{ m}^3/\text{d})(375 \text{ g/m}^3)/(10^3 \text{ g/1 kg})$$
$$= 8100 \text{ kg/d}$$

    b. TSS_M in influent:

$$TSS_M = (21{,}600 \text{ m}^3/\text{d})(400 \text{ g/m}^3)/(10^3 \text{ g/1 kg})$$
$$= 8640 \text{ kg/d}$$

    c. TSS_M after grit removal (influent to primary settling tanks):

$$TSS_M = (21{,}600 \text{ m}^3/\text{d})(360 \text{ g/m}^3)/(10^3 \text{ g/1 kg})$$
$$= 7776 \text{ kg/d}$$

2. Estimate the concentration of soluble BOD_C in the effluent using the following relationship:

    Effluent BOD_C = influent soluble BOD_C escaping treatment + BOD_C of effluent TSS_C

    a. Determine the BOD_C of the effluent TSS_C.

        i. Biodegradable portion of effluent TSS_C is 0.65 (22 g/m³) = 14.3 g/m³

        ii. UBOD of the biodegradable effluent TSS_C is [0.65(22 g/m³)](1.42 g/g) = 20.3 g/m³

        iii. BOD_C of effluent suspended solids = 20.3 g/m³ (0.68) = 13.8 g/m³

    b. Solve for the influent soluble BOD_C escaping treatment.

        20 g/m³ = S + 13.8 g/m³

        S = 6.2 g/m³

3. Prepare the first iteration of the solids balance. (In the first iteration, the effluent wastewater total suspended solids and the biological solids generated in the process are distributed among the unit operations and processes that make up the treatment system.)

    a. Primary setting

        i. Operating parameters:

            BOD_C removed = 33%

            TSS_C removed = 70% (see also Fig. 5–51)

        ii. BOD_M removed = 0.33(8100 kg/d) = 2700 kg/d

        iii. BOD_M to secondary = (8100 – 2700) kg/d = 5400 kg/d

        iv. TSS_M removed = 0.7(7776 kg/d) = 5443 kg/d

        v. TSS_M to secondary = (7776 − 5443) kg/d = 2333 kg/d

    b. Determine the volatile fraction of primary sludge.

        i. Operating parameters:

        Volatile fraction of TSS_C in influent = 67%

        Volatile fraction of grit = 10%

        Volatile fraction of incoming TSS_C discharged to the secondary process = 85%

        ii. Volatile suspended solids (VSS_M) in influent prior to grit removal = 0.67 (8640 kg/d) = 5789 kg/d

        iii. VSS_M removed in grit chamber = 0.10(8640 − 7776) kg/d = 86 kg/d

        iv. VSS_M in secondary influent, kg/d = 0.85(2333 kg/d) = 1983 kg/d

        v. VSS_M in primary sludge, kg/d = (5789 − 86 − 1983) kg/d = 3710 kg/d

        vi. Volatile fraction in primary sludge = [(3710 kg/d)/(5443 kg/d)](100) = 68.2%

    c. Secondary process

        i. Determine the secondary process operating parameters

$$\text{Mixed liquor } VSS_C = \frac{(Q)(Y)(S_o - S)\text{SRT}}{[1 + b(\text{SRT})] (V_r)}$$

$$= \frac{(21{,}600 \text{ m}^3/\text{d})(0.5)[(250 - 6.2) \text{ g/m}^3](10\text{d})}{[1 + (0.06\,\text{d}^{-1})(10\,\text{d})](4700 \text{ m}^3)}$$

$$= 3500 \text{ g/m}^3$$

$$\text{Mixed Liquor TSS}_C = \frac{\text{VSS}_c}{0.8} = \frac{(3500 \text{ g/m}^3)}{0.8} = 4375 \text{ g/m}^3$$

$$Y_{\text{obs}} = \frac{Y}{1 + b(\text{SRT})} = \frac{0.5}{1 + 0.06 \times 10} = 0.3125$$

ii. Determine the effluent mass quantities.

$$\text{BOD}_M = (21{,}600 \text{ m}^3/\text{d})(20 \text{ g/m}^3)/(10^3 \text{ g/1 kg}) = 432 \text{ kg/d}$$
$$\text{TSS}_M = (21{,}600 \text{ m}^3/\text{d})(22 \text{ g/m}^3)/(10^3 \text{ g/1 kg}) = 475 \text{ kg/d}$$

iii. Estimate the mass of volatile solids produced in the activated-sludge process that must be wasted. [The required value is computed using Eq. (8–14)].

$$P_{x,\text{VSS}} = Y_{\text{obs}}Q(S_o - S)/(10^3 \text{ g/1 kg})$$

$$= \frac{0.3125(21{,}600 \text{ m}^3/\text{d})[(250 - 6.2) \text{ g/m}^3]}{(10^3 \text{ g/1 kg})} = 1646 \text{ kg/d}$$

Note: The actual flowrate will be the primary influent less the flowrate of the primary underflow. However, the primary underflow is normally small and can be neglected. If the underflow is significant, the actual flowrate should be used to determine the volatile solids production.

iv. Estimate the TSS_M that must be wasted assuming the volatile fraction represents 0.80 of the total solids.

$$\text{TSS}_M = 1646/0.80 = 2057 \text{ kg/d}$$

Note: If it is assumed that the fixed solids portion of the influent suspended solids equals 0.15, the mass of fixed solids in the input from the primary settling facilities is equal to $0.15 \times 2333 = 350$ kg/d. This value can then be compared with the fixed solids determined in the above computations, which is equal to $2057 - 1646 = 411$ kg/d. The ratio of these values is 1.18[(411 kg/d)/(350 kg/d)]. Values that have been observed for this ratio vary from about 1.0 to 1.3; a value of 1.15 is considered to be the most representative.

v. Estimate the waste quantities discharged to the thickener. (It is assumed in this example that wasting is from the biological reactor.)

$$\text{TSS}_M = (2057 - 475) \text{ kg/d} = 1582 \text{ kg/d}$$

$$\text{Flowrate} = \frac{(1582 \text{ kg/d})(10^3 \text{ g/1 kg})}{(4375 \text{ g/m}^3)} = 362 \text{ m}^3/\text{d}$$

The assumed concentration value of MLSS of 4375 g/m³ in the aeration tank will increase when the recycled BOD_C and TSS_C are taken into consideration in the second and subsequent iterations of the mass balance.

d. Flotation thickeners
   i. Operating parameters:
      Concentration of thickened sludge = 4%
      Assumed solids recovery = 90%
      Assumed specific gravity of feed and thickened sludge = 1.0
   ii. Determine the flowrate of the thickened sludge.

$$\text{Flowrate} = \frac{(1582 \text{ kg/d})(0.9)}{(10^3 \text{ kg/m}^3)(0.04)} = 35.6 \text{ m}^3/\text{d}$$

   iii. Determine the flowrate recycled to the plant influent.

      Recycled flowrate = (362 – 35.6) m³/d = 326.4 m³/d

   iv. Determine the TSS_M to the digester.

      TSS_M = (1582 kg/d)(0.9) = 1424 kg/d

   v. Determine the TSS_M recycled to the plant influent.

      TSS_M = (1582 − 1424) kg/d = 158 kg/d

   vi. Determine the BOD_C of the TSS_C in the recycled flow.

$$\text{TSS}_C \text{ in recycled flow} = \frac{(158 \text{ kg/d})(10^3 \text{ g/1 kg})}{326 \text{ m}^3/\text{d}} = 485 \text{ g/m}^3$$

$$\text{BOD}_C \text{ of the TSS}_C = (485 \text{ g/m}^3)(0.65)(1.42)(0.68)$$
$$= 304.6 \text{ g/m}^3$$

$$\text{BOD}_M = (304.6 \text{ g/m}^3)(326 \text{ m}^3/\text{d})(1 \text{ kg/10}^3 \text{ g}) = 99 \text{ kg/d}$$

e. Sludge digestion

   i. Operating parameters:

      SRT = 20 d
      VSS destruction during digestion = 50%
      Gas production = 1.12 m³/kg of VSS destroyed
      BOD_C in digester supernatant = 1000 g/m³ (0.1%)
      TSS_C in digester supernatant = 5000 g/m³ (0.5%)
      TSS_C in digested sludge = 5%

   ii. Determine the total solids fed to the digester and the corresponding flowrate.

      TSS_M = solids from primary settling plus waste solids from thickener
      TSS_M = 5443 kg/d + 1424 kg/d = 6867 kg/d

$$\text{Total flowrate} = \frac{(5443 \text{ kg/d})}{0.06(10^3 \text{ kg/m}^3)} + \frac{(1424 \text{ kg/d})}{0.04(10^3 \text{ kg/m}^3)}$$

$$= (90.7 + 35.6) \text{ m}^3/\text{d} = 126.3 \text{ m}^3/\text{d}$$

   iii. Determine the VSS_M fed to the digester.

      VSS_M = 0.682(5443 kg/d) + 0.80(1424 kg/d)
      = (3712 + 1139) kg/d = 4851 kg/d

$$\text{Percent VSS}_M \text{ in mixture fed to digester} = \frac{(4851 \text{ kg/d})}{(6867 \text{ kg/d})}(100)$$

$$= 70.6\%$$

   iv. Determine the VSSM destroyed.

      VSS_M = 0.5(4851 kg/d) = 2426 kg/d

   v. Determine the mass flowrate to the digester.

      Primary sludge at 6% solids:

$$\text{Mass flow} = \frac{(5443 \text{ kg/d})}{0.06} = 90,717 \text{ kg/d}$$

Thickened waste-activated sludge at 4% solids:

$$\text{Mass flowrate} = \frac{(1424 \text{ kg/d})}{0.04} = 35,600 \text{ kg/d}$$

Total mass flowrate = $(90,717 + 35,600)$ kg/d = 126,317 kg/d

Note: The total mass flow can also be computed by multiplying the total flowrate to the digester by the density of the combined primary sludge and the thickened biosolids, if known.

vi. Determine the mass quantities of gas and sludge after digestion. Assume that the total mass of fixed solids does not change during digestion and that 50% of the volatile solids is destroyed.

Fixed solids = $\text{TSS}_M - \text{VSS}_M = (6867 - 4851)$ kg/d = 2016 kg/d
TSS_M in digested sludge = 2016 kg/d + 0.5(4851 kg/d) = 4441 kg/d

Gas production assuming that the density of digester gas is equal to 0.86 times that of air (1.204 kg/m³, see Appendix B):

Gas = (1.12 m³/kg)(0.5)(4851 kg/d)(0.86)(1.204 kg/m³) = 2813 kg/d

Mass balance of digester output:

Mass input = 126,317 kg/d
Less gas = − 2813 kg/d
Mass output = 123,504 kg/d (solids and liquid)

vii. Determine the flowrate distribution between the supernatant at 5000 mg/L and digested sludge at 5% solids. Let TSS_{SP} = kg/d of supernatant suspended solids.

$$\frac{\text{TSS}_{SP}}{0.005} = \frac{4441 - \text{TSS}_{SP}}{0.05} = 123,504 \text{ kg/d}$$

$\text{TSS}_{SP} + 444.1 - (0.1)\text{TSS}_{SP} = 617.5$ kg/d

$(0.9)\text{TSS}_{SP} = 173$ kg/d

$\text{TSS}_{SP} = 192$ kg/d

Digested solids = $(4441 - 192)$ kg/d = 4249 kg/d

$$\text{Supernatant flowrate} = \frac{(192 \text{ kg/d})}{0.005(10^3 \text{ kg/m}^3)} = 38.4 \text{ m}^3/\text{d}$$

$$\text{Digested sludge flowrate} = \frac{(4929 \text{ kg/d})}{0.05(10^3 \text{ kg/m}^3)} = 85 \text{ m}^3/\text{d}$$

viii. Establish the characteristics of the recycled flow.
Flowrate = 38.4 m³/d
BOD_C = (38.4 m³/d)(1000 g/m³)/(10³ g/1 kg) = 38 kg/d
TSS_M = (38.4 m³/d)(5000 g/m³)/(10³ g/1 kg) = 192 kg/d

f.   Sludge dewatering. (Note: In the analysis that follows, the weight of the polymer or other sludge-conditioning chemicals that may be added was not considered. In some cases, their contribution can be significant and must be considered.)

i. Operating parameters for centrifuge:
Sludge cake = 22% solids
Specific gravity of sludge = 1.06
Solids capture = 93%
Centrate BOD_C = 2000 mg/L

    ii. Determine the sludge-cake characteristics.

$$\text{Solids} = (4249 \text{ kg/d})(0.93) = 3952 \text{ kg/d}$$

$$\text{Volume} = \frac{(3952 \text{ kg/d})}{1.06 \,(0.22)(0.22)(10^3 \text{ kg/m}^3)} = 16.9 \text{ m}^3/\text{d}$$

    iii. Determine the centrate characteristics.

$$\text{Flowrate} = (85 - 16.9) \text{ m}^3/\text{d} = 68.1 \text{ m}^3/\text{d}$$
$$BOD_M \text{ (at 2000 g/m}^3) = (2000 \text{ g/m}^3)(68.1 \text{ m}^3/\text{d})(10^3 \text{ g/1 kg})$$
$$= 136 \text{ kg/d}$$
$$TSS_M = (4249 \text{ kg/d})(0.07) = 297 \text{ kg/d}$$

g. Prepare a summary table of the recycle flows and waste characteristics for the first iteration.

Operation	Flowrate, m³/d	BOD$_M$, kg/d	TSS$_M$, kg/d
Flotation thickener	326.0	99	158
Digester supernatant	38.4	38	192
Centrate	68.1	136	297
Totals	432.5	273	647[a]

[a] The volatile fraction of the returned suspended solids will typically vary from 50 to 75 percent. A value of 60 percent will be used for the computation in the second iteration.

4. Prepare the second iteration of the solids balance.
  a. Primary settling
    i. Operating parameters = same as those in the first iteration
    ii. TSS$_M$ and BOD$_M$ entering the primary tanks

$$TSS_M = \text{influent } TSS_M + \text{recycled } TSS_M$$
$$= 7776 \text{ kg/d} + 647 \text{ kg/d} = 8423 \text{ kg/d}$$
$$\text{Total } BOD_M = \text{influent } BOD_M + \text{recycled } BOD_M$$
$$= 8100 \text{ kg/d} + 273 \text{ kg/d} = 8373 \text{ kg/d}$$

    iii. BOD$_M$ removed = 0.33(8373 kg/d) = 2763 kg/d
    iv. BOD$_M$ to secondary = (8,373 − 2763) kg/d = 5610 kg/d
    v. TSS$_M$ removed = 0.7(8423 kg/d) = 5896 kg/d
    vi. TSS$_M$ to secondary = (8423 − 5896) kg/d = 2527 kg/d
  b. Determine the volatile fraction of the primary sludge and effluent suspended solids.
    i. Operating parameters:

Incoming wastewater = same as those for the first iteration
Volatile fraction of solids in recycle returned to headworks = 60%

    ii. Although the computations are not shown, the computed change in the volatile fractions determined in the first iteration is slight and, therefore, the values determined previously are used for the second iteration. If the volatile fraction of the return is less than about 50%, the volatile fractions should be recomputed.
  c. Secondary process
    i. Operating parameters = same as those for the first iteration and as follows:
Aeration tank volume = 4700 m³
SRT = 10 d
Y = 0.50 kg/kg
b = 0.06 d^{-1}

ii. Determine the BOD_C in the influent to the aeration tank.

Flowrate to aeration tank = influent flowrate + recycled flowrate
$$= (21{,}600 + 432.5) \text{ m}^3/\text{d} = 22{,}033 \text{ m}^3/\text{d}$$

$$BOD_C = \frac{(5610 \text{ kg/d})(10^3 \text{ g/1 kg})}{(22{,}032.5 \text{ m}^3/\text{d})} = 255 \text{ g/m}^3$$

iii. Determine the new concentration of mixed liquor VSS.

$$X_{VSS} = \frac{(Q)(Y)(S_o - S)\text{SRT}}{[1 + b(\text{SRT})](V_r)}$$

$$X_{VSS} = \frac{(22{,}035 \text{ m}^3/\text{d})(0.5)(255 - 6.2)(10 \text{ d})}{[1 + (0.06 \text{ d}^{-1})(10 \text{ d})](4700 \text{ m}^3/\text{d})} = 3648 \text{ g/m}^3$$

iv. Determine the mixed liquor suspended solids.

$$X_{SS} = \frac{X_{VSS}}{0.8}$$

$$X_{SS} = 3648/0.8 = 4560 \text{ g/m}^3$$

v. Determine the cell growth.

$$P_{x,VSS} = Y_{OBS} \, Q \, (S_o - S)/(10^3 \text{ g/1 kg})$$

$$= \frac{0.3125 \, (22{,}032.5 \text{ m}^3/\text{d})[(255 - 6.2 \text{ g/m}^3)]}{(10^3 \text{ g/1 kg})} = 1714 \text{ kg/d}$$

$$P_{x,TSS} = 1714/0.8 = 2143 \text{ kg/d}$$

vi. Determine the waste quantities discharged to the thickener.

Effluent TSS_M = 432 kg/d (specified in the first iteration)

Total TSS_M to be wanted to the thicker = (2143 − 432) kg/d
$$= 1711 \text{ kg/d}$$

$$\text{Flowrate} = \frac{(1711 \text{ kg/d})(10^3 \text{ g/1 kg})}{(4560 \text{ g/m}^3)} = 375 \text{ m}^3/\text{d}$$

d. Flotation thickeners

i. Operating parameters:

Concentration of thickened sludge = 4%
Assumed solids recovery = 90%
Assumed specific gravity of feed and thickened sludge = 1.0

ii. Determine the flowrate of the thickened sludge.

$$\text{Flowrate} = \frac{(1711 \text{ kg/d})(0.9)}{(10^3 \text{ kg/m}^3)(0.04)} = 38.5 \text{ m}^3/\text{d}$$

iii. Determine the flowrate recycled to the plant influent.

Recycled flowrate = (375 − 38.5) m^3/d = 336.5 m^3/d

iv. Determine the TSS_M to the digester.

TSS_M = (1711 kg/d)(0.9) = 1540 kg/d

v. Determine the TSS_M recycled to the plant influent.

TSS_M = (1711 − 1540) kg/d = 171 kg/d

    vi. Determine the BOD_C of the TSS_C in the recycled flow.

$$TSS_C \text{ in recycled flow} = \frac{(171 \text{ kg/d})(10^3 \text{ g/1 kg})}{(336.5 \text{ m}^3/\text{d})} = 508 \text{ g/m}^3$$

BOD_C of $TSS_C = (508 \text{ g/m}^3)(0.65)(1.42)(0.68) = 319 \text{ g/m}^3$
$BOD_M = (319 \text{ g/m}^3)(336.5 \text{ m}^3/\text{d})(10^3 \text{ g/1 kg})^{-1} = 107 \text{ kg/d}$

e. Sludge digestion

    i. Operating parameters = same as those in the first iteration
    ii. Determine the total solids fed to the digester and the corresponding flowrate.
       $TSS_M = TSS_M$ from primary settling plus waste TSSM from thickener
       $TSS_M = 5443 \text{ kg/d} + 1540 \text{ kg/d} = 6983 \text{ kg/d}$

$$\text{Total flowrate} = \frac{(5443 \text{ kg/d})}{0.06(10^3 \text{ kg/m}^3)} + \frac{(1540 \text{ kg/d})}{0.04(10^3 \text{ kg/m}^3)}$$

$$= (90.7 + 38.5) \text{ m}^3/\text{d} = 129.2 \text{ m}^3/\text{d}$$

    iii. Determine the total VSSM fed to the digester.

$$VSS_M = 0.682(5443 \text{ kg/d}) + 0.80(1540 \text{ kg/d})$$

$$= (3712 + 1232) \text{ kg/d} = 4944 \text{ kg/d}$$

$$\text{Percent VSS in mixture fed to digester} = \frac{(4944 \text{ kg/d})}{(6983 \text{ kg/d})}(100)$$

$$= 71.3\%$$

    iv. Determine the VSS destroyed.

       VSS destroyed = $0.5(4944 \text{ kg/d}) = 2472 \text{ kg/d}$

    v. Determine the mass flowrate to the digester.

       Primary sludge at 6% solids:

$$\text{Mass flowrate} = \frac{(5443 \text{ kg/d})}{0.06} = 90,717 \text{ kg/d}$$

       Thickened waste-activated sludge at 4% solids:

$$\text{Mass flowrate} = \frac{(1540 \text{ kg/d})}{0.04} = 38,500 \text{ kg/d}$$

       Total mass flowrate = $(90,717 + 38,500) \text{ kg/d} = 129,217 \text{ kg/d}$

    vi. Determine the mass quantities of gas and sludge after digestion. Assume that the total mass of fixed solids does not change during digestion and that 50% of the volatile solids is destroyed.

       Fixed solids = $TSS_M - VSS_M = (6983 - 4944) \text{ kg/d} = 2039 \text{ kg/d}$
       TSS in digested sludge = $2039 \text{ kg/d} + 0.5(4944) \text{ kg/d} = 4511 \text{ kg/d}$

       Gas production assuming that the density of digester gas is equal to 0.86 times that of air (1.204 kg/m^3):

       Gas = $(1.12 \text{ m}^3/\text{kg})(0.5)(4944 \text{ kg/d})(0.86)(1.204 \text{ kg/m}^3) = 2867 \text{ kg/d}$
       Mass balance of digester output:

Mass input $= 129{,}217$ kg/d
Less gas $= -2867$ kg/d
Mass output $= 126{,}350$ kg/d (solids and liquid)

vii. Determine the flowrate distribution between the supernatant at 5000 mg/L and digested sludge at 5 percent solids. Let $\text{TSS}_{sp} = $ kg/d of supernatant suspended solids.

$$\frac{\text{TSS}_{sp}}{0.005} = \frac{4441 - \text{TSS}_{sp}}{0.05} = 126{,}350 \text{ kg/d}$$

$$\text{TSS}_{sp} + 451.1 - (0.1)\text{TSS}_{sp} = 631.8 \text{ kg/d}$$

$$(0.9)\text{TSS}_{sp} = 180.7 \text{ kg/d}$$
$$\text{TSS}_{sp} = 201 \text{ kg/d}$$

Digested $\text{TSS}_M = (4511 - 201) \text{ kg/d} = 4310 \text{ kg/d}$

$$\text{Supernatant flowrate} = \frac{(201 \text{ kg/d})}{0.005(10^3 \text{ kg/m}^3)} = 40.2 \text{ m}^3/\text{d}$$

$$\text{Digested sludge flowrate} = \frac{(4310 \text{ kg/d})}{0.05(10^3 \text{ kg/m}^3)} = 86.2 \text{ m}^3/\text{d}$$

viii. Establish the characteristics of the recycled flow.

Flowrate $= 40.2 \text{ m}^3/\text{d}$
$\text{BOD}_M = (40.2 \text{ m}^3/\text{d})(1000 \text{ g/m}^3)/(10^3 \text{ g/1 kg}) = 40 \text{ kg/d}$
$\text{TBB}_M = (40.2 \text{ m}^3/\text{d})(5000 \text{ g/m}^3)/(10^3 \text{ g/1 kg}) = 201 \text{ kg/d}$

f. Sludge dewatering

i. Operating parameters for centrifuge $=$ same as those in the first iteration
ii. Determine the sludge-cake characteristics.

$$\text{TSS}_M = (4310 \text{ kg/d})(0.93) = 4008 \text{ kg/d}$$

$$\text{Volume} = \frac{(4008 \text{ kg/d})}{1.06(0.22)(10^3 \text{kg/m}^3)} = 17.2 \text{ m}^3/\text{d}$$

iii. Determine the centrate characteristics.

Flow $= (86.2 - 17.2) \text{ m}^3/\text{d} = 69 \text{ m}^3/\text{d}$
$\text{BOD}_M \text{ (at 2000 g/m}^3) = (2000 \text{ g/m}^3)(69 \text{ g/m}^3)/(10^3 \text{ g/1 kg}) = 138 \text{ kg/d}$
$\text{TSS}_M = (4310 \text{ kg/d})(0.07) = 302 \text{ kg/d}$

g. Prepare a summary table of the recycle flows and waste characteristics for the second iteration.

| Operation/process | Flow, m³/d | BOD$_M$, kg/d | TSS$_M$, kg/d | Incremental change from previous iteration | | |
				Flowrate, m³/d	BOD$_M$, kg/d	TSS$_M$, kg/d
Flotation thickener	326.0	99	158	10.5	8	13
Digester supernatant	38.4	38	192	1.8	2	9
Centrate	68.1	136	297	0.9	2	5
Totals	432.5	273	647[a]	13.2	12	27

5. Because the incremental change in the return quantities is less than 5 percent, the values summarized in the above table are acceptable for design. Given that the above computations would be done on a spreadsheet program, additional iterations could be made to obtain an incremental change of less than 1 percent. The flow, TSSM, and BODM values for the various processes from the second iteration are presented in following figure.

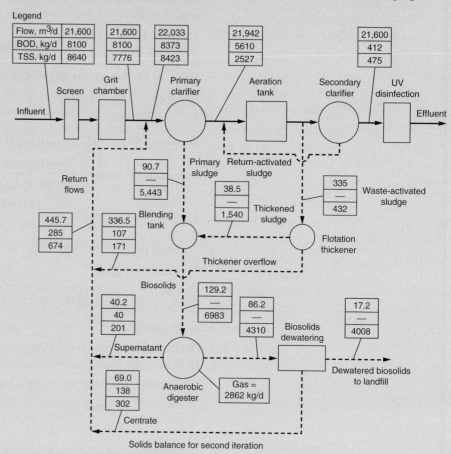

Solids balance for second iteration

Comment In this example, an iterative approach was used to illustrate the computational steps in preparing a solids mass balance. Sludge and biosolids solids balances can be prepared using a specially designed spreadsheet or a proprietary solids balance software program. In general, if the iterative computational procedure is used, similar to the method used in this example, it should be carried out until the incremental change in all of the return quantities from the previous iteration is equal to or less than 5 percent.

14–8 RESOURCE RECOVERY FROM SLUDGES AND BIOSOLIDS

Sludges and biosolids can serve as a source of nutrients that can be used as a fertilizer, as a feedstock for the production of energy, and in the fabrication of value-added products. As discussed in Sec. 13–2 in Chap. 13, biosolids that have been stabilized properly according to EPA Part 503 Regulations can be used for a variety of beneficial uses.

Significant fractions of nutrients in biosolids become solubilized when biosolids are stabilized by anaerobic digestion and return flow from the dewatering process (often called centrate, filtrate or sidestream) contains high concentrations of nutrients. Nutrients in the sidestream can be returned to the main liquid treatment trains or recovered through sidestream treatment. Recovery and use of nutrients in solids is discussed in this section, and recovery of nutrients in the liquid stream is discussed in Chap. 15. The recovery of energy from biosolids and sludge is considered in the following section and in Chap. 17.

Recovery of Nutrients

Among the most valuable nutrients contained in sludges and biosolids are organic nitrogen and phosphorous. The recovery of phosphorous and ammonia is usually from the liquid stream of the dewatering process, as discussed in detail in Chap.15. Following dewatering, Class A or B biosolids can be used beneficially in agriculture land application or non-agriculture land application. The principal beneficial uses of biosolids include (1) agricultural land application, (2) non-agricultural land application, (3) energy recovery and generation, and (4) commercial use. Agricultural and non-agricultural land application are examined below. Energy recovery and generation are discussed in Sec. 14–9.

Agricultural Land Application

Biosolids contain a range of valuable nutrients such as nitrogen, phosphorus, iron, calcium, magnesium and various other macro and micro nutrients which are essential for plant growth. Many nutrients found in biosolids are also essential components in the healthy diet of animals. Biosolids are usually applied to agricultural land at rates designed to supply crops with adequate nitrogen (see Sec. 14–9 for details regarding biosolids land application). Recycling biosolids to agricultural land enables farmers to improve the economics of crop production in lieu of the use of expensive chemical fertilizers and reduces greenhouse gas emission generated from producing chemical fertilizer and carbon sequestration. Class B or A biosolids from processes such as anaerobic digestion and lime stabilization discussed in Chap. 13 are applied to land for crop production as cake; as compost material (see Sec. 14–5); or as a granular material in pellets form produced from drying processes (see Sec. 14–3).

Non-Agricultural Land Applications

Non-agricultural uses of biosolids include reclaiming mining sites, land reclamation, landscaping, and forest crops. Biosolids have been used widely for repairing land damaged by mining such as surface mined areas, abandoned mine lands, and coal refuse piles. Combining mine soils with biosolids, increases the organic matter, the cation exchange capacity, and soil nutrient levels. In mine damaged lands, biosolids application controls pH, metal content, and fertilization of the soils. Reclaiming and improving disturbed and marginal soils is another beneficial use of biosolids. Several biosolids characteristics make this use very successful. The organic matter in biosolids improves the soil physical properties through improving soil granulation and increasing soil water holding capacity. Biosolids increase soil cation exchange capacity, supply plant nutrients, and buffer soil. Land application to forest land is also practiced. However, this use has been difficult to achieve due to difficulties in spreading biosolids evenly through heavily forested areas. Finally, the use of biosolids for horticulture and landscaping, such as golf courses, is similar to agricultural land application but not used as a fertilizer. Biosolids compost is most popular product for landscaping uses as compost is primarily a soil conditioner and not a fertilizer.

14-9 **ENERGY RECOVERY FROM SLUDGE AND BIOSOLIDS**

Sludges and biosolids are considered renewable energy resources as they contain organic material that has a fuel value that can be harnessed. Under a properly engineered and controlled environment, energy recovery and generation from biosolids is considered at the top of the hierarchy of beneficial use due to the increased cost of energy and the need to obtain green energy credit by municipal utilities. Sludges from wastewater can be processed to generate energy by (WERF, 2008b):

1. Producing methane through anaerobic digester
2. Thermally oxidize sludge
3. Producing syngas through gasification process and/or pyrolysis
4. Producing oil and liquid fuel

Each of these options along with the appropriate method for energy recovery is considered briefly below. Additional details on energy recovery are presented in Chap. 17.

Energy Recovery through Anaerobic Digestion

Production of methane rich gas by anaerobic digestion is one pathway used to recover energy from biosolids. Gas production, collection and use methods are discussed in Sec. 13–9. Combined heat and power (CHP) recovery from the biogas generated from anaerobic digestion of sludge is illustrated on Fig. 14–34. As shown on Fig. 14–34 , the waste heat can be used as an energy source to further dry and stabilize the cake solids. Drying processes are discussed in Sec. 14–3. The dried material can also be used beneficially in agricultural or non-agricultural land application. Another beneficial use of the dried material can be hauling to cement kilns to supplement coal as a fuel. The value of the dried material is usually compared to the market value of coal on a heat content basis. Energy can be further recovered from the dried material on site to provide energy needed for drying as discussed below.

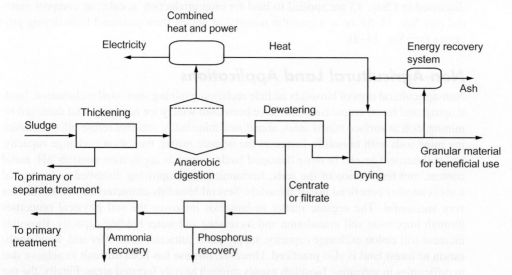

Figure 14-34

Schematic process flow diagram for the recovery of energy with anaerobic digestion. (The simultaneous recovery of nutrients is considered in Chap. 15.)

Table 14-23

Characteristics of different thermal processing technologies

Parameter	Combustion	Gasification	Pyrolysis
Temperature, °C (°F)	900–1100 (1650–2000)	590–980 (1100–1800)	200–590 (390–1100)
O_2 supplied	>Stoichiometric (excess air)	<Stoichiometric (limited air)	None
By-products	Flue gas (CO_2, H_2O) and ash	Syngas (CO, H_2) and ash	Pyrolysis gas, oils, tars and char

Energy Recovery by Thermal Oxidation

The three basic thermal processes are combustion or advanced thermal oxidation, gasification, and pyrolysis. All thermal processing of dried biosolids is followed by energy recovery methods. The difference between the three thermal processing technologies, in terms of operating temperature and oxygen requirements, and the main byproducts from each are summarized in Table 14–23. Thermal oxidation of sludge or biosolids is more common in the areas where ultimate disposal of sludge and biosolids is costly due to limited disposal sites. Sludges or biosolids to be incinerated, as described in Sec. 14–4, are typically dewatered to 15 to 30 percent dry solids. Further drying and combustion occurs simultaneously in the combustion reactor. The combustion process produces hot flue gas where the energy can be recovered for combustion air preheating and other energy needs or electrical production as shown previously on Fig. 14–25. The ash produced from the advanced thermal oxidation is an inert material that can be used in commercial applications such as cement making, asphalt, etc. Technologies to recover valuable material from the ash such as phosphorous are under development and once commercially developed may gain wide interest due to the potential market value of such resources. Excess heat from thermal oxidation of dewatered sludge has been used recently to generate steam for electricity generation.

Energy Recovery from Dried Material through Gasification and Pyrolysis

Gasification is an established process for converting organic materials to a fuel gas called synthetic gas or syngas, and has been practiced since the 1800s to generate fuel gas from coal and other biomass. Syngas is composed mainly of CO, CO_2, H_2, and CH_4 and has a low heating value of 4500–5500 kJ/m³ (120–150 BTU/ft³), which is approximately 25 percent of the heat value of biogas generated from anaerobic digestion. While gasification is still in the early phases of development for processing biosolids, there has been increased interest in applying this technology to sludges and biosolids. Sludges or biosolids have a higher ash content as compared to traditional organic materials used as a fuel source in gasification processes, which makes ash handling more difficult. Further details of gasification and energy recovery is considered in Chap. 17.

Pyrolysis is also an established technology used in the chemical industry to produce charcoal, activated carbon and methanol. Similar to gasification, pyrolysis at high temperatures generates a combustible gas, pyrolysis gas, with a low heating value but also can be used to generate char and oil. Pyrolysis is actually the first step that occurs in both gasification and combustion reactions. Pyrolysis of sludges and biosolids is still considered innovative technology and not widely practiced for municipal applications.

To effectively harness energy through a gasification or pyrolysis technology, most commercially available systems require dried biosolids with greater than 75 percent solids in granular form. Pelletization is not required; however, a certain degree of uniformity in the dried granular material along with low dust content is required. The required dryness depends on the technology. The energy required for drying is typically supplied by the thermal technology. Energy can also be recovered from waste heat, such as a combined heat and power (CHP) system, if available onsite.

Production of Oil and Liquid Fuel

Converting sludge to liquid and oil fuel is another use of sludge as energy source. However, the commercial application of such technologies is hindered by the high capital cost and requirement for large amount of sludge feed to lower the capital and operating costs.

14-10 APPLICATION OF BIOSOLIDS TO LAND

Land application of biosolids is defined as the spreading of biosolids on or just below the soil surface. Biosolids may be applied to (1) agricultural land, (2) forest land, (3) disturbed land, and (4) dedicated land disposal sites. In all four cases, the land application is designed with the objective of providing further biosolids treatment. Sunlight, soil microorganisms, and desiccation combine to further inactivate any residual pathogens and many toxic organic substances. Trace metals are trapped in the soil matrix and nutrients are taken up by plants and converted to useful biomass. In some cases, a geomembrane liner is installed below a dedicated land disposal area.

To qualify for application to agricultural and nonagricultural land, biosolids or material derived from biosolids must meet at least the pollutant ceiling concentrations, Class B requirements for pathogens, and vector attraction requirements. Bulk biosolids applied to lawns and home gardens and biosolids that are sold or given away in bags or containers must meet the Class A criteria and one of several available vector-attraction reduction processes.

Benefits of Land Application

The application of biosolids to land for agricultural purposes is beneficial because organic matter improves soil structure, tilth, water holding capacity, water infiltration, and soil aeration. Macronutrients (nitrogen, phosphorus, potassium) and micronutrients (iron, manganese, copper, chromium, selenium, and zinc) aid plant growth. Organic matter also contributes to the cation-exchange capacity (CEC) of the soil, which allows the soil to retain potassium, calcium, and magnesium. The presence of organic matter improves the biological diversity in soil and improves the availability of nutrients to the plants (Wegner, 1992). Nutrients in the biosolids also serve as a partial replacement for expensive chemical fertilizers.

Land application can also be of great value in silviculture and site reclamation. Forest utilization has been practiced extensively in the northwest, and biosolids application has been recognized as being beneficial to forest growth (WEF, 2010). Reclamation of disturbed land such as superfund sites has also been successful (Henry and Brown, 1997).

U.S. EPA Regulations for Beneficial Use and Disposal of Biosolids

As discussed in Sec. 13–2, the U.S. EPA published regulations for biosolids (sewage sludge is the term used in the regulations) use and disposal under the code of Federal Regulations (CFR), 40 CFR Part 503. For land application, the regulations provide numerical limits on 10 metals, management practice guidance, and requirements for monitoring, record keeping, and reporting. The regulations are summarized in Table 14–24 and discussed in the following paragraphs.

Table 14–24

U.S. EPA sludge regulations for land application

Classification	Class A: no restrictions[a] Class B: site restrictions
Management practices	See Table 14–25
Pathogen reduction alternatives	See Table 13–9
Vector attraction reduction	See Table 13–10
Site restrictions for Class B biosolids	See Table 14–26
Metal limits and loading rates	See Table 14–30

[a] Other than bag labeling (like a fertilizer).

Table 14–25

Land application management practices under U.S. EPA Part 503 rule[a]

For bulk biosolids[b]

- Bulk biosolids cannot be applied to flooded, frozen, or snow-covered agricultural land, forests, public contact sites, or reclamation sites in such a way that the biosolids enters a wetland or other waters of the United States (as defined in 40 CFR Part 122.2), except as provided in a permit issued pursuant to Section 402 (NPDES permit) or Section 404 (Dredge and Fill Permit) of the Clean Water Act, as amended.
- Bulk biosolids cannot be applied to agricultural land, forests, or reclamation sites that are 10 m or less from U.S. waters, unless otherwise specified by the permitting authority.
- If applied to agricultural lands, forests, or public contact sites, bulk biosolids must be applied at a rate that is equal to or less than the agronomic rate for the site. Biosolids applied to reclamation sites may exceed the agronomic rate if allowed by the permitting authority.
- Bulk biosolids must not harm or contribute to the harm of a threatened or endangered species or result in the destruction or adverse modification of the species' critical habitat when applied to the land. Threatened or endangered species and their critical habitats are listed in Section 4 of the Endangered Species Act. Critical habitat is defined as any place where a threatened or endangered species lives and grows during any stage of its life cycle. Any direct or indirect action (or the result of any direct or indirect action) in a critical habitat that diminishes the likelihood of survival and recovery of a listed species is considered destruction or adverse modification of a critical habitat.

For biosolids sold or given away in a bag or other container for application to the land[a]

- A label must be affixed to the bag or other container, or an information sheet must be provided to the person who receives this type of biosolids in another container. At a minimum, the label or information sheet must contain the following information:
  - The name and address of the person who prepared the biosolids for sale or give-away in a bag or other container.
  - A statement that prohibits application of the biosolids to the land except in accordance with the instructions on the label or information sheet.
  - An AWSAR (annual whole sludge application rate) for the biosolids that does not cause the annual pollutant loading rate limits to be exceeded.

[a] from U.S. EPA (1995).
[b] These management practices do not apply if the biosolids is of "exceptional quality."

Management Practices

Management practices that must be followed when biosolids are applied on land are specified in the Part 503 rule (see Table 14–25). The practices vary depending on whether the material that is applied is hauled in bulk or in individual bags.

Pathogen-Reduction Alternatives. As discussed in Chap. 13 and in Sec. 13–2, the Part 503 pathogen-reduction requirements for biosolids are divided into Class A and Class B categories (see Table 13–9). The goal of the Class A requirements is to reduce the pathogens in the biosolids (including Salmonella sp. bacteria, enteric viruses, and viable helminth ova) to below detectable levels. When this goal is achieved, Class A biosolids can be land applied without any pathogen-related restrictions on the site (U.S. EPA, 1995). The goal of the Class B requirements is to ensure that pathogens have been reduced to levels that are unlikely to pose a threat to public health and the environment under specific use conditions. Site restrictions on land application of Class B biosolids minimize the potential for human and animal contact with the biosolids until environmental factors have reduced pathogens to below detectable levels.

Vector Attraction Reduction. There are 10 potential vector attraction reduction measures that can be combined with pathogen-reduction alternatives for an acceptable land-application project using Class B biosolids (see Table 13–10). The list in Table 13–10 also includes some stabilization processes that reduce pathogens.

Site Restrictions for Class B Biosolids. Site restrictions, listed in Table 14–26, depend on the crops to be used and the contact control for animals and the public. Food crops and turf grass are given the longest time restrictions because of the potential for public exposure (U.S. EPA, 1995).

Table 14–26

Site restrictions for Class B biosolids[a]

Restrictions for the harvesting of crops and turf

- Food crops with harvested parts that touch the biosolids/soil mixture and are totally above ground shall not be harvested for 14 mo after application of biosolids.

- Food crops with harvested parts below the land surface where biosolids remains on the land surface for 4 mo or longer prior to incorporation into the soil shall not be harvested for 20 mo after biosolids application.

- Food crops with harvested parts below the land surface where biosolids remains on the land surface for less than 4 mo prior to incorporation shall not be harvested for 38 mo after biosolids application.

- Food crops, feed crops, and fiber crops, whose edible parts do not touch the surface of the soil, shall not be harvested for 30 d after biosolids application.

- Turf grown on land where biosolids is applied shall not be harvested for 1 y after application of the biosolids when the harvested turf is placed on either land with a high potential for public exposure or a lawn, unless otherwise specified by the permitting authority.

Restriction for the grazing of animals

- Animals shall not graze on land for 30 d after application of biosolids to the land.

Restrictions for public contact

- Access to land with a high potential for public exposure, such as a park or ball field, is restricted for 1 y after biosolids application. Examples of restricted access include posting with no trespassing signs or fencing.

- Access to land with a low potential for public exposure (e.g., private farmland) is restricted for 30 d after biosolids application. An example of restricted access is remoteness.

[a] From U.S. EPA (1995).

Exceptional Quality Biosolids. The category of "exceptional quality" biosolids has been defined as those biosolids that meet metal standards, Class A pathogen reduction standards, and vector reduction standards as defined in the Part 503 regulations.

Site Evaluation and Selection

A critical step in land application of biosolids is finding a suitable site. The characteristics of the site will determine the actual design and will influence the overall effectiveness of the land-application concept. The sites considered potentially suitable will depend on the land-application option or options being considered, such as application to agricultural lands, forest lands, etc. The site-selection process should include an initial screening on the basis of the factors and criteria described in the following discussion. For screening purposes, it is necessary to have at least a rough estimate of land-area requirements for each feasible option.

Ideal sites for land application of biosolids have deep silty loam to sandy loam soils, groundwater deeper than 3 m (10 ft), slopes at 0 to 3 percent; no wells, wetlands, or streams; and few neighbors. Site characteristics of importance are topography, soil characteristics, soil depth to groundwater, and accessibility and proximity to critical areas.

Topography. Topography is important as it affects the potential for erosion and runoff. Suitability of site topographies also depends on the type of biosolids and the method of application. As shown in Table 14–27, liquid biosolids can be spread, sprayed, or injected onto sites with rolling terrain up to 15 percent in slope. Dewatered sludge is usually spread on agricultural land that requires a tractor and spreader. Forested sites can accommodate slopes up to 30 percent if adequate setbacks from streams are provided.

Soil Characteristics. In general, desirable soil characteristics include (1) loamy soil, (2) slow to moderate permeability, (3) soil depth of 0.6 m (2 ft) or more, (4) alkaline or neutral soil pH (pH > 6.5), and (5) well drained to moderately well drained soil. Practically any soil can be adapted to a well-designed and well-operated system.

Soil Depth to Groundwater. A basic philosophy inherent in federal and state regulations is to design biosolids application systems that are based on sound agronomic principles, so that biosolids pose no greater threat to groundwater than current agricultural practices. Because the groundwater fluctuates on a seasonal basis in many soils, difficulties are encountered in establishing an acceptable minimum depth to groundwater. The quality of the underlying groundwater and the biosolids application option have to be considered carefully, especially where groundwater nondegradation restrictions apply.

Table 14–27

Typical slope limitations for land application of biosolids

Slope,%	Comment
0–3	Ideal; no concern for runoff or erosion of liquid or dewatered biosolids
3–6	Acceptable; slight risk of erosion; surface application of liquid or dewatered biosolids is acceptable
6–12	Injection of liquid biosolids required for general cases, except in closed drainage basin and/or when extensive runoff control is provided; surface application of dewatered biosolids is generally acceptable
12–15	No application of liquid biosolids should be made without extensive runoff control; surface application of dewatered biosolids is acceptable, but immediate incorporation into the soil is recommended
Over 15	Slopes greater than 15 percent are suitable only for sites with good permeability where the length of slope is short and where the area with steep slope is a minor part of the total application area

Generally, the greater the depth to the water table, the more desirable a site is for biosolids application. At least 1 m (3 ft) to groundwater is preferred for land-application sites. Seasonal water-table fluctuations to within 0.5 m (1.5 ft) of the surface can be tolerated. If the shallow groundwater is excluded as a drinking-water aquifer, the groundwater depth can be as shallow as 0.5 m before problems with trafficability of the soil arise. The presence of faults, solution channels, and other similar connections between soil and groundwater is undesirable unless the depth of overlying soil is adequate. When a specific site or sites are being considered for biosolids application, a detailed field investigation may be necessary to obtain the required groundwater information.

Accessibility and Proximity to Critical Areas.   Buffer zones or setbacks are needed to separate the active application area from sensitive areas such as residences, wells, roads, surface waters, and property boundaries. Local and state regulations often include minimum distances for setbacks depending on the method of application; example minimum setback distances used in California are listed in Table 14–28.

Design Loading Rates

Design loading rates for land application of biosolids can be limited by pollutants (heavy metals) or by nitrogen. The long-term loadings of heavy metals are based on U.S. EPA 503 regulations. The annual loading rate is usually limited by the nitrogen loading rate.

Nitrogen Loading Rates.   Nitrogen loading rates are set typically to match the available nitrogen provided by commercial fertilizers (Chang et al., 1995). Because municipal biosolids represent a slow-release organic fertilizer, a combination of ammonia and organic nitrogen must be made according to Eq. (14–6).

$$L_N = [(NO_3) + K_V(NH_4) + f_n(N_o)]F \qquad (14\text{–}6)$$

where L_N = plant available nitrogen in the application year, g N/kg (lb N/ton)

NO_3 = percent nitrate nitrogen in biosolids, decimal

Table 14–28

Typical setback distances for land application sites[a]

Setback from:	Minimum distance	
	ft	m
Property Boundaries	10	3
Domestic water supply wells	500	150
Nondomestic water supply wells	100	30
Public roads and onsite occupied residences	50	15
Surface waters (wetlands, creeks, ponds, lakes, underground aqueducts, and marshes)	100	30
Primary agricultural drainageways	33	10
Occupied nonagricultural buildings and offsite residences	500	150
Domestic water supply reservoir	400	120
Primary tributary to a domestic water supply	200	60
Domestic surface water supply intake	2500	750

[a] CSWRCB (2000).

$$K_v = \text{volatilization factor for ammonia loss}$$
$$= 0.5 \text{ for surface-applied liquid sludge}$$
$$= 0.75 \text{ for surface-applied dewatered sludge}$$
$$= 1.0 \text{ for injected liquid or dewatered sludge}$$
$$NH_4 = \text{percent ammonia nitrogen in sludge, decimal}$$
$$f_n = \text{mineralization factor for organic nitrogen}$$
$$= 0.5 \text{ for warm climates and digested sludge}$$
$$= 0.4 \text{ for cool climates and digested sludge}$$
$$= 0.3 \text{ for cold climates or composted sludge}$$
$$N_o = \text{percent organic nitrogen in sludge, decimal}$$
$$F = \text{conversion factor, 1000 g/kg of dry solids (2000 lb/ton)}$$

To use Eq. (14–6) requires knowledge of the method of application, the nitrogen content of the biosolids (nitrate, ammonia, and organic), the type of stabilization, and the type of climate. The use of the mineralization factors simplifies the previously used method of calculating the amount of organic nitrogen mineralized each year and adding up the total for an annual equivalent. The use of Eq. (14–6) is also appropriate if biosolids are applied to a single site once every 2 to 3 y.

The loading rate based on nitrogen loadings is then calculated from Eq. (14–7).

$$L_{SN} = \frac{U}{N_P F} \tag{14–7}$$

where L_{SN} = biosolids loading rate based on N, kg/ha·y (ton/ac·y)
  U = crop uptake of nitrogen, kg/ha (lb/ac) (see Table 14–29)
  N_p = plant available nitrogen in sludge, g/kg (lb/ton)
  F = conversion factor, 10^{-3} kg/g (1 lb/lb)

Loading Rates Based on Pollutant Loading. The pollutants of concern are those listed in Table 14–30. To calculate the biosolids loading rate based on pollutant loading use Eq. (14–8).

$$L_S = \frac{L_C}{CF} \tag{14–8}$$

where L_S = maximum amount of biosolids that can be applied per year, kg/ha·y (tons/ac·y)
  L_C = maximum amount of constituent that can be applied per year, kg/ha·y (lb/ac·y)
  C = pollutant concentration in biosolids, decimal (mg/kg)
  F = conversion factor, 10^{-6} kg/mg (2000 lb/ton)

Land Requirements. Once the maximum biosolids loading rate is determined [by comparing the values from Eqs. (14–7) and (14–8)], the field area can be calculated using Eq. (14–9).

$$A = \frac{B}{L_S} \tag{14–9}$$

where A = application area required, ha
  B = biosolids production, kg of dry solids/y

Table 14–29

Typical nitrogen uptake values for selected crops[a]

Crop	Nitrogen uptake		Crop	Nitrogen uptake	
	lb/ac·y	kg/ha·y		lb/ac·y	kg/ha·y
Forage crops			Tree crops		
Alfalfa	200–600	220–670	Eastern forests		
Brome grass	115–200	130–220	Mixed hardwoods	200	225
Coastal Bermuda grass	350–600	390–670	Red pine	100	110
Kentucky bluegrass	175–240	195–270	White spruce	200	225
Quack grass	210–250	235–280	Pioneer succession	200	225
Orchard grass	220–310	250–350	Aspen sprouts	100	110
Reed canary grass	300–400	335–450	Southern forests		
Ryegrass	160–250	180–280	Mixed hardwoods	250	280
Sweet clover[b]	155	175	Loblolly pine	200–250	225–280
Tall fescue	130–290	145–325	Lake states forest		
Field crop			Mixed hardwoods	100	110
Barley	110	120	Hybrid poplar	140	155
Corn	155–180	175–200	Western forest		
Cotton	65–100	70–110	Hybrid poplar	270	300
Grain sorghum	120	135	Douglas fir	200	225
Potatoes	200	225			
Soybeans	220	245			
Wheat	140	155			

[a] Adapted from U.S. EPA (1981).

[b] Legume crops can fix nitrogen from the air but will take up most of their nitrogen from applied wastewater.

Table 14–30

Metals concentrations and loading rates for land application of biosolids[a]

Pollutant	Ceiling concentration[b]		Cumulative pollutant loading rate[c]		Pollutant concentration for exceptional quality[d]		Annual pollutant loading rate[e]	
	lb/ton	mg/kg	lb/ac	kg/ha	lb/ton	mg/kg	lb/ac	kg/ha
Arsenic	0.15	75	37	41	0.08	41	1.78	2.0
Cadmium	0.17	85	35	39	0.08	39	1.70	1.9
Chromium[f]	—	—	—	—	—	—	—	—
Copper	8.60	4300	1338	1500	3.00	1500	66.91	75
Lead	1.68	840	268	300	0.60	300	13.38	15

(continued)

Table 14–30 (Continued)									
	Ceiling concentration[b]		**Cumulative pollutant loading rate[c]**		**Pollutant concentration for exceptional quality[d]**		**Annual pollutant loading rate[e]**	
Pollutant	lb/ton	mg/kg	lb/ac	kg/ha	lb/ton	mg/kg	lb/ac	kg/ha
Mercury	0.11	57	15	17	0.03	17	0.76	0.85
Molybdenum[f]	0.15	75	—	—	—	—	—	—
Nickel	0.84	420	374	420	0.84	420	18.74	21
Selenium	0.20	100	89	100	0.20	100	4.46	5.0
Zinc	15.00	7500	2498	2800	15.00	2800	124.91	140

[a] Adapted from Federal Register (1993).

[b] Dry weight basis, Table 1 from Part 503 regulations, instantaneous maximum.

[c] Dry weight basis, Table 2 from 503 regulations.

[d] Dry weight basis, Table 3 from 503 regulations, monthly average.

[e] Table 4 from 503 regulations.

[f] A February 25, 1994, Federal Register Notice deleted chromium; deleted the molybdenum values for Tables 2, 3, and 4; and raised the selenium value in Table 3 from 36 to 100.

EXAMPLE 14–4 **Metals Loadings in Land Application** A community has stockpiled biosolids in a storage lagoon. The lagoon needs to be cleaned and the biosolids disposed of to make room for a plant expansion. The metals concentrations (mg/kg) in the lagoon are as follows:

As = 45	Hg = 5
Cd = 30	Ni = 350
Cu = 1,200	Se = 15
Pb = 250	Zn = 3100

Determine if the biosolids are acceptable for land application.

Solution

1. Compare the concentrations for the above metals to the ceiling concentration (column 2) and the pollutant concentration for exceptional quality (column 4)
   a. All metals concentrations are under the ceiling limits in column 2. The biosolids are suitable for land application.
   b. Arsenic and zinc exceed the values for exceptional quality. Calculations of annual loadings are necessary.
2. Calculate the allowable annual biosolids loading rates, using Eq. (14–8), for the two metals using the annual pollutant loading rates in Table 14–30.
   a. Arsenic-based loading rate (L_C = 2 kg/ha·y)

$$L_S = \frac{L_C}{L_C(10^{-6})} = \frac{(2 \text{ kg/ha·y})}{(45 \text{ mg/kg})(1 \text{ kg/}10^6 \text{ mg})} = 44{,}444 \text{ kg/ha·y}$$

b. Zinc based loading rate ($L_C = 140$ kg/ha·y)

$$L_S = \frac{(140 \text{ kg/ha·y})}{(3100 \text{ mg/kg})(1 \text{ kg}/10^6 \text{ mg})} = 45{,}161 \text{ kg/ha·y}$$

3. Compare the whole biosolids loading rates to determine the limiting rate. The 44,444 kg/ha·y biosolids loading based on arsenic is limiting.

Comment Nitrogen loadings typically are more limiting than metals loadings. If the nitrogen loading rate exceeds 20 Mg/ha·y (U.S. EPA, 1995), then the arsenic loading rate will determine the whole biosolids loading rate.

Application Methods

Application methods for biosolids range from direct injection of liquid biosolids to surface spreading of dewatered biosolids. The method of application selected will depend on the physical characteristics of the biosolids (liquid or dewatered), site topography, and the type of vegetation present (annual field crops, existing forage crops, trees, or preplanted land).

Liquid or Thickened Biosolids Application. Application of biosolids in the liquid or thickened state is attractive because of its simplicity. Dewatering processes are not required, and the liquid or thickened biosolids can be transferred by pumping. Typical solids concentrations of liquid or thickened biosolids applied to land range from 1 to 8 percent. Liquid or thickened biosolids may be applied to land by vehicular application or by irrigation methods similar to those used for wastewater distribution.

Vehicular application may be by surface distribution, subsurface injection or incorporation. Limitations to vehicular application include limited tractability on wet soil and potential reduction in crop yields due to soil compaction from truck traffic. Use of high-flotation tires can minimize these problems.

Surface distribution may be accomplished by tank truck or tank wagon equipped with rear-mounted spreading manifolds or by tank trucks mounted with high-capacity spray nozzles or guns. Specially designed, all-terrain biosolids application vehicles with spray guns are ideally suited for biosolids application on forest lands. Vehicular surface application is the most common method used for field and forage croplands. The procedure used commonly for annual crops is to (1) spread biosolids prior to planting, (2) allow the biosolids to dry partially, and (3) incorporate the biosolids by disking or plowing. The process is repeated then after harvest.

Liquid biosolids can be injected below the soil surface by using tank wagon or tank trucks with injection shanks or incorporated immediately after surface application by using plows or disks equipped with biosolids distribution manifolds and covering spoons (see Fig. 14–35). Advantages of injection or immediate incorporation methods include minimization of potential odors and vector attraction, minimization of ammonia loss due to volatilization, elimination of surface runoff, and minimum visibility leading to better public acceptance. Injection shanks and plows are very disruptive to perennial forage crops or pastures. To minimize such effects, special grassland biosolids injectors have been developed (Crites and Tchobanoglous, 1998).

Figure 14–35

Land application of liquid sludge: (a) self contained vehicle used to haul and to inject the liquid sludge into the ground. Self-contained vehicles of the type shown are used for relatively small amounts of liquid sludge and (b) tractor equipped with subsurface liquid sludge injection tines. The liquid sludge to be injected is supplied by a hose connected to the injection device. The tethered sludge supply hose is dragged along by the tractor. The injected liquid sludge is disked in the ground using the tractor and disk such as shown on Fig. 14–36(b).

(a)

(b)

Irrigation methods include sprinkling and furrow irrigation. Typically, large-diameter, high-capacity sprinkler guns are used to avoid clogging problems. Sprinkling has been used mainly for application to forested lands and occasionally for application to dedicated disposal sites that are relatively isolated from public view and access. Sprinklers can operate satisfactorily on land too rough or wet for tank trucks or injection equipment and can be used throughout the growing season. Disadvantages to sprinkling include power costs of high-pressure pumps, contact of biosolids with all parts of the crop, possible foliage damage to sensitive crops, potential odors and vector attraction problems, and potentially high visibility to the public.

Furrow irrigation can be used to apply biosolids to row crops during the growing season. Disadvantages associated with furrow irrigation are localized settling of solids and the potential for ponding of biosolids in the furrows, both of which can result in odor problems.

Dewatered Biosolids Application. Application of dewatered biosolids to the land is similar to an application of semisolid animal manure. The use of conventional manure spreaders is an important advantage because farmers can apply biosolids on their lands with their own equipment. Typical solids concentrations of dewatered biosolids applied to land range from 20 to 30 percent. Dewatered biosolids are spread most commonly using tractor-mounted box spreaders or manure spreaders followed by plowing or disking into the soil (see Fig. 14–36). For high application rates bulldozers, loaders, or graders may be used. For forest application, a side-slinging vehicle has been tested that can apply dewatered biosolids up to 60 m (200 ft) (Leonard et al., 1992).

Figure 14–36

Land application of dewatered sludge: (a) typical example of vehicle used to apply dewatered sludge on the surface of the soil and (b) typical tractor and rotating two-way disk used to disk dewatered and/or liquid sludge into the ground.

(a)

(b)

Table 14–31

Criteria for dedicated land disposal (DLD) sites for biosolids[a]

Parameter	Unacceptable condition	Ideal condition
Slope	Deep gullies, slope >12%	<3%
Soil permeability	$>1 \times 10^5$ cm/s[b]	$\leq 10^{-7}$ cm/s[c]
Soil depth	<0.6 m (2 ft)	>3 m (10 ft)
Distance to surface water	<90 m (300 ft) to any pond or lake used for recreational or livestock purposes, or any surface water body officially classified under state law	>300 m (1000 ft) from any surface water
Depth to groundwater	<3 m (10 ft) to groundwater table (wells tapping shallow aquifers)[d]	>15 m (50 ft)
Supply wells	Within 300 m (1000 ft) radius	No wells within 600 m (2000 ft)

[a] From U.S. EPA (1983).

[b] Permeable soil can be used for DLD if appropriate engineering design preventing DLD leachate from reaching the groundwater is feasible.

[c] When low-permeability soils are at or too close to the surface, liquid disposal operations can be hindered due to water ponding.

[d] If an exempted aquifer underlies the site, poor quality leachate may be permitted to enter groundwater.

Application to Dedicated Lands

Disturbed land reclamation and dedicated land disposal are two types of high-rate land application. Disturbed land reclamation consists of a one-time application of 110 to 220 Mg/ha (50 to 100 dry tons/ac) to correct adverse soil conditions. Lack of soil fertility and poor physical properties can be corrected by biosolids application to allow revegetation programs to proceed. For disturbed land reclamation to be the sole avenue for biosolids reuse, a large area of disturbed land must be available on an ongoing basis. Dedicated land disposal requires a site where high rates of biosolids application are acceptable environmentally on a continuing basis. Biosolids for a dedicated land disposal (DLD) operation should meet at least Class B requirements.

Site Selection. Siting criteria for a dedicated land disposal (DLD) site are presented in Table 14–31. Major issues in DLD siting are nitrogen control and the avoidance of groundwater contamination. Groundwater contamination can be avoided by (1) locating sites remote from useful aquifers, (2) intercepting of leachate, and (3) constructing an impervious geological barrier. Low percolation rates and deep aquifers will substantially reduce or eliminate potential contamination effects.

Where groundwater nondegradation restrictions apply, it has been found that for most DLD sites it is less costly to excavate the site entirely, install a geomembrane liner, and replace the excavated material, than to dispose of the sludge by some other means (e.g., dewatering and landfilling). The limited amount, if any, of leachate collected from the liner is returned to the treatment plant for processing.

Loading Rates. Annual biosolids loading rates have ranged from 12 to 2250 tonne/ha (5 to 1000 tons/ac). The higher rates have been associated with sites that

- Receive dewatered biosolids
- Mechanically incorporate the biosolids into the soil

- Have relatively low precipitation
- Have no leachate problems because of site conditions or project design

Design loading rates for DLD can be estimated using Eq. (14–10).

$$L_S = \frac{E(TS)F}{100} - TS \qquad (14\text{–}10)$$

where L_S = annual biosolids loading rate, Mg/ha (ton/ac)
E = net evaporation rate from soil, mm/y (in./y)
TS = total solids content, percent by weight
F = conversion factor, 10 Mg/mm (113.3 ton/in.)
The net soil evaporation can be estimated from Eq. (14–11).

$$E = (f)E_L - P \qquad (14\text{–}11)$$

where E = net evaporation rate from soil, mm/y (in./y)
f = 0.7
E_L = pan evaporation rate, mm/y (in./y)
P = annual precipitation, mm/y (in./y)

It should be noted that infiltration into the soil is not considered in Eq. (14–10). If infiltration is allowed, the term E should be increased by the annual infiltration rate in mm/y.

Once the annual loading rate is calculated, the field area can be determined using Eq. (14–9) (dividing the biosolids production by the loading rate). Other area requirements include buffer zones, surface runoff control, roads, and supporting facilities.

Landfilling

Landfilling of biosolids in a monofill is covered under 40 CFR Part 503. Landfilling of biosolids in a sanitary landfill with municipal solid waste is regulated by the U.S. EPA under 40 CFR 258. If an acceptable site is convenient, landfilling can be used for disposal of biosolids, grit, screenings, and other solids. Stabilization may be required depending on state or local regulations. Dewatering of biosolids is usually required to reduce the volume to be transported and to control the generation of leachate from the landfill. In many cases, solids concentration is an important factor in determining the acceptability of biosolids in landfills. The sanitary landfill method is most suitable if it is also used for disposal of the other types of solid wastes. In a true sanitary landfill, the wastes are deposited in a designated area, compacted in place with a tractor or roller, and covered with a 350 mm (14 in.) layer of clean soil. With daily coverage of the newly deposited wastes, nuisance conditions, such as odors and flies, are minimized.

PROBLEMS AND DISCUSSION TOPICS

14-1 A wastewater treatment plant is producing 55,000 L/d of thickened biosolids containing 2.8 percent solids. A belt-filter press installation is to be designed based on a normal operation of 8 h/d and 5 d/wk, a belt-filter press loading rate of 280 kg/m·h, and the following data. Compute the number and size of the belt filters, and the expected solids capture, in percent. Determine the daily hours of operation required if a sustained 5-d peak solids load occurs.

1. Total solids in dewatered sludge = 26 percent.

2. Total suspended solids concentration in filtrate = 800 mg/L.

3. Washwater flowrate = 90 L/min per m of belt width.

4. Specific gravities of sludge feed, dewatered cake, and filtrate are 1.02, 1.08, and 1.01, respectively.

5. Use Fig. 3–14 to estimate the peaking factor for a sustained 5-d peak solids load

14–2 The ultimate elemental analysis of a dried sludge yields the following data:

Element	Percent
Carbon	52.1
Hydrogen	2.7
Oxygen	38.3
Nitrogen	6.9
Total	100.0

How many kg of air will be required per kg of sludge for its complete oxidation?

14–3 Compute the fuel value of the sludge from a primary settling tank having a composition (by weight) of 64.5 percent carbon, 8.5 percent hydrogen, 21.0 percent oxygen, and 4 percent sulfur.

14–4 Calculate the theoretical heat requirement to dry 1000 m^3/d of sludge cake from solid contents of 15, 20 or 25 percent (to be selected by instructor) to 92 percent. Assume feed sludge temperature at 20°C and final temperature at 100°C. Latent heat of evaporation is 2260 kJ/kg H_2O. Assume a specific gravity of the dewatered cake is 1.05. Assume dry solids heat capacity of 1.5 kJ/kg·°C. Estimate the fuel requirement assuming 5 percent heat loss and 85 percent heater efficiency.

14–5 A community of 25,000 persons has asked you to serve as a consultant on their sludge disposal problems. Specifically, you have been asked to determine if it is feasible to compost waste activated sludge with the community's solid waste. If this plan is not feasible, you have been asked to recommend a feasible solution. Currently the waste sludge from the WWTP is dewatered on a belt filter press. Assume the following data are applicable:

Solid waste data:
Waste production = 2 kg/person·d (wet basis)
Compostable fraction = 55%
Moisture content of compostable fraction = 22%

Sludge production:
Net sludge production = 0.12 kg/person·d (dry basis)
Concentration of sludge out of the belt filter press = 22%
Specific gravity of dewatered cake = 1.05

Compost:
Final moisture content of composted biosolids/solid waste mixture = 55%

14–6 A municipality with a population of 200,000 has hired you as a consultant to investigate alternative mechanical sludge dewatering options. The three alternatives to be investigated are belt press dewatering, centrifugation, and pressure filter press dewatering. The biosolids to be dewatered are stabilized by anaerobic digestion, are a mixture of primary sludge and waste activated sludge, and have a solids concentration of 5 percent. Ultimate disposal is by landfilling at a site located 50 kilometers from the treatment plant. Compare the various dewatering alternatives and recommend one. State the reasons for your recommendation. Should newer technologies such as rotary presses or screw presses be considered?

14-7 Prepare a solids balance for the peak loading condition for the treatment plant used in Example 14–3 for one of the data sets to be selected by instructor. Enter your final values on the solids balance figure in Example 14–3.

	Data set			
	1	2	3	4
Peak flowrate, m³/d	54,000	60,000	50,000	54,000
Average BOD at peak flowrate, mg/L	340	300	350	300
Average TSS at peak flowrate, mg/L	350	320	330	320
TSS after grit removal, mg/L	325	300	310	300

Use data given in Example 14–3 for other parameters.

14-8 Prepare a solids balance, using the iterative technique delineated in Example 14–3, for the following treatment process flow diagram and one of the following data sets to be selected by instructor. Also determine the effluent flowrate and suspended solids concentration.

	Data set			
	1	2	3	4
Influent characteristics				
Flowrate, m³/s	10,000	20,000	30,000	40,000
Suspended solids, mg/L	1000	350	400	300
Sedimentation tank				
TSS removal efficiency, %	75	60	65	60
Underflow TSS concentration, %	7	6.5	6	5.5
Specific gravity of sludge	1.1	1.1	1.1	1.1
Alum addition				
Dosage, mg/L	10	10	20	15
Chemical solution, kg alum/L of solution	0.5	0.5	0.5	0.5
Filters				
TSS removal efficiency, %	90	90	95	92
Washwater solids concentration, %	6	6	6.8	6.5
Specific gravity of backwash	1.08	1.08	1.089	1.085
Thickener				
Supernatant TSS, mg/L	400	300	200	250
Concentration of solids in underflow, %	12	8	9	8
Chemical addition				
Dosage, percent of underflow solids from thickener	0.8	1.0	1.0	1.0
Chemical solution, kg/L of solution	2.0	2.0	2.0	2.0
Filter press				
TSS concentration in filtrate, mg/L	200	300	250	200
Concentration of dewatered solids, %	40	38	42	42
Specific gravity of sludge cake	1.6	1.5	1.65	1.65

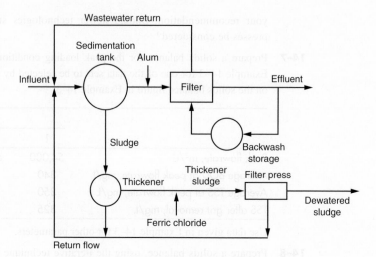

In preparing the solids balance assume that all of the unit operations respond linearly such that the removal efficiency for recycled solids is the same as that for the solids in the influent wastewater. Also assume that the distribution of the chemicals added to improve the performance of the filter and filter press is proportional to the total solids in the return flows and the effluent solids.

14-9 Assume the HHV value of biosolids as derived in Example 14–2, and compute the energy contents in biosolids with 25, 65 and 90 percent solid contents (to be selected by instructor). Compare the results with the heat requirements to vaporize water in the biosolids for each solids content level. Discuss supplemental energy requirements for the incineration process in relation to the solids contents of the feed biosolids.

14-10 Determine the dry sludge application rate for Reed canary grass on the basis of satisfying crop nitrogen uptake. Assume that biosolids containing 3 percent organic nitrogen by weight are applied to a soil that has an initial nitrogen content of zero. Use a mineralization rate of 30 percent for the first year, 15 percent for the second year, and 5 percent for the third and subsequent years.

14-11 Biosolids containing 50 ppm of cadmium on a dry basis are to be applied to land. If the limiting mass loading to the soil is set at 10 kg/ha, what would be the safe loading rate for 50 y of application?

14-12 Compare the advantages and disadvantages of land application of liquid biosolids to dewatered biosolids. Assume the land application site is located 15 km from the treatment plant in primarily an agricultural area and the biosolids are transported by truck. The biosolids are stabilized by anaerobic digestion and the liquid biosolids concentration is 6 percent and the dewatered solids concentration is 25 percent.

14-13 In problem 14–13, what would be the advantages and limitations of conveying the liquid biosolids by pipeline to the land application site? What types of facilities, i.e., structures, equipment, and vehicles, would be required and what are the operating and maintenance considerations?

The following two problems are best solved in groups of two or three.

14-14 A treatment plant is treating an average wastewater flowrate of 500,000 m³·d with 430 mg/L TSS, 345 mg/L VSS, and 335 mg/L BOD. The volatile part of TSS is 80 percent. The peaking factors are 1.3, 1.4, and 1.5 for max month, max week, and max day, respectively. The main treatment processes include primary clarifiers and activated sludge system for secondary treatment. Primary sludge is thickened via gravity thickener while waste activated sludge is thickened by gravity belt thickening. Combined thickened sludges are stabilized via anaerobic digestion. The primary clarifier is expected to remove 50 percent of the TSS and 30 percent of the BOD. Students are encouraged to solve this problem through establishing a spreadsheet mass and energy balance.

1. Determine the average amount of primary sludge produced in kg/d at a TS concentration of 1.5 percent.

2. Determine the average amount of WAS produced in kg/d assuming the BOD effluent from the secondary clarifier to be 10 mg/L and a sludge yield of 0.6. Assume the mean cell residence time or SRT of the secondary treatment is 10 d; endogenous decay rate, k_d, is 0.06 d^{-1}; the VS content in the mixed liquor suspended solids is 70 percent.

3. Determine the average amount of primary thickened sludge in kg/d at a TS concentration of 5 percent and assuming a solids capture of 95 percent in the gravity thickener.

4. Determine the average amount thickened WAS in kg/d at a solids concentration of 6 percent assuming a 95 percent capture rate in the gravity belt thickener. Assume the WAS from the secondary clarifier is 1 percent TS.

5. The two thickened sludges are mixed in a combined primary and activated sludge (CPAS) tank with a volume of 40 m^3. Determine the combined sludges percent TS content out of the CPAS, and percent volatile solids content out of the CPAS. Determine the HRT of the CPAS tank in min.

6. The anaerobic digestion complex achieves 50 percent volatile solids destruction. Digesters are cylindrical with cone bottom. The dimensions of each digester are 37 m diameter, 11 m side water depth, 1.2 m freeboard, and cone bottom slope of 1:10. Assuming all sludge filled digester volume is usable:

   a. Determine the number of digesters if the HRT of all digesters is not reduced below 15 d HRT to meet max month per MOP 8 design criteria. Assume the digesters are high rate complete mix with no decant.

   b. Determine the number of digesters to process annual average sludge production.

   c. Determine the annual average and max month VS loading to the digester complex in kg/m^3·d and compare it to typical design criteria

   d. Determine the amount of heat needed to maintain the mesophilic temperature within the digestion complex during the winter months assuming minimum income sludge temp to be 5°C and assuming no heat loss from the digesters.

   e. If you are the operations manager do you recommend processing the maximum day sludge in the digester complex? If not what do you recommend an operation strategy?

7. Determine the amount of biogas production in m^3/h assuming 0.95 m^3/kg VS destroyed.

8. The plant produces electricity from the biogas system through an internal combustion engine of 38 percent electrical efficiency and 40 percent thermal energy recovery.

   a. Determine the amount of electricity produced in KW if the generated biogas has 22,400 kJ/m^3.

   b. Is the recovered heat from the engine enough to supplement the energy used to keep the digesters heated during winter months?

9. If a portion of the biogas is used to supplement the energy needed for the digesters' heating through a boiler system with 80 percent energy efficiency prior to feeding the engine, determine the amount of electricity generated.

10. Provide your thoughts on which energy management system through comparing results from 8 and 9 above.

11. The biosolids produced from digestion process are dewatered via centrifuges. Determine the following:

   a. Dry solids loading to the centrifuge complex following digestion in kg/d

   b. Volatile solids content of the biosolids out of digestion in kg/d

   c. The percent solids total solids (%TS) content of the biosolids after digestion

   d. The volumetric flowrate of the biosolids in m³/h going to the centrifuges.

   e. Determine the amount of wet tonnes produced daily from the centrifuge and the concentration (%TS) and the centrate flowrate (m³/h) if the centrifuge complex is producing a solids cake of 23 percent with a 95 percent solids recovery. Assume the BOD concentration in the centrate is 1500 mg/L and is recycled to the primary clarifier.

REFERENCES

Abu-Orf, M. M., and S. K. Dentel (1997) "Polymer Dose Assessment Using the Streaming Current Detector," *Water Environ. Res.,* **69**, 6, 1075–1085.

Abu-Orf, M. M, P. Griffin, and S. K. Dentel (2001) "Chemical and Physical Pretreatment of ATAD for Dewatering," *Water Sci. Technol.,* **44**, 10, 309–314.

Abu-Orf, M. M., C. A. Walker, and S. K. Dentel (2003) "Centrate Viscosity for Continuous Monitoring of Polymer Feed in Dewatering Applications," *Adv. Environ. Res.,* **7**, 7, 687–694.

Abu-Orf M. M., and B. Ormeci (2005) A Protocol to Measure Network Strength of Sludges and Its Implication for Dewaterability, *J. Envir. Engrg., ASCE,* **131**, 1, 80–85.

Abu-Orf, M. M., N. Tepe, S. K. Dentel, R. Mahmudov, A. Tesfaye, and W. Smith (2009) "Is Your Plant Receiving the Same Product in Different Polymer Batches?" *Proceedings of the WEF Residuals and Biosolids Management Conference 2009,* Water Environment Federation, Alexandria, VA.

ASTM (2007) *Standard Practice for Proximate Analysis of Coal and Coke,* ASTM D3172–07, ASTM International, West Conshohocken, PA.

ASTM (2008) *Standard Practice for Coagulation-Flocculation Jar Test of Water,* ASTM D2035–08, ASTM International, West Conshohocken, PA.

ASTM (2009) *Standard Practice for Ultimate Analysis of Coal and Coke,* ASTM D3176–09, ASTM International, West Conshohocken, PA.

ASTM (2011) *Standard Test Method for Gross Calorific Value of Coal and Coke,* ASTM D5865 – 11, ASTM International, West Conshohocken, PA.

CSWRCB (2000) *General Waste Discharge Requirements for the Discharge of Biosolids to Land for Use as a Soil Amendment in Agricultural, Silvicultural, Horticultural, and Land Reclamation Activities,* Water Quality Order No. 2000-10-DWQ, California State Water Resources Control Board, Sacramento, CA.

Chang, A. C., A. L. Page, and T. Asano (1995) *Developing Human Health-Related Chemicals Guidelines for Reclaimed Wastewater and Sewage Sludge Applications in Agriculture,* World Health Organization, Geneva, Switzerland.

Cooper, P. F., G. D. Job, M. B. Green, and R. B.E. Shutes (1996) "Reed Beds and Constructed Wetland for Wastewater Treatment," *WRc Swindon,* ISBN 1.898920-27-3, Swindon, Wiltshire, England.

Crites, R.W., and G. Tchobanoglous (1998) *Small and Decentralized Wastewater Management Systems,* McGraw-Hill, New York.

Epstein, E. (1997) *The Science of Composting,* Technomic Publishing Co., Lancaster, PA.

Eschborn, R., M. M. Abu-Orf, D. Sarrazin-Sullivan (2011) "Integrating Elecgtrodewatering into Advanced Biosolids Processing – Where Does It Makes Sense?" *Proceedings of the WEF Residuals and Biosolids Management Conference,* Sacramento, CA. Water Environment Federation, Alexandria VA.

Gazbar, S., J. M. Abadie, and F. Colin (1994) "Combined Action of Electro-Osmotic Drainage and Mechanical Compression on Sludge Dewatering," *Water Sci. Technol.*, **30**, 8, 169–175.

Haug, R. T., F. M. Lewis, G. Petino, and W. J. Harnett (1993) "Explosion Protection and Fire Prevention at a Biosolids Drying Facility," *Proceedings of the 66th Annual Conference & Exposition*, Water Environment Federation, Alexandria, VA.

Henry, C., and S. Brown (1997) "Restoring a Superfund Site with Biosolids and Fly Ash," *Biocycle*, **38**, 11, 79–83.

Holcomb, S. P., B. Dahl, T. A. Cummings, and G. P. Shimp (2000) "Fluidized Bed Drying Replaces Incineration at Pensacola, Florida," *Proceedings of the 73rd Annual Conference & Exposition on Water Quality and Wastewater Treatment,* Water Environment Federation, Alexandria, VA.

Leonard, P., R. King, and M. Lucas (1992) "Fertilizing Forests with Biosolids; How to Plan, Operate, and Maintain a Long-Term Program," *Proceedings, The Future of Municipal Sludge (Biosolids) Management,* WEF Specialty Conference, 233–250, Water Environment Federation, Alexandria, VA.

Mahmoud, A., J. Olivier, J. Vaxelaire, and A. F. A. Hoadley (2010) "Electrical Field: A Historical Review of its Application and Contributions in Wastewater Sludge Dewatering," *Water Res.*, **44**, 8, 2381–2407.

Mujumdar, A. S. (ed.) (2007) *Handbook of Industrial Drying,* CRC Press, Taylor & Frances Group, Boca Raton, FL.

Novak, J. T., M. M. Abu-Orf, and C. Park (2004) "Conditioning and Dewatering of Digested Waste Activated Sludge," *J. Resid. Sci Technol.*, **1**, 1, 45–51.

Standard Methods (2012) *Standard Methods for the Examination of Water and Waste Water,* 21st ed., American Public Health Association, Washington, DC.

Tsang, K. R., and P. A. Vesilind (1990) "Moisture Distribution in Sludge," *Water Sci. Technol.*, **22**, 12, 135–142.

Tchobanoglous, G., H. Theisen, and S. Vigil (1993) *Integrated Solid Waste Management,* McGraw-Hill, New York.

U.S. EPA (1981) *Process Design Manual for Land Treatment of Municipal Wastewater,* EPA/625/1-81-013, Center for Environmental Research Information, U.S. Environmental Protection Agency, Cincinnati, OH.

U.S. EPA (1983) *Process Design Manual for Land Application of Municipal Sludge,* EPA 625/1-83-016, Center for Environmental Research Information, U.S. Environmental Protection Agency, Cincinnati, OH.

U.S. EPA (1987a) *Design Manual, Dewatering Municipal Wastewater Sludges,* EPA/625/1–87/014, Office of Research and Development, U.S. Environmental Protection Agency, Cincinnati, OH.

U.S. EPA (1987b) *Design Information Report—Design, Operational, and Cost Considerations for Vacuum Assisted Sludge Dewatering Bed Systems,* EPA-68-03-1821, EPA/600/J-87/078, Water Engineering Research Laboratory, U.S Environmental Protection Agency, Cincinnati, OH (see also *J. WPCF,* **59**, 4, 228–234).

U.S. EPA (1987c) *Design Information Report—Sidestreams in Wastewater Treatment Plants,* EPA-68-03-3208, EPA/600/J-87/004,Water Engineering Research Laboratory, U.S. Environmental Protection Agency, Cincinnati, OH (see also *J. WPCF,* **59**, 1, 54–61).

U.S. EPA (1989) *Summary Report, In-Vessel Composting of Municipal Wastewater Sludge,* EPA/625/8-89 /016, Center for Environmental Research Information, U.S. Environmental Protection Agency, Cincinnati, OH.

U.S. EPA (1995) *Process Design Manual—Land Application of Sewage Sludge and Domestic Septage,* EPA/625/R-95/001, Center for Environmental Research Information, U.S. Environmental Protection Agency, Cincinnati, OH.

U.S. EPA (2000) *Biosolids Technology Fact Sheet: Centrifuge Thickening and Dewatering,* EPA 832-F-00-053 , U.S. Environmental Protection Agency, Washington, DC.

WEF (1988) *Sludge Conditioning, Manual of Practice No. FD-14,* Water Environment Federation, Alexandria, VA.

WEF (2010) *Design of Municipal Wastewater Treatment Plants,* 5th ed., Manual of Practice no. 8, vol. 3, Chaps. 20–27, Water Environment Federation, Alexandria, VA.

WEF (2012) *Solids Process Design and Management,* Water Environment Federation, Alexandria, VA.

Wegner, G. (1992) "The Benefits of Biosolids from a Farmer's Perspective." 39–44 , *Proceedings The Future Direction of Municipal Sludge (Biosolids) Management,* Water Environment Federation. Specialty Conference, Portland, OR.

WERF (2003) *Identifying and Controlling Odor in the Municipal Wastewater Environment Phase 3: Biosolids Processing Modifications for Cake,* Project Number 03-CTS-9T. Water Environment Research Foundation, Alexandria, VA.

WERF (2008a) *Evaluation of Bacterial Pathogen and Indicator Densities After Dewatering of Anaerobically Digested Biosolids: Phases II and III,* Project Number 04-CTS-3T), Water Environment Research Foundation, Alexandria, VA.

WERF (2008b) State of the Science Report: Energy and Resource Recovery from Sludge.

Yoshida, H., (1993) "Practical Aspects of Dewatering Enhanced by Electroosmosis," *Drying Tech.,* **11**, 4, 787–814.

15

Plant Recycle Flow Treatment and Nutrient Recovery

WORKING TERMINOLOGY

Term	Definition
Acid absorption	The absorption of constituents from an air stream into an acid solution.
Air stripping	The use of air to remove (transfer) constituents from a liquid stream to an air stream.
Anaerobic ammonium oxidation (Anammox)	Biological oxidation of ammonia under an oxygen-free environment with nitrite as the electron acceptor.
AOB	Ammonia oxidizing bacteria.
Bioaugmentation	The introduction of a selected or engineered group of microorganisms to enhance treatment.
Crystallizers	A liquid-solid separation technique in which solid crystals are formed from constituents in solution.
Deammonification	A two-step biological process consisting of partial nitritation and the Anammox reaction.
Denitritation	Biological conversion of nitrite to nitrogen gas and other intermediate compounds containing nitrogen.
Denitratation	Biological conversion of nitrate to nitrite.
Fermentation	A biological process in which organic matter is converted under anaerobic conditions to volatile fatty acids.
Integrated sidestream-mainstream treatment	An integrated system is any process where the sidestream reactor waste solids are fed to the mainstream secondary process or mixed liquor solids are interchanged between the two processes.
Limit of technology	Represents the lowest achievable concentration of a specific constituent by the best available technology.
Nitratation	The conversion of nitrite to nitrate.
Nitritation	The conversion of ammonia to nitrite.
NOB	Nitrite oxidizing bacteria.
Sidestream	The collective term used to describe all recycle streams. Sidestreams derived from digestion are also known in some countries as liquor or reject water.

Term	Definition
Separate sidestream treatment process	Biological treatment process that is isolated from the mainstream treatment process and dedicated to treat sidestream flows.
Struvite	A precipitated compound form of magnesium ammonium phosphate hexahydrate, $MgNH_4PO_4 \cdot 6H_2O$.
Steam stripping	The use of steam to remove a constituent from a waste stream.

The separation of water from primary, secondary, combined, or digested sludges during solids processing generates a liquid stream, which has characteristics that prevent direct discharge of the stream with the wastewater treatment plant final effluent. At facilities that thicken primary and secondary waste sludges before aerobic or anaerobic digestion and dewater the digested solids, multiple recycle streams are generated, each with a different composition, flowrate, and impact on the treatment plant. Because anaerobic and aerobic digestion result in the release of soluble organic nitrogen-containing compounds, ammonium and orthophosphate into the bulk liquid, the post-digestion recycle stream generated by the dewatering of the digested solids will have elevated nutrient concentrations resulting in an increased nutrient loading to the primary and secondary treatment processes. In addition, the release of orthophosphate and ammonium during solids digestion often results in the formation of insoluble inorganic compounds such as magnesium ammonium phosphate, also known as struvite, which can cause operational and maintenance problems in mechanical dewatering equipment and pipes that convey the recycle stream.

Current practice at most wastewater treatment plants is to recycle these sidestreams to the head of the plant or directly to the secondary process for treatment. However, because these sidestreams can impact significantly the performance of the secondary treatment process, many treatment plants now treat these streams separately. Interest in reducing nutrient loadings from plant recycle streams rich in ammonium and phosphate through the implementation of dedicated or *sidestream* treatment processes has been increasing since the late 1980s due to more stringent effluent discharge limits on ammonium-N, total nitrogen (TN) and total phosphorus (TP) and a desire to reduce plant operating costs (energy, chemicals, and maintenance). In addition, the impact of suspended solids and colloidal material present in some recycle streams on treatment plant effluent quality must be considered.

The types of recycle streams commonly found in conventional wastewater treatment plants, their characteristics, and their potential impacts on the operation of wastewater treatment facilities are reviewed in this chapter. Following the review of types of recycle streams and their characteristics, the focus of the remainder of the chapter is on physio-chemical and biological treatment technologies that have been developed and implemented to treat separately the nutrient-rich streams, which are typically the primary recycle streams of concern for nutrient removal facilities. Typical process design and performance information are also provided.

15–1 SIDESTREAM IDENTIFICATION AND CHARACTERIZATION

Recycle sidestreams from the thickening and dewatering of raw and digested solids are typically identified with the particular process or mechanical equipment from which they originate. The composition of each recycle sidestream also varies depending on the source of the solids being thickened or dewatered. For the purpose of simplifying terminology, all recycle streams discussed in this chapter will be called *sidestream*, but specific terminology that allows for the distinction of one type of recycle stream from another will be retained as needed and used interchangeably with the general name. Common sources of sidestreams, typical flowrates, and typical sidestream characteristics [total suspended solids (TSS), total Kjeldahl nitrogen (TKN), ammonium nitrogen (ammonium-N), total phosphorus (TP) and orthophosphate (ortho-P)] are summarized in Table 15–1. Additional characteristics that distinguish one type of sidestream from another are discussed below.

Sidestreams Derived from Primary and Secondary Sludges

The soluble nutrient concentrations in sidestreams generated from the thickening of primary and waste activated sludges (gravity thickening; dissolved air flotation; centrifugation) and the filtration of secondary effluent generally represent the soluble composition of the liquids from which they originate. The partial fermentation of sludges that often occurs in gravity thickeners, will generate volatile fatty acids, decrease the alkalinity, and increase the ammonium and phosphate concentrations. The wide ranges for the ammonium and orthophosphate concentrations shown in Table 15–1 for these sidestreams represent the typical range of weak to untreated domestic wastewaters (Table 3–18) and nitrifying and non-nitrifying secondary treatment systems. Concentrations of TKN, TP, and BOD are strongly dependent on the TSS concentration. In general, the daily mass nutrient loadings contributed by these sidestreams to the secondary process are relatively minor in comparison to the raw influent or primary tank effluent. The major impact of their return to the primary and/or secondary treatment process is on the solids mass balance for the treatment plant, as discussed in Sec. 14–7 in Chap. 14. The solids impact is dependent on the solids capture efficiency of the thickening or dewatering process. For most treatment facilities, sidestream flowrates from the thickening of primary and secondary sludges and backwash from final filtration are continuous or near-continuous.

Sidestreams Derived from Fermented Primary and Digested Primary and Secondary Sludges

Sidestream characteristics resulting from the fermentation of primary sludge and the digestion of primary and waste activated sludges are significantly different than sidestreams generated from the thickening and dewatering of primary and waste activated sludges. The principal reason is the release of organic nitrogen-containing compounds, ammonium and phosphate into the bulk liquid. The soluble nutrient concentrations in these higher strength sidestreams are, as indicated in Table 15–1, dependent on the process from which they originate. For example, in the fermentation of primary sludge to generate volatile fatty acids (VFAs) for the enhancement of biological phosphorus removal, the ammonium and orthophosphate concentrations in the fermentate are far lower than the concentrations in the sidestreams resulting from aerobic and anaerobic digestion. Further, the ammonium and orthophosphate concentrations resulting from digestion are dependent on the feed solids concentrations to the digester and the volatile solids destruction efficiency, as reflected in the broad concentration ranges shown in Table 15–1. For example, if thermal hydrolysis is applied in the pretreatment of sludge before anaerobic digestion, the total solids (TS) concentration in the feed to the digesters will be 8 to 11 percent by weight. With the higher TS concentration and enhanced volatile solids destruction efficiency in the digester associated with this advanced digestion process, the digester ammonium-N concentration will be two to three times the concentration observed in typical conventional anaerobic digesters.

Table 15–1

Characteristics of various sidestreams from the thickening, stabilization and dewatering of raw and digested solids

Operation	Flowrate, Percent of plant influent	Value, mg/L						
		TSS[a]	BOD[a]	TKN	NH4-N	NOx	TP	Ortho-P
Gravity thickening supernatant:								
Primary sludge	2–3	80–350	100–400	19–70	12–45	0	4–11	3–8[e]
Primary sludge + waste activated sludge	3–5	100–350	60–400	20–70	8–45	0–8	4–15[b]	2–7[b,e]
Primary sludge fermentate, including elutriation water	3–4	700–900	2000–2500	80–120	60–100	0	10–20	5–15
Flotation thickening subnatant (waste activated sludge)	0.7–1	100–2500	50–1200	8–250	0–45	0–30	2–50	0.05–8
Centrifuge thickening centrate (waste activated sludge)	0.7–1	500–3000	170–3000	40–280	0–45	0–30	8–60	0.05–8
Screw press-filtrate + pressate (alkaline and heat stabilization for Class A)[d]	0.3–0.5	400–500	600–1300	120–250	10–20	0–5	6–14	< 1
Aerobic digestion supernatant (mesophilic; continuous and intermittent aeration)	0.1–0.5	100–10,000	100–1700	100–1200	20–400	0–400	200–350[b,c]	200[b,c]
Anaerobic digestion supernatant (two-stage, high-rate)	0.1–0.5	1000–11,500	500–5000	850–1800	800–1300	0	110–470[b,c]	100–350[b,c]
Centrifuge dewatering centrate:								
Two-stage, high rate anaerobic digestion	0.5–1	200–20,000	100–2000	810–2100	800–1300	0	100–550[b,c]	100–350[b,c]
Thermal hydrolysis + single stage mesophilic anaerobic digestion	0.2–0.5	1500–10,000	1500–3000	2200–3700	2000–3000	0	220–800[b,c]	200–700[b,c]
Belt-filter press filtrate: two-stage, high rate anaerobic digestion, including belt washwater	1–2	100–2000	50–500	410–730	400–650	0	50–200[b,c]	50–180[b,c]
Recessed-plate-filter press filtrate	0.5–1	50–1000	50–250		800–1300	0		100–350[b,c]
Sludge lagoon supernatant		5–200	100–200					
Sludge drying bed underdrainage	0.3–0.5	20–500	100–200		0–400	0–400	2–210	2–200
Composting leachate		500	2000					

(continued)

1663

Table 15–1 (Continued)

Operation	Flowrate, Percent of plant influent	Value, mg/L						
		TSS[a]	BOD[a]	TKN	NH4-N	NOx	TP	Ortho-P
Incinerator scrubber water		600–8000	30–80					
Depth filter washwater		100–1000	50–500					
Microscreen washwater		240–1000	100–500					
Carbon adsorber washwater		100–1000	50–400					
Dryer condensate								

[a] Adapted, in part, from U.S. EPA (1987b) and WEF (1998).

[b] Orthophosphate concentration does not include potential phosphorus release from waste activated sludges derived from plants operating with biological phosphorus removal.

[c] Orthophosphate concentration does not include reduction by chemical precipitation within the digester through the natural formation of salts such as hydroxyapatite and magnesium ammonium phosphate (struvite), through the addition of ferric or ferrous salts to the digester to control struvite formation in dewatering equipment and pipes conveying sidestream or for facilities that practice chemically enhanced primary or secondary treatment for phosphorus removal.

[d] Based on unpublished data collected from the stabilization and dewatering of waste activated sludge with the FKC Class A process at the Sequin, WA, wastewater treatment plant in July-August, 2011 (no primary sedimentation; oxidation ditch secondary process).

[e] Orthophosphate concentration does not reflect facilities that practice chemically enhanced primary treatment.

Nitrogen Content. In addition to a higher ammonium concentration, soluble organic nitrogen-containing compounds are released into the bulk liquid during digestion, accounting for roughly 10 percent of the soluble TKN. Of this soluble organic nitrogen fraction, approximately 50 percent is considered essentially non-biodegradable or recalcitrant dissolved organic nitrogen (rDON). Recalcitrant DON is typically less than 1 mg/L in plant effluents due to rDON in the raw influent and the production of soluble microbial products in the secondary treatment process. In general, the presence of rDON is not of concern for most nutrient removal plants. However, for limit of technology (LOT) plants with a low effluent TN limit (e.g., 3 mg/L), rDON becomes a larger fraction of the plant effluent TN. Typical digester sidestream will add approximately 0.2 mg N/L of rDON to the plant effluent, assuming no removal by adsorption onto the activated sludge in the secondary process or capture in primary tanks where chemically enhanced clarification is being performed. The concentration of rDON in digester sidestream will be higher if thermal hydrolysis is applied for sludge pretreatment before digestion; the concentration being dependent on the hydrolysis reactor temperature (Dwyer et al., 2008).

Phosphorus Content. The orthophosphate concentration range, as noted in Table 15–1, for the various digestion sidestreams does not include phosphate released by waste activated sludge from biological phosphorus removal processes, nor do the ranges account for reduction in phosphate due to precipitation within digester or in downstream mechanical equipment and pipes conveying sidestream. The values shown in Table 15–1 represent the stoichiometric amounts of phosphorus expected to be released into the bulk liquid for the various operating and performance conditions considered. For example, at facilities performing chemically enhanced primary treatment with ferric salts and anaerobically digesting the combined primary and waste activated sludges, the orthophosphate concentration in digester sidestream will typically be below 10 mg P/L due to precipitation or absorption onto iron floc in the digester. For facilities where ferric or ferrous salts are added directly to the anaerobic digester for control of hydrogen sulfide, partial removal of orthophosphate would also be expected. At facilities where no iron salts are added, partial precipitation of released phosphate as magnesium ammonium phosphate (struvite) at the digester outlet structure, in the biosolids dewatering process and in the pipes that convey digester sidestream is common. The amount that is formed is dictated by the ion molar concentrations of the struvite constituents and the elevated pH induced by CO_2 release from the bulk liquid at these locations.

Alkalinity Content. The alkalinity in sidestreams will vary depending on the source. The concentration of alkalinity in a sidestream is of importance in the operation of separate sidestream treatment processes. In sidestreams generated from fully aerated conventional low rate aerobic digesters, the alkalinity concentration is anticipated to be low due to the acidification of the digester sludge by nitrification. For aerobic digesters that operate with intermittent aeration to allow denitrification, the residual alkalinity in the digester will be slightly higher. In contrast, anaerobic digestion and autothermal thermophilic aerobic digestion (ATAD) sidestreams typically contain higher alkalinity concentrations, primarily in the form of bicarbonate. The cause of the high alkalinity concentrations is due to the retention of carbon dioxide in the digester bulk liquid to balance the positively charged ammonium ion at the typical pH range of the digesters (7.2 to 7.8). Therefore, the bicarbonate and ammonium-N concentrations will be equal on a molar basis. In terms of the standard measurement as $CaCO_3$, the alkalinity to ammonium-N mass ratio is 3.5 to 1 (kg $CaCO_3$/kg N).

Total Suspended Solids Content. Total suspended solids in digester sidestreams are composed largely of stabilized biologically inert solids with a relatively low volatile solids content (e.g., 65 percent). If the digester sidestream is sent to the primary sedimentation tanks, the TSS will settle readily and will be recycled to the solids processing train. If the sidestream is sent directly to the secondary treatment process, the solids will accumulate in the activated sludge. If the dewatering equipment is not working well, resulting in poor capture efficiency, the inert solids loading to the main plant can be substantial, potentially inducing foaming in the secondary process and effectively reducing the active fraction of the sludge in the secondary system.

Colloidal Material Content. Depending on the secondary treatment system configuration, digestion sidestreams, which can also contain colloidal material, can impact plant effluent quality. In addition to the potential for inorganic precipitate formation (e.g., struvite, hydroxyapatite), the presence of the colloidal material, can make the use of membrane-based technologies for separate treatment of digestion sidestream difficult due to membrane fouling, unless coagulation and filtration of these materials precedes the membrane process.

Temperature. Sidestream temperatures will vary depending on the source. The sidestream resulting from centrifugation of anaerobically-digested solids will typically have a temperature near the digester temperature (e.g., 30 to 35°C), depending on heat losses from a digested sludge holding tank and from the dewatering equipment. If a belt filter press is used for dewatering the same digested sludge, the temperature will typically be lower (e.g., 20 to 30°C) due to the inclusion of cooler belt washwater in the sidestream, unless warm filtrate is recycled and used as the washwater source. For aerobic digestion sidestreams, the temperature is also dependent on the type of process and the operating conditions. For example, a sidestream derived from ATAD process would have a higher temperature than a conventional non-insulated aerobic digester operating at a long hydraulic retention time. For the anaerobic sludge lagoons, the supernatant will be near ambient temperatures. The importance of temperature for the design of separate sidestream treatment processes is discussed in the individual sections dealing with physiochemical and biological treatment processes.

Flowrate. Sidestream flowrates from digestion processes, as reported in Table 15–1, are typically less than 1 percent of the daily average raw influent flow based on a continuous 7 d/wk operation. However, at facilities where digested sludge is not dewatered every day nor operate the dewatering equipment continuously (i.e., 24 h/d), the sidestream flow can result in a high instantaneous flow to the primary and secondary treatment processes if discharged directly without equalization. Estimation of peak sidestream flow is illustrated in Example 15–1.

EXAMPLE 15–1   **Estimate Peak Sidestream Flow**   The sidestream from an anaerobic digester contributes 0.7 percent of the daily average influent flow of 0.5 m³/s (11.4 Mgal/d), on a continuous basis, and is generated 5 d/wk, 8 h/d. What would be the instantaneous flow to the primary tanks? What percentage of the average influent flow would the instantaneous flow represent?

Solution

1. Determine the sidestream flowrate (SSF).

   $$SSF = (0.5 \, m^3/s)(0.007) = 0.0035 \, m^3/s$$

2. Determine the peak sidestream flowrate.

Peak SSF = $(0.0035 \text{m}^3/\text{s})[(7\text{d}/\text{wk})/(5\text{d}/\text{wk})][(24\text{h}/\text{d})/(8\text{h}/\text{d})]$
= $0.0147 \text{m}^3/\text{s}$

3. Determine the percentage of the flow represented by the instantaneous flow.

Percent of total flow = $[(0.0147 \text{ m}^3/\text{s})/(0.50 \text{ m}^3/\text{s})](100) = 2.94$

Comment   Although the facility likely has sufficient hydraulic capacity to receive this sidestream flow over the 8-h sludge dewatering period, the primary concern is the nutrient loading associated with this sidestream flow, particularly if the sidestream load occurs during the peak diurnal nutrient loading to the facility. The impact of the 8-h sidestream peak nutrient load can be dampened through flow equalization or by treating the sidestream load in a separate process to minimize the nutrient loading to the secondary process.

15–2 MITIGATING RECYCLE FLOWS AND LOADS

Recycle streams from sludge thickening, digestion, dewatering and storage processes may have negative impacts on the performance of the mainstream process. These impacts vary in degree and are specific to each facility. Summary information on the impact of recycle streams on the mainstream plant and potential mitigation measures is provided in Table 15–2.

Sidestream Pretreatment

Sidestreams will increase the solids load to the mainstream process, affecting solids inventory and the mixed liquor solids concentration., Some sidestreams, as noted previously, will also contain colloidal material which may impact plant effluent quality. Depending on the plant operating condition and performance, sidestream pretreatment to reduce its suspended solids and colloidal material content may be beneficial. The reduction of suspended solids and colloidal material in sidestreams is discussed in Sec. 15–3.

For treatment plants with stringent nutrient removal requirements, pretreatment of the nutrient-rich sidestreams prior to return to the mainstream process may be cost effective. Physiochemical and biological treatment options for these sidestreams are discussed in Sec. 15–4 through 15–11. As demonstrated in Example 15–1, the hydraulic load contributed by sidestream generated by dewatering operations only performed 7–8 h/d will increase the instantaneous hydraulic load through the mainstream facility. However, this hydraulic load increase is typically not a concern.

Equalization of Sidestream Flows and Loads

Digested sludge dewatering operations are commonly carried out as a batch operation, over several hours during the daytime hours five or six days per week. Consequently, nutrient-rich sidestream constituent loads to the mainstream plant will increase near instantaneously and will coincide with the diurnal influent load peak period. Where nitrification, denitrification, chemical phosphorus removal or biological phosphorus removal are practiced, the contribution of the sidestream nutrient load will increase peak air and chemical demands (external organic carbon, iron salts or alum), and may deteriorate plant effluent quality if control systems are not adequate to respond to the sudden and significant increase in nutrient loads. Nitrification may also be destabilized if the alkalinity concentration in the plant influent is relatively low and no supplemental alkalinity is available.

Table 15–2

Major impacts and potential mitigation measures for return flows from sludge and biosolids processing facilities[a]

Source of return flow	Impact	Process impacted	Mitigation measure
Sludge thickening	Effluent degradation by colloidal SS	Sedimentation	Add flocculent aid ahead of sedimentation tank
			Separately thicken primary and biological sludges
			Optimize gravity thickener dilution water
	Floating sludge	Sedimentation	Minimize gravity thickener detention time
			Remove sludge continuously and uniformly
	Odor release and septicity	Recycle point	Reduce gravity thickener detention time
			Return flows ahead of aerated grit chamber
			Provide odor containment, ventilation, and treatment (scrubber or biofilter)
		Biological	Return odorous flows to aeration tank
			Remove sludge continuously and uniformly
			Provide separate return flow treatment (with other recycle streams)
	Solids buildup	Sedimentation	Increase thickening unit operation time or capacity to maintain desired solids inventory in sedimentation units
		Biological	Remove sludge continuously and uniformly
			Include recycle loads in mass balance analysis
Sludge dewatering	Effluent degradation by colloidal suspended solids	Sedimentation	Optimize dewatering unit solids capture by improved sludge conditioning
			Add flocculent aid ahead of sedimentation tank
			Return centrate/filtrate to thickener
			Provide separate return flow treatment (with other recycle streams)
	Solids buildup	Sedimentation	Increase dewatering unit operation time or capacity to maintain desired solids inventory in sedimentation units
		Biological	Remove sludge continuously and uniformly
			Reduce trickling-filter recycle rate
			Include recycle loads in mass balance analysis
Sludge stabilization	Effluent degradation by excessive BOD load	Biological	Optimize supernant/decant removal, i.e., remove smaller amounts over a longer period of time, or reschedule removal to off-peak periods
			Provide separate return flow treatment
			Increase RBC speed
			Increase MLVSS in activated sludge system (decrease F:M ratio)
			Increase dissolved oxygen level in activated sludge process
	Effluent degradation by nutrients	Biological	Regulate digester supernatant/decant removal
			Thicken sludge before stabilization
			Provide separate return flow treatment
Washwater from depth filters	Hydraulic surges	Sedimentation	Provide backwash storage for flow equalization
			Schedule filter backwashing for off-peak periods

[a] Adapted, in part, from U.S. EPA (1987b).

To minimize these potential impacts, sidestream flow equalization can be employed. In larger facilities, where sludge is dewatered continuously, the benefit of sidestream equalization depends on the variability of the sidestream nutrient loads.

Equalization of sidestream flow is also commonly used to reduce the peak air demand in a biological pretreatment process. As discussed in Sec. 15–11, flow equalization will also have a direct impact on the sidestream reactor volume requirement. For batch treatment processes, where interruption in sidestream flow to the treatment process will occur intermittently during the treatment cycle, flow equalization is required if no release of raw sidestream to the mainstream plant is desired.

Equalization Volume Requirement. The general principles of flow and load equalization discussed in Sec. 3–7 in Chap. 3 apply to nutrient-rich sidestreams, but the basis for calculating the equalization volume requirement is slightly different. The equalization volume may be based on full equalization where sidestream is returned continuously to the mainstream plant or sent to a pretreatment process at a relatively stable, but adjustable, flowrate. The required equalization volume can be estimated using Eq. (15–1).

$$V = (N)(Q_{dw})(D_{dw}/7) \qquad (15\text{–}1)$$

Where V = equalization tank volume, m^3
  N = maximum number of consecutive days without dewatering, d
  Q_{dw} = average daily sidestream volume generated, m^3/d
  D_{dw} = number of days per week sludge is dewatered, d

Depending on the sidestream constituent concentrations, the alkalinity of the mainstream plant influent, and the treatment objectives of the mainstream plant, capturing sidestream in an equalization tank and returning the non-pretreated sidestream load to the mainstream plant in the off-peak hours when the facility influent flow and loads are at their minimum may be sufficient. However, for facilities that perform nitrate and phosphate removal, continuous return of the sidestream load at a steady rate is recommended. Where space constraints and tank costs are of concern, a smaller equalization tank volume may be used if the sidestream load variability does not impact plant performance. Dynamic process modeling is commonly used to assess the feasibility of using a smaller tank volume. Depending on the results of such an assessment, it may also be found that increasing the number of days each week or the hours per day that sludge is dewatered is the most cost-effective option.

Design Considerations. The features of a sidestream equalization tank vary depending on the design objectives. For an ideally equalized stream, mechanical mixing is provided to return a uniform composition to the mainstream plant. A mixing power input of 8 to 13 kW/10^3 m^3 (0.3 to 0.5 hp/10^3 ft^3) may be required if high suspended solids concentrations of suspended solids frequently occur ($>$ 1000 mg/L). The mixing power input may be decreased by 50 percent if high suspended solids concentrations are infrequent. Aeration is not provided typically, unless the tank is performing a dual function of equalization and biological treatment, in which case the tank design is dictated by the desired biological treatment performance.

Depending on the sidestream characteristics, the tank may be covered and equipped with an odor control device to minimize odorous emissions. Liquid level sensors with high and low alarms, an overflow to the plant drain system, a sloped floor with drain line, and a variable speed return pump or control valve with flow metering are common features in equalization tank design. Struvite formation in the tank is a potential concern and will impact the design of the mixer and selection of pipe or pipe liner material. The flexibility of adding iron salts to the tank to limit struvite formation may be a consideration. Configuring sidestream piping systems to avoid introduction or contact with air will also help limit struvite formation.

Equalization and TSS Reduction.   Equalization tanks for sidestream pretreatment processes often are used to reduce the TSS concentration if solids reduction is a requirement for the physiochemical or biological pretreatment process. In this design, mechanical mixing is not provided, the liquid level is greater and the operating liquid level range limited as to not disturb solids settling and thickening in the bottom section of the tank. Thickened solids are pumped intermittently to the solids processing facility. Because the tank is performing the dual function of sidestream storage and suspended solids reduction, the tank volume requirement may be greater than a tank only performing equalization, but a separate solids removal process is avoided.

Equalization and Biological Pretreatment.   Sidestream equalization and biological pretreatment can be accomplished within a single tank. In a process called storage and treat (SAT), the tank is designed as a sequencing batch reactor but has the flexibility to operate in a continuous overflow mode during short periods of peak sludge dewatering. Tank volume and aeration capacity are dictated by the treatment objectives, rather than the equalization requirement. At full scale, SAT has been used to provide partial treatment, resulting in pretreated sidestream containing a reduced ammonium concentration and a mixture of nitrite and nitrate (Laurich, 2004). Biological treatment of sidestream is discussed in Sec. 15–7 through 15–11.

EXAMPLE 15–2   Impact of Full Sidestream Equalization on Plant Influent Ammonium-N Concentration and Calculation of Equalization Tank Volume   Anaerobically digested sludge is normally dewatered from 8:00 a.m. to 4:00 p.m., 5 d/wk. Sidestream is produced at an average rate of 0.006 m^3/s during each 8-h period and contains an ammonium-N concentration of 1000 mg/L. The sidestream is sent to the facility's headworks. The longest number of consecutive days in which no sludge is dewatered is 3 d. For the average hourly facility influent flows and ammonium-N concentrations shown in the table below: (1) develop concentration-versus-time graphs for non-equalized and fully-equalized sidestream and (2) calculate the equalization tank volume required for full equalization.

	Given data	
Time period	Average flowrate during time period, m^3/s	Average NH_4^+-N concentration during time period, mg/L
M–1	0.275	20.0
1–2	0.220	18.8
2–3	0.165	17.9
3–4	0.130	20.2
4–5	0.105	17.3
5–6	0.100	15.6
6–7	0.120	15.3
7–8	0.205	13.4
8–9	0.355	19.6
9–10	0.410	27.0

(continued)

(*Continued*)

	Given data	
Time period	Average flowrate during time period, m³/s	Average NH₄⁺-N concentration during time period, mg/L
10–11	0.425	30.2
11–N	0.430	35.1
N–1	0.425	35.3
1–2	0.405	28.5
2–3	0.385	24.9
3–4	0.350	22.8
4–5	0.325	21.3
5–6	0.325	21.3
6–7	0.330	21.0
7–8	0.365	19.5
8–9	0.400	21.8
9–10	0.400	21.0
10–11	0.380	19.7
11–M	0.345	20.6
Average	0.307	

Note: m³/s × 35.3147 = ft³/s.

m³ × 35.3147 = ft³.

mg/L = g/ m³.

Solution

1. Calculate the fully equalized average sidestream flowrate.

Equalized flowrate = (0.006 m³/s)(3600 s/h)(8 h/d) (5 d/wk)/[(7 d/wk) (86,400 s/d)]

= 0.00143 m³/s

2. Using the non-equalized and fully-equalized sidestream flows, calculate the hourly average ammonium-N in the facility headworks.

	Plant Influent without sidestream		No Sidestream Equalization		Sidestream Equalization	
Time period	Average flowrate during time period, m³/s	Average NH₄⁺-N concentration during time period, mg/L	Sidestream flowrate, m³/s	Average ammonium-N concentration, mg/L	Sidestream flowrate, m³/s	Average ammonium-N concentration, mg/L
M–1	0.275	20.0		20.0	0.00143	25.1
1–2	0.220	18.8		18.8	0.00143	25.3
2–3	0.165	17.9		17.9	0.00143	26.5
3–4	0.130	20.2		20.2	0.00143	31.2

(*continued*)

(*Continued*)

Time period	Plant Influent without sidestream		No Sidestream Equalization		Sidestream Equalization	
	Average flowrate during time period, m³/s	Average NH$_4^+$-N concentration during time period, mg/L	Sidestream flowrate, m³/s	Average ammonium-N concentration, mg/L	Sidestream flowrate, m³/s	Average ammonium-N concentration, mg/L
4–5	0.105	17.3		17.3	0.00143	30.9
5–6	0.100	15.6		15.6	0.00143	29.9
6–7	0.120	15.3		15.3	0.00143	27.2
7–8	0.205	13.4		13.4	0.00143	20.3
8–9	0.355	19.6	0.006	36.5	0.00143	23.7
9–10	0.410	27.0	0.006	41.6	0.00143	30.5
10–11	0.425	30.2	0.006	44.3	0.00143	33.5
11–N	0.430	35.1	0.006	49.1	0.00143	38.4
N–1	0.425	35.3	0.006	49.4	0.00143	38.7
1–2	0.405	28.5	0.006	43.4	0.00143	32.1
2–3	0.385	24.9	0.006	40.5	0.00143	28.6
3–4	0.350	22.8	0.006	40.0	0.00143	26.9
4–5	0.325	21.3		21.3	0.00143	25.7
5–6	0.325	21.3		21.3	0.00143	25.7
6–7	0.330	21.0		21.0	0.00143	25.3
7–8	0.365	19.5		19.5	0.00143	23.4
8–9	0.400	21.8		21.8	0.00143	25.4
9–10	0.400	21.0		21.0	0.00143	24.6
10–11	0.380	19.7		19.7	0.00143	23.5
11–M	0.345	20.6		20.6	0.00143	24.7
Average	0.307					

3. Plot the resulting influent ammonium concentration.

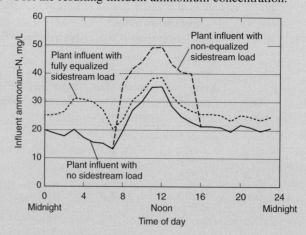

4. Calculate the equalization tank volume required for full equalization with 3 consecutive days with no sludge dewatering.

$$V = (N)(Q_{dw})(D_{dw}/7)$$

$$V = (3) [(0.006 \text{ m}^3/\text{s})(3600 \text{ s/h})] (8 \text{ h/d}) [(5 \text{ d/wk of sludge dewatering})/(7 \text{ d/wk})]$$
$$= 370 \text{ m}^3 (13,066 \text{ ft}^3)$$

15–3 REDUCTION OF SUSPENDED SOLIDS AND COLLOIDAL MATERIAL

The solids balance for a wastewater treatment plant, as illustrated in Sec. 14–7 in Chap. 14, is influenced by the TSS in the sidestreams derived from sludge thickening and biosolids dewatering processes. Sidestreams are commonly returned to the headworks where solids settle in primary sedimentation tanks. In plants that do not have primary sedimentation or return the sidestreams directly to the mainstream secondary process, the solids become integrated into the mixed liquor suspended solids. Because the majority of the post-digestion sidestream TSS are biologically inert, they will contribute directly to the secondary solids inventory and will exert only a minor oxygen demand on the process. The impact of these solids on the primary tank effluent or the secondary process operating conditions and performance varies with the sidestream solids loading rate, the operating conditions of the primary process, and the type of the secondary process employed.

Some sidestreams, particularly from post-digestion dewatering processes, contain colloidal matter that may impact the plant effluent quality, depending on the type of secondary process in the facility and its operating conditions. The particles will contribute to turbidity and may negatively impact effluent disinfection. Plant effluent colloidal particles are a particular concern for facilities that provide treated effluent for reuse.

Common practices used for the management of TSS and colloidal matter in thickening and dewatering sidestreams, at full scale, are discussed below.

Sidestreams Derived from Sludge Thickening

Thickeners that suffer from poor capture efficiencies will generate sidestreams with a high TSS concentration. The elevated sidestream solids may negatively impact the solids removal efficiency across the primary sedimentation tanks or contribute an unacceptable solids loading rate to a secondary process. Thickener operating conditions are commonly adjusted to improve solids capture to resolve this problem rather than installing a separate solids reduction process for the sidestream.

Sidestreams Derived from Biosolids Dewatering

Reducing the TSS concentration in post-digestion sidestream before the sidestream is sent to the mainstream plant is uncommon. If the dewatering sidestream solids concentration is frequently high and the solids returned to the mainstream plant prove detrimental, the dewatering process operating conditions must be adjusted to improve solids capture. A solids balance should be performed to assess its contribution to the mixed liquor solids inventory and concentration and determine if a sidestream solids reduction step would be beneficial. Life cycle cost analyses can be performed comparing operating costs of power, aeration, and chemicals in mainstream treatment with that of a sidestream pretreatment step.

Reduction of TSS is more common when sidestream is pretreated to reduce the nutrient load returned to the mainstream plant. The pretreatment requirement is specific to the physiochemical or biological process employed for nutrient reduction, as discussed in Sec. 15–4 through 15–11.

Process Options. Reduction of TSS before sidestream pretreatment has been accomplished in practice through gravity settling in sidestream equalization tanks. The suspended solids concentration in the clarified sidestream is typically sufficient for biological treatment processes that benefit from suspended solids less than 200 mg/L. If higher removal efficiency is required, separation processes such as Lamella inclined plate settlers or high rate clarification processes (see Sec. 5–7 in Chap. 5 for process descriptions) may be applied where chemically enhanced flocculation can be employed optimally to improve solids removal. Experience is largely limited to inclined plate settlers, without chemical addition, as the majority of the pretreatment processes applied at full scale do not require a low suspended solids concentration. Filtration is rarely used on sidestream due to the high fouling potential of the colloidal material and residual polymer and the formation of inorganic foulants such as struvite.

In belt filter presses, belt washwater contains the majority of the uncaptured solids and the filtrate has a TSS concentration less than 500 mg/L. Belt filter presses can be designed for separate collection of washwater and filtrate, potentially eliminating the need for a solids reduction step. Separate collection of washwater and filtrate results in higher nutrient concentrations in the filtrate, which will improve the nutrient removal efficiency in the pretreatment process. Some pretreatment processes may also benefit from the elevated temperature of the filtrate, which is no longer being cooled by washwater.

Removal of Colloidal Matter

Sidestreams derived from post-digestion dewatering processes commonly contain colloidal particles that may impact the facility effluent. Typically, the colloids are not removed by primary sedimentation. If chemically-enhanced primary treatment is employed, partial or complete capture of the colloidal particles may occur. In the absence of chemical addition, the colloidal particles will enter the secondary treatment process where some portion of the particles will be captured by the mixed liquor suspended solids. The degree of capture is dependent on the type of secondary process and its operating conditions. If a sidestream pretreatment process is employed for nutrient reduction, a portion of the colloidal particles may be removed before the stream is sent to the mainstream plant.

Removal of sidestream colloidal particles by chemically-enhanced flocculation in a separate pretreatment process is not a common practice. If colloidal particles derived from post-digestion dewatering sidestream are identified as a primary source of turbidity in a plant effluent, addressing the problem by sidestream pretreatment may be the most cost-effective option as the colloids are present at their highest concentration in the sidestream and the sidestream flow is less than 1 percent of the mainstream flowrate. A process configuration could consist of solids reduction by gravity settling in an unmixed sidestream equalization tank, followed by an advanced chemically-enhanced filtration process. Filtration options are presented in Chap. 11. Bench or pilot-scale testing is recommended to assess feasibility and chemical requirements.

15–4 PHYSIOCHEMICAL PROCESSES FOR PHOSPHORUS RECOVERY

Phosphate ore, which serves as the primary source of phosphorus for modern agricultural practices is a limited resource. Significant depletion of known global reserves had been projected to occur as early as the end of the 21st century (Cordell et al., 2009), but based on the most recent projections known reserves will be available for the next 300 to 400 years (Van Kauwenbergh, 2010). However, as demand for phosphorus increases and the exploration for new reserves, the mining of lower quality ores, and the adoption of more expensive processing equipment occur, the price of phosphorus will continue to escalate in the future. Consequently, the phosphorus in wastewater treatment facility sludges and return flows may be viewed increasingly as an asset that should be recovered and reused as fertilizer rather than a nutrient that must be treated and disposed.

The primary focus of this section is on phosphorus recovery processes that have been developed and demonstrated at a full-scale on nutrient-rich sidestreams and industrial wastewaters. Other technologies have been developed to recover phosphorus from sludge ash, but these processes are not presented here. The phosphorus recovery processes, based on crystallization, considered in this section include those for the recovery of magnesium ammonium phosphate (struvite) and calcium phosphate (hydroxyapatite). Descriptions of the following crystallization processes that have been demonstrated in full-scale facilities are presented in Table 15–3.

AirPrex® process
Cone-shaped fluidized bed crystallizer
Crystalactor®
NuReSys® process
Pearl® process
Phosnix® process
PHOSPAQ™ process

However, before discussing these processes in greater detail, it will be useful to consider the fundamental aspects of the crystallization process. The beneficial reuse of recovered nutrients as fertilizers is discussed in Sec. 15–6.

Description of the Crystallization Process

All of the technologies presented in Table 15–3 are based on the three fundamental stages that occur within the physical environment of the crystallizer (reactor): (1) supersaturated ion concentrations, (2) primary and secondary nucleation processes, and (3) crystal growth. The phosphate removal efficiency, crystal size distribution achieved within the reactor, and the purity of the final product are all influenced by the temperature, pH, ionic composition, and hydrodynamic conditions within the reactor. In the recovery of phosphate for reuse, a larger crystalline product is desired rather than an amorphous solid phase consisting of fine particles that are difficult to recover from the liquid and process to generate a reusable product.

Supersaturation. In supersaturated solutions, the product of the ion molar concentrations (expressed as activities) of the desired product exceeds the value of its solubility constant at given reaction conditions. As the degree of supersaturation increases, there is increased potential of forming fine particles through primary nucleation, not a desirable condition, as discussed below. Therefore, the supersaturated condition within a crystallizer is controlled to avoid fine particle formation and to provide sufficient driving force for mass transfer of ions to the surface of the growing crystals. Supersaturation is the driving force for subsequent processes of nucleation and crystal growth.

Nucleation. The process by which ions come together under supersaturated conditions and aggregate into a solid form, or "nuclei," according to their solubility at a given temperature and pH is known as *nucleation*. The nucleation process cycles through

Table 15–3

Processes for recovery of sidestream phosphorus as magnesium ammonium phosphate (struvite)

	Description
(a) AirPrex® process 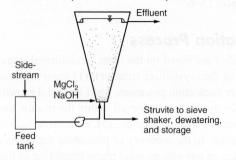	The AirPrex® process was, developed by Berliner Wasserbetriebe (Germany) in collaboration with the Berlin Institute of Technology. In this process, struvite is crystallized directly from the sludge stream from an anaerobic digester, rather than from sidestream, to prevent struvite formation in the sludge dewatering process. AirPrex® consists of a dual-stage aerated tank configuration, either as separate tanks or as a single tank with a dividing wall, with a hydraulic retention time (HRT) of approximately 8 h. An air-lift aeration design is used in each stage to induce sludge mixing and strip CO_2 to increase pH. Magnesium chloride is used as the magnesium source and is added to the first, second or both stages. As struvite forms and develops into particle sizes of sufficient settling velocity, it settles into the bottom conical section of each stage. The product is withdrawn intermittently or continuously from each stage and transferred by a screw conveyer to a sand washer. Washed product is stored wet or is subsequently dried. Aerated sludge overflows the second stage and is sent to a sedimentation vessel where additional struvite may be recovered or to the dewatering process. Processing the exhaust air through an odor control system may be required.
(b) Cone-shaped fluidized bed crystallizer 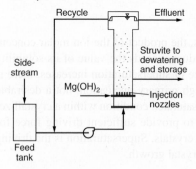	A cone-shaped fluidized bed crystallizer was developed by Multiform Harvest Inc. (USA). The crystallizer consists of a conical section and a solids-liquid separation zone located at the top. The dimensions of the conical section are selected to provide a desired range of superficial upflow velocities. The HRT is typically less than 1 h. As struvite crystals grow, they settle towards the bottom of the cone where they are removed intermittently, processed through a sieve shaker or drum screen, disinfected, and bagged for off-site processing. Magnesium chloride and sodium hydroxide are added through a proprietary injection system at the bottom of the cone to provide supersaturated conditions and to increase pH to the desired range.
(c) Crystalactor®	The Crystalactor® is a fluidized bed crystallizer developed by DHV (The Netherlands). The crystallizer consists of a cylindrical reactor with a solids-liquid separation zone located at the top. Effluent is recirculated to the bottom of the reactor where it is blended with sidestream and injected into the crystallizer through nozzles to achieve optimum cross-section liquid distribution. The effluent recirculation rate is adjusted to maintain a superficial upflow velocity in the range of 40 to 75 m/h (130 to 250 ft/h) in the reaction section of the vessel. The HRT based on sidestream flow is typically less than 1 h. Quartz sand is added initially as seed material to accelerate startup, but further sand addition is not required once struvite crystals form. As the pellets grow, they settle to the bottom of the crystallizer, where a portion of the pellets is removed at regular intervals, dewatered, and stored for offsite transport.

(continued)

Table 15–3 (*Continued*)

	Description
(d) NuReSys® process 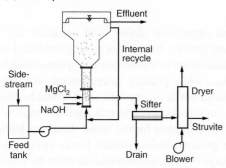	The NuReSys® process (NUtrient REcovery SYStem) was developed by Akwadok/NuReSys (Belgium). The process consists of a CO_2 stripping tank followed by a mechanically stirred crystallizer and a sedimentation zone. A HRT of 0.5 to 1 h is provided in the stripping tank where mechanical agitation and air are provided to strip CO_2 from the sidestream. Mixer speed and air flow are adjusted to control the pH to limit the formation of fine crystals in the stripping tank. In the crystallizer, mechanical stirring provides mixing and creates the hydrodynamic environment conducive to pelletized struvite formation. Stirrer speed and the product withdrawal rate are adjusted to provide the desired pellet size of the harvested product. Magnesium chloride is used as the magnesium source and NaOH is added to control the pH in the range of 8.1 to 8.3. A crystallizer HRT of 0.5 to 1 h is typical. Smaller crystals are settled in the sedimentation zone where they are returned to the crystallizer. Processing the exhaust air through an odor control system may be required.
(e) Pearl® process	The Pearl® process was developed at the University of British Columbia for the crystallization of magnesium ammonium phosphate and was introduced at full-scale by Ostara Nutrients Recovery Technologies Inc. (USA). The Pearl® reactor is a fluidized bed crystallizer with a segmented construction where the segment or zone diameter increases from the bottom of the reactor to the top to reduce the upflow liquid velocity incrementally and retain struvite crystals of various sizes within each zone. A liquid/solids separation section is located at the top of the reactor. Effluent is recirculated to the bottom of the reactor to maintain the upflow velocity profile within the desired range. The HRT based on sidestream flow, is typically less than 1 h. As struvite pellet diameter increases, the pellets gradually sink from one zone to the next. The final product is removed from the bottom zone, separated from the liquid by screening, dried, and bagged. The effluent recirculation rate and struvite retention time are adjusted to control the pellet size in the final product. The magnesium source is typically magnesium chloride. Sodium hydroxide is used to maintain the pH within the desired range.
(f) Phosnix® process 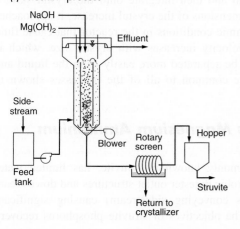	The Phosnix® crystallizer, developed by Unitika Ltd (Japan), consists of a cylindrical reaction zone with a conical bottom section and a larger diameter solids-liquid-gas separation section at the top. The crystallizer is aerated to provide mixing and to increase the pH by stripping CO_2 from the liquid. If required, the exhaust air is treated with an odor control system. The HRT in the reaction zone is less than 1 h. Magnesium hydroxide is typically used as the magnesium source and sodium hydroxide is added to control pH within the desired range. Larger struvite pellets that settle into the conical section of the reactor are pumped intermittently to a rotary drum screen. Liquid and associated smaller struvite particles that pass through the screens are sent to the mainstream plant or returned to the crystallizer to allow the fine particles to serve as seed material for struvite pellet growth.

(*continued*)

Table 15–3 (Continued)

	Description
(g) PHOSPAQ™ process 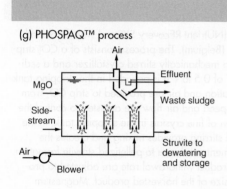	PHOSPAQ™ was developed by Paques (The Netherlands). The process consists of an aerated reaction zone and a proprietary solids-liquid-air separation device in the upper section of the vessel. An air-lift aeration design is used to (1) provide mixing, (2) strip CO_2 from the liquid to increase pH, and (3) provide dissolved oxygen (DO) for biological treatment. The HRT is approximately 5 to 6 h. Magnesium oxide is typically used as the magnesium source. Struvite is harvested from the bottom of the reactor by pumping struvite-rich mixed liquor through a hydrocyclone. The recovered product is dewatered in a screw press and transferred to a container as a 70 percent dry material. The separation device above the reaction zone allows smaller struvite particles and biomass to settle and return to the reaction zone. An odor control system may be required for the exhaust air.

aggregate formation and dissolution until an aggregated cluster is of sufficient size to remain stable and provide a surface for crystal growth. Nucleation is subcategorized into primary and secondary processes. Primary homogeneous nucleation is a spontaneous formation of nuclei from supersaturated solution, while primary heterogeneous nucleation occurs on foreign solid surfaces (interior reactor surfaces, crystals, sand, colloids, dust). Of the two forms of primary nucleation, the latter is dominant in real applications.

Secondary nucleation also occurs in two fundamental forms: "fluid shear" nucleation and "contact" nucleation. In the fluid shear process, hydrodynamic forces sweep small nuclei from the crystal surface or cause breakage of branched protrusions from the crystal, resulting in the generation of a new "seed" surface for nucleation and growth of new crystals. Contact nucleation is a result of crystal attrition where physical forces fracture crystals, increasing the number of crystals and surface area for growth.

Crystal Growth. The growth of crystals comprises the third primary stage of the crystallization process. Under supersaturated conditions, ions diffuse through the boundary layer near the surface of each crystal and then integrate onto the crystal surface through a complex mechanism. As the dimensions of the crystal increase, it approaches a terminal size dictated by the hydrodynamic conditions in the reactor that cause fluid shearing and crystal attrition. Settling velocity increases with crystal size, which is advantageous as it allows the product to be separated more easily from the liquid and other solids within the reactor, a feature common to all of the processes shown in Table 15–3.

Recovery of Phosphorus as Magnesium Ammonium Phosphate (Struvite)

Magnesium ammonium phosphate, commonly known as struvite, has limited water solubility and forms encrustations in anaerobic digester outlet structures and downstream processes (dewatering equipment, pipes conveying sidestream) causing significant operational and maintenance problems. The objectives of struvite phosphorus recovery processes are to form struvite and produce a crystallized product with sufficient purity and physical characteristics that it qualifies for reuse as a fertilizer. The value of struvite in the fertilizer market is discussed in Sec. 15–6.

Reaction Stoichiometry.   The chemistry of struvite formation is presented in detail in Sec. 6–5 in Chap. 6. The stoichiometry of the struvite precipitation reaction is shown in Eq. (6–25), repeated here for convenience.

$$Mg^{2+} + NH_4^+ + PO_4^{3-} + 6H_2O \rightarrow MgNH_4PO_4 \cdot 6H_2O \tag{6–25}$$

Typically, the product of the ion molar concentrations of the struvite constituents in the anaerobic digester bulk liquid is not high enough to induce a substantial level of struvite formation, thus limiting the phosphate recovery efficiency. To overcome this limitation in the processes described in Table 15–3, a magnesium salt is added to the crystallizer. As discussed below, the addition of magnesium is not the only requirement for successful production of struvite. The pH and hydrodynamic conditions within the crystallizer must be controlled to optimize phosphate removal and generate a product of sufficient size that can be separated easily from the liquid phase and other suspended solids.

Calcium Inhibition.   Calcium competes with magnesium for phosphate under conditions conducive to precipitation of calcium phosphates and magnesium ammonium phosphate. A Mg/Ca molar ratio of less than 2/1 results in a longer induction time and a lower struvite crystal growth rate. As the ratio decreases below 1/1, amorphous calcium phosphate formation is dominant (Le Corre et al., 2005).

Operational Considerations.   The principal operational requirements that must be considered in the application of phosphorus recovery as struvite include (1) pretreatment requirements, (2) pH and temperature control, (3) chemical requirements, (4) seed requirements, and (5) mixing and hydraulics. Each of these factors is considered in the following discussion.

Pretreatment Requirements. Sidestream pretreatment to reduce TSS and colloidal material has not been standard practice. Upflow fluidized bed crystallizers can tolerate sidestream TSS concentrations ranging from 1500 to 5000 mg/L, depending on the technology. The upflow velocities are high enough to prevent digested biological solids from settling. Based on operating experience with the AirPrex® process, struvite crystals can be produced and separated from anaerobic digester effluent (biosolids), albeit with a larger reactor.

pH and Temperature Control. Struvite solubility decreases with increasing pH, reaching a minimum solubility near pH 10.3 (see Sec. 6–5 in Chap. 6). However, in practice, struvite crystallizers do not operate at pH greater than 9.0. Typically, pH is controlled in the range of 8.0 to 8.8 to minimize the addition of base chemicals, limit the degree of supersaturation, and limit the potential formation of other solids such as calcium carbonate and calcium phosphate (hydroxyapatite). These solids would lead to greater impurity of the final crystalline product. A reduction in the phosphorus recovery efficiency may also occur due to the formation of fine amorphous calcium phosphate precipitate that will exit the crystallizer with the treated effluent. In the pH range of 8.0 to 8.8, phosphorus recovery efficiencies greater than 80 percent and the production of a highly pure struvite product have been demonstrated.

The formation of struvite at a pH less than 8.0 has also been demonstrated. In the AirPrex® process, where struvite is formed directly from anaerobic digester effluent (sludge), operating in a pH range of 7.2 to 7.4 results in a high phosphorus recovery efficiency and generation of struvite crystals of sufficient size that allows separation from the sludge in the conical section of the reactor (Nieminen, 2010). The crystal growth rate is reduced as the operating pH decreases, which results in a longer residence time in the crystallizer to achieve the same level of phosphorus recovery and crystal sizes. In sidestreams derived from the digestion of WAS from biological phosphorus removal processes, the orthophosphate concentration and the resulting level of supersaturation are sufficiently high that a pH greater than 8.0 may not be required to achieve the desired level of phosphorus recovery.

The strategies used in attaining the desired operating pH range varies by process as summarized in Table 15–3 and the sidestream characteristics. Aeration of the crystallizer has been shown to be an effective method for increasing pH through stripping of CO_2. Anaerobic digester effluent and sidestream are supersaturated with CO_2 due to the high CO_2 content of the digester gas. Once exposed to air under atmospheric conditions, CO_2 will diffuse from the liquid and pH will increase above 8.0. In the AirPrex®, PHOSPAQ™, and Phosnix® technologies described in Table 15–3, aeration is applied directly to the crystallizer to strip CO_2 from the liquid. Aeration also provides mixing and induces hydrodynamic conditions conducive to developing crystals over a range of sizes that allows separation and processing. In the NuReSys® process, CO_2 is stripped from the sidestream through aeration and mechanical agitation in a vessel preceding the crystallizer. The application of high speed mechanical agitation and aeration reduces energy consumption and limits the amount of air that may require treatment in an odor control process. Mixer speed and air flow are adjusted to limit the formation of fine struvite crystals before the sidestream enters the crystallizer.

The dominant forms of orthophosphate in an anaerobic digester and sidestream are HPO_4^{2-} and $H_2PO_4^-$, with PO_4^{3-}, the form required for struvite formation, existing at a low concentration. As shown in Eq. (15–2), as struvite forms, HPO_4^{2-} and $H_2PO_4^-$ shift towards PO_4^{3-}, resulting in proton release or an increase in acidity.

$$H_2PO_4^- \rightleftarrows HPO_4^{2-} + H^+ \rightleftarrows PO_4^{3-} + 2H^+ \tag{15–2}$$

If magnesium chloride is used as the source of magnesium, additional acidity is generated through the chloride ions remaining in solution. The acidity generated by these chemical reactions will be buffered by neutralization with bicarbonate, but if sufficient acidity is generated, pH will decrease and a target pH above 8.0 may not be achieved by aeration alone. Magnesium oxide or hydroxide may be selected as the magnesium source as an alternative to magnesium chloride to provide alkalinity or sodium hydroxide can be added for pH control. Depending on the amount of acidity created as a result of struvite formation, magnesium oxide (or hydroxide) and sodium hydroxide may be required to control the process.

Struvite solubility is a function of temperature, but temperature control is not practiced in the processes presented in Table 15–3. Cooling or heating the reactor to operate at an ideal temperature to maximize the phosphate removal efficiency is not justified economically.

Chemical Requirements. The source of magnesium is typically magnesium chloride, magnesium hydroxide, or magnesium oxide, which forms magnesium hydroxide upon contact with water. The choice of chemical for each process varies and is based on vendor or end user preference and chemical cost.

Magnesium chloride has the advantage of disassociating faster than magnesium hydroxide, leading to higher reaction rates. Consequently, an optimized chemical feed dispersion system is required for the reactor to prevent localized excessively high supersaturated conditions near the chemical addition point.

Magnesium oxide and hydroxide provide both magnesium and alkalinity to the crystallizer. The primary role of both compounds is to provide magnesium. The alkalinity provided by the chemical is advantageous, but may not increase the pH and stabilize it within the desired operating range. In this case, sodium hydroxide addition will be required for pH control. Magnesium oxide and hydroxide have limited water solubility and are fed to the crystallizer as suspensions in water. Consequently, the solubilization rate of the compounds may control the level of supersaturation and the crystal growth rate. Nucleation may also occur on the surface of undissolved reagent leading to reduced purity of the harvested product.

Magnesium is added to the process in molar excess to maintain a supersaturated condition for crystal growth. A molar ratio of Mg^{2+} to PO_4^{3-} in range of 1.1 to 1.6 has been used in practice for municipal sidestreams, with a value of 1.3 being typical. If excess magnesium is added, resulting in a higher level of supersaturation, excessive primary nucleation will occur, leading to the formation of small crystals that may not be retained in the crystallizer, depending on its hydraulic design. At the Mg^{2+}/PO_4^{3-} molar ratio and operating pH range used in crystallizing struvite from municipal sidestreams, inhibition by calcium and the formation of calcium phosphate is minimal, leading to a high recovery efficiency and high purity of the harvested product.

Sodium hydroxide is used typically to adjust and control pH, depending on the process technology and desired operating pH. Sodium hydroxide is preferred due to ease of handling, and it can be stored at concentrations as high as 50 percent by weight.

Chemical requirements are specific to the sidestream as operating conditions in the mainstream process and solids processing facility dictate the phosphate concentration in the digester and its variability. Pilot or demonstration-scale testing of the selected crystallization technology is also commonly performed to assess the chemical requirements and other operating conditions for design of the full-scale process.

Seed Requirements. A seed material such as sand or struvite crystals is often added to a crystallizer to rapidly start the process. After a certain level of crystal inventory is achieved in the reactor, the process is self-sustaining at the target operating conditions (e.g., level of supersaturation, pH) and further seed addition is not required.

Mixing and Hydraulic Requirements. The mixing and hydraulic conditions within the reactor have an impact on the size of the crystals harvested from the process. In the AirPrex®, PHOSPAQ™ and Phosnix® processes, an air-lift aeration design is used to strip CO_2 and induce a mixing pattern within the reactor, resulting in the development of crystal sizes sufficient for separation and processing. In the AirPrex® and Phosnix® reactors, a conical bottom provides an environment where the larger struvite crystals can separate and thicken prior to their removal. Alternatively, solids from the bottom of the reactor can be intermittently pumped through a hydrocyclone to recover the product for further processing, as done in the PHOSPAQ™ process.

In the Pearl® and Crystalactor® processes, effluent from the crystallizer is recycled to maintain the upflow liquid velocity within the desired range to develop crystals and a pelletized product with a specific range of sizes. If the size of the crystals or pellets harvested from the reactor is of less importance, effluent recirculation may not be required. In the cone-shaped fluidized bed crystallizer (Table 15–3), where no internal recirculation is applied, crystals develop of sufficient size that allows separation from the liquid. The

Figure 15–1

Examples of full scale struvite crystallizers: (a) Ostara Pearl® reactor, Tigard, OR and (b) Multiform Harvest cone-shaped reactor, Yakima, WA (courtesy of Multiform Harvest Inc.).

(a)

(b)

struvite is removed intermittently and processed offsite to generate product characteristics specific to a particular end use.

The specific geometry of the crystallizer may also be designed to create a range of upflow velocities that impact the characteristics of the harvested product and allow retention of smaller crystals, which continue to grow and sink into the lower sections of the crystallizer. The Pearl and cone-shaped crystallizers shown on Figs. 15–1(a) and (b) are examples of processes that create variable upflow velocities through an increasing cross-sectional diameter. In the Phosnix®, Crystalactor® and PHOSPAQ™ processes, the smaller crystals are retained in low upflow velocity zones or sedimentation zones in the upper section of the reactors and gradually return to the main reaction zone.

The hydrodynamic conditions required for crystallization and pellet formation can also be achieved through mechanical mixing. In the NuReSys® process, a mechanical mixer with a three-blade impeller has been used successfully to provide mixing and generate a pelletized final product.

Product Separation and Purification.   Crystals are allowed to grow to sizes that allow separation from the liquid and other suspended solids and are easy to process after harvesting (e.g., screening, washing). Separation and recovery of product with a mean crystal size of 0.2 mm or greater from the liquid has been demonstrated. Depending on the crystallizer design, the hydrodynamic conditions will enhance crystal agglomeration and form spherical pellets with a mean diameter up to 2–4 mm. The mean diameter of the product is controlled by adjusting the product harvesting rate to increase or decrease the product residence time in the crystallizer.

Processing requirements for the struvite product vary. The product may be concentrated by screening, disinfected with chlorinated water or heat, rinsed, dewatered, and dried. Product purity typically exceeds regulatory requirements and no further purification steps are required. The degree to which the final product is processed is dependent on the end user requirements. Examples of processing equipment for the pelletized product harvested from a Pearl crystallizer are shown on Fig. 15–2.

Struvite Phosphorus Recovery Limitations.   In practice, the orthophosphate is not recovered completely and a crystallizer effluent concentration below 5 mg P/L is not typical. For fluidized bed reactors with a relatively low hydraulic retention time (less than 1 hour), consistently achieving an effluent concentration below 10 mg P/L in these systems

Figure 15–2

Processing of pelletized struvite from a Pearl crystallizer: (a) drying; (b) product bagging.

(a)   (b)

requires greater chemical addition to maintain sufficient supersaturated conditions throughout the crystallizer as orthophosphate decreases. Based on economic considerations, i.e., chemical cost versus higher recovery efficiency, an effluent orthophosphate concentration in the range of 10 to 25 mg P/L is typical. In processes such as AirPrex® where the hydraulic retention time is much higher (e.g., 8 h), effluent concentrations below 10 mg P/L have been demonstrated, but at the expense of requiring a larger reactor.

Recovery of Phosphorus as Calcium Phosphate

Phosphorus can be recovered from sidestream flows through precipitation as calcium phosphate. A typical process flow diagram for the recovery of phosphorus as calcium phosphate is presented on Fig. 15–3. The process was developed initially and demonstrated successfully in The Netherlands in the 1980s (Piekema and Giesen, 2001). However, the rise in chemical costs has made this process generally uneconomical, especially as compared to the recovery of phosphorus as struvite. Although the recovery of phosphorus as calcium phosphate is now largely limited to industrial, food, and dairy applications, there are still wastewater applications. Phosphate recovery from mainstream processes with lime precipitation continues to be used in several facilities and is discussed following the discussion of the recovery of calcium phosphate.

Reaction Stoichiometry. Calcium phosphate can exist in several forms depending on pH and the ionic composition of the sidestream. The basic chemistry of phosphate precipitation with lime is presented in Sec. 6–5. The primary product generated during

Figure 15–3

Process flow diagram for the production of calcium phosphate from nutrient-rich sidestream.

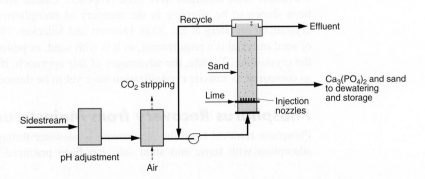

crystallization is hydroxyapatite $[Ca_5(PO_4)_3OH]$, the most thermodynamically stable form of calcium phosphate.

Operational Considerations. The principal operational requirements that must be considered in the application of phosphorus recovery as calcium phosphate include (1) pretreatment requirements, (2) pH and temperature control, (3) chemical requirements, and (4) seed requirements. Each of these factors is considered in the following discussion.

Pretreatment Requirements. Most high-strength sidestreams contain a high bicarbonate concentration that will cause significant formation of calcium carbonate precipitate in the crystallizer. To limit the formation of calcium carbonate, the sidestream is pretreated through acidification to a pH less than 5 with a strong inorganic acid, and the acidified stream is subjected to air stripping to remove CO_2. The pretreated stream is then treated with calcium hydroxide slurry to precipitate phosphate.

pH and Temperature Requirements. Calcium phosphate crystallization is performed over a pH range of 8.0 to 9.0. The optimum pH for a given application is developed through pilot-scale tests. In practice, the crystallizer temperature is not controlled, due to economic considerations.

Chemical Requirements. An inorganic acid is required for sidestream pretreatment. The amount of acid can be estimated preliminarily by assuming complete conversion of bicarbonate to CO_2. Quick lime (CaO) is used typically as the calcium source for the crystallizer and fed as slurry (forming calcium hydroxide upon addition to water). Lime is added to increase the pretreated sidestream pH from 5.0 or less to the crystallizer pH setpoint and provide the desired level of supersaturation in the crystallizer. A lime overdose in the range of 0.5 to 5.0 mM has been used in practice (Piekema and Giesen, 2001). The inorganic acid and lime requirements are determined through pilot-scale tests.

Seed requirements. Calcium phosphate crystallization through primary nucleation creates a fine microcrystalline product that is difficult to separate from water. However, in the presence of seed material such as sand and under well-controlled supersaturated conditions, nucleation will occur on the surface of the sand particles and crystal growth on the newly formed pellet is sustained. By controlling the hydrodynamic conditions within the fluidized-bed crystallizer, a pelletized product with a mean diameter of approximately 1-mm is harvested from the bottom of the crystallizer. As pelletized product is removed intermittently, virgin seed material is added. The superficial upflow velocity through the crystallizer is in the range of 40 to 75 m/h (130 to 250 ft/h).

To eliminate the need for inorganic carbon removal via acidification and air stripping, alternative seed materials have been proposed. Calcite and calcium silica hydrates have been shown to be effective in the recovery of phosphorus by means of hydroxyapatite crystallization (Berg et al., 2006; Donnert and Salecker, 1999). The intermittent addition of seed material is a requirement, as it is with sand, as pelletized product is removed from the crystallizer. To date, the advantages of this approach, if any, for phosphorus recovery as compared to struvite crystallization have yet to be demonstrated at full scale.

Phosphorus Recovery from Mainstream Processes

Phosphate removal from the mainstream wastewater through chemical precipitation and adsorption with ferric and alum salts has been practiced widely. The chemical sludge produced by this reaction is amorphous and is formed in primary sedimentation tanks or in the activated sludge process where it is intimately blended with primary sludge or activated sludge. The form of the chemical sludge and its intimate mixing with non-chemical sludges makes phosphate recovery more challenging. Technologies for recovering phosphate from sludges and incinerator ash have been developed, but are not discussed in this chapter.

Phostrip Process. One of the first processes used for the recovery of phosphate from the mainstream process is the Phostrip process, developed specifically for enhanced biological phosphorus removal (EBPR). A description of the Phostrip process is presented in Table 8–27 in Sec. 8–8. In the Phostrip process, a portion of the phosphorus-rich return activated sludge (20 to 40 percent) from the EBPR process is subjected to anaerobic conditions at a sludge retention time in the range of 12 to 20 h. Under these conditions, endogenous production of readily biodegradable COD (rbCOD) causes release or *stripping* of orthophosphate from the sludge to the bulk water. The phosphate *stripper* tank typically consists of a gravity thickener with thickened sludge recirculation to elutriate the released phosphorus from the sludge to the thickener overflow. For nitrifying mainstream processes, a prestripper tank is provided prior to phosphate stripping to denitrify the RAS flow and optimize phosphate release. Thickened RAS is returned to the mainstream process or a portion is sent to the solids processing facility as waste activated sludge (WAS).

Traditionally, stripper overflow has been subjected to chemical precipitation with lime at a pH of 9.0–9.5 to remove phosphate. Due to the presence of bicarbonate in the stripper overflow, calcium carbonate formation will also occur, resulting in a mixed solids composition. Typically, the chemical solids are removed separately or settled in the primary sedimentation tanks. Alternatively, the overflow from the stripper tank can be blended with sidestream from post-digestion dewatering and fed to a crystallization process to produce magnesium ammonium phosphate.

Release of Phosphate from Waste Activated Sludge. The Phostrip concept can be adapted for waste activated sludge derived from an EBPR process where WAS is subjected to anaerobic conditions to allow solids hydrolysis and fermentation, which results in the release of phosphate. Releasing phosphate from WAS prior to anaerobic digestion significantly reduces struvite formation in the digester and digester overflow. The WAS fermenter or stripper may be configured as a WAS thickener with internal thickened sludge recycle, similar to the Phostrip process. The thickener overflow is blended with post-digestion sidestream for struvite recovery. Magnesium is also released from the WAS at a mass ratio of approximately 0.25 g Mg/g PO_4^{3-}-P released, reducing the magnesium chloride or magnesium oxide/hydroxide requirement in the struvite crystallizer.

Enhanced Release of Phosphate from Waste Activated Sludge. The addition of primary sludge fermentate or acetic acid to the WAS stripper will enhance the phosphate release rate and reduce the sludge residence time in the stripper from 12–20 h (endogenous process) to 2–5 h. A VFA to VSS ratio of 0.02 to 0.04 g/g is required (Schauer et al., 2011; Corrado, 2009). In general, the phosphate release rate is insensitive to the VFA concentration in the WAS stripper. The optimal configuration for the enhanced release WAS stripping process is a complete mixed tank followed by a solids thickening step, as shown in the process flow diagram presented on Fig. 15–4. Depending on the operating conditions, the Phostrip thickener configuration with thickened sludge recycle may also provide enhanced phosphate release.

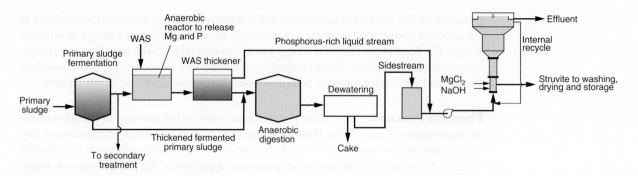

Figure 15–4

Process flow diagram for volatile fatty acid (VFA) enhanced stripping of phosphate from secondary waste sludge enriched with polyphosphate.

15–5 PHYSIOCHEMICAL PROCESSES FOR AMMONIA RECOVERY AND DESTRUCTION

Physiochemical processes for sidestream ammonium treatment are alternatives to biological treatment, which is the dominant method for ammonium removal in practice. Recovery of ammonia from wastewaters to produce aqueous ammonia or an ammonium salt (e.g., ammonium sulfate, ammonium nitrate) for use in industrial and agricultural applications has been of interest for decades, and a number of processes have been developed, demonstrated, and practiced at a full scale. Where ammonia reuse is not desired, a thermal catalytic destruction technology has also been developed and practiced at full scale where ammonia stripped from wastewater is catalytically converted to N_2 at high temperature.

Processes that have been demonstrated at a full scale and are currently used in recovering or destroying ammonia from wastewaters derived from industrial processes, municipal sludge digestion, landfill leachate treatment and animal manure digestion are described in this section. Technologies such as ion exchange and adsorption are rarely, if ever, practiced in the treatment of high strength wastewaters and are not considered in this section. Emerging technologies such as ammonia electrolysis to generate hydrogen for energy recovery (Vitse et al., 2005) and alternative ammonia stripping or volatilization technologies such as Vacuum Flash Distillation (Kemp et al., 2007) and Membrane Contactors (Membrana, 2007; du Preez et al., 2005) are also not considered as application at a full scale is limited or in development.

Recovery of Ammonia by Air Stripping and Acid Absorption

The recovery of ammonia from high strength wastewaters by air stripping–acid absorption technology has been used in both industrial and municipal applications. Most notable in the municipal sector is the process at the VEAS facility [3.5 m³/s (80 Mgal/d)] in Oslo, Norway, where ammonium sulfate was produced from 1996 to 1998, and ammonium nitrate has been produced since 1998 (Sagberg et al., 2006). While this technology has not been applied widely in North America, several processes have been in operation in Europe since the late 1980s for the recovery of ammonia from municipal digester sidestreams, manure digestion sidestreams, landfill leachate, and industrial wastewaters.

Although sulfuric acid is the least expensive and most commonly used, other types of acids can also be used:

- Phosphoric acid—produces mono-ammonium phosphate (MAP) or di-ammonium phosphate (DAP)
- Hydrochloric acid—produces ammonium chloride (NH_4Cl)
- Acetic acid—produces ammonium acetate ($NH_4C_2H_3O_2$)
- Nitric acid—produces ammonium nitrate (NH_4NO_3)

The selection of a specific acid depends on local or regional demand for the resulting product and if revenue could be realized by the sale of the product. However, ammonium sulfate is the dominant product of choice from air stripping–acid absorption processes currently in practice, largely driven by chemical cost and market demand.

Process Description.   The basic air stripping/acid absorption process flow diagram and an example of a full-scale process are shown on Fig. 15–5. The process is comprised of pH adjustment, TSS removal, a dual column air stripper–acid absorber system and chemical storage tanks with associated delivery systems and controls. As described in Sec. 11–10 in Chap. 11, stripping ammonia from wastewater with air requires ammonium to be converted to ammonia in the liquid phase through an increase in pH. The effects of temperature and pH on the percentage of ammonium in the form of ammonia for an aqueous ammonia solution are illustrated on Fig. 15–6. At the typical digestion sidestream temperature range of 25 to 35°C, a pH of 11 or higher is required to shift nearly 100 percent of the ammonium to ammonia. The ionic composition of the sidestream affects the chemical equilibrium illustrated on Fig. 15–6 due to non-ideal conditions created by

Figure 15–5

Air stripping–acid absorption process for recovery of ammonia and production of concentrated ammonium sulfate solution:
(a) process flow diagram and
(b) full-scale process at the VEAS wastewater treatment plant, Oslo, Norway (courtesy of Paul Sagberg and VEAS—Vestfjorden Avløpsselskap).

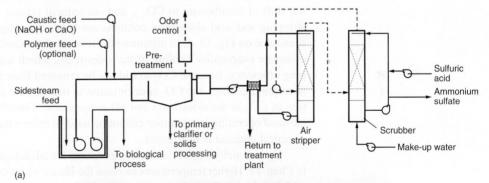

(a)

(b)

Figure 15–6

Ammonia-ammonium equilibrium as a function of pH and temperature.

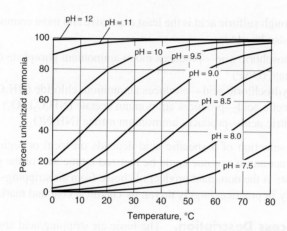

ion-ion interactions. The impact of non-ideal conditions on ammonium-ammonia equilibrium is accounted for through the use of monovalent and divalent activity coefficients as multipliers on the ion concentrations. The calculation of activity coefficients is described in detail in Sec. 2–2 in Chap. 2.

Temperature Impacts. From the plots on Fig. 15–6 it would appear that as sidestream temperature increases, the required operating pH would be lower, resulting in a lower alkaline chemical requirement. However, as noted in Sec. 15–1, ammonium exists in digestion sidestream as ammonium bicarbonate in the pH range of 7.0 to 8.0. A shift from ammonium to ammonia due to an increase in temperature would be accompanied by the shift of bicarbonate to $CO_{2,\,aq}$ and subsequent release of CO_2 into the gas phase. Air stripping and acid absorption columns are typically designed as a closed air system, as illustrated on Fig. 15–5, to minimize the release of contaminated air to the atmosphere and eliminate evaporative cooling of the sidestream, which would impact the ammonia stripping efficiency. Because CO_2 would not be removed from the air with the ammonia in the absorption column, the CO_2 concentration in the air loop increases, resulting in the retention of CO_2 in the sidestream and pH suppression. Therefore, for an enclosed system, an elevated operating temperature cannot be used to reduce the operating pH and the alkaline chemical demand is unaffected.

Operationally, higher column temperatures are advantageous, as illustrated on Fig. 11–62 in Chap. 11. Higher temperatures increase the Henry's Law coefficient for ammonia, creating a higher driving force for mass transfer in the air stripping column. In addition, ammonia diffusivities in water and air increase with temperature, further increasing the mass transfer rate, mathematically expressed as the overall mass transfer coefficient, K_L, in the stripping column design calculations shown in Sec. 11–10 in Chap. 11. In total, higher operating temperatures have the net effect of lowering the air flowrate required to achieve the same ammonia removal efficiency (see Fig. 11–63). A reduction in the air requirement has the further effect of reducing the diameters of the stripping and absorption columns.

Alkaline Chemical Demand. Caustic soda (sodium hydroxide) and lime have been used in practice for adjusting the pH of digestion sidestream. The two primary reactions that govern the caustic soda requirement are

$$NH_4HCO_3 + NaOH \rightarrow NH_3 + H_2O + NaHCO_3 \tag{15-3}$$

$$NaHCO_3 + NaOH \rightarrow Na_2CO_3 + H_2O \tag{15-4}$$

Due to the solubility of sodium carbonate, the pH-adjusted sidestream contains a significant buffering capacity, allowing the air stripping column to operate at the desired pH throughout the column depth. For wastewater treatment facilities that require alkalinity addition to sustain nitrification, the sidestream processed by air stripping–acid absorption where caustic soda is used for pH adjustment can provide a portion or all of the alkalinity requirement for the secondary treatment system. Because the cost of caustic soda is a significant portion of the operating cost for an air stripping–acid absorption process, assuming partial or total credit for the alkalinity provided to the secondary treatment system is an important economic consideration.

Alternatively, lime (CaO) can be used for pH adjustment. The primary benefit of lime is the significantly lower cost compared to caustic soda. However, as described in Sec. 6–3 in Chap. 6, lime removes carbonate alkalinity through the precipitation of calcium carbonate. If a beneficial use of the calcium carbonate cannot be found (e.g., beneficially land applied as *Farmer's Lime*), disposal of the solids may result in an additional operating cost, offsetting the cost advantage lime has over caustic soda. The substantial loss in buffering capacity results in a decreasing pH gradient through the depth of the air stripping column as ammonia is stripped from the sidestream, reducing the effectiveness of the lower section of the column. The use of lime has the added disadvantage of increasing the air stripping column fouling rate, thereby increasing the frequency at which the air stripping column is removed from service for cleaning. Lime is also more difficult to handle in comparison to caustic soda.

Solids Removal for Enhanced Air Stripping.

Digestion sidestreams contain TSS (see Table 15–1), which can cause significant fouling in the air stripping column. In addition, during pH adjustment, inorganic precipitates will form such as calcium carbonate, although the solids mass generated by pH adjustment with caustic soda will be far less than the amount of chemical sludge generated by lime. To reduce the negative impact of the TSS on the air stripping column, a solids removal step is recommended such as an inclined plate settler or a high rate clarification technology (see Sec. 5–7 in Chap. 5 for process descriptions). If lime is used for pH adjustment, the solids removal process is essentially conventional cold lime softening, which is commonly practiced in water treatment.

Air Stripping and Acid Absorption Column Operations.

The basic design approach for determining the dimensions and packing depth of the air stripping column were introduced in Sec. 11–10 in Chap. 11. The design of the acid absorption column is more complicated because, in addition to mass transfer, a chemical reaction is occurring in the acidic ammonium salt solution recirculated through the column and heat is being generated, which affects the liquid and air temperatures in the column. An advanced chemical process model such as ASPEN+ or a similar modeling software package are typically employed to calculate the scrubbing efficiency, packing depth, column diameter, and temperature of the air returned to the air stripping column. Alternatively, an equipment provider may have a proprietary design method based on their experience that is specific to the type of mass transfer media being used in the column.

As illustrated on Fig. 15–5, the two columns contain randomly placed plastic medium and a closed air system where air blown through the stripper column, countercurrent to the sidestream flow, is sent to the absorber and air from the top of the absorber flows to the inlet of the fan. Sizing of the two columns, the fan and product recirculation pump is dependent on the sidestream flow and ammonia concentration, the desired sidestream ammonia removal efficiency and the operating conditions in the two columns (pH, temperature). Typically, the diameters of the stripper and absorption columns are equivalent, but the height of the absorption column and its associated packing depth is roughly 80 percent of stripping column. As the operating temperature increases beyond 40°C, the absorption column height and packing depth will increase and approach those of the stripping column. The air stripping–acid absorption process is typically controlled by pH (pretreated sidestream and recirculated product solution in absorption column) and the density of the recirculated product solution.

In the absorption column, acid is dosed into the product solution recirculation loop to maintain an acidic pH, which provides sufficient driving force for absorption. Real time measurement of the recirculated product solution density is used to fine tune pH control. The product solution is removed continuously from the bottom of the column. For the production of ammonium sulfate, the water content of the 93 percent sulfuric acid (by weight) typically used in larger applications is insufficient to meet the water requirement of a 40 percent (by weight) ammonium sulfate solution. Therefore, water is added continuously to the absorption column, primarily controlled by the density of the recirculated solution. The addition of concentrated sulfuric acid and the subsequent formation of ammonium sulfate results in significant chemical heat generation. A portion of this heat can be used to elevate the temperature of both columns by allowing the temperature of the absorption column to increase, which transfers heat to the air being recirculated back to the stripper column. Less chemical heat is generated with the use of concentrated nitric acid to produce ammonium nitrate.

The direct addition of low pressure waste steam to the stripper column can also be done as higher operating temperatures reduce the fan capacity and the column diameters up to a temperature limit of approximately 70°C. Beyond this temperature, the ammonia removal efficiency in the absorption column will begin to deteriorate and this is the temperature limit for the fiberglass reinforced plastic used for column construction.

Economic Considerations. As discussed in the overview of the air stripping–acid absorption process, concentrated inorganic acid is used in the absorption column and caustic soda or lime is used for pH adjustment of the digestion sidestream. Because these chemicals represent a major operating cost, the future pricing trends of these chemicals are critically important information to obtain to assess the economic viability of this process, in comparison to biological treatment. To offset the costs of the caustic soda or lime and the acid, selling the product to generate revenue will improve the economics of ammonia recovery. The use of ammonium sulfate or ammonium nitrate as a fertilizer is discussed further in Sec. 15–6.

Recovery of Ammonia by Steam Stripping

The use of steam to volatilize ammonia from water is practiced at several industrial installations, however, the implementation of steam stripping for municipal sidestreams has been limited. The only performance data reported are from pilot or demonstration-scale studies (Teichgräber and Stein, 1994; Gopalakrishnan et al., 2000). An ammonia concentration of 100 mg N/L in the steam-stripped sidestream is the practical limit for the process with energy consumption and associated operating cost being the limiting condition.

Process Description. As shown on Fig. 15–7, the steam stripping process consists of contacting sidestream with low pressure steam in a packed column with random-dumped media. At an operating temperature in the range of 95 to 100°C, ammonium bicarbonate is decomposed thermally into ammonia and carbon dioxide in the column and the dissolved gasses are subsequently stripped from the sidestream into the vapor phase.

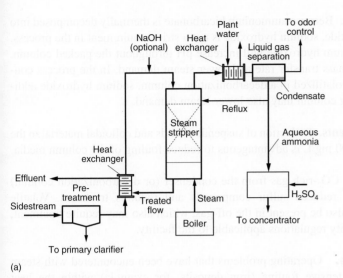

(a)   (b)

Figure 15-7

Steam stripping process for recovery of ammonia from sidestream for production of aqueous ammonia or ammonium sulfate: (a) process flow diagram and (b) view of pilot-plant facility (courtesy of The New York City Department of Environmental Protection).

The stripped sidestream is used to preheat the sidestream influent to reduce energy consumed by the process.

The vapor phase from the column is cooled in a condenser to a two-phase mixture using plant effluent as single-pass cooling water. The liquid and gas phases are separated and a portion of the liquid is returned to the stream stripping column ("reflux"). The remainder of the ammonia-rich liquid is processed further to produce concentrated aqueous ammonia or the liquid is neutralized with sulfuric acid to produce an ammonium sulfate or ammonium nitrate solution, which is further processed to generate a concentrated solution for reuse as a fertilizer. Due to the high volatility of carbon dioxide at the condenser temperature, the majority of the carbon dioxide remains in the gas phase, limiting the reformation of ammonium bicarbonate in the liquid. The gas is highly odorous and requires treatment.

Volatilizing CO_2 from the preheated sidestream in a "decarbonization" packed column prior to the steam stripping column may be beneficial. By removing the majority of the CO_2 before the stripping column, a pH of 9.5 to 9.9 can be sustained throughout the full packing depth resulting in a higher mass transfer rate and lower steam demand (Teichgräber and Stein, 1994). Steam is added to the decarbonization column to increase the sidestream temperature and enhance CO_2 volatilization, but the steam volume requirement does not induce significant ammonia stripping.

Energy Requirements. Approximately 0.15 to 0.18 kg of low pressure steam is required per kg of sidestream to achieve a stripped sidestream ammonia concentration of approximately 100 mg N/L, if energy is recovered from the stripped sidestream to preheat the sidestream influent. The steam demand will be lower if CO_2 is removed from the preheated sidestream in a decarbonization step. Additional energy is consumed to process the low strength aqueous ammonia to a higher concentration or to produce higher strength ammonium sulfate solution for reuse.

Chemical Requirements. Because ammonium bicarbonate is thermally decomposed into ammonia and carbon dioxide, sodium hydroxide is not a strict requirement in the process. However, addition of sodium hydroxide will increase pH throughout the packed column, which will increase the mass transfer rate and lower steam demand. In the process configuration where CO_2 is volatilized in a decarbonization column, sodium hydroxide addition to the steam stripping column may also lower steam demand.

Pretreatment Requirements. Reduction of suspended solids and colloidal material in the sidestream to less than 100 mg/L is advantageous to reduce fouling of the column media.

Off-gas Treatment. The CO_2-rich gas from the condenser (or decarbonization column) contains highly odorous reduced sulfur compounds and requires treatment. Volatile organic compounds will also be present in the off-gas, which also may require treatment, depending on the air quality regulations applicable to the facility.

Operating Problems.   Operating problems that have been encountered with steam stripping include: (1) extensive fouling (iron deposits, for example) within the heat exchanger and in the stripper due to the presence of waste constituents at elevated temperatures, (2) maintaining the required pH for effective stripping, (3) controlling the steam flow, and (4) maintaining the stripping tower temperature. Because of the importance of temperature, steam stripping should be carried out in enclosed facilities. Spiral wound heat exchangers have proven to be effective. In Europe, an acid wash is used to clean the piping, the heat exchanger, and the stripping column.

Air Stripping with Thermocatalytic Destruction of Ammonia

As an alternative to capturing the stripped ammonia to produce a fertilizer, the ammonia-laden air from the stripping column can, as shown on Fig. 15–8, be subjected to thermo-catalytic oxidation (TCO). In TCO up to 98 percent of the ammonia is oxidized selectively to N_2 (~95 percent selectivity) with the balance of the oxidation products primarily being N_2O or NO, depending on the catalyst temperature. This process has not been applied to municipal sidestreams, but has been used for landfill leachate treatment applications where waste heat is available from the combustion of landfill gas to meet the thermal demand of the stripping column (Organics Limited, 2009).

Figure 15–8

Schematic of catalytic reactor used for the thermocatalytic destruction of ammonia in air.

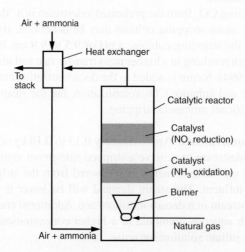

The air stripping column is operated similarly to the column in the air stripping–acid absorption process, only the air is single-pass ambient air that must be heated via steam injection to compensate for evaporative cooling of the column to maintain the desired temperature. In full-scale applications on landfill leachate, sufficient waste heat is available to increase the stripping column temperature high enough to induce thermal decomposition of ammonium bicarbonate, similar to the steam stripping process (Organics Limited, 2009). Consequently, the caustic soda requirement is greatly reduced or eliminated and the air requirement for stripping is reduced.

The ammonia-laden air is then subjected to oxidation at temperatures in the range of 288 to 316°C (550 to 600°F) with a catalyst typically composed of mixtures of transition metals or transition metal oxides on silica or alumina supports and doped with a noble metal such as platinum. The reaction is exothermic (generates heat), reducing or eliminating the natural gas input to the oxidizer required to sustain the target temperature, depending on the ammonia concentration in the air. The primary reactions of interest in this process are:

$$NH_3 + 0.75 O_2 \rightarrow 0.5 N_2 + 1.5 H_2O \tag{15-5}$$

$$NH_3 + O_2 \rightarrow 0.5 N_2O + 1.5 H_2O \tag{15-6}$$

$$NH_3 + 1.25 O_2 \rightarrow NO + 1.5 H_2O \tag{15-7}$$

Temperatures above 350°C (660°F) are avoided to limit the formation of NO_2. Even when operating within the desired temperature range, a certain portion of the ammonia will convert to NO, which must be reduced by selective catalytic reduction (SCR). Unoxidized ammonia will react with NO in the downstream SCR, but there is likely a need for supplemental addition of urea or anhydrous ammonia to maximize NO removal. Exhaust from the TCO/SCR process is used to preheat the ammonia-laden air to the TCO.

The application of air stripping/TCO is limited to cases where waste heat is available from gas engine exhaust or can be generated from otherwise unused digester gas. For leachate treatment applications, the energy demand has been estimated at 450 MJ per m^3 of leachate (Organics Limited, 2009). Depending on the ammonia concentration in the inlet to the TCO, the injection of natural gas to the TCO may also be required to maintain the temperature within the required range. An energy balance across the stripping/TCO system and a life cycle cost analysis is required to assess the economic viability of the process.

15–6   BENEFICIAL USE OF RECOVERED PHOSPHATE AND AMMONIUM PRODUCTS

Phosphate and ammonia recovered from sidestream can be reused as fertilizers or for other industrial applications. The principal products recovered from sidestreams include:

1. Magnesium ammonium phosphate hexahydrate (struvite)
2. Calcium phosphate (hydroxyapatite)
3. Ammonium sulfate
4. Ammonium nitrate

The beneficial use of these products is considered in this section.

Magnesium Ammonium Phosphate Hexahydrate (Struvite)

Struvite has been recognized as a fertilizer for over 150 years and is considered a slow-release fertilizer due to its low water solubility. The low dissolution rate into the soil limits high soluble nutrient concentrations around the root structure and the occurrence of "fertilizer burn," and minimizes the loss of nutrients due to surface runoff and groundwater percolation.

Struvite is only a minor contributor to the overall slow-release fertilizer market, dominated by slowly soluble urea-aldehyde reaction products and polymer and sulfur-coated controlled-release fertilizers. Due to high manufacturing costs, slow-release fertilizers have been predominately used in high valued-added applications in non-agricultural markets such as nurseries, greenhouses and golf courses. Struvite is currently available in a blend with potassium magnesium phosphate under the tradename of MagAmp or MagAmp®-K, typically defined as a 7 percent N, 40 percent P_2O_5, 6 percent K_2O fertilizer with 12 percent Mg.

Demand for slow-release fertilizers has been increasing world-wide to improve fertilizer utilization efficiency and limit the release of nutrients into water bodies. Consequently, struvite recovered from municipal sidestreams is being viewed increasingly as an asset rather than a nutrient that must be treated and disposed. As presented in Sec. 15–4, several struvite crystallization and recovery processes are in full-scale operation. Product purity reported from these facilities has been high, generally greater than 99 percent, with low heavy metal concentrations and coliform counts after the product is washed and decontaminated (Nawa, 2009; Baur et al., 2011; Moerman, 2011). Purity requirements and product accreditation vary by country, state or region, but the product purities have not limited their introduction into regional markets. The product is not considered an organic fertilizer, as it is derived from a wastewater treatment facility.

The recovered struvite may be sold directly by the municipality to a fertilizer blender who markets slow-release fertilizers, but it is a common business model by struvite crystallizer technology providers to assume this responsibility as part of their contract with the municipality. The fertilizer company will blend the struvite with other chemicals to create the desired nutrient blend for a specific application. The fertilizer application may also dictate the required physical characteristics of the product. For example, a pelletized product with a large mean diameter may be specified for a particular application. Consequently, the crystallizer operating conditions and processing requirements for the harvested product will be adjusted to generate a product that complies with this specification. Alternatively, the harvested product will be processed offsite to generate the desired physical characteristics.

Calcium Phosphate (Hydroxyapatite)

Although the recovery of phosphorus from municipal sidestreams as hydroxyapatite has fallen out of favor due to chemical pretreatment costs and the more favorable economics of struvite recovery, hydroxyapatite will continue to be favorable for certain industrial and dairy waste streams. If the use of calcite and calcium silica hydrate as seed materials for hydroxyapatite crystallization is proven to be successful at a full-scale, the production of hydroxyapatite may become a viable alternative to struvite production from municipal sidestreams.

As demonstrated in The Netherlands in the 1990s, hydroxyapatite can be crystallized from sidestream using a seed material such as sand to generate a granular product of high purity and low heavy metals content (Piekema and Giesen, 2001). The sand content of the harvested product is 5 percent by weight or less. The product can be used as feed stock by a phosphate rock processor to generate other phosphate compounds for the fertilizer market such as $Ca(HPO_4^{2-})_2$ (*Superphosphate*), it can also be blended with other nutrients to produce a formulated fertilizer or applied directly as a slow-release fertilizer. Similar to struvite, the pelletized product can be easily dewatered and stored.

Ammonium Sulfate

The primary use of ammonium sulfate is as a fertilizer, but there is no established market for ammonium sulfate produced from wastewater treatment plants. However, there is an established market for ammonium sulfate as a byproduct of various industries, primarily from the production of nylon. The product is typically marketed worldwide and produced in a crystallized form due to customer preference and to reduce shipping costs. Ammonium sulfate can be used in a direct application, it can be blended in custom fertilizer solutions, and it can be blended with biosolids as discussed below. For example, ammonium sulfate has been blended with urea ammonium nitrate (UAN) to produce a fertilizer solution that will increase the sulfur content of the solution fertilizer blend. While the use of sulfur-containing fertilizers are increasing due to the substantial reduction in atmospheric sulfur deposition in certain regions, the use of nitrate and anhydrous ammonia based fertilizers are decreasing due to security issues, ammonia volatilization, and a general shift to liquid fertilizers.

Use of Ammonium Sulfate (AS). Although used throughout the year, the highest use of AS (50–75 percent of annual usage) will typically occur in the late spring and early summer. Smaller amounts would be used in late summer and in the fall. For the remainder of the year (approximately six months), AS must be stored. Typically, a fertilizer blender/distributor will have storage capability, thus avoiding the need to install storage tanks at the wastewater treatment facility.

Blending Ammonium Sulfate with Biosolids. If biosolids are to be reused beneficially, the 40 percent by weight AS solution can be mixed with the dewatered solids to enhance N and S content. However, there are several potential concerns with this approach:

1. Blending the AS solution with the biosolids will the increase the mass of the biosolids and its associated hauling costs.
2. Where biosolids are land applied at an agronomic rate based on nitrogen content, the acreage needed will be larger as the nutrient content will be higher.
3. Increased potential for hydrogen sulfide odors: Because the solution has a high level of sulfate, there is a potential to generate hydrogen sulfide if the mixed biosolids becomes anaerobic during storage.

Marketing Ammonium Sulfate. The ammonium sulfate supply chain contains three distinct groups: manufacturer, blender/distributor, and applier/end-user. A treatment facility would sell their product, as a manufacturer, to a blender/distributor for them to either incorporate into a fertilizer blend or sell it as is.

The most critical issue with marketing AS for the fertilizer suppliers and farmers is its quality and consistency. The fertilizer/chemical suppliers generally will not accept the AS unless the quality and consistency are known and meet their specifications. However, product quality is anticipated to be higher than the standard specifications based on current experience with ammonium nitrate (Sagberg, 2006) and ammonium sulfate (ThermoEnergy, 2009), where the products were found to have low levels of heavy metals (<1 ppm and largely attributed to the quality of the commercial grade acids used in the process) and low total organic carbon (TOC < 50 ppm). Typically, TOC content is tightly specified for ammonium nitrate and monitored to limit explosion potential, but this would not be a factor with ammonium sulfate. Based on the data from operating processes, the TOC in the product is most likely methylamine, which is stripped from the digestion sidestream and absorbed into the product solution along with the ammonia.

Ammonium Nitrate

Ammonium nitrate is an important fertilizer used extensively throughout the world and is commercially available in both dry and liquid forms. Because ammonium nitrate can be used in explosives, the handling and purchase of dry ammonium nitrate is strictly regulated to prevent misuse. This regulation would not apply to "liquid" form, and liquid usage is increasing substantially. However, most nitric acid capacity is "captive," i.e., manufactured for internal use, and, therefore, the price and availability are uncertain. Based on experience in Norway, the fertilizer manufacturer would supply the nitric acid to the wastewater treatment plant and take the ammonium nitrate product; however, this arrangement results in a less cost competitive environment compared to ammonium sulfate production as the fertilizer manufacturer has more extensive control of the acid supply and purchase of the ammonium nitrate.

15–7 BIOLOGICAL REMOVAL OF NITROGEN FROM SIDESTREAMS

Sidestreams resulting from sludge and biosolids processing are most commonly treated in the mainstream plant. Typically, the sidestreams are returned to the headworks, the inlet of primary sedimentation tanks or directly to a location near or in the secondary treatment process (e.g., channels conveying primary effluent or RAS to the activated sludge reactors), depending on the nutrient, BOD and suspended solids loadings associated with the sidestream, and physical constraints such as the plant piping configuration and the location of the dewatering process relative to the process units associated with the main liquid treatment train.

Treatment of the nitrogen-rich sidestreams derived from the dewatering of digested solids, a principal focus of this chapter, can occur in separate treatment processes or in a treatment processes that are integrated with the mainstream treatment process. Both separate and integrated treatment processes are introduced and discussed in this section. The biological treatment processes used for the removal of nitrogen, introduced in this section, are described in greater detail in the subsequent three sections.

Nitrogen Removal Processes

Inorganic nitrogen can be removed biologically by three general processes:

1. Nitrification-denitrification
2. Nitritation-denitritation
3. Partial nitritation-anaerobic ammonium oxidation (deammonification)

Although these processes have been described previously in Chaps. 7 through 10 for the treatment of wastewater, the discussion in this chapter deals with the application of these processes for treatment of sidestreams containing high concentrations of ammonium. For the purpose of comparison, the pathways involved in each of these processes are illustrated on Fig. 15–9 and described below.

Nitrification-Denitrification. In the nitrification-denitrification process, as illustrated on Fig 15–9(a), ammonium is first oxidized to nitrite (nitritation) and subsequently to nitrate (nitratation). Ammonia oxidizing bacteria (AOB) and nitrite oxidizing bacteria (NOB) are responsible for the two steps in the nitrification process. In the denitrification process nitrate is first reduced to nitrite and subsequently to nitrogen gas. As shown on Fig. 15–9(a), oxygen must be added to complete the oxidation of ammonia to nitrate and a carbon source must be available to complete the reduction of nitrate to nitrogen gas.

Nitritation-Denitritation. In the nitritation-denitritation process, as illustrated on Fig 15–9(b), ammonium is first oxidized to nitrite (nitritation). In the next step, nitrite, under anoxic conditions, is reduced to nitrogen gas (denitritation). Short-circuiting the nitrification-denitrification pathway, as illustrated on Fig. 15–9(b), through the restriction

Figure 15–9

Biological reaction pathways for ammonium oxidation and inorganic nitrogen removal: (a) nitrification-denitrification, (b) nitritation-denitritation; and (c) deammonification.

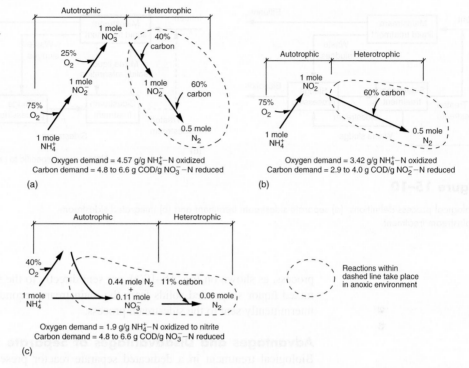

Oxygen demand = 4.57 g/g NH_4^+–N oxidized
Carbon demand = 4.8 to 6.6 g COD/g NO_3^- –N reduced

(a)

Oxygen demand = 3.42 g/g NH_4^+–N oxidized
Carbon demand = 2.9 to 4.0 g COD/g NO_2^-–N reduced

(b)

Oxygen demand = 1.9 g/g NH_4^+–N oxidized to nitrite
Carbon demand = 4.8 to 6.6 g COD/g NO_3^-–N reduced

(c)

Reactions within dashed line take place in anoxic environment

or prevention of nitrite oxidation to nitrate, reduces the stoichiometric oxygen demand and associated aeration energy by 25 percent. In the subsequent anoxic step, the heterotrophic bacteria require 40 percent less degradable organic carbon to reduce nitrite to nitrogen gas. Reducing the aeration power and chemical requirements for sidestream treatment has been the primary driver in the development of alternative advanced biological treatment processes.

Partial Nitritation-Anaerobic Ammonium Oxidation (Deammonification).
In the deammonification process, as illustrated on Fig 15–9(c), a portion of the ammonium is first oxidized to nitrite (partial nitritation). In the next step of the process, ammonia and nitrite are converted to nitrogen gas and nitrate under oxygen-free conditions (anaerobic ammonium oxidation) by a special group of autotrophic bacteria collectively known as anaerobic ammonium oxidizers (Anammox). As illustrated on Fig. 15–9(c), deammonification further reduces the organic carbon requirement.

Process Design Considerations.
Although there are distinct differences in the three different nitrogen removal processes, there is commonality in the design approaches used to determine aeration, chemical, heat removal and tank volume requirements. A unified section on process design for the three major types of sidestream processes is presented in Sec. 15–11. Additional information on the specific design requirements for proprietary processes is provided in the process descriptions in Sec. 15–8 through 15–10.

Separate Treatment Processes for Nitrogen Removal

A sidestream treatment process is considered *separate* from the mainstream plant if the mixed liquor suspended solids in the sidestream reactor are isolated from the mainstream secondary treatment process [see Fig. 15–10(a)]. Waste solids from the sidestream treatment

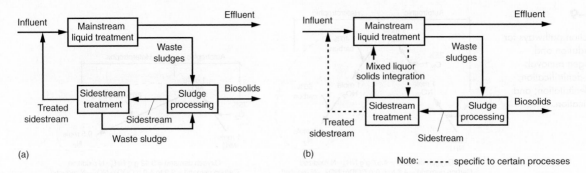

Figure 15–10

Biological process definitions: (a) separate sidestream treatment and (b) integrated sidestream-mainstream treatment.

process, as shown on Fig. 15–10(a), are sent directly to the solids processing train and no mixed liquor suspended solids from the mainstream secondary process are constantly or intermittently sent to the sidestream process.

Advantages and Disadvantages of Separate Sidestream Treatment.

Biological treatment in a dedicated separate reactor presents two primary advantages. First, the nutrient and particulate loadings associated with the sidestream are greatly diminished, resulting in a stream that is less likely to impact the performance of the main treatment plant. Second, treatment in a dedicated reactor provides an opportunity to operate under conditions that can limit the final nitrification product to nitrite. This mode of operation results in lower oxygen and COD requirements for inorganic nitrogen removal (Sec. 15–9); it also provides an ideal environment for the growth of bacteria that possess unique characteristics and biochemistry that can be exploited to further reduce the cost of inorganic nitrogen removal (Sec. 15–10).

A separate sidestream treatment process may also be advantageous or cost-effective for facilities where a required marginal reduction in the plant effluent total nitrogen (TN) concentration or loading can be achieved by treating the sidestream alone and bioaugmentation via an integrated sidestream treatment process does not provide any distinct advantages in terms of overall plant performance (integrated systems are described below). The cost effectiveness of separate sidestream treatment, in comparison to alternative secondary process upgrades that achieve the same improvement in TN removal, is dependent on site specific factors unique to the specific facility, e.g., available footprint for plant expansion, constructability of the various plant upgrade options.

Historically, many separate sidestream treatment processes were designed to provide only nitrification. In these cases, the mainstream plant has limited nitrification capacity and full or partial nitrification of the sidestream ammonium load was the most cost effective option.

From an operations perspective, a separate sidestream treatment system is an additional process at the facility that requires operational oversight and maintenance, which may not be desirable. Also, some separate sidestream nitrification-denitrification processes require the addition of an alkalinity source, a supplemental COD source or both. Chemical consumption would only be a disadvantage if other sidestream nitrification-denitrification options do not require chemical addition for the facility to achieve the desired effluent discharge quality.

Impact of Other Treatment Processes. However, despite the isolation of the sidestream mixed liquor solids from the mainstream secondary process, the sidestream treatment process can be affected by the operating conditions of the other process units in the facility. For example, if the operation of the anaerobic digesters is disturbed or mechanical or performance problems are encountered with the post-digestion dewatering process, the quality and quantity of the sidestream may change, potentially resulting in a perturbation in the sidestream treatment process operating conditions and performance. However, as discussed in Sec. 15–2, the inclusion of sidestream equalization and solids removal will minimize the impact on the downstream process. Because the quality of the effluent from the sidestream treatment process is typically not sufficient to allow blending with the mainstream secondary effluent, the treated sidestream is sent to the mainstream secondary process for further treatment.

Integrated Sidestream-Mainstream Treatment and Bioaugmentation

A sidestream treatment process is defined as "integrated" when mixed liquor suspended solids in the mainstream secondary and sidestream processes are interchanged or nitrifier-enriched waste solids from the sidestream process are fed to the mainstream secondary process to induce bioaugmentation. An integrated configuration, as illustrated on Fig. 15–10(b), would be represented by a flow scheme where a portion of a mainstream return activated sludge (RAS) is fed to the sidestream reactor or, similarly, the sidestream is fed to a RAS reaeration tank. An integrated system would also be represented by any process where sidestream reactor waste solids are fed to the mainstream secondary process.

Advantages and Disadvantages of Integrated Sidestream Treatment.
The main advantage of integrating the sidestream and mainstream processes is the enhancement and stabilization of mainstream nitrification with nitrifier-enriched mixed liquor solids from the sidestream reactor. The optimum configuration is dependent on the mainstream process configuration, operating conditions, and the desired plant effluent quality. Operating costs associated with chemical addition are also a consideration during process selection. A disadvantage of integrating the sidestream and mainstream processes is the introduction of additional inert solids into the secondary process mixed liquor; however, if the sidestream is equalized and the majority of the solids are removed and recycled to the solids processing train, the effect of the sidestream on the mainstream process solids balance is greatly diminished.

Augmentation of Mainstream Treatment Process. The augmentation of the mainstream process with nitrifier-enriched sidestream mixed liquor solids has been documented, yet a unified mechanistic model that can be used to predict the impact of bioaugmentation for all process configurations has not yet been developed. In general, the effectiveness of nitrifying biomass grown in a sidestream reactor in enhancing nitrification performance in the mainstream secondary process may be greatly impacted by the bulk liquid environment in which these organisms grow (e.g., osmotic pressure, temperature, pH, ionic composition and strength, substrate concentration). In integrated configurations where the sidestream is fed to a RAS reaeration tank, the nitrifying bacteria grow within the flocculated mixed liquor suspended solids in the RAS reaeration tank at the same temperature and nearly identical bulk liquid conditions as the mainstream secondary process; therefore, complete retention of nitrifier activity is anticipated as the nitrifier-enriched mixed liquor passes from the reaeration tank to the activated sludge tanks.

In process configurations where the difference in the bulk liquid sidestream and mainstream process environments are significant, the sidestream reactor operating conditions allow the growth and domination of specific types of nitrifying organisms over others, yet when placed in the mainstream environment, are at a competitive disadvantage to the nitrifying organisms that dominate under the mainstream operating conditions. Intuitively, the bioaugmentation effect would be lower with these configurations in comparison to the configurations where sidestream is substantially diluted into a RAS stream and only increase as the sidestream and mainstream processes are increasingly integrated through interchange of mixed liquor solids. However, an exact quantification of the effect has not been thoroughly demonstrated over a wide range of operating conditions.

Bioaugmentation Effect of Integrated Sidestream Treatment. The bioaugmentation effect of the sidestream mixed liquor solids is also impacted by the nitrifier mass discharge rate to the mainstream process. Therefore, operating conditions that lead to a reduced nitrifier mass in the sidestream process will reduce the mass in the waste sludge. For example, a higher operating temperature in the sidestream process will increase the decay rate of the nitrifying organisms. Enhanced decay also occurs as the sidestream process SRT is allowed to increase well above the minimum SRT required to sustain the desired sidestream treatment performance. In sidestream processes that operate at elevated temperatures and a high SRT, the combined effect of these two conditions will diminish the nitrifier mass available for bioaugmentation. Based on these observations operating the sidestream treatment reactor at (1) a temperature as close as practically possible to the mainstream process, (2) with a bulk liquid environment similar to the mainstream process and (3) at a low SRT that provides the desired performance will result in optimum retention of nitrifier activity and, hence, the greatest bioaugmentation effect.

15–8 NITRIFICATION AND DENITRIFICATION PROCESSES

Several biological treatment processes have been developed for sidestream treatment where the end product of ammonium oxidation is primarily nitrate and the nitrate is subsequently denitrified in part or in whole, depending on the treatment objective. The processes are subdivided into two categories: separate treatment and integrated sidestream-mainstream treatment, according to the definitions provided in Sec. 15–7. The implementation of one type of process over the other is dependent on the type of mainstream secondary process, its operating conditions, plant effluent quality objectives and economic considerations. The purpose of this section is to consider the application of nitrification and denitrification processes for the treatment of sidestreams.

Fundamental Process Considerations

To understand the nitrification-denitrification process, it is useful to consider (1) the process biology, kinetics, and stoichiometry, (2) the alkalinity requirements, (3) the importance of inorganic carbon, and (4) the need for degradable organic carbon.

Process Biology, Kinetics, and Stoichiometry. The microbiology, basic biochemical reaction stoichiometries and autotrophic growth kinetics associated with the oxidation of ammonium to nitrite and nitrite to nitrate are presented in Sec. 7–9 in Chap. 7. Although kinetic rates in sidestream biological treatment processes are influenced by the same environmental conditions as the mainstream nitrification processes, differences between ammonia oxidizing bacteria (AOB) and nitrite oxidizing bacteria (NOB) growth kinetics in the sidestream and mainstream environments must be considered for sidestream process design and operation. Specifically, the key differences are the impact of temperature on growth and decay, nitrous acid and free ammonia inhibition of AOB and NOB populations, and the effect of bicarbonate concentration on the autotrophic growth rates.

Biological denitrification reaction stoichiometry and kinetics are presented in Sec. 7–10 in Chap. 7. The pertinent stoichiometric reactions involved in nitrification-denitrification process, repeated here for convenience, are as follows.

Nitrification. The principal nitrification reactions are as follows.
Biological conversion of ammonium to nitrite

$$2NH_4^+ + 3O_2 \rightarrow 2NO_2^- + 4H^+ + 2H_2O \tag{7–88}$$

Biological conversion of nitrite to nitrate

$$2NO_2^- + O_2 \rightarrow 2NO_3^- \tag{7–89}$$

Total oxidation reaction:

$$NH_4^+ + 2O_2 \rightarrow NO_3^- + 2H^+ + H_2O \tag{7–90}$$

If cell tissue is neglected, the amount of alkalinity required to carry out the ammonium oxidation reaction is given by the following reaction, obtained by rewriting Eq. (7–90) as follows.

$$NH_4^+ + 2HCO_3^- + 2O_2 \rightarrow NO_3^- + 2CO_2 + 3H_2O \tag{7–91}$$

When cell mass synthesis is included in the overall oxidation of ammonium to nitrate, Eq. (7–91) becomes Eq. (7–93). The cell mass yield is based on yields of 0.12 g VSS/g NH_4-N and 0.04 g VSS/g NO_2-N for the nitritation and nitratation reactions, respectively.

$$NH_4HCO_3 + 0.9852NaHCO_3 + 0.0991CO_2 + 1.8675O_2 \rightarrow$$
$$0.01982C_5H_7NO_2 + 0.9852NaNO_3 + 2.9232H_2O + 1.9852CO_2 \tag{7–93}$$

Denitrification. The amount of biodegradable organic compound required to reduce nitrate to nitrogen gas is dependent on the carbon source as illustrated by the following equations where the organics in wastewater, methanol, or acetate are used for nitrate reduction.

Wastewater:

$$C_{10}H_{19}O_3N + 10NO_3^- \rightarrow 5N_2 + 10CO_2 + 3H_2O + NH_3 + 10OH^- \tag{7–110}$$

Methanol:

$$5CH_3OH + 6NO_3^- \rightarrow 3N_2 + 5CO_2 + 7H_2O + 6OH^- \tag{7–111}$$

Acetic Acid:

$$5CH_3COOH + 8NO_3^- \rightarrow 4N_2 + 10CO_2 + 6H_2O + 8OH^- \tag{7–112}$$

The above equations do not reflect the actual carbon requirement for denitrification as they do not include cell mass synthesis. When expressed as COD, the mass ratio of biodegradable COD to nitrate-N taking into account the biomass yield is given by Eq. (7–126).

$$\frac{bsCOD}{NO_3\text{-}N} = \frac{2.86}{1 + 1.42Y_n} \tag{7–126}$$

Where 2.86 = oxygen equivalent of nitrate-N (g O_2/g NO_3-N)

Y_n = net biomass yield, as defined by Eq. (7–121).

$$Y_n = \frac{Y}{1 + b(\text{SRT})} \qquad (7\text{–}121)$$

Where b = heterotrophic anoxic decay rate

Based on these equations, the observed COD/N ratio is dependent on the SRT of the reactor and the value of b, which, in turn, is dependent on the reactor temperature. The reader should refer to Secs. 7–9 and 7–10 for more detailed discussions of the nitrification and denitrification reactions, respectively.

Alkalinity Requirements. The alkalinity requirements are given by Eqs. (7–91) and (7–93). Referring to Eq. (7–91), two moles of bicarbonate are required to neutralize the acidity generated per mole of ammonium oxidized during nitrification. When cell growth is included [see Eq. (7–93)], 1.98 moles of bicarbonate are required for acid neutralization and 0.099 moles of inorganic carbon are required for cell growth in the complete nitrification of one mole of ammonium-N. If nitrate is denitrified completely, 50 percent of the alkalinity destroyed during nitrification is recovered (see Sec. 7–10 in Chap. 7), resulting in a net alkalinity reduction of 1 mole of bicarbonate per mole of ammonium-N nitrified and denitrified (3.57 g $CaCO_3$/g NH_4-N).

For sidestreams derived from anaerobic digestion and ATAD processes, the alkalinity is primarily in the form of bicarbonate and is typically equal to the ammonium concentration on a molar basis, providing only one half of the alkalinity required for complete nitrification. The remaining alkalinity demand can be satisfied through the addition of an external alkalinity source (e.g., caustic soda; see Sec. 15–11), dilution of the sidestream into another stream with sufficient alkalinity (e.g., return activated sludge) or by generating alkalinity via denitrification. In the absence of sufficient alkalinity, a stoichiometric amount of the sidestream ammonium-N will be oxidized to a mixture of nitrite and nitrate in accordance with the available alkalinity.

The addition of a COD source, to enhance denitrification and generate alkalinity to support nitrification, may not eliminate the need for an external alkalinity source. For example, if ferric or ferrous chloride is added to the digesters to control struvite formation or to the sidestream to control struvite formation in the pipe or channel conveying sidestream, a reduction in sidestream alkalinity will occur due to the acidity associated with these chemicals. Therefore, in subsequent sidestream treatment, the addition of supplemental alkalinity may be required depending on the treatment objective.

Importance of Inorganic Carbon. The role of inorganic carbon in nitrifier growth is often ignored in mainstream nitrification processes where bicarbonate and CO_2 are readily available due to the degradation of abundant organic carbon compounds in the plant influent. However, the residual inorganic carbon concentration in separate sidestream reactors has an impact on the ammonium removal rate and removal efficiency due to the high autotrophic growth rates in the reactor. In high rate nitritation-denitritation sidestream treatment processes operated at high temperature (greater than 30°C) at a full-scale, the effect of inorganic carbon concentration on the autotrophic growth rate was not found to follow a conventional Monod kinetic form, but is represented by a logistic function ("S-curve") as defined by Eq. (15–8). Monod kinetic terms for dissolved oxygen, substrate concentration, free ammonia inhibition, and nitrous acid inhibition along with the decay rate and Arrhenius temperature function are not included in Eq. (15–8) for simplicity.

$$\mu_n = \mu_m \frac{e^{[(HCO_3^- - k)/a]}}{e^{[(HCO_3^- - k)/a]} + 1} \tag{15–8}$$

Where μ_n = growth rate of nitrifying bacteria, g new cells/g cells·d

μ_m = maximum specific growth rate of nitrifying bacteria, g new cells/g cells·d

HCO_3^- = bicarbonate concentration, mM

k = saturation constant, mM

a = constant (estimated value of 0.83 based on Wett and Rauch, 2003)

While the value of the saturation constant, k, for mainstream processes is 0.5 mM or less, the value has been estimated to be around 4 mM in the warm, high growth rate environment of a separate sidestream reactor (Wett and Rauch, 2003). Consequently, for sidestream biological treatment processes, the residual inorganic carbon concentrations in the reactor must be considered carefully when selecting the reactor operating conditions that provide optimum performance. Bicarbonate and CO_2 exist in equilibrium at concentrations dictated by the reactor bulk liquid conditions, such as temperature, pH, and calcium concentration (formation of calcium carbonate). As the reactor pH decreases below 7, more bicarbonate shifts to $CO_{2,aq}$, which is removed subsequently from the reactor via air stripping. Therefore, under increasingly acidic conditions in a sidestream biological reactor, a restricted nitrifier growth rate may occur due to an inorganic carbon limitation.

Need For Degradable Organic Matter. Depending on the COD source (e.g. methanol, glycerol, volatile fatty acids, municipal wastewater), the observed sludge yield under anoxic conditions is typically in the range of 0.28 to 0.4 g VSS /g COD consumed, resulting in a degradable COD to nitrate-N ratio of 4.8 to 6.6 g/g [see Eq. (7–126)]. In a typical high strength sidestream, the ratio of degradable COD to TKN is less than 1. Thus, the available degradable COD for a sidestream treatment system is insufficient where a high denitrification efficiency is desired. Supplementation of the COD with commercially available organic carbon sources is an option; however, the use of COD sources within the facility (e.g primary sludge; endogenous decay of secondary mixed liquor solids) is desired as the purchase of a commercial COD source increases operating cost.

Treatment Processes

Over the past 20 years, a number of separate and integrated nitrification-denitrification treatment process configurations have been developed or implemented for sidestream treatment. The principal processes are

BAR/R-D-N process

InNitri® process

ScanDeNi® process

Sequencing batch reactor

Summary information on these processes, including process flow diagrams, is provided in Table 15–4.

In addition to the process configurations described in Table 15–4, trickling filters have been used for sidestream treatment in a limited number of facilities, but typically their application has occurred where the mainstream process has been upgraded from trickling filters to a suspended growth activated sludge system, resulting in the availability of a

Table 15–4

Description of separate and integrated nitrification–denitrification processes for sidestream treatment

Process	Description
(a) BAR and R-D-N process	BioAugmentation Reaeration (BAR) or Regeneration-Denitrification-Nitrification (R-D-N) process consists of nitrification of the sidestream in a plug-flow return activated sludge (RAS) reaeration tank. In a typical design, the sidestream is mixed with the entire RAS flow at the head of the tank. An anoxic zone can be provided at the head of the tank with a HRT of 1 h to promote partial denitrification of the RAS and suppress odors associated with the sidestream. The aerobic zones that follow typically have a total hydraulic retention time (HRT) of 2 h to allow complete ammonium oxidation. A final anoxic zone may be provided at the end of the tank with a 1 h HRT to further promote endogenous denitrification if needed. Process modeling is used to refine the reaeration tank volume requirement. Addition of the entire RAS flow dilutes the sidestream ammonium-N and other constituents by 50 to 100-fold, creating a mixed liquor environment for nitrification that is similar to that in the mainstream activated sludge reactors. RAS alkalinity is typically sufficient to meet the alkalinity demand for full nitrification of the sidestream ammonium load. Ideally, mechanical mixers are used in the anoxic zone(s), but for large plants coarse bubble aeration has proven to be a cost-effective method of providing mixing with limited impact on denitrification.
(b) InNitri® process	The InNitri® process was developed with the key objective of bioaugmenting the mainstream process with nitrifier-rich waste sludge. The InNitri process is typically designed in a Modified Ludzack-Ettinger (MLE) configuration with a separate gravity clarification step. An external carbon source is added to promote denitrification and generate alkalinity to enhance the nitrification efficiency. The internal recycle rate from the aerobic to the anoxic zone depends on the desired total inorganic nitrogen (TIN) removal efficiency. An external alkalinity source may be added if a high nitrification efficiency is required to meet the treatment objective. Alternatively, the InNitri reactor may be operated entirely in an aerobic mode, which would require external alkalinity addition to achieve a high nitrification efficiency. Primary or plant effluent is fed to the reactor to prevent the operating temperature from exceeding 38°C or to maintain the reactor within a specific temperature range. To maximize the amount of nitrifier mass sent to the mainstream secondary process, the InNitri sidestream reactor is operated at an aerobic SRT in the range of 3 to 5 d, which provides stable sidestream performance, but limits the loss of nitrifier mass through decay.

(continued)

Table 15–4 (Continued)	
Process	**Description**

(c) ScanDeNi® process

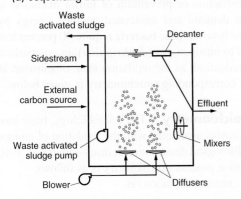

Similar to the BAR/R-D-N process, the ScanDeNi® process nitrifies the sidestream nitrogen load in a RAS reaeration tank. Unlike the other processes, a post anoxic zone is provided and an external carbon source is added to enhance denitrification. No internal recycle within the reaeration tank is provided since the RAS has sufficient alkalinity for complete nitrification. The design HRT of the anoxic zone will be impacted by the selected external carbon as the denitrification rates will vary with the source. The ScanDeNi process was developed to provide a means of nitrifying and denitrifying a sidestream ammonium load and provide a denitrified RAS flow to the mainstream process where biological phosphorus removal is performed.

(d) Sequencing Batch Reactor (SBR)

Sequencing Batch Reactors (SBR) have been used most commonly for separate sidestream treatment. In the SBR configuration, sidestream is fed continuously during the *React* period or rapidly fed to the reactor at the beginning of the SBR cycle in a defined anoxic *Fill* period. Intermittent aeration is applied at defined time intervals to provide aerobic and anoxic periods for nitrification and denitrification. Typically, the total aerobic time is two-thirds of the *React* period. If a high ammonium removal efficiency is desired, an external alkalinity source is provided. If high ammonium and inorganic nitrogen removal efficiencies are desired, an external carbon source or a facility carbon source such as primary sludge/WAS is added for denitrification and to generate alkalinity for nitrification. Primary or plant effluent is fed to the SBR for temperature control as needed. At the end of the *React* period, the suspended solids are settled (*Settle* period), treated sidestream is decanted to the mainstream plant (*Decant* period) and waste sludge is pumped to the solids processing train. The total SRT is typically 10 d or higher.

trickling filter for sidestream treatment. Performance is largely limited to partial nitrification and a dilution water source (e.g., primary effluent) and a high internal recycle rate are required to maintain the desired hydraulic loading rate. Trickling filters are not considered further in this chapter.

Other attached growth process configurations such as moving bed biofilm reactors and rotating biological contactors, historically, have not been used for nitrification-denitrification of sidestreams, but have been proven effective for deammonification of sidestreams, as presented in Sec. 15–10. Submerged attached growth processes such as fluidized bed reactors and biological aerated filters have also not been applied full-scale for sidestream treatment. However, in pilot studies with submerged reactors with a sand medium nitritation-denitritation with methanol as the supplemental carbon source has been demonstrated successfully. The application of attached growth systems is considered further in Sec. 15–10.

Finally, activated sludge reactors have been applied successfully for the nitrification of digester sidestreams (Jeavons et al., 1998), but these systems have been decommissioned.

15-9 NITRITATION AND DENITRITATION PROCESSES

The development of the nitritation-denitritation processes in the 1990s was driven by the desire to reduce the energy and chemical requirements for high strength sidestream nitrogen removal. Nitritation-denitritation processes that have been implemented in full-scale facilities are described in this section along with general design information. Separate and integrated sidestream-mainstream processes and those that produce a mixture of nitrite and nitrate during the ammonium oxidation step are also included in this section.

Fundamental Process Considerations

To understand the nitritation-denitritation process it is useful to consider (1) the biological pathways, (2) process biology, kinetics, and stoichiometry, (3) the alkalinity requirements, (4) the need for biodegradable organic carbon, and (5) the operating modes that limit the oxidation of nitrite to nitrate.

Biological Pathways. As illustrated on Fig. 15–9(b), short-circuiting the nitrification-denitrification pathway through the restriction or prevention of nitrite oxidation to nitrate, reduces the stoichiometric oxygen demand and associated aeration energy by 25 percent. In the subsequent anoxic step, the heterotrophic bacteria require 40 percent less degradable organic carbon to reduce nitrite to nitrogen gas (*denitritation*), in comparison to the organic carbon required for nitrate reduction, as the oxidation state of nitrogen in nitrite ($+3$) is lower than nitrate ($+5$). The corresponding stoichiometry is given below.

Process Biology, Kinetics, and Stoichiometry. The microbiology, basic biochemical reaction and autotrophic growth kinetics associated with the oxidation of ammonium to nitrite are presented in this Section. Biological denitritation is discussed in Sec. 7–10 in Chap. 7. The pertinent nitritation-denitritation process stoichiometry is as follows.

Biological conversion of ammonium to nitrite (nitritation) is

$$NH_4HCO_3 + 0.9852NaHCO_3 + 0.07425CO_2 + 1.4035O_2 \rightarrow$$
$$0.01485C_5H_7NO_2 + 0.9852NaNO_3 + 2.9406H_2O + 1.9852CO_2 \qquad (7–92)$$

The amount of biodegradable organic compound required to reduce nitrite to nitrogen gas is dependent on the carbon source as illustrated by the following reactions based on municipal wastewater, methanol, and acetate.

Wastewater:

$$C_{10}H_{19}O_3N + 16.66NO_2^- + 0.33H_2O \rightarrow 10CO_2 + NH_3 + 8.33N_2 + 16.66OH^- \qquad (15–9)$$

Methanol:

$$5CH_3OH + 10NO_2^- \rightarrow 5N_2 + 5CO_2 + 5H_2O + 10OH^- \qquad (15–10)$$

Acetic Acid:

$$5CH_3COOH + 13.33NO_2^- \rightarrow 10CO_2 + 3.33H_2O + 13.33OH^- \qquad (15–11)$$

The equations given above can be compared to the nitrate reduction reactions in Eq. (7–110) through (7–112) given in Sec. 15–8. Comparing the corresponding equations, it can be seen that the mass of a specific organic compound can reduce a greater amount of inorganic nitrogen if nitrite is not allowed to oxidize to nitrate during the ammonium oxidation step in the process.

Similar to denitrification, the equations above do not reflect the actual carbon requirement for denitritation, as the reactions do not account for cell mass synthesis. When expressed on a COD basis, the actual biodegradable COD requirement for denitritation takes into account the effect of biomass yield and is calculated as follows.

$$\frac{bsCOD}{NO_2\text{-}N} = \frac{1.71}{1 - 1.42Y_n} \tag{15–12}$$

Where 1.71 = oxygen equivalent of nitrite (g O_2/g NO_2-N)
Y_n, = net biomass yield, calculated with Eq. (7–121), presented in Sec. 15–8.

Alkalinity Requirement. Alkalinity destruction during the complete oxidation of ammonium to nitrate occurs during the nitritation step. Therefore, the alkalinity demand for acid neutralization in a nitritation process is identical to a process performing complete nitrification. The alkalinity requirement for complete ammonium removal must be provided through an external alkalinity source or alkalinity generation through nitrite reduction. Providing the alkalinity requirement by diluting the sidestream into another plant stream (raw influent, primary tank effluent; plant effluent) is not a viable option, as dilution creates a growth environment where restriction of nitrite oxidation to nitrate is difficult to control. As discussed in Sec. 15-8, considering the impact of bicarbonate concentration on the autotrophic growth rate and the potential reduction in alkalinity through the use of chemicals (e.g. iron salts) upstream of the sidestream treatment process, the addition of supplemental alkalinity may be required depending on the treatment objective.

Need For Degradable Organic Matter. Despite a reduction in the organic carbon requirement, the amount of degradable organic carbon in a typical sidestream is insufficient to allow complete denitritation. An external organic carbon source or a carbon source from within the plant such as primary solids is required. In the absence of sufficient denitritation, an external alkalinity source is required for a high nitritation efficiency or incomplete ammonium oxidation to nitrite will occur.

Restriction of Nitrite Oxidation and the Impact of Nitrite Accumulation. Preventing or limiting NOB growth in a nitritation-denitritation process can be accomplished through four primary mechanisms: (1) low aerobic SRT at a reactor temperature greater than 20°C, (2) intermittent aeration at low dissolved oxygen (DO) concentration, (3) free ammonia inhibition, and (4) free nitrous acid inhibition, where the latter two inhibitory effects are dependent on reactor pH. The role of one or more of these mechanisms to restrict NOB growth is discussed in the process descriptions given in Table 15–5. For further discussion of ammonium and nitrite oxidizing bacteria growth kinetics and conditions that restrict NOB growth, refer to Sec. 7–9 in Chap. 7.

The accumulation of nitrite to high concentrations can result in a reduction in the ammonium oxidation rate through free nitrous acid inhibition. Therefore, inclusion of denitritation in the treatment process is preferred for very high strength sidestreams, or the sidestream can be diluted with primary or plant effluent, with the restriction that excessive dilution may result in sidestream reactor conditions that limit the control of nitrite oxidation to nitrate. Where a high residual nitrite concentration is present in the sidestream reactor effluent, the effluent should be discharged to an anoxic zone in the mainstream secondary process to prevent further oxidation of the nitrite to nitrate and to minimize the organic carbon demand for inorganic nitrogen removal.

Table 15–5
Description of nitritation–denitritation processes for sidestream treatment

Process	Description
(a) BABE® process 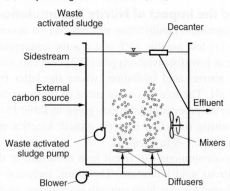	The BABE® (Biological Augmentation Batch Enhanced) process is an integrated sidestream-mainstream configuration designed to provide a source of nitrifier-enriched sludge for bioaugmentation of the mainstream process, but is operated under conditions that limit the oxidation of nitrite to nitrate (Berends et al., 2005). The BABE reactor can be a sequencing batch reactor (SBR) operated with intermittent aeration, a plug-flow reactor with external clarification where alternating anoxic and aerobic zones are provided or a plug-flow reactor in a MLE configuration. Regardless of the reactor configuration, the distinguishing feature of the BABE process is the addition of a portion of mainstream return activated sludge (RAS) (< 10 percent of the total RAS flow) to the reactor, which serves to integrate the BABE reactor with the mainstream process and control the BABE reactor temperature at or below 25°C. Sidestream ammonium is oxidized to a mixture of nitrite and nitrate, which are subsequently reduced through endogenous denitrification with the RAS solids and the addition of an external carbon source. Free ammonia inhibition and transient anoxia are believed to be the primary mechanisms that limit nitrite oxidizing bacteria (NOB) growth in the BABE reactor. Waste sludge from the BABE reactor is sent to the mainstream process. The reactor volume requirement and the RAS flowrate (or daily volume) required for the BABE process are dependent on the overall facility inorganic nitrogen removal objective and typically developed through process modeling.
(b) Sequencing Batch Reactor (SBR)	Nitritation-denitritation can be achieved in a SBR configuration through pH-controlled or time-based intermittent aeration with the DO concentration controlled at 1 mg/L or less during the aerated periods (Wett, 1998). An external carbon source or a facility carbon source such as primary sludge is added during the anoxic phases to promote denitritation. Continuously feeding sidestream to the SBR during the *React* phase is typically practiced in full-scale systems. However, for sidestreams with a COD/N ratio approaching 1, an intermittent feeding strategy may be beneficial to minimize the external carbon demand. The SBR typically operates at temperatures above 30°C due to biological heat generation and may require heat removal through an external mixed liquor cooling loop or through the addition of dilution water to maintain the reactor temperature below 38°C. A total solids retention time (SRT) in range of 5 to 10 d is typical.

(continued)

Table 15–5 (*Continued*)	
Process	**Description**
(c) SHARON® process 	The SHARON® (Stable reactor system for High Ammonia Removal Over Nitrite) process consists of anoxic-aerobic continuous stirred tank reactors (CSTRs) in a series configuration, without suspended solids retention. The process is operated in the temperature range of 35 to 38°C. The design SRT of the aerobic zone is 1.5 d under this elevated operating temperature and controlled at this value through intermittent aeration, which allows growth of ammonia oxidizing bacteria, but provides a washout rate that prevents the growth of nitrite oxidizing bacteria (Hellinga, 1998). The design anoxic SRT is typically 0.75 d. The anoxic and aerobic volumes are based on the sidestream design flow and their respective design SRT values. If the sidestream is diluted with primary and plant effluent, the diluted flowrate serves as the basis for the anoxic and aerobic zone volumes. An internal recycle flow of 13 times the feed flow (undiluted or diluted sidestream) is provided to supply nitrite to the anoxic zone. An external carbon source is supplied to the anoxic zone for denitritation and to generate alkalinity to support nitritation. The addition of an external alkalinity source may also be required to sustain the target inorganic nitrogen removal efficiency. Heat generated by the biological reactions is removed through an external mixed liquor cooling loop, which can also serve as a method for heat addition if heat losses during the winter period are excessive. Highly concentrated sidestreams are typically diluted with primary or plant effluent to lower the ammonium concentration to 1500 mg N/L or less.

Treatment Processes

Nitritation-denitritation processes have also been developed and implemented at a full-scale for separate sidestream treatment. Process development has also led to integrated sidestream-mainstream configurations where the oxidation of nitrite to nitrate is not controlled completely, but nitrite remains the dominant product that accumulates during the ammonium oxidation step. The principal separate and integrated processes are

> BABE® process
>
> Sequencing batch reactor
>
> SHARON® process

Summary information on these processes, including process flow diagrams, is provided in Table 15–5.

15–10 PARTIAL NITRITATION AND ANAEROBIC AMMONIUM OXIDATION (DEAMMONIFICATION) PROCESSES

Further reduction in aeration energy and chemical demand for sidestream inorganic nitrogen removal can be achieved by implementing processes that perform partial nitritation and support the growth and enrichment of Anammox bacteria. Anammox bacteria are autotrophic organisms capable of oxidizing ammonium under anoxic conditions, using nitrite as the electron acceptor. The development and implementation of processes designed to perform sequential partial nitritation and anaerobic ammonium oxidation ("Deammonification") has accelerated since the first full-scale systems were commissioned in Germany and The Netherlands in 2001–2002.

In the initial development of deammonification processes, researchers found that stable deammonification can be achieved within a single reactor where the two primary reactions occur in a suspended biomass or a biofilm (Kuai and Verstraete, 1998; Olav Sliekers et al., 2002; Siegrist et al., 1998; Seyfried et al., 2001). Two early process names (1) Completely Autotrophic Nitrogen removal Over Nitrite (CANON) and (2) Oxygen-Limited Autotrophic Nitrification Denitrification (OLAND) soon fell out of use, as specific trademarked process names became the norm. The processes described in this section are summarized in Table 15–6.

Fundamental Process Considerations

To understand deammonification processes and the advantages they present in reducing the cost of removing inorganic nitrogen from sidestreams, the biology, kinetics and reaction stoichiometry of Anammox bacteria, and their syntrophic relationship with aerobic ammonium oxidizing bacteria are summarized briefly below.

Process Biology, Kinetics, and Stoichiometry.
Anammox organisms have been detected in marine and fresh water environments, soils, sediments, wetlands, and of particular interest for the discussion of sidestream treatment, in wastewater treatment plants (Kuenen, 2008; Van Hulle et al., 2010). Activated sludges from conventional nitrification/denitrification plants have been used for the startup of bench, pilot, and full-scale processes that incorporate the Anammox reaction (Fux et al., 2002; Third et al., 2005; van der Star et al., 2007).

The biochemistry and growth kinetics of the anammox bacteria are provided in Chaps. 7 and 8. As discussed in these chapters and summarized below, anammox bacteria have a maximum specific growth rate that is approximately one-tenth of the growth rate of aerobic nitrifying organisms. Therefore, a long SRT is a key feature in the design of all deammonification processes. Additional environmental factors that control their growth are also summarized below. The stoichiometry of the partial nitritation and Anammox reactions is given below.

Partial Nitritation.
If ammonium is oxidized partially to nitrite (*partial nitritation*) according to the stoichiometric reaction given by Eq. (7–92), the stoichiometric ratio of ammonium and nitrite needed for the anammox reaction is obtained as shown in Eq. (15–13). The neutralization of the acidity produced by the nitritation reaction with bicarbonate alkalinity is included in the partial nitritation stoichiometry for discussion purposes.

$$2.34NH_4^+ + 1.87O_2 + 2.66HCO_3^- \rightarrow$$
$$0.02C_5H_7NO_2 + NH_4^+ + 1.32NO_2^- + 2.55CO_2 + 3.94H_2O$$
(15–13)

Anammox Reaction.
Under oxygen-free conditions anammox bacteria will oxidize ammonium using nitrite as the electron acceptor. The principal stoichiometric reaction, including cell mass synthesis, is shown in Eq. (15–14) (Strous et al., 1998).

$$NH_4^+ + 1.32NO_2^- + 0.066HCO_3^- + 0.13H^+ \rightarrow$$
$$1.02N_2 + 0.26NO_3^- + 0.066CH_2O_{0.5}N_{0.15} + 2.03H_2O$$
(15–14)

Table 15-6

Description of partial nitritation—anaerobic ammonium oxidation (*deammonification*) processes

Process	Description
(a) ANITA™Mox–Single stage moving bed biofilm reactor process 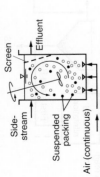	ANITA™Mox is a single-stage deammonification moving bed bioreactor (MBBR) system that is continuously aerated with a variable dissolved oxygen (DO) setpoint in the range of 0.5 to 1.5 mg/L, based on online reactor ammonium and nitrate measurements. DO setpoint adjustment is required to control nitrite oxidizing bacteria (NOB) growth and provide stable ammonium removal. Due to continuous aeration, mechanical mixing is only required during startup and periods of low ammonia loads. The fraction of the reactor volume occupied by the media does not exceed 50 percent to prevent insufficient mixing and media movement throughout the reactor volume. The AnoxKaldnes plastic media with an active specific area of 500 m²/m³ or higher have been used in this process.
(b) DeAmmon® moving bed biofilm reactor process 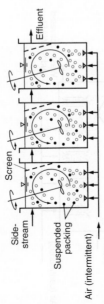	DeAmmon® consists of a single or dual train reactor system with three stages per reactor (designed by Purac). The stages are operated in series, but piping flexibility is provided to allow parallel operation. Kaldnes (AnoxKaldnes/Veolia) K1 media has been used typically for biofilm support (active surface area of 500 m²/m³). Each stage is aerated intermittently to provide aerobic and anoxic periods for the partial nitritation and anammox reactions, respectively, and continuously mixed with mechanical mixers. The duration of the aerobic and anoxic periods are dependent on the ammonium loading to the system and the removal rate. Aeration and anoxic times of 20 to 50 min and 10 to 20 min, respectively, are typical. A DO concentration of 3 mg/L during the aeration periods is targeted, but higher concentrations are avoided to prevent the potential for NOB growth and to limit anammox inhibition. Typically, the fraction of the reactor volume occupied by the media does not exceed 50 percent to prevent insufficient mixing and media movement throughout the reactor volume.
(c) DEMON® Sequence Batch Reactor (SBR) 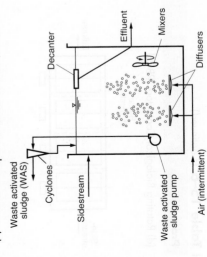	The DEMON® SBR is a suspended growth reactor operated by pH-controlled or time-based intermittent aeration to provide aerobic periods for the partial conversion of ammonium to nitrite and anoxic periods for the anammox reaction (Wett, 2006). The peak DO concentration during each aerobic phase is controlled at approximately 0.3 mg/L to provide selective pressure against NOB growth and allow a rapid transition to an anoxic condition after the air is shut off to promote an optimal environment for the anammox bacteria. To minimize the impact of nitrite on anammox activity, the aeration period in each cycle is typically around 10 to 15 min (anoxic period is typically 5 to 10 min). If pH is used to control the aeration cycle, a pH interval of 0.01 or 0.02 units is applied. The mean operating pH is typically maintained above 6.8 to minimize inorganic carbon loss through air stripping of CO_2. The SBR is continuously fed during the *React* phase of each SBR cycle. Because the anammox bacteria grow in the SBR is in a dense granulated form, the waste sludge is pumped through hydrocyclones to separate the anammox granules from the remaining flocculated solids and return them to the reactor. Consequently, the solids retention time (SRT) of the anammox bacteria approaches 40 to 50 d and the SRT of the remaining solids [ammonia oxidizing bacteria (AOB), heterotrophs, inert solids] is maintained around 10 d, which provides further selective pressure against NOB growth and more stable performance over a range of loading conditions (Wett et al., 2010).

(continued)

| Table 15-6 (*Continued*)

Process	Description
(d) Rotating Biological Contactors (RBCs) 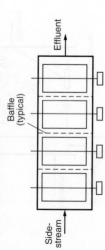	Rotating Biological Contactors (RBC) have been applied for leachate treatment and were found to promote the development of deammonification as the dominant nitrogen removal pathway (Seyfried et al., 2001). The deammonifying RBCs vary in configuration and media material, but demonstrated an average surface specific deammonification rate of approximately 2.5 g N/m²·d with peak rates up to 4.8 g N/m²·d. The typical RBC bulk liquid DO concentration is 1 mg/L. For leachates rich in degradable organic carbon, biological removal of the carbon is required before the deammonification RBCs. Deammonification performance was stable at operating temperatures less than 20°C and a nitrogen removal efficiency of 70 percent was reported at temperatures as low as 10°C (Seyfried, 2002). Information is not readily available on submergence depth or rotational velocity required to induce or optimize deammonification performance.
(e) Single-Stage ANAMMOX® process 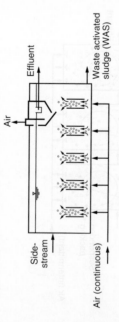	The single-stage ANAMMOX® reactor (designed by Paques BV, The Netherlands) was initially based on an air lift design where air is continuously applied to the bases of multiple riser tubes, resulting in upward liquid-solids movement through the risers and downward movement of liquid-solids outside the risers after gas disengagement. The hydrodynamic conditions favor the development of a thick well-granulated biomass where the AOB and anammox populations grow synergistically within the same granules. In a later development of this process, sludge granulation was shown to be stable without the need for riser tubes, simplifying the design. A proprietary gas-liquid-solids separator(s) located in the upper section of the reactor provides separation of the liquid-solids from the exhaust air and a settling zone for the granulated solids, which return to the main reaction zone. Pretreatment of the sidestream to remove denser inert solids is advisable to prevent their accumulation in the reactor. Online measurement of DO, nitrite and ammonium are used to control the process.
(f) Terra-N® process 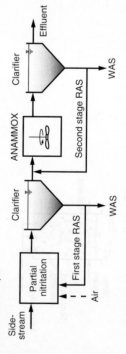	The Terra-N® MBBR process was developed by SÜD-Chemie/Clariant GmbH (Munich, Germany) where bentonite is used as the support media for biofilm growth (TERRANA® product; mean particle diameter in the range of 25 to 45 μm). Due to the rapid settling rate and compaction of the bentonite particles, gravity clarification is applied for separation of the media and biofilm from the bulk liquid. The process is designed in a two-stage configuration, with each stage consisting of a completely mixed reactor and a gravity clarifier (continuously aerated partial nitritation stage followed by a completely mixed anoxic stage for the anammox reaction), or as a single-stage SBR with intermittent aeration. In the SBR and the partial nitritation stage of the two-stage configuration, bentonite is added to a concentration of 10 to 12 g/L. With biomass attachment, the total suspended solids concentration is typically 15 to 20 g/L. In the anammox stage of the two-stage configuration, granulation of the anammox bacteria eliminates the need for a support media, although the addition of bentonite is not detrimental. Second-stage biomass concentrations of 5 to 7 g/L have been reported (Clariant/SÜD Chemie, 2012).

Deammonification can be accomplished in a two-stage system consisting of partial nitritation followed by the anammox reaction. In the first stage, a CSTR is operated at an aerobic SRT of approximately 1.5 d (based on the SHARON® concept) at a temperature above 30°C, without alkalinity or organic carbon addition, to convert approximately 50 percent of the sidestream ammonium to nitrite. The partially nitritated sidestream is fed to the second stage ANAMMOX® reactor (provided by Paques BV, The Netherlands) operated in an upflow configuration. Under a high upflow superficial liquid velocity, anammox bacteria form dense granules with settling velocities greater than 100 m/h, which allows the development of a sludge bed in the lower section of the reactor with a solids concentration as high as 5 to 7 percent. Flocculated particles with lower densities and settling velocities are washed out. Solids separation between the two reactors is recommended to prevent the accumulation of denser inert solids in the ANAMMOX® reactor sludge bed. Internal liquid mixing within the ANAMMOX® reactor is provided by a gas-lift mechanism where nitrogen gas produced by the sludge bed is collected through a proprietary gas collection system located above the mid-depth point of the tank, which conveys the gas to a central riser pipe, inducing gas-lift of liquid to a gas-liquid separa-tor located on top of the reactor. De-gassed liquid is returned to the bottom of the tank to increase the superficial liquid velocity in the sludge bed and dilute the partially nitritated sidestream entering the reactor.

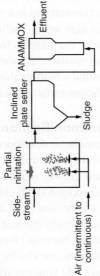

(g) Two-stage SHARON®-ANAMMOX® process

Because the Anammox bacteria use inorganic carbon as their source of carbon for growth and nitrite is used as the electron acceptor, inorganic nitrogen is removed without the addition of an organic carbon substrate. As illustrated in Eq. (15–14), nitrate is a product of the reaction; hence, some organic carbon is required for complete inorganic nitrogen removal. For every kg of ammonium-N that is removed by the two-step deammonification pathway [see Fig. 15–9(c)], 0.11 kg of nitrate-N is produced. Consequently, deammonification reduces the organic carbon demand to about 11 percent of the carbon demand required by heterotrophic reduction of nitrate to nitrogen gas in a nitrification-denitrification process. The corresponding oxygen requirement for deammonification is also reduced to approximately 40 percent of the oxygen required for complete oxidation of ammonium to nitrate by the two-step aerobic autotrophic pathway.

Alkalinity Requirement.   Bicarbonate in a typical high strength sidestream is equal to the ammonium concentration on a molar basis. As shown in Eqs. (15–13) and (15–14) above, the total amount of bicarbonate required to neutralize acidity produced by ammonium oxidation to nitrite and to support autotrophic cell growth will exceed the amount available in the sidestream. Nitrate reduction in the sidestream reactor, in part or entirely, depending on the degradable COD in the sidestream, will generate bicarbonate alkalinity, but the amount is insufficient to provide the balance required to satisfy the overall bicarbonate stoichiometric demand. In practice, deammonification processes typically achieve sufficiently high ammonium removal efficiencies that no external alkalinity source is added (greater than 80 percent removal, with 90 to 95 percent being typical). If ferrous or ferric chloride is added to the digesters or other locations to control struvite formation, the addition of alkalinity to the deammonification system to compensate for the alkalinity destroyed by iron salt addition may be justified or required if a high inorganic nitrogen removal efficiency is required.

Solids Retention Time.   Anammox organisms have a maximum specific growth rate that is less than one-tenth the rate of aerobic nitrifying bacteria (see also Table 7–13 in Chap. 7). Hence, a SRT greater than 20 d is required to sustain these organisms in a biological reactor. Fortunately, anammox bacteria produce an excessive amount of exocellular polymeric substances (Cirpus et al., 2006), which results in tightly bound aggregation and granule formation as the organisms are concentrated in a suspended growth reactor. The formation of a granulated biomass results in high solids settling rates and a high solids specific gravity which can be exploited to separate the granules from flocculated solids. These features allow process configurations that can easily retain the anammox bacteria within the processes. Integration of the organisms within a biofilm also occurs readily once the environment conducive to their growth and sufficient surface area are provided.

Implementation of Deammonification Processes and Challenges.   The first deammonification processes were developed for high strength sidestreams where the warm sidestream temperatures are advantageous in enhancing the growth rate of the anammox organisms. For suspended growth sequencing batch reactors, intermittent aeration combined with a low dissolved oxygen (DO) concentration (0.3 mg/L) during the aerobic period of each aeration cycle has proven to be an effective way to provide conditions for the growth of aerobic ammonia oxidizing and anammox bacteria while restricting the growth of aerobic nitrite oxidizing bacteria (Wett, 2007).

Nitrite accumulation in biofilms can also be induced by operating the attached growth reactor at a reduced bulk liquid DO concentration, providing an environment within the biofilm where the Anammox organisms can thrive. Sustaining the growth of aerobic ammonia oxidizing and anammox bacteria within a single reactor, while preventing growth of aerobic nitrite oxidizing bacteria, has also been demonstrated with continuously aerated gas-lift reactors where the hydrodynamic conditions induced by the aeration system creates a granulated sludge, which can be retained to a high concentration (Olav Sliekers et al., 2003; Abma et al., 2010). With a low DO concentration, aerobic and anoxic zones within the granulated sludge provide environments for aerobic ammonia oxidizer and anammox growth.

Although the anammox reaction is attractive from the perspective of reducing the plant operating cost associated with inorganic nitrogen removal, incorporating anammox bacteria within a biological reactor system has presented a number of technical challenges. First and foremost, due to their low growth rate, a long reactor startup period is required in comparison to a nitrification-denitrification or nitritation-denitritation system and a long recovery period is also needed in the event of a significant reduction in anammox activity. However, seeding a new installation with anammox-enriched biomass has been demonstrated to facilitate startup (Wett, 2006; Abma et al., 2007; Schneider et al., 2009; Christensson et al., 2011). Due to the increasing number of deammonification plants, seed availability is increasing. The degree to which startup can be hastened depends on ammonium load to the startup plant and mass of seed that can be obtained.

The anammox bacteria are also sensitive to nitrite, which can cause irreversible loss of activity. Various levels of nitrite exposure, both in terms of concentration and exposure time, and the corresponding loss of activity have been reported (Fux et al., 2004; Wett et al., 2007), which indicate that nitrite toxicity is a function of both concentration and exposure time. Exposure to oxygen also results in the inhibition of anammox activity, but the effect is reversible. Finally, anammox organisms compete with NOB for nitrite; therefore, the reactor operating conditions must be well-controlled to limit NOB growth or provide an environment where anammox bacteria have a competitive advantage.

Treatment Processes

Process configurations designed for deammonification include

> ANITA™Mox moving-bed biofilm reactor process
>
> DeAmmon® moving-bed biofilm reactor process
>
> DEMON® suspended growth sequencing batch reactor
>
> Rotating biological contactors
>
> Single-stage ANAMMOX® process
>
> Terra-N® moving-bed biofilm reactor process
>
> Two-stage SHARON®-ANAMMOX® process

Summary information on these processes, including process flow diagrams, is provided in Table 15–6.

15–11 PROCESS DESIGN CONSIDERATIONS FOR BIOLOGICAL TREATMENT PROCESSES

Process design considerations for the biological systems described in Sec. 15–8 through 15–10 are presented in this section. The principal design considerations include (1) sidestream characteristics and treatment objectives (2) design loadings and load equalization,

(3) sidestream pretreatment, (4) reactor volume requirements, (5) aeration system design, (6) sludge retention time and mixed liquor suspended solids concentration, (7) chemical addition (organic carbon requirements for denitrification and denitritation; alkalinity), (8) operating temperature and pH, and (9) heat removal. An example problem is presented following the discussion of the design considerations, to illustrate their application.

Sidestream Characteristics and Treatment Objectives

The characteristics of high strength sidestream, as noted in Sec. 15–1, are dependent on the process from which the sidestream is derived and the process operating conditions. Therefore, when designing a process for sidestream treatment, the characteristics of the specific sidestream should be measured. The sidestream parameters that should be considered are summarized in Table 15–7. Because nitrogenous oxygen demand is an important process design parameter, both soluble and total TKN measurements are included in sidestream characterization, to estimate the biodegradable organic nitrogen concentration. However, a significant fraction of the organic TKN in sidestreams may be comprised of recalcitrant dissolved organic nitrogen (rDON) and nitrogen associated with biologically inert suspended solids. Typically, about 50 percent of the soluble organic TKN (soluble TKN minus ammonium-N) in sidestream derived from anaerobic digestion is biodegradable. Of the particulate TKN, typically 10 to 15 percent of TKN fraction is biodegradable under aerobic conditions.

If the sidestream is to be generated in the future as a result of the implementation of a new solids processing system, a bench or pilot-scale sludge digestion study is recommended so that the soluble constituents in the sidestream reflect the conditions in the

Table 15–7

Parameters to consider when designing a sidestream treatment system

Parameter	Unit	Remarks
Flowrate (average/minimum/maximum)	m³/d	
Dewatering operation frequency	d/wk	
Dewatering operation hours	h/d	
Temperature (average/minimum/maximum)	°C	
Total suspended solids	mg/L	Dependent on pretreatment requirements
Volatile suspended solids	mg/L	Dependent on pretreatment requirements
Alkalinity	mg/L as CaCO₃	
Total cBOD₅ᵃ	mg/L	
Soluble cBOD₅ᵃ	mg/L	0.45μm membrane filtered
Ammonium-N	mg N/L	Typically 90 to 95 percent of soluble TKN
Soluble TKNᵇ	mg/L	0.45μm membrane filtered
Total TKNᶜ	mg/L	
Soluble ortho-P	mg/L	

ᵃ A degradable-COD to cBOD5 ratio of 1.5 g/g can be assumed to estimate the degradable particulate and soluble COD fractions.

ᵇ Difference between the soluble TKN and ammonium-N is the soluble organic N of which roughly 50 percent is typically considered biodegradable for sidestream derived from anaerobic digestion of primary and secondary sludges.

ᶜ Difference between the total TKN and soluble TKN is the particulate TKN of which roughly 10 to 15 percent is typically considered biodegradable for sidestream derived from anaerobic digestion of primary and secondary sludges.

digestion process and the characteristics of the sludges and biosolids. The TSS concentration in the sidestream is estimated by the capture efficiency range anticipated in the full-scale dewatering process. In the absence of such tests, the ammonium-N and other constituent concentrations can be estimated based on digestion process modeling.

Treatment objectives for the sidestream process will differ with each facility and are dependent on the facility effluent permit limits on nutrients and the costs associated with the energy and chemical requirements. Selection of a treatment option is also dependent on the potential benefits associated with bioaugmentation of the mainstream plant. Typically, the separate and integrated processes being considered are subjected to a life cycle cost assessment to identify the most cost effective option.

Design Loading and Load Equalization

The sidestream nitrogen and degradable carbon mass loading rates will affect the process volume and the aeration and chemical requirements. For separate treatment processes that employ solids separation for SRT control, the sidestream TSS loading rate may also impact the design as the majority of the solids are biologically inert and will accumulate in the reactor. Therefore, the maximum oxygen demand and solids loading rate are used commonly for reactor design.

The peak sidestream loading rate for reactor design is dependent on the operation of the biosolids dewatering process. Dewatering facilities at smaller treatment plants are commonly operated during daytime hours for 5 or 6 d/wk. At larger facilities, the dewatering operation typically is continuous. Where biosolids dewatering is intermittent, the implementation of flow and load equalization is a cost-effective option for reducing the volume of the treatment process and the aeration energy associated with the peak oxygen demand. Equalization of sidestream flows and loads were discussed previously in Sec. 15–2.

Sidestream Pretreatment

The majority of the separate sidestream treatment processes described in Sec. 15–8 to 15–10 will benefit from a reduction in the sidestream TSS concentration before treatment. Most of the sidestream TSS are biologically inert and will, therefore, accumulate in the sidestream reactor, potentially elevating the solids concentration to a level that can negatively impact the solids-liquid separation process and the ability to control the SRT within the desired range. Pretreatment requirements are dependent on historical sidestream TSS concentration data and the specific process being used or considered for sidestream treatment. Dynamic process modeling of many of the sidestream processes is an effective tool in assessing the need for pretreatment for existing sidestream where sufficient TSS data are available. Pretreatment options were considered in Sec. 15–3.

Reduction in the sidestream TSS to an average concentration of 200 mg/L or less is typically sufficient. For the DEMON deammonification SBR where hydrocyclones are used to separate anammox granules from other suspended solids, removal of dense particulate matter is particularly important because the hydrocyclones will also separate this material from the waste sludge and retain it in the reactor. Removal of large inert debris is critical for MBBR systems that use screens for media retention as the debris will accumulate in the reactor over time.

Sidestream solids removal is typically not a requirement for the SHARON nitritation-denitritation process as solids retention is not performed. However, if the sidestream solids concentration is high (> 2000 mg/L) periodically, the removal of solids may prove beneficial in reducing oxygen demand in the aerated zone. For integrated sidestream-mainstream processes, a solids balance across the facility should be conducted to assess the impact of sidestream solids on the mainstream process.

Sidestream Reactor Volume

The conditions that define the minimum reactor volume requirement vary with the type of sidestream treatment process and are dictated by the treatment objective, the maximum achievable oxygen transfer rate, and the biological kinetic rates, as well as other considerations such as solids-liquid separation. The volume requirement for a separate sidestream reactor is commonly estimated through the use of a specific nitrogen loading rate on the system, expressed as the nitrogen loading per unit reactor volume per day (as kg N/m^3·d) or, in the case of fixed-film systems per unit area of active media surface area per day (g N/m^2·d). The maximum nitrogen loading rate (kg N/d) is selected commonly for design. The specific loading rate for a given type of process is often developed through pilot or demonstration-scale studies and full-scale operating experience. Typical values for the specific loading rate for the processes discussed in Sec. 15–8 through 15–10 are provided in Table 15–8.

A more detailed analysis is required to determine the volume requirement for a SBR or a suspended growth process with external clarification. For SBRs, the minimum volume is dictated largely by the aeration system and its ability to provide sufficient oxygen to satisfy the peak oxygen demand imposed by the sidestream (design condition) during the aerobic periods of the *react* phase. Separation of solids from the treated liquid is an additional design consideration. The reactor volume determined through the aeration system analysis must be sufficient to ensure that settled sludge inventory is not disturbed during the decanting of the treated sidestream, which further ensures that the desired SRT can be maintained. Where the anoxic reaction is occurring in a separate stage, the volume requirement depends on the reaction kinetics and is determined through process modeling or pilot studies.

For the integrated sidestream-mainstream processes, process modeling is commonly used to determine the reactor volume requirement. The volume will be influenced significantly by the desired level of inorganic nitrogen removal across the facility and the mainstream operating conditions, which affect the biological kinetic rates in the reactor or zones nitrifying and denitrifying the sidestream nitrogen load. Typical loading rates or hydraulic retention times for the various processes are also presented in Table 15–8.

Where a mainstream process of a facility does not have sufficient capacity to treat sidestream and unacceptable deterioration of the facility effluent quality would result, division of the sidestream reactor volume into multiple tanks may be a design requirement, giving the facility operator the flexibility to remove a reactor from service for maintenance and inspection. Process modeling is often used to evaluate these operating scenarios.

Aeration System

The application of oxygen demand in the design of the aeration system is specific to the type of sidestream process. The oxygen demand is comprised of the oxygen requirement for the carbonaceous oxygen demand and the oxidation of ammonium to nitrite or nitrate. Typically, complete conversion of ammonium to nitrite or nitrate is assumed in the calculation of the design oxygen demand for nitritation-denitritation and nitrification-denitrification processes, respectively. Where processes have been shown to produce a mixture of nitrite and nitrate, a conservative design approach is to provide sufficient aeration capacity for complete oxidation of ammonium to nitrate. For deammonification, the nitrogenous oxygen demand is based on the stoichiometric conversion of ammonium to nitrite that provides complete ammonium removal across the process.

Aeration Requirements for Sequencing Batch Reactors.

For sequencing batch reactors where aerobic and anoxic reactions are occurring within the same reactor through alternating aerobic and anoxic phases, the temporal distribution of oxygen demand

Table 15-8

Typical design and control parameters for the biological sidestream treatment processes

Process	Key design parameters	HRT, h Ammonium-N loading rate (ALR), kg N/m³·d Media loading rate, g N/m²·d	SRT[r], d	Process control parameters
Nitrification and denitrification processes				
BAR and R-D-N process (RAS reaeration tank)	SRT (total), HRT (reaeration tank)	HRT = 2 (aerobic)[b] 1–2 (anoxic)	8–15 (SS+MS, total)	DO, SRT
InNitri or short SRT process (2-stage)	SRT, ALR	ALR = 0.4–0.5[c]	3–5 (SS, total)	DO, external Alk or COD add., temp., SRT
ScanDeNi process (RAS reaeration tank)	SRT (total), HRT (reaeration tank)	HRT = 1–2 (aerobic)[b] 1–2 (anoxic)	8–15 (SS+MS, total)	DO, SRT, external COD add.
Sequence Batch Reactor (SBR)	SRT, ALR	ALR = 0.3–0.4[d]	10–15 (SS, total)	DO, external Alk or COD add., temp, SRT
Nitritation and denitritation processes				
BABE process (SBR)	SRT (total), ALR, temp.	ALR = 0.4	4–8 (SS+MS, total)	RAS flow to BABE reactor, temp. (25°C, max), DO
Sequencing batch reactor	SRT, ALR	ALR = 0.4–0.6[d]	5–10 (total)	DO (<1 mg/L); external Alk/ COD add.
SHARON Process	SRT (aerobic, anoxic), temp.	ALR < 0.7[e]	1.5/0.75[e] (aerobic/anoxic)	Temp (35–38 °C); aerobic SRT; external COD add.
Partial nitritation and anaerobic ammonium oxidation processes				
ANITA™Mox	Media surface area, ALR	ALR = 0.7–1.2[f]	> 20	DO (0.5–1.5 mg/L)
DeAmmon®	Media surface area, ALR	ALR = 0.6–0.8[g]	> 20	DO, aerobic-anoxic cycle times
DEMON® SBR	SRT, ALR	ALR = 0.7–1.2[h]	40–50 (Anammox granules) 10–15 (floc)	pH interval (0.01–0.02 s.u.); DO (0.3 mg/L); SRT
Rotating biological contactor	Media surface loading rate	loading rate = 2.3–2.8 g N/m²·d[i]	> 20	DO (1 mg/L)
Single-stage ANAMMOX®	SRT, ALR, air-lift aeration design	ALR = 2.0	>20	DO, SRT, temp.
Terra-N® two-stage with intermediate clarification	Bentonite concentration, ALR, clarifier overflowrate	ALR = 1.2–2.1 (partial nitritation stage)[j] 1.2–2.1 (Anammox stage)[j]	> 20	DO

(continued)

Table 15-8 (Continued)

Process	Key design parameters	HRT, h Ammonium-N loading rate (ALR), kg N/m³·d Media loading rate, g N/m²·d	SRT[b], d	Process control parameters
Terra-N®-single-stage	Bentonite concentration, ALR	ALR = 0.25–0.7 (1.5)[k]	> 20	DO
Two-stage SHARON®-ANAMMOX®	SRT, ALR, upflow liquid velocity	ALR < 0.7 (SHARON); 3–10 (ANAMMOX)	1.5 (SHARON, aerobic); > 20 (ANAMMOX)	Aerobic SRT, temp (SHARON) Upflow liquid velocity, SRT (ANAMMOX)

[a] Notations: SS = sidestream reactor(s); MS = mainstream secondary reactor(s); total = sum of the aerobic and anoxic SRT values.

[b] Aerobic and anoxic hydraulic retention times and RAS reaeration tank volume based on the maximum RAS flow, typically a 100% return rate, at the minimum operating temperature. HRT values shown are typical. Tank geometry is plug flow and the volume is dependent on the desired nitrification performances in the RAS reaeration tank and the mainstream plant at the minimum operating temperature. Modeling is typically used in this assessment and the autotrophic growth rates are adjusted to approximately 0.5–0.6 d⁻¹ (20°C) to account for the effect of low pH on the growth rate.

[c] Loading rate based on total system volume with an aerobic/anoxic volume ratio of 2:1 and a maximum OUR of 150 mg/L·h.

[d] Maximum ammonium removal rate based on three 8-h cycles per day with a react period of 6 h per cycle, an intermittent aeration cycle of 66 percent aerobic–34 percent anoxic, a maximum total OUR of 150 mg/L·h.

[e] Maximum specific rate based on total system volume with aerobic/anoxic ratio of 2:1 and a maximum OUR of 150 mg/L·h. Aerobic and anoxic SRTs based on influent flowrate, e.g. if dilution water is added, the volume requirements are based on the diluted sidestream flowrate.

[f] Loading rate dependent on the type of plastic media employed and its effective surface area for biofilm growth. A removal rate up to 1.2 kg N/m³·d with AnoxKaldnes BiofilmChip™M (effective surface area of 1200 m²/m³ of media; 40 percent media fill volume) was reported by Christensson et al. (2011) and Lemaire et al. (2011).

[g] AnoxKaldnes K1 media, effective surface area for biofilm = 500 m²/m³ of media; deammonification rate = 1.5 to 2 g N/m² of effective surface area per day (Plaza et al., 2011).

[h] Demonstrated range of loading rates. Design loading rate is typically 0.7 kg N/m³·d.

[i] Loading rates correspond to average nitrogen removal rates demonstrated by RBC systems treating landfill leachate. Peak removal rates up to 4 g/m²·d have been demonstrated (Siegrist et al., 1998; Seyfried, 2002).

[j] Loading rates demonstrated for retrofitted tanks (Clariant/SÜD Chemie, 2012). Partial nitritation and anammox reactor volumes are typically equal. Proprietary clarifier design-clarifier surface overflowrate and typical solids return flowrate not disclosed by vendor.

[k] Range of values demonstrated for retrofitted tanks. Value in parenthesis is considered by the vendor to be the maximum design loading rate for a new tank (Clariant/SÜD Chemie, 2012).

has an effect on reactor volume and blower capacity requirements. For example, a SBR operating at three cycles per day with a combined *settle*, *decant* and *idle* time of 1.5 h/cycle, the total *react* time over the day is 19.5 h. If the reactor is operated with intermittent aeration during the *react* periods with an aerated:anoxic time ratio of 2/1, the total aeration time is 13 h/d. Therefore, the minimum reactor volume and the design of the aeration system are based on the daily oxygen demand being satisfied over the total aerated time period. Allocation of carbonaceous oxygen demand in the aerated and anoxic periods is dependent on if the reactor is fed continuously during the *react* period, fed rapidly at the beginning of each SBR cycle or fed intermittently during the anoxic phases of the aeration cycle. Process modeling is often used in assessing the total oxygen demand for design.

Type of Aeration System.   Because the development of advanced biological treatment processes has been driven partially by the desire to reduce aeration energy, fine and ultra-fine bubble air diffusion systems are common in suspended growth reactors. To achieve a more compact reactor design, maximum floor coverage applicable to the type of diffuser is applied. A design oxygen uptake rate (OUR) of 150 mg/L·h is used commonly for fine and ultrafine bubble diffused air systems. Diffuser selection takes into account the material of construction as some manufacturers specify a maximum wastewater temperature limit for their diffusers (e.g., 30°C), which may be lower than the anticipated average operating temperature of the reactor.

Alpha and fouling factors (see Sec. 5–11 in Chap. 5) are developed through testing or provided by the process technology or diffuser vendor based on their experience with similar applications. The utilization of high purity oxygen is uncommon, but is applicable to separate and integrated sidestream-mainstream systems. High purity oxygen would satisfy a considerably higher OUR, which may allow a reduction in the reactor volume, depending on other operating conditions such as SRT, mixed liquor suspended solids concentration and kinetic rates.

In MBBR systems that employ plastic media, medium to coarse bubble stainless steel air diffusers are standard practice to reduce maintenance requirements and the risk of having to remove the media from the reactor to replace diffusers. The dissolved oxygen concentration requirement is specific to the process and typical values are provided in Table 15–8.

Sludge Retention Time and Mixed Liquor Suspended Solids Concentration

The design SRT is specific to each type of sidestream process. Typical SRT ranges employed in the processes described in Sec. 15–8 through 15–10 are presented In Table 15–8. For suspended growth separate reactors where solids-liquid separation is employed, the suspended solids concentration is typically kept below 4000 mg/L (at the maximum liquid level in a SBR), depending on the settling and compaction characteristics of the solids. Consequently, the impact of the sidestream TSS load on the reactor solids concentration should be examined to determine if the process would benefit from a sidestream pretreatment step where the suspended solids concentration is reduced.

Chemical Requirements

Alkalinity and/or organic carbon addition is often required for many of the nitrification-denitrification and nitritation-denitritation processes, especially if a high ammonium or inorganic nitrogen removal efficiency is desired. Deammonification processes generally do not require any chemical addition, but in certain cases alkalinity addition may prove beneficial. Some of the key considerations for each chemical category are summarized below.

Alkalinity. The importance of alkalinity for ammonium oxidation and the stoichiometric requirements are discussed in Sec. 15–8. As described in Sec. 15–1 and 15–8, the alkalinity in high strength sidestream typically provides up to 50 percent of the total alkalinity required for complete ammonium oxidation. Therefore, an external alkalinity source is added or alkalinity is generated through nitrate or nitrite reduction with an external organic carbon source. In the integrated sidestream-mainstream systems where a portion of the mainstream RAS is fed to the sidestream reactor, RAS will provide a fraction of the alkalinity requirement, reducing the external alkalinity demand. For the BAR/R-D-N/ScanDeNi processes where sidestream is diluted into the entire RAS stream in a RAS reaeration tank, the RAS alkalinity is typically sufficient to meet the alkalinity requirement.

Alkalinity Addition. Under the warm, high rate growth conditions in a separate sidestream reactor, inorganic carbon can impact the autotrophic growth rate at concentrations that would not impact their growth rate under mainstream process conditions (see discussion in Sec. 15–8). As shown in Eq. (15–8), as the bicarbonate concentration approaches the saturation coefficient value, which has been estimated at 4 mole/L (200 mg/L as $CaCO_3$). Concomitantly, the AOB growth rate is reduced, resulting in a reduction in the ammonium removal efficiency in systems operated at or near their design sidestream loading rate. External alkalinity addition has been shown to increase the nitrogen removal in deammonification pilot reactors, reducing the residual ammonium concentration (Yang et al., 2011). The improvement in reactor performance through alkalinity addition may be justified, depending on an economic analysis in which the cost of alkalinity addition is compared to the cost associated with treating the residual ammonium in the mainstream plant (e.g., aeration energy, organic carbon requirement).

Sources of Alkalinity. Caustic soda is the source of alkalinity used most commonly, due to ease of handling and availability. Magnesium hydroxide and sodium carbonate can also be used. Lime is generally avoided due to its limited water solubility and the potential formation of insoluble calcium carbonate. The selection of alkalinity source may be dependent on the regional market conditions for these chemicals.

Organic Carbon. The organic carbon requirement for design is based on complete nitrite or nitrate removal and the type of carbon source. A discussion of commercially available carbon sources is presented in Sec. 8–7 in Chap. 8. Although methanol and glycerol are the carbon sources used most commonly in sidestream treatment, any readily biodegradable carbon source can be used so long as nitrification reaction is not inhibited as a consequence of its addition. The COD demand for nitrate and nitrite reduction can be estimated using Eqs. (7–126) and (15–12), respectively. The observed sludge yields required in these two equations can also be estimated or the COD-to-N ratio can be developed through pilot testing.

Due to the cost of purchasing, storing, and handling an external carbon source, particularly if the carbon source is classified as a hazardous compound (e.g., methanol), a source of organic carbon from within the wastewater treatment facility may be preferred. Fermentation of degritted primary and secondary sludges are established methods for producing readily biodegradable COD in the form of volatile fatty acids to enhance nutrient removal in the mainstream process as described in Sec. 8–7 in Chap. 8. However, there are practical limitations to using primary sludge fermentate in separate sidestream reactors, as the rbCOD is dilute and typically separated from the residual primary solids by elutriation. Primary sludge and combined primary and secondary sludges have been used as the carbon source for nitrite and nitrate reduction (Wett et al., 1998; Bowden et al., 2012).

In both cases, pretreatment of the sludges through hydrolysis and fermentation may enhance the utilization of the carbon for nitrite and nitrate removal. Screened and degritted sludge is also preferred to limit the introduction of inert materials and debris into the side-stream reactor. Diverting a portion of the plant sludge to the sidestream process results in an incremental decrease in anaerobic digester gas production, but such a diversion may be acceptable, if the sludge has greater economic value as a carbon source for nitrogen removal.

Operating Temperature and pH

The operating temperature of the separate sidestream reactor should not exceed 38°C to avoid inhibiting the ammonium oxidation rate. The lower temperature limit is specific to the process and the desired performance of the system. In general, separate sidestream reactors do not operate at temperatures below 20°C due to the sidestream temperature and biological heat generation. Integrated sidestream-mainstream processes where the side-stream is blended with a portion or the entire mainstream RAS will operate over a broad temperature range due to seasonal changes in the plant influent temperature.

Operating pH

The operating pH of separate sidestream reactors is typically above 6.8 to avoid a pH-limited autotrophic growth rate and limit free nitrous acid inhibition, if nitrite is accu-mulating to a high concentration. In the RAS reaeration tank of integrated sidestream-mainstream processes (BAR/R-D-N/ScanDeNi), the pH typically decreases to or below 6.5. The impact of lower pH on nitrification performance is accounted for typically in the design by reducing the autotrophic maximum specific growth rates by 40 to 50 percent when modeling the process to determine the reaeration tank volume requirement or to predict performance with an existing tank.

Energy Balance to Determine Reactor Cooling Requirements

The biological reactions associated with nitrification-denitrification, nitritation-denitritation and deammonification are exothermic. Due to the high ammonium concentration and tem-perature of sidestreams derived from anaerobic sludge digestion, biological heat genera-tion may increase the separate sidestream reactor temperature beyond 38°C, impairing process stability and performance. Therefore, a heat balance is required using the follow-ing information to determine if heat removal or the addition of dilution water is required to maintain the sidestream reactor temperature within the desired range.

1. Sidestream flow and constituent concentrations
2. Sidestream enthalpy at reactor inlet
3. Biological heat generation rates
4. Heat losses due to evaporative cooling (aeration and induced by air movement across the open reactor surface), radiation and solar heat transfer and conductive heat trans-fer through reactor floor and walls
5. Treated sidestream enthalpy at reactor temperature
6. Mechanical energy inputs (blower compression energy, mechanical mixers)

Heat of reactions associated with the principal biological reactions are provided in Table 15–9. The values reported in Table 15–9 take into account cell mass growth. The influent and treated sidestream enthalpies are approximated by the enthalpies of pure water at their respective temperatures.

Table 15–9

Heat of reaction at standard conditions for biological reactions

Reaction	Heat of reaction at 25°C and 1 atm[a]
Nitritation[b]: $NH_4HCO_3 + 1.5O_2 + HCO_3^- \rightarrow NO_2^- + 2CO_2 + 3H_2O$	−14.3 MJ/kg-N
Nitrification[b]: $NH_4HCO_3 + 2O_2 + HCO_3^- \rightarrow NO_3^- + 2CO_2 + 3H_2O$	−21.8 MJ/kg-N
Deammonification[b]: $NH_4HCO_3 + 0.85O_2 + 0.11HCO_3^- \rightarrow 0.44N_2 + 0.11NO_3^- + 1.11CO_2 + 2.56H_2O$	−18.6 MJ/kg-N
Denitritation (external rbCOD): $COD + a\ NO_2^- + b\ CO_2 \rightarrow a\ HCO_3^- + 0.5b\ N_2 + c\ C_5H_7NO_2 + d\ H_2O$	$= [-17.0 + (25.5 \times Y_H^c)]$ MJ/kg-COD[d]
Denitritation (primary/secondary sludges): $COD_{VSS} + e\ NO_2^- + f\ CO_2 \rightarrow e\ HCO_3^- + 0.5f\ N_2 + g\ NH_4HCO_3 + h\ H_2O$	
Primary: $C_{4.66}H_{7.2}N_{0.21}O_{2.06}$	−23.9 MJ/kg-COD
Secondary: $C_5H_7NO_2$	−21.8 MJ/kg-COD
Denitrification (external rbCOD): $COD + i\ NO_3^- + j\ CO_2 \rightarrow i\ HCO_3^- + 0.5i\ N_2 + k\ C_5H_7NO_2 + k\ H_2O$	$= [-13.6 + (20.7 \times Y_H^c)]$ MJ/kg-COD[d]
Denitrification (primary/secondary sludges): $COD_{VSS} + l\ NO_3^- + m\ CO_2 \rightarrow l\ HCO_3^- + 0.5l\ N_2 + n\ NH_4HCO_3 + o\ H_2O$	
Primary: $C_{4.66}H_{7.2}N_{0.21}O_{2.06}$	−14.1 MJ/kg-COD
Secondary: $C_5H_7NO_2$	−14.3 MJ/kg-COD

[a] Heat of reaction values calculated by:

$$\Delta H^\circ = \left(\sum_i \nu_i \overline{H}_{f,i}^\circ \right)_{products} - \left(\sum_i \nu_i \overline{H}_{f,i}^\circ \right)_{reactants}$$

Where ν_i = the stoichiometric coefficient

$\overline{H}_{f,i}^\circ$ = standard heat of formation of species i at standard conditions of 25°C and 1 atm pressure.

Standard heats of formation for bicarbonate, nitrite and nitrate based on aqueous sodium salts (Green and Perry, 2007; Haynes, 2012). Heats of formation were not adjusted from standard conditions to typical sidestream reactor conditions. Heat of formation for cell mass estimated at −258 kJ/mole based on a calorimetric higher heat value of −24 kJ/g-VSS (−2712 kJ/mole). Heat of formation for primary sludge estimated at −290 kJ/mole based on a calorimetric higher heat value of −26 kJ/g-VSS (−2574 kJ/mole).

[b] Due to the low autotrophic cell mass yield under the typical sidestream reactor operating conditions, cell mass yield is ignored in the stoichiometric equation and calculation of the heat of reaction.

[c] Y_H defined as the net cell mass yield, mg VSS/mg COD.

[d] Equation based on heats of reaction for methanol, ethanol, glycerol and acetic acid.

Separate sidestream reactors are covered or uncovered depending on the climatic conditions and the desired reactor temperature range. If heat removal is a daily requirement, the reactors may be uncovered to promote evaporative cooling to minimize the design heat load or eliminate the need for heat removal. The heat load selected for design is typically based on summer climatic conditions and a low wind velocity, which corresponds to the lowest heat loss rate to the surrounding environment. The heat transfer calculations presented in Sec. 13–9 in Chap. 13 for anaerobic digesters are applicable for covered sidestream reactors constructed of concrete. For smaller reactors constructed of steel, the heat transfer coefficient estimation methods presented by Kumana and Kothari (1982) are recommended. The correlations presented by Al-Shammiri (2002) and

Bansal and Xie (1998) are recommended for estimating evaporation cooling induced by air movement across an open reactor surface.

EXAMPLE 15–3   **Estimating Reactor Volumes and Chemical Requirements for Nitritation-Denitritation and Deammonification Processes**   For the facility described in Example 15–2, calculate the reactor volume and chemical requirements for (a) nitritation-denitritation SBR and (b) deammonification SBR processes for separate sidestream treatment. Assume the following design conditions apply:

1. Average equalized sidestream flow = 124 m^3/d
2. Peak equalized sidestream flow = 149 m^3/d
3. Dilution water requirement for reactor cooling during peak summer conditions
   a. Nitritation-denitritation = 30 m^3/d
   b. Deammonification = 15 m^3/d
4. Equalized sidestream ammonium concentration = 1000 g N/m^3 (valid for all flows)
5. Equalized sidestream TSS concentration < 200 g/m^3
6. Design maximum OUR = 150 $g/m^3 \cdot h$
7. Carbonaceous OUR = 3% of the nitrogenous OUR for nitritation-denitritation
8. Carbonaceous OUR = 6% of the nitrogenous OUR for deammonification
9. External organic carbon source: methanol
   a. Concentration = 100%
   b. Specific gravity = 790 kg/m^3
   c. COD/mass = 1.5 g/g
10. Biomass yield on methanol = 0.28 g VSS/g COD
11. Nitritation-denitritation and deammonification SBR basis: Three cycles per day with each cycle consisting of:
    a. 6 h react
    b. 1 h settle
    c. 1 h decant
12. Intermittent aeration during react period: 66% aerobic and 34% anoxic

Solution Part A—Nitritation-denitritation SBR

1. Determine nitritation-denitritation reactor volume.
   a. Determine design oxygen demand.

   Design oxygen demand = nitrogenous oxygen demand + carbonaceous oxygen demand at the maximum load

   Nitrogenous oxygen demand = (149 m^3/d)(1 kg N/m^3)(3.43 kg O_2/kg N)
   = 511 kg O_2/d

   Carbonaceous oxygen demand = (0.03)(511 kg O_2/d) = 15 kg O_2/d

   Total oxygen demand = 511 kg O_2/d + 15 kg O_2/d = 526 kg O_2/d

   b. Determine the aerobic reaction time.

   Total aerobic time/d = (3 cycles/d)(6-h react/cycle)(0.66-h aerobic/h react)
   = 12-h aerobic/d

   c. Determine AOR during aeration period.

   AOR during each aeration period = (526 kg O_2/d)/(12 h/d) = 43.8 kg O_2/h

   d. Determine the required reactor volume.

   Using the oxygen requirement computed above, reactor volume at minimum liquid level is:

$$\text{Reactor volume at minimum liquid level} = \frac{\text{AOR}}{\text{Design OUR}}$$

$$= \frac{(43.8 \text{ kg } O_2/h)}{(150 \text{ g } O_2/m^3 \cdot h)} \times \frac{10^3 \text{ g}}{1 \text{ kg}}$$

$$= 292 \text{ m}^3$$

   Maximum reactor volume corresponds to the reactor volume at the minimum liquid level plus the additional volume needed for the maximum hydraulic load per cycle.

$$\text{Maximum hydraulic load/cycle} = \frac{(\text{Maximum hydraulic load/d})}{(\text{Number of cycles/d})}$$

$$= \frac{(\text{maximum sidestream flow}) + (\text{maximum dilution water})}{(\text{Number of cycles/d})}$$

$$= \frac{[(149 \text{ m}^3/d) + (30 \text{ m}^3/d)]}{(3 \text{ cycles/d})}$$

$$= 60 \text{ m}^3/\text{cycle}$$

   Reactor volume at maximum liquid level = 292 m³ + 60 m³ = 352 m³

2. Determine methanol requirement. Assume 100% oxidation of ammonium to nitrite and 100% reduction of nitrite to N_2.
   a. Determine COD requirement for denitritation.
      Using Eq. (15–12) and assuming a biomass yield = 0.28 g VSS/g COD,

      COD/N = 1.71/(1 − 1.42 × 0.28) = 2.85 kg COD/kg NO_2 N

   b. Calculate the annual ammonium loading.

$$\text{Annual ammonium loading} = (1000 \text{ g N/m}^3)(124 \text{ m}^3/d)(365 \text{ d/y})$$

$$= (1 \text{ kg N/m}^3)(124 \text{ m}^3/d)(365 \text{ d/y})$$

$$= 45,260 \text{ kg N/y}$$

   c. Determine methanol consumption.

      From the problem statement, COD of the methanol is 1.5 kg COD/kg-methanol.

      Average annual methanol consumption

$$= (45,260 \text{ kg N/y})[(2.85 \text{ kg COD/kg N})/(1.5 \text{ kg COD/kg methanol})]$$

$$= 85,994 \text{ kg methanol/y}$$

      Specific gravity of methanol is 790 kg/m³, therefore annual methanol consumption is:

$$\text{Annual methanol consumption, m}^3/\text{y} = (85,994 \text{ kg methanol/y})(1/790 \text{ kg/m}^3)$$

$$= 109 \text{ m}^3/\text{y}$$

Solution, Part B— Deammonification SBR

1. Determine deammonification reactor volume.
   a. Determine design oxygen demand.

      From Eq. (15–13), the oxygen requirement for complete deammonification = 1.87 kg O_2/kg NH_4 N.

$$\text{Maximum nitrogenous oxygen demand} = (149 \text{ m}^3/\text{d})(1000 \text{ g N/m}^3)(1.87 \text{ kg O}_2/\text{kg N})$$
$$= 279 \text{ kg O}_2/\text{d}$$

Maximum carbonaceous oxygen demand $= 6\%$ of nitrogenous oxygen demand.

Maximum carbonaceous oxygen demand $= (0.06)(279 \text{ kg O}_2/\text{d}) = 17 \text{ kg O}_2/\text{d}$

Total maximum oxygen demand $= 273 \text{ kg O}_2/\text{d} + 17 \text{ kg O}_2/\text{d} = 290 \text{ kg O}_2/\text{d}$

b. Determine the time of aerobic reaction.

$$\text{Total aerobic time/d} = (3 \text{ cycles/d})(6\text{-h react/cycle})(0.66\text{-h aerobic/h react})$$
$$= 12\text{-h aerobic/d}$$

c. Determine AOR during aeration period.

AOR during each aeration period $= 290 \text{ kg O}_2/\text{d} \div 12 \text{ h/d} = 24 \text{ kg O}_2/\text{h}$

d. Determine the reactor volume requirement.
Using the oxygen requirement obtained from Step (c), reactor volume at minimum liquid level is calculated as:

$$\text{Reactor volume at minimum liquid level} = \frac{\text{AOR}}{\text{Design OUR}}$$
$$= \frac{(24 \text{ kg O}_2/\text{h})(10^3 \text{ g/1 kg})}{(150 \text{ g O}_2/\text{m}^3\cdot\text{h})}$$
$$= 160 \text{ m}^3$$

Maximum reactor volume corresponds to the reactor volume at the minimum liquid level plus the additional volume needed for the maximum hydraulic load per cycle.

$$\text{Maximum hydraulic load/cycle} = \frac{(\text{Maximum hydraulic load/d})}{(\text{Number of cycles/d})}$$
$$= \frac{(\text{maximum sidestream flow}) + (\text{maximum dilution water})}{(\text{Number of cycles/d})}$$
$$= \frac{[(149 \text{ m}^3/\text{d}) + (15 \text{ m}^3/\text{d})]}{(3 \text{ cycles/d})}$$
$$= 55 \text{ m}^3/\text{cycle}$$

Reactor volume at maximum liquid level $= 160 \text{ m}^3 + 55 \text{ m}^3 = 215 \text{ m}^3$, or 61% of the volume requirement for nitritation-denitritation.

2. Determine methanol requirement.
The deammonification process, as shown in Eq. (15–14), does not require an external carbon source. Therefore methanol requirement $= 0 \text{ m}^3/\text{y}$

Comment   Although Deammonification does not require an external organic carbon source, the nitrate product from the anammox reaction would have to be reduced with a carbon source to yield an equivalent inorganic nitrogen removal efficiency. In the calculations presented above it was assumed that the maximum OUR of the aeration system controls the design reactor volume, which is typical of a conventional digester sidestream. In the case where sidestream TSS is consistently and considerably higher and no reduction is provided, the accumulation of inert suspended solids will increase the reactor MLSS concentration and may result in a need for a greater reactor volume. For this situation, reduction of the side-stream TSS may be cost effective.

PROBLEMS AND DISCUSSION TOPICS

15-1 Alkaline stabilization of combined primary and waste activated solids is currently practiced at a wastewater treatment facility, but there is a plan to replace this system with a single-stage mesophilic anaerobic digestion process. Centrifugation has been selected for biosolids dewatering. In the absence of pilot study data, a preliminary estimate of the daily sidestream volume and characteristics are required to assess the nutrient and solids loads returned to the mainstream nutrient removal process. Using the following data and assumptions, estimate the future daily sidestream volumes and concentrations of soluble TKN, total TKN, soluble orthophosphate, and total suspended solids at the average and peak thickened sludge loads to the proposed digestion and dewatering processes. In the calculation of the orthophosphate concentration, assume no precipitation of phosphate. The contribution of centrifuge wash-water to the sidestream flow can also be ignored.

Thickened combined raw sludge data

Parameter	Unit	Value
Daily volume, average—maximum two-week	m³/d	530–700
Total solids concentration (applicable for all flowrates)	%	4.5
Volatile fraction	%	78
Volatile fraction nitrogen content	%	6.5
Volatile fraction phosphorus content	%	1.5
Specific gravity		1.02

Single-stage mesophilic digestion performance criteria

Parameter	Unit	Value
Temperature	°C	35
Volatile solids destruction efficiency at minimum digester SRT	%	45
Volatile solids destruction efficiency at average digester SRT	%	50
Digested volatile solids nitrogen content	%	6.5
Digested volatile solids phosphorus content	%	1.5

Biosolids dewatering performance criteria (applicable for all sludge loading rates)

Parameter	Unit	Value
Cake solids concentration	%	22
Solids capture efficiency	%	95

15-2 Digested primary and waste activated sludge is dewatered six day per week and eight hours per day. The average sidestream flow, ammonium-N concentration and soluble orthophosphate concentration are shown below:

Parameter	Unit	Value
Flowrate during biosolids dewatering,	m³/h	83
Ammonium-N concentration	mg/L	1050
Orthophosphate-P concentration	mg/L	190
Maximum period without dewatering	d	2

Using the given information, (a) calculate the tank volume required for full equalization of the sidestream so that the equalized flow is returned continuously at constant flowrate to the mainstream plant and (b) the required volume if the equalized sidestream is to be returned to the mainstream plant seven days per week between the hours of 10 p.m. and 6 a.m.

15-3 The sidestream described in Problem 15-2 is discharged to the inlet of the primary clarifiers in the mainstream process. Ferric chloride is applied to the primary clarifiers at an average Fe/P mass ratio of 2 kg/kg to reduce the orthophosphate concentration in the primary tank effluent. Using the given data,

a. Estimate the volume of concentrated ferric chloride solution required to precipitate the orthophosphate-P contributed by sidestream to the primary tank influent. The physical property data for the ferric chloride solution is shown below:

Parameter	Unit	Value
FeCl$_3$ concentration, percent by weight	%	37
Specific gravity of concentrated solution		1.4

b. The implementation of a struvite crystallization process for sidestream pretreatment is being considered. Assuming negligible loss of phosphate during to precipitation in the equalization tank, a target soluble orthophosphate-P concentration of 15 mg/L in the crystallizer effluent and 100 percent recovery of the crystallized product, estimate the average daily mass of struvite that can be potentially harvested from this process. An air-dried product in the form of a hexahydrate salt should be assumed in the calculation.

15-4 For the sidestream described in Problem 15–2, a process consisting of struvite crystallization followed by deammonification in a SBR has been proposed for pretreatment before discharge to the mainstream plant. Using the struvite crystallization criteria provided in Problem 15–3(b),

a. Calculate the ammonium-N concentration in the struvite crystallizer effluent.

b. Estimate the deammonification SBR volume requirement. The following conditions apply:

  i. Sidestream flow will be equalized as calculated in Problem 15–2(a) and this equalized flow will serve as the design basis.

  ii. An energy balance has revealed that sufficient heat losses from the equalization tank, struvite crystallizer and SBR will occur to maintain the SBR temperature below 38°C during the peak summer conditions; thus, no dilution water is required.

  iii. Fine bubble diffusers will be used and can satisfy a maximum OUR of 150 mg/L·h.

  iv. Carbonaceous OUR is estimated to be 8 percent of the nitrogenous OUR.

  v. Intermittent aeration is applied with each aeration cycle time consisting of 66 percent aerobic and 34 percent anoxic.

  vi. Three SBR cycles per day with each cycle consisting of 6-h *react*, 1-h *settle* and 1-h *decant*.

  vii. Assume an ammonia removal efficiency of 100 percent as the design basis.

15-5 A sidestream derived from mesophilic anaerobic digestion of primary and waste activated sludge is produced at a equalized daily volume of 600 m^3/d and contains an ammonium concentration of 900 mg N/L. For the three biological treatment options, nitrification-denitrification, nitritation-denitritation and deammonification,

a. Calculate the daily biological heat generated by the three processes in megajoules/day using the following performance conditions:

  i. Ammonium-N removal efficiency = 95% in all processes

  ii. NOx-N removal efficiency = 95% for denitrification and denitritation.

iii. Denitrification of the nitrate produced by the anammox reaction with sidestream degradable carbon is not significant; thus, this reaction can be excluded from the heat calculation.

iv. Net sludge yield, $Y_H = 0.2$ g VSS/g COD and is applicable to denitrification and denitritation.

b. Determine if the biological heat generated in the three processes is sufficient to increase the reactor temperature beyond a maximum limit of 38°C. If the temperature will exceed 38°C, determine the fraction of the biological heat in each process that must be removed or absorbed via dilution water addition. The equalized sidestream has a peak temperature of 35°C.

15-6 For the sidestream and the process performance criteria described in Problem 15-5, calculate the daily mechanical mixing and aeration energy consumption (kWh/d) for nitrification-denitrification, nitritation-denitritation, deammonification or all three processes (instructor's preference). The following information should be used for the calculations:

Sidestream Characteristics

Parameter	Unit	Value
Equalized flowrate for design	m³/d	600
Ammonium-N concentration	mg/L	900
Degradable COD concentration[a]	mg/L	200

[a] sum of the readily and complex biodegradable COD.

A sequencing batch reactor is selected for design. The SBR will be fed continuously during the *react* period of each SBR cycle. The following operating and design conditions apply to all three processes:

Parameter	Unit	Value
Number of SBR cycles per day	–	3
Settle period duration	h/cycle	1
Decant period duration	h/cycle	1
Aerobic fraction of react period	%	66
Average reactor temperature	°C	34
Idle period duration	h/cycle	0
Maximum OUR for design	mg/L·h	150
Maximum sidewater depth	m	7
Ammonia and degradable COD removal efficiencies for SBR design	%	100
Actual ammonia removal efficiency	%	90
Actual degradable COD removal efficiency	%	95
Net heterotrophic yield, Y_H	gVSS/gCOD	0.2

Based on a heat loss analysis for open top concrete reactors, the following average daily dilution water requirements were estimated for each process:

Process	Unit	Value
Nitrification-denitrification	m³/d	200
Nitritation-denitritation	m³/d	100
Deammonification	m³/d	0

Fine bubble membrane disk diffusers and submersible mixers are selected for design.

Parameter	Unit	Value
Mech. mixing intensity (nitrification-denitrification)[a]	W/m^3	4
Mech. mixing intensity (nitritation-denitritation)[a]	W/m^3	4
Mech. mixing intensity (deammonification)[a]	W/m^3	6
Mixer total efficiency (electrical + mechanical)	%	84
Operating DO conc. (nitrification-denitrification)	mg/L	2.0
Operating DO conc. (nitritation-denitritation)	mg/L	0.5
Operating DO conc. (deammonification)	mg/L	0.3
Alpha factor, α	—	0.5
Fouling factor, F	—	0.85
Beta factor, β	—	0.95
Temperature correction factor, θ	—	1.024
Site barometric pressure	kPa	99.97
Distance from floor to membrane surface	m	0.25
Pressure drop (blower inlet)	kPa	1.7
Pressure drop (piping, valves, diffusers)	kPa	12
Standard oxygen transfer efficiency, SOTE[b]	%/m	6
Average ambient air temperature	°C	20
Blower mechanical efficiency	%	75
Blower motor electrical efficiency	%	90

[a] Mixing intensity at maximum liquid level. The submersible mixers only operate during the anoxic periods and when they operate, the mixers run at a constant speed regardless of the liquid depth.

[b] SOTE per meter of diffuser submergence. For simplification, assume a linear relationship with depth.

15-7 A wastewater treatment facility requires a capacity expansion to accommodate the projected growth in the service population. The following processes will be added to the facility:

   i. Primary sedimentation tanks
   ii. Gravity thickening of primary sludge
   iii. Dissolved air flotation for thickening of waste activated sludge
   iv. Mesophilic anaerobic digestion of combined primary and waste activated sludges
   v. Screw press for dewatering digested sludge

Using the plant information given below:

a. Calculate the flow and suspended solids concentrations for the gravity thickener overflow and the dissolved air flotation subnatant.

b. Calculate the pressate flow and concentrations of soluble TKN, total suspended solids and soluble phosphorus.

c. Calculate the percent contribution of the pressate soluble TKN and soluble phosphorus to the primary effluent TKN and TP loads to the secondary process.

Future plant influent flowrate and characteristics:

Parameter	Unit	Value
Average daily flowrate	m³/d	26,500
COD	mg/L	580
Carbonaceous BOD	mg/L	275
Total suspended solids	mg/L	290
Volatile suspended solids	mg/L	226
TKN	mg/L	40
Ammonium-N	mg/L	23
Total phosphorus	mg/L	7
Ortho-phosphorus	mg/L	3.6
Particulate TKN/VSS ratio	—	0.04
Particulate P/VSS ratio	—	0.015
Particulate COD/VSS ratio	—	1.6

Future solids removal, thickening and digester performance criteria:

Parameter	Unit	Value
Gravity thickener TSS capture efficiency	%	93
Thickened primary sludge percent solids	%	6
DAF TSS capture efficiency	%	95
Thickened waste activated sludge percent solids	%	5
Primary tank TSS removal efficiency	%	60
Primary tank underflow percent solids	%	1
Primary tank cBOD removal efficiency	%	30
Screw press TSS capture efficiency	%	95
Digested sludge cake percent solids	%	25
Digester volatile solids destruction	%	50
Nitrogen content of digested VSS	%	6
Phosphorus content of digested VSS	%	1.8

Future secondary process performance and solids production:

Parameter	Unit	Value
SRT	d	12
cBOD removal efficiency	%	98
Observed yield	gVSS/gBOD	0.55
Volatile content of WAS TSS	%	80
WAS MLSS concentration	mg/L	7500
Nitrogen content of WAS VSS	%	9.5
Phosphorus content of WAS VSS	%	2
Filtered plant effluent TSS concentration	mg/L	< 2

15-8 Using Eqs. (7-96) in Sec. 15-8 and (15-11) in Sec. 15-9, estimate the amount of methanol that can be saved with the conversion of a sidestream treatment process from nitrification-denitrification to nitritation-denitritation. Express the results as kilograms of methanol saved per kilogram of NOx-N-removed.

15-9 For the sidestream described in Problem 15-2, and assuming a sidestream alkalinity of 3750 mg/L as $CaCO_3$, estimate the soda ash (Na_2CO_3) dosing rate required for complete nitrification of the sidestream ammonium-N in a separate reactor performing only nitrification. Describe how the soda ash dosing requirement would change if the sidestream treatment process is modified from nitrification-denitrification to nitritation-denitritation.

REFERENCES

Abma, W., C. E. Schultz, J. W. Mulder, M. C. M. van Loosdrecht, W. R. L. van der Star, M. Strous, and T. Tokutomi (2007) "The Advance of Anammox," *Water 21*, **36**, 2, 36–37.

Abma, W. R., W. Driessen, R. Haarhuis, and M. C. M. van Loosdrecht (2010) "Upgrading of Sewage Treatment Plant by Sustainable and Cost-Effective Separate Treatment of Industrial Wastewater," *Water Sci. Technol.*, **61**, 7, 1715–1722.

AI-Shammiri, M. (2002) "Evaporation Rate as a Function of Water Salinity," *Desalination*, **150**, 2, 189–203.

Bansal, P. K., and G. Xie (1998) "A Unified Empirical Correlation for Evaporation of Water at Low Air Velocities," *Int. Comm. Heat Mass Transfer*, **25**, 2, 183–190.

Baur, R., N. Cullen, and B. Laney (2011) "Nutrient Recovery: One Million Pounds Recovered-With Benefits," *Proceedings of the Water Environment Federation 84th Annual Conference and Exposition*, San Diego, CA.

Berends, D., S. Salem, H. van der Roest, and M. C. M. van Loosdrecht (2005) "Boosting Nitrification with the BABE Technology," *Water Sci. Technol.*, **52**, 4, 63–70.

Berg, U., M. Schwotzer, P. Weidler, and R. Nüesch (2006) "Calcium Silicate Hydrate Triggered Phosphorus Recovery–An Efficient Way to Tap the Potential of Waste- and Process Waters as Key Resource," *Proceedings of the Water Environment Federation 79th Annual Conference and Exposition*, Dallas, TX.

Bowden, G., D. Lippman, B. Dingman, and E. Casares (2012) "Case Study in Optimizing the Use of Existing Infrastructure and Plant Carbon Sources to Reduce the Effluent Total Nitrogen: Upgrade of the Tapia Water Reclamation Facility," *Proceedings of the Water Environment Federation 85th Annual Conference and Exposition*, New Orleans, LA.

Christensson, M., S. Ekström, R. Lemaire, E. Le Vaillant, E. Bundgaard, J. Chauzy, L. Stålhandske, Z. Hong, and M. Ekenberg (2011) "ANITA™Mox–A Biofarm Solution for Fast Start-Up of Deammonifying MBBRs," *Proceedings of the Water Environment Federation 84th Annual Conference and Exposition*, San Diego, CA.

Cirpus, I. E. Y., W. Geerts, J. H. M. Hermans, H. J. M. Op den Campa, M. Strous, J. G. Kuenen, and M. S. M. Jetten (2006) "Challenging Protein Purification from Anammox Bacteria," *Int. J. Biol. Macromol.*, **39**, 1–3, 88–94.

Clariant/SÜD Chemie (2012) Terra-N® performance data, facilities list and operating conditions provided to author by personal correspondence.

Cordell, D., J-O. Drangert, and S. White (2009) "The story of phosphorus: Global food security and food for thought," *Global Env. Change*, **19**, 2, 292–305.

Corrado, M. (2009) "Reducing Struvite Formation Potential in Anaerobic Digesters by Controlled Release of Phosphate from Waste Activated Sludge," M.S. Thesis, Univ. of Wisconsin, Madison, WI.

Donnert, D., and M. Salecker (1999) "Elimination of Phosphorus from Waste Water by Crystallisation," *Env. Technol.*, **20**, 7, 735–742.

du Preez, J., B. Norddahl, and K. Christensen (2005) "The BIOREK® Concept: A Hybrid Membrane Bioreactor Concept for Very Strong Wastewater," *Desalination*, **183**, 1–3, 407–415.

Dwyer, J., D. Starrenburg, S. Tait, K. Barr, D. Batstone, and P. Lant (2008) "Decreasing activated sludge thermal hydrolysis temperature reduces product colour, without decreasing degradability," *Water Res., 42*, 18, 4699–4709.

Fux, C., M. Boehler, P. Huber, I. Brunner, and H. Siegrist (2002) "Biological Treatment of Ammonium-Rich Wastewater by Partial Nitritation and Subsequent Anaerobic Ammonium Oxidation (ANAMMOX) in a Pilot Plant," *J. Biotech., 99*, 3, 295–306.

Fux, C., V. Marchesi, I. Brunner, and H. Siegrist (2004) "Anaerobic Ammonium Oxidation of Ammonium-Rich Waste Streams in Fixed-Bed Reactors," *Water Sci. Technol., 49*, 11–12, 77–82.

Fux C., S. Velten, V. Carozzi, D. Solley, and J. Keller (2006) "Efficient and Stable Nitritation and Denitritation of Ammonium-Rich Sludge Dewatering Liquor Using an SBR with Continuous Loading," *Water Res., 40*, 14, 2765–2775.

Giesen, A. (2009) "P Recovery with the Crystalactor® Process," Presentation in *BALTIC 21 Phosphorus Recycling and Good Agricultural Management Practice,* September 28–30, 2009, Berlin, Germany.

Gopalakrishnan, K., J. Anderson, L. Carrio, K. Abraham, and B. Stinson (2000) "Design and Operational Considerations for Ammonia Removal from Centrate by Steam Stripping," *Proceedings of the Water Environment Federation 73rd Annual Conference and Exposition,* Los Angeles, CA.

Green, D. W., and R. H. Perry (2007) *Perry's Chemical Engineers' Handbook,* 8th ed., McGraw-Hill, New York.

Haynes, W. M. (2012) *CRC Handbook of Chemistry and Physics,* 93rd ed., Taylor & Francis, Inc., Florence, KY.

Hellinga, C., A. A. J. C. Schellen, J. W. Mulder, M. C. M. van Loosdrecht, and J. J. Heijnen (1998) "The SHARON Process: An Innovative Method for Nitrogen Removal from Ammonium-Rich Wastewater," *Water Sci. Technol., 37*, 9, 135–142.

Jeavons, J., L. Stokes, J. Upton, and M. Bingley (1998) "Successful Sidestream Nitrification of Digested Sludge Liquors," *Water Sci. Technol., 38*, 3, 111–118.

Kemp, P., M. Simon, and S. Brown (2007) "Ammonium/Ammonia Removal from a Stream," U.S. Patent 7270796.

Kuai, L., and W. Verstraete (1998) "Ammonium Removal by the Oxygen-Limited Autotrophic Nitrification-Denitrification System," *Appl. Env. Microbiol., 64*, 11, 4500–4506.

Kuenen, J. G. (2008) "Timeline: Anammox Bacteria: From Discovery to Application," *Nature Rev. Microb., 6*, 4, 320–326.

Kumana, J. D., and S. P. Kothari (1982) "Predict storage-Tank Heat Transfer Precisely," *Chem. Eng.-New York, 89*, 6, 127–132.

Laurich, F. (2004) "Combined Quantity Management and Biological Treatment of Sludge Liquor at Hamburg's Wastewater Treatment Plants—First Experience in Operation with the Store and Treat Process," *Water Sci. Technol., 50*, 7, 49–52.

Le Corre, K. S., E. Valsami-Jones, P. Hobbs, and S. A. Parsons (2005) "Impact of calcium on struvite crystal size, shape and purity," *J. Crystal Growth, 283*, 3–4, 514–522.

Lemaire, R., I. Liviano, S. Esktröm, C. Roselius, J. Chauzy, D. Thornberg, C. Thirsing, and S. Deleris (2011) "1-Stage Deammonification MBBR Process for Reject Water Sidestream Treatment: Investigation of Start-Up Strategy and Carriers Design," *Proceedings of the Water Environment Federation, Nutrient Recovery and Management 2011 Conference,* Miami, FL.

Membrana (2007) "Successful Ammonia Removal from Wastewater Using Liqui-Cel® Membrane Contactors at a European Manufacturing Facility," Technical Brief no. 43, revision 2, www.membrana.com or www.liqui-cel.com.

Moerman W. H. M. (2011) "Full Scale Phosphate Recovery: Process Control Affecting Pellet Growth and Struvite Purity," *Proceedings of the Water Environment Federation, Nutrient Recovery and Management 2011 Conference,* Miami, FL.

Nawa, Y. (2009) "P-Recovery in Japan - the PHOSNIX Process," Poster from *BALTIC 21 Phosphorus Recycling and Good Agricultural Management Practice,* Berlin, Germany.

Nieminen, J. (2010) "Phosphorus Recovery and Recycling from Municipal Wastewater Sludge," M. S. Thesis, Aalto Univ., Finland.

Olav Sliekers, A., N. Derwort, J. L. Campos Gomez, M. Strous, J. G. Kuenen, and M. S. M. Jetten (2002) "Completely Autotrophic Nitrogen Removal Over Nitrite in One Single Reactor," *Water Res.,* **36**, 10, 2475–2482.

Olav Sliekers, A., K. A. Third, W. Abma, J. G. Kuenen, and M. S. M Jetten (2003) "CANON and Anammox in a Gas-Lift Reactor," *FEMS Microbiol. Let.,* **218**, 2, 339–344.

Organics Limited (2009) "Thermally-Driven Ammonia Strippers: Ammonia Stripping with Waste Heat," Data Sheet ODSP09, www.organics.com.

Piekema, P., and A. Giesen (2001) "Phosphate recovery by the crystallization process: experience and developments, " *Proceedings of the Second International Conference on Recovery of Phosphate from Sewage and Animal Wastes,* Noordwijkerhout, The Netherlands.

Plaza, E., S. Stridh, J. Örnmark, L. Kanders, and J. Trela (2011) "Swedish Experience of the Deammonification Process in a Biofilm System," *Proceedings of the Water Environment Federation, Nutrient Recovery and Management 2011 Conference,* Miami, FL.

Sagberg, P., P. Ryrfors, and K. G. Berg (2006) "10 Years of Operation of an Integrated Nutrient Removal Treatment: Ups and Downs. Background and Water Treatment," *Water Sci. Technol.,* **53**, 12, 83–90.

Schauer, P., R. Baur, J. Barnard, and A. Britton (2011) "Increasing Revenue While Reducing Nuisance Struvite Precipitation: Pilot Scale Testing of the WASSTRIP Process," *Proceedings of the Water Environment Federation, Nutrient Recovery and Management 2011 Conference,* Miami, FL.

Schneider Y., M. Beier, and K. Rosenwinkel (2009) "Impact of Seeding on the Start-up of the Deammonification Process with Different Sludge Systems," *Proceedings of the Second IWA Specialized Conference on Nutrient Management in Wastewater Treatment Processes,* Krakow, Poland.

Seyfried, C., A. Hippen, C. Helmer, S. Kunst, and K. H. Rosenwinkel (2001) "One-Stage Deammonification: Nitrogen Elimination at Low Costs," *Water Sci. Technol.: Water Supply,* **1**,1, 71–80.

Seyfried, C. (2002) "Deammonification: A Cost-Effective Treatment Process for Nitrogen-Rich Wastewaters," *Proceedings of the Water Environment Federation 75ᵗʰ Annual Conference and Exposition,* Chicago, IL.

Siegrist, H., S. Reithaar, and P. Lais (1998) "Nitrogen Loss in a Nitrifying Rotating Contactor Treating Ammonium Rich Leachate Without Organic Carbon," *Water Sci. Technol.,* **37**, 4–5, 589–591.

Strous, M., J. J. Heijnen, J. G. Kuenen, and M. S. M. Jetten (1998) "The Sequencing Batch Reactor as a Powerful Tool for the Study of Slowly Growing Anaerobic Ammonium-Oxidizing Microorganisms," *Appl. Microbiol. Biotechnol.,* **50**, 5, 589–596.

Teichgräber, B., and A. Stein (1994) "Nitrogen Elimination from Sludge Treatment Reject Water: Comparison of Steam Stripping and Denitrification Processes," *Water Sci. Technol.,* **30**, 6, 41–51.

ThermoEnergy (2009) "Ammonium Recovery Process (ARP) Ammonium Sulfate Purity Study, 26ᵗʰ Ward ARP Project," Report Issued to the New York City Department of Environmental Protection, Worcester, MA.

Third, K. A., J. Paxman, M. Schmid, M. Stous, M. S. M. Jetten, and R. Cord-Ruwisch (2005) "Enrichment of Anammox from Activated Sludge and its Application in the Canon Process," *Microbial Ecol.,* **49**, 2, 236–244.

U.S. EPA (1987) *Design Information Report-Sidestreams in Wastewater Treatment Plants,* EPA-68-03-3208, EPA/600/J-87/004, Water Engineering Research Laboratory, U.S. Environmental Protection Agency, Cincinnati, OH (see also *J. WPCF,* **59**, 1, 54–61).

van der Star, W., W. Abma, D. Blommers, J. W. Mulder, T. Tokutomi, M. Strous, C. Picioreanu, and M. van Loosdrecht (2007) "Startup of Reactors for Anoxic Ammonium Oxidation: Experiences from the First Full-Scale Anammox Reactor in Rotterdam," *Water Res.,* **41**, 18, 4149–4163.

van Hulle, S. W. H., H. J. P. Vandeweyer, B. D. Meesschaert, P. A. Vanrolleghem, P. Dejans, and A. Dumoulin (2010) "Engineering Aspects and Practical Application of Autotrophic Nitrogen Removal from Nitrogen Rich Streams," *Chem. Eng. J.,* **162**, 1, 1–20.

van Kauwenbergh, S. (2010) "World Phosphate Rock Reserves and Resources," Technical Bulletin T-75 International Fertilizer Development Center, Muscle Shoals, AL.

Vitse, F., M. Cooper, and G. G. Botte (2005) On the use of ammonia electrolysis for hydrogen production, *J. Power Sources, 142*, 1–2, 18–26.

Wett, B., R. Rostek, W. Rauch, and K. Ingerle (1998) "pH-Controlled Reject-Water-Treatment," *Water Sci. Technol., 37*, 12, 165–172.

Wett, B., and W. Rauch (2003) "The Role of Inorganic Carbon Limitation in Biological Nitrogen Removal of Extremely Ammonia Concentrated Wastewater," *Water Res., 37*, 5, 1100–1110.

Wett, B. (2006) "Solved Upscaling Problems for Implementing Deammonification of Rejection Water," *Water Sci. Technol., 53*, 12, 121–128.

Wett, B. (2007) "Development and Implementation of a Robust Deammonification Process," *Water Sci. Technol., 56*, 7, 81–88.

Wett, B., I. Takacs, S. Murthy, M. Hell, G. Bowden, A. Deur, and M. O'Shaughnessy (2007) "Key Parameters for Control of DEMON Deammonification Process," *Water Pract., 1*, 5, 1–11.

Wett, B., M. Hell, G. Nyhuis, T. Puempel, I. Takacs, and S. Murthy (2010) "Syntrophy of Aerobic and Anaerobic Ammonia Oxidizers," *Water Sci. Technol., 61*, 8, 1915–1922.

Yang, J., L. Zhang, Y. Fukuzaki, D. Hira, and K. Furukawa (2011) "The Positive Effect of Inorganic Carbon on Anammox Process" *Proceedings of the Water Environment Federation, Nutrient Recovery and Management 2011 Conference,* Miami, FL.

16

Air Emissions from Wastewater Treatment Facilities and Their Control

WORKING TERMINOLOGY

Term	Definition
Absorption	The process by which atoms, ions, molecules, and other constituents are transferred from one phase and are distributed uniformly in another phase (see also adsorption).
Adsorption	The process by which atoms, ions, molecules, and other constituents are transferred from one phase and accumulate on the surface of another phase (see also absorption).
Air stripping	The removal of volatile and semi-volatile contaminants from a liquid by passing air and liquid counter currently through a packed tower.
Buffer zone	An area around a facility which serves to diminish the impact of any odors emitted from the facility. Trees are sometimes planted at the periphery of buffer zones to further reduce the impact of odors.
Catalytic incineration	Controlled process used to oxidize VOCs with the help of a catalyst such as platinum and palladium.
Chemical scrubber	A reactor used to provide contact between air, water, and chemicals, if used, to provide oxidation or entrainment of odorous compounds.
Biofilter	Open or closed packed-bed filters used for the removal of odors biologically. In open biofilters, gases to be treated move upward through the filter bed. In closed biofilters, the gases to be treated are either blown or drawn through the packing lateral.
Biotrickling filter	Similar to biofilters, with the exception that moisture is provided continuously or intermittently over the packing. Liquid is recirculated and nutrients are often added.
Digester gas	Gas produced from anaerobic digestion of sludges. Also often referred to as biogas. Digester gas typically contains 60 percent or higher methane gas by volume and can be used as a fuel source.
Gas stripping	The purposeful introduction of air or other gases to transfer volatile constituents such as VOCs and odors from a liquid phase to a gaseous phase.
Global warming potential (GWP)	A measure of relative effects of a gas to trap heat in the atmosphere in reference to that of carbon dioxide.
Greenhouse gases	Gases that have been identified as contributing to global warming.
Mass transfer	The transfer of material from one homogeneous phase to another; aeration, gas stripping, and adsorption are examples of mass transfer.
Mechanical aerators	Devices used to agitate water to promote mixing with atmospheric air.
Mixing	The agitation of a liquid-solids suspension for the purpose of blending the mixture and keeping solids in suspension, entraining gases, or for accelerating a chemical reaction.
Odor threshold	The concentration at which an odor is detectable by human sense of smell.
Off-gas	The gaseous emission from a process; off-gas may be odorous and/or contain greenhouse gases and VOCs.
Stripping tower	A closed vertical reactor used to bring about the transfer of VOCs from a liquid phase to a gaseous phase.
Thermal oxidation	Controlled process used to oxidize VOCs at high temperatures.
Vapor phase adsorption	Process whereby hydrocarbons and other compounds arc are adsorbed selectively on the surface or such materials as activated carbon, silica gel, or alumina.
Volatile organic compounds (VOCs)	A term often used generically to mean total organic carbon; in the context of air quality the term means total nonmethane hydrocarbons.
Volatilization	The release of VOCs from water surface to the atmosphere.

The treatment of wastewater results in the release of a variety of air emissions, many of which are odorous and/or may contain air pollutants. Activities that result in emissions of air pollutants to the outdoor atmosphere may require approval of the state or regional environmental agency or the U.S. Environmental Protection Agency (U.S. EPA). The extent of the air permitting requirements is based on a number of factors such as the nature of the air pollutant, the quantity of emissions, the air quality in the vicinity of the facility, the existing air emissions and emission sources, the state air permitting requirements, and emissions control regulations mandated by the governing agency and federal regulations. The governing agency may be at the state, region, city, or tribal level. Air permitting is typically at the federal and state level, but authority may vary. As used herein, "state" refers to agencies other than the U.S. EPA that have jurisdiction to implement air permitting under the Clean Air Act (CAA) at the state, regional, city, or tribal level. Topics discussed in this chapter include (1) types of emissions, (2) regulatory requirements, (3) odor management, (4) volatile organic carbon emissions and their control, (5) emissions from combustion of gases and solids, and (6) emission of greenhouse gases.

16–1 TYPES OF EMISSIONS

Typical air emissions from wastewater treatment plants that are subject to regulation under the CAA are provided in Table 16–1. Some states require that additional classes of compounds (e.g., odors and greenhouse gases) be calculated at these facilities. For example, a recently promulgated regulation by the U.S. EPA requires that the total emissions from all sources of greenhouse gases be calculated and compared to regulatory thresholds. Certain processes within wastewater treatment plants have qualified for exemptions to the rule due to the "biogenic" nature of their emissions (i.e., non-fossil fuel origin). Emission of greenhouse gases is discussed further in Sec 16–6. Sources of emissions within odor management is particularly important for the operation of wastewater treatment facilities, and it is discussed in detail in Sec 16–3.

16–2 REGULATORY REQUIREMENTS

The CAA is the basis for federal and state regulations, which are codified in Subchapter C, Air Programs, of Title 40 of the Code of Federal Regulations. The CAA and amendments establish air permitting programs at the federal level and provide a regulatory framework for state and regional regulations. Permit authority may be at the federal, state, or regional level. Emissions of air pollutants fall into three general categories: criteria pollutants, non-criteria pollutants, and hazardous air pollutants (HAPs). Criteria pollutants have an associated air quality standard. The key air quality regulations cover broad categories, as described herein.

Ambient Air Quality and Attainment Status

The U.S. EPA has established National Ambient Air Quality Standards (NAAQS) to protect public health and public welfare. Primary standards are based on observable human health responses and are set at levels that provide an adequate margin of safety for sensitive segments of the population. Secondary standards are intended to protect public welfare interests such as structures, vegetation, and livestock. States may also establish ambient air quality standards that are more stringent than the federal standards and may retain a revoked federal standard.

Table 16–1

Typical air pollutants associated with wastewater treatment plants

Air pollutant	Source(s)
Criteria pollutants	
Carbon monoxide (CO)	Incomplete combustion, partial oxidation of organic material
Nitrogen dioxide (NO_2)	Combustion processes
Sulfur dioxide (SO_2)	Combustion processes
Total suspended particulate matter (TSP)	Combustion processes, material handling, aggregate handling, other process sources
Respirable particulate matter with a diameter of up to 10 microns (PM_{10})	Combustion processes, material handling, aggregate handling, other process sources
Fine particulate matter with a diameter of up to 2.5 microns ($PM_{2.5}$)	Combustion processes, material handling, aggregate handling, other process sources
Ozone (O_3):	Generated at ground level through photochemical oxidation of precursors NOx and VOC
Oxides of nitrogen (NOx)	Combustion processes
Volatile organic compounds (VOC)	Combustion processes, organic material storage and use, other process sources
Lead (Pb)	Combustion processes
Non-criteria pollutants	
Hydrogen sulfide (H_2S)	Process sources, anaerobic reduction of sulfur compounds
Methane (CH_4)	Anaerobic digestion, combustion processes
Carbon dioxide (CO_2)	Combustion processes, anaerobic digestion
Ammonia (NH_3)	Solids processing/polymer breakdown
Nitrous oxide (N_2O)	Biological nitrogen removal systems
Hazardous air pollutants (HAPs)	
Toluene, benzene, xylene, etc	Influent constituents from industrial sources
Methanol	Nutrient removal systems
Trimethylammine and dimehtylammine	Solids processing/polymer breakdown
Carbon disulfide and carbonyl sulfide	Total reduced sulfur compounds from sulfate oxidation
Formaldehyde and hexane	Combustion by-products from combined heat and power/boilers/process heaters/etc.
Chlorine (Cl_2)	Used in chlorination processes
Mercury, other heavy metals and polycyclic organic compounds (POM)	Wastewater sludge incinerators

The CAA requires the U.S. EPA and states to identify by category the ambient air quality compliance status for specific geographic regions (air quality control regions or portions thereof). Areas may be designated "attainment" or "nonattainment" of the NAAQS based on monitoring data, or "unclassifiable" if insufficient ambient monitoring data exists; "unclassifiable" areas are generally considered as "attainment" areas. The attainment status of the existing/new source location determines the applicability of pre-construction permitting programs.

Table 16-2

Permitting programs and control technologies

Requirements	Regulations, rules and technologies
Preconstruction and operating permitting	Nonattainment new source review (NNSR or NANSR)
	Prevention of significant deterioration (PSD)
	Title V operating permits
	Minor source permitting programs
Stationary source control permitting	New Source Performance Standards (NSPS per 40 CFR Part 60)
	National Emission Standards for Hazardous Air Pollutants (NESHAPs per 40 CFR Parts 61 and 63)
	Reasonably Available Control Technology (RACT)

Preconstruction and Operating Permitting Programs

New sources and modifications/reconstructions of existing sources that result in air pollution emissions (above specified thresholds) will be subject to air permitting requirements. Air permitting programs are listed in Table 16–2. The applicability of nonattainment new source review (NNSR) and prevention of significant deterioration (PSD) depends upon the nature of the project, the air quality designation in the vicinity of the project, and the quantity of annual emissions. Both are preconstruction approval programs applicable to major projects, and both prohibit commencement of construction until the preconstruction approval is issued. If major source or major modification criteria are met, NNSR may be applicable to a nonattainment pollutant (and precursor pollutants) while PSD may be applicable to an attainment pollutant. If NNSR is applicable, then the lowest achievable emission rate (LAER) control technology that is technically feasible must be incorporated in the project design, regardless of cost.

If PSD is applicable, then the best available control technology (BACT) must be incorporated into the project design. Factors such as cost, energy requirements, and other environmental consequences/benefits are part of a BACT analysis. The U.S. EPA may delegate authority to approve NNSR and PSD projects to state agencies; alternatively, it may approve state regulations that meet U.S. EPA requirements (e.g., approval of State Implementation Plan). Within the PSD program, more stringent protection of air quality is afforded to "Class I" areas, which include national parks, wildlife areas, and other designated areas. Impacts to visibility and ecology must also be minimized for the project to obtain an air permit to construct.

The CAA Amendments of 1990 added the Title V operating permit program, which established state and federal permitting procedures for major facilities that allow for the consolidation of facility specific requirements into a single, federally-enforceable permit. These requirements include emission limitations, work practice standards, monitoring requirements, recordkeeping requirements, and submittal/notification requirements. Applicable requirements may be based on regulations or preconstruction approval conditions. Title V permitting programs are usually administered at the state level. Minor source permitting programs apply to sources that are not subject to Title V operating permits. These requirements are established by state agencies.

Stationary Source Control Technology Requirements

Stationary source emissions control technology requirements are listed in Table 16–2. Unlike the control technology requirements of the NNSR and PSD preconstruction programs, regulations such as new source performance standards (NSPS) and national emission standards for hazardous air pollutants (NESHAPs) apply to equipment based on the function and category of the equipment. The NSPS apply to new, modified, or reconstructed sources that meet the applicability criteria specific to the source type. The NESHAP requirements apply primarily to major sources of HAPs, but some requirements are applicable to "area sources" (minor sources) of HAPs. Reasonably available control technology (RACT) broadly refers to the control technology requirements specified by state regulations that establish the minimum level of emissions control. RACT requirements can be applied retroactively to a source and are generally more stringent in nonattainment areas than in attainment areas.

16–3 ODOR MANAGEMENT

The potential release of odors is a major concern of the public relative to modifying existing wastewater treatment facilities and constructing new facilities. Thus, the control of odors has become a major consideration in the design and operation of wastewater collection, treatment, and disposal facilities, especially with respect to the public acceptance of these facilities. In many instances, projects have been rejected because of the fear of potential odors. In several states, wastewater management agencies are now subject to fines and other legal action over odor violations. In view of the importance of odors in the field of wastewater management, the following topics are considered in this section: (1) the types of odors encountered, (2) the sources of odors, (3) measurement of odors, (4) the movement of odorous gases, (5) strategies for odor control, (6) odor control methods, and (7) the design of odor control facilities.

Types of Odors

For humans, the importance of odors at low concentrations is related primarily to the psychological stress the odors cause, rather than to the harm they do to the body. The principal types of odors encountered in wastewater management facilities are reported in Table 16–3. With few exceptions, odorous compounds typically contain either sulfur or nitrogen. The characteristic odor of organic compounds containing sulfur is that of decayed organic material. Of the odorous compounds reported in Table 16–3, the rotten egg smell of hydrogen sulfide is the odor encountered most commonly in wastewater management facilities. As noted in Chap. 2, gas chromatography has been used successfully for the identification of specific compounds responsible for odors. Unfortunately, this technique has not proved as successful in the detection and quantification of odors derived from wastewater collection, treatment, and disposal facilities, because of the many compounds that may be involved. It should be noted that at higher concentrations, many of the odorous gases (e.g., hydrogen sulfide) can, depending on exposure, be lethal.

Sources of Odors

The principal sources of odors in wastewater management facilities and the relative potential for release of odor are presented in Table 16–4. Minimization of odors from these sources is the concern of odor management.

Wastewater Collection Systems. The principal sources of odorous compounds in collection systems are from (1) the biological conversion, under anaerobic conditions, of organic matter containing nitrogen and sulfur, and (2) the discharge of industrial wastewater that may contain odorous compounds or compounds that may react with compounds in the wastewater to produce odorous compounds. Odorous gases released to the sewer atmosphere can accumulate and be released at air release valves, cleanouts, access ports (i.e., manholes), and house vents.

Table 16–3

Odor thresholds of odorous compounds and their characteristics associated with wastewater management

Odorous compound	Chemical formula	Molecular weight	Odor threshold, ppm,[a]	Characteristic odor
Ammonia	NH_3	17.0	46.8	Pungent, irritating
Chlorine	Cl_2	71.0	0.314	Pungent, suffocating
Chlorophenol	ClC_6H_4OH	128.51	0.00018	Medicinal odor
Crotyl mercaptan	$CH_3\text{-}CH{=}CH\text{-}CH_2\text{-}SH$	90.19	0.000029	Skunk like
Dimethyl sulfide	$CH_3\text{-}S\text{-}CH_3$	62	0.0001	Decayed cabbage
Diphenyl sulfide	$(C_6H_5)_2S$	186	0.0047	Unpleasant
Ethyl mercaptan	$CH_3CH_2\text{-}SH$	62	0.00019	Decayed cabbage
Ethyl sulfide	$(C_2H_5)_2SH$	91.9	0.000025	Nauseating odor
Hydrogen sulfide	H_2S	34	0.00047	Rotten eggs
Indole	C_8H_6NH	117	0.0001	Fecal, nauseating
Methyl amine	CH_3NH_2	31	21.0	Putrid, fishy
Methyl mercaptan	CH_3SH	48	0.0021	Decayed cabbage
Skatole	C_9H_9NH	132	0.019	Fecal odor, nauseating
Sulfur dioxide	SO_2	64.07	0.009	Pungent, irritating
Thiocresol	$CH_3\text{-}C_6H_4\text{-}SH$	124	0.000062	Skunk like, irritating
Trimethyl amine	$(CH_3)_3N$	59	0.0004	Pungent, fishy

[a] Parts per million by volume.

Table 16–4

Sources of odor in wastewater management systems[a]

Location	Source/cause	Odor potential
Wastewater collection system		
Air release valves	Accumulation of odorous gases released from wastewater	High
Cleanouts	Accumulation of odorous gases released from wastewater	High
Access ports (manholes)	Accumulation of odorous gases released from wastewater	High
Industrial wastewater discharges	Odorous compounds may be discharged to wastewater collection system	
Raw wastewater pumping station	Wetwell/septic raw wastewater, solids and scum deposits	High
Wastewater treatment facilities		
Headworks	Release odorous gases generated in the wastewater collection system due to turbulence in hydraulic channels and transfer points	High
Screening facilities	Putrescible matter removed by screening	High
Preaeration	Release of odorous compounds generated in wastewater collection system	High

(continued)

| **Table 16-4** (Continued)

Location	Source/cause	Odor potential
Grit removal	Organic matter removed with grit	High
Flow equalization basins	Basin surfaces/septic conditions due to accumulation of scum and solids deposits	High
Septage receiving and handling facilities	Odorous compounds can be released at septage receiving stations, especially when septage is being transferred	High
Sidestream returns[b]	Return flows from biosolids processing facilities	High
Primary clarifiers	Effluent weirs and troughs/turbulence that releases odorous gases. Scum-either floating or accumulated on weirs and baffles/putrescible matter. Floating sludge/septic conditions	High/moderate
Fixed film processes (trickling filters or RBCs)	Biological film/septicity due to insufficient oxygen, high organic loading, or plugging of trickling filter medium; turbulence causing release of odorous material	Moderate/high
Aeration basins	Mixed liquor/septic return sludge, odorous sidestream flows, high organic loading, poor mixing, inadequate DO, solids deposits	Low/moderate
Secondary clarifiers	Floating solids/excessive solids retention	Low/moderate
Sludge and biosolids facilities		
Thickeners, solids holding tanks	Floating solids; weirs and troughs/scum and solids septicity due to long holding periods, solids deposits, and temperature increases; odor release by turbulence	High/moderate
Aerobic digestion	Incomplete mixing in reactor	Low/moderate
Anaerobic digestion	Leaking hydrogen sulfide gas/upset conditions, high sulfate content in solids	Moderate/high
Sludge storage basins	Lack of mixing, formation of scum layer	Moderate/high
Mechanical dewatering by belt filter press, recessed plate filter press, or centrifuge	Cake solids/putrescible matter; chemical addition, ammonia release	Moderate/high
Sludge loadout facilities	Release of odors during the transfer of biosolids from storage to transfer facilities	High
Composting facilities	Composting solids/insufficient aeration, inadequate ventilation	High
Alkaline stabilization	Stabilized solids/ammonia generation resulting from reaction with lime	Moderate
Incineration	Air emissions/combustion temperature is not high enough to destroy all organic substances	Low
Sludge drying beds	Drying solids/excess putrescible matter due to insufficient stabilization	Moderate/high

[a] Adapted in part from WEF (1996a).

[b] Sidestreams could include digester decant, dewatering return flows, or backwash water (see Chap. 15).

Figure 16-1

Typical examples of odor management for preliminary treatment processes: (a) bar screen in enclosed building with odor control facilities and (b) enclosed grit processing facilities.

 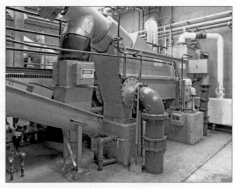

(a)   (b)

Wastewater Treatment Facilities.   In considering the potential for the generation and release of odors from treatment plants, it is common practice to consider the liquid and solids processing facilities separately. The headworks and preliminary treatment operations have the highest potential for release of odor, especially for treatment plants that have long collection systems where anaerobic conditions can develop (see Fig. 16–1). Sidestream discharges including return flows from filter backwashing and from sludge and biosolids processing facilities are often a major source of odors, especially where these flows are allowed to discharge freely into a control structure or mixing chamber.

Sludge and Biosolids Handling Facilities.   Typically, the most significant sources of odors at wastewater treatment plants are sludge thickening facilities, anaerobic digesters, and sludge loadout facilities. The highest potential for odor release occurs when unstabilized sludge is handled (e.g., turned, spread, or stored).

One of the major contributors to odor in solids processing and an important item to consider in treatment plant design is shear. Shear is the cutting or tearing of solids by shear stress. When solids undergo mixing by either high shear dewatering or conveyance equipment, particle size reduction occurs, and odor production increases. Solids that exit a dewatering facility can be sheared enough to release odors. The major mechanism appears to be the release of proteinaceous biopolymer. Once released, these proteins are degraded, liberating a number of odorous compounds, but mostly mercaptans. The increase in solution protein also makes dewatering more difficult. The solution proteins can be "coagulated" by addition of polymer, but the synthetic polymers are degraded and the protein becomes degradable. The synthetic polymer can also generate methylamines when degraded (Novak, 2001; Murthy, 2001).

Trimethylamine (TMA) is present in the liquid phase in many anaerobically digested sludges. Trimethylamine, like ammonia, is soluble below pH 9, but above this pH level is a gas, which can be released into the air. Adding lime to digested sludge for odor control may, in fact, enhance the release of odors by converting TMA to a gas (Novak, 2001; Murthy, 2001). Some plants may be unable to land apply dewatered sludge because of the increased odor production. Thus, in evaluating processing and disposal options, the ramifications of odor generation and control have to be evaluated carefully.

Measurement of Odors

The sensory (organoleptic) measurement of odors by the human olfactory system is used most often to detect odors emanating from wastewater treatment facilities. Detection of odors by the human olfactory system and instrumental methods is described in Chap. 2.

Although there are several different ways of assessing the impact of odors, the primary method at wastewater treatment facilities is dilution-to-threshold ratio (D/T). The D/T ratio is a measure of the number of dilutions of fresh air needed to render the odorous ambient air nondetectable. The higher the D/T value, the greater amount of fresh air needed to render the odor nondetectable. The D/T values, determined by an odor panel, are used as a regulatory standard for assessing off-site odor impacts. Typical D/T values vary from 1 to 50, with a median value of 5. The location where D/T impacts are measured varies with the governing jurisdiction, but typically is at the location of the receptor of odor, e.g., property fence line, residence, community, park, etc. Modeling, discussed subsequently, is used to predict the D/T values at the receptor location using source-specific D/T values.

There are two challenges with the measurement of odors; one challenge is that the detection threshold varies between individuals, and the second challenge is that odorous compounds tend to be mixed together, such as hydrogen sulfide and methyl mercaptan. The first challenge can be overcome through the use of standard methods for measuring odor, such as ASTM E679–04 where a specialized device presents different D/T mixtures to subjects for testing. The second challenge can only be overcome through the use of an odor panel, as the impact of individual odors cannot be summed. Typical probability distributions of threshold odor levels for different odorous compounds, as reported in the literature, are illustrated on Fig. 16–2. As shown on Fig. 16–2, the range of threshold odor values can vary by as much as seven orders of magnitude.

Odor Dispersion Modeling

Odor dispersion modeling is used to assess the impact from an odor source at the receptor location and the type and the extent of the odor management facilities must be implemented. Typically, the odor dispersion model is run over a set time period, and a probability of exceeding a given D/T value at the receptor location is estimated. Input to the model must be obtained from the emission source, and usually these samples must be sent for analysis by an odor panel. The model run time corresponds to the amount of time a receptor can be exposed to an odor. Average expose times range between 3 and 60 min. Depending on the governing jurisdiction or agency, the odor D/T value at the receptor location must be lower than the established regulatory D/T value (e.g., 5), 98 to 100 percent of the time.

Movement of Odors from Wastewater Treatment Facilities

Under quiescent meteorological conditions (i.e., calm winds and low atmospheric mixing), odorous gases that develop at treatment facilities tend to hover over the point of generation (e.g., sludge thickening facilities, sludge storage lagoons), because the odorous gases are more dense than air. Depending on the local meteorological conditions, it has been observed that odors may be measured at undiluted concentrations at great distances from the point of generation. The following events appear to happen: (1) in the evening or early morning hours, under quiescent meteorological conditions, a cloud of odors will develop over the wastewater treatment unit prone to the release of odors; and (2) the concentrated cloud of odors can then be transported (i.e., pushed along), without breaking up, over great distances by the weak evening or early morning breezes, as they develop. In some cases, odors have been detected at distances of up to 25 km from their source. This transport phenomenon has been termed the puff movement of odors (Tchobanoglous and Schroeder, 1985). The puff movement of odors was first described by Wilson (1975). Air dispersion modeling can be used to predict whether quiescent conditions will tend to persist over prolonged periods of time. The most common method used to mitigate the effects of the

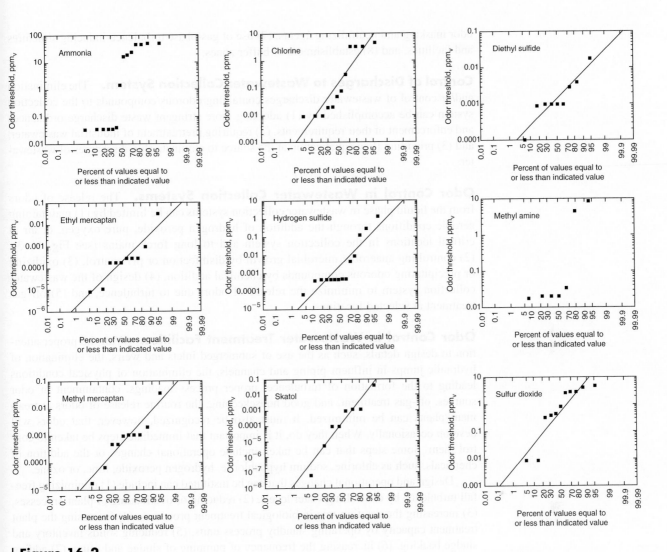

Figure 16–2

Log probability plots of odor threshhold odor values reported in the literature for a variety of odorous compounds.

odor puff is to install barriers to induce turbulence, thus breaking up and dispersing the cloud of concentrated odors, and/or to use wind generators to maintain a minimum velocity across the source.

Strategies for Odor Management

Strategies for the management and control of odors are presented and discussed below. An overview of some of the methods used to control and treat odorous gases is presented in the following section. Where chronic odor problems occur at treatment facilities, approaches to solving these problems may include (1) control of odor-causing wastewaters discharged to the collection system and treatment plant that creates odor problems, (2) control of odors generated in the wastewater collection system, (3) control of odors generated in wastewater treatment facilities, (4) installation of odor containment and treatment facilities, (5) application of chemicals to the liquid (wastewater) phase, (6) use of odor masking and neutralizing agents, (7) use of gas-phase turbulence-inducing structures and facilities, and (8) establishment of buffer zones.

Control of Discharges to Wastewater Collection System.   The elimination and/or control of wastewater discharges containing odorous compounds to the collection system can be accomplished by (1) adopting more stringent waste discharge ordinances and enforcement of their requirements, (2) requiring pretreatment of industrial wastewater, and (3) providing flow equalization at the source to eliminate slug discharges of wastewater.

Odor Control in Wastewater Collection Systems.   The release of odors from the liquid phase in wastewater collection systems can be limited by (1) maintaining aerobic conditions through the addition of hydrogen peroxide, pure oxygen, or air at critical locations in the collection system and to long force mains (see Fig. 16–3), (2) controlling anaerobic microbial growth by disinfection or pH control, (3) oxidizing or precipitating odorous compounds by chemical addition, (4) design of the wastewater collection system to minimize the release of odors due to turbulence, and (5) off-gas treatment at selected locations.

Odor Control in Wastewater Treatment Facilities.   With the proper attention to design details, such as the use of submerged inlets and weirs, the elimination of hydraulic jumps in influent piping and channels, the elimination of physical conditions leading to the formation of turbulence, proper process loadings, containment of odor sources, off-gas treatment, and good housekeeping, the routine release of odors at treatment plants can be minimized. It must also be recognized, however, that odors will develop occasionally. When they do, it is important that immediate steps be taken to control them. Some steps that can be taken include operational changes or the addition of chemicals, such as chlorine, sodium hypochlorite, hydrogen peroxide, lime, or ozone.

Design and operational changes that can be instituted can include (1) minimizing free-fall turbulence by controlling water levels, (2) reducing of overloading of plant processes, (3) increasing the aeration rate in biological treatment processes, (4) increasing the plant treatment capacity by operating standby process units, (5) reducing solids inventory and sludge backlog, (6) increasing the frequency of pumping of sludge and scum, (7) adding chlorinated dilution water to sludge thickeners, (8) controlling the release of aerosols, (9) increasing the frequency of disposal of grit and screenings, (10) cleaning odorous accumulations more frequently, and (11) containment, ventilation, and treatment of odorous gases.

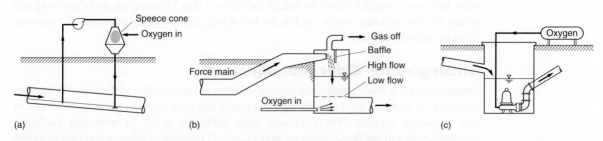

Figure 16–3

Typical uses of commercial oxygen in wastewater collection systems for odor control: (a) sidestream oxygenation and reinjection of wastewater into a gravity sewer, (b) Injection of oxygen into a hydraulic fall and (c) Injection of oxygen into two-phase flow in force main. (From Speece et al., 1990.)

Figure 16–4

Typical odor containment facilities at wastewater treatment plants: (a) and (b) covered primary sedimentation tanks, (c) covered trickling filter, (d) cover over the solids contact portion of a trickling filter solids contact biological treatment process, (e) covered primary sludge fermentor, and (f) view inside of an enclosed sludge thickener.

(a)   (b)

(c)   (d)

(e)   (f)

Odor Containment.   Odor containment includes the installation of covers, collection hoods, and air handling equipment for containing and directing odorous gases to disposal or treatment systems. In cases where the treatment facilities are close to developed areas, it has become common practice to enclose or cover treatment units such as the bar screens and grit processing units [see Fig. 16–1(a) and (b)], primary clarifiers [see Figs. 16–4(a) and (b)], trickling filters [see Fig. 16–4(c)], (d) biological treatment processes [see Fig. 16–4(d)], (e) sludge fermentors [see Fig. 16–4(e)], sludge thickeners [see Fig. 16–4(f)], sludge-processing facilities, and sludge-loadout facilities. Where covers are used, the trapped gases must be collected and treated. The specific method of treatment will depend on the characteristics of the odorous compounds. Typical containment alternatives for the control of odor emissions from wastewater management facilities are reported in Table 16–5.

Chemical Additions to Wastewater for Odor Control.   Odors can be eliminated in the liquid phase through the addition of a variety of chemicals to achieve (1) chemical oxidation, (2) chemical precipitation, and (3) pH control. The most common oxidizing chemicals that can be added to wastewater include oxygen, air, chlorine, sodium

Table 16–5

Odor containment and process alternatives for the control of odors the emission from wastewater management facilities

Source	Suggested control strategies
Wastewater sewers	Seal existing access ports (i.e., manholes). Eliminate the use of structures that create turbulence and enhance volatilization.
Sewer appurtenances	Isolate and cover existing appurtenances.
Pump stations	Vent odorous gases from wet-well to treatment unit. Use variable-speed pumps to reduce the size of the wet-well.
Bar racks	Cover existing units. Reduce headloss through bar racks.
Comminutors	Cover existing units. Use inline enclosed comminutors.
Parshall flume	Cover existing units. Use alternative measuring device.
Grit chamber	Cover existing aerated grit chambers. Reduce turbulence in conventional horizontal-flow grit chambers; cover if necessary. Avoid the use of aerated grit chambers.
Equalization basins	Cover existing units. Use submerged mixers, and reduce air flow.
Primary and secondary sedimentation tanks	Cover existing units [see Figs. 16–4(a) and (b)]. Replace conventional overflow weirs with submerged weirs.
Biological treatment	Cover existing units. Use submerged mixers and reduce aeration rate.
Transfer channels	Use enclosed transfer channels.

hypochlorite, potassium permanganate, hydrogen peroxide, and ozone. While all of these compounds will oxidize hydrogen sulfide (H_2S) and other odorous compounds, their use is complicated by the chemical matrix in which the odorous gases exist. The only way to establish the required chemical dosages for the removal of chemical compounds is through bench- or pilot-scale testing.

Odorous compounds can also be reduced by precipitation. For example, ferrous chloride and ferrous sulfate can be used for the control of H_2S odors by precipitation of the sulfide ion as ferrous sulfide. As with the oxidation reactions, the required chemical dosage can be determined only through bench- or pilot-scale testing. The release of H_2S can also be controlled by increasing the pH value of the wastewater. Increasing the pH of the wastewater results in reduced bacterial activity and also shifts the equilibrium so that the sulfide ion is present as HS^-. With most of the odor control methods involving the addition of chemicals to wastewater, some residual product is formed that must ultimately be dealt with. Shock treatment involving the addition of sodium hydroxide (NaOH) can be used to reduce microbial slimes in sewers. The high pH also reduces sulfide (S^{2-}) formation. Additional details on chemical addition may be found in Rafson (1998).

Use of Odor Masking and Neutralization. On occasion, chemicals have been added to wastewater or offgases to mask an offensive odor with a less offensive odor. Masking chemicals are based on essential oils with the most common aromas being vanilla, citrus, pine, or floral (Williams, 1996). Typically, enough masking chemical is added to wastewater to overpower the offensive odor. Masking chemicals, however, do not modify or neutralize the offensive odors. Neutralization involves finding chemical compounds that can be combined with the odorous gases in the vapor state so that the

Figure 16–5

High barrier fence placed around sludge holding lagoons to induce air turbulence and mixing, and thus limit the release of odors off the treatment plant site (a) aerial view of sludge storage basins (coordinates N 38.439, W121,480) and (b) view of high barrier fence.

(a)   (b)

combined gases cancel each other's odor, produce an odor of lower intensity, or eliminate the odorous compounds. Although odor masking and neutralization are viable options for short-term management of odor problems, the key to long-term odor management is to identify the source of the odors and implement corrective measures.

Use of Turbulence-Inducing Structures and Facilities for Odor Dispersion. In a number of wastewater treatment plants, physical facilities used to induce atmospheric turbulence have been constructed specifically for the purpose of gas-phase odor reduction. The high barrier fence [3.7 m (12 ft)] shown on Fig. 16–5 surrounds sludge-storage lagoons. Operationally, any odorous gases that develop under quiescent conditions over the lagoons are diluted as they move away from the storage lagoons, due to the local turbulence induced by the barrier. Trees are also used commonly to dilute odorous gases by inducing turbulence (i.e., the formation of eddies) and mixing. Trees are also known to help purify the air as a result of respirometric activity.

Use of Buffer Zones. The use of buffer zones can also help in reducing the impact of odors on developed areas. Typical buffer zone distances used by regulatory agencies are presented in Table 16–6. If buffer zones are used, odor studies should be conducted that identify the type and magnitude of the odor source, meteorological conditions, dispersion characteristics, and type of adjacent development. Trees that grow rapidly are often planted at the periphery of the buffer zones to further reduce the impact of odors.

Odor Treatment Methods

The general classification of odor treatment methods is presented in Table 16–7, along with typical applications in wastewater management. Odor treatment methods are designed either to treat the odor producing compounds in the wastewater stream or to treat the foul air. Most of the methods in Table 16–7 are meant to be used to treat the foul air (i.e., the gas phase). As noted above, to control the release of odorous gases from treatment facilities, it has become more common to cover wastewater treatment processes (see Fig. 16–4).

The principal methods used to treat odorous gases include the use of (1) chemical scrubbers, (2) activated-carbon absorbers, (3) vapor-phase biological treatment processes (i.e., compost filters), (4) treatment in conventional biological treatment processes, and (5) thermal processes. Each of these methods is discussed below. The specific method of odor control and treatment that should be applied will vary with local conditions. However, because odor control measures are expensive, the cost of making process changes or

Table 16–6

Methods to control and treat odorous gases found in wastewater management systems[a]

Method	Description and/or application
Physical methods	
Adsorption on activated carbon	Odorous gases can be passed through beds of activated carbon to remove odors. Carbon regeneration can be used to reduce costs. Additional details may be found in Chap. 11.
Adsorption on sand, soil, or compost beds	Odorous gases can be passed through sand, soil, or compost beds. Odorous gases from pumping stations may be vented to the surrounding soils or to specially designed beds containing sand or soils. Odorous gases collected from treatment units may be passed through compost beds.
Combustion	Gaseous odors can be eliminated by combustion at temperatures varying from 650 to 815°C (1200 to 1500°F). Gases can be combusted in conjunction with treatment plant solids or separately in a fume incinerator.
Containment	Installation of covers, collection hoods, and air handling equipment for containing and directing odorous gases to disposal or treatment systems.
Dilution with odor-free air	Gases can be mixed with fresh air sources to reduce the odor unit values. Alternatively, gases can be discharged through tall stacks to achieve atmospheric dilution and dispersion.
Masking agents	Perfume scents can be sprayed in fine mists near offending process units to overpower or mask objectionable odors. In some cases, the odor of the masking agent is worse than the original odor. Effectiveness of masking agents is limited.
Oxygen injection	The injection of oxygen (either air or pure oxygen) into the wastewater to control the development of anaerobic conditions has proven to be effective.
Scrubbing towers	Odorous gases can be passed through specially designed scrubbing towers to remove odors. Some type of chemical or biological agent is usually used in conjunction with the tower.
Thermal oxidation	Combustion of off-gases at temperatures from 800 to 1400°C will eliminate odors. Lower temperatures (400 to 800°C) are used with catalytic incineration.
Turbulence inducing facilities	Use of wind breaks, such as high fences and trees, and propeller fans.
Chemical oxidation	Oxidizing the odor compounds in wastewater is one of the most common methods used to achieve odor control. Chlorine, ozone, hydrogen peroxide, and potassium permanganate are among the oxidants that have been used. Chlorine also limits the development of a slime layer.
Chemical precipitation	Chemical precipitation refers to the precipitation of sulfide with metallic salts, especially iron.
Scrubbing with various alkalies	Odorous gases can be passed through specially designed scrubbing towers to remove odors. If the level of carbon dioxide is high, costs may be prohibitive.
Biological methods	
Activated sludge aeration tanks	Odorous gases can be combined with the process air for activated sludge aeration tanks to remove odorous compounds.
Biological conversion	Biological processes in the wastewater can reduce odors by converting malodorous constituents through oxidation.
Biological stripping towers	Specially designed towers can be used to strip odorous compounds. Typically, the towers are filled with plastic packing of various types on which biological growths can be maintained.
Compost filters	Gases can be passed through biologically active beds of compost to remove odors.
Sand and soil filters	Gases can be passed through biologically active beds of compost to remove odors.
Trickling filters	Odorous gases can be passed through existing trickling filters to remove odorous compounds.

[a] Adapted in part from U.S. EPA (1985).

Table 16–7

Suggested minimum buffer distances from treatment units for odor containment[a,b]

Treatment process unit	Buffer distance	
	ft	m
Sedimentation tank	400	125
Trickling filter	400	125
Aeration tank	500	150
Aerated lagoon	1000	300
Sludge digester (aerobic or anaerobic)	500	150
Sludge handling units	1000	300
Open drying beds	500	150
Covered drying beds	400	125
Sludge holding tank	1000	300
Sludge thickening tank	1000	300
Vacuum filter	500	150
Wet air oxidation	1500	450
Effluent recharge bed	800	250
Secondary effluent filters		
Open	500	150
Enclosed	200	75
Advanced wastewater treatment		
Tertiary effluent filters		
Open	300	100
Enclosed	200	75
Denitrification	300	100
Polishing lagoon	500	150
Land disposal	500	150

[a] Source: New York State Department of Environmental Conservation.
[b] Actual buffer distance requirements will depend on local conditions.

modifications to the facilities to eliminate odor development should always be evaluated and compared to the cost of various alternative odor control measures before their adoption is suggested.

Chemical Scrubbers. The basic design objective of a chemical scrubber is to provide contact between air, water, and chemicals (if used) to provide oxidation or entrainment of the odorous compounds. The principal wet scrubber types, as shown on Fig. 16–6, include single-stage countercurrent packed towers, countercurrent spray chamber absorbers, and cross-flow scrubbers. In most single-stage scrubbers, such as shown on Fig. 16–7, the scrubbing fluid (usually sodium hypochlorite) is recirculated. The commonly used oxidizing scrubbing liquids are sodium hypochlorite, potassium permanganate, and hydrogen peroxide solutions. Because of safety and handling issues, chlorine gas is not used commonly in scrubbing applications at wastewater treatment facilities. Sodium hydroxide is also used in scrubbers where H_2S concentrations in the gas phase are high.

Figure 16-6

Typical wet scrubbers systems for odor control: (a) countercurrent packed tower, (b) spray chamber absorber, and (c) cross-flow scrubber.

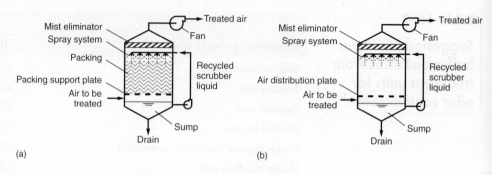

(a)

(b)

(c)

Figure 16-7

Typical sodium hypochlorite scrubber used to treat the odors from the trickling filters shown on Fig. 16-4(c).

Chemical Scrubbing Reactions for Hydrogen Sulfide. Typical simplified scrubbing reactions for H$_2$S with chlorine, sodium hypochlorite, potassium permanganate, and hydrogen peroxide are as follows:

With chlorine

$$H_2S + 4Cl_2 + 4H_2O \rightarrow H_2SO_4 + 8HCl \tag{16-1}$$
(34.06) (4 × 70.91)

$$H_2S + Cl_2 \rightarrow S°\downarrow + 2HCl \tag{16-2}$$
(34.06) (70.91)

With sodium hypochlorite

$$H_2S + 4NaOCl + 2NaOH \rightarrow Na_2SO_4 + 2H_2O + 4NaCl \tag{16-3}$$
(34.06) (4 × 74.45)

$$H_2S + NaOCl \rightarrow S°\downarrow + NaCl + H_2O \tag{16-4}$$
(34.06) (74.45)

With potassium permanganate

$$3H_2S + 2KMnO_4 \rightarrow 3S + 2KOH + 2MnO_2 + 2H_2O \text{ (acidic pH)} \tag{16-5}$$
(3 × 34.06) (2 × 142.04)

$$3H_2S + 8KMnO_4 \rightarrow 3K_2SO_4 + 2KOH + 8MnO_2 + 2H_2O \text{ (basic pH)} \tag{16-6}$$
(3 × 34.06) (8 × 142.04)

With hydrogen peroxide

$$H_2S + H_2O_2 \rightarrow S°\downarrow + 2H_2O \text{ (pH < 8.5)} \tag{16-7}$$
(34.06) (34.0)

In the reaction given by Eq. (16–3), 8.74 mg/L of sodium hypochlorite is required per mg/L of hydrogen sulfide, or 9.29 mg/L if the hydrogen sulfide is expressed as sulfide. In addition, in the reaction given by Eq. (16–3), 2.35 mg/L of sodium hydroxide (caustic) will be required per mg/L of hydrogen sulfide to make up for alkalinity consumed in the reaction. In practice, the required sodium hypochlorite dosage for the reaction given by Eq. (16–3) will vary from 8 to 10 mg/L per mg/L of H_2S. For the reaction given by Eq. (16–4), 2.19 mg/L of sodium hypochlorite is required per mg/L of hydrogen sulfide.

When potassium permanganate is used, the reactions that occur are typically various combinations of the reactions given by Eqs. (16–5) and (16–6). Reaction products that can occur, depending on the local wastewater chemistry, include elemental sulfur, sulfate, thionates, dithionates, and manganese sulfide. Stoichiometrically, about 2.8 and 11.1 mg/L of $KMnO_4$ are required for each mg/L of H_2S oxidized as given by Eqs. (16–5) and (16–6), respectively. However, based on operating data from actual field installations, about 6 to 7 mg/L $KMnO_4$ are required for each mg/L of H_2S oxidized. Potassium permanganate is generally used in smaller installations because of the cost (U.S. EPA, 1985; WEF, 1995).

In the reaction given in Eq. (16–7), 1.0 mg/L of hydrogen peroxide is required for each mg/L of sulfide expressed as hydrogen sulfide. In practice, the required dosage can vary from 1 to 4 mg/L per mg/L of H_2S. Because the systems used to carry out the reactions defined by Eqs. (16–1) through (16–7) are complex, especially where competing reactions may occur, the proper dosage should be established by site-specific testing.

Hypochlorite scrubbers can be expected to remove oxidizable odorous gases when other gas concentrations are minimal. Typical removal efficiencies for single-stage scrubbers are reported in Table 16–8. In cases where the concentrations of odorous components in the exhaust gas from the scrubbers are above desirable levels, multistage scrubbers (see Fig. 16–8) are often used. In the three-stage scrubber shown on Fig. 16–8, the first stage is a pretreatment stage used to raise the pH so that a portion of the odorous gases (e.g., hydrogen sulfide) is reduced before treatment with chlorine in the second and third stages. The reaction that occurs in the first stage of a three-stage unit can be represented as follows.

$$H_2S + 2NaOH \rightarrow Na_2S + 2H_2O \tag{16–8}$$

To reduce maintenance problems due to precipitation, it is recommended that a low hardness (less than 50 mg/L as $CaCO_3$) be used for the makeup water.

Chemical Scrubbing of Ammonia and/or Amine Compounds. In situations where there is a potential for the release of high concentrations of ammonia or amine compounds, such as from solids handling and lime stabilization processes, an additional tower may be needed. The removal of ammonia or amine compounds is accomplished with an

Table 16–8

Effectiveness of hypochlorite wet scrubbers for removal of several odorous gases[a]

Gas	Expected removal efficiency, %	
	Range	Typical
Hydrogen sulfide	90–99	98
Ammonia	90–99	98
Sulfur dioxide	90–96	95
Mercaptans	85–92	90
Other oxidizable compounds	70–90	85

[a] Adapted in part from U.S. EPA (1985).

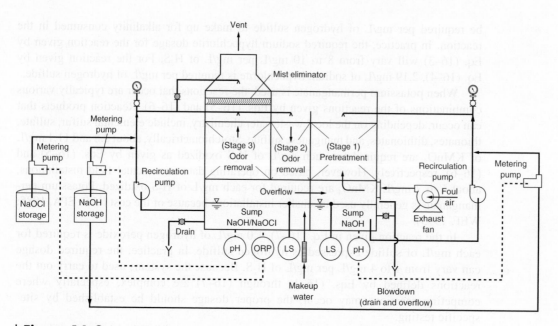

Figure 16-8

Three-stage odor control process flow diagram. (From Lo/Pro Systems, Inc.)

acid/base reaction with sulfuric acid. Typically, the pH of recirculating sulfuric acid solutions is around 4.0 to 6.0 and the following reaction occurs:

$$2NH_3 + H_2SO_4 \rightarrow (NH_4)_2SO_4 \qquad (16\text{-}9)$$

Some of the spent scrubbing liquid that contains the solid ammonium sulfate is discharged from the scrubber. The waste spent scrubbing fluid is typically returned to the treatment plant headworks. The remaining scrubbing liquid is recirculated back into the tower with additional sulfuric acid added to maintain the proper pH of the scrubbing liquid.

Activated Carbon Adsorbers. Activated carbon adsorbers are used commonly for odor control (see Fig. 16–9). The rate of adsorption for different constituents or compounds will depend on the nature of the constituents or compounds being adsorbed (nonpolar versus polar). It has also been found that the removal of odors depends on the

Figure 16-9

Use of activated carbon for odor control: (a) schematic of typical downflow activated carbon reactor and (b) view of multiple activated carbon odor control reactors.

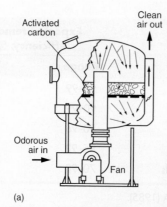

(a)   (b)

concentration of the hydrocarbons in the odorous gas. Typically, hydrocarbons are adsorbed preferentially before polar compounds such as H_2S are removed (note activated carbon is nonpolar). Thus, the composition of the odorous gases to be treated must be known if activated carbon is to be used effectively. Because the life of a carbon bed is limited, carbon must be regenerated (see Sec. 11–7 in Chap. 11) or replaced regularly for continued odor removal. To prolong the life of the carbon, two-stage systems have been used, with the first stage being a wet scrubber followed by activated-carbon adsorption.

Vapor-Phase Biological Treatment Processes. The two principal biological processes used for the treatment of odorous gases present in the vapor phase are (1) biofilters and (2) biotrickling filters (Eweis et al., 1998). The use of microbial growths for the treatment of odors was the subject of an early patent by Pomeroy (1957), one of the important early researchers in the area of odor management in wastewater collection and treatment facilities.

Biofilters. Biofilters are packed-bed filters. In open biofilters [see Fig. 16–10(a)], the gases to be treated move upward through the filter bed. In closed biofilters [see Fig. 16–10(b)], the gases to be treated are either blown or drawn through the packing material. As the odorous gases move through the packing in the biofilter, two processes occur simultaneously: sorption (i.e., absorption/adsorption) and bioconversion. Odorous gases are absorbed into the moist surface biofilm layer and the surfaces of the biofilter packing material. Microorganisms, principally bacteria, actinomycetes, and fungi, attached to the packing material, oxidize the absorbed/adsorbed gases and renew the treatment capacity of the packing material. Moisture content and temperature are important environmental conditions that must be maintained to optimize microorganism activity (Williams and Miller, 1992a, 1992b; Yang and Allen, 1994; Eweis et al., 1998). Although compost biofilters are used commonly, one drawback is the large surface area (footprint) required for these units.

Figure 16–10

Typical packed bed biofilters:
(a) open bed type and
(b) enclosed reactor type.

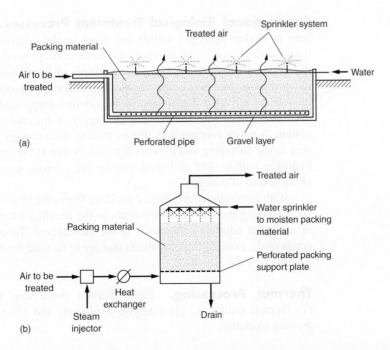

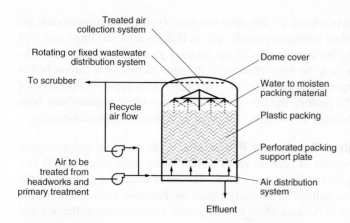

Figure 16–11

Typical covered biological stripping tower used for odor control (a) schematic and (b) view of multiple biological stripping towers.

Biotrickling Filters. Biotrickling filters are essentially the same as biofilters with the exception that moisture is provided continuously or intermittently by applying (typically spraying) a liquid (e.g., treated effluent) over the packing (see Fig. 16–11). The liquid is recirculated and nutrients are often added. Because water is lost in the gas leaving the filter, makeup water must be provided. Similarly, because of the accumulation of salts in the recycled water, a blowdown stream is required. Compost is not a suitable packing material for biotrickling filters because water will accumulate within the compost, thereby limiting the free movement of air within the filter. Typical packing materials include Pall rings, Raschig rings, lava rock, and granular activated carbon (Eweis et al., 1998; see also Sec. 11–8, Gas Stripping, in Chap. 11).

Conventional Biological Treatment Processes. The ability of microorganisms to oxidize hydrogen sulfide and other similar odorous compounds dissolved in the liquid under aerobic conditions is the basic concept used for the treatment of odors in liquid-based systems. The two principal types of conventional liquid-based systems used in wastewater treatment plants are the activated sludge process and the trickling filter process. In the activated sludge process, the odorous compounds are introduced into the aeration basin either with the existing air supply or injected separately through a manifold system. A major concern with this method of odor management is the high rate of corrosion in the air piping and blowers that occurs due to the presence of moist air containing hydrogen sulfide. The ability to transfer the odorous gaseous compounds to the liquid phase is also of concern.

With conventional uncovered trickling filters the major issues are how to transfer the air containing the odorous compounds to the trickling filter and how to avoid the release of untreated odorous compounds to the atmosphere. To control the release of odorous compounds, existing trickling filters that are to be used for odor control are almost always covered [see Fig. 16–4(c)].

Thermal Processing. Three thermal processing techniques have been used: (1) thermal oxidation, (2) catalytic oxidation, and (3) recuperative and regenerative thermal oxidation.

The Oxidation Process. The oxidation of CH_4 (methane) and H_2S can be used to illustrate the basic principle of all three thermal processes.

$$CH_4(gas) + 2O_2(gas) \rightarrow CO_2(gas) + 2H_2O(vapor) + heat \tag{16–10}$$

$$H_2S(gas) + 2O_2(gas) \rightarrow H_2SO_4(vapor) \tag{16–11}$$

If the gas to be combusted does not liberate enough heat to sustain the combustion process, it is usually necessary to use an external fuel source such as fuel oil, natural gas, or propane. Unfortunately, because of the low concentrations of odorous combustible gases in most waste streams, sustainable thermal oxidation is seldom possible, and large amounts of fuel are typically required to maintain the combustion temperatures needed to eliminate odors.

Thermal Oxidation. Thermal oxidation is used, more commonly, for concentrated waste streams. The flaring of odorous gases is a relatively crude form of thermal combustion [see Fig. 16–12(a)]. Depending on the design of the combustion facility, incomplete combustion can occur due to variations in gas flow. For this method of odor control to be sustainable, the waste gas must typically contain 50 percent of the fuel value of the gas stream to be combusted.

Catalytic oxidation. A flameless oxidation that occurs in the range from 310 to 425°C (600 to 800°F) in the presence of a catalyst is defined as catalytic oxidation [see Fig. 16–12(b)]. Common catalysts include platinum, palladium, and rubidium. The decrease in temperature as compared to complete thermal oxidation reduces the energy requirements significantly. However, because the catalysts can become fouled, the gas to be

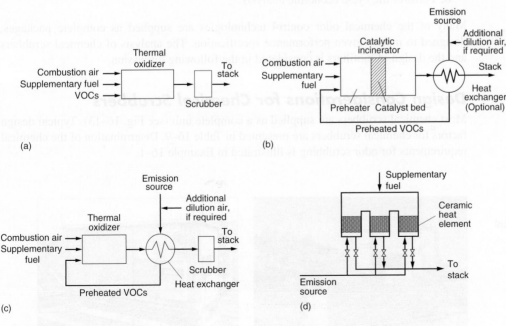

(a)   (b)

(c)   (d)

Figure 16–12

Schematic diagrams of thermal processes for treatment of VOCs: (a) thermal oxidation, (b) catalytical oxidizer, (c) recuperative thermal oxidizer, and (d) regenerating thermal oxidizer.

oxidized must not contain particulate material or constituents that will result in a residue. Additional information on the physical facilities used for thermal processing of VOCs may be found in Sec. 16–4.

Recuperative and Regenerative Thermal Oxidation. This process involves preheating the odorous gases before passing them into the combustion chamber so that complete oxidation can be achieved. Combustion occurs at temperatures in the range from 425 to 760°C (800 to 1400°F). Recuperative and regenerative thermal oxidation processes are used to reduce fuel consumption by preheating the incoming air, especially in large installations. In recuperative oxidizers, thin wall tubes are used to transfer heat recovered from exhaust air to the incoming air [see Fig. 16–12(c)]. In regenerative oxidizers, ceramic packing material is used to capture the heat from the hot exhaust gas and subsequently to release it to the incoming air. To maintain optimal heat recovery, the exhaust and incoming air are cycled through the packing material so that the incoming air is always passed through the hottest packing material. Typically, three stages of packing material are used in regenerative oxidizers [see Fig. 16–12(d)].

Selection and Design of Odor Control Facilities

The following steps are involved in the selection and design of odor control and treatment facilities:

1. Determine the characteristics and volumes of the gas to be treated.
2. Define the exhaust requirements for the treated gas.
3. Evaluate climatic and atmospheric conditions.
4. Select one or more odor control and treatment technologies to be evaluated.
5. Conduct pilot tests to determine design criteria and performance.
6. Perform life cycle economic analysis.

Many of the chemical odor control technologies are supplied as complete packages, designed to meet a given performance specification. The analysis of chemical scrubbers and the design of biofilters is considered in the following discussion.

Design Considerations for Chemical Scrubbers

Most chemical scrubbers are supplied as a complete unit (see Fig. 16–13). Typical design factors for chemical scrubbers are presented in Table 16–9. Determination of the chemical requirements for odor scrubbing is illustrated in Example 16–1.

Figure 16–13

Typical self contained chemical odor stripping unit.

(a)   (b)

Table 16–9

Typical design factors for chemical scrubbers[a]

Item	Unit	Value
Packing depth	m	1.8–3
Gas residence time in packing	s	1.3–2.0
Scrubbant flowrate	kg H_2O/kg air flow	1.5–2.5
	L/s per m^3/s air flow	2–3
Makeup water flow	L/s per kg sulfide at pH 11	0.075
	L/s per kg sulfide at pH 12.5	0.004
pH	unitless	11–12.5
Temperature	°C	15–40
Caustic usage	kg NaOH/kg sulfide	2–3

[a] Adapted in part from from WEF (1995), Devinny et al. (1999).

EXAMPLE 16–1   **Chemical Requirements for Odor Scrubbing**   Hydrogen sulfide is to be scrubbed from a waste airstream using sodium hypochlorite. Determine the chemical (i.e., sodium hypochlorite and caustic) and water requirements for the following conditions.

1. Waste airstream flowrate = 1000 m^3/min
2. H_2S concentration in waste stream = 20 ppm$_v$ at 20°C
3. Specific weight of air = 0.0118 kN/m^3 at 20°C
4. Density of air = 1.204 kg/m^3 at 20°C (see Sec. B-3 in Appendix B)
5. Assume liquid to gas ratio for scrubber = 1.75
6. Density of 50 percent NaOH solution = 1.52 kg/L

Solution

1. Determine the volume occupied by one mole of a gas at a temperature of 20°C and a pressure of 1.0 atm using Eq. (2–44).

$$V = \frac{nRT}{P}$$

$$V = \frac{(1 \text{ mole})(0.082057 \text{ atm·L/mole·K})[(273.15 + 20)\text{K}]}{1.0 \text{ atm}}$$

$$= 24.055 \text{L} \qquad \text{use } 24.1 \text{L}$$

2. Estimate the sodium hypochlorite requirement.
    a. Using Eq. (2–45), convert the H_2S concentration from ppm$_v$ to g/m^3.

$$20 \text{ ppm}_v = \left(\frac{20 \text{ m}^3}{10^6 \text{ m}^3}\right)\left[\frac{(34.8 \text{ g/mole } H_2S)}{(24.1 \times 10^{-3} \text{ m}^3/\text{mole of } H_2S)}\right]$$

H_2S concentration = 28.3×10^{-3} g/m^3

    b. Determine the amount of H_2S that must be treated per day.

$$(1000 \text{ m}^3/\text{min}) \times (28.3 \times 10^{-3} \text{ g/m}^3)(1440 \text{ min/d})(1 \text{ kg}/10^3 \text{ g}) = 40.8 \text{ kg/d}$$

    c. Estimate the sodium hypochlorite dose. From Eq. (16–3), 8.74 mg/L of sodium hypochlorite are required per mg/L of sulfide, expressed as hydrogen sulfide.

$$NaOCl_2 \text{ required per day} = (40.8 \text{ kg/d})(8.74) = 356.6 \text{ kg/d}$$

3. Estimate the water requirement for the scrubbing tower.
    a. Determine the mass air flowrate.

$$(1000 \text{ m}^3/\text{min})(1.204 \text{ kg/m}^3) = 1204 \text{ kg/min}$$

    b. Determine the water flowrate.

$$(1204 \text{ kg/min})(1.75) = 2107 \text{ kg/min} = 2.1 \text{ m}^3/\text{min}$$

4. Determine the amount of sodium hydroxide (caustic) that must be added to replace the alkalinity consumed in the reaction.
    a. From the reaction given by Eq. (16–3), 2.35 mg/L of NaOH is required for each mg/L of H_2S removed.
    b. Determine the amount of NaOH required.

$$NaOH = (40.8 \text{ kg/d})(2.35) = 95.9 \text{ kg/d}$$

    c. Determine the volume of NaOH required. The amount of caustic per liter at 50% NaOH is

$$NaOH = (1.52 \text{ kg/L})(0.50) = 0.76 \text{ kg/L}$$

$$\text{Volume of NaOH} = \frac{(95.9 \text{ kg/d})}{(0.76 \text{ kg/L})} = 126.2 \text{ L/d}$$

Comment   The water requirement for the scrubbing tower will be specified initially by the scrubber supplier and field adjusted based on the results of pilot-plant studies and past operating experience. If the wastewater has sufficient alkalinity, it may not be necessary to add sodium hydroxide.

Design Considerations for Odor Control Biofilters

Important design considerations for biofilters include (1) the type and composition of the packing material, (2) facilities for gas distribution, (3) maintenance of moisture within the biofilter, and (4) temperature control. Each of these topics is considered below. Design and operational parameters are presented and discussed following the discussion of the above topics. Additional details on biofilters may be found in van Lith (1989), Allen and Yang (1991, 1992), WEF (1995), Eweis et al., (1998), Devinny et al. (1999), and WEF (2004).

Packing Material.   The requirements for the packing used in biofilters are (1) sufficient porosity and near-uniform particle size, (2) particles with large surface areas and significant pH-buffering capacities, and (3) the ability to support a large population of microflora (WEF, 1995). Packing materials used in biofilters include compost, peat, and a variety of synthetic mediums. Although soil and sand have been used in the past, they are less used today because of excessive headloss and clogging problems (Bohn and Bohn, 1988). Bulking materials used to maintain the porosity of compost and peat biofilters include perlite, Styrofoam™ pellets, wood chips, bark, and a variety of ceramic and plastic materials. A typical recipe for a compost biofilter is as follows (Schroeder, 2001):

Compost = 50 percent by volume

Bulking agent = 50 percent by volume

1 meq $CaCO_3$/g of packing material by weight

Figure 16–14

Definition sketch for packed bed biofilters: (a) open bed and (b) trench type.

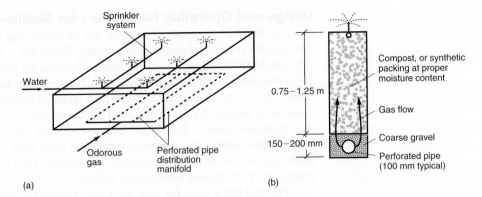

Optimal physical characteristics of a packing material include a pH between 7 and 8, air-filled pore space between 40 and 80 percent, and organic matter content of 35 to 55 percent (Williams and Miller, 1992a, 1992b). When compost is used, additional compost must be added periodically to account for the loss due to biological conversion. Bed depths of up to 1.8 m (6.0 ft) have been used. However, because most of the removal takes place in the first 20 percent of the bed, the use of deeper beds is not recommended.

Gas Distribution.   An important design feature of a biofilter is the method used to introduce the gas to be treated. The most commonly used gas distribution systems include (1) perforated pipes, (2) prefabricated underdrain systems, and (3) plenums. Perforated pipes are usually placed in a gravel layer below the compost (see Fig. 16–14). Where perforated pipes are used, it is important to size the pipe so that it performs as a reservoir and not a manifold to assure uniform distribution (Crites and Tchobanoglous, 1998). A variety of prefabricated underdrain systems are available that allow for the movement of gas upward through the compost bed and allow for the collection of drainage. Air plenums are used to equalize the air pressure to achieve uniform flow upward through the compost bed. The height of air plenums will typically vary from 200 to 500 mm.

Moisture Control.   Perhaps the most critical item in the successful operation of a biofilter is to maintain the proper moisture within the filter bed. If the moisture content is too low, biological activity will be reduced. If the moisture content is too high, the flow of air will be restricted and anaerobic conditions may develop within the bed. Also, biofilters tend to dry out unless moisture or humidity is added. The optimal moisture content is between about 50 and 65 percent, defined as follows:

$$\text{Moisture content, \%} = \left(\frac{\text{mass of water}}{\text{mass of water + mass of dry packing}} \right) \times 100 \qquad (16\text{–}12)$$

Moisture can be supplied by adding water to the top of the bed (usually by spraying) or by humidifying the incoming gas in a humidification chamber. The relative humidity of the gas entering the biofilter should be 100 percent at the operating temperature of the biofilter (Eweis et al., 1998). The liquid application rate for biotrickling filters is typically about 0.75 to 1.25 $m^3/m^2{\cdot}d$.

Temperature Control.   The operating temperature range for biofilters is between 15 and 45°C, with the optimal range being between 25 and 35°C. In cold climates, biofilters must be insulated, and the incoming gas must be heated. Where the incoming gas is warmer, it may have to be cooled before being introduced to the biofilter. Operation at higher temperatures (e.g., 45 to 60°C) is often possible, as long as the temperature remains relatively constant.

Design and Operating Parameters for Biofilters.   The sizing of biofilters is typically based on a consideration of the gas residence time in the bed, the unit air loading rate, and the constituent elimination capacity. Terms that will be encountered in the literature and the relationships commonly used to describe the performance of bulk media filters are summarized in Table 16–10. The empty bed residence time (EBRT) [see Eq. (16–13)], used to define the relationship between the volume of the contactor and the volumetric gas flowrate, is similar to Eq. (11–62) used for the analysis of activated-carbon systems. The true residence time is determined by incorporating the porosity, a [see Eq. (16–14)]. Surface and volumetric mass loading rates are often used to define the operation of bulk media filters. The elimination capacity, as given by Eq. (16–20), is used to compare the performance of different odor control systems.

The residence time for foul air from wastewater treatment facilities is typically between 15 and 60 s, and surface loading rates have ranged up to 120 m³/m²·min for H_2S concentrations up to 20 mg/L. Constituent elimination rates are determined experimentally and are usually reported as a function of the constituent loading rate (e.g., mg H_2S/m^3·h). An essentially linear, 1 to 1, constituent elimination rate up to a critical loading rate has been observed for hydrogen sulfide and other odorous compounds

Table 16–10

Parameters used for the design and analysis of bulk media filters[a]

Parameter		Definition
Empty bed residence time		EBRT = empty bed residence time, h
$EBRT = \dfrac{V_f}{Q}$	(16–13)	V_f = total volume of filter bed contactor, m³
		Q = volumetric flowrate, m³/h
Actual residence time in filter		RT = residence time, h, min, s
$RT = \dfrac{V_f \times \alpha}{Q}$	(16–14)	a = porosity of filter bed contactor
		SLR = surface loading rate, m³/m²·h
Surface loading rate		A_f = surface area of filter bed contactor, m²
$SLR = \dfrac{Q}{A_f}$	(16–15)	VLR = volumetric loading rate, m³/m³·h
		RE = removal efficiency, %
Surface mass loading rate		C_o = influent gas concentration, mg/L
		EC = elimination capacity, g/m³·h
$SLR_m = \dfrac{Q \times C_o}{A_f}$	(16–16)	C_e = effluent gas concentration, mg/L
Volume loading rate		
$VLR = \dfrac{Q}{V_f}$	(16–17)	
Volume mass loading rate		
$VLR_m = \dfrac{Q \times C_o}{V_f}$	(16–18)	
Removal efficiency		
$RE = \dfrac{C_o - C_e}{C_o} \times 100$	(16–19)	
Elimination capacity		
$EC = \dfrac{Q(C_o - C_e)}{V_f}$	(16–20)	

[a] Adapted in part from from Eweis et al. (1998); Devinny et al. (1999).

Figure 16–15

Typical odor elimination capacity versus applied load.

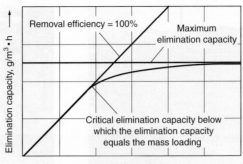

(see Fig. 16–15). Yang and Allen (1994) have reported a linear 1 to 1 elimination rate for H₂S, with loading rates for compost filters up to a maximum value of about 130 g S/m³·h, beyond which the elimination rate becomes essentially constant at a rate of 130 g S/m³·h with increased loading. It should be noted that H₂S is eliminated easily as it passes through a biofilter.

Typical design criteria for biofilters are presented in Table 16–11. Typical biofilters are shown on Fig. 16–16. Some states regulate the design of compost biofilters including

Table 16–11

Typical design factors for biofilters[a]

| | | Type of biofilter | |
Item	Units	Biofilter	Biotrickling filter
Oxygen concentration	Parts oxygen/ parts oxidizable gas	100	100
Moisture			
Compost filter	%	50–65	50–65
Synthetic media	%	55–65	55–65
Temperature, optimum	°C	15–35	15–35
pH	unitless	6–8	6–8
Porosity	%	35–50	35–50
Gas residence time	s	30–60	30–60
Depth of medium	m	1–1.25	1–1.25
Inlet odorous gas concentration	g/m³	0.01–0.5	0.01–0.5
Surface loading rate	m³/m²·h	10–100[b]	10–100[b]
Volume loading rate	m³/m³·h	10–100	10–100
Liquid application rate	m³/m²·d		0.75 to 1.25
Elimination capacity			
H₂S (in compost filter)	mg/m³·h	80–130	80–130
Other odorous gases	mg/m³·h	20–100	20–100
Back pressure, maximum	mm of water	50–100	50–100

[a] Adapted in part from van Lith (1989), Yang and Allen (1994), WEF (1995), and Devinny et al. (1999).

[b] Loading rates as high as 500 m³/m²·h have been reported, depending on the compound and its concentration.

Figure 16–16

Typical bulk biofilters for odor control: (a) compost biofilter (b) gravel-type biofilter.

(a)

(b)

loading rates, biofilter emission rates, odor-sampling procedures, and setbacks from property lines. A typical odor-emission limit at the surface of the biofilter is 50 dilutions to threshold [see Eq. (2–52) in Chap. 2)] (Finn and Spencer, 1997). The design of a compost biofilter for the elimination of hydrogen sulfide is illustrated in Example 16–2.

EXAMPLE 16–2 Design of a Compost Biofilter for Odor Control   Determine the size of compost biofilter needed to scrub the air from a 100 m³ enclosed volume using the design criteria given in Table 16–11. Also estimate the mass of the buffer compound needed to neutralize the acid formed as a result of treatment within the filter. Assume 12 air changes per hour are needed. Assume a bed porosity of 40 percent. Will the volume selected be adequate if the air contains 40 ppm$_v$ of H$_2$S in addition to other odorous constituents? Assume an elimination rate of 65 g S/m³·h, which incorporates a factor of safety of 2 as compared to the maximum rate given in Table 16–11. The temperature of the air is 20°C.

Solution

1. Estimate the airflow to be scrubbed.
    Flow = volume/time
    Flow = (100 m³)(12 changes per h) = 1200 m³/h
2. Select a loading rate from Table 16–11; use 90 m³/m²·h.
3. Select a filter-bed depth from Table 16–11; use 1.0 m.
4. Calculate the area needed for the filter bed.
    Area = gas flow/loading rate
    Area = (11200 m³/h)/(90 m³/m²·h)
    Area = 13.3 m²
5. Check the empty bed residence time using Eq. (16–13).

    $$\text{EBRT} = \frac{V_f}{Q} = \frac{(13.3 \text{ m}^2)(1 \text{ m})}{(1200 \text{ m}^3/\text{h})}$$
    $$= 0.011 \text{ h} = 39.9 \text{ s (OK } 39.9 \text{ s} > 30 \text{ s)}$$

6. Determine if the volume of the biofilter determined in Step 5 is adequate to treat the H$_2$S.

a. Determine the concentration of H_2S in g/m^3 using Eq. (2–45). From Example 16–1 the volume of gas occupied by one mole of a gas at a temperature of 20°C and a pressure of 1.0 atm is 24.1 L. Thus, the concentration of H_2S is

$$g/m^3 = \left(\frac{40\ L^3}{10^6\ L^3}\right)\left[\frac{(34.08\ g/mole\ H_2S)}{(24.1 \times 10^{-3}\ m^3/mole\ of\ H_2S)}\right]$$

$$= 0.057\ g/m^3$$

b. Determine the mass loading rate of S^{2-} in g S/h.

$$M_s = \left(\frac{1200\ m^3}{h}\right)\left(\frac{0.057\ g\ H_2S}{m^3}\right)\left(\frac{32\ g\ S^{2-}}{34.08\ g\ H_2S}\right)$$

$$= 64.2\ g\ S^{2-}/h$$

c. Determine the required volume assuming an elimination rate of 65 g S/m³·h.

$$V = \frac{(64.2\ g\ S/h)}{(65\ g\ S/m^3\cdot h)} = 0.99\ m^3$$

Because the volume of the bed (13.3 m³) is significantly greater than the required volume, the removal of H_2S will not be an issue.

7. Determine the mass of the buffer compound needed to neutralize the acid formed as a result of treatment within the filter.

a. Determine the mass of H_2S in kg applied per year.

$$H_2S,\ kg/year = \frac{(1200\ m^3/h)(0.057\ g/m^3)(24\ h/d)(365\ d/y)}{(10^3\ g/1\ kg)}$$

$$= 599.2\ kg/y$$

b. Determine the mass of buffer compound required. Assume the following equation applies:

$$H_2S + Ca(OH)_2 + 2O_2 \rightarrow Ca_2SO_4 + 2H_2O$$

34.06   74.08

Thus, about 2.18 kg of $Ca(OH)_2$ (74.08/34.06) will be required per kg of H_2S. If the compost biofilter has a useful life of 2 y, then a total of 2457 kg of $Ca(OH)_2$ equivalent will be required to be added to the bed. Typically, 1.25 to 1.5 times as much are added. The buffer compound is mixed in with the compost and the bulking agent.

Comment   Based on the results of the computation carried out in Step 6, it is clear why compost and soil filters are so effective in the elimination of H_2S.

16–4   CONTROL OF VOLATILE ORGANIC CARBON EMISSIONS

Many of the organic priority pollutants of concern in wastewater treatment are, as noted in Chap. 2, also classified as volatile organic compounds (VOCs). At some wastewater-treatment facilities, volatile organic compounds (VOCs) such as trichloroethylene (TCE) and 1,2-dibromo-3-chloropropane (DBCP) have been detected in wastewater.

The uncontrolled release of such compounds that now occurs in wastewater collection systems and wastewater-treatment plants is of concern because (1) once such compounds are in the vapor state they are much more mobile and, therefore, more likely to be released to the environment, (2) the presence of some of these compounds in the atmosphere may pose a significant public health risk, and (3) they contribute to a general increase in reactive hydrocarbons in the atmosphere, which can lead to the formation of ground-level ozone. The physical properties of selected VOCs, the mechanisms governing the release of these compounds, the locations where the release of these compounds is most prevalent, and the methods of controlling the discharge of these compounds to the atmosphere are discussed in this section.

Physical Properties of Selected VOCs

The physical properties of selected VOCs are presented in Table 16–12. Organic compounds that have a boiling point less than or equal to 100°C and/or a vapor pressure greater than 1 mm Hg at 25°C are generally considered to be volatile organic compounds (VOCs). For example, chloroethene (vinyl chloride), which has a boiling point of −13.9°C and a vapor pressure of 2548 mm Hg at 20°C, is an example of an extremely volatile organic compound.

Emission of VOCs

The release of VOCs in collection systems and at treatment plants, especially at the headworks, is of particular concern with respect to the health of the collection system and treatment plant workers. The principal mechanisms governing the release of VOCs in wastewater collection and treatment facilities are (1) volatilization and (2) gas stripping. These mechanisms and the principal locations where VOCs are released are considered in the following discussion.

Volatilization.   The release of VOCs from wastewater surfaces to the atmosphere is termed volatilization. Volatile organic compounds are released because they partition between the gas and water phase until equilibrium concentrations are reached (Roberts et al., 1984). The mass transfer (movement) of a constituent between these two phases is a function of the constituent concentration in each phase relative to the equilibrium concentration. Thus, the transfer of a constituent between phases is greatest when the concentration in one of the phases is far from equilibrium. Because the concentration of VOCs in the atmosphere is extremely low, the transfer of VOCs usually occurs from wastewater to the atmosphere. However, because of the dynamic nature of the flows found within a wastewater treatment plant equilibrium conditions rarely exist, VOCs are either volatilized and/or degraded biologically.

Gas Stripping.   Gas stripping of VOCs occurs when a gas (usually air) is entrained temporarily in wastewater or is introduced purposefully to achieve a treatment objective. When gas is introduced into a wastewater, VOCs are transferred from the wastewater to the gas. The forces governing the transfer between phases are the same as described above. For this reason, gas (air) stripping is most effective when contaminated wastewater is exposed to contaminant-free air. In wastewater treatment, air stripping occurs most commonly in aerated grit chambers, aerated biological treatment processes, and aerated transfer channels. Specially designed facilities (e.g., stripping towers) for gas stripping are considered in Sec. 11–10 in Chap 11 and Sec 15–5 in Chap. 15.

Table 16–12

Physical properties of selected volatile and semi-volatile organic compounds[a,b]

Compounds	mw	mp, °C	bp, °C	vp, mm Hg	vd	sg	Sol., mg/L	Cs, g/m³	H, m³·atm/mol	log K_{ow}
Benzene	78.11	5.5	80.1	76	2.77	.8786	1780	319	5.49×10^{-3}	2.1206
Chlorobenzene	112.56	−45	132	8.8	3.88	1.1066	500	54	3.70×10^{-3}	2.18–3.79
o-Dichlorobenzene	147.01	18	180.5	1.60	5.07	1.036	150	N/A	1.7×10^{-3}	3.3997
Ethylbenzene	106.17	−94.97	136.2	7	3.66	0.867	152	40	8.43×10^{-3}	3.13
1,2-Dibromoethane	187.87	9.8	131.3	10.25	0.105	2.18	2699	93.61	6.29×10^{-4}	N/A
1,1-Dichloroethane	98.96	−97.4	57.3	297	3.42	1.176	7840	160.93	5.1×10^{-3}	N/A
1,2-Dichloroethane	98.96	−35.4	83.5	61	3.4	1.25	8690	350	1.14×10^{-3}	1.4502
1,1,2,2-Tetrachloroethane	167.85	−36	146.2	14.74	5.79	1.595	2800	13.10	4.2×10^{-4}	2.389
1,1,1-Trichlorethane	133.41	−32	74	100	4.63	1.35	4400	715.9	3.6×10^{-3}	2.17
1,1,2-Trichloroethane	133.4	−36.5	133.8	19	N/A	N/A	4400	13.89	7.69×10^{-4}	N/A
Chloroethene	62.5	−153	−13.9	2548	2.15	0.912	6000	8521	6.4×10^{-2}	N/A
1,1-Dichloroethene	96.94	−122.1	31.9	500	3.3	1.21	5000	2640	1.51×10^{-2}	N/A
c-1,2-Dichloroethene	96.95	−80.5	60.3	200	3.34	1.284	800	104.39	4.08×10^{-3}	N/A
t-1,2-Dichloroethene	96.95	−50	48	269	3.34	1.26	6300	1428	4.05×10^{-3}	N/A
Tetrachloroethene	165.83	−22.5	121	15.6	N/A	1.63	160	126	2.85×10^{-2}	2.5289
Trichloroethene	131.5	−87	86.7	60	4.54	1.46	1100	415	1.17×10^{-2}	2.4200
Bromodichloromethane	163.8	−57.1	90	N/A	N/A	1.971	N/A	N/A	2.12×10^{-3}	N/A
Chlorodibromomethane	208.29	<−20	120	50	N/A	2.451	N/A	N/A	8.4×10^{-4}	N/A
Dichloromethane	84.93	−97	39.8	349	2.93	1.327	20000	1702	3.04×10^{-3}	N/A
Tetrachloromethane	153.82	−23	76.7	90	5.3	1.59	800	754	2.86×10^{-2}	2.7300
Tribromomethane	252.77	8.3	149	5.6	8.7	2.89	3130	7.62	5.84×10^{-4}	N/A
Trichloromethane	119.38	−64	62	160	4.12	1.49	7840	1027	3.10×10^{-3}	1.8998
1,2-Dichloropropane	112.99	−100.5	96.4	41.2	3.5	1.156	2600	25.49	2.75×10^{-3}	N/A
2,3-Dichloropropene	110.98	−81.7	94	135	3.8	1.211	insol.	110	N/A	N/A
t-1,3-Dichloropropene	110.97	N/A	112	99.6	N/A	1.224	515	110	N/A	N/A
Toluene	92.1	−95.1	110.8	22	3.14	0.867	515	110	6.44×10^{-3}	2.2095

[a] Data were adapted from Lang (1987).

[b] All values are reported at 20°C.

Note: mw = molecular weight, mp = melting point, bp = boiling point, vd = vapor density relative to air, s_g = specific gravity relative to water, Sol = solubility, C_s = saturation concentration in air, H = Henry's Law Constant, log K_{ow} = logarithm of the octanol-water partition coefficient.

Locations Where VOCs Are Emitted. The principal locations where VOCs are emitted from wastewater collection and treatment facilities are summarized in Table 16–13. The degree of VOC removal at a given location will depend on local conditions. Mass transfer is considered in the following section.

Table 16–13

Sources, methods of release, and control of VOCs from wastewater facilities

Source	Method of release	Suggested control strategies
Domestic, commercial, and industrial discharges	Discharge of small amounts of VOCs in liquid wastes	Institute active source control program to limit the discharge of VOCs to municipal sewers
Wastewater sewers	Volatilization from the surface enhanced by flow induced turbulence	Seal existing manholes. Eliminate the use of structures that create turbulence and enhance volatilization
Sewer appurtenances	Volatilization due to turbulence at junctions, etc., volatilization and air stripping at drop manholes and junction chambers	Isolate and cover existing appurtenances
Pump stations	Volatilization and air stripping at influent wet-well inlets	Vent gases from wet-well to VOC treatment unit Use variable-speed pumps to reduce size of wet-well
Bar racks	Volatilization due to turbulence	Cover units, reduce headloss through bar racks
Comminutors	Volatilization due to turbulence	Cover units, use inline enclosed comminutors
Parshall flume	Volatilization due to turbulence	Cover units, use alternative measuring device
Grit chamber	Volatilization due to turbulence in horizontal-flow grit chambers Volatilization and air stripping in aerated grit chambers Volatilization in vortex-type grit chambers	Cover aerated and vortex-type grit chambers Reduce turbulence in horizontal-flow grit chambers; cover if necessary
Equalization basins	Volatilization from surface enhanced by local turbulence Air stripping where diffused air is used	Cover units, use submerged mixers. Reduce air flow
Primary and secondary sedimentation tanks	Volatilization from surface. Volatilization and air stripping at overflow weirs, in effluent channel, and at other discharge points	Cover tanks, replace overflow weirs with drops with submerged launders
Biological treatment	Air stripping in diffused-air activated sludge. Volatilization in activated sludge processes with surface aerators. Volatilization from surface enhanced by local turbulence. Volatilization from trickling filters	Cover units, in activated sludge systems, use submerged mixers and reduce aeration rate
Transfer channels	Volatilization from surface enhanced by local turbulence. Volatilization and air stripping in aerated transfer channels	Use enclosed transfer channels
Digester gas	Uncontrolled release of digester gas. Discharge of incompletely combusted or incinerated digester gas	Controlled thermal incineration, combustion, or flaring of digester gas

Mass Transfer Rates for VOCs

The mass transfer of VOCs can, for practical purposes, be modeled using the following equation (Roberts et al., 1984, and Thibodeaux, 1979):

$$r_{VOC} = -(K_L a)_{VOC}(C - C_s) \tag{16-21}$$

where r_{VOC} = rate of VOC mass transfer, mg/m³·h
$(K_L a)_{VOC}$ = overall VOC mass transfer coefficient, 1/h
C = concentration of VOC in liquid, mg/m³
C_s = saturation concentration of VOC in liquid, mg/m³

Due to chemical handling and analytical requirements, measuring $K_L a_{VOC}$ is much more difficult than measuring $K_L a_{O_2}$. Therefore, a practical approach is to relate the $K_L a_{VOC}$ to the $K_L a_{O_2}$. The following equation is used to relate the mass transfer coefficients as a function of the VOC and O_2 diffusion coefficients in water:

$$(K_L a)_{VOC} = (K_L a)_{O_2}\left(\frac{D_{VOC}}{D_{O_2}}\right)^n \tag{16-22}$$

where $(K_L a)_{VOC}$ = system mass transfer coefficient, T^{-1} (1/h)
$(K_L a)_{O_2}$ = system oxygen mass transfer coefficient, T^{-1} (1/h)
D_{VOC} = diffusion coefficient of VOC in water, $L^2 T^{-1}$ (cm²/s)
D_{O_2} = diffusion coefficient of oxygen in water, $L^2 T^{-1}$ (cm²/s)
n = coefficient

Diffusion coefficient values for different compounds can be obtained from Schwarzenbach et al. (1993) or in other handbooks. It should be noted that there is often considerable variation in the values reported in the literature. However, based on the results a variety of experimental investigations of the relationship between $K_L a_{VOC}$ and $K_L a_{O_2}$ it has been found that Eq. (16–22) is generally applicable, and that the value for n varies depending on whether the gas/liquid transfer is accomplished by surface aeration, diffused aeration, or a packed column air stripper, and the power intensity of the gas transfer device (Roberts and Dandliker, 1983; Matter-Muller et al., 1981; Hsieh et al., 1993; Libra, 1993; and Bielefeldt and Stensel, 1999). For practical power intensities of less than 100 W/m³ a reasonable value of n is 0.50 for packed columns and mechanical aeration and 1.0 for diffused aeration. For higher power intensities the work of Hsieh et al. (1993) should be consulted. The $K_L a_{VOC}$ value has also been found to be essentially the same in wastewater as in tap water (Bielefeldt and Stensel, 1999).

Mass Transfer of VOCs from Surface and Diffused-Air Aeration Processes

The amount of VOCs released from a complete-mix reactor used for the activated-sludge process will depend on the method of aeration (e.g., surface aeration or diffused aeration).

Complete-Mix Reactor with Surface Aeration.

A materials balance for the stripping of a VOC written around a complete-mix reactor is as follows, assuming no other removal mechanisms for the VOC compound such as biodegradation or solids sorption are applicable.

1. General word statement:

$$\begin{pmatrix}\text{Rate of accumulation} \\ \text{of VOC within} \\ \text{the system boundary}\end{pmatrix} = \begin{pmatrix}\text{rate of flow of} \\ \text{VOC into the} \\ \text{system boundary}\end{pmatrix} - \begin{pmatrix}\text{rate of flow of} \\ \text{VOC out of the} \\ \text{system boundary}\end{pmatrix} + \begin{pmatrix}\text{amount of VOC} \\ \text{removed through} \\ \text{system boundary} \\ \text{by stripping}\end{pmatrix} \tag{16-23}$$

2. Simplified word statement:

Accumulation = inflow-outflow + decrease due to stripping $\hspace{2cm}$ (16–24)

3. Symbolic representation:

$$\frac{dC}{dt}V = QC_i - QC_e + r_{\text{VOC}}V \hspace{2cm} (16\text{–}25)$$

where dC/dt = rate of change in VOC concentration in reactor
$\hspace{1.5cm}V$ = volume of complete mix reactor, L^3 (m^3)
$\hspace{1.5cm}Q$ = liquid flowrate, L^3T^{-1} (m^3/s)
$\hspace{1.5cm}C_i$ = concentration of VOC in influent to reactor, ML^{-3} (mg/m^3)
$\hspace{1.5cm}C_e$ = concentration of VOC in effluent from reactor, ML^{-3} (mg/m^3)
$\hspace{1.5cm}r\text{VOC}$ = rate of VOC mass transfer, $ML^{-3}T^{-1}$ ($mg/m^3 \cdot h$)

Substituting for rVOC from Eq. (16–21) and τ for V/Q yields

$$\frac{dC}{dt} = \frac{C_i - C_e}{\tau} + [-(K_L a)_{\text{VOC}}(C_e - C_s)] \hspace{2cm} (16\text{–}26)$$

If steady-state conditions are assumed and it is further assumed that C_s is equal to zero, then the amount of VOC that can be removed by surface aeration is given by the following expression:

$$1 - \frac{C_e}{C_i} = 1 - [1 + (K_L a)\tau]^{-1} \hspace{2cm} (16\text{–}27)$$

If a significant amount of the VOC is adsorbed or biodegraded, the results obtained with the above equation will be overestimated. The above analysis can also be used to estimate the release of VOCs at weirs and drops by assuming the time period is about 30s.

Complete Mix Reactor with Diffused-Air Aeration. The corresponding expression to Eq. (16–24) for a complete-mix reactor with diffused-air aeration is developed by a mass balance on the VOC compound. At steady state the VOC in equals the VOC out, and the corresponding mass balance is

$$\frac{\text{Inflow in}}{\text{liquid stream}} = \frac{\text{outflow}}{\text{liquid stream}} + \frac{\text{outflow}}{\text{in exit gas}}$$

$$QC_i = QC_e + Q_g C_{g,e} \hspace{2cm} (16\text{–}28)$$

where Q = liquid flowrate, L^3T^{-1} (m^3/s)
$\hspace{1.5cm}C_i$ = VOC concentration in influent, ML^{-3} (mg/m^3)
$\hspace{1.5cm}C_e$ = VOC concentration in effluent, ML^{-3} (mg/m^3)
$\hspace{1.5cm}Q_g$ = gas flowrate, L^3T^{-1} (m^3/s)
$\hspace{1.5cm}C_{g,e}$ = VOC concentration in exit gas, ML^{-3} (mg/m^3)

The general expression for the removal of VOC by gas sparging through a liquid is (Bielefeldt and Stensel, 1999)

$$Q_g C_{g,e} = Q_g H_u C_e (1 - e^{-\phi}) \hspace{2cm} (16\text{–}29)$$

where H_u = Henry's law constant, dimensionless
$\hspace{1.5cm}\phi$ = VOC saturation parameter defined as

$$\phi = \frac{(K_L a)_{\text{VOC}} V}{H_u Q_g} \hspace{2cm} (16\text{–}30)$$

Eq. (16–29) can be rearranged as follows:

$$Q(C_i - C_e) = H_u C_e(1 - e^{-\phi})$$ (16–31)

Solving Eq. (16–31) for C_e/C_i yields

$$\frac{C_e}{C_i} = \left[1 + \frac{Q_g}{Q}H_u(1 - e^{-\phi})\right]^{-1}$$ (16–32)

and the fraction removed is given by

$$1 - \frac{C_e}{C_i} = 1 - \left[1 + \frac{Q_g}{Q}H_u(1 - e^{-\phi})\right]^{-1}$$ (16–33)

The application of the above equations is illustrated in Example 16–3.

EXAMPLE 16–3   **Stripping of Trichloroethene in the Activated Sludge Process**   Determine the amount of trichloroethene (TCE) that can be stripped in a complete-mix activated sludge reactor equipped with a diffused-air aeration system. Assume the following conditions apply:

1.   Wastewater flowrate = 4000 m³/d
2.   Aeration tank volume = 1000 m³
3.   Depth of aeration tank = 6 m
4.   Air flowrate = 50 m³/min at standard conditions
5.   Oxygen mass transfer rate, $(K_L a)_{O_2}$ = 6.2/h
6. $H_{TCE} = 1.17 \times 10^{-2}$ m³·atm/mol (see Table 16–10)
7. $n = 1.0$
8.   Temperature = 20°C
9.   Oxygen diffusivity = 2.11 × 10⁻⁵ cm²/s
10. Trichloroethene diffusivity ~ 1.0 × 10⁻⁵ cm²/s

Solution

1.   Determine the quantity of air referenced to the mid-depth of the aeration tank, which represents the depth for an average bubble size. Using the universal gas law, the air flowrate at mid-depth (3 m) is:

$$Q_g = (50 \text{ m}^3/\text{min})\frac{(10.33 \text{ m})}{(10.33 \text{ m} + 3 \text{ m})} = 38.7 \text{ m}^3/\text{min}$$

Note: 10.33 = standard atmospheric pressure expressed in m of H_2O.

2.   Determine the air/liquid ratio.

$$Q = \frac{(4000 \text{ m}^3/\text{d})}{(1440 \text{ min/d})} = 2.78 \text{ m}^3/\text{min}$$

$$\frac{Q_g}{Q} = \frac{(38.7 \text{ m}^3/\text{min})}{(2.78 \text{ m}^3/\text{min})} = 13.9$$

3.   Estimate the mass transfer coefficient for TCE using Eq. (16–22).

$$(K_L a)_{VOC} = (K_L a)_{O_2}\left(\frac{D_{VOC}}{D_{O_2}}\right)^n = (6.2/\text{h})\left[\frac{(1.0 \times 10^{-5} \text{ cm}^2/\text{s})}{(2.11 \times 10^{-5} \text{ cm}^2/\text{s})}\right]^{1.0}$$

$$(K_L a)_{VOC} = 2.94 \text{ /h} = 0.049/\text{min}$$

4. Determine the dimensionless value of the Henry's law constant for TCE using Eq. (2–51).

$$H_u = \frac{H}{RT}$$

$$H_u = \frac{0.0117}{0.000082057 \times (273 + 20)} = 0.487$$

5. Determine the saturation parameter ϕ using Eq (16–30).

$$\phi = \frac{(K_L a)_{VOC} V}{H_u Q_g}$$

$$\phi = \frac{(0.049/min)(1000 \text{ m}^3)}{(0.228 \times 38.7 \text{ m}^3/min)} = 5.55$$

6. Determine the fraction of TCE removed from the liquid phase using Eq. (16–33).

$$1 - \frac{C_e}{C_i} = 1 - \left[1 + \frac{Q_g}{Q}(H_u)(1 - e^{-\phi})\right]^{-1}$$

$$1 - \frac{C_e}{C_i} = 1 - [1 + 13.9(0.487)(1 - e^{-5.55})]^{-1}$$

$$1 - \frac{C_e}{C_i} = 1 - 0.13 = 0.87 \text{ or } 87\%$$

Comment   The computations presented in this example are based on the assumption that the concentration of TCE in the influent is not being reduced by diffusion and turbulence in the headworks, primary sedimentation tank or by adsorption or biological degradation in the aeration tank.

Control Strategies for VOCs

Volatilization and gas stripping are, as noted previously, the principal means by which VOCs are released from wastewater-treatment facilities. In general, it can be shown that the release of VOCs from open surfaces is quite low compared to the release of VOCs at points of liquid turbulence and by gas stripping. Thus, the principal strategies for controlling the release of VOCs, as reported in Table 16–13, are (1) source control, (2) elimination of points of turbulence, and (3) the covering of various treatment facilities. Three serious problems associated with the covering of treatment facilities are (1) treatment of the off-gases containing VOCs, (2) corrosion of mechanical parts, and (3) provision for confined space entry for personnel for equipment maintenance.

Treatment of Off-Gases

The off-gases containing VOCs from covered treatment facilities will have to be treated before they can be discharged to the atmosphere. Options for the off-gas treatment include (see Fig. 16–17) (1) vapor-phase adsorption on granular activated carbon or other VOC selective resins, (2) thermal incineration, (3) catalytic oxidation, (4) combustion in a flare, (5) biofiltration, and (6) combustion in a boiler or process heater (U.S. EPA, 1986; WEF, 1997). The application of these processes will depend primarily on the volume of air to be treated and the types and concentrations of the VOCs contained in the airstream. The first four of these off-gas treatment processes are considered in greater detail in the

Figure 16–17

Options for treating off-gases containing VOCs. (Adapted from Eckenfelder, 2000.)

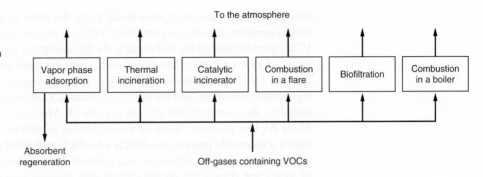

following discussion. Biofiltration was discussed previously in Sec. 16–3. A boiler or process heater is used only where a combustion process is included as part of the plant facilities.

Vapor-Phase Adsorption. Adsorption is the process whereby hydrocarbons and other compounds are adsorbed selectively on the surface of materials such as activated carbon, silica gel, or alumina. Of the available adsorbents, activated carbon is used most widely. The adsorption capacity of an adsorbent for a given VOC is often represented by adsorption isotherms that relate the amount of VOC adsorbed (adsorbate) to the equilibrium pressure (or concentration) at constant temperature. Typically, the adsorption capacity increases with the molecular weight of the VOC being adsorbed. In addition, unsaturated compounds are generally adsorbed more completely than saturated compounds, and cyclical compounds are adsorbed more easily than linearly structured materials. It should be noted that careful evaluation of the adsorption media should be undertaken to be assured that the compound being adsorbed does not react with the adsorbent. Also, the adsorption capacity is enhanced by lower operating temperatures and higher concentrations. VOCs characterized by low vapor pressures are more easily adsorbed than those with high vapor pressures (U.S. EPA, 1986).

Steps in VOC Adsorption Process. The two main steps involved in the VOC adsorption process are (1) the continuous adsorption in multiple beds (see Fig. 16–18) and (2) batch regeneration of the adsorbent. For control of continuous emission streams, at least one bed

Figure 16–18

Gas phase carbon adsorption and regeneration system for the treatment of VOCs in off-gas.

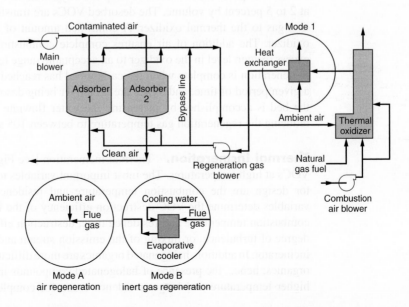

remains online in the adsorption mode while the other is being regenerated. In a typical batch operation, the off-gas containing VOCs is passed through the carbon bed where the VOCs are adsorbed on the bed surface. As the adsorption capacity of the bed is approached, traces of VOCs appear in the exit stream, up to the level where the breakthrough point of the bed has been attained. The off-gas is then directed to a parallel bed containing regenerated adsorbent, and the process continues. Concurrently, the saturated bed is regenerated by the passage of hot air (see Fig. 16–18, Mode A), hot inert gases (see Fig. 16–18, Mode B), low-pressure steam, or a combination of vacuum and hot gas. Because adsorption is a reversible process, the VOCs adsorbed on the bed can be desorbed by supplying heat (equivalent to the amount of heat released during adsorption). Small residual amounts of VOCs are always left on the carbon bed, because complete desorption is technically difficult to achieve and economically impractical. Regeneration with hot air and a hot inert gas is considered in the following discussion.

Hot Air Regeneration. When the VOCs are either nonflammable or have a high ignition temperature and thus do not pose a risk of carbon fires, hot air regeneration is used. A portion of the hot flue gas in the oxidizer is mixed with ambient air to cool the gas to below 180°C (350°F). The regeneration gas is driven upflow (or countercurrent to adsorption flow) through the GAC adsorber. As the temperature of the carbon bed rises, the desorbed organics are transferred to the regeneration gas stream. The regeneration gas containing the desorbed VOCs is sent directly to the thermal oxidizer where the VOCs are oxidized. After the bed has been maintained at the desired regeneration temperature for a sufficient period of time, regeneration is ended. The bed is then cooled to approximately ambient temperature by shutting off the hot regeneration gas and continuing to pass ambient air through the carbon bed. The regeneration and cooling times are predetermined based on the amount of carbon in the adsorber and the expected loading on the carbon (U.S. EPA, 1986).

Inlet Gas Regeneration. Where the VOCs contained in the off-gas include compounds such as ketones and aldehydes that may pose fire risks at elevated temperatures in the presence of oxygen, inert gas regeneration is used. A relatively inert gas can be obtained by passing a portion of the hot flue gas from the thermal oxidizer through an evaporative cooler. It is possible, therefore, to keep the oxygen concentration in the regeneration gas at 2 to 5 percent by volume. The desorbed VOCs are transferred along with the regeneration gas to the thermal oxidizer. A controlled amount of secondary air is added to the oxidizer. The addition of air ensures complete combustion of the VOCs but limits the excess oxygen level in the oxidizer to an acceptable range (e.g., 2 to 5 percent by volume). Regeneration is complete when the carbon bed has reached the necessary temperature for a given period of time, and VOCs are no longer being desorbed from the bed. Cooling of the bed is accomplished by increasing the water flowrate to the evaporative cooler and reducing the regeneration gas temperature to between 105 and 120°C (220 and 250°F).

Thermal Incineration. Thermal incineration [see Fig. 16–12(a)] is used to oxidize VOCs at high temperatures. The most important variables to consider in thermal incinerator design are the combustion temperature and residence time because these design variables determine the VOC destruction efficiency of the incinerator. Further, at a given combustion temperature and residence time, destruction efficiency is also affected by the degree of turbulence, or mixing of the emission stream and hot combustion gases, in the incinerator. In addition, halogenated organics are more difficult to oxidize than unsubstituted organics; hence, the presence of halogenated compounds in the emission stream requires higher temperature and longer residence times for complete oxidation. When emission streams treated by thermal incineration are dilute (i.e., low heat content), supplementary fuel is required to maintain the desired combustion temperatures. Supplementary fuel requirements may be reduced by recovering the energy contained in the hot flue gases from the incinerator. Also, depending on the byproducts of incineration (e.g., HCl, H$_2$SO$_4$, HF), it may be necessary to provide an acid gas scrubber.

Catalytic Oxidation.   In catalytic oxidation [see Fig. 16–12(b)], VOCs in an emission stream are oxidized with the help of a catalyst. A catalyst is a substance that accelerates the rate of a reaction at a given temperature without being appreciably changed during the reaction. Catalysts typically used for VOC oxidation include platinum and palladium; other formulations are also used, including metal oxides for emission streams containing chlorinated compounds. The catalyst bed (or matrix) in the oxidizer is generally a metal mesh-mat, ceramic honeycomb, or other ceramic matrix structure designed to maximize catalyst surface area. The catalysts may also be in the form of spheres or pellets. Before passing through the catalyst bed, the emission stream is preheated, if necessary, in a natural gas-fired preheater (U.S. EPA, 1986).

The performance of a catalytic oxidizer is affected by several factors including (1) operating temperature, (2) space velocity (reciprocal of residence time), (3) VOC composition and concentration, (4) catalyst properties, and (5) presence of catalyst poisons or inhibitors in the emission stream. In catalytic incinerator design, the important variables are the operating temperature at the catalyst bed inlet and the space velocity. The operating temperature for a particular destruction efficiency is dependent on the concentration and composition of the VOC in the emission stream and the type of catalyst used (U.S. EPA, 1986). As with incineration, it may be necessary to provide an acid gas scrubber.

Combustion in a Flare.   Flares, another type of thermal incineration commonly used for disposal of waste digester gas, can be used to destroy most VOCs found in off-gas streams. Flares can be designed and operated to handle fluctuations in emission VOC content, inerts content, and flowrate. Several different types of flares are available, including steam-assisted, air-assisted, and pressure-head flares. Steam-assisted flares are employed in cases where large volumes of waste gases are released. Air-assisted flares are generally used for moderate off-gas gas flows. Pressure-head flares are used for small gas flows.

16–5 EMISSIONS FROM THE COMBUSTION OF GASES AND SOLIDS

Various types of fuels are used at wastewater treatment facilities to generate heat and electricity onsite to supplement energy supplied from outside sources and to operate equipment. As discussed in Chaps. 13 and 14, part or all of gases and solids generated at wastewater treatment facilities can also be used as fuels within the wastewater treatment facility. Fuel sources, combustion systems used at wastewater treatment plants and their emissions, and flaring of excess digester gas are discussed in this section.

Sources of Fuels

Fuels used for the operation of a wastewater treatment plant include various grades of fuel oils, natural gas, and digester gas. The types of fuels used will depend on the types of combustion systems used at a treatment facility. Emissions of certain pollutants will vary with the quality of the fuel and the type and condition of the combustion system. Typical fuels used at a wastewater treatment plant are listed in Table 16–14.

Fuel type	**Characteristics**
No. 2 fuel oil[a]	A distillate fuel oil that has a distillation temperature of 640 degrees Fahrenheit at the 90-percent recovery point. It is used in atomizing type burners for domestic heating or for moderate capacity commercial/ industrial burner units.
No. 4 fuel oil[a]	A distillate fuel oil made by blending distillate fuel oil and residual fuel oil stocks. It is used extensively in industrial plants and in commercial burner installations that are not equipped with preheating facilities. It also includes No. 4 diesel fuel used for low- and medium-speed diesel engines.
Natural gas[b]	Typical composition of natural gas before refining is 70 to 90 percent methane, up to 20 percent of ethane, propane, and butane, and up to 8 percent carbon dioxide. Sulfur content is typically low but unprocessed natural gas may have higher sulfur content.
Digester gas	Typically contains 50 to 65 percent methane. Siloxanes in the digester gas can be detrimental to the combustion system. The level of cleaning required for use depends on the types of the combustion system. Emission of sulfur oxides depends on the sulfur contents in the digester gas.

Table 16–14

Fuels used commonly at wastewater treatment plants

[a] From U.S. Energy Information Administration.

[b] Typical composition from Natural Gas Supply Association (naturalgas.org).

Combustion Systems Used at Wastewater Treatment Plants

Major combustion systems used at wastewater treatment plants include boilers and power generators. In some cases, solids are incinerated within the treatment facility (see Sec. 14–4 in Chap. 14). Flares are used to combust excess digester gas.

Boiler. Boilers used at wastewater treatment facilities use fuel oil, natural gas, digester gas, or a combination of these fuels to produce hot water or steam [see Fig. 16–19(a)]. The main purposes of using boilers are to provide heat for maintaining temperature of anaerobic digesters and to provide heat and hot water in the buildings. In addition, boilers are used to produce process steam for use in sludge drying and to treat biosolids thermally in various processes. The use of steam boilers is less common, especially at smaller wastewater treatment facilities, as additional treatment of the boiler feed water may be required. Combustion of a fuel is controlled to heat water to a desired temperature, and hot water or steam is circulated for various purposes.

Reciprocating Engine. Internal combustion reciprocating engines are used most commonly for the generation of electricity [see Fig. 16–19(b)]. Reciprocating engines can also be used to operate pumps, gas compression, or chillers. Reciprocating engines are categorized as spark ignition or compression ignition. Diesel fuel is used for compression ignition engines, or the engines can be set up to use natural gas with a small amount of diesel to ignite with compression. Emission of nitrogen oxides (NOx) is correlated with the combustion temperature as well as nitrogen in the fuel (typically a concern with liquid fuels), and NOx removal may be necessary if the emission exceeds the permit level. Emission of particulate matters and carbon monoxide (CO) are generally correlated with incomplete combustion of fuels.

Figure 16–19

Combustion systems used commonly at wastewater treatment facilities: (a) view of heat recovery boiler and (b) view of internal combustion engine.

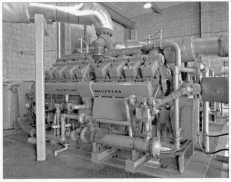

(a)   (b)

Gas Turbine. In gas turbines, compressed fuel and air is combusted in a combustion chamber, and the expanded hot exhaust gas is used to spin the power turbine. Because gas turbines generally operate at a significantly higher temperature than reciprocating engines, the emission of NOx has to be controlled. The use of gas turbines is increasing with the use of combined heat and power (CHP) system at wastewater treatment facilities [see Fig. 17–8(d) in Chap. 17]. Steam is generated using a heat exchanger that removes excess energy from turbine exhaust gases. The steam produced in this manner can be used to replace boiler-produced steam for heating or used in a steam turbine to produce additional electricity.

Emissions of Concern from Combustion Sources

The major emissions of concern from combustion sources are NOx, CO, particulate matter (PM) and sulfur dioxide (SO_2). VOCs are generally emitted in small quantities from the incomplete combustion of fuel and are only a concern when a significant combustion equipment is being used. In addition, combustion emissions may contain hazardous air pollutants (HAPs) in small quantities such as formaldehyde, and any metals contained in the fuel.

Fuel gas such as natural gas and digester gas is oxidized completely to carbon dioxide (CO_2) when a combustion temperature is at or above 850°C (1560°F) and the combustion time is 0.3 s or longer. Lower combustion temperatures, shorter combustion times, and incomplete mixing of fuel and air can cause the incomplete oxidation of the fuel resulting in the production of CO. Carbon monoxide emission is controlled by adjusting the operating conditions, or by the use of a post-combustion catalytic oxidation system. Emission of VOC is essentially the emission of unburned hydrocarbons in the fuel as a result of incomplete combustion.

At combustion temperatures above 1200°C, oxidation of the nitrogen gas contained in the combustion air will occur, and NOx will be formed. Particulate matter emissions from fuel combustion include both organic and inorganic fractions. The organic fractions of particulate matter include by-products of incomplete combustion, such as polycyclic organic matter. Inorganic fractions include metals as well as acid mists. Emissions of sulfur dioxide occur as a result of the oxidation of sulfur contained in the fuel. Both liquid and gaseous fuels contain sulfur both in elemental as well as in organic fractions.

Emission Factor. An emission factor is a representative value that is used to relate the quantity of a pollutant released to the atmosphere by a specific activity associated with the release of that pollutant (U.S. EPA, 1995). Pollutant emissions are estimated using Eq. (16–34):

$$E = (A)(EF)(1 - ER/100) \tag{16–34}$$

where, E = emissions

A = activity rate, e.g., rate of the fuel consumption

EF = emission factor for given activity

ER = overall emission reduction efficiency for given activity, %

Activity rate is a general term for the rate of the fuel consumption when emissions from a combustion system are being considered. For example, if there is no device to destroy or capture the pollutant, the overall emission reduction efficiency (ER) is zero. Thus, without emission reduction Eq. (16–34) is simply the product of the rate of fuel consumption and the emission factor. As noted above, emission factors are available for specific emission generating activities. The U.S. EPA has been compiling the emission factors in AP 42, Compilation of Air Pollutant Emission Factors (U.S. EPA, 1995). Emission factors for criteria pollutants from fuel oil and natural gas combustion are reported in Table 16–15. The overall emission reduction is a product of destruction or removal efficiency and the capture efficiency of the control system such as cyclones, scrubbers, and catalytic reduction systems. It should be noted that the estimate based on the emission factor from AP42 is usually conservative. For more accurate estimate, emission data from the equipment supplier should be referenced.

Control of Nitrogen Oxides. Methods to control the emission of NOx include control of combustion temperature and residence time, and the use of catalytic reduction systems. Combustion temperature and residence time may be controlled by adjusting air to fuel ratio or by injecting water or steam. The latter is used mainly for the gas turbine systems. It should be noted that the use of dilution air could result in the emission of CO, and the operating conditions must be controlled to minimize the emission of both NOx and CO. Catalytic reduction systems include selective catalytic reduction with the use of ammonia or urea, and non-selective catalytic reduction. Non-selective catalytic reduction is applicable only for the combustion system with exhaust oxygen levels of 4 percent or less. In selective catalytic reduction, nitrogen oxides, ammonia (or urea) and oxygen react in presence of the catalyst to form nitrogen gas (N_2) and water (H_2O). Selective catalytic reduction is used for both reciprocating engines and gas turbines.

Flaring of Digester Gas

Digester gas, a gas generated from anaerobic digestion of sludges, typically contains 55 to 65 percent methane and can be used as a fuel source to generate heat and electricity. When a wastewater treatment facility does not utilize all digester gas, the excess gas must be fared (see Fig. 16–20).

Combustion of methane gas can be expressed as

$$CH_4 + 2O_2 \rightarrow CO_2 + 2H_2O \tag{16–35}$$

Air contains approximately 21 percent oxygen and requiring 2 moles of oxygen to completely oxidize 1 mole of methane. Therefore, a theoretical ratio of methane to air to achieve complete combustion of methane is approximately 9.5. If digester gas contains 60 percent methane, the theoretical digester gas to air ratio is 5.7. Practically, excess air with a well controlled flaring system is necessary to achieve effective combustion of digester gas. Flaring of digester gas will result in emissions described previously in Table 16–1.

Table 16-15

Emission factors for criteria pollutants from fuels used commonly at wastewater treatment facilities[a]

Category	Unit	CO	NOx	SO$_2$[b]	PM (filterable)	PM (condensable)	Lead
Boilers, larger than 106 GJ/h (100 MBtu/h)							
Natural gas							
Uncontrolled[c]	kg/m^3 (lb/10^6 ft^3)	1344 (84)	3040 (190)	9.6 (0.6)	30.4 (1.9)	91.2 (5.7)	0.08 (0.0005)
Controlled, low NOx burner	kg/m^3 (lb/10^6 ft^3)	1344 (84)	2240 (140)	9.6 (0.6)	30.4 (1.9)	91.2 (5.7)	0.08 (0.0005)
Controlled, flue gas recirculation	kg/m^3 (lb/10^6 ft^3)	1344 (84)	1600 (100)	9.6 (0.6)	30.4 (1.9)	91.2 (5.7)	0.08 (0.0005)
Fuel oil No. 2	kg/m^3 (lb/10^3 gal)	0.60 (5)	2.88 (24)	17.0S (142S)	0.24 (2)	0.16 (1.3)	
Fuel oil No. 2, low NOx burner, flue gas recirculation	kg/m^3 (lb/10^3 gal)	0.60 (5)	1.20 (10)	17.0S (142S)	0.24 (2)	0.16 (1.3)	
Fuel oil No. 4, normal firing	kg/m^3 (lb/10^3 gal)	0.60 (5)	5.63 (47)	18.0S (150S)	0.24 (2)	0.16 (1.3)	
Fuel oil No. 4, tangential firing	kg/m^3 (lb/10^3 gal)	0.60 (5)	3.83 (32)	18.0S (150S)	0.24 (2)	0.16 (1.3)	
Boilers, smaller than 106 GJ/h (100 MBtu/h)							
Natural gas							
Uncontrolled	kg/1 m^3 (lb/10^6 ft^3)	1344 (84)	1600 (100)	9.6 (0.6)	30.4 (1.9)	91.2 (5.7)	0.08 (0.0005)
Controlled, low NOx burner	kg/m^3 (lb/10^6 ft^3)	1344 (84)	800 (50)	9.6 (0.6)	30.4 (1.9)	91.2 (5.7)	0.08 (0.0005)
Controlled, flue gas recirculation	kg/m^3 (lb/10^6 ft^3)	1344 (84)	512 (32)	9.6 (0.6)	30.4 (1.9)	91.2 (5.7)	0.08 (0.0005)
Fuel oil No. 4, oil fired	kg/m^3 (lb/10^3 gal)	0.60 (5)	2.40 (20)	18.0S (150S)	0.84 (7)	0.16 (1.3)	

(continued)

Table 16–15 (Continued)

Category	Unit	CO	NOx	SO$_2^b$	PM (filterable)	PM (condensable)	Lead
Gas turbine							
Natural gas							
Uncontrolled	kg/GJ	0.19	0.74	2.19S	0.0042	0.011	
	(lb/MMBtu)	(0.082)	(0.32)	(0.94S)	(0.0019)	(0.0047)	
Water steam injection	kg/GJ	0.070	0.30	2.19S	0.0042	0.011	
	(lb/MMBtu)	(0.030)	(0.13)	(0.94S)	(0.0019)	(0.0047)	
Lean premix	kg/GJ	0.035	0.23	2.19S	0.0042	0.011	
	(lb/MMBtu)	(0.015)	(0.099)	(0.94S)	(0.0019)	(0.0047)	
Digester gas, uncontrolled	kg/GJ	0.040	0.37	2.19S	0.0042	0.011	
	(lb/MMBtu)	(0.017)	(0.16)	(0.94S)	(0.0019)	(0.0047)	
Natural gas fired reciprocating engines							
4-stroke, lean burn, 90–105% load	kg/GJ	0.737	9.49	5.58×10^{-4}	7.71×10^{-5}	9.91×10^{-3}	
	(lb/MMBtu)	(0.317)	(4.08)				
4-stroke, lean burn, <90% load	kg/GJ	1.30	1.97	5.58×10^{-4}	7.71×10^{-5}	9.91×10^{-3}	
	(lb/MMBtu)	(0.557)	(0.847)				
Large diesel (>600hp) and dual fuel reciprocating engines							
Large diesel engines, uncontrolled	kg/GJ	2.0	7.4	2.35S$_1$	0.14	0.018	
	(lb/MMBtu)	(0.85)	(3.2)	(1.01S$_1$)	(0.062)	(0.0077)	
Large diesel engines, controlled	kg/GJ	2.0	4.4	2.35S$_1$			
	(lb/MMBtu)	(0.85)	(1.9)	(1.01S$_1$)			
Dual fuel engines[d]	kg/GJ	2.70	6.3	0.12S$_1$ + 2.08S$_2$			
	(lb/MMBtu)	(1.16)	(2.7)	(0.05S$_1$ + 0.895S$_2$)			

[a] From U.S. EPA (1998, 1999).

[b] S = weight percent of sulfur in the fuel oil.

[c] Values for post New Source Performance Standard (NSPS).

[d] S$_1$ = weight percent of sulfur in the fuel oil, S$_2$ = weight percent of sulfur in natural gas.

Figure 16–20

Flares for the combustion of excess digester gas: (a) ground effect flare and (b) open air flare.

Similar to the internal combustion systems, the emissions from flaring of digester gas will depend on the completeness of the combustion and the combustion temperature. Digester gas to air ratio should be adjusted according to the quality of the digester gas, but typically 150 to 200 percent of excess air in addition to the stoichiometric air requirement is necessary to achieve sufficient oxidation of methane while minimizing the generation of NOx (IEA Bioenergy, 2000). To minimize emission of CO or NOx, the flare temperature should be controlled between 850 to 1200°C. Technical specification of flaring systems must also meet the Code of Federal Regulations, CFR 40, 60.18, and 60.31.

EXAMPLE 16–4   Calculation of NO$_x$ Emissions from a Natural Gas-Fired Boiler Using U.S. EPA's AP42 compilation of air emission factors, estimate NOx emissions from a natural-gas fired boiler. Use the following data to calculate the emissions.

Item	Unit	Basis
Boiler heat input rating	52,753 MJ/h (50.0 MMBtu/h)	Vendor or Manufacturer
Boiler fuel	Natural Gas	
Natural gas heat content	39.1 MJ/m³ (1050 MMBtu/MMscf)	U.S. EPA AP42ᵃ, Appendix A
Boiler operation	8760 h/y	
Emission factor	1600 kg/m³ (100 lb/MMscf)	U.S. EPA AP42ᵃ, 1.4–5, Small Boilers, Uncontrolled

ᵃ Latest version of AP42 sections can be found in: http://www.epa.gov/ttnchie1/ap42/.

Solution

1. Determine the amount of fuel that is combusted over a a period of a year. The amount of fuel is given by

    Amount of fuel = (Boiler heat input rating / fuel heat content) × h of operation/y

    $$(52,753 \text{ MJ/h}) / (39.1 \text{ MJ/m}^3) \times 8760 \text{ h/y}$$

    $$= 11.82 \times 10^6 \text{ m}^3\text{/y of natural gas consumed}$$

2. Using the emission factor found in U.S. EPA AP42 (see Table 16–15), estimate the anticipated emissions:

Emissions = fuel consumed × emission factor

Emissions = $(11.82 \times 10^6 \text{ m}^3/\text{y}) \times (1600 \text{ kg/m}^3) \times (1 \text{ tonne}/10^3 \text{ kg})$

= 18.9 tonne of NOx emissions/y

Comment As reported in Table 16–15, the emission factor for low NOx burner is 800 kg/10^3 m^3 (50 lb/10^6 ft^3), which is half of uncontrolled small natural gas boilers, and the emission factor is 512 kg/10^3 m^3 (32 lb/10^6 ft^3), less than one-third, with a flue gas recirculation.

16–6 EMISSION OF GREENHOUSE GASES

Recognizing the evidence and projected impacts of the emission of greenhouse gases (GHG), their reduction and/or elimination has become an important element of wastewater management. Assessment of GHG emissions is important in establishing priorities for future capital investments. In planning wastewater treatment facilities, reduction of GHG emissions is often considered in the environmental assessment of the triple bottom line (TBL) approach (see Chap. 18). Because there was no standard protocol for the measurement of GHG emissions until recently and the methodology can affect the outcome of the assessment, it is useful to first review the GHG measurement framework and protocols. After the review of the framework and protocols, opportunities for the reduction of GHG emissions from wastewater treatment plant are discussed.

Framework for Greenhouse Gases Reduction

The United Nations Framework Convention on Climate Change (UNFCCC), first signed in Rio de Janeiro in June 1992, was the first major world-wide recognition of the challenges posed by climate change. The Kyoto Protocol, adopted in 1997, is the first international agreement that set binding targets for industrialized countries towards actions to reduce greenhouse gas emissions. Following the Kyoto Protocol, Intergovernmental Panel on Climate Change (IPCC) issued National Greenhouse Gas Inventories Guidelines (updated most recently in 2006). Protocols developed thereafter to measure the GHG emissions in smaller scales generally follow the greenhouse gases specified in the Kyoto Protocol. In the United States, US EPA issued the Mandatory Reporting of Greenhouse Gases Rule (74 FR 56260) in 2008.

Assessment Protocols

The protocol entitled "GHG Protocol Corporate Accounting and Reporting Standard," often referred simply as the *GHG Protocol*, developed by the World Resources Institute (WRI) and the World Business Council for Sustainable Development (WBCSD) is consistent with the IPCC Guidelines and accepted widely as the standard protocol to quantify GHG emissions from a specific business(WRI and WBCSD, 2004). In the United States, the Local Government Operations (LGO) protocol was developed based on the *GHG Protocol* in partnership with the California Air Resources Board (ARB), California Climate Action Registry (CCAR), ICLEI-Local Government for Sustainability, in collaboration with the Climate Registry and dozens of stakeholders (ARB et al., 2010). In the LGO protocol, procedures to calculate GHG emissions from government-owned facilities including water and wastewater treatment facilities are identified. The procedure to calculate GHG emissions from a wastewater treatment plant is illustrated in Example 16–4. It should be noted that these protocols are being updated to reflect the latest scientific findings, and latest publications of relevant protocols should be consulted.

Greenhouse Gases.   Based on the Kyoto Protocol, the following six greenhouse gases are usually considered in the assessment: carbon dioxide (CO_2), methane (CH_4), nitrous oxide (N_2O), hydrofluorocarbons (HFCs), perfluorocarbons (PFCs), and sulfur hexafluoride (SF_6). Because each greenhouse gas has a different level of impact on retaining heat in the atmosphere, the amount of GHG is usually reported as carbon dioxide equivalent, using global warming potentials (GWPs) determined for each greenhouse gas (see Table 16–16). For wastewater treatment facilities, only CO_2, CH_4 and N_2O are usually assessed because emissions of HFCs, PFCs, and SF_6 are generally negligible.

GHG Emissions Categories.   Generally, GHG emissions are categorized in three categories:

  Scope 1: All direct GHG emissions
  Scope 2: Indirect emissions associated with the consumption of purchased or
                acquired electricity, steam, heating, or cooling
  Scope 3: All other indirect emissions not covered in Scope 2

In the LGO protocol, Scopes 1 and 2 are considered mandatory for reporting, while Scope 3 is stated as optional. In the LGO protocol it is also recommended that CO_2 emissions from the combustion of biomass (or biomass-based fuels including digester gas) be quantified, but not included in the Scope 1 emissions, and reported separately (LGOP, 2008) as "biogenic" emissions.

The protocol, PAS 2050, prepared by the British Standards Institution (BSI), is another approach that can be used for the assessment of greenhouse gas emissions (BSI, 2011). The PAS 2050 protocol is based on the life cycle assessment, and encompasses the boundary to the Scope 3 in the WRI/WBSCD protocol.

Nitrous Oxide Emission.   Because of its global warming potential (about 310 times the effect of carbon dioxide), emission of nitrous oxide from wastewater treatment facilities could potentially be significant relative to the emission of other greenhouse gases.

Table 16–16

Fuels used commonly at wastewater treatment plants[a]

Greenhouse gas	Global warming potential (GWP)
Carbon dioxide (CO_2)	1
Methane (CH_4)	21
Nitrous oxide (N_2O)	310
Hydrofluorocarbons (HFCs)	12–11,700
Perfluorocarbons (PFCs)	6500–9200
Sulfur hexafluoride (SF_6)	23,900

[a] From LGO Protocol (2010). GWP values vary for HFCs and PFCs depending on the specific chemical compounds.

Currently, the protocols described above use an approximate estimate of nitrous oxide emission from wastewater treatment facilities based on the nitrogen loading to the plant or population served, types of treatment, and types of receiving waters. The emission is considered as a direct emission from the facility and included in the Scope 1 item. As discussed in Chap. 7, the mechanism of nitrous oxide emission from wastewater treatment facility is highly dependent on the operating conditions, making it difficult to make accurate estimates. Based on the protocol, nitrous oxide emission could make up almost one third, or more, of the total GHG emissions from a wastewater treatment facility.

EXAMPLE 16–5 **Calculation of GHG Emissions from a Wastewater Treatment Plant** Calculate the GHG emissions from the wastewater treatment plant. The treatment plant is designed for BOD removal and nitrification/denitrification, and effluent is discharged to an estuary. The treatment plant serves a combination of residential and industrial/commercial customers. The data required to complete the calculations for the treatment plant were collected and summarized as below. Report Scope 1 and Scope 2 emissions, and report biogenic emissions separately.

Item	Unit	Value
Energy use		
Electricity	kWh/y	14,100,000
Natural gas	m³/y	17,300
Fuel oil #2	m³/y	390
Digester gas production	m³/y	1,047,900
Digester gas used	m³/y	755,000
Digester gas flared	m³/y	290,500
Digester gas vented	m³/y	2400
Plant performance		
Annual average flowrate	m³/d	100,000
Average influent ammonium concentration	mg/L	18
Average influent total nitrogen concentration	mg/L	32
Average effluent ammonium concentration	mg/L	1.5
Average effluent total nitrogen concentration	mg/L	8.3
Other required information		
Population served	persons	430,000
Methane content in digester gas	%	60
Carbon dioxide in digester gas	%	35
Energy content in natural gas	GJ/m³	0.0383
Energy content in fuel oil #2	GJ/m³	38.47
Energy content in digester gas	GJ/m³	0.0224

(continued)

(*Continued*)

Item	Unit	Value
Emission factors		
Electricity	g CO_2e/kWh	720
Natural gas		
CO_2	g/GJ	50,253
CH_4	g/GJ	0.948
N_2O	g/GJ	0.0948
Fuel oil #2		
CO_2	g/GJ	70,100
CH_4	g/GJ	2.844
N_2O	g/GJ	0.569
Digester gas		
CO_2	g/GJ	49,353
CH_4	g/GJ	3.033
N_2O	g/GJ	0.597

Solution

1. Calculate Scope 1 emissions which Include emissions from stationary combustion of natural gas, fuel oil, fugitive emission of digester gas, and emissions associated with wastewater treatment process:

   a. Emissions from natural gas

   CO_2 emission = (Natural gas use, m^3/y)(energy content, GJ/m^3)(emission factor)

         = (17,300 m^3/y)(0.0383 GJ/m^3)(50,253 g/GJ)(1 tonne/10^6 g)

         = (33,297,135 g/y)(1 tonne/10^6 g)

         = 33.3 tonne/y

   CH_4 emission = (Natural gas use, m^3/y)(energy content, GJ/m^3)(emission factor)

         = (17,300 m^3/y)(0.0383 GJ/m^3)(0.948 g/CH_4/GJ)(1 tonne/10^6 g)

         = (628.1g CH_4/y)(1 tonne/10^6 g)

         = 0.000628 tonne CH_4/y

   N_2O emission = (Natural gas use, m^3/y)(energy content, GJ/m^3)(emission factor)

         = (17,300 m^3/y)(0.0383 GJ/m^3)(0.095 g N_2O/y)(1 tonne/10^6 g)

         = (628.1g N_2O/y)(1 tonne/10^6 g)

         = 0.0000628 tonne N_2O/y

   Total emissions = (CO_2 emission) GWP_{CO_2}

         + (CH_4 emission) GWP_{CH_4}

         + (N_2O emission) GWP_{N_2O}

       = 33.3 × 1.0 + 0.000628 × 21 + 0.0000628 × 310

       = 33.3 tonne CO_2 e/y

b. Emissions from fuel oil #2

CO_2 emission = (Fuel oil use, m^3/y)(energy content, GJ/m^3)(emission factor)

$$= (390 \text{ m}^3/\text{y})(38.47 \text{ GJ/m}^3)(70,100 \text{ g CO}_2/\text{GJ})(1 \text{ tonne}/10^6 \text{ g})$$

$$= (1,051,731,330 \text{ g/y})(1 \text{ tonne}/10^6 \text{ g})$$

$$= 1051.7 \text{ tonne/y}$$

CH_4 emission = (Fuel oil use, m^3/y)(energy content, GJ/m^3)(emission factor)

$$= (390 \text{ m}^3/\text{y})(38.47 \text{ GJ/m}^3)(2.844 \text{ kg CH}_4/\text{GJ})(1 \text{ tonne}/10^6 \text{ g})$$

$$= (42,669.4 \text{ g CH}_4/\text{y})(1 \text{ tonne}/10^6 \text{ g})$$

$$= 0.0427 \text{ tonne CH}_4/\text{y}$$

N_2O emission = (Fuel oil use, m^3/y)(energy content, GJ/m^3)(emission factor, kg-N_2O/GJ)

$$= (390 \text{ m}^3/\text{y})(38.47 \text{ GJ/m}^3)(0.569 \text{ kg N}_2\text{O/GJ})(1 \text{ tonne}/10^6 \text{ g})$$

$$= (8536.9 \text{ g N}_2\text{O/y})(1 \text{ tonne}/10^6 \text{ g})$$

$$= 0.00854 \text{ tonne N}_2\text{O/y}$$

Total emissions = (CO_2 emission) GWP_{CO_2}
+ (CH_4 emission) GWP_{CH_4}
+ (N_2O emission) GWP_{N_2O}

$$= 1050.7 \times 1.0 + 0.0427 \times 21 + 0.00854 \times 310$$

$$= 1054.2 \text{ tonne CO}_2 \text{ e/y}$$

c. Emissions from digester gas used. Note CO_2 emission from digester gas combustion is considered "biogenic." Biogenic emissions are often not required to be reported as part of the GHG emission but to be reported separately.

CO_2 emission = (Digester gas use, m^3/y)(energy content, GJ/m^3)(emission factor)

$$= (755,000 \text{ m}^3/\text{y})(0.0224 \text{ GJ/m}^3)(49,353 \text{ kg-CO}_2/\text{GJ})(1 \text{ tonne}/10^6 \text{ g})$$

$$= (834,657,936 \text{ g/y})(1 \text{ tonne}/10^6 \text{ g})$$

$$= 834.7 \text{ tonne/y (biogenic)}$$

CH_4 emission = (Digester gas use, m^3/y)(energy content, GJ/m^3)(emission factor)

$$= (755,000 \text{ m}^3/\text{y})(0.0224 \text{ GJ/m}^3)(3.033 \text{ kg CH}_4/\text{GJ})(1 \text{ tonne}/10^6 \text{ g})$$

$$= (51,294 \text{ g CH}_4/\text{y})(1 \text{ tonne}/10^6 \text{ g})$$

$$= 0.0513 \text{ tonne-CH}_4/\text{y}$$

N_2O emission = (Digester gas use, m^3/y)(energy content, GJ/m^3)(emission factor)

$$= (755,000 \text{ m}^3/\text{y})(0.0224 \text{ GJ/m}^3)(0.597 \text{ kg N}_2\text{O/GJ})(1 \text{ tonne}/10^6 \text{ g})$$

$$= (10,096 \text{ g N}_2\text{O/y})(1 \text{ tonne}/10^6 \text{ g})$$

$$= 0.0101 \text{ tonne N}_2\text{O/y}$$

Total emissions = (CO_4 emission) GWP_{CH_4}

$$+ (N_2O \text{ emission}) \, GWP_{N_2O}$$

$$= 0.0513 \times 21 + 0.0101 \times 310$$

$$= 4.21 \text{ tonne } CO_2 \text{ e/y}$$

Biogenic emissions = (CO_2 emission) GWP_{CO_2}

$$= 834.7 \times 1.0$$

$$= 834.7 \text{ tonne } CO_2 \text{ e/y}$$

d. Emissions from digester gas flared. In the LOG Protocol, digester gas flaring is assumed to leave 1 percent of the methane gas within the digester gas due to incomplete combustion, and no nitrous oxide emission is assumed from the digester gas flaring. The approach taken by the LOG protocol is followed in this example. Similarly to combusted digester gas, CO_2 emission is considered "biogenic."

CO_2 emission = (Digester gas flared, m^3/y)(energy content, GJ/m^3)(emission factor)

$$= (290,500 \text{ m}^3\text{/y})(0.0224 \text{ GJ/m}^3)(49,353 \text{ kg } CO_2\text{/GJ})(1 \text{ tonne/}10^6 \text{ g})$$

$$= (834,657,936 \text{ g/y})(1 \text{ tonne/}10^6 \text{ g})$$

$$= 321.15 \text{ tonne/y (biogenic)}$$

CH_4 emission = (Digester gas flared, m^3/y)(methane content)
$\times$ (incomplete combustion)(mass of methane, g/m^3)

$$= (290,500 \text{ m}^3\text{/y})(0.60)(0.01)(656 \text{ g/m}^3)(1 \text{ tonne/}10^6 \text{ g})$$

$$= (1,143,408 \text{ g } CH_4\text{/y})(1 \text{ tonne/}10^6 \text{ g})$$

$$= 1.143 \text{ tonne } CH_4\text{/y}$$

Total emissions = (CH_4 emission) GWP_{CH_4}

$$= 1.143 \times 21$$

$$= 24.0 \text{ tonne } CO_2 \text{ e/y}$$

Biogenic emissions = (CO_2 emission) GWP_{CO_2}

$$= 321.15 \times 1.0$$

$$= 321.2 \text{ tonne } CO_2 \text{ e/y}$$

e. Emissions from digester gas vented. Of the digester gas vented to the atmosphere, 60 percent is methane, to be reported as Scope 1 emission. Carbon dioxide (35 percent) is not counted in the GHG inventory.

CH_4 emission = (Digester gas vented, m^3/y)(methane content)
(mass of methane, g/m^3)

$$= (2400 \text{ m}^3\text{/y})(0.60)(656 \text{ g/m}^3)(1 \text{ tonne/}10^6 \text{ g})$$

$$= (944,640 \text{ g-}CH_4\text{/y})(1 \text{ tonne/}10^6 \text{ g})$$

$$= 0.945 \text{ tonne-}CH_4\text{/y}$$

$$\text{Total emissions} = (\text{CH}_4 \text{ emission}) \text{GWP}_{\text{CH}_4}$$

$$= 0.945 \times 21$$

$$= 19.8 \text{ tonne CO}_2 \text{ e/y}$$

f. Determine process N_2O emissions from WWTP with nitrification/denitrification.
In the LGO Protocol, process N_2O emission from WWTP with nitrification/denitrification is estimated using an emission factor of 7 g-N_2O/person/y. Using the given data, the emission is estimated as

$$\text{Process N}_2\text{O emission} = [(P_{\text{total}} \times F_{\text{ind-com}})EF_{\text{nit/denit}}(1 \text{ tonne}/10^6 \text{ g})]\text{GWP}_{\text{N}_2\text{O}}$$

$$= [(430,000 \times 1.25) \times 7 \times 10^{-6}] \times 310$$

$$= 1166.4 \text{ tonne CO}_2 \text{ e/y}$$

where, P_{total} = total population served by the treatment plant
$F_{\text{ind-com}}$ = factor for industrial and commercial co-discharge waste into the sewer system = 1.25 for WWTPs
$EF_{\text{nit/denit}}$ = emission factor = 7 g-N_2O/person·y
$\text{GWP}_{\text{N}_2\text{O}}$ = global warming potential for N_2O = 310

g. Determine process N_2O emissions from effluent discharge.
In the LGO Protocol, process N_2O emission from wastewater discharge is estimated based on the measured average total nitrogen discharge. The emission factor for N_2O from effluent discharge is 0.005 kg N_2O-N/kg-N discharged in the effluent. Using the given data, the N_2O emission is estimated as follows.

$$\text{Process N}_2\text{O emission} = (\text{N load})(EF_{\text{effluent}})(365.25)(44/28)(10^{-6})\text{CWP}_{\text{N}_2\text{O}}$$

$$= [(8.3 \text{ g-N/m}^3)(100,000 \text{ m}^3/\text{d})](0.005 \text{ kg N}_2\text{O-N/kg-N})$$
$$(365.25 \text{ d/y})(44 \text{ g-CO}_2/28 \text{ g-N})(1 \text{ tonne}/10^6 \text{ g})(310)$$

$$= (2,381,952 \text{ g/y})(1 \text{ tonne}/10^6 \text{ g})(310)$$

$$= 738.4 \text{ tonne CO}_2\text{e/y}$$

h. Prepare a summary of Scope 1 emissions.

$$\text{Total Scope 1 emissions} = 33.3 + 1054.2 + 4.21 + 24.0 + 19.8 + 1166.4$$
$$+ 738.4 = 3040 \text{ tonne CO}_2\text{e/y}$$

$$\text{Biogenic emissions} = 834.7 + 321.2 = 1156 \text{ tonne CO}_2\text{e/y}$$

2. Calculate Scope 2 emissions.
Emissions associated with electricity:

$$\text{CO}_2 \text{ emission} = (\text{Electricity use, kWh/y})(\text{emission factor, g CO}_2\text{e/kWh})$$

$$= 14,100,000 \times 720 \text{ g/y}$$

$$= (10,152,000,000 \text{ g/y})(1 \text{ tonne}/10^6 \text{ g})$$

$$= 10,152 \text{ tonne/y}$$

3. Prepare a summary of the total and biogenic emissions.

Total emissions = Scope 1 emissions + Scope 2 emissions

$$= 3040 + 10{,}152 = 13192 \text{ tonne } CO_2e/y$$

Total biogenic emissions = 1156 tonne CO_2e/y

Comment The calculations shown in this example are based on the LGO Protocol, presented in SI units. It should be noted that nearly two-thirds of Scope 1 emissions is attributed to nitrous oxide emissions, and it constitutes about 14 percent of the total emissions.

Opportunities for GHG Reduction at Wastewater Treatment Facilities

Opportunities typically investigated for the reduction of GHG emissions from wastewater treatment facilities are summarized in Table 16–17. Many of the GHG reduction opportunities are related directly to reduction of energy use, discussed further in Chap. 17. Other opportunities are related to shifting the energy sources to those with lower GHG emissions. Examples of opportunities for GHG reductions are discussed briefly in the following subsections. It should be noted that there are numerous options, most of them site-specific, to achieve energy conservation and GHG emission reduction. It is also important to note that a thorough evaluation of existing conditions must be made before options for GHG reduction are evaluated. In conducting the evaluation of existing conditions and opportunities for GHG reduction, latest publications from U.S. EPA, WEF and other sources should be consulted.

Control of Dissolved Oxygen. Aeration of activated sludge reactors is the most significant process element when evaluating energy use as it takes nearly half of the total process-related energy at a typical conventional secondary treatment facility (see Chaps. 4 and 17). Control of dissolved oxygen not only helps reduce energy use, it also helps improve the treatment performance. The use of automated DO control with online DO analyzers has become a common practice. Other process upgrades to improve DO control and associated energy efficiencies include the improvements in blowers and diffusers (see Chap. 5).

Treatment Process Modification. In selecting a treatment process, or planning a process upgrade and modifications, considerations should be given to the treatment processes that require less energy to achieve the treatment objective. For example, by controlling the nitrogen removal process to nitritation and denitritation (see Chap. 7, 8 and 15), oxygen requirements for nitrogen removal could be reduced by 25 percent. Generation and recovery of energy from wastewater and waste energy within the treatment facility is also a significant consideration in GHG reduction as energy recovered from wastewater constituents (e.g., in the form of digester gas) is considered as biogenic and not counted in the Scope 1 and Scope 2 emissions. Greenhouse gas emissions associated with the use of chemicals of fossil fuel sources are usually counted as a Scope 3 emission. However, CO_2 emissions from the use of fossil fuel-origin methanol for denitrification may be counted as process-oriented Scope 1 emission (ICLEI, 2012).

Management of Nitrous Oxide Emission. Emission of nitrous oxide is estimated with assumptions that may not reflect actual emissions from the treatment processes and receiving waters, and it constitutes a significant fraction of total GHG emissions from wastewater treatment facilities. Depending on the mode of operation and

Table 16–17

Considerations in greenhouse gas reduction options for wastewater treatment facilities

Unit process	Description
General	• Selection of lower GHG emission energy sources (e.g., digester gas or natural gas vs. fuel oil)
	• Waste energy recovery within wastewater treatment plant (see Chap. 17)
Inlet/preliminary treatment	• Pump efficiency improvement
	• Flow equalization
Primary treatment	• Improvements in solids removal
Secondary treatment and sidestream treatment	• Use of diffusers with high oxygen transfer efficiencies
	• Use of blowers with high energy efficiencies
	• Selection of blowers or combination of blowers sized to allow operation in high efficiency range over full range of possible air requirements
	• Use of energy efficient mixing system for activated sludge anoxic/anaerobic zones
	• Control of aeration with DO monitoring
	• Control of aeration with NH_4-N monitoring
	• Selection of biological process requiring less oxygen (e.g., nitritation/denitritation, partial nitritation/deammonification) (see Chaps. 7, 8, 9 and 15)
	• Process configuration and control to minimize generation of N_2O (see Chap. 7)
Sludge processing and biosolids utilization/disposal	• Optimization of sludge thickening and dewatering
	• Elimination of unaccounted and un-combusted methane emissions
	• Complete utilization of digester gas
	• Enhancement in digester gas production (see Chap. 13)
	• Use of waste heat for pre-heating of sludge drying (see Chap. 14)
	• Heat recovery from sludge incineration systems (see Chap. 14)
	• Onsite energy and heat production from digester gas and biosolids (see Chaps. 14 and 17)
Disinfection	• Use of high-output lamps for UV disinfection
Advanced treatment	• Use of membrane systems with lower energy requirements
	• Energy recovery from residual pressure in the membrane treatment system

control, emission of nitrous oxide from biological treatment process could increase significantly. Generation of nitrous oxide during biological treatment is discussed in Sec. 7–12 in Chap. 7.

Use of Digester Gas.   Even though the use of digester gas for boilers and power generation has been a common practice at many wastewater treatment facilities, not all facilities with anaerobic sludge digestion utilize all of the digester gas. Because methane has 21 times higher global warming potential than carbon dioxide, one of the major GHG emission sources can be the release of uncombusted digester gas into the atmosphere. By simply flaring the digester gas, the GHG emission is reduced by 21 fold, and because the CO_2 released from the flare is originated from biological sources, it is counted as a "biogenic" emission, which is reported separately, but not included in the GHG emissions inventory. By maximizing the use of digester gas as a fuel source, part of the GHG emissions associated with the use of electricity and fuels from other sources could be avoided. The use of digester gas and management of energy use are discussed further in Chap. 17.

PROBLEMS AND DISCUSSION TOPICS

16-1 Verify that for each mg/L of H_2S removed with chlorine 10.87 mg/L of alkalinity as $CaCO_3$ will be required.

16-2 Determine the amount of hydrogen peroxide (H_2O_2) required for the oxidation of hydrogen sulfide H_2S.

16-3 Using the following half reaction for permanganate (MnO_4^-), estimate the amount of permanganate that would be required per day to oxidize 100 ppm_v of H_2S from an foul air stream with a flowrate of 1500, 2000, 1800 or 2200 m^3/min (flowrate to be selected by instructor).

$$MnO_4^- + 4H^+ + 3e^- \rightarrow MnO_2(s) + 2H_2O$$

16-4 Determine the amount of ferrous sulfate ($FeSO_4$) that would be required to remove 150 mg/L of H_2S from digester supernatant. Assume the sulfide ion in H_2S will be converted to ferrous sulfide in an exchange reaction.

16-5 Four different waste air streams have been sampled and the results are summarized below. For one of these waste air streams (to be selected by instructor), determine the chemical requirements. Sodium hypochlorite and sodium hydroxide are to be used in the chemical scrubber.

Item	Unit	Plant			
		1	2	3	4
Air waste stream flowrate	m^3/min	1000	2500	3200	1800
H_2S concentration	ppm_v	75	45	65	35
Liquid to gas ratio	kg/kg	1.85	2.0	2.1	1.9
Temperature	°C	28	33	30	25
Density of 50 percent NaOH solution	kg/L	1.52	1.52	1.52	1.52

16-6 Using the design criteria given in Table 16–11, determine the size of compost filter needed to scrub 65 ppm_v H_2S from foul air at a flowrate of 1500, 1880, 2100 or 2300 m^3/min (value to be selected by instructor). Also estimate the mass of the buffer compound needed to neutralize the acid formed as a result of treatment within the filter. Assume a packed bed porosity of 43 percent. The temperature of the foul air is 20°C.

16-7 Using U.S. EPA's AP42 compilation of air emission factors, estimate (CO, NOx or SO_2 emissions, to be selected by instructor) from a dual-fuel reciprocating engine. The data used to calculate the emissions are summarized below and emission factors are reported in Table 16–15.

Item	Unit
Engine power rating	2386 kW (3200 bhp)
Fuel	Fuel oil #2, natural gas
Fuel oil #2 heat content	38.47 GJ/m^3
Natural gas heat content	0.0383 GJ/m^3
Average load by fuel oil	35 percent
Engine operation	8640 h/y

16–8 In Example 16–4, part of the digester gas was flared and vented. Consider the situation where all unutilized digester gas in Example 16–4 is utilized to reduce natural gas consumption, assuming the equipment at the treatment plant is capable to switch the use between natural gas and digester gas. Calculate: (a) the amount of natural gas consumption saved per year, and (b) reduction in the total GHG emissions.

REFERENCES

Allen, E. R., and Y. Yang (1991) "Biofiltration Control of Hydrogen Sulfide Emissions," *Proceedings of the 84th Annual Meeting of the Air and Waste Management Association,* Vancouver, BC, Canada.

Allen, E. R., and Y. Yang (1992) "Operational Parameters for the Control of Hydrogen Sulfide Emissions Using Biofiltration," *Proceedings of the 85th Annual Meeting of the Air and Waste Management Association,* Kansas City, MO.

ARB, CCAR, ICLEI, and the Climate Registry (2010) *Local Government Operations Protocol for the Quantification and Reporting of Greenhouse Gas Emissions Inventories,* Version 1.1, California Air Resources Board, California Climate Action Registry, ICLEI – Local Governments for Sustainability, and the Climate Registry, Sacramento, CA.

Bielefeldt, A. R., and H. D. Stensel (1999) "Treating VOC-Contaminated Gases in Activated Sludge: Mechanistic Model to Evaluate Design and Performance," *Environ. Sci. Technol.,* **33**, 18, 3234–3240.

Bohn, H. L., and R. K. Bohn (1988) "Soil Beds Weed Out Air Pollutants," *Chem. Engnr.,* **95**, 6, 73–76.

BSI (2011) *Specification for the Assessment of the Life Cycle Greenhouse Gas Emissions of Good and Services,* PAS 2050:2011, British Standards Institution, London.

Crites, R. W., and G. Tchobanoglous (1998) *Small and Decentralized Wastewater Management Systems,* McGraw-Hill, New York.

Devinny, J S., M. A. Deshusses, and T. S. Webster (1999) *Biofiltration For Air Pollution Control,* Lewis Publishers, Boca Raton, FL.

Eweis, J. B., S. J. Ergas, D. P. Y. Chang, and E. D. Schroeder (1998) *Bioremediation Principles,* McGraw-Hill, Boston, MA.

Finn, L., and R. Spencer (1997) "Managing Biofilters for Consistent Odor and VOC Treatment," *BioCycle,* **38**, 1, 40–44.

Hsieh, C., K. S. Ro, and M. K. Stenstrom (1993) "Estimating Emissions of 20 VOCs II, Diffused Aeration," *J. Environ. Engr.,* **119**, 6, 1099–1118.

ICLEI (2012) *U.S. Community Protocol for Accounting and Reporting of Greenhouse Gas Emissions,* V1.0, ICLEI - Local Government for Sustainability USA, Oakland, CA.

IEA Bioenergy (2000) *Biogas Flares: State of the Art and Market Review,* Topic Report of the IEA Bioenergy Agreement Task 24 – Biological Conversion of Municipal Solid Waste, *AEA Technology Environment,* Culham, Abingdon, Oxfordshire, UK.

Libra, J. A. (1993) "Stripping of Organic Compounds in an Aerated Stirred Tank Reactor;" Fortschc.-ber. VD1 Rhhe 15, Nr. 102, VDI-Verlag, Düsseldorf, Germany.

Matter-Muller, C., W. Gujer, and W. Giger (1981) "Transfer of Volatile Substances from Water to the Atmosphere," *Water Res.,* **15**, 11, 1271–1279.

Murthy, S. (2001) Personal communication.

Novak, J. T. (2001) Personal communication.

Pomeroy, R. D. (1957) "Deodorizing Gas Streams by the Use of Microbiological Growths," U.S. Patent No. 2,793,096.

Rafson, H. J. (ed.) (1998) *Odor and VOC Control Handbook,* McGraw-Hill, New York.

Roberts, P. V., and P.G. Dandliker (1983) "Mass Transfer of Volatile Organic Contaminants from Aqueous Solution to the Atmosphere During Surface Aeration," *Environ. Sci. Technol.,* **17**, 8, 484–489.

Roberts, P. V., C. Munz, P. G. Dandliker, and C. Matter-Muller (1984) *Volatilization of Organic Pollutants in Wastewater Treatment-Model Studies,* EPA-600/S2-84-047.

Schroeder, E. D. (2001) Personal Communication, Department of Civil and Environmental Engineering, University of California at Davis, Davis, CA.

Schwarzenbach, R. P., P. M. Gschwend, and D. M. Imboden (1993) *Environ. Org. Chem.*, Wiley, New York.

Tchobanoglous, G., and E. D. Schroeder (1985) *Water Quality: Characteristics, Modeling, Modification,* Addison-Wesley Publishing Company, Reading, MA.

Thibodeaux, L. J. (1979) *Chemodynamics: Environmental Movement of Chemicals in Air, Water, and Soil,* Wiley, New York.

U.S. EPA (1985) *Design Manual, Odor and Corrosion Control in Sanitary Sewerage Systems and Treatment Plants,* EPA/625/1–85/018, U.S. Environmental Protection, Agency Washington, DC.

U.S. EPA (1986) *Handbook: Control Technologies for Hazardous Air Pollutants,* EPA/625/6–86/014, U.S. Environmental Protection Agency, Research Triangie Park, NC.

U.S. EPA (1995) *Compilation of Air Pollutant Emission Factors,* AP-42, 5th ed., U.S. Environmental Protection Agency, Research Triangle Park, NC.

U.S. EPA (1998) *Compilation of Air Pollutant Emission Factors,* AP 42, 5th ed., Vol. I, Chap. 1: External Combustion Sources, Sec 1.4, Natural Gas Combustion, U.S. Environmental Protection Agency, Research Triangle Park, NC.

U.S. EPA (1999) *Compilation of Air Pollutant Emission Factors,* AP 42, 5th ed., Vol. I, Chap. 1: External Combustion Sources, Sec 1.3, Fuel Oil Combustion, U.S. Environmental Protection Agency, Research Triangle Park, NC.

U.S. EPA (2010) *Evaluation of Energy Conservation Measures for Wastewater Treatment Facilities,* EPA 832-R-10–005, Office of Wastewater Management, U.S. Environmental Protection Agency, Washington, DC.

van Lith, C. (1989) "Design Criteria for Biofilters," Proceedings of the 82th Annual Meeting of the Air and Waste management Association, Anaheim, CA.

WEF (1995) *Odor Control in Wastewater Treatment Plants,* WEF Manual of Practice No. 22, Water Environment Federation, Alexandria, VA.

WEF (1997) *Biofiltration: Controlling Air Emissions through Innovative Technology,* Project 92-VOC-1, Water Environment Federation, Alexandria, VA.

WEF (2004) *Control of Odors and Emissions from Wastewater Treatment Plants,* WEF Manual of Practice No. 25, Water Environment Federation, Alexandria, VA.

Williams, D. G. (1996) *The Chemistry of Essential Oils,* Michelle Press, Dorset, England.

Williams, T. O., and F. C. Miller (1992a) "Odor Control Using Biofilters," *BioCycle,* **33**, 10, 72–77.

Williams, T. O., and F. C. Miller (1992b) "Biofilters and Facilities Operations," *BioCycle,* **33**, 11, 75–79.

Wilson, G. (1975) Odors: Their Detection and Measurement, EUTEK Process Development and Engineering, Sacramento, CA.

Yang, Y., and E. R. Allen (1994) "Biofiltration Control of Hydrogen Sulfide: Design and Operational Parameters," *J. AWWA,* **44**, 7, 863–868.

17

Energy Considerations in Wastewater Management

WORKING TERMINOLOGY

Term	Definition
Biogas	A generic term used for the gas produced during anaerobic treatment of organic compounds. Typical composition of biogas generated by anaerobic digestion of sludges is 60 to 65 percent methane, and approximately 30 percent carbon dioxide. The term is used interexchangeably with digester gas.
Bomb calorimeter	A device in which a known mass of sample is combusted and the released energy is measured.
Carnot cycle	A reversible power cycle consisting of two isothermal processes and two isentropic processes without internal heat transfer.
Coefficient of performance (COP)	A ratio of the energy extracted from an energy recovery process to the energy input to the energy recovery process such as heat pump.
Convection	Heat transferred by the movement of gas or liquid.
Combined heat and power (CHP)	A system used to generate both electricity and heat. Also called cogeneration.
Electrical efficiency	The percentage of energy that is converted into electricity relative to the total energy input.
Enthalpy	A thermodynamic property of a system expressed as the sum of the internal energy and the product of pressure and volume.
Exothermic reaction	A chemical reaction resulting in a release of energy from the molecules.
Fuel cell	A device used to generate electricity from the reaction of hydrogen and oxygen gas. In wastewater treatment facilities, methane in biogas can be used to generate hydrogen. Recoverable thermal energy can also be generated with a fuel cell.
Heat exchanger	A device that allows the extraction of heat through a refrigerant with external power input.
Heat pump	A device used to transfer heat from one body to another using a refrigerant and energy from an external power source.
Higher heating value (HHV)	The total amount of energy released by complete combustion of a unit mass of a substance. Latent heat of vaporization is counted in HHV.
Integrated resource recovery (IRR)	A management approach in which the recovery of energy, water, and other resources is achieved by integrating the management of various waste streams such as wastewater, solid waste, and others.
Latent heat	Heat released or absorbed during the change in phases of a chemical substance (e.g., from liquid to gas).
Lower heating value (LHV)	The total amount of energy released by complete combustion of a unit mass of a substance, less the latent heat of vaporization of the water vapor formed by the combustion.

Term	Definition
Organic Rankine process	A process to generate electricity using low grade heat such as waste heat from a stationary combustion system and with an organic fluid that changes phases at a lower temperature than water.
Sensible heat	Energy transferred to a substance that results in a change in temperature of the substance.
Specific heat	The amount of heat necessary for the temperature of a unit mass of a material to increase by one unit.
Syngas	Syngas, or synthesis gas, is a mix of hydrogen, carbon monoxide, and other gaseous compounds generated as a result of the incomplete oxidation of organic materials under restricted oxygen environment. Syngas can be used as a fuel gas or to generate liquid fuel through the Fischer-Tropsch process.
Thermal efficiency	The ratio of net usable thermal output to the energy input, expressed in percentage. Also known as total system efficiency.

In the past, due to the relatively low and stable cost of fossil fuels and electricity, the use, recovery, and management of energy was not typically emphasized in the design and operation of wastewater management facilities. With increasing energy costs, uncertainties about future fossil fuel supplies, and increasing awareness of the impacts of greenhouse gas emissions, the efficient management of energy is now of greater concern with both private and public entities. Recognizing the importance of energy in the implementation of wastewater treatment facilities, and opportunities to recover and utilize energy from various sources within the treatment facilities, the focus of this chapter is on the subject of energy recovery and utilization. Topics considered in this chapter include: (1) need for energy management, (2) energy in wastewater, (3) fundamentals of heat balance, (4) energy usage in treatment plants, (5) recovery and utilization of chemical energy (6) recovery and utilization of heat energy, (7) recovery and utilization of hydraulic potential energy, (8) energy management, and (9) future opportunities for alternative wastewater treatment processes. The recovery of nutrients is considered in Chaps. 14 and 15.

17–1   FACTORS DRIVING ENERGY MANAGEMENT

The principal driving forces for achieving more efficient management of energy in wastewater treatment are

1. Potential for energy cost savings, including opportunity to become a net energy supplier
2. Potential for improved energy supply reliability
3. Considerations for sustainability, including the greenhouse gas reduction goals put forth by local, state, and federal governmental agencies

Potential for Energy Cost Savings

The operation of wastewater treatment facilities, as discussed in Chap. 4, depends largely on the use of energy resources to bring about various reactions. Therefore it is important to appraise energy requirements to better manage the energy usage, which constitutes the second largest expenditure in the operation of wastewater treatment facilities after labor costs. Examples of energy cost saving opportunities include the use of energy efficient equipment, process control for optimized energy use, and selection of energy sources including pricing negotiation.

Energy Supply Reliability

Reliability of energy supply for wastewater treatment facilities is an important consideration because of the potential for unforeseeable events such as an area-wide blackout of electrical power supply and disruption in energy supply after natural disasters. Generally, wastewater treatment facilities are equipped with emergency generators to operate critical elements of the treatment facility during disruption of energy supply, but few treatment plants are able to operate the entire treatment process with the emergency power supply. In recent years, it has been recognized that wastewater theoretically contains more energy than that required for treatment. It is also recognized that wastewater treatment plants could become net exporters of energy if the energy contained in incoming wastewater could be recovered effectively. Becoming self sufficient in the production and utilization of energy would significantly improve treatment plant reliability, even during power outages. As discussed in this chapter, however, inherent inefficiencies with energy recovery process are significant challenges. The use of anaerobic digestion and combustion of biogas, for example, usually achieves one third or less of typical energy requirements for the conventional wastewater treatment facility. Some of the technologies for energy recovery and utilization considered in this chapter are relatively new to the wastewater field, but many of them have been used in various industrial and commercial applications.

Considerations for Sustainability

In addition to the need to reduce costs for purchasing energy from outside sources, the reduction of greenhouse gas (GHG) emissions has become one of key factors affecting decisions on wastewater treatment processes and equipment selection. As discussed in Sec. 4–1, many existing treatment plants built during the 1970s and 1980s will need to be upgraded to meet increasingly stringent discharge limits. In planning treatment plant upgrades, reduction of GHG emissions (see Chap. 16) is often highlighted in the objectives of the project. Even though justifications for funding tend to focus on the immediate cost savings, when similar alternatives are compared, the options which address GHG reduction goals and other environmental impacts are considered favorably. As wastewater treatment facilities are considered more frequently to be an integral part of a regional energy management scheme, various alternative organic wastes such as food waste and grease, could be brought to the treatment plant to enhance energy production (Chap. 14).

17-2 ENERGY IN WASTEWATER

The energy contained in wastewater is comprised of (1) chemical energy, (2) thermal energy, and (3) hydraulic energy. Chemical energy is the energy contained in organic molecules which can be released by chemical reactions. Thermal energy is the heat retained in wastewater. Hydraulic energy of wastewater fluid is the sum of the gravitational potential energy due to elevation head, energy associated with pressure head, and kinetic energy embodied in the wastewater flows as velocity head. Each of the three forms of energy in wastewater is considered in the following discussion.

Chemical Energy

Wastewater contains organic and inorganic molecules, and exothermic reactions of these constituents will result in a release of chemical energy retained in the molecules. A majority of the chemical energy in wastewater is contained in organic compounds measured as COD, even though some inorganic constituents including ammonia also contain chemical energy that could be extracted. In wastewater treatment, part of the chemical energy is removed from the liquid stream in the form of sludge during preliminary and primary treatment. During the biological treatment process, some of the chemical energy is transformed into biomass and reaction products such as carbon dioxide and methane, or released as heat through metabolism of microorganisms. In sludge processing, part of the chemical energy may be recovered in the form of methane gas and utilized as an energy source.

Energy Contents of Wastewater Constituents. The energy content of wastewater, as described in Chaps. 2 and 14, can be estimated from an elemental analysis of the organic constituents in wastewater using the following empirical expression, which is a modified form of the DuLong formula developed by Channiwala (1992), repeated here from Chap. 2 for convenience.

$$\text{HHV (MJ/kg)} = 34.91\,C + 117.83\,H - 10.34\,O - 1.51\,N + 10.05\,S - 2.11A \quad (2\text{–}66)$$

Where C, H, O, N, S, and A are the weight fraction of carbon, hydrogen, oxygen, nitrogen, sulfur, and ash, respectively. Note that higher heating value (HHV) is the total amount of energy released by complete combustion of a unit mass of a substance when water remains as liquid. In actual combustion systems, the combustion temperature is well above 100°C and water will be vaporized, absorbing the latent heat of vaporization. The total amount of energy released from a complete combustion of unit mass of a substance, considering the heat required for vaporization of water is the lower heating value (LHV). For example, LHV of natural gas is typically 10 percent lower than the HHV. While Eq. (2–66) can be used to estimate the energy content, experimental data on chemical energy in wastewater samples are collected using a bomb calorimeter.

For simple molecules, chemical energy can be calculated based on the enthalpy of reaction. The enthalpy of reaction is defined by Hess' Law as the difference between the sum of enthalpy of formation of reactants and the sum of enthalpy of formation of all products:

$$H_{\text{reaction}} = \sum H^o_{f\,\text{products}} - \sum H^o_{f\,\text{reactants}} \quad (17\text{–}1)$$

where H_{reaction} = enthalpy of reaction
    H^o_f = enthalpy of formation

The reaction is an exothermic reaction when the enthalpy of reaction is negative. Enthalpy of formation of chemical compounds found commonly in wastewater is shown in Table 17–1, and a more complete list can be found in references in chemical engineering. Calculation of the heat of reaction is illustrated in Example 17–1.

EXAMPLE 17–1   Calculation of Enthalpy of Reaction   Calculate the enthalpy of reaction for the oxidation of methane.

The enthalpy of formation for each reactant and product at 25°C is

$CH_{4(g)}$	−74.6 kJ/mole
$O_{2(g)}$	0 kJ/mole
$CO_{2(g)}$	−393.5 kJ/mole
$H_2O_{(l)}$	−285.8 kJ/mole

Solution

1. The oxidation of methane can be expressed as follows.

$$CH_4 + 2O_2 \rightarrow CO_2 + 2H_2O$$

2. Determine the enthalpy for the oxidation of methane using Eq. (17–1).

$$H_{reaction} = \sum H^o_{f\,products} - \sum H^o_{f\,reactants}$$

$$H_{reaction} = [(H^o_{fCO_2}) + 2(H^o_{fH_2O})] - [(H^o_{fCH_4}) + 2(H^o_{fO_2})]$$

$$= [(-393.5) + 2(-285.8)] - [(-74.6) + 2(0)]$$

$$= -890.5 \text{ kJ/mole}$$

Comment   Because the enthalpy of reaction was calculated for 25°C, water is in the liquid phase and the calculated value is the higher heating value at 25°C. The volume of one mole of methane at 25°C at atmospheric pressure is approximately 24L. Assuming 65 percent methane content by volume in digester gas, 1 m³ of digester gas contains approximately 27 moles of methane. Thus, the HHV in digester gas can be estimated as 890 (kJ/mole) × 27 (mole/m³) = 24,112 kJ/m³ (~647 Btu/ft³). Stationary combustion systems are operated at a much higher temperature and the latent heat of water vaporization must be accounted for by estimating the lower heating value.

Table 17–1

Enthalpy of formation for selected chemical compounds at 25°C[a]

Substance	State[b]	ΔH^o_f, kJ/mole	Substance	State[b]	ΔH^o_f, kJ/mole
Ca^{2+}	aq	−542.8	H_2O	g	−241.8
$CaCO_3$	s	−1206.87	HS^-	aq	−17.6
$Ca(OH)_2$	s	−986.6	H_2S	g	−20.6
$CaSO_4$	s	−1434.5	H_2S	aq	−39.3
CH_4	g	−74.6	H_2SO_4	l	−814
CH_3CH_3	g	−84.67	Mg^{2+}	aq	−466.9
CH_3COOH	aq	−488.4	$Mg(OH)_2$	s	−924.5
CH_3COO^-	aq	−486.0	Na^+	aq	−240.1
$C_6H_{12}O_6$	s	−1275	NH_3	g	−45.9
Cl_2	g	0	NH_3	aq	−80.83
Cl_2	aq	−23.4	NH_4^+	aq	−132.5
Cl^-	aq	−167.20	NO_2^-	aq	−104.6
CO_2	g	−393.51	NO_3^-	aq	−207.4
CO_2	aq	−412.92	O_2	g	0
CO_3^{2-}	aq	−677.10	O_2	aq	−11.71
HCO_3^-	aq	−692.0	OH^-	aq	−230.0
H_2CO_3	aq	−699.0	S^{2-}	aq	30.1
H_2O	l	−285.8	SO_4^{2-}	aq	−909.3

[a] From Sawyer et al. (2003).
[b] g = gas, aq = aqueous solution, l = liquid.

For larger and more complex molecules in wastewater, theoretical values cannot be calculated with Eq. (17–1) because the enthalpy of formation values for many of the reactants are not available. There have been attempts to find correlations between COD and chemical energy in wastewater, and the values ranging between 14.7 and 17.8 kJ/gCOD have been reported for untreated domestic wastewater (Shiraz and Bagley, 2004; Heidrich et al., 2011). However, as described by Heidrich et al. (2011), the amount of energy per gram COD can vary widely. Typical energy content in primary sludge has been reported to be between 23,000 and 29,000 kJ/kg dry solids (see Table 14–17 in Chap. 14). Determination of energy in primary sludge is illustrated in Example 17–2.

EXAMPLE 17–2 | **Chemical Energy in Sludge Removed by Primary Clarification** The flowrate to a wastewater treatment plant is 1000 m³/d. The average total suspended solids concentration in the wastewater is 720 g/m³. Using the typical energy contents in primary sludge reported in Table 14–17, estimate the chemical energy removed by primary clarification assuming 50 percent solids removal.

Solution
1. Calculate TSS loading.

   TSS loading = 1000 m³/d × 720 g/m³ = 720,000 g/d = 720 kg/d

2. Estimate the amount of chemical energy removed by primary clarification.

   TSS removed = (720 kg/d) × 0.5 = 360 kg/d.

   Chemical energy in the primary sludge is 23,000 to 29,000 kJ/kg. Assuming 26,000 kJ/kg,

   Chemical energy removed = (360 kg/d) × (26,000 kJ/kg) = 9,360,000 kJ/d = 9.36 GJ/d

Comment | The amount of energy that can be recovered and utilized will depend on the solids processing, energy recovery and energy use efficiencies, as discussed later in this Chapter. It should be noted that the loss of dissolved and volatile organic compounds during sample preparation have been recognized and high variability in values between samples have been reported (Shiraz and Bagley, 2004; Heidrich et al., 2011).

Energy Content of Ammonia. As described in Sec. 15–5, thermal oxidation of ammonia is an exothermic reaction, and combustion of ammonia will release energy. The heat balance based on the heat of formation at 25°C at 1 atm is expressed as:

$$NH_3 + 0.75\ O_2 \rightarrow N_2 + 1.5\ H_2O \qquad \Delta H° = -317\ kJ/mole \qquad (17–2)$$

$$NH_3 + O_2 \rightarrow 0.5\ N_2O + 1.5\ H_2O \qquad \Delta H° = -276\ kJ/mole \qquad (17–3)$$

$$NH_3 + 1.25\ O_2 \rightarrow NO + 1.5\ H_2O \qquad \Delta H° = -227\ kJ/mole \qquad (17–4)$$

Anhydrous ammonia has been used as a fuel in the past and is currently the subject of extensive research as a potential alternative fuel source. Even though the heat content is significantly lower than that of methane (802.6 kJ/mole or 50,163 kJ/kg), there is a potential to recover ammonia from nitrogen-rich waste streams and use the energy to supplement other fuel sources.

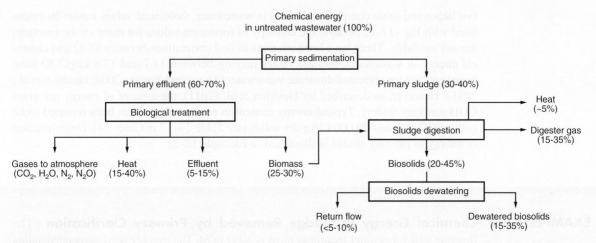

Figure 17-1

Fate of chemical energy in wastewater treatment with activated sludge and anaerobic sludge digestion. A range is shown for the percentages as they vary with the treatment technologies and wastewater characteristics.

Distribution of Chemical Energy during Wastewater Treatment. The distribution and fate of chemical energy in conventional wastewater treatment is illustrated conceptually on Fig. 17-1. Transformation of chemical energy occurs primarily during two major treatment processes: biological treatment of liquid stream, and treatment and processing of sludge. Even though each wastewater has different characteristics and distribution of energy varies with treatment processes, it is worth noting that a considerable amount of chemical energy is sent to the biological treatment process where additional energy is added to the process to convert chemical energy into CO_2, H_2O, N_2, N_2O, heat, and other byproducts that cannot be utilized as an energy source.

Heat generated from exothermic reactions can be significant in processes such as autothermal aerobic digestion (ATAD) and sidestream treatment involving oxidation and reduction of highly concentrated organic and nitrogenous compounds. In these processes, a heat balance must be prepared to determine if cooling of the treatment unit process or dilution of the treated stream is necessary (see Chap. 15) to maintain a desired range of operating temperature. However, in normal domestic wastewater, the heat generated from biochemical reactions is not significant enough to require heat balance evaluation.

Part of the chemical energy is retained in biomass, which can be transformed into an energy source such as biogas and syngas through sludge processing (see Chaps. 13 and 14). With a conventional treatment system, only a fraction of chemical energy that reaches solids processing can be recovered. Historically, wastewater treatment systems were not designed with the intent to maximize the flow of chemical energy to the processes capable of energy recovery.

Thermal Energy

The thermal energy in wastewater is in the form of temperature. When the temperature of a liquid or gas changes from T_1 to T_2, the change in thermal energy is expressed as:

$$Q = mc\Delta T \tag{17-5}$$

where, Q = change in heat content in wastewater flowrate, kJ/h
$\quad m$ = mass flowrate of water, kg/h
$\quad c$ = specific heat of a substance, kJ/kg·°C
$\quad \Delta T$ = temperature change, °C

Thermal energy can be recovered from water (e.g., wastewater effluent) or heated air, such as exhaust from unit processes involving combustion of fuel (e.g., engine generators, boilers, incinerators). It is common to utilize the excess heat in air and/or water for various uses within the treatment facility including digester heating, solids drying, hot water supply, and space heating. Excess heat can also be provided to users outside of the facility, if the amount of heat produced at the plant is large enough.

When the temperature of the wastewater is significantly different from the ambient temperature or the temperature of other streams at the treatment facility, there is a potential that heat can be transferred in a form that could be used for other purposes. Heat added to (or lost from) wastewater can be approximated by assuming that the specific heat of wastewater is equal to that of water at the given temperature (see Chap 2). The specific heat of water, c_w, is 4.1816 kJ/kg·°C at 20°C, and even though it varies with temperature the difference is within 1 percent over the range between 0 to 100°C. For example, if the temperature of wastewater is raised from 21 to 23°C and the flowrate is 160,000 L/h, the heat added to wastewater is

$$Q = mc_w\Delta T$$
$$= 160,000 \times 4.18 \times 2$$
$$= 1,337,600 \text{ kJ/h}$$

Hydraulic Energy

Exclusive of chemical and heat energy, wastewater fluid can also contain energy in the form of elevation head, h_e, (the relative position of the influent to effluent free water surface), pressure head, h_p, (as in pressurized processes such as reverse osmosis) and, velocity head h_v, (associated with the kinetic energy of the moving fluid). These forms of energy are usually quantified in terms of the Bernoulli equation. It should be noted that each of the three terms represents a linear dimension. The energy involved, expressed in kJ or other appropriate units, is obtained by taking into account the corresponding mass of the fluid. The power, expressed in W, kJ/h, or other appropriate units, is the energy per unit time.

Elevation Head, h_e. Most conventional wastewater treatment plants are designed to allow wastewater to flow by gravity from the headworks to the receiving waters. To minimize the power requirements, the hydraulic profile is set to minimize the excess head at the end of the treatment process. Because the majority of wastewater treatment plants are located adjacent to the receiving water body, the plant is often designed to have minimal head at the discharge point.

Pressure Head, h_p. Some wastewater treatment processes such as reverse osmosis operate under pressurized conditions. The pressure head is expressed as p/γ, where γ is a specific weight of the fluid. The recovery of energy from reverse osmosis is considered in Sec. 11–7 in Chap. 11.

Velocity Head, h_v. Because the wastewater is moving through the plant, it also contains kinetic energy, expressed as $v^2/2g$. In general, the velocity head contribution to the total energy is relatively small.

Determination of Total Fluid Head. If H_t, represents the total head transferred to (+) or from (−) the fluid (e.g., in a pump, fan, or turbine) then application of the law of conservation of energy between any two points can be expressed as follows:

$$(h_e + h_p + h_v)_1 \pm H_t = (h_e + h_p + h_v)_2 + \text{losses} \tag{17–6}$$

The losses in Eq. (17–6) represent the head that has been transformed into nonrecoverable forms of energy (e.g., heat or noise). The general expression for an incompressible liquid may be rewritten as

$$\pm H_t = (p_2/\gamma - p_1/\gamma) + (v_2^2/2g - v_1^2/2g) + (z_2 - z_1) + h_L \tag{17–7}$$

where H_t = total head transferred to/received from fluid, m (ft)

p_1, p_2 = pressure, kN/m^2 (lb$_f$/in.^{2})

γ = specific weight of water, kN/m^3 (lb/ft^3)

v = velocity of water, m/s (ft/s)

g = acceleration due to gravity, 9.81 m/s^2 (32.2 ft/s^2)

z_1, z_2 = height above any assumed datum plane, m (ft)

h_L = headloss, m (ft)

Conversion of Fluid Potential Energy to Other Forms of Energy. The fluid energy as given by Eq. (17–7) can be converted to electrical power for a given flowrate using the following expression:

$$P_e = \rho Q g H_t \eta_t \eta_e \tag{17–8}$$

where, P_e = electrical power obtained, W

ρ = density of wastewater, kg/m^3

Q = flowrate, m^3/s

g = acceleration due to gravity, 9.81 m/s^2

η_t = efficiency of mechanical device (e.g., Pelton Wheel, reverse pump, pump, etc.) expressed as a fraction, dimensionless

η_e = efficiency of electrical conversion device expressed as a fraction, dimensionless

Efficiency factors for power generation systems are given in Sec. 17–7. The application of Eqs. (17–7) and (17–8) is demonstrated in Example 17–3.

EXAMPLE 17–3 **Determine Hydraulic Energy of Wastewater Discharge** Consider a wastewater treatment facility located near a coastline, with an effluent discharge point 3 m above mean sea level and an average effluent flowrate of 4 ML/d. Treated effluent flowing out of the chlorine contact tank has a velocity of 0.5 m/s. Calculate the potential energy contained in the effluent. Estimate the actual electrical energy output that can be obtained if a hydraulic turbine generator is located at sea level. Assume a turbine generator combination is to be used with an overall efficiency on 40 percent, The wastewater temperature is 20°C.

Solution

1. Determine the potential and velocity energy.

   a. The potential energy is equal to

   $$(z_2 - z_1) = [0 - 3 \text{ m})] = -3 \text{ m}$$

   b. The velocity energy is equal to

   $$(v_2^2/2g - v_1^2/2g) = 0 - (0.5 \text{ m/s})^2/2g = -0.0127 \text{ m}$$

2. Calculate the potential mechanical energy that can be transferred using Eq. (17–7) neglecting any losses.

   $$H_t = (z_2 - z_1) + (v_2^2/2g - v_1^2/2g) = -3 \text{ m} + (-0.0127) \text{ m} = -3.0127 \text{ m}$$

Because the sign is negative, energy can be produced.

3. Calculate the electrical energy that can be produced with a turbine generator combination using Eq. (17–8) with a combined efficiency factor η and neglecting any other losses.

$$
\begin{aligned}
P_e &= \rho Q g H_i \eta \\
&= (1000 \text{ kg/m}^3)(4000 \text{ m}^3/\text{d})(9.81 \text{ m/s}^2)(3.0127 \text{ m})(0.40) \\
&= 47.3 \times 10^6 \text{ kg·m}^2/\text{s}^2\text{·d} \\
&= 47.3 \text{ MJ/d}
\end{aligned}
$$

Comment   Depending on the receiving water, the hydraulic potential that could be made available for power generation may vary with tides, seasons or other factors altering the surface level. Recovery of hydraulic potential is further discussed in Sec. 17–7.

17–3 **FUNDAMENTALS OF A HEAT BALANCE**

In the following sections, the various forms of energy utilized in wastewater treatment and the management of that energy usage in wastewater treatment facilities are introduced and discussed. To understand the issues involved, it will be useful to review the basic concept involved in the preparation of heat balances.

Concept of a Heat Balance

The concept of heat balance is based on the first law of thermodynamics, i.e., enthalpy is conserved. The mathematical approach is similar to the mass balance calculation described in Chap. 1. For a given system boundary, the general heat balance analysis is given by:

1. General word statement:

Rate of accumulation of heat within the system boundary	=	rate of flow of heat into the system boundary	−	rate of flow of heat out of the system boundary	+	rate of generation of heat within the system boundary	(17–9)
(1)		(2)		(3)		(4)	

2. The corresponding simplified word statement is

$$
\text{Accumulation} = \text{inflow} - \text{outflow} + \text{generation} \tag{17–10}
$$

3. Assuming there is no change in mass flowrate and specific heat of the mass, symbolic representation of heat balance is (see Fig. 17–2):

$$
\Delta H = mcT_o + Q_1 - mcT_e - Q_2 + Q_r \tag{17–11}
$$

where ΔH = change of enthalpy within the system boundary

m = influent mass flowrate

c = specific heat of the mass

T = temperature

Q_1, Q_2 = heat added to or lost from the system boundary, such as added energy through mechanical mixing, or heat loss through the reactor wall

Q_r = heat generated/absorbed within the system boundary, such as heat generated /absorbed during due to chemical reaction and latent heat of vaporization

Figure 17-2

Conceptual diagram of heat balance.

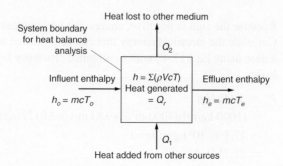

Preparation of a Heat Balance

The preparation of a heat balance is similar to the preparation of a mass balance in that all sources of heat that enter, exit, or are released or taken up within the system boundary must be considered. In most cases, however, internal energy within chemical compounds is not considered except for the energy released into the system in the form of thermal energy. The steps given below should be followed in the preparation of heat balances, as the techniques involved are being mastered.

1. Prepare a simplified schematic or flow diagram of the system or process.
2. Draw a system boundary to define the limits over which the heat balance is to be applied. Proper selection of the system boundary is important because, in many situations, it may be possible to simplify the heat balance computations.
3. List all of the pertinent data and assumptions that will be used in the preparation of the heat balance on the schematic or flow diagram.
4. List all of the rate expressions for the biological or chemical reactions that occur in the process.
5. Select a convenient basis on which the numerical calculations will be based.
6. Solve the balance equations.

The application of the heat balance calculation is illustrated in Example 17–4.

EXAMPLE 17–4 **Heat Balance Calculation** Evaluate the heat balance for an anaerobic digester, using the following process information. How much additional heat will be required to maintain the digester temperature? Assume that the sludge in the reactor is well mixed and has a uniform temperature and specific heat. Ignore heat in the digester gas.

Process information	Unit	Value
Incoming sludge flowrate	m³/d	100
Incoming sludge temperature	°C	10
Volume of sludge in the digester	m³	2000
Liquid temperature in the digester	°C	32
Specific heat of sludge in the digester	kJ/kg·°C	4.2
Heat loss by conduction[a]	kJ/d	1.9×10^6
Heat added by sludge mixing	kJ/d	negligible

[a] Conduction losses occur through the digester walls above and below the ground level, the bottom, and the top, depending on the design of the digester.

Solution

1. Describe the heat balance.

   Accumulation/loss of heat = (heat in incoming sludge) + (heat added by sludge mixing) − (heat in effluent sludge) − (heat loss by conduction through the wall)

2. Develop a symbolic representation from Eq. (17–3).

$$\Delta H = mcT_o + Q_1 - mcT_e - Q_2 + Q_r$$

3. Using the given data and assuming specific gravity of water is 1.0, solve for ΔH.

$$\Delta H = (100\,\text{m}^3/\text{d})(10^3\,\text{kg/m}^3)(4.2\,\text{kJ/kg})(10\,^\circ\text{C}) + 0 - (100\,\text{m}^3/\text{d})(10^3\,\text{kg/m}^3)$$
$$(4.2\,\text{kJ/kg})(32\,^\circ\text{C}) - (1.9 \times 10^6\,\text{kJ/d}) + 0$$
$$= 4.2 \times 10^6 - 13.44 \times 10^6 - 1.9 \times 10^6\,\text{kJ/d}$$
$$= -11.14 \times 10^6\,\text{kJ/d}$$

   To maintain the heat balance, an additional 11.14×10^6 kJ/d, or 11.14 GJ/d of heat must be added to the digester.

Comments For simplicity, heat in the digester gas was not included in the calculation. Even though a single number was assumed in this example to account for conduction losses, evaluation of the heat loss from the digester tank is a significant part of the heat balance calculation around the digester (see Chap. 13).

In normal municipal wastewater treatment, the heat balance around the liquid treatment train is important primarily to maintain temperature for biological treatment. However, it is sometimes observed that the temperature of mixed liquor is slightly higher than the incoming wastewater. The elevated temperature is generally attributed to biological oxidation of organics and nitrogen which is an exothermic reaction as discussed in Sec. 17–5 and Sec. 15–6 in Chap. 15. The aeration system will also contribute to the heat balance in bioreactors as compressed air, at an elevated temperature, is introduced into the bioreactor. Mixing in open activated sludge tanks can add heat (if done with the aeration system), but mixing also enhances the removal of heat liquid by evaporative cooling as moist air exits from the reactor. In sidestream treatment, heat generated from oxidation and reduction of nitrogen is significant and the heat balance must be evaluated to maintain the operating temperature with dilution water or a heat removal mechanism (see Sec. 15–11). Excess heat in the incoming wastewater and treated effluent can be recovered for beneficial purposes as discussed in Sec. 17–6.

17–4 ENERGY USAGE IN WASTEWATER TREATMENT PLANTS

The energy contained in wastewater and the fundamentals of the heat balance used to assess the potential for the recovery and utilization of energy were introduced in the previous sections. To achieve efficient use of energy and utilization of energy extracted from the treatment facility, it is useful to review the usage of energy in wastewater treatment plants. The review of energy usage in wastewater treatment plants is also important because the cost for energy ranges between 15 and 40 percent of the total operation and maintenance costs for wastewater treatment (WEF, 2009), the second highest after labor costs. The topics considered in this section include (1) types of energy sources, (2) energy use for wastewater treatment, (3) energy usage by various treatment processes, and (4) advanced and new wastewater treatment technologies.

Types of Energy Sources Used at Wastewater Treatment Facilities

The energy required to operate a wastewater treatment facilities is supplied by a number of different sources, but primarily by electricity. Electricity is used to operate motors for blowers, pumps and other facilities with moving parts. Electricity is also used for instrumentation and control equipment, lighting, and cooling and heating of buildings with air conditioning equipment. Fuel oils and natural gas are used primarily to operate boilers for heating, and to operate a stationary combustion system to produce electricity. Emergency generators are typically operated by fuel oils. The heat generated from the combustion system can also be utilized (combined heat and power, CHP). In some cases, reciprocating dual fuel engines are used to drive pumps or blowers. Digester gas is used commonly for both boilers and stationary combustion systems for electricity generation. Wind and solar electric power generation systems are used in some facilities, but typically for a minor power demand. The use of electricity from onsite power generation and the use of renewable energy such as solar, wind, tidal, and other energy has been considered and implemented in a number of wastewater treatment facilities (U.S. EPA, 2006a). However, the contribution of these energy sources to the total energy consumption has been minor due primarily to relatively low and stable cost of electricity.

Energy Use for Wastewater Treatment

At the present time, as discussed in Chap. 1, nearly all of the publicly owned wastewater treatment plants in the United States provide secondary or higher levels of treatment. Treatment plants with more stringent treatment limits generally require greater amounts of electric energy per volume of water treated. Plants that have biological treatment for nutrient removal and filtration use on the order of 30 to 50 percent more electricity for aeration, pumping, and solids processing than conventional activated sludge treatment (EPRI, 1994). Wastewater reclamation plants with advanced treatment processes described in Chap. 11 also require significantly more energy to operate.

Energy Use by Individual Treatment Processes

The energy requirements of wastewater treatment systems depend on the flowrate, the characteristics of the incoming raw wastewater, and the treatment process employed. Treatment processes and equipment requiring electric energy in a municipal wastewater treatment plant are presented in Table 17–2. Various types of electric motor-driven equipment are involved in these operations and processes including pumps, blowers, mixers, sludge collectors, and centrifuges. In conventional secondary treatment, most of the electricity is used for (1) biological treatment by either the activated sludge process that requires energy for aeration blowers or trickling filters that require energy for influent pumping and effluent recirculation; (2) pumping systems for the transfer of wastewater, liquid sludge, biosolids, and process water; and (3) equipment for the processing, dewatering, and drying of residuals and biosolids.

Typical energy requirements for individual treatment processes are reported in Table 17–3. The electrical energy required for wastewater treatment vary widely but typically between 950 MJ/10^3 m^3 and 2900 MJ/10^3 m^3 (between 1000 and 3000 kWh/Mgal) for most treatment facilities (AWWARF, 2007). The amount of electricity consumed varies with plant size, influent and effluent characteristics, pumping requirements for influent

Table 17–2

Commonly used electric motor driven equipment used in wastewater treatment

Process or operation	Commonly used electric motor driven equipment
Pumping and preliminary treatment	Chemical feeders for prechlorination, influent pumps, screens, screenings press, grinders and macerators, blowers for preaeration and aerated grit chambers, grit collectors, grit pumps, air lift pumps
Primary treatment	Flocculators, clarifier drives, sludge and scum pumps, blowers for channel aeration
Secondary (biological) treatment	Blowers for channel and activated sludge aeration, mechanical aerators, trickling filter pumps, trickling filter distributors, clarifier drives, return and waste activated sludge pumps
Disinfection	Chemical feeders, evaporators, exhaust fans, neutralization facilities, mixers, injector water pumps, UV lamps
Advanced wastewater treatment	Blowers for nitrification aeration, mechanical aerators, mixers, trickling filter pumps, pumps for depth filters, blowers for air backwash, pumps for membrane filtration
Solids processing	Pumps, grinders, thickener drives, chemical feeders, mixers for anaerobic digesters and blending tanks, aerators for aerobic digesters, centrifuges, belt presses, heat dryer drives, incinerator drives, conveyors
Ancillary systems	
Odor control	Odor control fans, chemical feeders
Process water	Pumps
Plant air	Compressors

and effluent, energy requirements for odor control, and type of treatment system employed. However, generally energy use per volume tends to be lower at larger treatment plants, and activated sludge processes tend to require more energy than trickling filters (see Fig. 17–3). A typical percentage distribution of energy use in a conventional activated sludge treatment plant is illustrated on Fig. 17–4.

The operational requirements for wastewater collection and treatment systems are correlated with the wastewater load (see Fig. 3–6 in Chap. 3). If a diurnal electricity demand curve were developed for the treatment facilities, it would be of a similar shape to the flowrate and loading curves shown on Fig. 3–6. Therefore the peak energy demand at a wastewater treatment facility would likely occur from midday to the early evening hours when peak electricity consumption occurs in the community.

Advanced and New Wastewater Treatment Technologies

With the introduction of new technologies for wastewater treatment, energy usage requirements will change, particularly where effluent from a conventional treatment process is treated further for reuse applications. As reported in Table 17–3, higher levels of treatment, or new technologies whose operation is based on electric energy, i.e., membrane treatment, UV disinfection, and advanced oxidation, tend to require more energy to operate.

Table 17–3

Typical energy consumption of various treatment processes on wastewater treatment[a]

Technology	Energy consumption[b]	
	kWh/10³ gal	kWh/m³
Conventional secondary treatment WWTP[c]	**0.38 to 0.67**	**0.10–0.18**
Wastewater influent pumping	0.12–0.17	0.032–0.045
Screens	0.001–0.002	0.0003–0.0005
Grit removal (aerated grit removal)	0.01–0.05	0.003–0.013
Trickling filters	0.23–0.35	0.061–0.093
Trickling filter-solids contact	0.35	0.093
Activated sludge for BOD removal	0.53–4.1	0.14
Activated sludge with nitrification/denitrification	0.87–0.88	0.23
Membrane bioreactor	1.9–3.8	0.5–1.0[d]
Return sludge pumping	0.03–0.05	0.008–0.013
Secondary settling	0.013–0.015	0.003–0.004
Dissolved air flotation	0.12–0.15	0.03–0.04
Tertiary filtration (depth filtration)	0.1–0.3	0.03–0.08
Tertiary filtration (surface filtration)		
Chlorination (sodium hypochlorite)	0.001–0.003	0.0003–0.0008
UV (ultraviolet) disinfection	0.05–0.2	0.01–0.05
Microfiltration/ultrafiltration	0.75–1.1	0.2–0.3
Reverse osmosis (without energy recovery)	1.9–2.5	0.5–0.65
Reverse osmosis (with energy recovery)	1.7–2.3	0.46–0.6
Electrodialysis (TDS range 800–1200 mg/L)	4.2–8.4	1.1–2.2
UV photolysis with O_3 or H_2O_2 (advanced oxidation)[e]	0.2–0.4	0.05–0.1
Sludge pumping	0.003	0.0008
Gravity thickening	0.001–0.006	0.0003–0.0016
Aerobic digestion	0.48–1.2	0.13–0.32
Mesophilic anaerobic digestion (primary plus waste activated sludge)[f]	0.35–0.6	0.093–0.16
Mesophilic anaerobic digestion with thermal hydrolysis pretreatment (primary plus waste activated sludge)[f]	0.58–0.6	0.015–0.02
Sludge dewatering (centrifuge)	0.02–0.05	0.005–0.013
Sludge dewatering (belt filter press)	0.002–0.005	0.0005–0.0013

[a] Adapted in part from Burton (1996).

[b] Energy requirement per unit volume of wastewater treated.

[c] For treatment, not including conveyance. From Global Water Research Coalition (2008).

[d] From Krzeminski et al. (2012).

[e] For RO permeate.

[f] Energy recovery is not counted. Including electrical power and heating requirements.

Figure 17–3

Comparison of electrical energy used for different types of treatment processes as a function of flowrate. (From Burton, 1996.)

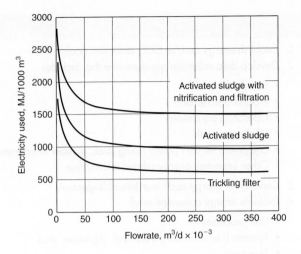

Figure 17–4

Distribution of energy usage in a typical wastewater treatment plant employing the activated-sludge process. (From EPRI, 1994.)

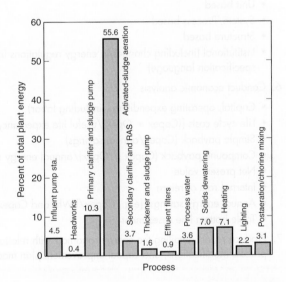

17–5 ENERGY AUDITS AND BENCHMARKING

An energy audit is a procedure and methodology used to identify energy conservation opportunities. The performance of an energy audit is an important first step to identify inefficiencies with existing plant operation and opportunities to improve energy efficiency. Different levels of energy audits can be performed varying from a preliminary "walk-through" to obtain an overview of principal equipment to a detailed process audit in which the energy used by each unit processes is evaluated (EPRI, 1994). In general, one of the initial tasks in an energy audit is the evaluation of the unit processes known to consume a significant amount of energy, such as influent pumping, aeration for activated sludge plants, pumping systems at trickling filter plants, and sludge processing units. A typical procedure used to conduct an energy audit is presented in Table 17–4. Common energy audit recommendations are reported in Table 17–5.

Table 17–4

Typical procedure for energy audit

1. Establish energy and/or GHG reduction goals
2. Develop data collection questionnaire that includes

   - Energy demands
   - Process description
   - Operating parameters
   - Discharge goals
   - Equipment list with power rating, size, number in operation, etc.
   - Other important energy related information

3. Conduct an energy audit tour (visual inspection)
4. Evaluate energy consumption of:

   - Unit processes
   - Systems (i.e. secondary treatment, digestion, etc.)
   - Structures

5. Develop a matrix of energy conservation measures (ECMs)

   - Unit based
   - System/Process based
   - Structure based
   - Institutional (including changes in energy regulations including required design specification language)

6. Conduct economic analysis

   - Capital, operating expenditures (including labor)
   - Life cycle costs [(Capex + Opex)/useful life expectancy]
   - Simple payback (Capex/annual savings)
   - Compound payback (Capex + Opex)/annual energy savings
   - Net present value
   - Internal rate of return
   - Annual energy savings, [e.g., Capex/kWh, and Capex/MT $CO_2(e)$ avoided]
   - Triple bottom line assessment

7. Produce report detailing the energy roadmap with relations to operational modifications, process modifications, and capital improvement plan modifications

Benchmarking Energy Usage

Benchmarking of energy use at a wastewater treatment plant is a fundamental and essential tool in assessing energy usage and conservation opportunities. Energy benchmarking is used to estimate the energy consumption reduction potential and to provide a basis for identifying increases or decreases in energy efficiencies as new processes, technologies, and energy conservation measures are employed. Benchmarking the energy usage of a wastewater treatment plant involves the comparison of energy consumption to a so called "standard" wastewater treatment plant. Because the design flowrates, influent loadings, process steps, operational modes, and treatment technologies can all impact energy consumption and vary significantly from plant to plant, the energy consumption must first be normalized to allow a comparison between treatment plants with different characteristics. For example, two physically identical plants with the same flowrate, but different BOD loadings will have different benchmark results when based on flow but should have more similar results when based on BOD loading.

Table 17-5

Summary of common audit recommendations for energy savings at several U.S. wastewater treatment plants[a]

1. Install adjustable speed drives on pumps and blowers for variable flowrate operations
2. Install DO monitoring and control in aeration tanks
3. Conduct periodic pumps tests and repair or replace inefficient pumps
4. Operate emergency generators to reduce peak hour power demand
5. Install CHP when replacing emergency generators
6. Install electric load monitoring devices
7. Install capacitors to improve power factor
8. Operate less reactors during prolonged under-loading conditions
9. Change or reduce pumping operations
10. Reduce odor control/ventilation areas where possible
11. Install motion sensors for lighting in areas not occupied frequently
12. Control heating and cooling in areas that are not occupied all day
13. Replace oversized motors
14. Change selected operations to off-peak periods

[a] Adapted in part from Burton (1996).

Benchmarking Protocol

In the protocol published by AWWARF (2007) the parameters that appear to have the greatest impact on energy consumption were identified based on the analysis of data collected from 266 wastewater treatment plants throughout the United States. Using multi-linear log-regression, an empirical model was developed for benchmarking energy in wastewater treatment plants. This method is based on the multi-parameter benchmark score method used by U.S. EPA Energy Star rating for buildings (U.S. EPA, 2007) with minor modifications.

In the AWWARF protocol, six parameters were identified as key variables affecting the energy use, including (1) daily average flowrate, (2) design flowrate, (3) influent and (4) effluent BOD concentrations, (5) fixed versus suspended media, and (6) conventional treatment versus biological nutrient removal. Using multi-parameter log-regression analysis, the following wastewater treatment plant energy use model was developed:

$$\ln(E_s) = 15.8741 \qquad (17\text{-}12)$$
$$+ 0.8944 \times \ln(\text{influent average Mgal/d})$$
$$+ 0.4510 \times \ln(\text{influent BOD mg/L})$$
$$- 0.1943 \times \ln(\text{effluent BOD mg/L})$$
$$- 0.4280 \times \ln(\text{influent average flowrate / influent design flowrate} \times 100)$$
$$- 0.3256 \times (\text{trickling filter? Yes-1, No-0})$$
$$+ 0.1774 \times (\text{nutrient removal? Yes-1, No-0})$$

where E_s = source energy use estimated from the model, kBtu/y (defined below)

Because the model was developed for treatment plants in the United States, the units used for the development of the model are U.S. customary units. An SI version is not available at the present time.

Figure 17-5

Benchmarking score plot for wastewater treatment plant energy use. (Adapted from AWWARF, 2007.)

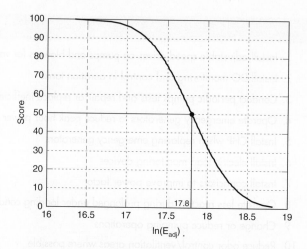

Performance Rating. The value obtained from Eq. (17–12) is an estimate of the energy usage at a specific facility. To compare the estimated performance of the specific facility to other facilities, the ratio of the predicted performance to the average performance derived from the complete data set is used. The range of $\ln(E_s)$ values in the complete data set used to derive Eq. (17–12) ranges from approximately 16 to 19.6 with a mean value (50th percentile) of 17.8 (see Fig. 17–5). Thus the adjustment factor, F_{adj}, used to normalize the individual energy use factors is given as the following expression.

$$F_{adj} = \ln(E_s) / 17.8 \tag{17-13}$$

In effect the adjustment factor allows a utility to assess its benchmark score by its location on the distribution curve.

Conversion of Energy Usage Data to Source Energy Use. The source energy use is defined as the amount of raw fuel energy that is required to operate the facility (U.S. EPA, 2011). Whereas the energy usage data obtained at the wastewater treatment facility is the amount of energy measured at the point of use. The difference between the two values is due to the energy losses during production, transmission, and delivery. To allow for a consistent comparison, conversion factors (source energy factor, F_s) between the actual energy use and the source energy use were developed based at a national average, as presented in Table 17–6. It should be noted that the energy generated onsite from renewable energy sources such as the use of digester gas or solar power is not included in the source energy use calculation. The value

Table 17-6

Source energy factor used for the benchmarking of wastewater treatment plant energy usage[a]

Energy Source	US customary units		SI units	
	Unit	Value	Unit	Value
Electricity	kBtu/kWh	11.1	kBtu/kWh	11.1
Natural Gas	kBtu/therm	102.5	kBtu/MJ	0.97
Fuel Oil	kBtu/gal	141	kBtu/L	37.25
Propane	kBtu/gal	91	kBtu/L	24.04
Digester Gas	kBtu/ft³	0.6	kBtu/m³	21.2

[a] Adapted from AWWARF (2007).

for digester gas is included in Table 17–6 as a reference, and the use of the value is discussed in Example 17–5. Actual source energy use, E_{as}, is calculated as

$$E_{as} = \sum E_u \cdot F_s \qquad (17\text{–}14)$$

where, E_{as} = Actual source energy use
E_u = Energy use measured at the point of use (= data from the treatment plant)
F_s = Source energy factor

The source energy factor will convert all energy use into the source energy use, and also convert the unit into kBtu/y, by which energy usage from various energy sources could be added to determine the total source energy use.

Calculation of Adjusted Energy Use. The adjustment factor is used to calculate the adjusted energy use factor, a normalized log-value of the energy use:

$$\ln(E_{adj}) = \ln(E_{as}) / F_{adj} \qquad (17\text{–}15)$$

The adjusted energy use factors allow comparison between treatment facilities with different treatment levels and wastewater characteristics. The benchmark score can be obtained from Fig. 17–5, or the value for each score can be looked up from the AWWARF report (AWWARF, 2007). The score corresponds to the percentile in terms of normalized energy as compared to other studied plants. Benchmarking is illustrated in Example 17–5.

EXAMPLE 17–5 **Benchmarking of Energy Use at a Wastewater Treatment Plant** A BNR activated sludge process is used to treat the wastewater from a community. Based on the plant data given below, calculate the source energy usage and compare with the actual energy use to determine the benchmark score. How does the score changes if the use of energy generated onsite from digester gas was counted in the source energy use?

Item	Unit	Value
Energy use		
Electricity	kWh/y	14,100,000
Natural gas	m³/y	17,300
Fuel oil #2	m³/y	390
Digester gas production	m³/y	1,047,900
Digester gas used	m³/y	755,000
Digester gas flared	m³/y	290,500
Digester gas vented	m³/y	2400
Plant performance		
Annual average flowrate	m³/d	100,000
Average influent BOD concentration	mg/L	180

(continued)

(*Continued*)

Item	Unit	Value
Average influent ammonium concentration	mg/L	18
Average influent total nitrogen concentration	mg/L	32
Average effluent BOD concentration	mg/L	4
Average effluent ammonium concentration	mg/L	1.5
Average effluent total nitrogen concentration	mg/L	8.3
Other required information		
Design flowrate	m³/d	180,000
Population served	persons	430,000

Solution

1. Calculate the natural log of the source energy use value using Eq. (17–12) and the given data:

$$
\begin{aligned}
\ln(E_s \text{ kBtu/y}) = {} & 15.8741 \\
& + 0.8944 \times \ln\{[100,000 \ (\text{m}^3/\text{d})]/[3785 \ (\text{m}^3/\text{Mgal})]\} \\
& + 0.4510 \times \ln(180) \\
& - 0.1943 \times \ln(4) \\
& - 0.4280 \times \ln(100,000/180,000 \times 100) \\
& - 0.3256 \times (0) \\
& + 0.1774 \times (1) \\
= {} & 15.8741 + 2.9284 + 2.3420 - 0.2694 - 1.7194 - 0 + 0.1774 \\
= {} & 19.33
\end{aligned}
$$

2. Calculate the adjustment factor from the value obtained in Step 1 using Eq. (17–13):

Adjustment factor = 19.33 / 17.8 = 1.086

3. Calculate the natural log of the source energy use value using the energy usage data from the wastewater treatment plant and source energy factor in Table 17–6.
   a. Calculate actual source energy use for the energy from outside sources (not including energy generated from digester gas) using Eq. (17–14).

$$
\begin{aligned}
E_{as} \text{ (no digester gas)} = {} & (14,100,000 \ \text{kWH/y})(11.1 \ \text{kBtu/kWH}) \\
& + (17,300 \ \text{m}^3/\text{y})(35.31 \ \text{ft}^3/1 \ \text{m}^3)(1.025 \ \text{kBtu/ft}^3) \\
& + (390 \ \text{m}^3/\text{y})(264.2 \ \text{gal}/1 \ \text{m}^3)(141 \ \text{kBtu/gal}) \\
= {} & 156,510,000 + 626,135 + 14,528,358 \\
= {} & 171,664,493 \ \text{kBtu/y}
\end{aligned}
$$

   b. Calculate source energy use for the energy including the energy generated onsite from digester gas.

$$
\begin{aligned}
\text{Digester gas energy} = {} & (755,000 \ \text{m}^3/\text{y})(35.31 \ \text{ft}^3/1 \ \text{m}^3)(0.6 \ \text{kBtu/ft}^3) \\
= {} & 15,995,430 \ \text{kBtu/y}
\end{aligned}
$$

$$
\begin{aligned}
E_{as} \text{ (with energy from digester gas)} = {} & 171,664,493 \ \text{kBtu/y} + 15,995,430 \ \text{kBtu/y} \\
= {} & 187,659,923 \ \text{kBtu/y}
\end{aligned}
$$

4. Convert the source energy usage calculated in Step 3 to the natural log of the adjusted energy use using the adjustment factor from Step 2 and Eq. (17–15).

a. Calculate $\ln(E_{adj})$ not including energy generated from digester gas.

$$\ln(E_{adj}) = \ln(171{,}664{,}493 \text{ kBtu/y}) \,/\, 1.086 = 18.96 \,/\, 1.086$$

$$= 17.46$$

b. Calculate $\ln(E_{adj})$ including energy generated from digester gas.

$$\ln(E_{adj}) = \ln(187{,}659{,}923 \text{ kBtu/y}) \,/\, 1.086 = 19.05 \,/\, 1.086$$

$$= 17.54$$

5. Using Fig. 17–5, find the benchmark score with and without digester gas usage.

Score without counting the use of energy generated from digester gas = 78

Score counting the use of energy generated from digester gas = 72

Comment In this example, the treatment plant's normalized percentile score for energy usage, without including the use of energy generated from digester gas as specified in the protocol, is 78, or slightly better than the average of the treatment facilities used for the development of the model. Including the use of energy generated onsite from digester gas will lower the benchmark score to 72. In other words, if this treatment plant was generating digester gas but not utilizing it (i.e., flaring it) and depending solely on energy from outside sources, this plant would have scored 72 on benchmarking. Therefore the use of energy generated onsite from digester gas will significantly improve the benchmark score.

17–6 RECOVERY AND UTILIZATION OF CHEMICAL ENERGY

Recovery and utilization of chemical energy involves transformation of wastewater constituents containing chemical energy into fuel, and the use of the fuel for beneficial purposes. In some cases, pretreatment of the fuel is necessary before it is used. Recovery of chemical energy has been practiced at wastewater treatment facilities primarily by producing digester gas (biogas) from sludges with anaerobic sludge digestion, and digester gas has been used widely for boilers and other combustion systems to supplement other energy sources. Dried biosolids have also been used as an energy source where sludge incineration is practiced.

Fuels Derived from Wastewater

The types of fuels derived from wastewater constituents can be categorized as (1) gaseous fuels, (2) solids, and (3) liquids/oils. The gaseous fuels include biogas from anaerobic digestion and syngas from gasification. The solids include primary sludge, waste secondary sludge, and stabilized biosolids. Liquid fuels and oils could be produced from the solid contents of wastewater, but the generation and utilization of liquid fuels and oils derived from wastewater constituents is not common.

Generally, chemical energy in wastewater is extracted from the solid contents, and by means of a physical/chemical process or a biological treatment process. In anaerobic treatment processes (see Chap. 10), some dissolved organics are converted biologically to methane, but the use of anaerobic treatment processes for liquid treatment in municipal wastewater has been uncommon. In conventional municipal wastewater treatment facilities with aerobic biological treatment, only a fraction of the dissolved chemical energy is assimilated into biomass which is subsequently collected for the solids processing and converted into biogas through anaerobic digestion.

Biogas. The method used most commonly to recover energy from the solid contents of wastewater is to produce methane through anaerobic digestion (see Chap. 13). Typical production of digester gas through an anaerobic biological process varies between 0.75 and 1.12 m^3/kg volatile solids destroyed (12 to 18 ft^3/lb VSS). Typically, digester gas contains 55 to 70 percent methane, 30 to 40 percent CO_2, and small amounts of N_2, H_2, H_2S, water vapor, and other gases. The energy content of digester gas is typically in the range of 22 to 24 MJ/m^3 (600 to 650 Btu/ft^3) in HHV. The methane gas content depends primarily on the pH of the digester as it affects the amount of CO_2 that is released to the gas phase. The theoretical biogas production reaction, as described in Chap. 10, is

$$C_vH_wO_xN_yS_z + \left(v - \frac{w}{4} - \frac{x}{2} + \frac{3y}{4} + \frac{z}{2}\right)H_2O \rightarrow$$

$$\left(\frac{v}{2} + \frac{w}{8} - \frac{x}{4} - \frac{3y}{8} - \frac{z}{4}\right)CH_4 + \left(\frac{v}{2} - \frac{w}{8} + \frac{x}{4} + \frac{3y}{8} + \frac{z}{4}\right)CO_2 + yNH_3 + zH_2S$$

$$(10\text{--}4)$$

As shown in Eq. (10–4), ammonia and hydrogen sulfide are produced, and other volatile compounds are also produced during anaerobic digestion. Because some of the compounds could be detrimental to the combustion system, gas cleaning may be required before collected biogas is used for the combustion system. The cleaned biogas can also be used to generate electricity using a reciprocating engine, gas turbine, microturbine, or fuel cell. The use of biogas for boilers typically does not require gas cleaning. The biogas conveyance and storage system must be kept in a positive pressure to prevent accidental mixing of biogas with air, which could result in explosion.

Syngas. Syngas is a mixture of gases comprised mainly of CO, H_2, CO_2, and CH_4 generated through a gasification process (see Chap. 14) and has a lower heating value (LHV) ranging between 4 and 15 MJ/m^3 with a typical range in the wastewater application between 4.5 and 5.5 MJ/m^3 (120–150 BTU/ft^3). Energy content of syngas varies widely and depends on the gasification process technology and its operating conditions such as amount of air and moisture injected to the gasification process. The unit processes to produce syngas are described in Sec. 14–9 in Chap. 14.

Syngas can be oxidized immediately, or cleaned and used for internal combustion systems in a two stage system. Cleaned syngas could be processed further using a catalytic Fisher-Tropsch (FT) process to a liquid fuel, which can be used in an internal combustion engine-generator, boiler, or fuel cell. Liquid fuel can also be used for the production of various chemicals (Valkenburg et al., 2008). Limited information is available regarding the production rate of syngas, as it varies widely with the gasification process, characteristics of the feed solids, and the operating conditions used. In a full-scale installation of fluidized bed gasification process in Balingen, Germany, 1000 kg of dewatered sludge with 32 percent dry solid content was dried to produce 400 kg of dried sludge with 80 percent solids content, which generated to 510 m^3 of syngas and 160 kg of mineral granule (WERF, 2008).

Solid Fuels. Solid contents in wastewater are largely organic compounds. Depending on the water content of the sludges or biosolids separated from wastewater they can be incinerated without additional fuels. Typical heating values of sludges and biosolids are reported in Table 14–17 in Chap. 14, and theoretical values could be derived from Eq. (2–66) if the chemical composition is known (see Sec. 17–2).

Liquid Fuel and Oils.   Technologies are available to produce liquid fuel and oils from the solid constituents present in wastewater. For example, syngas could be converted into a liquid fuel through the Fisher-Tropsch (FT) process, and char and oil are generated from pyrolysis of solid contents. However the production of oil and liquid fuel from wastewater solid contents has not been implemented in full-scale.

Energy Recovery from Gaseous Fuels with Engines and Turbines

Reciprocating engines, gas turbines, and microturbines are the principal technologies used to generate electricity from the combustion of gaseous fuels derived from waste-water. Even though it is practiced less commonly, pumps and blowers can be operated with a direct drive from the engines fueled by gaseous fuels. Typical energy recovery systems with engines and turbines include, as illustrated on Fig. 17–6, a gas generation process, gas holding vessel, compressors, gas cleaning, an engine/turbine, emission control (see Chap. 16), and a waste heat recovery system. Fuel cells are also used to generate electricity from biogas but by a different mechanism. Thus fuel cells are considered separately in this section.

Exhaust heat from these combustion processes can be used to heat water for building heating, heat anaerobic digesters, or to provide hot water supply, as discussed further in Sec. 17–6. It should be noted that the use of biogas in boilers and engines will result in the emission of flue gas, which may be subject to strict regulations to minimize air pollution. Emissions from boilers and engines are considered in Sec. 16–5 in Chap. 16.

Figure 17–6

Typical process for the recovery and utilization of biogas generated by anaerobic digestion of sludges: (a) process flow diagram, (b) egg shape anaerobic digester, (c) view of dual fuel type reciprocating engine, and (d) view of heat recovery boiler.

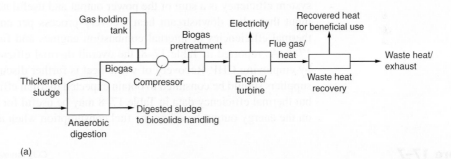

(a)

(b)

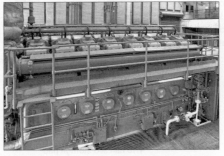

(c)

(d)

Gas Pretreatment Requirements. Constituents in biogas and syngas that affect the operation of equipment to generate power and heat include hydrogen sulfide (H_2S), siloxanes, carbon dioxide (CO_2), and moisture (H_2O). Syngas also contains particulate matters that may need to be removed. Hydrogen sulfide (H_2S) is an odorous gas and highly corrosive. At relatively low concentrations H_2S is hazardous to human health. The immediately dangerous to life or health (IDLH) value used by the National Institute for Occupational Safety and Health (NIOSH) for H_2S is 100 ppm (CDC, 1994). An H_2S value of 100 ppm is often exceeded in biogas generated from anaerobic digestion. Siloxanes are known to cause detrimental impacts to the combustion systems. Water condensation occurs in the biogas transfer pipelines and a compressor, and enhances corrosion by hydrogen sulfide. To minimize these adverse impacts, digester gas is often cleaned before combustion. Carbon dioxide may be removed only when biogas is to be sold as natural gas (sometimes referred to as bio-methane), thus CO_2 removal is rarely practiced at a wastewater treatment facility unless there is an incentive to sell bio-methane instead of using the biogas within the facility. When biogas is mixed with natural gas at a treatment facility, it is common to mix natural gas with air to reduce the heat content to the level of biogas, instead of removing CO_2 from the biogas to raise the biogas heat content to that of natural gas. Because the air added to the natural gas is not sufficient to cause explosion, the removal of oxygen is not required for the natural gas/air mixing process.

Typical treatment systems for biogas and syngas are shown on Fig. 17–7. The types of gas cleaning technologies used for digester gas and syngas are reported in Table 17–7. Description of each gaseous compound and its treatment is described in detail in Sec. 13–9 in Chap. 13. In many cases, not all of these gas cleaning processes are necessary, and the cleaning requirements are determined according to the gas quality requirements for the end use.

Total System Efficiency. The total system efficiency, also known as thermal efficiency, is defined as the ratio of the total energy output, either work, heat, or electricity, to the total energy input. For an engine generator without heat recovery, total system efficiency is the electrical power output per fuel energy input. In a CHP system, the total system efficiency is a sum of the power output and useful net thermal output per total fuel input through a downstream heat recovery process per energy input by a fuel. Typical thermal efficiencies of internal combustion engines and fuel cell are presented in Table 17–8. As reported in Table 17–8, the overall thermal efficiency is improved significantly by employing CHP. Recovery of waste heat is further discussed in Sec. 17–6. Equipment suppliers should be consulted to obtain expected thermal efficiency for specific equipment, but thermal efficiency data in Table 17–8 may be useful for making preliminary estimates on the energy output based on the fuel consumption when accurate data on energy output

Figure 17–7

Typical gas cleaning process flow diagrams: (a) for biogas from anaerobic digestion and (b) syngas from gasification.

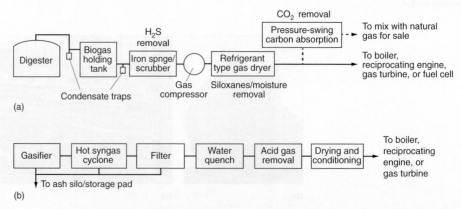

Table 17–7

Major contaminants in digester gas, and treatment method for their removal[a]

Gaseous compound	Typical concentration in digester gas, mg/L[b]	Treatment methods
Hydrogen sulfide (H_2S)	1000–2000	Adsorption Scrubber Chemical addition to digester
Siloxanes (in silica)	0.3–11	Refrigeration/drying Scrubber Adsorption (activated carbon, proprietary media)
Water vapor (H_2O)	Saturation	Solid desicants Liquid absorption Refrigeration/drying
Carbon dioxide (CO_2)	30 to 35%	Scrubber. CO_2 is removed only for the application to produce natural gas quality gaseous fuel and the application has been limited

[a] Compiled from various sources.

[b] Unless otherwise noted.

are not available. When digester gas is used for engines and boilers, the actual heat content of digester gas should be measured to accurately project the expected energy output.

Reciprocating Engine Generators.   Reciprocating engines and gas turbines are used widely to produce electricity on site at wastewater treatment facilities using recovered chemical energy in the form of digester gas [see Figs. 17–8(a) and (c)]. Both spark ignition engines (Otto-cycle engines) and compression ignition engines (diesel-cycle engines) have been used with biogas. When compression ignition engines are used, a supplemental fuel oil will be added to induce the ignition. The typical size of the reciprocating engines ranges between 20 kW and 6 MW (see Table 17–8).

Gas Turbines.   Gas turbines used at wastewater treatment facilities have a wind range of capacity [see Fig. 17–8(d)], in the range of 1 to 250 MW. Their electrical efficiency is slightly lower than reciprocating engines, varying from 30 and 40 percent. As shown in Table 17–8, overall thermal efficiency can be in the same range with reciprocating engines when it is used for CHP because exhaust gas from gas turbines has a higher temperature and can be used for boilers and a wide range of heat recovery processes. However, reciprocating engines have been used more commonly at wastewater treatment facilities because the reciprocating engines generally have a wider operational range than the gas turbines and the energy output could be adjusted more easily in response to the diurnal energy demand variations.

Microturbines.   A microturbine is a smaller gas turbine, often packaged by the burbine suppliers. The typical size of microturbines used at wastewater treatment facilities range from 30 to 250 kW. In the typical microturbines, a recuperation cycle is used to utilize the heat in the exhaust gas to preheat the combustion gas. Remaining heat is recovered to produce hot water or connected to other thermal energy recovery devices. Microturbines are suited for distributed generation applications to supplement a specific electrical load, or to be used in parallel to serve large loads.

Table 17–8

Devices used to recover energy from digester gas and syngas

Device	Typical efficiency[a], %	Typical efficiency with CHP, %	Gas cleaning requirements	Typical size, kW	Conversion of energy
Reciprocating engine	25–50	70–80	siloxiane	20–6000	electricity, mechanical power, waste heat
Gas turbine (simple cycle)	25–40	70–80	siloxiane, H_2S	1000–250,000	electricity, mechanical power, waste heat
Gas turbine (combined cycle)	40–60	70–80	siloxiane, H_2S	1000–250,000	electricity, mechanical power, residual heat to produce steam for more electricity generation
Micro-turbine	25–35	70–85	siloxiane, H_2S	30–250	electricity, mechanical power, waste heat
Sterling engine	~30	~80	no cleaning required		electricity, mechanical power, waste heat
Fuel cell	40–60	70–80	siloxiane, H_2S, H_2O	200–3000	electricity, waste heat
Boiler	80–90+	-	typically no cleaning required		steam, hot water

[a] Efficiencies based on the manufacturers' rating for a new system, not including heat recovery.

Sterling Engine. Sterling engine is an external combustion engine and a wide variety of fuels can be used without the level of gas cleaning required for internal combustion engines. Electrical efficiency is approximately 30 percent, and a total system efficiency can be up to 80 percent when heat recovery is included.

Energy Recovery from Gaseous Fuels with Boilers

Boilers are used at wastewater treatment facilities to generate hot water or steam for steam turbines, space heating, and hot water supply. Heating requirements for wastewater treatment processes include heating of anaerobic digester and building heating and various solids handling processes such as sludge pretreatment and sludge drying (see Chaps. 13 and 14). Boilers are used also for heating of administration and control building, as well as hot water supply. Typically gas pretreatment is not included when biogas or syngas is used only for the boilers, and it is designed to maximize the thermal efficiency of the boiler, even though further recovery of heat from the waste heat is possible (see Fig. 17–9).

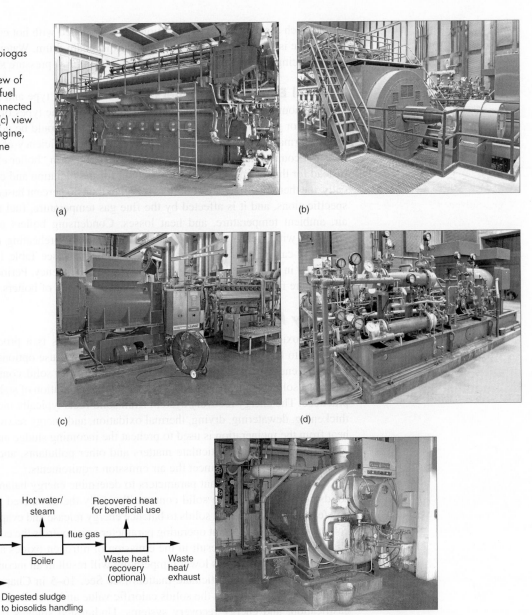

Figure 17–8

Typical devices utilizing biogas generated at wastewater treatment facilities: (a) view of large reciprocating dual fuel engine, (b) generator connected to engine shown on (a), (c) view of alternative dual fuel engine, and (d) view of gas turbine generator.

Figure 17–9

Energy recovery system utilizing biogas in a dual-fuel boiler: (a) process flow diagram and (b) view of boiler.

Types of Boilers.   The typical boilers used commonly at wastewater treatment facilities are hot water boilers to provide building heating, heating for treatment processes, and hot water supply. The boilers producing steam are also used especially when a high temperature heat source is necessary for sludges and biosolids processing, such as thermal hydrolysis. Boilers can generally be categorized as fire tube type and water tube type. In fire tube boilers, hot gases flow through the tubes to heat water that surrounds the tubes to generate hot water and steam. Fire tube boilers are suited for smaller, low pressure (less than 17 bar or 250 lb$_f$/in.^{2}) steam generation. In water tube boilers, water runs through the tubes which are placed in a combustion chamber, filled with hot combustion gases. Water in the tube is heated and produces steam in the steam drum. Water tube boilers are suited for producing higher (higher than 17 bar or 250 $lb_f/in.^2$) pressure steam.

Thermal Efficiency of Boilers.   Selection of the type and size of boilers depends on the amount and quality of heat to be delivered and the types of fuel to be used. Either dual-fuel or gas boilers could be used with biogas. It should be noted that sometimes the term "thermal efficiency" is used specifically for the efficiency of the heat exchanger but does not count radiation and convection losses. The term "boiler efficiency" may be used instead for the thermal efficiency to account for the radiation and convection losses. Typically, the thermal efficiency of boilers is higher than 80 percent based on the manufacturers' specifications, and it is affected by the flue gas temperature, fuel type, amount of excess air, ambient temperature, and heat losses. Condensing boilers are the type of boilers equipped with preheating of the cold feed water. By preheating the feed water, overall efficiency can be improved to greater than 90 percent (see Table 17–8). Build up of soot or scaling in boilers will result in reduced thermal efficiency. Periodic cleaning and maintenance are necessary to maintain the thermal efficiency of boilers.

Energy Recovery from Solid Fuels

Thermal oxidation (i.e., incineration) of solid contents is a process typically used by medium- to large-size plants with limited disposal or reuse options (Chap. 14). To maximize the energy recovery from thermal oxidation, the solid contents are used directly without a solid stabilization process even though incineration of stabilized biosolids is also practiced. The energy recovery system from solid fuels typically includes sludge/biosolids thickening, dewatering, drying, thermal oxidation, and energy recovery system. Part of the heat from the incineration is used to preheat the incoming sludge and air (see Fig. 17–10). The flue gas contains particulate matters and other pollutants, and appropriate emission control must be placed to meet the air emission requirements.

The two most important parameters to determine energy balance are dewatered cake solids content and volatile solid content. Typically the dewatered solids content must be greater than 22–28 percent solids to balance energy release and evaporative cooling. Solids content is also an important operating parameter as it controls the combustion temperature. Higher temperature will result in the emission of nitrogen oxides and formation of amalgam from the ash, whereas lower temperature will result in an incomplete combustion and increased emission of particulate matters (see Sec. 16–5 in Chap. 16). The overall heat balance can vary based on the solids calorific value and overall efficiencies of the drying, gasification, and energy recovery systems. Undigested sludges typically have a higher volatile content so supplemental fuel use can be eliminated at a lower solids content. Anaerobically digested biosolids contain lower calorific values per unit dry material as some of the volatile material is consumed and converted to biogas. Digested sludge with chemical addition within the treatment system generally contains lower calorific value

Figure 17–10

Typical process flow diagram for the recovery of chemical energy by combustion of sludges and biosolids.

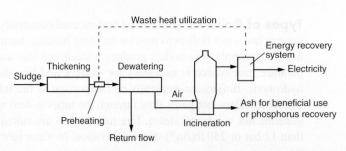

than digested sludge without chemical precipitation (Barber, 2007). Because the calorific value of digested biosolids is lower, a higher dewatered solid content is required to achieve an energy neutral drying and gasification process.

Combustion of Sludge and Biosolids.   Combustion of solid and semi-solid sludge and biosolids has been implemented in some parts of world where land application and/or landfill disposal is not economical. The main types of combustors used for biosolids combustion include multiple hearth, fluidized bed, and electric infrared. Devices used to recover energy from wastewater solids are summarized in Table 17–9.

Combustion Stoichiometry.   The stoichiometric expression for the complete combustion of sludges and biosolids, discussed in Sec. 14–4 in Chap. 14 and repeated here for convenience, is:

$$C_aO_bH_cN_d + (a + 0.25c - 0.5b)O_2 \rightarrow aCO_2 + 0.5cH_2O + 0.5dN_2 \qquad (14\text{–}3)$$

If it is assumed that dry air contains 23.15 percent of oxygen by weight, the amount of air required for complete oxidation is approximately 4.3 times the calculated amount of required oxygen. In addition, excess air is required to induce near sufficient turbulence and mixing in the combustion chamber. For the incineration of dried sludges and biosolids, typically about 50 percent of excess air is required. Excess air will also affect the combustion temperature, and the air to fuel ratio must be controlled to maximize the efficiency of combustion and minimize the emission of pollutants such as nitrogen oxides, carbon monoxide, volatile organic compounds and other potentially hazardous compounds (see Chaps. 14 and 16).

Table 17–9

Devices used to recover energy from wastewater solid contents[a]

Device	Output[a]	Remarks
Fluidized bed incinerator	Waste heat	• Used widely for dewatered sludge incineration • Typical feed solids content 20 to 35 percent
Multiple heath incinerator	Waste heat	• Used widely for dewatered sludge incineration • Typical feed solids content 20 to 35 percent
Fluidized bed gasifier	Syngas, waste heat	• Uniform product • Require high solids content (>85 percent) • Operating temperature 700 to 900°C
Fixed bed updraft gasifier	Syngas, waste heat	• Relatively low solids content is acceptable (70 to 80 percent) • Relatively high tar content in syngas • Operating temperature ~1000°C
Fixed bed downdraft gasifier	Syngas, waste heat	• Require high solids content (>80 percent) • Relatively low tar content in syngas • Operating temperature 900 to 1000°C • Unconverted carbon 4 to 7 percent
Entrained flow gasifier	Syngas, waste heat	• Relatively low solids content is acceptable • Low tar and CO_2 • Low CH_4

[a] Adapted in part from DOE (2002) and McKendry (2002).

[b] Depending on the heat balance, waste heat could be used to operate turbine to generate electricity and steam or hot water.

Heat Released from Combustion.   Part of the heat released from the combustion of solid materials is stored in the combustion product to raise the temperature of the material, and partly transferred by convection, conduction, and radiation to the wall of the combustion system. Depending on the dewatering and drying process, sludges and biosolids may contain a significant amount of water which will be vaporized in the combustion chamber, taking up latent heat of evaporation. As discussed in Chap. 14, the heat required to maintain the combustion system at the final temperature is expressed as:

$$Q = \Sigma(Q_S) + Q_e + Q_L = \Sigma C_P W_s(T_2 - T_1) + W_W \lambda + Q_L \qquad (14\text{-}4)$$

where Q = total heat, kJ (Btu)
$\quad Q_s$ = sensible heat in the ash, kJ (Btu)
$\quad Q_e$ = latent heat, kJ (Btu)
$\quad Q_L$ = heat loss
$\quad C_p$ = specific heat for each category of substance in ash and flue gases, kJ/kg·°C (Btu/lb·°F)
$\quad W_s$ = mass of each substance, kg (lb)
$\quad T_1, T_2$ = initial and final temperatures
$\quad W_w$ = mass of water, kg (lb)
$\quad \lambda$ = latent heat of evaporation, kJ/kg (Btu/lb)

By comparing the required heat and heat content in the combusted material as calculated by Eq. (2–66), additional energy requirement, or the excess heat that could be recovered and utilized, can be calculated. In Example 17–6, a heat balance for combustion of biosolids is presented.

EXAMPLE 17–6   **Computation of Heat Balance for Combustion of Biosolids**   Dewatered biosolids with the following characteristics is to be combusted in an incinerator. Using the information, given below, on the operating conditions for the incinerator and related process data, determine the air requirement and develop a heat balance. Determine heat content of the flue gas after it was used to preheat the inlet air. For simplicity, assume all combustible materials listed below are oxidized to CO_2, H_2O, N_2 and SO_2, and ignore SO_2 in heat balance calculations.

Biosolids composition and elemental analysis

Element	Percent of total weight
Combustible	
Carbon	14.2
Hydrogen	1.0
Oxygen	5.2
Nitrogen	0.3
Sulfur	0.4
Inerts	8.9
Water	70
Total	100

Operating conditions

Item	Unit	Value
Ambient air temperature	°C	20
Relative humidity at 20°C	%	50
Target combustion temperature	°C	850
Heat loss	% of gross heat input	0.5

Thermodynamic properties

Component	Unit	Value
Specific heat of water	kJ/kg·°C	4.19
Specific heat of biosolids	kJ/kg·°C	1.26
Specific heat of dry air	kJ/kg·°C	1.01
Specific heat of water vapor	kJ/kg·°C	1.88
Specific heat of the ash	kJ/kg·°C	1.05
Latent heat of water evaporation	kJ/kg	2257

Air composition at 20°C and 50 percent humidity

Gas	Percent
N_2	77.1
O_2	20.7
CO_2	0.04
H_2O	1.17
Other gas (ignore in calculations)	1.00

Enthalpy of ideal gas, kJ/mole

	20°C	850°C
N_2	8.53	34.2
O_2	8.54	35.7
CO_2	9.18	49.6
H_2O	9.74	41.1

Solution

1. Estimate the air requirement.
    a. Estimate the stoichiometric requirement to completely oxidize the sludge, assuming CO_2, H_2O, N_2, and SO_2 are the only combustion products. Set up a computation table to determine the moles of oxygen and kg of air required per kg of biosolids.

Component	Weight fraction	Atomic weight, kg/mole	Atomic weight units[a], mole/kg	O_2 required[b], mole	Combustion reaction and products	Product gas formed, mole/kg
Carbon	0.142	0.012	11.83	11.83	$C + O_2 \rightarrow CO_2$	11.83
Hydrogen	0.010	0.001	10.00	2.50	$4H + O_2 \rightarrow 2H_2O$	5.0
Oxygen	0.052	0.016	3.25	−1.63	$2O \rightarrow O_2$	0
Nitrogen	0.003	0.014	0.214	–	$2N \rightarrow N_2$	0.11
Sulfur	0.004	0.0321	0.125	0.062	$S + O_2 \rightarrow SO_2$	0.12
Water	0.70	0.018			H_2O (vapor)	38.9
Inerts	0.086					
Total	1.00			12.83		56.0

[a] Atomic weight unit = weight fraction/atomic weight.

[b] Moles required = atomic weight unit × mole of O_2 required per atom being oxidized based on the combustion reaction. For oxygen, O_2 saved due to oxygen in the biosolids is recorded, indicated with a negative sign.

From the calculations above, 1 kg of biosolids will require 12.8 moles of O_2. Oxygen content in the air = 20.7 percent by volume at 20°C and 50 percent humidity, and assuming mole fraction = volume fraction, moles of air required is:

Air required = 12.8/0.207 = 62.0 mole air/kg biosolids

b. Determine the amount of gas generated from combustion of 1 kg biosolids.
   For CO_2, 1 mole C is converted to 1 mole CO_2. From the summary above, CO_2 formed is 11.8 mole/kg biosolids (assuming complete combustion). Similarly, H_2O from hydrogen, N_2 from nitrogen, and SO_2 from sulfur are calculated and summarized in the table. Note that the water content in biosolids (70 percent) also becomes water vapor.

2. Develop a heat balance for various quantities of excess air for a unit mass of biosolids. Consider 0, 50, and 100 percent excess air and prepare a computation table to summarize the results. Ignore SO_2 for the rest of the calculations.
   a. Determine air flows.
      From Step 1, stoichiometric air requirement is 62.0 mole/kg biosolids
      For 50 percent excess air: 62.0 × 1.5 = 93.0 mole/kg biosolids
   b. Determine heat content of added air.
      From air composition and enthalpy data given in the problem statement, heat content of added air at 20°C without excess air is

      H = [(8.53 × 0.771 + 8.54 × 0.207 + 9.18 × 0.0004 + 9.74 × 0.0117 (kJ/mole)] × 62.0
      mole air/kg biosolids = 524 kJ/kg biosolids = 0.524 MJ/kg biosolids

      Heat content with 50 and 100 percent excess air is calculated as summarized in the computation table.
   c. Determine the heat content in biosolids at 20°C.
      Solid content of biosolids is 30 percent and specific heat of dry biosolids is given in the problem statement. Thus heat content is 0.30 × 1.26 × 20 = 7.56 kJ/kg biosolids. Water content of biosolids is 70 percent. Thus, the heat content is 0.70 × 4.19 × 20 = 58.7 kJ/kg biosolids. Total heat content is 7.6 + 58.7 = 66.3 kJ/kg biosolids = 0.066 MJ/kg biosolids.

    d. Determine the flue gas composition for 0, 50, and 100 percent excess air flows and calculate heat content at 850°C. Calculations for 0 percent excess air flow is shown as an example.

      i. Determine gas composition.

        N_2: (N_2 from N in biosolids) + (N_2 in the air) = 0.11 (mole/kg biosolids) + 62.0 (mole air/kg biosolids) × 0.771 = 47.9 mole/kg biosolids

        O_2: With no excess air, all oxygen in the air is used. For 1 kg of biosolids, 12.8 moles of oxygen is used

        CO_2: 11.8 (mole/kg biosolids) + 62.0 (mole air/kg biosolids) × 0.0004 = 11.9 mole/kg-biosolids

        H_2O: [5.00 + 38.9 (mole/kg biosolids)] + 62.0 (mole air/kg biosolids) × 0.011 = 44.6 (mole/kg biosolids)

        Similarly, the flue gas composition with excess air is calculated and the results are summarized in the table.

      ii. Calculate the heat content using the data given in the problem statement.

        N_2: 47.9 × 34.2 = 1638 kJ/kg biosolids

        O_2: 0 × 35.7 = 0 kJ/kg biosolids

        CO_2: 11.9 × 49.6 = 587 kJ/kg biosolids

        H_2O: 44.6 × 41.1 = 1832 kJ/kg biosolids

        Total = 1638 + 0 + 587 + 1832 = 4057 kJ/kg biosolids = 4.057 MJ/kg biosolids

    e. Calculate the heat content remaining in the ash.

      Assuming complete combustion, ash = inert. Heat content is

      0.089 × 1.05 × 850 = 79.4 kJ/kg biosolids = 0.079 MJ/kg biosolids

    f. Estimate the heat released from combustion of biosolids, assuming complete combustion.

      Using Eq. (2–66) and elemental analysis data given in the problem statement, estimate the heat contents of the solids:

      HHV (MJ/kg) = 34.91 C + 117.83 H − 10.34 O − 1.51 N + 10.05 S − 2.11 A

$$HHV \text{ (MJ/kg)} = 34.91 \times 0.142 + 117.83 \times 0.010 - 10.34 \times 0.052$$
$$-1.51 \times 0.003 + 10.05 \times 0.004 - 2.11 \times 0.089$$
$$= 5.446 \text{ MJ/kg-biosolids)}$$

    g. Estimate evaporative cooling from the vaporization of water in biosolids and water formed by combustion of biosolids.

      From 1 kg of biosolids, 43.9 moles of water are formed. Latent heat of vaporization is 2257 kJ/kg. Therefore, latent heat of vaporization associated with every kg of biosolids is

      [(43.9 × 18)/1000] × 2257 = 1783 kJ/kg-biosolids = 1.783 MJ/kg-biosolids.

    h. Estimate the heat loss. Assume 0.5 percent of gross heat input.

      Gross heat input = (heat of biosolids) + (heat of inlet air) + (heat of combustion) − (heat loss from evaporation). Note that latent heat of vaporization should be subtracted from the gross heat input as the heat of combustion calculated in Step f above is HHV.

      Gross heat loss = (0.524 + 0.066 + 5.446 − 1.783) × 0.005 = 0.021 MJ/kg-biosolids

3. Evaluate the heat balance to determine the air flow to maintain the operating temperature at 850°C.

From the summary table below, energy balance is barely positive with stoichiometric air flow, and the balance is negative with excess air flow. If the heat balance is linear to the air flow, heat balance is exactly zero at 4.3 percent excess air.

	Unit[a]	Stoichiometric air flow	With 50% excess air	With 100% excess air
Air added	mole/kg biosolids	60.3	90.4	120.5
Flue gas composition				
N_2	mole/kg biosolids	46.6	69.8	93.0
O_2	mole/kg biosolids	0	6.2	12.5
CO_2	mole/kg biosolids	11.4	11.5	11.5
H_2O	mole/kg biosolids	45.1	45.5	45.9
Heat content of added air at 20°C	MJ/kg biosolids	0.524	0.786	1.05
Heat content of biosolids at 20°C	MJ/kg biosolids	0.066	0.066	0.066
Flue gas heat content at 850°C	MJ/kg biosolids	4.057	5.119	6.180
Ash heat content at 850°C	MJ/kg biosolids	0.079	0.079	0.079
Energy released from combustion	MJ/kg biosolids	5.446	5.446	5.446
Heat loss by evaporation of water	MJ/kg biosolids	1.783	1.783	1.783
System heat loss	MJ/kg biosolids	0.021	0.023	0.024
Net energy balance		0.095	−0.7	−1.5

[a] Units are per kg biosolids on a wet basis.

Note: Because values were calculated on a spreadsheet and rounded, some values may not match exactly with manual calculations.

4. Determine the water content that will allow self-sustained combustion. From Step 3, the heat generated by combustion of chemical contents in the biosolids is barely sufficient to maintain the heat balance. Therefore in this example, 30 percent solids, 70 percent water content was the limit to sustain combustion. At water content of 71 percent, the heat balance is −0.12 MJ/kg biosolids. It is important to note that at stoichiometric air flow, the heat content in the flue gas is approximately 4.1 MJ/kg biosolids. The heat in the flue gas could be used to preheat biosolids and inlet air or to generate electricity and/or hot water.

Comment   The dewatered sludge solid content assumed in this example is on the higher end of the typical water content with conventional centrifuge. Even though the calculations used in this example may not be completely representative of the actual combustion system as ideal conditions were assumed, they are helpful in illustrating how to conduct a preliminary assessment of the dewatering/drying requirements and operational conditions such as excess air requirement and operating temperature. In biosolids incineration facilities, inlet air is often preheated with flue gas to save the use of supplemental fuel. In a typical incineration facility, a solid content of 26 to 28 percent is considered the threshold needed to sustain combustion without supplemental fuel, but no significant excess energy will be available for other purposes. During the plant start up, supplemental fuel must be used to raise the temperature of the reactor.

Energy Recovery from Syngas

There are two approaches to utilize energy from syngas. One is to clean the syngas and use it for conventional boilers and engine generators, often referred to as two-stage gasification. The other approach is to use the syngas directly in the thermal oxidation chamber, referred to as close-coupled gasification.

Two Stage Gasification.   In two stage gasification systems, as depicted on Fig. 17–11(a), the syngas produced from gasifying the dried biosolids is cleaned and the cleaned syngas can be used as a fuel source for an internal combustion engine. Cleaning the syngas is required to remove sulfur, siloxanes, and other contaminants such as tar that could damage the engine. The syngas cleaning process is not fully developed for the application of biosolids and currently considered in the innovative phase. Syngas cleaning, however, is commercially practiced in the coal industry, but on a much larger scale.

Close-Coupled Gasification.   Close-coupled gasification, illustrated on Fig. 17–11(b) does not require syngas cleaning and instead the syngas is thermally oxidized. Syngas oxidation generates high temperature, approximately 980°C (1800°F), flue gas which can be used for thermal heat recovery. The energy recovered from the flue gas can be used as the energy source to dry to the biosolids to the desired dryness and thus minimize or eliminate the need for fossil fuels (e.g., natural gas or fuel oil). The hot flue gas can also be used as an energy source for generating electricity through the use of boiler and steam turbine or an organic Rankine cycle (ORC) engine. Electricity generation with close-coupled gasification is practiced commonly on other types of biomass, however, this system is not common for biosolids because it is generally more economical to use the energy to offset the drying energy requirement.

Energy Recovery with Fuel Cell

Fuel cell systems used at wastewater treatment facilities utilize methane gas generated from anaerobic digestion. In the fuel cell system, methane gas is used to generate hydrogen as expressed by Eqs. (17–16) and (17–17):

$$CH_4 + 2H_2O \rightarrow CO + 3H_2 \tag{17–16}$$

$$2H_2O + CO \rightarrow H_2 + CO_2 \tag{17–17}$$

Figure 17–11

Recovery and utilization of chemical energy by gasification: (a) two-stage gasification and (b) closed-couple gasification.

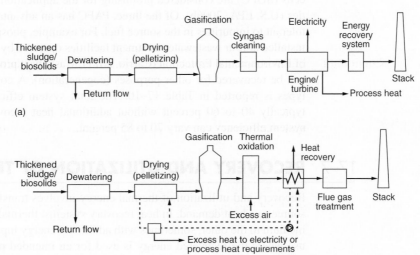

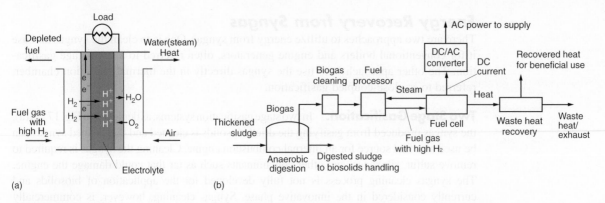

Figure 17–12

Fuel cell energy recovery system: (a) schematic illustration of fuel cell (adapted from U.S. EPA, 2006b) and (b) typical process flow diagram of a fuel cell system used at a wastewater treatment facility.

Hydrogen gas is then introduced to the anode side of the fuel cell system, where electrons are released and protons move within the fuel cell to the cathode. On the cathode side of the fuel cell, protons react with oxygen to produce water [see Fig. 17–12(a)].

Fuel cells are in general sensitive to impurity in the source fuel, and pretreatment is necessary to remove at least hydrogen sulfide (H_2S) and halides before digester gas can be sent to fuel cells. Fuel-cell systems also require essentially non-detect for siloxanes. Cleaned gas is mixed with steam to induce the reactions shown in Eqs. (17–16) and (17–17), and introduced into the fuel cell stack. Steam is recycled from the fuel cell where protons (H^+) and oxygen react on the cathode side of the fuel cell to produce heat and water. Electricity generated from a fuel cell is in direct current (DC), and is converted to alternating current (AC) for use in AC circuits. A typical process flow diagram is shown on Fig. 17–12(b).

Among a number of fuel cell types being used or studied, three fuel cell types: phosphoric acid fuel cells (PAFC), molten carbonate fuel cells (MCFC), and solid-oxide fuel cells (SOFC) are considered promising for the application at wastewater treatment facilities (U.S. EPA, 2006b). Of the three, PAFC has an advantage that the system is relatively tolerant to impurities in the source fuel. For example, phosphoric acid fuel cells have been installed at four wastewater treatment facilities operated by the New York City Department of Environmental Protection (Carrio, 2011). Waste heat produced from the fuel cell system can be recovered for other purposes (cogeneration). A comparison of the three fuel cell types is reported in Table 17–10. The total system efficiency of a fuel cell system is typically 40 to 60 percent without additional heat recovery. With heat recovery, total system efficiency can vary 70 to 85 percent.

17–7 RECOVERY AND UTILIZATION OF THERMAL ENERGY

Recovery and utilization of thermal energy involves transferring heat energy from a heat source to a heat demand. In heat recovery systems, thermal energy is exchanged from one medium to the other, sometimes with additional energy input, and conveyed to the point of use where the recovered energy is used for an intended purpose, such as space heating, heating of an anaerobic digester, or electrical power generation.

Table 17–10

Fuel cell systems used at wastewater treatment facilities[a]

System	Typical power output, kW	Efficiency, %	Advantage	Disadvantage
Alkaline (AFC)	10–100	60	• High performance due to fast cathode reaction in alkaline electrolyte • Low cost components	• Sensitive to CO_2 in fuel and air • High electrolyte management needs
Phosphoric acid (PAFC)	100–400	40	• Suitable for combined heat and power due to high operating temperature • Tolerant to fuel impurities	• Expensive catalyst (Pt) • Long start-up time • Low current and power
Molten carbonate (MCFC)	300–3000	45–50	• High efficiency compared to PAFC • Fuel flexibility • A variety of catalysts could be used • Suitable for combined heat and power due to high operating temperature	• Susceptible to high temperature corrosion and breakdown of cell components • Long start-up time • Low power density

[a] Adapted from DOE (2011).

Sources of Heat

Major sources of thermal energy in wastewater treatment facilities include heat from the combustion processes, as discussed in Sec. 17–5, and sensible heat in wastewater. Excess heat from engine generators and boilers has been used widely to heat water for building heating, anaerobic digesters, or hot water supply. In some wastewater treatment plants, lower quality heat such as the residual heat from the exhaust air stream in heat recovery ventilation systems and the heat in treated wastewater effluent has been recovered and used for various purposes.

Combined Heat and Power (CHP) System. In a CHP system, or cogeneration system, both electricity and usable heat are generated. Typically thermal energy is transformed into hot water or steam and used for space heating, digester heating, drying, and other purposes. The typical range of total system efficiency from internal combustion engines without CHP is between 25 and 50 percent, as reported in Table 17–8, depending on the type of engine and operational conditions, whereas the total system efficiency with CHP can be in the range of 70 to 85 percent.

Low Grade Waste Heat. Low grade waste heat is the heat either coming directly out of combustion processes or after the heat has been recovered from the heat source by other heat recovery systems, and considered not sufficient to generate steam or high temperature hot water (80 to 90°C) directly. The low grade waste heat may have the temperature from 30°C or lower (e.g., wastewater effluent) to as high as 230°C (e.g., exhaust air from existing heat recovery devices), depending on the heat recovery system considered

(DOE, 2008). Devices such as heat pumps and organic Rankine cycle engines have been used to recover energy from low grade waste heat.

Other Sources of Heat. Digested sludge, process air exiting bioreactors, and air in the building exhaust also contain heat that can be used for preheating incoming sludge or process air. Sidestream treatment processes can be a potential source of excess heat as temperature of sidestream tends to be higher than normal wastewater and due to high concentrations of nitrogen, the heat generated from oxidation and reduction of nitrogen is typically significant enough to require dilution of the sidestream or heat removal through a heat exchanger (see Chap. 15). However, the feasibility of heat recovery and its use at a specific treatment facility has to be evaluated as the cost savings from the use of recovered heat may not be justifiable based on the installed cost of the heat recovery equipment and conveyance of recovered heat to the point of heat demand or use.

Demands for Heat

Heating requirements in wastewater treatment facilities are primarily for sludge and biosolids processing, building heating, and generation of hot water for various uses. Heating requirements in solids processing include anaerobic digester heating to maintain the operating temperature and for drying sludges and biosolids. Effluent disinfection by pasteurization, a relatively new development for wastewater applications, also requires significant amount of heat (see Sec. 12–10 in Chap. 12).

Heating and Cooling of Buildings. Typically, heating, ventilation, and air conditioning (HVAC) requirements are assessed during the design of building structures and a dedicated HVAC system is installed to maintain the climatic condition in the buildings. For high level assessment, the estimates on heating demand may be developed from published energy data. More detailed heating demand estimates are typically developed using microclimate-specific energy models. Major heating requirements must be identified and heat recovery considered during the design phase when heat supply for major facilities is considered. Smaller heat recovery systems may be considered separately for specific heating needs and available heat. In rare cases, excess heat at a wastewater treatment plant may be exported to outside of the facility, either to a specific building or to be integrated with a district heat system. Typically, however, the amount and quality of heat is not significant enough for energy input to district heating systems and the use is limited to within the treatment facility as the piping distances are much shorter.

Digester Heating. In conventional mesophilic anaerobic digestion, the heating of the digester is the most significant heating demand. The heating demand in sludge processing will increase significantly with thermophilic anaerobic digestion and also when thermal hydrolysis is applied for the pretreatment of the sludge. Heating of the feed sludge up to the operating temperature is the largest heating demand, and additional energy may be required to maintain the digester temperature, depending on the heat loss from the digester walls, floor, and the roof (see Example 13–7 in Chap. 13).

Drying. Sludge and biosolids drying is an energy intensive operation, and it can raise the treatment facility's energy demand significantly. Heating demand is determined based on the temperature of incoming sludges/biosolids and initial and final water contents. The heat transfer rate is estimated from the heat-transfer coefficient, contact area of heat source, and the temperature difference between the sludge and heating medium as expressed in Eqs. (14–6) and (14–7) (See Sec. 14–3 in Chap. 14). Depending on the heat recovery system employed, the heat required for drying can be supplied by combusting the dried sludge. In Example 17–6, the heat balance for a sludge drying process is presented.

EXAMPLE 17-7   Computation of Heat Balance for Sludge Drying Process   Two different sludges and anaerobically digested biosolids are to be dried and subsequently combusted to generate heat for a sludge drying process. Heat content of the sludges and biosolids are presented below. Assume the sludge drying process requires 3.5 MJ to remove 1 kg of water from sludge, and dried cake has 90 percent solids content. Determine the initial water content above which heat content of the solids is sufficient for the drying process. Assume the heat loss through the transfer of heat from the combustion process to sludge drying process is inclusive of the heating requirement (3.5 MJ/kg). Latent heat of water vaporization is 2.257 MJ/kg.

Type of feed	Heat content, MJ/kg
Primary sludge	11.6
Waste activated sludge	14.0
Anaerobically digested biosolids	16.3

Solution

1. Calculate the mass of wet cake for 1000 kg dry solids and solids contents between 15 and 35 percent.
   For 15 percent solids content, the mass of wet cake is

   Wet cake $= (1000 \text{ kg})/0.15 = 6666.7$ kg

   Set up a spreadsheet to calculate the mass of wet cake between 15 and 35 percent solids content.

2. Calculate the amount of water to be evaporated and the heat required to dry wet cake from the initial solids content to 90 percent.

   Mass of 90 percent solids content cake $= (1000 \text{ kg})/0.90 = 1111.1$ kg

   The amount of water to be evaporated for 15 percent cake is

   Water to be evaporated $= 6666.7 - 1111.1 = 5555.6$ kg

   Heat required to evaporate $= 5555.6 \times 3.5 = 19,444$ MJ

   Similarly, calculate the amount of water to evaporate for the range of solids content and compute the heat required to evaporate water.

3. Calculate heat generated from combustion of the 90 percent solids.
   For primary sludge, 1000 kg of dry solids will generate $1000 \times 11.6 = 11,600$ MJ
   The remaining water (10 percent, or 111.1 kg) will be evaporated.

   The latent heat of vaporization is $111.1 \times 2.257 = 250.8$ MJ

   Heat generated from the combustion of 90 percent solids cake $= 11,600 - 250.8 = 11,349$ MJ.

4. Calculate the net heat balance and plot the results for the three types of sludge for the range of solids content.
   For primary sludge with 15 percent solids content, the heat balance is

   Net heat generated $= 11,349 \text{ MJ} - 19,444 \text{ MJ} = -8905$ MJ.

   Similarly, calculate the heat balance and plot the results for primary sludge, waste activated sludge, and digested sludge. The result is shown on the following plot.

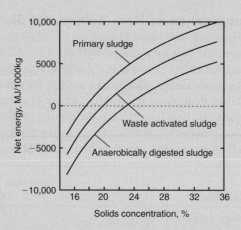

Using the solver function on the spreadsheet program, the solids contents to achieve zero net energy are 20.0, 19.8, and 17.6 percent for primary sludge, waste activated sludge, and digested sludge, respectively.

Comment   Depending on the initial solids concentration and efficiency of the solids combustion and drying process, heat generated from combustion may be sufficient to operate the drying process without supplemental fuel. Detailed evaluation of the thermal efficiency was not included in this example, but a simplified evaluation is often used for a preliminary assessment.

Devices for Waste Heat Recovery and Utilization

Suitable heat recovery applications are determined based on the thermal energy contents and the type of output (e.g., electricity, hot water, hot air, cold water, cold air). Examples of devices used at wastewater treatment facilities to recover and utilize heat include heat exchangers, heat pumps, and heat absorption chillers. Organic Rankine cycle engines have also been used to recover energy from waste heat to produce electricity.

Heat Exchanger.   A heat exchanger is a device that transfers heat from one source (hotter fluid) to another (colder fluid) through a conductive material that separates fluids with different temperatures. Heat exchangers are used widely at wastewater treatment plants, especially where anaerobic digesters are used to stabilize sludges and produce digester gas. Exhaust heat from the water jacket of engine generators is often used for digester heating and heating of buildings. Various types of heat exchangers have been installed at wastewater treatment plants. Common heat exchanger types are (1) coil heat exchangers, (2) plate heat exchangers, (3) shell-and-tube heat exchangers [see Fig. 17–13(a)], and (4) spiral heat exchangers [see Fig. 17–13(b)].

A simplified schematic of a shell-and-tube heat exchanger is shown on Fig. 17–14. In simpler forms, heat exchanger configurations can be categorized as cocurrent [see Fig. 17–14(a)] and countercurrent [see Fig. 17–14(b)]. Overall heat transfer with a shell-and-tube heat exchanger is expressed as

$$Q = UAF\Delta T \tag{17–18}$$

where Q = total heat load transferred, MJ/h (Btu/h)
   U = heat transfer coefficient, MJ/m²·h·°C (Btu/ft·h·°F)
   A = surface area, m² (ft²)

Figure 17-13

Typical heat exchangers used at wastewater treatment facilities: (a) schematic of shell-and-tube heat exchanger for the recovery of heat from engine exhaust, (b) view of a shell-and-tube heat exchanger, (c) schematic of spiral heat exchanger for sludge heating, and (d) view of inside of a spiral heat exchanger.

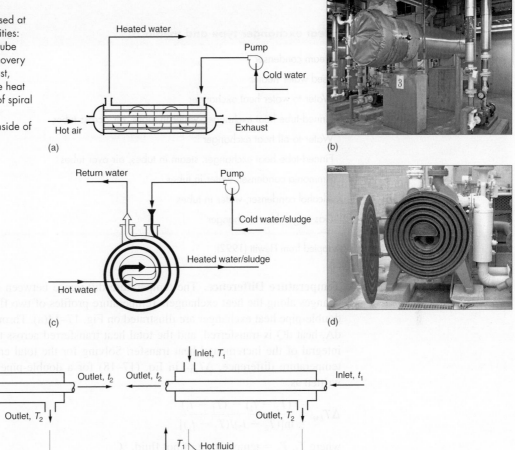

Figure 17-14

Simplified diagram of heat exchangers and corresponding temperature gradients: (a) cocurrent type and (b) counter current type.

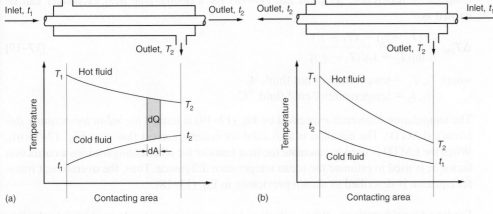

F = correction factor for specific flow arrangements within the heat exchanger, unitless

ΔT = mean temperature difference, °C (°F)

Heat Transfer Coefficient. The heat transfer coefficient depends on the conduction heat transfer coefficients for the materials used, and to a lesser degree the heat convection resistance (Holman, 2009). A detailed analysis of heat exchangers can be found in references on chemical engineering (for example, Green and Perry, 2007). The typical range of heat transfer coefficients for various materials are reported in Table 17–11. The correction factor, F, for various heat exchanger configurations could be found in references on heat exchangers, such as Holman (2009), Kuppan (2000), and others.

Table 17–11

Typical heat transfer coefficients for various materials[a]

Heat exchanger type and media	Heat transfer coefficient, U W/m²·K
Steam condenser	1100–5600
Feed water heater	1100–8500
Water to water heat exchanger	850–1700
Finned-tube heat exchanger, water in tubes, air across tubes	25–55
Water-to-oil heat exchanger	110–350
Finned-tube heat exchanger, steam in tubes, air over tubes	28–280
Ammonia condenser, water in tubes	850–1400
Alcohol condenser, water in tubes	255–680
Gas-to-gas heat exchanger	10–40

Adapted from Hewitt (1992).

Temperature Difference. The temperature difference between hot fluid and cold fluid changes along the heat exchanger. Temperature profiles of two fluids in a cocurrent flow double-pipe heat exchanger are illustrated on Fig. 17–14(a). Through an incremental area, dA, heat dQ is transferred, and the total heat transferred across the heat exchanger is an integral of the incremental heat transfer. Solving for the total energy transfer, the mean temperature difference, ΔT_{LM}, in Eq. (17–18) for a double-pipe heat exchanger can be written as

$$\Delta T_{LM} = \frac{(T_2 - t_2) - (T_1 - t_1)}{\ln[(T_2 - t_2)/(T_1 - t_1)]} \qquad (17\text{–}19)$$

where T_1, T_2 = temperature of hot fluid, °C
$\quad\quad\;\; t_1$, t_2 = temperature of cold fluid, °C

The temperature difference expressed by Eq. (17–19) is termed *log mean temperature difference* (LMTD). The equation is also valid for counter current flow [see Fig. 17–14(b)]. When the LMTD is used to calculate the heat transfer for other configurations a correction factor, F, is used to estimate the mean temperature difference. Thus, the overall heat transfer equation is described as shown previously in Eq. (17–18).

Fouling Considerations. When selecting heat exchangers, the designer must take into account the fouling factors which will impede the heat transfer. Consideration of the fouling factor, is a critical design factor, especially when the fluid is not within a closed loop and is not a clean fluid. Generally, selection of a heat exchanger would be determined by the cost and specifications by manufacturers providing packaged heat exchangers. The characteristics of the fluid from which heat is recovered, including the type of fluid, temperature, any constituents in the fluid, and desired heat recovery are put together and manufacturers are consulted. Critical information that would determine the feasibility of heat recovery and the type of heat exchanger includes the heat transfer requirements, capital cost, physical size, pressure drop across the heat exchanger, and ease of cleaning. Design of a heat exchanger that meets the heat transfer requirements will depend on the relative weights of cost versus pressure drop and physical size limitations (Holman, 2009). The use of Eq. (17–18) and Table 17–11 would be useful in conducting a preliminary assessment before detailed information is collected for design.

Heat Pump. The heat pump is a device that uses a refrigerant to take heat from one source typically at a lower temperature and transfer it to another medium which usually has a higher temperature (DOE, 2003). A simplified schematic is shown on Fig. 17–15. Briefly, the low temperature-low pressure refrigerant is vaporized using a heat source. The vaporized refrigerant is compressed to a high pressure-high temperature vapor by a compressor which requires energy (electricity). It then goes through another heat exchanger to transfer the latent heat of vaporization to a heating medium as the refrigerant is condensed to liquid. The refrigerant that is condensed to liquid still retains relatively high temperature. In the expansion valve, the temperature and pressure of the refrigerant is lowered. The low pressure, low temperature refrigerant goes back to the first step to be vaporized by taking up heat from wastewater effluent. The heat transferred to a medium could be used for heating of buildings, water, digesters, and other purposes.

When wastewater is used as a heat source, usually an intermediate medium such as propylene glycol is used to take heat from wastewater by a heat exchanger suitable for wastewater, and glycol is used to operate the heat pump. The intermediate step is recommended to avoid fouling of the heat exchanger within the heat pump. The use of a heat pump for heating can also provide a side benefit by lowering the temperature of the effluent to be discharged to receiving water that is sensitive to temperature difference. The most common use of heat pump at wastewater treatment facilities is for space-heating and ventilation preheating.

The same principle for a heat pump could also be used on an opposite direction to cool a medium. A refrigerator is an example of a cooling cycle heat pump. In a warm climatic condition, wastewater can be used as a heat sink (i.e., heat taken out for cooling is disposed) in a heat pump operation to provide space cooling. In a large scale system, both heating and cooling cycles could be used so that the heat pump system could be utilized throughout a year. A district energy system installed in Tokyo is such an example (Funamizu et al., 2001). It is a key design consideration to determine if the heat pump system would be used only for heating, or used for both heating and cooling.

Heat Balance. Referring to Fig. 17–15, the heat transferred from the effluent, Δh_{in}, J/s, is

$$\Delta h_{in} = m_o C_o (T_1 - T_2) \tag{17–20}$$

where m_o = mass flowrate of effluent, kg/s
$\quad\ C_o$ = specific heat of heat source, J/kg·°C
$\quad\ T_1$ = heat source temperature entering the heat pump, °C
$\quad\ T_2$ = heat source temperature exiting the heat pump, °C

Figure 17–15

Schematic illustration of a heat pump system.

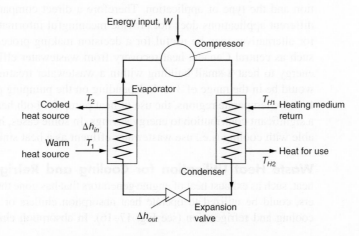

The heat transferred from the heat pump to the heating water, Δh_{out}, is

$$\Delta h_{out} = m_H C_H (T_{H2} - T_{H1}) \tag{17-21}$$

where m_H = mass flowrate of heating medium, kg/s
T_{H1} = heating medium temperature entering the heat pump, °C
C_H = specific heat of heating medium, J/kg·°C
T_{H2} = heating medium temperature exiting the heat pump, °C

Energy loss through the piping, h_c, is

$$h_c = h_f + h_a = g m_e H(2cL) + UA\Delta T \tag{17-22}$$

where, h_f = energy loss due to friction headloss, J/s
h_a = heat loss through the pipe, J/s
g = acceleration of gravity, m/s^2
H = headloss per unit pipe length, m/m
c = friction loss per unit pipe length, m/m
L = distance between heat pump and heat delivery point, m
Other terms were defined previously [see Eq. (17–18)].

If the electricity used to compress the refrigerant is W_1 (J/s) and the compressor efficiency is E_c, the energy balance is as follows:

$$m_o C_o (T_1 - T_2) + E_c W = m_H C_H (T_{H2} - T_{H1}) + h_c \tag{17-23}$$

where E_c = compressor efficiency, unitless
W = energy input, J/s

Coefficient of Performance. Coefficient of performance (COP) is an expression used to quantify heat pump performance in terms of the heat delivered from a unit of energy input.

$$COP = h_h / h_w \tag{17-24}$$

where, COP = coefficient of performance, unitless
h_h = heat output from the heat pump, Joule (or J/s)
h_w = energy input to the heat pump, Joule (or J/s)

Using the terms shown previously, COP can also be written as

$$COP = \frac{m_H C_H (T_{H2} - T_{H1})}{W} = E_c W + C_o (T_2 - T_1) - h_c/W \text{ (use the same format as the term before)} \tag{17-25}$$

The level of COP achievable for the specific application will depend on the climatic condition and the type of application. Therefore a direct comparison of COP values between different applications does not provide meaningful information, but the estimate of COP for alternatives may be useful for a decision making process. In relatively cold climate, such as central Canada, heat recovery from wastewater effluent could provide sufficient energy to heat a small building within a wastewater treatment facility, and typical COP would be in the range of 3 to 4, depending on the pumping power and fluid temperatures. In more temperate regions, the use of heat pump for both heating and cooling can provide a significant contribution to energy savings. In some cases, higher COP values are achievable with cooling (i.e., use wastewater effluent as a heat sink).

Waste Heat Utilization for Cooling and Refrigeration. Low grade waste heat, such as exhaust heat of engine generators that has gone through a series of heat exchangers, could be utilized to operate heat absorption chillers or other technologies to provide cooling and refrigeration (see Fig. 17–16). In absorption chillers, lithium bromide is used

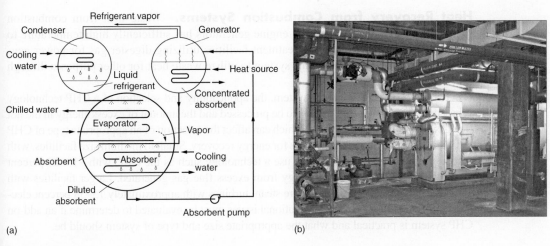

(a)   (b)

Figure 17-16

Chiller using low grade heat: (a) schematic of chiller composed of generator, condenser evaporator and absorber and (b) view of typical chiller (courtesy of Philadelphia Water Department).

typically to absorb the vaporized water (refrigerant) to create a vacuum environment. Water absorbed in the lithium bromide solution is vaporized with the low grade heat, and concentrated lithium bromide solution is recirculated. Water vapor is sent to a condenser, and sent back to the vacuum environment, where it boils at a low temperature and evaporative cooling is used to produce chilled water. The quality of energy source suitable for the absorption chillers varies with the chiller types, but suitable heat source is typically between 70 and 130°C.

The performance of absorption chiller or other cooling and refrigerating systems can be quantified in a similar manner as the heat pump, in terms of COP. The range of COP depends on the energy source and the cooling requirement. When hot water is used as a heat source, the COP value can be 4.0 or higher. Because the design of chillers will involve evaluation of the heat source, chilled water temperature, cooling water temperature and other site specific conditions, manufacturers of chillers should be consulted and a computer modeling is used to determine the sizing and operating conditions.

Organic Rankine Cycle Engine. The organic Rankine cycle (ORC) engine is an electrical power generator utilizing an organic refrigerant. The mechanism of ORC engines is similar to that of the heat pump, and a generator is used in place of the expansion valve. The ORC is capable of generating electricity from waste heat as low as 67°C (150°F), and the efficiency is generally 10 to 20 percent (DOE, 2008).

Design Considerations for Thermal Energy Recovery Systems

Evaluation of the characterization of waste heat is an important step to determine the feasibility of thermal energy recovery and the selection of the recovery systems. Important parameters that must be determined include (DOE, 2008):

1. Heat quantity
2. Heat temperature/quality
3. Composition
4. Minimum allowed temperature
5. Operating schedules, availability, and other logistics
6. Diurnal and seasonal variations in the parameters identified above

Heat Recovery from Combustion Systems. Waste heat from combustion systems such as exhaust heat from engine generators has sufficiently high temperature to generate hot water or steam. At treatment facilities utilizing digester gas for power generation, it is common to design the system to utilize waste heat for other purposes through a CHP system (see Chap. 13).

For a biosolids incineration system, the applicability and selection of a CHP technology is based on the amount of biosolids to be processed and the amount of excess energy in the hot flue gas generated. Another factor which can affect the selection of an appropriate type of CHP is the size of the facility and the goals for energy recovery. Small to medium size facilities, with a goal of producing electricity may use a technology such as the ORC with ~10–20 percent electrical efficiency to recover energy from excess flue gas generated. Larger facilities with similar goals may select high pressure steam turbines with approximately 15–38 percent electrical efficiency. Economic and operational issues must be evaluated to determine if an add-on CHP system is practical and what the appropriate size and type of system should be.

Heat Recovery from Wastewater. The temperature of wastewater is typically higher than that of potable water, and the variation in wastewater temperature is smaller than that of ambient temperature (see Fig. 17–17) throughout a year. The heat from wastewater is a reliable source of thermal energy for beneficial uses during the colder season, and it can be used as a heat sink during the warmer season. Depending on the size of the treatment facility, thermal energy can be used to supplement heating requirements within the wastewater treatment facility, or integrated into a district heating/cooling system. Because the available heat in wastewater is low, a heat pump is used to extract heat from wastewater (Pallio, 1977). Depending on the availability of heat, heating of a specific building within a wastewater treatment facility may be considered, or it may be integrated into the centralized heating system for the entire treatment plant. The estimation of heat to be recovered from wastewater is illustrated in Example 17–7. Heat recovery from raw wastewater in the collection system or at the beginning of the treatment facility can also be considered. When heat is recovered from raw wastewater, the impact of lowered wastewater temperature on the treatment performance must be evaluated (Wanner et al., 2005).

In North America, heating or cooling capacity of air conditioning equipment is often measured in terms of the tons of refrigeration. One ton of refrigeration is approximately 12,000 Btu/h or 3.517 kW. Similarly, the power requirement for the heat extraction is often expressed in brake horse power (bhp) per ton of refrigeration. One horse power (hp) is approximately 0.745 kW. Thus, if a power requirement for a heat pump is rated as 1.0 bhp/ton, then 0.745 kW is required to provide 3.517 kW of heat extraction.

Figure 17–17

Seasonal variation of ambient temperature and wastewater temperature in north eastern United States.

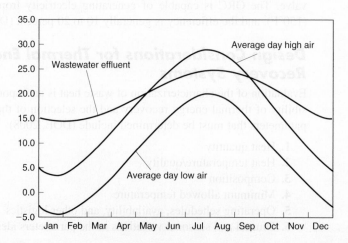

EXAMPLE 17–8   **Heat Pump for Effluent Heat Recovery**   A water to water heat pump is to be used to provide heating in a building at a wastewater treatment plant. The heat source is treated effluent. The lowest effluent temperature is 12°C, and temperature drop of the effluent in the shell and tube heat exchanger is assumed to be 4°C. A 40 percent propylene glycol mixture is used as an intermediate heat transfer medium to extract heat from the effluent, and the heat pump will have an 8°C entering temperature and a 5°C leaving fluid temperature. The peak heating requirement for the building was estimated to be 360 kW.

A heat pump supplier was contacted and a heat pump was selected requiring 0.222 kW electrical input per kW heat output to extract 300 kW from the effluent.

1. Calculate the COP for the heat pump, and the heat output and the electrical power input in kW of the heat pump. For simplicity, assume compressor efficiency = 1
2. Three circulation pumps, 14, 11, and 15kW, are used to transfer effluent to the heat exchanger and to pump the glycol on each side of the heat pump system. Calculate the overall system COP including pumping power.
3. Given the energy to be extracted from wastewater and assumed temperature drop, determine the wastewater flowrate that needs to be transferred to the heat pump system.

Solution   1. Calculate the COP for the heat pump using Eq. (17–24).

Power input = 0.222 kW per 1 kW of output

$$COP = \frac{1.0}{0.222} = 4.5$$

Using Eq. (7–25) and ignoring the heat loss from the transfer of the heating medium,

$$4.5 = \frac{W + m_oC_o(T_1 - T_2)}{W} = \frac{W + 300}{W}$$

Power input = W = 86 kW

Using Eq. (17–23),

Total power output = 300 kW + 86 kW = 386 kW

2. Calculate the COP for the entire system.

$$COP = \frac{386\,kW}{(86\,kW + 14\,kW + 11\,kW + 15\,kW)}$$
$$= 3.06$$

3. Calculate the wastewater flowrate to be transferred to the heat pump system.

300 kW = 300 kJ/s

Specific heat of water = 4.2 kJ/kg·°C

Assume density of water = 1.0

$$\text{Wastewater flowrate required, m}^3\text{/s} = \frac{(300\,kJ/s)}{(4.2\,kJ/kg\cdot°C)(1000\,kg/m^3)(4°C)}$$
$$= \frac{300}{4.2\cdot1000\cdot4} = 0.0179\ m^3/s = 1.07\ m^3/min$$

Comment   Additional energy required to convey (i.e., pump) wastewater and heated water or glycol can be significant if the heat extracted by a heat pump is used for a large area or outside of the wastewater treatment facility (e.g., for district heating system).

17–8 RECOVERY AND UTILIZATION OF HYDRAULIC POTENTIAL ENERGY

Utilization of hydraulic potential energy involves conversion of hydraulic potential energy into electricity, and transferring of hydraulic pressure in a stream to different parts of the treatment process. The purpose of this section is to introduce (1) the type of devices used to recover energy from hydraulic potential, (2) the hydraulic energy recovery applications adopted by wastewater treatment facilities, and (3) considerations for the selection and design of hydraulic energy recovery systems.

Type of Hydraulic Potential Energy Recovery Devices

Hydraulic turbines for electrical power generation are the primary machines used for the recovery of hydraulic potential energy from the main liquid process flow. Devices to utilize the pressure head or kinetic energy of the stream are usually integrated into a specific energy-intensive unit process to reduce the energy requirement of the process. Electrical power generated from hydraulic potential energy can be sent to the electrical distribution system within the treatment facility to supplement the electricity drawn from the electrical grid or directly connected to the grid and sold to the electric utility.

Hydraulic Turbines. Turbines are turned by water flow, and the movement of the turbine is directed to the attached generator to produce electricity. Hydraulic turbines to generate electricity are essentially pumps in reverse. An electrical generator connected to the turbine produces electricity rather than an electric motor which consumes electricity to move water. The type of turbine follows the flow and head characteristics of the related pumps. The head and flow ranges for most wastewater treatment facilities would place the hydroturbines generators in the small (< 5 MW) or micro (< 250kW) range.

The two types of hydraulic turbines are: (1) reaction turbines, and (2) impulse turbines. In the reaction turbines, water pressure is applied to the face of the runner blades to turn the turbine. Radial (Francis) turbines, and axial type (Kaplan and propeller) turbines are the common reaction turbines. In the impulse turbines, water entering the turbine has sufficient velocity (i.e., kinetic energy), and turn the turbine as a high velocity water hits the buckets on the periphery of the runner. Common impulse turbines include Pelton turbines [see Figs. 17–18(a) and (b)], Turgo turbine, and cross-flow turbines (ESHA, 2004). Impulse turbines exhaust to atmosphere and are not suitable for energy recovery from force mains or where the continuity of the water column must be maintained through the turbine to capture the available head. The application ranges for various types of hydraulic turbines are graphically presented on Fig. 17–19. The Pelton and Turgo are the best fit for low flow and high head conditions. The axial flow Kaplan-type turbines are suitable for low head, high flow conditions that would be typical of wastewater treatment plant effluent streams. Crossflow turbines are used most commonly for small scale applications, and suitable for low head applications.

Devices for the Transferring of Pressure Head. Hydraulic turbochargers and pressure exchangers used for the nanofiltration and reverse osmosis systems are the examples of devices used to reduce energy requirements by transferring hydraulic potential energy from a stream to the other part of the process. Hydraulic turbochargers utilize the hydraulic head in the membrane reject stream to boost the inlet pressure for the membrane modules. The use of energy recovery devices in membrane treatment systems are discussed in detail in Sec. 11–6 in Chap. 11.

Figure 17–18

Examples of hydraulic turbines: (a) schematic of a pelton wheel turbine, (b) view of pelton turbine used with raw wastewater for power generation in Jordan, (c) schematic of Francis turbine, and (d) schematic of crossflow turbine. (Adapted from BHA, 2005.)

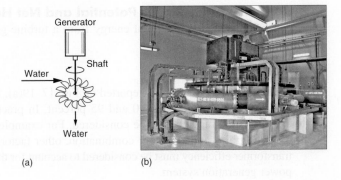

(a) (b)

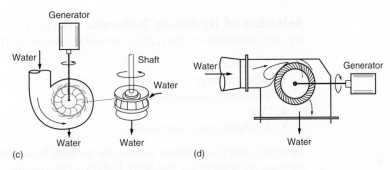

(c) (d)

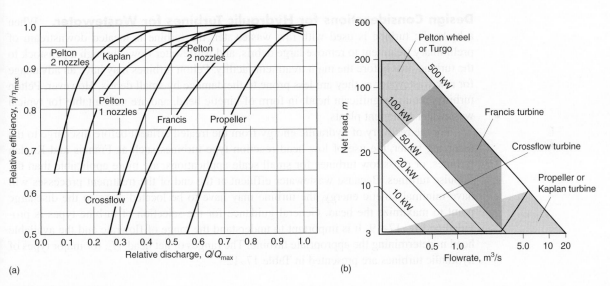

(a) (b)

Figure 17–19

Guidelines for the selection of hydraulic turbine: (a) typical turbine efficiency for various turbine types, and (b) range of operation for hydraulic turbine based on the net head and flowrate (adapted from the BHA, 2005, and ESHA, 2009).

Application of Hydraulic Energy Recovery Devices

The two most common applications to recover hydraulic potential energy from wastewater treatment facilities are (1) recovery of hydraulic potential at the influent or effluent of the wastewater treatment plant and (2) recovery of residual head from high-pressure membrane treatment processes.

Recovery of Hydraulic Potential and Net Head. The hydraulic energy that can be converted to electrical energy from a turbine-generator system is defined previously as.

$$P_e = \rho Q g \Delta H \eta_i \eta_e \qquad (17\text{-}8)$$

Typical turbine efficiency is reported on Fig. 17–19(a). Typically, generator efficiency for new generators is between 90 and 95 percent. In practical applications, additional efficiency factors may need to be considered. For example, where power is to be generated using a turbine and generator combination, other factors such as penstock efficiency and transformer efficiency must be considered to account for the efficiency for the entire electrical power generation system.

Selection of Hydraulic Turbines. The key criteria in selecting the type, geometry, and dimensions of the turbine include the following (ESHA, 2004):

1. Available head
2. Flowrate range
3. Variation in tailwater elevation
4. Cost
5. Available space and access

Generally, impulse turbines are suitable for high head, low flowrate streams, and reaction turbines are suitable for low head, high flowrate streams.

Design Considerations for Hydraulic Turbines for Wastewater. When a hydraulic turbine is used with raw wastewater, the turbine is located downstream of preliminary treatment to remove large debris, and wastewater flows through a penstock to the turbine, just before the main treatment facility. Pelton turbines may have an advantage for raw wastewater as they are less prone to the damage by small debris. However, Pelton turbines require significant head, in form of kinetic head, and are not suitable for typical wastewater treatment plants.

For the recovery of hydraulic energy from the treated effluent before discharge to the receiving water, the use of low head reaction type turbines such as Francis and Kaplan turbines, or crossflow turbines for small scale applications, are more applicable than the impulse turbines. Because wastewater effluent at the end of the treatment processes will have minimal kinetic energy, the turbine may have to be located close to the discharge point to maximize the head. General guidance for the selection of turbine types is provided on Fig. 17–19. It is important to understand the range of flowrate and the available head in determining the appropriate turbine. The ranges of net head, H_n, for major types of hydraulic turbines are presented in Table 17–12.

Table 17–12

Typical range of net head for major hydraulic turbine types[a]

Turbine type	Range of net head, H_n, m
Kaplan and propeller	2–40
Francis	25–350
Pelton	50–1300
Crossflow	2–250

[a] Adapted from ESHA (2004).

Recovery of hydraulic head from wastewater plant effluent flow will not fit the typical range of turbines used for hydroelectric generating facilities. The recovery of energy from a large WWTP (432 ML/d) with 7 meters of available head is illustrated in Example 17–9.

EXAMPLE 17–9   Power Generation from Wastewater Discharge Location   Estimate the power that can be generated by a hydro-power generation unit located at the effluent discharge for the following conditions. Also, select the type of turbine that would be most suitable the given conditions.

Parameter	Unit	Value
Available head	m	7
Flowrate	m³/s	5
Penstock efficiency	%	0.92
Turbine efficiency	%	0.8
Generator efficiency	%	0.92
Transformer efficiency	%	0.95

Solution

1. Estimate the the power that can be generated using Eq. (17–8).

$$P_e = (1000)(5)(9.81)(7)(0.8)(0.92) = 2.53 \times 10^5 \text{ W} = 253 \text{ kW}$$

2. Selection of turbine type.
   From Fig. 17–19(b), for the flow and available head, a Kaplan or propeller turbine could be used.

Comment   Consideration should be given to looking at the hydroturbine as a pump in reverse and coupling the appropriate pump to a generator. The reverse pump may not be as efficient as a turbine designed specifically for hydropower, but it may be difficult to find a hydroturbine in the flow and head ranges seen at wastewater treatment facilities. Note that flywheels are often used to prevent freewheeling of the turbine if the connection to the electrical system is interrupted.

Use of Residual Pressure Head in Treatment Processes

For industrial and indirect/direct potable reuse applications, reverse osmosis has become a common treatment unit process used to remove dissolved constituents that are otherwise difficult to remove. In water reuse applications, typical feed water pressure for the RO unit is in the range of 12 to 18 bar, and the concentrate (brine) will retain the pressure ranging 10 to 16 bar. Because transfer of the brine usually does not require as much head to the discharge point or to the brine storage, the residual pressure is often used to recover energy. Devices used for the recovery of energy from the residual pressure sources were discussed previously in Sec. 11–6. Typically, the energy recovery from RO units are considered economically viable for larger treatment plants as the energy savings is close to proportional to the flowrate whereas the additional costs for the energy recovery devices are not proportional to the size of the units.

17–9 ENERGY MANAGEMENT

Important energy management considerations include maximizing the use of available energy as discussed in previous sections, reducing the use of energy at wastewater treatment facilities, increasing the energy production onsite by utilizing waste from other sources, and utilizing renewable energy such as solar or wind power. In addition, a wide range of management goals must be evaluated, including (NYSERDA, 2010)

1. Improving energy efficiency and managing total energy consumption
2. Controlling peak demand for energy
3. Managing energy cost volatility
4. Improving energy reliability

In addition to these management goals, greenhouse gas reduction has become a management goal for wastewater treatment facilities. Assessment of greenhouse gas emissions is described in Sec. 16–6 in Chap. 16. Topics considered in this section include (1) process optimization and modification for energy saving, (2) process modification for increased energy production, (3) peak flowrate management, and (4) selection of energy sources. It should be noted that the topics covered in this section are illustrative of what can be done to improve energy management, and are not meant to be inclusive of all energy management opportunities. Additional information on energy management can be found in a number of publications including WEF (2009), and U.S. EPA (2008, 2010, 2012).

Process Optimization and Modification for Energy Saving

Opportunities to improve energy efficiency in existing wastewater treatment facilities are identified through the energy audit described in Sec. 17–4. Typically, the evaluation following the energy audit involves process optimization and modification. Examples of process optimization and modifications for energy saving are summarized in Table 17–13. Energy saving may be achieved through operational changes to existing processes or a modification to processes and equipment. Even though the opportunities for energy saving from operational changes are limited, they can be made with little or no capital cost. Process modifications and the use of energy efficient equipment could make improvements in energy management, but may require significant capital expenditures. A life cycle cost analysis should be conducted for the shortlisted options to determine if the energy savings generated by making the change justify the capital cost of the change. To illustrate the energy reduction through process modifications, energy saving opportunities with large pumps and activated sludge aeration systems are considered below.

Major Pumps at Wastewater Treatment Facilities.
Many wastewater treatment facilities require raw wastewater be pumped at the headworks to provide sufficient hydraulic head for wastewater to flow through the rest of the liquid treatment processes by gravity. These pumps, often identified as *main sewage pumps (MSPs)*, are typically the largest pumps at a wastewater treatment plant. The MSPs are required to handle a wide range of flowrates through diurnal variations of the flowrate as well as the wet weather flows. Treatment facilities designed for nutrient removal typically have significant mixed liquor or return activated sludge (WAS) flows. In MBR systems, the recycle flow is often 6 times the influent wastewater flowrate to maintain high MLSS concentrations (see Chap. 8). Energy efficiency in these major pumps can affect the overall energy efficiency of the treatment facility significantly.

Table 17–13

Example of process optimization and modifications to manage energy consumption

Operational modification	Expected outcome	Potential issues
Main sewer pump control	Reduced energy requirement	Modification to the pumps may be necessary
Wastewater loading distribution by flow equalization	Reduced peak-hour energy use	
DO control in activated sludge	Reduced power consumption	Usually require online monitoring of DO and other parameters
	Increased process performance reliability	Increased monitoring requirements
		Potential process upset if online analyzers are not working properly
Conversion of nitrification to nitritation or simultaneous nitrification/denitrification	Reduced oxygen requirement	precise process control is required for nitritation
Use of energy efficient diffusers	Reduced air flow requirement	Air flow distribution should be checked
Use of energy efficient blowers	Reduced energy requirement	Some energy efficient blowers may not be suitable for deep reactors due to pressure head limitations
Use of energy efficient UV lamps	Reduced energy requirement	
Energy recovery system for RO	Reduced energy requirement	
Use of off-peak hours for biosolids dewatering	Reduced peak-hour energy use	

The overall efficiency of a pumping system is defined as

$$E = E_p \times E_m \times E_c \qquad (17\text{–}26)$$

where E = overall efficiency
E_p = pump efficiency
E_m = motor efficiency
E_c = control efficiency

The relative efficiency of each of the three elements identified in Eq. (17–26) is reported in Table 17–14. Pump and motor efficiencies depend on the selection of specific models, whereas the control efficiency can vary depending on the type of the control device used. The same principle applies to the control of blowers as described in the following discussion.

Variable Frequency Drive. The flow control efficiency depends on the type of control used, and the use of variable frequency drives (VFDs) is the most energy efficient, corresponding to the higher end of the range of control efficiency reported in Table 17–14. Operationally, VFDs manipulate the frequency of the alternating current to vary the speed

Table 17–14

Pump systems used for wastewater treatment and typical efficiency[a]

Pump system component	Range, %	Typical value, %
Pump	30–75	60
Flowrate control	20–98	60
Motor	85–95	90
Overall efficiency	5–80	

[a] Adapted from U.S. EPA (2010).

of a motor. With a VFD, the pump can continue to operate along the same pump system curve while shifting the performance curve [see Fig. 17–20(a)]. Throttling and bypass controls artificially shift the system curve while maintaining the power input, thereby reducing the actual flow delivered from the pump [see Fig. 17–20(b-d)]. Stop/start control does not use part of the output that is available (U.S. EPA, 2010).

Aeration Control for the Activated Sludge Process. Energy demand associated with aeration of the activated sludge process could be managed by (1) controlling the aeration rate to match actual oxygen requirements, (2) using energy efficient and appropriately sized blowers, (3) using diffusers with higher oxygen transfer efficiencies for

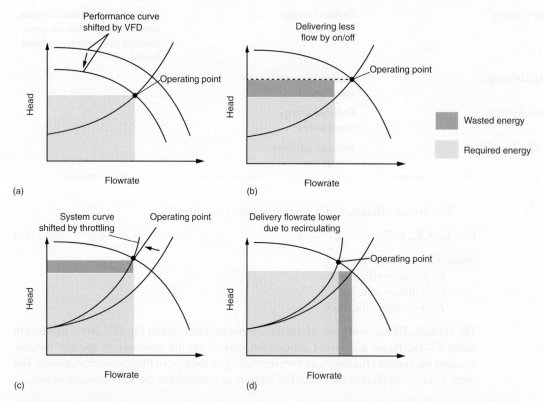

Figure 17–20

Pump efficiency with various control methods: (a) with VFD control, (b) energy loss due to intermittent pumping, (c) energy loss due to throttling, and (d) energy loss due to recirculation. (Adapted from U.S. DOE, 2006.)

Figure 17–21

Examples of aeration blowers for activated sludge process: (a) view of large centrifugal blower driven by 1120 kW (1500 hp) electric motor, and (b) view of single stage centrifugal blower with adjustable inlet guide vanes, (c) schematic of high speed turbo blower (adapted from APG Neuros), and (d) view of high speed turbo blower installation in self contained sound-dampening enclosure.

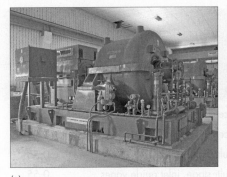

(a)

(b)

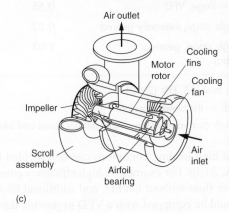

(c)

(d)

the given locations and other conditions, and (4) controlling organic loading entering the activated sludge process. Minimizing energy loss from inefficient piping of air supply system can also contribute to the overall energy demand, but the contribution is generally minor compared to other factors (WEF/ASCE 2010).

Excessive DO concentration in the aerobic zone of BNR processes is not only wasteful of energy but can result in the carryover of DO into anoxic or anaerobic zones. Excessive DO can also result in biological oxidation of organic carbon meant for denitrification. A deoxigenation zone is often allocated to avoid the DO intrusion into anoxic and anaerobic zones.

Blowers. The types of blowers used commonly in North America include single stage centrifugal, multi stage centrifugal, rotary lobe positive displacement, and high speed single stage turbo blowers (see Fig. 17–21 and Fig. 5–72 in Sec. 5–11 in Chap. 5). High speed turbo blowers are relatively new technology that utilizes air bearings or magnetic bearings to achieve high speed. Comparison of the blower types and blower efficiencies are summarized in Table 17–15. It should be noted the technologies for blowers are advancing quickly and most recent information should be consulted to determine the applicability of the turbo blowers. Airflow control devices include inlet dampers, inlet guide vanes at the inlet to the blower, and outlet dampers. Inlet guide vanes are used most commonly with centrifugal blowers and are more energy efficient than dampers or throttling. Inlet guide vanes create swirls to the airstream entering the blower, lowering the load on the blower, and are an effective airflow control between 80 and 100 percent of full flow. Another method of airflow control is fan speed adjustment, typically by VFD (DOE, 1989). Similar to the pumps, the use of VFD is an energy-efficient control of the air flowrate. The turn-down ratio, the ratio between the lowest operable flowrate to the maximum flowrate, for the specific blower must be evaluated so that the air flowrate for both

Table 17–15

Types of blowers and typical blower efficiency[a]

Blower type	Range of discharge pressure Bar	Nominal blower efficiency %	Nominal turndown % of rated flow[b]
Positive displacement, VFD	0.55–1.03	45–65	50
Positive displacement hybrid, VFD	0.20–1.5	70–80	25
Centrifugal, multi-stage, throttled	0.55	60–70	60
Centrifugal, multi-stage, inlet guide vanes	0.55	60–70	60
Centrifugal, multi-stage, VFD	0.55	60–70	50
Centrifugal, single stage, internally geared[c]	0.83	75–80	45
Centrifugal, single stage, gearless (high-speed turbo)	1.03	75–80	45–50

[a] Adapted, in part, from U.S. EPA (2010).
[b] The value depends on the blower pressure.
[c] Constant Speed with dual vane system, Outlet diffuser vanes and inlet guide vanes.

low loading and high loading conditions can be provided by the selected blower and the VFD (U.S. EPA, 2010). For example, a high efficiency constant speed blower may be used to cover the base flow without a VFD, and additional blowers, either the same size or a smaller size, could be equipped with a VFD to provide the required range of air flow.

Aeration Diffusers. Oxygen transfer efficiencies of various diffuser types were discussed in Sec. 5–11 in Chap. 5. Most activated sludge facilities designed and constructed in the 1970s and 1980s and earlier employed coarse bubble diffusers or surface aerators. As these facilities approach their design life and design capacity, their replacement with higher

EXAMPLE 7–10 **Energy Saving by DO Control** Using the complete-mix activated sludge process to treat 22,700 m³/d of primary effluent for COD removal without nitrification as shown in Example 8–3 in Chap. 8, estimate the reduction in power demand by changing the DO setpoint from 3.5 mg/L to 2.0 mg/L.

Use the following design information extracted from Example 8–3 and supplemental information necessary to determine the energy demand.

Parameter	Unit	Value
Required oxygen transfer rate, OTR_f	kg/h	111.3
Alpha factor, a	–	0.50
Fouling factor, F	–	0.90
Beta factor, β	–	0.95
Dissolved oxygen surface saturation concentration at 12°C, C_{st}	mg/L	10.78
Dissolved oxygen surface saturation concentration at standard temperature (20°C), C^*_{s20}	mg/L	9.09

(continued)

(*Continued*)

Parameter	Unit	Value
Pressure correction factor, P_b/P_s		0.94
Steady-state DO saturation concentration, $C^*_{\infty 20}$		10.64
Empirical temperature correction factor, θ		1.024
Temperature of mixed liquor	°C	12
Ambient temperature	°C	15
Oxygen in the air	kg/kg air	0.232
Oxygen transfer efficiency, OTE	%	25
Universal gas constant	kJ/kmole	8.314
Blower inlet absolute pressure	kPa	101.3
Blower discharge absolute pressure	kPa	121.5

For simplicity, assume the same oxygen transfer efficiency of 25 percent for both DO concentrations.

Solution

1. Determine the SOTR for DO in aeration basin = 3.5 mg/L using Eq. (5–70).

$$\text{SOTR} = \left[\frac{\text{OTR}_f}{(\alpha)(F)}\right]\left[\frac{C^*_{\infty 20}}{(\beta)(C_{st}/C^*_{s20})(P_b/P_s)(C^*_{\infty 20}) - C}\right][(1.024)^{20-t}]$$

Using the data given above, SOTR is calculated as

$$\text{SOTR} = \left[\frac{(111.3 \text{ kg/h})}{(0.50)(0.90)}\right]\left\{\frac{10.64}{\left[0.95\left(\frac{10.78}{9.09}\right)(0.94)(10.64) - 3.5\right]}\right\}(1.024^{20-12}) = 409.6 \text{ kg/h}$$

2. Determine air flowrate in terms of kg/min.

$$\text{Air flowrate, kg/min} = \frac{(\text{SOTR kg/h})}{[(E)(60 \text{ min/h})(0.232 \text{ kg O}_2/\text{kg air})]}$$

$$= \frac{409.6}{(0.25)(60)(0.232)} = 117.7 \text{ kg/min}$$

3. Calculate an estimated power requirement using Eq. (5–77a) in Chap. 5.

$$P_w = \frac{wRT_1}{28.97\,n\,e}\left[\left(\frac{p_2}{p_1}\right)^n - 1\right]$$

Using the data give above, P_w is calculated as

$$P_w, \text{kW} = \frac{(117.7/60)(8.314)(273.15 + 15)}{28.97(0.283)(0.85)}\left[\left(\frac{121.5}{101.3}\right)^{0.283} - 1\right]$$

$$= 35.6 \text{ kW}$$

4. Repeat Steps 1 through 3 for DO = 2.0mg/L.

$$\text{SOTR} = \left[\frac{(111.3 \text{ kg/h})}{(0.50)(0.90)}\right]\left\{\frac{10.64}{\left[0.95\left(\frac{10.78}{9.09}\right)(0.94)(10.64) - 2.0\right]}\right\}(1.024^{20-12}) = 343.3 \text{ kg/h}$$

$$\text{Air flowrate, kg/min} = \frac{(\text{SOTR kg/h})}{[(\text{OTE})(60 \text{ min/h})(0.232 \text{ kg O}_2/\text{kg air})]}$$

$$= \frac{343.3}{(0.25)(60)(0.232)} = 98.6 \text{ kg/min}$$

$$P_w, \text{kW} = \frac{(98.6/60)8.314 \cdot (273.15 + 15)}{28.97 \cdot 0.283 \cdot 0.85}\left[\left(\frac{121.5}{101.3}\right)^{0.283} - 1\right]$$

$$= 29.8 \text{ kW}$$

5. Compare the energy demand at DO setpoint at 3.5 and 2.0 mg/L.
    From Step 3 and Step 4, the energy requirement for aeration is reduced from 35.6 to 29.8 kW by reducing the DO setpoint from 3.5 to 2.0 mg/L. The reduction in energy requirement is 16.2 percent.

Comment   The energy demand estimate with Eq. (5–77) is for a single-stage centrifugal blower and assuming the air flowrate is controlled by VFD. The estimated energy saving cannot be recognized if the air flow is controlled with throttling and other methods which do not reduce energy consumption with reduced air flowrates.

oxygen transfer efficiency aerators will, in most cases, save energy. When such a retrofit/upgrade is considered, it is also important to review and make necessary improvements in the air piping, air flow meters, and the DO monitoring and control system.

Process Modification for Increased Energy Production

Increased energy production can be achieved by improving the removal of organic material from wastewater prior to biological secondary treatment for anaerobic digestion, and by improving the volatile solids destruction in the anaerobic digesters. In recent years, the use of waste products from other sources has been studied for the enhanced energy production at wastewater treatment facilities. Enhanced energy production with waste products from other sources such as food waste, fats, oil and grease, and waste from industrialized livestock operations can be implemented at a wastewater treatment facility. Depending on the scale of the energy production, it may be possible to produce more energy than the treatment facility consumes (net energy positive). The use of renewable energy generated onsite, such as solar and wind power, has been implemented on a smaller scale, however, the contribution of these energy sources is relatively minor in comparison to the total energy demand at a wastewater treatment facility.

Additional Removal of Organic Matter from Wastewater.   One effective way to remove additional organic matter from wastewater is through primary effluent filtration. In recent studies, it has been possible to remove up to 70 percent of the TSS remaining in primary effluent. Primary effluent filtration is considered in Sec. 5–9 in Chap. 5. Another technology that has been tested and found to remove a significant fraction of COD and TSS from raw wastewater is charged bubble flotation, also considered in Sec. 5–9 in Chap. 5, in which air bubbles are coated with a polymer. In the coming years it is anticipated that a variety of new technologies will be developed to capture the organic matter in wastewater before aerobic oxidation. Removal of organic matter at the primary treatment can also be improved by chemically enhanced primary treatment (CEPT) but care must be taken in evaluating the effectiveness of CEPT for the enhanced energy recovery because the increase in inorganic precipitates in the sludge could potentially lower the efficiency of biogas production with anaerobic digestion. Other design considerations with enhanced removal of organic matter include the impacts of iron on UV disinfection (when iron salts are used for precipitation), and carbon requirements for denitrification in BNR processes.

Process Modifications to Enhance Digester Gas Production. It is often recognized that older anaerobic digesters are not operated and maintained to maximize the production of biogas. Two common issues are insufficient digester heating and insufficient digester mixing. Another common loss of energy is due to leakage in biogas piping and storage facilities. Improvements in the operating conditions of the existing anaerobic digesters and biogas capturing could improve the amount of energy available for beneficial use. In addition to these operational and maintenance issues, methods to enhance solids loading and digester performance are discussed in detail in Sec. 13–9 in Chap. 13.

Use of Waste Products from Other Sources. Anaerobic digesters are often designed with multiple tanks but not all tanks are built in the beginning, and additional space is allocated for the later stage of the design horizon. Even with the phasing of the digester expansion, anaerobic digesters often operate at below the design capacity. The excess capacity available in most anaerobic digesters can be utilized to increase the digester gas production. The use of food waste, fat, and grease for enhanced production of biogas has been studied extensively in recent years, based on the fact that enhanced biogas production could be implemented with minimal modifications to the existing anaerobic digestion processes. The use of food waste, fat, and grease for the enhanced production of biogas is considered in Chaps. 10 and 13.

Use of Renewable Energy Sources. Considerations for the use of renewable energy sources such as wind and solar energy have become a common practice in many buildings but implementation at wastewater treatment facilities has been relatively minor, typically due to space limitations. It is considered more commonly when a building or the facility is pursuing a sustainability certificate, such as a LEED (leaders in energy and environmental design) program, but often these energy sources could fill only a small fraction of the total energy demand at a treatment plant. Common usage of renewable energy at a wastewater treatment facility includes the use of solar panels for lighting, and the use of solar energy for the production of hot water.

Peak Flowrate Management (Peak Energy Usage)

The pricing structure for the electricity for large power users such as wastewater treatment facilities typically includes a demand charge based on the peak equipment power demand, and energy charge based on the amount of energy consumed. The pricing for the energy, in most cases, is different during peak demand hours. In some regions, a spot market is adopted for the electricity pricing. When considering energy management in terms of energy cost savings, one of the effective measures is to shift the use of electricity during the hours with lower energy charge. The shift could be made, for example, by operating sludge dewatering only during off-peak hours.

Flow equalization can be an effective measure to divert part of the wastewater entering the facility during the peak power demand (highest power cost) hours and treat it during off-peak hours. Peak flowrate management not only reduces the use of power during peak demand hours, but also allows for more uniform flow and loading to the treatment processes which will also increase the energy efficiency for most electrically driven equipment. Flow equalization of return flows for treatment in off hours will also reduce peak demand. Flow equalization of return flows is considered in Chap. 15.

Opportunities exist for managing energy use in wastewater treatment by employing the concept of demand-side management. The electric power industry has long recognized the importance of integrating traditional supply-side planning with demand-side management to reduce peak demand. The goal of demand-side management is to change the electrical load characteristics (the amount of energy used at different times of the day) by improving energy efficiency and managing equipment operation. In demand-side management, it is also recognized that continued load growth will occur as the systems expand to meet new domestic and industrial wastewater collection and treatment requirements.

Selection of Energy Sources

Wastewater treatment agencies typically sign contracts with an electrical power supplier and suppliers of fuels (e.g., natural gas, fuel oils and other fuels) to set the pricing. The pricing for electrical power can be a based on a fixed plus usage fee, which can be a fixed rate or dynamic rate. When the electrical power charge is based on the dynamic pricing, a significant cost saving can be expected by implementing peak flowrate management with flow equalization, by which aeration power demand could be shifted to off-peak hours, or during the time electrical power cost is lower.

The mechanism of fuel cost contracts may affect the selection and design of mechanical equipment as different types of contracts become available for the wastewater treatment facilities. For example, a 20-y contract for natural gas at a fixed price can lead to the selection of direct drive motors fueled by natural gas over electrical motors. The fixed price contract can be compared to the purchase of electrical power which can be unpredictable.

17–10 FUTURE OPPORTUNITIES FOR ALTERNATIVE WASTEWATER TREATMENT PROCESSES

The principal sources of energy in wastewater have been identified and discussed previously in Sec. 17–2. The means used currently to recover and utilize energy from the available sources has been discussed in Secs. 17–5 through 17–7. The purpose of this section is to highlight briefly some innovations that are being developed that will make it possible to (1) recover additional energy from the various sources, (2) reduce energy usage through the implementation of different biological treatment technologies, and (3) reduce energy usage though the implementation of alternative treatment process options. Energy reduction through the use of improved process equipment is discussed in the previous section dealing with energy management.

Enhanced Energy Recovery of Particulate Organic Matter

Currently, conventional primary clarification achieves about 50 percent removal of the applied TSS and 30 to 40 percent of incoming COD. The TSS and COD not removed by sedimentation must be treated in the downstream biological treatment process. If all of the particulate COD could be removed before biological treatment, additional energy could be recovered from the material removed and the amount of energy need for carbonaceous oxidation could be reduced.

The TSS and associated COD can be removed by depth, surface, or membrane filtration. Primary effluent filtration (PEF) has been studied since the late 1970s (Matsumoto et al.,1980, 1982). Using PEF in combination with primary sedimentation, removal rates of TSS and COD around 90 and 60 percent, respectively, can be achieved. The PEF process was not adopted widely because the cost of energy at the time the process was first studied was less than $0.03/kWh and the recovery of energy was not an issue. With the current interest in the recovery of energy, the desire to reduce energy usage, and the development of a number of new filtration technologies, there is a resurgence of interest in PEF. A major advantage of PEF is that a separate waste stream is not created, because the solids removed by filtration, either continuously or intermittently, can be mixed with the primary sludge. Primary sludge from PEF or CEPT processes is converted readily to biogas using anaerobic digestion. Alternatively, the solids can be diverted to a fermenter for the production of volatile fatty acids for use in the enhanced removal of phosphorus (see Sec. 8–8 in Chap. 8).

Reduced Energy Usage in Biological Treatment

As discussed in Chap. 7 and more extensively in Chap. 15, new biological processes are being developed that offer potential savings in energy and chemical consumption, especially with respect to the removal of nitrogen. For example, the use of nitritation/denitritation and partial nitritation and deammonification processes can be used to reduce both oxygen and carbon demand. Nitrogen removal through nitritation/denitritation would require 25 percent less oxygen and 40 percent less external carbon source than conventional nitrification/denitrification processes. With partial nitritation and deammonification, the requirement for oxygen is 60 percent less and demand for an external carbon source is nearly 90 percent less than the conventional nitrification and denitrification (see Chap. 15).

When partial nitritation and deammonification is used for sidestream treatment and the main treatment process is also designed for nutrient removal, the use of external carbon in the sidestream treatment may be eliminated and the remaining nitrate treated in the main stream process, which will simplify the chemical dosing system for the sidestream treatment. While the application of this process is predominantly for the treatment of recycle flow with high ammonium concentration, studies are ongoing to apply the process for the treatment of main stream wastewater at ambient temperatures (Al-Omari et al., 2012).

Reduced Energy Usage through the Use of Alternative Treatment Processes

Future treatment process flow diagrams will incorporate alternative biological processes as well treatment without biological processes to achieve more effective reduction and utilization of the energy in wastewater. Examples of alternative approaches are illustrated on Fig. 17–22. With advances in treatment technologies, a number of alternative approaches are expected to emerge.

Anaerobic Treatment at Ambient Temperature.
Anaerobic processes which were discussed in detail in Chap. 10 are in general less energy intensive than aerobic processes as they do not require aeration for the bulk removal of bCOD. Anaerobic processes also generate biogas containing methane which could be recovered and used for heat and power generation. In the future, it is anticipated that treatment process flow diagrams will be developed for the anaerobic treatment of wastewater at ambient temperature (McCarty, 2011).

Use of Trickling Filters for Treatment of Filtered Wastewater.
Another example of an alternative treatment process flow diagram involves the use of a cloth screen for primary treatment followed by a filtration step as discussed above. The filtered effluent could be applied directly to a tricking filter in a single stage, or two stage process where nutrient removal is an issue. A sedimentation tank would not be needed following the trickling filter as the small amount of residual solids could be removed, if needed, with a high-rate filtration process (Koltz, 1985).

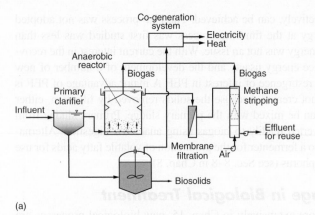

(a)

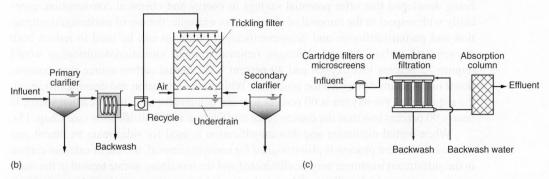

(b)   (c)

Figure 17-22

Process flow diagrams of alternative wastewater treatment processes: (a) anaerobic treatment at ambient temperature (adapted from McCarty, 2011), (b) trickling filter for treatment of filtered wastewater, and (c) membrane absorption process.

Membrane-Absorption Process. It is also anticipated that processes will be developed to treat wastewater that will not employ biological treatment. Filtered wastewater, as discussed above, could be processed further with a variety of different membranes. Any residual organic matter passing the membrane could be removed by adsorption. In turn, the adsorbed organic matter could be digested or processed thermally (Adams et al., 2011).

Prospects for the Future

The key to thinking about alternative treatment processes is that the technologies available currently, or under development, will make it possible to alter the characteristics of wastewater to optimize any given treatment process while maximizing the recovery and utilization the energy in wastewater.

PROBLEMS AND DISCUSSION TOPICS

17-1 Review three current (since 2005) peer-reviewed articles on co-digestion of wastewater sludge and food waste. Summarize the key findings with respect to increased gas production. Discuss the obstacles/challenges with implementation of co-digestion based on your review.

17–2 Given the following summary information for a wastewater treatment plant, determine the benchmark score for this treatment plant using the AWWARF methodology described in Sec. 17–4. Calculate the benchmark score for 30, 60, and 100 percent use (value to be selected by instructor) of biogas to offset natural gas use. Assume LHV of 35 MJ/m³ for natural gas, 22 MJ/m³ for digester gas.

Parameter	Unit	Value
Flowrate (annual average/design capacity)	10^3 m³/d	24/30
Influent BOD	mg/L	120
Influent ammonium	mg N/L	24
Influent total nitrogen	mg N/L	36
Effluent BOD	mg/L	6.2
Effluent ammonium	mg N/L	2.1
Effluent total nitrogen	mg N/L	5.5
Energy use		
Electricity	kWh/y	3,600,000
Biogas production	m³/y	605,000
Biogas used	m³/y	0
Natural gas	m³/y	372,700

17–3 Using the data presented in Example 17–5 for a wastewater treatment facility, calculate the benchmarking score when all biogas is utilized to generate electricity and heat is recovered to offset natural gas use. Assume typical energy efficiencies for CHP as reported in Table 17–8.

17–4 A wastewater treatment facility receives raw wastewater at an average flowrate of 7500 m³/d, with a velocity of 1.5 m/s, and a pressure head of 2.5 bar. Identify the type of hydro-turbine that could be used to generate electricity using the incoming raw wastewater, and estimate the potential power generation. Use Fig. 3–14 to estimate the peaking factor and Fig. 17–19 to estimate the average efficiency at the average flowrate when the turbine is sized for the peak flowrate. Discuss potential operational issues in using raw wastewater for hydraulic turbine.

17–5 Calculate the heat balance for sludge combustion assuming the following solids composition (to be selected by instructor). Use the procedure presented in Example 17–6.

Element	Sample constituent composition, % by wt		
	1	2	3
Combustible			
Carbon	13.3	18.9	15.0
Hydrogen	0.9	1.3	1.2
Oxygen	4.9	6.9	5.3
Nitrogen	0.3	0.4	0.3
Sulfur	0.4	0.5	0.4
Inerts	8.3	11.9	7.8
Water	72	60	70
Total	100	100	100

17-6 Determine the coefficient of performance for the heat pump as well as the entire system and the wastewater flowrate necessary to provide the heating requirement based on the information summarized in the table below (to be selected by instructor).

Parameter	Unit	Value
Heating requirement (value to be selected by instructor)	kW	(1) 200 (2) 240 (3) 280
Heat pump electrical input	kW per kW output	0.24
Wastewater temperature	°C	15
Temperature drop through heat exchanger	°C	5
Total power requirements for pumps (value to be selected by instructor)	kW	(1) 40 (2) 45 (3) 50

17-7 For Problem 17–6 above, discuss benefits and potential issues to extract heat from raw wastewater in the collection system, instead of the wastewater effluent, for district heat supply in a small commercial area.

17-8 For Example 17–10 and Example 8–3 in Chap. 8, the treatment process operated at the design capacity is upgraded with a primary effluent filtration device between primary clarifiers and activated sludge process, and BOD loading to the activated sludge process was reduced by 20, 30, or 40 percent (value to be selected by instructor). Calculate the air requirement and aeration power demand for the same COD removal without nitrification process. Assume the DO set point is 2.0 mg/L.

17-9 The wastewater treatment plant in Example 8–3, designed for BOD removal, is currently operating at 60 percent of its capacity. The municipality is experiencing lower population growth in the service area as compared to the projected growth when the treatment plant was designed and built. As part of an integrated resource recovery scheme being studied by the municipality, you are asked to determine if implementing the use of kitchen food waste grinders in residential homes will result in saving energy. It is projected that installation of food waste grinders will result in 600, 1000, or 1200 kg/d (value to be selected by instructor) increase in biodegradable COD loading, of which 75 percent will be captured as primary sludge, and 20 percent will be treated in the activated sludge process. Primary sludge is transferred to anaerobic digestion to produce biogas.

a. Determine the aeration power demand at 60 percent loading.

b. Estimate the increased aeration power demand if the use of food waste grinders was implemented.

c. Determine the increase in digester gas production if food waste grinders were implemented (assume anaerobic digester has sufficient capacity)

d. Assuming 45 percent engine efficiency and 93 percent generator efficiency, determine whether the implementation of food waste grinders will result in overall energy saving.

For simplicity, ignore the increased biomass and biogas production from WAS and the corresponding energy savings from heat recovery in the calculations.

17-10 A wastewater pump station is used to transfer 20,000 m³/d screened raw wastewater from a wet well to an aerated grit chamber. The pump station contains three pumps. Two pumps are used continuously and one pump serves as a standby, with an on/off control. Discuss potential changes to the pump system that can be made for energy conservation and explain how the savings are achieved.

REFERENCES

Adams, R. M., H. L. Leverenz, and G. Tchobanoglous (2011) "Re-Orienting Municipal Wastewater Management System for Energy Reduction and Energy Production," *Proceedings of the WEF 83rd ACE,* Los Angeles, CA.

Al-Omari, A., M. Han, B. Wett, N. Dockett, B. Stinson, S. Okogi, C. Bott, and S. Murthy (2012) "Main-Stream Deammonification Evaluation at Blue Plains Advanced Wastewater Treatment Plant (AWTP)," *Proceedings of the WEF 84th AEC,* New Orleans, LA.

AWWARF (2007) *Energy Index Development for Benchmarking Water and Wastewater Utilities,* AWWA Research Foundation, Denver, CO.

Barber, W. P. F. (2007) "Observing the Effects of Digestion and Chemical Dosing on the Calorific Value of Sewage Sludge," *Proceedings of the IWA Specialist Conference: Moving Forward Wastewater Biosolids Sustainability,* Moncton, Canada.

BHA (2005) *A Guide to UK Mini-Hydro Developments,* The British Hydropower Association, Dorset, UK.

Burton, F. L. (1996) *Water and Wastewater Industries: Characteristics and Energy Management Opportunities,* CEC Report 106941, Electric Power Research Institute, St Louis, MO.

Carrio, L. (2011) "Combined Heat and Power at NYCDEP Wastewater Treatment Plants," U.S. Environmental Protection Agency Combined Heat and Power Partnership Webinar, *Strengthening Critical Infrastructure: Combined Heat and Power at Wastewater Treatment Facilities,* U.S. Environmental Protection Agency, Washington, DC.

CDC (1994) *Chemical Listing and Documentation of Revised IDLH Values,* Center for Disease Control, Washington, DC.

Channiwala, S. A. (1992) *On Biomass Gasification Process and Technology Development-Some Analytical and Experimental Investigations,* Ph.D. Thesis, Indian Institute of Technology, Department of Mechanical Engineering, Bombay (Munbai), India.

DOE (1989) *Improving Fan System Performance: A Sourcebook for Industry,* U.S. Department of Energy, Washington, DC.

DOE (2002) *Benchmarking Biomass Gasification Technologies for fuels, Chemicals and Hydrogen Production,* U.S. Department of Energy National Energy Technology Laboratory, Morgantown, WV.

DOE (2003) *Industrial Heat Pumps for Steam and Fuel Savings,* U.S. Department of Energy, Industrial Technology Program, Energy Efficiency and Renewable Energy, Washington, DC.

DOE (2006) *Improving Pumping System Performance: A Sourcebook for Industry,* 2nd Ed., U.S. Department of Energy, Washington, DC.

DOE (2008) *Waste Heat Recovery: Technology and Opportunities in U.S. Industry,* U.S. Department of Energy, Industrial Technologies Program, Washington, DC.

DOE (2011) *Comparison of Fuel Cell Technologies,* Department of Energy, Energy Efficiency and Renewable Energy Fuel Cell Technologies Program, Washington, DC.

EPRI (1994) *Energy Audit Manual for Water and Wastewater Facilities,* Electric Power Research Institute, St. Louis, MO.

ESHA (2004) *Guide on How to Develop a Small Hydropower Plant,* European Small Hydropower Association, Brussels.

ESHA (2009) "Guide on How to Develop a Small Hydropower Plant: Information from the European Small Hydro Association (ESHA)," *Energize,* June, 16–20.

Funamizu, N., M. Iida, Y. Sakakura, and T. Takakuwa (2001) "Reuse of Heat Energy in Wastewater: Implementation Examples in Japan," *Water Sci. Technol.,* **43,** 10, 277–285.

Green, D., and R. Perry (2007) *Perry's Chemical Engineers' Handbook,* 8th Ed., McGraw-Hill, New York.

Heidrich, E. S., T. P. Curtis, and J. Dolfing (2011) "Determination of the Internal Chemical Energy of Wastewater," *Environ. Sci. Technol.,* **45,** 2, 827–832.

Hewitt, G. F. (1992) *Handbook of Heat Exchanger Design,* Begell House, Inc., New York.

Holman, J. P. (2009) *Heat Transfer,* Tenth Ed., McGraw-Hill, New York.

Koltz, J. K., (1985) *Treatment of an Altered Waste by Trickling Filters,* M.Sc Thesis, University of California at Davis, Davis, CA.

Krzeminski, P., J. H. J. M. van der Graaf, and J. B. van Lier (2012) "Specific Energy Consumption of Membrane Bioreactor (MBR) for Sewage Treatment," *Water Sci. Technol.,* **65**, 2, 380–392.

Kuppan, T. (2000) *Heat Exchanger Design Handbook,* CRC Press, Boca Raton, FL.

Matsumoto, M. R., T. M. Galeziewski, G. Tchobanoglous, and D. S. Ross (1980) "Pulsed Bed Filtration of Primary Effluent," Proceedings of the Research Symposium, 53rd Annual Water Pollution Control Federation Conference, 1–21, Las Vegas, NV.

Matsumoto, M. R., T. M. Galeziewski, G. Tchobanoglous, and D. S. Ross (1982) "Filtration of Primary Effluent," J. WPCF, 54, 12, 1581–1591.

McCarty, P. L., J. Bae, and J. Kim (2011) "Domestic Wastewater Treatment as Net Energy Producer – Can This be Achieved?" *Environ. Sci. Technol.,* **45**, 17, 7099–7106.

McKendry, P. (2002) "Energy Production from Biomass (Part 3): Gasification Technologies," *Bioresource Technol.,* **83**, 1, 55–63.

NYSERDA (2010) *Water and Wastewater Energy Management: Best Practices Handbook,* New York State Energy Research and Development Authority, Albany, NY.

Pallio, F. S. (1977) "Energy Conservation and Heat Recovery in Wastewater Treatment Plants," *Water Sewage Works,* **12**, 2, 62–65.

Sawyer, C. N., P. L. McCarty, and G. F. Parkin (2003) *Chemistry for Environmental Engineering and Science,* 5th Ed., McGraw-Hill, New York.

Shizas, I., and D. M. Bagley (2004) "Experimental Determination of Energy Content of Unknown Organics in Municipal Wastewater Streams," *J. Energy Eng.,* **130**, 2, 45–53.

U.S. EPA (2006a) *Auxiliary and Supplemental Power Fact Sheet: Fuel Cells,* EPA832-F-05–012, Office of Water, U.S. Environmental Protection Agency, Washington, DC.

U.S. EPA (2006b) *Auxiliary and Supplemental Power Fact Sheet: Viable Sources,* EPA832-F-05–009, Office of Water, U.S. Environmental Protection Agency, Washington, DC.

U.S. EPA (2007) *ENERGY STAR® Performance Ratings Technical Methodology for Wastewater Treatment Plant,* U.S. Environmental Protection Agency (www.energystar.gov), Washington, DC.

U.S. EPA (2008) *Ensuring a Sustainable Future: An Energy Management Guidebook for Wastewater and Water Utilities,* U.S. Environmental Protection Agency, Washington, DC.

U.S. EPA (2010) *Evaluation of Energy Conservation Measures for Wastewater Treatment Facilities,* EPA 832-R-10–005, Office of Wastewater Management, U.S. Environmental Protection Agency, Washington, DC.

U.S. EPA (2011) *ENERGY STAR Performance Ratings Methodology for Incorporating Source Energy Use,* Air and Radiation, U.S. Environmental Protection Agency, Washington, DC.

U.S. EPA (2012) *Energy Management System Manual Wastewater Treatment Utility: A Supplement to the EPA Energy Management Guidebook for Drinking Water and Wastewater Utilities (2008),* U.S. Environmental Protection Agency, Washington, DC.

Valkenburg, C., M. A. Gerber, C. W. Walton, S. B. Jones, B. L. Thompson, and D. J. Stevens (2008) *Municipal Solid Waste (MSW) to Liquid Fuels Synthesis, Volume 1: Availability of Feedstock and Technology,* Pacific Northwest National Laboratory, Prepared for U.S. Department of Energy, Washington, DC.

Wanner, O., V. Panagiotidis, P. Clavadetscher, and H. Siegrist (2005) "Effect of Heat Recovery from Raw Wastewater on Nitrification and Nitrogen Removal in Activated Sludge Plants," *Water Res.,* **39**, 19, 4725–4734.

WEF (2009) *Energy Conservation in Water and Wastewater Treatment Facilities,* WEF Manual of Practice, No. 32, Prepared by the Energy Conservation in Water and Wastewater Treatment Task Force of the Water Environment Federation, Water Environment Federation, Alexandria, VA.

WEF/ASCE (2010) *Design of Municipal Wastewater Treatment Plants,* 5th Ed., WEF Manual of Practice No. 8, ASCE Manuals and reports on Engineering Practice No. 76, Water Environment Federation, Alexandria, VA.

WERF (2008) *State of Science Report: Energy and Resource Recovery from Sludge,* Prepared for the Global Water Research Coalition, Water Environment Research Foundation, Alexandria, VA.

18

Wastewater Management: Future Challenges and Opportunities

WORKING TERMINOLOGY

Term	Definition
Climate change	Significant change in measures of climate such as temperature, precipitation or wind, lasting for an extended period of time (decades or longer).
Decentralized (satellite) treatment system	System used for the treatment of wastewater located close to the point of reuse. Satellite treatment plants generally do not have solids processing facilities; solids are returned to the collection system for processing in a central treatment plant located downstream. Three types of satellite systems are identified: (1) interception type, (2) extraction type, and (3) upstream type.
Direct potable reuse	The introduction of purified water either directly into the potable water supply distribution system downstream of a water treatment plant, or into the raw water supply immediately upstream of a water treatment plant.
Histogram	A graphical representation of the frequency with which an event occurs over a range of different conditions.
Indirect potable reuse	The planned incorporation of purified water into a raw water supply, such as in potable water storage reservoirs or a groundwater aquifer, resulting in mixing, dilution, and assimilation, thus providing an environmental buffer.
Linear correlation	The linear relationship between two variables.
Low impact design (LID)	A design approach that protects surface and ground water quality, maintains the integrity of aquatic living resources and ecosystems, and preserves the physical integrity of receiving streams with designs that are in harmony with nature.
Natural disasters	Extreme sudden natural events such as floods, tornadoes, hurricanes, heat waves, or droughts that injure people and damage property.
NPDES	The National Pollution Elimination Discharge System (NPDES) was established based on uniform technological minimums with which each point source discharger has to comply.
Pareto analysis	A statistical decision making technique to select a limited number of tasks or variables that will result in a significant overall impact.
Pilot plant studies	Studies conducted at test beds at a scale larger than bench-scale, to establish the suitability of a process in the treatment of a specific wastewater under specific environmental conditions and to obtain data that can be used for full-scale design and operation.
Population demographics	The study of populations based on statistics such as economics; migration patterns; and birth, deaths, and disease.
Potable reuse, direct	See Direct potable reuse.
Potable reuse, indirect	See Indirect potable reuse.
Privatization	Private sector ownership and operation of facilities and services used by government entities in performing their public function.
Satellite treatment system	See decentralized treatment.
Sustainability	The principle of optimizing the benefits of a present system without diminishing the capacity for similar benefits in the future.
Triple bottom line	Project analysis in which profit (economics), people (social) and planet (environment) are considered equally.
Uncontrollable events	Events in nature beyond the control of humans, such as tornadoes and hurricanes.
Unintended consequences	Consequences (outcomes) that are not anticipated or intended by a particular action.
Variability, inherent	Based on the laws of chance, all physical, chemical, and biological treatment processes exhibit some measure of variability with respect to the performance that can be achieved. Variability is inherent in biological treatment processes.
Wet-weather flow	The runoff that results when it rains or snows.

The concepts and elements involved in the general design of wastewater treatment plants have been presented and discussed in the previous chapters. The purpose of this chapter is to consider the various concepts and design elements in light of the challenges and opportunities for wastewater management in the future. The topics considered are (1) a general discussion of some important future challenges and sustainability issues, (2) the impact of global and uncontrollable events, (3) upgrading existing treatment plant performance through process optimization, (4) upgrading existing treatment plant performance through process modification, and (5) the management of wet-weather flows. Important future challenges and sustainability issues and the impact of global and uncontrolled events are discussed first because improvements to existing plants as well as the design of new plants must be undertaken in light of these challenges and issues. Wherever possible, reference will be made to the pertinent section and/or chapter where additional information can be found on the individual topics discussed in this chapter.

18-1 FUTURE CHALLENGES AND OPPORTUNITIES

In the twentieth century the primary focus of wastewater treatment was on the removal and treatment of settleable and floatable solids, organic matter expressed as biochemical oxygen demand (BOD), total suspended solids (TSS), and pathogenic microorganisms (see Chap. 1). Late in the twentieth century, nutrient removal and odors also became issues, and controlled, non-potable use of reclaimed water became a common practice in many parts of the world (Asano et al., 2007). In the twenty-first century, as a result of numerous environmental issues and events, a paradigm shift has occurred in how wastewater is viewed. In the twenty-first century, wastewater is no longer viewed as a waste requiring disposal, but as a "renewable recoverable source of energy, resources, and potable water" (Tchobanoglous, 2010; Tchobanoglous et al., 2011). In light of this view of wastewater, it is appropriate to consider briefly some challenges as well as the opportunities that will become increasingly important in the future, including: (1) asset management, (2) the need to design wastewater treatment plants (WWTPs) for the recovery of energy and resources, (3) the need to produce effluent suitable for the production of potable water, (4) the implementation of decentralized (satellite) treatment systems, (5) integrated wastewater management, (6) the use of low impact development, and (7) the use of the triple bottom line for project evaluation. It should be noted that the challenges discussed below are beyond the need to meet more stringent discharge standards for trace organics and residuals processing as discussed in Chap. 4. The impact of global and uncontrollable events on the future of wastewater management is considered in the following section.

Asset Management

With aging infrastructure and reduced funding for repair and replacement, wastewater agencies are exploring a number of different techniques that can be used to prioritize future capital expenditures that will yield the greatest value added for the consumer. Asset management, in its many forms, is one of the techniques now being investigated and applied. The U.S. EPA defines asset management (AM) as "Managing infrastructure assets to minimize the total cost of owning and operating them, while delivering the service levels customers desire" (U.S. EPA, 2012a). The U.S. EPA further reports (U.S. EPA, 2012b) the implementation of AM based decisions have the potential to save 20 to 30 percent of future lifecycle costs within U.S. wastewater utilities and defined strategies and procedures have reportedly shown operational cost savings and more efficient working practices.

Objectives of Asset Management. The principal objective of AM is to develop sufficient information in an organized fashion to support strategic decision making. All agencies conduct some form of AM, but since the late 1990's there has been an increased emphasis on the adoption of more focused approaches that encourage the understanding and reporting of all aspects of the lifecycle of an asset. From the initial conceptual need through to complete implementation, owners can draw from a wide range of proven techniques and available technologies to assist in the design, operation, and maintenance of new and replacement assets. The application of these techniques and technologies rely on the collection and analysis of data that enable the owner to analyze a range of performance metrics to support strategic decision making.

Practice of Asset Management. To date, the most widespread development and implementation of AM has been in Australia, New Zealand, and the UK where the performance of all aspects of the water and wastewater industries are more closely regulated. From the experience gained in these countries, it is clear that although it takes a number of years to see true financial returns, there are very real benefits by adopting more advanced AM techniques. While regulatory requirements in the United States do not require the implementation of AM, many agencies including WEF, AWWA, and the Association of Metropolitan Water Agencies (AMWA) have taken a lead role in promoting AM. The U.S. EPA continues to blend education of the principles and tools, with a formal need for implementation as a component of compliance regulations for individual agencies and owners.

Asset Management Methodologies. There are many AM techniques and technologies and it is beyond the scope of this text to discuss each of them adequately. Numerous publications are available to provide guidance but the fundamental core is in understanding the required performance from an asset and being able to operate and maintain the asset cost-effectively. In general, AM techniques include (1) developing an asset inventory, (2) assessment of the condition of asset, (3) determining the level of service to be provided, (4) identifying the critical asset to sustain the performance, (5) determining the cost for the entire life cycle of the asset, and (6) determining the best long-term strategy. Agencies that have a mature understanding of AM will likely have a better understanding of the condition and remaining life of their infrastructure and of the best maintenance methodologies. This level of knowledge can prove to be beneficial as agencies, with such an approach, are considered to be well managed businesses. Information on AM is also needed to justify funding either to customers, financial institutions, or within their own organization.

Ramifications of Asset Management. While much of the current focus of AM is related to existing infrastructure, it is clear that the lifecycle of every asset is significantly influenced by the decisions that are made at the concept and design stages. The designer has the opportunity to determine the optimum performance of the asset while considering criteria relevant to operations and maintenance. The correct combination of these interrelated factors will enable the end user to derive the most benefit with assets that provide the anticipated levels of service with reduced operations, maintenance, and energy costs.

While the principles of AM are considered by many to be the most effective way to manage infrastructure, there are many organizations in the United States that are still in the early stages of developing suitable approaches. Organizations that have adopted more advanced AM methodologies are however seeing benefits both in terms of performance levels and financial management. It is clear, therefore, that a more defined implementation of AM should be considered for all aspects of infrastructure management from the development and justification of need, to the ultimate disposal.

Design for Energy and Resource Recovery

The chemical and heat energy content of wastewater has been delineated and discussed previously in Chaps. 2, 14, and 17. As noted in the earlier discussions, wastewater treatment plants could potentially become net exporters of energy, and especially so if external sources of energy contained in food waste and fats, oils, and grease are included. The challenge in the future is how to extract the energy in wastewater most effectively. For example, food waste could be ground up in kitchen food waste grinders and transported to the wastewater treatment facilities in the collection system, or it could be intercepted at various upstream locations and extracted from the wastewater using a micro- or cloth-screen such as described in Sec. 5–9 in Chap. 5. The solids removed from the wastewater could be placed directly in an anaerobic digester. Alternatively, conventional aerobic treatment processes could be replaced with ambient-temperature low hydraulic retention time anaerobic treatment processes (McCarty, 2011). Heat recovered from wastewater could be used for drying screenings as well as in other applications, especially in the processing of biosolids. The key concept here is to think about how the characteristics of wastewater could be altered to enhance the recovery of energy from wastewater.

In the future, the recovery of resources from wastewater will occur simultaneously with the recovery of energy. To date, the removal of nitrogen and phosphorus has received the greatest attention as nitrogen and phosphorus discharge standards have become more stringent. The option of recovering, rather than simply removing, these constituents has became economically feasible, especially from return flows. Biological phosphorus removal was considered in Sec. 8–8 in Chap. 8. The recovery of nitrogen and phosphorus in the form of struvite is considered in Sec. 6–5 in Chap. 6 and Sec. 15–4 in Chap. 15. The recovery of nitrogen as ammonium sulfate is considered in Sec. 15–5 in Chap. 15. The recovery of resources from fly ash following combustion is considered in Sec. 14–4 in Chap. 14. The recovery of nutrients including nitrogen, phosphorus, and potassium from urine (see Table 3–15 in Chap. 3) is another resource recovery opportunity that has received considerable attention, especially in Europe and Australia. What role urine separation will play in the United States remains to be seen. Clearly, finding the optimum cost and energy-effective approach for the recovery of resources, coupled with the recovery of energy and potable water from wastewater will be a major challenge in the future.

Design of Wastewater Treatment Plants for Potable Reuse

As a result of population growth, urbanization, and climate change, public water supplies are becoming stressed, and the chances of tapping new water supplies for metropolitan areas are getting more difficult, if not impossible. As a consequence, existing and new water supplies must go further. One way to achieve this objective is by increased water reuse, particularly in supplementing municipal water supplies by means of indirect or direct potable reuse (IPR or DPR) (Leverenz et al., 2011; Tchobanoglous et al., 2011). As a result of the development and demonstration of full-scale advanced treatment processes (see Fig. 18–1), the use of purified water that has been recovered from municipal wastewater directly for potable purposes is now considered to be a viable alternative (NRC, 2012). It is also recognized that there is a continuum of possibilities. The challenge for the future is how to design or upgrade treatment plants so that the effluent produced will be

Figure 18–1

Advanced water treatment facility at the Orange County Water District, Fountain Valley, CA: (a) schematic flow diagram for 2.65×10^5 m³/d (70 Mgal/d) advanced water treatment facility, (b) microfilters, (c) cartridge filters, (d) reverse osmosis module, and (e) advanced UV oxidation reactors.

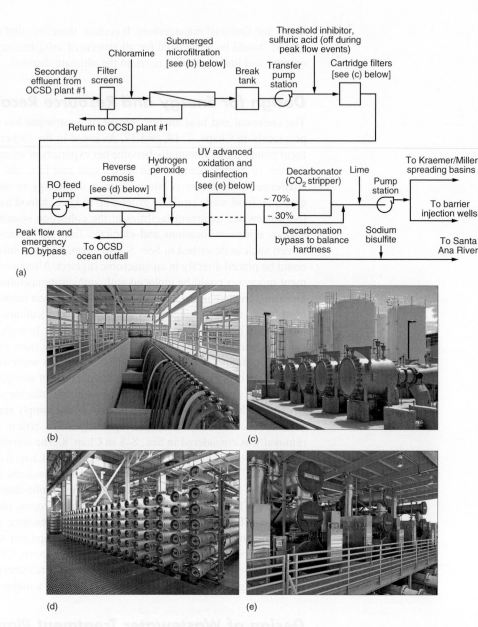

(a)

(b)

(c)

(d)

(e)

suitable for the production of purified water. Technical issues involved in the implementation of potable reuse are identified in Table 18–1.

When considering wastewater treatment plants for potable reuse, a key question is, what should the treatment plant of the future look like? In examining existing treatment process flow diagrams for IPR and DPR, it can be concluded that the production of purified water for DPR was an afterthought. Basically, additional unit processes were tacked on to the end of the existing secondary treatment process flow diagrams to remove specific compounds. However, at some point in the future there will need to be a complete rethinking of urban infrastructure to obtain the highest levels of performance and reliability. For water and wastewater systems, the advanced infrastructure model will likely include decentralization (also known as distributed), remote management, resource recovery, source separated waste streams, and application of specific optimization of water quality.

Table 18–1

Technical issues in the implementation of potable reuse

Consideration	Comments / questions
Source control	• Identification of constituents that may be difficult to remove (depends on technologies used) • Development of baseline sources and concentrations of selected constituents • Define the improvements that need to be made to existing source control programs where DPR is to be implemented
Influent monitoring	• Development of influent monitoring systems, including constituents, parameters, and monitoring recommendations • Investigate potential benefits of various influent monitoring schemes that may be used for early detection of constituents • Consideration of how influent monitoring data could be used to adapt treatment operations depending on variable influent characteristics
Flow equalization	• Determination of the optimum location and type (inline or offline) in secondary treatment process with respect to enhanced reliability and removal of trace constituents • Determination of optimum size of flow equalization before advanced treatment • Quantify the benefits of flow equalization on the performance and reliability of biological and other pretreatment processes
Wastewater treatment	• Quantify benefits of optimizing conventional (primary, secondary, and tertiary) processes to improve overall reliability of entire system • Quantify the benefits of complete nitrification or nitrification and denitrification on the performance of membrane systems used for DPR applications
Performance monitoring	• Determine monitoring schemes to document reliability of treatment performance for each unit process and validate end-of-process water quality
Analytical/monitoring requirements	• Selection of constituents and parameters that will require monitoring, including analytical methods, detection limits, quality assurance/quality control methods, and frequency • Determination of how monitoring systems should be designed in relation to process design • Development of appropriate monitoring systems for use with alternative buffer designs
Advanced wastewater treatment (water purification)	• Develop baseline data for treatment processes employing reverse osmosis. OCWD can be used as a benchmark • Development of alternative treatment schemes with and without demineralization that can be used for water purification • Quantify benefit of second stage (redundant) reverse osmosis
Engineered storage buffer	• Development of sizing guidelines based principally on existing analytical, detection, and monitoring capabilities to assess technical and economic feasibility of utilizing engineered storage buffer • Characterize the impact of existing monitoring response times on the safety and economic feasibility of implementing an engineered storage buffer
Balancing mineral content	• Development of recommendations for balancing water supply mineral content in consideration of site-specific factors, such as magnesium and calcium • Determination of potential impacts of various water chemistries on infrastructure and public acceptance • Development of specifications for chemicals used for balancing water quality

(continued)

Table 18–1 (Continued)

Consideration	Comments / questions
Blending	• Development of guidance on what level of blending, if any, is required based on the quality of the purified water and alternative water sources • Investigation of the significance of and rationale for blend ratios in terms of engineered buffer, protection of public health, public acceptance, and regulatory acceptance • Investigation of potential impacts of purified water on drinking water distribution system, e.g., corrosion issues, water quality impacts, etc.
Emergency facilities	• Stand-by power systems in the event of power loss or other emergency • Availability of all replacement parts and components that would be required in the event of a process breakdown • Process redundancy so that treatment trains can be taken offline for maintenance • Facilities for the by-pass or discharge of off-spec water in the event that the water does not meet the established quality requirements
Pilot testing	• Utilization of a review panel for advice and recommendations on the design, operation, monitoring plan for a project's pilot system to ensure that it will be representative of the proposed full-scale system • Development of monitoring protocol for collection of baseline data for "raw" water input to AWT pilot plant; how much testing and for what duration (e.g., 6 mo to 1 y). • Development of pilot study design so that results can be used to assess reliability with proposed source water

From Tchobanoglous et al., (2011).

What is needed is the development of integrated water management systems in which new wastewater treatment plants are planned and designed from the ground up to optimize treatment performance with respect to the production of purified water, along with the recovery of energy and resources.

Decentralized (Satellite) Wastewater Treatment

In most collection and treatment systems, wastewater is transported through the collection system to a centralized treatment plant located at the downstream end of the collection system near the point of dispersal (disposal) to the environment. Because centralized wastewater collection systems are generally arranged to route wastewater to these remote locations for treatment, water reuse in urban areas is often inhibited by the lack of a dual distribution system (i.e., purple pipe). The infrastructure costs for storing and transporting reclaimed water to the points of use are often prohibitive, thus making reuse uneconomic. An alternative to the conventional approach of transporting reclaimed water from a central treatment plant is the concept of decentralized (satellite) treatment at upstream locations with localized reuse and/or the recovery of wastewater solids.

Decentralized wastewater systems are used to treat wastewater at or near the point of waste generation and reuse (see Fig. 18–2). Decentralized treatment plants generally do not have solids processing facilities; solids are returned to the collection system for processing in a central treatment plant located downstream. Individual decentralized systems can be used for water reclamation and reuse for applications such as landscape irrigation, toilet flushing, cooling applications, and water features. Use of decentralized systems is predicated on the assumption that the existing collection system can be utilized for the

Figure 18–2

Schematic illustration of four types of satellite water reclamation systems: (a) interception type where wastewater to be reclaimed and recycled is intercepted before discharge to a centralized collection system, (b) extraction type (i.e., sewer mining) in which wastewater is extracted (i.e., pumped) from a centralized collection system for local reuse, (c) upstream type for treatment and reuse for a remote community or development with solids discharged to a centralized collection system, and (d) satellite system for individual home.

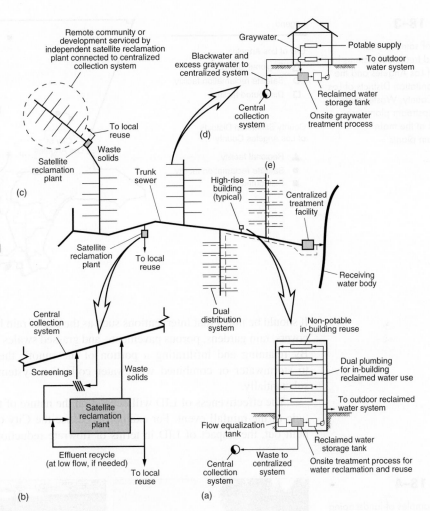

transport of solids and reduced flow. Onsite reclamation systems may obviate the need for large-scale dual piping systems, which are generally prohibitively expensive in urbanized areas and reduce the need to expand existing treatment plants to meet future growth projections. Two notable examples of the use of decentralized wastewater treatment systems are in the Los Angeles, CA area as illustrated on Fig. 18–3. Both the City of Los Angeles and the County Sanitation District's of Los Angeles County upstream treatment plants discharge screenings and biological solids to the large treatment plants located near the ocean where the treated effluent is discharged. The implementation of decentralized wastewater treatment will require a new approach to the management and reuse of wastewater (Tchobanoglous and Leverenz, 2013).

Low Impact Development

Low impact development (LID)(also known as low impact design) is a concept that has been applied to the management of wet-weather flow. With respect to the management of wet-weather flows, the goal of LID is to control both rainfall and stormwater runoff, at or near the source. Both engineered and vegetated natural systems are utilized to filter the rainfall runoff and replenish groundwater locally through infiltration (see Fig. 18–4).

Figure 18-3

Diagram of satellite and centralized treatment systems in the City of Los Angeles and the County Sanitation Districts of Los Angeles County. Waste solids from the upstream plants are processed at the main downstream plants.

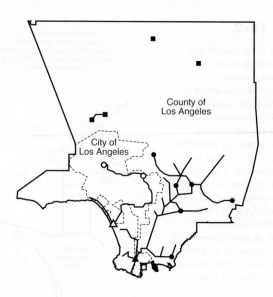

Legend

City of Los Angeles

△ Regional facility
○ Satellite reclamation facility
□ Distributed facility

County Sanitation Districts
of Los Angeles County

▲ Regional facility
● Satellite reclamation facility
■ Distributed facility

County of
Los Angeles

City of
Los Angeles

It should be noted that interventions such as the use of rain barrels, cisterns, vegetated roof covers, rain gardens, porous pavements, and grassed swales are all part of the LID strategy. By retaining and infiltrating a portion of the runoff, the quantity of flow discharged to stormwater or combined wastewater collection systems can potentially be reduced substantially.

The effectiveness of LID will depend on the nature of the community and the magnitude of the rainfall event. For example, because the City of San Francisco is essentially built out, the impact of LID in terms of flowrate reduction has been relatively minimal

Figure 18-4

Typical examples of landscaping designed to limit the effects of wet weather flows: (a) swale, (b) around treatment unit, (c) open area at treatment plant, and (d) roadway runoff capture.

(a)

(b)

(c)

(d)

Figure 18–5

Relationship of the elements involved in the triple bottom line method of analysis.

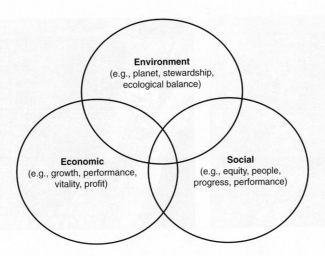

(e.g., on the order of two to five percent). Yet in other cities, flowrate reductions as high as 20 percent have been reported. With some of the recent record (extreme) rainfall events, the impact of LID has been difficult to assess. The challenge is how to integrate LID most effectively into the management of wastewater. Some notable examples in the form of case studies have been reviewed by the U.S. EPA (2010).

Triple Bottom Line

The *triple bottom line* refers to a method of analysis in which engineers are encouraged to consider social and environmental bottom lines in addition to the financial bottom line in arriving at the most sustainable solution (see Fig. 18–5). While the concept is laudable, in practice it has proven to be difficult to implement uniformly. Assessing the pluses and minuses of the financial aspects of a project has proven to be much easier than assessing the social and environmental pluses and minuses. In practice, to implement social and environmental bottom lines effectively, dollar values must be assigned to the corresponding benefits and drawbacks. If dollar values cannot be assigned, social and environmental concerns must be considered even though they will have little practical impact on project implementation beyond the normal concern for these issues incorporated into any project analysis. The challenge moving forward is how to best incorporate the concepts embodied in the triple bottom line analysis in project planning, design, and implementation.

18–2   IMPACT OF POPULATION DEMOGRAPHICS, CLIMATE CHANGE AND SEA LEVEL RISE, UNCONTROLLABLE EVENTS, AND UNINTENDED CONSEQUENCES

In addition to the identifiable causes of the variability observed in the treatment of wastewater as described in Chap. 3 and 4, a number of other global and local events now routinely impact the design and operation of wastewater treatment plants (WWTPs). Four such event categories are considered in the following discussion: (1) the impact of population demographics, (2) the impact of climate change and sea level rise, (3) the impact of uncontrollable events, and (4) the impact of the law of unintended consequences. Consideration of these topics is especially important in meeting the challenges and sustainability issues identified in the previous section.

(a)

(b)

Figure 18-6

Impact of urbanization on wastewater treatment plants: (a) inland location, Sacramento, CA (coordinates 38.439 N, 121.480 W, view at altitude 10 km) and (b) coastal location, Los Angeles (coordinates 33.923 N, 118.429 W, view at altitude 2.75 km).

Impact of Population Demographics

Population demographics will continue to impact existing and future of WWTPs in a variety of ways including the impact of urban spread and urbanization along coastal areas.

Impact of Urban Spread and Higher Density Housing.   Since the early twentieth century, treatment plants have typically been located at some remote location, distant from the city they serve, where the wastewater would flow by gravity and near some water body that could be used as to receive the treated wastewater. In some cases, pumping was required. What has happened in the 50 to 100 intervening years is that urban spread has essentially encircled most of these early WWTPs (see Fig. 18–6). Urbanization, especially near the boundaries of the WWTPs, has often resulted in a number of diverse complaints about the WWTPs, including odors, noise, excess birds, unsightly vistas, and truck traffic, among others. As a result, a number of corrective measures have had to be implemented to deal with citizen complaints. Perhaps the most common intervention has been to cover open treatment tanks and to install odor management facilities (see Fig. 18–7). Scheduling truck deliveries in off hours has been used in a number of locations. Moving forward, care must be taken to identify and deal with citizens issues and concerns both in building new WWTPs and in upgrading existing WWTPs.

Along with urbanization, higher density housing will be necessary to accommodate anticipated population growth, especially along coastal areas as described below. Although, higher density housing poses a variety of infrastructure challenges, it also offers new opportunities to implement cost-effective decentralized wastewater treatment plants and localized reuse. As discussed previously, treatment plants located in high-density urban areas could be either of the interception or extraction type, as illustrated on Fig. 18–2. In many cases, it may make more sense to integrate extraction type treatment facilities in upscale apartment buildings, as now done in New York City (e.g., the apartment buildings in the Battery Park area of New York City are classic examples). The challenge is to develop a rational cost-effective plan for co-locating high-density housing and wastewater treatment plants.

(a)

(b)

Figure 18–7

Typical odor control facilities for new and existing wastewater treatment plants: (a) covered primary sedimentation facilities (b) compost filter for odor control for odorous gases from covered primary sedimentation tanks.

Urbanization Along Coastal Areas.   Currently, about 50 percent of the population of the United States lives within 80 km (50 mi.) of a coastal area; about the same ratio applies worldwide. Worldwide it is projected that up to 60 percent of the world's population will live near coastal areas by 2025 to 2030. The implications for water resources management of such a population shift from rural to coastal areas are significant. For example, withdrawing water from inland areas, transporting it to coastal population centers, treating and using it once, and then discharging it to coastal waters is, in the long run, unsustainable. Clearly, wastewater must be reused if the accumulation of large populations along coastal areas is to remain a viable option.

Although irrigation with treated wastewater has been occurring for decades, it is reaching logistical and economic constraints. In general, agricultural irrigation with reclaimed water is not feasible for most cities due to the long distance between the large sources of recycled water (e.g., coastal cities) and the major agricultural demand (rural areas). Further, the cost and disruption to construct a separate pipe system to convey recycled water back to agricultural areas and the need to provide winter water storage facilities are significant impediments for agricultural reuse. Thus, if significant amounts of wastewater are to be reused, the solution is to implement either IPR or the DPR of purified water in the existing water distribution system (see Fig. 18–8). In the future, because it is inevitable that DPR will become part of the water management portfolio, it is important that water and wastewater agencies begin to develop the necessary information that will allow DPR to become a reality (Haarhoff and van der Merwe, 1995; Leverenz et al., 2011; Tchobanoglous et al., 2011).

Impact of Climate Change and Sea Level Rise

Climate change and sea level rise have already had an impact on wastewater management facilities, but moving forward, even greater impacts should be anticipated and must be accounted for in the planning, design, and implementation of wastewater management facilities. Examples of the impacts of climate change and sea level rise that must be considered are reviewed in the following discussion.

Climate Change.   The most immediate impacts of climate change are reflected in increased temperature, increased evaporation rates, earlier snowmelt, and reduced or increased

Figure 18-8

Definition sketch for direct and indirect potable reuse. The bold solid line corresponds to a system in which an engineered storage buffer is used to replace the environmental buffer used for indirect potable reuse. The bold dashed line corresponds to a DPR system in which an engineered storage buffer is not used. (Adapted from Tchobanoglous et al., 2011.)

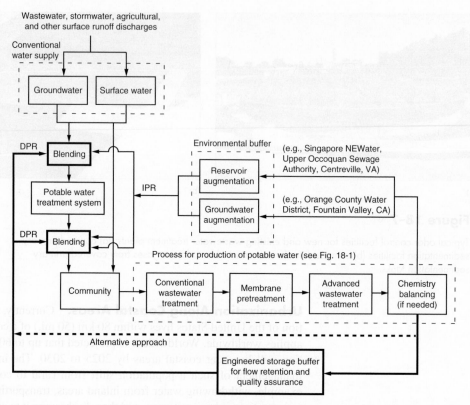

rainfall events. Although the number of rainfall events has decreased in a number of locations, the intensity of rainfall events has increased. While the impacts of climate change on the water sector are perhaps most severe, the impacts of extreme rainfall events on wastewater management systems have also been significant, including (1) extreme flooding events, exceeding the capacity of the existing collection systems; (2) increased pumping costs; (3) the discharge of untreated stormwater runoff; (4) damage to collection system infrastructure; (5) flooding of wastewater treatment plants; (6) washout of biological treatment process at treatment plants with limited storage capacity in the collection system; and (7) flows beyond the capacity of the disinfection facilities. The impact of extreme increases in rainfall may necessitate significant increases in peak capacity and/or in system retention. Because such rainfall events can be expected to continue in the future, planning efforts must be undertaken to assess how to best adapt to these changes. Increasing temperatures are also of concern because as the wastewater in the collection system gets warmer, effluent temperature TMDL values may be exceeded. A discussion of the potential costs of adaptation of water and wastewater infrastructure to climate change may be found in NACWA (2009).

Sea Level Rise. Coupled with the effects of climate change, sea level rise has already impacted a number of facilities located on or near coastal areas. A significant impact of sea level rise is that it contributes to increased stormwater flooding along coasts and tidal rivers and estuaries. When tidal surges occur, the discharge from treatment plants located along the path of the surge is often blocked, leading to the release of untreated wastewater. Localized flooding in low lying areas is also exacerbated because stormwater tide-gates cannot open. Because sea level rise also increases the level of sea-water intrusion along with the hydrostatic pressure, the design of facilities for coastal areas must account for these occurrences. Corrective measures that have been considered include the construction of levees and sea walls, raising the elevation of equipment prone to flooding, relocating stormwater discharge locations, pumping effluent through existing gravity outfalls, improvements to the collection system to reduce extraneous flows and, in the extreme, relocating wastewater management facilities.

Impact of Uncontrollable Events

In addition to the future challenges identified in Sec. 18–1 and the impact of demographics, climate change, and sea level rise discussed previously, wastewater treatment plants are subject to the effects of uncontrollable events such as natural disasters and the price of chemicals and supplies.

Natural Disasters. Natural disasters are sudden events such as hurricanes, floods, cyclones, earthquakes, and brushfires caused by natural phenomena that result in the loss of life and extensive property damage. Natural disasters that come to mind readily with respect to their impact on wastewater management facilities are Hurricane Katrina, which occurred in 2005, and the resulting damage to the wastewater management facilities of the city of New Orleans, Louisiana; the string of earthquakes that struck New Zealand in 2011; and Superstorm Sandy that struck the East coast of the United States in 2012. The effects of Katrina were magnified by the tidal surge which breached the levees that were supposed to protect against flooding. The lesson from Katrina and Superstorm Sandy is that when thinking about natural disasters with respect to the construction of new facilities or upgrading existing facilities, the unthinkable must be thought.

Chemical Costs. The impact of chemical costs that are difficult to control, especially for relatively small wastewater treatment plants, must be considered carefully. In many communities, treatment processes have been abandoned because of increases in the cost of chemicals. Because chemical costs are generally beyond the control of small municipalities, it is important to consider designs that will minimize the need for chemicals.

Impact of the Law of Unintended Consequences

Unintended consequences are outcomes that are not anticipated or intended by a particular action. Even beyond all of the factors considered previously, unintended consequences must be anticipated, as resources and cost for wastewater management become more restrictive. The field of environmental engineering is littered with monuments to the law of unintended consequences. Some examples are considered below.

Treatment Plant Siting. The unintended consequences of locating treatment plants near coastal areas, as described previously, is a prime example of an unintended consequence. When the treatment plants were located originally, little or no thought was given to sea level rise or the subsequent development that would amplify the impact of tidal surges (see Fig. 18–9). To mitigate the unintended consequences will now require the construction of expensive levees or a sea wall and/or the installation of well points to depress the groundwater level or even the possibility of having to relocate one or more of the treatment plants, an unbelievably costly undertaking.

Location of Stormwater Storage Basins. Another example of an unanticipated consequence is related to the stormwater storage basins, located around the periphery of the City of San Francisco. When the storage basins were designed and constructed, the storage basins were equipped with discharge weirs that were located so that they would

Figure 18-9

Flooding at the NYC DEP Bronx WPCP on March 2001. (Courtesy of CU-CCSR and NASA-GISS.)

only be breeched under an extreme rainfall event when the storage capacity of the basins was exceeded (see Fig. 18–10). However, with sea level rise and the subsequent development which channelized the slough to which the basins discharge, flow from tidal surges now overtops the weirs, allowing the seawater to enter the storage basins. The presence of sea water containing high sulfate concentrations has led to excessive hydrogen sulfide corrosion. To remedy the situation, the overflow weirs will have to be raised and pumps will be needed to pump the excess flow. Here again, the designers did not consider the potential impacts of sea level rise or that the subsequent development would amplify the impact of tidal surges.

Water Conservation.   In the twenty-first century, water conservation has become a goal for most water and wastewater management agencies. Water conservation is an important element of Leadership in Energy and Environmental Design (LEED) certification. In simple terms LEED is a rating system for buildings. Points can be accumulated for features such as energy conservation, water conservation, water reuse, and mitigation of stormwater. The impact of water conservation, as discussed in Chap. 3, has been significant with respect to wastewater flowrates and constituent concentrations. In the past, per capita wastewater flowrates greater than 450 L/capita·d (120 gal/capita·d) were common. In the not-so-distant future, it is reasonable to assume the per capita flowrates could decrease to below 150 L/capita·d (40 gal/capita·d). Such a decrease would have a

Figure 18-10

Impact of sea level rise on the operation of stormwater storage basins in San Francisco, CA. (Courtesy of City of San Francisco, CA.)

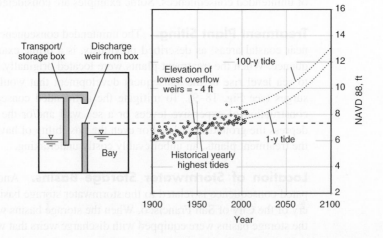

Figure 18–11

Use of a pipe within a pipe to accommodate reduced wastewater flows from residences.

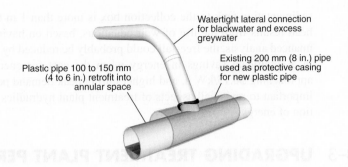

Watertight lateral connection for blackwater and excess greywater

Plastic pipe 100 to 150 mm (4 to 6 in.) retrofit into annular space

Existing 200 mm (8 in.) pipe used as protective casing for new plastic pipe

significant impact on wastewater collection systems with respect to the release of odors, grease accumulations, and hydrogen sulfide corrosion.

Most wastewater collection systems were designed to be self-cleaning at flowrates on the order of 380 to 450 L/capita·d (100 to 150 gal/capita·d). At reduced flowrates, wastewater solids and grease tend to accumulate and undergo anaerobic decomposition. In general, the release of odors locally has not been a significant problem. However, the transport of hydrogen sulfide by surface water friction has resulted in a significant increase in the rate of corrosion of downstream facilities. In some cities the rate of corrosion has increased by a factor of eight as compared to historic rates of corrosion. Grease accumulations resulting in flow blockages are another important issue associated with reduced wastewater flowrates.

In the future, with reduced rainfall and drought conditions associated with climate change, hydrogen sulfide corrosion and grease accumulation problems will continue to increase and exacerbate the problem of collection system maintenance. One mitigation measure that has been proposed is a pipe within a pipe (see Fig. 18–11). In such an arrangement, a smaller diameter plastic pipe would be placed within an existing collection system. Regardless of the mitigation measures, water conservation will continue to have a number of unintended consequences for the operation and maintenance of wastewater collection systems. Developing workable solutions will be a major change for wastewater management agencies.

Treatment Plant Hydraulics. In the past, when energy was $0.02–0.03/kWh, little attention or effort was devoted to minimizing treatment plant hydraulic headlosses. For example, as illustrated on Fig. 18–12, the water free-fall over the weirs in the primary

Figure 18–12

Examples of excessive headloss at a primary clarifier launder discharge box.

(a)

(b)

sedimentation tank to the collection box is more than 1 m (3.25 ft). Such a large loss of head results from the use of weir equations, based on having a free nappe. With a more nuanced analysis, the free-fall could probably be reduced by 50 percent, or more, resulting in a significant savings in energy when the cost for energy in many location is now approaching $0.15/kW h, and higher during peak demand periods. In the future, it will be important to revisit all aspects of treatment plant hydraulics to achieve a effective utilization of energy.

18–3 UPGRADING TREATMENT PLANT PERFORMANCE THROUGH PROCESS OPTIMIZATION AND/OR OPERATIONAL CHANGES

Establishing proper control of treatment processes was always considered a primary duty of treatment plant operating personnel. In the past, the criterion of good control and operation was producing an effluent with water quality indicators that on the average did not exceed limits established by a regulatory authority, typically prescribed in a NPDES permit. In the twenty-first century, the operational challenge for existing WWTPs will be to improve treatment performance to meet more stringent discharge requirements, to meet new constituent discharge requirements as discussed in Sec 4–2 in Chap. 4, to minimize the use of resources, and to meet some of the challenges identified in Secs. 18–1 and 18–2.

In recent years, many secondary treatment plants were converted to biological nutrient removal and water reclamation facilities. Some plants are even required to produce an effluent suitable for the subsequent production of potable water. In a number of cases, because of funding limitations for plant expansion, plants operate in excess of their design capacity. At the same time, more emphasis is being placed on operational reliability, and, in some cases, plants are cited and fined for even the slightest violation of water quality standards. Finally, because of the need to improve plant efficiency and reduce capital and operating cost, privatization of facilities construction and operation has entered the wastewater field, resulting in cost competition between public and private operating entities.

Process Optimization

As a result of the social, economic, and technological changes discussed previously and throughout this book, the wastewater treatment plant operating staff is faced with many operational challenges. Often, these challenges can be addressed by optimizing process parameters and operational procedures, modernizing facilities, and retrofitting existing equipment and processes. Development of such improvements usually requires the use of sophisticated tools and protocols, rather than just intuitive approaches. Analysis of plant operating data is a first step in facility evaluation and performance optimization. Utilization of several methods can simplify this analysis including the use of (1) histograms, (2) linear correlation, (3) online process monitoring, (4) computer models, and (5) pilot scale testing. Each of these methods is described in Table 18–2 and expanded upon below. Plant optimization through process modification is considered in the following section.

Use of Histograms. A histogram is a graphical representation of the frequency with which an event occurs over a range of different values for the same variable. An example is the number of times in a year the TSS in the secondary effluent is from 0 to 9.99, 10 to 19.9, 20 to 29.9, and greater than 30 mg/L. Histograms are particularly useful if the

Table 18–2

Methods used to evaluate process performance

Method	Description
Histograms	Graphs displaying frequency of occurrence of parameters such as wastewater characteristics, flow, and cost of chemicals and electricity
Linear correlation	A statistical method used to evaluate data from historical records such as flowrates and water quality parameters
Online process monitoring	Instruments are used continuously or intermittently to record and track important operating parameters such as flowrates, dissolved oxygen concentrations, chlorine residual concentrations, and tank levels. Data from such monitoring can be used to identify trends in operation so that process changes can be implemented before a problem occurs
Computer models	Computer modeling is a useful tool to simulate existing process operations and the effects of possible changes such as modifications to operating strategies or the addition of new equipment or processes
Pilot scale testing	Pilot scale testing is useful in evaluating the performance of new or alternative technologies and in developing criteria that can be used for the design of full scale facilities

frequency of occurrence of undesirable values, such as effluent TSS, is high. Depending on the frequency of occurrence and critical nature of the parameter, significant changes may have to be made to the process operation or perhaps to the process unit infrastructure to achieve correction. Infrequent occurrences of undesirable values often mean that only minor adjustments to operation and maintenance practices may be necessary to achieve the desired results.

Another application of histograms is in establishing the priorities for the tasks that have to be performed to avoid an undesirable event. This type of histogram, called a Pareto chart, represents a graph displaying relative contribution of small problems causing one larger problem. The Pareto chart was named after Italian economist Vilfredo Pareto (1848–1923) who made the famous observation that 80 percent of the world wealth was owned by only 20 percent of people. This principle is valid in many situations and can be equated to process operations, as 80 percent of process problems are due to 20 percent of the causes. Pareto analysis is used to identify those 20 percent that led to majority of process troubles. Analysis includes following four steps:

1. Identifying the causes of a particular problem
2. Determining the frequency of each cause
3. Calculating the percentage of each cause of the total occurrences
4. Plotting the percentage in descending order

Use of a Pareto chart in analyzing the occurrence of tank overflows due to specific causes, such as instrumentation and electrical instrumentation failures, is illustrated on Fig. 18–13.

Use of Linear Correlation. If the relationship between two variables is linear, a linear *correlation* is said to exist. The strength of the relationship is defined by the correlation coefficient, which can vary between -1 to $+1$. Linear correlation is often used to evaluate relationship between a parameter that needs to be optimized and other variables

Figure 18-13

Pareto plot used to illustrate relative contribution of various causes to overall failure.

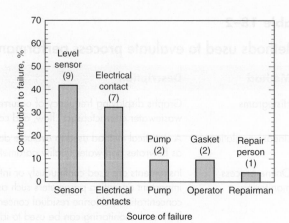

that may have some effect on that parameter. Many models used in process control are linear because they are well understood and are simple to analyze and solve. Usually, linear correlation is written as follows:

$$y = a_0 + a_1 x_1 \qquad (18\text{--}1)$$

where y = parameter to be optimized
$\quad a_0, a_1$ = coefficients
$\quad x_1$ = variable

The higher the probability that coefficient a_1 is not zero, the higher is the effect of the variable x_1 on the parameter y. This probability can be calculated using various statistical software packages. Usually, a 95 percent probability that a coefficient a_1 is not equal to zero is considered an indication that a variable x_1 has an effect on the optimized parameter y. The sign of the a_1 coefficient is used to determine in which direction the variable affects an optimized parameter. It is also important to note that lack of correlation between an optimized parameter and a variable does not necessarily mean that they are not related in general. In some cases, an appropriate relationship may be more complex than just a linear relationship (for example, the relationship may be log-normal as discussed in the Sec. 12–9 on UV disinfection), or the effect of their relationship was minimal within examined ranges.

Use of Online Process Monitoring. Since the 1950s, many industries have been using a variety of statistical analyses for making continuous process improvements. In the wastewater industry, statistical process control is not as popular because most effluent water quality characteristics are not distributed normally (i.e., bell-shaped). As a result, the methods required for implementation of statistical control for municipal wastewater treatment processes have to be more sophisticated than ones used in other industries.

Conventional analysis of historical data includes the analysis of data obtained through grab and composite sampling. Such data often do not reflect full dynamics of a treatment process. Online instrumentation, however, can be used to provide information that is more representative of the process dynamics. An example of online monitoring of dissolved oxygen (DO) for process control is shown on Fig. 18–14. The DO values vary significantly from the target value of 2 mg/L, indicating for much of the time the wastewater is overaerated. Installation of automatic DO control can prevent over- as well as underaeration resulting in savings in aeration energy and improved process control.

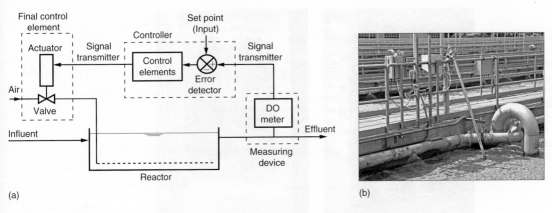

(a)

(b)

Figure 18–14

Dissolved oxygen (DO) control in an activated sludge biological treatment process: (a) general structure and components of an oxygen control loop and (b) typical installation of online meter used to monitor DO in activate sludge aeration tank.

The use of online ammonium analyzers and nitrate analyzers have been increasing at the WWTPs with biological nutrient removal and addition of external carbon for denitrification. The analyzers are sometimes used to control the dosing of the external carbon in combination with the process control algorithm. In some treatment plants, online phosphorus analyzers have been used to control chemical dosing for the chemical phosphorus removal process.

Use of Computer Models. Computer modeling of treatment processes is an effective tool for development of facility improvements because it is now possible to understand the dynamics of treatment processes and to optimize operational parameters. The effect of changes in the operational modes on process performance can also be analyzed. For example, the results of computer modeling of the activated sludge process and the hydraulic modeling of clarifiers can be used to evaluate whether an existing activated sludge plant, designed for conventional treatment (i.e., BOD and TSS removal) can be modified to remove nutrients or/and to treat higher flow than it was designed for. Computer modeling is also very helpful in the evaluation of retrofit options, such as the installation of baffles in aeration tanks and clarifiers, changing of media size in filters, and so forth. In many cases, computer modeling can simplify physical modeling. The use of modeling in activated sludge design is discussed in Sec. 8–5 in Chap. 8.

To provide reliable results, a computer model requires calibration to verify the values that have been chosen for the coefficients included in the model. These coefficients often depend on wastewater characteristics and the design features of the particular process. Calibration involves modifying the coefficient values until model output matches the plant performance data collected using online water quality and flow monitors, composite samplers, and other data collection equipment. Sometimes to improve the quality of the model calibration, the process and equipment need to be operated under extreme conditions. Such experiments are called *stress testing* and require careful planning and execution to avoid threatening plant reliability. Poor planning of stress testing at the Chernobyl nuclear power station, for example, caused the worst nuclear disaster in the history of mankind.

Figure 18–15

Full and bench scale pilot plant facilities: (a) reverse osmosis module shown above the first row of modules is used to test different RO membranes in a side-by-side comparison with existing membrane modules and (b) laboratory scale reactors set up in the temperature control incubator at 25°C to study the anammox process.

(a)　　　　　　　　　　　　　　(b)

Use of Pilot-Scale Testing.　A full-scale treatment process is the ultimate tool for evaluation of proposed improvements [see Fig. 18–15(a)]. In recent years, more use has been made of laboratory (bench) and pilot-scale testing of proposed improvements before they are implemented at full scale. Pilot scale testing was discussed previously in Sec. 4–2 in Chap. 4; laboratory-scale biological test reactors are shown on Fig. 18–15(b). Pilot scale units can be operated under a variety of different conditions, and the quality of effluent and other parameters can be compared between old and new modes of operation. Such piloting not only provides the most accurate information about the benefits of proposed process improvements, but also provides an opportunity for training the operating staff in the operation of new processes.

Operational Changes to Improve Plant Performance

In some cases, process modifications can be made at little or no cost. Some of the common problems faced in the operation of treatment facilities relate to the changing nature of the wastewater to be treated, variations in flows and loads, utilizing and maintaining each process component to achieve its maximum capability, and maintaining quality control so that treated effluent and plant residuals meet exacting reuse or permit requirements. Examples of operational changes that can be used to resolve plant performance issues are presented in Table 18–3. An example would be the turndown of the aeration system in an activated sludge plant during low flow periods to reduce over-aeration and to save on energy costs.

Where a more comprehensive approach is needed, a capacity evaluation may be required to determine the capability of the critical elements in the treatment process. For example, short-circuiting may exist in the primary clarifiers resulting in the carryover of settleable material to the biological treatment units, thus causing overloading conditions and increasing the cost of treatment. Stress testing of the process components may be necessary to determine the limits of operating capability of the key components. In other cases, tracer testing to document hydraulic flow patterns (see Appendix H and I) or measuring dissolved oxygen transfer rates to determine aeration efficiency may be employed. These types of tests can also be used to determine if new physical facilities are required.

Table 18–3

Examples of operational changes that can be used to improve plant performance

Issue	Possible remedial action	See Sec. no.
General plant		
Odors from open tanks and channels	Reduce turbulence by controlling water levels to eliminate free falls and splashing	
	Add chemicals (such a chlorine, ferric chloride, or hydrogen peroxide) to influent wastewater	6–2, 6–3
	Modify process loading	
Wide influent flow variations	Conduct collection system I/I investigation to identify sources of extraneous flow	
Short circuiting in clarifiers and chlorine contact tanks	Conduct tracer tests	12–6
	Conduct stress tests	
Headworks		
Odorous grit	Adjust air flowrates in aerated grit chambers to obtain cleaner grit	
	Add lime to dewatered grit	
Odors and vectors in headworks	Cover grit and screenings receptacles	16–3
	Add lime to dewatered grit and screenings	
Inadequate grit removal	Analyze channel and grit chamber hydraulics	5–5
	Adjust flow distribution to grit chamber	
	Add temporary baffles to prevent short circuiting	
	Adjust airflowrate in aerated grit chambers	
Grit deposition in channels	Modify/adjust channel flow-through velocity	
Primary clarifiers		
Poor solids removal	Check for short circuiting/modify baffling	
	Improve flow distribution	5–6
	Add chemicals to influent	6–3
	Reduce return flows from other processes	15–2
Low solids concentration in primary sludge	Modify sludge pumping rate/install timers	
	Increase sludge blanket depth	
Aeration tanks		
Low DO	Determine DO profile in tank and adjust air flowrate	
	Conduct oxygen transfer test	5–11
	Assess diffuser fouling/clean fouled diffusers	
	Change conventional plug flow operation to step feed (if possible)	
	Check wastewater characteristics (rbCOD and nbVSS)	8–2
High DO	Turn off aeration equipment during low flow and loading periods	
	Install timers to control blower or mechanical aerator operation	
Filamentous organisms in mixed liquor	Examine mixed liquor microscopically to identify types of organisms	8–3
	Increase sludge wasting	
	Chlorinate return sludge	

(continued)

Table 18-3 (*Continued*)

Issue	Possible remedial action	See Sec. no.
Process is nitrifying	Check SRT	
Process is not nitrifying	Check SRT, alkalinity, and temperature	
Low pH	Check if nitrification is occurring	8–6
	Add alkalinity	8–6
Foaming	Identify nature of foam	8–3
	Change MLSS concentration	8–5
	Add defoaming agent to spray water	
Nocardia foaming	Use dilute chlorine solution spray on foam	
	Reduce oil and grease discharges to collection system	
Trickling filters		
Poor BOD removal	Reduce dosing rate	9–2
Solids washout at peak flows	Reduce recirculation rate	
Biological nitrogen removal process		
Inadequate removal	Check rbCOD, MLSS, and temperature/revise feed	
	Increase external carbon feed	
Final clarifiers		
Bulking sludge	Increase dissolved oxygen concentration	
	Increase F/M ratio	
	Modify return and waste activated sludge pumping rates	
	Chlorinate influent wastewater	
	Chlorinate return activated sludge	
Rising sludge	Increase return activated sludge pumping rate to reduce sludge blanket depth in clarifier	
	Conduct state point analysis of final clarifiers	8–10
Poor solids separation	Perform state point analysis	8–10
Disinfection		
High coliform count in effluent	Improve chlorine mixing	12–6
	Conduct tracer tests to determine short circuiting	12–6
	Check process for is partial nitrification	8–6
Solids processing		
Low solids concentration from dissolved air flotation thickener	Check air-solids ratio	
	Reduce solids loading rate	
	Add/increase polymer feed	
Poor anaerobic digester performance	Change frequency of solids feeding	13–9
	Increase concentration of feed solids	
	Check adequacy of mixing	
	Remove sludge and grit deposits	
	Increase SRT	

(*continued*)

Table 18–3 (Continued)		
Issue	**Possible remedial action**	**See Sec. no.**
Poor aerobic digester performance	Check temperature/adjust SRT	13–9
	Increase concentration of feed solids	
	Check adequacy of mixing	
	Increase DO	
	Check pH/adjust alkalinity	
Odors from composting operations	Increase aeration by air addition or frequency of turning	14–5
Poor compost quality	Perform materials balance/adjust feed composition	14–5
Excessive moisture in compost mixture	Change compost mixture by adding amendment or bulking agent	14–5
	Improve sludge dewatering operations	
Processing of return flows	Use flow equalization to input return flows more uniformly over the day or in the evening hours when excess aeration capacity is available	15–2
	Consider installation of separate treatment facilities for return flows.	15–3
	Consider the recovery of nutrients	15–4, 15–5

18–4 UPGRADING TREATMENT PLANT PERFORMANCE THROUGH PROCESS MODIFICATION

In many cases, to meet some of the many challenges identified in Secs. 18–1 and 18–2, it will be necessary to upgrade existing treatment plants. Upgrading of a wastewater treatment plant can encompass a large number or only a few factors. The scope of this text cannot include identifying and discussing all of the factors that may go into the decision to upgrade a facility; however, some common factors that necessitate upgrading existing treatment facilities include the need to:

1. Improve treatment plant performance
2. Reduce chemical, energy, and maintenance costs
3. Meet more stringent discharge requirements
4. Meet additional treatment capacity needs for population growth
5. Meet new constituent removal requirements for potable reuse
6. Meet new energy and resource recovery objectives

Because these factors have been discussed throughout this textbook, they are not discussed further, but are presented as a basis for understanding the material presented in this section. The various issues related to upgrading existing treatment plants are divided into two categories: (1) upgrading of physical facilities for liquid and solids treatment and (2) potential process modifications for meeting new requirements for constituent removal.

Upgrading Physical Facilities

Most treatment plants contain all of the essential elements necessary to meet treatment goals, but in some instances, the capacity of some of the components is underutilized or overloaded, hydraulic bottlenecks exist that constrain effective and efficient operation, and inadequacies in facilities design affect plant operations and maintenance. Some common issues related to upgrading liquid treatment and solids processing facilities are discussed below.

Upgrading Liquid Treatment Facilities. Upgrading existing facilities may be necessary to mitigate existing operating problems. An example might be the carryover of shredded screenings and hair from the screening process that is causing plugging of membrane filters. Replacement of coarse bar racks or comminutors may be necessary to reduce screenings passthrough and filter plugging. Covering primary sedimentation tanks to limit the release of odors and installing odor control facilities is a common plant upgrade (see Fig. 18–7). Another example is the installation of equalization basins or sidestream treatment processes to accommodate wet-weather flows. Examples of issues relating to the upgrading of physical facilities for the treatment of wastewater (liquid treatment facilities) are summarized in Table 18–4.

In some cases, a capacity evaluation is required in which a full-scale test program is required to determine the capability of the critical elements in the treatment process. For example, short-circuiting may exist in the primary clarifiers resulting in the carryover of settleable material to the biological treatment units, thus causing overloading conditions or increasing the cost of treatment. Stress testing of the process components may be necessary to determine the limits of operating capability of the key components. In other cases, tracer testing to document hydraulic flow patterns or measuring dissolved oxygen transfer rates to determine aeration efficiency may be employed.

Upgrading Solids Processing Facilities. Although a major focus of treatment plant design and operation is on the liquid treatment facilities because of the standards for treated effluent reuse and disposal, solids processing facilities are receiving increased attention because of the potential to recover energy and resources. Solids processing is often the most vexing problem for many plants because of operational difficulties, increasingly stringent requirements for reuse, and limited options for disposal. However, some of the new technologies described in this text can be used to upgrade the design and operation of solids processing facilities. Example upgrade options are described in Table 18–5. Frequent problems are often associated the return flows and loads from solids processing facilities such as thickening and dewatering. However, some of the new technologies described in this text can be used to upgrade the design and operation of solids processing facilities (see Chaps. 13 and 14). The separate treatment of return flows is considered in Chap. 15.

Upgrading to Meet New Constituent Removal Requirements

As discussed in Chaps. 1, 4, 8, 13, and 15 and in earlier sections of this chapter, standards for constituent removal have changed in recent years and will continue to change as more scientific information is developed and as the reuse of treated wastewater and biosolids becomes increasingly important. Examples of many of the current and future issues and upgrade options in resolving these issues are summarized in Table 18–6.

18–5 MANAGEMENT OF WET-WEATHER FLOWS

Although the management of wet-weather flows has, to date, received considerable attention, it is anticipated that even greater attention will be devoted to this aspect of wastewater management in the future. Sanitary sewer overflows (SSOs) are currently regulated under the Clean Water Act, with the U.S. EPA taking the position that all SSOs are illegal discharges and must be eliminated. It is widely recognized, however, that no matter how well a sanitary sewer collection system is operated and maintained, occasional unintentional discharges occur from almost every system. To date, efforts to develop and implement a national policy to regulate the control of SSOs have not been successful.

Table 18–4

Examples of upgrading of physical liquid treatment facilities to improve plant performance

Issue	Remedial action/upgrade option	See sec. no.
General plant		
Odors	Cover structures	16–3
	Add odor collection and treatment system	
	Reduce turbulence by eliminating free falls and sharp bends	
	Add chemical feed facilities	
Wide influent flow variations	Add upstream flow equalization	3–7
	Install variable speed drives on pumps	
	Install small capacity pumps for low flows	
Flow control/ distribution	Improve flow splitting	
	Add metering	
Return flows from sludge processing facilities upset biological process	Provide flow storage/equalization	15–2
	Provide sidestream treatment of return flows	15–3
	Modify operations/upgrade solids processing facilities to reduce load	
Headworks		
Inadequate screenings removal	Modify/replace screens to prevent screenings carryover	5–1
	Install fine screens	
	Replace comminutors	
Odorous, wet screenings	Install screenings press	5–2
	Replace screens with macerators	
	Enclose and ventilate screening equipment	16–3
Odorous grit	Install grit washer	
	Enclose and ventilate grit equipment	16–3
Inadequate grit removal	Add permanent baffles to prevent short circuiting	
	Replace/upgrade grit removal equipment	5–5
Primary clarifiers		
Inadequate solids removal in primary clarifiers	Add chemical treatment and flocculation	6–2, 6–3
	Add high rate clarification	5–7
	Install baffles at effluent weirs	
Aeration tanks		
Low DO	Install DO probes for DO monitoring	
	Replace coarse bubble with fine pore diffusers	5–11
	Change diffuser placement to a grid pattern	
High DO	Install variable speed drives on centrifugal blowers to provide turndown capability	5–11
	Install inlet guides vanes on centrifugal blowers to provide turndown capability	
	Install variable speed drives on positive displacement blowers	
	Install timers and two-speed motors on mechanical aerators	
	Install automatic DO control system	

(continued)

| **Table 18-4** (Continued) |

Issue	Remedial action/upgrade option	See sec. no.
Unbalanced DO profile in plug flow aeration tanks	Change to step feed process	8-9
	Add DO control system	
Solids deposition	Increase mixing capacity	5-3
Nocardia foaming	Add selector	8-4
Trickling filter		
Plugging and ponding of rock filters	Install plastic packing	9-2
Odors and poor BOD removal	Increase airflow by improving natural draft or adding ventilation system	9-2
Biological treatment system		
Insufficient reactor and solids separation capacity	Add chemical treatment	6-2, 6-3
	Add high rate clarification to reduce loading on biological treatment system	5-7
	Add membrane bioreactor	8-12
Solids washout from high flow-rates	Add flow equalization	3-7
	Add high rate clarification process for excess flows	5-7
	Use contact stabilization process	
Secondary clarifiers		
Inadequate solids separation in secondary clarifiers	Modify flow distribution	8-11
	Modify circular clarifier center feedwell	8-11
	Add flocculating center feedwell	8-11
	Install baffles at effluent weirs	
	Add tube or plate settlers	5-4
	Modify effluent weir configuration	
	Modify sludge collector to improve solids withdrawal	
Disinfection		
Inadequate chlorine disinfection	Add/replace chlorine mixers	12-6
	Add/modify baffles to reduce short circuiting in chlorine contact tank	
	Add chlorine residual control system	
TSS in effluent	Add depth filters before disinfection	
Excessive chlorine residual	Add chlorine residual analyzer and automatic control system	12-6
	Add dechlorination facilities	
	Replace chlorination system with UV	

SSO Policy Issues

In 1994, a stakeholder process was initiated to consider technical and policy issues pertaining to SSOs. An SSO Subcommittee was formed to evaluate the need for national consistency in regulating SSOs, and defining public policy issues related to collection system operation and maintenance as well as public health and environmental impacts related to SSOs. SSO Subcommittee efforts continued until 1999 and several basic principles were identified as suggested NPDES Permit requirements. In 2001, a Notice of Proposed Rulemaking (NPRM)

Table 18-5

Examples of upgrading physical solids treatment facilities to improve plant performance

Issue	Remedial action/upgrade option	See sec. no.
Thickening		13–6
Low solids concentration in primary sludge	Add gravity thickening	
	Add co-settling thickening	
Insufficient gravity thickening of waste activated sludge	Use alternative thickeners (dissolved air flotation, or centrifuge)	
Alkaline stabilization		13–8
Odor and vector problems in dewatered sludge	Add post-lime stabilization	
Anaerobic digesters		13–9
Excessive hydraulic loading	Add sludge thickening prior to digestion	
Inadequate mixing	Upgrade digester mixing system	
	Install egg-shaped digester	
Poor digestion of mixed primary and biological sludge	Install separate digesters	
Inadequate solids destruction	Install two-phased anaerobic digestion process	
Aerobic digestion		13–10
Insufficient pathogen removal	Increase SRT by adding thickening or additional aerobic digester capacity	
	Add ATAD process	
Inadequate mixing	Increase mixing energy	
Composting		
Excessive plastics and inert material in product to be reused	Install fine screens in plant influent	5–1
	Install sludge screens	13–5
Dewatering and Drying		14–2, 14–3
Excessive water in dewatered sludge cake	Add sludge thickeners	
	Install high solids centrifuge dewatering	
	Install filter press	
	Add solar drying beds	
	Add heat dryers	
Sludge Lagoons and drying beds		
Odors	Construct turbulence inducing structures	16–3
Land application of biosolids		14–10
Excessive attraction of vectors	Modify preapplication treatment methods or method of biosolids application	
Excessive pathogen levels	For Class A biosolids, use one of the six prescribed alternative treatment alternatives	
	For Class B biosolids, use of three prescribed alternative monitoring or treatment alternatives	

Table 18–6

Potential process modifications for meeting new standards for constituent removal

Issue	Remedial action/upgrade option	See Sec. no.
TSS discharge standards	Investigate alternative solids separation facilities	5–9
	Chemical treatment to enhance settling	6–2, 6–3
	Addition of tube or plate settlers to final clarifiers	5–7
	Addition of depth filtration	11–5
	Addition of surface filtration	11–6
	Addition of membrane separation	8–12, 11–7
BOD/COD standards	Investigate alternative treatment facilities	
	Supplemental chemical treatment	6–6
	Nitrification	7–9, 8–6
	Combined aerobic biological processes	9–3
	Membrane biological reactors	8–12
	Adsorption	11–9
	Advanced oxidation	6–8
Removal of nitrogen and phosphorus	Investigate alternative removal facilities	
	Chemical treatment for phosphorus removal	6–4
	Activated sludge selector	8–4
	Suspended growth processes	
	Nitrification	8–6
	Nitrogen removal	8–7
	Phosphorus removal	6–4, 8–8
	Attached growth processes	9–7
	Ammonia stripping of digester supernatant	15–5
	Ion exchange for nitrogen removal	11–11
New disinfection standards	Add depth filtration (prior to disinfection)	11–5, 11–6
	Improve chlorine mixing and dispersion	12–6
	Add dechlorination system	12–5
	Replace chlorination with UV	12–9
VOC emission requirements	Investigate alternative advanced treatment systems	
	Adsorption	16–4
	Air stripping	16–4
	Advanced oxidation	6–8
Removal of residual solids for water reuse	Investigate alternative advanced treatment systems	
	Depth filtration	11–5
	Surface filtration	11–6
	Microfiltration	11–7
	Activated carbon adsorption	11–9
	Ion exchange	11–11
	Advanced oxidation	6–8

(continued)

Table 18–6 (Continued)		
Issue	**Remedial action/upgrade option**	**See Sec. no.**
Removal of trace constituents	Investigate alternative treatment systems	
	Chemical precipitation and oxidation	6–6, 6–8
	Microfiltration and/or reverse osmosis	11–7
	Microfiltration and/or reverse osmosis with UV oxidation	6–8, 11–7
Part 503 biosolids regulations for Class A land application	Investigate alternative processes to further reduce pathogens (PFRP) including	14–5
	Thermophilic aerobic digestion	13–9
	Composting	14–5
	Heat drying	14–3
	Heat treatment	14–3
Part 503 biosolids regulations for Class B land application	Investigate alternative processes to significantly reduce pathogens (PSRP) including	
	Lime stabilization	13–8
	Anaerobic digestion	13–9
	Aerobic digestion	13–10
	Composting	14–5
	Air drying	14–3

intended to further development of these basic principles was signed by the EPA Administrator but rulemaking efforts never advanced.

SSO Guidance

In 2005 EPA issued a guidance document that contains most of what was intended to be in the original rule relative to these basic principles, which have become known as CMOM – Capacity, Management, Operations, and Maintenance. Individual EPA Regional offices have required implementation of CMOM principles through the NPDES Permit process. During 2010, a series of listening sessions were held by EPA to obtain public input for use by EPA in considering whether and how to modify NPDES Regulations as they pertain to municipal sanitary sewer systems and SSOs. EPA is currently preparing a summary of the input received.

Wet-Weather Management Options

Wet weather treatment systems and processes are designed similar to conventional wastewater treatment processes in many respects. The key difference between wet-weather treatment and conventional wastewater treatment is the highly variable, intermittent nature of wet weather flow as compared to the continuous (24 h/d, 365 d/y) operation of a conventional wastewater treatment process. To withstand the rigors of constant ON/OFF, wet-dry cycles and widely varying flows and loads, wet weather treatment processes must be designed to be simple, rugged, and reliable. In assessing the applicability of any technology, system, or process for wet weather treatment, the designer must be assured that it can withstand the highly variable operating conditions outlined previously. A number of treatment technologies are identified and discussed in Table 18–7. Each of the processes

Table 18–7

Representative wet-weather treatment processes in various combinations

Unit processes	Description
Screening and disinfection 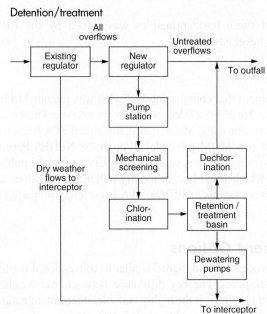	Screening and disinfection facilities combine these two unit processes to control the two most prevalent constituents in CSOs in terms of contribution to water quality impacts—floatable debris and pathogens. Screening equipment must be selected to withstand repeated wet-dry cycles, and screens that avoid moving parts below the water surface are preferred. Screen opening sizes can range from approximately 12 to 20 mm (0.5 to 0.75 in.) down to 3 to 4 mm (0.1 to 0.15 in.). Screenings are normally containerized for disposal following each storm event but, in some cases, can also be macerated and introduced into a sewer tributary to a downstream WWTP. Disinfection is normally accomplished using sodium hypochlorite due to the proven nature of high-rate hypochlorite disinfection and the ability of the equipment to start and stop automatically in response to storm events. Hypochlorite contact times of 15 min are common, but times as short as five min have been used in combination with high-intensity, induction mixers to introduce the hypochlorite into the screened flow. When required, dechlorination is accomplished by applying liquid sodium bisulfite following disinfection

Detention/treatment

Detention/treatment facilities employ gravity sedimentation (see Chap. 5) in combination with offline storage to reduce impacts of CSOs. Unit processes typically include: influent screening (to capture floating pollutants and remove large objects), gravity sedimentation, sodium hypochlorite disinfection, and de-chlorination using liquid sodium bisulfite (when required). High-rate disinfection practices, using induction mixers, are customarily used. High-rate disinfection practices improve disinfection efficiency, which is important as disinfection takes place concurrent with sedimentation. Ancillary processes typically include detention basin dewatering and flushing systems, and odor control. Dewatering and flushing systems are automated typically to reduce staffing requirements, with flushing gates and tipping buckets are used commonly to scour settled solids from the tank bottom following each storm event.

Side water depths ranging from 3 to 6 m (10 to 20 ft), and surface overflowrates (SORs) on the order of 180 to 240 m/d (4500 to 6000 gal/ft²·d) are used in conjunction with the peak design flow to size detention basins. These high SORs take into account the transient nature of wet weather flows, meaning that the peak design flow may occur for only minutes during each storm event that causes the facility to activate, and that for much of the storm the actual SOR is much lower. In addition to pollutant removal achieved by gravity sedimentation, a portion of the influent flow remains in the basin following each storm event. On an annual basis, the percent removal achieved by this captured volume is significant due to the prevalence of small storms that can be largely captured in the detention basin.

Chemically-enhanced detention treatment is an emerging wet weather treatment process that involves the addition of a coagulant (typically a metal salt) and flocculent (polymer) to improve solids removal efficiency. When chemically-enhanced detention treatment is used, high-rate disinfection follows in a separate contact basin.

(continued)

Table 18–7 (Continued)

Unit processes	Description
Swirl/vortex Existing regulator → New regulator → Untreated overflows → To outfall All overflows Pump station → Dechlorination Mechanical screening → Chlorine contact tank Dry weather flows to interceptor Swirl / vortex separators → Chlorination Underflow pumps To interceptor	Swirl/vortex treatment uses a combination of gravitational and rotational forces to enhance liquid-solids separation. Flow is introduced tangentially into a circular basin with specific geometry to direct the flow on a long spiral path. While some rotational motion occurs, gravitational forces acting on the solids as they travel along the spiral flow path are primarily responsible for solids separation. Solids that settle are concentrated at the bottom of the unit and from there are conveyed as a slurry to a sewer or interceptor and on to the WWTP. Swirl/vortex units are equipped with integral screens and/or baffles to capture floating pollutants. For configurations that do not include an integral screen, a separate screening process is provided typically upstream of the swirl/vortex unit. High-rate disinfection, using sodium hypochlorite, is often performed concurrent with swirl/vortex treatment, followed by de-chlorination using liquid sodium bisulfite (when required). Provisions to dewater the unit following each storm event are normally provided, but automated flushing mechanisms and odor control systems are uncommon. There are three specific geometric configurations commonly used for swirl/vortex treatment units: one developed by the U.S. EPA, which is in the public domain, and two proprietary configurations. Details on the geometry of the U.S. EPA design are found in EPA-R2—72-008. Each swirl/vortex unit operates on the basic principles outlined above. It is imperative to characterize the particle settling velocity specific to the wet weather flow to be treated using swirl/vortex technology. The use of settling column tests is highly recommended as a basis for establishing an appropriate surface overflowrate (SOR). For removals that approach a primary sedimentation level of efficiency (TSS removals of 50 percent or more) SORs on the order of 300 to 400 m/d (7500 to 10,000 gal/ft^2·d) are typical. Higher SORs, on the order of 600 to 1200 m/d (15,000 to 30,000 gal/ft^2·d), can be used with recognition that a swirl/vortex unit sized at those SORs will generally only remove floatable pollutants and heavier solids (grit). Chemically-enhanced swirl/vortex treatment is an emerging wet weather treatment process that involves the addition of a coagulant (typically a metal salt) and flocculent (polymer) to improve solids removal efficiency. When chemically-enhanced swirl/vortex treatment is used, high-rate disinfection follows in a separate contact basin. *(continued)*

Table 18-7 (Continued)	
Unit processes	**Description**

Ballasted flocculation

Ballasted flocculation is a physical-chemical treatment process that uses a recycled media and chemical addition to improve the settling characteristics of solids. By agglomerating and increasing the specific gravity of the solids particles, effective solids separation can be accomplished at surface over-flowrates (SORs) many times higher than with conventional sedimentation, ranging from 0.80 to 3.25 m/min (20 to 80 gal/ft²·min). Appurtenant unit processes include fine screening (screens with clear openings of 6 mm (0.25 in.) or smaller are generally required), and may include grit removal. A separate grit removal process is typically included for wet weather applications with higher activation frequencies and volumes to mitigate contamination of the ballast with grit particles. Because coagulant (typically a metal salt) and flocculent (polymer) are added to enhance solids separation, disinfection and de-chlorination (if necessary) are provided downstream of ballasted flocculation. Because ballasted flocculation produces a high-quality effluent, UV disinfection provides an alternative to sodium hypochlorite disinfection. When used for wet weather treatment the ballasted flocculation process train incorporates an automated shut-down procedure to partially drain the system and recycle the ballast to the influent end of the process.

There are two ballasted flocculation process systems that have been used successfully for wet weather treatment. One uses recycled sludge from the process as ballast and the other uses a fine "microsand," which is recovered using hydrocyclones. Both processes involve dosing the influent flow with coagulant upstream of initial rapid mixing, then adding polymer and ballast to increase the size and specific gravity of solids in the flow. As the flow passes through a series of mixing compartments the mixing intensity decreases, allowing the particles to agglomerate. Solids separation usually takes place in a circular clarifier equipped with plate or tube settlers to provide a greater effective surface area for settling. While there are many functional similarities between the two processes, both have specific advantages and disadvantages which must be carefully weighed relative to each unique application.

Storage basins

Storage basins are used to hold excess wet weather flow temporarily during and shortly after storm events, and can be installed at the WWTP or at satellite locations. These basins fill during periods when conveyance and/or treatment capacity is not available and are drained when capacity is available. Basin volume is generally determined by defining the excess (or overflow) volume corresponding to a design storm event and annual performance is checked using a collection system model. Ancillary processes typically include detention basin dewatering and flushing systems, and odor control. Influent screens are normally provided to prevent large objects (sticks, bricks, etc) from entering the basin. Because the purpose of these screens is protection from overflows rather than treatment, screen openings are relatively large 25 to 75 mm (1 to 3 in.), which mitigates the quantities of screenings to be handled at satellite locations.

(continued)

Table 18–7 (Continued)	

Unit processes	Description
Tunnel storage Existing regulators → New regulator (All overflows / Untreated overflows / To outfall) Drop shafts (To outfall) De-aeration chambers Horizontal connectors (drift tunnels) Tunnel storage → Dewatering pumps Dry weather flows to interceptor To interceptor	Tunnel storage is used to temporarily hold excess wet weather flow during and shortly after storm events in the same manner as storage basins. While tunnel storage often mitigates peak flows to be handled at a WWTP, it is typically implemented in the collection system. Tunnel volume is generally determined by defining the excess (or overflow) volume corresponding to a design storm event and annual performance is checked using a collection system model. Ancillary processes typically include a tunnel dewatering system and odor control. Storage tunnels are normally designed to be self-cleansing but, if low velocities are expected, flushing systems can be included. A means of preventing large objects from entering the tunnel, such as a coarse bar rack or baffle arrangement can be devised, and screens sized with 25 to 75 mm (1 to 3 in.) clear openings to prevent large objects (sticks, bricks, etc) from entering the dewatering pump station at the downstream end of the tunnel are normally provided.

described in Table 18–7 has performed satisfactorily under various wet-weather operating conditions. The challenge moving forward is how to integrate the technologies identified and others into existing wastewater management programs.

DISCUSSION TOPICS

18–1 Review three articles on asset management for wastewater treatment facilities and discuss the benefits and potential risks in implementing asset management as discussed in the articles.

18–2 A large development is planned at the outskirt of a city with 60,000 population, and population is expected to increase by 20,000 in the next 10 y. Currently the city has one wastewater treatment facility, which is nearing its plant capacity. Discuss the potential advantages and/or disadvantages of considering decentralized wastewater management based on the geographic and climatic characteristics of your location.

18–3 Assuming that the intensity of wet-weather events has continued to increase, what measures would you propose investigate to alleviate the potential washout of solids at a treatment plant with shallow secondary clarifiers [3 m (10 ft) side water depth].

18–4 Discuss advantages and disadvantages of converting a combined wastewater collection system into separate wastewater and stormwater collection systems.

18–5 Discuss advantages and disadvantages of converting separate wastewater and stormwater collection systems into a combined wastewater collection system.

18–6 Discuss the benefits and drawbacks of either of the actions proposed in Problems 18–4 and 18–5 with respect to the management of stormwater.

REFERENCES

Asano, T., F. L. Burton, H. Leverenz, R. Tsuchihashi, and G. Tchobanoglous (2007) *Water Reuse: Issues, Technologies, and Applications,* McGraw-Hill, New York.

Haarhoff, J., and B. van der Merwe (1995) "Twenty-Five Years of Wastewater Reclamation in Windhoek, Namibia," in A. Angelakis, T. Asano, E. Diamadopoulos, and G. Tchobanoglous (eds.), *Proc. 2nd International Symposium on Wastewater Reclamation and Reuse,* 29–40.

Leverenz, H. L., G. Tchobanoglous, and T. Asano (2011) "Direct Potable Reuse: A Future Imperative," *J. Water Reuse and Desal.,* 1, 1, 2–10.

McCarty, P. L. (2011) *Back to the Future—Towards Sustainability in Water Resources.* Presented at 84th Annual Water Environment Federation Technical Exhibition And Conference, Los Angeles, CA.

NACWA (2009) *Confronting Climate Change: An Early Analysis of Water and Wastewater Adaptation Costs,* National Association of Clean Water Agencies and Association of the Metropolitan Water Agencies, Washington, DC.

NRC (2012) *Water Reuse: Potential for Expanding the Nation's Water Supply Through Reuse of Municipal Wastewater,* National Research Council, National Academies Press, Washington, DC.

Tchobanoglous, G. (2010) *Wastewater Management in the Twenty-First Century.* Presented at the College of Engineering Distinguished Speaker Series. University of Miami, Miami, FL.

Tchobanoglous, G., H. Leverenz, M. H. Nellor, and J. Crook (2011) *Direct Potable Reuse: A Path Forward,* WateReuse Research and WateReuse California, Washington, DC.

Tchobanoglous, G., and H. Leverenz (2013) The Rationale for Decentralization of Wastewater Infrastructure, Chap. 8, in T.A. Larson, K.M. Udert, and J. Lienert (eds.) *Wastewater Treatment: Source Separation and Decentralisation,* IWA Publishing, London.

U.S. EPA (1994) *Water Quality Standards Handbook,* 2nd ed., EPA-823-B-94-005a, U.S. Environmental Protection Agency, Washington, DC.

U.S. EPA (2010) *Green Infrastructure Case Studies: Municipal policies for Managing Stormwater with Green Infrastructure,* EPA-841-F-10-004, Office of Wetlands, Oceans and Watersheds, U.S. Environmental Protection Agency, Washington, DC.

U.S. EPA (2012a) website: http://www.epa.gov/owm/assetmanage/, accessed in August 2012.

U.S. EPA (2012b) website: http://www.epa.gov/owm/assetmanage/assets_training.htm, accessed in August 2012.

Conversion Factors

Appendix A

Table A–1

Unit conversion factors, SI units to U.S. customary units and U.S. customary units to SI units

		To convert, multiply in direction shown by arrows			
SI unit name	**Symbol**	→	←	**Symbol**	**U.S. customary unit name**
Acceleration					
meters per second squared	m/s²	3.2808	0.3048	ft/s²	feet per second squared
meters per second squared	m/s²	39.3701	0.0254	in./s²	inches per second squared
Area					
hectare (10,000 m²)	ha	2.4711	0.4047	ac	acre
square centimeter	cm²	0.1550	6.4516	in.²	square inch
square kilometer	km²	0.3861	2.5900	mi²	square mile
square kilometer	km²	247.1054	4.047×10^{-2}	ac	acre
square meter	m²	10.7639	9.2903×10^{-2}	ft²	square foot
square meter	m²	1.1960	0.8361	yd²	square yard
Energy					
kilojoule	kJ	0.9478	1.0551	Btu	British thermal unit
joule	J	2.7778×10^{-7}	3.6×10^{6}	kW·h	kilowatt-hour
joule	J	0.7376	1.356	ft·lb$_f$	foot-pound (force)
joule	J	1.0000	1.0000	W·s	watt-second
joule	J	0.2388	4.1876	cal	calorie
kilojoule	kJ	2.7778×10^{-4}	3600	kW·h	kilowatt-hour
kilojoule	kJ	0.2778	3.600	W·h	watt-hour
megajoule	kJ	0.3725	2.6845	hp·h	horsepower-hour
Force					
newton	N	0.2248	4.4482	lb$_f$	pound force
Flowrate					
cubic hectometers per day	hm³	264.1720	3.7854×10^{3}	Mgal/d	million gallons per day
cubic meters per day	m³/d	264.1720	3.785×10^{-3}	gal/d	gallons per day
cubic meters per day	m³/d	2.6417×10^{-4}	3.7854×10^{3}	Mgal/d	million gallons per day

(continued)

1901

| **Table A-1** (Continued) |

		To convert, multiply in direction shown by arrows			
SI unit name	**Symbol**	$\rightarrow$	$\leftarrow$	**Symbol**	**U.S. customary unit name**
cubic meters per second	m³/s	35.3147	2.8317×10^{-2}	ft³/s	cubic feet per second
cubic meters per second	m³/s	22.8245	4.3813×10^{-2}	Mgal/d	million gallons per day
cubic meters per second	m³/s	15850.3	6.3090×10^{-5}	gal/min	gallons per minute
liters per second	L/s	22,824.5	4.3813×10^{-2}	gal/d	gallons per day
liters per second	L/s	2.2825×10^{-2}	43.8126	Mgal/d	million gallons per day
liters per second	L/s	15.8508	6.3090×10^{-2}	gal/min	gallons per minute
Length					
centimeter	cm	0.3937	2.540	in.	inch
kilometer	km	0.6214	1.6093	mi	mile
meter	m	39.3701	2.54×10^{-2}	in.	inch
meter	m	3.2808	0.3048	ft	foot
meter	m	1.0936	0.9144	yd	yard
millimeter	mm	0.03937	25.4	in.	inch
Mass					
gram	g	0.0353	28.3495	oz	ounce
gram	g	0.0022	4.5359×10^{2}	lb	pound
kilogram	kg	2.2046	0.45359	lb	pound
megagram (10³ kg)	Mg	1.1023	0.9072	ton	ton (short: 2000 lb)
megagram (10³ kg)	Mg	0.9842	1.0160	ton	ton (long: 2240)
Power					
kilowatt	kW	0.9478	1.0551	Btu/s	British thermal units per second
kilowatt	kW	1.3410	0.7457	hp	horsepower
watt	W	0.7376	1.3558	ft-lb$_f$/s	foot-pounds (force) per second
Pressure (force/area)					
Pascal (newtons per square meter)	Pa (N/m²)	1.4504×10^{-4}	6.8948×10^{3}	lb$_f$/in.²	pounds (force) per square inch
Pascal (newtons per square meter)	Pa (N/m²)	2.0885×10^{-2}	47.8803	lb$_f$/ft²	pounds (force) per square foot
Pascal (newtons per square meter)	Pa (N/m²)	2.9613×10^{-4}	3.3768×10^{3}	in. Hg	inches of mercury (60°F)
Pascal (newtons per square meter)	Pa (N/m²)	4.0187×10^{-3}	2.4884×10^{2}	in. H$_2$O	inches of water (60°F)
kilopascal (kilonewtons per square meter)	kPa (kN/m²)	0.1450	6.8948	lb$_f$/in.²	pounds (force) per square inch
kilopascal (kilonewtons per square meter)	kPa (kN/m²)	0.0099	1.0133×10^{2}	atm	atmosphere (standard)

(continued)

| **Table A-1** (*Continued*)

	To convert, multiply in direction shown by arrows				
SI unit name	**Symbol**	→	←	**Symbol**	**U.S. customary unit name**
Temperature					
degree Celsius (centigrade)	°C	$1.8(°C) + 32$	$0.0555(°F) - 32$	°F	degree Fahrenheit
degree kelvin	K	$1.8(K) - 459.67$	$0.0555(°F) + 459.67$	°F	degree Fahrenheit
Velocity					
kilometers per second	km/s	2.2369	0.44704	mi/h	miles per hour
meters per second	m/s	3.2808	0.3048	ft/s	feet per second
Volume					
cubic centimeter	cm^3	0.0610	16.3781	$in.^3$	cubic inch
cubic hectometer (100 m × 100 m × 100 m)	hm^3	8.1071×10^2	1.2335×10^{-3}	ac·ft	acre·foot
cubic hectometer	hm^3	264.1720	3.7854×10^3	Mgal	million gallons
cubic meter	m^3	35.3147	2.8317×10^{-2}	ft^3	cubic foot
cubic meter	m^3	1.3079	0.7646	yd^3	cubic yard
cubic meter	m^3	264.1720	3.7854×10^{-3}	gal	gallon
cubic meter	m^3	8.1071×10^{-4}	1.2335×10^3	ac·ft	acre·foot
liter	L	0.2642	3.7854	gal	gallon
liter	L	0.0353	28.3168	ft^3	cubic foot
liter	L	33.8150	2.9573×10^{-2}	oz	ounce (U.S. fluid)

Table A–2

Conversion factors for commonly used wastewater treatment plant design parameters

	To convert, multiply in direction shown by arrows		
SI units	$\rightarrow$	$\leftarrow$	**U.S. units**
g/m^3	8.3454	0.1198	lb/Mgal
ha	2.4711	0.4047	ac
hm^3	264.1720	3.785×10^3	Mgal
kg	2.2046	0.4536	lb
kg/ha	0.8922	1.1209	lb/ac
$kg/kW \cdot h$	1.6440	0.6083	$lb/hp \cdot h$
kg/m^2	0.2048	4.8824	lb/ft^2
kg/m^3	8345.4	1.1983×10^{-4}	lb/Mgal
$kg/m^3 \cdot d$	62.4280	0.0160	$lb/10^3 ft^3 \cdot d$
$kg/m^3 \cdot h$	0.0624	16.0185	$lb/ft^3 \cdot h$
kJ	0.9478	1.0551	Btu
kJ/kg	0.4299	2.3260	Btu/lb
kPa (gage)	0.1450	6.8948	$lb_f/in.^2$ (gage)
kPa Hg (60 °F)	0.2961	3.3768	in. Hg (60 °F)
kW/m^3	5.0763	0.197	$hp/10^3$ gal
$kW/10^3 m^3$	0.0380	26.3342	$hp/10^3 ft^3$
L	0.2642	3.7854	gal
L	0.0353	28.3168	ft^3
$L/m^2 \cdot d$	2.4542×10^{-2}	40.7458	$gal/ft^2 \cdot d$
$L/m^2 \cdot h$	0.5890	1.6978	$gal/ft^2 \cdot d$
$L/m^2 \cdot min$	0.0245	40.7458	$gal/ft^2 \cdot min$
$m^3/m^2 \cdot min$	24.5424	4.0746×19^{-2}	$gal/ft^2 \cdot min$
$L/m^2 \cdot min$	35.3420	0.0283	$gal/ft^2 \cdot d$
m	3.2808	0.3048	ft
m/h	3.2808	0.3048	ft/h
m/h	0.0547	18.2880	ft/min
m/h	0.4090	2.4448	$gal/ft^2 \cdot min$
$m^2/10^3 m^3 \cdot d$	0.0025	407.4611	$ft^2/Mgal \cdot d$
m^3	1.3079	0.7646	yd^3
$m^3/capita$	35.3147	0.0283	$ft^3/capita$
m^3/d	264.1720	3.785×10^{-3}	gal/d
m^3/d	2.6417×10^{-4}	3.7854×10^3	Mgal/d
m^3/h	0.5886	1.6990	ft^3/min
$m^3/ha \cdot d$	106.9064	0.0094	$gal/ac \cdot d$
m^3/kg	16.0185	0.0624	ft^3/lb
$m^3/m \cdot d$	80.5196	0.0124	$gal/ft \cdot d$
$m^3/m \cdot min$	10.7639	0.0929	$ft^3/ft \cdot min$

(continued)

Table A–2

(*Continued*)

To convert, multiply in direction shown by arrows			
SI units	**→**	**←**	**U.S. units**
$m^3/m^2 \cdot d$	24.5424	0.0407	$gal/ft^2 \cdot d$
$m^3/m^2 \cdot d$	0.0170	58.6740	$gal/ft^2 \cdot min$
$m^3/m^2 \cdot d$	1.0691	0.9354	$Mgal/ac \cdot d$
$m^3/m^2 \cdot h$	3.2808	0.3048	$ft^3/ft^2 \cdot h$
$m^3/m^2 \cdot h$	589.0173	0.0017	$gal/ft^2 \cdot d$
m^3/m^3	0.1337	7.4805	ft^3/gal
$m^3/10^3\ m^3$	133.6805	7.4805×10^{-3}	$ft^3/Mgal$
$m^3/m^3 \cdot min$	133.6805	7.4805×10^{-3}	$ft^3/10^3\ gal \cdot min$
$m^3/m^3 \cdot min$	1,000.0	0.001	$ft^3/10^3\ ft^3 \cdot min$
Mg/ha	0.4461	2.2417	ton/ac
mm	3.9370×10^{-2}	25.4	in.
ML/d	0.2642	3.785	Mgal/d
ML/d	0.4087	2.4466	ft^3/s

Table A–3

Abbreviations for SI units

Abbreviation	SI unit
°C	degree Celsius
cm	centimeter
g	gram
g/m^2	gram per square meter
g/m^3	gram per cubic meter (= mg/L)
ha	hectare (= 100 m × 100 m)
hm^3	cubic hectometer (= 100 m × 100 m × 100 m)
J	Joule
K	Kelvin
kg	kilogram
kg/capita · d	kilogram per capita per day
kg/ha	kilogram per hectare
kg/m^3	kilogram per cubic meter
kJ	kilojoule
kJ/kg	kilojoule per kilogram
kJ/kW · h	kilojoule per kilowatt-hour
km	kilometer (= 1000 m)
km^2	square kilometer
km/h	kilometer per hour
km/L	kilometer per liter
kN/m^2	kiloNewton per square meter

(*continued*)

Table A–3

(Continued)

Abbreviation	SI unit
kPa	kiloPascal
ks	kilosecond
kW	kilowatt
L	liter
L/s	liters per second
m	meter
m^2	square meter
m^3	cubic meter
mm	millimeter
m/s	meter per second
mg/L	milligram per liter ($= g/m^3$)
m^3/s	cubic meter per second
MJ	megajoule
N	Newton
N/m^2	Newton per square meter
Pa	Pascal (usually reported as kilopascal, kPa)
W	Watt

Table A–4

Abbreviations for US customary units

Abbreviation	US Customary Units
ac	acre
ac-ft	acre foot
Btu	British thermal unit
Btu/ft^3	British thermal unit per cubic foot
d	day
ft	foot
ft^2	square foot
ft^3	cubic foot
ft/min	feet per minute
ft/s	feet per second
ft^3/min	cubic feet per minute
ft^3/s	cubic feet per second
°F	degree Fahrenheit
gal	gallon
gal/ft^2·d	gallon per square foot per day
gal/ft^2·min	gallon per square foot per minute
gal/min	gallon per minute

(continued)

Table A–4

(*Continued*)

Abbreviation	US Customary Units
h	hour
hp	horsepower
hp-h	horsepower-hour
in.	inch
kWh	kilowatt-hour
lb_f	pound (force)
lb_m	pound (mass)
lb/ac	pound per acre
lb/ac·d	pound per acre per day
lb/capita·d	pound per capita per day
lb/ft^2	pound per square foot
lb/ft^3	pound per cubic foot
lb/in^2	pound per square inch
lb/yd^3	pound per cubic yard
Mgal/d	million gallons per day
mi	mile
mi^2	square mile
mi/h	mile per hour
min	minute
mo	month
ppb	part per billion
ppm	part per million
s	second
ton (2000 lbm)	ton (2000 pounds mass)
wk	week
y	year
yd	yard
yd^2	square yard
yd^3	cubic yard

Physical Properties of Selected Gases and the Composition of Air

Appendix B

B-1 PHYSICAL PROPERTIES OF SELECTED GASES

Table B-1

Molecular weight, specific weight, and density of gases found in wastewater at standard conditions (0°C, 1 atm)[a]

Gas	Formula	Molecular weight	Specific weight, lb/ft³	Density, g/L,
Air	–	28.97	0.0808	1.2928
Ammonia	NH_3	17.03	0.0482	0.7708
Carbon dioxide	CO_2	44.00	0.1235	1.9768
Carbon monoxide	CO	28.00	0.0781	1.2501
Hydrogen	H_2	2.016	0.0056	0.0898
Hydrogen sulfide	H_2S	34.08	0.0961	1.5392
Methane	CH_4	16.03	0.0448	0.7167
Nitrogen	N_2	28.02	0.0782	1.2507
Oxygen	O_2	32.00	0.0892	1.4289

[a] Adapted from Perry, R. H., D. W. Green, and J. O. Maloney: Perry's (eds) (1984) *Chemical Engineers' Handbook*, 6th ed., McGraw-Hill Book Company, New York.

B-2 COMPOSITION OF DRY AIR

Table B-2

Composition of dry air at 0°C and 1.0 atmosphere[a]

Gas	Formula	Percent by volume[b,c]	Percent by weight
Nitrogen	N_2	78.03	75.47
Oxygen	O_2	20.99	23.18
Argon	Ar	0.94	1.30
Carbon dioxide	CO_2	0.03	0.05
Other[d]	–	0.01	–

[a] Note: Values reported in the literature vary depending on the standard conditions.

[b] Adapted from North American Combustion Handbook, 2nd ed., North American Mfg., Co., Cleveland, OH.

[c] For ordinary purposes air is assumed to be composed of 79 percent N_2 and 21 percent O_2 by volume.

[d] Hydrogen, Neon, Helium, Krypton, Xenon.

Note: Molecular weight of air = $(0.7803 \times 28.02) + (0.2099 \times 32.00) + (0.0094 \times 39.95) + (0.0003 \times 44.00) = 28.97$ (see Table B-1 above).

B–3   DENSITY OF AIR AT OTHER TEMPERATURES

In SI units

The following relationship can be used to compute the density of air, ρ_a, at other temperatures at atmospheric pressure.

$$\rho_a = \frac{PM}{RT}$$

where P = atmospheric pressure, $1.0132^5 \times 10^5$ N/m^2

M = molecular weight of air (see Table B-1), 28.97 g/ g mole

R = universal gas constant, 8314 N · m/(mole air · K)

T = temperature, K (273.15 + °C)

For example, at 20°C, the density of air is:

$$\rho_{a,20°C} = \frac{(1.01325 \times 10^5 \text{ N/m}^2)(28.97 \text{ g/ mole air})}{[8314 \text{ N} \cdot \text{m/(mole air} \cdot \text{K)}][(273.15 + 20)\text{K}]}$$

$$= 1.204 \times 10^3 \text{ g/m}^3 = 1.204 \text{ kg/m}^3$$

In U.S. customary units

The following relationship can be used to compute the specific weight of air, γ_a, at other temperatures at atmospheric pressure.

$$\gamma_a = \frac{P(144 \text{ in}^2/\text{ft}^2)M}{RT}$$

where P = atmospheric pressure, 14.7 lb/in^2

M = molecular weight of air (see Table B-1), 28.97 lb/ lb mole air

R = universal gas constant, 1544 ft · lb/(lb mole air · °R)

T = temperature, °R (460 + °F)

For example, at 68°F, the specific weight of air is:

$$\gamma_a = \frac{(14.7 \text{ lb/in}^2)(144 \text{ in}^2/\text{ft}^2)(28.97 \text{ lb/lb mole air})}{[1544 \text{ ft} \cdot \text{lb/(lb mole air} \cdot °\text{R)}][(460 + 68)°\text{R}]} = 0.0752 \text{ lb/ft}^3$$

B–4   CHANGE IN ATMOSPHERIC PRESSURE WITH ELEVATION

In SI units

The following relationship can be used to compute the change in atmospheric pressure with elevation.

$$\frac{P_b}{P_s} = \exp\left[-\frac{gM(z_b - z_a)}{RT}\right]$$

where P_b = pressure at elevation z_b, N/m^2

P_s = atmospheric pressure at sea level, 1.01325×10^5 N/m^2

g = acceleration due to gravity, 9.81 m/s^2

M = molecular weight of air (see Table B-1), 28.97 g/mole air

z_b = elevation b, m

z_a = elevation b, ft

R = universal gas constant, 8314 N $\cdot$ m/(mole air $\cdot$ K)

T = temperature, K (273.15 + °C)

In U.S. customary units

The following relationship can be used to compute the change in atmospheric pressure with elevation.

$$\frac{P_b}{P_s} = \exp\left[-\frac{gM(z_b - z_a)}{g_c RT} \right]$$

where P_b = pressure at elevation z_b, lb/in^2

P_s = atmospheric pressure at sea level, lb/in^2

g = acceleration due to gravity, 32.2 ft/s^2

M = molecular weight of air (see Table B–1), 28.97 lb$_m$/lb mole air

z_b = elevation b, ft

z_a = elevation b, ft

g_c = 32.2 ft $\cdot$ lb$_m$/lb $\cdot s^2$

R = universal gas constant, 1544 ft $\cdot$ lb/(lb mole air $\cdot$ °R)

T = temperature, °R (460 + °F)

The principal physical properties of water are summarized in SI units in Table C–1 and in U.S. customary units in Table C–2. They are described briefly below (Vennard and Street, 1975; Webber, 1971).

C–1 SPECIFIC WEIGHT

The specific weight of a fluid, γ, is its weight per unit volume. In SI units, specific weight is expressed in kilonewtons per cubic meter (kN/m^3). The relationship between γ, ρ, and the acceleration due to gravity g is $\gamma = \rho g$.

C–2 DENSITY

The density of a fluid, ρ, is its mass per unit volume. In SI units density is expressed in kilograms per cubic meter (kg/m^3). For water, ρ is 1000 kg/m^3 at 4°C. There is a slight decrease in density with increasing temperature.

C–3 MODULUS OF ELASTICITY

For most practical purposes, liquids may be regarded as incompressible. The bulk modulus of elasticity, E, is given by

$$E = \frac{\Delta p}{(\Delta V/V)}$$

where Δp is the increase in pressure, which when applied to a volume V, results in a decrease in volume ΔV. In SI units, the modulus of elasticity is expressed in kilonewtons per meter squared (kN/m^2).

C–4 DYNAMIC VISCOSITY

The viscosity of a fluid, μ, is a measure of its resistance to tangential or shear stress. In SI units, the dynamic viscosity is expressed in Newton seconds per square meter (N·s/m^2).

C–5 KINEMATIC VISCOSITY

In many problems concerning fluid motion, the viscosity appears with the density in the form μ/ρ, and it is convenient to use a single term, ν, known as the kinematic viscosity. In SI units, the kinematic viscosity is expressed in meters squared per second (m^2/s). The kinematic viscosity of a liquid diminishes with increasing temperature.

C–6 **SURFACE TENSION**

The surface tension of a fluid, σ, is the physical property that enables a drop of water to be held in suspension at a tap, a glass to be filled with liquid slightly above the brim and yet not spill, or a needle to float on the surface of a liquid. The surface-tension force across any imaginary line at a free surface is proportional to the length of the line and acts in a direction perpendicular to it. In SI units, surface tension per unit length is expressed in Newtons per meter (N/m). There is a slight decrease in surface tension with increasing temperature.

C–7 **VAPOR PRESSURE**

Liquid molecules that possess sufficient kinetic energy are projected out of the main body of a liquid at its free surface and become vapor. In a system open to the atmosphere, the vapor pressure, p_v, is the partial pressure exerted by the liquid vapor in the atmosphere. In a closed system, the vapor molecules are in equilibrium with the liquid; the pressure exerted by the vapor molecules is known as the saturated vapor pressure. In SI units, vapor pressure is expressed in kilonewtons per square meter (kN/m^2).

REFERENCES

Vennard, J.K., and R.L. Street (1975) *Elementary Fluid Mechanics*, 5th ed., Wiley, New York.

Webber, N.B. (1971) *Fluid Mechanics for Civil Engineers*, SI ed., Chapman and Hall, London.

Table C–1

Physical properties of water (SI units)[a]

Temp-erature, °C	Specific weight, γ, kN/m³	Density[b], ρ, kg/m³	Modulus of elasticity[b], $E/10^6$, kN/m²	Dynamic viscosity, $\mu \times 10^3$, N·s/m²	Kinematic viscosity, $\nu \times 10^6$, m²/s	Surface tension[c], σ, N/m	Vapor pressure, p_v, kN/m²
0	9.805	999.8	1.98	1.781	1.785	0.0765	0.61
5	9.807	1000.0	2.05	1.518	1.519	0.0749	0.87
10	9.804	999.7	2.10	1.307	1.306	0.0742	1.23
15	9.798	999.1	2.15	1.139	1.139	0.0735	1.70
20	9.789	998.2	2.17	1.002	1.003	0.0728	2.34
25	9.777	997.0	2.22	0.890	0.893	0.0720	3.17
30	9.764	995.7	2.25	0.798	0.800	0.0712	4.24
40	9.730	992.2	2.28	0.653	0.658	0.0696	7.38
50	9.689	988.0	2.29	0.547	0.553	0.0679	12.33
60	9.642	983.2	2.28	0.466	0.474	0.0662	19.92
70	9.589	977.8	2.25	0.404	0.413	0.0644	31.16
80	9.530	971.8	2.20	0.354	0.364	0.0626	47.34
90	9.466	965.3	2.14	0.315	0.326	0.0608	70.10
100	9.399	958.4	2.07	0.282	0.294	0.0589	101.33

[a] Adapted from Vennard and Street (1975).
[b] At atmospheric pressure.
[c] In contact with the air.

Table C–2

Physical properties of water (U.S. customary units)[a]

Temp-erature, °F	Specific weight, γ, lb/ft³	Density[b], ρ, slug/ft³	Modulus of elasticity[b], $E/10^3$, lb$_f$/in.²	Dynamic viscosity, $\mu \times 10^5$, lb·s/ft²	Kinematic viscosity, $\nu \times 10^5$, ft²/s	Surface tension[c], σ, lb/ft	Vapor pressure, p_v, lb$_f$/in.²
32	62.42	1.940	287	3.746	1.931	0.00518	0.09
40	62.43	1.940	296	3.229	1.664	0.00614	0.12
50	62.41	1.940	305	2.735	1.410	0.00509	0.18
60	62.37	1.938	313	2.359	1.217	0.00504	0.26
70	62.30	1.936	319	2.050	1.059	0.00498	0.36
80	62.21	1.934	324	1.799	0.930	0.00492	0.51
90	62.11	1.931	328	1.595	0.826	0.00486	0.70
100	62.00	1.927	331	1.424	0.739	0.00480	0.95
110	61.86	1.923	332	1.284	0.667	0.00473	1.27
120	61.71	1.918	332	1.168	0.609	0.00467	1.69
130	61.55	1.913	331	1.069	0.558	0.00460	2.22
140	61.38	1.908	330	0.981	0.514	0.00454	2.89
150	61.20	1.902	328	0.905	0.476	0.00447	3.72
160	61.00	1.896	326	0.838	0.442	0.00441	4.74
170	60.80	1.890	322	0.780	0.413	0.00434	5.99
180	60.58	1.883	318	0.726	0.385	0.00427	7.51
190	60.36	1.876	313	0.678	0.362	0.00420	9.34
200	60.12	1.868	308	0.637	0.341	0.00413	11.52
212	59.83	1.860	300	0.593	0.319	0.00404	14.70

[a] Adapted from Vennard and Street (1975).

[b] At atmospheric pressure.

[c] In contact with the air.

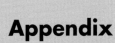
The statistical analysis of wastewater flowrate and constituent concentration data involves the determination of statistical parameters used to quantify a series of measurements. Commonly used statistical parameters and graphical techniques for the analysis of wastewater management data are reviewed below.

D–1   COMMON STATISTICAL PARAMETERS

Commonly used statistical measures include the mean, median, mode, standard deviation, and coefficient of variation, based on the assumption that the data are distributed normally. Although the terms just cited are the most commonly used statistical measures, two additional statistical measures are needed to quantify the nature of a given distribution. The two additional measures are the coefficient of skewness, and coefficient of kurtosis. If a distribution is highly skewed, as determined by the coefficient of skewness, normal statistics cannot be used. For most wastewater data that are skewed, it has been found that the log of the value is normally distributed. Where the log of the values is normally distributed, the distribution is said to be log normal. The common statistical measures used for the analysis of wastewater management data (Eqs. D–1 through D–9) are summarized in Table D–1.

D–2   GRAPHICAL ANALYSIS OF DATA

Graphical analysis of wastewater management data is used to determine the nature of the distribution. For most practical purposes, the type of the distribution can be determined by plotting the data on both arithmetic- and logarithmic-probability paper and noting whether the data can be fitted with a straight line. The three steps involved in the use of arithmetic, and logarithmic-probability paper are as follows.

1. Arrange the measurements in a data set in order of increasing magnitude and assign a rank serial number.
2. Compute a corresponding plotting position for each data point using Eqs. (D–10) and (D–11).

$$\text{Plotting position } (\%) = \left(\frac{m}{n+1} \right) \times 100 \tag{D–10}$$

where m = rank serial number
n = number of observations

The term $(n + 1)$ is used to correct for a-small-sample bias. The plotting position represents the percent or frequency of observations that are equal to or less than the indicated value. Another expression often used to define the plotting position is known as Blom's transformation:

$$\text{Plotting position } (\%) = \frac{m - 3/8}{n + 1/4} \times 100 \tag{D–11}$$

3. Plot the data on arithmetic- and logarithmic-probability paper. The probability scale is labeled "Percent of values equal to or less than the indicated value."

Table D–1

Statistical parameters used for the analysis of wastewater management data[a]

Parameter	Definition
	Terms
Mean value	$\bar{x}$ = mean value
$$\bar{x} = \frac{\sum f_i x_i}{n} \qquad \text{(D–1)}$$	f_i = frequency (for ungrouped data $f_i = 1$)
	x_i = the mid-point of the ith data range (For ungrouped data x_i = the ith observation)
Standard Deviation	n = number of observations (Note $\sum f_i = n$)
$$s = \sqrt{\frac{\sum f_i(x_i - \bar{x})^2}{n - 1}} \qquad \text{(D–2)}$$	s = standard deviation
	C_v = coefficient of variation, percent
Coefficient of variation	α_3 = coefficient of skewness
	α_4 = coefficient of kurtosis
$$C_v = \frac{100\,s}{\bar{x}} \qquad \text{(D–3)}$$	M_g = geometric mean
	s_g = geometric standard deviation
Coefficient of skewness	$P_{15.9}$ and $P_{84.1}$ = values from arithmetic or logarithmic probability plots at indicated percent values, corresponding to one standard deviation
$$\alpha_3 = \frac{\sum f_i(x_i - \bar{x})^3/(n-1)}{s^3} \qquad \text{(D–4)}$$	
Coefficient of kurtosis	**Median value**
	If a series of observations are arranged in order of increasing value, the middlemost observation, or the arithmetic mean of the two middlemost observations, in a series is known as the median.
$$\alpha_4 = \frac{\sum f_i(x_i - \bar{x})^4/(n-1)}{s^4} \qquad \text{(D–5)}$$	
	Mode
Geometric mean	The value occurring with the greatest frequency in a set of observations is known as the mode. If a continuous graph of the frequency distribution is drawn, the mode is the value of the high point, or hump, of the curve. In a symmetrical set of observations, the mean, median, and mode will be the same value. The mode can be estimated with the following expression. Mode = 3(median) $-$ 2($\bar{x}$).
$$\log M_g = \frac{\sum f_i(\log x_i)}{n} \qquad \text{(D–6)}$$	
Geometric standard deviation	
$$\log s_g = \sqrt{\frac{\sum f_i(\log^2 x_g)}{n-1}} \qquad \text{(D–7)}$$	**Coefficient of skewness** When a frequency distribution is asymmetrical, it is usually defined as being a skewed distribution.
Using probability paper	**Coefficient of kurtosis**
$$s = P_{84.1} - \bar{x} \text{ or } P_{15.9} + \bar{x} \qquad \text{(D–8)}$$	Used to define the peakedness of the distribution. The value of the kurtosis for a normal distribution is 3. A peaked curve will have a value greater than 3 whereas a flatter curve it will have a value less than 3.
$$s_g = \frac{P_{84.1}}{M_g} = \frac{M_g}{P_{15.9}} \qquad \text{(D–9)}$$	

[a] Adapted from Metcalf & Eddy (1991) and Crites and Tchobanoglous (1998).

If the data, plotted on arithmetic-probability paper, can be fit with a straight line, then the data are assumed to be normally distributed. Significant departure from a straight line can be taken as an indication of skewness. If the data are skewed, logarithmic probability paper can be used. The implication here is that the logarithm of the observed values is normally distributed. On logarithmic-probability paper, the straight line of best fit passes through the geometric mean, M_g, and through the intersection of $M_g \times s_g$ at a value of 84.1 percent and M_g/s_g at a value of 15.9 percent. The geometric standard deviation, s_g, can be determined using Eq. D–9 given in Table D–1. The use of arithmetic- and logarithmic-probability paper is illustrated in Example D–1.

EXAMPLE D–1   Statistical Analysis of Wastewater Flowrate Data.   Using the following weekly flowrate data obtained from an industrial discharger for a calendar quarter of operation, determine the statistical characteristics and predict the maximum weekly flow-rate that will occur during a full year's operation.

Week No.	Flowrate, m³/wk	Week No.	Flowrate, m³/wk
1	2900	8	3675
2	3040	9	3810
3	3540	10	3450
4	3360	11	3265
5	3770	12	3180
6	4080	13	3135
7	4015		

Solution

1. Plot the flowrate data using the log/probability method.
   a. Set up a data analysis table with three columns as described below.
      i. In column 1, enter the rank serial number starting with number 1
      ii. In column 2, arrange the flowrate data in ascending order
      iii. In column 3, enter the probability plotting position

Rank serial no., m	Flowrate, m³/wk	Plotting position,ᵃ %
1	2900	7.1
2	3040	14.3
3	3135	21.4
4	3180	28.6
5	3265	35.7
6	3360	42.9
7	3450	50.0
8	3540	57.1
9	3675	64.3
10	3770	71.4
11	3810	78.6
12	4015	85.7
13	4080	92.9

ᵃ Plotting position = $[m/(n + 1)]100$.

b. Plot the weekly flowrates expressed in m^3/wk versus the plotting position. The resulting plots are presented below. Because the data fall on a straight line on both plots, the flowrate data can be described adequately by either distribution. This fact can be taken as indication that the distribution is not skewed significantly and that normal statistics can be applied.

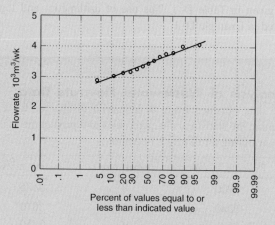 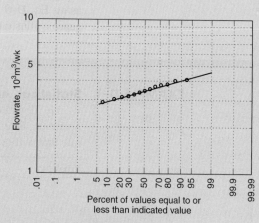

2. Determine the statistical characteristics of the flowrate data.
   a. Set up a data analysis table to obtain the quantities needed to determine the statistical characteristics.

Flowrate, m^3/wk	$(x - \bar{x})$	$(x - \bar{x})^2$	$(x - \bar{x})^3$ 10^{-6}	$(x - \bar{x})^4$ 10^{-9}
2900	−578	334,084	−193	11,161
3040	−438	191,844	−84	3680
3135	−343	117,649	−40	1384
3180	−298	88,804	−26	789
3265	−213	45,369	−9.6	206
3360	−118	13,924	−1.6	19.4
3450	−28	784	−0.02	0.06
3540	62	3844	0.24	1.48
3675	197	38,809	7.6	151
3770	292	85,264	25	727
3810	332	110,224	37	1215
4015	537	288,369	155	8316
4080	602	362,404	218	13,134
45,220		1,681,372	88.62	40,784

b. Determine the statistical characteristics using the parameters given in Table D–1.
   i. Mean

$$\bar{x} = \frac{\sum x}{n}$$

$$\bar{x} = \frac{45,220}{13} = 3478 \ m^3/wk$$

ii. Median (the middle-most value)

Median = 3450 m³/wk (see data table above)

iii. Mode

Mode = 3(Median) − 2($\bar{x}$) = 3(3450) − 2(3478) = 3394 m³/wk

iv. Standard deviation

$$s = \sqrt{\frac{\sum(x - \bar{x})^2}{n - 1}}$$

$$s = \sqrt{\frac{1{,}681{,}372}{12}} = 374.3 \text{ m}^3/\text{wk}$$

v. Coefficient of variation

$$C_V = \frac{100s}{\bar{x}}$$

$$C_V = \frac{100(374.3)}{3478} = 10.8\%$$

vi. Coefficient of skewness

$$\alpha_3 = \frac{[\sum(x - \bar{x})^3/(n - 1)]}{s^3}$$

$$\alpha_3 = \frac{(88.62 \times 10^6/12)}{(374.3)^3} = 0.141$$

vii. Coefficient of kurtosis

$$\alpha_4 = \frac{[\sum(x - \bar{x})^4/(n - 1)]}{s^4}$$

$$\alpha_4 = \frac{(40{,}784 \times 10^9/12)}{(374.3)^4} = 1.73$$

Reviewing the statistical characteristics, it can be seen that the distribution is somewhat skewed (α_3 = 0.141 versus 0 for a normal distribution) and is considerably flatter than a normal distribution would be (α_4 = 1.73 versus 3.0 for a normal distribution).

3. Determine the probable annual maximum weekly flowrate.

a. Determine the probability factor:

$$\text{Peak week} = \frac{m}{n + 1} = \frac{52}{52 + 1} = 0.981$$

b. Determine the flowrate from the figure given in Step 1b at the 98.1 percentile:

Peak weekly flowrate = 4500 m³/wk

Comment   The statistical analysis of data is important in establishing the design conditions for wastewater treatment plants. The application of statistical analysis to the selection of design flowrates and mass loadings rates is considered in the Sec. 3–6 in Chap. 3.

Dissolved Oxygen Concentration in Water as a Function of Temperature, Salinity, and Barometric Pressure

Appendix E

Table E–1

The air solubility of oxygen in mgIL as functions of temperature and elevation in meters for 0–1800 m[a]

Temp., °C	Elevation above sea level, m									
	0	200	400	600	800	1,000	1,200	1,400	1,600	1,800
0	14.621	14.276	13.94	13.612	13.291	12.978	12.672	12.373	12.081	11.796
1	14.216	13.881	13.554	13.234	12.922	12.617	12.32	12.029	11.745	11.468
2	13.829	13.503	13.185	12.874	12.57	12.273	11.984	11.701	11.425	11.155
3	13.46	13.142	12.832	12.53	12.234	11.945	11.663	11.387	11.118	10.856
4	13.107	12.798	12.496	12.201	11.912	11.631	11.356	11.088	10.826	10.57
5	12.77	12.468	12.174	11.886	11.605	11.331	11.063	10.801	10.546	10.296
6	12.447	12.174	11.866	11.585	11.311	11.044	10.782	10.527	10.278	10.035
7	12.138	11.851	11.571	11.297	11.03	10.769	10.514	10.265	10.022	9.784
8	11.843	11.562	11.289	11.021	10.76	10.505	10.256	10.013	9.776	9.544
9	11.559	11.285	11.018	10.757	10.502	10.253	10.01	9.772	9.54	9.314
10	11.288	11.02	10.759	10.504	10.254	10.011	9.773	9.541	9.315	9.093
11	11.027	10.765	10.51	10.26	10.017	9.779	9.546	9.319	9.098	8.881
12	10.777	10.521	10.271	10.027	9.789	9.556	9.329	9.107	8.89	8.678
13	10.536	10.286	10.041	9.803	9.569	9.342	9.119	8.902	8.69	8.483
14	10.306	10.06	9.821	9.587	9.359	9.136	8.918	8.705	8.498	8.295
15	10.084	9.843	9.609	9.38	9.156	8.938	8.724	8.516	8.313	8.114
16	9.87	9.635	9.405	9.18	8.961	8.747	8.538	8.334	8.135	7.94
17	9.665	9.434	9.209	8.988	8.774	8.564	8.359	8.159	7.963	7.772
18	9.467	9.24	9.019	8.804	8.593	8.387	8.186	7.99	7.798	7.611
19	9.276	9.054	8.837	8.625	8.418	8.216	8.019	7.827	7.639	7.455
20	9.092	8.874	8.661	8.453	8.25	8.052	7.858	7.669	7.485	7.304
21	8.914	8.7	8.491	8.287	8.088	7.893	7.703	7.518	7.336	7.159
22	8.743	8.533	8.328	8.127	7.931	7.74	7.553	7.371	7.193	7.019
23	8.578	8.371	8.169	7.972	7.78	7.592	7.408	7.229	7.054	6.883
24	8.418	8.214	8.016	7.822	7.633	7.449	7.268	7.092	6.92	6.752
25	8.263	8.063	7.868	7.678	7.491	7.31	7.132	6.959	6.79	6.625
26	8.113	7.917	7.725	7.537	7.354	7.175	7.001	6.83	6.664	6.501
27	7.968	7.775	7.586	7.401	7.221	7.045	6.873	6.706	6.542	6.382
28	7.827	7.637	7.451	7.269	7.092	6.919	6.75	6.584	6.423	6.266
29	7.691	7.503	7.32	7.141	6.967	6.796	6.63	6.467	6.308	6.153
30	7.559	7.374	7.193	7.017	6.845	6.677	6.513	6.353	6.196	6.043
31	7.43	7.248	7.07	6.896	6.727	6.561	6.399	6.241	6.087	5.937
32	7.305	7.125	6.95	6.779	6.612	6.448	6.289	6.133	5.981	5.833
33	7.183	7.006	6.833	6.665	6.5	6.339	6.181	6.028	5.878	5.731
34	7.065	6.89	6.72	6.553	6.39	6.232	6.077	5.925	5.777	5.633
35	6.949	6.777	6.609	6.445	6.284	6.127	5.974	5.825	5.679	5.536
36	6.837	6.667	6.501	6.338	6.18	6.025	5.874	5.727	5.583	5.442
37	6.727	6.559	6.395	6.235	6.078	5.926	5.776	5.631	5.489	5.35
38	6.62	6.454	6.292	6.134	5.979	5.828	5.681	5.537	5.396	5.259
39	6.515	6.351	6.191	6.035	5.882	5.733	5.587	5.445	5.306	5.171
40	6.412	6.25	6.092	5.937	5.787	5.639	5.495	5.355	5.218	5.084

[a] From Colt, J. (2012) *Dissolved Gas Concentration in Water: Computation as Functions of Temperature, Salinity and Pressure*, 2nd ed., Elsevier, Boston, MA.

Table E–2

Standard air saturation concentration of oxygen as a function of temperature and salinity in mg/L. 0–40 g/kg (Seawater. I atm moist air)[a]

Temp., °C	Salinity, g/kg								
	0.0	5.0	10.0	15.0	20.0	25.0	30.0	35.0	40.0
0	14.621	14.120	13.635	13.167	12.714	12.276	11.854	11.445	11.050
1	14.216	13.733	13.266	12.815	12.378	11.956	11.548	11.153	10.772
2	13.829	13.364	12.914	12.478	12.057	11.649	11.255	10.875	10.506
3	13.460	13.011	12.577	12.156	11.750	11.356	10.976	10.608	10.252
4	13.107	12.674	12.255	11.849	11.456	11.076	10.708	10.352	10.008
5	12.770	12.352	11.946	11.554	11.174	10.807	10.451	10.107	9.774
6	12.447	12.043	11.652	11.272	10.905	10.550	10.205	9.872	9.550
7	12.138	11.748	11.369	11.002	10.647	10.303	9.970	9.647	9.335
8	11.843	11.465	11.098	10.743	10.399	10.066	9.743	9.431	9.128
9	11.559	11.194	10.839	10.495	10.162	9.839	9.526	9.223	8.930
10	11.288	10.933	10.590	10.257	9.934	9.621	9.318	9.024	8.739
11	11.027	10.684	10.351	10.028	9.715	9.411	9.117	8.832	8.556
12	10.777	10.444	10.121	9.808	9.505	9.210	8.925	8.648	8.379
13	10.536	10.214	9.901	9.597	9.302	9.016	8.739	8.470	8.209
14	10.306	9.993	9.689	9.394	9.108	8.830	8.561	8.299	8.046
15	10.084	9.780	9.485	9.198	8.920	8.651	8.389	8.135	7.888
16	9.870	9.575	9.289	9.010	8.740	8.478	8.223	7.976	7.736
17	9.665	9.378	9.099	8.829	8.566	8.311	8.064	7.823	7.590
18	9.467	9.188	8.917	8.654	8.399	8.151	7.910	7.676	7.448
19	9.276	9.005	8.742	8.486	8.237	7.996	7.761	7.533	7.312
20	9.092	8.828	8.572	8.323	8.081	7.846	7.617	7.395	7.180
21	8.914	8.658	8.408	8.166	7.930	7.701	7.479	7.262	7.052
22	8.743	8.493	8.250	8.014	7.785	7.561	7.344	7.134	6.929
23	8.578	8.334	8.098	7.868	7.644	7.426	7.215	7.009	6.809
24	8.418	8.181	7.950	7.726	7.507	7.295	7.089	6.888	6.693
25	8.263	8.032	7.807	7.588	7.375	7.168	6.967	6.771	6.581
26	8.113	7.888	7.668	7.455	7.247	7.045	6.849	6.658	6.472
27	7.968	7.748	7.534	7.326	7.123	6.926	6.734	6.548	6.366
28	7.827	7.613	7.404	7.201	7.003	6.811	6.623	6.441	6.263
29	7.691	7.482	7.278	7.079	6.886	6.698	6.515	6.337	6.164
30	7.559	7.354	7.155	6.961	6.773	6.589	6.410	6.236	6.066
31	7.430	7.230	7.036	6.847	6.662	6.483	6.308	6.138	5.972
32	7.305	7.110	6.920	6.735	6.555	6.379	6.208	6.042	5.880
33	7.183	6.993	6.807	6.626	6.450	6.279	6.111	5.949	5.790
34	7.065	6.879	6.697	6.521	6.348	6.180	6.017	5.857	5.702
35	6.949	6.768	6.590	6.417	6.249	6.085	5.925	5.769	5.617
36	6.837	6.659	6.486	6.316	6.152	5.991	5.834	5.682	5.533
37	6.727	6.553	6.383	6.218	6.057	5.899	5.746	5.597	5.451
38	6.620	6.450	6.284	6.122	5.964	5.810	5.660	5.514	5.371
39	6.515	6.348	6.186	6.027	5.873	5.722	5.575	5.432	5.292
40	6.412	6.249	6.090	5.935	5.784	5.636	5.492	5.352	5.215

[a] From Colt, J. (2012) *Dissolved Gas Concentration in Water: Computation as Functions of Temperature, Salinity and Pressure*, 2nd ed., Elsevier, Boston, MA.

The chemical species that comprise the carbonate system include gaseous carbon dioxide [$(CO_2)g$], aqueous carbon dioxide [$(CO_2)_{aq}$], carbonic acid [H_2CO_3], bicarbonate [HCO_3^-], carbonate [CO_2^{-2}], and solids containing carbonates. In waters exposed to the atmosphere, the equilibrium concentration of dissolved CO_2 is a function of the liquid phase CO_2 mole fraction and the partial pressure of CO_2 in the atmosphere. Henry's law (see Chap. 2) is applicable to the CO_2 equilibrium between air and water; thus

$$x_g = \frac{P_T}{H} p_g \qquad\qquad (F-1)$$

where x_g = mole fraction of gas in water, mole gas/mole water

$$= \frac{\text{mole gas}(n_g)}{\text{mole gas}(n_g) + \text{mole water}(n_w)}$$

P_T = total pressure, usually 1.0 atm

H = Henry's law constant, $\dfrac{\text{atm (mole gas/mole air)}}{\text{(mole gas/mole water)}}$

p_g = mole fraction of gas in air, mole gas/mole of air (Note: The mole fraction of a gas is proportional to the volume fraction.)

The concentration of aqueous carbon dioxide is determined using Eq (F–1). At sea level, where the average atmospheric pressure is 1 atm, or 101.325 kPa, carbon dioxide comprises approximately 0.03 percent of the atmosphere by volume (see Appendix B). Values of the Henry's law constant for CO_2 as a function of temperature are given in Table F–1. The values in Table F–1 were computed using Eq. (2–48) and the data given in Table 2–7 in Chap. 2.

Table F–1

Henry's Law constant for CO_2 as a function of temperature

T, °C	H, atm
0	794
10	1073
20	1420
30	1847
40	2361
50	2972
60	3691

Aqueous carbon dioxide [$(CO_2)_{aq}$] reacts reversibly with water to form carbonic acid.

$$(CO_2)aq + H_2O \rightleftarrows H_2CO_3 \qquad\qquad (F-2)$$

The corresponding equilibrium expression is

$$\frac{[H_2CO_3]}{[CO_2]} = K_m \tag{F–3}$$

The value of K_m at 25°C is 1.58×10^{-3}. Note that K_m is unitless. Because of the difficulty of differentiating between $(CO_2)_{aq}$ and H_2CO_3 in solution and the observation that very little H_2CO_3 is ever present in natural waters, an effective carbonic acid value $(H_2CO_3^*)$ is used which is defined as:

$$H_2CO_3^* \rightleftarrows (CO_2)_{aq} + H_2CO_3 \tag{F–4}$$

Because carbonic acid is a diprotic acid it will dissociate in two steps - first to bicarbonate and then to carbonate. The first dissociation of carbonic acid to bicarbonate can be represented as

$$H_2CO_3^* \rightleftarrows H^+ + HCO_3^- \tag{F–5}$$

The corresponding equilibrium relationship is defined as

$$\frac{[H^+][HCO_3^-]}{[H_2CO_3^*]} = K_{a1} \tag{F–6}$$

The value of first acid dissociation constant K_{a1} at 25°C is 4.467×10^{-7} mole/L. Values of K_{a1} at other temperatures are given in Table F–2, which is repeated here from Table 6–16 in Chap. 6.

The second dissociation of carbonic acid is from bicarbonate to carbonate as given below

$$HCO_3^- \rightleftarrows H^+ + CO_3^{2-} \tag{F–7}$$

The corresponding equilibrium relationship is defined as:

$$\frac{[H^+][CO_3^{2-}]}{[HCO_3^-]} = K_{a2} \tag{F–8}$$

The value of the second acid dissociation constant K_{a2} at 25°C is 4.477×10^{-11} mole/L. Values of K_{a2} at other temperatures are given in Table F–2.

Table F–2

Carbonate equilibrium constants as function of temperature[a]

Temperature, °C	Equilibrium constant[b]	
	$K_{a1} \times 10^7$	$K_{a2} \times 10^{11}$
5	3.020	2.754
10	3.467	3.236
15	3.802	3.715
20	4.169	4.169
25	4.467	4.477
30	4.677	5.129
40	5.012	6.026

[a] Adapted from Table 6–20 in Chap. 6.
[b] The reported values have been multiplied by the indicated exponents.
Thus, the value K_{a2} at 20°C is equal to 4.169×10^{-11}.

The distribution of carbonate species as function of pH is illustrated on Fig. F–1.

Figure F–1

Log concentration versus pH diagram for a 10^{-3} molar solution of carbonate at 25°C. By sliding the constituent curves up or down, pH values can be obtained at different concentration values. (LD)

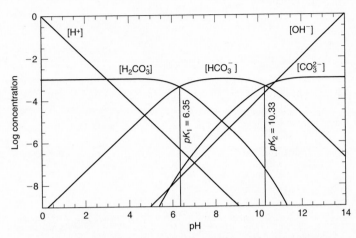

To illustrate the use of the data presented in Tables F–1 and F–2, it will be helpful to estimate the pH of a water, assuming the atmosphere above the water contains 0.03 percent CO_2 by volume (see Appendix B), the bicarbonate (HCO_3^-) concentration in the water is 610 mg/L, and the temperature of the water is 20°C.

1. Use Eq. (F–1) to determine concentration of H_2CO_3 in the water. The value of Henry's constant from Table F–1 is 1420 atm, thus

$$x_{H_2CO_3} = \frac{P_T}{H}p_g = \frac{(1\ \text{atm})(0.00030)}{1420\ \text{atm}} = 2.113 \times 10^{-7}$$

Because one liter of water contains 55.6 mole [1000 g/(18 g/mole)], the mole fraction of H_2CO_3 is equal to:

$$x_{H_2CO_3} = \frac{\text{mole gas}\,(n_g)}{\text{mole gas}\,(n_g) + \text{mole water}\,(n_w)}$$

$$2.113 \times 10^{-7} = \frac{[H_2CO_3]}{[H_2CO_3] + 55.6\ \text{mole/L}}$$

Because the number of moles of dissolved gas in a liter of water is much less than the number of moles of water,

$$[H_2CO_3] \approx (2.113 \times 10^{-7})(55.6\ \text{mole/L}) \approx 11.75 \times 10^{-6}\ \text{mole/L}$$

2. Use Eq. (F–6) to determine the pH of the water. The value of K_{a1} at 20°C from Table F–2 is 4.169×10^{-7}, thus

$$[H^+] = \frac{K_{a1}[H_2CO_3^*]}{[HCO_3^-]}$$

Substitute known values and solve for $[H^+]$

$$[H^+] = \frac{(4.169 \times 10^{-7})(11.75 \times 10^{-6}\ \text{mole/L})}{[(610\ \text{mg/L})/61{,}000\ \text{mg/mole})]} = 4.90 \times 10^{-10}$$

$$pH = -\log[H^+] = -\log[4.90 \times 10^{-10}] = 9.31$$

From Fig. F–1, the amount of carbonate present is essentially non-measurable.

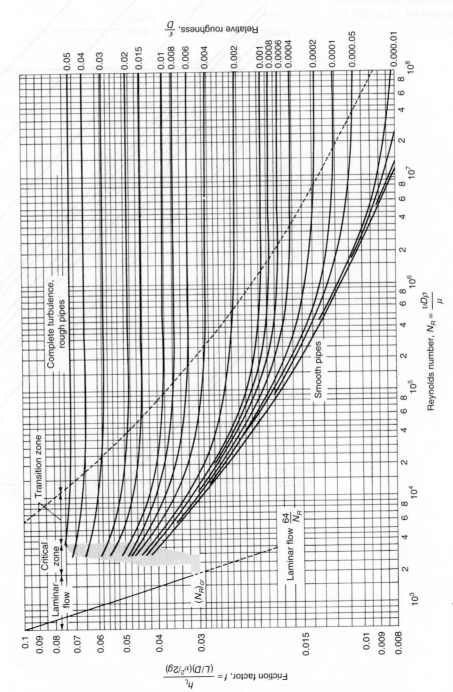

Figure G–1

Moody diagram for friction factor in pipes versus Reynolds number and relative roughness. [From Moody, L.F. (1944) Friction Factors for Pipe Flow, Transactions American Society of Civil Engineers vol. 66, p. 671.]

Figure G-2

Moody diagram for relative roughness as a function of diameter for pipes constructed of various materials. [Adapted from Moody, L.F. (1944) Friction Factors for Pipe Flow, Transactions American Society of Civil Engineers vol. 66, p. 671.]

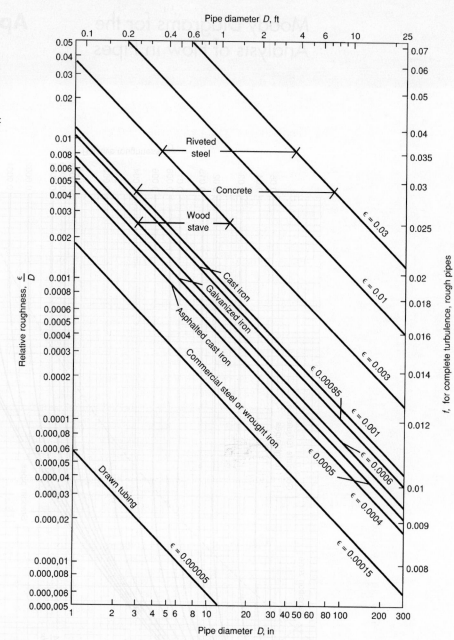

The analysis of reactor hydraulic performance using tracers is considered in this appendix. Important applications of tracer studies include the assessment of: (1) short circuiting in sedimentation tanks and biological reactors, (2) the effective contact time in chlorine contact basins, (3) hydraulic approach conditions in UV reactors, and (4) patterns in constructed wetlands and other natural treatment systems. Tracer studies are also of critical importance in assessing the degree of success that has been achieved with corrective measures. Topics considered in this appendix include: (1) factors leading to nonideal flow in reactors, (2) important characteristics of tracers, (3) analysis of tracer response curves, and (4) practical interpretation of tracer measurements. The types of tracers used and the conduct of tracer tests are discussed in Sec. 12-3 in Chap. 12. The discussion of nonideal flow in this section will also serve as an introduction to the modeling of nonideal flow considered in Appendix I.

H–1 FACTORS LEADING TO NONIDEAL FLOW IN REACTORS

Nonideal flow occurs when a portion of the flow which enters the reactor during a given time period arrives at the outlet, in less than the theoretical detention time, ahead of the bulk flow which entered the reactor during the same time period. The theoretical detention time, τ, is defined as V/Q, where V is the volume and Q is the flowrate in consistent units. Nonideal flow is often identified as short circuiting. Factors leading to nonideal flow in reactors include:

1. Temperature differences. In complete-mix and plug-flow reactors, nonideal flow (short circuiting) can be caused by density currents due to temperature differences. When the water entering the reactor is colder or warmer than the water in the tank, a portion of the water can travel to the outlet along the bottom of or across the top of the reactor without mixing completely [see Fig. H–1(a)].

2. Wind driven circulation patterns. In shallow reactors, wind circulation patterns can be set up that will transport a portion of the incoming water to the outlet in a fraction of the actual detention time [see Fig. H–1(b)].

3. Inadequate mixing. Without sufficient energy input, portions of the reactor contents may not mix with the incoming water [see Fig. H–1(c)].

4. Poor design. Depending on the design of the inlet and outlet of the reactor relative to the reactor aspect ratio, dead zones may develop within the reactor which will not mix with the incoming water.

5. Axial dispersion in plug-flow reactors [see Fig. H–1(d)]. In plug-flow reactors the forward movement of the tracer is due to advection and dispersion. *Advection* is the term used to describe the movement of dissolved or colloidal material with the current velocity. For example, in a tubular plug-flow reactor (e.g., a pipeline), the early arrival of the tracer at the outlet can be reasoned partially by remembering that the velocity distribution in the pipeline will be parabolic. *Dispersion* is the term used to describe the axial and longitudinal transport of material brought about by velocity differences and dispersion. The distinction between molecular diffusion and dispersion is considered further in Appendix I.

Figure H-1

Definition sketch for short circuiting caused by (a) density currents caused by temperature differences, (b) wind circulation patterns, (c) inadequate mixing, (d) fluid advection and dispersion.

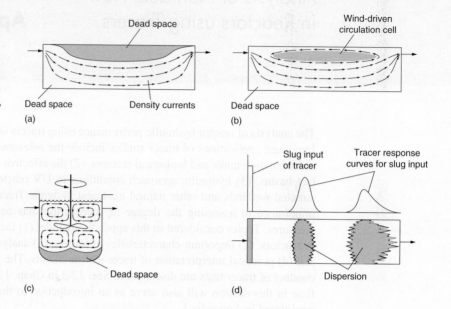

Ultimately, the inefficient use of the reactor volume due to short circuiting resulting from any of the factors described above, can result in reduced treatment performance. The subject of short circuiting in a series of complete-mix reactors was examined extensively in an early paper by MacMullin and Weber (1935); Fitch (1956) and Morrill (1932) examined the effects of short circuiting on the performance of sedimentation tanks.

H–2 TYPES OF TRACERS

Over the years, a number of tracers have been used to evaluate the hydraulic performance of reactors. Important characteristics for a tracer include (adapted in part from Denbigh and Turner, 1984):

1. The tracer should not affect the flow (should have essentially the same density as water when diluted).
2. The tracer must be conservative so that a mass balance can be performed.
3. It must be possible to inject the tracer over a short time period.
4. The tracer should be able to be analyzed conveniently.
5. The molecular diffusivity of the tracer should be low.
6. The tracer should not be adsorbed onto or react with the exposed reactor surfaces.
7. The tracer should not be adsorbed onto or react with the particles in wastewater.

Dyes and chemicals that have been used successfully in tracer studies at wastewater treatment plants are discussed in Sec. 12–6 in Chap. 12. In addition to the tracers considered in Sec. 12-3, lithium chloride is used commonly for the study of natural systems. Sodium chloride, used extensively in the past, has a tendency to form density currents unless mixed completely. Sulfur hexafluoride gas (SF_6) is used most commonly for tracing the movement of groundwater.

H–3   ANALYSIS OF TRACER RESPONSE CURVES

Because of the complexity of the hydraulic response of full-scale reactors, tracer response curves are used to analyze the hydraulics of reactors. Typical examples of tracer response curves are shown on Fig. H–2. Tracer response curves, measured using a short-term and continuous injection of tracer, are known as C (concentration versus time) and F (fraction of tracer remaining in the reactor versus time) curves, respectively. The fraction remaining is based on the volume of water displaced from the reactor by the step input of tracer. The terms used to characterize tracer response curves, the analysis of concentration versus time, and the development of residence time distribution (RTD) curves are described below.

Terms Used to Characterize Tracer Response Curves

Over the years, a number of different symbols and numerical values, as reported in Table H–1, have been used to characterize output tracer curves. The relationship of the terms in Table H–1 and a typical tracer response curve are illustrated on Fig. H–3.

Concentration Versus Time Tracer Response Curves

As noted previously, tracer response curves measured using a short-term or continuous injection of tracer are known as "C" curves (concentration versus time). To characterize C curves, such as those shown on Fig. H–3, the mean value is given by the centroid of the distribution. For C curves, the theoretical mean residence time is determined as follows.

$$\bar{t}_c = \frac{\displaystyle\int_0^\infty tC(t)dt}{\displaystyle\int_0^\infty C(t)dt} \tag{H–1}$$

where $\bar{t}_c$ = mean residence time derived from tracer curve, T

t = time, T

$C(t)$ = tracer concentration at time t, ML^{-3}

The variance, σ_c^2 used to define the spread of the distribution, is defined as

$$\sigma_c^2 = \frac{\displaystyle\int_0^\infty (t - \bar{t})^2 C(t)dt}{\displaystyle\int_0^\infty C(t)dt} = \frac{\displaystyle\int_0^\infty t^2 C(t)dt}{\displaystyle\int_0^\infty C(t)dt} - (\bar{t}_c)^2 \tag{H–2}$$

Figure H–2

Typical tracer response curves: (a) two different types of circular clarifiers (adapted from Dague and Baumann, 1961) and (b) open channel UV disinfection system (courtesy of Andy Salveson).

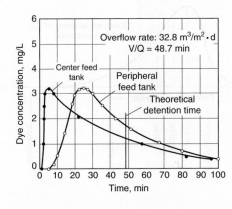

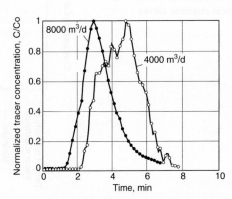

Table H–1

Various terms used to describe the hydraulic performance of reactors used for wastewater treatment[a]

Term	Definition
τ	Theoretical hydraulic residence time (V, volume/Q, flowrate)
t_i	Time at which tracer first appears
t_p	Time at which the peak concentration of the tracer is observed (mode)
t_g	Mean time to reach centroid of the RTD curve
t_{10}, t_{50}, t_{90}	Time at which 10, 50, and 90 percent of the tracer had passed through the reactor
t_{90}/t_{10}	Morrill Dispersion Index, MDI
$1/MDI$	Volumetric efficiency as defined by Morrill (1932)
t_i/τ	Index of short circuiting. In an ideal plug-flow reactor, the ratio is one, and approaches zero with increased mixing
t_p/τ	Index of modal retention time. Ratio will approach 1 in a plug-flow reactor, and 0 in a complete-mix reactor. For values of the ratio greater than or less than 1.0 the flow distribution in the reactor is not uniform
t_g/τ	Index of average retention time. A value of one would indicate that full use is being made of the volume. A value of the ratio greater than or less than 1.0 indicates the flow distribution is not uniform
t_{50}/τ	Index of mean retention time. The ratio t_{50}/τ, is a measure of the skew of the RTD curve. In an effective plug-flow reactor, the RTD curve is very similar to a normal or Gaussian distribution (U.S. EPA, 1986). A value of t_{50}/τ, less than 1.0 corresponds to an RTD curve that is skewed to the left. Similarly, for values greater than 1.0 the RTD curve is skewed to the right
$t/\tau = \theta$	Normalized time, used in the development of the normalized RTD curve

[a] Adapted, in part, from Morrill (1932) and U.S. EPA (1986).

It will be recognized that the integral term in the denominator in Eqs. (H–1) and (H–2) corresponds to the area under the concentration versus time curve. It will also be recognized that the mean and variance are equal to the *first* and *second moments* of the distribution about the y axis.

Figure H–3

Definition sketch for the parameters used in the analysis of concentration versus time tracer response curves.

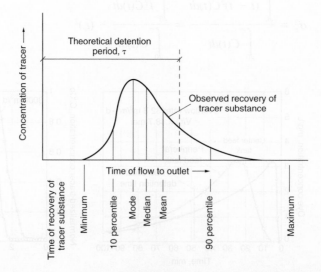

If the concentration versus time tracer response curve is defined by a series of discrete time step measurements, the theoretical mean residence time is typically approximated as

$$\bar{t}_{\Delta c} \approx \frac{\sum t_i C_i \Delta t_i}{\sum C_i \Delta t_i} \tag{H-3}$$

where $\bar{t}_{\Delta c}$ = mean detention time based on discrete time step measurements
t_i = time at ith measurement, T
C_i = concentration at ith measurement, ML^{-3}
Δt_i = time increment about C_i, T

The variance for a concentration versus time tracer response curve, defined by a series of discrete time step measurements, is defined as

$$\sigma_{\Delta c}^2 \approx \frac{\sum t_i^2 C_i \Delta t_i}{\sum C_i \Delta t_i} - (\bar{t}_{\Delta c})^2 \tag{H-4}$$

Where $\sigma_{\Delta C}^2$ = variance based on discrete time measurements, T^2. The application of Eqs. (H–3) and (H–4) is illustrated in Example 12–8 in Chap. 12. Additional details on the analysis of tracer response curves may be found in Levenspiel (1998).

Residence Time Distribution (RTD) Curves

To standardize the analysis of output concentration versus time curves a single reactor for a pulse input of tracer, such as shown on Figs. H–2 and H–3, the output concentration measurements are often normalized by dividing the measured concentration values by an appropriate function such that the area under the normalized curve is equal to one. The normalized curves are known, more formally, as *residence time distribution (RTD) curves* (see Fig. H–4). When a pulse addition of tracer is used, the area under the normalized curve is known as an E curve (also known as the exit age curve). The most important characteristic of an E curve is that the area under the curve is equal to one, as defined by the following integral.

$$\int_0^\infty E(\bar{t}_{\Delta c})\, dt = 1 \tag{H-5}$$

Figure H–4

Normalized residence time distribution curves. The curve on the bottom is known as the exit age curve, identified as the "E curve." The curve on top is the cumulative residence time curve, identified as the "F curve."

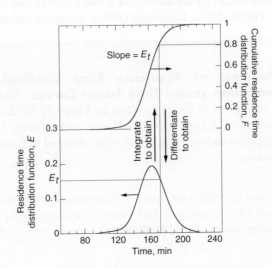

where $E(t)$ is the residence time distribution function. The $E(t)$ value is related to the $C(t)$ value as follows:

$$E(t) = \frac{C(t)}{\displaystyle\int_0^\infty C(t)dt} \tag{H-6}$$

As in Eqs. (H–1) and (H–2), the integral term in the denominator in Eqs. (H–6) corresponds to the area under the concentration versus time curve. Applying Eq. (H–6) to the expression, obtained previously, for the complete-mix reactor, the exit age curve $E(t)$ for a complete-mix reactor is obtained as follows:

$$E(t) = \frac{C(t)}{\displaystyle\int_0^\infty C(t)\,dt} = \frac{C_o e^{-t/\tau}}{\displaystyle\int_0^\infty C_o e^{-t/\tau}\,dt} = \frac{e^{-t/\tau}}{\tau} \tag{H-7}$$

and the corresponding value based on normalized time, $\theta = t/\tau$, is

$$E(\theta) = \tau E(t) = e^{-\theta} \tag{H-8}$$

The mean residence time for the $E(t)$ curve, given by Eq. (H–8), can be derived by applying Eq. (H–23); the resulting expression is given by

$$t_m = \frac{\displaystyle\int_0^\infty t E(t)dt}{\displaystyle\int_0^\infty E(t)dt} = \int_0^\infty t E(t)dt \tag{H-9}$$

In a similar manner when a step input is used, the normalized concentration curve is known as the cumulative residence time distribution curve and is designated as the F curve. The F curve is defined as

$$F(t) = \int_0^t E(t)dt = 1 - e^{-t/\tau} \tag{H-10}$$

where $F(t)$ is the cumulative residence time distribution function. As shown on Fig. H–4, the $F(t)$ curve is the integral of the $E(t)$ curve while the $E(t)$ curve is the derivative of the $F(t)$ curve. In effect, $F(t)$ represents the amount of tracer that has been in the reactor for less than the time t. The development of E and F RTD curves is illustrated in Example H–1. Additional details on the analysis of E and F curves may be found in Denbigh and Turner (1984), Fogler (2005), Levenspiel (1998), and in the chemical engineering literature.

EXAMPLE H–1 **Development of Residence Time Distribution (RTD) Curves from Concentration versus Time Tracer Curves** Use the concentration versus time tracer data given in Example 12–8 in Chap. 12 to develop E and F residence time distribution (RTD) curves for the chlorine contact basin. Using the E curve, compute the residence time and compare to the value obtained in Example 12–8. Plot the resulting E and F curves.

Solution

1. Using the values for time and concentration from Example 12–8, set up a computation table to calculate $E(t)$ values. The computation table is shown below. The $E(t)$ values are calculated by finding the sum of the $C \times \Delta t$ values, which corresponds to

the area under the C (concentration) curve, and dividing the original concentrations by the sum as illustrated below.

Area under C curve = $\Sigma C \Delta t$

$$E(t) = \frac{C}{\Sigma C \Delta t}$$

a. For each time interval, multiply the concentration by the time step ($t = 8$ min) and obtain the sum of the multiplied values. As shown in the computation table, the sum (which is the approximate area under the curve) is 817.778 μg/L-min.

b. Calculate the $E(t)$ values by dividing the original concentration values by the area under the curve.

c. Confirm that the $E(t)$ values are correct by multiplying each one by the time step, and calculating the sum. According to Eq. (H–6) written in summation form, the sum should be 1.00.

$$\Sigma E(t) \Delta t = \Sigma \left(\frac{C \Delta t}{\Sigma C \Delta t} \right) = \Sigma \left(\frac{C}{\Sigma C} \right) = 1$$

Time, t, min	Conc., C, μg/L	$C \times \Delta t$ μg/L·min	$E(t)$, min^{-1}	$E(t) \times \Delta t$, unitless	$t \times E(t) \times \Delta t$, min
88	0.000	0.000	0.00000	0.00000	0.000
96	0.056	0.445	0.00007	0.00054	0.077
104	0.333	2.666	0.00041	0.00326	0.333
112	0.556	4.445	0.00068	0.00544	0.627
120	0.833	6.666	0.00102	0.00815	0.960
128	1.278	10.222	0.00156	0.01250	1.638
136	3.722	29.778	0.00455	0.03641	5.005
144	9.333	74.666	0.01141	0.09130	13.133
152	16.167	129.336	0.01977	0.15816	24.077
160	20.778	166.224	0.02541	0.20326	32.512
168	19.944	159.552	0.02439	0.19510	32.794
176	14.111	112.888	0.01726	0.13804	24.358
184	8.056	64.445	0.00985	0.07880	14.573
192	4.333	34.666	0.00530	0.04239	8.141
200	1.556	12.445	0.00190	0.01522	3.040
208	0.889	7.111	0.00109	0.00870	1.830
216	0.278	2.222	0.00034	0.00272	0.518
224	0.000	0.000	0.00000	0.00000	0.000
Total	102.222	817.778	-	1.0000	163.616

^{a} ppb = parts per billion

2. Determine the mean residence time using the following summation form of Eq. (H–28).

$$\bar{t} = \Sigma (t) E(t) \Delta t$$

The required computation is presented in the final column of the computation table given above. As shown, the computed mean residence time (163.6 min) is essentially the same as the value (163.4 min) determined in Example 12–8 in Chap.12.

3. Develop the F RTD curve. The values for plotting the F curve are obtained by summing cumulatively the $E(t)\Delta t$ values to obtain the coordinates of the F curve.

Time t, min	$E(t)$, min^{-1}	$E(t) \times \Delta t$, unitless	Cumulative total, F, unitless
88	0.00000	0.00000	0.0000
96	0.00007	0.00054	0.0005
104	0.00041	0.00326	0.0038
112	0.00068	0.00544	0.0092
120	0.00102	0.00815	0.0174
128	0.00156	0.01250	0.0299
136	0.00455	0.03641	0.0663
144	0.01141	0.09130	0.1576
152	0.01977	0.15816	0.3158
160	0.02541	0.20326	0.5191
168	0.02439	0.19510	0.7142
176	0.01726	0.13804	0.8522
184	0.00985	0.07880	0.931
192	0.00530	0.04239	0.9734
200	0.00190	0.01522	0.9886
208	0.00109	0.00870	0.9973
216	0.00034	0.00272	1.0000
224	0.00000	0.00000	1.0000
Total		1.0000	

4. Plot the resulting E and F curves. The plot of the E (column 3) and F (column 4) curves using the values determined above is shown below:

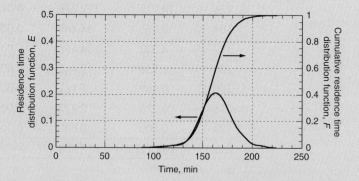

Comment   The use of normalized RTD curves to obtain coefficients of dispersion is included in the discussion of the Hydraulic Characteristics of Nonideal Reactors. Additional details may be found in Denbigh and Turner (1984) and Levenspiel (1998).

Practical Interpretation of Tracer Measurements

In 1932, based on his studies of sedimentation basins, Morrill (1932) suggested that the ratio of the 90 percentile to the 10 percentile value from the cumulative tracer curve could be used as a measure of the dispersion index, and that the inverse of the dispersion index is a measure of the volumetric efficiency (Morrill, 1932). The dispersion index as proposed by Morrill is given by

$$\text{Morrill Dispersion Index, MDI} = \frac{P_{90}}{P_{10}} \tag{H--11}$$

where P_{90} = 90 percentile value from log-probability plot

P_{10} = 10 percentile value from log-probability plot

The percentile values are obtained from a log-probability plot of the time (log scale) versus the cumulative percentage of the total tracer which has passed out of the basin (on probability scale). The value of the MDI for an ideal plug-flow reactor is 1.0 and about 22 for a complete-mix reactor. A plug-flow reactor with an MDI value of 2.0 or less is considered by the U.S. EPA to be an effective plug-flow reactor (U.S. EPA, 1986). The volumetric efficiency is given by the following expression:

$$\text{Volumetric efficiency, \%} = \frac{1}{\text{MDI}} \times 100 \tag{H--12}$$

The determination of the Morrill dispersion index and the volumetric efficiency for the analysis of the flow pattern in a chlorine contact basin is illustrated in Example 12–8 in Chap. 12.

REFERENCES

Denbigh, K.G., and J.C.R. Turner (1984) *Chemical Reactor Theory: An Introduction*, 2nd ed., Cambridge, New York.

Fitch, E.B. (1956) "Effect of Flow Path Effect on Sedimentation," *Sewage and Industrial Wastes*, **28**, 1, 1–9.

Folger, H.S. (2005) *Elements of Chemical Reaction Engineering*, 4th ed., Prentice Hall, Upper Saddle River, NJ.

Levenspiel, O. (1998) *Chemical Reaction Engineering*, 3rd ed., John Wiley & Sons, Inc., New York.

MacMullin, R.B., and M. Weber, Jr. (1935) "The Theory of Short-Circuiting in Continuous Mixing Vessels in Series and Kinetics of Chemical Reactions in such Systems," *J. Am. Inst. Chem. Engrs.*, **31**, 409–458.

Morrill, A.B. (1932) "Sedimentation Basin Research and Design," *J. AWWA*, **24**, 9, 1442–1463.

U.S. EPA (1986) *Design Manual, Municipal Wastewater Disinfection*, EPA/625/1-86/021, U.S. Environmental Protection Agency, Cincinanati, OH.

Modeling Nonideal Flow in Reactors

Appendix I

The hydraulic characteristics of nonideal reactors can be modeled by taking dispersion into consideration. For example, if dispersion becomes infinite, the plug-flow reactor with axial dispersion is equivalent to a complete-mix reactor. Both the plug-flow reactor with axial dispersion and complete-mix reactors in series are considered in the following discussion. However, before considering nonideal flow in reactors it will be helpful to examine the distinction between the diffusion and dispersion as applied to the analysis of reactors used for wastewater treatment.

I–1 THE DISTINCTION BETWEEN MOLECULAR DIFFUSION AND DISPERSION

In addition to short-circuiting caused by the use of improper design of reactor inlets and outlets, inadequate mixing, thermal and density currents, and diffusion and dispersion can also result in nonideal flow. The distinction between the diffusion and dispersion is as follows.

DIFFUSION

Under quiescent flow conditions (i.e., no flow), the mass transfer of material is brought about by *molecular diffusion*, in which dissolved constituents and/or small particles move randomly. This random motion is known as *Brownian Motion*. Further, it should be noted that molecular diffusion can occur under either laminar of turbulent flow conditions, as it does not depend on the bulk movement of a liquid. The transfer of mass by molecular diffusion in stationary systems can be represented by the following expression, known as Fick's first law:

$$r = -D_m \frac{\partial C}{\partial x} \tag{I–1}$$

where r = rate of mass transfer per unit area per unit time, $ML^{-2}T^{-1}$

D_m = coefficient of molecular diffusion in the x direction, L^2T^{-1}

C = concentration of constituent being transferred, ML^{-3}

In the chemical engineering literature the symbol "J" is used to denote mass transfer in concentration units whereas the symbol "N" is used to denote the transfer of mass expressed as moles. The negative sign in Eq. (I–1) is used to denote the fact that diffusion takes place in the direction of decreasing concentration (Shaw, 1966). Adolf Fick (1829–1901), a physician and physiologist, derived the first, and second, laws of diffusion in the 1850s by direct analogy to the equations used to describe the conduction of heat in solids as proposed by Fourier (Crank, 1957). Determination of numerical values for the coefficient of molecular diffusion is illustrated in Sec. 1–10 in Chap. 1.

DISPERSION

The transfer of a constituent from a higher concentration to a lower concentration (e.g., blending) brought about by eddies formed by turbulent flow or by the shearing forces between fluid layers is termed *dispersion*. Under this definition eddies can vary in size from microscale to macroscale to large circulation patterns in the oceans. While microscale transport can only be brought about by molecular diffusion, macroscale transport is brought about by both molecular diffusion and dispersion (Crittenden et al., 2012). Under turbulent flow conditions, the longitudinal spreading of a tracer is caused by dispersion, in which case the coefficient of molecular diffusion term, D_m, in Eq. (I–1) is replaced by the "coefficient of dispersion", D.

While the magnitude of the molecular diffusion depends primarily on the chemical and fluid properties, turbulent or eddy diffusion and dispersion depend primarily on the flow regime. Typical observed ranges for the coefficient of molecular diffusion and dispersion are reported in Table I–1. In all cases, it is important to remember that regardless of whether the coefficient of molecular diffusion or the coefficient of turbulent or eddy diffusion, or dispersion is operative, the driving force for mass transfer is the concentration gradient.

I–2 PLUG-FLOW REACTOR WITH AXIAL DISPERSION

In the following analysis only the one-dimensional problem is considered. However, it should be noted that all dispersion problems are three dimensional, with the dispersion coefficient varying with direction and the degree of turbulence. Using the relationship given above [Eq. (I–1)] and referring to Fig. I–1(c), the one-dimensional materials mass balance for the transport of a conservative dye tracer by advection and dispersion is:

Accumulation = inflow − outflow

$$\frac{\partial C}{\partial t}A\Delta x = \left(vAC - AD\frac{\Delta C}{\Delta x}\right)\bigg|_x - \left(vAC - AD\frac{\Delta C}{\Delta x}\right)\bigg|_{x+\Delta x} \tag{I–3}$$

where $\partial C/\partial t$ = change in concentration with time, $ML^{-3}T^{-1}$, (g/m$^3 \cdot$ s)
  A = cross-sectional area in x direction, L^2, (m^2)
  Δx = differential distance, L, m
  C = constituent concentration, ML^{-3}, (g/m^3)
  D = coefficient of axial dispersion, L^2T^{-1}, (m^2/s)
  v = average velocity in x direction, LT^{-1}, (m/s)

In Eq. (I–3) the term vAC represents the transport of mass due to advection and the term $AD(\Delta C/\Delta x)$ represents the transport brought about by dispersion. Taking the limit of Eq. (I–3) as Δx approaches zero results in the following two expressions:

$$\frac{\partial C}{\partial t} = -D\frac{\partial^2 C}{\partial x^2} - v\frac{\partial C}{\partial x} \tag{I–4}$$

Table I–1			
Typical range of values for molecular diffusion and dispersion[a]	**Coefficient**	**Symbol**	**Range of values, cm²/s**
	Molecular diffusion	D_m	$10^{-10} - 10^{-7}$
	Dispersion	D	$10^{-3} - 10^{0}$

[a] Adapted from Schnoor (1996), Shaw (1966), and Thibodeaux et al. (2012).

Figure I-1

Views of plug-flow reactors and definition sketch: (a) and (b) views of plug-flow activated sludge process reactors and (c) definition sketch for the hydraulic analysis of a plug-flow reactor with (1) advection only and (2) with advection and axial dispersion.

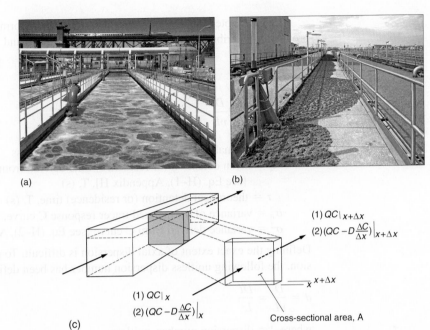

(a)   (b)

(1) $QC|_{x+\Delta x}$
(2) $(QC - D\frac{\Delta C}{\Delta x})|_{x+\Delta x}$

$x+\Delta x$

(1) $QC|_x$
(2) $(QC - D\frac{\Delta C}{\Delta x})|_x$

Cross-sectional area, A

(c)

$$\frac{\partial C}{\partial t} = -D\frac{\partial^2 C}{\partial x^2} - \frac{\partial C}{\partial \tau} \qquad (I\text{-}5)$$

In Eq. (I–5) the hydraulic detention time $\partial\tau$ has been substituted for the term $\partial x/v$. Equation (I–5) has been solved for small amounts of axial dispersion (see below). The solution for a unit pulse input leading to symmetrical output tracer response curves for small amounts of axial dispersion is given by Levenspiel (1998):

$$C_\theta = \frac{1}{2\sqrt{\pi(D/vL)}} \exp\left[-\frac{(1-\theta)^2}{4(D/vL)}\right] \qquad (I\text{-}6)$$

where C_θ = normalized tracer response, C/C_o, unitless
$\qquad \theta$ = normalized time, t/τ, unitless
$\qquad t$ = time, T, (s)
$\qquad \tau$ = theoretical detention time, V/Q, T, (s)
$\qquad D$ = coefficient of axial dispersion, L^2T^{-1}, (m²/s)
$\qquad v$ = fluid velocity, LT^{-1}, (m/s)
$\qquad L$ = characteristic length, L, (m)

The solution given by Eq. (I–6) is for what is known as a *closed system* in which it is assumed that there is no dispersion upstream or downstream of the boundaries of the reactor (e.g., a reactor with inlet and outlet weirs). A large reactor, such as a rectangular sedimentation basin, fed by a small diameter pipe, is an example of a closed system. If a tracer were added to the flow in the small diameter pipe, the tracer would be transported by advection with little or no dispersion. The same situation exists in the discharge pipe from the reactor. Nevertheless, for small amounts of dispersion, Eq. (I–6) can be used to approximate the performance of an open or closed reactor, regardless of the boundary conditions.

It will be noted that Eq. (I–6) has the same general form as the equation for the normal probability distribution. Thus, the corresponding mean and variance are

$$\bar{\theta} = \frac{\bar{t}_c}{\tau} = 1 \tag{I–7}$$

$$\sigma_\theta^2 = \frac{\sigma_c^2}{\tau^2} = 2\frac{D}{vL} \tag{I–8}$$

where $\bar{\theta}$ = normalized mean detention time, $\bar{t}_c/\tau$, unitless

$\bar{t}_c$ = mean detention (or residence) time derived from C curve [see Eq. (H–1), Appendix H], T, (s)

τ = theoretical detention (or residence) time, T, (s)

σ_θ^2 = variance of normalized tracer response C curve, T^2, (s^2)

σ_c^2 = variance derived from C curve [see Eq. (H–2), Appendix H], T^2, (s^2)

Defining the exact extent of axial dispersion is difficult. To provide an estimate of dispersion, the following unitless dispersion number has been defined:

$$d = \frac{D}{vL} = \frac{Dt}{L^2} \tag{I–9}$$

where d = dispersion number, unitless

D = coefficient of axial dispersion, L^2T^{-1}, (m^2/s)

v = fluid velocity, LT^{-1}, (m/s)

L = characteristic length, L, (m)

t = travel time (L/v), T, (s)

Normalized effluent concentration versus time curves, obtained using Eq. (I–6), for a plug-flow reactor with limited axial dispersion for various values of the dispersion number are shown on Fig. I–2.

When dispersion is large, the output curve becomes increasingly nonsymmetrical, and the problem becomes sensitive to the boundary conditions. In environmental problems, a wide variety of entrance and exit conditions are encountered, but most can be considered approximately open; that is, the flow characteristics do not change greatly as the boundaries are crossed. The solution to Eq. (I–5) for a unit pulse input in an *open system* with larger amounts of dispersion is as follows (Fogler, 2005; Levenspiel, 1998):

$$C_\theta = \frac{1}{2\sqrt{\pi\theta(D/vL)}} \exp\left[-\frac{(1-\theta)^2}{4\theta(D/vL)}\right] \tag{I–10}$$

Figure I–2

Typical concentration versus time tracer response curves for a plug flow reactor with small amounts of axial dispersion subject to a pulse (slug) input of tracer.

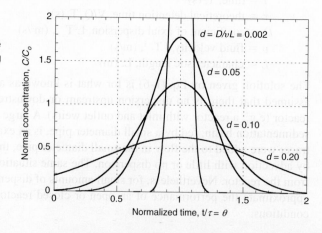

The corresponding mean and variance are:

$$\bar{\theta} = \frac{\bar{t}_c}{\tau} = 1 + 2\frac{D}{vL} \tag{I–11}$$

$$\sigma_\theta^2 = \frac{\sigma_c^2}{\tau^2} = 2\frac{D}{vL} + 8\left(\frac{D}{vL}\right)^2 \tag{I–12}$$

where the terms are as defined previously for Eqs. (I–7), (I–8) and (I–9).

The mean as given by Eq. (I–11) is greater than the hydraulic detention time because of the forward movement of the tracer due to dispersion. Effluent concentration versus time curves, obtained using Eq. (I–12), for a plug-flow reactor with significant axial dispersion for various dispersion factors are shown on Fig. I–3. A more detailed discussion of closed and open reactors may be found in Fogler (2005) and Levenspiel (1998).

In the literature, the inverse of Eq (I–9), as given below, is often identified as the Peclet number of longitudinal dispersion (Kramer and Westererp, 1963).

$$P_e = \frac{vL}{D} = \frac{1}{d} \tag{I–13}$$

In effect, the Peclet number represents the ratio of the mass transport brought about by advection and dispersion. If the Peclet number is significantly greater than one, advection is the dominant factor in mass transport. If the Peclet number is significantly less than one, dispersion is the dominant factor in mass transport. Although beyond the scope of this presentation, it can also be shown that the number of complete-mix reactors in series required to simulate a plug-flow reactor with axial dispersion is approximately equal to the Peclet number divided by 2. Thus, for a dispersion factor of 0.025, the Peclet number is equal to 40 and the corresponding number of reactors in series needed to simulate the dispersion in a plug-flow reactor is equal to 20. This relationship will be illustrated in the following discussion dealing with complete-mix reactors in series. The Peclet number is also used to define transverse diffusion in packed bed plug-flow reactors (Denbigh and Turner, 1984).

For practical purposes, the following dispersion values can be used to assess the degree of axial dispersion in wastewater treatment facilities.

No dispersion	$d = 0$ (ideal plug-flow)
Low dispersion	$d = <0.05$
Moderate dispersion	$d = 0.05$ to 0.25
High dispersion	$d = >0.25$
	$d \to \infty$ (complete-mix)

Figure I–3

Typical concentration versus time tracer response curves for a plug flow reactor with large amounts of axial dispersion subject to a pulse (slug) input of tracer.

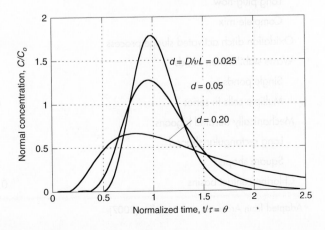

Typical dispersion numbers determined for actual treatment facilities are given in Table I–2. The considerable range in the dispersion numbers reported in Table I–2 for individual treatment processes is due, most often, to one or more of the following factors (Arceivala and Asolekar, 2007):

1. The scale of the mixing phenomenon
2. Geometry (i.e., aspect ratio) of the unit
3. Power input per unit volume (i.e., mechanical and pneumatic)
4. Type and disposition of the inlets and outlets of the treatment units
5. Inflow velocity and its fluctuations
6. Density and temperature differences between the inflow and the contents of the reactor.

Because of the wide range of dispersion numbers that can result for individual treatment processes, special attention must be devoted to the factors cited above in the design of treatment facilities. Determination of the dispersion number and the coefficient of dispersion using the tracer response curves is illustrated in Example I–1.

Evaluation of the coefficient of axial dispersion D for existing facilities is done experimentally using the results of tracer tests, as discussed previously. Because the systems encountered in wastewater treatment are large, experimental work is often difficult and expensive, and is, unfortunately, usually after the fact. To take into account axial dispersion in the design of treatment facilities both scaled models and empirical relationships have been developed for a variety of treatment units including oxidation ponds (Polprasert and Bhattasrai, 1992) and chlorine contact basins (Crittenden et al., 2012). An approximate value of D for water for large Reynolds numbers is (Davies, 1972)

$$D = 1.01\nu N_R^{0.875} \qquad (I-14)$$

where D = coefficient of dispersion, L^2T^{-1}, (m²/s)
  ν = kinematic viscosity, L^2T^{-1}, m²/s (see Appendix C)
  N_R = Reynolds number, unitless
   = $4\upsilon R/\nu$
  υ = velocity in open channel, LT^{-1}, (m/s)
  R = hydraulic radius = area/wetted perimeter, L, (m)

Table I–2

Typical dispersion numbers for various wastewater treatment facilities [a]

Treatment facility	Range of values for dispersion number
Rectangular sedimentation tanks	0.2–2.0
Activated sludge aeration reactors	
Long plug-flow	0.1–1.0
Complete-mix	3.0–4.0+
Oxidation ditch activated sludge process	3.0–4.0+
Waste stabilization ponds	
Single ponds	1.0–4.0+
Multiple cells in series	0.1–1.0
Mechanically aerated lagoons	
Long rectangular shaped	1.0–4.0+
Square shaped	3.0–4.0+
Chlorine contact basins	0.02–0.004

[a] Adapted from Arceivala and Asolekar (2007).

Values for N_R found in open channel flow in wastewater treatment plants are typically in the range from 10^3 to 10^4. The corresponding values for D range from 0.0004 to 0.003 m²/s (4 to 30 cm²/s).

EXAMPLE I–1   **Determination of the Dispersion Number and the Coefficient of Dispersion from Concentration versus Time Tracer Response Curves**   Use the concentration versus time tracer data given in Example 12-8 in Chap. 12 to determine the dispersion number and the coefficient of dispersion for the chlorine contact basin described in Example 12–8. Compare the value of the coefficient of dispersion computed, using the tracer data, to the value computed using Eq. (I–14). Assume the following data are applicable:

1. The flowrate at the time when the tracer test was conducted = 240,000 m³/d
2. Number of chlorine contact basin = 4
3. Number of channels per contact basins = 13
4. Channel dimensions
   a. Length = 36.6 m
   b. Width = 3.0 m
   c. Depth = 4.9 m
5. Temperature = 20°C

Solution

1. From Example 12-8, the mean value and variance for the tracer response C curve are:
   a. Mean, $\tau_{\Delta c}$ = 2.7 h
   b. Variance, $\sigma^2_{\Delta c}$ = 280.5 min²
2. Determine the theoretical detention time for the chlorine contact basin using the given data.

$$\tau = \frac{4 \times 13 \times (36.6 \text{ m} \times 3.0 \text{ m} \times 4.9 \text{ m})}{(240,000 \text{ m}^3/\text{d})}$$

3. Determine the normalized mean detention time using Eq. (I–7). Use the approximate value of $\bar{t}_{\Delta c}$ for $\bar{t}_c$.

$$\bar{\theta}_{\Delta c} = \frac{\bar{t}_{\Delta c}}{\tau} \approx \frac{2.7}{2.8} \approx 0.96 \approx 1.0$$

4. Determine the dispersion number using Eq. (I–8). Use the approximate value of $\sigma^2_{\Delta c}$ for σ^2_c.

$$\sigma^2_{\Delta c} = \frac{\sigma^2_{\Delta c}}{\tau^2} \approx 2\frac{D}{vL} = 2d$$

$$d \approx \frac{1}{2}\frac{\sigma^2_{\Delta c}}{\tau^2} \approx \frac{1}{2}\frac{280.5 \text{ min}^2}{(167.9 \text{ min})^2} = 0.00498$$

5. Using Eq. (I–9), determine the coefficient of dispersion.

$$D = d \times v \times L$$

$$v = \frac{Q}{A} = \frac{(240,000 \text{ m}^3/\text{d})}{(4 \times 13 \times 3.0 \text{ m} \times 4.9 \text{ m})} = 314.0 \text{ m/d} = 0.00363 \text{ m/s}$$

$$D = (0.00498)(0.00363 \text{ m/s})(36.6 \text{ m}) = 6.62 \times 10^{-4} \text{ m}^2/\text{s}$$

6. Compare the value of the coefficient of dispersion computed in Step 5 with the value computed using Eq. (I–14).

$$D = 1.01\nu N_R^{0.875}$$

a. Compute the Reynolds number

$$N_R = 4vR/\nu$$
$$\nu = 1.002 \times 10^{-6} \text{ m}^2/\text{s}$$
$$N_R = \frac{(4)(0.00363 \text{ m/s}) \left[(3.0 \text{ m} \times 4.9 \text{ m})/(2 \times 4.9 \text{ m} + 3.0 \text{ m})\right]}{(1.002 \times 10^{-6} \text{ m}^2/\text{s})} = 1664$$

b. Compute the coefficient of dispersion

$$D = (1.01)(1.002 \times 10^{-6})(1664)^{0.875} = 6.66 \times 10^{-4} \text{ m}^2/\text{s}$$

Comment

Based on the computed value of the dispersion number (0.00498), the chlorine contact basin would be classified as having low dispersion (i.e., $d = < 0.05$). The coefficient of dispersion determined using the results of the tracer study and Eq. (I–14) are remarkably close, given the nature of such measurements.

I–3 COMPLETE-MIX REACTORS IN SERIES

When varying amounts of axial dispersion are encountered, the flow is sometimes identified as *arbitrary flow*. The output from a plug-flow reactor with axial dispersion (arbitrary flow) is often modeled using a number of complete-mix reactors in series, as outlined below. In some situations, the use of a series of complete-mix reactors may have certain advantages with respect to treatment. To understand the hydraulic characteristics of reactors in series (see Fig. I–4), assume that a pulse input (i.e., a slug) of tracer is injected into the first reactor in a series of equally sized reactors so that the resulting instantaneous concentration of tracer in the first reactor is C_o. The total volume of all the reactors is V, the volume of an individual reactor is V_i, and the hydraulic residence time V_i/Q is τ_i. The effluent concentration from the first reactor as is given by the following equation [see Eq. (1–13) in Chap. 1].

$$C_1 = C_o e^{-t/\tau_i} \tag{I–15}$$

Writing a materials balance for the second reactor results in the following:
Accumulation = inflow − outflow

$$\frac{dC_2}{dt} = \frac{1}{\tau_i}C_1 - \frac{1}{\tau_i}C_2 \tag{I–16}$$

Figure I–4

Definition sketch for the analysis of complete-reactors in series.

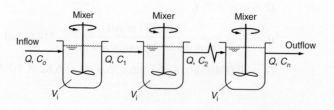

Rearranging the terms in Eq. (I–16) and substituting Eq. (I–15) for C_1 results in

$$\frac{dC_2}{dt} + \frac{C_2}{\tau_i} = \frac{C_o}{\tau_i} e^{-t/\tau_i} \qquad (\text{I–17})$$

The general non-steady-state solution for Eq. (I–17) is obtained by first noting that Eq. (I–17) has the form of the standard first-order linear differential equation. The solution procedure outlined in Eqs. (I–18) through (I–23) involves the use of an integrating factor. It should be noted that Eq. (I–17) can also be solved using the separation of variables method. The solution to Eq. (I–17) is included here because these types of equations are encountered frequently in the field of environmental engineering and in this text. The first step in the solution is to rewrite Eq. (I–17) in the form

$$C'_2 + \frac{C_2}{\tau_i} = \frac{C_o}{\tau_i} e^{-t/\tau_i} \qquad (\text{I–18})$$

where C'_2 is used to denote the derivative dC_2/dt. In the next step, both sides of the expression are multiplied by the integrating factor $e^{\beta t}$, where $\beta = (1/\tau_i)$.

$$e^{\beta t}(C'_2 + \beta C_2) = e^{\beta t}(\beta C_o e^{-\beta t}) = \beta C_o \qquad (\text{I–19})$$

The left hand side of the above expression can be written as a differential as follows:

$$(C_2 e^{\beta t})' = \beta C_o \qquad (\text{I–20})$$

The differential sign is removed by integrating the above expression

$$C_2 e^{\beta t} = \beta C_o \int dt \qquad (\text{I–21})$$

Integration of Eq. (I–21) yields

$$C_2 e^{\beta t} = \beta C_o t + K \text{ (constant of integration)} \qquad (\text{I–22})$$

Dividing by $e^{\beta t}$ yields

$$C_2 = \beta C_o t e^{-\beta t} + K e^{-\beta t} \qquad (\text{I–23})$$

But when $t = 0$, $C_2 = 0$ and K is equal to 0. Thus,

$$C_2 = C_o \frac{t}{\tau_i} e^{-t/\tau_i} \qquad (\text{I–24})$$

Following the same solution procedure, the generalized expression for the effluent concentration for the ith reactor is

$$C_i = \frac{C_o}{(i-1)!} \left(\frac{t}{\tau_i}\right)^{i-1} e^{-t/\tau_i} \qquad (\text{I–25})$$

The effluent concentration from each of four complete-mix reactors in series is shown on Fig. I–5.

The corresponding expression based on the overall hydraulic residence time τ, where τ is equal to $n\tau_i$ is

$$C_i = \frac{C_o}{(i-1)!} \left(\frac{nt}{\tau}\right)^{i-1} e^{-nt/\tau} \qquad (\text{I–26})$$

Figure I–5

Effluent concentration curves for each of four complete-mix reactors in series subject to a slug input of tracer into the first reactor of the series.

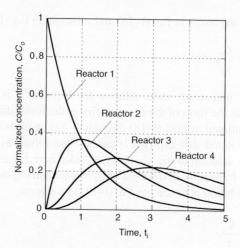

Fraction Remaining

Equation (I–25) can also be used to obtain the fraction of tracer remaining in a series of complete-mix reactors at any time t. The fraction of tracer remaining, F, at time t, is equal to

$$F = \frac{C_1 + C_2 + \cdots + C_n}{C_o} \tag{I–27}$$

Using Eq. (I–25) to obtain the individual effluent concentrations, the fraction remaining in four equal sized reactors in series is given by

$$F_{4R} = \frac{C_o e^{-4t/\tau} + C_o(4t/\tau)\, e^{-4t/\tau} + (C_o/2)\,(4t/\tau)^2 e^{-4t/\tau} + (C_o/6)\,(4t/\tau)^3 e^{-4t/\tau}}{C_o}$$

$$F_{4R} = \left[1 + (4t/\tau) + \frac{(4t/\tau)^2}{2} + \frac{(4t/\tau)^3}{6}\right] e^{-4t/\tau} \tag{I–28}$$

The fraction of a tracer remaining in a series of 2, 4, 6 and 75 complete-mix reactors in series is given on Fig. I–6.

Comparison of Nonideal Plug-Flow Reactor and Complete-Mix Reactors in Series

In many cases it will be useful to model the performance of plug-flow reactors with axial dispersion, as discussed previously, with a series of complete-mix reactors in series. To obtain

Figure I–6

Fraction of tracer remaining in a system comprised of reactors in series.

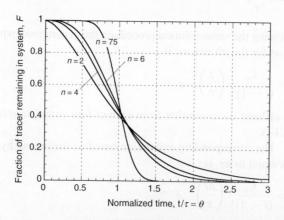

Figure I-7

Effluent tracer concentration curves for reactors in series, subject to a slug input of tracer into the first reactor of the series. Concentration values greater than one occur because the same amount of tracer is placed in the first reactor in each series of reactors.

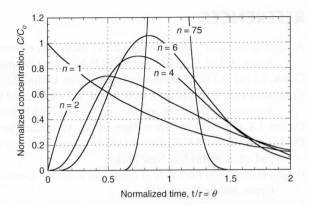

the normalized residence time distribution (RTD) curve for n reactors in series, Eq. (I–26) can be written as follows by noting that the total volume is nV_i and $\tau = n\tau_i$:

$$E(\theta) = \frac{C_n}{C_o/n} = \frac{n}{(n-1)!}(n\theta)^{n-1}e^{-n\theta} \tag{I-29}$$

In effect, in Eq. (I–29) it is assumed that the same amount of tracer is always added to the first reactor in series. Effluent residence time distribution curves, obtained using Eq. (I–57), for 1, 2, 4, 6, and 75 reactors in series are shown on Fig. I–7. It is interesting to note that a model comprised of four complete-mix reactors in series is often used to describe the hydraulic characteristics of constructed wetlands. As shown on Fig. I–7, the concentration increases as the number of reactors in series increases because the same amount of tracer is used regardless of the number of reactors in series. It should also be noted that the F curves shown on Fig. I–6 can also be obtained by integrating the E curves given on Fig. I–7.

A comparison of the residence time distribution curves obtained for a plug-flow reactor, with a dispersion number of 0.05, and to the residence time distribution curves obtained for six, eight and ten complete-mix reactors in series is shown on Fig. I–8. As shown, all three of the complete-mix reactors in series can be used, for practical purposes, to simulate a plug-flow reactor with a dispersion factor of 0.05. As noted previously, the number of reactors in series needed to simulate a plug-flow reactor with dispersion is approximately equal to the Peclet number divided by 2. Thus, for a dispersion factor of 0.05, the Peclet number is equal to 20 and the corresponding number of reactors in series needed to simulate the dispersion in a plug-flow reactor is equal to 10. As shown on Fig. I–8, the response curves computed using ten reactors in series and Eq. (I–10) with a dispersion number, $d = 0.05$ are essentially the same.

Figure I-8

Comparison of effluent response curves for a plug-flow reactor with a dispersion factor of 0.05 and reactor systems comprised of six, eight, and ten reactors in series subject to a pulse input of tracer.

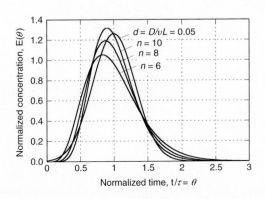

REFERENCES

Arceivala, S.J., and S.R. Asolekar (2007) *Wastewater Treatment for Pollution Control and Reuse*, 3rd ed.,Tata McGraw-Hill Publishing Company Limited, New Delhi, India.

Crank, J. (1957) *The Mathematics of Diffusion*, Oxford University Press, London.

Crittenden, J.C., R.R. Trussell, D.W. Hand, K.J. Howe, and G. Tchobanoglous (2012) *Water Treatment: Principles and Design*, 3rd ed., John Wiley & Sons, Inc, New York.

Davies, J.T. (1972) *Turbulence Phenomena*, Academic Press, New York.

Denbigh, K.G., and J.C.R. Turner (1984) *Chemical Reactor Theory: An Introduction*, 2nd ed., Cambridge, New York.

Folger, H.S. (2005) *Elements of Chemical Reaction Engineering*, 4th ed., Prentice Hall, Upper Saddle River, NJ.

Kramer, H., and K.R. Wsetererp (1963) *Elements of Chemical Reactor Design and Operation*, Academic press, Inc., New York.

Levenspiel, O. (1998) *Chemical Reaction Engineering*, 3rd ed., John Wiley & Sons, Inc., New York.

Polprasert C., and K.K. Bhattarai (1985) Dispersion Model for Waste Stabilization Ponds, *J. Environ. Eng. Div.*, ASCE, **11**, EE1, 45–58.

Schnoor, J.L. (1996) *Environmental Modeling*, A Wiley-Interscience Publication, John Wiley & Sons, Inc., New York.

Shaw, D.J (1966) *Introduction to Colloid and Surface Chemistry*, Butterworth, London.

Thibodeaux, L.J., J.L. Schnoor, and A.J.B. Zehnder (2012) *Environmental Chemodynamics: Movement of Chemicals in Air, Water, and Soil*, Wiley Interscience, New York.

Wallis, P. M., 1391, 1397
Wallis-Lage, C., 1038, 1039
Waltrip, G. D., 868
Walton, C. W., 1820, 1864
Waltz, T., 1488
Wang, H., 1447
Wang, J. F., 1046
Wang, K. Y., 1082
Wang, L., 667, 686
Wang, S., 647, 800
Wang, Y. T., 612
Wang, X., 1046
Wanner, J., 733, 734, 740
Wanner, O., 612, 1015, 1844, 1864
Warakomski, A., 999
Ward, B. B., 621, 629, 631, 641, 644, 646, 647
Ward, R. W.. 1374
Warnecke, F., 571, 651
Washington State Department of Health
  (WDOH), 138, 181
Water Environment Federation (WEF), 234,
  261, 274, 296, 297, 304, 311, 406, 429, 454,
  482, 549, 611, 723, 890, 915, 945, 946, 953,
  954, 955, 958, 959, 960, 961, 963, 966, 970,
  972, 973, 974, 978, 979, 982, 983, 999, 1003,
  1005, 1006, 1008, 1016, 1020, 1027, 1030,
  1030, 1032, 1034, 1035, 1038, 1451, 1470,
  1489, 1490, 1493, 1494, 1495, 1496, 1498,
  1500, 1501, 1502, 1503, 1504, 1509, 1517,
  1527, 1529, 1532, 1533, 1538, 1539, 1540,
  1541, 1542, 1544, 1545, 1546, 1549, 1551,
  1744, 1755, 1761, 1762, 1765, 1774, 1791,
  1795, 1809, 1850, 1852, 1864
Water Environment Research Foundation
  (WERF), 619, 626, 860, 1034, 1041, 1820, 1864
Water Pollution Control Federation (WPCF), 304,
  393, 395, 430, 438, 440, 454, 912, 957, 1058
Waterbury, J. B., 619, 631, 688
Watson, H. E., 1301
Watson, R. S., 969, 978, 979, 981
Wattie, E., 1320, 1322
Watts, R. J., 664
Wayner, W. J., 1240
WCPH, *see* Western Consortium for Public
  Health
WDOH, *see* Washington State Department of
  Health
Weand, B. L., 469, 472, 505, 547
Webb, R. I., 641, 644, 645, 688
Webber, N. B., 1913, 1914
Weber, M., Jr., 1932, 1939
Webster, T. S., 1761, 1762, 1764, 1765, 1794
WEF, *see* Water Environment Federation
Wegner, G., 1640, 1658
Wehner, J. F., 50, 56
Wehner, M., 526, 549
Weidler, P., 1684
Weil, I., 1322
Weissenbacher, N., 647

Welander, T., 637, 685, 850, 1002
Wellinger, A., 1539, 1540
Wells, G. F., 619, 622, 631
Wells, W. N., 861
Weltevrede, R., 1046
Wenzel, L. A., 1289
Wentzel, M. C., 587, 627, 687
Werner, A., 628, 685
Werzernak, C. T., 623
West, P. M., 156, 181
Westererp, K., R., 52, 56, 1945, 1952
Westerhoff, H. V., 647, 684
Westerhoff, P., 503, 549
Western Consortium for Public Health
  (WCPH), 282, 286, 304
Wett, B., 624, 626, 627, 628, 641, 645, 647,
  687, 1090, 1703, 1708, 1711, 1714, 1715,
  1722, 1859, 1863
Wetterau, G. D., 1208
Whalen, T., 1278
Whang, L. M., 631, 652
Wheelis, M. L., 140, 141, 145–147, 180, 181
White, G. C., 1313, 1315, 1329, 1339, 1341,
  1368, 1374, 1375
White, S., 1674
Whitley, R., 1404
Whitman, W. C., 411, 413, 414, 417, 453
Wicht, H., 646
Wienberg, H. S., 1335
Wiesner, M., 1202
Wilderer, P. A., 1046
Wilf, M., 1196
Wilhelm R. F., 50, 56
Williams, D. G., 1750, 1795
Williams, T. O., 1757, 1763, 1795
Williamson, K., 612, 613, 614, 1045
Wilson, A. W., 626, 627, 711, 754
Wilson, C. A., 1090
Wilson, G., 1746, 1795
Wilson, G. G., 366, 367, 368, 369, 454
Wilson, M. F., 161, 179
Wilson, T. E., 895, 1532
Winkler, M.-K. H., 644
Winogradsky, M. S., 619
Winslow, C. E. A., 1064
Winslow, D., 1036, 1037
Witzgall, R., 737, 741, 1005
Woese, C., 140, 181
Wohlfarth, G., 665, 687
Wolf, D. C., 674
Wolfner, J. P., 473, 547
Wong, J., 1211, 1212, 1290
Wong, P., 1033
Wong, M. T., 652
Wong, T., 1429, 1432
Wong, T. H. F., 380, 454
Wong-Chong, G. M., 630, 631
Wood, D. K., 735
Woods, K., 1392

WPCF, *see* Water Pollution Control Federation
Wurhmann, K., 841

Xie, G., 1725
Xin, G. X., 741

Yamamoto, K., 705
Yamashita, K., 657, 688, 1084
Yanez-Noguez, I., 1382
Yang, J., 1722
Yang, Q., 647
Yang, W., 705, 915, 917
Yang, X., 800
Yang, Y., 1757, 1762, 1765, 1794, 1795
Yaun, Z., 645, 646, 647, 684, 688, 1046, 1058
Ye, L., 619, 631, 645, 647, 688
Yeates, C., 571, 651
Yendt, C. M., 664, 667, 669
Yeung, C. H., 619, 631
Ydstebo, L., 737
Yilmaz, G., 1046, 1058
Yoo, H., 1223
Yoshida, H., 1585, 1658
Yoshika, N., 364, 454
Young, J. C., 122, 126, 127, 181, 1026, 1065,
  1067, 1169
Young, K. W., 364, 453
Young, M. Y., 672, 687
Young, P., 859
Young, T. M., 494, 495
Yu, H. Q., 623, 686, 1096
Yu, R., 634, 637, 647, 684, 850
Yuan, Z. G., 629, 652

Zacheis, A., 1178, 1180
Zahlier, J. D., 661, 685
Zaki, W. N., 1140
Zandvoort, M. H., 1080
Zanoni, A. E., 129, 181
Zart, D., 634, 647, 684
Zeeman, G., 1082, 1096, 1097
Zehnder, A. J. B., 1942, 1952
Zhang, L., 110, 111, 178, 1083, 1722
Zhang, T., 619, 631, 652
Zhao, Z. G., 1046
Zheng, P., 644
Zhou, S., 1077
Zhu, G., 628
Ziarkowski, S.. 1271
Ziemke, N., 1065, 1067
Zimmer, J. L., 1392
Zinder, S. H., 658
Zitomer, D., 1539
Zuckut, S. W., 619, 685
Zumft, W. G., 587
Zwillinger, D., 282

Because a number of the subjects covered in this text can be referenced (i.e., indexed) under different alphabetical listings, it has been necessary to develop an approach to limit the degree of duplication, yet not affect the utility of the index. The approach used is as follows. Each subject with multiple subentries is indexed in detail under one letter of the alphabet. Where the same subject is indexed under another letter of the alphabet, inclusive page numbers are given and a *See also* citation is given to the location where the subject is indexed in detail. For example, Activated carbon adsorption, 1138–1162, is followed by (*See also* Adsorption). Where a subject is indexed in detail (e.g., Biochemical oxygen demand) and a commonly used abbreviation (e.g., BOD) is also cited under the same letter of the alphabet, the abbreviated entry is followed by a *See* citation. For example, the index entry BOD is followed by (*see* Biochemical oxygen demand). Where an abbreviation occurs under a different letter of the alphabet than where the detailed citation is given, inclusive page numbers are given followed by the *See also* citation. For older or unused terms, the *see* citation is used to direct the reader to the appropriate term used.

To access the number of data tables in the textbook more easily, an index entry followed by the capital letter T in parenthesis [e.g., (T)] is used to denote a data table related to the subject matter.